ECHOLOCATION
in Bats and Dolphins

ECHOLOCATION
in Bats and Dolphins

EDITED BY

Jeanette A. Thomas,

Cynthia F. Moss, *and*

Marianne Vater

THE UNIVERSITY OF CHICAGO PRESS

Chicago and London

JEANETTE THOMAS is professor of biology at Western Illinois University. She is the coeditor of four books, including *Sensory Abilities of Aquatic Mammals* and *Marine Mammal Sensory Systems.*

CYNTHIA MOSS is professor of psychology a the University of Maryland. She is the coeditor of *Neuroethological Studies on Cognitive and Perceptual Processes.*

MARIANNE VATER is professor in and chair of the Department of Physiology and Cell Biology at the University of Potsdam, Germany.

The University of Chicago Press, Chicago 60637
The University of Chicago Press, Ltd., London
© 2004 by The University of Chicago
All rights reserved. Published 2004
Printed in the United States of America

12 11 10 09 08 07 06 05 04   1 2 3 4 5
ISBN: 0-226-68446-6 (cloth)
ISBN: 0-226-79599-3 (paper)

Library of Congress Cataloging-in-Publication Data

Echolocation in bats and dolphins / edited by Jeanette A. Thomas, Cynthia F. Moss, and Marianne Vater.
    p.   cm.
    Includes bibliographical references (p.   ).
    ISBN 0-226-68446-6 (cloth : alk. paper) —
    ISBN 0-226-79599-3 (pbk. : alk. paper)
    1. Bats—Physiology.   2. Dolphins—Physiology.
3. Echolocation (Physiology)   I. Thomas, Jeanette A.
II. Moss, Cynthia.   III. Vater, Marianne.

QL737.C5 E28 2003
573.8'719—dc21   2002010606

⊗ The paper used in this publication meets the minimum requirements of the American National Standard for Information Sciences—Permanence of Paper for Printed Library Materials, ANSI Z39.48-1992.

To my father, L. Keith Thomas, for providing me with a childhood on a farm and instilling an appreciation for nature—JAT

To my family, students, and mentors, for immeasurable support and inspiration—CFM

To my mother and teachers—MV

Contents

....

Part Three | PERFORMANCE AND COGNITION IN ECHOLOCATING MAMMALS

Part Four | ECOLOGICAL AND EVOLUTIONARY ASPECTS OF ECHOLOCATING MAMMALS

Part Five | ECHOLOCATION THEORY, ANALYSIS TECHNIQUES, AND APPLICATIONS

Part Six | POSSIBLE ECHOLOCATION ABILITIES IN OTHER MAMMALS

Preface

The editors gratefully acknowledge the contributions by our many colleagues involved in the writing of this book. A scientific organizing committee (Paul Nachtigall from the University of Hawaii, Darlene Ketten from Woods Hole Institute of Oceanography, Patrick Moore from SPAWARS Systems Center, Brock Fenton from York University, Cynthia Moss from the University of Maryland, Jeanette Thomas from Western Illinois University, and Marianne Vater from Potsdam University) planned a conference on biological sonars held 27 May to 2 June 1998 in Carvoeiro, Portugal. The meeting was hosted by Zoomarine in Albufeira, Portugal. Elio Vicente and Pedro Lavia, from Zoomarine, and Manuel dos Santos, from Institute Superior de Psicologia, served on a committee that organized local accommodations and travel.

Five days of meetings included 55 oral and 37 poster presentations by international experts studying animals that echolocate, primarily bats and dolphins. Scientists from the United Kingdom, Canada, the United States, Germany, Japan, Portugal, Holland, China, Russia, Denmark, Sweden, Israel, Switzerland, and Poland participated.

This meeting followed a series of conferences on echolocation in animals held at Frescati in 1966, at Jersey in 1979, and at Helsingør in 1986. Interactions among scientists in the bat and dolphin scientific community during the three previous Biosonar Conferences largely set the direction of research on biological sonars. Examination of the books from these previous conferences (*Animal Sonar Systems: Biology and Bionics* by Busnel, 1967; *Animal Sonar Systems* by Busnel and Fish, 1980; and *Animal Sonar: Processes and Performance* by Nachtigall and Moore, 1988) demonstrates the continuity of research efforts over time, the impact of new directions of research, and the importance of coordinating research between scientists studying bats and dolphins. A special banquet was held honoring researchers who have attended all biosonar conferences: William E. Evans, Bertel Møhl, Hans-Ulrich Schnitzler, Ronald J. Schusterman, Nobua Suga, and William Watkins. Further recognition was given to Kenneth Norris and Donald Griffin for their special contributions to the field of echolocation in animals.

Often bioacoustic researchers identify a natural adaptation or ability possessed in an echolocating mammal, perfected over evolutionary time, which can offer direct applications to humanmade sonar systems. Publishing a book from this conference is the best and quickest way to assemble the most recent research findings from each contributor into one single source. Participants divided into six working groups at the meeting to develop the format and outline contributions for *Echolocation in Bats and Dolphins*. Lee Miller from Odense University obtained funds from the Office of Naval Research in Europe to support this publication. The Office of Naval Research in Washington, D.C., under the direction of Robert Gisiner, also subsidized the costs of equipment, commodities, and labor associated with the publication of this book. A grant from the National Science Foundation to Cynthia F. Moss provided additional support for editorial work.

ECHOLOCATION
in Bats and Dolphins

A Comparison of the Sonar Capabilities of Bats and Dolphins

Whitlow W. L. Au

Introduction

Despite three previous international animal sonar symposiums (1966, 1979, and 1986), scientists have expended very little effort in comparing the sonar capabilities of bats and dolphins. Part of the reason for this has to do with the vast differences between the two orders, their environments, and the type of echolocation tasks they perform. Therefore, researchers in both fields tend to concentrate on vastly different sonar problems. However, over these many years some echolocation experiments have been similar enough to permit some quantitative comparisons of the capabilities of both species.

The comparison that will be performed in this chapter is not meant to be some kind of competition between bats and dolphins, but is undertaken with the goal of securing a deeper understanding of the differences in both sonar systems. Bats are small flying animals, weighing no more than 30 grams, whereas dolphins are much larger animals, weighing from tens of kilograms to several hundred kilograms. Bats use their sonar system at short ranges of up to approximately 3–4 m (see Popper and Fay 1995), whereas dolphins can detect small targets at ranges varying from a few tens of meters for the harbor porpoise (Kastelein et al. 2000) to over a hundred meters for bottlenose and other large dolphins (Au 1993). Many bats hunt for insects that dart rapidly to and fro in rapid sequences that are very different than the escape behavior of fish chased by dolphins. Finally, the speed of sound in air is about one fifth that in water, so that the information transfer rate during sonar transmission for bats is much slower than for dolphins. These and many other differences in environment and prey have led to the evolution of totally different types of sonar systems, which naturally makes direct comparisons between species difficult at best and perhaps irrelevant at worst.

Several similar experiments performed with bats and dolphins permit comparative results for six different echolocation tasks: (1) target detection in a "noiseless" environment, (2) target detection in noise, (3) range difference discrimination, (4) differential range difference discrimination, (5) target localization, and (6) auditory integration time for echoes. However, before proceeding, it is important to briefly discuss the vastly different types of echolocation signals used by bats and dolphins.

ECHOLOCATION SIGNALS OF BATS

Echolocation signals used by bats are brief sounds varying in duration from 0.3 to 300 ms and in frequency from 12 to 200 kHz (Neuweiler 1990). The structure of bat echolocation sounds is varied and diverse, being both species and situation specific (Pye 1980). In most species the sounds consist of either frequency-modulated (FM) components alone or a combination of a constant-frequency (CF) component coupled with FM components. Echolocation signals typically consist of the following elements or of combinations of them emitted as single or multiple harmonics: (1) downward FM sweep with linear or exponential time course (FM-down); (2) CF tone or shallowly modulated tonal element; or (3) upward FM sweep with linear or curved time course (FMup), which only occurs in combination with other sound elements (Neuweiler 1990). FM-only signals are brief in duration, varying from 0.5 to 10 ms. The sweep is usually downward. CF signals are either short in duration, varying from 1 to 10 ms, or quite long, varying from 10 to 100 and sometimes to 300 ms.

Orientation signals in the time domain with their representative spectrograms are shown in the top panel of fig. 1 for three species of bats. The orientation sounds emitted by four bat species, each representing a different family, are shown in the bottom panels of fig. 1. These examples indicate that the bandwidth of the FM signals can be very wide, extending over an octave. The FM signals used by bats are Doppler tolerant (Altes and Titlebaum 1970), so they are not affected significantly by either their own motion or that of the prey. On the other hand, long-CF signals used by some bats like *Rhinolophus* are affected by the velocity of both bat and prey so that echoes can carry Doppler information.

Bats typically adjust their emission rate so that echoes are returned before the next signal is transmitted. As bats fly closer to their prey, the two-way travel time decreases, and some bats correspondingly reduce the duration of their echolocation signals in order avoid overlap between the outgoing signal and the returning echo

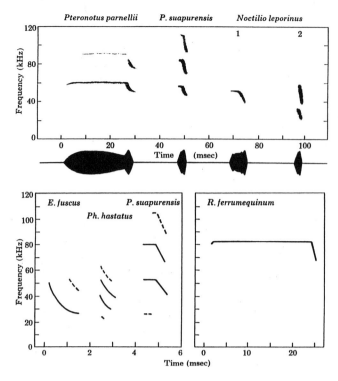

Fig. 1. Orientation echolocation signals of six species of bats (from Simmons, Howell, and Suga 1975)

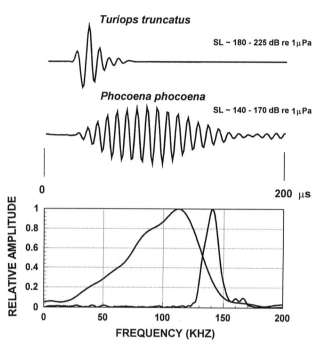

Fig. 2. Examples of the echolocation signals of a *T. truncatus* (a whistling dolphin) and a *P. phocoena* (a nonwhistling porpoise)

(Kick and Simmons 1984). Conversely, when some bats echolocate on rather distant targets (greater than about 3 m), they will increase the duration of the signals to transmit higher energy signals toward a target.

ECHOLOCATION SIGNALS OF DOLPHINS

The echolocation signals emitted by dolphins tend to fall into two broad categories. Dolphins that typically emit whistle signals also emit brief broadband echolocation clicks having between four and eight cycles and durations of 40–70 μs (see fig. 2). Most dolphins fall into this class (Au 1993). The signals do not change much in duration and shape. The center frequency is a function of the intensity of the projected signal varying almost linearly with the level of the signal (Au et al. 1995). High-intensity signals often have center frequencies of 100 kHz and higher, whereas very low intensity signals often have center frequencies between 30 and 60 kHz.

Dolphins and porpoises that do not emit whistle signals emit narrowband echolocation signals with at least 12 cycles and a duration generally greater than 100 μs (Au 1993). Among odontocetes that emit signals in this category are the harbor porpoise, *Phocoena phocoena;* finless porpoise, *Neophocoaena phocaenoides;* Commerson's dolphin, *Cephalorhynchus commersonnii;* Hector's dolphin, *Cephalorhynchus hectori;* Dall's porpoise, *Phocoenoides dalli;* and the pygmy sperm whale, *Kogia* sp. (see Au 2000). Whether riverine dolphins whistle or not is still an open question, although Ding, Würsig, and Leatherwood (2001) have evidence that *Inia geoffrensis*

produce whistles. Riverine dolphins emit echolocation signals that would fall in the short duration, broadband category. An example of typical signals used by a *Tursiops truncatus* (a whistling dolphin) and a *P. phocoena* (a nonwhistling dolphin) is shown in fig. 2, along with the range of peak-to-peak source levels (sound pressure level 1 m from the animal).

Maximum Target Detection Range

TARGET DETECTION IN A QUIET ENVIRONMENT

Kick and Simmons (1984) measured the target detection capabilities of two *Eptesicus fuscus* using nylon spheres with diameters of 0.48 cm and 1.91 cm as targets. The performance of one bat with the 1.91 cm sphere is shown in fig. 3. The 75% correct response threshold range was 5.1 m. (The noise level was too low to be measured.)

Kastelein et al. (2000) measured the target detection capability of a *P. phocoena* in a quiet environment using water-filled stainless-steel spheres of 5.08 cm and 7.62 cm diameters. The porpoise correct detection performance for the 5.08 cm sphere is also shown in fig. 3. The correct detection threshold was 15.5 m. The bat results of Kick and Simmons 1984 were for all correct response trials that included target present as well as target absent trials. The porpoise results of Kastelein et al. 2000 were only for correct target present trails. A 75% correct performance level is approximately equivalent to a 50% correct detection threshold. One of the reasons why the threshold target detection ranges are so different for the bat and porpoise can be attributed to the vast difference in the acoustic absorption losses in air and water. At a frequency of 50 kHz, the absorption loss in air is about 2 dB/m compared with 0.01 dB/m in saltwater. The bat uses signals with duration of about 8 ms, which is considerably longer than the 125–150 μs for the porpoise. Because of the long duration of the projected signal, the bat actually projects slightly more energy than the porpoise. This will become obvious in the following analysis.

The performance of both species can be compared by calculating the received energy in the echoes at each detection threshold range. The energy of the echo (EE) in dB from a target at range R is equal to

$$EE = SE - 40 \log R - 2\alpha(f)R + TS_E \qquad (1)$$

where SE is the source energy flux density, 40 log R is the two-way spreading loss, $\alpha(f)$ is the absorption coefficient of the medium (which is frequency dependent), and TS_E is the target strength of the target in terms of energy.

The source energy flux density in dB re 1 J/m² (J stands for joules) can be expressed as

$$SE = 10 \log\left(\frac{1}{\rho c}\int_0^T p_s^2(t)\, dt\right) \qquad (2)$$

where $p_s(t)$ is the instantaneous acoustic pressure of the source at the reference range of 1 m, ρ is the density, and c is the sound speed of the medium. The effects of the high absorption of airborne sounds on the echoloca-

tion signals used by bats can be determined by combining the source pressure waveform with the absorption term, letting

$$p_s'(t) = \Im^{-1}\{P_S(f)e^{-2\alpha(f)R}\} \qquad (3)$$

where $P_S(f) = \Im[p_S(t)]$ is the Fourier transform of the source pressure waveform. The modified source energy flux density in dB corresponding to eq. 3 is

$$SE' = 10 \log\left(\frac{1}{\rho c}\int_0^T p_s'^2(t)\, dt\right)$$
$$= 10 \log\left(\frac{1}{\rho c}\int_0^T \Im^{-1}\{P_S(f)e^{-2\eta(f)R}\}^2\, dt\right) \qquad (4)$$

where $\alpha(f)$ in dB/m was converted in Neper/m. Kick (1982) reported that the average peak-to-peak sound pressure level (SPL) of the emitted signal of *Eptesicus* was 109 dB re 20 μPa at 10 cm, which is equivalent to 115 dB re 1 μPa at 1 m. One of these long echolocation signals of *E. fuscus* is shown in the top panel of fig. 4. The signal waveform after propagating 10.2 m, corresponding to the target threshold range of 5.1 m (shown

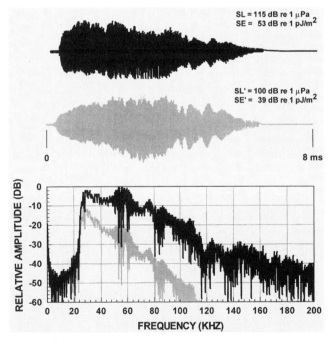

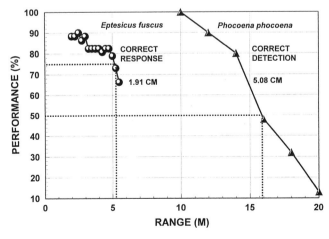

Fig. 3. Target detection performance of the big brown bat, *E. fuscus*, with a 1.91 cm sphere (adapted from Kick 1982); and a harbor porpoise, *P. phocoena*, with a 5.08 water-filled stainless-steel sphere (from Kastelein et al. 2000). The 75% correct response threshold for the bat is shown along with the 50% correct detection threshold for the porpoise.

Fig. 4. The effects of absorption losses on the shape of the echolocation signal of a bat after propagating to a target at 5.1 m and back. The top panel is the emitted signal (courtesy of James Simmons); the middle panel is the reflected signal taking into account the absorption losses; and the bottom panel contains the spectrum of the emitted and reflected echolocation signal. The peak-to-peak SPL is shown above each signal, along with the energy flux density in dB re 1 pJ/m2, where p stands for pico and is equal to 10^{-12}.

in the bottom panel of fig. 4), is the frequency spectrum of both the emitted and reflected signal at the bat. It is obvious from both the time waveform and the frequency spectra in fig. 4 that the higher-frequency components are attenuated more than the low-frequency ones. The relatively high absorption of the high-frequency components of the signal can also be seen in the envelope of the signal waveform. Such changes in the signal shape due to absorption will not occur for the porpoise signal since the signal has such a narrow bandwidth that the absorption across the bandwidth of the signal is nearly constant.

Eq. 1 can now be written as

$$EE = SE - 40 \log R + TS_E \qquad (5)$$

As demonstrated in Urick 1983, the target strength of a sphere where there is little or no penetration of the acoustic signal into the sphere is given by the equation

$$TS = 20 \log(a/2) \qquad (6)$$

where a is the radius of the sphere in meters. For a sphere of 1.91 cm diameter, the target strength is equal to -46.4 db. As fig. 5 shows, $SE' = 39.0$ dB re 1 pJ/m^2 so that the echo energy for the bat detection case is equal to

$$EE = -[7.4 + 40 \log R] \ (\text{dB re 1 pJ/m}^2) \qquad (7)$$

Eq. 1 can be used directly for the harbor porpoise in order to determine the energy flux density in the echo at any target range. The average peak-to-peak SPL of the emitted signal for the animal was 157.2 dB re 1 μPa (Au, Lammers, and Aubauer 1999). Inserting the waveform of fig. 2 having a peak-to-peak SPL of 157.2 dB into eq. 2, we find that $SE = 43.8$ dB re 1pJ/m^2, which is slightly lower than the 53 dB for the bat. Since the bandwidth of the *Phocoena* signal is not very large, we can use a value of 0.038 dB/m for $\alpha(f)$ for the average peak frequency of 127.5 kHz (Au, Lammers, and Aubauer 1999). The target strength, measured using a simulated *Phocoena* echolocation signal, was found to be -36.3 dB (Kastelein et al. 2000). Therefore the energy flux density of the echo at the *Phocoena* threshold is equal to

$$EE = 7.5 - 40 \log R - 0.08 \ R \ (\text{dB re 1 pJ/m}^2) \qquad (8)$$

The animals' performance data shown in fig. 3 were used along with eqs. 7 and 8 to estimate the energy flux density necessary for any performance level; the results are shown in fig. 5. The curves indicate that at the bat's 75% correct performance threshold, an echo energy flux density of -35.7 dB was required, and at the porpoise's 50% correct detection threshold, an energy flux density of -40.7 dB was required. This suggests that the *P. phocoena* is slightly more sensitive than *E. fuscus* by ap-

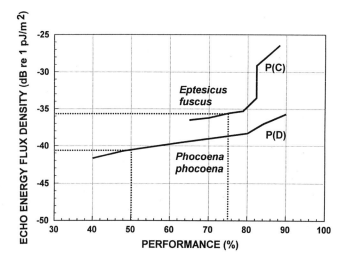

Fig. 5. Estimate of echo energy flux density required by *E. fuscus* and *P. phocoena* in order to achieve a particular performance level

proximately 5 dB. This difference is relatively small and may be due to the difference in experimental methodology or the selection of the source levels used to derive eqs. 7 and 8. For all practical purposes, it would not be too far-fetched to state that the echo detection sensitivity of *P. phocoena* and *E. fuscus* is about the same.

TARGET DETECTION IN A NOISY ENVIRONMENT

Three different types of target detection experiments were performed in order to determine the target detection capabilities of *T. turncatus* in noise. The maximum detection range of two *Tursiops* was determined in Kaneohe Bay by Murchison (1980) using a 2.54 cm diameter solid steel sphere and by Au and Snyder (1980) using a 7.62 cm diameter water-filled sphere. The noise produced by snapping shrimp was high enough to mask the dolphins (Au and Snyder 1980). The 50% detection threshold ranges for the 2.54 cm and 7.62 cm diameter spheres were 73 m and 113 m, respectively. Both these results are consistent if the difference in the target strength of the two spheres is considered.

The target detection capability of *Tursiops* in noise was measured by two other techniques. A target was positioned at a fixed range, and the dolphin's capability of detecting it was measured as a function of the level of a wideband masking noise (Au and Penner 1981; Au, Moore, and Pawloski 1988; and Turl, Penner, and Au 1987). In another experiment an electronic echo generator was used to simulate a phantom target at 20 m with the level of the echo made progressively smaller as the echolocating dolphin performed a detection task in a fixed noise field (Au, Moore, and Pawloski 1988). The results of the three different methods of measuring a dolphin's target detection capability were very similar, after echo energy-to-noise ratio (E_e/N) for the range de-

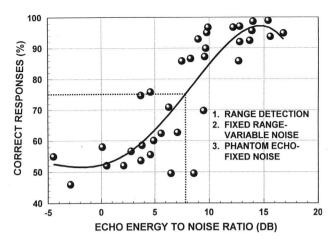

Fig. 6. *T. truncatus* target detection performance as a function of the echo energy-to-noise ratio (from Au 1993)

tection data was determined using the transient form of the sonar equation (Au 1993).

The results of the three different types of experiments are shown in fig. 6, with correct responses plotted against the ratio of the echo energy flux density over noise spectral density (Au 1993). In determining the echo energy, the maximum source level per trial adjusted downward by 2.9 dB (Au and Pawloski 1989) was used. The 75% correct response threshold occurred at approximately 7.8 dB.

Møhl (1986) conducted an echo detection in noise experiment with a Pipistrelle bat (*Pipistrellus pipistrellus*). A phantom target electronic playback system was used, playing back a previously recorded echo after a specific delay whenever the bat emitted a sonar signal. The delay was chosen so that the bat would receive the phantom echo 8 ms after it emitted a signal. The microphone to measure the bat's sonar signal and the playback speaker were located about 25 and 85 cm from the bat's platform. A constant amplitude broadband noise was played back with the phantom echo. The animal's performance was measured as the amplitude of the echo signal was varied. At threshold, the ratio of the total echo over the noise spectrum density was 50 dB. Møhl attributed this high E/N ratio to the effects of clutter caused by echoes produced by the speaker (front face diameter of 15 mm) used to project the echo and noise.

Troest and Møhl (1986) also measured the echo detection in noise capabilities of three serotine bats (*Eptesicus serotinus*), using the same phantom target system of Møhl (1986) and a variable echo amplitude. The delay between the emission of a sonar signal and the reception of an artificial echo was reduced to 3.2 ms in this study. The microphone and speaker were located at a distance of 22 cm and 88 cm, respectively, from the bat's platform. The average echo energy-to-noise ratio at threshold was found to be 36 dB. However, the presence

of clutter from the speaker, which trailed the echo by 2 ms, may have affected the bats' performance.

The echo energy-to-noise ratio of 7.8 dB at threshold for the three *T. truncatus* (fig. 6) is considerably smaller than the 36 and 50 dB measured for the bat experiments at the University of Aarhus, Denmark. However, the results of the bat experiments were contaminated by clutter caused by the playback speakers. Simmons (personal comm., 1999) used phantom echoes to measure the detection performance of *E. fuscus* in noise with minimal extraneous clutter. The 75% correct performance threshold occurred for a detection index d' (see Au 1993, p. 166) of about 5.5, which corresponds to an E/N of 11.8 dB. This E/N is much smaller than that obtained by Møhl (1986) and Troest and Møhl (1986) and only 4 dB higher than the bottlenose dolphin.

Target Range Difference Discrimination

Target range difference discrimination experiments with echolocating bats have been conducted by a number of investigators. Simmons (1973) used four different species of bats: *Eptesicus, Phyllostomus, Pteronotus* sp., and *Rhinolophus*. The bats were required to lie on a starting platform and determine which one of two triangular planar targets was closer to the platform. One of the targets was kept at a fixed range from the bat's platform and the other at a closer variable range. The difference in distance between the nearer and farther targets was decreased from 10 cm down to zero in steps of 5 or 10 mm. The targets were separated by an azimuth of 40°. The results of the experiment with *E. fuscus* are shown in fig. 7a with the fixed target located at a range of either 30 or 60 cm. The bat had the same threshold of 1.3 cm for both fixed target ranges of 30 and 60 cm. Simmons (1973) also used phantom electronic targets to mimic the experiment with real targets and obtained similar results. Denzinger and Schnitzler (1994) performed a similar phantom electronic target experiment using four *E. fuscus* and obtained nearly identical threshold results. Denzinger and Schnitzler also studied the effect of the echo amplitude on the bat's ability to discriminate range differences and found an optimal echo level of −28 dB SPL. The bats' performance dropped off when the signal levels were both lower than and greater than −28 dB.

Simmons (1969) speculated that the bat may have a coherent or matched-filter receiver, since the range discrimination capability of a coherent receiver should not depend on range for high signal-to-noise conditions. Simmons's support of a possible matched-filter model was strengthened by theoretical predictions of the bat's performance based on the envelope of the autocorrelation function of the emitted sonar signal that is depicted in fig. 7b. The theoretical curve matched the bat's performance curve very well.

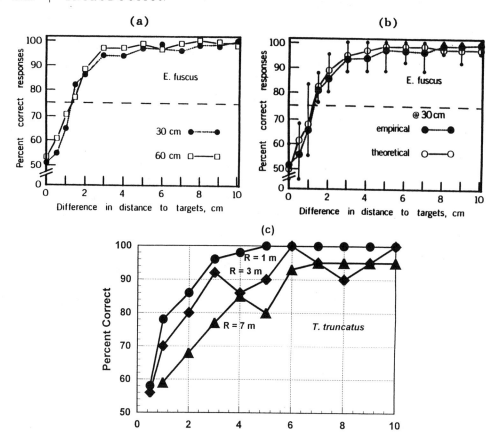

(a) **(b)**

(c)

Fig. 7. (*a*) Average target range difference discrimination performance of *E. fuscus* at absolute ranges of 30 cm (for eight bats) and 60 cm (for three bats); (*b*) average target range difference discrimination performance of eight *Eptesicus* at an absolute range of 30 cm. The "empirical" curve shows the actual performance of the bats, and the "theoretical" curve shows the performance predicted from the envelope of the autocorrelation function of the bat's target-ranging sonar signal. (*c*) The performance of *T. truncatus* as a function of the difference in the relative target range (from Simmons 1973).

The corresponding experiment to determine the range difference discrimination capability of an echolocating dolphin was conducted by Murchison (1980). The dolphin was trained to wear rubber eyecups and station in a chin cup that could swivel from side to side, and echolocate two identical 7.62 cm polyurethane foam spheres separated in azimuth by 40°. The dolphin's task was to determine which of the spheres was closer by touching the paddle on the same side of the center line as the closer target. The animal's performance as a function of the difference in the relative target range, when the closer target was 1, 3, and 7 m from the front of the chin cup, is shown in fig. 7c. The 75% correct response thresholds were at ΔR of 0.9, 1.5, and 3 cm for absolute target ranges of 1, 3, and 7 m. Unlike the bat *E. fuscus,* the dolphin's performance varied as a function of the distance of the closer target.

The performance curves of fig. 7 suggest that the bat and dolphin range difference discrimination capabilities are similar. However, the difference in the speed of sound in air and under water affecting the two-way travel time of the sonar signal should be considered. The bat results of Simmons 1973 for *Eptesicus,* Ayrapet'yant and Konstantinov 1974 for *Rhinolophus,* and Surlykke and Miller 1985 for *Pipistrellus* are plotted in fig. 8, along

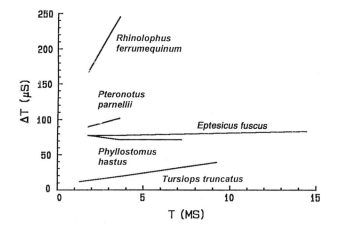

Fig. 8. Graphs of the arrival time difference discrimination threshold in bats and dolphins as a function of the two-way travel time for a sonar signal to leave and return to an animal after reflecting off a target

with the dolphin results of Murchison 1980. The curve for the dolphin indicates that *T. truncatus* can discriminate arrival time differences of echoes as small as about 12 μs. The arrival time difference discrimination capability increases linearly with range or total elapsed travel time to about 40 μs for a travel time of 9.2 ms. *E. fuscus,* on the other hand, can resolve arrival time differences of about 77 μs for absolute travel times between 2.8 and 14.6 ms.

Differential Range Difference Discrimination

HOLE-DEPTH DIFFERENCE AND TWO-WAVE-FRONT EXPERIMENTS WITH BATS

Simmons et al. (1974) performed an experiment to determine the differential range difference discrimination capability of an echolocating bat. The specific task was to consider two echoes, each with two highlights or glints: one with a highlight separation of τ and another with a highlight separation of $\tau + \Delta\tau$. The parameter of interest is $\Delta\tau$, the differential time difference that is directly related to range difference. The stimuli were provided by two Plexiglas plates, each with holes drilled to different depths. Each Plexiglas target measured 7.3 × 7.3 × 2.2 cm thick. Twenty-four holes in a 5 × 5 array with the center hole missing were drilled into each plate. One plate, with holes drilled to a depth of 8.0 mm, was considered the standard or correct target. The other plates, with holes of smaller depths—6.5, 7.0, 7.2, 7.6, and 8.0 mm—were considered the comparison or negative targets. The standard and comparison plates were separated by an angle of 40° and mounted 30 cm away from the bat's starting position. Simmons et al. (1974) trained two *E. fuscus* to echolocate on the standard and a comparison plate and indicate which plate was the standard with the 8.0 mm holes. The performance curves for the bats in the hole-depth discrimination experiment are shown in fig. 9. The 75% correct re-

sponse threshold for bat D was 0.62 mm; that of bat J was 0.88 cm. The differential time difference discrimination can be estimated by considering the reflection process for the standard and comparison targets.

Habersetzer and Vogler (1983) performed a similar hole-depth experiment using *Myotis myotis* and two reference hole depths: 8 mm and 4 mm. They found that the bat's discrimination threshold was about 1 mm for the 8 mm reference depth and about 0.8 mm for the 4 mm reference depth. These values are in general agreement with the values obtained by Simmons et al. (1974).

Schmidt (1988, 1992) took a different approach to studying the time difference discrimination threshold of an echolocating bat. The bat *Megaderma lyra* was trained to discriminate between echoes of two phantom targets consisting of replicas of the bat's echolocation signal. The standard phantom target produced an echo consisting of two replicas of the bat's signal, separated by a time τ_S, and the comparison target produced an echo with a time separation of $\tau_S + \Delta\tau$ between each replica. Schmidt's results, plotted in terms of the calculated difference in the differential distance between the standard and comparison stimuli that $\Delta\tau$ represented, are shown in fig. 9, along with the hole-depth results of Simmons et al. (1974). The *Myotis* in the Schmidt experiment had thresholds of −0.18 and 0.22 mm, which were much better than the *Eptesicus* in the experiment of Simmons et al. (1974). The difference between the results of Schmidt (1988, 1992) and Simmons et al. (1974) may be attributed to a difference in methodology rather than a difference in species—for example, the hole-depth results of Simmons et al. (1974) with *Eptesicus* were comparable to the hole-depth results of Habersetzer and Vogler (1983) with *Myotis*.

Simmons et al. (1974) suggested that the bats were able to perform the hole-depth discrimination by listening for differences in the ripple spectra of the echoes from the standard and comparison plates. Ripples in the frequency spectra of the echoes are the result of having correlated echo highlights or glints formed by the acoustic signal reflecting off the front face of the plates and off the back face, corresponding to the depth of the holes. The time separation between acoustic energy reflecting off the front face and bottom of a hole will be

$$\tau_S = \frac{2d_S}{c_a} \qquad \tau_C = \frac{2(d_S - \Delta d)}{c_a} \qquad (9)$$

where c_a is the speed of sound in air (approximately 344 m/s); d_S is the depth of the standard (8 mm); and Δd is the hole-depth difference in the comparison plate. From eq. 9, the differential time difference discrimination threshold is

$$\Delta\tau = \tau_S - \tau_C = \frac{2\Delta\tau}{c_a} \qquad (10)$$

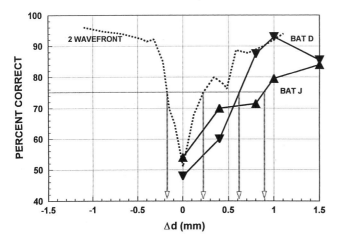

Fig. 9. Differential range difference results for echolocating bats. The dark curves are the hole-depth discrimination performance of two *Eptesicus*, bat D and bat J, as a function of the difference between the hole depths (after Simmons et al. 1974). The lighter curve is the two-wave-front result of Schmidt (1988, 1992).

Let an echo, $e(t)$, be expressed by the sum of two reflected signals such that

$$e(t) = s(t) + as(t - \tau) \tag{11}$$

where τ is the time separation between the echo off the front and back of the hole face, and a is the amount the back of the hole echo is lower in amplitude than the front face echo. If $S(f)$ is the Fourier transform of $s(t)$, then the Fourier transform of eq. 11 can be expressed as (see Au 1993)

$$|\Im[e(t)]| = \sqrt{1 + a^2 + 2a\cos(2\pi f\tau)}|S(f)| \tag{12}$$

The ripple in the spectrum comes from the argument of the cosine term, which will cause the cosine to vary between ± 1 with frequency f. The peaks in the spectrum will occur when $\cos(2\pi f\tau) = 1$, and nulls will occur when $\cos(2\pi f\tau) = -1$.

Examples of the spectra of an echo reflecting off the plates with the standard hole depth of 8 mm and comparison plates with hole depths of 7.6 and 7.2 mm in the experiment of Simmons et al. (1974) are shown in fig. 10. For the plate with the standard hole depth of 8 mm, $\tau = 46.5$ μs, and from eq. 14 a null should occur at 32.3 kHz for $n = 2$. For the comparison plate with a hole depth of 7.6 mm, which was below the threshold of both bats, $\tau = 44.2$ μs and the first null should be at 33.9 kHz. For the comparison plate with a hole depth of 7.2 mm, which was above the threshold of both bats, $\tau = 41.9$ μs, and the first null should be at 35.8 kHz. Therefore, the difference in the frequencies of the first nulls between the standard plate and the plate with the 7.6 mm deep holes was 1.6 kHz; the difference in the standard plate and the

plate with the 7.2 mm deep holes was 3.5 kHz. The bats performed below the 75% threshold with the 7.6 mm holes and above the threshold with the 7.2 mm holes. Therefore, if the bats were using a difference in the ripple pattern of the standard and comparison plates, the frequency difference threshold should be between 1.6 and 3.5 kHz. The two-wave-front stimuli used by Schmidt (1988, 1992) will have similar ripple patterns as the echoes in the experiments of Simmons et al. (1974) and Habersetzer and Vogler (1983). It is impossible to determine from these experiments whether the bats were using time-domain or frequency-domain cues.

CYLINDER WALL THICKNESS DIFFERENCE: DOLPHIN

A dolphin echolocation experiment that is analogous to the hole-depth and two-wave-front experiment with bats is the wall thickness discrimination experiment of Au and Pawloski (1990). In this experiment a dolphin's capability to discriminate wall thickness differences in metallic cylinders by echolocation was determined. The major echo components from the cylinders were a reflection from the front face and a second reflection from the inside back face of the cylinder, which is analogous to the reflection from the front face of the Plexiglas slab and a second reflection from the face at the bottom of the drilled holes in the bat experiment.

On any given trial the dolphin was required to echolocate on a standard target and a comparison target separated by an azimuth of 22° and located 8 m from the dolphin's hoop station. The dolphin was required to touch a paddle located on the same side of the center line as the standard target. The standard target was constructed from aluminum with a 3.81 cm outer diameter (OD), a 6.35 mm wall thickness, and a length of 12.7 cm. Comparison targets with wall thicknesses both thinner and thicker than the standard but of the same OD and length were used. The dolphin's performance as a function of the difference in wall thickness between the standard and comparison targets is shown in fig. 11. The 75% correct response threshold corresponded to a wall thickness difference of 0.23 mm for the thinner targets and 0.27 mm for the thicker targets. Au and Pawloski (1990) examined the targets acoustically and found that the components of the echo from the front surface and the inside back surface of the cylinders were prominent, and the time interval between the first and second reflections was sufficiently different for the standard and comparison targets for the dolphin to discriminate between them. An example of the echoes from the standard cylinder and the comparison cylinder having a wall thickness difference of −0.3 mm are shown in fig. 12. The difference in the highlight separation time between both targets was approximately 0.6 μs. This difference in the time separation between the first and second highlights for the standard and comparison targets also caused the spectrum of the echo from one target to shift

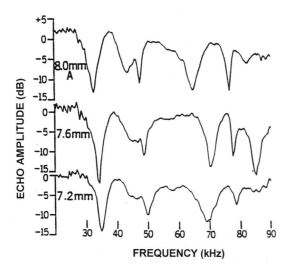

Fig. 10. Spectra of echoes from the plate with a standard hole depth of 8 mm and comparison plates with hole depths of 7.6 and 7.2 mm (from Simmons et al. 1974)

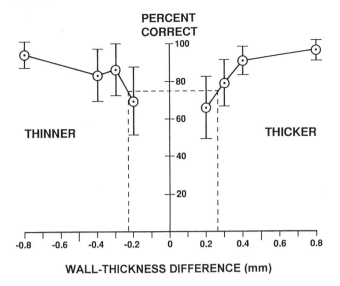

PERCENT CORRECT

THINNER   THICKER

WALL-THICKNESS DIFFERENCE (mm)

Fig. 11. Dolphin wall thickness discrimination performance as a function of wall thickness difference (from Au and Pawloski 1990)

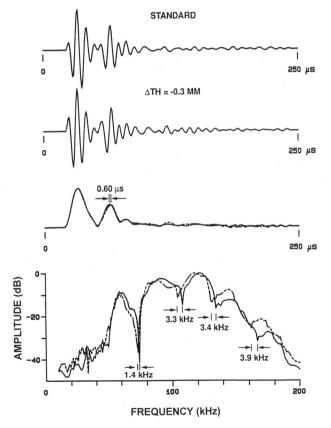

STANDARD

0   250 μS

ΔTH = -0.3 MM

0   250 μS

0.60 μs

0   250 μS

Fig. 12. Echo waveform, waveform envelope, and frequency spectrum for the standard and the comparison target having a wall thickness difference of -0.3 mm. The dashed curves for the envelope and frequency spectrum are for the comparison target (from Au and Pawloski 1990).

in frequency in comparison to the other, as can be seen in the spectral plots.

The amount of shift in spectrum between the standard and the -0.3 mm and -0.2 mm comparison target were 3.2 and 2.2 kHz, respectively. The wall thickness difference thresholds of -.27 mm and 0.23 mm are better than the 0.6 to 1 mm for the bats in the hole-depth experiment, but slightly worse than the 0.18 mm and 0.22 mm for the bats in the two-wave-front experiment. However, any comparison of results should be performed on the basis of time difference discrimination, because of the difference in the sound velocity of air and water. The time difference in the arrival of the first two echo components can be determined by considering the two-way propagation of a signal through the wall of the cylinder and through the water medium to the back wall. This time difference can be expressed as

$$\tau = \frac{2th}{c_{alum}} + \frac{2(O.D. - 2th)}{c_o} \tag{13}$$

where *th* is the wall thickness of the cylinder, c_{alum} is the speed of sound in the aluminum, *O.D.* is the outer diameter, and c_o is the speed of sound of water. The difference in the time difference in the arrival of the back wall echo from one cylinder to another can be determined by applying eq. 13 to each cylinder and subtracting the results to get

$$\Delta\tau = \left| 2\frac{\Delta th}{c_1} - 4\frac{\Delta th}{c_o} \right| \tag{14}$$

where $\Delta th = th_S - th_C$, and the subscript *S* refers to the standard cylinder and subscript *C* to a comparison cylinder. Fig. 13 depicts the bats and dolphin's performance as a function of the difference in the highlight separation time for the standard and comparison targets. As shown in fig. 13, the dolphin's threshold occurred at differential highlight separation times of -0.52 and 0.60 μs, compared with -1.00 and 1.30 μs for the bats in the two-wave-front experiment and 3.58 μs for bat D in the hole-depth experiment.

Target Localization

Lawrence and Simmons (1982) examined the angular resolution capability of two *E. fuscus* in the vertical plane using two arrays of horizontal rods with two rods in each array, positioned along an arc. The rods in one array were separated in vertical angle by 6.5°, with the rods in the other array separated by a greater vertical angle. The bats were required to indicate which array had a smaller angular separation between the rods. The results of their experiment in the vertical plane are shown in fig. 14a. At the 75% correct response threshold, the bats could resolve angular separations of 3-3.5°.

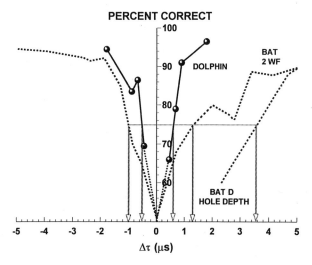

Fig. 13. Echolocation discrimination performance as a function of the differential time difference τ_s between two stimuli, each with two highlights. The highlights of the standard stimulus are separated by a time τ_s; the highlights of the comparison stimulus are separated by $\tau_s + \Delta\tau$. The dolphin results are from Au and Pawloski 1990; bat D results are from Simmons et al. 1974; and the two-wave-front results (2 WF) are from Schmidt 1988, 1992.

Lawrence and Simmons (1982) demonstrated that the tragus of the bat's external ear played a major role in assisting the bats to perform the vertical localization task. When they glued the tragus of each ear forward and down to the adjacent fur, the bats' performance fell significantly.

Simmons et al. (1983) conducted a similar experiment in order to investigate the angular resolution capabilities of bats in the horizontal plane. They once again used two sets of targets, each consisting of an array of vertical rods placed in an arc 44 cm from the bat's observation platform. In one case, an array of two rods separated by a small angle was presented with another array of two rods separated by a different small angle.

Each array was spaced randomly apart by an azimuth of 40–48°. Three *E. fuscus* were first trained to locate the array of rods that were separated by an angle of 6.5° versus the pair of rods separated by 13°. The smaller angle was then progressively increased in 1° increments, and the bats were required to locate the array with the smaller angle between the two rods. Later, an array of five rods was used. The results of horizontal angle discrimination experiment are shown in fig. 14b. The 75% correct response threshold occurred at an angular separation of 1.5°.

Bel'kovich, Borisov, and Gurevch (1970), using a *Delphinus delphis,* performed a similar experiment as that discussed for the bat using two arrays of two cylinders each. However, the cylinders were not positioned along an arc, which probably introduced time-delay cues caused by the targets being at slightly different distances from the dolphin, making the results of 0.028° suspect. Such a cue would not be present if the target were positioned along an arc. The passive sound localization study can provide a hint to the dolphin's sonar angle discrimination. Renaud and Popper (1975) measured a minimum audible angle of 0.7° in the vertical plane and 0.9° in the horizontal plane for *Tursiops* when the stimulus was a broadband click signal with a peak frequency of 64 kHz. Their data suggest that dolphins may have vertical and horizontal angular resolution capabilities slightly better than those of bats. An echolocating dolphin may have an even keener angular resolution because of its narrow transmission beam.

Auditory Integration Time for Echoes

Au, Moore, and Pawloski (1988) measured the integration time of an echolocating *T. truncatus* by using a phantom echo generator that captured each of the dolphin's echolocation signals and returned echoes consisting of a replica of the projected signal, delayed by a time corresponding to a target range of 20 m. They used a staircase

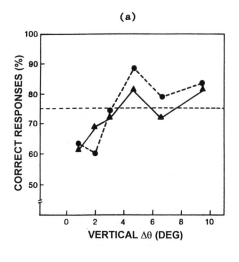

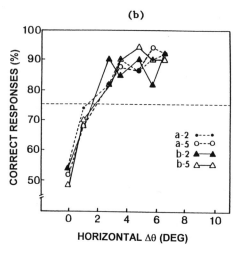

Fig. 14. Results of angular resolution experiments with echolocating *E. fuscus* (*a*) in the vertical plane (from Lawrence and Simmons 1982), and (*b*) in the horizontal plane with arrays of two and five rods (from Simmons et al. 1983)

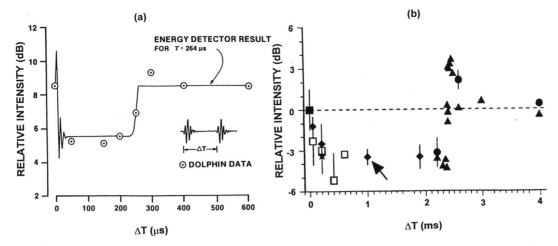

Fig. 15. Relative staircase threshold intensity for a single-click echo and double-click echoes, with varying values for the separation time between clicks for (*a*) *T. truncatus* (after Au et al. 1988), and (*b*) for four *E. fuscus* and one *V. murinus* (after Surlykke and Bojesen 1996)

procedure to measure the dolphin's detection threshold. First the dolphin's threshold was measured when only one click was returned to the dolphin for each click transmitted. Then two replicas of the transmitted signal were returned to the dolphin for each transmitted signal. The time between the double click was a variable, and the animal's threshold was measured as a function of the separation time, Δt, between the two clicks in the phantom echo. The results of their integration time measurement are shown in fig. 15a. The animal's threshold decreased by 3 dB when the echo contained two replicas of the transmitted signal for Δt less than 250 μs. When Δt increased to 300 μs and greater, the animal's detection threshold returned to the single-click threshold, indicating that the integration time for detecting broadband echo signals was between 250 and 300 μs. The curve for a ideal energy detector having an integration time of 264 μs best fitted the dolphin's data. This integration time is considerably shorter than the tens of ms obtained by Johnson (1968) using pulse pure-tone stimuli. The results also suggest that dolphins may process broadband transient signals differently than narrowband pure-tone signals. Bullock and Ridgway (1972) found that different areas of a dolphin's brain are used to process broadband clicks and narrowband pure-tone signals. The 264 μs integration time also corresponded with the backward-masking threshold reported by Vel'min and Dubrovskiy (1975) and Moore et al. (1984) for echolocating *T. truncatus*.

Surlykke and Bojesen (1996) performed a similar integration time measurement with four big brown bats (*E. fuscus*) and one parti-colored bat (*Vespertilio murinus*). The major difference in the experiment of Surlykke and Bojesen (1996) and Au, Moore, and Pawloski (1988) was in the composition of the phantom echo. Sur-

lykke and Bojesen (1996) used broadband clicks with a peak frequency close to 40 kHz as echoes rather than a replica of the transmitted signal, whereas Au, Moore, and Pawloski (1988) used a replica of the transmitter signal. Two target ranges of 51 and 85 cm were simulated.

The results of Surlykke and Bojesen (1996) are shown in fig. 15b. When double clicks were used for the echoes, the bats' relative threshold decreased by 3 dB from their single-click threshold in a similar manner as the dolphin. When the separation time between clicks increased to about 2.41 ms, the threshold began to increase; when the separation time reached 2.5 ms, the threshold was essentially the same as the single-click threshold. Therefore, Surlykke and Bojesen (1996) estimated the bats' integration time to be about 2.4 ms. This is about an order of magnitude lower than integration times measured with pulse pure-tone signals, and it is also an order of magnitude greater than the integration time for the dolphin. The 2.4 ms integration time is consistent with the backward-masking results of Møhl and Surlykke (1989). Surlykke and Bojesen (1996) suggested that perhaps the integration time they measured was somehow adapted to the 2 ms duration signals the bats were typically emitting.

Discussion and Conclusions

The maximum target detection range experiments suggest that bats and dolphins have similar detection sensitivities when the results are analyzed in terms of the amount of energy in the echoes necessary to achieve comparable performance levels. However, the jury is still out in regard to the detection of target echoes in noise. The experiments performed at the University of Aarhus in Denmark with bats (Møhl 1986; Troest and

Møhl 1986) were severely affected by clutter echoes from the loudspeaker projecting the phantom echoes, which caused the bats to have a rather high echo signal-to-noise ratio compared to experiments with dolphins (Au 1993).

When comparing the detection sensitivity of bats and dolphins, extreme care must be taken in choosing the signals to use in calculating the echo energy at the animals' threshold. Dolphins and bats typically vary their signal levels when performing a detection task in an experimental setting. Møhl and Surlykke (1989) found variations in the energy flux density of *E. fuscus* on the order of 6–8 dB. Bottlenose dolphins often emitted click trains with SL varying over 10–15 dB (from the minimum to the maximum), with the average SL usually about 5–6 dB below the largest SL (Au 1993). Au, Lammers, and Aubauer (1999) found that a *P. phocoena* typically had an average SL of about 8–10 dB below the level of the maximum SL in a click train. This question of choosing an appropriate value of SL to use in estimating the energy in an echo for *Tursiops* was addressed by Au, Moore, and Pawloski (1988), who found that the most appropriate SL should be 2.9 dB below the maximum SL in a click train. Whether or not such a rule of thumb can be applied to bats is not readily apparent. The results shown in fig. 8 were determined with a 2.9 dB correction to the energy-to-noise ratio calculated using the maximum SL signal per trial. Perhaps the best method with the least amount of ambiguity in making a comparison between bats and dolphins is to use the signal with the largest amplitude per trial to calculate the maximum echo energy available to the subject on any given trial. Another method is to use all signals with amplitude within 3 dB of the largest signal.

It should not be too surprising that the echo detection sensitivities of bats and dolphins are comparable from both an evolutionary and an ecological perspective. Both species require a sensitive auditory system to detect minute echoes in seeking prey, and it is possible that their auditory systems have evolved to be among the most sensitive in the animal kingdom. Although the prey and environment are vastly different for both species, their auditory systems rely on sensing the energy in echoes when detecting targets (Au, Moore, and Pawloski 1988; Surlykke and Bojesen 1996).

Nor is it very surprising that dolphins have better time resolution capabilities, whether resolving absolute time differences or differential time differences. The echo integration time for *Tursiops* is approximately $\frac{1}{10}$ that of *Eptesicus,* suggesting that the dolphin auditory system responds to acoustic stimuli much more quickly than the bat. The 264 μs integration time suggests that the bottlenose dolphin is sensitive to features in echoes within the integration time window. The bat, on the other hand, would be sensitive to echo features within a larger integration window. We can also compare the range resolution property of the bat and dolphin sonar signal by using the Woodward time resolution constant (Woodward 1953), which is

$$\Delta\tau = \frac{\int_0^\infty |S(f)|^4 \, df}{\left[\int_0^\infty |S(f)| \, df\right]^2} \tag{15}$$

where $S(f)$ is the Fourier transform of the signal. Using eq. 15 for the *Tursiops* signal shown in fig. 2, we get $\Delta\tau = 14\ \mu$s, while $\Delta\tau = 21\ \mu$s represents a typical *Eptesicus* sonar signal having a duration of 2.6 ms. Therefore, if an ideal detector is used to process dolphin and bat signals, better time resolution would be possible with dolphin signals.

An animal's accuracy in judging the difference in the range of two targets will depend on the accuracy and stability of its internal clock and the time resolution capability of its transmitted signal. The results of Simmons (1973), in which *E. fuscus* has essentially the same accuracy in determining the nearer of two targets being independent of the absolute range of the target, suggests that the bat has an extremely accurate and stable internal clock. O'Neill and Suga (1982) found time delay neurons in the auditory cortex of the mustached bat, *Pteronotus parnellii rubigninosus,* that were highly tuned to precise delays between the time of signal emission and echo reception. Although the dolphin signal has a better time resolution property than the bat signal, the internal clock of the dolphin is either not as stable or as accurate as the bat. The range difference discrimination experiment of Murchison (1980) showed that the accuracy of the *T. truncatus* diminished with range.

The smaller time resolution constant associated with the dolphin's echolocation signal is consistent with the dolphin having a much lower differential time difference threshold than the bat. However, this difference between the two species is much smaller if the wall thickness results for *Tursiops* are compared with the two-wave-front results for *Eptesicus*. However, it might have been easier for the bat to obtain time resolution cues from the two-wave-front stimulus than from echoes in the hole-depth experiments of both Simmons et al. (1974) and Habersetzer and Vogler (1983). The 7.3 cm × 7.3 cm plates used for targets by Simmons et al. (1974) subtends an angle of approximately 13.8°, so that the signal path to the center of the target is shorter than the path to the edge of the target by about 0.5 ms. The second echo component consists of the sum of the echoes from the bottom of each hole, each arriving at the bat at slightly different times. This causes the second echo component to be less distinctive compared to the case where all the echoes arrive at exactly the same time.

The phantom echo experiment of Schmidt (1988, 1992) probably presented much cleaner and distinctive stimuli to the bat than the plates with holes drilled in them. The same argument can be applied to the cylinder wall thickness experiment with the dolphin (Au and Pawloski 1990), since there are other echo components than those from the front and inside back wall of the cylinders used. A two-wave-front experiment with a dolphin using a phantom echo generator may well provide smaller differential time differences than the cylinder wall thickness experiment.

Although spatial resolution could not be compared directly between bats and dolphins, the passive sound localization capability of *T. truncatus* was found to be better than the echolocation angular resolution of *E. fuscus*. The head of the dolphin is considerably larger than that of a bat, so that the ears of the dolphin also have a larger separation distance between them. Sound localization in mammals is based on time difference cues at low frequency and intensity difference cues at high frequency. Its larger head gives the dolphin a distinct advantage over the bat. Angular resolution for an echolocating dolphin should be even better than its passive capability, since the narrow transmission beam of the dolphin will also be a positive factor for finer angular resolution.

All of the experiments discussed above dealt with essentially stationary targets. However, many bats forage on insects whose rapidly fluttering wings enable them to fly quickly back and forth. The wing beat of some insects can undergo several cycles within the time duration of a bat echolocation signal, causing the amplitude of the echo to be amplitude modulated. The echoes of a fluttering insect are shown in fig. 16, which demonstrates that not only will the echoes be amplitude modulated within a pulse, but there will be fluctuation in the echoes from pulse to pulse. Therefore, the bat's auditory system may be more attuned to fluctuations in echoes from a moving target rather than fine differences in echo parameters that can be achieved with stationary targets. Dolphins, on the other hand, use much shorter pulses, so that amplitude modulation within a pulse caused by motion of a fish would not be present. Also, because the speed of sound is 4.5 times faster than in air, the echoes from a fish may not be very different over a number of signal emissions, especially when the dolphin is within several meters of the prey.

Unlike dolphins, some bats can sense a Doppler shift in echoes. Tonal CF signals used by rhinolophid bats, hipposiderid bats, and the mormoophid bat can encode velocity information (Schnitzler 1984). For a bat flying toward a stationary target or toward a prey flying away from it, there will be a positive Doppler shift in the echoes, and the bat will lower its emission frequency to compensate for the Doppler shift (Schnitzler 1984). Dolphin echolocation signals are very Doppler tolerant—either their movement or a prey's movement will not cause the signal to experience any Doppler effects.

The echolocation capabilities of bats and dolphins have evolved to be extremely well adapted for survival in their respective environments and prey type. Only two types of echolocation signals seem to be used by different dolphin species, whereas the difference in echolocation signals used by different species of bats can be very great. Bats also have a very plastic system in which

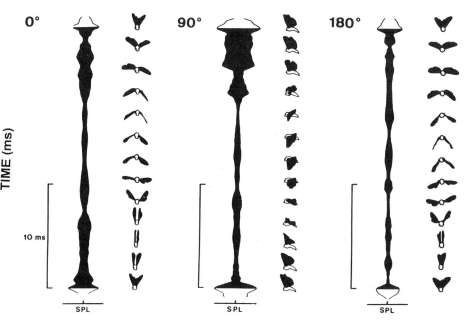

Fig. 16. Amplitude-modulated echo of a flying insect oriented at three different angles (from Schnitzler et al. 1983)

RELATIVE AMPLITUDE

the signal characteristics can change considerably during different phases of a prey pursuit sequence. This does not seem to be the case with dolphins—although very little is known about how dolphins use echolocation when foraging for prey. Nevertheless, their excellent sonar systems enable both bats and dolphins to thrive—except in situations where they are encroached upon by humans.

Literature Cited

ALTES, R. A., and E. L. TITLEBAUM. 1970. Bat signals as optimally Doppler tolerant wave forms. *Journal of the Acoustical Society of America* 48:1014–1020.

AU, W. W. L. 1993. *The sonar of dolphins.* New York: Springer-Verlag.

———. 2000. Echolocation in dolphins. Pp. 364–408 in *Hearing by whales and dolphins,* ed. W. W. L. Au, A. N. Popper, and R. R. Fay. New York: Springer-Verlag.

AU, W. W. L., and D. A. PAWLOSKI. 1989. A comparison of signal detection between an echolocating dolphin and an optimal receiver. *Journal of Comparative Physiology A* 164:451–458.

———. 1990. Cylinder wall thickness discrimination by an echolocating dolphin. *Journal of Comparative Physiology A* 172:41–47.

AU, W. W. L., and K. J. SNYDER. 1980. Long-range target detection in open waters by an echolocating Atlantic bottlenose dolphin (*Tursiops truncatus*). *Journal of the Acoustical Society of America* 68:1077–1084.

AU, W. W. L., M. O. LAMMERS, and R. AUBAUER. 1999. A portable broadband data acquisition system for field studies in bioacoustics. *Marine Mammal Science* 15:526–531.

AU, W. W. L., P. W. B. MOORE, and D. PAWLOSKI. 1986. Echolocation transmitting beam of the Atlantic bottlenose dolphin. *Journal of the Acoustical Society of America* 80:688–691.

———. 1988. Detection of complex echoes in noise by an echolocating dolphin. *Journal of the Acoustical Society of America* 83:662–688.

AU, W. W. L., J. L. PAWLOSKI, P. E. NACHTIGALL, M. BLONZ, and R. C. GISINER. 1995. Echolocation signals and transmission beam pattern of a false killer whale (*Pseudorca crassidens*). *Journal of the Acoustical Society of America* 98:51–59.

AYRAPET'YANT, E., and A. I. KONSTANTINOV. 1974. *Echolocation in nature.* Leningrad: Nauka. (English translation, Arlington: Joint Publication Research Service.)

BEL'KOVICH, V. M., V. I. BORISOV, and V. S. GUREVCH. 1970. Angular resolution by echolocation by *Delphinus delphis.* Pp. 66–67 in the Proceedings of the Twenty-third Science-Technology Conference, Leningrad.

BULLOCK, T. H., and S. H. RIDGWAY. 1972. Evoked potentials in the central auditory systems of alert porpoises to their own and artificial sounds. *Journal of Neurobiology* 3:79–99.

DENZINGER, A., and H.-U. SCHNITZLER. 1994. Echo SPL influences the ranging performance of the big brown bat, *Eptesicus fuscus. Journal of Comparative Physiology A* 175:563–571.

DING, W., B. WÜRSIG, and S. LEATHERWOOD. 2001. Whistles of boto, *Inia geoffrensis,* and tucuxi, *Sotalia fluviatilis. Journal of the Acoustical Society of America* 109:407–411.

HABERSETZER, J., and B. VOGLER. 1983. Discrimination of surface-structured targets by the echolocating bat, *Myotis myotis,* during flight. *Journal of Comparative Physiology A* 152:275–282.

JOHNSON, C. S. 1968. Relation between absolute threshold and duration of tone pulse in the bottlenose porpoise. *Journal of the Acoustical Society of America* 43:757–763.

KASTELEIN, R. A., W. W. L. AU, H. T. RIPPE, and N. M. SCHOONEMAN. 2000. Target detection by an echolocating harbour porpoise (*Phocoena phocoena*). *Journal of the Acoustical Society of America* 49:359–375.

KICK, S. A. 1982. Target detection by the echolocating bat *Eptesicus fuscus. Journal of Comparative Physiology A* 145:431–435.

KICK, S. A., and J. A. SIMMONS. 1984. Automatic gain control in the bat's sonar receiver and the neuroethology of echolocation. *Journal of Neuroscience* 4:2725–2737.

LAWRENCE, B. D., and J. A. SIMMONS. 1982. Measurements of atmospheric attenuation at ultrasonic frequencies and the significance for echolocation by bats. *Journal of the Acoustical Society of America* 71:585–590.

MØHL, B. 1986. Detection by a pipistrelle bat of normal and reversed replica of its sonar pulses. *Acustica* 61:75–82.

MØHL, B., and A. SURLYKKE. 1989. Detection of sonar signals in the presence of pulses of masking noise by the echolocating bat, *Eptesicus fuscus. Journal of Comparative Physiology A* 165:119–124.

MURCHISON, A. E. 1980. Maximum detection range and range resolution in echolocating bottlenose porpoise

(*Tursiops truncatus*). Pp. 43–70 *Animal sonar systems,* ed. R. G. Busnel and J. F. Fish. New York: Plenum Press.

NEUWEILER, G. 1990. Auditory adaptations for prey capture in echolocating bats. *Physiological Reviews* 70:615–641.

POPPER, A. N., and R. R. FAY, eds. 1995. *Hearing by bats.* New York: Springer-Verlag.

PYE, J. D. 1980. Echolocation signals and echoes in air. Pp. 309–353 in *Animal sonar systems,* ed. R. G. Busnel and J. F. Fish. New York: Plenum Press.

RENAUD, D. L., and A. N. POPPER. 1975. Sound localization by the bottlenose porpoise. *Tursiops truncatus. Journal of Experimental Biology* 63:569–585.

SCHMIDT, S. 1988. Evidence for a spectral basis of texture perception in bat sonar. *Nature* 331:617–619.

———. 1992. Perception of structured phantom targets in the echolocating bat, *Megaderma lyra. Journal of the Acoustical Society of America* 91:2203–2223.

SCHNITZLER, H. U., D. MENNE, R. KOBER, and K. HEBLICH. 1983. The acoustical image of fluttering insects in echolocating bats. *Neuroethology and Behavioral Physiology* 236–250.

SIMMONS, J. A. 1969. Depth perception by sonar in the bat *Eptesicus fuscus.* Ph.D. dissertation, Princeton University.

———. 1973. The resolution of target range by echolocating bats. *Journal of the Acoustical Society of America* 54(1):157–173.

SIMMONS, J. A., D. J. HOWELL, and N. SUGA. 1975. Information content of bat sonar echoes. *American Scientist* 63:204–215.

SIMMONS, J. A., S. A. KICK, B. D. LAWRENCE, C. HALE, C. BARD, and B. ESCUDIE. 1983. Acuity of horizontal angle discrimination by the echolocating bat, *Eptesicus fuscus. Journal of Comparative Physiology A* 153:321–330.

SIMMONS, J. A., W. A. LAVENDER, B. A. LAVENDER, C. A. DOROSHOW, S. W. KIEFER, R. LIVINGSTON, A. C. SCALLET, and D. E. CROWLEY. 1974. Target structure and echo spectral discrimination by echolocating bats. *Science* 186:1130–1132.

SURLYKKE, A., and O. BOJESEN. 1996. Integration time for short broad band clicks in echolocating FM-bats (*Eptesicus fuscus*). *Journal of Comparative Physiology A* 178:235–241.

SURLYKKE, A, and L. A. MILLER. 1985. The influence of arctiid moth clicks on bat echolocation: Jamming or warning? *Journal of Comparative Physiology A* 156:831–843.

TROEST, N., and B. MØHL. 1986. The detection of phantom targets in noise by serotine bats: Negative evidence for the coherent receiver. *Journal of Comparative Physiology A* 159:559–567.

TURL, C. W., R. H. PENNER, and W. W. L. AU. 1987. Comparison of target detection capabilities of the beluga and bottlenose dolphin. *Journal of the Acoustical Society of America* 82:1487–1491.

VEL'MIN, V. A., and N. A. DUBROVSKIY, N. A. 1975. On the analysis of pulsed sounds by dolphins. *Dokl. Akademy Nauk. SSSR* 225:470–473.

WOODWARD, P. M. 1953. Probability and information theory with application to radar. New York: Pergamon Press.

PART ONE

Echolocation Signal Production, Feedback, and Control Systems

{ **1** } INTRODUCTION TO PART ONE

Vocal Control and Acoustically Guided Behavior in Bats

Gerd Schuller and *Cynthia F. Moss*

Introduction

Echolocating animals transmit sonar signals and use information contained in the returning echoes to determine the position, size, and other features of objects. This active system allows echolocating bats and dolphins to forage, avoid obstacles, and orient in the absence of light. Since echolocating animals probe the environment with the acoustic signals that drive behavior, detailed studies of sound production and feedback control are central to understanding the process of echolocation. This chapter reviews studies of sonar signal production and feedback control in echolocating bats.

Acoustically Guided Behavior in Echolocating Bats

Field and laboratory studies of foraging behavior in bats reveal a host of adaptive motor responses to dynamic acoustic input from sonar echoes (e.g., Griffin 1958). These adaptive behaviors include changes in the aim of the bat's head and external ears, the direction of the flight path, and the features of the sonar signals. The aim accuracy of the bat's head is approximately 1–2°, and adjustments in the flight path are both rapid and agile (Webster 1963a, 1963b; Wilson and Moss, chapter 3, this volume). As a bat flies toward an insect target, the features of the sonar vocalizations change (see fig. 1.1), contributing to a complex set of adaptive behaviors in response to dynamic acoustic information (Griffin 1958).

Sonar signal production determines the acoustic energy that drives the animal's behavior. That is, the timing, frequency content, and duration of sonar signals used to ensonify the environment directly influence the information available to the bat's acoustic imaging system. In turn, the bat's perception of the environment that builds upon information contained in sonar echoes influences the bat's motor behaviors, including vocal control. By detailing the features of the bat's vocal behavior that enter into this audiomotor feedback system, we develop a better understanding of the underlying neural and perceptual processes.

Adaptive vocal behavior exhibited by bats falls broadly into two categories: velocity-dependent adjustments in sound frequency and range-dependent adjustments in sound duration, bandwidth, and repetition rate. Adjustments in sound frequency are most salient in species of CF-FM bats, and adjustments in sound duration and repetition rate are most salient in FM bats; however, CF-FM bats also show changes in temporal patterning with closing target distance, and FM bats show changes in the bandwidth of their signals during target approach (Schnitzler and Henson 1980). Details of adaptive vocal behaviors are discussed below.

CF-FM BATS

As a bat flies toward a target, its relative velocity introduces Doppler shifts in the returning echoes. Some species of CF-FM bats lower the frequency of their sonar vocalizations to compensate for echo Doppler shifts, receiving echoes at a relatively constant reference frequency (see fig. 1.2; Schnitzler 1968; Schuller, Beuter, and Rübsamen 1975). The CF-FM bat's Doppler-shift compensation thus serves to cancel echo frequency shifts due to its own flight velocity and isolates Doppler shifts in echoes that come from fluttering insect prey. Laboratory experiments demonstrated that CF-FM bats do not

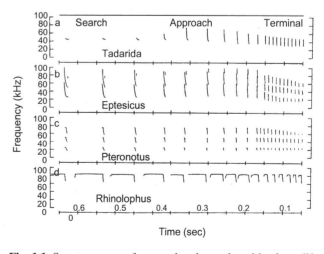

Fig. 1.1. Spectrograms of sonar signals produced by four different bat species as they advance from the search to approach and finally to the terminal phase of insect pursuit. Note that signal duration decreases and repetition rate increases as the bat gets closer to contact with the insect prey. (Adapted from Simmons, Fenton, and O'Farrell 1979.)

A.

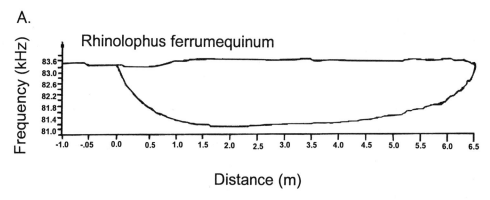

Distance (m)

B.

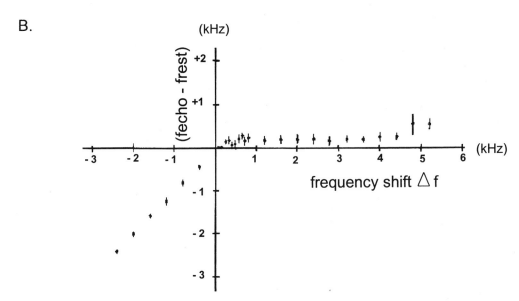

Fig. 1.2. (*A*) *Stabilization of the echo frequency within the acoustic fovea by a feedback system:* A horseshoe bat, emitting 83.3 kHz, flies from a starting point (0.0 on abscissa) to a landing post (6.5 m distance). The emitted frequency before takeoff is 83.3 kHz (ordinate). As soon as the bat starts flying, the emitted frequency (lower trace) is lowered in such a way that the echo frequency (upper trace) is constant within the foveal frequency. (After Schnitzler 1968.)

(*B*) *Capacity of the feedback system:* A horseshoe bat receives artificially frequency shifted echoes (*f*, abscissa). The bat reacts by lowering the emitted sound frequency so that the echo frequency is kept constant and close to the frequency emitted in nonflying situations ($f_{echo} - f_{rest}$ ordinate). Frequency shifts of more than 5–6 kHz and shifts below the f_{rest} (negative quadrant) are not compensated for. (From Schuller, Beuter, and Schnitzler 1974; adapted from Neuweiler, Bruns, and Schuller 1980.)

compensate for Doppler shifts that arise from fast movement of single targets (e.g., insect prey) in the bat's environment (e.g., von der Emde and Menne 1989).

Doppler-shift compensation first was discovered in the greater horseshoe bat, *Rhinolophus ferrumequinum* (Schnitzler 1968), a species that uses CF-FM signals up to more than 100 ms in duration. This behavior has been demonstrated also in other species, like *Pteronotus p. parnellii* (Schnitzler 1970) and in hipposiderid bats (Habersetzer, Schuller, and Neuweiler 1984). Accuracy and consistency of compensation performance are different among species.

The echo reference frequency to which bats compensate is species-specific and characteristic for each individual (e.g., approximately 83 kHz in *R. ferrumequinum* (Neuweiler, Bruns, and Schuller 1980). The auditory systems of Doppler-shift compensating bats are highly specialized in the relevant frequency range for fine frequency analysis (e.g., *R. ferrumequinum* can detect frequency shifts as small as 30–60 Hz, Schnitzler and Flieger 1983). Details of auditory specialization for Doppler-shift compensation are described elsewhere in this volume. The co-evolution of echolocation sound production and auditory system in Doppler-compensating CF-FM bats renders these species particularly well suited to use Doppler information to evaluate target movement. These bats can use their ability to analyze very small frequency and amplitude modulations superimposed on the carrier frequency to discriminate between insect species on the basis of flutter rate and spectral cues (von der Emde and Menne 1989).

FM Bats

In FM bats, the most notable changes in sonar vocalizations occur with a reduction in target range. The FM bat increases the signal repetition rate, decreases the signal duration, and modifies the signal bandwidth as it approaches a sonar target (Webster 1963a, 1963b). Similar to the CF-FM bat, the FM bat's echolocation behavior depends on an audio-vocal feedback system, but the most dramatic changes in vocalizations occur with a reduction in target echo delay, rather than with changes in relative velocity. Each frequency in the FM sound provides a marker for arrival time (Moss and Schnitzler 1989). Bats that use broadband FM signals are well suited to detect small changes in echo delay, which is the bat's cue for target distance (Simmons 1973). Indeed, one FM bat species, *Eptesicus fuscus,* detects differences in echo arrival time smaller than 0.5 μs, corresponding to differences in target range less than 0.1 mm (Moss and Schnitzler 1989; Simmons, Moss, and Ferragamo 1990).

Vocalization in the Developing Bat

The development of vocal signals has been studied in bats of several different families, including Vespertilionidae (e.g., *Antrozous pallidus* by Brown, Grinnell, and Harrison 1978; *E. fuscus* and *Myotis lucifugus* by Gould 1971; *Nycticeius humeralis* by Scherrer and Wilkinson 1993; *Pipistrellus pipistrellus* by Jones, Hughes, and Rayner 1991; Mollossidae (*Tadarida brasiliensis*) by Gelfand and McCracken 1986); Noctilionidae (*Noctilio albiventris*) by Brown, Brown, and Grinnell 1983; Phyllostomidae (*Phyllostomus discolor*) by Esser and Schmidt 1989; Rhinolophidae (*R. ferrumequinum*) by Konstantinov and Makarov 1987; *R. ferrumequinum* Nippon by Matsumara 1979; *Rhinolophus rouxi* by Rübsamen 1987; and Hipposideridae (*Hipposideros speoris*) by Habersetzer and Marimuthu 1986. In the early postnatal period, the vocal repertoire of an infant bat differs from that of an adult (Brown and Grinnell 1980). During postnatal development, a bat's vocalizations rise in frequency, decrease in duration, and become increasingly stereotyped (e.g., Gould 1971; Konstantinov and Makarov 1987; Matsumura 1979; Brown, Brown, and Grinnell 1983; Habersetzer and Marimuthu 1986; Rübsamen 1987; Moss 1988; Jones, Hughes, and Rayner 1991). See, for example, fig. 1.3.

The ontogenetic changes in frequency, bandwidth, sweep rate, and duration of the sounds produced by bats

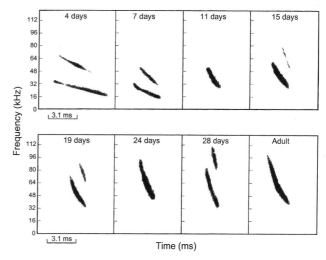

Fig. 1.3. Spectrograms of FM sounds produced by *M. lucifugus* ranging in age from 4 to 28 days. Data for an adult are shown for comparison. (Adapted from Moss et al. 1997.)

can reflect maturation of the larynx, its muscles and its innervation (Gould 1975). Maturation of the respiratory organs and their coordination with vocal organ musculature, along with the development of central motor circuits controlling vocalizations, also are important to the ontogeny of sonar signal production in bats; however, the relative contribution of each has not been studied developmentally.

The change in vocalization frequency occurs with an ontogenetic increase in auditory responses to high-frequency sounds (e.g., Konstantinov 1973; Brown, Grinnell, and Harrison 1978; Rübsamen 1987; Rübsamen, Neuweiler, and Marimuthu 1989). Developmental data on auditory function in young bats come from neurophysiological measures, typically pure-tone evoked responses from the inferior colliculus in anesthetized animals; auditory evoked responses from several bat species show that the onset of hearing generally occurs by the second postnatal week (Konstantinov 1973; Brown, Grinnell, and Harrison 1978; Rübsamen 1987). Neurophysiological data indicate that the bat's auditory system develops to process the ultrasound frequencies present in the adult echolocation signal; however, they do not provide direct measures of echo information processing during development.

Research on developing horseshoe bats demonstrates that early deafening (three to five weeks) affects the constant-frequency (CF) component of the echolocation pulses (Rübsamen 1987). The frequency shifted by between +4 kHz and −14 kHz from preoperational conditions, and the intensity of the first and third harmonics increased considerably. The latter is due to a mismatch of vocal production and filter characteristics of the supralaryngeal filter. While disruption of auditory feedback markedly influenced the development of the normal "adult" vocal frequency pattern, the prevention of normal vocalization by cutting the laryngeal nerves had no influence on the normal development of cochlear tuning and the responses of the auditory system. The control of echolocation pulses seems to be under auditory feedback control throughout postnatal development, whereas maturation of the auditory periphery and its frequency tuning appear to be innate processes.

Vocal Control

SOUND PRODUCTION

The peripheral vocal apparatus of the bat is a "basic" mammalian larynx, and the mechanisms for the production of calls are essentially the same as those found in other mammals. The spectral composition and the temporal structure of the bat's vocal signals are determined largely at the level of the larynx. The supralaryngeal transmission pathway has relatively fixed filtering properties that allow only moderate modulations of the spectral content of the vocalizations. Some bat species emit vocalizations through the nostrils, and the morphology of the transmission pathway determines the transmission filter characteristics. Modulation of the emitted calls in nose-leaf vocalizers therefore is more restricted than in open-mouth vocalizers. As described above, the duration of echolocation calls spans from very short (fractions of milliseconds) to comparatively long (several tens of milliseconds). In bats using FM calls that sweep through an octave in only a few milliseconds, the laryngeal musculature controlling sound frequency must deploy its activity in very short periods. In bats using long CF-FM signals, the emitted constant-frequency component is maintained for tens of milliseconds, immediately followed by a rapid modulation of frequency. The spectral composition of these signals requires very fast and precise neural control over the laryngeal musculature controlling frequency. The fundamental glottal pulse frequencies produced by bats generally are high in comparison to other mammals.

In addition, the repetition rate of vocalizations ranges from 5 to almost 200 sounds per second. At slow sound production rates, the bat typically emits a single vocalization for each expiratory cycle. At higher rates, the bat's vocalizations are toggled on and off within milliseconds, resulting in a burst of calls during a single expiration. Concurrent with sonar signal production is the contraction of middle-ear muscles that temporarily influences hearing sensitivity (Jen, Ostwald, and Suga 1978).

The dynamic control of vocal parameters in bats requires a laryngeal apparatus that produces high-frequency signals with very rapid spectral and sound pressure changes in the range of milliseconds. Of the many chiropteran species, the larynx has been studied in detail in only a few (Fischer and Gerken 1961; Griffiths 1983; Denny 1976). For a review of sound-production mechanisms in bats, see Suthers 1988.

It should be emphasized that there is no typical "bat larynx," but that different species show modifications of a basic mammalian plan (see fig. 1.4). The size of the bat larynges relative to body size generally is large compared to other mammals. In most bat species that emit echolocation pulses through the nostrils, the larynx fits into the nasal part of the pharynx, thus separating the trachea and the mouth mechanically (Denny 1976), which allows for feeding and vocalizing in parallel. Generally, the supraglottal tube is short in bats, and the epiglottis is in contact with the palate or within the nasopharynx. This is also the case in some other mammals, but not in humans.

In many bat species, the laryngeal cartilage (thyroid, cricoid, and arytenoid) shows ossification at an early age. Calcification and ossification in the human larynx do not typically occur before the third decade and are considered to be degenerative processes. The ossification of laryngeal cartilage found in bats probably provides rigid scaffolding for spanning the strongly developed intrinsic laryngeal musculature. Ossification is different among species: early ossification appears in Rhinolophidae and Emballonuridae, but it is less common in Phyllostomatidae (Denny 1976); usually it is limited to locations where muscles are attached.

The combination of aerial living and echolocation demands high metabolic consumption, along with high respiratory activity for flight and high laryngeal activity for vocal production. Vocalization occurs principally with an increase of resistance in the respiratory tract. The need to breathe and vocalize at the same time resulted in glottic modifications in bats. Enlargements in the region of the posterior commissure in the glottis permits respiration during adduction of the cords, and thus vocalization takes place during effective expiration (Denny 1976). Whether the large tracheal air sacs in bats, which emit long constant-frequency sonar signals, play some essential role within this context or serve different purposes remains a matter of discussion (Suthers 1988).

INNERVATION

The intrinsic laryngeal musculature is innervated by two laryngeal nerve branches, the recurrent (or inferior) laryngeal nerve and the superior laryngeal nerve. Both have their somata in the nucleus ambiguus (NA), located in the medulla oblongata. The recurrent laryngeal nerve ipsilaterally contacts all muscles except the cricothyroid muscle, for which motor innervation originates exclusively from the external branch of the superior laryngeal nerve. The only bilaterally supplied muscle (through the recurrent laryngeal nerve) is the inter-arytenoid muscle, the relaxation of which leads to adduction of the vocal folds. Bilateral, but not unilateral, denervation of the recurrent laryngeal nerves resulted in suffocation of the horseshoe bat, perhaps due to bilateral innervation of the inter-arytenoid muscle (Schuller and Suga 1976).

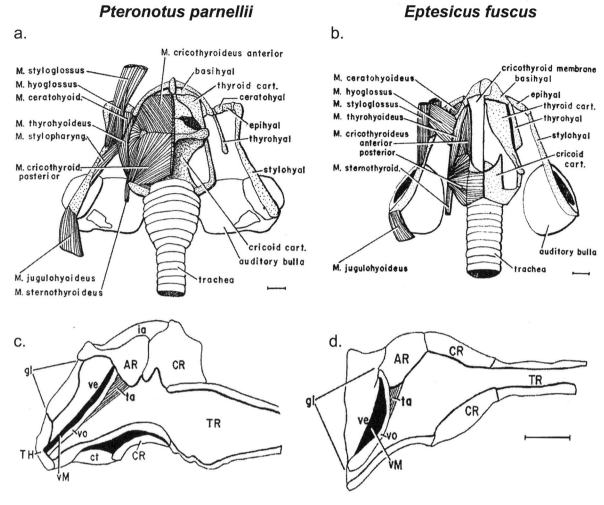

Fig. 1.4. Ventral view of the larynges of (*a*) *Pteronotus parnellii* and (*b*) *Eptesicus fuscus.* Internal lateral view of the larynges of (*c*) *P. parnellii,* and (*d*) *E. fuscus.* (Adapted from Griffiths 1983.)

In addition, the branches of the laryngeal nerves also have vaso- and secretomotor fibers of vegetative origin that carry sensory information from laryngeal tissue to the central nervous system. Both aspects have drawn little attention for study in the bat. The laryngeal nerve supply in mammals is not uniform, but several types of nerve communications between recurrent and superior laryngeal nerve branches have been described. The reported patterns range from no connection between the two nerves to various communicating patterns between the recurrent branch and the interior superior branch. Two species of Chiroptera have been investigated in this respect and showed the pattern of complete separation of the nerve branches (Bowden 1974).

Comparison of Final Common Pathway for Vocalization in Bats with Other Mammals

As noted above, somata of the laryngeal nerve fibers are located in the nucleus ambiguus (NA), a rostrocaudally elongated structure in the medulla oblongata. Nerve transections show the functional involvement of the superior laryngeal nerve in the control of vocal frequency, as well as the importance of the recurrent laryngeal nerve for the temporal control of call production (Rübsamen and Schuller 1981). Tracer experiments revealed a partial topographic separation of motoneuron pools in the NA, with superior laryngeal nerve motoneurons represented more rostrally and recurrent laryngeal nerve motoneurons located more caudally (Schweizer, Rubsamen, and Rühle 1986; Kobler 1983). This organization in bats compares well with that in other mammals.

The control of laryngeal muscles is only one of many preconditions for vocal production, and it must be coordinated with the control of respiratory and pharyngeal muscular activity (see fig. 1.5, Lancaster, Henson, and Keating 1995). Holstege (1989) proposed a final common pathway for vocalizations in mammals, coordinated by activity in the nucleus retroambigualis that lies caudal to the NA. His neuroanatomical investigations in

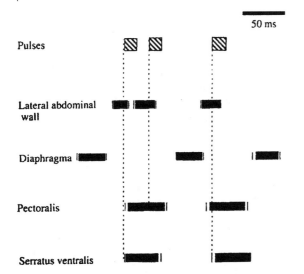

Fig. 1.5. Composite diagram depicting the relative timing characteristics of respiratory and flight muscle activities and biosonar vocalizations during flight in bats. Error bars represent standard error of the mean. (Adapted from Lancaster, Henson, and Keating 1995.)

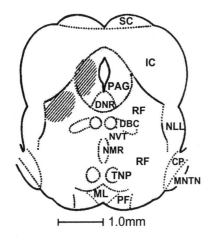

Fig. 1.6. Schematic of cross section of the midbrain of the FM bat, *Myotis austroriparius*. Stippled areas denote regions in which electrical stimulation evokes vocalizations. (Adapted from Suga and Yajima 1988.)

the cat showed that the retroambigual nucleus has connections to respiratory centers and the NA as a vocal motoneuron pool, and it receives input from the lateral periaqueductal gray (PAG) and adjacent tegmental areas (Holstege 1989). In the rat, cat, and monkey, the PAG is considered the most important relay station for the control of vocalizations in the midbrain. Stimulation in the PAG evokes vocalizations (Jürgens 1994), while bilateral lesion of this region leads to mutism. Stimulation in the PAG and the adjacent tegmental areas in bats also evokes emission of vocalizations (see fig. 1.6; Suga et al. 1973; Schuller and Radtke-Schuller 1990). However, the electrical current levels required to elicit vocal-

izations from the PAG in bats is higher than several distinct, highly specific low-threshold areas in more lateral regions of the mesencephalon. Furthermore, there are differences in the vocal signals elicited by microstimulation of the PAG and other low-threshold vocal motor structures (see below).

Differences in the stimulation paradigms and in the conditions for eliciting vocal responses at different sites are important and must be considered carefully. In bats, short bursts of electrical pulses lasting up to some tens of ms at vocalization-specific sites (midbrain, not PAG) induce single calls or short (two to five) series of calls. The latency is well defined and short (several tens of ms) and the stimulation-response ratio is clearly one-to-one. In the nonchiropteran mammals, stimulation in the PAG commonly consisted of a series of electrical pulses at a given frequency, lasting for several seconds that provoked prolonged call sequences. Thus, in the latter case, the stimulation appears to induce vocal behavior in the animal in a more general sense and not with a distinct one-to-one relationship of stimulus and vocal response. In general, the indicated threshold currents (peak current applied) for eliciting vocalizations in the PAG and adjacent tegmental regions are higher than in vocalization-specific areas in the bat (by a factor of about three).

Several questions, therefore, arise in the context of electrical stimulation for eliciting vocalizations. Are the mechanisms to evoke vocal responses in the PAG and other vocalization-specific sites in the midbrain functionally different? Are the stimulation sites part of different systems influencing the vocal behavior? Can the pathway originating from the PAG, as determined by limbic influences, be distinguished from a pathway that sets predominantly the physical preconditions of vocal response?

To date, answers to these questions are inconclusive. It seems, however, that a hierarchical organization of the vocalization pathway leading from the PAG to the retroambigual nucleus, as proposed for other mammals, does not adequately describe the complexity of the descending vocalization system at the brainstem level in echolocating bats. The connections from the PAG to the NA have been established in the bat, using retrograde marking after tracer injections (horseradish peroxidase, or HRP) in the NA (Rübsamen and Schweizer 1986). However, the only explicit demonstration of a connection targeting the NA proper was from the cuneiform nucleus with WGA-HRP injection in this nucleus after functional identification via electrical stimulation (Schuller and Radtke-Schuller 1988). The study of bats lacks conclusive demonstrations of vocal connections reaching the NA and more caudal structures immediately adjacent to the NA. To date, there has been no demonstration of a separable retroambigual structure in the bat. It is therefore unclear whether a final common pathway controlling vocalizations in the bat is organized similarly to other mammals. Therefore, tracing of the connections with the NA and adjacent regions using reciprocal tracer methods and concurrent physiological measures of the vocal relevance of these connections is needed.

Control Levels Converging on the Final Common Pathway for Vocalization: Respiratory Control and Vocalization

Vocalizations in bats are linked closely to respiratory activity and other motor behaviors, such as pinna movements, wing beats, or middle-ear muscle contraction (Jen, Ostwald, and Suga 1978; Lancaster 1994; Valentine, Sinha, and Moss in press; Wilson and Moss, chapter 3, this volume). These motor activities are controlled and coordinated precisely. The neural pathways involved in coordinating respiration, vocalization, middle-ear muscle contraction, pinna movements, and wing beats have not been investigated in the bat in detail. Only pieces of information form parts of that puzzle.

Temporal structuring of outgoing vocalizations depends predominantly on the activation of laryngeal muscles in precise temporal coordination with expiration. The recurrent laryngeal nerve, which innervates all laryngeal muscles (except the cricothyroid muscle), is the main laryngeal afferent involved in the timing of vocalizations (Schuller and Suga 1976; Rübsamen and Schuller 1981). The motoneurons of the recurrent laryngeal nerve are found predominantly in the caudal portions of the NA (Schweizer, Rübsamen, and Rühle 1986). There are some indications in the greater horseshoe bat that different portions of the NA are connected differentially to structures involved in the control of respiration; however, the functional connection of subdivisions of the NA has not been studied in detail (Rübsamen and Schweizer 1986). Neurophysiological measurements in the horseshoe bat revealed neurons that are exclusively active during a particular phase of respiration were found, as well as neurons that are exclusively active during vocalization and silent during expiration. Researchers also identified a class of neurons that are active during both respiration and vocalization, but no clear topographical arrangement could be determined (Rübsamen and Betz 1986; Rübsamen and Schweizer 1986).

The intensity of vocalization is not a parameter determined by laryngeal control exclusively, but it demonstrates most impressively how a vocal parameter is determined by expiratory control. The building of subglottic pressure and thus the availability of respiratory volume determines the vocal intensity over long calls or during bursts of multiple vocalizations (e.g., final buzz). It is unknown whether and where there is a unique structure providing the final common pathway for vocalization in the bat, as in the cat or monkey (Holstege 1989). This topic is addressed in following sections.

General Vocal-Motor System

Motoneurons controlling the laryngeal and respiratory musculature are modulated by premotor interneurons in the brainstem that receive input from various sources. Tegmental areas project to the trigeminal, facial, hypoglossal, and ambiguus nuclei, whereas premotor neurons projecting to respiratory motor centers are located in the medulla. In the cat, the rat, and the monkey, the nucleus retroambigualis, an area posterior to the NA, plays a coordinating role as center of the common final pathway, just before the control level of the laryngeal and respiratory motoneurons for the vocalization pathway (Holstege 1989). In the bat, this portion of the reticular formation surrounding the NA caudally has neither been distinguished cytoarchitectonically or its afferent connections established differentially. The afferent and efferent connections to the NA has, however, been studied extensively in two bat species: the horseshoe bat (Rübsamen and Schweizer 1986) and the mustached bat (Kobler 1983).

A number of brainstem areas potentially project to the region of the NA; however, the relevance of these afferent projections to the NA for vocal control has been studied in only a few sites. In the horseshoe bat, retrograde labeling from the NA was found in very lateral portions of the PAG and the adjacent tegmentum, the cuneiform nucleus. Tracer injections centered in the cuneiform nucleus, but possibly encroaching the lateral PAG, yield anterograde labeling in the NA. There have not been tracer injections into the PAG proper in this bat species yet, so the connection from the PAG has not been explicitly shown and could result from tracer uptake by fibers of passage in the NA. In the mustached bat, the connection from the PAG to NA was shown by retrograde labeling concentrated to the medial divisions (adjacent to the aqueduct) and not encroaching to the lateral adjacent tegmental regions (Kobler 1983). There is no reciprocal demonstration of this connection in this bat species. The sparse knowledge on the afferent projections to NA in the two species, besides being partly contradictory, shows that the NA neurons receive potentially modulatory input from tegmental or periaqueductal regions, which are not mediated by a retroambigual area (see fig. 1.7).

In monkeys, the PAG constitutes the most important brain center for vocal coordination. This was demonstrated in numerous reports (Jürgens 1994) that showed that the influence of anterior structures on vocal utterances are obligatorily mediated by the PAG. Electrical stimulation of PAG subdivisions further supports the notion that there exists a local differentiation with respect to different communication call structures.

The involvement of PAG in vocal control has not been demonstrated in the bat as conclusively as in the monkey. Electrical stimulation in the periaqueductal re-

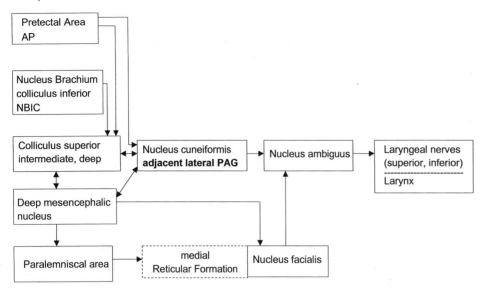

Fig. 1.7. Brainstem areas where vocalization can be elicited by electrical microstimulation, shown in the left column. The most relevant descending connections to structures of the vocal pathway are presented. (Adapted from Schuller 1998.)

gion elicited vocalizations in bats (Suga et al. 1973), although it is not clear how much the laterally adjacent tegmental structures were involved in the vocal activation in these studies.

Electrical stimulation in the lateral regions of the PAG and the cuneiform nucleus in the horseshoe bat elicits vocalizations (Schuller and Radtke-Schuller 1990). However, electrical stimulation in these structures always elicits general arousal or body movements after short periods of current delivery. The latency of the response is long and less consistent compared to stimulation in other vocally responsive regions of the midbrain (e.g., the paralemniscal area). This clearly indicates a functional involvement of these two regions in vocal control. On the other hand, the nonvocal reactions further suggest their involvement in the motivational control of vocalizations.

The PAG constitutes a part of the descending vocal pathway and receives input from limbic structures that presumably play a role in emotional behavior. An important afferent connection originates in the anterior part of the cingulate cortex and was demonstrated in the bat as a functionally important structure for vocalization. Electrical stimulation of this site in the limbic cortex induces vocalizations with relatively long latencies; the locus of stimulation determines the frequency of the onset of the vocalization sequence (Gooler and O'Neill 1987, see fig. 1.8). There is no evidence to date that the emotional motor system in bats (anterior cingulate cortex, PAG, etc.) is organized differently than in other mammals.

Vocal utterances are themselves the result of a complex coordination of many muscles in the laryngeal, pharyngeal, and respiratory musculature (NA, ncl. hypoglossus, ncl. accessorius, spinal nerves to C1–C5, T1–T14). On the other hand, a number of coordinated motor events accompany vocalizations in parallel. The most important examples are movements of the pinnae and/or the noseleaf, middle-ear muscle contractions, and co-

ordination with head and wing movements. The neural substrate of this coordinated control of motor behaviors has been only partially detailed. The premotor and motor nuclei for the separate efferent control circuits (facial nucleus, trigeminal nucleus, motor division) were studied in the general mammalian system and, to a lesser extent, in bats. However, the circuitry for the temporal coordination and integration of activities resulting in complex behavioral patterns is largely unknown.

Higher Level Influences on Vocal Production

Very recently the idea of separate processing pathways in the auditory forebrain has gained more support from imaging studies, as well as from neurophysiological and neuroanatomical investigations (Rauschecker et al. 1997). According to this concept, and analogous to the visual system, structural analysis of sound would be carried out predominantly in the ventral stream, whereas the spatial information on sound source location would be processed in a dorsal stream. The question arises whether and where information on the spatial location and acoustic structure are integrated for perceptions that guide behavioral responses.

There are considerable data on the cortical processing of complex sound patterns in the CF-FM bat, as well as anatomical data on the structure and connectivity of forebrain areas (see chapters 24–32 of this volume). Thus, the connection between the auditory thalamus and cortex clearly is organized in parallel pathways, distinguishing the afferent influx of the dorsal region from the ventral regions of the auditory cortex (Radtke-Schuller, chapter 30, this volume). The physiological responses clearly are different and in many respects specialized in dorsal fields of the bat cortex, in comparison to the more temporal (primary) areas.

The extralemniscal central acoustic tract contributes additional input to frontal regions of the forebrain,

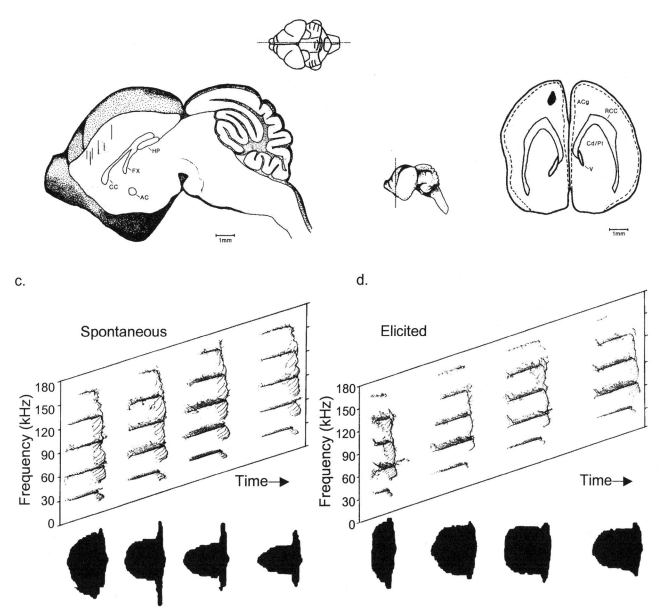

a.

b.

c.

Spontaneous

Frequency (kHz)
180
150
120
90
60
30
0

Time→

d.

Elicited

Frequency (kHz)
180
150
120
90
60
30
0

Time→

Fig. 1.8. Sites in the mustached bat forebrain from which biosonar vocalizations were elicited by electrical microstimulation. The horizontal and lateral views of the brain (insets) indicate the levels of the respective sections. (*a*) The penetrations from which electrical stimulation elicited vocalizations are shown in this sagittal section by vertical lines located anterior and dorsal to the corpus callosum. (*b*) A transverse section taken from a different animal shows an HRP deposit (dark spot) midway along the rostrocaudal extent, which indicates the mediolateral position of the stimulation sites. (*AC,* anterior commissure; *Acg,* anterior cingulate cortex; *CC,* corpus callosum; *Cd/Pt,* caudate/putamen; *FX,* fornix; *HP,* hippocampus; *RCC,* radiations of the corpus callosum; *V,* lateral ventricle.) (*c* and *d*) Spectrograms (*top*) and associated time waveforms (*bottom*) of spontaneous and electrically elicited vocal pulses. The relative amplitude of harmonics is represented by the peaks of the fast Fourier transforms (FFTs) in the waterfall spectrographic display. Each FFT contains 25% new information. These signals show the characteristics common to biosonar vocalizations emitted spontaneously or elicited by microstimulation. The vocalizations typically begin with a short spectral broadening, followed by a long CF component and a terminal FM sweep. The fifth harmonic is down by 14 dB due to attenuation in the recording instrumentation. (Adapted from Gooler and O'Neill 1987.)

which have connections to the acoustic areas (Esser et al. 1997; Casseday et al. 1989). A considerable investigation should be carried out to study the details of physiological, anatomical, and connectional properties of this frontal cortical region. The idea that this cortical region represents a level of integration of the parallel auditory streams and is at the same time involved in audiomotor feedback control is compelling.

Audiomotor Systems in the Echolocating Bat

In this section the central neural mechanisms that support vocal control in echolocating bats are considered. The focus is on five brain regions: the central nucleus of the acoustic tract (NCAT), the superior colliculus (SC), the paralemniscal area (PLA), the rostral pole of the inferior colliculus (ICrp), and the pontine gray (PG). All regions show at least one of the following features: vocalization can be elicited by electrical or pharmacological stimulation; and the region has efferent or afferent connections with premotor structures involved in vocal control. The main characteristics of the neural response characteristics in these regions and their possible relationship to vocal control are described below.

ACTIVATION/DEACTIVATION STUDIES

Brain structures involved in the control of motor responses or behavioral patterns can be determined by showing that their activation is necessary and/or sufficient for eliciting the behavior or that their damage or absence suppresses or disturbs relevant motor response. Among a variety of methods, electrical or pharmacological microstimulation and focal lesioning have proven to be valuable tools in this respect. However, the drawbacks of these methods (estimation of effective stimulation area, secondary effects, etc.) must be taken into account when interpreting the data. The conditions to evoke behaviors—for example, vocal responses by electrical stimulation—are not uniform in all relevant brain regions and could point to differences in the functional involvement of the specific areas.

With regard to electrically elicited vocalizations, three different patterns have been distinguished (Radtke-Schuller and Schuller 1990). In the first response pattern, the stimulation triggers sonar vocalizations at low-threshold currents (less than 20 μA) without any concurrent limb or body movements. The evoked echolocation calls are indistinguishable from spontaneously emitted calls. Stimulus and vocalizations show mostly a one-to-one relationship at a relatively constant response latency. In the second response pattern, the electrical stimulus influences the pattern of the vocalizations. Stimulation can cause distortions in the temporal or, in rare cases, in the spectral composition of the calls. In the third response pattern, vocalizations are not elicited in a one-to-one relationship to the stimulus at a distinct latency, but start after prolonged repetition of electri-

cal stimuli. The evoked vocal responses are not synchronized to the stimulation rhythm and persist for some time even after the electrical stimulation has been switched off.

The first stimulus response pattern ("trigger type") has been encountered in five areas in the horseshoe bat's brain (Schuller and Radtke-Schuller 1990), presumably premotor parts of the descending vocal motor system. These areas are (1) the paralemniscal tegmental area (PLA) located rostrally and medially to the dorsal nucleus of the lateral lemniscus (Pillat and Schuller 1998); (2) the dorsolateral parts of the mesencephalic reticular formation, corresponding to the deep mesencephalic nucleus in the rat; and (3) the intermediate and deep layers of the SC (also *Eptesicus fuscus;* see Valentine and Moss 1998), which are enormously hypertrophied in the bat relative to the visual input layers (*Eptesicus fuscus,* Covey, Hall, and Kobler 1987; *Rhinolophus rouxi,* Reimer 1989). Two further areas showing this response to electrical stimulation are the nucleus of the brachium of the inferior colliculus (NBIC) and the pretectal area (AP) at the transition between the SC and medial geniculate body (Schuller and Sripathi 1999). The shortest latencies, around 25 ms, were found in the PLA; whereas the latencies in the other regions could be longer than 100 ms.

In addition, microstimulation in these brain regions generally evokes coordinated pinna movements, another orienting behavior tightly coupled to echolocation (Metzner 1996; Schuller 1998; Valentine and Moss 1998; Valentine, Sinha, and Moss in press). In all these areas, it is sufficient to stimulate unilaterally to elicit calls. However, there is no evidence that microstimulation parameters (intensity, rate, or timing) or the locus of stimulation in the PLA or SC have a direct influence on the features of the bat's sonar vocalizations. This suggests that activity in these brain regions does not modulate the final motor pattern for echolocation sound production. These areas seem to gate vocal emissions under specific conditions. It is still unclear under which conditions the activity of these areas is necessary for the emission of echolocation calls.

Lesioning studies, however, raise questions about the importance of the PLA in Doppler-shift compensation (DSC) behavior. Bilateral lesions of the PLA of the CF-FM bat, *Rhinolophus rouxi,* do not disrupt DSC behavior (Pillat and Schuller 1998). These findings suggest that PLA is not the sensorimotor interface for echolocation in bats. Manipulation of GABAergic activity in the PLA, however, affects vocal parameters (Schuller and Sripathi 1999). These results provide evidence for modulatory influences of the PLA on vocal control. It appears that this brain region could play a role in a larger circuit for audio-vocal feedback control but is not essential to the modulation of echo-dependent vocal behavior.

In posterior parts of the pretectal area, at the transition between the SC and medial geniculate body, vocal responses are elicited at low-threshold currents with electrical stimulation. The region receives prominent afferent input from the auditory pathway—that is, the dorsal field of the auditory cortex, the inferior colliculus (central and rostral pole nucleus), and the nucleus of the central acoustic tract. Input from nonauditory structures originates in the nucleus ruber, the deep mesencephalic nucleus, and the lateral nuclei of the cerebellum. Efferent connections project back to thalamic targets (zona incerta and ncl. reticularis thalami), to the nucleus ruber, the cuneiform nucleus, and distinct areas of the pontine gray. The connectivity pattern of this region suggests an important functional role of the pretectal area for acoustically guided behavior in bats. However, such function must be demonstrated by showing direct effects of manipulations of this region on vocal characteristics.

Interestingly, the PAG, an area commonly associated with vocal behavior in mammals, and the cuneiform nucleus (CUN) do not appear to be directly involved in the control of sonar vocalization parameters in the horseshoe bat (Schuller and Radtke-Schuller 1990). The cuneiform nucleus is located lateroventrally to the PAG and medioventrally to the colliculi.

While microstimulation in the lateral parts of the PAG and in the cuneiform nucleus elicits vocalizations in the bat, the vocalizations obtained at low stimulation currents (below 20 μA) are normal echolocation calls. However, the one-to-one relationship between stimulation and calls is absent. Vocalizations also persist for some time after stimulation ceases. The vocal responses are always accompanied by arousal of the animal, which increases further with persisting stimulation.

Given its anatomical connections, the cuneiform nucleus is probably one of the relays mediating vocal control information from higher brain levels to the level of the laryngeal motoneurons. It has direct descending access to the laryngeal motor nucleus in the medulla, and it receives input from four of the five regions in which vocalizations could be elicited electrically (but not influenced in their spectral or temporal characteristics by the stimulation parameters or location within the structure). The arousal evoked by electrical stimulation of CUN suggests that this nucleus may modulate affective components of vocalizations. The fact that the structures projecting to the CUN do not directly influence spectrotemporal parameters of vocalizations upon stimulation offers further support to the notion that this nucleus might not be directly involved in vocal parameter control.

The pathway involving the cuneiform nucleus could be important for gating the production of vocalizations under distinct behavioral situations, but it does not represent a direct vocal control interface. This suggests that the PAG and CUN could be involved in activating vocal production in particular emotional states, rather than

in shaping the signals used for echolocation. In this respect, the cuneiform nucleus and adjacent periaqueductal regions in the bat compare functionally to the periaqueductal gray in primates, which is considered to be an important relay station of the descending vocalization system (Jürgens 1994).

Brainstem areas that have immediate influence on the spectral parameters of vocalization are not well defined in the bat. Loci, where electrical stimulation led to distortion of vocal parameters and thus influenced the vocal output directly, were found mostly in the vicinity of vocal areas in the lateral tegmental area and in the lateral pontine regions (Schuller and Radtke-Schuller 1990). Spectral and temporal distortion of calls can be due either to direct influence on premotor or laryngeal motor neuron pools, on descending fibers or to a temporal mismatch of respiratory and vocal control resulting from the electrical stimulation. In the latter case, the electrically initiated laryngeal response for vocalization can fall into a period of inhalation instead of exhalation, and the lack of exhalation volume can distort the calls.

The lateral pontine area yielding distorted calls and arousal is the nucleus of the central acoustic tract (NCAT), with neurons tuned to the constant-frequency component of the bat's echolocation signal. The NCAT projects to putative audiomotor control structures (see fig. 1.9) and could be a strong candidate for audio-vocal functions (Casseday et al. 1989, Behrend and Schuller, chapter 2, this volume).

In the nucleus of the brachium of the inferior colliculus (NBIC), located in a rostrolateral position to the in-

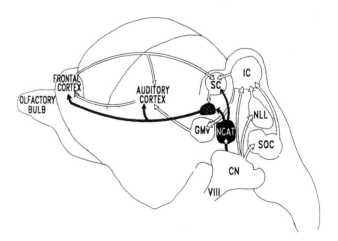

Fig. 1.9. Schematic diagram of the projections from the nucleus of the acoustic tract (NCAT) to the midbrain superior colliculus (SC), supragenicular nucleus (SG), auditory cortex, and frontal cortex. Also shown are projections from the cochlear nucleus (CN) to the superior olivary complex (SOC), nuclei of the lateral leminiscus (NLL) and to the inferior colliculus (IC), which also projects to the SC and the medial geniculate body (GMv). The schematic also shows descending projections from the frontal cortex to the auditory cortex and SC. (Adapted from Casseday et al. 1989.)

ferior colliculus, is another brain area in which vocal responses can be evoked electrically at low stimulation current. Lesioning of this structure has marked influences on Doppler-shift compensation behavior, even though the resting frequency is unimpaired (Schuller and Sripathi 1999). Connectivity and physiological response properties of this brain region are not available in the bat; and the afferent connections of this area have yet to be studied in detail.

Microstimulation in cortical brain regions also elicits sonar vocalizations without distortion of the spectrotemporal signal patterns. In the anterior cingulate cortex, the locus of microstimulation influences the features of the sonar vocalizations (Gooler and O'Neill 1987), in that the frequency of electrically elicited calls depends on the locus of cortical stimulation. In other mammals including primates, the anterior cingulate cortex is an important forebrain structure controlling vocal utterance (Jürgens and Pratt 1979; see also fig. 1.8).

Auditory Responses in Putative Prevocal Areas

Information carried by echoes guides vocal behavior of the echolocating bat. As described above, the CF-FM bat adjusts the frequency of sonar emissions to compensate for echo Doppler shifts introduced by flight velocity, achieving a relatively constant echo return at the reference frequency to which it is maximally sensitive (Schnitzler 1968).

The FM bat adjusts the repetition rate, duration, and bandwidth of sonar signals with changing target distance, using echo delay information to avoid overlap of outgoing sounds and returning echoes (Schnitzler and Kalko 1999; see also fig. 1.1).

The CF-FM bat tolerates overlap of outgoing sounds and returning echoes, because the sonar vocalizations and returning echoes stimulate separate frequency channels in the bat's auditory system. While the Doppler-shifted echoes can differ from the sonar vocalizations by less than 0.5%, such frequency changes in the signals are perceptually distinct to the CF-FM bat, due to specializations in its auditory system (Neuweiler, Bruns, and Schuller 1980). A detailed review of auditory information processing in the bat sonar receiver is presented by Suga et al., chapter 24, this volume. Herein, the auditory input to putative prevocal areas is considered, with particular emphasis on the auditory coding of information needed to guide vocal control of echolocation signal production.

Doppler-Shift Compensation

CF-FM bats respond to Doppler shifts by adjusting sonar vocalization frequency to produce a relatively constant echo return Doppler-shift compensation (Schnitzler 1968) that isolates spectral changes in returning echoes

from the wing-beat patterns of insect prey. As noted above, relative target velocity resulting in frequency changes of <0.5% is perceptually salient to some species of CF-FM bats and can trigger subsequent behavior. The neural basis for fine frequency discrimination in CF-FM bats relies on very sharp tuning in the range of a CF-FM bat's echo reference frequency (Q10 dB values up to 400) (Neuweiler, Bruns, and Schuller 1980) and can be traced to mechanical specializations of the cochlea. Doppler-compensating CF-FM bats show in addition an expanded representation of frequencies at and a few kHz above the resting frequency throughout the auditory system (Neuweiler, Bruns, and Schuller 1980). Thus CF-FM bats are very sensitive to small Doppler shifts in sonar echoes, which in turn support Doppler-shift compensation behavior in which the bat adjusts sonar vocalizations to stabilize returning echo frequency (see fig. 1.2).

The Doppler-shift compensation system in the horseshoe bat and in the mustached bat is a model example of an audiomotor feedback system and lends itself to investigating the mechanisms and underlying structural organization of information transfer from the auditory system to the vocal control system. There is evidence that neurons in the paralemniscal tegmental area (PLA) of the Doppler-shift compensating horseshoe bat (*R. rouxi*) are involved in audio-vocal interactions. Metzner (1989, 1993, 1996) found a population of audio-vocal neurons whose responses were influenced by sonar signal production (see fig. 1.10). He located neurons in the rostral region of the PLA that were active before and/or during vocalizations with little or no response to auditory stimuli (vocal neurons). Neurons in more caudal regions were active during or after sonar vocalizations and also to acoustic stimuli (auditory or audio-vocal neurons). Most audio-vocal neurons had best frequencies in the band relevant for DSC and exhibited a large activity change within a small frequency increase (see below, auditory responses in "nonprimary" auditory areas). The concurrent occurrence of vocal, audio-vocal, and auditory neurons in this restricted area suggests a putative role of the PLA in the interaction of auditory and vocal systems. Suga and Yajima (1988) also described audio-vocal responses in the midbrain of the mustached bat.

The nucleus of the central acoustic tract (NCAT) in the horseshoe bat, *R. rouxi,* consists largely of neurons with best frequencies at and in a narrow frequency band above the bat's resting frequency with very narrow tuning (Schuller, Covey, and Casseday 1991). More than 60% of the neural population is devoted to processing sound frequencies in the region of the constant-frequency (CF) portion of the bat's echolocation call. Such overrepresentation of the CF frequencies hints to an involvement of the region in the analysis of spectral fine structure of the echo. On the other hand, these neu-

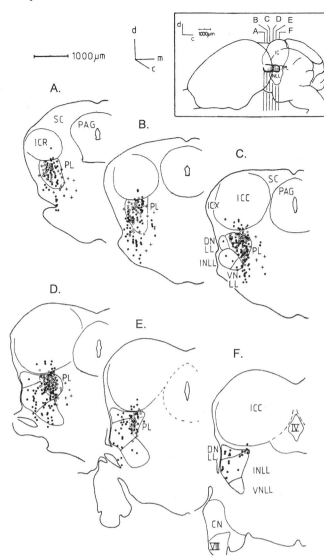

Fig. 1.10. The recording sites of audio-vocal neurons in the paralemniscal tegmental area (PL), superior colliculus (SC), central nucleus of the inferior colliculus (ICC), external nucleus of the inferior colliculus (ICX), dorsal nucleus of the lateral lemniscus (DNLL), inferior nucleus of the lateral leminscus (INLL), ventral nucleus of the lateral lemniscus (VNLL), periaquiductal gray (PAG), and cochlear nucleus (CN). (Adapted from Metzner 1993.)

rons report with high fidelity any shifts of frequencies above the resting frequency and therefore could play an important role in Doppler-shift compensation (Behrend and Schuller 2000). The transmission delay is short, as NCAT neurons receive direct input from the anterioventral cochlear nucleus (AVCN) (Casseday et al. 1989). The NCAT's connection to the pretectal area could rapidly transmit the appropriate acoustic information to accurately coordinate with vocalizations, respiration, and pinna movements (Casseday et al. 1989; Huffman and Henson 1990; Schuller, Covey, and Casseday 1991). Manipulation of this region by stimulation or lesion induces changes of acoustically guided behavior, but further in-

vestigation is needed to develop a complete understanding of the functional role of the NCAT in audiomotor control.

Spatially Guided Behaviors

Central to the control of sonar signal production in the FM bat is target distance information. As described above, the echolocating bat estimates target range from the time delay between sonar emissions and returning echoes. Psychophysical experiments suggest that CF-FM and FM bats both use the FM component of the sonar signal for target distance measurement (Simmons 1973). The neural basis for target ranging has been studied extensively, and researchers hypothesize that echo-delay-tuned neurons play a central role in distance measurement by echolocating bats. Echo-delay-tuned neurons, found in the auditory brainstem (Mittman and Wenstrup 1995; Dear and Suga 1995; Valentine and Moss 1997), thalamus (Olson and Musil 1992), and cortex (O'Neill and Suga 1982; Wong and Shannon 1988; Dear et al. 1993), show very weak responses to single FM sounds. However, these neurons respond vigorously to pairs of FM sounds (a simulated pulse and a weaker echo) separated by a delay. Typically, echo-delay-tuned neurons show facilitated responses to simulated pulse-echo pairs over a delay range of several milliseconds. The delay to which an echo-delay-tuned neuron shows the largest response is referred to as the best delay (BD). In CF-FM bats, BD is organized topographically in the auditory cortex (O'Neill and Suga 1982); but in FM bats, there is no evidence for an orderly representation of BD (Dear et al. 1993; Wong and Shannon 1988). (The neural basis for target ranging in echolocating bats is described in detail in part 2 of this volume.)

In the SC of the FM bat, *Eptesicus fuscus*, about 33% of the auditory neurons sampled showed echo-delay tuning, a response characteristic that could be used by the bat to guide vocal behavior. Best delays for the population of echo-delay-sensitive SC units were largely between 8 and 16 ms, corresponding roughly to 1.4–2.7 m target distance. Responses of echo-delay-tuned neurons showed selectivity to echo arrival time that was often tagged to the azimuth of stimulation. A majority of echo-delay-tuned neurons exhibited facilitation along the delay axis only from a restricted azimuth within the cell's two-dimensional receptive field (Valentine and Moss 1997).

In the horseshoe bat, relatively few neurons in the SC showed responses to vocalizations, which in most cases resembled responses to acoustic stimuli mimicking vocalizations (Reimer 1991). About one third of the neurons active during vocalization could not be driven by comparable acoustic stimuli. Vocally active neurons in the SC did not discharge prior to the emission of the calls but always with a latency of some milliseconds after the start of the vocalization. Latency to vocalization often was shorter than to passive acoustic stimulation. The auditory response to pure tones in the intermediate and deep layer SC neurons was primarily tuned to frequencies at and above the resting frequency of the bat, whereas the remaining sound frequencies were underrepresented. The frequency tuning was relatively narrow with Q10 dB values often above 80, in contrast to broad frequency-tuning measurements in cats. In CF bats narrowband noise was less effective as a stimulus than pure tones.

More than two thirds of the SC neurons in the horseshoe bat showed a binaural response, which was in most cases characterized by contralateral excitation and ipsilateral inhibition of the response. The inhibition was effective for ipsilateral interaural intensity differences greater than 10 dB on the ipsilateral side, so that signals from frontal directions were processed best. This finding is consistent with the importance of target reflections from straight ahead for echolocation. The SC showed a topographical trend for interaural intensity differences with representation of more ipsilateral positions in medial portions and more contralateral positions in lateral parts of the nucleus. The SC seems to be more involved in functional mechanisms for directional encoding and control of orientation in space (e.g., pinna orientation) than in audio-vocal feedback control. The SC could participate in the temporal coordination of vocal utterances and pinna and orientation movements (Valentine and Moss 1998). The rostral pole nucleus of the inferior colliculus (ICrp) is important for audiomotor interaction on the basis of anatomical connections with various nuclei of the ascending auditory system and premotor structures. The auditory responses of ICrp neurons are grouped into roughly two areas: a medial tonotopically organized region and a lateral nontonotopic region with many complex response patterns and strong inhibitory influences. Throughout the ICrp an overrepresentation of the frequencies of the final frequency-modulated part of the bat's echolocation call can be found (Prechtl 1995). This predominance of neural processing in the FM frequency range suggests that audiomotor functions of the ICrp are related to acoustically guided behavior in which the final FM portion is of major importance. Thus, the ICrp could mediate in the temporal coordination of vocalizations and pinna movements, but it is most probably not important for Doppler-shift compensation. Somatosensory projections to the ICrp and the external nucleus of the IC hint to the integrative functions of these marginal IC subnuclei for multisensory information processing.

The pretectal area receives very dense auditory projections from almost all levels of the auditory pathway (NCAT, IC, dorsal field of the dorsal auditory cortex) and exhibits projections to premotor areas. Therefore, it could be an important structure for audiomotor behavior. However, information on auditory responses of pretectal neurons is not available to date.

Conclusions

The echolocating bat uses dynamic acoustic information about the position and velocity of a target to make rapid and fine adjustments in the features of sonar vocalizations. Echo-dependent adjustments of the frequency, duration, and timing of echolocation calls require a sensorimotor interface, where acoustic information carried by sonar echoes is used to guide motor commands for vocal production and control. Investigations into the neural mechanisms of audio-vocal interaction have demonstrated that several nuclei on different brain levels show anatomical connections and functional response characteristics that would suggest their role in audio-vocal feedback for echolocation in bats. Certainly, the view that distinct structures mediating the transformation of Doppler-frequency shifts into a motor signal for lowering the emitted call frequency, as in the seemingly simple Doppler-shift compensation system, must be modified. Instead, we propose an interwoven network of audio-vocal control on many different brain levels. Further research to clarify the contribution of each potential interfacing brain structure and the neuroanatomical details of the circuitry is needed to come closer to an understanding of audio-vocal control mechanisms in echolocating bats. Comprehension of audio-vocal mechanisms in such a sophisticated hearing-oriented system like echolocation will advance our knowledge of audio-vocal interactions in less specialized mammals.

Acknowledgments

The authors thank Kari Bohn and Amy Kryjak for their valuable assistance preparing the figures and copyediting the manuscript. C. F. Moss gratefully acknowledges support from the National Science Foundation for this project (NYI-POWRE supplement). Support was granted to Gerd Schuller by the Deutsche Forschungsgemeinschaft (SCHU390/5–1,2,3).

{ 2 }

New Aspects of Doppler-Shift Compensation

in the Horseshoe Bat, *Rhinolophus rouxi*

Oliver Behrend and *Gerd Schuller*

Introduction

The rufous horseshoe bat (*Rhinolophus rouxi*) has evolved a sophisticated biosonar system for obstacle avoidance and foraging. Its species-specific orientation signal contains up to three harmonics, and the pulse consists of three major elements (Neuweiler et al. 1987): a short upward frequency-modulated (FM) sweep at the beginning of the call is followed by a constant-frequency (CF) part of 20–60 ms; the call ends with a downward FM sweep of about 1–3 ms, with a variable bandwidth up to 16 kHz. The constant-frequency element of the prominent second harmonic (CF_2) varies from specimen to specimen between 70 kHz and 84 kHz. The "personal" CF_2 of an individual horseshoe bat shows only small variations in a resting specimen (± 50 Hz; Schnitzler 1968).

The horseshoe bat's individually tuned vocalization apparatus is matched by a specialized auditory system. A sharply tuned mechanical filter is implemented in the cochlea (Vater, Feng, and Betz 1985). Based on the expanded cochlear representation, central auditory neurons also overrepresent a narrow frequency range of about 1.5 kHz above the CF_2. The neurons feature extremely sharp tuning and very low thresholds ("auditory fovea"; Schuller and Pollak 1979). The individualized foveal filters render the foraging bat's echolocation clutter resistant; periodic movement of insect wings imprint brief frequency and amplitude modulations ("glints") onto the pure-tone carrier of the echoes. These glints indicate potential prey within the echo clutter reflected from static foliage and twigs. Field studies and behavioral experiments show indeed that echolocating horseshoe bats only detect wing-beating insects (Trappe 1982).

However, for a flying bat echoes are additionally Doppler shifted as a whole in consequence of relative velocities. Frequency shifts up to +3 kHz occur for flight speeds ≤6 m/s. Thus, during flight a vocal compensation system is engaged to keep the echoes in the foveal range. The bat compensates frequency shifts in echoes of emitted pulses by lowering the frequency of subsequent calls. This behavior is known as Doppler-shift compensation (DSC; Schnitzler 1968). That is, the flying bat centers the echoes' CF_2 in the acoustic fovea, at a "reference frequency" (RF) about 50–300 Hz above its individual resting frequency (Schuller, Beuter, and Schnitzler 1974).

A remaining question, after 30 years of research on DSC, is whether bats indiscriminately compensate frequency-shifted echoes from any direction or select echoes for DSC (e.g., from straight ahead). The results show that DSC is influenced by binaural analysis.

Materials and Methods

SURGICAL PROCEDURES

Consistently compensating bats were selected for the experiments. Surgery was performed under halothane anesthesia. After incision of the skin along the skull midline, the skull surface was exposed and freed from tissue at its anterior part for the fixation of a small tube. The tube was attached to the skull with cyan acrylic adhesive and dental cement and allowed a reproducible fixation of the animal's head in an animal holder (an exact description of the procedure is given in Schuller, Radtke-Schuller, and Betz 1986). Three animals were unilaterally deafened by removing the middle-ear structures and puncturing the first or second cochlear turn with a cannula (0.4 mm). The animals' sense of balance and their feeding and cleaning behavior were unimpaired after surgery. All three animals' cochleae were examined for proper cannula penetration after the end of the experiments.

PHYSIOLOGICAL VERIFICATION OF DEAFENING

The success of the unilateral deafening procedure was monitored by recording local acoustically evoked potentials at a constant position either in the ipsilateral nucleus of the central acoustic tract or the contralateral ventral nucleus of the lateral lemniscus before and after surgery in two animals. Foveal pure tones were binaurally presented by earphones at sound pressure levels of -10 to $+50$ dB relative to the excitatory threshold of intact animals before and after deafening. The positioning of the recording electrodes was determined by a stereotactic method described in Schuller, Radtke-Schuller, and Betz 1986. Verification of the stereotactic coordinates was achieved after termination of the experiment by reconstruction of WGA-HRP injections at the recording position.

DOPPLER-SHIFT COMPENSATION
INDUCED ON THE PENDULUM

Three bats were tested for Doppler-shift compensation behavior on a pendulum prior to surgical preparation. For DSC testing, bats were placed in a Styrofoam body mold with their head protruding. On the swing, the bat's head was positioned 5 cm behind the pendulum's front edge and echolocation calls were picked up as described below. The pendulum, 2 m in length, swung through an arc of 110°; its lower vertex was 39 cm above the ground. On completion of the first forward swing, the distance between the bat and a wooden wall that served as a sound reflection device was 38 cm. To induce DSC, the swing was displaced to a height of 1 m above the lower vertex and then released. Acceleration in the first forward swing leads to a maximum speed of 4.4 m/s at the midpoint of the swing. The peak velocity caused Doppler shifts of about 2 kHz, depending on the individual's resting frequency.

COMPENSATION OF EXPERIMENTAL FREQUENCY SHIFTS
IN THE PLAYBACK CONFIGURATION

Six animals were tested for compensation of artificial frequency shifts in the playback configuration. Bats were immobilized in the setup with the technique described above. Recorded orientation calls (see below) were shifted electrically in frequency by a double-heterodyning technique (Schuller, Beuter, and Schnitzler 1974). The frequency shifts were varied sinusoidally (modulation frequency 0.1 Hz; modulation depth 0 to +2 kHz). The procedure does induce a delay of approximately 1 ms in the playback but not any other changes in the spectral composition of the sounds (Schuller 1986). To elicit DSC, the frequency-shifted echolocation pulses were presented to the animal via ultrasonic loudspeaker (distance 19 cm) or earphones (Schuller 1997). The playback was attenuated (Adret AP401) relative to the sound pressure level of the emitted calls. Call intensities vary by ±3 dB around 108 dB SPL at a distance of 10 cm from the bat (Pietsch and Schuller 1987). Therefore, the intensities of the playback varied in the same range but had a defined attenuation relative to the emitted call intensity. The determination of thresholds was carried out in 6 to 10 dB steps.

CALL REGISTRATION AND DSC EVALUATION

A ¼″ Brüel & Kjær (B&K) type 4135 microphone was placed 6 cm in front of the bat's head. Echolocation calls were picked up under this condition during the pendulum swing sequences, as well as in the playback setup. The frequency course of echolocation calls was visually monitored using a frequency to voltage converter connected to an oscilloscope, and the constant-frequency part of the echolocation call's second harmonic (CF_2) was recorded with a Watanabe Mark VII thermorecorder. Accuracy of the frequency recording was ± 50 Hz or ± 0.07%. On a second channel of the recording device, call intensities (AC/DC converter Phillips PM 5171 provided a proportional voltage to SPL) were recorded. They varied between 105 and 112 dB SPL. The position of the pendulum or the phase of the sinusoidally modulated playback signal was registered simultaneously on a third channel of the recording device.

Only CF_2 was evaluated and used for the description of DSC performance. As a measure for DSC performance, the compensation of the maximum Doppler shift during the swing or playback cycles was used. The compensation value was calculated out of 4–20 cycles (mean with standard deviation).

Results

PENDULUM-INDUCED DSC OF HORSESHOE BATS
BEFORE AND AFTER UNILATERAL DEAFENING

Three animals were tested for DSC induced on a swing before and after unilateral deafening. At foveal frequencies, postoperational thresholds for excitatory auditory responses in the brainstem increased by at least 50 dB. After surgery, basic call parameters like duration, intensity, and frequency course remained within control levels. However, the animal's compensation of Doppler shifts was severely degraded. In our experiments, maximal DSC was achieved most consistently during the second forward swing of the pendulum, causing a maximum Doppler shift of about +1.8 kHz (depending on the individual CF_2 of an animal). Therefore, the second swing cycle was used to compare the DSC performance in three individual bats before and after unilateral deafening (fig. 2.1). The mean value of maximum compensation amplitude during the second forward swing was calculated. These "monaural" horseshoe bats compensated the maximum Doppler shift induced by top speed during the second forward swing by only 28–48% compared to controls. The capacity for DSC did not recover to control levels during the observed period of 24 hours after surgery. These experiments indicated that maximal DSC requires binaural echo input.

In an echolocating, nonflying horseshoe bat, the resting frequency of call sequences varies by only 0.1% (Schnitzler 1968). After unilateral deafening, two out of three animals showed no changes in the resting frequency. The third animal was unable to maintain resting frequency after unilateral deafening—that is, the postoperational resting frequency deviated by −0.8 kHz to +1.3 kHz from the preoperational one. High deviations in the remaining DSC performance of this animal (fig. 2.1, animal A; SD = 130%; not shown) were observed.

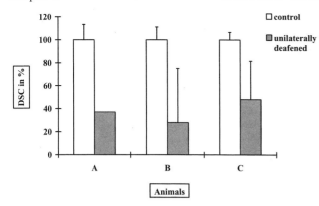

Fig. 2.1. Three bats' (A, B, and C) maximum Doppler-shift compensation (DSC) during the second forward swing in pendulum swing sequences. DSC is compared before and after unilateral deafening. Bars represent means (±SD) of maximum compensation during 10–20 swings. Maximum DSC before unilateral deafening is set to 100% for each specimen. Error of frequency recording is ±50 Hz. DSC performance after unilateral deafening decreased to 37% in animal A. This animal was unable to maintain the resting frequency after surgery. A high standard deviation in DSC performance after surgery (SD = 130%; not shown) is due to this lack of frequency control. The DSC performance after unilateral deafening of animals B and C decreased to 28% and 48%, respectively.

BINAURAL INFLUENCE ON DSC OF HORSESHOE BATS IN THE PLAYBACK CONFIGURATION

To mimic Doppler shifts, frequencies of recorded sequences of echolocation calls were sinusoidally (0.1 Hz) shifted between 0 and +2 kHz and played back via earphones to the animals. Presented with an interaural intensity difference (IID) of zero, the playback intensity threshold to induce compensation of the experimental frequency shifts varied individually between 58 and 95 dB SPL. In the given example, the artificial shift was fully compensated when echoes were played back at binaural levels ≥ 58 dB SPL (fig. 2.2A). A binaural playback presentation at 58 dB SPL was considered as the intensity threshold for full compensation in this specimen.

In sharp contrast to binaural echo presentation, no horseshoe bat compensated for exclusively monaural playbacks even if presented at intensities of 28 dB above the binaural playback intensity threshold for compensation (fig. 2.2B).

Our results indicated that DSC performance depends on the IIDs of the echoes. Playback presentation of experimentally frequency-shifted echolocation calls with stepwise increased IIDs resulted in gradually decreasing DSC performance in four animals. None of the tested bats fully compensated for the experimental frequency shifts when the IID of the playback exceeded 20 dB (fig. 2.3).

Discussion

By Doppler-shift compensation, flying horseshoe bats stabilize the echoes of orientation calls in a narrow frequency band. That is, DSC establishes a constant "echo background" for the echolocating bat, in spite of variations in ground speed (Schnitzler 1973). Since the "background frequencies" match the foveal hearing range, foraging bats benefit from enhanced glint detection (Schnitzler and Ostwald 1983; Schuller 1984). This study demonstrates that, in addition to ground speed, binaural cues also affect DSC. Considerable consequences are illuminated below.

That IIDs are essential for the orientation of flying horseshoe bats was indicated by the results of Flieger and Schnitzler (1973). The authors showed that a monaural echo attenuation of 15–25 dB by unilateral ear plugs was detrimental to obstacle avoidance in *Rhinolophus ferrumequinum*. Plugging both ears, however, neutralized the experimental IID to zero and induced an almost complete recovery of obstacle avoidance ability. Our results led us to speculate whether the impact of IIDs on obstacle avoidance was related to poor DSC performance.

In a free-field setup, horseshoe bats compensating for experimentally shifted echoes with superimposed narrowband noise signals compensate for the upper edge of the noise band. Thus, the highest frequency in the echo is stabilized at the reference frequency if echo information is ambiguous (Neumann and Schuller 1991). That is, the DSC system might preferentially compensate for frontal echoes that display the maximal frequency shift during flight. However, a bat turning its head relative to the flight direction would listen into "silence," regardless of sonar echoes, since the highest echo Doppler shift still is tied to the flight direction. Integrating head-related cues like IIDs could add directional flexibility to the DSC system.

In day roosts, thousands of bats are echolocating simultaneously with high pulse rates and high intensities (Neuweiler et al. 1987). Given the limited variation in individual CF$_2$ frequencies, an IID-sensitive DSC system could reduce the number of calls of other colony members mistakenly compensated.

Generally, but not principally, playback has to be relatively intense to elicit DSC. We consider that the additive experimental frequency shift of the orientation call does not perfectly mimic a Doppler-shifted echo; moreover, facilitating inputs from, for example, the vestibular system that might support DSC during flight or on the swing are missing (Gaioni, Riquimaroux, and Suga 1990). Hence, a low level of attention could produce high individual differences in DSC of playbacks, whereas a high level of attention might facilitate DSC performance of bats on the swing. This interpretation matches a bat's relatively good DSC performance on the swing,

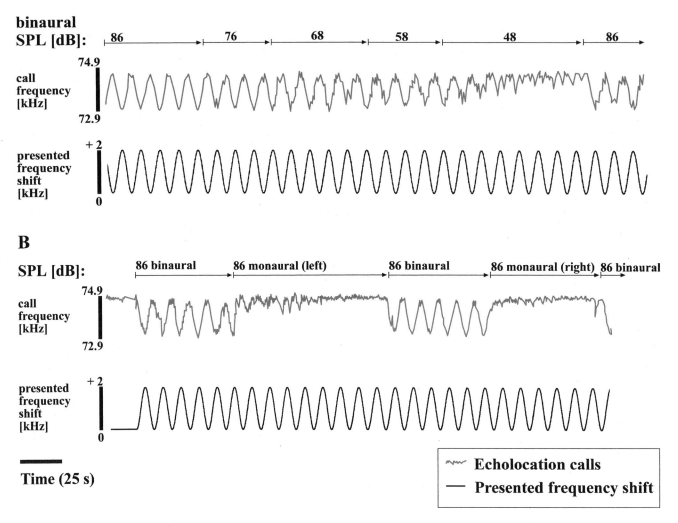

Fig. 2.2. Compensation of experimental frequency shifts in playbacks presented by earphones to a single horseshoe bat. The dotted line indicates call frequencies; the numbers above the arrows indicate playback intensity, which was constant over a period corresponding to the arrow's length. Frequencies of the vocalizations are shifted sinusoidally between 0 and +2 kHz. Modulation frequency is 0.1 Hz (solid lines). *A:* Threshold for compensation of binaural playback with an IID of zero: experimental frequency shifts are completely compensated down to a binaural playback intensity level of 58 dB SPL. *B:* Monaural playback fails to induce compensation, whereas experimental frequency shifts of the playback are fully compensated if presented binaurally. Vocalizations of the horseshoe bats were played back to the animal either monaurally or binaurally 28 dB above the binaural DSC threshold (i.e., 86 dB SPL).

even when unilaterally deafened, compared to poor DSC of playbacks with high IIDs. However, DSC thresholds as low as 50 to 60 dB SPL were observed in the playback configuration for individual animals, indicating that low intensities—comparable to free-field echoes—are sufficient for compensation of experimental frequency shifts.

The main conclusion we draw from the deafening experiments is that monaural input is not sufficient for full DSC. At least within the observed period of 24 hours after unilateral surgery, no recovery occurred. Whether animals are capable of adapting to the monaural acoustic situation to regain full DSC performance during a longer survival time poses an interesting question. So far, it seems that binaural inputs improve the DSC performance dramatically in agreement with the earphone experiments.

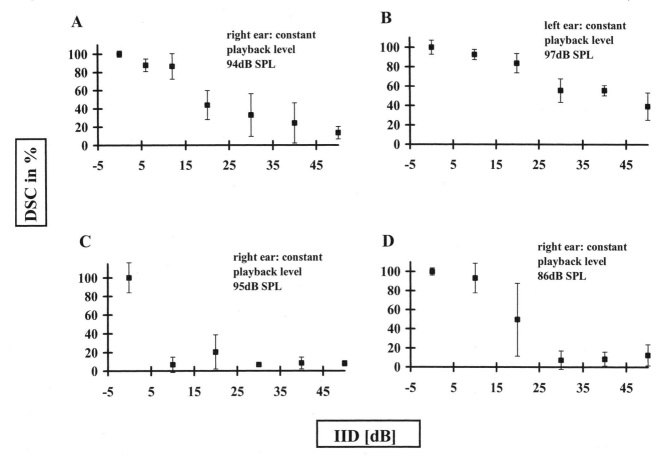

Fig. 2.3. Four bats (A, B, C, and D) show IID sensitivity compensating for experimental frequency shifts in playbacks of echolocation calls by earphones. Data points are means of the maximum compensation of a +2 kHz shift during 4–10 compensation cycles. Error bars show SD. Compensation performance of a +2 kHz shift in the playback presented with an IID = 0 is set to 100% for each specimen. Playback is unilaterally presented with constant SPL. Contralateral playback intensity is stepwise decreased starting with IID = 0. The individual DSC performance decreases with IIDs >12 dB (animal A); >20 dB (animal B); ≥10 dB (animal C); and ≥20 dB (animal D).

Nevertheless, the IID sensitivity of DSC in the earphone playback setup could not be supported by free-field playback of frequency-shifted echoes from different azimuth angles. We consider the possibility that the bats adapt to IIDs in the frequency-shifted playback under free-field conditions. In the free field, the partially variable physical directionality of sound emission and pinnae transfer functions might blur the binaural characteristics of DSC. However, during earphone presentation and after complete unilateral deafening, none of these adaptations work. The search for the neural interface that controls DSC will certainly be influenced by the finding that binaural information has to be processed to maintain proper audio-vocal coupling.

Acknowledgments

This study was supported by DFG/SFB 204 "Gehör" and Graduiertenkolleg "Sensorische Interaktion in biologischen und technischen Systemen" München. The Department of Wildlife Conservation of the Government of Sri Lanka has contributed to this research by providing permission to capture and export specimens from Sri Lanka. Principles of laboratory animal care were followed and experiments were conducted under the regulations of the current version of German Law on Animal Protection (approval 211–2531–17/94, Institute 06, Reg. Oberbayern).

{ 3 }

Sensory-Motor Behavior of Free-Flying FM Bats during Target Capture

Willard W. Wilson and *Cynthia F. Moss*

Introduction

Insectivorous bats face the formidable task of localizing and intercepting small, rapidly flying insect prey while the bat itself is on the wing. This task is successfully accomplished hundreds of times over a single evening in capture sequences lasting less than one second (Griffin, Webster, and Michael 1960). To understand the dynamic sensory processing and motor control underlying this complex task, our research focused on vocal-motor behavior of free-flying FM bats during sonar-guided localization and interception of targets.

In the laboratory, we recorded and analyzed the flight and vocal behavior of bats trained to intercept stationary tethered insects. The stationary-target paradigm has the advantage that the bat has complete control of its behavior during capture and does not need to compensate for the movement of the target. This advantage might reveal fundamental capture strategies used by bats during more complex captures of naturally moving insects.

We studied two species of bat, *Myotis septentrionalis* and *Eptesicus fuscus*. Both species use broadband frequency-modulated (FM) signals for echolocation and hunt insects on the wing. However, *M. septentrionalis* also gleans insects from surfaces (Miller and Treat 1993) and can eliminate sonar vocal emissions during the final portion of gleaning captures (Faure, Fullard, and Dawson 1993). In addition, *M. septentrionalis* is smaller in size (Kalcounis and Brigham 1995) and exhibits greater maneuverability in the lab. These factors suggest that *M. septentrionalis* may show greater flexibility in echolocation behavior than *E. fuscus*. Here, we compare the coordinated vocal-motor behaviors of these two bat species in an insect-capture task and consider whether sonar-target localization and motor planning early in the capture sequence determine behavior in the final stages of the sequence.

Materials and Methods

Five echolocating, big brown bats (*E. fuscus*) and six northern long-eared bats (*M. septentrionalis*) were trained to capture tethered mealworms in a large flight chamber (6.4 × 7.3 × 2.5 m) lined with acoustical foam. Mealworms were suspended by monofilament line within a 5.3 m diameter target area in the center of the room. On each trial, bats were released from a consistent location, facing away from the target. They typically searched the room in a circular or figure-8 path for several seconds before localizing and capturing the target. Once bats achieved a consistent capture rate of nearly 100%, we carried out high-speed audio and video recording of their capture behavior.

Experiments were carried out using long wavelength lighting (Plexiglas no. 2711 and Bogen filter no. 182) to preclude visual orienting by the bat (Hope and Bhatnagar 1979). Two Gen-locked, high-speed video cameras (Kodak MotionCorder, 640 × 240 pixels, typically 240 Hz frame rate and 1/240 s shutter speed) recorded target position, bat flight path, and target-capture behavior. A spatial calibration frame (2.2 × 1.9 × 1.6 m) was placed in the center of the room and filmed by both cameras prior to each session. These images were used in later calculation of the three-dimensional position of the bat, target, and microphones. We placed the target at random locations throughout the target area but recorded only captures that were within the calibrated portion of this space.

Echolocation signals were transduced using two ultrasonic microphones (UltraSound Advice SM1, spectral variation ± 5.5 dB from 10 to 100 kHz) placed within the calibrated area and recorded on the direct channels of a high-speed tape recorder (Racal Store-4, 30 ips). An additional FM channel recorded TTL synchronization pulses corresponding to the start of each video frame and gated to the end of video acquisition. The S/N ratio of the audio recordings was good during captures within the calibrated area, but it deteriorated at greater distances and when the bat was oriented away from the microphone.

Analysis

A commercial motion analysis system (Peak Motus) was used to digitize selected points in each frame for both camera views and to calculate the three-dimensional location of marked points. The accuracy of the system typically was within ± 0.5% (approximately ± 1.5 cm for the largest calibrated area). All audio channels were digitized at ¼ recording speed (National Instruments AT-MIO-16-1, effective sampling rate 240 kHz/channel). Custom software trimmed the digitized audio data to correspond to the first and last frame of video acquisition, and output files were exported to a digital signal-processing program (Sona-PC, Waldman). The time and frequency structure of the first harmonic of the emissions was determined from sonograms (approximately 80 μs time resolution, 470 Hz frequency resolution) and downloaded to a spreadsheet for further analysis. Off-line analyses correlated audio and video data and tracked changes in vocal-motor behavior with the relative position of the sonar target.

Results

Behaviors used to capture stationary tethered targets were similar to interfemoral membrane and wingtip catches used by vespertilionid bats capturing flying insects in the field (Webster and Griffin 1962). Both *E. fuscus* and *M. septentrionalis* captured tethered targets using a coordinated movement of the head, wing, and body to scoop the insect into the uropatagium (a pouch formed by the membrane between the tail and hind legs).

Biosonar behavior is divided into *search, approach,* and *terminal* phases (Griffin, Webster, and Michael 1960), based on the acoustic emission patterns of the bat as it approaches a target. During captures of stationary tethered insects, the vocal behavior of *E. fuscus* exhibited these characteristic changes in spectrum, interpulse interval (IPI), and pulse duration, culminating in a "terminal buzz" with an average minimum IPI of approximately 6.0 ms. The substrate-gleaning behavior used by *M. septentrionalis* lacks a terminal phase (Faure, Fullard, and Dawson 1993). However, while capturing stationary targets in the laboratory, *M. septentrionalis* also exhibited an approach and terminal phase in its vocal emission pattern (average minimum IPI of approximately 5.2 ms). This suggests that the stationary-target capture paradigm taps into the natural behaviors used for aerial hawking captures by both species. One striking difference in the vocal emission patterns of the two species (fig. 3.1) was that the terminal buzz was much longer for *E. fuscus* (mean ± s.e.m. = 229.9 ± 12.1 ms) than for *M. septentrionalis* (134.5 ± 3.3 ms). Note also that the final echolocation pulse in the terminal buzz occurred before target contact, preceding contact on the average by 30 ms for *E. fuscus* and 25 ms for *M. septentrionalis*. A similar termination of the buzz just prior to contact has been reported in other species (e.g., Kalko 1995).

We observed stereotyped patterns in the IPI of both species while approaching and capturing stationary targets. Fig. 3.1 shows IPI functions for seven trials in which an individual *E. fuscus* intercepted tethered targets (3.1A) and seven trials for an individual *M. septen-*

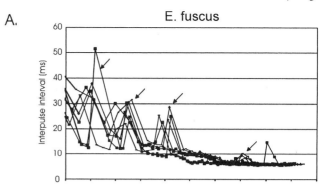

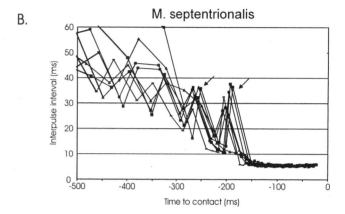

Fig. 3.1. Interpulse interval (IPI) functions for (*A*) seven trials in an individual *E. fuscus* and (*B*) seven trials for an individual *M. septentrionalis* while intercepting stationary tethered targets. The terminal buzz was defined as those IPIs less than twice the observed minimum IPI for each trial, which was shorter in *M. septentrionalis* (average onset 160 ms) than in *E. fuscus* (average onset 250 ms). Arrows indicate consistent increases in the IPI functions of both species (PEGs).

trionalis (3.1B). In each sequence, the target was at a different point in the room and the bat approached the target from a different location. However, peaks in the IPI functions, or pulse emission gaps (PEGs), occurred at consistent times during the late approach stage of the vocal sequence (arrows). PEGs occurred at about 450, 375, 300, and 150 ms before target contact in the example shown for *E. fuscus* and at about 250 and 200 ms before contact in the examples for *M. septentrionalis*. Note that the final PEG for *E. fuscus* occurred during the longer terminal buzz of this species. Similar breaks in vocal emission during the terminal phase have been reported by Schnitzler et al. (1987) in *Pipistrellus kuhli*.

In addition, both species appear to use a stereotypical pattern in their wing-beat cycle during the last approximately 400 ms of the capture sequence. Fig. 3.2 shows a raster plot of wing-beat timing relative to target contact for 13 target captures from *E. fuscus* and 13 target-capture trials from *M. septentrionalis*. In general, wing-beat frequency was lower for *E. fuscus* than for *M. septentrionalis*—13 Hz and 14.2 Hz, respectively.

Willard W. Wilson and Cynthia F. Moss

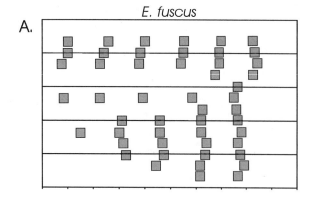
E. fuscus

A.

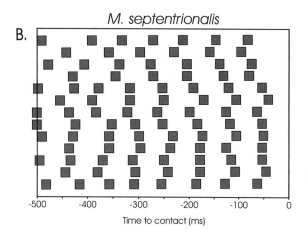
M. septentrionalis

B.

-500   -400   -300   -200   -100   0
Time to contact (ms)

Fig. 3.2. Raster plot of wing-beat timing relative to target contact for (*A*) 13 target captures from an individual *E. fuscus* and (*B*) 13 trials in an individual *M. septentrionalis.* Single trials are shown along each row, with each symbol representing the timing of the bottom of a wing stroke. We used a different lens for the video recordings of the *E. fuscus* and the bat was in view for a shorter period before capture, decreasing the number of data points per trial.

However, for the *E. fuscus* shown in fig. 3.2A, wing beats do not appear to be randomly distributed; rather, they exhibited a consistent pattern across trials, suggesting that the bat used stereotypical wing-beat patterns to approach the target. The different pattern in the top four trials and the bottom nine trials suggests that *E. fuscus* might use two different, but stereotyped, strategies to capture stationary targets. For *M. septentrionalis* (fig. 3.2B), a columnar pattern also emerged about 300–400 ms before target contact, again regardless of target position or angle. While larger sample sizes would permit a more rigorous analysis, these data, particularly for *E. fuscus,* show that wing gait is not randomly distributed in the last 300–400 ms before contact.

Fig. 3.3 shows the timing of pulse emission gaps relative to wing-beat cycle in an individual *E. fuscus.* Almost all PEGs occurred during the second half of the wing-beat cycle, during the downward phase of the wing

stroke. These findings are consistent with work in *Phyllostomus hastatus* (Suthers, Thomas, and Suthers 1972), demonstrating that inspiration occurs during the downward portion of the wing cycle and exhalation on the upward phase. This relationship might be expected to produce gaps in vocal output synchronized with wing cycle, as has been observed in other species (Schnitzler and Henson 1980). Therefore, the stereotypy of the wing cycle before target capture demonstrated above could produce consistent respiratory patterns and, consequently, pulse emisssion gaps in the late approach phase of target capture.

A third consistent motor pattern is the bat's control of its range relative to the target during the final stages of capture. Fig. 3.4 shows the distance of *E. fuscus* (fig. 3.4A) and *M. septentrionalis* (fig. 3.4B) to the target as a function of time. Both species control their velocity relative to the target in the final 300–400 ms before contact. *E. fuscus* approached the target at about 3.2 m/s during the final portion of the capture. For *M. septentrionalis,* the relative velocity during this same period was about 2 m/s. Lee et al. (1995) calculated the function τ (estimated time of contact based on distance and velocity) for two *E. fuscus* catching stationary tethered targets. This estimate was accurate throughout stationary-target captures (their fig. 3E and 3F), suggesting that a constant capture velocity was maintained in these sequences as well (Lee et al. 1992).

The beginning of terminal buzz for *M. septentrionalis*

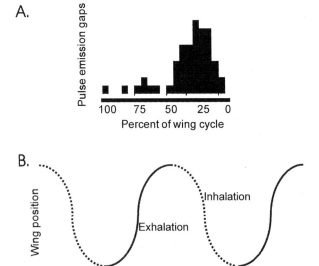

A.

B.

Fig. 3.3. *A:* Timing of pulse emission gaps at a function of wing cycle (defined as the time between the bottom of wing strokes) in an individual *E. fuscus.* The majority of PEGs were during the second half of the wing cycle, or during the downward portion of the stroke. *B:* The sinusoid shows the relationship between wing cycle and respiratory cycle (Suthers, Thomas, and Suthers 1972) on the same time scale.

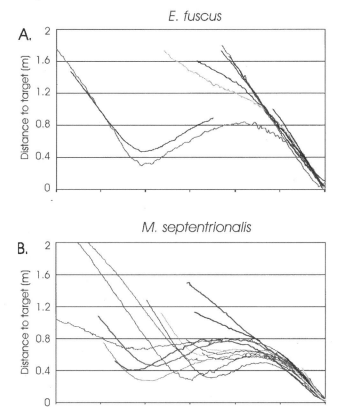

Fig. 3.4. Target range as a function of time to contact for (*A*) an individual *E. fuscus* and (*B*) three *M. septentrionalis* as determined from three-dimensional reconstruction of the flight path during target capture. A consistent linear portion and slope is present in each of these curves immediately before contact, indicating that a consistent relative velocity is maintained in the final stage of the capture sequences. Note that *M. septentrionalis* is capable of higher relative velocities than used during the final portion of the capture, exhibiting relative velocities of over 3 m/s in the early portion of some sequences.

averaged about 160 ms before contact (fig. 3.1B), well after the beginning of the constant relative velocity phase. Likewise, the beginning of the terminal buzz for *E. fuscus* averaged about 250 ms, also after constant relative velocity was initiated. Given the relative velocities of the two species and the timing of their terminal buzz, the distance from the target at buzz onset works out to about 32 cm for *Myotis septentrionalis,* similar to that observed by Webster (1962) in *Myotis lucifugus.* However, *Eptesicus fuscus* began the buzz at a much greater range, about 82 cm. Based on the maximum pulse emission rate of around 167 Hz for *E. fuscus* and 192 Hz for *M. septentrionalis,* the two species travel about 1.9 and 1.0 cm/pulse between pulses, respectively, during the buzz. Therefore, the rate at which acoustic information is acquired by *M. septentrionalis* is higher than for *E. fuscus,* measured in terms of both time and distance.

Discussion

We showed that acoustic information available to the bat prior to the onset of the terminal buzz is sufficient not only to induce changes in vocal emission pattern, but also to adjust wing gait and relative velocity. Our working hypothesis is that both *E. fuscus* and *M. septentrionalis* generate a motor plan for target capture about 400 ms prior to contact, presumably to coordinate scooping the target into the interfemoral membrane with a favorable wing-beat phase. Preparatory adjustments in tail position before contact have been observed in *Myotis lucifugus* (Webster and Griffin 1962) and in several pipistrelle species (Kalko 1995). These behaviors suggest that bats may calculate the time to contact to initiate preparatory movements at the correct point in the capture sequence. Webster and Griffin (1962) also argued that calculating a motor plan before target contact would be necessary to position the bat at the location of a moving target. Other investigators have suggested that echolocating bats might update spatial information on an echo-to-echo basis, orienting toward the last known target position until collision (e.g., Masters, Moffat, and Simmons 1985). If the motor patterns shown above are in fact anticipatory adjustments for capture at a favorable wing-beat phase, this would indicate that the bat can calculate location and wing phase at least 400 ms in the future. However, the observed pattern might also emerge as a result of the bat controlling its relative velocity or direction to the target (i.e., without planning per se). Experiments that might shed light on this problem would determine if the stereotyped motor behaviors are also present during capture of a moving target and when the bat must either predict, or adjust for, changing spatial relationships introduced by the target.

The pulse emission gaps demonstrated above are similar to those shown by Turl and Penner (1989) in the beluga, *Delphinapterus leucas.* PEGs in the sonar emissions of stationary beluga were longer than the two-way travel time of the signal to the target, thus allowing echoes to return without overlapping with outgoing sonar pulses—a potentially confounding factor in echo information processing. Fig. 3.5A shows how echoes returning from targets within an "inner window" can overlap with the end of the outgoing pulse. Echoes returning from targets outside an "outer window" can temporally overlap with later sonar emissions.

Fig. 3.5B shows the IPI and sonar pulse duration for a stationary-target capture by *E. fuscus.* The ranges within an "overlap-free" window based on these acoustic data are shown in fig. 3.5C. The actual target range during this trial, determined from three-dimensional reconstruction, is within the overlap-free window except during the final 60 ms (18 cm), where overlap occurs in the inner window. Note that the PEGs at approximately 450, 380, and 300 ms before target contact have a large

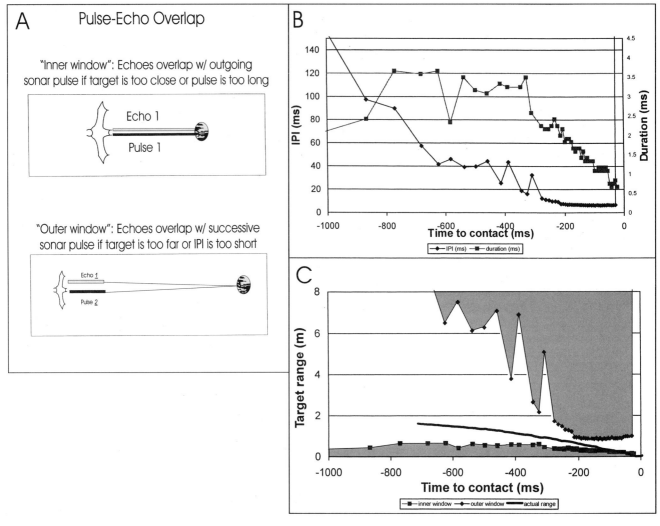

Fig. 3.5. *A, inner window:* Pulse-echo overlap can occur when targets are closer than the two-way travel time of a given pulse. The inner window is determined by the sonar pulse duration. *A, outer window:* Echoes from targets farther than the "outer window" can temporally overlap with later sonar emissions. The outer window is determined primarily by IPI and duration of the echolocation pulses. *B:* IPI and sonar pulse duration for a representative stationary-target capture by *E. fuscus* showing PEGs at about 450, 380, and 300 ms before target contact. *C:* Target ranges between the inner and outer window shown for acoustic data in *B.* Shaded areas show target ranges at which pulse/echo overlap is possible. Distance of the bat to the target for this trial is also shown.

influence on the outer window, producing periodic increases in the maximum target distance without overlap. These periodic increases in IPI might influence the perceptual system of the echolocating bat. For example, brief increases in the operating range of the biosonar system caused by PEGs might be useful in localizing more distant targets, allowing the bat to establish figure/ground relationships before the effective range of the system is diminished during the terminal buzz (fig. 3.5C).

The common features in the motor patterns of *E. fuscus* and *M. septentrionalis* described above are produced by modifying underlying behaviors that differ in detail.

The most striking differences between the two species are the emission rate during the terminal buzz, the terminal buzz duration, and the relative velocity during the final stages of target capture. The difference in emission rate (or IPI) between *E. fuscus* and *M. septentrionalis* might be explained by size differences between the two species. For example, smaller pipistrelle species have vocal emission patterns with shorter IPIs (Kalko 1994). Moreover, given the coupling between wing-beat cycle and vocalization and the higher wing-beat rate in *M. septentrionalis,* one might expect higher vocalization rates in this species.

The difference in terminal buzz duration could be explained by different target detection ranges in the two species, with a later detection in *M. septentrionalis* giving rise to a shorter terminal buzz. However, we showed that stereotypical wing-beat and velocity control emerge at roughly the same time in the capture sequence of both species. This suggests that the later terminal buzz in *M. septentrionalis* is not due to later detection. Estimates of the vocal emission levels suggest that the two species could receive echoes of similar intensities (Kick 1982; Miller and Treat 1993), which also argues against the idea of late detection in *Myotis septentrionalis*. Terminal buzz duration is labile within a number of species, varying dramatically from trial to trial (Schnitzler et al. 1987). Here, we showed differences in the buzz duration of two bat species capturing targets under identical conditions, suggesting that species-specific characteristics of biosonar systems also influence terminal buzz duration.

It is difficult to interpret any relative advantage of the more rapid approach and longer terminal buzz used by *E. fuscus* or the slower approach and shorter terminal buzz of *M. septentrionalis*. This is due, in part, to having little information about the function of the terminal buzz. The higher emission rate of the terminal buzz must be useful in increasing both the temporal and spatial information flow in the biosonar system. This might help the bat in fine control of its flight behavior before contact, especially if control of flight behavior is accomplished by monitoring acoustic variables related to relative velocity and location (Lee et al. 1995). From our data, it appears that *M. septentrionalis* puts a premium on the information conveyed by this acoustic behavior. Both the lower flight speed and higher emission rate of *M. septentrionalis* served to increase the pulse density per unit distance during the terminal buzz to twice that of *E. fuscus*. This would provide more detailed information during the final few hundred milliseconds before contact, and it also might compensate for the shorter buzz duration in this species.

In summary, we made a direct comparison of the vocal-motor capture behavior of two FM bat species capturing targets under identical conditions. These comparisons revealed a set of stereotypical behaviors used by echolocating bats for capturing insect prey. Early motor behaviors included patterning of the wing-beat cycle, a coincident pattern in biosonar vocalization pattern, and a period of constant relative velocity. These behaviors emerged about 300–400 ms before target contact in both species and could be evidence for an early motor plan for target capture. Future experiments will explore the question of motor planning further, focusing on the influence of target movement on these behaviors.

Acknowledgments

This research was supported by grants 5 T32 DC-00046-02, NIDCD, and 1 F32 MH11489-01, NIMH, to Willard Wilson, and grant 1 R01 MH56366, NIMH, and IBN-9258255 NSF Young Investigator Award to Cynthia Moss. We thank Amy Kryjak, Myriam Tron, Pete Abrams, Melonie Mavilia, Harry Erwin, and others for assistance with data collection and analysis.

{ 4 }

Biosonar Pulse Production in Odontocetes:
The State of Our Knowledge

Ted W. Cranford and *Mats Amundin*

Introduction

The story of the evolution and diversification of odontocetes is primarily one about the development of echolocation/biosonar. The development of echolocation was arguably the preeminent adaptive advantage that allowed odontocetes to diversify into the oceans of the world. The apparatus that supports the biosonar function can be divided into sound-generation and sound-reception components, one no less important than the other. In this chapter, we focus on toothed whale sound-generation components. In some species, these tissue components constitute a system that is capable of producing sounds that approach the finite limit of acoustic intensity in water (Griffin 1980) and may border on the potential to cause tissue damage (Norris and Møhl 1983).

In this summary of odontocete biosonar signal generation, we outline the current primitive understanding of this system and attempt to draw some comparisons with the system used by bats for biosonar signal generation. Since so much more is known about sound production in bats, it will hardly be an equivalent comparison, but we can highlight some of the advantages and limitations inherent in the generalized sound-generation system of each group.

Research into Source(s) of Sonar Sounds in Odontocetes

Debate over the location of an odontocete sonar source has been lively over the past forty years (see reviews in Norris et al. 1961; Norris 1964, 1975; Popper 1980; Amundin 1991b; Cranford 2000). At the same time, there has been little or no discussion of the number of possible sonar sources in the odontocete head. No doubt both aspects of the issue are important, and addressing one could shed light on the other. A review of the literature suggests that previous workers primarily approached the sound-generation question from the view of locating any source rather than considering how many there might be. The development of medical devices for the study of human physiology has opened a number of possible avenues for the study of marine mammal physiology, in particular for investigating sonar signal generation (Cranford 1988, 1992b; Cranford et al. 1997a, 1997b).

The early work on odontocete sonar source location indicated that it resides in the nasal apparatus (Norris et al. 1961; Norris 1975). The X-ray cinematography work by Norris et al. (1971) showed that the larynx moves forward during a pressurization phase, just prior to an intricate set of activities in the nasal apparatus concurrent with sound generation. Electromyography and pressure event studies conducted by Ridgway et al. (1980), as well as by Amundin and Andersen (1983), provided unequivocal evidence that sonar sounds were being generated somewhere in the nasal apparatus, the "forehead" of dolphins and porpoises.

Probing with a finger in the blowhole of a phonating harbor porpoise revealed strong tissue vibrations in the right side of the main air passage, approximately at the level of the entrance to the vestibular sac, but not in the left side (Amundin and Andersen 1983). Observations using video endoscopy at the same location in bottlenose dolphins (Cranford et al. 1997a) confirmed the earlier palpation results. Ultrasonic imaging experiments conducted by Mackay and Liaw (1981) demonstrated that the tissues of the nasal apparatus, as well as the nasal air spaces themselves, were moving during the process of generating sonar signals. That both tissue and air spaces are moving during sound generation is a fact that is often lost these days. We suggest that this seemingly innocuous detail is important for understanding the sound-generation process from interpretations of

anatomic and physiologic measurements. The implication is that the sound-generation process is initiated pneumatically and results in signals that are largely born from, and propagated in, tissues (Amundin 1991a).

Ridgway and his colleagues (Ridgway et al. 1980; Ridgway and Carder 1988; Ridgway 1990) and Amundin and Andersen (1983) pioneered the use of internal electronic probes to gather physiologic information from inside the dolphin's head during echolocation. They began using electromyography and pressure catheters. Recently, Cranford et al. (1997b) pushed ahead to include high-speed video endoscopy to study internal physiologic events during biosonar signal generation. They found activity at the phonic lips is synchronous with pulse production. In addition, substituting a helium-oxygen gas mixture in place of normal sound generation air separated critical components of echolocative pulses in both toothed whales (Amundin 1991a) and bats (Suthers and Durrant 1980; D. J. Hartley and Suthers 1990).

Most of the pertinent direct work into odontocete sound generation over the last two decades has focused on localizing and characterizing the nasal sound source(s) by attempting to measure physiologic parameters or to find the acoustic limits of echolocation behavior. Once the sounds have been generated they must be propagated through the head. The acoustic results of the sound propagation process also can shed light on the nature of the source of the sound.

Sound-Generation Anatomy in Odontocetes

One major research thrust that grew out of dolphin biosonar research was based upon curiosity about how, and anatomically where, these animals produced their echolocation sounds. Reviews of this subject can be found in Norris 1964; Norris 1969; Norris 1975; Mead 1975; Popper 1980; and Cranford 2000. The search for the anatomic site of sound generation was perhaps the most intractable obstacle in understanding the odontocete biosonar puzzle.

The search for the source of odontocete biosonar signals began in earnest once Norris and his colleagues (1961) suggested that sounds might originate within the forehead, actually part of the hypertrophied nasal complex. This notion was supported by the early work of Evans and Prescott (1962), but contradicted by the work of Purves (1966). These studies marked the beginning of a fervent debate that emerged in the 1960s and has simmered in the literature until recently (Cranford, Amundin, and Norris 1996; Cranford et al. 2000; Cranford 2000).

Historically, the assumption was that dolphins generated their sonar signals from the larynx, as do bats and most other mammals (Schevill and Lawrence 1956; Purves 1966). This assumption was challenged with keen observations using a blindfolded dolphin and an elegantly simple experimental design (Norris et al. 1961). Their observations were subsequently buttressed by the work of Diercks et al. (1971), Norris et al. (1971), Norris and Harvey (1974), Ridgway et al. (1980), Ridgway and Carder (1988), Mackay and Liaw (1981), Amundin and Andersen (1983), and Amundin (1991a). It was clear from these studies that the source of odontocete sonar signals would be found in the nasal apparatus, although the exact location was the subject of debate for more than two decades.

The task of precisely locating the sonar signal source within the nasal apparatus has been the focus of much anatomic work (Evans and Maderson 1973; Mead 1975; Green, Ridgway, and Evans 1980; Amundin, Kallin, and Kallin 1988; Cranford 1988; Amundin and Cranford 1990; Amundin 1991b; Cranford, Amundin, and Norris 1996). Cranford (1988), Amundin and Cranford (1990), and Cranford (1992b) used modern medical imaging devices to discover specific structural complexes in each nasal passage, which they identified as candidates for the sound-generation sites. Each structural monkey lips and dorsal bursae (MLDB) complex is composed of the same basic tissue structures in whichever species it is found, and it apparently varies primarily in size and position in three-dimensional space (Cranford, Amundin, and Norris 1996).

The anatomic and geometric details of these structural complexes can be found in a few papers (Cranford 1988; Amundin and Cranford 1990; Cranford, Amundin, and Norris 1996; Cranford 1999b). The precise roles of the various components of the MLDB complex in the sonar signal generation process are not well understood. It is likely that the fatty dorsal bursae, stout blowhole ligament, slender bursal cartilages, and dense connective tissue theca all have important roles in the formation of sonar signals (Cranford, Amundin, and Norris 1996). The current balance of evidence suggests that the fatty dorsal bursae within the "phonic lips" (Cranford et al. 1997b) play a central role in the process of sound generation (Cranford 2000; Cranford et al. 2000). In fact, a strong case can be made that the fatty dorsal bursae are the vibratory sources within the phonic lips of each MLDB complex (Cranford, Amundin, and Bain et al. 1987; Cranford 1990; Cranford 1992b; Cranford, Amundin, and Norris 1996; Cranford et al. 1997a).

The phonic lips (Cranford 2000) in dolphins are located just proximal to the vestibular sacs, the first enlarged air space inside the blowhole. These phonic lips can be distinguished by a suite of structural peculiarities. There are unique histological differences between the phonic lips and the surrounding epithelium (Degollada, García-Hartmann, and Cranford 1998), which may or may not be analogous to those reported for bats (see below). Histologically, the lips are heavily keratinized and thickened with slightly raised and flattened promon-

tories. There are also minute furrows or grooves that run across the lips in the direction of airflow (Cranford 1992b; Cranford, Amundin, and Norris 1996; Cranford et al. 1997b). Interestingly, the peculiar structure of the phonic lips is similar in dolphins and sperm whales (Norris and Harvey 1972; Cranford 1992b; Cranford 1999b; Cranford 2000), suggesting that it is common across all odontocetes.

Studies carried out primarily by Cranford et al. show that size, shape, material composition, and the intricate structure of the MLDB complex, as well as movement patterns during the sound-generation process, all indicate that this is the location of sound generation (Cranford, Amundin, and Norris 1996; Cranford et al. 1997a; Cranford 2000; Cranford et al. 2000). The proposition that these MLDB complexes contain the source of sonar signals now has some experimental support (Aroyan et al. 1992; Cranford et al. 1997a; Aroyan et al. 2000; Cranford et al. 2000). These recent results corroborate those of earlier anatomic (Mead 1975; Cranford 1992b) and experimental (Norris et al. 1971; Ridgway et al. 1980, Amundin and Andersen 1983) work.

Generally, the primary (spiracular) air space conducts the airstream dorsally from the bony nasal passages and past the phonic lips of the MLDB complex. Between the superior bony nares and the phonic lips, an opening in the posterior wall of the spiracular cavity (inferior vestibule) gives off a short air passage that rises behind the posterior part of the MLDB complex. The inferior vestibule then empties into the nasofrontal sacs, peculiar blind-ended sacs that partially encircle the spiracular cavity (Dormer 1974; Mead 1975; Amundin and Cranford 1990; Amundin 1991b). When the nasofrontal sacs are pressurized, they may also form a pneumatic lock to help prevent the intrusion of water from above (Lawrence and Schevill 1956).

The sound-generation process is apparently initiated in dolphins by moving the larynx anterodorsally using the gular musculature. At the same time, the palatopharyngeal muscle complex pulls the epiglottal spout of the larynx into the inferior bony nares and forms a tight seal between them. There is an interesting anatomic parallel with rhinolophid and hipposiderid bats, which emit their sonar pulses through the nasal openings or nostrils. In these forms the epiglottis is similarly inserted into the nasopharyngeal opening of the soft palate, forming a tight laryngonasal junction and isolating the mouth and buccal cavity from the nasal vocal tract (Matsumura 1979).

The action of the palatopharyngeal muscle complex upon the epiglottis acts to pressurize the nasopharyngeal air space. The pressurized air is then metered past the nasal plugs (Norris et al. 1971; Ridgway et al. 1980; Amundin and Andersen 1983; Ridgway and Carder 1988) and the phonic lips (within the MLDB complex). This action presumably sets the phonic lips and associated complexes into vibration, completing the process of sonar signal generation (Norris et al. 1971; Ridgway et al. 1980; Amundin and Andersen 1983; Ridgway and Carder 1988; Amundin 1991b; Cranford 1992b; Cranford et al. 1997b). The constriction in the spiracular passageway at the phonic lips probably facilitates the process of setting the system into oscillation (Dubrovskiy and Giro, chapter 10, this volume), which may then be sustained or adjusted.

There are two basic anatomic configurations of the proposed sound-generation anatomy in odontocetes (Cranford, Amundin, and Norris 1996). A unilateral configuration is found in sperm whales (Physeteridae), including *Physeter catodon* and both species of *Kogia*. All other odontocetes (nonphyseterids) have a bilateral configuration with a propensity toward directional asymmetry (right side larger than the left). So it appears as though sperm whales (physeterids) have one MLDB complex (the right side) that has become extremely hypertrophied (Cranford 1999b), while all nonphyseterid odontocetes have two MLDB complexes, one associated with each of two nasal passages. It is likely that these structural complexes are associated with sound generation in *all* odontocetes (Cranford 1992b; Cranford, Amundin, and Norris 1996). In addition, the particular configuration apparently has implications for the structure of the sounds that are generated from them.

Possible Sonar Pulse Production Mechanism(s) in Odontocetes

The sound-generation mechanism in dolphins is less well known than it is in bats. Generally, dolphins produce echolocation signals within the large soft-tissue nasal complex superior to the skull and project the signals out through the fatty forehead tissues (the melon) into the water. Over the past 40 years three basic proposals have been put forward to explain the mechanism of sound generation in odontocetes (Cranford 2000). It is fair to conclude that we do not currently have definitive evidence that would allow us to distinguish between the "cavitation" and "mechanical" proposals. However, it does seem clear that the stridulation proposal can be set aside for the time being (Amundin and Andersen 1983; Amundin 1991a; Cranford 2000).

STRIDULATION

This sound-generation mechanism was originally proposed by Evans and his colleagues (Evans 1973; Evans and Maderson 1973). According to their "friction/ stiction" proposal, the nasal plug is moved against the caudal wall of the main air passage and, by means of a series of small friction-based stops followed by slippage between the surfaces, produces vibrations in the nasal-plug tissues. This mechanism can be compared with the squeaking sound sometimes produced by chalk against

a blackboard. The appeal of this mechanism is found in the argument for efficient energy conversion and power output. Despite these appealing aspects, there does not appear to be a structural or behavioral basis in delphinids to support it. In fact, high-speed video endoscopy of the ventral surface of the nasal plug during sound production shows no such movements in the nasal plugs (Cranford et al. 1997a), directly contradicting the "friction/stiction" proposal.

CAVITATION

Perhaps the most intriguing alternative proposal for the sound-generation mechanism in dolphins is one that involves cavitation. It was suggested to us a couple of decades ago by K. S. Norris, but the idea has never been explicitly tested. (These suggestions were offered during a lecture course on the biology of marine mammals offered at U.C. Santa Cruz in 1986. Documentation and audio records of these lectures can be found at the McHenry Library Special Collections on that campus.) Interestingly, the seduction of a cavitation mechanism seems to be the same as it was for stridulation—that is, an efficient way to produce high intensity signals. Goodson, Flint, and Cranford (chapter 11, this volume) briefly put forth a notion for how a cavitation-based sonar signal generation mechanism might work, but currently no direct evidence supports this conjecture.

The current thinking for how such a cavitation mechanism might work is that during a low-pressure phase, bubbles of the same size are formed within the fluid layer that bathes the surfaces of the phonic lips. Bubble formation is then followed immediately by simultaneous collapse. The simultaneous collapse of numerous minute bubbles could produce a brief and powerful pulse, which could then be transmitted, through the surrounding fluid, into the adjacent tissues. The potential for the release of high-intensity signals via this cavitation mechanism is appealing, but the seemingly extraordinary requirements of uniform bubble size (during formation) followed by their simultaneous collapse should cause us to approach this idea with a healthy dose of skepticism. Because the current force of evidence does not strongly support or discount the cavitation notion, any rigorous evaluation of the idea must await experimentation.

PNEUMATIC-MECHANICAL

It would seem that the most parsimonious choice for the sound-generation mechanism is a simple, pneumatically driven mechanical tissue generator (Amundin 1991a, 1991b; Cranford 1992a; Cranford, Amundin, and Norris 1996; Cranford 2000). In contrast to a cavitation-based mechanism, the pneumatic-mechanical process is one in which no extraordinary physiological circumstances are required. In this case, the sound-generation tissues are set into vibration, after which the system achieves a state of "relaxed oscillation," sometimes called "self-oscillation" (Dubrovskiy and Giro, chapter 10, this volume). The tissue-borne (echolocation) signals could be generated when tissues impact upon one another or when the change in acceleration is greatest. This is, in a manner, similar to the action that occurs in the lips of a trumpet player (Chen and Weinreich 1996; Copley and Strong 1996), except that the important vibrations are those that are propagated in the lips and surrounding tissues of the odontocete forehead rather than those in the air spaces. Alternatively, echolocation signals might instead be generated during the deformation or bending of the bursal cartilages, small cartilaginous blades found embedded in the tissue just behind each fatty posterior dorsal bursa in all nonphyseterid odontocetes examined by Cranford (1992b).

Both the cavitation and mechanical models have a pneumatic driving force in common; they differ primarily in the details of the events taking place at the phonic lips. Whether or not the airstream is distributed across the full width of the lips during phonation is unclear at this time, but some structures and behaviors indicate that the direction of airflow is perpendicular to the axis of the lips (Cranford et al. 1997b). The resistive forces are apparently applied both with air pressure in the nasofrontal sacs and by the nasal musculature, particularly those acting upon the lateral aponeuroses of the spiracular cavity.

It has been proposed that pneumatic action forces the phonic lips apart, in the absence of cavitation, and air flows across the lips into the vestibular sacs, creating a sudden pressure drop (by Bernoulli or other fluid dynamic forces) that pulls the lips together again and creates a pulse. This simple mechanism is supported, to some degree, by the modeling work of Dubrovskiy and (chapter 10, this volume), and it explains the consistent low-frequency preamble to dolphin sonar pulses as described by Dubrovskiy and Zaslavskiy (1975).

Frequency Distribution in Delphinid Sonar Signals

From a signal generation standpoint, there are interesting correlates between anatomical geometry and peak frequency or waveform type (Cranford 1992a). That is, if the anatomy of sound generation is bilaterally symmetrical or unilateral, the signals produced tend to have a unimodal spectral peak, a narrow bandwidth, and a polycyclic waveform. If the sound-generation anatomy is moderately asymmetric (bilaterally), then the signal generation process seems to be capable of supporting two peaks in the frequency spectrum, a correspondingly broad bandwidth, and an oligocyclic waveform (Cranford 1992b; Au et al. 1995; Cranford 2000; Cranford et al. 2000).

In bats, "vocal tract resonances and filters can provide a means of regulating the harmonic emphasis and spec-

tral content of the sonar signal" (Suthers 1988, p. 35). A number of microchiropterans that produce lengthy CF components have been tested with light and heavy gas mixtures (D. J. Hartley and Suthers 1990). These and similar studies indicated that significant filtering can take place as a result of vocal tract resonances (Roberts 1973; Suthers 1994).

It is intriguing to ask whether the melon, air sacs, and other associated structures in the odontocetes perform some sort of "vocal tract filtering" that could result in polycyclic waveforms. Two species that produce polycyclic waveform pulses are the porpoise, *Phocoena phocoena* (Møhl and Andersen 1973), and the dolphin, *Cephalorhynchus commersonii* (Kamminga 1988). There are some anatomical similarities, as well as many differences, in nasal morphology between these two species. The differences are particularly striking with regard to the floor of the vestibular sacs. According to Goodson (1997), the peculiar corrugated folds in the floor of the vestibular sacs in *P. phocoena* may play an important role in shaping the signal into an almost "monochromatic" polycyclic sonar pulse (Kamminga 1988).

Goodson, Flint, and Cranford (chapter 11, this volume) buttress their proposed mechanism for vocal tract filtering with a sophisticated computer model that should not be dismissed lightly. Their hypothesis is contradicted by the fact that *P. phocoena* and *C. commersonii* produce virtually identical, polycyclic signals—yet *C. commersonii* lacks the corrugated folds in the floor of the vestibular sac (Amundin and Cranford 1990). It is of course possible that these polycyclic signals might be produced by a different mechanism in each species, although we must approach this explanation with some trepidation since it appears to violate the principal of parsimony. An example of an alternative mechanism might be as simple as shifting the phase of actuation between multiple sound sources to produce the desired effect in the interference beam. It is also possible that polycyclic waveforms might have arisen by two different phylogenetic pathways, particularly if selective pressures have promoted convergence in these dolphin and porpoise species.

Contrary to the difference in the anatomy of the vestibular sacs, however, there are striking similarities found in the MLDB complexes in *P. phocoena* and *C. commersonii*. The fatty bodies within the MLDB complexes are almost equal in size on both sides in the species that produce polycyclic waveforms (Cranford 1992a); whereas in species that produce oligocyclic signals (e.g., *Tursiops truncatus*), the complex on the right side is about twice as large as the complex on the left side. One could argue that finding general similarities in the central components of the sound generator across a range of odontocetes should be given more weight than finding unique differences in ancillary structures whose functions are thought to be auxiliary.

Sound-Generation Comparisons between Bats and Dolphins

For the purposes of the comparisons in this section, "bats" will refer to microchiropterans (to the exclusion of megachiropterans) unless otherwise stated. The most important perspective to retain when considering the differences between bat and dolphin sonar signal sources is that while bats employ the traditional mammalian location (the larynx) for producing sounds, dolphins have usurped and completely revamped the nasal apparatus for producing sonar sounds. As a consequence, there are no homologous relationships between the structures used for sonar signal generation in bats and dolphins (Cranford, Amundin, and Norris 1996). The de novo development of a highly complex nasal apparatus in odontocetes, particularly since it contains a plethora of lipid compounds that are apparently toxic to normal metabolic pathways, suggests that enormous importance can be assigned to the function(s) of this nasal contrivance. It appears as though the relative size of the nasal apparatus, and perhaps the importance of its function(s), has reached a zenith in the sperm whale (Cranford 1999b).

In comparing the sonar systems of bats and dolphins, it is essential to first consider the density differences of the mediums in which the signals are at first produced and finally propagated. Water is a thousand times the density of air. Consequently, for any given displacement of the medium, the vibrations that ultimately cause movement of water molecules will require proportionately more energy than those responsible for moving air molecules. The propagation characteristics are also different in the two mediums: sound will travel five times faster and considerably farther in water than in air.

While bats and dolphins propagate their signals into drastically different physical mediums, there are some similarities in their sound-generation systems. Bats and toothed whales both power their sound-generation systems pneumatically and produce their signals by pushing air past tissue gates or valves.

So each group uses moving air and has usurped portions of the respiratory system to generate sonar signals. Both bats and dolphins use air that is forced through the sound-generation apparatus in the same direction as respiratory expiration (expiratory airflow). Bats use subglottic pressure (Fattu and Suthers 1981), while dolphins use intranasal pressure (Ridgway et al. 1980; Amundin and Andersen 1983). Bats essentially have an unlimited source of air, while odontocetes must recycle the air from a single breath they take at the water surface and then compensate for the compressive effects of hydrostatic pressure throughout the dive cycle. This necessarily limits the air volume available for producing sound in odontocetes, which in turn may restrict the number of possible clicks emitted, or the mechanism by which they are generated, particularly at depth.

Microchiropteran bats generate ultrasonic echolocation signals in the larynx and project them through the mouth and/or the nasal openings (Griffin 1958). Their pressure gate is the vestibular fold of the glottis; the vibratory source is thought to be the extremely thin vocal membranes that project from the vocal folds, which are immediately inferior to the vestibular fold (Fattu and Suthers 1981; Suthers 1988). Sound pressure level is correlated with subglottic pressure, but frequency does not appear to be tightly correlated with subglottic pressure. In bats, the cricothyroid muscles are instrumental in providing the tension on the vocal membranes that determine frequency composition.

The process is a great deal different in the megachiropteran echolocators like the fruit bat, *Rousettus agyptiacus,* which generates clicks by the lateral action of the tongue against the side of the gums and teeth of the lower jaws (Kuzler 1960). In fact, the process of sonar signal generation in microchiropterans is arguably more similar to that in echolocating birds than to that in megachiropterans. In echolocating birds, such as oilbirds and grey swiftlets, the syringeal membranes interact with the flowing airstream to produce clicks (Suthers and Hector 1982, 1985).

Since bats use subglottic pressure, then factors like lung compliance and mechanical advantage created through the mechanical relationships of muscles and bones accomplish abdominal and thoracic pressure fluctuations that are important contributors to acoustic output. The work of Fattu and Suthers (1981) suggests that the resistive and vibratory functions are attributable to separate components in the larynx of echolocating bats. They proposed that the muscular vestibular fold constitutes the glottis and serves in the vocal tract's resistive function. This muscular hump is lined with specialized epithelial cells that are peculiar to that region of the vocal tract. As Fattu and Suthers assert,

phonation is assumed to begin when subglottal air pressure becomes great enough to overcome the glottal adducting forces. At this point the glottis opens slightly causing an initial abrupt pressure drop. Air rushing though the glottis creates Bernoulli or other aerodynamic forces, which combine with the elastic recoil and tension in the vocal membranes, drawing them medially and setting them in oscillation. The frequency of the waveform generated within the larynx is determined primarily by the tension exerted by the cricothyroid muscles on the vocal membranes, while its intensity depends primarily on the amplitude of membrane vibrations. (1981, p. 472)

Unfortunately, a similarly detailed understanding of the vibratory mechanism in the phonic lips of odontocetes has not been clearly delineated. Endoscopic studies indicate that the mechanism in dolphins may be similar to one of three basic models that have been used to explain the behavior of a human trumpet player's lips—the so-called lip reed (Chen and Weinreich 1996, Copley

TABLE 4.1. Acoustic comparisons between delphinids and microchiropteran bats.

	Tursiops truncatus	*Phocoena phocoena*	**Microchiropterans**
Intensity (energy flux density)	Very high (-20.8 dB re 1 joule/m^2)	High (-72 dB re 1 joule/m^2)	High (-66.4 dB re 1 joule/m^2) *Eptesicus serotinus*
Directivity	Very high (-3 dB beamwidth 10–50° or better)	Very high (-3 dB beamwidth 10–12°)	High (species dependent) (-3 dB beamwidth 20–40°)
Frequency	High (30–150 kHz)	Very high 120–140 kHz	High (12–200 kHz)
Time frequency	Very short duration signals, medium or broad bandwidth (time-bandwidth product close to 1)	Very short duration signals, narrow bandwidth (time-bandwidth product close to 1)	Highly regular broadband FM sweeps and/or extremely narrow bandwidth CF signals

Source: Compiled from Au 1993; Kamminga 1988; Neuweiler 1990; Troest and Møhl 1986; Tougaard 2000; D. J. Hartley and Suthers 1989, 1990.

and Strong 1996). With the current state of our understanding of the dolphin system, it is not possible to determine whether the sound-generation mechanism is more like the "swinging door" model, the "sliding door" model, or some hybrid of the two. Future research efforts might consider computer simulations or analytical models that could shed light on this issue (Adachi and Sato 1995, 1996). One analytical model that has great potential for describing the vibrational behavior of the dolphin's sound-generation elements is the asymmetric vocal-fold model developed by Steinecke and Herzel (1995).

Profound constraints are thus placed on the internal physiology of the beast by the physics of the external environment, but the coupling of the sound-generation apparatus to the propagation medium is also significant. It would seem that there is a near perfect coupling of the gaseous sound-generation medium to the gaseous sound-propagation medium in bats. Bats set air into vibration inside their bodies and then channel those airborne vibrations into the gaseous environment. Odontocetes also have an excellent impedance match between the tissues of signal generation/propagation and the aqueous environment (Litchfield, Karol, and Greenberg 1973, 1978; Norris and Harvey 1974; Litchfield et al. 1979). As a result, tissue-borne vibrations can be efficiently coupled between tissue and water.

The comparative story may reveal major differences if we consider the energetic costs of producing echolocation signals—which, as a percentage of the daily energy budget, might be greater in bats than in dolphins. Even though signal production requires greater absolute energy from a dolphin, the cost, relative to body size, may actually be greater for a bat. The absolute costs should be higher for odontocetes due primarily to acoustic impedance differences in the sound propagation media between air (bats) and water (dolphins). In addition, the total power available is partially a function of sound production muscle mass, which is comparatively smaller in bats than in dolphins. Alternatively, bats have coopted

flight muscles for the sound-generation process (Lancaster and Speakman 1999), which lowers or eliminates the energetic costs of making sonar sounds during flight (Speakman and Racey 1991; Lancaster 1993).

Differential constraints placed on sound generation in bats and dolphins are indicated by comparing signal levels. This exercise is somewhat complicated by the impedance differences in their respective sound propagation mediums and the different standard references used in those media. Au takes these factors into account and calculates the comparative values for us (Au 1993, pp. 247–248). The sound intensity, expressed as energy flux density (-66.4 dB re 1 joule/m^2), achieved by a roosting bat, *Eptesicus serotinus* (see table 4.1), is truly remarkable when we consider its 25 g body mass (Troest and Møhl 1986). Biosonar sounds from bottlenose dolphins are 45 dB (3.5×10^4) greater than those in the bat (Au 1993), but the dolphin's body mass is about 10^4 times greater than the bat (Read et al. 1993). By comparison, the harbor porpoise, *P. phocoena*, is 2000 times more massive than the bat, yet it appears to produce signals of only -72 dB re 1 joule/m^2 (i.e., 6 dB weaker). Other signal comparisons between bats and odontocetes are also instructive (see table 4.1).

Research has tied the bat's sound-generation system to the flight and respiratory muscles (Lancaster 1993; Lancaster 1994; Lancanster, Henson, and Keating 1995), which would dramatically reduce the specific costs of acoustic power output during flight (Speakman et al., chapter 49, this volume). Even so, Fenton suggests that the acoustic properties of the gaseous media combined with body size, locomotor considerations, and sound pressure levels suggest that bats may be acoustically (energy) limited (Fenton 1984; Fenton et al. 1995).

Even the smallest odontocetes are orders of magnitude more massive than the largest microchiropterans, and this difference is directly proportional to muscle mass and to the acoustic power output. Consequently, odontocete echolocation may not be constrained by energetic concerns. This is not to suggest that sonar signal generation is untaxed in odontocetes, only that it is not likely to be a significant part of a daily energy budget.

A survey of sonar signal values across the Odontoceti and Microchiroptera would cover a broad range of acoustic characteristics. The specific values are undoubtedly shaped by differences in body size, anatomy, various ecological pressures, phylogenetic inertia, and certainly physical constraints of the propagating medium.

Undoubtedly, more comparisons between bats and dolphins could shed light on the limitations and advantages driving the evolution of each group's sound-generation apparatus. Other important comparisons might include the size of the animal in relation to the speed of locomotion and the speed of sound in the medium, the wavelength of sound at peak frequency and prey size, or the sensitivity of prey and predators to the biosonar signals. These and other important considerations form a basis for intriguing questions that could become fodder for future investigations.

Feedback and Control Systems in Odontocetes

Control of sonar signal intensity is certainly influenced by air pressure in the bony nasal passages. Currently, we do not know if there is a minimum pressure to signal onset—but it is apparently controlled by means of muscular action (Ridgway et al. 1980; Amundin and Andersen 1983).

Tension on the vibratory elements and control of the mode of vibration in odontocetes may be accomplished largely by the action of a family of nasal muscles with fan-shaped origins on the skull (Lawrence and Schevill 1956; Mead 1975; Heyning 1989). Most of these nasal muscles act through their insertion upon the connective tissue aponeurosis along each lateral margin of the spiracular cavity. It is clear that their action exerts tension primarily along the axis of the phonic lips, more or less laterally with respect to the long axis of the body (Cranford 1992b). This action, along with that provided by the muscles intrinsic to the nasofrontal sacs (which are posterior to the spiracular cavity), and the air pressure in the nasofrontal sacs may function to adjust or control pulse repetition rate. In fact, pressurization of the nasofrontal sacs could function as a fluid-based spring constant and may be one of the parameters that could be used to adjust the mode of vibration. The details of the relationships among sound pressure level, pulse repetition rate, and intranasal pressure are currently being teased apart by Cranford and colleagues using pressure catheters and endoscopes (2000).

One intriguing experiment (Moore and Pawloski 1990) probed the limits of biosonar signal intensity and frequency composition in a bottlenose dolphin. They easily trained a bottlenose dolphin to control the intensity of outgoing signals but had more difficulty training them to control the frequency content of the clicks.

Their results suggest that there may be constraints on the dolphin's ability to control frequency composition. They concluded that such constraints might imply limits imposed by the sound-generation system.

The miniaturization of electronic devices has spawned a new age of physiologic and psychoacoustic research into dolphin sonar (Sigurdson 1997a, 1997b) and will no doubt lead to other innovations. Sigurdson (1997a) trained dolphins to carry a broadband, computerized recording unit with a hydrophone mounted in front of the rostrum, in the center core of the sound beam. These dolphins used sonar while swimming and were directed to search, detect, and report targets in the water column, on the bottom, or buried in the sediment. Under these circumstances, Sigurdson's preliminary analysis suggests that dolphins have some control over the frequency content of their sonar clicks (Sigurdson, pers. comm., 1999).

Sigurdson demonstrated that bottlenose dolphins often switch between two separate and disparate peak frequencies within the clicks of the sonar beam. The peak frequency switched between two narrow ranges, 110–130 kHz and 40–60 kHz. Switching between the two peak ranges can occur quickly, over the span of one or a few clicks in a train; or peaks may remain stable throughout a train of clicks, with one or both ranges emphasized. These results corroborate the earlier results of Moore and Pawloski (1990).

Demonstration of a dolphin's ability to produce bimodal click spectra with significant energy in both or either of these frequency ranges was noted previously for the false killer whale (*Pseudorca crassidens*) (Au et al. 1993a, 1995). Another consistent frequency shift was noted by Au et al. (1985) when white whales (*Delphinapterus leucas*) were moved from a relatively quiet environment in San Diego Bay to the considerably more noisy surroundings of Kaneohe Bay, Oahu, Hawaii. Au and his colleagues noted that this probably occurred in response to the prominent sounds of snapping shrimp (*Synalpheus parneomeris*) effectively masking the lower, 40–60 kHz frequency range. Similar results were reported for bottlenose dolphins in Hawaii (Au et al. 1974), although the same animals had not been recorded in the same two circumstances as for the white whales.

There are some anomalous instances in which a frequency plateau appears between the two predominate peaks. At first it might seem that this plateau is not easily explained, particularly if we assume that dolphins have two sonar signal generators, each with its own spectral signature. Upon further consideration, it might be possible to interpret this feature as the result of an interaction between the two sources. A potentially similar (heterodyne) mechanism has been suggested to occur in the syrinx of oscine birds (Nowicki and Capranica 1986).

The mechanisms behind the extent and degree of frequency control are currently not well understood and should be examined. In addition, we know very little about innervation or central nervous system control over the sound-generation system in odontocetes when compared to the expansive knowledge of the bat biosonar system.

Research Goals

A number of research goals will expand our knowledge of the dolphin biosonar system. A forum such as this will not support enumeration of all the intriguing avenues of study. Consequently, the remaining sections will present topics considered to be the most pressing or whose impact promises to be the greatest.

SOUND-GENERATION MECHANISM(S)

A detailed look at the basics of the sonar signal generation mechanism should be initiated. This is particularly important for resolving or settling the controversy between proposals for the cavitation and the pneumatic-mechanical mechanisms. Structural acoustics, which investigates the kinematics of the sound-generation apparatus, and numerical analysis, which can be used to calculate the resultant sound field from the displacement curves, are promising avenues for study.

NUMBER OF SOUND SOURCES

Perhaps foremost among the many remaining questions surrounding odontocete sonar signal generation is figuring out exactly how many sonar signal generators exist in the odontocete nasal region. The underlying assumption of most previous odontocete biosonar research has been that a single sound source, independent of location, sufficiently explains the observed acoustical phenomena.

There is considerable indirect evidence (Cranford 1992a; Cranford 1992b; Cranford, Amundin, and Norris 1996; Cranford 2000) and recently some direct evidence (Cranford 1999; Cranford et al. 2000) that there are at least two separately controllable pulse generators in the heads of bottlenose dolphins. We cannot currently show that both pulse generators are involved in the production of pulses used for sonar, although this seems a reasonable assumption. If we find that dolphins possess multiple sonar signal generators, we may need to reconsider our views of the basic characteristics and limits upon the signals themselves, in addition to the range and variety of functions within the generation apparatus.

Multiple sonar signal sources in odontocetes could affect emission patterns and the resultant sonar beam, particularly if the sources can be activated with variable phase relationships. In bats, the sonar signals are most often emitted through the open mouth, but some bats emit sounds through the nostrils, often with the aid of the elaborate facial adornments called noseleafs. These noseleafs affect the transmission beam patterns (D. J. Hartley and Suthers 1987).

Frequency filtering occurs within the vocal tract in multiple groups of microchiropterans (D. J. Hartley and Suthers 1990). Nasal emission pathways in bats may function as a frequency filter and perhaps also to increase directionality through interference patterns between the emissions from each nostril acting as two separate sound sources (Pye 1988). There is no evidence to suggest that bats can control the emissions from each nostril separately, but there is indirect evidence that dolphins may be able to do so, as they probably possess (at least) two sonar signal generators (Cranford 1992b, 1992a).

In the bats, since presumably the nostrils are simply two emission points for sound emanating from the same source (the larynx), the effect of interference on directionality depends upon the spacing between the nostrils, which should be fixed, and the wavelength of the sound emitted. Thus, in an FM bat, the directionality could be expected to change during the course of the signal (Pye 1988).

The multiple sonar source analogy might be more correctly applied to the avian sound production system (Suthers 1990; R. S. Hartley and Suthers 1990). In these groups the syrinx comprises two medially connected parts (Brackenbury 1980; Bradbury and Vehrencamp 1998). In echolocators, such as oilbirds and swiftlets, for example, the syrinx is divided into two semi-syrinxes, located along each primary bronchus. The activity of each semi-syrinx influences the activity of the other—that is, they are not independent of each other, and both sides contribute to the sonar sounds (Suthers 1988; Suthers and Hector 1988). In the dolphin, the two sources should be able to act independently (Cranford et al. 2000), but the degrees of freedom remain to be determined.

One primary message that can be gleaned from this chapter is that our current understanding of sonar signal generation in odontocetes is incomplete. The corollary is that a more complete picture is within our grasp through the use of new technological advances properly applied. The realization of this promise will undoubtedly cause a change in our basic view of the functional capabilities and limitations of toothed whale underwater biosonar and the implications for their place in the ecological milieu.

Acknowledgments

We are sincerely grateful to Rodrick Suthers (Indiana University) and Winston Lancaster (California State University, Sacramento), who reviewed earlier drafts of this chapter and made suggestions that improved it tremendously. Despite their expert assistance, we accept full responsibility for any errors, omissions, or inaccuracies.

{ 5 }

Comparison of Click Characteristics among Odontocete Species

Koji Nakamura and *Tomonari Akamatsu*

Introduction

Most odontocete species possess active sonar capabilities, and their echolocation signals (clicks), although superficially similar, appear to contain species-specific characteristics (Kamminga, Stuart, and Silber 1996). The differences between these sonar signals are believed to be caused by morphological differences associated with the sound generation. However, the sound sources and the precise sound-production mechanisms of echolocation signals have proven to be localized. Historically, there were three alternative hypotheses for the basic sound source mechanism: the larynx (Purves and Pilleri 1983), the nasal plug (Evans and Maderson 1973), and the monkey lips and dorsal bursae (MLDB) (Cranford, Amundin, and Norris 1996). Some evidence from experiments with live, phonating dolphins supports the MLDB hypothesis (Cranford et al. 1998 [the endoscope study, presented at the World Marine Mammal Science Conference]).

The dorsal bursae, proposed by Cranford et al. (1996) to be the sound source of odontocete echolocation signal, "are small ellipsoid fat bodies, encapsulated by a thin connective tissue sheath, or pouch, arranged in bilateral pairs" (225). These four small fat bodies are found near the posterodorsal terminus of the melon, embedded in the anterior and posterior walls of the spiracular cavity. One pair of dorsal bursae is located to the right of the midline and the other to the left. Cranford et al. (1996) suggested that there is a homologous relationship between the MLDB structure of every extant odontocete superfamily, and they described the detailed differences in the size, shape, degree of asymmetry, and geometric configuration of these structures in a number of species.

Thomas et al. (1988) and Au et al. (1995) showed that spectrum of pulsed signals produced by a false killer whale (*Pseudorca crassidens*) had bimodal peaks of frequency at <70 kHz and >70 kHz. Cranford et al. (1996) referred to these results and suggested that "these observations are consistent with the bilateral generation of odontocete biosonar signals, where one MLDB complex is approximately twice the size of the other" (276).

In this study, the acoustic characteristics of clicks produced by six odontocete species were compared with consideration for the dorsal bursae and the nasal structures, as described by Cranford et al. (1996).

Materials and Methods

DOLPHINS

Profiles of the six odontocete species are shown in table 5.1. The dolphins were filmed with an 8 mm camcorder while swimming toward the hydrophone; only the on-axis echolocation signals were used in the analysis. The size of the experimental tanks were larger than the

TABLE 5.1. Summary of basic data of six odontocete species: *Delphinus delphis* (Dd); *Lagenorhynchus obliquidens* (Lo); *Tursiops truncatus* (Tt); *Pseudorca crassidens* (Pc); *Lipotes vexillifer* (Lv); *Phocoena phocoena* (Pp). BL: body length; BW: body weight; KSW: Kamogawa Sea World (Japan); IHCAS: Institute of Hydrobiology, the Chinese Academy of Science (China); OA: Otaru Aquarium (Japan).

Species	Name	Sex	BL (cm)	BW (kg)	Site	Tank volume
Dd	Kapera	F	213	99	KSW	$12.0 \times 8.0 \times 3.4$ m^3
Lo	Arow	M	233	137	KSW	$19.0 \times 14.0 \times 3.5$ m^3
Lo	Sam	M	219	123	KSW	$19.0 \times 14.0 \times 3.5$ m^3
Lo	Hokuto	M	205	106	KSW	$19.0 \times 14.0 \times 3.5$ m^3
Tt	Slim	F	294	287	KSW	$13.0 \times 12.5 \times 3.0$ m^3
Tt	Orino	F	259	186	KSW	$13.0 \times 12.5 \times 3.0$ m^3
Tt	Karis	F	246	214	KSW	$13.0 \times 12.5 \times 3.0$ m^3
Tt	Uran	F	200	105	KSW	$13.0 \times 12.5 \times 3.0$ m^3
Pc	Kool	F	330	151	KSW	10.0 m diameter, 2.5 m depth
Lv	QiQi	M	215	125	IHCAS	13.0 m diameter, 4.0 m depth
Pp	Otaru No. 1	M	160	55	OA	$23.0 \times 9.0 \times 2.5$ m^3

sound traveling distance during pulse duration of each click. Hence, the reverberation could be separated from direct path signal in the time domain.

RECORDING AND ANALYSIS EQUIPMENT

Two hydrophones (a B&K type 8103 hydrophone, sensitivity −211 dB re 1 V/μPa + 2/− 9 dB, up to 180 kHz, or an OKI ST8004, sensitivity −220 dB re : 1 V/μPa + 3/− 2 dB, up to 200 kHz), two preamplifiers with 1 kHz high-pass filter (a B&K type 2635 or an OKI ST-80B), and a DAT recorder (SONY PCHB 244, sampling rate 384 kHz) were used to record the dolphin clicks. Using 10-cycle sinusoidal test signals with different frequencies, the frequency response of the data recorder was determined to be flat from DC to 147 kHz within 3 dB. Though the Nyquist frequency of the data recorder was 192 kHz, the upper limit of the recordable frequency was 147 kHz, due to an anti-aliasing filter built in the data recorder. As Mitson (1990) reported, clicks may have higher frequency components than this recordable range. We could not analyze if high-frequency components exist because they were eliminated from the data by the anti-aliasing filter.

Odontocete echolocation signals are composed of a series of clicks. The waveform of each individual click in a train tends to be very repetitive and stereotyped (Au et al. 1974). Thus the acoustic characteristics of clicks in the same click train may not be independent of one another. For the independent sampling of sound data, only the highest amplitude click was chosen from each click train. The digitized data sampled by LeCroy Model 9304AM, with a 16-bit A/D converter operating at 25 MHz, was transferred to a PC (NEC PC9821 with Windows 95). Cool Edit software was used to determine the peaks in the frequency spectrum (PF5) and to measure the pulse duration (PD) of a click. The PD was defined by the time elapsed during more than half amplitude of pulse relative to the maximum amplitude, to avoid subjective decision by human observation. The beginning and the end of a click is usually hard to determine because the ratio of the signal and noise was comparable at both edges of a click.

Results

ACOUSTIC CHARACTERISTICS OF CLICKS

The typical waveforms and frequency spectrum of clicks of the six species are depicted in fig. 5.1. The waveform characteristics were visibly different among species; the harbor porpoise's waveform was particularly distinct. Typically, the oscillations defining the envelope of the harbor porpoise's clicks increased in amplitude for the first 5 cycles and decayed exponentially (polycyclic waveform). Spectrally, the harbor porpoise's signal peaked between 120 and 140 kHz. The typical waveforms of clicks from the other species examined were less

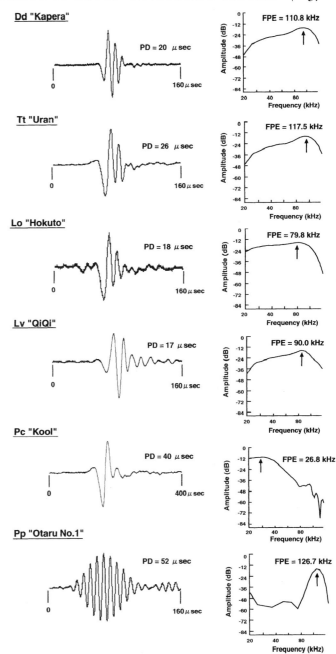

Fig. 5.1. Examples of the clicks produced by six odontocete species. The typical waveform is shown on the left and the frequency spectrum on the right. For abbreviations of odontocete species, see table 5.1.

than 3–5 cycles, with the first cycle achieving maximum amplitude (oligocyclic waveform). Their frequency spectrum was much broader than that of harbor porpoise, containing energy between 20 and 120 kHz. Therefore, it was easy to recognize differences in clicks between the harbor porpoise and the other species merely from the waveform characteristics and frequency spectrum.

Scatter plots of the peak frequency versus the pulse duration of clicks for six odontocete species are presented in fig. 5.2. Acoustic characteristics differences

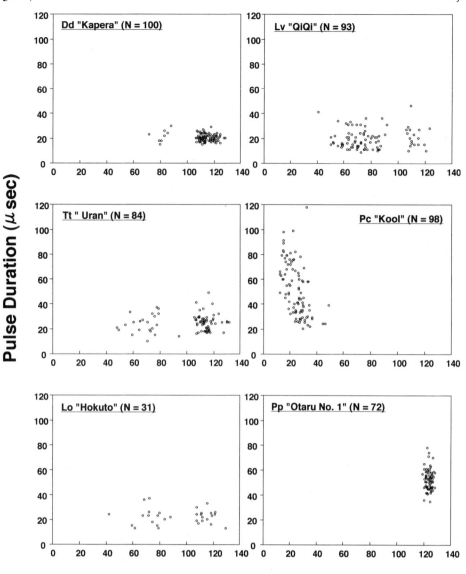

Frequency of Peak Energy (kHz)

Fig. 5.2. Scatter plots of the frequency of peak energy versus the pulse duration of clicks for six odontocete species. For abbrevations of odontocete species, see table 5.1.

among these species are recognizable from scatter plots of these two acoustic parameters. The delphinid species (the common dolphin, the bottlenose dolphin, the Pacific white-sided dolphin) and the baiji exhibit a bimodal double peak in their power spectrum at <100 kHz and >100 kHz, with shorter pulse duration (approximately 10–40 μs). The peak frequency of the false killer whale's clicks was lower (12.6–49.5 kHz) and the duration was longer (20–142 μs). The harbor porpoise produced a narrower frequency range and longer duration clicks than all the other species except for the false killer whale.

CLUSTER ANALYSIS

Cluster analysis was conducted using 16 acoustic parameters (mean, standard deviation, maximum, and minimum of PD and PFS) of the clicks of all six species. Different sets of analyses were done for the mean and standard deviation of the PD and PFS at low (<100 kHz) and high (>100 kHz) peak frequency of the click spectrum for each species.

The false killer whale clicks were classified using the different clusters from the other species. The euclidean distance between the false killer whale and the other species was 1.2734 (fig. 5.3). When the harbor porpoise was classified, its euclidean distance to the other species was found to be 0.4181. The delphinid species formed one cluster, with *Delphinus delphis* being slightly different. The four bottlenose dolphins, the three Pacific white-sided dolphins, and the baiji could not be differentiated into independent clusters.

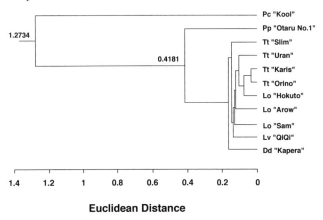

Fig. 5.3. Similarity of six odontocete species' clicks. Cluster analysis was conducted using 16 acoustic parameters (mean, standard deviation, maximum, and minimum PD and PFS) of clicks.

Discussion

DIFFERENCES AMONG DELPHINID AND PHOCOENID SPECIES

In the present study, the waveforms of six odontocete species were classified into two types: oligocyclic and polycyclic waveforms. The posterodorsal terminus of the melon of delphinid species (bottlenose dolphin and common dolphin) is attached directly to the dorsal bursae (Cranford, Amundin, and Norris 1996). Thus the impact vibration at the dorsal bursae may be transmitted directly into the melon. However, the posterodorsal terminus of the melon of the phocoenid species (the harbor porpoise and Dall's porpoise) does not attach to the dorsal bursae (Cranford, Amundin, and Norris 1996), and thus the impact vibration at the dorsal bursae will not be transmitted directly into the melon.

The floor of the vestibular sacs in the harbor porpoise have several deep, transverse ridges (Curry 1992). Goodson (1997) examined the dimensions of these ridges and suggested a passive signal processing mechanism to explain the formation of the phocoenid polycyclic waveform and narrow frequency band of the high-frequency clicks.

The waveform characteristics of odontocete species may be determined by both the configuration of the posterodorsal terminus of the melon and the structure of the nasal sacs surrounding the dorsal bursae.

DIFFERENCES AMONG DELPHINID SPECIES

The double frequency peak structure of clicks produced by the bottlenose dolphins and the common dolphin suggest that there may be two different sound sources. Cranford et al. (1997a) reported that the right dorsal bursae is twice as long as that of the left in these two species. Although the MLDB structure of the baiji is not known, Ness (1967) reported that the skull of this species is characteristically asymmetrical, and in this respect it is similar to that of the delphinids. The double peak spectrum of the baiji clicks is also similar to that of the delphinid species, suggesting the dorsal bursae may be asymmetrical. Dorsal bursae in the harbor porpoises are paired with similar dimensions and produce single peak spectrum clicks (Amundin 1991b). The peak spectrum structure is consistent with the dimension of the right and left pair of dorsal bursae in these three odontocete species.

Although the dorsal bursae of the Pacific white-sided dolphin revealed symmetrical dimensions for the right and left pairs (Cranford, Amundin, and Norris 1996), all three Pacific white-sided dolphins in this study produced clicks with double peak spectrum. This species could thus have a different frequency tuned mechanism between the dorsal bursae and the melon.

In this study, the echolocation signals of the false killer whale had peak frequencies ranging from 12.6 to 49.5 kHz. Thomas et al. (1988) also showed that the clicks of the false killer whale in a tank experiment had a low-frequency range (20–65 kHz) with peak-to-peak source level between 144 and 152 dB re 1 μPa at 1 m. The received sound pressure level (rms) of our false killer whale was much lower, varying between 122.1 and 135.0 dB re 1 μPa at 1 m, probably a result of the closed environment.

PHYLOGENETIC CONSTRAINTS AND CONVERGENCE OF ODONTOCETE SPECIES CLICKS

The present study showed that clicks of six odontocete species could be clearly classified into three clusters (Globicephalidae, Phocoenidae, and Delphinidae). The several specimens of bottlenose dolphins and Pacific white-sided dolphins could not be separated into independent clusters for each species. The echolocation signals of odontocete species seem to be exposed to some phylogenetic constraints at the family level.

However, the baiji could be included in the Delphinidae cluster. The echolocation signal of the baiji may thus have been convergently evolved to a delphinid type. In addition, the convergence can be seen in the echolocation signal of the Commerson's dolphin, *Cephalorhynchus commersonii,* and the Hector's dolphin, *Cephalorhynchus hectori,* two delphinid species with echolocation clicks very similar to the phocoenid clicks (Kamminga and Wiersma 1981). Mead (1975) reported that the posterodorsal terminus of the melon of Hector's dolphin was unusual in not extending into the dorsal bursae region as in most other delphinids. The posterodorsal terminus of the melon of the Commerson's dolphin also seems not to extend into the dorsal bursae region (Amundin 1991b). These anatomical differences may affect the click waveform of two Cephalorhynchus species, such as proposed by Goodson (1997) as a phocoenid waveform formation hypothesis. However,

Amundin (1991b) showed that the Commerson's dolphin does not have the vestibular sac ridges as seen in the *Phocoena*. We will have to find another explanation for the formation of the narrowband, polycyclic click of the Jacobita and *Phocoena*.

The echolocation signals of bats differ between species, and these differences seem to reflect the differences in echolocation ability and habitat choice (Neuweiler 1984). The clicks of odontocetes also seem to reveal differences in echolocation ability and habitat (Evans, Awbrey, and Hackbarth 1988). For the correct understanding of the differences in the acoustic characteristics of clicks in different species and taking evolutionary adaptation aspects into consideration, we will have to exclude phylogenetic constraints. The present study is the basis for future research.

Acknowledgments

We thank the Institute of Hydrobiology, The Chinese Academy of Science, Otaru Aquarium, and Kamogawa Sea World. We are also grateful to W. Ding (Institute of Hydrobiology, The Chinese Academy of Science), T. Tobayama (Kamogawa Sea World) for their kind help. K. Kagoshima (Otaru Aquarium) and H. Katsumata (Kamogawa Sea World) worked with us and greatly supported our experiments. M. Amundin provided constructive criticism on this manuscript.

{ 6 }

Structure of Harbor Porpoise (*Phocoena phocoena*) Acoustic Signals with High Repetition Rates

Willem C. Verboom and *Ronald A. Kastelein*

Introduction

Since July 1993, a systematic study on harbor porpoise (*Phocoena phocoena*) acoustic signals was conducted in the Netherlands. The first general results were published in Verboom and Kastelein (1995, 1997). Studied were animals kept for veterinary treatment in the Cetacean Rehabilitation Centre at Harderwijk, after being stranded on the Dutch coast. So far, the general conclusion of this acoustic study has been that harbor porpoises produce acoustic signals covering a very broad frequency range (<100 Hz to >160 kHz). These signals fall into three ranges: (1) low frequency (LF), <100 Hz to 10 kHz; (2) midfrequency (MF), 10 kHz to 100 kHz; and (3) high frequency (HF), 100 kHz to >160 kHz. The various signal ranges probably each have a specific function. Harbor porpoise echolocation clicks have fundamental frequencies of between 1.4 kHz and 2.5 kHz, and above 100 kHz. Porpoises frequently produce social and communicative signals, such as vocalizations sounding like grunts, bleats, and whoops, in the LF range up to 2 kHz as well (Verboom pers. obs.). Short duration communicative signals, including sine wave signals, have frequencies of below 10 kHz, most commonly below 2 kHz. Social signals up to 2 kHz can be described as grunts, whoops, and bleats. HF echolocation clicks are similar in frequency spectrum and waveform; LF signals are more variable. Busnel and Dziedzic (1967) showed that the click repetition rate in harbor porpoise signals depends on the distance between the animal's head and the ensonified object. High-repetition-rate trains are used when the observation distance is only a few centimeters.

This chapter describes harbor porpoise click trains with pulse repetition frequencies (PRFs) >50 Hz in terms of frequency content, click duration, and variability in click repetition rate. Corresponding data of trains with a lower PRF (<50 Hz) were published by Verboom and Kastelein (1997). Furthermore, a comparison is made of echolocation clicks emitted voluntarily (at animal's initiative) and when stimulated, during a discrimination experiment, as well as the relation between observation distance (distance snout to object) and PRF. So far, only Busnel and Dziedzic (1967) have published harbor porpoise PRF data; and we are unaware of any published detailed data on PRFs of other species.

Materials and Methods

Acoustic signals with high repetition rates were recorded from porpoise PpSH030, a 2–3-year-old male, housed in an indoor oval pool (8.6 × 6.3 m, water depth 1.3 m). Recordings were made in two different conditions, while the porpoise swam freely in the pool and acoustically scanned the environment. In the first experiment, the porpoise approached a hydrophone hanging in the middle of the pool (mid–water depth) and seemed to "attack," as if it were prey, while echolocating (voluntarily) in the direction of the hydrophone. During signal registration, the distance between the tip of the snout and the hydrophone varied; minimum distance was <10 cm, usually approximately 4 cm. Sometimes the porpoise touched the hydrophone with the side of its body while passing. In the second experiment, a box filled with sand was placed on the floor of the pool, with a hydrophone 3 cm above the sand. During these recordings, the animal searched, as part of a discrimination experiment, for objects buried in the sand and was rewarded for a correct hit (for details, see Kastelein et al. 1997). When stimulated to scan the sand, the tip of the snout was approximately 1 cm above the sediment. The distance between the hydrophone and the center of the melon was 10–20 cm.

Recording equipment consisted of a hydrophone and a tape recorder system (system response 0–160 kHz). Spectral analysis was carried out with a computerized analysis system, based on fast Fourier transformation (FFT). The analysis system frequency range was 0–22 kHz. By reducing the playback speed by a factor of 8, the virtual analysis frequency range was 0–176 kHz. Duration of the LF and HF clicks was determined by filtering the broadband signals (LF clicks at 0.5–3 kHz band pass; HF clicks at 90 kHz high pass) and plotting the time display. The duration of HF clicks was particularly difficult to determine, because very often the acoustic environment influenced the waveform. For instance, reflections could interfere with the emitted signal, thus influencing the waveform of the emitted pulse. Reverberations could "lengthen" the click, hampering the exact determination of duration.

Results

CLICK REPETITION RATE

Harbor porpoise click repetition rates per second vary from a few clicks up to 600. Examples of the PRF in four high-repetition-rate click trains, taken at random, are shown in fig. 6.1. The PRF varies between 295 and 480 Hz (signal *A*), between 400 and 475 Hz (signal *B*), and between 320 and 383 Hz (signal *C*). Signal *D* has a larger range, between 328 and 508 Hz. During the study, PRFs up to 590 Hz were measured. The figure thus demonstrates the PRF variability in click trains.

CLICK DURATION

The mean duration (+ standard deviation) of LF and HF clicks in four trains are shown in table 6.1. These recordings were made while PpSH030 was scanning the sand. Trains were selected with PRFs varying between 56 and 352 Hz.

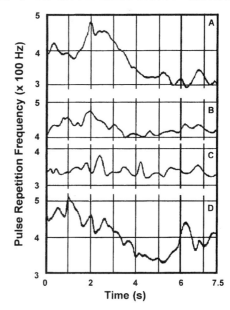

Fig. 6.1. Harbor porpoise pulse repetition frequencies recorded when the animal was scanning the sand in search for a hidden object. The repetition frequency range is between 300 and 500 Hz. Four examples are shown here, each with duration of 7.5 s.

FREQUENCY CONTENT

Fig. 6.2 depicts a three-dimensional waterfall display containing 10 clicks from a train with a PRF of 590 Hz. Typical spectral components in a harbor porpoise click train—HF-component, MF-component, and LF-component—are also shown.

Discussion

CLICK REPETITION RATE

Porpoise echolocation PRFs are low (range 1–50 Hz, typically 20 Hz) when the animal swims "at ease," without "interesting" objects in the water. Also when echolocating over large distances, the PRF is low. Miller et al. (1998) found that the PRF of PpSH030, during accurate target detection between 12 and 20 m, was very stable (18–20 Hz). Au et al. (2000) determined, again for the same animal, a PRF of between 22 and 33 Hz for an observation distance of 7.5 m. Twenty Hz seems to be the PRF for relatively large detection distances. When there is a certain reason or when the animal's attention is drawn by a specific event/object at very short distance, PRF increases to above 500 Hz, but it never exceeds 650 Hz.

The present study showed that PRFs vary between 300 and 590 Hz for observation distances of <10 cm. This confirms observations of Busnel and Dziedzic (1967) that harbor porpoise PRFs increase from 20 Hz to roughly 600 Hz when a prey item, or another interesting object, was observed at short distance. They observed the following relationship between PRF and observation distance (for 53 measurements on 2 animals): 416–640 Hz for an observation distance of 2–10 cm and

Table 6.1. Mean duration (+ standard deviation) of LF and HF clicks in 4 click trains, and click interval time (*n* = number of clicks in train, PRF = average PRF in train of *n* clicks).

Train	Number of clicks	PRF (Hz)	Mean duration ± SD LF clicks (ms)	Mean duration ± SD HF clicks (μs)	Click interval time (ms)
1	40	352	no LF component	117.8 ± 16.4	2.8
2	10	250	2.37 ± 0.52	91.6 ± 8.2	4.0
3	17	135	3.05 ± 0.34	116.4 ± 10.7	7.4
4	17	56	2.20 ± 0.21	265.0 ± 57.5	17.9

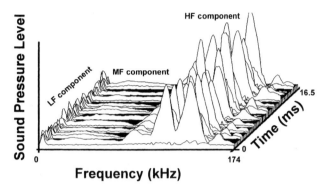

Fig. 6.2. A waterfall display of a series of 10 clicks: PRF = 590 Hz, linear time scale 0–16.5 ms, linear frequency range 0–174 kHz, logarithmic level scale (not calibrated). The LF component peaks at about 2.5 kHz. The HF component may consist of a number of energy peaks (up to 9).

64–192 Hz for 20–34 cm (in animal 1); and 375–515 Hz for 0–11 cm and 80–160 Hz for 11–37 cm (in animal 2). Note that these researchers gave no definition for the observation distance. From their tabled observations we derived the average PRF per observation distance and the total (linear) average PRFs. The average PRF was 494 Hz at distances <10 cm and 114 Hz at distances around 30 cm. Busnel and Dziedzic (1967) gave a gradually decaying curve as the relationship between PRF and observation distance. Our study showed that harbor porpoises seem to have a number of "favorite" PRFs, used frequently, depending on the observation distance. This would mean that the relationship is a curve decaying not gradually, but rather in a few steps. A relation could be (1) PRFs around 500 Hz for close distance (0–10 cm), (2) PRFs around 100 Hz for larger distances (20–40 cm), and (3) a PRF of around 20 Hz for distances larger than 10 m and for swimming in an inattentive state. When visibility is good and the animals are in a familiar area, PRF is reduced to around 1 Hz.

Click Duration

In this study, mean duration of the LF clicks varied between 2.20 and 3.05 ms. Busnel and Dziedzic (1967) found a duration for LF clicks of 1–3 ms (mean value 1.2 ms), and Schevill, Watkins, and Ray (1969) observed 0.5–5.0 ms. For PRFs <35 Hz, Verboom and Kastelein (1997) found a range of 2.5–4.5 ms. From these observations, it can be concluded that the range of LF click

duration is between 0.5 and 5.0 ms. Mean duration of the HF clicks varied between 91.6 and 265.0 μs. Average duration at a PRF of 56 Hz in our study was 265 μs, but at higher PRFs duration decreased to around 100 μs. Undisturbed HF click shape typically contained 12–14 sine waves, corresponding with an average duration of approximately 100 μs. This suggests that the duration of HF clicks is influenced by the PRF: increasing click duration at decreasing PRF.

Click Interval

The (minimum) observation distance in the present study was <0.1 m, which means a two-way propagation time of 130 μs (or less), being equal or less than the duration of the HF click. This would make an accurate determination of the object distance impossible. Therefore, it is hypothesized that under these circumstances the porpoise only echolocates to determine the presence, but not the distance, of an object. At short distances, and if the object was not hidden in the sediment, the actual distance determination could be by vision.

Table 6.1 shows the mean LF click duration and click interval. Minimum time between the end of a click and the start of the next click is 1.6 ms (related to a PRF of 250 Hz). The click train with a PRF of 352 Hz did not contain LF clicks. Absence of LF clicks has been seen frequently. A PRF of 352 Hz corresponds with an interval of 2.8 ms, roughly equal to the mean LF click duration found in this study. This suggests that when PRF increases and the click duration approaches the interval duration, (1) the porpoise switches off the LF component (as in this study); (2) the porpoise reduces the LF click duration (Schevill, Watkins, and Ray [1969] found 0.5 ms as minimum click duration); or (3) the LF clicks touch each other and a continuous LF sine wave is emitted (Verboom and Kastelein [1995] reported that in high-PRF trains the HF clicks can be superimposed on a continuous LF wave).

Possibly the study gives an impression of the minimum time a porpoise needs to distinguish two successive clicks. Because approximately 600 Hz is the maximum PRF for harbor porpoise echolocation signals, the minimum interval between two HF clicks is 1.6 ms. At high PRFs the duration of HF clicks is 0.1 ms, which means that the ""dead" time between two clicks is 1.5 ms. It might be hypothesized that this dead time is the time a porpoise need to process acoustic information. Above it is stated that also for LF clicks the dead time is in the same order, namely 1.6 ms. Busnel and Dziedzic (1967) found a 0.6 ms dead time in combination with a LF click duration of 1.2 ms and an 1.8 ms interval (560 Hz PRF). In comparison, humans distinguish pulses with a minimum interval of 1.1 ms (Busnel and Dziedzic 1967).

HIGH- AND MID-FREQUENCY COMPONENTS

HF clicks and MF components, observed in this study, were similar to those found in earlier studies, for instance in click series with a low repetition rate (Verboom and Kastelein 1997).

LOW-FREQUENCY COMPONENTS

During click trains with a high repetition rate, harbor porpoises do not always emit LF clicks. When emitted, LF clicks were of the usual damped sine type. In combined LF/HF clicks, the HF component is usually emitted during the first sine wave of the LF component, mostly during the first half of the LF cycle. Sometimes the HF click is emitted during the second half of the cycle.

SEDIMENT PENETRATING SIGNALS

A comparison of signals emitted voluntarily and those recorded during the discrimination experiment (Kastelein et al. 1997), when the porpoise was stimulated to echolocate, showed no significant differences. The sediment penetrating signals do not deviate from other echolocation signals. Because LF signals were not always emitted, the porpoise used HF echolocation clicks (>100 kHz) to scan the sand. One would expect,

however, that LF signals would be used for sediment penetration, due to their lower signal losses.

LF AND HF GENERATOR

The LF and HF components are probably not produced by the same sound generator, because during HF click trains LF signals (echolocation clicks or noiselike social or communicative signals) are switched on and off randomly, without influencing the HF clicks. Also, the difference in click duration (LF 2.5 ms versus HF 150 μs) might indicate that harbor porpoises have a generator for LF clicks and another for HF clicks. However, they appear to be coupled, because emission of the HF click coincides with the emission of the first LF click cycle. Furthermore, it appears that nonecholocation signals are also emitted without affecting the echolocation signals. Because there is a large variety of social signals (very low frequencies, continuous sine waves, etc.) and the LF component can be superimposed on the social signal, possibly even a third generator exists and produces the social signals.

Acknowledgments

We gratefully thank Teun van den Dool (TNO Institute of Applied Physics, the Netherlands) for his technical assistance; the editors of this book, Jeanette Thomas, Cynthia Moss, and Marianne Vater, for their thoughtful and constructive reviews of the paper; and two anonymous reviewers and Nancy Vaughan (Bristol University, UK) for their helpful comments, which significantly improved this chapter.

{ 7 }

Echolocation in the Risso's Dolphin, *Grampus griseus:* A Preliminary Report

Jennifer D. Philips, Paul E. Nachtigall, Whitlow W. L. Au, Jeffrey L. Pawloski, and *Herbert L. Roitblat*

Introduction

The Risso's dolphin (*Grampus griseus*) is a member of the family Delphinidae, usually occurring in tropical and temperate waters near continental slopes and deep-water canyons (Leatherwood et al. 1980). Partly due to its infrequent occurrence near areas of common boating

activities and because few individuals of this species are or ever have been held in captivity, relatively little is known about its biology, behavior, or natural history.

Among other aspects of its biology and behavior, very little is known about the echolocation system of the Risso's dolphin. Though the species is presumed capable of producing and using echolocation, based partly on its membership in a family and order of other echo-locating mammals, there is no experimental evidence to demonstrate it.

Suggestive evidence does, however, support the assumption that the Risso's dolphin echolocates. First, individuals of the species often produce apparent sonar clicks while free swimming, both in captivity and in the wild (e.g., Corkeron and Van Parijs 2001; Au 1993). Second, the species possesses anatomical features in its forehead (e.g., the phonic lips–dorsal bursae complex; Cranford, Amundin, and Norris 1996) found in other echolocating species that are thought to be important for the production of echolocation clicks. And third, the species is capable of hearing high frequencies to 80 kHz, as are other odontocetes (Nachtigall et al. 1995). Nachtigall et al. (1995) proposed that a high-frequency hearing system has evolved along with an active high-frequency echolocation system in dolphins. Despite suggestive evidence, however, there is still no experimentally convincing substantiation that Risso's dolphins actually do echolocate.

The Risso's dolphin has unique characteristics of its biology and behavior relative to other odontocetes that may be important determiners of its sonar system and echolocation capabilities. For example, the species has a unique, vertically indented forehead (Leatherwood, Reeves, and Foster 1983), unlike the smoothly rounded forehead of all other known echolocating odontocetes (fig. 7.1). The function of such a groove (and whether or not the underlying melon is also indented) is unknown. Since the curvature of the surface of the forehead and the shape and density features of the melon are thought to be acoustically important to the propagation of the outgoing sonar pulse (e.g., Litchfield et al. 1979), an indented forehead could have a significant effect on the sonar beam produced by this species.

Fig. 7.1. View of the Risso's dolphin, showing the vertically indented forehead. The Risso's dolphin is the only odontocete to possess such a vertical groove; its function is unknown.

Also, the Risso's dolphin feeds nearly exclusively on cephalopods (e.g., Clarke and Pascoe 1985; Cockcroft, Haschick, and Klagers 1993). The consumption of cephalopods by this species is particularly interesting considering the likely function of dolphin sonar in searching for and localizing prey. According to Foote (1980), reflection of the incident sonar signal by the gaseous swim bladder in many fish species accounts for more than 90% of the sonar energy reflected from a fish target. Cephalopods, by contrast, do not have swim bladders and, therefore, have measured acoustic target strengths of levels 18 to 25 dB lower than the target strengths measured for swim-bladder-possessing fish species (MacLennan and Simmonds 1992). The cephalopod probably, therefore, presents a considerably more difficult sonar target than the swim-bladder-possessing fish preyed upon by most currently studied echolocating dolphin species. The unique problems faced by the Risso's dolphin using echolocation to locate its cephalopod prey may have led to specializations in its sonar system relative to that of fish-eating dolphins.

This chapter reports preliminary results from an ongoing investigation of echolocation in a captive Risso's dolphin.

Materials and Methods

EXPERIMENTAL DESIGN

The subject of this preliminary study was an older, female Risso's dolphin named Hana, housed at the Marine Mammal Research Program (MMRP) facilities at the Hawaii Institute of Marine Biology in Kaneohe Bay, Oahu, Hawaii. She was maintained on a diet of primarily squid. Her age was unknown, though numerous markings, white overall coloration, and the worn-down appearance of her teeth suggested that she was mature and perhaps already relatively old at the time of her 1989 capture. This study was conducted in April 1998.

A history of attempts to train this dolphin (as well as a second Risso's dolphin) to perform a trained echolocation task from an underwater stationary hoop had previously been met with some difficulty. As a result, the decision was made to remove the hoop station and allow the animal to swim freely in her pen while performing a very basic echolocation task. The animal's task was to search for, swim to, and physically touch a 15 cm air-filled, plastic, oval-shaped target with her forehead/melon. She was trained to voluntarily accept suction cups over her eyes to ensure that she could not use vision to locate the target.

At the start of each trail, the animal (wearing eye-cups) was positioned at the far end of the pen (range to target = 6.5–7.5 m). The target was placed into the water to approximately 1 m depth and held in place by the trainer. This signaled the animal to search for and touch the target. After touching the target, the animal was rewarded with a squid. Incorrect trials (i.e., trials in which the animal did not touch the target) were not rewarded. The target was then removed from the water, signaling the animal to return to the opposite end of the pen. The next trial was begun as the target was reinserted into the water.

After the difficulties encountered with the hoop-stationed experiment previously attempted, the goal of this preliminary investigation was only to establish that the dolphin had the capability of performing a very basic echolocation task when her vision was occluded. Direct efforts were not made, therefore, to completely control every cueing factor associated with the target, such as the noise it might have made going into the water or the precise depth or location at which it was held during trials. The target was, however, quietly placed into the water when the animal was at the opposite side of the pen, and the trainer attempted to hold it in place and not move it during trials.

The target was present in all trials (again, efforts were not made in this preliminary investigation to control the animal's response bias by balancing the number of target-present and target-absent trials). A total of 50 trials were conducted over 2 sessions.

SIGNAL ACQUISITION

Clicks emitted by the dolphin during her approach were acquired using a four-hydrophone symmetrical-star aligned array (fig. 7.2). The array was placed in the water at 1.0 m depth (relative to the center hydrophone, H1) and positioned in front of or just behind the plastic target so that clicks emitted by the animal during her approach would be optimally directed for acquisition. Clicks were sampled at 500 kHz, 256 points per click, using dual, simultaneous GAGE 1210, 12 bit, ± 5V A/D boards operated by a portable PC (486DX-33). A full technical description of the design of the array and detailed localization computations are available in this volume (Schotten et al., chapter 54, this volume).

There were particular advantages to using the star array, rather than a single hydrophone, to acquire the dolphin's clicks in this study. Sonar clicks are highly directional, and they appear distorted and lower in frequency and amplitude, relative to the source clicks, when acquired off the axis of the transmission beam (Au, Moore, and Pawloski 1986). Because the animal was allowed to swim freely while performing this detection task, her position relative to the acquisition hydrophone was variable and the acquisition of off-axis clicks was,

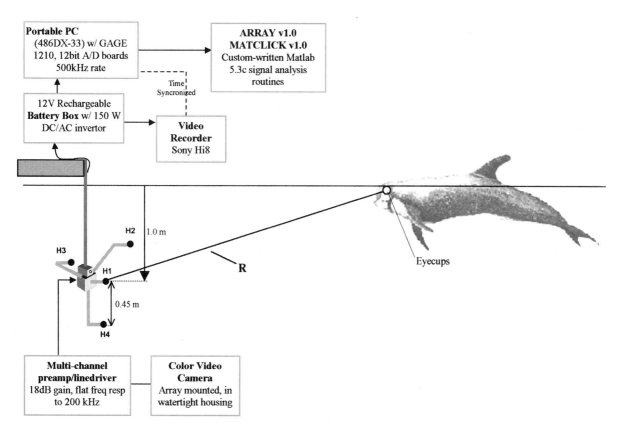

Fig. 7.2. Four-hydrophone symmetrical-star aligned array and data acquisition and analysis system. Range, R, is calculated by signal time-of-arrival differences and is the distance (m) from H1 (center, trigger hydrophone) to the dolphin. The dolphin is shown at a typical approach position.

therefore, likely. This study's hydrophone array provided the means with which to view a single click as acquired from four positions around the animal. Comparing the relative amplitudes of the four acquired clicks, it was possible to determine which click was acquired from nearest the center of the transmission beam (i.e., the clicks with the highest amplitudes). Clicks acquired closer to or along the beam axis could thus provide a more accurate description of the source click emitted by this animal. Additionally, time of arrival differences at the four hydrophones on the array could be used to calculate the range to the animal (R) and, therefore, the source level (SL) of the acquired clicks.

A small color video camera mounted onto the array was time-synchronized with the click acquisition system ($\pm$ 50 ms) and provided a real-time video recording of the approaching animal for each trial.

SIGNAL ANALYSIS

Digitized clicks were analyzed using custom-written Matlab 5.3c signal analysis routines. Raw click trains for each trial were first reviewed to assess relative click amplitude information. Then the individual clicks acquired by the array were examined. When an emitted click was acquired by all four hydrophones on the array (i.e., a "valid click set"), the source range (R, m) was calculated. Sound pressure level (SPL, dB re 1 μPa), source level (SL, dB re 1 μPa 1m), peak frequency (f_p, kHz), center frequency (f_0, kHz), 3 dB frequency bandwidth (BW, kHz), and duration (τ, μsec) were calculated for the clicks of each valid click set. The time-synchronized video recordings of the dolphin during each trial were reviewed after data collection along with the acoustic data to qualitatively describe the position of the dolphin relative to the array (including depth, range, and angle of head) in relation to the properties of the clicks acquired at the same approximate moment.

Results

The dolphin performed the target detection task with little difficulty. She successfully searched out and touched the target in 100% of trials, despite being blindfolded.

The animal emitted echolocation clicks immediately as each trial began and as she approached the target. She typically remained at or very near the surface of the water with the upper portion of her forehead out of the water during her approach and dove down to touch the target at ranges of only 2 m or less. The trial depicted in fig. 7.3 represents a typical example of the four click trains acquired by the array during one trial. Typically, the highest amplitude clicks were acquired by the deepest hydrophones on the array (H1 and H4). In general, it appeared from video recordings of the animal during trials that the highest amplitude clicks (i.e., on-axis clicks) were acquired by the deeper hydrophones even when the dolphin scanned from the surface of the water

with her head relatively level with the longitudinal axis of her body and not tilted downward.

Two examples of valid click sets are given in fig. 7.4. The figure includes a visual still frame of the position of the animal at the approximate time of acquisition of the click set, a plot of the time-domain waveforms for each hydrophone, the corresponding frequency spectrum for each waveform, and calculated amplitude and time-frequency characteristics of each click.

Maximum source levels of 208 dB re 1 μPa 1m were calculated, with mean source level for the trial depicted in fig. 7.3 of 193 dB re 1 μPa 1m (table 7.1). Maximum peak frequency and center frequency acquired were 50 kHz and 75 kHz, respectively. Mean peak and center frequency were 24 kHz and 40 kHz, respectively (table 7.1). The 3 dB bandwidth varied considerably, ranging from 15 to 35 kHz. Click duration ranged from 30 to 100 μsec and averaged 56 μsec (SD = 18 μsec). Greater than 80% of the clicks acquired during the trial depicted in fig. 7.3, for example, had durations between 35 and 75 μsec. Overall, the characteristics of the clicks emitted by this dolphin were very similar to the sonar clicks emitted by other echolocating delphinids (e.g., *Tursiops truncatus, Pseudorca crassidens*).

Discussion

Because the dolphin was able to locate and touch the plastic target consistently, despite being blindfolded, it is strongly suggested that she used echolocation to perform the task. While we do not conclude that echolocation has been definitively demonstrated in this species, these data do provide the first analysis of the clicks emitted by an individual of the species while successfully performing an apparently echolocation-dependent task. Overall, the clicks appeared similar to those of other echolocating delphinids (see Au 1993), with source levels above 200 dB re 1 μPa 1m, peak frequencies up to 50 kHz, and mean click durations of 56 μsec.

The surface-based swim pattern employed by the dolphin during her search for the target was interesting in view of the finding that very high amplitude, high-frequency clicks were more often acquired by the deepest hydrophones on the array. The dolphin was also apparently able to emit sonar clicks into the water even when the upper portion of her indented forehead was above the surface. These findings suggest that the vertical sonar transmission beam of the Risso's dolphin may project downward from the lower portion of the dolphin's forehead by 20° or more, relative to the longitudinal axis of the dolphin's body. Detailed measurements of the vertical transmission beam pattern will be necessary to test this hypothesis.

We do not yet know the function of the vertical indentation of the Risso's dolphin's forehead or how the echolocation system of this species might be specialized for cephalopod consumption. Indeed, although a good

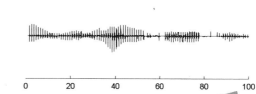

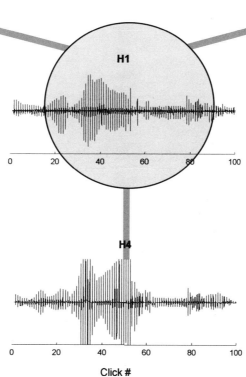

Click #

Fig. 7.3. Click trains acquired by each of the four array hydrophones in a single trial, shown in array configuration layout. Vertical pulses along the *x*-axis represent echolocation clicks (the time intervals between clicks were not acquired, so successive clicks are plotted immediately following the previous), with relative amplitudes being represented by the relative heights of pulses. This trial represents a typical click acquisition pattern, with the highest amplitude clicks acquired at the deepest hydrophones on the array (H1 and H4). Higher amplitude clicks are considered to have been acquired from nearer the 0° propagation axis of the signal transmission beam.

TABLE 7.1. Summary of click amplitude and frequency characteristics (mean ± SD) for all valid click sets in the trial depicted in fig. 7.3 ($n = 74$ clicks for each hydrophone). H1:H4 = mean of the highest amplitude click from each click set, $n = 74$; SPL = sound pressure level, dB re 1μPa; SL = source level, dB re 1μPa 1m; f_p = peak frequency, kHz; f_0 = center frequency, kHz.

	SPL (dB)	SL (dB)	f_p (kHz)	f_0 (kHz)
Hydrophone 1	182.50 ± 5.7	188.95 ± 8.2	19.32 ± 8.2	36.32 ± 10.7
Hydrophone 2	176.98 ± 4.8	183.45 ± 7.8	19.25 ± 7.1	33.31 ± 8.7
Hydrophone 3	179.11 ± 6.4	185.59 ± 9.3	20.19 ± 21.3	36.09 ± 12.7
Hydrophone 4	182.92 ± 8.2	189.40 ± 10.8	24.54 ± 7.5	34.67 ± 10.3
H1:H4	185.86 ± 6.8	192.26 ± 9.6	23.56 ± 9.6	39.73 ± 12.2

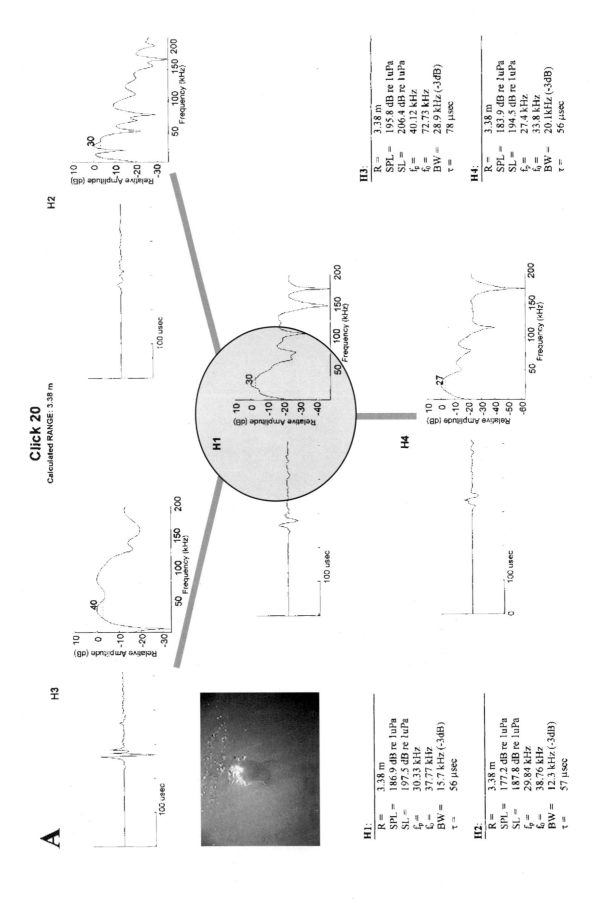

Click 20

Calculated RANGE: 3.38 m

H2

H3

H1

H4

A

H1:
R = 3.38 m
SPL = 186.9 dB re 1uPa
SL = 197.5 dB re 1uPa
f_p = 30.33 kHz
f_0 = 37.77 kHz
BW = 15.7 kHz (-3dB)
τ = 56 μsec

H2:
R = 3.38 m
SPL = 177.2 dB re 1uPa
SL = 187.8 dB re 1uPa
f_p = 29.84 kHz
f_0 = 38.76 kHz
BW = 12.3 kHz (-3dB)
τ = 57 μsec

H3:
R = 3.38 m
SPL = 195.8 dB re 1uPa
SL = 206.4 dB re 1uPa
f_p = 40.12 kHz
f_0 = 72.73 kHz
BW = 28.9 kHz (-3dB)
τ = 78 μsec

H4:
R = 3.38 m
SPL = 183.9 dB re 1uPa
SL = 194.5 dB re 1uPa
f_p = 27.4 kHz
f_0 = 33.8 kHz
BW = 20.1kHz (-3dB)
τ = 56 μsec

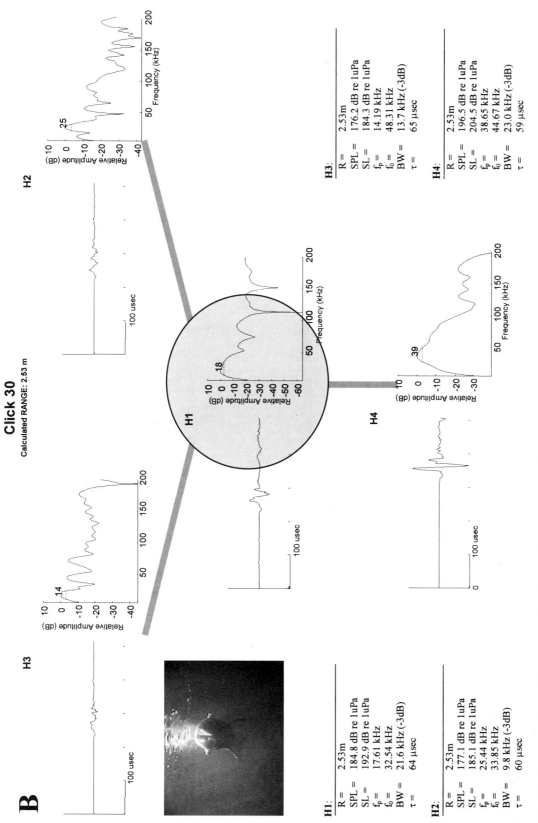

Click 30

Calculated RANGE: 2.53 m

H3:
R =	2.53m
SPL =	176.2 dB re 1uPa
SL =	184.3 dB re 1uPa
f$_p$ =	14.19 kHz
f$_0$ =	48.31 kHz
BW =	13.7 kHz (-3dB)
τ =	65 μsec

H4:
R =	2.53m
SPL =	196.5 dB re 1uPa
SL =	204.5 dB re 1uPa
f$_p$ =	38.65 kHz
f$_0$ =	44.67 kHz
BW =	23.0 kHz (-3dB)
τ =	59 μsec

H1:
R =	2.53m
SPL =	184.8 dB re 1uPa
SL =	192.9 dB re 1uPa
f$_p$ =	17.61 kHz
f$_0$ =	32.54 kHz
BW =	21.6 kHz (-3dB)
τ =	64 μsec

H2:
R =	2.53m
SPL =	177.1 dB re 1uPa
SL =	185.1 dB re 1uPa
f$_p$ =	25.44 kHz
f$_0$ =	33.85 kHz
BW =	9.8 kHz (-3dB)
τ =	60 μsec

Fig. 7.4. Two examples of click sets acquired during the trial depicted in fig. 7.3: click 20 (*A*) and click 30 (*B*). Signal waveforms are plotted at left and the frequency spectra are plotted at right for each click. A visual still frame shows the position of the animal at the approximate time of click acquisition. R = range, m; SPL = sound pressure level, dB re 1μPa; SL = source level, dB re 1μPa 1m; f$_p$ = peak frequency, kHz; f$_0$ = center frequency, kHz; BW = 3 dB bandwidth, kHz; τ = duration, μsec.

deal is known about the general system of echolocation in dolphins, much about the natural functional use of echolocation by dolphins in the wild—such as, for example, how and when dolphins use echolocation to forage or how their sonar is adapted to the type of prey they consume—remains to be discovered. The study of echolocation in species like the Risso's dolphin that are biologically and behaviorally different from those more typically studied could be useful in providing a more comprehensive understanding of dolphin echolocation.

Acknowledgments

The authors thank Dr. Roland Aubauer and Michiel Schotten for their work in developing the symmetrical-star array system and Dave Lemonds for invaluable analysis assistance. This research was funded by the Biosonar Program of the Office of Naval Research, grant N00014-99-1-0800, to Paul Nachtigall and Whitlow Au. The support of scientific officers Harold Hawkins, Robert Gisiner, and Teresa McMullen is gratefully acknowledged.

{ 8 }

Echolocation and Social Signals from White-Beaked Dolphins, *Lagenorhynchus albirostris,* Recorded in Icelandic Waters

Marianne H. Rasmussen and *Lee A. Miller*

Introduction

The white-beaked dolphin, *Lagenorhynchus albirostris* (Gray 1846), is distributed in the North Atlantic. They are seen both in small groups from two to five individuals or in large groups of up to 1500 animals. White-beaked dolphins are curious animals and often bow ride and swim near boats. They are very social and exhibit different aerial displays like tail walking, jumping, tail splashing, and twisting in the air. They feed on herring (*Clupea harengus*), Atlantic cod (*Gadus morhua*), haddock (*Melanogrammus aeglefinus*), whiting (*Merlangius merlangus*), sandeel (*Ammodytes tobianus*), and squid (Evans 1980). They associate with gannets (*Sula bassana*), fin whales (*Balaenoptera physalus*), minke whales (*Balaenoptera acutorostrata*), and humpback whales (*Megaptera novaeangliae*).

Mitson (1990) reported that acoustic emissions from white-beaked dolphins have significant energy at frequencies around 305 kHz. (They used a passive sonar centered on 305 kHz with a bandwidth of 40 kHz.) Mitson (1990) recorded signals from white-beaked dolphins as far away as 70 m. No audiogram exists from a white-beaked dolphin, but Tremel et al. (1998) measured an audiogram from a Pacific white-sided dolphin, *Lagenorhynchus obliquidens*. This audiogram was very similar to those known from other odontocetes and showed best hearing sensitivities from 4 to 128 kHz.

Echolocation of the bottlenose dolphin, *Tursiops truncatus,* has been intensively studied (Au 1993). It uses echolocation clicks with a peak frequency (frequency of maximum amplitude in the spectrum) of 120 kHz when recorded in open waters (Au 1993). Other characteristics of their sonar clicks are center frequency (of the spectral energy density) at 100 kHz; 3 dB bandwidth of 41 kHz (−3 dB to each side of the peak frequency), rms bandwidth of 22 kHz, and typical duration of 40 to 70 μs. The sonar beam is very directional, becoming narrower at higher frequencies. The 3 dB beam bandwidth is directed between 5 and 10° above the longitudinal axis of the animal (Au 1993).

The aim of this study was to describe the whistles and presumed echolocation clicks from white-beaked dolphins. We found that the clicks from white-beaked dolphins resemble those from the bottlenose dolphins and that they contain energy above 250 kHz, supporting the results of Mitson (1990). We also found that white-beaked dolphin whistles contain ultrasonic components.

Material and Methods

Field recordings of echolocation and social signals were made in the waters off southwest Iceland, not far from Keflavik harbor, 64°00.33′ N, 22°33.30′ W. Dolphins were located about one to three nautical miles offshore, where the water was 35 to 37 m deep. We recorded echolocation and social signals using Brüel & Kjær (B&K)

type 8103 hydrophones, an Etec amplifier, and a Racal high-speed tape recorder set at 30 in/s or 60 in/s. The recording system was calibrated up to 300 kHz. The dolphins were 1–10 m away during recordings of echolocation signals. Some social signals were recorded with a monitoring hydrophone or a B&K 8103 and recorded on a Sony digital cassette DAT recorder. Recordings were made with one hydrophone over each side of a 10 m fiberglass motorboat at a depth of 4 m. The echolocation clicks were sampled at 705.6 kHz and the social signals at 88.2 kHz. The frequency and time characteristics of the signals were analyzed using standard methods with special software (S.B. Pedersen) and BatSound (Pettersson Elektronik AB, Uppsala, Sweden). Spectrograms of whistles were generated using a sample rate of 88.2 kHz and consecutive 512 point FFTs with Hann windows (85% overlap). Whistles were categorized based on duration and frequency contours, similar to the method used by Moore and Ridgway (1995).

Results

A total of 1500 clicks were analyzed. These were taken from 16 trains or bursts presumably produced by individual dolphins. Click intervals varied from 2.8 to 40.0 ms (mean click interval $\pm$ SD was 5.8 $\pm$ 5.3 ms). In the sequence shown in fig. 8.1, dolphins were about 5 m from the boat when they dove and swam straight toward the hydrophone. The sequence was probably produced by one individual. The highest repetition rate (357 clicks/s) was found in this sequence, which contained 490 clicks (fig. 8.1A). Click interval varied from 2.8 to 6.8 ms, and the power spectra of these clicks had peak frequencies of 115 $\pm$ 3 kHz ($n = 100$). However, some clicks could have a lower amplitude peak at 250 kHz; these clicks were probably directed at the hydrophone (fig. 8.1B). Other properties of the clicks were center frequency at 82 $\pm$ 4 kHz, 3 dB bandwidth at 70 $\pm$ 12 kHz, and rms bandwidth at 36 $\pm$ 2 kHz. The duration

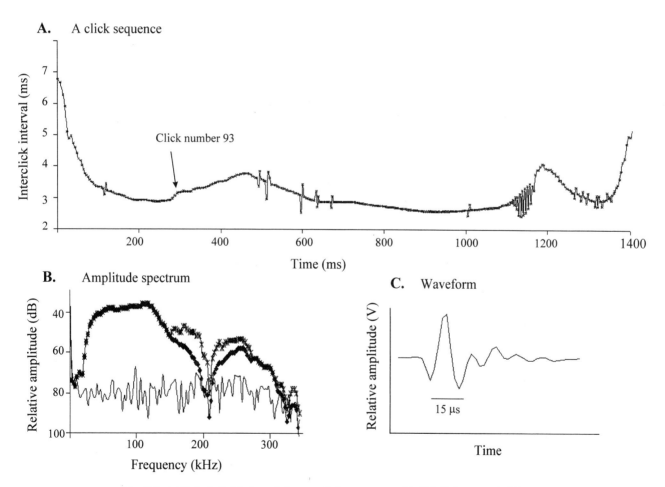

Fig. 8.1. *A:* Plot of click intervals from a click sequence with 490 clicks recorded from a wild white-beaked dolphin, *L. albirostris,* in Icelandic waters, September 1997. *B:* Amplitude spectrum of click number 93, which is found 290 ms from the start of the recording (—♦— shows the spectrum recorded with a B&K 8103; —×— shows the corrected spectrum according to B&K calibrations; and —— shows the noise floor). *C:* Waveform of click number 93.

of a single click in this sequence was between 10 and 30 μs (fig. 8.1C).

We analyzed 1536 whistles. These social sounds were only recorded when dolphins were showing aerial displays, when a group of 30 dolphins swam around a humpback whale, or when a diver was in the water. Usually, many dolphins whistled at the same time. We never recorded whistles when the dolphins were feeding or traveling. Most recordings were made using a DAT, but high-speed recordings are included (345 whistles out of 1536 whistles). Whistles ranged from 3 kHz up to 35 kHz. The duration of a single whistle was from 3 ms to 1.2 s. Whistles were classified into 22 categories (table 8.1). Whistles from categories 13, 14, 15, 17, and 20 were not found in the high-speed recordings. Fig. 8.2

shows spectrograms of some whistles from category 1, "long up-sweep," and category 6, "long down-up."

Discussion

Spectral and temporal parameters of the clicks from the white-beaked dolphin, *L. albirostris,* resembled those of clicks produced by the bottlenose dolphin, *T. truncatus* (Au 1993). Therefore, we assumed that white-beaked dolphins use clicks for echolocation. We observed waxing and waning in the amplitude of the power spectrum above 100 kHz for signals in a long sequence (fig. 8.1A). This is probably due to the narrower beamwidth for high-frequency signals and to the dolphin turning its head or body. Only clicks projected directly at the hydrophone contained high frequencies (Au 1993). Commonly clicks had a second, high-frequency region between 200 and 250 kHz (fig. 8.1B), which confirms earlier studies reporting very high frequencies in signals recorded from the white-beaked dolphin (Mitson 1990).

Why should the white-beaked dolphin have a second, high-frequency peak in its sonar signals? One reason could be that *L. albirostris* are mainly feeding on sand eels (*Ammodytes* sp.) around Iceland during the summer. Sand eels have very small otoliths, between 0.5 and 4.0 mm in length, which correlate to a fish length between 41.6 and 201.7 mm (Härkönen 1986). Sand eels also have no swim bladder (Reay 1986). Using high-frequency sonar up to 300 kHz gives a wavelength of 5 mm (approximately 1500 m/s), which improves the chance of detecting a sand eel. Another explanation might be found in the mechanism of sound production. According to Cranford, Amundin, and Norris (1996), sound is produced by the dorsal bursae complex. CT scans of the head of *L. obliquidens* were not very different from those of *L. albirostris,* but quite different from those of *T. truncatus* (Cranford, Amundin, and Norris 1996). The most striking difference is two enlarged fatty basins located between the melon and the dorsal bursae complex seen in *L. obliquidens.* The height, width, and length of the dorsal bursae are slightly larger for *L. obliquidens* than for other dolphin species. There is less skull asymmetry in *L. obliquidens* than in *T. truncatus.* We do not know whether these morphological differences affect sonar signals.

White-beaked dolphin whistles contained higher frequencies than reported from other dolphin species. Although we used a DAT to record most whistles, some were recorded on a Racal tape recorder (frequency range of up to 150 kHz). An example of high-frequency whistles is shown in fig. 8.2, where the frequency range is between 8 and 35 kHz. Most recordings of dolphin whistles are made with audio equipment and thus limited to about 20 kHz. Herzing (1996) described signature whistles from the spotted dolphin (*Stenella frontalis*) to between 4 and 18 kHz. Steiner (1981) studied whistles for free-ranging *T. truncatus, L. acutus, Globi-*

TABLE 8.1 Whistle repertoire of the white-beaked dolphin, *L. albirostris.* Most recordings were made in Icelandic waters in June and September 1997 using a DAT, but high-speed recordings are also included (345 whistles out of a total of 1536 whistles). Whistles from categories 13, 14, 15, 17, and 20 were not found in the high-speed recordings. The start and end frequencies are mean values.

Categories	Spectrogram	Number	Mean duration (s)	Frequency range (kHz)	Start frequency (kHz)	End frequency (kHz)
1	/	268	0.26	3.0 – 29.1	9.9	13.0
2	\	72	0.20	3.7 – 23.3	10.4	8.4
3	,	86	0.06	4.8 – 18.9	9.5	12.5
4	،	45	0.07	3.5 – 18.8	10.7	8.9
5	∧	139	0.27	5.4 – 20.4	9.9	10.3
6	∪	124	0.17	4.3 – 33.6	12.0	13.9
7	∧	145	0.07	3.6 – 29.8	10.2	10.4
8	∨	62	0.07	4.0 – 23.3	10.8	11.7
9	—	21	0.33	3.4 – 18.5	9.9	10.0
10	-	42	0.07	5.2 – 26.6	9.7	9.8
11	⌐	12	0.32	6.7 – 35.0	9.6	12.1
12	⌐	7	0.34	6.0 – 17.7	10.9	9.1
13	∿	309	0.32	5.1 – 16.9	10.3	16.1
14	⌐	28	0.16	6.2 – 17.5	10.6	8.9
15	⌣	10	0.26	6.9 – 18.9	8.5	11.0
16	⌐	11	1.05	5.6 – 20.1	11.1	8.7
17	∿∿	15	0.73	4.8 – 14.5	10.0	10.5
18	⌐	13	0.22	6.6 – 20.1	11.8	15.5
19	∼	34	0.32	5.1 – 18.4	10.2	12.0
20	⌢	12	0.24	4.2 – 16.4	11.1	10.4
21	⌒	17	0.68	5.2 – 16.3	7.6	7.3
22	others	64		4.6 – 23.5		

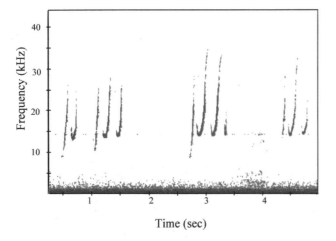

Fig. 8.2. Spectrogram of whistles from wild *L. albirostris,* recorded in Icelandic waters in September 1997. Whistles are from category 1, "long up-sweep" and category 6, "long down-up." Note that some whistles contain frequencies up to 35 kHz.

cephala malaena, Stenella plagiodon, and *Stenella longirostris* and reported frequencies up to 20 kHz for these dolphins. Ding, Würsig, and Evans (1995) described geographic population differences of whistles for free-ranging bottlenose dolphins. The frequency response for all recordings ranged from about 100 Hz to about 15 kHz. Recently, Au, Lammers, and Aubauer (1999) reported ultrasonic harmonic components above 70 kHz in the whistles of spinner dolphins, *Stenella longirostris,* recorded with broadband equipment.

Whistle contours shown in table 8.1 are similar to those of other dolphin species—for example, the bottlenose dolphin in captivity (McCowan and Reiss 1995) or free-ranging dolphins (Steiner 1981). Moore and Ridgway (1995) divided whistles from captive and free-ranging common dolphins, *Delphinus delphis,* into eight categories. The whistles from white-beaked dolphins fell into similar categories (categories 1, 2, 3, 6, 10, 20, and 21 in table 8.1), but we could classify additional categories. The more varied whistle repertoire from the white-beaked dolphin could reflect the large sample size (1536 whistles) compared to 186 whistles from free-ranging common dolphins (Moore and Ridgway 1995). Also free-ranging dolphins could use a greater variety of whistles, perhaps reflecting a richer variety of social interactions in the field. But some variety could stem from the fact that dolphins mimic one another (Tyack 1997). Attributing the whistles from free-ranging dolphins to specific social situations is difficult—that is, who is whistling and what provoked it are unknown.

Acknowledgments

This study was supported by the Danish National Research Foundation, Danish Research Council for Natural Sciences, and Dansk-Islandsk Fond. Thanks to Börje Wijk at Chalmers tekniska Högskola, Göteborg, Sweden for lending us the Racal recorder and to North Sailing in Husavik for their hospitality. We are grateful to Simon Boel Pedersen for software development and helpful discussions concerning signal analysis, to Brüel & Kjær for providing a special calibration chart, and to David Helweg for kindly donating a monitoring hydrophone. Special thanks go to Dolphin and Whale Spotting in Keflavik, to Helga Ingimundardottir for her hospitality, and to the captain, Olafur Björnsson, for use of the boat *Hnoss*. The studies were done in cooperation with the Marine Research Institute in Reykjavik, with special thanks to Gisli Vikingsson and the staff of the Whale Department for their helpfulness.

{ 9 }

Acoustic Properties of Echolocation Signals by Captive Pacific White-Sided Dolphins (*Lagenorhynchus obliquidens*)

Margaret Fahner, Jeanette Thomas, Ken Ramirez, and *Jeff Boehm*

Introduction

Pacific white-sided dolphins (*Lagenorhynchus obliquidens*) inhabit temperate waters of the Pacific Ocean and are found mainly offshore along the continental shelf (Leatherwood et al. 1984). They are highly gregarious, travel in groups of up to hundreds of individuals, and feed on a variety of schooling fish and squid (Jones 1981). Stacey and Baird (1990) reported that Pacific white-sided dolphins often swim in mixed species herds.

Pacific white-sided dolphins have been collected for research and public display since 1966 (Leatherwood et al. 1984). Norris (1968) first demonstrated that this species could echolocate by training the dolphin to wear eyecups and avoid obstacles as it swam freely in a pool. Evans (1973) first recorded and documented the acoustic properties of echolocation signals from a captive, adult female Pacific white-sided dolphin. He reported a peak frequency from 60 kHz to 80 kHz, pulse duration of 0.25 to 1.00 ms, and a source level of 170 dB re 1 μPa. Evans also discovered that this same female could discriminate objects using echolocation. Hatakeyama et al. (1994) reported that captive Pacific white-sided dolphins could detect nylon monofilament gillnets using echolocation. To date, details about the acoustic properties of biosonar signals for wild Pacific white-sided dolphins are not available.

In this study, we further characterized the acoustic

properties of sonar pulses and trains for the sparsely studied Pacific white-sided dolphin. Most captive studies on echolocation data come from a single animal; however, we collected biosonar data from four dolphins. Our study also took advantage of existing trained behaviors to collect data quickly, rather than requiring months of special training for echolocation tests. The main objectives were to (1) document acoustic properties of echolocation signals in this species, (2) investigate variations in acoustic properties of pulses and trains among individual dolphins, and (3) examine signal differences when dolphins echolocated on a cylindrical versus a spherical target.

Materials and Methods

The subjects were four adult Pacific white-sided dolphins (one male and three females) housed at the John G. Shedd Aquarium in Chicago, Illinois. All dolphins were 8–11 years old, collected from the wild as juveniles, and part of the same social group at the aquarium since 1994. Echolocation sessions took place in a small, concrete pool ($8 \times 5.5 \times 3$m) within the oceanarium that was out of public view (fig. 9.1). Each dolphin was alone in the test pool for trials. To isolate the test dolphin acoustically (to the best of our abilities), the remaining five dolphins and five belugas (*Delphinapterus leucas*) were involved in a training session in another pool farthest from the test pool. As part of basic training at the Shedd Aquarium, each dolphin was trained to recognize and station with the rostrum on its own uniquely shaped target (i.e., circle, square, star, or cross sign). For these tests, a pole attached to the walkway at one end of the pool placed the dolphin's shape at 1.5 m under water (fig. 9.1).

In our study, dolphins did not wear eyecups and were not trained for a go/no-go or match-to-sample test paradigm (as in other echolocation studies). Rather, we assumed that when given a hand-cue to swim toward their shape the dolphin would submerge and echolocate on any novel object in the path to its shape.

Sessions were conducted during November and December 1997. Each of four dolphins participated in two sessions lasting approximately 20 to 25 minutes. In each session, the dolphin had 10 trials with a cylindrical target, 10 trials with a spherical target, and 10 trials with only the hydrophone (the control). Trial types were randomized over the whole session. The spherical target was a stainless-steel, water-filled sphere 7.5 cm in diameter (target strength approximately −34 dB). The cylindrical target was a hollow aluminum cylinder with a 6.65 mm wall thickness, 3.81 cm outer diameter, and 12.7 cm length (target strength approximately −20 dB).

At the beginning of a trial, a target was lowered in the middle of the pool and directly in front of the hydrophone using a pulley. If the trial was a control, the target was quietly removed, leaving only the hydrophone in the dolphin's path. When lowered, the target and hydrophone were 1.0 m apart and directly in-line with the dolphin's shape; all were at a 1.5 m depth.

Trainers, on opposite sides of the pool, initiated a trial by giving a hand-cue for the dolphin to submerge and swim down the middle of the pool toward its shape, thus encountering a hydrophone and/or target along the path. Once a dolphin reached the shape, it stationed on the shape, and then was cued to return to the first trainer for reinforcement. One researcher controlled the pulley and target type and narrated events into a handheld cassette recorder. An assistant controlled the tape recorder

Fig. 9.1. Experimental setup

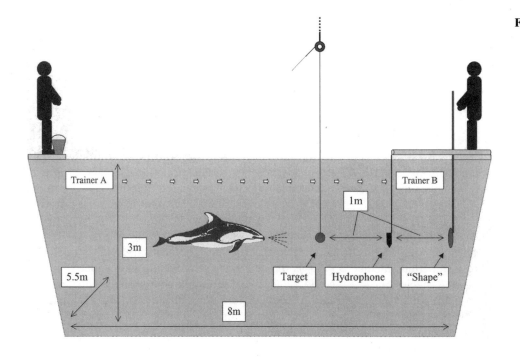

and noted events on a data sheet, including in which trials the dolphin swam directly toward the target and hydrophone.

The hydrophone was an Ithaco model 601 (sensitivity −169 dB re 1 μPa). Recordings were made using a Telex 434 reel-to-reel recorder with a Krohn-hite 3901 HP/LP two-channel filter (10 dB gain). The system frequency response was linear to 100 kHz ± 3 dB.

Analysis of pulses and trains was conducted using the PC-based SIGNAL software and a LeCroy 9310A digital oscilloscope (analysis range linear to 100 kHz). To avoid problems with off-axis signals, only pulse trains produced by a dolphin swimming directly toward the hydrophone were analyzed. For each pulse, the waveform, peak frequency, bandwidth (at −3 dBm), peak amplitude (dB re 1 μPa), and pulse duration were documented. For each train of pulses, the train duration, number of pulses in a train, interpulse intervals, and presence of echoes were examined. One-way ANOVA on variables was conducted to examine signal differences among dolphins and among targets. Two-way ANOVA was conducted to examine signal characteristics dependent on the interactions between specific dolphins and target types. All statistical tests were conducted at α = 0.05 level.

Results

PULSE CHARACTERISTICS

Among echolocation trains from all four dolphins, 172 pulses were recorded directly in-line with the hydrophone and analyzed. Comparing all dolphins in all tasks, pulse duration ranged from 40 to 71 μs, peak frequencies fell between 50 kHz and 80 kHz, bandwidth (at −3 dBm) ranged from 9.5 kHz to 36 kHz, and peak amplitudes were from 149 dB to 157 dB re 1 μPa (table 9.1).

The waveform was simple and no evidence of doublette pulses was found (fig. 9.2).

One-way ANOVA (table 9.1) showed that among the four dolphins, there were significant differences in all pulse parameters: duration (F = 6.568, df = 3, P =

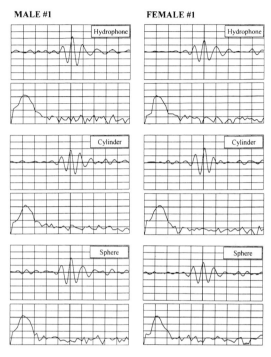

Fig. 9.2. Representative oscillogram of pulses of Pacific white-sided dolphins echolocating on a sphere, cylinder, or hydrophone: male #1; female #1. Power spectrum is a zoom of the pulse only. *Top waveform:* x-axis is time (μs), 20 μs per division; y-axis is amplitude (mV), 20 mV per division. *Bottom power spectrum:* x-axis is frequency (kHz), 50 kHz per division; y-axis is amplitude (dBm), 13.6 dBm per division.

TABLE 9.1. Summary statistics for echolocation pulses from Pacific white-sided dolphins. Highlighted areas indicate significant differences from one-way ANOVA tests.

	Pulse duration (μs)				Peak frequency (kHz)				−3dBm bandwidth			Peak amplitude		
	Minimum	Maximum	Mean	Standard deviation	Minimum	Maximum	Mean	Standard deviation	Minimum	Maximum	Mean	Minimum	Maximum	Mean
All Pulses (n = 172)	39.7	70.9	52.8	5.1	50.0	80.0	57.3	7.4	9.5	35.8	22.7	149	157	152
Male #1 (n = 45)	46.5	67.4	54.4	4.5	50.0	70.0	54.4	6.2	9.5	29.8	21.2	149	154	152
Female #1 (n = 27)	46.7	60.3	54.1	6.0	50.0	70.0	57.1	7.6	10.0	30.3	21.8	151	157	153
Female #2 (n = 45)	39.7	70.9	50.2	5.9	50.0	80.0	61.1	6.0	12.8	35.8	24.9	150	154	151
Female #3 (n = 55)	43.2	64.6	53.1	4.5	50.0	80.0	56.5	7.9	12.0	33.3	22.6	149	156	153
Hydrophone (n = 70)	46.1	70.9	53.0	3.8	50.0	70.1	56.4	6.8	12.0	35.8	23.9	149	156	152
Cylinder (n = 50)	42.7	67.4	52.4	6.3	50.0	70.1	59.2	6.6	12.8	33.3	23.1	149	157	152
Sphere (n = 52)	39.7	64.6	53.0	5.3	50.0	80.0	56.6	8.6	9.5	33.8	20.7	150	156	152

Table 9.2. Summary statistics for echolocation pulses from Pacific white-sided dolphins. Highlighted areas indicate significant differences from two-way ANOVA tests.

	Pulse duration (μs)				Peak frequency (kHz)				−3 dBm bandwidth			Peak amplitude		
	Minimum	Maximum	Mean	Standard deviation	Minimum	Maximum	Mean	Standard deviation	Minimum	Maximum	Mean	Minimum	Maximum	Mean
All Pulses (n = 172)	39.7	70.9	52.8	5.1	50.0	80.0	57.3	7.4	9.5	35.8	22.7	149	157	152
M #1/Hydrophone (n = 20)	46.5	54.2	50.9	2.0	50.0	70.0	57.4	7.1	22.5	29.8	26.2	149	153	152
M #1/Cylinder (n = 5)	50.4	67.4	58.7	7.8	50.0	60.1	58.0	5.4	14.8	21.5	18.3	151	154	152
M #1/Sphere (n = 20)	50.9	61.4	56.7	2.6	50.0	60.0	50.5	2.2	9.5	25.5	16.9	150	154	153
F #1/Hydrophone (n = 10)	51.3	58.4	55.4	2.4	50.0	70.0	54.0	6.9	13.8	30.3	21.6	151	155	153
F #1/Cylinder (n = 10)	47.5	58.5	53.9	4.1	50.0	70.0	60.0	9.4	13.5	29.8	24.7	151	157	154
F #1/Sphere (n = 7)	46.7	60.3	52.5	5.3	50.1	61.1	57.2	4.9	10.0	25.3	17.9	152	155	153
F #2/Hydrophone (n = 20)	46.1	70.9	52.7	5.5	50.1	70.1	60.6	6.0	13.0	35.8	24.4	150	154	152
F #2/Cylinder (n = 20)	42.7	61.3	48.9	5.8	50.0	70.1	60.4	4.9	12.7	33.3	24.9	150	153	151
F #2/Sphere (n = 5)	39.7	49.4	45.2	3.8	60.0	80.0	66.0	8.9	20.5	33.8	27.0	150	154	152
F #3/Hydrophone (n = 20)	48.4	59.1	54.1	2.8	50.0	60.0	52.5	4.4	12.0	30.8	22.4	150	156	154
F #3/Cylinder (n = 15)	46.9	62.4	53.9	5.6	50.0	70.0	57.3	7.0	16.3	33.3	21.2	149	153	152
F #3/Sphere (n = 20)	43.2	64.6	51.5	4.8	50.0	80.0	60.0	9.7	14.5	29.5	23.9	150	156	152

0.000), peak frequency (F = 7.175, df = 3, P = 0.000), 3 dBm bandwidth (F = 4.030, df = 3, P = 0.000), and peak amplitude (F = 9.551, df = 3, P = 0.000). In comparison, only the 3 dBm bandwidth was affected by the type of target—for example, there were significant differences among pulses only in bandwidth while echolocating on a cylinder, sphere, or the hydrophone alone (F = 5.715, df = 2, P = 0.004).

A two-way ANOVA (table 9.2) comparing subjects by target type showed significant differences in all pulse parameters (fig. 9.3): duration (F = 7.002, df = 6, P = 0.000), peak frequency (F = 5.197, df = 6, P = 0.00), 3 dBm bandwidth (F = 7.084, df = 6, P = 0.000), and peak amplitude (F = 4.501, df = 6, P = 0.000). However, there were no discernible trends in these differences.

Train Characteristics

Forty-three pulse trains were recorded clearly as the dolphins swam directly in-line with the hydrophone. Echoes were identifiable in some recordings and typically were attenuated and out-of-phase with the original pulse. Pulse trains lasted from 0.03 to 3.81 s (table 9.3). The number of pulses per train varied from 3 to 54. Time between pulses was from 2.9 ms to 1198 ms (table 9.3). In most cases, pulses accelerated in repetition rate as the dolphin approached a target, but occasionally they were even or irregular.

One-way ANOVA (table 9.3) showed a significant difference in train duration and the number of pulses among dolphins. Female 3 had much longer trains and a greater number of pulses (mean = 2.38 s and 37 pulses) versus the pooled average of the other three dolphins

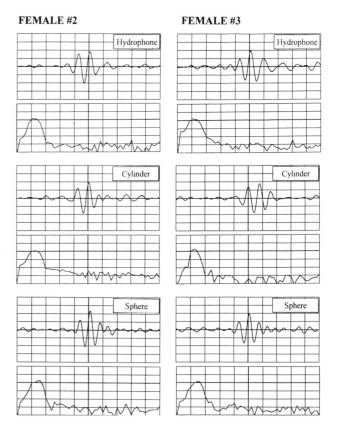

Fig. 9.3. Representative waveform and power spectrum of two female Pacific white-sided dolphin echolocation pulses on a hydrophone versus a cylindrical or a spherical target. Power spectrum is a zoom of the pulse only. *Top waveform: x*-axis is time (μs), 20 μs per division; *y*-axis is amplitude (mV), 20 mV per division. *Bottom power spectrum: x*-axis is frequency (kHz), 50 kHz per division; *y*-axis is amplitude (dBm), 13.6 dBm per division.

TABLE 9.3. Summary statistics for echolocation trains from each Pacific white-sided dolphin and each target type. Highlighted areas indicate significant differences from one-way ANOVA tests.

	Train duration (s)					Number of pulses					Interpulse interval (ms)				
	Number of cases	Minimum	Maximum	Mean	Standard deviation	Number of cases	Minimum	Maximum	Mean	Standard deviation	Number of cases	Minimum	Maximum	Mean	Standard deviation
All Trains	43	0.03	3.81	1.39	1.11	43	3	54	21	15	770	2.9	1197.9	62.9	104.9
Male #1	13	0.03	2.50	0.95	0.85	13	3	27	15	8	169	2.9	981.1	61.7	124.6
Female #1	9	0.19	3.08	0.95	0.86	9	4	35	11	10	87	9.5	740.4	82.1	120.8
Female #2	7	0.15	1.90	0.80	0.61	7	6	32	16	8	104	6.2	588.6	49.1	75.5
Female #3	14	0.50	3.81	2.38	1.06	14	11	54	37	14	410	10.9	1197.9	62.9	98.4
Hydrophone	13	0.03	3.81	1.47	1.49	13	3	52	22	18	250	2.9	718.5	58.7	92.4
Cylinder	17	0.11	3.08	1.07	0.83	17	4	46	20	12	321	5.8	740.4	52.1	77.3
Sphere	13	0.36	3.53	1.74	0.94	13	5	54	22	15	199	17.7	1197.9	85.8	146.9

TABLE 9.4. Summary statistics for echolocation trains from each Pacific white-sided dolphin by target type. Highlighted areas indicate significant differences from two-way ANOVA tests.

	Train duration (s)					Number of pulses					Interpulse interval (ms)				
	Number of cases	Minimum	Maximum	Mean	Standard deviation	Number of cases	Minimum	Maximum	Mean	Standard deviation	Number of cases	Minimum	Maximum	Mean	Standard deviation
All Trains	43	0.03	3.81	1.39	1.11	43	3	54	21	15	770	2.9	1197.9	62.9	104.9
M #1/Hydrophone	3	0.03	0.28	0.15	0.13	3	3	8	6	3	13	2.9	104.8	28.1	27.7
M #1/Cylinder	5	0.11	1.02	0.05	0.34	5	13	2	17	3	81	5.8	228.1	30.4	36.6
M #1/Sphere	5	1.09	2.50	1.87	0.51	5	5	27	17	9	75	18.3	984.1	101.3	175.5
F #1/Hydrophone	4	0.19	1.03	0.74	0.39	4	6	16	10	5	37	9.5	545.4	80.5	114.5
F #1/Cylinder	3	0.29	3.08	1.45	1.45	3	4	35	15	17	39	12.4	740.4	77.1	119.8
F #1/Sphere	2	0.36	0.88	0.62	0.37	2	6	10	8	3	11	17.7	414.1	105.0	151.9
F #2/Hydrophone	1	0.15	0.15	0.15	—	1	22	22	22	—	21	6.2	9.1	7.4	1.0
F #2/Cylinder	4	0.20	1.90	0.89	0.73	4	6	32	16	11	59	29.1	588.6	60.1	90.4
F #2/Sphere	2	0.72	1.21	0.96	0.35	2	14	16	15	1	24	33.6	229.9	58.4	54.1
F #3/Hydrophone	5	1.37	3.81	3.10	0.99	5	18	52	41	14	179	20.7	718.5	62.4	93.5
F #3/Cylinder	5	0.50	2.00	1.55	0.60	5	11	46	30	13	142	10.9	595.7	54.4	71.5
F #3/Sphere	4	1.22	3.53	2.53	1.00	4	19	54	40	15	89	19.3	1197.9	77.7	137.1

(mean = 0.95 s and 14 pulses). Only interpulse interval was significantly different among target types, with the spherical target prompting the longest interval between echolocation signals.

A two-way ANOVA on train duration detected a significant interaction between the specific dolphin and type of target (table 9.4). Female 3 had noticeably longer train duration than the other three dolphins. However, a similar test using the number of pulses and interpulse interval did not detect a significant interaction between a specific dolphin and target type.

Discussion

Echolocation signal characteristics vary among species of marine mammals. These differences could be related to anatomy and physiology, ecology, ambient noise levels, or type of task. Echolocation pulses from the four captive Pacific white-sided dolphins had a simple waveform, similar to pulses from bottlenose dolphins (*Tursiops truncatus*), killer whales (*Orcinus orca*), and false killer whales (*Pseudorca crassidens*) (Au 1993). Rasmussen and Miller (chapter 8, this volume) reported a comparable simple waveform for a congener, the wild white-beaked dolphins (*Lagenorhynchus albirostris*). In contrast, species in the family Phocoenidae and in the delphinid genus, *Cephalorhynchus,* emit complex, single, or double pulses (Dawson 1988; Au 1993). Belugas produce both simple pulses and packets of pulses (Au 1993).

There are five other species within the genus *Lagenorhynchus* (*L. acutus, L. albirostris, L. australis, L. cruciger,* and *L. obscurus*), but echolocation data are only available for the white-beaked dolphin, *L. albirostris* (Mitson 1990; Mitson and Morris 1988; Rasmussen and Miller, chapter 8, this volume). Wild white-beaked dolphins emit pulses recorded at extremely high frequencies near 305 kHz (Mitson 1990; Mitson and Morris 1988). Rasmussen and Miller (chapter 8) recorded peak frequencies of 115 ± 3 kHz. No captive discrimination studies have been conducted on this species; however, Rasmussen and Miller (chapter 8) suspect that *L. albirostris* also uses clicks for echolocation due to the similarity of the waveform with that of the well-studied bottlenose dolphin. Cranford, Amundin, and Norris (1996) made CT scans of *L. obliquidens* and *L. albirostris* and found similarities between their skulls and tissue structures. So results of echolocation studies on one species likely are a good predictor of the abilities of the other species.

Tremel et al. (1998) studied the underwater hearing of a captive female Pacific white-sided dolphin and found a typical mammalian, U-shaped audiometric curve, with the greatest sensitivities from 4 kHz to 128 kHz. The sonar pulses of the Pacific white-sided dolphins in our study fell within this range of best frequencies for underwater hearing. Pulse peak frequencies fell between 50 kHz and 80 kHz—similar, but more broadband, than the peak frequencies of 60 kHz to 80 kHz reported by Evans (1973) for this species. Our peak-frequency data were comparable to echolocation pulses from killer whales, common dolphins, false killer whales, and pilot whales (*Globicephala* spp.). However, peak frequencies from these *L. obliquidens* signals were lower than pulses from belugas, bottlenose dolphins, any porpoise in the family Phocoenidae, and members of the genus *Cephalorhynchus* (Richardson et al. 1995).

Odontocetes in reverberant pools have different signal characteristics than their wild counterparts, often being lower in amplitude and lower in peak frequency. Pulses from *Tursiops truncatus, Delphinapterus leucas,* and *Pseudorca crassidens* measured in a pool (Au 1993) all had lower peak frequencies (30 kHz to 60 kHz) than in an open water, high-ambient noise environment (100 kHz to 130 kHz). The low-ambient noise environment in the Shedd Aquarium pools could have facilitated dolphins using lower peak frequencies (Tremel et al. 1998) than might be used in the wild.

Cranford, Amundin, and Norris (1996) studied the skull and tissue structures of a variety of odontocetes

and suggested that the sound-production mechanisms of *L. obliquidens* and *L. albirostris* are similar. In our study, pulses of *L. obliquidens* had much lower peak frequencies and lacked a second high-frequency component, as seen in *L. albirostris* (Mitson 1990; Mitson and Morris 1988; Rasmussen and Miller, chapter 8). Cranford, Amundin, and Norris (1996) suggested, based on anatomical data, that *L. obliquidens* should have a dual frequency component in their echolocation pulses. Additionally, Nakamura and Akamatsu (chapter 5, this volume) studied three captive Pacific white-sided dolphins and recorded peak frequencies up to the 120 kHz range. Our study, and perhaps that of Evans (1973), was limited by equipment from detecting a second frequency component over 100 kHz. Hearing studies by Tremel et al. (1998) on this species only tested up to 140 kHz, so if a secondary area of sensitivity occurs in this species it would be well above this frequency.

Pulse duration in our study was from 40 to 70 μs, compared with Evans's (1973) pulse durations of 0.25 to 1.00 ms. Pulse durations reported herein are near the typical length of echolocation pulses (50–200 μs) for odontocetes reported by Richardson et al. (1995).

Peak amplitude averaged 152 dB re 1 μPa, lower than 170 dB reported for the same species by Evans (1973). Source levels over 200 dB re 1 μPa have been reported in belugas, false killer whales, and bottlenose dolphins (Richardson et al. 1995). However, echolocating cetaceans typically do not produce their maximum signal level when echolocating in a small, reverberant pool (Au 1993). As in other echolocating odontocetes, the amplitude and frequency of an echolocation signal are codependent—when one decreases, so does the other. So it is not surprising that both source level and peak frequency were lower in our study animals. Data on the source level of pulses of wild Pacific white-sided dolphins presently are unavailable; however, we would not be surprised if pulses in the wild are both higher in frequency and in amplitude. Studies on source levels of echolocation are best conducted in quiet field situations. The ideal study of echolocation in this species would be in an open ocean environment using extremely broadband sensing and recording equipment. Clearly, more work is needed on both hearing and echolocation abilities in this species.

Richardson et al. (1995) reported three types of echolocation trains in odontocetes: (1) orientation clicks, with long interclick intervals used to scan the environment at a distance; (2) discrimination clicks, with brief interclick intervals and an accelerating repetition rate; and (3) nonfunctional, collateral acoustic behavior or a partial pulse train. Pulse trains from our Pacific white-sided dolphins were probably the second type, with an accelerating repetition rate and brief interpulse intervals used to detect a target or the hydrophone.

In our study, target type seemed to have little influence on pulse duration, peak frequency, or peak amplitude of outgoing pulses. Only interpulse intervals and -3 dBm bandwidth were significantly different among target types. Different targets should provide different highlight structures in the echo. Unfortunately, low source levels of outgoing pulses and the short distance to the target (i.e., short signal-to-echo delay) made detection and detailed examination of echoes nearly impossible. The highlight structure of Pacific white-sided dolphin echoes off different targets should be studied in a larger pool or semicaptive environment.

Perhaps the most important finding in our study were the significant differences in echolocation signals among individual dolphins. Pulse duration, peak frequency, peak amplitude, -3 dBm bandwidth, train duration, and number of pulses were considerably different among the four individual dolphins conducting the same echolocation task. All had unique sonar characteristics—even though they were of comparable age, echolocating in an identical environment, performing the same task, and part of the same social group. Some differences could relate to the fact that the dolphins also could see the targets, so their pulse trains might not reflect a typical search pattern used by dolphins with eyecups. Sonar differences discernible among individual dolphins could be from slight differences in sound-generating mechanisms, resulting in unique sounds. The individualized echolocation signals could help identify specific dolphins. Alternatively, slight differences in pulses by individuals could assist in discrimination of one's own echo among those from a group of echolocating dolphins exploring the environment. Little is known about the "etiquette" of echolocation in a group of dolphins. Many questions remain to be answered. Is there a social rule concerning who echolocates? How do individual dolphins interpret the many echoes they receive while swimming in a group? Can dolphins eavesdrop and take advantage of the echoes produced by a neighbor dolphin? How do dolphins avoid masking one another's signals? These issues likely will receive considerable attention in future research on dolphin echolocation.

Acknowledgments

We give many thanks to Andrew Bielecki, Joseph Coelho, Jaqueline Connour, Megan Cope, Anne and Ty Fahner, Dave Hurd and LeCroy Corporation, Mersedeh Jalili, Mercury Metalcraft Corporation, Amy Nester, Kerri Still, Joe Utas, and Shannon Weight for their time and assistance. We especially acknowledge the marine mammal staff at the Shedd Aquarium for sharing their time and resources and for providing endless support for this study. Funding was granted from the Western Illinois University Research Council and the John G. Shedd Aquarium Research and Conservation Council.

{ 10 }

Modeling of the Click-Production Mechanism in the Dolphin

Nikolai A. Dubrovskiy and *Ludmila R. Giro*

Experiments with echolocating dolphins were conducted to determine a plausible click source location. An analysis of recorded signals confirmed that the source of the echolocation pulses is situated near the right nasal plug. A physical and a mathematical model are proposed to explain how dolphins can produce short, broadband, and powerful echolocation clicks. The physical model comprises a rubber tube and a rubber ring put on the underwater portion of the tube. This ring can block the air passage through the tube at some access air pressure. By increasing this pressure, one can put the model in a typical self-oscillation mode of vibrations. The displacement of the ring surface has a typical triangular shape. Acoustic clicks similar to those observed in dolphins are locked at the bends in the displacement curve. The mathematical model helps reveal the origin of the acoustic pressure clicks and their locking at bends of the displacement curve.

The question of how dolphins produce echolocation clicks has attracted much attention for more than three decades (Dubrovskiy and Zaslavskiy 1975; Giro and Dubrovskiy 1975; Dormer 1979; Ridgway et al. 1980; Ridgway and Carder 1988; Au 1993; Cranford 1992b; Cranford, Amundin, and Norris 1996). In spite of the considerable progress in understanding many important features of the click-production mechanism, there is still no definite explanation, to our knowledge, regarding how some biological structures in the dolphin can produce very short (15 μs), broadband (50 kHz), and powerful (226 dB re 1 μPa at 1 m) echolocation pulses.

In this chapter, we present an approach aiming to model the physical mechanisms underlying echolocation click production in the dolphin. Three points will be considered: (1) acoustic experiments and computations intended to estimate the position of the click source in the dolphin head, (2) a physical and a mathematical model for simulation of the physical mechanism responsible for the echolocation click production, and (3) comparison of the model predictions with actual parameters of the echolocation clicks.

This chapter is based mostly on unpublished technical reports (1975–1987) and a doctoral thesis (Giro 1987, supervised by Dubrovskiy).

Plausible Location of the Echolocation Click Source in the Dolphin Head

A special experimental setup was developed to determine the acoustic pressure distribution near the head of an echolocating dolphin. A Black Sea dolphin (*Tursiops truncatus*) was trained to put his head on a firmly fixed stand underwater and remain motionless for several dozens of seconds, while emitting echolocation pulses aimed at a small spherical target. Echolocation clicks were picked up by several broadband acoustic sensors suspended approximately 1 m from the dolphin's head and were recorded on a movie camera in front of the screen of a multitrace oscilloscope. The latter was triggered by a click from a reference sensor. Positions of the dolphin's head and sensors were recorded in the horizontal plane by another movie camera, synchronized with performance of the first camera. By analyzing pictures taken by the cameras it was possible to determine the positions of the dolphin's head and sensors, as well as delays in click arrivals at different sensors when echolocation pulses triggered both cameras and the oscilloscope (fig. 10.1).

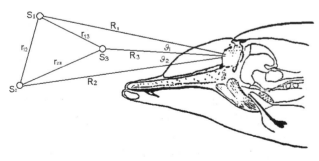

Fig. 10.1. System of spherical coordinates used for determining the click source location

A set of equations was derived to calculate an unknown location of the click source for every click recorded. For the spherical coordinates,

$$(r_{12})^2 = (R_1)^2 + (R_2)^2 - 2R_1R_2\cos(\theta_1 - \theta_2)$$
$$(r_{13})^2 = (R_1)^2 + (R_3)^2 - 2R_1R_3\cos\theta_1$$
$$(r_{23})^2 = (R_2)^2 + (R_3)^2 - 2R_2R_3\cos\theta_2$$
$$T_{12}c = R_1 - R_2$$
$$T_{13}c = R_1 - R_3 \tag{10.1}$$

Here R_1, R_2, R_3 are unknown distances between a click source position and locations of the sensors S_1, S_2, S_3. Values r_{12}, r_{13}, r_{23} are distances between the sensors S_1, S_2, and S_3; θ_1 and θ_2 are unknown spherical angles (fig. 10.1). Values T_{12} and T_{13} are delays in click arrivals to the sensors; c is the speed of sound in the seawater.

In deriving this set of equations we postulated that acoustic impedances of tissue over the skull bone (muscles, fatty substance of the melon, etc.) equal that of the seawater. We also disregarded delays caused by possible click diffraction over the air-filled sacks.

There are five unknown values (R_1, R_2, R_3, θ_1, θ_2) and five independent equations. The roots of these equations were calculated with some approximation using specially developed software.

Sixteen different configurations of the sensor locations in the space around the dolphin's head were explored in the experiment, and many dozens of clicks were recorded at each sensor configuration. Results of these measurements and computations suggest that the acoustic source of the echolocation pulses was situated within the compact volume with linear dimension 4.0 cm in close proximity of the right nasal plug.

A Physical Model for Simulation of the Click-Production Mechanism

From early observations, we infer that the physical mechanism of sound production, both for clicks and whistles, is driven by air pressure created either by the lungs or air-sac muscles. The essential element of the sound-production mechanism is formation of an air jet (for a short period of time in case of the click generation) through some passages, which can be partially or completely blocked by sphincters or plugs. Some experimenters have concluded that the air jet is actually involved in sound production (Norris 1969; Norris et al. 1971; Ridgway et al. 1980; Amundin 1991b; Cranford 1992b).

Similarities with vowel production by vocal cords in humans have also been observed (Giro 1987). Constant force caused by air pressure creates oscillations of the vocal cords due to its internal compliant and inertial (mass) properties, and the period of these self-oscillations determines the tonal quality of vowel. It equals approximately 10 ms (100 oscillations per second in a human's voice). Similarly, the period of self-oscillations in the click-production mechanism in the dolphin corresponds to click repetition rate. Understanding the origin of energy and the "self-oscillation" mode of vibrations of some biological structures responsible for click-train production has left unresolved the basic question of how to create very brief, broadband, and powerful clicks.

As shown in fig. 10.2, we implemented a simple physical model of a nasal passage, consisting of a rubber tube opened at one end and attached to the reductor of a high-pressure gas bottle on the other end. A sensor (1) with a flat frequency response from 0 to 40 kHz measured the displacement of a rubber ring (2), put on the underwater portion of the tube (3). This ring blocked the air passage at some small access pressure from the reductor. By increasing this pressure, one can put the model in a typical self-oscillation mode of vibrations, provided that the elastic characteristics of the tube and the ring are chosen properly. Acoustic signals emitted by a model were picked up by a broadband hydrophone (Brüel & Kjær type 8103).

Let us consider how the model generates the self-oscillation mode of vibrations. Due to blockage by the ring, the air cannot flow through the tube at the very beginning of access pressure application (fig. 10.2A). At this stage, the air pressure forces inside the tube (expanding forces) are still smaller than the compliant forces of the ring compressing the tube from outside (compressing forces). When these antagonistic forces are equal, or expanding forces become greater than compressing ones, a small opening occurs inside the tube and the air begins to flow through the opening (fig. 10.2B). According to Bernoulli's law, the total air pressure inside the tube, p_t, is composed of two mutually complimentary portions: a dynamic one, $p_d = \rho v^2/2$, where ρ is the air density and v is the speed of airflow, and the static one, p_s—that is, $p_t = p_s \pm p_d$. When air is not moving—that is, $v = 0$ and $p_d = 0$—then $p_t = p_s$. But any flow of air immediately creates dynamic pressure, thereby reducing the static pressure p_s, which becomes $p_s < p_t$, and

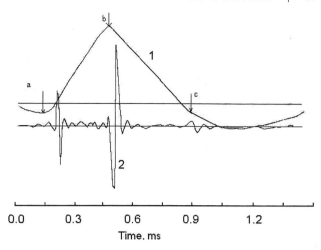

Fig. 10.3. Physical model of the acoustic source. The triangular curve (1) represents displacement of the rubber ring surface; curve (2) represents the pressure signature recorded by a hydrophone in water. Three noticeable acoustic clicks are locked to bends *a*, *b*, and *c* in the displacement curve. Time scale is 150 μs/div.

the outside compressing forces from the ring again block the tube, and a new cycle begins. By varying the rate of access pressure increase, stiffness of the ring, or the tube itself, one can easily control duration of the cycle.

Fig. 10.3 demonstrates a displacement curve of the ring (1), picked up by the sensor, and the acoustic pressure signature (2) recorded in the water by the hydrophone at about 1 m from the ring. The displacement curve has two portions: the first positive portion corresponds to expansion of the ring surface and has a triangular shape and duration of about 700 μs. The second negative portion has a lower peak value and duration of about 600 μs. Note that the shape of the displacement curve is determined by the specific set of the model parameters. We associate the positive triangular portion of the curve with opening of the tube and passage of air jet. The positive slope of the triangular portion of the curve corresponds to the widening of the opening. At this time, the expanding forces are bigger than the compressing forces from the rubber ring. These forces are equal at the moment corresponding to the peak of the displacement curve and hence the widest opening of the tube. Then compressing forces again become greater and the opening begins to close. The negative portion of the displacement curve demonstrates the additional compression of the ring and the tube walls. This joint compression of the tube and the ring is due to inertia of the ring and the tube.

There are three noticeable acoustic pressure clicks, with the second one of the highest peak value. Duration of the clicks is within a time interval of 50–80 μs, which correlates with the duration range of other dolphin clicks (Dubrovskiy and Zaslavskiy 1975; Au 1993). The

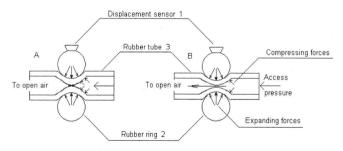

Fig. 10.2. Physical (pneumatic) model of the acoustic source of probing pulses: (1) displacement sensor, (2) rubber ring, compressing (blocking) the air passage, and (3) rubber tube. *A:* Compressing forces of the ring block the tube. *B:* The tube is open due to expanding forces of excess pressure, and air jet is coming through a small opening.

position of these clicks along the time scale is locked to the most pronounced changes (bends) in the displacement curve. The first click is locked to the beginning of the displacement raise; the second click is locked to the peak; and the third click is locked to a bend of the curve (caused by closing the passage and further compression of the tube walls and the ring due to inertia).

A Mathematical Model of the Click-Production Mechanism

To explain the origin of the acoustic pressure clicks and their locking at the bends of the displacement curve in our physical model, a simple mathematical model was developed (fig. 10.4). Let us imagine that the whole compressing ring has no weight—except for a small portion, to which we will assign a spherical shape to simplify further calculations. Assuming that both outside compressing and inside expanding forces are applied only to this selected sphere, the sphere will move in accord with the triangular displacement curve. The linearized Euler's (acoustic) equation states that

$$\rho \frac{d^2\xi}{dt^2} + \text{grad } p = 0 \qquad (10.2)$$

where ξ is the displacement, ρ is density and p is the acoustic pressure—that is, the acoustic pressure gradient is proportional to the second time derivative of the displacement. Returning to the triangular curve, observed in the physical model of the acoustic source (fig. 10.4), one can easily see that the first and the second derivatives of the displacement should be most pronounced only near the displacement curve bends. This is because the first derivative for linear portion of the displacement curve (constant slope portion) is constant, and hence the second derivative is equal to zero. The complex (i.e., containing real and imaginary parts)

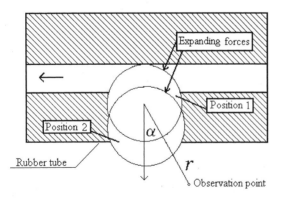

Fig. 10.4. Schematic diagram showing oscillations of the selected sphere. In position 1, the sphere blocks air jet; position 2 corresponds to the presence of air jet at maximum displacement of the selected sphere.

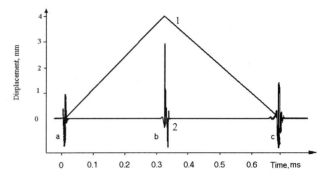

Fig. 10.5. A mathematical model of the acoustic source of probing pulses. Curve 1 is the displacement of the selected oscillating sphere; curve 2 is the calculated sound pressure in water: the "sensor-source" distance is 1 m, maximum peak-to-peak sound pressure of click *b* is 28200 Pa (209) dB; duration of displacement triangle 1 is 700 μs; duration of acoustic clicks *a*, *b*, and *c* is within 15–50 μs.

acoustic pressure caused by a sphere oscillation displacement (an acoustic dipole) can be represented in the frequency domain as follows (Isakovich 1973):

$$p(r, t) = \rho\omega^2 a^3 \xi(\omega)$$
$$\cdot \frac{(ikr - 1)\cos\alpha \exp(-ika)\exp(ikr - i\omega t)}{[2 - 2ika - (ka)^2]r^2} \qquad (10.3)$$

where *a* is the radius of the selected sphere, *r* is a distance from the center of the sphere to the observation point, *k* is a wave number $\omega = 2\pi f$, α is an angle between direction of oscillations and direction to the observation point, and *i* is an imaginary unit ($i = \sqrt{-1}$). Expression for the real part of the acoustic pressure can be simplified for low-frequency range $ka < 1$ (eq. 10.4):

$$p(r, t) = \rho c \frac{(ka)^2 a\xi(\omega)\cos(kr - \omega t)}{2r} \qquad (10.4)$$

Acoustic pressure grows proportional to $(ka)^2$ at low-frequency range. It means that the proposed mechanism of click emission is effective only at high frequencies, for $(ka)^2$ has negligible value at low frequencies where $ka < 1$. The acoustic pressure $p(r, t)$ is very sensitive to the size of the oscillating sphere *a*. It depends also on the displacement spectral density $\xi(\omega)$.

Pressure signature of radiated acoustic click $p(r, t)$ is computed by means of the inverse Fourier transform of the click complex amplitude spectrum. A program was created for calculating of the click amplitude spectrum density (ASD) and acoustic pressure signature $p(r, t)$ in the water for arbitrary displacement $\xi(t)$ of the sphere. As an example, the computed acoustic pulses for triangular displacement curve are shown in fig. 10.5. Indeed, these pulses are locked to the bends of the displacement curve, similar to what we observed in our physical model.

Discussion

POSITION OF THE CLICK SOURCE

Our findings support the view that structures situated in close proximity to the right nasal plug are responsible for echolocation click production. The new element of these findings is that they were derived from straightforward acoustic measurements in the echolocating dolphin. These findings are correlated with observations of Norris et al. (1971); Mead (1972); Dormer (1979); and Amundin and Andersen (1983). Based on palpation experiments, Dormer (1979) and Amundin and Andersen (1983) concluded that clicking apparently takes place on the right side of the nasal plug. Norris et al. (1971), Mead (1972), and Dormer (1979) proposed a bilateral sound generator with clicks assigned to the right side of the air passage and whistles to the left.

However, our finding does not imply that only the right side of the air passage can be assigned for click production or only the left side of the air passage can be assigned for whistle production. Rather, because the suggested physical mechanism can produce both clicks and whistles, both left and right air passages may be sources of clicks and whistles.

PHYSICAL AND MATHEMATICAL MODELING OF THE CLICK-PRODUCTION MECHANISM

The main finding of our work consists in formulation of the hypothesis on how the click source functions. Feasibility of the suggested mechanism can be proven by comparing the predicted characteristics of the model source with actual characteristics of the dolphin click source.

Duration of the click

Our physical and mathematical models provide a click duration that is well within the limits recorded in the *Tursiops truncatus* and the other species of dolphins (Au 1993; Bel'kovich and Dubrovskiy 1976). Fig. 10.4 shows that duration of the main click in the physical model was close to 60 μs. It is reasonable to assume that a more careful choice of the elastic and mass properties of rubber, as well as the dimension of the model components, would provide a wide variety in click duration. Our mathematical model predicted click duration from 30 to 40 μs (fig. 10.5) at $a = 5$ mm and $\xi_0 = 5$ mm.

Composition of the echolocation emission

Our models also explain the presence of additional pulses, which can be considered as "forerunners" or "afterrunners"—that is, arriving ahead (Dubrovskiy and Zaslavskiy 1975) of the main (high peak value) click or accompanying it (Au et al. 1974) (fig. 10.6). It is evident from our models that forerunners and afterrunners have the same origin as the main echolocation click. They differ from it by the peak value and the waveform.

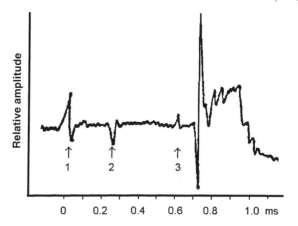

Fig. 10.6. The pressure signature of a dolphin probing pulse: first, second, and third "forerunners"; a main echolocation pulse; and a low-frequency component. Time delays between the forerunners and the main pulse are approximately 700 μs, 380 μs, and 100 μs (Dubrovskiy and Zaslavskiy 1975).

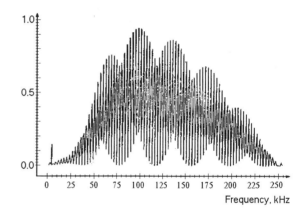

Fig. 10.7. Amplitude spectrum of the acoustic pressure caused by a single triangular displacement of an oscillating sphere

The forerunners and afterrunners manifest the physics of the sound-production mechanism. However, the presence of these pulses does not refute the idea of multiple sources of sound in the dolphin.

Amplitude spectrum of the click

Fig. 10.7 demonstrates ASD of three clicks occurring at the bends of the computed triangular displacement curve shown in fig. 10.5. The ASD curve has oscillations with periods equal to the inverse values of the time delays between these three pulses. Two inferences can be drawn from this computation: the spectrum of clicks is wide (up to 80 kHz at −3 dB point) and its peak is situated near 100 kHz.

ASD of the clicks obtained from the physical model (fig. 10.3) are not so wide because the bends in the displacement curve are not as abrupt as in the strict mathematical triangle (fig. 10.5). The bandwidth of clicks obtained in the physical model was estimated to be about

20 kHz. Again, the proper choice of the elastic and mass properties of rubber should be made as well as the dimension of the model elements.

Peak value of the click

Observed peak values of clicks for *T. truncatus* are within the limits from 153 to 226 dB re 1 μPa at 1 m (Bel'kovich and Dubrovskiy 1976; Au 1993). Peak acoustic pressure calculated using the mathematical model depends on radius of the "selected" sphere a, an amplitude of displacement ξ_0, and duration of the displacement triangle T. By increasing the radius of the selected sphere from 1 mm to 5 mm, while keeping constant the duration of the displacement $T = 700$ μs and $\xi_0 = 5$ mm, it was possible to increase the peak pressure from 152 dB to 214 dB re 1 μPa at 1 m. Obtained acoustic pressure peak values are well within peak values observed in experiments with dolphins.

Feasibility of adaptation of click parameters

The proposed physical and mathematical models of the dolphin click source provide possibilities for adaptation of click parameters. Variations of radius a and the displacement amplitude ξ_0 change the click waveform, peak value, duration, and frequency bandwidth. The smaller the radius, the longer the duration of the click and the smaller its peak value. The maximum of the ASD shifts toward high frequencies with decreasing radius a.

How can the dolphin change the radius of a selected sphere? It is feasible that the dolphin can somehow vary a portion of lip, muscle, or plug involved in the dipole-like oscillations of the sphere. We observed many times that the same animal (Black Sea *T. truncatus*) easily changed the type of clicks from a "standard" one containing from two to four "half-waves" to a "tone burst" type containing four and more periods of oscillation.

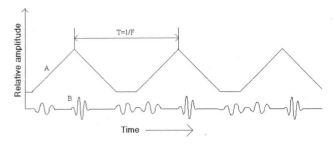

Fig. 10.8. Schematic diagram of the displacement curve (*A*) and quasi-periodic sequence of whistlelike pulses (*B*)

Are the physical mechanism of whistle production and the mechanism of the click production different?

Recently Murray, Mercado, and Roitblat (1998), based on observation of the entire continuum (click train to whistle) in single vocalizations, suggested that a single production mechanism can potentially produce all of the observed signal types. We contend that the suggested physical mechanism can produce both clicks and whistles. In order to emit whistles the dolphin has to balance compressive and expanding forces (fig. 10.2A). Such a balance can be established at different absolute values of these forces. The higher the absolute values of expanding and compressive forces, the higher are repetition rates of pulses and the intensity of radiation. Near the balance, displacement triangles can follow with a short or no pause (fig. 10.8). The source could create the basic sequence of displacement triangles in this case. This sequence can be made quasi periodic and lead to the formation of a quasi-harmonic structure in the emitted signal, similar to what is observed in the frequency structure of whistles. It is also similar to the process of vowel phonation by humans.

{ 11 }

The Harbor Porpoise *(Phocoena phocoena)* — Modeling the Sonar Transmission Mechanism

A. David Goodson, James A. Flint, and *Ted W. Cranford*

Introduction

Morphologically the head of the harbor porpoise, *Phocoena phocoena,* although smaller and lacking the obvious extended rostrum of the bottlenose dolphin, *Tur-* *siops truncatus,* contains the same identifiable internal structures. Both odontocetes employ an active sonar to detect and intercept prey, but the significant differences in their respective echolocation pulse shapes and power spectra require some explanation.

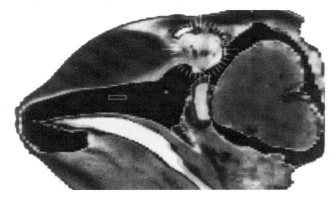

Fig. 11.1. X-ray CT scan showing a vertical tomographic reconstruction (right parasaggital section) from the head of a harbor porpoise (*Phocoena phocoena*) (see Cranford, Amundin, and Norris 1996 for a more detailed description of the morphology)

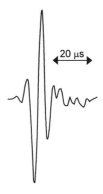

Fig. 11.2. Wideband echolocation pulse from a bottlenose dolphin

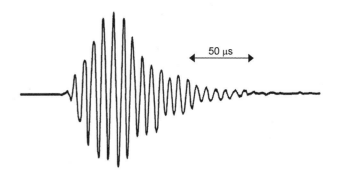

Fig. 11.3. Narrowband echolocation pulse from a harbor porpoise observed on axis

Some insuperable difficulties restrict attempts to measure acoustic signals propagating inside a biological organism, such as an odontocete's head, and most direct measurements can only be made on the emerging signals. However, given accurate dimensions and detailed knowledge of the local sound velocity characteristics for the materials involved, computer models now can construct and predict wave propagation to explain waveform development inside such complex structures. Two quite different approaches are discussed herein to illustrate how the separate parts of the harbor porpoise echolocation mechanism interact. With the current advances in three-dimensional tissue imaging using magnetic resonance (MR) and X-ray computed tomography (CT) (see fig. 11.1), together with high-quality, full bandwidth, on-axis recordings of the sonar signals produced by an animal, such computer models can provide the foundation for a better understanding of the way this biological sonar evolved.

Harbor Porpoise Sonar Pulse Formation

The pulse emissions of the harbor porpoise are extended in duration and more oscillatory in nature, and the energy spectrum confined within a much narrower bandwidth, than those used by most dolphins (see figs. 11.2 and 11.3). The small body size, typically around 1.5 m length, appears to impose some physical restrictions to the available acoustic power. Typically, these animals emit source levels (SL) of 150–160 dB re 1 μPa at 1 m (Akamatsu et al. 1994; Goodson, Kastelein, and Sturtivant 1995; Goodson and Sturtivant 1996), some 50 dB less than can be achieved by a bottlenose dolphin (Au 1980). The actual sound source for echolocation signals in dolphins has been localized to the paired structural complex known as monkey lip and dorsal bursae (MLDB), which includes the fatty bursae cantantes (Cranford 1992b; Cranford, Amundin, and Norris 1996).

These inconspicuous fatty bursae are a few centimeters below the blowhole and affiliated with the left and right nasal passages, displaced to each side of the midline. In some dolphins, the right MLDB complex is significantly larger than that of the left. In the harbor porpoise, the two MLDB complexes are small and more closely matched in size. However, when considered mechanically, these structures are similar and can reasonably be expected to function in the same way and initiate similar impulsive acoustic events.

Air Sacs Forming Reflective Baffles

If unconstrained by discontinuities in the surrounding tissue, the pulse of acoustic energy generated by an MLDB complex should radiate omnidirectionally, as the dimensions of this structure are close to the wavelengths being generated. In fact, the pulses generated are forced to propagate forward by the air-filled nasal diverticuli surrounding the source and in particular by the pair of saddlelike vestibular air sacs that roof-over the sound-generation tissues just below the blowhole. The inner tissue/air interface of these surrounding air sacs baffles most of the radiating sound energy, restricting its escape to the forward direction and hence into the melon. Very little high-frequency energy radiates from the head in any other direction, although the small fraction of energy that passes into these air sacs will be reradiated omnidirectionally from their outer surfaces at frequencies defined by the volume resonances of these cavities.

The Melon Beam Former

The melon structure forming the large forepart of the head comprises specialized lipid tissues. These triglycerol and wax esters are distributed in cross section within this organ. Shorter-chain fatty-acid fractions are found nearer the central sound axis and medium-chain material, with increasing density, toward the periphery. As sound velocity is affected by the material's density and compressibility, this structure can be expected to modify a propagating wave front, ensuring that the parts of the wave traveling closer to the periphery move faster than at the center. This low-loss wave guide effect further narrows the projected −3 dB beamwidth of the emerging sonar pulse and, by increasing the propagation velocity in the outer parts, effectively flattens the expanding wave front toward a plane wave condition. This is a desirable focusing effect that will improve the radiated pattern within the conventional near field. In all the small odontocetes the echolocation mechanism could be formed from these three closely coupled structures: the MLDB complex providing the initial acoustic stimulus; the tissue channel, bounded by air and connective tissue structures, restricting the propagation of the expanding sound pulse to the forward direction; and the fatty tissue "lens" furnishing the final "beam forming" before transferring the shaped pulse into the water.

Narrowband versus Wideband Signal Formation

The vestibular air sacs wrapped around the acoustic channel can be considered as a very short "megaphone," and the reverberation of the initial impulse will naturally extend the duration of the pulse by adding internal reflections. If this constraining sac floor/air boundary expands rapidly in the forward direction, then the emerging pulse acquires minimal reverberation and retains a wideband characteristic.

If the channel is longer and expands more slowly, then the reverberation added to the emerging pulse will be greater and the pulse length consequently extended. The rate of expansion and length of this constraining channel therefore influences both the wave shape and the bandwidth of the emerging pulse. The presence of several deep transverse folds within the floor of the harbor porpoise's vestibular sacs are a feature not seen in the larger dolphins. However, this acoustic boundary, which is closely associated with the propagation of sound within the *P. phocoena* head, is markedly different from the smooth sac floor structure of delphinid morphology. A simple computer model has been used to examine how the expanding acoustic impulse may be influenced

by such deep tissue corrugations (Goodson and Datta 1995; Goodson 1997). The deeply folded floor appears to form a succession of acoustic delays, with each fold returning the energy captured by it after a finite delay determined by its depth. In combination the delays introduced by each succeeding fold can select and enhance a single frequency, creating a tuned baffle effect. This model describes a very compact passive mechanism, which can explain the transition from an initial wideband stimulus to the narrowband, single mode, wave shape emitted by the harbor porpoise.

The Signal Generation Hypothesis

Although morphologically the head of harbor porpoise contains the same identifiable homologous internal structures as the bottlenose dolphins, there are significant differences in the respective echolocation signal wave shapes and power spectra. The MLDB complex is believed to be activated pneumatically by increased air pressure applied to the lower side of the MLDB complex—air that is contained and compressed in the bony nasal passages above the laryngeal spout. The mating surfaces are lubricated and maintained in a wetted condition by a serous fluid produced by glands directly below these structures (Degollada, García-Hartmann, and Cranford 1998; Evans and Maderson 1973). The action of forcing these closely fitting smooth wetted surfaces apart, under the combined action of muscle tension and compressed air, will cause numerous vacuum bubbles to nucleate across this surface as the liquid film ruptures prior to actual separation. Once physical separation breaks the seal, the vacuum bubbles expanding within the serous fluid will collapse, accelerated in their destruction by the compressed air applied from the nares. It is suggested that the shock wave generated by the simultaneous extinction of these bubbles, rather than a tissue-on-tissue impact, provides the actual energy release in the form of a very fast rising edge pressure pulse. This energy-release mechanism by exploiting the physical phenomena known as cavitation (Leighton 1994) can generate a high-frequency, high-energy peak in the signal spectrum in a coherent and repeatable form. Since the liquid film will be violently disrupted by this process, this could explain the spray of droplets ejected into the upper airway as the lips open during the formation of acoustic pulses (Cranford et al. 1997a, 1997b). This proposed cavitation mechanism for initiating the pulse appears consistent with recent high-speed video endoscopy observations by Cranford et al. (1997a). They found the opening of these phonic lips, along with signal generation, and the expulsion of a burst of fine droplets into the upper airway all occur within a 2.5 ms video-defined window. An alternative hypothesis (Cranford, Amundin, and Norris 1996; Cranford 2000), based on exactly the same structures, suggests that the initial impulse is generated by the rapid closure of the mating anterior and posteria phonic lips of the MLDB complexes. This sound-generation mechanism may also be consistent with the endoscopic observations of Cranford et al. (1997a) given perfect simultaneous mechanical contact over the mating area of the lips. As the impulse duration will control the spectral energy peak in the emerging signal, this event should not extend for more than half a cycle—that is, a period of <3.5 μs for a 140 kHz porpoise signal. Some limiting effects due to the mass and inertia of the moving parts should be considered, as different pulse repetition rates may modify the achievable closure rates. However, with either interpretation, the acoustic energy released by the MLDB complex is efficiently coupled directly to the tissues that conduct this toward the melon.

Beam Forming

The melon is a complex connective tissue and lipid structure that forms the bulk of the odontocete forehead and is usually described in the literature as an acoustic lens. This description, however, can be misleading since the optical analogy suggests that refraction occurring at the surface boundaries is responsible for shaping the projected beam. The term beam implies a continuous illumination analogy suggestive of a vehicle headlight. In reality, the projected sounds emerge as a sequence of short duration pulses that propagate into the water directly ahead of the animal. The expanding conical volume of water ensonified in this way represents the beam and targets inside this limited zone return echoes that will be detected after a range related time delay. The air sacs of the nasal region improve the effective source level (SL) by confining the angular dispersion of the pulse energy at its source into a narrow angle, while ensuring that little off-axis side-lobe energy is radiated. Reverberation of side-lobe energy can become a serious limiting factor in very shallow water and any off-axis radiation limits performance when operating very close to either the sea surface or the sea bottom. The use of a narrow projected beam with low side-lobe radiation therefore produces a gain in target detection performance over an equivalent omnidirectional power source. However, if a very narrow beam is selected, then more transmissions must be made to search a given volume of water. The time taken to do this will be greater and some moving targets could escape detection. These conflicting pressures appear to favor solid-angle beamwidths of around 8–10° in a number of odontocete and human-made sonar systems concerned with midwater target search and detection tasks.

Fish Prey Detection— Some Acoustics Constraints

In the context of simple fish detection, several interrelated parameters need to be considered:

Choice of operating frequency

High-frequency, short (approximately 1 cm) wavelengths are essential to produce detectable echoes from small prey. Also, seastate and molecular agitation noise masking will be minimum at frequencies from 50 kHz to 150 kHz. Losses due to increased absorption at these frequencies are more than compensated for by the improved signal to seastate-noise ratio.

Source level and bandwidth

Higher power transmissions normally extend detection ranges unless conditions become reverberation limited, as in shallow water or when working very close to the bottom. Short wideband signals return more information from a target and aid recognition and target classification. Narrowband signals return less target information but improve signal-to-noise performance. The use of extended tonal signals suitable for Doppler processing has not been reported in small cetaceans.

Beamwidth

Sonar systems designed to search in three dimensions must compromise between the projected beam angle and the maximum target detection range, as a moving target limits the available detection time. A narrow sonar beam requires more transmission periods to search a volume than a wide beam; but, for the same power, the narrow beam achieves longer range detection for a given size target.

The harbor porpoise sonar operates at much lower power levels than a bottlenose dolphin, SLs around 150–160 dB$_{rms}$ (re 1 μPa at 1m) being typical; these animals do not appear to transmit at levels in excess of 170 dB. Their power spectrum is centered around 140 kHz, although it seems that this may fall toward 120 kHz with increasing age and body size. The -3 dB bandwidth of the porpoise pulse is around 13 kHz (Goodson, Kastelein, and Sturtivant 1995). These values contrast with data reported for the bottlenose dolphin where maximum SLs in excess of 220 dB$_{rms}$ re 1 μPa at 1m have been quoted and where their bimodal spectrum characteristic can exploit an adaptable bandwidth in excess of 100 kHz (Sigurdson 1997a).

Some care is required when comparing marine mammal source sevels as dB$_{peak\text{-}to\text{-}peak}$ values are sometimes quoted in the literature. The internationally accepted 1 μPa reference level used in underwater acoustics should not be confused with the 20 μPa usage, dB(A), traditionally employed in airborne mammalian audiology (Chapman and Ellis 1998).

More information about a target's characteristics are obtained from wideband ensonification, but the signal-to-noise ratio, in an uncorrelated system, favors a narrowband emission. Most small cetaceans experience difficulty when trying to break up very large prey, and all seem to prefer to swallow a fish whole (Bloom 1991). Preferred prey sizes for the harbor porpoise are there-

A. David Goodson, James A. Flint, and Ted W. Cranford

fore appreciably smaller than those of a dolphin. Given that the maximum SL produced by the porpoise could be body-size related and that echoes from smaller prey are necessarily much weaker than for large ones, it seems that the harbor porpoise is quite severely restricted by both signal-to-noise ratio and by the small target strength (TS) of its prey. From sonar equation predictions and from observations of interclick intervals, it seems unlikely to be searching for prey more than 30 m ahead (Goodson and Sturtivant 1996).

Modeling the Effect of a Reverberant Delay-Line Air Sac Structure

Reverberation that occurs as the initial pulse propagates between the vestibular sacs can be shown to have an additive effect at a single frequency (f) when the depth of the folded tissue forming the sac floor returns reflections from the tissue/air boundary to the center of the sound channel with a $1/f$ delay. If the adjacent folds are formed at one-wavelength (λ) intervals, then each contributes in turn, superimposing the energy returned to the center channel to the reverberation of its predecessor (figs. 11.4, 11.5, and 11.6). Because some of the energy must be lost at the tissue/air boundary, this reverberant process dies away with an exponential tail. Typically, the amplitude of the porpoise pulse increases during the first five cycles; coincidentally, there are usually five transversal folds formed in the floor of the vestibular sacs. This process was first modeled by varying the reflection coefficient (a) and iteratively substituting different values. This model accounted for the fraction of energy lost at the tissue/air interface (Goodson and Datta 1995). The best fit to the duration and shape of the far-field waveform was obtained when this boundary reflection loss was 16%. This simple model assumed that each delay line contributed equally and that each generates a one-cycle delay. The summation of the energy

trapped and returned by each successive tissue fold and the resulting amplitude distribution in the emerging pulse can be described by the following formula:

$$x_L(n) = \begin{cases} U_0 \dfrac{(1 - a^n)}{(1 - a)} & 1 \le n \le m \\ U_0 \dfrac{a^{n-m}(1 - a^m)}{(1 - a)} & n > m \end{cases} \tag{11.1}$$

letting $U_0 = 1$, $m = 5$, $a = 0.9|0.8|0.7| \ldots |0.5$. The amplitude of the output waveform results from the sum of the reverberant contributions from m delay lines (sac-

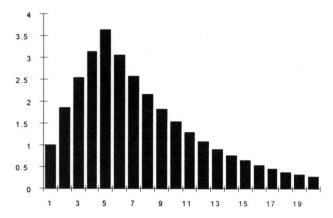

Fig. 11.5. The computed output amplitude for each cycle of the signal in the delay-line model. The best fit to the real signal amplitude data occurred when the reflectance factor (a) was 0.84, indicating 16% of the energy was lost through the tissue/air boundaries.

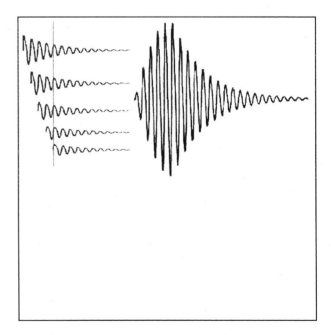

Fig. 11.6. Overlaying the reverberations of a single wideband initiating impulse (contributed by each sac floor fold) with successive $1/f$ delays generated a porpoiselike pulse envelope at a single frequency.

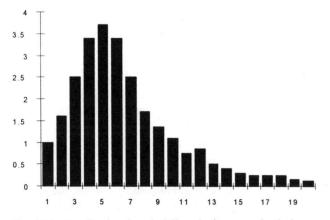

Fig. 11.4. Amplitude of each full cycle (measured relative to the first) from a real porpoise echolocation pulse recorded on-axis

floor corrugations) at each of *n* cycles, where *a* is the tissue/air boundary reflection coefficient.

This model shows that the resulting waveform envelope is sensitive to the proportion of the signal energy lost into the air sac with the best fit occuring at around 16% (fig. 11.5).

Transmission Line Modeling (TLM)

TLM is an established time-domain numerical technique for examining wave propagation through and around complex structures. TLM is a full-field solution method akin to the more well-known finite difference numerical methods (Aroyan et al. 1992). The strength of TLM lies in its computational simplicity and ability to retain a close link between the model and the physical situation under investigation.

Originally the method was developed for solving two-dimensional electromagnetic problems in waveguides (Johns and Beurle 1971). More recently, a number of authors have applied the method to solving the wave equation in acoustics (Orme, Jones, and Arnold 1988; Pomeroy, Williams, and Blanchfield 1991). In essence, TLM is a discretised version of Huygens' principle, which states that every point along a propagating wave front acts as an isotropic secondary radiator. Each secondary source acts individually as a spherical wavelet transmitter, and so the wave propagates by the net effect. To simulate this scattering process in a discretised model, it is necessary to divide the space into incremental steps on a Cartesian grid. Each block of space bounded by this division is modeled by a node that, by taking account of the signals arriving via the mesh connections, computes the local absorption and transmission loss and passes the result to immediate neighbors. The instantaneous pressure and particle velocities are then available at every point in the grid, and the process is repeated to obtain values at subsequent discrete time intervals.

In the original formulation, applications were restricted to materials with uniform propagation velocity; however, later work (Saleh and Blanchfield 1990) extended the method to inhomogeneous problems in acoustics. The ability to model large data sets in TLM with spatially varying sound velocity has not been exploited until recently (Flint, Goodson, and Pomeroy 1997). The odontocete melon is a suitable candidate for modeling with TLM, since its dimensions, the sound velocities involved, and the frequencies of interest are within the capabilities of current desktop computers.

For modeling propagation within the melon, the injected waveform must be defined very accurately, as the iterative computation must be carried out at mesh node increments of $<0.1 \lambda$.

Although the simple delay-line computer model creates a wave shape with the required porpoiselike pulse envelope, the mathematical description (eq. 11.1) syn-

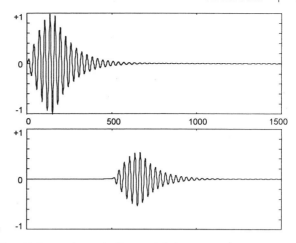

Fig. 11.7. Mathematically modeled porpoise waveform injected into the melon TLM model together with the output signal emerging for comparison. The time axis is scaled in iteration steps.

thesizes too coarse a waveform for direct injection to the TLM mesh. For the study of propagation within the melon, a higher resolution signal was synthesized by a curve-fitting formula (eq. 11.2) matched to an actual porpoise sonar pulse waveform. This use of a far-field measured waveform as the injected signal would be difficult to justify if the melon structure propagated the signal with significant boundary reflections, because multipath interference will modify the output. In fact, the waveform output by the model showed that no distortion occurs and indicated that the melon does not exploit reflection pathways. The use of this mathematical function ensured that dispersive artifacts, created in TLM when the 0.1 λ mesh criteria is not met, were avoided and the waveform was synthesized with a known frequency characteristic—that is to say, within the TLM mesh bandwidth.

The discrete waveform generated by this function for use in the melon modeling simulation is shown in fig. 11.7. The fundamental frequency is dictated by the simulated center frequency of the porpoise (120–140 kHz), and this is enveloped with a sinusoid ($0 < \theta < \pi/4$) at the peak amplitude and by an exponential decay:

$$x_H(n) = \begin{cases} U_0 \sin\left(\dfrac{n\pi}{5\sqrt{2}}\right).\sin\left(\dfrac{n\pi}{142}\right) & 0 \leq n \leq 70 \\ U_0 \sin\left(\dfrac{n\pi}{5\sqrt{2}}\right).\exp\left(\dfrac{71-n}{57}\right) & n > 70 \end{cases}$$

$$(11.2)$$

The first sinusoid term in each part of this waveform definition also contains a scaling factor to account for the fact that in this particular form of two-dimensional TLM, waves travel in the model at $1/\sqrt{2}$ times the speed of pulses. More details and justification can be found in Saleh and Blanchfield (1990). Further time scaling was necessary in the actual model because the injection point in the melon tissue occured in a region of low sound velocity.

Limitations of the TLM Model

This model used a subset of a digital three-dimensional CT scan of the head of a (1.565 m body length) male harbor porpoise. The selected data correspond to a thin section through the melon (right parasagittal) aligned to pass through the right MLDB complex (Cranford, Amundin, and Norris 1996, 255–259). The small (2 cm) offset from the median just missed the maximum vertical dimension of the melon and a part of the lowest velocity material at the core.

The melon tissue model was treated in isolation, and the adjacent blubber, connective tissue, muscle, and bony rostrum below were removed. As will be evident from the model, very little injected energy actually impinges on these boundaries, and this approximation appears to be valid. In fact, any energy reaching the porous oil-saturated rostrum is likely to be absorbed rather than reflected, as the mixed density material has obvious similarities to some engineered anechoic structures.

The relative density of the fatty tissues obtained from the CT scan data was linearly mapped to sound velocity. The melon material was assumed to be isothermal, and the effect of the temperature gradient that must occur at the tissue/water interface was ignored.

The maximum value for c was matched to seawater, and the slowest sound velocity quantified was matched to the sound velocity measurements of Norris and Harvey (1974) for melon core material.

Results

The point at which sound enters the melon from the air-sac-bounded channel defined the injection aperture; from the tomographic images this measurement was determined to be in the order of 1.5 λ (at 130 kHz).

Initial tests carried out at both 1 λ and 2 λ appeared suboptimum. At 1 λ, the expanding waveform was not fully captured, and the emerging pulse did not achieve the maximum amplitude. At 2 λ, the sound propagates into the melon without the main lobe (-3 dB) dimensions initially filling the structure, and more significant side lobes were created.

At 1.5 λ, the injected signal expands following the overall shape of the structure; the side lobes formed were very small.

This model was then run and a time sequential set of images captured for replay on the computer display as a slow-motion video, allowing the propagating wave front to be examined dynamically as it passed through each part of the melon.

Fig. 11.8 shows the density distribution in the melon tissue mapped from the digital CT scan slice, fig. 11.1,

provided by Cranford. (In the later figures these density pixels are outlined to delineate their positions.)

Fig. 11.9 shows a captured frame at $T_0 + 36\ \mu s$ (T_0 corresponds to the time of signal injection unto the melon). The signal has reached its maximum amplitude and the leading edge traversed more than one third of the structure; the graded index velocity characteristic of the melon then started to flatten the expanding wave front. Three waveform sampling positions, outside the melon, are shown in this image.

Fig. 11.10 shows a captured frame at $T_0 + 75\ \mu s$. The signal is starting to emerge into the water and the flattened wave front quite apparent.

Fig. 11.11 shows a captured frame at $T_0 + 120\ \mu s$. The bulk of the echolocation signal has emerged into the water, and the effect of the low-velocity melon material situated asymmetrically below the center axis in the forepart of the melon is refracting the emerging wave front so that the projection axis is now shifted closer to horizontal.

Fig. 11.12 shows the emerging signal waveforms sampled across a wave front at three sampling points just outside the melon structure: (*A*) amplitude above the projection axis is -3 dB; (*B*) amplitude on the projection axis (reference) is 0 dB; and (*C*) amplitude below the projection axis is -2.1 dB.

The respective wave shapes differ slightly along their time axes; however, there is significant vertical asymmetry in the formed beam above and below the peak amplitude axis. The associated power spectrum of the emerg-

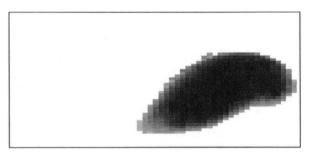

Fig. 11.8. Melon tissue CT slice showing lipid density distribution

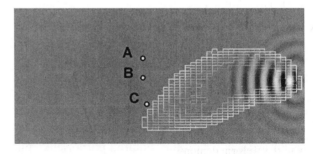

Fig. 11.9. Propagation of the injected wave at $T_0 + 36\ \mu s$

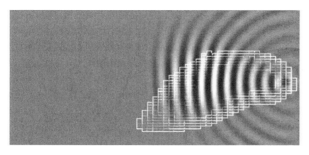

Fig. 11.10. Propagation of the injected wave at $T_0 + 75\ \mu s$

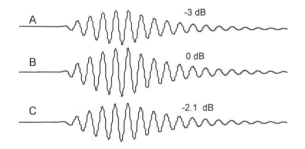

Fig. 11.11. Propagation of the injected wave at $T_0 + 120\ \mu s$

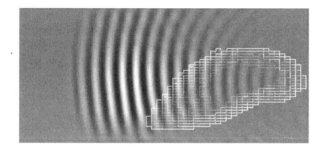

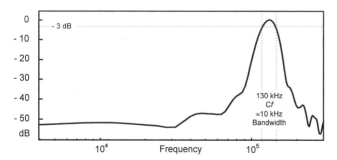

Fig. 11.12. Output signal waveforms sampled at the marked positions in fig. 11.9 as they emerged from the model: (*A*) above transmission axis (-3 dB); (*B*) on the transmission axis (0 dB); (*C*) below the transmission axis (-2.1 dB). The vertical beamwidth is estimated to be 18°, but for more accurate predictions the size of the mesh in this model needs to be increased to permit the signal to propagate further downrange. The spectrum of the *B* waveform demonstrates the bandwidth of the synthesized signal.

ing B waveform (Hamming window) indicates that the synthesized signal has a nominal 10 kHz bandwidth.

Discussion

The use of computer modeling to examine complex bioacoustic structures with graded sound velocity characteristics is a practical technique, and the TLM modeling method described is within the capabilities of most current desktop computers. These particular examples were prepared using a Sun Ultra 10, but the software was also compiled to run successfully in a Windows-NT environment. The two-dimensional mesh defining the model space involved 10^5 nodes, and the time sequence required to propagate the signal through this space involved 10^3 steps or some 10^8 node iterations. With each nodal calculation taking around 3 μs, the model ran in approximately 5 minutes.

The visualization provided by the TLM method showed how signal propagation proceeds through the melon and provided useful insights to the more subtle features of the porpoise's sonar system.

The melon model shows that the graded index wave guide effect very efficiently captures and beam-forms the pulse energy without exploiting internal reflections or causing significant waveform distortion. This conclusion provides support for the assumption that the signal is already in its final extended oscillatory form as it enters the melon from the air sac bounded channel.

In the harbor porpoise, the assertion that the initial impulse is provided by a single-step pressure function, generated as the phonic lips separate, requires further investigation. A modification to this hypothesis should be considered in which, on occasion, both MLDB complexes are fired with a small time delay between them. If this double stimulus can be proved to occur, then it may explain the lack of a low-frequency communication whistle signal in the harbor porpoise, as the left MLDB complex appears to be reserved for this function in the bottlenose dolphin (Goodson, Farooq, and Datta 2001).

The deeply folded floor of the vestibular air sacs can explain the translation of the initial stimulating impulse into the polycyclic waveform characteristic of porpoise signals (Goodson 1997). With accurate tissue dimensions, this structure can also be modeled using the TLM method, but to do this effectively the boundary dimensions will need to be determined with more precision than has been possible to date.

The length and rate at which the air-sac-bounded channel expands toward the melon contributes to the duration of the reverberant condition following the pulse and hence the resulting wave shape. The effect of the short megaphone-like section between the MLDB complexes and the melon is worth careful investigation. *Cephalorhynchus commersonii*, a very small delphinid species occupying a similar ecological niche in the Southern Hemisphere, also produce the polycyclic waveform characteristic of porpoises. It is possible that a porpoise-like pulse could be produced by animals without a folded vestibular sac floor, provided that the bounded channel is extended in length and expands at a slower rate. The structures unique to porpoises, therefore, could have evolved from a dolphinlike configuration by compressing a much longer channel into a short one while retaining the narrowband signal-to-noise advantage. If this folded structure functions in the porpoise as described, then the tuned baffle filter effect will only efficiently transmit one frequency. If the source level is raised or lowered, then the energy peak contained in the stimulating impulse may change in frequency and fail to match this tuned structure. This in turn could provide an explanation for the similarity of SLs reported for different harbor porpoises in a variety of conditions.

The TLM modeling approach allowed hypotheses to be tested using dimensions and data obtained from dissections, X-ray CT, and perhaps in the future from magnetic resonance imaging. More importantly, the TLM method can be used to model structures that include subwavelength dimensions and variable sound velocity media—problems that invalidate the application of traditional ray tracing methodologies at this scale. To be effective, the signal bandwidth must be carefully defined and the mesh dimensions selected so that the 0.1λ sampling criteria is met at the highest frequencies and in the lowest velocity material. The mesh resolution and the size of the field modeled has a direct effect on the computational load; there are practical limitations associated with attempts to examine high-frequency signals propagating in much larger fields.

This first application of TLM modeling to bioacoustic structures highlights some features where a graded index wave guide appears to outperform the conventional beam-forming approaches of humanmade systems. In particular, the method offers better control of side lobe energy while maintaining the beam-forming capability of the much larger exit aperture. The graded velocity structure of the melon reduces the near/far field transition distance by producing a pronounced flattening effect on the emerging wave front, and this could actually occur within the wave guide structure. Conventional transducer systems, with otherwise similar characteristics, suffer severely degraded performance when a sonar target enters the fluctuating near-field diffraction zone close to the source. The porpoise has evidently evolved an effective solution to this problem and has been observed operating its sonar at extremely short range (Lockyer et al. 1999). The benefits of maintaining consistent target ensonification during the final phase of prey capture appear substantial. Other more subtle advantages, such as achieving a lower hydrodynamic profile by bending the emerging beam, will be addressed in the three-dimensional tlm models planned for future investigations.

Acknowledgments

Discussions with John Sigurdson and other colleagues (SPAWAR Biosciences group, San Diego) and with Mats Amundin (Kolmården Djurpark in Sweden) provided valuable feedback and encouragement. The original dissections that initiated this study of the harbor porpoise were carried out by Harry Ross, Vic Simpson, and Paul Jepson during the UK (Department of the Environment) harbor porpoise strandings autopsy workshops. Harderwijk Marine Mammal Park in The Netherlands and the Fjørd and Bælt Center in Denmark provided access to record the sonar signals of harbor porpoises.

Sekharjit Datta and Simon Pomeroy (Loughborough University) helped with the modeling and reviewed the manuscript. This work was supported by the European Commission DG XIV and in the UK by the Ministry of Agriculture Fisheries and Food, the Department of the Environment, and by the Engineering and Physical Sciences Research Council.

Part One / Literature Cited

ADACHI, S., and M. SATO. 1995. Time-domain simulation of sound production in the brass instrument. *Journal of the Acoustical Society of America* 97:3850–3861.

———. 1996. Trumpet sound simulation using a two-dimensional lip vibration model. *Journal of the Acoustical Society of America* 99:1200–1209.

AKAMATSU, T., Y. HATAKEYAMA, T. KOJIMA, and H. SOEDA. 1994. Echolocation rates of two harbor porpoises (*Phocoena phocoena*). *Marine Mammal Science* 10:401–411.

ALTES, R. A., and E. L. TITLEBAUM. 1970. Bat signals as optimally Doppler tolerant wave forms. *Journal of the Acoustical Society of America* 48:1014–1020.

AMUNDIN, M. 1991a. Helium effects on the click frequency spectrum of the harbor porpoise, *Phocoena phocoena*. *Journal of the Acoustical Society of America* 90:53–59.

———. 1991b. Sound production in odontocetes with emphasis on the harbour porpoise *Phocoena phocoena*. Ph.D. dissertation, University of Stockholm, Sweden.

AMUNDIN, M., and S. H. ANDERSEN. 1983. Bony nares air pressure and nasal plug muscle activity during click production in the harbour porpoise, *Phocoena phocoena,* and the bottlenose dolphin, *Tursiops truncatus. Journal of Experimental Biology* 105:275–282.

AMUNDIN, M., and T. W. CRANFORD. 1990. Forehead anatomy of *Phocoena phocoena* and *Cephalorhynchus commersonii:* 3-dimensional computer reconstructions with emphasis on the nasal diverticula. Pp. 1–18 in *Sensory abilities of cetaceans: Laboratory and field evidence,* ed. J. A. Thomas and R. A. Kastelein. New York: Plenum Press.

AMUNDIN, M., E. KALLIN, and S. KALLIN. 1988. The study of the sound production apparatus in the harbour porpoise, *Phocoena phocoena* and the jacobita, *Cephalorhynchus commersoni,* by means of serial cryo-microtome sectioning and 3-D computer graphics. Pp. 61–66 in *Animal sonar: Processes and performance,* ed. P. E. Nachtigall and P. W. B. Moore. New York: Plenum Press.

AROYAN, J. L., T. W. CRANFORD, J. KENT, and K. S. NORRIS. 1992. Computer modeling of acoustic beam formation in *Delphinus delphis. Journal of the Acoustical Society of America* 92:2539–2545.

AROYAN, J. L., M. A. MCDONALD, S. C. WEBB, J. A. HILDEBRAND, D. CLARK, J. T. LAITMAN, and J. S. REIDENBERG. 2000. Acoustic models of sound production and propagation. Pp. 409–469 in *Hearing by whales and dolphins,* ed. W. W. L. Au, A. N. Popper, and R. R. Fay. New York: Springer-Verlag.

AU, W. W. L. 1980. Echolocation signals of the Atlantic bottlenose dolphin (*Tursiops truncatus*) in open waters. In *Animal sonar systems,* ed. R.-G. Busnel and J. F. Fish. New York: Plenum Press.

———. 1993. *The sonar of dolphins.* New York: Springer-Verlag.

———. 2000. Echolocation in dolphins. Pp. 364–408 in *Hearing by whales and dolphins,* ed. W. W. L. Au, A. N. Popper, and R. R. Fay. New York: Springer-Verlag.

AU, W. W. L., and D. A. PAWLOSKI. 1989. A comparison of signal detection between an echolocating dolphin and an optimal receiver. *Journal of Comparative Physiology A* 164:451–458.

———. 1990. Cylinder wall thickness discrimination by an echolocating dolphin. *Journal of Comparative Physiology A* 172:41–47.

AU, W. W. L., and R. H. PENNER 1981. Target detection in noise by echolocating Atlantic bottlenose dolphins. *Journal of the Acoustical Society of America* 70:687–693.

AU, W. W. L., and K. J. SNYDER. 1980. Long-range target detection in open waters by an echolocating Atlantic bottlenose dolphin (*Tursiops truncatus*). *Journal of the Acoustical Society of America* 68:1077–1084.

AU, W. W. L., M. O. LAMMERS, and R. AUBAUER. 1999. A portable broadband data acquisition system for field studies in bioacoustics. *Marine Mammal Science* 15:526–531.

AU, W. W. L., P. W. B. MOORE, and D. PAWLOSKI. 1986. Echolocation transmitting beam of the Atlantic bottlenose dolphin. *Journal of the Acoustical Society of America* 80:688–691.

———. 1988. Detection of complex echoes in noise by an echolocating dolphin. *Journal of the Acoustical Society of America* 83:662–688.

AU, W. W. L., D. A. CARDER, R. H. PENNER, and B. L. SCRONCE. 1985. Demonstration of adaptation in beluga whale echolocation signals. *Journal of the Acoustical Society of America* 77:726–730.

AU, W. W. L., R. W. FLOYD, R. H. PENNER, and A. E. MURCHISON. 1974. Measurement of echolocation signals of the Atlantic bottlenose dolphin, *Tursiops truncatus* (Montagu), in open waters. *Journal of the Acoustical Society of America* 56:1280–1290.

AU, W. W. L., J. L. PAWLOSKI, T. W. CRANFORD, R. C. GISINER, and P. E. NACHTIGALL. 1993a. Transmission beam pattern of a false killer whale. *Journal of the Acoustical Society of America* 93:2358–2359.

AU, W. W. L., J. L. PAWLOSKI, P. E. NACHTIGALL, M. BLONZ, and R. C. GISINER. 1995. Echolocation signals and transmission beam pattern of a false killer whale (*Pseudorca crassidens*). *Journal of the Acoustical Society of America* 98:51–59.

AU, W. W. L., J. L. PAWLOSKI, P. E. NACHTIGALL, T. W. CRANFORD, and R. C. GISINER. 1993b. Echolocation signals and transmission beam pattern of a false killer whale (*Pseudorca crassidens*). In Proceedings of the Tenth Biennial Conference on the Biology of Marine Mammals, Society for Marine Mammalogy, Galveston, Texas.

AYRAPET'YANT, E., and A. I. KONSTANTINOV. 1974. *Echolocation in nature.* Leningrad: Nauka. (English translation, Arlington: Joint Publication Research Service.)

BAIRD, R. W., and P. J. STACEY. 1991. Status of the Risso's dolphin, *Grampus griseus,* in Canada. *Canadian Field Naturalist* 105:233–242.

BEHREND, O., and G. SCHULLER. 2000. The central acoustic tract and audio-vocal coupling in the horseshoe bat, *Rhinolophus rouxi. European Journal of Neuroscience* 12:4268–4280.

BEL'KOVICH V. M., and N. A. DUBROVSKIY. 1976. *Sensornye osnovy orientatsii kitoobraznykh.* Moscow: Nauka. (English translation, 1977: Sensory basis of cetacean orientation. Arlington: Joint Publications Research Service L/7157.)

BEL'KOVICH, V. M., V. I. BORISOV, and V. S. GUREVCH. 1970. Angular resolution by echolocation by *Delphinus delphis.* Pp. 66–67 in Proceedings of the Twenty-third Science-Technology Conference, Leningrad.

BLOOM, P. 1991. The diary of a wild bottlenose dolphin (*Tursiops truncatus*) resident off Amble on the North Northumberland coast of England from April 1987 to January 1991. *Aquatic Mammals* 173(3): 103–119.

BOWDEN, R. E. M. 1974. Innervation of intrinsic laryngeal muscles. Pp. 370–391 in *Ventilatory and phonatory control systems,* ed. B. Wyke. London: Oxford University Press.

BRACKENBURY, J. 1980. Respiration and production of sounds by birds. *Biological Reviews* 55:363–378.

BRADBURY, J. W., and S. L. VEHRENCAMP. 1998. *Principals of animal communication.* Sinauer Associates, Inc., Sunderland.

BROWN, P. E., and A. D. GRINNELL. 1980. Echolocation ontogeny in bats. Pp. 355–377 in *Animal sonar systems,* ed. R.-G. Busnel and J. F. Fish. New York: Plenum Press.

BROWN, P. E., T. W. BROWN, and A. D. GRINNELL. 1983. Echolocation, development and vocal communication in the lesser bulldog bat, *Noctilio albiventris. Behavioral Ecology and Sociobiology* 6:211–218.

BROWN, P. E., A. D. GRINNELL, and J. B. HARRISON. 1978. The development of hearing in the pallid bat, *Antrozous pallidus. Journal of Comparative Physiology A* 126:169–182.

BULLOCK, T. H., and S. H. RIDGWAY. 1972. Evoked potentials in the central auditory systems of alert porpoises to their own and artificial sounds. *Journal of Neurobiology* 3:79–99.

BUSNEL, R.-G., and A. DZIEDZIC. 1967. Résultats métrologiques expérimentaux de l'echolocation chez le Phocaena phocaena et leur comparaison avec ceux de certaines chauves-souris. Pp. 307–335 in *Animal sonar systems: Biology and bionics,* ed. R.-G. Busnel. Laboratoire de Physiologie Acoustique, INRA-CNRZ, Jouy-en-Josas, France.

BUSNEL, R.-G., and J. F. FISH, eds. 1980. *Animal sonar systems.* NATO Advanced Study Institute Series A: Life Sciences. New York: Plenum Press.

CASSEDY, J. H., J. B. KOBLER, S. F. ISBEY. and E. COVEY. 1989. Central acoustic track in an echolocating bat: An extralemniscal auditory pathway to the thalamus. *Journal of Comparative Neurology* 287:247–259.

CHAPMAN, M. F., and D. D. ELLIS. 1998. The elusive decibel: Thoughts on sonars and marine mammals. *Canadian Acoustics/Acoustique Canadienne* 26(2): 29–31.

CHEN, F., and G WEINREICH. 1996. Nature of the lip reed. *Journal of the Acoustical Society of America* 99:1227–1233.

CLARKE, M. R., and P. L. PASCOE. 1985. The stomach contents of a Risso's dolphin (*Grampus griseus*) stranded at Thurlestone, South Devon. *Journal of the Marine Biological Association of the United Kingdom* 65:663–665.

COCKCROFT, V. G., S. L. HASCHICK, and N. T. W. KLAGERS. 1993. The diet of Risso's dolphin, *Grampus griseus* (Cuvier, 1812), from the east coast of South Africa. *Zeitschrift für Säugetierkunde* 58:286–293.

CONNOR, R. C., and R. A. SMOLKER. 1996. "Pop" goes the dolphin: A vocalization male bottlenose dolphins produce during courtships. *Behaviour* 133:643–662.

COPLEY, D. C., and W. J. STRONG. 1996. A stroboscopic study of lip vibrations in a trombone. *Journal of the Acoustical Society of America* 99:1219–1226.

CORKERON, P. J., and S. M. VAN PARIJS. 2001. Vocalizations of eastern Australian Risso's dolphins, *Grampus griseus. Canadian Journal of Zoology/Revue Canadienne de Zoologie* 79:160–164.

COVEY, E., W. C. HALL, and J. B. KOBLER. 1987. Subcortical connections of the superior colliculus in the mustache bat, *Pteronotus parnellii. Journal of Comparative Neurology* 263(2): 179–197.

CRANFORD, T. W. 1988. The anatomy of acoustic structures in the spinner dolphin forehead as shown by X-ray computed tomography and computer graphics. Pp. 67–77 in *Animal sonar: Processes and performance,* ed. P. E. Nachtigall and P. W. B. Moore. New York: Plenum Press.

———. 1990. Morphology of the odontocete nose: Components of a biosonar signal generator. *American Zoologist* 30(4): abstract 500.

———. 1992a. Directional asymmetry in odontocetes. *American Zoologist* 32:140.

———. 1992b. Functional morphology of the odontocete forehead: Implications for sound generation. Ph.D. dissertation, University of California, Santa Cruz.

———. 1999a. Evidence for multiple sonar signal generators in odontocetes. P. 40 in *Proceedings of the Thirteenth Biennial Conference on the Biology of Ma-*

rine Mammals. The Society for Marine Mammalogy Conference, Maui, Hawaii.

———. 1999b. The sperm whale's nose: Sexual selection on a grand scale? *Marine Mammal Science* 15: 1134–1158.

———. 2000. In search of impulse sound sources in odontocetes. Pp. 109–156 in *Hearing by whales and dolphins,* ed. W. W. L. Au, A. N. Popper, and R. R. Fay. New York: Springer-Verlag.

CRANFORD, T. W., M. AMUNDIN, and D. E. BAIN. 1987. A unified hypothesis for click generation in odotocetes. P. 14 in *Proceedings of the Seventh Biennial Conference on the Biology of Marine Mammals.* The Society for Marine Mammalogy, Miami, Florida.

CRANFORD, T. W., M. AMUNDIN, and K. S. NORRIS. 1996. Functional morphology and homology in the odontocete nasal complex: Implications for sound generation. *Journal of Morphology* 228:223–285.

CRANFORD, T. W., W. G. VAN BONN, S. H. RIDGWAY, M. S. CHAPLIN, and J. A. CARR. 1997b. Functional morphology of the dolphin's biosonar signal generator studied by high-speed video endoscopy. *Journal of Morphology* 232(3): 243.

CRANFORD, T. W., W. G. VAN BONN, M. S. CHAPLIN, J. A. CARR, D. A. CARDER, T. KAMOLNICK, and S. H. RIDGWAY. 1997a. Visualizing dolphin sonar signal generation using high-speed video endoscopy. *Journal of the Acoustical Society of America* 102(5): 3123.

CRANFORD, T. W., W. G. VAN BONN, M. S. CHAPLIN, J. A. CARR, D. A. CARDER, T. KAMOLNICK, and S. H. RIDGWAY. 1988. High-speed video endoscopy of delphinid sonar signal generators. World Marine Mammal Science Conference—Joint Meeting in Monaco of the Society for Marine Mammalogy and European Cetacean Society. P. 30 in *European Research on Cetaceans,* ed. P. G. Evans and E. C. M. Parsons. Valencia: European Cetacean Society.

CRANFORD, T. W., W. G. VAN BONN, S. H. RIDGWAY, M. S. CHAPLIN, J. A. CARR, D. A. CARDER, T. KAMOLNICK, and S. H. RIDGWAY. 1998. High-speed video endoscopy of delphinid sonar signal generators. P. 30 in *Abstracts of Biological Sonar Conference,* ed. J. Thomas. Moline: Western Illinois University.

CRANFORD, T. W., W. R. ELSBERRY, D. J. BLACKWOOD, J. A. CARR, T. KAMOLNICK, M. TODD, W. G. VAN BONN, D. A. CARDER, S. H. RIDGWAY, D. M. BOZLINSKI, and E. C. DECKER. 2000. Physiological evidence for two independent sonar signal generators in the bottlenose dolphin. *Journal of the Acoustical Society of America* 108:2613.

CURRY, B. E. 1992. Facial anatomy and potential function of facial structures for sound production in the

harbor porpoise (*Phocoena phocoena*) and Dall's porpoise (*Phocoenoides dalli*). *Canadian Journal of Zoology* 70:2103–2114.

DAWSON, S. M. 1988. The high frequency sounds of free-ranging Hector's dolphins, *Cephalorhynchus hectori. International Whaling Commission,* special issue, 9: 339–344.

DEAR, S. P., and N. SUGA. 1995. Delay-tuned neurons in the midbrain of the big brown bat. *Journal of Neurophysiology* 73(3): 1084–1100.

DEAR, S. P., J. FRITZ, T. HARESIGN, M. FERRAGAMO, and J. A. SIMMONS. 1993. Tonotopic and functional organization in the auditory cortex of the big brown bat, *Eptesicus fuscus. Journal of Neurophysiology* 70(5): 1988–2009.

DEGOLLADA, E., M. GARCÍA-HARTMANN, and T. W. CRANFORD. 1998. Histological structure of the sound generation complex in delphinoid cetaceans. World Marine Mammal Science Conference—Joint Meeting in Monaco of the Society for Marine Mammalogy and European Cetacean Society. P. 298 in *European Research on Cetaceans,* ed. P. G. Evans and E. C. M. Parsons. Valencia: European Cetacean Society.

DENNY, S. P. 1976. The bat larynx. Pp. 346–370 in *Scientific foundations of otolaryngology,* ed. R. Hinchcliffe and D. F. Harrison. London: Heinemann (Med Books Ltd.).

DENZINGER, A., and H.-U. SCHNITZLER. 1994. Echo SPL influences the ranging performance of the big brown bat, *Eptescicus fuscus. Journal of Comparative Physiology A* 175:563–571.

———. 1998. The possible role of flow field information in CF-bats: Field studies. P. 11 in *Abstracts of Biological Sonar Conference,* ed. J. Thomas. Western Illinois University, Moline.

DIERCKS, K. T., R. T. TROCHTA, C. F. GREENLAW, and W. E. EVANS. 1971. Recording and analysis of dolphin echolocation signals. *Journal of the Acoustical Society of America* 49:1729–1732.

DING, W., B. WÜRSIG, and W. E. EVANS. 1995. Whistles of bottlenose dolphins: Comparisons among populations. *Aquatic Mammals* 21:65–77.

DORMER, K. J. 1974. The mechanism of sound production and measurement of sound processing in delphinid cetaceans. Ph.D. dissertation, University of California, Los Angeles.

———. 1979. Mechanism of sound production and air recycling in delphinids: Cineradiographic evidence. *Journal of the Acoustical Society of America* 65:229–239.

DOS SANTOS, M. E., G. CAPORIN, H. O. MOREIRA, A. J. FERREIRA, and J. L. B. COELHO. 1990. Acoustic behavior in a local population of bottlenose dolphins. Pp. 585–598 in *Sensory abilities of cetaceans: Laboratory and field evidence,* ed. J. A. Thomas and R. A. Kastelein. New York: Plenum Press.

DUBROVSKIY, N. A., and L. R. GIRO. 1999. A plausible mechanism of acoustic click production in the dolphin. *Journal of the Acoustical Society of America* 105:1263.

DUBROVSKIY, N. A., and G. L ZASLAVSKIY. 1973. Time structure and directionality of sound emission by the bottlenose dolphin. Pp. 56–59 in Proceedings of the Eighth All-Union Acoustics Conference, Moscow, Russia (in Russian).

———. 1975. Role of the skull bones in the space-time development of the dolphin echolocation signal. *Sov. Phys. Acoust.* 21:255–258.

ESSER, K. H., and U. SCHMIDT. 1989. Mother-infant communication in the lesser spear-nosed bat: *Phyllostomus discolor* (Chiroptera, Phyllostomidae)—evidence for acoustic learning. *Ethology* 82(2): 156.

ESSER, K. H., C. J. CONDON, N. SUGA, and J. S. KANWAL. 1997. Syntax processing by auditory cortical neurons in the CM-FM area of the mustached bat *Pteronotus parnellii*. *Proceedings of the National Academy of Sciences, United States* 94:14019–14024.

EVANS, P. G. H. 1980. Cetaceans in British waters. *Mammal Review* 10:1–52.

EVANS, W. E. 1973. Echolocation by marine delphinids and one species of fresh-water dolphin. *Journal of the Acoustical Society of America* 54:191–199.

EVANS, W. E., and P. F. MADERSON. 1973. Mechanisms of sound production in delphinid cetaceans: A review and some anatomical considerations. *American Zoologist* 13:1205–1213.

EVANS, W. E., and J. H. PRESCOTT. 1962. Observations of the sound capabilities of the bottlenose porpoise: A study of whistles and clicks. *Zoologica* 47:121–128.

EVANS, W. E., F. T. AWBREY, and H. HACKBARTH. 1988. High frequency pulses produced by free ranging Commerson's dolphin (*Cephalorhynchus commersonii*) compared to those of phocoenids. *International Whaling Commission,* special issue, 9:173–181.

FATTU, J. M., and R. A. SUTHERS. 1981. Subglottic pressure and the control of phonation by the echolocating bat, *Eptesicus*. *Journal of Comparative Physiology A* 143:465–475.

FAURE, P. A., J. H. FULLARD, and J. W. DAWSON. 1993. The gleaning attacks of the Northern long-eared bat, *My-otis septentrionalis,* are relatively inaudible to moths. *Journal of Experimental Biology* 178:173–189.

FENTON, M. B. 1984. Echolocation: Implications for ecology and evolution of bats. *Quart. Rev. Biol.* 59:33–53.

FENTON, M. B., D. AUDET, M. K. OBRIST, and J. RYDELL. 1995. Signal strength, timing and self-deafening: The evolution of echolocation in bats. *Paleobiology* 21: 229–242.

FISCHER, H., and H. GERKEN. 1961. Le larynx de la chauve-souris *(Myotis myotis)* et le larynx human. *Annals Oto-laryngology* 78:577–585.

FISCHER, H., and H. J. VOEMEL. 1961. Der Ultraschallapparat des Larynx von *Myotis myotis. Gegenbaurs Jahrbuch der Morphologie und Mikroskopischen Anatomie* 102:200–226.

FLIEGER, E., and H.-U. SCHNITZLER. 1973. Ortungsleistungen der Fledermaus *Rhinolophus ferrumequinum* bei ein- und beidseitiger Ohrverstopfung. *Journal of Comparative Physiology A* 82:93–102.

FLINT, J. A., A. D. GOODSON, and S. C. POMEROY. 1997. Visualising wave propagation in bio-acoustic lens structures using the transmission line modelling method. *Proceedings of the Institute of Acoustics, Underwater Bio-Sonar and Bioacoustics Symposium* 19(9): 29–38.

FOOTE, K. G. 1980. Importance of the swimbladder in acoustic scattering by fish: A comparison of gadoid and mackerel target strengths. *Journal of the Acoustical Society of America* 67:2084–2089.

GAIONI, S. J., H. RIQUIMAROUX, and N. SUGA. 1990. Biosonar behavior of mustached bats swung on a pendulum prior to cortical ablation. *Journal of Neurophysiology* 64:1801–1817.

GELFAND, D. L., and G. F. MCCRACKEN. 1986. Individual variation in the isolation calls of Mexican free-tailed bat pups (*Tadarida brasiliensis mexicana*). *Animal Behavior* 34:1078–1086.

GIRO, L. R. 1987. A mechanism of high frequency echolocation clicks creation and formation in the dolphin. Ph.D. dissertation, Andreyev Acoustics Institute, Moscow (in Russian).

GIRO, L. R., and N. A. DUBROVSKIY. 1975. Possible role of the pericranial diverticula in the production of dolphin echolocation signals. *Soviet Physics–Acoustics* 20(5): 428–430.

GOODSON, A. D. 1997. A narrow band bio-sonar: Investigating echolocation in the harbour porpoise, *Phocoena phocoena.* Pp. 19–28 in *Underwater bio-sonar and bioacoustics symposium,* vol. 19 (9). Institute of Acoustics, Loughborough.

GOODSON, A. D., and S. DATTA. 1995. Investigating the sonar signals of the harbour porpoise, *Phocoena phocoena.* Proceedings of the Silver Jubilee National Symposium on Acoustics, *Journal of the Acoustical Society of India* 23 (4): 205–211.

GOODSON, A. D., and C. R. STURTIVANT. 1996. Sonar characteristics of the harbour porpoise *Phocoena phocoena. ICES Journal of Marine Science* 53(2): 465–472.

GOODSON, A. D., O. FAROOQ, and S. DATTA. 2001. Phase and amplitude changes in echolocation signals from the harbour porpoise. *Proceedings of the Institute of Acoustics, 2nd Symposium on Underwater Bio-sonar and Bioacoustic Systems* 23(4): 133–140.

GOODSON, A. D., R. A. KASTELEIN, and C. R. STURTI-VANT. 1995. Source levels and echolocation signal characteristics of juvenile harbour porpoises (*Phocoena phocoena*) in a pool. Pages 41–53 in *Harbour porpoises: Laboratory studies to reduce bycatch,* ed. P. E. Nachtigall, J. Lien, W. W. L. Au, and A. J. Read. Woerden, The Netherlands: De Spil Publishers.

GOOLER, D. M., and W. E. O'NEILL. 1987. Topographic representation of vocal frequency demonstrated by microstimulation of anterior cingulate cortex in the echolocating bat, *Pteronotus parnelli. Journal of Comparative Physiology A* 161 (1987): 283–294.

GOULD, E. 1971. Studies of maternal-infant communication and development of vocalizations in the bat *Myotis* and *Eptesicus. Communication and Behavioural Biology* 5:263–313.

———. 1975. Experimental studies of the ontogeny of ultrasonic vocalizations in the bats. *Developmental Psychobiology* 8(4): 333–346.

GRAY, J. E. 1846. On the British cetacea. *Annals and Magazine of Natural History* 17:82–85.

GREEN, R. F., S. H. RIDGWAY, and W. E. EVANS. 1980. Functional and descriptive anatomy of the bottle-nosed dolphin nasolaryngeal system with special reference to the musculature associated with sound production. Pp. 199–238 in *Animal sonar systems,* ed. R.-G. Busnel and J. F. Fish. New York: Plenum Press.

GRIFFIN, D. R. 1958. The laryngeal mechanisms for the production of high frequency sound. Pp. 110–119 in *Listening in the dark.* New Haven: Yale University Press.

———. 1980. Early history of research on echolocation. Pp. 1–10 in *Animal sonar systems,* ed. R.-G. Busnel and J. F. Fish. New York: Plenum Press.

GRIFFIN, D. R., F. A. WEBSTER, and C. R. MICHAEL. 1960. The echolocation of flying insects by bats. *Animal Behavior* 8:141–154.

GRIFFITHS, T. A. 1983. Comparative laryngeal anatomy of the big brown bat, *Eptesicus fuscus,* and the mustached bat, *Pteronotus parnellii. Mammalia* 47: 377–394.

HABERSETZER, J., and G. MARIMUTHU. 1986. Ontogeny of sounds in the echolocating bat, *Hipposideros speoris. Journal of Comparative Physiology A* 158:247–257.

HABERSETZER, J., and B. VOGLER. 1983. Discrimination of surface-structured targets by the echolocating bat, *Myotis myotis,* during flight. *Journal of Comparative Physiology A* 152:275–282.

HABERSETZER, J., G. SCHULLER, and G. NEUWEILER. 1984. Foraging behavior and Doppler shift compensation in echolocating hipposiderid bats, *H. bicolor and H. speoris. Journal of Comparative Physiology A* 155: 559–567.

HÄRKÖNEN, T. 1986. *Guide to the otoliths of the bony fishes of the Northeast Atlantic.* Sweden: Danbiu Aps.

HARTLEY, D. J., and R. A. SUTHERS. 1987. The sound emission pattern and the acoustical role of the noseleaf in the echolocating bat, *Carollia perspicillata. Journal of the Acoustical Society of America* 85:1348–1351.

———. 1989. The sound emission pattern of the echolocating bat, *Eptesicus fuscus. Journal of the Acoustical Society of America* 85:1348–1351.

———. 1990. Sonar pulse radiation and filtering in the mustached bat, *Pteronotus parnellii rubiginosus. Journal of the Acoustical Society of America* 87:2756–2772.

HARTLEY, R. S., and R. A. SUTHERS. 1990. Lateralization of syringeal function during song production in the canary. *Journal of Neurobiology* 21:1236–1248.

HATAKEYAMA, Y., K. ISHII, T. AKAMATSU, H. SOEDA, T. SHIMAMURA, and T. KOJIMA. 1994. A review of studies on attempts to reduce the entanglement of the Dall's porpoise, *Phocoenoides dalli,* in the Japanese salmon gillnet. *Report to the International Whaling Commission,* special issue, 15:549–563.

HERZING, D. L. 1996. Vocalizations and associated underwater behavior of free-ranging Atlantic spotted dolphins, *Stenella frontalis,* and bottlenose dolphins, *Tursiops truncatus. Aquatic Mammals* 22:61–79.

HEYNING, J. E. 1989. Comparative facial anatomy of beaked whales (Ziphiidae) and a systematic revision among the families of extant Odontoceti. *Contributions in Science, Los Angeles County Museum of Natural History* 405:1–64.

HOLSTEGE, G. 1989. Anatomical study of the final common pathway for vocalization in the cat. *Journal of Comparative Neurology* 284:242–252.

HOPE, G. M., and K. P. BHATNAGAR. 1979. Electrical response of bat retina to spectral stimulation: Comparison of four microchiropteran species. *Experientia* 35:1189–1191.

HUFFMAN, R. F., and O. W. HENSON. 1990. The descending auditory pathway and acousticomotor systems—connections with the inferior colliculus. *Brain Research Review* 15(3): 295–324.

ISAKOVICH, M. A. 1973. *Obshchaya akustika (Fundamentals of acoustics)*. Moscow, Russia: Nauka Publishing House (in Russian).

JEN, P. H. S., J. OSTWALD, and N. SUGA. 1978. Electrophysiological properties of the acoustic middle ear and laryngeal muscle reflexes in the awake echolocating FM-Bats, *Myotis lucifugus. Journal of Comparative Physiology A* 124:61–73.

JENNINGS, R. 1982. Pelagic sightings of Risso's dolphin, *Grampus griseus,* in the Gulf of Mexico and Atlantic Ocean adjacent to Florida. *Journal of Mammalogy* 63:522–523.

JOHNS, P. B., and R. L. BEURLE. 1971. Numerical solution of 2-dimensional scattering problems using a transmission line matrix. *Proceedings of the IEE* 118(9): 1203–1208.

JOHNSON, C. S. 1968. Relation between absolute threshold and duration of tone pulse in the bottlenose porpoise. *Journal of the Acoustical Society of America* 43:757–763.

JONES, G., P. M. HUGHES, and J. M. V. RAYNER. 1991. The development of vocalizations in *Pipistrellus pipistrellus* (Chiroptera: Vespertilionidae) during postnatal growth and the maintenance of individual vocal signatures. *Journal of Zoology* 225:71–84.

JONES, R. E. 1981. Food habits of smaller marine mammals from Northern California. *Proceedings of the California Academy of Sciences* 42(16): 409–43.

JÜRGENS, U. 1994. The role of the periaqueductal grey in vocal behaviour. *Behavioural Brain Research* 62(2): 107–117.

JÜRGENS, U., and D. PLOOG. 1981. On the neural control of mammalian vocalization. *Trends in the Neurosciences* 6:135–137.

JÜRGENS, U., and R. PRATT. 1979, The cingular vocalization pathway in the squirrel monkey. *Experimental Brain Research* 34:499–510.

KALCOUNIS, M. C., and R. M. BRIGHAM. 1995. Intraspecific variation in wing loading affects habitat use by little brown bats (*Myotis lucifugus*). *Canadian Journal of Zoology* 73:89–95.

KALKO, E. K. V. 1994. Coupling of sound emission and wingbeat in naturally foraging European pipistrelle bats (Microchiroptera: Vespertilionidae). *Folia Zoologica* 43:363–76.

———. 1995. Insect pursuit, prey capture, and echolocation in pipistrelle bats (Microchiroptera). *Animal Behavior* 50:861–880.

KAMMINGA, C. 1988. Echolocation signal types of odontocetes. Pp. 9–22 in *Animal sonar: Processes and performance,* ed. P. E. Nachtigall and P. W. B. Moore. New York: Plenum Press.

KAMMINGA, C., and H. WIERSMA. 1981. Investigations on cetacean sonar II: Acoustical similarities and differences in odontocete sonar signals. *Aquatic Mammals* 8:41–62.

KAMMINGA, C., A. C. STUART, and G. K. SILBER. 1996. Investigations on cetacean sonar XI: Intrinsic comparison of the wave shapes of some members of the Phocoenidae family. *Aquatic Mammals* 22:45–55.

KASTELEIN, R. A., W. W. L. AU, H. T. RIPPE, and N. M. SCHOONEMAN. 2000. Target detection by an echolocating harbour porpoise (*Phocoena phocoena*). *Journal of the Acoustical Society of America* 49:359–375.

KASTELEIN, R. A., N. M. SCHOONEMAN, W. W. L. AU, W. C. VERBOOM, and N. VAUGHAN. 1997. The ability of a harbour porpoise (*Phocoena phocoena*) to discriminate between objects buried in sand. Pp. 329–342 in *The biology of the harbour porpoise,* ed. A. J. Read, P. R. Wiepkema, and P. E. Nachtigall. Woerden, The Netherlands: De Spil Publishers.

KICK, S. A. 1982. Target detection by the echolocating bat *Eptesicus fuscus. Journal of Comparative Physiology A* 145:431–435.

KICK, S. A., and J. A. SIMMONS. 1984. Automatic gain control in the bat's sonar receiver and the neuroethology of echolocation. *Journal of Neuroscience* 4:2725–2737.

KOBLER, J. B. 1983. The nucleus ambiguus of the bat, *Pteronotus parnellii:* Peripheral targets and central inputs. Ph.D. dissertation, University of North Carolina at Chapel Hill.

KONSTANTINOV, A. I. 1973. Development of echolocation of bats in postnatal ontogenesis. *Periodicum Biologorum* 75:13–19.

KONSTANTINOV, A. I., and A. K. MAKAROV. 1987. Development of echolocation in the ontogenesis of the greater horseshoe bat, *Rhinolophus ferrumequinum. Journal Evolution Biochemistry Physiology* 23(1): 80–87.

KÖSSL, M., and M. VATER. 1995. Cochlear structure and function in bats. Pp. 191–234 in *Hearing by bats,* ed. A. N. Popper and R. R. Fay. New York: Springer-Verlag.

KUZLER, E. 1960. Physiolgisahe und morphologische untersuchungen uber die erzeugung der orientierungslaute von flughunden der gattung *Rousettus. Zeitschrift für Vergleichende Physiologie* 43:231–268.

LANCASTER, W. C. 1993. Saving energy by flying: The economy of echolocation in flight. *Bat Research News* 33:22.

———. 1994. Morphological and physiological correlates of biosonar vocalizations in bats. Ph.D. dissertation, University of North Carolina, Chapel Hill.

LANCASTER, W. C., and J. R. SPEAKMAN. 1999. Respiratory muscle recruitment in echolocation: interspecific variation and implications for efficiency. *Bat Research News* 40:178.

LANCASTER, W. C., O. W. HENSON, and A. W. KEATING. 1995. Respiratory muscle activity in relation to vocalization in flying bats. *Journal of Experimental Biology* 198:175–191.

LAWRENCE, B., and W. E. SCHEVILL. 1956. The functional anatomy of the delphinid nose. *Bulletin of the Museum of Comparative Zoology* (Harvard) 114:103–151.

LAWRENCE, B. D., and J. A. SIMMONS. 1982a. Measurements of atmospheric attenuation at ultrasonic frequencies and the significance for echolocation by bats. *Journal of the Acoustical Society of America* 71: 585–590.

———. 1982b. Echolocation in bats: The external ear and perception of the vertical positions of targets. *Science* 218:481–483.

LEATHERWOOD, S., R. R. REEVES, and L. FOSTER. 1983. *The Sierra Club handbook of whales and dolphins.* Singapore: Tien Wah Press.

LEATHERWOOD, S., W. F. PERRIN, V. L. KIRBY, C. L. HUBBS, and M. DALHEIM. 1980. Distribution and movements of Risso's dolphins, *Grampus griseus,* in the Eastern North Pacific. *Fisheries Bulletin* 77:951–963.

LEATHERWOOD, S. J., R. R. REEVES, A. E. BOWLES, B. S. STEWART, and K. R. GOODRICH. 1984. Distribution, seasonal movements, and abundance of Pacific whitesided dolphins in the eastern North Pacific. *Scientific Reports of Whale Research Institute* 35:128–157.

LEE, D. N., J. A. SIMMONS, P. A. SAILLANT, and F. BOUFFARD. 1995. Steering by echolocation: A paradigm of ecological acoustics. *Journal of Comparative Physiology A* 176:347–354.

LEE, D. N., F. R. VAN DER WEEL, T. HITCHCOCK, E. MATEJOWSKY, and J. D. PETTIGREW. 1992. Common principle of guidance by echolocation and vision. *Journal of Comparative Physiology A* 171:563–571.

LEIGHTON, T. 1994. *The acoustic bubble.* San Diego: Academic Press.

LITCHFIELD, C., C. KAROL, and A. J. GREENBERG. 1973. Compositional topography of melon lipids in the Atlantic bottlenose dolphin (*Tursiops truncatus*): Implications for echolocation. *Marine Biology* 23:165–169.

———. 1978. Compositional topography of melon lipids in the Atlantic bottlenose dolphin *Tursiops truncatus:* Implications for echo-location. *Marine Biology* 28:165–169.

LITCHFIELD, C., R. KAROL, M. E. MULLEN, J. P. DILGER, and B. LUTHI. 1979. Physical factors influencing refraction of the echolocative sound beam in delphinid cetaceans. *Marine Biology* 52:285–290.

LOCKYER, C., M. AMUNDIN, G. DESPORTES, A. D. GOODSON, and F. LARSEN. 1999. EPIC: Elimination of harbour porpoise incidental catch. Report to the European Commission DG XIV. Special Study Project DG XIV 97/0006:1–18.

MACKAY, R. S., and C. LIAW. 1981. Dolphin vocalization mechanisms. *Science* 212:676–678.

MACLENNAN, D. N., and E. J. SIMMONDS. 1992. *Fisheries acoustics.* London: Chapman and Hall.

MARTEN, K., K. S. NORRIS, P. W. B. MOORE, and K. A. ENGLUND. 1988. Loud impulse sounds in odontocete predation and social behavior. Pp. 567–579 in *Animal sonar: Processes and performance,* ed. P. E. Nachtigall and P. W. B. Moore. New York: Plenum Press.

MASTERS, W. M., A. J. M. MOFFAT, and J. A. SIMMONS. 1985. Sonar tracking of horizontally moving targets by the big brown bat *Eptesicus fuscus. Science* 228: 1331–1333.

MATSUMURA, S. 1979. Mother-infant communication in a horseshoe bat (*Rhinolophus ferrumequninum nippon*): Development of vocalization. *Journal of Mammalogy* 60:76–84.

McCOWAN, B., and D. REISS. 1995. Quantitative comparison of whistle repertoires from captive adult bottlenose dolphins (Delphinidae, *Tursiops truncatus*): A re-evaluation of the signature whistle hypothesis. *Ethology* 100:194–209.

MEAD, J. G. 1972. Anatomy of the external nasal passage and facial complex in the Delphinidae (Mammalia: Cetacea). Ph.D. dissertation, University of Chicago.

———. 1975. Anatomy of the external nasal passages and facial complex in the Delphinidae (Mammalia: Cetacea). *Smithsonian Contributions to Zoology* 207:1–72.

METZNER, W. 1989. A possible neuronal basis for Doppler-shift compensation in echolocating horseshoe bats. *Nature* 341:529–532.

———. 1993. An audio-vocal interface in echolocating horseshoe bats. *Journal of Neuroscience* 13(5): 1899–1915.

————. 1996. Anatomical basis for audio-vocal integration in echolocating horseshoe bats. *Journal of Comparative Neurology* 368(2): 252–269.

MILLER, L. A., and A. E. TREAT. 1993. Field recordings of echolocation and social signals from the gleaning bat *Myotis septentrionalis. Bioacoustics* 5:67–87.

MILLER, L. A., T. KIRKETERP, J. TEILMANN, B. MØHL, and R. A. KASTELEIN. 1998. Target detection by the harbour porpoise, *Phocoena phocoena.* P. 346 in *Proceedings Göttingen Neurobiological Conference,* vol. 2, ed. N. Elsner and R. Wehner.

MITSON, R. B. 1990. Very-high-frequency acoustic emissions from the white-beaked dolphin (*Lagenorhynchus albirostris*). Pp. 269–281 in *Sensory abilities of cetaceans: Laboratory and field evidence,* ed. J. A. Thomas and R. A. Kastelein. New York: Plenum Press.

MITSON, R. B., and R. J. MORRIS. 1988. Evidence of high-frequency acoustic emissions from the white-beaked dolphin (*Lagenorhynchus albirostris*). *Journal of the Acoustical Society of America* 83:825–826.

MITTMANN, D. H., and J. J. WENSTRUP. 1995. Combination-sensitive neurons in the inferior colliculus. *Hearing Research* 90(1–2): 185–191.

MØHL, B. 1986. Detection by a pipisrelle bat of normal and reversed replica of its sonar pulses. *Acustica* 61:75–82.

MØHL, B., and S. ANDERSEN. 1973. Echolocation: High-frequency component in the click of the harbour porpoise (*Phocoena phocoena,* L.). *Journal of the Acoustical Society of America* 54:1368–1372.

MØHL, B., and A. SURLYKKE. 1989. Detection of sonar signals in the presence of pulses of masking noise by the echolocating bat, *Eptesicus fuscus. Journal of Comparative Physiology A* 165:119–124.

MOORE, P. W. B., and D. A. PAWLOSKI. 1990. Investigations on the control of echolocation pulses in the dolphin (*Tursiops truncatus*). Pp. 305–316 in *Sensory abilities of cetaceans: Laboratory and field evidence,* ed. J. A. Thomas and R. A. Kastelein. New York: Plenum Press.

MOORE, P. W. B., R. W. HALL, W. A. FRIEDL, and P. E. NACHTIGALL. 1984. The critical interval in dolphin echolocation: What is it? *Journal of the Acoustical Society of America* 76:314–317.

MOORE, S. E., and S. H. RIDGWAY. 1995. Whistles produced by common dolphins from the Southern California Bight. *Aquatic Mammals* 21:55–63.

Moss, C. F. 1988. Ontogeny of vocal signals in the big brown bat, *Eptesicus fuscus.* NATO Life Science Series A, 156:115–120.

MOSS, C. F., and H.-U. SCHNITZLER. 1989. Accuracy of target ranging in echolocating bats: Acoustic information processing. *Journal of Comparative Physiology A* 165(3): 383–393.

MOSS, C. F., D. REDISH, C. GOUNDEN, and T. H. KUNZ. 1997. Ontogeny of vocal signals in the little brown bat, *Myotis lucifugus. Animal Behaviour* 54:131–141.

MURCHISON, A. E. 1980. Detection range and range resolution of porpoise. Pp. 43–70 in *Animal sonar systems,* ed. R.-G. Busnel and J. F. Fish. New York: Plenum Press.

MURRAY, S. O., E. MERCADO, and H. L. ROITBLAT. 1998. Characterizing the graded structure of false killer whale (*Pseudorca crassidens*) vocalizations. *Journal of the Acoustical Society of America* 104:1679–1688.

NACHTIGALL, P. E., and P. W. B. MOORE, eds. 1988. *Animal sonar: Processes and performance.* New York: Plenum Press.

NACHTIGALL, P. E., W. W. L. AU, J. L. PAWLOSKI, and P. W. B. MOORE. 1995. Risso's dolphin (*Grampus griseus*) hearing thresholds in Kaneohe Bay, Hawaii. Pp. 49–53 in *Sensory systems of aquatic mammals,* ed. R. A. Kastelein, J. A. Thomas, and P. E. Nachtigall. Woerden, The Netherlands: De Spil Publishers.

NESS, A. R. 1967. A measure of asymmetry of the skulls of odontocete whales. *Journal of Zoology* 153:209–221.

NEUMANN, I., and G. SCHULLER. 1991. Spectral and temporal gating mechanisms enhance the clutter rejection in the echolocating bat, *Rhinolophus rouxi. Journal of Comparative Physiology A* 169:109–116.

NEUWEILER, G. 1984. Foraging, echolocation and audition in bats. *Naturwissenschaften* 71:446–455.

————. 1990. Auditory adaptations for prey capture in echolocating bats. *Physiological Reviews* 70:615–641.

NEUWEILER, G., V. BRUNS, and G. SCHULLER. 1980. Ears adapted for the detection of motion, or how echolocating bats have exploited the capacities of the mammalian auditory system. *Journal of the Acoustical Society of America* 68(3): 741–753.

NEUWEILER, G., W. METZNER, R. RÜBSAMEN, M. ECKRICH, and H. H. COSTA. 1987. Foraging behavior and echolocation in the rufous horseshoe bat (*Rhinolophus rouxi*) of Sri Lanka. *Behavioral Ecology and Sociobiology* 20:53–67.

NORRIS, K. S. 1964. Some problems of echolocation in cetaceans. Pp. 317–336 in *Marine bio-acoustics,* ed. W. N. Tavolga. New York: Pergamon Press.

————. 1968. The evolution of acoustic mechanisms in odontocete cetaceans. Pp. 297–324 in *Evolution and environment,* ed. E. T. Drake. New Haven: Yale University Press.

———. 1969. The echolocation of marine mammals. Pp. 391–423 in *The biology of marine mammals,* ed. H. T. Andersen. New York: Academic Press.

———. 1975. Cetacean biosonar: Part 1—Anatomical and behavioral studies. Pp. 215–234 in *Biochemical and biophysical perspectives in marine,* ed. D. C. Malins and J. R. Sargent. New York: Academic Press.

NORRIS, K. S., and G. W. HARVEY. 1972. A theory for the function of the spermaceti organ of the sperm whale (*Physeter catodon* L.). Pp. 397–417 in *Animal orientation and navigation,* S. R. Galler, K. Schmidt-Koenig, G. J. Jacobs, and R. E. Belleville. (NASA Special Publication 262). Washington, D.C.: NASA Scientific and Technical Office.

———. 1974. Sound transmission in the porpoise head. *Journal of the Acoustical Society of America* 56:659–664.

NORRIS, K. S., and B. MØHL. 1983. Can odontocetes debilitate prey with sound? *American Naturalist* 122:85–104.

NORRIS, K. S., W. E. EVANS, and R. N. TURNER. 1967. Echolocation in an Atlantic bottlenose porpoise during discrimination. Pp. 409–437 in *Animal sonar systems: Biology and bionics,* ed. R.-G. Busnel. New York: Plenum Press.

NORRIS, K. S., K. J. DORMER, J. PEGG, and G. T. LIESE. 1971. The mechanism of sound production and air recycling in porpoises: A preliminary report. Pp. 113–129 in Proceedings of the Eighth Conference on the Biological Sonar of Diving Mammals. Stanford Research Institute, Menlo Park, California.

NORRIS, K. S., J. H. PRESCOTT, P. V. ASA-DORIAN, and P. PERKINS. 1961. An experimental demonstration of echolocation behavior in the porpoise, *Tursiops truncatus* (Montagu). *Biological Bulletin* 120:163–176.

NORRIS, K. S., B. WÜRSIG, R. S. WELLS, and M. WÜRSIG. 1994. *The Hawaiian spinner dolphin.* Berkeley and Los Angeles: University of California Press.

NOWICKI, S., and R. R. CAPRANICA. 1986. Bilateral syringeal interaction in vocal production of an oscine bird sound. *Science* 231:1297–1299.

OLSON, C. R., and S. Y. MUSIL. 1992. Topographic organization of cortical and subcortical projections to posterior cingulate cortex in the cat—evidence for somatic, ocular, and complex subregions. *Journal of Comparative Neurology* 324(2): 237–260.

O'NEILL, W. E., and N. SUGA. 1982. Encoding of target range and its representation in the auditory cortex of the mustached bat. *Journal of Neuroscience* 2(1): 17–31.

ORME, E. A., P. B. JONES, and J. M. ARNOLD. 1988. A hybrid modelling technique for underwater acoustic scattering. *International Journal of Numerical Modelling: Electronic Networks, Devices and Fields* 1: 189–206.

PIETSCH, G., and G. SCHULLER. 1987. Auditory self-stimulation by vocalization in the CF-FM bat, *Rhinolophus rouxi. Journal of Comparative Physiology A* 160:635–644.

PILLAT, J., and G. SCHULLER. 1998. Audiovocal behavior of Doppler-shift compensation in the horseshoe bat survives bilateral lesion of the paralemniscal tegmental area. *Experimental Brain Research* 119:17–26.

POMEROY, S. C., H. R. WILLIAMS, and P. BLANCHFIELD. 1991. Evaluation of ultrasonic inspection and imaging systems for robotics using TLM modelling. *Robotica* 9 :283–290.

POPPER, A. N. 1980. Sound emission and detection by delphinids. Pp. 1–52 in *Cetacean behavior: Mechanisms and function,* ed. L. M. Herman. New York: John Wiley & Sons.

POPPER, A. N., and R. R. FAY, eds. 1995. *Hearing by bats.* New York: Springer-Verlag.

PRECHTL, H. 1995. Senso-motorische Wechselwirkung im auditorischen Mittelhirn der Hufeisennasen-Fledermaus *Rhinolophus rouxi.* Ph.D. dissertation, University of Munich, Germany.

PURVES, P. E. 1966. Anatomical and experimental observations on the cetacean sonar system. Pp. 197–270 in *Animal sonar systems: Biology and bionics,* ed. R.-G. Busnel. New York: Plenum Press.

PURVES, P. E., and G. E. PILLERI. 1983. *Echolocation in whales and dolphins.* New York: Academic Press.

PYE, J. D. 1980. Echolocation signals and echoes in air. Pp. 309–353 in *Animal sonar systems,* ed. R.-G. Busnel and J. F. Fish. New York: Plenum Press.

———. Noseleaves and bat pulses. Pp. 791–796 in *Animal sonar: Processes and performance,* ed. P. E. Nachtigall and P. W. B. Moore. New York: Plenum Press.

RAUSCHECKER, J. P., B. TIAN, T. PONS, and M. MISHKIN. 1997. Serial and parallel processing in rhesus monkey auditory cortex. *Journal of Comparative Neurology* 382:89–103.

READ, A. J., R. S. WELLS, A. A. HOHN, and M. D. SCOTT. 1993. Patterns of growth in wild bottlenose dolphins, *Tursiops truncatus. Journal of Zoology* (London) 231:107–123.

REAY, P. J. 1986. Ammodytidae. Pp. 945–950 in *Fishes of the North-eastern Atlantic and the Mediterranean,* vol. 2, ed. P. J. Whitehead, M. L. Bauchot, J. C. Hureau, J. Nielsen, and E. Tortonese. UNESCO. Bungay, United Kingdom: The Chaucer Press Ltd.

REIMER, K. 1989. Retinofugal projections in the rufous horseshoe bat, *Rhinolophus rouxi. Anatomical Embryology* 180:89–98.

———. 1991. Auditory properties of the superior colliculus in the horseshoe bat, *Rhinolophus rouxi. Journal of Comparative Physiology A* 169:719–728.

RENAUD, D. L., and A. N. POPPER. 1975. Sound localization by the bottlenose porpoise. *Tursiops truncatus. Journal of Experimental Biology* 63:569–585.

RICHARDSON, J. W., C. R. GREENE JR., C. I. MALME, and D. H. THOMSON. 1995. *Marine mammals and noise.* New York: Academic Press.

RIDGWAY, S. H. 1990. Dolphin sound production: physiologic, diurnal, and behavioral correlates. *Journal of the Acoustical Society of America* 74 Suppl. 1:S73.

RIDGWAY, S. H., and D. A. CARDER. 1988. Nasal pressure and sound production in an echolocating white whale, (*Delphinapterus leucas*). Pp. 53–60 in *Animal sonar: Processes and performance,* ed. P. E. Nachtigall and P. W. B. Moore. New York: Plenum Press.

RIDGWAY, S. H., D. A. CARDER, R. F. GREEN, A. S. GAUNT, S. L. L. GAUNT, and W. E. EVANS. 1980. Electromyographic and pressure events in the nasolaryngeal system of dolphins during sound production. Pp. 239–250 in *Animal sonar systems,* ed. R.-G. Busnel and J. F. Fish. New York: Plenum Press.

ROBERTS, L. H. 1973. Cavity resonances in the production of orientation cries. *Periodicum Biologorum* 75:27–32.

RÜBSAMEN, R. 1987. Ontogenesis of the echolocation system in the rufous horseshoe bat, *Rhinolophus rouxi* (audition and vocalization in early postnatal development). *Journal of Comparative Physiology A* 161:899–904.

RÜBSAMEN, R., and M. BETZ. 1986. Control of echolocation pulses by neurons of the nucleus ambiguus in the rufous horseshoe bat, *Rhinolophus rouxi.* I. Single unit recordings in the ventral motor nucleus of the laryngeal nerves in spontaneously vocalizing bats. *Journal of Comparative Physiology A* 159:675–687.

RÜBSAMEN, R., and G. SCHULLER. 1981. Laryngeal nerve activity during pulse emission in the CF-FM bat, *Rhinolophus ferrumequinum. Journal of Comparative Physiology A* 143:323–327.

RÜBSAMEN, R., and H. SCHWEIZER. 1986. Control of echolocation pulses by neurons of the nucleus ambiguus in the rufous horseshoe bat, *Rhinolophus rouxi.* II. Afferent and efferent connections of the motor nucleus of the laryngeal nerves. *Journal of Comparative Physiology A* 159:689–699.

RÜBSAMEN, R., G. NEUWEILER, and G. MARIMUTHU. 1989 Ontogenesis of tonotopy in inferior colliculus of a hipposiderid bat reveals postnatal shift in frequency-place code. *Journal of Comparative Physiology A* 165:755–769.

SALEH, H. M., and P. BLANCHFIELD. 1990. Analysis of acoustic radiation patterns of array transducers using the TLM method. *International Journal of Numerical Modelling: Electronic Networks, Devices and Fields* 3:39–56.

SCHERRER, J. A., and G. S. WILKINSON. 1993. Evening bat isolation calls provide evidence for heritable signatures. *Animal Behaviour* 46(5): 847–860.

SCHEVILL, W. E., and B. LAWRENCE. 1956. Food-finding by a captive porpoise (*Tursiops truncatus*). *Breviora* (Museum of Comparative Zoology) 53:1–15.

SCHEVILL, W. E., W. A. WATKINS, and C. RAY. 1969. Click structure in the porpoise, *Phocoena phocoena. Journal of Mammalogy* 50:721–728.

SCHMIDT, S. 1988. Evidence for a spectral basis of texture perception in bat sonar. *Nature* 331:617–619.

———. 1992. Perception of structured phantom targets in the echolocating bat, *Megaderma lyra. Journal of the Acoustical Society of America* 91:2203–2223.

SCHNITZLER, H.-U. 1968. Die Ultraschall-Ortungslaute der Hufeisen-Fledermaeuse (Chiroptera: Rhinolophidae) in verschiedenen Orientierungssituationen. *Zeitschrift für Vergleichende Physiologie* 57:376–408.

———. 1970. Echoortung bei der Fledermaus *Chilonycteris rubiginosa. Zeitschrift Vergleichende Physiologie* 68:25–38.

———. 1973. Control of Doppler shift compensation in the greater horseshoe bat, *Rhinolophus ferrumequinum. Journal of Comparative Physiology A* 82:79–92.

———. 1984. The performance of bat sonar systems. Pp. 211–224 in *Localization and orientation in biology and engineering,* ed. V. Vrju and H.-U. Schnitzler. Springer-Verlag, Berlin.

SCHNITZLER, H.-U., and O. W. HENSON. 1980. Performance of airborne animal sonar systems. Pp. 109–181 in *Animal sonar systems,* ed. R.-G. Busnel and J. F. Fish. New York: Plenum Press.

SCHNITZLER, H.-U., and E. KALKO. 1999. Roosting and foraging behavior of two neotropical gleaning bats, *Tonatia silvicola* and *Trachops cirrhosus* (Phyllostomidae). *Biotropica* 31(2): 344–353.

SCHNITZLER, H.-U., and J. OSTWALD. 1983. Adaptations for the detection of fluttering insects by echolocation in horseshoe bats. Pp. 801–827 in *Advances in vertebrate neuroethology,* ed. J. P. Ewert, R. R. Capranica, and D. J. Ingle. New York: Plenum Press.

SCHNITZLER, H.-U., E. KALKO, L. MILLER, and A. SUR-LYKKE. 1987. The echolocation and hunting behavior of the bat, *Pipistrellus kuhli. Journal of Comparative Physiology A* 161(2): 267–74.

SCHNITZLER, H.-U., D. MENNE, R. KOBER, and K. HEB-LICH. 1983. The acoustical image of fluttering insects in echolocating bats. *Neuroethology and Behavioral Physiology* 236–250.

SCHULLER, G. 1984. Natural ultrasonic echoes from wing beating insects are encoded by collicular neurons in the CF-FM bat, *Rhinolophus ferrumequinum. Journal of Comparative Physiology A* 155:121–128.

———. 1986. Influence of echolocation pulse rate on Doppler shift compensation control system in the greater horseshoe bat. *Journal of Comparative Physiology A* 158:239–246.

———. 1997. A cheap earphone for small animals with good frequency response in the ultrasonic frequency range. *Journal of Neuroscience Methods* 71:187–190.

———. 1998. Neural mechanisms of vocal control in bats. In *Neural control of mammalian vocalization,* ed. S. Brudzynski. On-line Proceedings of the 5th Internet World Congress on Biomedical Sciences, 1998, at McMaster University, Canada.

SCHULLER, G., and G. POLLAK. 1979. Disproportionate frequency representation in the inferior colliculus of Doppler-compensating greater horseshoe bats: Evidence for an acoustic fovea. *Journal of Comparative Physiology A* 132:47–54.

SCHULLER, G., and S. RADTKE-SCHULLER. 1988. Midbrain areas as candidates for audio-vocal interface in echolocating bats. Pp. 93–98. In *Animal sonar: Processes and performance,* ed. P. E. Nachtigall and P. W. B. Moore. New York: Plenum Press.

———. 1990. Neural control of vocalization in bats: Mapping of brainstem areas with electrical microstimulation eliciting species-specific echolocation calls in the rufous horseshoe bat. *Experimental Brain Research* 79:192–206.

SCHULLER, G., and K. SRIPATHI. 1999. Audio-motor control in horseshoe bats. P. 62 in *Advances in ethology* 34, ed. S. Sridhara. Wien: Blackwell Science Berlin.

SCHULLER, G., K. BEUTER, and R. RÜBSAMEN. 1975. Dynamic properties of the compensation system for Doppler shifts in the bat, *Rhinolophus ferrumequinum. Journal of Comparative Physiology A* 97: 113–125.

SCHULLER, G., K. BEUTER, and H.-U. SCHNITZLER. 1974. Response to frequency-shifted artificial echoes in the bat *Rhinolophus ferrumequinum. Journal of Comparative Physiology A* 89:275–286.

SCHULLER, G., E. COVEY, and J. H. CASSEDAY. 1991. Auditory pontine grey: Connections and response properties in the horseshoe bat. *European Journal of Neuroscience* 3:648–662.

SCHULLER, G., S. RADTKE-SCHULLER, and M. BETZ. 1986. A stereotaxic method for small animals using experimentally determined reference profiles. *Journal of Neuroscience Methods* 18:339–350.

SCHWEIZER, H., R. RÜBSAMEN, and C. RÜHLE. 1986. Localization of brain stem motoneurons innervating the laryngeal muscles in the rufous horseshoe bat, *Rhinolophus rouxi. Brain Research* 230:41–50.

SIGURDSON, J. E. 1997a. Analyzing the dynamics of dolphin biosonar behavior during search and detection tasks. Pp. 123–132 in *Proceedings of the Institute of Acoustics, underwater bio-sonar and bioacoustics symposium* vol. 19(9).

———. 1997b. Biosonar dynamics of the bottlenose dolphin in VSW search and detection tasks. *Journal of the Acoustical Society of America* 102:3123.

SIMMONS, J. A. 1969. Depth perception by sonar in the bat *Eptesicus fuscus.* Ph.D. dissertation, Princeton University.

———. 1973. The resolution of target range by echolocating bats. *Journal of the Acoustical Society of America* 54(1): 157–173.

SIMMONS, J. A., M. B. FENTON, and M. J. O'FARRELL. 1979. Echolocation and pursuit of prey by bats. *Science* 203:16–21.

SIMMONS, J. A., D. J. HOWELL, and N. SUGA. 1975. Information content of bat sonar echoes. *American Scientist* 63:204–215.

SIMMONS, J. A., C. F. MOSS, and M. FERRAGAMO. 1990. Convergence of temporal and spectral information into acoustic images of complex sonar targets perceived by the echolocating bat, *Eptesicus fuscus. Journal of Comparative Physiology A* 166:449–470.

SIMMONS, J. A., S. A. KICK, B. D. LAWRENCE, C. HALE, C. BARD, and B. ESCUDIE. 1983. Acuity of horizontal angle discrimination by the echolocating bat, *Eptesicus fuscus. Journal of Comparative Physiology A* 153:321–330.

SIMMONS, J. A., W. A. LAVENDER, B. A. LAVENDER, C. A. DOROSHOW, S. W. KIEFER, R. LIVINGSTON, A. C. SCALLET, and D. E. CROWLEY. 1974. Target structure and echo spectral discrimination by echolocating bats. *Science* 186:1130–1132.

SPEAKMAN, J. R., and P. A. RACEY. 1991. No cost of echolocation for bats in flight. *Nature* 350:421–423.

STACEY, P. J., and R. W. BAIRD. 1990. Status of the Pacific white-sided dolphin, *Lagenorhynchus obliquidens,* in Canada. *Canadian Field Naturalist* 105(2): 219–232.

STEINECKE, I., and H. HERZEL. 1995. Bifurcations in an asymmetric vocal-fold model. *Journal of the Acoustical Society of America* 97:1874–1884.

STEINER, W. W. 1981. Species-specific differences in pure tonal whistle vocalizations of five Western North Atlantic dolphin species. *Behavioral Ecology and Sociobiology* 9:241–246.

SUGA, N., and Y. YAJIMA. 1988. Auditory-vocal integration in the midbrain of the mustached bat: Periaqueductal grey and reticular formation. Pp. 87–107 in *The physiological control of mammalian vocalization,* ed. J. D. Newman. New York: Plenum Press.

SUGA, N., P. SCHLEGEL, T. SHIMOZAWA, and J. SIMMONS. 1973. Orientation sounds evoked from echolocating bats by electrical stimulation of the brain. *Journal of the Acoustical Society of America* 54(3): 793–797.

SURLYKKE, A., and O. BOJESEN. 1996. Integration time for short broad band clicks in echolocating FM-bats (*Eptesicus fuscus*). *Journal of Comparative Physiology A* 178:235–241.

SURLYKKE, A, and L. A. MILLER. 1985. The influence of arctiid moth clicks on bat echolocation; jamming or warning? *Journal of Comparative Physiology A* 156: 831–843.

SUTHERS, R. A. 1988. The production of echolocation signals by bats and birds. Pp. 23–45 in *Animal sonar: Processes and performance,* ed. P. E. Nachtigall and P. W. B.Moore. New York: Plenum Press.

———. 1990. Contributions to birdsong from the left and right sides of the intact syrinx. *Nature* 347:473–477.

———. 1994. Variable asymmetry and resonance in the avian vocal tract: A structural basis for individually distinct vocalizations. *Journal of Comparative Physiology A* 175:457–466.

SUTHERS, R. A., and G. E. DURRANT. 1980. The role of the anterior and posterior cricothyroid muscles in the production of the echolocative pulses by Mormoopideae. Pp. 995–997 in *Animal sonar systems,* ed. R.-G. Busnel and J. F. Fish. New York: Plenum Press.

SUTHERS, R. A., and D. H. HECTOR. 1982. Mechanism for the production of echolocating clicks by the grey swiftlet, *Collocalia spodiopygia. Journal of Comparative Physiology A* 148:457–470.

———. 1985. The physiology of vocalization by the echolocating oilbird, *Steatornis caripensis. Journal of Comparative Physiology A* 156:243–266.

———. 1988. Individual variation in vocal tract resonance may assist oilbirds in recognizing echos of their own sonar clicks. Pp. 87–92 in *Animal sonar: Processes and performance,* ed. P. E. Nachtigall and P. W. B. Moore. New York: Plenum Press.

SUTHERS, R. A., S. P. THOMAS, and B. J. SUTHERS. 1972. Respiration, wing-beat and ultrasonic pulse emission in an echo-locating bat. *Journal of Experimental Biology* 56:37–48.

TAVOLGA, W. N., ed. 1967. *Marine bio-acoustics.* New York: Pergamon Press.

THOMAS, J., M. STOEMER, C. BOWER, L. ANDERSON, and A. GARVER. 1988. Detection abilities and signal characteristics of echolocating false killer whale (*Pseudorca crassidens*). Pp. 323–328 in *Animal sonar: Processes and performance,* ed. P. E. Nachtigall and P. W. B. Moore. New York: Plenum Press.

TOUGAARD, J. 2000. Using echoes in orientation: Why are odontocetes (and bats) something special? In *Fourteenth Annual Conference of the European Cetacean Society,* vol. 14. Cork, Ireland: European Cetacean Society.

TRAPPE, M. 1982. Verhalten und Echoortung der grossen Hufeisennase beim Insektenfang. Ph.D. dissertation, University of Tübingen, Germany.

TREMEL, D. P., J. A. THOMAS, K. T. RAMIREZ, G. S. DYE, W. A. BACHMAN, A. N. ORBAN, and K. K. GRIMM. 1998. Underwater hearing sensitivity of a Pacific white-sided dolphin, *Lagenorhynchus obliquidens. Aquatic Mammals* 24:63–69.

TROEST, N., and B. MØHL. 1986. The detection of phantom targets in noise by serotine bats: Negative evidence for the coherent receiver. *Journal of Comparative Physiology A* 159:559–567.

TURL, C. W., and R. H. PENNER. 1989. Differences in the echolocation click patterns of the beluga (*Delphinapterus leucas*) and the bottlenose dolphin (*Tursiops truncatus*). *Journal of the Acoustical Society of America* 86:497–502.

TYACK, P. L. 1997. Development and social functions of signature whistles in bottlenose dolphins, *Tursiops truncatus. Bioacoustics* 8:21–46.

URICK, R. J. 1983. *Principles of underwater sound.* 3rd ed. New York: McGraw-Hill.

VALENTINE, D. E., and C. F. MOSS. 1997. Spatially selective auditory responses in the superior colliculus of the echolocating bat. *Journal of Neuroscience* 17(5): 1720–1733.

———. 1998. The sensorimotor integration in bat sonar. Pp. 220–230 in *Bat biology and conservation.* Washington, D.C.: Smithsonian Institution Press.

VALENTINE, D. E., S. R. SINHA, and C. F. MOSS. 2002. Orienting responses and vocalizations produced by microstimulation in the superior colliculus of the echolocating bat, *Eptesicus fuscus*. *Journal of Comparative Physiology A* 188:89–108.

VATER, M., A. S. FENG, and M. BETZ. 1985. An HRP-study of the frequency-place map of the horseshoe bat cochlea: Morphological correlates of the sharp tuning to a narrow frequency band. *Journal of Comparative Physiology A* 157:671–686.

VEL'MIN, V. A., and N. A. DUBROVSKIY, N. A. 1975. On the analysis of pulsed sounds by dolphins. *Dokl. Akademy Nauk. SSSR* 225:470–473.

VERBOOM W. C., and R. A. KASTELEIN. 1995. Acoustic signals by Harbour porpoises (*Phocoena phocoena*). Pp. 1–40 in *Harbour porpoises: Laboratory studies to reduce bycatch*, ed. P. E. Nachtigall, J. Lien, W. W. L. Au, and A. J. Read. Woerden, The Netherlands: De Spil Publishers.

———. 1997. Structure of harbour porpoise (*Phocoena phocoena*) click train signals. Pp. 343–362 in *The biology of the harbour porpoise*, ed. A. J. Read, P. R. Wiepkema, and P. E. Nachtigall. Woerden, The Netherlands: De Spil Publishers.

VON DER EMDE, G., and D. MENNE. 1989. Discrimination of insect wingbeat-frequencies by the bat, *Rhinolophus ferrumequinum*. *Journal of Comparative Physiology A* 164:663–371.

WEBSTER, F. A. 1962. Mobility without vision by living creatures other than man. Pp. 110–127 in *Proceedings of the Mobility Research Conference*, America Foundation for the Blind.

———. 1963a. Active energy radiating systems: The bat and ultrasonic principles in acoustical control of airborne interceptions by bats. *Proceedings of the International Congress on Technology and Blindness*, vol. 1.

———. 1963b. Bat-type signals and some implications. Pp. 378–408 in *Human factors in technology*, ed. E. Bennett, J. Degan, and J. Spiegel. New York: McGraw-Hill.

WONG, D., and S. L. SHANNON. 1988. Functional zones in the auditory cortex of the echolocating bat, *Myotis lucifugus*. *Brain Research* 453:349–352.

WOODWARD, P. M. 1953. *Probability and information theory with application to radar*. New York: Pergamon Press.

PART TWO

Auditory Systems in Echolocating Mammals

The Ears of Whales and Bats

Marianne Vater and *Manfred Kössl*

Introduction

Biosonar as an acoustic imaging system was invented independently by two unrelated mammalian taxa that drastically differ in body plan, evolution, and lifestyle: bats and dolphins (Pierce and Griffin 1938; Griffin 1958; Kellogg and Kohler 1952; McBride 1956). Both groups apply sonar during foraging and obstacle avoidance, but there are many differences in characteristics and capabilities of the two sonar systems (Au 1993; Au, introduction, this volume), the most obvious being that they operate in media with different physical characteristics: water and air.

It is of considerable interest and the central goal of this chapter to compare the principles of construction and function of the peripheral stages of the sonar receiver—the ear—in bats and dolphins. Numerous studies on both taxa have been devoted to the structure and function of the sound-conducting apparatus (outer and middle ear), as well as to the functional design of the auditory portion of the inner ear, the cochlea. Our comparative discussion of this work elucidates parallel and divergent evolutionary trends in ear function as adaptations to different auditory worlds imaged by aquatic and terrestrial sonars. It also illustrates both the constraints and potentials in mammalian ear design as a function of different environments and ecological niches. Since the sensory capacities of a particular species ultimately can only be understood in context with its evolutionary history, lifestyle, and vocal characteristics, we address these briefly before discussing the specifics of ear structures and mechanisms.

Bats and dolphins descended from land-dwelling ancestors with air-adapted ears. The water to land transition by the first tetrapods required the evolution of a sound-conducting apparatus that compensates for the difference between the low acoustic impedance of air and the high acoustic impedance of cochlear fluids. Bats kept the mammalian heritage of terrestrial ears. To operate successfully in a nocturnal niche, the acoustic sense gained high priority and was optimized for sensitive perception of the echoes from biosonar signals. Reentry to the aquatic environment by the ancestors of cetaceans posed the problem of impedance matching again; their terrestrial ears were maladapted to function under water. They needed to rematch the properties of their ears to the ancestral vertebrate medium. Water is about a thousand times denser than air and incompressible. Sound travels approximately five times faster than in air; the acoustic impedance of soft body tissues almost matches that of the surrounding medium, thus allowing sound energy to flow from water to body tissue with little energy loss. Specifically, the "acoustic transparency" of the cetacean body necessitated the evolution of structural specializations to channel acoustic energy selectively to the inner ear.

ANCESTRY AND LIFESTYLE

The land-water transition of the cetacean ancestor occurred sometime during the Paleocene (50–60 million years ago) and was performed either by amphibious carnivores or by early ungulates (Millinkovitsch et al. 1993). The two lines of extant cetaceans were established in the late Eocene; the Odontoceti (toothed whales) and the Mysticeti (baleen whales). This change in lifestyle to exploit the resources of an aquatic environment imposed new constraints and led to profound modifications in virtually all parts of cetacean anatomy (see Wartzok and Ketten 1999; Oelschläger 1990). Since acoustic energy propagates in water more efficiently than almost any other form of energy, the sense of hearing gained priority over other sensory systems (see Wartzok and Ketten 1999; Ketten 1992, 1997). It serves short- and long-distance communication tasks, as well as active acoustic imaging of the surround by echolocation. Extant odontocetes comprise more than 65 species of efficient carnivores that vary in size from 1 to 30 m. All odontocete species tested to date employ echolocation (see Au 1993; Au, introduction, this volume; Ketten 1998; for further data on lifestyle, see Herzing and Dos Santos, chapter 53, this volume). Mysticeti (11 species) are typically large pelagic planktivores and are not believed to echolocate. Many species produce infrasonic signals for long-range communication and maybe navigation (Clark and Ellison, chapter 73, this volume).

All Chiroptera are small mammals (weight 4–150 g in Microchiroptera and 20 g to 1.5 kg in Megachiroptera, Altringham 1999) that probably descended from small gliding nocturnal insectivores sometime in the late Cretaceous or early Paleocene. The ancestral forms are not documented in fossil records, and neither the relationship between the two groups of extant Chiroptera (Megachiroptera and Microchiroptera) nor the relation to other mammalian groups is clear. In fact, our inability to unequivocally link bats to any known mammalian group is indicative of their very early origin in mammalian history. The oldest complete fossil bat (*Icaronycteris index*) dates 50 million years ago and looks remarkably similar to modern microbats (Novacek 1985). Analysis of the oldest fossil bats, plus well-conserved fossils from oil pits in Messel (Germany) dating 45 million years (Habersetzer and Storch 1992), suggests that these species already possessed well-developed echolocation systems (for hypotheses on the evolution of echolocation, see Schnitzler, Kalko, and Denzinger, chapter 44, this volume). All bats are capable of active flight. Megabats (166 extant species) typically do not echolocate (with the exception of the Egyptian flying fox) and are exclusively frugivorous and restricted to the Old World tropics. The cosmopolitan extant Microchiroptera (759 species) exploit very different ecological niches ranging from frugivorous/nectarivorous species to insectivores, carnivores, and sanguivores. In adaptation to foraging on different prey in different habitats, elaborate echolocation systems were developed (see Denzinger, Kalko, and Jones, chapter 42, this volume; Fenton 1995; Neuweiler 1990).

Sonar

Dolphins generally use brief clicks for echolocation. Their sound production mechanisms are not understood fully, but in contrast to all other mammals, theirs involve highly specialized nasal-pharyngeal mechanisms rather than the larynx (Mackay and Liaw 1981; Cranford and Amundin, chapter 4, this volume). Click duration across species is 70–250 μs, and the signals are typically broadband with species-specific peak frequencies (Au 1993). Two acoustic categories have been defined: type I, with frequencies of maximum energy >100 kHz; and type II, with peak spectra below 80 kHz (Ketten 1984; Wartzok and Ketten 1999). These types correlate with habitat; smaller riverine and inshore species living in environments with high object density belong to type I and emit sonar clicks of relatively long duration (125–250 μs) and short wavelength (120–160 kHz in *Phocoena,* up to 200 kHz in *Inia geoffrensis*). Near- or offshore species use longer wavelength signals (40–70 kHz) suitable for detection of relatively large, distant targets. (For further details and references, see Au 1993; Au, introduction, this volume; Ketten 1998.)

Click-based sonar is uncommon among bats. Only megabat species of the genus *Rousettus* use tongue clicks for obstacle avoidance (see Fenton 1995). The echolocation signals of all microbats are produced by laryngeal mechanisms (see Schuller and Moss, chapter 1, this volume) and are more complex and varied in adaptation to different lifestyles and hunting habitats (Fenton 1995; Pye 1980; Schnitzler and Kalko 1998). Duration always exceeds 0.3 ms and reaches 100 ms in some species. The calls consist of either broadband components (brief downward FM sweep), pure tones (constant-frequency [CF] signals) or quasi CF, or a combination of narrowband and broadband components (see Denzinger, Kalko, and Jones, chapter 42, this volume). Call frequencies are species characteristic. The lowest sonar frequencies with the signal sweeping from 14.5 to 8.6 kHz are found in *Euderma maculatum;* the highest signal frequencies are employed by a small African hippossiderid bat (*Cloeotis percivali*) that emits a CF-signal component at 212 kHz (Fenton and Bell 1981). Unique in nature, Doppler-sensitive biosonar evolved in certain bats that routinely use bimodal calls consisting of a long CF component combined with brief FM sweeps (see Schnitzler and Henson 1980; Schuller and Moss, chapter 1, this volume; Behrend and Schuller, chapter 2, this volume). The long tonal component carries information on relative velocity and is used to detect fluttering prey insects in cluttered habitat.

Hearing Characteristics

A common trait in odontocetes and bats is their exquisite high-frequency hearing. The sonar signals typically coincide with the best hearing sensitivity of the particular species, but hearing also extends beyond the peak range related to communication sounds and prey-generated and environmental noise. The upper frequency limits of hearing are comparable in both groups and typically range from about 100 kHz to about 150 kHz (see Au 1993) and are thus higher than in most nonecholocating mammals. Inferred from species-specific sound emission frequencies, bats probably hold the record for high-frequency hearing at 212 kHz (Fenton and Bell 1981); at the other end of the acoustic spectrum, mysticetes probably hold the record for infrasound hearing at 20 Hz or lower (see Ketten 1997; Clark and Ellison, chapter 73, this volume). The audiograms of dolphins are the typical mammalian U-shaped broadband curves and basically resemble those obtained in bats that employ broadband sonar. Maximum sensitivity typically includes the frequency range of the dominant biosonar signals. Unique among mammals, CF-FM bats possess exceptionally sharp-tuned hearing within the narrow frequency range of CF signal, with filter qualities improved by at least a factor of 20 over those in other frequency ranges or in other vertebrates. The use of Doppler-sensitive sonar correlates with the evolution of the most sharply tuned cochlea filter mechanisms known in nature (Pollak, Henson, and Novick 1972; see Kössl and Vater 1995).

This chapter summarizes studies on ear design and function in dolphins and bats. We address the following issues: (1) design of the sound-conducting apparatus of

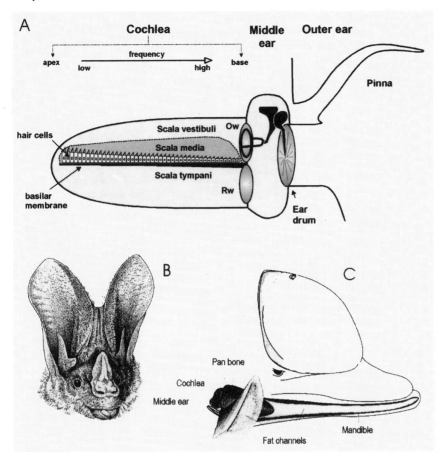

Fig. 12.1. *A:* Mammalian ear model (rw: round window; ow: oval window). *B:* Portrait of a gleaning bat (*Macroderma gigas*). Note the large pinnae (from Altringham 1999). *C:* Sound-conducting pathways to the dolphin middle and inner ear. The head is drawn semitransparent to show the fat channels (a lateral trumpet-shaped channel and the mandibular channel), the pan bone, and the location of the middle ear and cochlea (after Ketten 1997).

the ear in underwater and airborne echolocators, (2) basic principles in cochlea design in echolocating mammals, (3) task-related cochlear adaptations in different habitats and echolocical niches, and (4) basic cochlear filter mechanisms involved in ultrasound processing and unique filter mechanisms in Doppler-shift-compensating bats.

The Mammalian Ear Model

The ear consists of a sound-conducting apparatus (outer ear and middle ear) and a sound-reception apparatus (the auditory part of the inner ear, the cochlea). A functional model of the composite structures of the ear of terrestrial mammals is presented in fig. 12.1a. The pinnae act as directional antennae and mechanical amplifiers. Sound passes through the external auditory meatus and sets the tympanic membrane of the air-filled middle-ear cavity in motion. This motion is transferred to the cochlea via the ossicular chain of the middle ear. The middle ear acts as an impedance-matching device that counteracts the transmission loss between air and cochlear fluids. Additionally, it acts as a filter with characteristic resonance frequencies and thus profoundly influences the hearing range and the best frequencies of hearing (see Rosowski 1992, 1994). The cochlea per-

forms mechanoelectrical transduction and frequency analysis (see Dallos 1992).

The following chapters show that selection pressures imposed by life in different media have caused highly divergent constructions of the sound-conducting pathways in bats and dolphins. Yet the basic cochlear design appears evolutionarily quite conservative, and several parallel traits can be defined among airborne and marine echolocators.

The Sound-Conducting Apparatus in Bats and Dolphins

BATS

Large mobile pinnae are one of the most conspicuous external features of bats (fig. 12.1b). Pinnae shape and size vary considerably among species, and their construction to some extent reflects differences in echolocation behavior. The pinnae of bats act as mobile directional acoustic antennae (Guppy and Coles 1988; Obrist et al. 1993) analogous to magnifying glasses. Moving them changes the position of the acoustic axis, the focus. Pinna properties influence absolute hearing sensitivity due to their frequency-dependent amplification characteristics. Furthermore, they play a central role in sound localization since the pinnae gain also depends on the position of the sound source in azimuth and elevation. The directionality of sound emissions and the sound-receiving apparatus combine to produce a steep increase in total directionality of the sonar in front of the bat. The mechanical characteristics of the pinnae depend on their dimensions (i.e., length, areas of outer and inner openings) relative to wavelength of sound. In many bats, the pinnae dimensions correlate with the wavelength of a species' characteristic sonar signals (Guppy and Coles 1988; Obrist et al. 1993). A match in mechanical "tuning" of the external ear (i.e., maximal pinnae gain, directionality pattern, and maximal inter-aural intensity differences) and biosonar signals is most obvious in species that emphasize a narrow spectral band in echolocation calls. In *Rhinolophus rouxi,* there is a pinna gain of 14–20 dB and a maximum interaural intensity difference of 34 dB (Obrist et al. 1993) at the dominant CF-signal component of about 75 kHz. Exceptionally large pinnae are found among gleaning bats that rely on passive listening to noise generated by prey while foraging. In those bats, maximum pinna gain (5–22 dB, depending on species) occurs at lower frequencies (5–15 kHz, depending on species) than in bats that depend on echolocation for foraging. In megadermatid bats, an increase in auditory sensitivity by up to 25 dB SPL is produced by passive properties of the outer ear in the frequency range covered by prey-generated noise (Guppy and Coles 1988). There is no evidence for functional pinna specializations in megabats (Obrist et al. 1993).

The middle-ear design in bats follows the "general scheme" superimposed by several adaptations for high-frequency hearing. The microtype middle ear of bats (Fleischer 1978) is composed of low mass and high stiffness elements. These traits make the middle ear well adapted for transmission of ultrasonic frequencies, with rather poor low-frequency transfer characteristics (Fleischer 1978; Rosowski 1992, 1994). Bat middle ears show several characteristic design features. First are their small dimensions. The volume of the middle-ear cavity is small (7.1 mm^3 in *Rhinolophus*), and the bulla sits on top of a very large cochlea. The tympanic membrane is small and thin; its area amounts to about 2 mm^2 in *Rhinolophus* (Fleischer 1978) and in *Eptesicus* (Manley, Irvine, and Johnstone 1972). The area of the stapes footplate is 0.23 mm^2 in *Rhinolophus*. Although absolute dimensions are smaller in bats than in terrestrial mammals of larger size and good low-frequency hearing, the tympanic membrane-to-footplate area ratio is similar (Rosowski 1992). A second characteristic is the bats' decoupling of the middle ear from skull bones. The middle ear/cochlea complex is attached to the skull only via ligaments, which provides an effective isolation from bone-conducted sounds. Third, the middle-ear ossicles are small and stiffened in bats. The malleus is much larger than the incus. The manubrium of the malleus is

attached to the tympanic membrane, and the greatly expanded processus gracilis is fused with the tympanic bone (Fleischer 1978). Fleischer proposed that this feature is characteristic for mammals with good high-frequency hearing and insensitivity to frequencies below 1 kHz.

Measurements of middle-ear transfer functions are available for two species of bats (*Eptesicus:* Manley, Irvine, and Johnstone 1972; *Rhinolophus ferrumequinum:* Wilson and Bruns 1983) and corroborate the notion of increased middle-ear stiffness in adaptation for ultrasonic hearing. Optimal efficiency is at 25 kHz in *Eptesicus* and 55 kHz in *Rhinolophus,* with no indication of sharp middle-ear resonance at the CF frequency of 83 kHz.

DOLPHINS

The sound-conducting pathways of dolphins are unique among mammals (see Ketten 1997, 2000). The multiple specializations represent adaptations to rapid diving and long periods of submersion in water and are related to the fact that the acoustic impedance of water and soft body tissues are very similar. Thus, tissue-borne sound can act directly on the bony shelves of middle and inner ears. Dolphins and other cetaceans lack pinnae. Their external auditory meatus is rudimentary and plugged with tissue debris and waxy cerumen (Reysenbach de Haan 1956). The narrow, blocked meatus probably does not serve as an efficient acoustic pathway to the middle ear. The currently most accepted theory for sound conduction is via the lower jaw combined with soft tissues pathways that serve as acoustic wave guides (Norris 1980; Norris and Harvey 1974, 1980; Brill et al. 1988; Ketten 1994; see Ketten 1997, 2000). The transmission line includes an "acoustic window" made of an oval bulk of fatty tissue that overlies the thinned posterior portion of the mandible ("pan bone") and, according to recent data obtained with magnetic resonance imaging, several fat-filled channels to the tympanic periotic bone (Ketten 1994). The fatty tissues are made of "acoustic fat" (Varnassi and Malins 1971) and represent low impedance sound pathways. This system may function as a sound-focusing device toward the tympanic bulla (Norris 1980) and impose a directionality in sound reception with maximal sensitivity restricted to a narrow beam in the forward direction—that is, the direction of expected echoes (Bullock et al. 1968). This effect, which may be combined with a "preamplification" of sound, is analogous to pinna function in terrestrial mammals. But further research is clearly necessary to fully comprehend the complexity and function of the sound-conducting apparatus.

Cetaceans follow the mammalian bauplan of the middle ear: there is an equivalent to the tympanic membrane, a three-ossicular chain, and two middle-ear muscles, but comparative anatomy demonstrates that their whole middle ear is a highly derived structure (Fleischer 1978). The middle ears of mysticetes and odontocetes are characterized by the following features: (1) The middle-ear ossicles of both groups are massive and made of very dense mineralized bone (Fleischer 1978; Nummela et al. 1999a, 1999b); in odontocetes, their joints are stiffened with membranous sheets and ligaments. (2) The processus gracilis of the malleus is tightly attached to or even fused with the tympanic bone. (3) The tympanic membrane forms an elongated conical structure (tympanic conus of Reysenbach de Haan 1956) with a ligamentous attachment to the malleus. In sperm whales and ziphiids, the tympanic membrane is replaced by a thin bony tympanic plate (Fleischer 1978). (4) The middle-ear epithelium of both groups is thick and highly vascular. (5) The middle-ear cavity is air filled (Ketten 1994, 2000). (6) The eustachian tubes are large in both groups and serve in pressure equilibration. (7) The bony shells of middle and inner ears are combined in the tympanic-periotic complex that is positioned on the ventral side of the skull close to the lower jaw joint. The lateral aspect of the tympanic portion is in direct contact with soft tissue sound-conducting pathways from the lower jaw. The periotic bone is attached to the skull by ligaments. The whole complex is acoustically well isolated from the sound-production apparatus, which is located in the upper skull region.

Understanding middle-ear mechanisms in cetaceans is limited by the lack of direct experimental data: there are no measurements of vibration or transfer characteristics, and quantitative anatomical data were only recently integrated in models of the odontocete middle ear (Hemilä, Nummela, and Reuter 1999). The only physiological study of middle-ear function (McCormick et al. 1970, 1980) showed that dampening of the ossicular chain and bulla caused a drop in CM potentials to high-frequency stimulation, whereas disconnecting the malleus had little effect. Early hypotheses on middle-ear function (McCormick et al. 1970, 1980; Fleischer 1978) agree that cochlear excitation in cetaceans is produced by a pistonlike motion of the stapes in the oval window of the cochlea (as it is in terrestrial ears). McCormick et al. (1970, 1980) proposed functional principles similar to those involved in translatory bone conduction by terrestrial ears (i.e., tissue-borne sound causes the otic capsule to vibrate relative to the footplate of the stapes; the ossicular chain was viewed as a stiff, inert system). Fleischer (1978) emphasized vibrations of the tympanic conus transmitted to the cochlea by complex relative movements of the middle-ear bones.

Resent research (Nummela et al. 1999a) views the thin portions of the tympanic wall (referred to as "tympanic plate") as the whale analogue of the tympanic membrane of terrestrial mammals and interprets the tympanic conus (the whale homologue of the tympanic membrane) as nonfunctional in sound conduction. Hemilä, Nummela, and Reuter (1999) modeled the odontocete middle ear as a serial arrangement of two lever systems, the first of which is formed by the tympanic plate and the tympanic bone. The processus gracilis of the malleus transmits the amplified vibration to the second lever system, formed by the malleus-incus complex. Rotary movement of the malleus-incus complex pushes the stapes and causes it to move with increased velocity. These authors view the odontocete middle ear as a device for amplification of particle velocity—a completely different task in impedance matching than in terrestrial middle ears, which amplify pressure due to the large ratios of tympanic membrane area and the oval window and reduce particle velocity in the ossicle hinge system.

Cochlea Anatomy in Bats and Dolphins and Hearing Correlates

While selection pressure has sculptured the sound-conducting apparatus of outer and middle ear of bats and toothed whales to requirements of function in different media, the sense organ itself, the cochlea (fig. 12.2a–d), presents remarkably conservative overall structural arrangements: it is coiled, contains three fluid-filled scalae (scala tympani, scala media, and scala vestibuli), and houses all characteristic cellular and extracellular elements known from other mammals (Lim 1986; Echteler, Fay, and Popper 1994; Slepecky 1996). Several structural features have been identified that correlate with hearing range and/or behavior and habitat, and thus allow educated guesses on audition in extinct as well as in extant species where direct measures of hearing capabilities are lacking or impossible to obtain. Furthermore, a set of structural specializations appears unique for the cochlea of Doppler-shift-compensating bats.

GENERAL ORGANIZATION OF ORGAN OF CORTI

There are only few anatomical studies of the organ of Corti in odontocetes (*Tursiops truncatus*: Wever et al. 1971a, 1971b, 1971c; *Lagenorhynchus*: Wever et al. 1972; see Ketten 1997, 2000). Its ultrastructure has not been studied in detail due to the great difficulties in collecting specimens and typically rather long postmortem times prior to tissue fixation. Information on the mysticete cochlea is even sparser (see Ketten 2000). Several species of bats have been studied with both light-microscopic and ultrastructural techniques (*Rhinolophus*: Bruns 1976a, 1976b, 1980; Bruns and Schmieszek 1980; Bruns and Goldbach 1980; Vater and Lenoir 1992; Vater, Lenoir, and Pujol 1992; *Pteronotus*: Henson and Henson 1988, 1991; Henson, Henson, and Goldman 1977; Zook and Leake 1989; *Hipposideros*: Dannhof and Bruns 1991; *Tadarida*: Vater and Siefer 1995). The organization of the organ of Corti in bats and dolphins is compared in fig. 12.2a–d.

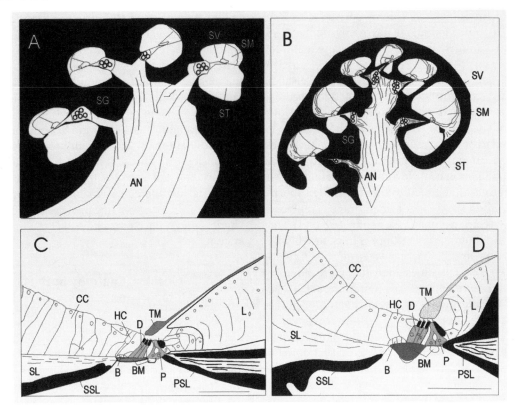

Fig. 12.2. Cross section of cochlear coils in (*A*) a dolphin (*Tursiops truncatus,* after Wever et al. 1971a); and (*B*) a bat (*Hipposideros lancadiva*). Details of the organ of Corti in (*C*) a dolphin (*Tursiops truncatus,* after Wever et al. 1971a); and (*D*) a bat (after Dannhof and Bruns 1991). AN: auditory nerve; B: Boettcher cells; BM: basilar membrane; CC: Claudius cells; D: Deiters cells; HC: Hensens cells; L: spiral limbus; PSL: primary spiral lamina; SG: spiral ganglion; SL: spiral ligament; SSL: secondary spiral lamina; TM: tectorial membrane. Inner and outer hair cells are drawn with black filling pattern; supporting cells of the organ of Corti (Deiters cells and pillar cells) are drawn with gray filling pattern. (Calibration bars: 100 μm.)

Within the prototypic mammalian arrangement of cellular and extracellular components of the hearing organ (Lim 1986), bats and dolphins have convergently evolved features that likely represent specializations for ultrasonic hearing. Shared anatomical traits of "high-frequency" cochleae include (1) a prominently thickened narrow basilar membrane (BM) that is anchored on the modiolar side by the bony primary osseous lamina and on the abmodiolar side by the secondary osseous spiral lamina; (2) miniaturized outer hair cells (OHCs); (3) sturdy and mechanically reinforced supporting cells (pillar and Deiters cells); (4) presence of Böttcher cells; and (5) hypertrophied Claudius cells.

SENSORY CELLS AND INNERVATION

Inner hair cells (IHCs) serve as classical mechanoreceptive cells and relay the information to the central auditory system. OHCs serve both as mechanoelectrical and electromechanical transducers by virtue of the ability for fast active contractions, thus pumping energy into the traveling wave (Brownell et al. 1985; Lim 1986; Dallos and Evans 1995; Holley 1996; Slepecki 1996). In both taxa, there is a single row of IHCs throughout the cochlea, and there are three rows of OHCs throughout most of the cochlea. An irregular presence of a fourth row of OHCs was noted in upper apical turn of dolphin (Wever et al. 1971c), similar to irregularities in OHC arrangements in the apex of several mammalian species (rodents, human). Such irregularities were so far noted only in one bat species (*Tadarida*). Basoapical gradients in OHC morphology play an integral role in cochlear frequency representation (Pujol et al. 1992; Dannhof, Roth, and Bruns 1991; Vater, Lenoir, and Pujol 1992). The OHCs are remarkably short and basoapical gradients in OHC length are very shallow in both taxa. In the cochlea of dolphins (*Tursiops truncatus*), OHCs measure 8 μm in the basal coil and lengthen to values of only 17 μm in the apical coil (Wever et al. 1971c). In bats, the gradients in OHC length along the cochlear spiral amount to 8–15 μm in *Hippossideros* (Dannhof and

Bruns 1991); and 12–17 μm in *Pteronotus* (Vater and Kössl 1996). This compares to 20–80 μm in guinea pigs (Pujol et al. 1992). Ultrastructural data on OHC and their stereocilia are not available in dolphins. At the high-frequency end of the bat cochlea, the length of OHC stereocilia amounts to only 0.8 μm, a value that likely presents the lower functional limit in stereocilia design (Vater and Lenoir 1992). Bat OHCs conserve the intricate subcortical lattice and cisternae system of mammalian OHC that has been indirectly linked with their fast active motility (Holley 1996).

The pioneering studies in dolphins reported an unusual high afferent innervation of the cochlear sensory epithelium by spiral ganglion cells. The ganglion cell-to-hair cell ratios amount to 5.4 to 1 in *Tursiops* (Wever et al. 1971c) and 4 to 1 in *Lagenorhynchus* (Wever et al. 1972), which compares to only 2 to 1 in humans. This suggests that odontocetes require additional neuronal channels for transmission of information contained in echolocation signals (Wever et al. 1972). The values in dolphins are matched and even exceeded by those reported for bats with FM calls: 5.5 to 1 in *Trachops cirrhosus* (Bruns, Burda, and Ryan 1988), 3.8 to 1 in *Taphozous* (Burda, Fiedler, and Bruns 1988), and even 15 to 1 in *Myotis* (Ramprashad et al. 1978). The highest densities typically occur in the midbasal turn.

CF-FM bats exhibit the largest variations in afferent innervation density along the cochlear spiral (M. M. Henson 1973; Bruns and Schmieszek 1980; Zook and Leake 1989), which correlate with profound specializations of the passive hydromechanical apparatus of the cochlea (i.e., morphology of BM and tectorial membrane [TM]; see Kössl and Vater 1995; Vater, chapter 13, this volume). The afferent innervation is densest in those cochlear regions that process the dominant components of the echolocation signal.

In all bat species studied, the lateral olivocochlear system, which provides descending efferent control at the level of the IHCs, is present. However, among bats there are significant and not yet fully understood variations in the expression of the medial olivocochlear system, which supplies the OHCs. While mustached bats possess one efferent synapse per OHC throughout the cochlea (with the largest terminals found in those regions that process the biosonar signal components) (Xie, Henson, and Bishop 1993), horseshoe bats completely lack an efferent supply of the OHC (Bruns and Schmieszek 1980; Bishop and Henson 1988; Vater, Lenoir, and Pujol 1992). Physiological data obtained in mustached bats suggest a tonic efferent control of cochlear mechanics by the medial efferent system, which dampens the vibration of the cochlear partition and can provide a protection from overdrive by loud vocalizations or noise (O. W. Henson et al. 1995; Xie and Henson 1998). To our knowledge, there are no published data on the efferent innervation of the cochlea in dolphins.

SUPPORTING CELLS, AND EXTRASENSORY CELLS

The morphology of supporting cells (pillar and Deiters cells) in the basal turn of both bats and dolphins is indicative of mechanical reinforcement and stiffening of the organ of Corti necessary for processing high frequencies (Reysenbach de Haan 1956; Wever et al. 1971a, 1971b, 1972; Vater, Lenoir, and Pujol 1992). In contrast, the support cells of the mysticete cochlea resemble those of humans, with no obvious overall specializations (see Ketten 1992). Böttcher cells are found throughout most of the cochlear coils in bats and dolphins, except at the very apex. Claudius cells are huge in the basal cochlear turn of echolocators, but the functional significance of this feature is unknown.

BASILAR MEMBRANE AND ANCHORING SYSTEM

The BM plays an integral role in mechanical frequency analysis of the mammalian cochlea (Békésy 1960). Typically, BM thickness and width vary inversely from base to apex, thus establishing a stiffness gradient along the cochlear duct that (in addition to other gradients in the design of the Organ of Corti such as OHC length) is the basis for the orderly representation of frequency (tonotopy) along the sensory epithelium. High frequencies are represented at the cochlear base, where the BM is thickest and most narrow—that is, stiffest. Low frequencies are represented apically, where the BM is more compliant due to an increase in width and decrease in stiffness. The exact course of the cochlear frequency map was obtained in three bat species in physiological studies combined with tracing the innervation of the IHCs (*Rhinolophus*: Vater, Feng, and Betz 1985; *Pteronotus*: Kössl and Vater 1985a; *Tadarida*: Vater and Siefer 1995). Direct frequency mapping of the dolphin cochlea is not available, but anatomical features of the BM were used to estimate the frequency representation (Ketten 1994). Such approaches are very useful, since data on BM morphology and anchorage can be obtained in postmortem tissue, and a comparison of BM morphology within and between cetaceans and bats gives important insights on hearing capabilities.

The edges of the BM are anchored firmly by ossified inner and outer spiral laminae throughout most of the basal turn in odontocetes (Ketten 1984, 1997) and bats (e.g., Bruns 1980; Dannhof and Bruns 1991; figs. 12.2 and 12.3). Ossified outer laminae are considered typical for high-frequency ears and provide firm anchoring sites for the BM; they are most elaborate in Rhinolophoidea, where they form tear-shaped protrusions into the scala tympani throughout the basal turn. In mysticetes, the outer laminae are vestigial (Ketten 1992, 2000).

BM length ranges from 22.5 mm in small dolphins to 64.7 mm in the fin whale (*Balaenoptera physalus*) and from 6 to 20 mm in echolocating bats (see Ketten 1997; Kössl and Vater 1995). BM length in mammals typically scales with body size, but there is no straightforward

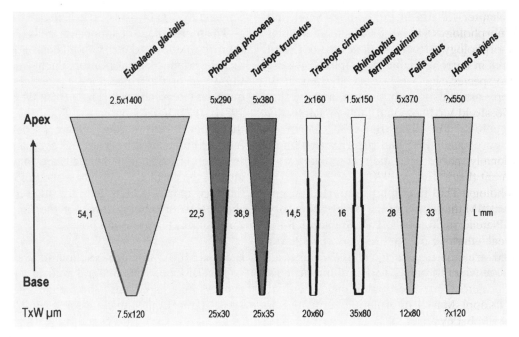

Fig. 12.3. Comparison of basilar membrane dimensions in whales, bats, cats, and humans (drawn after data summarized in Ketten 1997; Echteler, Fay, and Popper 1994). The regions of strengthened mechanical anchorage of the BM are indicated with thickened lines for bats and dolphins. T: BM thickness; W: BM width; L: absolute basoapical length of BM.

simple relation between hearing range and BM length across all mammalian species (Ketten 1994, 1997, 2000; Echteler, Fay, and Popper 1994). It is noteworthy that cochlear size and BM length in echolocating bats are greater than in nonecholocating terrestrial mammals of comparable body mass (Habersetzer and Storch 1992; Fiedler 1983), thus reflecting the dominance of the acoustic sense in behavior. Within Microchiroptera, the longest BMs are found in those species that employ Doppler-sensitive sonar. This elongation of BM relates to the implementation of an auditory fovea for processing the CF-call component (see Kössl and Vater 1995) and occurred in convergent evolution within the related Old World genera *Rhinolophus* and *Hipposideros,* and the New World species *Pteronotus parnellii.*

BM width and BM thickness are useful functional measures to estimate the stiffness gradients along the basoapical course of the BM, and a schematic comparison of BM dimensions in whales, bats, and terrestrial nonecholocators is given in fig. 12.3. In cetaceans, BM width increases 10- to 14-fold in basoapical direction and BM thickness decreases 5- to 6-fold. In humans, similarly directed gradients are established, but width only changes fivefold and thickness only twofold. In most bats, BM width only changes by a factor of about 2, and thickness decreases by a factor of 5–17 (see Kössl and Vater 1995; Echteler, Fay, and Popper 1994). The quotient of BM thickness and BM width has been used as an estimate of BM stiffness (Bruns 1980; Echteler, Fay, and Popper 1994; Ketten 1992, 1994) and provides a

hint to judge high- and low-frequency capacities of the respective species; but the exact cutoff frequencies cannot be predicted from these data. A typical odontocete BM has a nearly square cross section at the cochlear base and thickness-to-width (T/W) ratios range between 0.50 and 0.83, with the highest values found in high-frequency small species (Ketten 1994, 1997, 2000). In contrast, mysticete basal T/W ratios are much smaller and amount to only 0.06. Thus, it can be expected that the high-frequency limits are well below those of dolphins (Ketten 1994, 1997, 2000). Basal T/W ratios in bats are lower than those in odontocetes, with the highest ratios found in *Hipposideros bicolor* and *Rhinolophus ferrumequinum* (0.32 and 0.33, respectively), but these two species differ in upper hearing limit by almost one octave (Bruns 1976b; Dannhof and Bruns 1991). The apical T/W ratios in odontocetes range between 0.012 and 0.017 and are similar to those obtained in bats. With 0.001–0.002, the apical T/W ratios in mysticetes are much lower than those obtained in odontocetes and in terrestrial mammals, and they may be indicative of hearing capability in the infrasonic range (Ketten 1992, 1994).

As judged from published data, the basoapical gradients in BM dimensions are unspecialized in whales (Ketten 1992) and in the cochleae of most FM bats that have been studied (see Kössl and Vater 1995; Vater 1998). Focal transitions in BM dimensions bordered by plateau regions of BM dimensions are only observed in highly specialized cochleae of CF bats (see Kössl and Vater

1995; Vater, chapter 13, this volume). The discontinuities in BM morphology together with specialized features in TM morphology have been linked with enhanced filter properties in narrow frequency bands around the individual and species-characteristic CF-signal component in Doppler-shift-compensating bats (Vater and Kössl 1996; Kössl and Vater 1996a, 1996b; Vater, chapter 13, this volume; Kössl, Föller, and Faulstich, chapter 14, this volume). These mechanical filters are unique in nature. Exceptionally sharp-tuned responses arise from a cochlear segment that is located apical to the transition in BM morphology. This segment is characterized by almost constant BM dimensions and increased innervation density. Plateau regions in BM dimensions correspond to foveal—that is, expanded frequency—representations and are not limited to highly specialized CF-FM bats (see Vater 1998). Auditory fovea are also found in the cochlea of the mole rat (Müller et al. 1992), the barn owl (Köppl, Manley, and Gleich 1993), and in one species of FM bat (Vater and Siefer 1995), although with distinctly smaller mapping coefficients (maximally 5–11 mm/octave as compared to 70–150 mm/octave in CF-FM bats). These lesser foveae are not endowed with strongly enhanced tuning capacities. (For a more detailed comparative discussion see Kössl 1997.)

Cochlear Traits in Relation to Habitat Use and/or Sonar

Does cochlear anatomy reflect lifestyle or sonar systems? Are there convergent trends in bats and whales? In whales, three categories of cochlear design were defined based on a set of morphological measures (Ketten 1984, 1992; Wartzok and Ketten 1999), such as ratios of BM thickness and width in basal cochlear turn (basal ratios), the extent of secondary osseous spiral laminae, and spiral geometry. These categories correlate with habitat (riverine/inshore versus offshore/open sea—i.e., cluttered versus noncluttered acoustic environment) and frequency of the sonar signal. Type I cochlea are characterized by high basal ratios. Long secondary osseous spiral laminae and few coils are typical for inshore phocoenids and riverine dolphins, which emit the highest frequency signals and have the highest upper limit of hearing. Type II cochlea have basal ratios between 0.5 and 0.7, more coils, and well-developed secondary osseous spiral laminae. These are typical for offshore delphinids with lower echolocation signals. Type M cochlea have low basal and very low apical ratios, a residual spiral laminae, and a broad cochlear spiral with a steeper pitch than in odontocetes. These characterize baleen whales and indicate infrasonic capabilities.

Evidence obtained in CF-FM bats shows that cochlear dimensions scale with body size and signal frequency (Habersetzer and Storch 1992; Vater and Duifhuis 1986; Francis and Habersetzer 1998). It remains to be demonstrated if similar principles hold for bats with broadband sonar. Of central importance, however, is the existence of two separate traits in cochlear anatomy that relate to the type of sonar and sharpness of peripheral auditory filters. (1) Bats relying on broadband sonar possess a cochlea that exhibits nonspecialized tuning. Their cochlear anatomy is characterized by basic specializations for ultrasonic hearing (small dimensions of OHCs, sturdy supporting cells) and typically exhibits nonspecialized gradients in BM and TM morphology (see Vater 1998). This basic cochlea design is used by species widely differing in hunting strategy and diet, such as gleaning bats (*Megaderma:* Fiedler 1983; *Trachops:* Bruns, Burda, and Ryan 1988), open-air foragers (*Taphozous:* Fiedler 1983), or species that hunt along vegetation (*Pteronotus quadridens:* Vater, chapter 13, this volume). An interesting case within the group of bats characterized by nonspecialized peripheral tuning is *Tadarida.* In this species, specialized BM morphology creates an auditory fovea tuned to the lower end of the echolocation signal range; but, significantly, TM morphology is nonspecialized in this species (Vater and Siefer 1995; Vater 1998). (2) Doppler-shift-compensating bats, routinely employing long CF-FM signals, possess exceptionally sharp cochlear tuning at the CF frequencies. Cochlear anatomy features an auditory fovea with distinct specizations in both BM and TM morphology, combined with distinct variations in afferent innervation pattern (see Kössl and Vater 1995, 1996a, 1996b; Vater and Kössl 1996). These traits are unique and were convergently evolved in New World mustached bats and Old World horseshoe bats.

In conclusion, there clearly is a parallel evolution of general cochlear specializations for ultrasonic hearing in bats and dolphins. Parallels are found in miniaturization of OHCs, sturdy supporting cells, and a thickened narrow BM with bony anchorage. Judged from published data on basoapical gradients in BM morphology (Ketten 1992), the bauplan of the dolphin cochlea resembles that found in generalized bats with broadband sonar: there is no indication of stepwise focal changes in BM width or thickness. Is there a dolphin analogue to Doppler-shift-compensating bats? From physical and physiological properties of dolphin sonar such a convergence is not expected: brief clicks are Doppler-tolerant signals, and there is no evidence for enhanced cochlear tuning to a narrow preferred frequency range in any dolphin species studied to date (Au 1993; Au, introduction, this volume). Interestingly, in the basal cochlear turn of *Phocoena,* Ketten (1997) observed an unusual BM fine structure that resembles the specialized arrangements so far only found in CF-FM bats (Bruns 1976b, 1980; Vater and Kössl 1996). Although this represents an anatomical trait shared with CF-FM bats, there is not yet sufficient evidence to postulate a functional convergence. Further anatomical and physiological studies clearly are necessary to elucidate the functional role of cochlear specializations in phocoenids.

Filter Mechanisms of the Cochlea

Since there are no physiological data on cochlear function in cetaceans we only discuss here cochlear mechanisms in bats. A unique feature of the mammalian cochlea is active mechanical amplification of low-level signals by motile outer hair cells (Dallos 1992; Dallos and Evans 1995; Ashmore 1987; Brownell et al. 1985). The cochlear amplifier increases hearing sensitivity by at least 40 dB and, most important, enhances considerably the sharpness of traveling waves on the basilar membrane—and hence the quality of cochlear frequency filtering. According to current models of cochlear function, force generation by outer hair cells has to work on a cycle-by-cycle basis; the speed of hair cell contraction and elongation should match the frequency of the processed sound. This requirement would imply that the cochlear amplifier in echolocating mammals adapted to ultrasonic frequencies has to employ cellular motility, which is much faster than that known in any other cellular systems. Until recently, it was argued that the cochlear amplifier has a principal upper frequency limit close to 25 kHz (Gale and Ashmore 1997), and therefore animals with a hearing range beyond would have to use different mechanisms. However, the shape and level of dependence of neuronal tuning curves obtained from bats for frequencies above this value is remarkably similar to those at low frequencies and in other mammals (e.g., Suga and Jen 1977). Thus, there is no evidence that echolocating mammals would not use the mammalian cochlear amplifier. Data from ultrasonic otoacoustic emissions (Kössl and Vater 1985b; Kössl 1994a, 1994b; Kössl, Föller, and Faulstich, chapter 14, this volume) and recent laser Doppler measurements in nonecholocating mammals (Frank, Hemmert, and Gummer 1999) support this view. In the guinea pig, high-frequency force production by OHCs works up to frequencies of about 80 kHz (Frank, Hemmert, and Gummer 1999)

The only—but significant—deviation from normal mammalian cochlear function is found in CF-FM bats with a Doppler-sensitive sonar system. To resolve fine frequency modulations in the CF echoes caused by wing beats of prey insects, their cochlea is capable of extremely sharp tuning to the CF frequencies. In rhinolophid and hipposiderid bats, and in the New World mustached bat, the sharpness of neuronal tuning throughout the ascending auditory system can reach Q10 dB values of 200–600 at the dominant CF_2 frequency in comparison to maximum values of about 20 in other frequencies or in other mammals (see Kössl and Vater 1995). Enhanced tuning originates in the cochlea and can be measured on the level of cochlear mechanics in the form of sharp otoacoustic emissions (OAE) or basilar membrane tuning curves (fig. 12.4). (For measurements of

OAE, see Kössl et al., chapter 14, this volume.) The quality of tuning goes far beyond the capabilities of the normal cochlear amplifier (Kössl and Russell 1995; Russell and Kössl 1999). Anatomical adaptations of the TM and BM as well as of supporting structures (but not of outer hair cells) are present at specific cochlear regions, located basally to the CF_2 frequency place. In the mustached bat, strong cochlear resonance, which is evident in a prolonged ringing of the cochlear microphonic potentials (Suga and Jen 1977; Henson, Schuller, and Vater 1985; fig. 12.4), is associated with the frequency of sharpest tuning. There is evidence that the resonance is created by a double-resonator system consisting of specialized tectorial and basilar membranes (Russell and Kössl 1999). The most important component of this system seems to be an unusually shaped tectorial membrane (Vater and Kössl 1996; Kössl and Vater 1996a, 1996b). Other features of cochlear resonance are acoustical reflections and reverberations at sharp discontinuities of tectorial and basilar membrane in the form of standing waves along cochlear length (Russell and Kössl 1999).

An increase in frequency tuning due to resonance inevitably should lead to a deterioration of temporal acuity, as indicated by the prolonged rise and fall times of the cochlear microphonic response in the mustached bat. In the same species, neuronal temporal acuity does not seem to be grossly affected by the cochlear resonance—which opens the question of whether temporal acuity at CF_2 is improved by massive neuronal parallel processing due to the foveal representation of this frequency. In rhinolophid CF-FM bats, enhanced cochlear tuning at CF_2 is comparable to that of the mustached bat and also employs a resonator. However, cochlear microphonic responses are devoid of the typical ringing (Henson, Schuller, and Vater 1985), and it is unknown how cochlear mechanics can generate an extremely narrow filter without time-smearing effects in these bats.

Open Questions

Several fundamental questions on ear function in cetaceans are still unanswered (see also Popper, Hawkins, and Gisiner 1997; Ketten 1998). These concern the exact operation principles of the sound-conducting apparatus in general and in species-specific specializations. Because of great difficulties in working with these animals, for both practical and ethical reasons, there are almost no direct investigations of cochlear function and significant gaps in analysis of cochlear structure. In the long run, the use of recent advances in imaging techniques with higher resolution capabilities, investigations of further anatomical material from stranded specimens, and the application of noninvasive techniques to obtain hearing correlates can help to fill the gaps in knowledge. Significantly, in the case of cetaceans, this scientific basis is needed to assess the impact on behavior and survival of

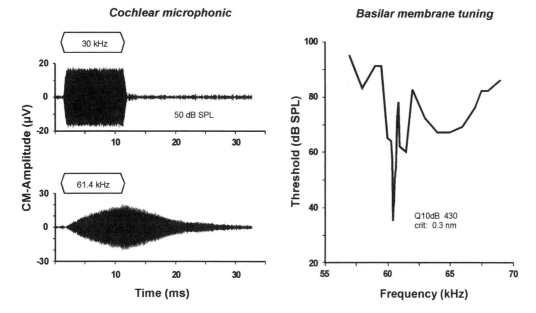

Cochlear microphonic   **Basilar membrane tuning**

Fig. 12.4. Cochlear microphonic responses in mustached bats to low-frequency stimulation (30 kHz, *top left*) and to pure-tone signals at the resonance frequency of the ear (*bottom left*). Note differences in amplitude scale and pronounced ringing of CM to 61.4 kHz; basilar membrane tuning curve in basal turn of mustached bat cochlea (*right*), obtained in laser interferometry with a threshold criterion of 0.3 nm BM vibration (after Kössl and Russell 1995).

processes such as normal age-related hearing loss, and hearing losses caused by human impact, such as medical treatment and environmental noise.

Knowledge of normal anatomy and function, and its intra- and interspecific variability, is essential to understand the sensory adaptations of a species to its characteristic environmental niche. In bats and dolphins, mechanical processing in the cochlea is probably at the physical limits of the design of a mechanoreceptive

sense organ, thus offering a chance to understand the restraining physical principles. However, to advance in this respect, both direct measurements and sophisticated cochlear modeling is required.

Acknowledgments

We thank Jeanette Thomas for critically reading the first version of this manuscript.

{ 13 }

Cochlear Anatomy Related to Bat Echolocation

Marianne Vater

Introduction

Cochlear technology in echolocating bats exploits the common mammalian design with focus on hydromechanical processing of ultrasound (see Kössl and Vater 1995). Prominent cochlear adaptations for process-

ing certain echolocation signal components are found in Doppler-shift-compensating bats that emit bimodal echolocation calls, composed of constant-frequency (CF) and frequency-modulated (FM) components (see Schnitzler and Kalko 1998). CF-FM bats of the genus *Rhinolophus* and one species of mormoopid, *Pteronotus parnellii,* have convergently evolved enhanced cochlear tuning within the frequency range of the second harmonic, constant-frequency component (CF$_2$). This frequency range occupies a large part of the region within the upper basal turn of the cochlea that is characterized by a high innervation density. Additionally, *P. parnellii* features enhanced tuning to the third harmonic CF component (CF$_3$) within a cochlear region of normal frequency mapping. Specialized frequency processing involves anatomical specializations of the basilar membrane (BM) (see Kössl and Vater 1995) and, according to recent results (Henson and Henson 1991; Vater and Kössl 1996), the tectorial membrane (TM). Among mormoopids, Doppler-sensitive sonar is a fairly recent evolutionary event, and emerged only once in *P. parnellii;* other related species employ FM-based sonar and possess broadly tuned hearing curves (Kössl et al., chapter 14, this volume).

The differences in sonar type and cochlear processing found within the genus *Pteronotus,* and the convergent evolution of Doppler-sensitive sonar in Old World horseshoe bats and New World mustached bats offer a unique opportunity to study evolutionary strategies used for engineering optimally adapted cochlear filter mechanisms. To address the question of how the structural design of the cochlea is related to a particular type of sonar, this study further investigated cochlear structure in two closely related species of the genus *Pteronotus:* the CF-FM bat *P. parnellii* and the FM bat *P. quadridens;* and two nonrelated species that have developed Doppler-sensitive sonar by convergent evolution: the Old World *Rhinolophus rouxi* and the New World *P. parnellii.*

Materials and Methods

Five cochleae from five different individuals of each species (*R. rouxi, P. parnelli, P. quadridens*) were processed for light microscopy and transmission electronmicroscopy (for details see Vater and Kössl 1996). Briefly, deeply anesthetized bats (Pentobarbital 60mg/100g) were decapitated, and the cochleae quickly removed. After opening the round and oval windows, and drilling a small hole into the cochlear apex, the cochleae were perfused through the scalae with 2.5% glutaraldehyde in phosphate buffer (0.1 M). After decalcification in 7.5% EDTA for 32 h, cochlear half turns were separated with a razor blade. The specimens were rinsed in buffer, postfixed in 1% osmium tetroxide, dehydrated, and embedded in epoxy resin (Durcopan). BM length was measured from photomicrographs of the half turns. Serial semi-thin sections were cut in the radial plane, mounted on glass slides, and stained with Richardson blue. Ultrathin sections were taken from remounted semi-thin sections at defined positions of the cochlea spiral. Dimensions of BM and TM were measured from semi-thin

sections using image analysis software. The ultrastructural composition of the TM was analyzed in electron photomicrographs at × 30,000 final magnification.

Results

The cochleae of the CF-FM bats, *P. parnelli* and *R. rouxi,* are considerably larger than the cochlea of the FM bat *P. quadridens* (fig. 13.1, top) and the cochlea of *P. macleayii* (O. W. Henson 1970). This enlargement is due to the implementation of an auditory fovea for processing CF$_2$ (Kössl and Vater 1995). Cochlea shape in *P. parnellii* is highly unusual and considerably different from the related FM bat, *P. quadridens.* Thus, within the same genus, a striking evolutionary change has occurred that concerns both external and internal morphology.

Innervation density of the basal cochlear turn, as revealed by surface preparations (fig. 13.1, bottom), appears homogenous in the FM bat cochlea. In both CF-FM bats, however, there are pronounced regional specific changes in innervation density. In *P. parnellii,* two regions of maximal innervation density are separated by a sparsely innervated (SI) zone (see also Henson and Henson 1991). The basally located maximum in innervation density represents the higher harmonics of the echolocation signal; the apically located maximum of innervation density is the site of origin of auditory nerve fibers sharply tuned to the CF$_2$ range. In *R. rouxi,* the SI zone encompasses the lower basal turn; the maximum of innervation density in the upper basal turn represents the CF$_2$ range (see Kössl and Vater 1995).

In the FM bat, BM thickness gradually decreases to-

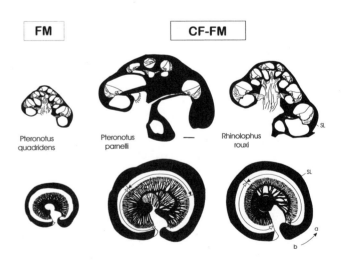

Fig. 13.1. Schematic illustrations of cross sections through the cochlea of FM and CF-FM bats (*top*) and innervation patterns of the basal cochlear turn (*bottom*). Solid arrowheads indicate the extent of the SI zone; open arrowheads indicate the regions of maxima in innervation density. SL: spiral ligament; b: basal; a: apical. Calibration bar: 200 μm.

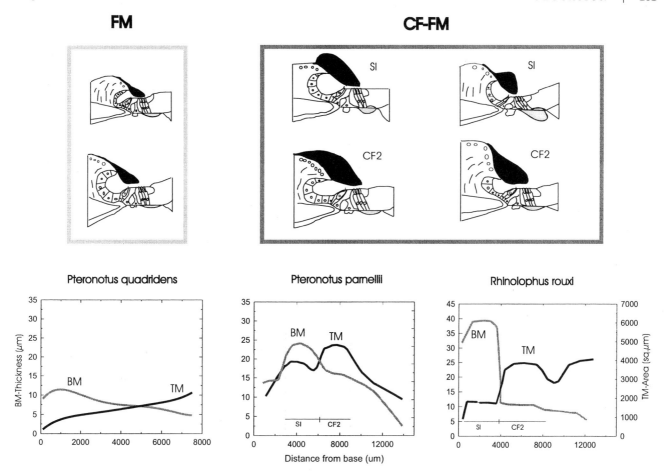

Fig. 13.2. Basoapical gradients in BM and TM dimensions and the appearance of the organ of Corti at different levels of the cochlear basal turn in the FM bat and the two CF-FM bats (after Vater 1997).

ward the apex, and the cross-sectional area (thus mass) of the TM gradually increases. Furthermore, TM shape and TM attachment sites to the spiral limbus change gradually in basoapical direction (fig. 13.2). The basoapical changes in dimensions of the BM and TM in nonrelated CF-FM bats follow common specialized traits that distinguish them from bats with other types of sonar. In both CF-FM bats, BM thickness is at maximum in the SI zone and sharply declines toward a lower plateau value in the CF_2 region. Cross-sectional area of the TM is at an absolute (*P. parnellii*) or relative maximum (*R. rouxi*) within the CF_2 region. The transitions in TM dimensions, attachment, and morphology coincide with stepwise changes in BM morphology.

As in other mammals (Hasko and Richardson 1988), the TM of bats is composed of two types of protofibrils: (1) thick unbranched type A protofibrils that form bundles; and (2) thin branched type B protofibrils that form the matrix. Several subregions of the TM can be defined according to location and ultrastructural composition, as illustrated for *P. quadridens* in fig. 13.3. The

main body of the TM can be divided into a core and a mantle region. The core region consists of bundles of predominantly radially oriented A protofibrils that are embedded in a loose matrix of B protofibrils. The mantle region consists of densely packed A protofibrils coursing parallel to the scala media surface of the TM. The cover net occupies the scala media side of the mantle and the marginal zone. Its electrondense appearance is created by a densely packed matrix. The marginal zone furthermore incorporates patches of longitudinally coursing A protofibrils. The limbal zone of the TM is located above the attachment region of the TM with the spiral limbus. A significant part of the limbal zone is composed of a meshwork of thick A protofibrils separated by electronlucent areas. The subsurface of the TM above the OHC domain is covered by an electrondense matrix region (Kimuras membrane), which carries the imprints of the tallest stereovilli of the outer hair cells. Above the inner hair cells, an electrondense matrix forms a distinct projection, the Hensens stripe (Vater, Lenoir, and Pujol 1992). A further region of

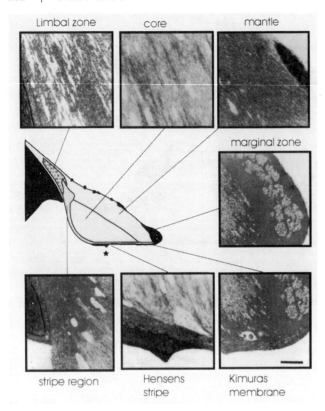

Limbal zone   core   mantle

marginal zone

stripe region   Hensens stripe   Kimuras membrane

Fig. 13.3. Subregions and ultrastructure of the TM in *P. quadridens*. Star: Hensens stripe; calibration bar: 1 μm.

densely packed A protofibrils is sandwiched between the matrix zones of the TM subsurface and the core region. It continues as a "stripe" toward the attachment of the TM with the spiral limbus.

These ultrastructurally defined subregions are present in the TM of all species (fig. 13.4), but to a different extent. The enlargement of the main body of the TM in both species of CF-FM bats, as compared to the FM bat, is created predominantly by hypertrophy of core and mantle zones. A unique feature of the TM of *P. parnellii* is the regional specific enlargement of the limbal zone, particularly the area that contains a crisscross pattern of type A protofibrils. The highly unusual shape of the TM in the SI zone is mostly created by this ultrastructurally defined region. In horseshoe bats, by contrast, a compatible subregion of the limbal zone was only observed apically to the CF_2 region and occupied only a small area. A unique specialization of the TM in horseshoe bats is the arrangement of the cover net, which forms a continuous layer on the scala media surface of the TM throughout the basal turn. A common derived character of the TM within the genus *Pteronotus* is the presence of patches of longitudinally directed A protofibrils within the marginal zone of the TM. This feature is lacking in *Rhinolophus* (this study) and *Tadarida* (Vater, unpublished).

Discussion

Two salient findings will be discussed in the context of the evolution of specialized cochlear filtering and micromechanical models of the cochlea: (1) the basoapical gradients in BM and TM morphology differ among related bats that employ different types of sonar; and (2) in both nonrelated CF-FM bats, there are specialized basoapical gradients and sharp transitions in the dimensions of BM and TM, the two key structures involved in passive hydromechanical processing.

The gradual basoapical change in BM and TM dimensions in mormoopid bats, which employ broadband sonar—the FM bats *P. quadridens* (this study), *P. macleai*, and *Mormops blainvillii* (Vater unpublished)—corresponds to the pattern typically observed in nonecholocating mammals, but it is scaled for the demands of high-frequency hearing. Although short CF components can precede the FM component of the echolocation signal (Kössl, Foeller, and Faulstich, chapter 14, this volume), neither mechanical measurements of cochlear function (Kössl, Foeller, and Faulstich, chapter 14, this volume) or anatomical data (this study) provide evidence for a preadaptation or gradual transition of the cochlear machinery from broadband to narrowband processing within the same genus. Nevertheless, the special organization of the marginal zone of the TM is a common derived feature within the genus *Pteronotus*. These findings suggest that the morphological specializations giving rise to enhanced cochlear tuning in *P. parnellii* arose as a recent, perhaps sudden, event on an evolutionary time scale—a cochlear big bang. The similarities in specialized longitudinal gradients in organization of the organ of Corti in the two nonrelated CF-FM bats argues for their origin by a similar modification of early developmental programs. In addition to a prolonged growth phase of the cochlear basal turn, changes seem to have occurred at the level of regulatory systems that control both spatiotemporal expression of secretory activity in cells that produce the extracellular matrix of BM and TM and afferent innervation patterns. Changes at these critical levels that still leave the cochlear machinery functional are rare. They could have occurred only twice in evolution, and by fortuitous coincidence created the most sharply tuned cochlear processors known to nature, which turned out to be advantageous for the lifestyle of hunting fluttering prey in dense bushes by means of a Doppler-sensitive sonar. This key evolutionary adaptation ignited a cascade of events at the level of sensory and motor control systems of the brain and emerged at an early point in time within the Old World genus *Rhinolophus*, leading to a radiation into more than 80 species, and more recently within the mormoopids with only one species and several subspecies employing Doppler-sensitive sonar.

The finding of specializations at the level of the TM in

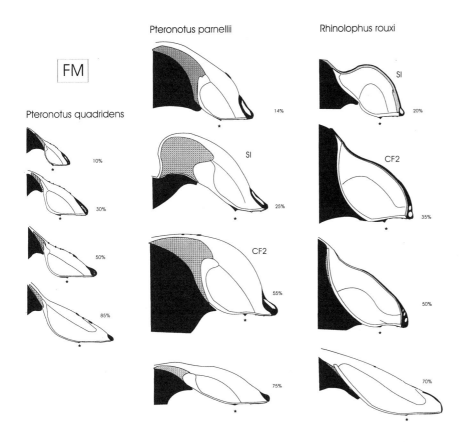

Fig. 13.4. Comparison of TM-ultrastructure in FM and CF-FM bats. Schematic cross sections are arranged in basoapical sequence from top to bottom; % distance from base is indicated. Stars: Hensens stripe; calibration bar: 1 μm.

CF-FM bats is particularly interesting, since recent micromechanical cochlear models view the TM as a spring-mass system rather than a rigid beam (Allen and Fahey 1993; Zwislocki and Cefaratti 1989). The resonance properties of this spring-mass system are thought to contribute to frequency processing. The TM acts as a second filter superimposed on BM mechanics and serves to sharpen the frequency response. In support of this passive mechanical second filter hypotheses, there are regional specific differences and stepwise changes in shape, geometry, and attachment of the TM that coincide with functionally specialized cochlear segments in both CF-FM bats out of all mammals studied. The morphological trends are similar in both species, but there are differences in TM dimensions and BM dimensions. In particular, the specialized shape of the TM in SI zone of the mustached bat has no counterpart in the horseshoe bat cochlea, and this could represent one critical feature that underlies the differences in resonance properties of the cochlear partition, as evidenced in recordings of cochlear potentials and otoacoustic emissions (Henson, Schuller, and Vater 1985; Kössl 1994a, 1994b; Kössl, Foeller, and Faulstich, chapter 14, this volume). The finding of differences in ultrastructure of the TM reflects the origin of specializations for sharp tuning by convergent (parallel) evolution. Whether these differences have consequences for mechanical function and vibration modes of the TM is an open question.

Acknowledgments

The author thanks Gesa Thies for excellent technical assistance. This work was supported by Sonderforschungsbereich 204 (Gehör).

{ 14 }

Otoacoustic Emissions and Cochlear Mechanisms in Echolocating Bats

Manfred Kössl, Elisabeth Foeller, and *Michael Faulstich*

Introduction

In bats, the requirements of echolocation have produced different adaptations of the mechanics of the inner ear. To compare the functional characteristics of bat cochleae in species that either use broadband frequency-modulated (FM) or CF-FM calls, we recorded otoacoustic emissions (OAEs), which are a consequence of active and nonlinear mechanical amplification by outer hair cells and also reflect specialized passive resonances in the cochlea. After the discovery of OAEs by Kemp (1978), their measurement became an important non-invasive method to study frequency processing in the mammalian cochlea and its pathology.

Spontaneous otoacoustic emissions (SOAEs) are the most direct proof that sound energy is actively generated by hearing organs. SOAEs are sinusoidal oscillations within a narrow frequency range. It is likely that their generation requires both active energy production, which in the case of mammals may be due to motile outer hair cells (OHCs), and structural discontinuities, which lead to the preferred emission at a specific frequency. They are emitted by ears of primates, dogs, and some sauropsid species (Probst, Lonsbury-Martin, and Martin 1991; Manley, Gallo, and Köppl 1996). They are hardly ever found in rodents or bats. The only bat species where pronounced SOAEs can be measured for periods of a few days is the mustached bat, *Pteronotus parnellii* (fig. 14.1A). Here, the SOAEs have a sound level up to 40 dB SPL at about 62 kHz, which is close to the dominant second harmonic, constant-frequency component (CF_2) of this bat's echolocation call. At this frequency, cochlear tuning is enhanced due to the presence of structural discontinuities in the tectorial membrane (TM) and basilar membrane (BM) (Vater and Kössl 1996; Kössl and Vater 1996b). These morphological specializations establish a passive mechanical resonator system, which greatly amplifies and sharpens cochlear responses to 62 kHz (see below). The mechanical feedback of the resonator also seems to enhance the output of spontaneous OHC activity and hence leads to large SOAEs.

Evoked OAEs can be induced by short tone bursts or clicks and then recorded at the eardrum as echolike phe-

nomena. The evoked OAEs also can be measured during presentation of a continuous pure-tone sweep. In this case, they appear as maxima and minima in the sound level and phase recorded at the eardrum, due to interference between the tone stimulus and the OAEs. In the mustached bat, such stimulus-frequency OAEs (SFOAEs) again are found close to 62 kHz (fig. 14.1B). Their maximum level approaches 70 dB SPL (Kössl and Vater 1985b), in contrast to maximum levels of about 20 dB SPL found in other mammals. SFOAEs can convert to SOAEs of the same frequency and vice versa. This emphasizes that the underlying mechanisms are the same and only differ in the degree of damping or, if described as feedback system, in the loop gain.

Distortion-product OAEs (DPOAEs) differ from the above-mentioned OAEs in that they are a direct consequence of the nonlinear mechanical properties of the cochlea. In the mammalian cochlea, the sensitivity to low sound levels is actively increased by nonlinear mechanical amplification. The basis for this "cochlear amplifier" are motile OHCs that change the shape of their cell body and thus generate force in response to their own receptor potential. Of course, the mechanical input to the OHCs strongly depends on properties of BM and TM and, as in the mustached bat, on resonant characteristics of both structures. Therefore, DPOAEs also reflect passive cochlear processes. DPOAEs are usually measured by using two continuous pure tones for stimulation; while these tones are present, DPOAEs appear as two-tone distortion peaks in frequency spectra recorded at the eardrum (fig. 14.1C). The advantage of the DPOAE method is that by choosing the stimulus frequencies, mechanical processing can be probed along the whole length of the cochlea, over the full hearing range of the animal. In all bat species investigated so far, pronounced DPOAEs can be measured. They reach frequencies >100 kHz. This gives a strong indication that the mammalian cochlear amplifier remains functional at the highest frequencies.

Materials and Methods

Measurement of OAEs is possible in an awake or lightly anaesthetized bat whose head is gently immobilized by

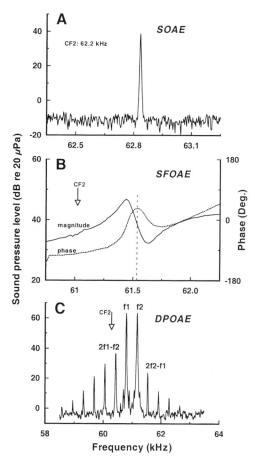

Fig. 14.1. Comparison of different types of OAEs in *P. parnellii. A:* Spectrum of a spontaneous OAE at 62.85 kHz, measured with a microphone positioned close to the eardrum. The dominant frequency of the emitted call (CF₂) was measured in the nonflying bat ("resting frequency"). *B:* Stimulus-frequency evoked OAE, measured during stimulation with a pure-tone sweep. The resulting frequency response recorded at the tympanum shows the characteristic sequence of a sound pressure maximum and minimum associated with phase changes, which are due to interference between the OAE and the stimulus. The transition between level maximum and minimum, where maximum phase change occurred, was used to define the SFOAE frequency (vertical line). *C:* Spectrum of distortion-product otoacoustic emissions. Two pure tones of different frequency (f1, f2) were the stimuli; all other peaks are generated by the ear's nonlinear properties. The frequency difference between the two stimuli was adjusted to induce maximum distortion at a frequency of 2f1–f2.

means of a mouth holder and an elastic strap. The recording system in our study consisted of a Brüel & Kjær (B&K) type 4135 microphone connected to a B&K type 2660 preamplifier and a B&K type 2610 measuring amplifier. The microphone was built into an acoustic coupler whose tip was custom-made to fit the individual bat's ear canal dimensions. The coupler tip was positioned 100–200 µm from the eardrum. Two D/A output

ports of a Microstar DAP 3200/e400 digital signal processing board, sampling at 250–400 kHz, produced the sinusoidal stimuli, which were fed into separate speakers. The microphone response was fed into an A/D input channel of the same board, and FFT analysis (Hanning windows) was performed to derive the amplitude and phase of the DPOAEs. DPOAEs could be recorded with this setup up to 160 kHz. Data presented here were measured at frequencies up to 100 kHz. (For further details of the measurement setup, see Kössl 1994a.)

The phase and group delay of the 2f1–f2 DPOAE were measured by keeping the f2 stimulus at a constant frequency and stepwise varying the f1 frequency. As a result, the 2f1–f2 frequency was swept across the frequency range tested. The measured phase of the 2f1–f2 DPOAE (φ_{2f1-f2}) was corrected for the stimulus phase (Mills and Rubel 1997). The corrected DPOAE phase angle (Φ_{2f1-f2}) is given by $\Phi_{2f1-f2} = \varphi_{2f1-f2} - (2\varphi_1 - \varphi_2)$, where φ_1 and φ_2 are measured phase angles for the f1 and f2 stimuli. The corrected DPOAE phase angles for successive f1 and DPOAE frequencies were unwrapped by a computer program and displayed. The group delay of the 2f1–f2 DPOAE (T_{2f1-f2}) was calculated from the phase change ($\Delta\Phi_{2f1-f2}$) versus frequency change (Δf_{2f1-f2}) of successive data points (Mills and Rubel 1997), where $T_{2f1-f2} = -\Delta\Phi_{2f1-f2} / \Delta f_{2f1-f2}$.

Results and Discussion

MECHANICAL SENSITIVITY OF THE COCHLEA
OF FM AND CF-FM BAT SPECIES

When cochlear sensitivity in a certain frequency region is high, lower levels of the appropriate frequency are required to induce active and nonlinear amplification and to evoke DPOAEs. This implies that it is possible to construct threshold curves of cochlear mechanics from DPOAE measurements. To measure hearing threshold curves on the basis of DPOAE recordings, we used the following procedures: The threshold was determined for different f2 frequencies within the hearing range of a bat. The reference to f2 was chosen, since DPOAE-suppression data (Brown and Kemp 1984) indicate that the 2f1–f2 DPOAE is generated close to the f2 place in the cochlea. For each f2 frequency, the optimum frequency ratio f2/f1 that evokes maximum DPOAE amplitudes at low stimulus levels was determined by varying f1. F1 was adjusted to this optimum ratio, and then DPOAE level growth functions were recorded by presenting a stepwise increase of both stimulus levels. From the growth functions, the f2 stimulus level that is sufficient to induce a small DPOAE of −10 dB SPL was interpolated and taken as threshold value. Such iso-response thresholds are relative curves. In most mammals we have investigated so far, the DPOAE thresholds run parallel to neuronal or behav-

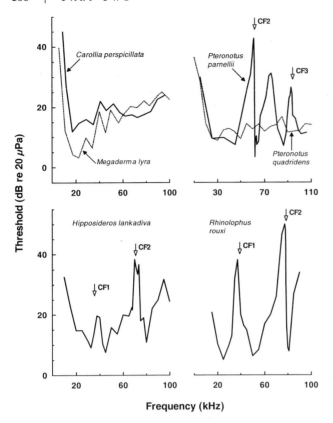

Fig. 14.2. DPOAE-threshold curves in different bat species. The level of the f2 stimulus sufficient to induce a 2f1–f2 DPOAE of −10 dB SPL (threshold criterion) is plotted as a function of f1 (*M. lyra* and *C. perspicillata* adapted from Kössl 1992; *R. rouxi* adapted from Kössl 1994a) or as a function of f2 (other species). The frequency of the second stimulus was adjusted according the best ratio f2/f1.

ioral threshold curves and are usually 10–30 dB higher, depending on the threshold criterion (e.g., Kössl 1992).

In bat species, like *Megaderma lyra* or *Carollia perspicillata* (fig. 14.2), that use broadband FM signals for echolocation, the mechanical threshold curves are characterized by a steep, low-frequency cutoff at 10–15 kHz and good sensitivity for the entire range of higher frequencies tested. Minimum thresholds are found between 20 and 40 kHz. This frequency range coincides either with communication calls or, in the case of the gleaning bat *M. lyra,* with noise produced by prey animals.

Whereas local threshold variations are below 5–10 dB in FM-bats, in *Pteronotus parnellii* and *Rhinolophus rouxi* there are pronounced threshold variations of about 40 dB in the range of the second harmonic CF components of their echolocation calls (fig. 14.2). These bats use 20–50 ms long CF components in their calls (long CF-FM calls) and keep the frequency of the returning CF echo constant by Doppler compensation. The emitted CF2 component coincides with the upper range of this threshold maximum. During flight, when the CF2 frequency is lowered to compensate the Dopp-

ler shift of echoes, the CF2 component is moved into the center of the threshold maximum (O. W. Henson et al. 1987). A few hundred Hz above this maximum, a sharp threshold minimum is found close to the frequency of Doppler-shifted CF2 echoes. In *P. parnellii*, the threshold minimum coincides exactly with the SOAE or SFOAE frequency. In both species sharpest neuronal tuning is found close to the threshold minimum. It appears as if a filter whose main role is to remove mechanical energy at the call frequency reduces cochlear response to the call, thereby allowing the bats to focus on the perception of Doppler-shifted echoes. In *P. parnellii* there is a second threshold maximum close to CF3, whereas in *R. rouxi* a second maximum is found close to CF1.

It is quite remarkable that such threshold features are completely absent in *Pteronotus quadridens* (fig. 14.2), a close relative of *P. parnellii*. *P. quadridens* uses FM calls or quasi–CF-FM calls where a 1–2 ms constant-frequency component at about 80 kHz precedes the FM component. A comparable picture is evident from cochlear anatomy: profound morphological adaptations that dominate the cochlea of *P. parnellii* are absent in *P. quadridens* (Vater, chapter 13, this volume). This difference would make it likely that both species are separated by a greater evolutionary distance. However, both species belong to the same genus and are quite similar to each other in general morphological and neuroanatomical properties. Therefore, it appears that the evolutionary development of cochlear function adapted to process CF calls happened quite recently or abruptly in *P. parnellii*. When comparing the threshold curve of the two *Pteronotus* species, it is obvious that the most dramatic change lies in the threshold maxima close to the CF call frequencies.

Whereas CF-specific cochlear adaptations are found only in one species of New World bat, *P. parnellii*, the Old World rhinolophid bats all have anatomical cochlear specializations. The species for which OAE data are available, *Rhinolophus rouxi*, shows threshold variations comparable to those of *P. parnellii*. It is likely that rhinolophids appeared much earlier in evolution than *P. parnellii*, and therefore a larger radiation and the formation of a large number of species could occur.

Bats from the family of Old World hipposiderids are closely related to the rhinolophids and use short CF-FM calls of about 10 ms. The CF components of the calls are less constant in frequency than those of the rhinolophids but are clearly narrower in frequency than quasi-CF components (see above), and Hipposiderids are able to Doppler compensate. The cochlear threshold curve of *Hipposideros lankadiva* shows a threshold maximum with a depth of about 20 dB close to the second harmonic CF frequency (fig. 14.2). This could indicate that hipposiderids represent an evolutionary stage between the long CF-FM bats and FM bats.

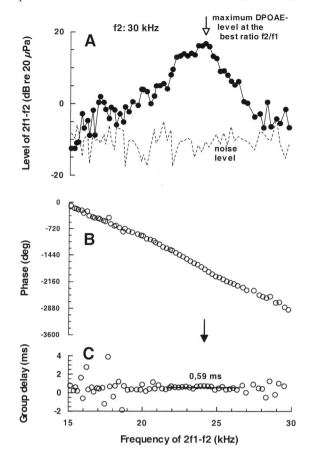

Fig. 14.3. Example for the calculation of the best ratio f2/f1 and the DPOAE group delay from *Hipposideros lankadiva*. *A:* Level of the 2f1–f2 DPOAE during stimulation with constant f2 at 30 kHz and 30 dB SPL (level of f1: 40 dB SPL, frequency varied). *B:* Phase of the 2f1–f2 DPOAE corrected for 360° phase transitions. *C:* Group delay calculated from *B;* the straight line gives a frequency range for which the DPOAE level was at least 12 dB above noise level. Within this frequency range, the phase values were used to calculate an average group delay (0.59 ms).

COCHLEAR FREQUENCY SEPARATION

When applying two pure-tone stimuli to the cochlea to induce DPOAEs, the frequency separation between f1 and f2 is an important parameter affecting DPOAE level. When both stimuli are far apart, there is no overlap in the corresponding traveling waves in the cochlea; as a consequence, no distortions are generated. Maximum DPOAE levels are induced for a specific frequency separation, defined by a best ratio f2/f1, which in noncholocating mammals ranges between 1.1 and 1.4, depending on the f2 frequency. Fig. 14.3A shows the DPOAE levels measured in *H. lankadiva* for a constant f2 of 30 kHz during stepwise changes of the f1 frequency. For an f2/f1 ratio, corresponding to a 2f1–f2 frequency of about 24 kHz, a pronounced maximum of the 2f1–f2 DPOAE is apparent. The best ratios most likely are the consequence of the action of a secondary cochlear fil-

ter whose morphological substrate is probably the TM (Brown, Gaskill, and Williams 1992). This secondary filter allows optimum throughput of DPOAE frequencies that correspond to the best ratio. The best ratio of 1.25, often found in mammals, indicates that the TM is tuned to frequencies roughly half an octave below the characteristic frequency of the BM. In bats, the best ratios usually vary between about 1.05 and 1.3 and are comparable to those of other mammals (fig. 14.4A: 20–30 kHz, 90–95 kHz). A remarkable deviation is found in the CF$_2$-frequency range of CF-FM bats. In *P. parnellii*, the best ratios are as small as 1.0005 (fig. 14.4A), corresponding to a stimulus frequency separation of 30 Hz for a f2 frequency of about 62 kHz. At this frequency, BM and TM are probably tuned to almost the same frequency and neuronal Q10 dB values (characteristic frequency of a tuning curve/bandwidth of curve 10 dB above threshold), as well as Q10 dB values of BM tuning, reaching values of 600 in *P. parnellii* in comparison to 10–30 in most other mammals. An exceptional frequency resolution of the auditory system of *P. parnellii* is also evident from behavioral data (Keating et al. 1994). In *R. rouxi* and *H. lankadiva*, close to CF$_2$ the neuronal Q10 dB values go up to about 400 and 200, respectively. Correspondingly, small best ratio values are found (fig. 14.4A). This indicates that the best ratio can be used as a noninvasive measure of tuning sharpness, and furthermore, that enhanced tuning is indeed generated by mechanical filter structures in the cochlea. In *P. parnellii* and *R. rouxi* (Vater, chapter 13, this volume), the shape of the TM changes dramatically and abruptly in the respective frequency regions of the cochlea. This is a strong indication that the TM is critically involved in producing enhanced tuning in CF bats.

COCHLEAR RESONANCE

The enhanced tuning and the distinct threshold variations in the CF$_2$ range are the consequence of specialized cochlear mechanics that lead to strong mechanical resonance. In *P. parnellii*, the resonance is obvious from ringing of cochlear microphonic potentials and from the emission of SOAEs and SFOAEs. Responsible for cochlear resonance are not OHC specializations but the design of their supporting structures. The hammerlike shape of the TM (Vater, chapter 13, this volume) in the cochlear region just basally to the 62 kHz place on the BM allows a strong resonance in the 60 kHz range (Steele 1997). This specialized TM is tuned to 62 kHz (Kössl and Vater 1996b) and standing-wave–like reflections at structural discontinuities in this region are thought to enhance the amplitude of resonance considerably. Fine-tuning and further sharpening of the resonance may arise from interaction of TM and BM with the OHCs positioned strategically to serve as critical control element.

In *R. rouxi* and *H. lankadiva*, SOAEs are absent; and

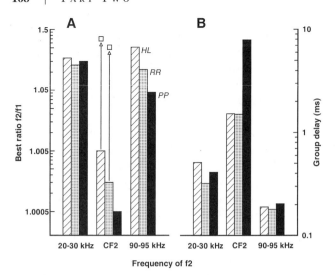

Fig. 14.4. *A:* Comparison of the best ratio f2/f1 and *B:* The group delay of the 2f1–f2 DPOAE at CF₂ and below/above CF₂ in *H. lankadiva* (*HL*), *R. rouxi* (*RR*), and *P. parnellii* (*PP*). In *R. rouxi* and *H. lankadiva*, close to CF₂, the best ratios are either very small (bars) or reach large values of up to 1.4 (open squares).

SFOAEs, as well as cochlear microphonic ringing, are small or below a detectable level (Henson, Schuller, and Vater 1985; Kössl 1994a). This implies that the underlying resonator is more strongly damped in the latter two species. In addition, in these species the best ratio measurements at or slightly below CF₂ reveal secondary maxima of the DPOAE level at large ratios (open squares in fig. 14.4A), which correspond to 2f1–f2 frequencies just above CF1. In *H. lankadiva*, such large ratios often induce the maximal DPOAE level. The fact that in these cases the resonator is not as sharply restricted to a single frequency as that of *P. parnellii* also could be due to increased damping. Anatomically, there is a significant difference between *P. parnellii* and the other two bat species in the BM of the cochlear region that is responsible for the resonance. In the genera *Rhinolophus* and *Hipposideros*, BM thickness reaches 30–35 µm (Bruns 1976a; Dannhof and Bruns 1991), compared to about 25 µm in *P. parnellii* (Vater and Kössl 1996). In addition, the region of maximum BM thickness extends over a much longer longitudinal cochlear distance than in *P. parnellii*. Taken together, these results indicate that the BM could be responsible for damping resonant oscillations created mainly by the TM, which should help to keep the resonator system stable.

COCHLEAR GROUP DELAY

Cochlear resonators obviously enhance the mechanical frequency resolution at CF₂; and, by absorbing input energy and thus acting as antiresonators, they also could produce the threshold maximum just below CF₂. However, in addition they should lead to a severe deteriora-

tion of the temporal acuity of cochlear processing due to the time required to build up the resonance and the long-lasting ringing after the end of acoustic stimulation. As a consequence, the generation of DPOAEs at CF₂ should need a longer delay than at other frequencies.

Cochlear delay can be determined as onset delay or as group delay. The group delay is measured on the basis of the phase lag of the response during ongoing stimulation and can easily be calculated using the phase of DPOAEs. Fig 14.3B shows the phase of the 2f1–f2 DPOAE that corresponds to the level data in fig. 14.3A. The corresponding delay that would produce the observed phase change is displayed in fig. 14.3C (see "Materials and Methods"). The resulting group delay takes into account both the forward and reverse traveling times of the main wave fronts in the cochlea. The group delay can be used to assess relative changes of temporal processing between different cochlear regions. Generally, in mammals the group delay increases with decreasing frequency (Mills and Rubel 1997). In the CF bats, this relationship is apparent when comparing the range of 20–30 kHz with that of 90–95 kHz (fig. 14.4B). Prolonged group delays at lower frequencies are most likely a consequence of both the longer traveling times required to induce a mechanical response at the cochlear apex and the time required to generate the DPOAEs. Close to CF₂, the regular mammalian gradient of decreasing group delays with increasing frequencies is interrupted; and here the absolute maxima of group delay are found, as predicted from the presence of resonators. The largest group delay values of about 8 ms occur in *P. parnellii*. In the other two species the group delays are smaller (1.5–2 ms), but still elevated. Because these measurements were made of only three individuals of *P. parnellii* (range of group delay values: 7.1–9.2 ms) and one individual *of R. rouxi* and *H. lankadiva*, additional data must be obtained before a detailed assessment can be made.

Conclusions

OAE measurements provide a noninvasive tool to probe cochlear mechanics in awake bats. Larger spontaneous OAEs or OAEs evoked by single tones are only found in the mustached bat. However, in all bat species DPOAEs that depend on nonlinear properties of the mammalian cochlear amplifier can be easily measured and are therefore the best method for assessing cochlear mechanics. The properties of DPOAEs also reflect macromechanical adaptations of cochlear structures, which produce passive resonances.

By comparing DPOAEs from FM bats and the three CF-FM bat species *P. parnellii*, *R. rouxi*, and *H. lankadiva*, it becomes obvious that the CF-FM bats have pronounced functional adaptations in the range around their dominant echolocation frequencies where neuronal tuning is sharpest: (1) Maximum DPOAE levels are evoked by minimal frequency separation between the two stimuli. This indicates a high mechanical frequency resolution. (2) Pronounced maxima and minima occur in auditory threshold curves. The maxima coincide with the frequency of the emitted call, the minima with that of Doppler-shifted echoes. (3) Cochlear group delays calculated from the phase of DPOAEs are very large, indicative of a mechanical resonator.

In *P. parnellii*, the resonance is less damped than in *R. rouxi* and *H. lankadiva* and closer to instability as becomes evident from the large spontaneous OAEs. In close relatives of *P. parnellii* (e.g., in the FM bat *P. quadridens*), they are completely absent.

The above-mentioned functional specializations are less obvious in *H. lankadiva* than in the other two CF-FM bat species. DPOAE-threshold variations and best ratio measurements, as a measure of cochlear frequency separation, indicate that in *H. lankadiva* specialized cochlear processing of CF frequencies is less pronounced. The question of whether *H. lankadiva* represents an evolutionary stage in between pure FM bats and CF bats is still open.

Acknowledgments

This study was supported by the DFG, KO 987.

{ 15 }

The Effect of Projector Position on the Underwater Hearing Thresholds of Bottlenose Dolphins (*Tursiops truncatus*) at 2, 8, and 12 kHz

Carolyn E. Schlundt, Donald A. Carder, and *Sam H. Ridgway*

Introduction

Many cetaceans have exceptional hearing and sound-production capabilities (Nachtigall 1986) that they use to forage, communicate, and navigate (Green et al. 1994). Investigations into their keen sense of hearing have spanned more than four decades. There is growing concern that anthropogenic (humanmade) noise in the marine environment may interfere with these abilities so crucial to the animals' survival (Richardson et al. 1995). This concern has focused especially on noise generated at lower frequencies, since many of the sources of interest emit louder sounds in this range (Greene 1995). Only recently have data been available to allow confident predictions regarding the parameters of sound that should be of concern for cetaceans (Finneran et al. 2000; Schlundt et al. 2000).

A key assumption in these studies, and in any studies relating to cetacean hearing, is that the audiograms available for several marine mammal species, on which predictions are based, reflect the most accurate range and sensitivity of the animal's hearing. Very often, behaviorally produced audiograms, that are referred to as the standard for that species, were based on one subject. Moreover, the experimental conditions used to behaviorally examine hearing sensitivity have rarely been modified between investigators. Studies suggest that hearing ability in some marine mammals is dependent on the location of the sound source (Au and Moore 1984; Bullock et al. 1968). Even so, the projector in hearing studies that produced the audiograms considered to be the standard for the species usually emitted sounds from a single position relative to the animal.

In 1967, C. S. Johnson produced the first complete dolphin audiogram, still considered the standard to which acoustic thresholds are compared today. To date, behavioral audiograms are available for several species of odontocetes. Most of these studies employed a single subject. (Richardson 1995 provides good reviews of some of them.) Results from each of these studies have produced audiograms with the familiar U-shaped audiometric curve, differing primarily in the high-frequency cutoff, which has been specific to the single subject of each species tested.

For the most part, methods for behaviorally estimating thresholds in the audiogram studies are similar in three respects. First, stimuli were typically presented using a staircase psychophysical procedure. Second, animal responses were observed using a go/no-go test paradigm (see Au 1993). Third, the test environment and

Carolyn E. Schlundt, Donald A. Carder, and Sam H. Ridgway

equipment configuration were similar among investigators. Test subjects, stationed under water, listened for continuous-wave tones that were projected from a sound source located directly in front of them along the longitudinal axis.

An alternative to the go/no-go test paradigm was developed to investigate acoustic response times (Ridgway et al. 1991) and to screen hearing in several animals rapidly. Similar to the method of free response described by Egan, Greenberg, and Schulman (1961), its application to behavioral testing of echolocation and hearing ability in dolphins and white whales has been described in Finneran et al. (2000) and Schlundt et al. (2000).

Several studies report that dolphins have frequency-dependent sensitivity for sound reception at various sites around the head. Bullock et al. (1968) described frequency dependence and angular sensitivity in the horizontal and vertical planes in their study using evoked potentials on several different dolphin species. They suggest that the pattern of angular sensitivity results from the use of the lower jaw as the primary pathway for sound reception to the ears (see Brill and Harder 1991).

Au and Moore (1984) used behavioral methods to investigate the receiving beam patterns of a *T. truncatus* in both the horizontal and vertical planes. Like Bullock et al. (1968), their results showed that beam patterns were frequency dependent, becoming increasingly narrower as frequency increased in both planes. They also found that angular sensitivity in the vertical plane was not symmetrical, finding that hearing ability dropped off less rapidly at angles tested below the animal's head than those tested above. They attribute this asymmetry in the vertical plane to a shadowing effect by the upper jaw and melon (if one is to believe that sounds are best received through the lower jaw).

A third study indirectly indicated frequency dependence based on the direction from which the sound source produced tones. Ridgway et al. (2001) noticed that white whales had markedly lower thresholds at 500 Hz and higher thresholds at 30 kHz during deep-water, open-ocean hearing tests than in earlier tests with the same whales in San Diego Bay. In the earlier tests, the projecting transducer was positioned in front of the whale. In the deep-water tests, the projector delivered test tones from below.

No studies have compared different projector positions on the same animal during hearing test procedures. It is possible that the results of audiograms produced using behavioral methods would be different if the projector were located in different positions. Moreover, it is likely that the effect of projector position on hearing thresholds will differ, depending on the test frequency.

The present study provides the first direct comparison of hearing thresholds in dolphins using behavioral methods when the projector emits test tones from two different positions within each test session. Thresholds were estimated for two dolphins at 2 kHz, 8 kHz, and 12 kHz, with the same projecting hydrophone in two positions using the staircase psychophysical procedure for stimulus presentation and our free response method. It was believed that thresholds estimated with the projector at one position would differ from those estimated with the projector at the other position and that thresholds at either location would differ, depending on frequency. It was expected that thresholds at 2 kHz would be lower when the projector was positioned directly below the animal's ears, and that thresholds at 12 kHz would be lower when the projector was positioned directly in front of the animal along its longitudinal axis. By testing a third intermediate frequency, we hoped to gain information regarding the frequency around which thresholds would not differ between projector positions. If correct, results would indicate that the low-frequency portion of cetacean audiograms might be reevaluated based on projector position.

Materials and Methods

STUDY ANIMALS

The subjects were two male Atlantic bottlenose dolphins (*Tursiops truncatus*), ages 20 years (BUS) and 36 years (BEN). Both animals were in good health at the time of this study and were monitored regularly by an on-site marine mammal veterinarian in accordance with applicable federal regulations. The study followed a protocol approved by the Institutional Animal Care and Use Committee under guidelines of the Association for the Accreditation of Laboratory Animal Care.

APPARATUS

The animals were tested in 12 × 12 m netted enclosures in San Diego Bay. An underwater hearing test station constructed of polyvinyl chloride (PVC) tubing was suspended from the deck surrounding the enclosure (fig. 15.1). The station supported a biteplate on which the dolphin stationed during test sessions. The biteplate ensured that the dolphin's position was fixed 2 m below the surface. Attached to the station was a PVC pipe molded to the shape of an arc bent to a soft 90° angle. A PVC "car" with a diameter just wider than the molded pipe was placed around the arc. Using a pulley arrangement, the car slid easily along the length of the arc and could be consistently locked into two positions at either end, in front and below the biteplate. A test-tone projector was attached to the car (see photos in fig. 15.1). At the in-front position, the projector was directly in front of the longitudinal axis of the animal 1.5 m from a line intersecting the animal's ears (position A in fig. 15.1). At the below position, the projector was 1.5 m directly below the line intersecting the ears (position B in fig. 15.1).

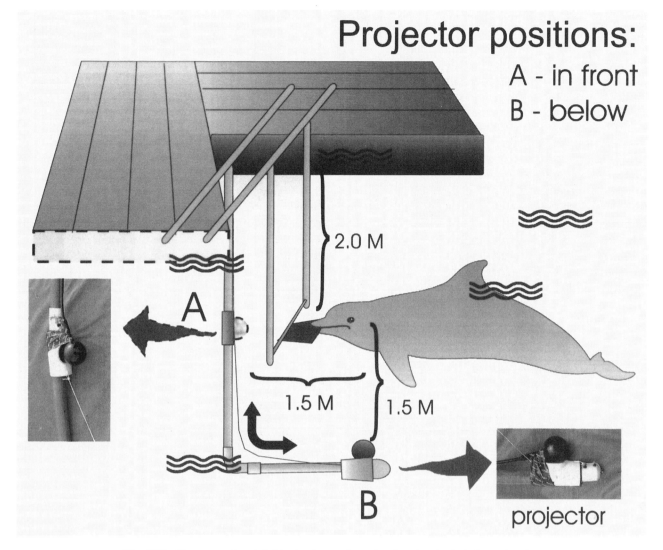

Fig. 15.1. Diagram of a dolphin in the underwater hearing test station, showing the arc on which the projector could be moved between two alternate positions during a session, in front of (projector position A) and below (projector position B) the dolphin.

PROCEDURES

Session overview

Dolphins were trained for hearing testing using standard operant conditioning techniques (Skinner 1961; Schusterman 1980). On a trainer's hand signal, the dolphin dove to the underwater station and positioned on the biteplate. Once on station, the animal listened for a variable number of test tones presented using an up-down staircase psychophysical procedure. If the animal heard a tone (hit) and responded with a whistle, the next tone presentation decreased by 4 dB. If the animal did not hear a tone (miss) and remained quiet, the next tone increased by 2 dB. Animals had to respond to a tone after 50 ms and within 2 s of the tone's onset to be recorded as a hit. In previous experiments, dolphin re-

sponse times to acoustic stimuli were sufficiently fast to fall well within this window (Ridgway et al. 1991). Responses that fell temporally past this window or within the first 50 ms of tone onset were recorded as false alarms. Additionally, variable-length catch trials (no tone) up to 75 s were introduced randomly and constituted at least 25% of the total number of trials in any session. During catch trials, no tones were presented and the dolphin was expected to remain quiet (a correct rejection). Responses made during catch-trial dives were recorded as false alarms.

Stimulus presentation and threshold estimation

Each animal participated in up to three sessions per day. Within a session, the projector was moved between the two positions for each frequency tested. The orders of projector position and frequency presentation were counterbalanced across sessions. The test tones were 250 ms in duration, including a 5 ms rise and fall times. Tones were presented at randomized intervals of 4.5–7 s in 0.1 s increments. Trainers were blind as to when the next tone in a series was coming. Each tone projection in a dive to the underwater station was considered a separate trial. On an average dive, the dolphin remained on the underwater station approximately 1 min for 12 consecutive trials. Two to four dives generally were sufficient to estimate threshold for a given frequency at one projector position.

The amplitude level of the first tone projected at any frequency was at least 20 dB above the animal's estimated threshold. The initial five tones cued the dolphin as to which frequency was being tested. The suprathreshold tones also provided the dolphin with a "warm-up" and enabled the trainers to evaluate the dolphin's attentiveness. Correct responses to at least 80% of these warm-up tones were obligatory for a session to continue.

Threshold for a given frequency and projector position was defined as the sound pressure level linear average (dB re 1 μPa [rms]) of the first 10 response/no-response reversal points (see Au 1993), following the five warm-up trials. The average of threshold estimations from three sessions in which a frequency was tested at both positions constituted the dolphin's ultimate threshold for that frequency and projector position.

Ambient noise

Human-generated masking noise was not used in this study; however, thresholds estimated in this study should not be considered nor compared to absolute thresholds. Indeed, thresholds presented here were masked by ambient bay noise. Consequently, it was necessary to monitor ambient noise levels in the bay at all times during testing. If at any time the noise level in the bay either exceeded or fell below that of the typical test levels ($\pm$5 dB), testing was suspended until ambient noise levels returned to within testing criteria.

Equipment and Calibration

The behavioral testing of hearing sensitivity was accomplished using computer-controlled methods. Test tones consisted of gated sinusoids created in custom software and output using a D/A converter on a Tucker-Davis Technologies QDA2 multifunction board. The output of the D/A board was filtered through two Ithaco filters (models 4302 and 4212), driven by a BGW power amplifier (Performance Series 2 model) and delivered via an International Transducer Corporation piezoelectric ceramic projector (model 1001).

Test tones emitted from the projector were calibrated at both projector positions twice daily, before and after sessions. A calibrated Brüel & Kjær (B&K) type 8103 hydrophone monitored and measured tones, back-

ground noise, and subject responses. The monitoring hydrophone was calibrated daily using a B&K calibrator (type 4223). During calibration, the monitor hydrophone was placed in the same position as the midpoint between the dolphin's ears when the animal was on the underwater station during a session. Test tones projected at various levels were picked up by the hydrophone, fed through a B&K charge amplifier (type 2635), and measured with a Hewlett-Packard (HP) signal analyzer (type 3561A). Tone projections at the underwater station were accurate within $\pm$1 dB. During sessions, the monitoring hydrophone was positioned just above the subject's blowhole and served both to receive the maximum amplitude of the animal's whistle responses and to monitor background noise levels. Input to the hydrophone was fed into the B&K charge amplifier. Background noise was displayed on the HP signal analyzer.

Results

Threshold Analysis

Thresholds (reported in dB re 1 μPa [rms]) for each dolphin at each frequency broken down by projector position and session are shown in table 15.1. Among-session consistency of thresholds for each dolphin was better than expected without the use of projected masking noise. For BEN, all thresholds estimated across the three sessions for each frequency and projector position were within 3 dB of one another with the exception of those at 8 kHz in the below position, where only two of the three were within 3 dB. For BUS, at least two of the three thresholds estimated across the three sessions for each frequency and projector position were within 3 dB of one another, with widest range being 6 dB at 12 kHz in the in-front position. There also was close agreement between animals. Across sessions, the mean hearing thresholds at 2 kHz, 8 kHz, and 12 kHz when the projector was in the in-front position were 109 dB, 90 dB, and 85 dB for BEN, and 104 dB, 89 dB, and 84 dB for BUS, respectively. Mean thresholds at 2 kHz, 8 kHz, and

TABLE 15.1. Thresholds (dB re 1 μPa [rms]) for each dolphin as a function of projector position, frequency, and session. Thresholds are based on the mean of 10 reversals.

Frequency	Session	BEN In front	BEN Below	BUS In front	BUS Below
2 kHz	1	111	97	102	94
	2	108	95	104	94
	3	108	94	106	96
8 kHz	1	90	93	90	90
	2	90	88	87	91
	3	89	93	91	87
12 kHz	1	84	96	84	94
	2	84	95	88	94
	3	86	97	82	94

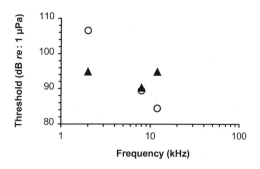

Fig. 15.2. Estimated thresholds at the three test frequencies when the test-tone projector was positioned in front (open circles) and below (closed triangles) the dolphin. Data shown are collapsed across both subjects.

12 kHz when the projector was in the below position were 95 dB, 91 dB, and 96 dB for BEN, and 94 dB, 89 dB, and 94 dB for BUS, respectively. The means of the estimated thresholds at each frequency per projector position collapsed across animals and sessions are shown in fig. 15.2. Hearing sensitivity at different frequencies in these dolphins was dependent on whether the projector delivering test tones was in front of or below the animal. At the lowest frequency tested, 2 kHz, hearing sensitivity was best when the projector was 1.5 m below the animal's ears. At the highest frequency tested, 12 kHz, hearing sensitivity was best when the projector was 1.5 m in front of the animal's ears. The 8 kHz hearing thresholds did not differ with projector position.

RESPONSE BIAS

The number of false alarms committed by each subject was consistently low throughout testing. No more than three false alarms were recorded for either animal during threshold estimation at any frequency/projector position combination. On most occasions, none were committed. Across sessions, the number of false alarms at 2 kHz, 8 kHz, and 12 kHz when the projector was in the in-front position were 1, 2, and 0 for dolphin BEN, and 0, 1, and 0 for dolphin BUS, respectively. The number of false alarms at 2 kHz, 8 kHz, and 12 kHz when the projector was in the below position were 2, 1, and 2 for BEN, and 4, 2, and 3 for BUS, respectively. A change in response bias did not account for the differences in thresholds at the two projector positions.

AMBIENT NOISE

Continuous 40-average samples of noise measurements (dB re 1 μPa^2/Hz) were taken before, during, and after sessions to monitor ambient levels. Based on numerous measurements such as these, noise levels in San Diego Bay averaged 85 dB, 79 dB, and 76 dB at 2 kHz, 8 kHz, and 12 kHz, respectively, during test sessions over the course of this study. No differences in average back-

ground noise levels within the test frequencies between projector positions were observed.

Discussion

The objective of this study was to determine the effect of projector position on dolphin hearing thresholds in a background of ambient bay noise. Behavioral conditioning techniques were used to estimate underwater hearing thresholds in two bottlenose dolphins when the position of the test-tone projector was alternately moved between two positions, in front of and below the animal. Table 15.1 and fig. 15.2 summarize the results. These results—the first information comparing thresholds estimated from different projector positions on the same animal during hearing test procedures—indicate that dolphin acoustic thresholds were dependent on the location from which sound is projected. Thresholds at 2 kHz were lower when the projector was positioned directly below the animal's ears, and thresholds at 12 kHz were lower when the projector was positioned directly in front of the animal along its longitudinal axis. Testing at the third, intermediate, frequency successfully provided information regarding the frequency around which thresholds do not differ between projector positions. The most complete cetacean audiogram should reflect thresholds that take into account the relationship between frequency and orientation of the sound source.

These results are consistent with the Au and Moore (1984) receiving beam study and the Bullock et al. (1968) evoked potential study, which show broader spatial hearing sensitivity at lower frequencies when tested in the vertical plane. They also support the observation that white whales had lower thresholds at frequencies below 3 kHz during deep-water, open-ocean hearing tests when the projector delivered test tones from below than in earlier tests with the same whales when the projector delivered test tones from in front (Ridgway et al. 2001). The findings of the present study are of particular interest given the growing concern over the potentially harmful effects on marine mammals of high levels of low-frequency humanmade noise being introduced in the open ocean. Ultimately, the results suggest that the low-frequency portion of the cetacean audiogram might be reevaluated based on projector position. The implications of such a reevaluation extend to the areas of absolute thresholds and critical ratios.

While it is not unusual to see low numbers of false alarms in animals tested at our facility using the free response method, the number of false alarms committed by each dolphin presented here indicated that the subjects were extremely conservative. It is worth noting that the intent of this study was not to establish absolute thresholds nor to compare these thresholds to ones obtained previously. Rather, our intent was to show that there is an effect of projector position on hearing sensi-

Carolyn E. Schlundt, Donald A. Carder, and Sam H. Ridgway

tivity and that current audiograms may not reflect the most accurate picture of cetacean hearing ability. Even if the thresholds presented here are conservatively high, they cannot explain the difference seen between thresholds for the same frequency. While both animals, BUS in particular, tended to commit more false alarms when the projector was positioned below them, it is important to note that these false alarms were relatively evenly distributed across the three test frequencies. Since there was no marked difference in either animal's likelihood to commit false alarms between projector positions, particularly within each test frequency, it is most likely that the difference in thresholds was due to the change of the location of the sound source and not to a change in response bias. Careful monitoring of background noise in the bay environment, varying of the order of both frequency delivery and projector position, and employing two subjects who behaved very similarly also support the findings that the differences in hearing thresholds were due only to the change in projector position.

There are at least two theories on how sounds are received by the dolphin. One is that the pathway for sound reception travels through the external auditory meatus and the auditory canal to receptor cells in the cochlea (Bel'kovich and Solntseva 1970). The other is that the pathways for sound reception are through the animal's lower jaw (Norris 1969). The latter theory has since become generally accepted through supporting physiological (McCormick et al. 1980) and behavioral (Brill and Harder 1991) research. In the deep-water open-ocean hearing tests with white whales, Ridgway et al. (2001) suggest that sound is conducted to the ear through their head tissues without requiring the usual eardrum–ossicular-chain amplification of the aerial middle ear, because whale hearing was not attenuated by an increase in depth. This finding also supports the idea that sound is conducted through the lower jaw, especially since the ear lies behind the lower jaw.

In fact, both theories are probably correct. Some studies suggest that the dolphin hearing ability depends on the interaction of multiple reception pathways and that each pathway is frequency dependent. Renaud and Popper (1975) indicated that the dolphin uses the region around the external auditory meatus to detect sounds at low frequencies, but uses the lateral sides of the lower jaw to detect sounds about 20 kHz. Brill et al. (1988) suggested that the lower jaw was the best reception site for signals above 20–30 kHz, but that the area around the meatus was better for signals below that.

The dolphin's sensitivity for low-frequency sounds, especially when the source comes from below, may enhance its ability to detect some prey and predators. While this speculative explanation for the adaptation of frequency dependence on directional reception for predator/prey identification is interesting, the fact remains that most marine mammals do not hear very well at lower frequencies. If cetaceans were to rely even moderately on low-frequency hearing to detect predators and prey, one would expect to see an evolved capacity for hearing in the lower range. Therefore a second, more realistic—albeit less exciting—explanation may be that the adaptation of frequency dependence on directional reception is simply a by-product of the adaptation for an advanced capacity in high-frequency hearing.

Acknowledgments

This work was supported by the Office of Naval Research and conducted in partial fulfillment of the requirements for the degree of master of arts. We thank Michelle Reddy for graphic design, layout, and production; Wesley R. Elsberry for software development; Jack Gonzales for underwater station and arc construction; Tricia Kamolnick, Jennifer Carr, Monica Chaplin, and Mark Todd for animal training; and Eric Jensen for veterinary support.

{ 16 }

Central Auditory Processing of Temporal Information
in Bats and Dolphins

Zoltan M. Fuzessery, Albert S. Feng, and *Alexander Supin*

The biosonar systems of bats and dolphins are remarkable in their capacity to extract precise spectral and temporal information from echoes of emitted pulses. These animals use acoustic information to image the external world with an accuracy that challenges or exceeds the capacity of the visual system. One example is the precision with which such systems can resolve target distance through the encoding of time delays between the pulse and echo with a resolution of better than 1 μs. Moreover, biosonar can obtain information that is invisible to the eye, as evidenced by the dolphin's ability to discriminate the internal structure and composition of a metal cylinder. Studies of the central auditory systems that process biosonar, particularly those of the microchiropteran bats, have greatly advanced our understanding of the fundamental neural mechanisms that underlie feature extraction in all mammalian species. This is largely due to the fact that biosonar is an active process in which a species generates much of its own acoustic environment. The emitted pulses have species-specific spectral and temporal characteristics, affording the researcher an immediate insight into the finite arrays of sounds that their auditory systems are designed to process. This has proven an invaluable tool for elucidating the nature of the neural circuitry that serves the processing of complex, behaviorally relevant sounds. This chapter introduces a series of reports on the neural mechanisms in the bat auditory brainstem that process the spectrotemporal features of both echolocation and nonecholocation sounds and summarizes recent advances in the processing of temporal acoustic information in dolphins.

Microchiropteran Bats

Studies in the last decade have focused on the questions of how and where neuronal selectivity for the spectrotemporal attributes of biosonar signals are created in the central auditory system (see also reviews by Casseday et al. 1995; Covey and Casseday 1999). These include selectivity for sound duration, pulse-echo delay, and the rate and direction of amplitude and frequency modulations. Several common themes have emerged. One is that the central auditory system may use similar types of synaptic integration "modules" to shape to neuronal se-

lectivity for different stimulus parameters. This integration is based on the timing of converging excitatory and inhibitory inputs. Excitatory postsynaptic events must coincide to excite a neuron, and this coincidence occurs only when a sound contains appropriate spectrotemporal characteristics. A second finding is that inhibition may contribute to excitation. A sound may generate an inhibitory postsynaptic potential that suppresses the neuron over the duration of the sound, but at the cessation of the sound, the release from inhibition results in an excitatory rebound—which, if it coincides with delayed excitatory potentials, drives the neuron to firing threshold. Inhibition thus shapes response selectivity not only by preventing a neuron from responding outside its range of selectivity; it also provides excitatory drive within the range of selectivity.

A second aspect of temporal processing addressed is the effect of the stimulus rate on response selectivity. Researchers typically examine selectivity with stimulus rates that are slow enough to isolate the effects of sequential stimuli. However, this is not how bat auditory systems operate in the real world. Bats will increase their rates of biosonar emission as high as 200 pulses per second when pursuing insects. Neurons can behave quite differently under these conditions. Their selectivity for various sound attributes can sharpen considerably, suggesting that our prevailing image of the precision of bat auditory systems may be founded on paradigms that do not adequately simulate the environment in which they function.

Finally, the extent of the absolute temporal resolution of the bat auditory system is addressed in terms of sensitivity to interaural time differences (ITDs). Most bats have not evolved a sensitivity to this binaural cue due to their high audible ranges and small interaural distances that restrict available ITDs to tens of microseconds. However, gleaning bats that listen for low-frequency sounds made by their prey may use ITDs to locate prey. One gleaning bat has been found to have the sharpest sensitivity to ITDs yet reported in a mammal, sufficient to resolve small available ITDs. Such findings, coupled with temporal acuity evident at the behavioral level, suggest that we have much to learn about the temporal resolving powers of bat auditory systems.

Processing of Sound Duration

Bat auditory systems have neurons tuned to the frequency, intensity, and pulse-echo delays of their biosonar signals. Recent studies demonstrate that they are also tuned to sound duration. Duration-selective neurons were first reported in the frog auditory system (Potter 1965; Narins and Capranica 1980; Hall and Feng 1986). This selectivity approximates the durations of communication signals, suggesting they serve as filters for signal detection. In bats, duration-selective neurons are abundant (Pinheiro, Wu, and Jen 1991; Casseday, Ehrlich, and Covey 1994; Ehrlich, Casseday, and Covey 1997; Galazyuk and Feng 1997; Fuzessery and Hall 1999) in the inferior colliculi (IC) and auditory cortices of three species (*Eptesicus fuscus, Myotis lucifugus,* and *Antrozous pallidus*). They prefer short durations (1–20 ms) of the biosonar signals of these species, and act as short-pass or band-pass duration filters. They may also serve to filter out irrelevant sounds on the basis of duration and to link images acquired from successive echoes. In *E. fuscus,* inferior colliculus neurons with similar best duration and frequency tuning have a wide range of response latencies (Ehrlich, Casseday, and Covey 1997). Echoes from sequential sonar pulses of similar duration and frequency can potentially arrive simultaneously at higher levels of the auditory system. This convergence in time can link the information obtained from successive echoes, providing a bat with a continuous perceptual image of a target.

Duration selectivity is not restricted to neurons that serve biosonar. In the pallid bat (*A. pallidus*), a gleaner that detects and locates terrestrial prey using passive sound localization, a large percentage of neurons tuned below the biosonar spectrum also exhibit a selectivity for short durations of <10 ms (Fuzessery and Hall 1999). A filtering function can also be suggested. The pallid bat attends to short, prey-generated, noise transients while hunting (Fuzessery et al. 1993). These duration-selective neurons may filter these transients from background noise.

The mechanisms underlying duration selectivity have received considerable attention. This selectivity appears to be created in the IC because it is not observed at lower system levels (Gooler and Feng 1992; Casseday, Ehrlich, and Covey 1994) and because the selectivity is often eliminated by blockade of GABAergic and glycinergic inhibitions (Casseday, Ehrlich, and Covey 1994; Jen and Feng 1999; Fuzessery and Hall 1999). These findings suggest that the timing of converging inhibitory and excitatory inputs to IC neurons is crucial in creating this selectivity. The temporal firing patterns of some duration-selective neurons in the IC also implicate the importance of the timing of these converging inputs. The response latencies of some duration-selective neurons are coupled to stimulus offset, not onset, indicating that they are excited by events that occur at the cessa-

tion of the sound (Casseday, Ehrlich, and Covey 1994; Ehrlich, Casseday, and Covey 1997; Fuzessery and Hall 1999). A patch clamp study found that some duration-selective neurons receive a short-latency inhibition input that precedes excitation and respond only when the inhibition has decreased over time (Casseday, Ehrlich, and Covey 1994). These two findings led to the formulation of a coincidence detection model with three components (Casseday, Ehrlich, and Covey 1994). The first is a short-latency, sustained inhibition. The second is an excitatory event at the end of the sound, perhaps due to an excitatory rebound from inhibition. The third is a delayed excitation that is triggered by sound onset. The second and third components must coincide to maximally excite the neuron. Because the second component has a variable latency that depends on sound duration, and the third component has a fixed latency, coincidence occurs only over a limited time window, resulting in duration selectivity.

This coincidence model is attractive for explaining the selectivity of neurons that respond at the end of the sound. However, other neurons selective for short-duration sounds have responses that are linked to stimulus onset. If their inhibitory component is eliminated by blocking receptors for inhibitory transmitters, these neurons lose their selectivity for short durations (Fuzessery and Hall 1999). Such neurons may derive their duration selectivity from two components: an early inhibition that persists while the sound is on and a delayed excitatory input. If the sound is long enough, the inhibition will encroach on the delayed excitatory event, and the response will be suppressed. This model suggests that the inhibition must not coincide with the excitation and that a rebound from inhibition does not contribute to excitation. This mechanism provides a simple but effective mechanism for creating short-pass neurons, with the favored duration being determined by the arrival time of the delayed excitatory input. The significance of the differences in the behaviors of duration-selective neurons is that it suggests that a given form of selectivity may be created by more than one synaptic mechanism, and that several mechanisms may exist both within and between species.

Processing of Frequency-Modulated (FM) Signals

Most bats produce sonar signals consisting entirely of a downward FM sweep. Even the signals of constant-frequency (CF) bats end with such a sweep. Frequency modulations are also present in intraspecific communication signals, and in echoes of CF emissions due to the dynamics of insect wing beat. The literature on the mechanisms underlying FM selectivity in the bat auditory system is extensive, providing evidence that the rate and direction of frequency modulations can be encoded by the timing of converging synaptic inputs.

Neurons that respond exclusively to single downward

FM sweeps (termed "FM specialists") have been reported only in echolocation bats. Selectivity for FM sweeps appears to first emerge at the level of the IC. The percentage of FM specialists varies considerably among species. Interestingly, the species relying most heavily on echolocation to capture prey are not those that show the greatest selectivity for FM sweeps. A comparative study of *M. lucifugus, M. yumanensis,* and *Plecotus townsendii* reported that 3% of recorded neurons are FM specialists (Suga 1965, 1969). In two mollosid bats (Vater and Schlegel 1979) and the *P. parnellii* (O'Neill 1985), bats that rely heavily on echolocation to find prey, the percentage of FM specialists is <2%. The percentage is higher (approximately 14%) in the *E. fuscus,* also an echolocating specialist (Casseday and Covey 1992). Interestingly, in *A. pallidus,* a species relying on passive listening to find prey instead of biosonar, 31% of neurons in the biosonar region of the IC responded exclusively to downward FM sweeps (Fuzessery 1994). This seems counterintuitive if one assumes that species that rely most heavily on biosonar will have more neurons that respond exclusively to these echoes. One explanation for these results is that *A. pallidus,* while hunting, must process both biosonar to avoid obstacles and passive listening to find prey. The neural pathways for these two functions are anatomically and functionally segregated in the IC (Fuzessery 1994, 1997). The large percentage of FM specialists may enhance this functional segregation by responding only to biosonar echoes. Indeed, these FM specialists are suppressed by other sounds that a pallid bat is likely to encounter (Fuzessery 1994; Fuzessery and Hall 1996). Such suppression may play a role in focusing the bat's attention between competing streams of information while the bat is hunting.

An unexpected form of selectivity found in *P. parnellii* is a preference for upward FM sweeps (Gordon and O'Neill 1998). These neurons are located in the external nucleus of the IC and tuned to the CF portion of the 60 kHz harmonic. Most (85%) show a preference for upward over downward FM sweeps. It is unclear what functions these neurons serve; they may encode upward FM sweeps in the species communication signals, or the upward sweeping components of sinusoidal FM in echoes produced by the wing beats of insects.

Neuronal selectivity for sinusoidal FM (SFM) signals has also been observed in the IC (Schuller and Pollak 1979; Casseday, Covey, and Grothe 1997; Koch and Grothe 1998). Such neurons can be selective for the direction, rate, or depth of SFMs, and they may not respond at all to other signals (tones or noise) with similar spectra. They presumably serve to detect acoustic "glints" that are imparted on echoes by the wing beats of flying insects. A pronounced form of SFM selectivity is observed in the IC of *E. fuscus* (Casseday, Covey, and Grothe 1997), in which neurons require more than one modulation cycle to respond maximally, or to respond at all. These neurons appear to be integrating information over several cycles before responding. This "buildup" response pattern has not been found at lower levels of the auditory brainstem (Grothe 1994; Huffman, Argeles, and Covey et al. 1998), suggesting that it is created in the IC.

The mechanisms that create selectivity for the direction of FM sweeps have been studied extensively. While they differ in detail, most rely on the timing of excitatory and inhibitory inputs tuned to different frequencies. One simple model postulates that the locations of inhibitory flanks around excitatory tuning curves can predict selectivity for FM sweep direction. If the FM sweep first crosses the inhibitory flank, the inhibitory input will arrive first and suppress the neuron's response (Suga 1965). However, it has also been reported that inhibitory flanks do not predict directional selectivity—since even FM sweeps that originate within the excitatory tuning curve, but proceed in the nonpreferred direction, can suppress the response (Fuzessery 1994). However, Gordon and O'Neill (1998) cautioned that such results can depend on how the inhibitory frequency domains are determined. If inhibitory domains are measured by a simultaneous presentation of excitatory and inhibitory tones, they may not predict directional selectivity. However, if a tone that was not inhibitory during simultaneous presence precedes the excitatory tone in the same manner that would occur during a nonpreferred FM sweep direction, it may suppress the response, indicating that the timing of inhibitory and excitatory inputs can be critical in shaping directional selectivity.

Other proposed mechanisms do not rely on excitatory/inhibitory interactions. Covey and Casseday (1999) suggest that coincident multiple excitatory inputs could produce the same selectivity if this coincidence could occur only in one sweep direction. A similar but more anatomically elaborate model (Fuzessery and Hall 1996, adapted from Rall 1964) relies on the spatiotemporal integration of excitatory inputs. If the inputs to a dendrite are arranged tonotopically, and the timing of inputs reflect the sequence of frequencies in an FM sweep, then a summation of excitatory graded potentials in the direction of the spike initiation zone will be greater in one direction than the other.

A second issue is the mechanism that creates neurons that respond exclusively to FM sweeps, and not to single tones within the sweep or spectrally identical band-pass noise. This type of selectivity (like others discussed in this chapter) is probably created at the level of the IC, because it has not been observed at lower levels of the auditory system (e.g., Huffman, Argeles, and Covey 1998) and because blocking receptors for inhibitory transmitters eliminates this response exclusivity (Fuzessery and Hall 1996). The latter finding suggests that inhibitory inputs to IC neurons are essential for this selectivity at the level of the IC. One simple mechanism that could explain these results is that the elimination of a sustained inhibition allows sounds (e.g., tones, noise) that normally provide only subthreshold excitation to now cause the neuron to fire.

Finally, what might create the selectivity of SFM neurons that require more than one modulation cycle to "build up" to a maximum response? A proposed mechanism is another variant of the coincidence model (Covey and Casseday 1999). Suppose an SFM-selective neuron were selective for the direction of frequency modulation, and excited on the downward modulation phase and inhibited on the upward phase. The first downward sweep would generate an excitatory input capable of eliciting a submaximal response. The next downward phase would evoke inhibitory input. If there were an excitatory rebound from this inhibition that coincided with the excitatory input produced by the next downward phase, the response would increase. The need for coincidence could also shape a neuron's selectivity for the rate of SFM (Koch and Grothe 1998; Burger and Pollak 1998).

Processing of Pulse-Echo Delay

Echolocating bats assess target distance by computing the delay between the emitted pulse and returning echo. This computation is accomplished by neurons that are selective to particular pulse-echo delays, in some species, by the delays between specific harmonics of the pulse and echo. In the auditory cortex, these neurons are partially (in FM bats) or completely (in CF-FM bats) segregated from the tonotopic region (O'Neill and Suga 1979; Sullivan 1982; Wong and Shannon 1988; Dear et al. 1993; Schuller, O'Neill, and Radtke-Schuller 1991). In the *P. parnellii*, the delay-tuned neurons are topographically organized into orderly representations of time delays across the rostrocaudal axis (O'Neill and Suga 1979). There are three parallel axes in which the timing of the first FM harmonic (FM_1) of the emitted pulse is compared with arrival of each of three higher harmonics of the FM components of the echo (FM_2, FM_3, FM_4).

How and where this computational map of target distance originates has received considerable attention. Neurons sensitive to pulse-echo combinations presented at specific delays and intensities were first discovered in the intercollicularis nucleus of *E. fuscus* (Feng, Simmons, and Kick 1978), a region between the inferior and superior colliculi. The location of these neurons, off the primary ascending auditory pathway, raises the question of whether they contribute to the delay sensitivity in the auditory cortex. More recently, delay-sensitive neurons were observed near the brachium of the inferior colliculus of the same species (Dear and Suga 1995), suggesting that these neurons may be part of a parallel pathway that leads to the processing of target distance information in the superior colliculus.

Along the primary auditory pathway in *P. parnellii*, delay-tuned neurons are abundant in the IC (Mittman

and Wenstrup 1995; Yan and Suga 1996b). This is the lowest auditory center where delay sensitivity has been found, and is likely where it is created. As in the medial geniculate body (MGB) (Olsen and Suga 1991) and the auditory cortex (O'Neill and Suga 1979), these neurons are sensitive to delays between FM_1 and one of the higher FM harmonics. The distribution of best delays (1–24 ms) is also similar in these three auditory centers. Whether there is a sharpening of delay tuning at higher levels of the system is not fully resolved. Portfors and Wenstrup (chapter 20, this volume) found little difference in the sharpness of delay tuning and the degree of response facilitation evoked by appropriate combinations of tone frequency, intensity, and delay in populations of IC and MGB neurons. However, Yan and Suga (1996b) reported that there is a considerable sharpening of delay tuning in the MGB, and they suggest a hierarchical model for encoding target distance.

The origin of the topographically organized delay map in the *P. parnellii* auditory system is also unresolved. Yan and Suga (1996b) report a systematic representation of the best delays as well as response latencies in the ICC neurons along the dorsoventral axis; both the best delay and response latency progressively decrease ventrally. In contrast, Portfors and Wenstrup (chapter 20, this volume) observed a topographic change in response latency in IC, but no orderly delay map. However, both studies support the idea that there is a gradual refinement in topographical representation of echo delays form the IC to the auditory cortex.

The mechanisms that underlie delay tuning in the IC of *P. parnellii* appear to arise from converging inputs from several lower brainstem nuclei, and not from interactions among IC neurons (Wenstrup et al., chapter 21, this volume). Injection of tract tracers near delay-sensitive neurons in the FM_2–FM_4 regions of the IC labels cells in the cochlear nuclei, the superior olivary complex, and the lateral lemniscal nuclei—but not in the FM_1 region of the IC. This indicates that inputs from lower level neurons tuned to the different FM harmonics converge in the IC.

Delay tuning may arise from a coincidence mechanism similar to those that have been previously described. Excitatory events generated by the FM_1 pulse and higher harmonic echoes must occur at the same time. Therefore, the excitatory event generated by the emitted pulse must somehow be delayed. There is evidence that this delayed excitation is generated by a rebound from inhibition (Wenstrup et al., chapter 21, this volume). This postinhibitory rebound must coincide with a late excitation generated by the echo to produce a maximum response. The duration of the inhibition, and the arrival time of the late excitatory input, will determine the delay selectivity of the neuron. The role of inhibition was determined by blocking receptors for glycine and GABA, the two major inhibitory transmitters in the IC. Blockade of the glycine receptor, but not the GABAa receptor, eliminates the facilitated response that normally occurs over a narrow range of pulse-echo delays, providing evidence that postinhibitory rebound contributes to excitation to shape neuronal selectivity. This result is also unusual in that it implicates a specific inhibitory neurotransmitter in creating this biosonar-specific form of a response selectivity.

Processing of Amplitude-Modulated (AM) Signals

A dynamic object such as a flying insect not only produces frequency (FM) but also amplitude (AM) modulations in the echo. For CF-FM bats, their echolocation signals are sufficiently long and can provide a full AM cycle that contains information of the wing-beat pattern and frequency within a single echo. In contrast, the echolocation signals of FM bats are short, and therefore they must integrate across a series of echoes to extract the same information. The precision of this temporal processing may depend on the rate of sonar emission. The emission rate is actively controlled during the pursuit of prey, from 5 to 200 pulses per second. A number of recent studies examined how neurons in the upper levels of the auditory system encode the AM across sound pulses. These studies revealed that neurons in the IC, medial geniculate body, and auditory cortex are sensitive to high-frequency trains of sound pulses, but only when their amplitudes vary (Condon, White, and Feng 1994; Condon et al. 1997; Llano and Feng 1999). In contrast, when the amplitudes of pulses within a train is constant, as expected of echoes from stationary background targets, neurons respond poorly. This sensitivity to changes in pulse amplitudes may serve to draw attention to dynamic targets such as flying insects.

In addition to the sensitivity to AM pulse trains, neurons show enhanced selectivity for changes in AM rates embedded in a pulse train (Condon, White, and Feng 1994; Condon et al. 1997). In the IC and cortex, neurons respond best to a specific range of AM rates. Rates to which each neuron responds maximally differs, so presumably a population of differentially tuned neurons confers the ability to discriminate insects on the basis of wing-beat frequency.

Neural representations of sinusoidal amplitude modulations (SAM) within a single prolonged signal have been extensively investigated but mostly in CF-FM bats, in part because the sonar signals of these bats are naturally long (Schuller and Pollak 1979; Vater 1982; Lesser et al. 1990; Yang and Pollak 1997; and see the review in Covey and Casseday 1999). Neurons in the auditory brainstem, from cochlear nucleus to IC, are selective for the AM rate. Their best modulation rates vary widely, from a few Hz up to 800 Hz. This range is far greater than the range observed in other mammals. Interest-

ingly, similar results are observed from brainstem neurons in FM bats (Condon et al. 1994; Grothe, Park, and Schuller 1997). In the IC of the *E. fuscus,* neurons with different temporal firing patterns show differential sensitivities to SAM signals. The mechanisms that underlie SAM-rate selectivity are unclear, but it appears that they differ at various levels of the ascending auditory system. In the lower brainstem, alternating cycles of phase-locked excitatory and inhibitory inputs may shape the modulation rate selectivity if early excitatory inputs and delayed inhibitory inputs begin to overlap with increasing modulation rates (Grothe 1994). This mechanism is supported by the finding that blockade of the inhibitory inputs indeed allows a neuron to respond to higher SAM rates. However, this is not the case in the IC (Burger and Pollak 1998), which suggests that a different mechanism may regulate this form of response selectivity.

Temporal Dynamics of Auditory Processing

Bats typically increase the rate of sonar emission when they confront challenging tasks such as capturing flying insects or avoiding obstacles. Early studies by Wong and colleagues (Wong, Maekawa, and Tanaka 1992; Tanaka and Wong 1993) observed that the temporal tuning property of cortical neurons is also dependent on the rate of stimulation. More recently, Galazyuk and Feng (1997a) and Galazyuk, Llano, and Feng (2000) reported that the rate of acoustic stimulation can alter neural processing of sound amplitude and frequency of IC neurons. About 50% of IC neurons in *M. lucifugus* exhibit dramatic increases in selectivity for the pulse frequency and intensity when the repetition rate of sound pulses is increased. In response to AM pulse trains with a dynamically changing rate—where the rate increases progressively from low to high within the train resembling bat's sonar emission pattern when pursing a flying insect—these IC neurons respond only to sound pulses within very narrow frequency and intensity ranges. These ranges of selectivity are far narrower than those derived from the great majority of studies that use much slower stimulus presentation rates, preventing one response from affecting the next. It seems that results obtained from physiologically isolated stimuli cannot predict the behavior of neurons under natural listening conditions. Future studies should direct attention to examining the selectivity of neurons under more natural conditions, with focus on the idea that response selectivity for a particular acoustic parameter may be context dependent, and depend on the rate of stimulation.

Processing of Interaural Time Differences (ITDs)

Echolocating bats were assumed to use only interaural intensity differences (IIDs) to acquire spatial information because of their high-frequency audible ranges and because their small head size and small interaural distance seems to preclude the use of ITDs. Indeed, bats that rely on echolocation for both obstacle avoidance and prey capture have little or no neuronal sensitivity to small ITDs created by their heads (Harnischfeger, Neuweiler, and Schlegel 1985; Pollak 1988; Grothe and Park 1995). However, gleaning bats, which hunt on terrestrial prey, appear to be an exception, since most rely on passive sound localization at the lower end of the audible range to detect and locate prey. The one gleaner thus far studied for ITD processing, *A. pallidus,* has binaural comparators that can resolve small ±70 μs ITDs (Fuzessery 1997; Fuzessery and Lohuis, chapter 18, this volume). This ITD sensitivity sharpens progressively from the level of the auditory midbrain to the cortex, suggesting that this binaural cue is important in the acquisition of spatial information. The most sensitive of these binaural neurons show a 100% change in response magnitude over an ITD range of less than 100 μs. This is an order of magnitude greater than the temporal resolution of neurons tuned to pulse-echo delays, and it hints at the absolute temporal acuity of bat auditory systems yet to be discovered. Given that the response selectivity of neurons increases when presented with dynamically changing sounds that more accurately represent real-world conditions, the ITD sensitivity of the *A. pallidus* auditory system may be more acute when tested under conditions that more accurately simulate a bat moving through space in search of prey.

In summary, recent studies of the bat auditory system suggest that many of the species-specific forms of response selectivity required of echolocators are created at the level of the IC through the timing of converging excitatory and inhibitory inputs. The exact nature of the synaptic mechanisms that create these selectivities remains elusive, despite the arsenal of techniques brought to bear on the issue. However, the emerging principle is one in which the auditory system may use a relatively small repertoire of integration "modules" to create neuronal selectivity for a number of signal attributes used to acquire information about the external environment.

Temporal Resolution and Processing of Temporal Information in Locating Odontocetes

Compared with bats, studies of hearing in echolocating odontocetes (dolphins) are much more restricted in terms of experimental approaches. Invasive experiments, such as single-neuron recordings, are not used for ethical reasons. Hearing abilities and mechanisms must therefore be studied with psychophysical (behavioral), noninvasive evoked-potential (EP), or morphological techniques. The information on hearing mechanisms in dolphins is therefore less well developed. The specific roles of the various levels of the central auditory system in the temporal processing of acoustic information are not fully understood. Nevertheless, many intriguing features of temporal processing in the dolphin's auditory system have been found in recent years.

A remarkable feature of the dolphin's auditory system is its high temporal resolution. Temporal resolution may be described in terms of temporal integration time: the shorter the integration time, the better the temporal resolution. Psychophysical data indicate that the dolphin auditory system has a very short integration time compared with humans. For example, detection thresholds of sound pulse pairs decrease by 3 dB when interpulse intervals are shorter than 0.2–0.3 ms. This temporal processing corresponds to an energy detector with an integration time between 0.2 and 0.3 ms (Au, Moore, and Pawloski 1988; Au 1990). Experiments with discrimination between pulse pairs have also shown that discrimination ability changes markedly when this interpulse interval becomes shorter, suggesting that pulses may merge into a unified acoustic image because they fall within the integration time (Dubrovskiy 1990). Backward masking in dolphins also occurs at same interpulse intervals of up to 0.2–0.3 ms (P. Moore et al. 1984; Dubrovskiy 1990). Thus, the integration times of dolphins are an order of magnitude shorter than that of humans (B. Moore et al. 1988; Plack and Moore 1990).

It should be noted that in some experimental paradigms, intervals can be analyzed when they are markedly shorter than 0.2–0.3 ms. Dolphins can discriminate pulse pairs with intervals as short as 0.05–0.2 ms, with discrimination thresholds on the order of a few microseconds (Vel'min and Dubrovskiy 1976; Dubrovskiy 1990). However, this discrimination may be due to spectral, rather than temporal, processing in the auditory system. Two sound pulses separated by a very short interval merge into an "acoustic whole" and are perceived as a single click. The frequency spectrum of this click depends on the interval. It contains "ripples"—that is, periodically alternating spectral peaks and valleys. The ripple spacing is the reciprocal of the delay. Thus, the spectrum structure can be a cue for discrimination pulse pairs with different delays. This discrimination mechanism does not have the low limit of interpulse delay; on the contrary, the shorter delay, the more rough the ripple spectrum pattern, and the more discernable to spectral difference. This discrimination mechanism must not be considered temporal processing. Of course, from a mathematical point of view, any signal processing can be described either in temporal or in spectral domains, so temporal and frequency processing cannot be contrasted. However, from a neurophysiological perspective, signal processing based either on temporal patterns of neuronal firing (temporal processing) or on frequency tuning of peripheral auditory filters (frequency processing) is readily contrasted.

Moreover, these data must not be confused with data

on the temporal summation limit. Studies of the temporal summation for tone pulses of various frequencies have shown that the summation (i.e., a threshold decrease with pulse prolongation) occurs at pulse durations up to dozens of milliseconds (C. S. Johnson 1968b, 1991). This summation time is about two orders of magnitude longer than the integration time of 0.2–0.3 ms indicated above, but actually there is no discrepancy between these data. The temporal resolution is dictated by the shortest possible integration time, whereas the temporal summation experiments reveal the longest possible one. Approximately the same ratio between these two parameters (1–2 orders of magnitude) exists, for example, in the human's auditory system, although both of these times are much longer than in dolphins: a few ms and 200–300 ms respectively (see Green 1984; de Boer 1984).

EP data has confirmed the very high temporal resolution of the dolphin's hearing obtained with psychophysical methods. When presented with paired click stimulation, EP recovery time in dolphins was several times shorter than in other mammals: a just-detectable response to the second click is observed at intervals as short as 0.2–0.3 ms (Supin and Popov 1985; Popov and Supin 1990a). However, these studies have shown that processes dictating the temporal resolution are more complicated than simple temporal integration within a certain time limit. Although just-detectable EPs appeared at very short intervals, complete recovery required interstimulus intervals of an order of a few ms. Moreover, the complete-recovery time increased with stimulus level up to 10 ms (Supin and Popov 1995a).

Very similar results were obtained by another popular method of measurement of temporal resolution: gap-in-noise detection in conjunction with EP technique. Threshold gap duration (when a just-detectable EP appeared) was as short as 0.1 ms. However, as EP amplitude increased with intensity, gap detection increased to 10 ms (Supin and Popov 1995b; Popov and Supin 1997). These results cannot be explained by any simple model based on temporal integration times of 0.2–0.3 ms.

Double-click EP data can be easily explained if the nonlinear transform of the afferent inflow in the auditory system is taken into account. EP amplitude in dolphins is roughly proportional to the decibel—that is, the amplitude depends nonlinearly on sound pressure or power (Popov and Supin 1990a, 1990b). This dependence obviously reflects a general intensity-response relation known as Weber's or Stevens' laws. Therefore EP recovery functions apparently reflect the integration time course via the nonlinear intensity-to-amplitude relation. From this perspective, data obtained in the double-click and gap-detection experiments in dolphins were transformed into temporal transfer function (Supin and Popov 1995a, 1995b; Popov and Supin 1997). This procedure produces a rather complicated form of the temporal transfer function. Its initial part is really very short (half-decay time about 0.3 ms) and in good agreement with psychophysical data. However, aside from this initial part, the function contains a long "tail" with a slope of 11–12 dB per time doubling (35–40 dB per time decade). Although this tail is of very low level, it determines response recovery in double-click and gap-detection experiments, since the nonlinear transform makes this low-level tail significant.

During echolocation, a dolphin probably hears its own emitted sound pulse and after a short delay hears the echo—that is, the situation is similar to that in double-click experiments. Therefore, some features of response recovery in the double-click experiments may be important to explain echolocation mechanisms. In particular, when a double-click stimulus increases in level (keeping ratio of two pulses constant), the recovery time enlarges. As a result, the absolute amplitude of the second response is almost independent of stimulus level. That is, an increase of the second click is compensated by the deeper suppression by increased first click (Supin and Popov 1995a). If the two pulses were the emitted pulse and echo, this suggests that the response to the echo may be independent of emitted pulse level, since a stronger emitted pulse results in a stronger echo. These two factors compensate each other. When the first of a pair of clicks becomes stronger than the second, the recovery time increases by approximately 10 times per 40 dB-level difference (Popov and Supin 1990a). Again suppose that the two pulses are the emitted pulse and echo. If the emitted sound spreads as spherical waves, its level decreases with distance with a rate of 20 dB per distance decade. The returning echo from a small target decreases in level with the same rate. The resulting dependence of echo level on distance is 40 dB per distance decade. On the other hand, the echo delay is proportional to the distance. Thus, the ratio between the echo delay and level is 40 dB per time decade. It is just the same ratio as between the level difference and recovery time. As a result, the response to the echo may be (within a certain range of distances) independent of distance: with distance increase, weakening the echo is compensated by more complete recovery after the emitted pulse. Thus, the model based on the nonlinear transform of input signal and subsequent integration by the temporal transfer function explains these properties of the recovery process.

As a rule, dolphins emit echolocation pulses as long trains rather than short pulses at high repetition rate. Therefore, the ability of the dolphin's auditory system to respond to rhythmic pulse sequences is important for understanding the temporal processing of signals. To some extent, this ability can be predicted based on the temporal transfer function and integration time in the system found in double-click and gap-detection experiments.

Psychophysical studies have mainly examined threshold dependence on pulse rate as a manifestation of temporal summation. Thresholds fall slowly with pulse rate increase (about 1 dB per rate doubling) until the rate is lower than 3000–4000 Hz. Beyond this rate, the slope of the summation effect increases abruptly up to about 3 dB per rate, and then doubles, indicating energy summation (Dubrovskiy 1990). These data indicate the same integration time of 0.2–0.3 ms.

Responses to rhythmic sounds were also studied with the EP method. An EP sequence can faithfully represent rather high rates of sound clicks, up to a few dozen Hz. Decreases in EP amplitude occur only at stimulation rates more than 100 Hz, although small evoked potentials were detectable even at rates up to a few hundred Hz (Ridgway et al. 1981; Popov and Supin 1990a, 1990b). This ability to represent rhythmic stimuli is very high compared to that of other mammals and humans. For example, in humans, EPs follow rhythmic sound pulsations up to 50–70 Hz (Rees, Green, and Kay 1986; Kuwada, Batra, and Maher 1986), in accordance with psychophysical studies that also show a cutoff frequency of about 50 Hz (Zwicker 1952; Viemeister 1979; Eddins 1993).

However, an EP can follow higher stimulation rates in response to sinusoidally amplitude-modulated sounds, up to 600–1000 Hz, with EPs detectable at rates up to 1700 Hz (Dolphin 1995; Dolphin, Au, and Nachtigall 1995; Supin and Popov 1995c). Thus, estimates of reproducible rates differed several times. Why would results differ when pulses and sinusoidally amplitude modulated signals are used? This issue is described in Supin and Popov, chapter 22, this volume. When the adaptation is reduced by using short pulse trains, the ability of EP to represent stimulation rate is the same as in conditions of sinusoidally modulated sounds: cutoff frequency was about 1700 Hz. This cutoff frequency corresponds to an integration time of 0.3 ms, in good agreement with other data reviewed above. Thus, in conditions of prolonged stimulation, only the short initial part of the temporal transfer function plays a significant role. The low-level tail of the transfer function markedly influences only responses appearing after a certain time of silence.

Up to this point, the temporal processing in the dolphin's auditory system was considered irrespectively of frequency analysis: the stimuli used in the experiments described above were either broadband (clicks in double-click and rhythmic-click studies, noise in gap-detection measurements) or tones of a certain frequency (sinusoidally amplitude-modulated stimuli). Further studies should consider temporal processing in conditions of interaction between different frequency channels in the auditory system. One such study found an effect termed "paradoxical lateral suppression." Briefly, the effect consists of a suppression of the EP ability to reproduce high-rate modulations of a certain carrier frequency when another carrier frequency is presented simultaneously. The paradoxical feature of this effect is that it appears when the suppressing carrier is 10–40 dB lower in level than the probe carrier. Thus, this effect has nothing to do with a regular masking. Only rhythmic EPs at high-rate amplitude modulations of the probe are deeply suppressed in these conditions; responses to single clicks, pips, sound onsets, and so on are not influenced by such weak sounds. Thus, the effect reflects some kind of regulation of temporal processing (enabling or disabling the ability to produce high-rate responses) in one frequency channel by a signal addressing another channel. Considering that dolphins emit pulses at rather high rates, it can be assumed that this effect influences the echo perception.

In summary, investigations of temporal processing in the dolphin's auditory system have resulted in several speculations regarding the role of temporal processing in perception and analysis of echo signals. At present, it is impossible to determine whether these speculations are accurate. One way to test their validity may be EP recording during natural echolocation in dolphins. Such recordings would directly demonstrate how the brain responds to echo signals as a function of target distance, emitted pulse level, pulse repetition rate, and so on. While this method would be difficult, it is not impossible.

{ 17 }

Midbrain Integrative Mechanisms and Temporal Pattern Analysis in Echolocating Bats

Ellen Covey

Introduction

TEMPORAL PATTERNS AND ECHOLOCATION

For all echolocating bats and dolphins, spectrotemporal acoustic patterns generated during active echolocation contain crucial information about the environment. Because bats and dolphins use somewhat different echolocation strategies, the specific patterns of sound that must be processed by their nervous systems necessarily differ. Nevertheless, it is likely that many general features of echolocation signals are important for all animals that echolocate, and that different species share common neural mechanisms for processing time-distributed auditory information.

The big brown bat, *Eptesicus fuscus,* is a typical insectivorous echolocator. Its biosonar calls are frequency-modulated (FM) signals that sweep from high to low frequency over a few milliseconds. The calls vary systematically according to the bat's foraging behavior, and echoes are modified in predictable ways by objects in the environment (e.g., Simmons 1989).

Information about the spectrotemporal characteristics of sound is represented in different ways at different levels of the mammalian nervous system. Neurons at early stages of the auditory pathway respond with discharge patterns that directly reflect the real-time characteristics of the stimulus, whereas neurons at progressively higher levels integrate information over longer periods of time and can provide filtering for biologically important features or patterns of sound (e.g., Casseday and Covey 1996). This chapter focuses on experimental evidence in *Eptesicus,* showing how the representation of sound parameters changes at successive levels of the ascending brainstem auditory system, with the result that many midbrain neurons in echolocating bats are tuned to parameters such as duration (Casseday, Ehrlich, and Covey 1994; Ehrlich, Casseday, and Covey 1997), amplitude (Casseday and Covey 1992), and pulse-echo delay (Feng, Simmons, and Kick 1978; Mittmann and Wenstrup 1995).

Materials and Methods

We have used several different approaches to address the question of how neural inputs are integrated at the level of the inferior colliculus (IC) to produce tuning to biologically relevant sound features. One approach involves blocking of inhibitory neurotransmitters at the IC neuron by iontophoretic application of bicuculline (a blocker of GABAa receptors) or strychnine (a blocker of glycine receptors). These methods are described in detail elsewhere (B. R. Johnson 1993; Casseday, Ehrlich, and Covey 1994, 2000). A second approach uses in vivo whole-cell patch clamp recording to examine the interaction of excitatory and inhibitory postsynaptic currents at the level of the IC neuron. This method is described in detail by Covey, Kauer, and Casseday (1996). All experiments were performed in awake, tranquilized animals.

Results

DURATION TUNING

At the auditory periphery of all mammals, afferent nerve fibers respond to sounds of all durations with a primary-like discharge pattern—that is, a peak firing rate elicited by the onset of a sound—followed by a discharge at a lower rate that is maintained throughout the sound's duration (e.g., Kiang et al. 1965). At the level of the cochlear nucleus, superior olivary complex, and nuclei of the lateral lemniscus, neurons respond with a variety of discharge patterns that range from a single action potential at sound onset to a sustained train of action potentials that persists for a time equal to the duration of the sound (e.g., Neuweiler and Vater 1977; Vater 1982; Haplea, Covey, and Casseday 1994; Covey and Casseday 1991; Covey, Vater, and Casseday 1991). Although discharge patterns vary among cells, each cell responds in the same way to an appropriate stimulus, regardless of duration. In contrast, approximately one third to one half of neurons in the inferior colliculus of *Eptesicus* show short-pass, long-pass, or band-pass tuning to duration (Ehrlich, Casseday, and Covey 1997). These neurons have, in effect, converted a temporal representation of duration to a spatial representation. That is, the duration of a sound is represented in terms of which neurons are active rather than for how long neurons with sustained discharge patterns are active.

Because there is good evidence that IC neurons receive convergent excitatory and inhibitory synaptic in-

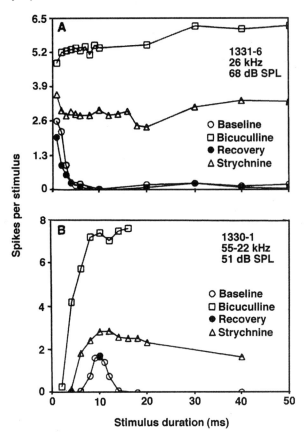

Fig. 17.1. Responses of two duration-tuned neurons as a function of sound duration under baseline conditions (open circles), during iontophoretic application of bicuculline (open squares), during iontophoretic application of strychnine (open triangles), and after recovery from the effects of drug application (solid circles). The legend at the upper right indicates unit number, sound frequency, and sound level. The unit in panel *A* was tested with a pure tone; the unit in panel *B* responded only to downward frequency-modulated sweeps, so the frequency range given indicates the upper and lower limits of the sweep. Both units were tested at 20 dB above threshold (modified from Casseday, Ehrlich, and Covey 1994).

puts, we tested the hypothesis that duration tuning arises through interactions between these two classes of input by blocking GABAergic and glycinergic inhibition at IC neurons from which we recorded sound-evoked responses.

Fig. 17.1 shows the effect of this procedure on responses of two neurons, one that responded best to a pure tone with a duration of about 1 ms (fig. 17.1A) and one that responded best to an FM sweep with a best duration of about 10 ms (fig. 17.1B). The neuron in fig. 17.1A appeared to respond to all durations shorter than about 5 ms, but not to longer durations—that is, it had short-pass filter characteristics. The neuron in fig. 17.1B responded to sounds longer than 4 ms and shorter than 15 ms—that is, it exhibited band-pass filtering for duration. Application of bicuculline, a blocker of GABAa

receptors, to both neurons completely eliminated their duration tuning. Blocking of GABAergic inhibition resulted in increased spike counts at all stimulus durations due to the fact that the evoked spike train lasted longer than under normal conditions. For both neurons, in the presence of bicuculline, spike counts increased with duration up to some point at which an asymptote was reached. At this point, further increases in duration did not evoke more spikes. This finding suggests that the excitatory input to both neurons was not sustained for the entire stimulus duration but was instead a transient burst consisting of multiple spikes. Across the population of duration-tuned neurons that were tested with bicuculline, duration tuning was completely eliminated in 75%; duration tuning was broadened in an additional 10% (Casseday, Ehrlich, and Covey 2000). Normally, responses of many duration-tuned neurons were clearly correlated with the offset of the stimulus. When inhibition was blocked, the responses of duration-tuned neurons were always correlated in time with stimulus onset.

Strychnine, a blocker of glycine, was also effective in eliminating duration tuning, although the effects generally were not as pronounced as those seen with bicuculline. For both neurons in fig. 17.1, strychnine eliminated the suppression of responses to long durations that was seen in the absence of drugs. Spike counts increased only slightly, suggesting that GABAergic, rather than glycinergic, inhibition is responsible for canceling the latter part of the excitatory input, thereby reducing the excitatory response to one, or at most a few, spikes. The band-pass neuron in fig. 17.1B remained unresponsive to durations of <4 ms, so it only lost its low-pass characteristics. Application of strychnine completely eliminated duration tuning in slightly over 60% of duration-tuned neurons and caused it to broaden in another 12%. These results indicate that for most duration-tuned neurons, this form of stimulus selectivity arises at the level of the IC and directly involves both GABAergic and glycinergic inhibition.

Additional insight into the sequence of events that creates duration tuning comes from experiments in which we recorded postsynaptic currents evoked by sounds of different durations, as shown in fig. 17.2 (Casseday, Ehrlich, and Covey 1994; Covey, Kauer, and Casseday 1996). For the neuron illustrated, the first evoked synaptic input generated an outward current, indicating that it was inhibitory. This was true regardless of the duration of the sound. At a duration of 5 ms, however, the initial inhibitory event was followed by a net inward current, signifying either a depolarizing excitatory input or a rebound from inhibition. At durations of 10–20 ms, the initial inhibitory event was still present, but it was followed by a depolarizing current large and rapid enough to evoke action potentials. Because the latency of the peak of the excitation increased as a function of duration, it appeared locked in time to the offset of the

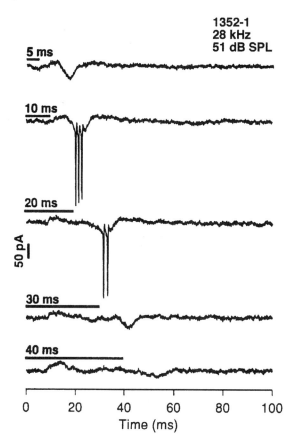

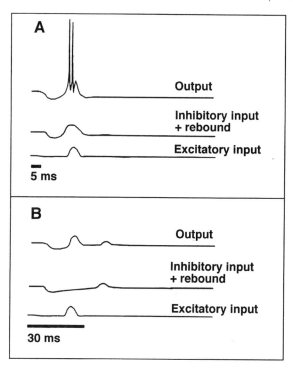

Fig. 17.2. Responses of a neuron to sounds of different durations recorded using the in vivo whole-cell patch clamp recording technique (Covey, Kauer, and Casseday 1996). Each trace represents one response of the neuron to the duration indicated by the stimulus bar and text at the upper left of the trace; recordings were performed in voltage clamp mode so that an upward deflection represents outward current (indicative of inhibitory input) and a downward reflection represents inward current (indicative of excitatory input). Because IC neurons have extensive dendritic trees, voltage clamp was not always complete, hence the action potentials fired during large, rapid inward currents (modified from Covey, Kauer, and Casseday 1996).

Fig. 17.3. A model illustrating how duration tuning is created by the interplay between excitation and inhibition. *A:* response of a hypothetical neuron to a 5 ms stimulus; *B:* response of the same hypothetical neuron to a 30 ms stimulus. The heavy bar at the lower left of each panel represents the stimulus duration. The upper trace in each panel represents the membrane potential of the neuron as it would appear in a standard intracellular recording. The middle trace represents a sustained, onset-evoked inhibitory input to the neuron, followed by a rebound at sound offset. The lower trace represents a transient, onset-evoked excitatory response, the latency of which is longer than that of the inhibitory input. When the duration of the sound is such that the onset excitation coincides with the rebound from inhibition, the membrane potential is depolarized to threshold and action potentials result. When the duration is such that the onset excitation and the rebound from inhibition occur at different times, threshold is not reached, and no action potentials are produced (modified from Covey and Casseday 1999).

sound. At durations longer than 20 ms, action potentials were no longer produced, the excitatory event that immediately followed the initial inhibition was very small, and the excitatory event correlated with sound offset became progressively smaller as duration increased, suggesting that it represented a rebound from inhibition that was evoked by sound onset and gradually decayed.

From these results, it appears that IC neurons with band-pass duration tuning acquire this property through interaction of several different excitatory and inhibitory inputs, offset in time from one another as illustrated in the model shown in fig. 17.3. In this model, the IC neuron receives transient excitatory input at a fixed latency following the sound onset. This input is rendered sub-

threshold by a sustained, onset-evoked inhibitory input, the beginning of which slightly precedes the onset-evoked excitation. At the end of the inhibitory period, there is another period of depolarization, probably due to a rebound from sustained, slowly decaying inhibition. If the duration of a sound is such that the offset-evoked rebound coincides with the onset-evoked subthreshold excitation, the cell reaches threshold and produces one or more action potentials. Because different IC neurons receive onset-evoked excitatory input with different latencies, different neurons respond maximally to different durations, resulting in a range of "best durations" across the population. This range of "best durations" corresponds closely to the range of biosonar call durations emitted by the bat during a typical foraging sequence (Simmons 1989; Ehrlich, Casseday, and Covey 1997). Because the duration of the bat's calls varies systematically during a foraging sequence, a consequence of duration tuning is that different populations of IC neurons are active during different stages of foraging. This finding raises the hypothesis that the different duration-specific neural populations have different patterns of output, each adapted for the behavioral patterns unique to a specific stage of foraging.

AMPLITUDE TUNING

At the level of the auditory nerve of all mammals, fibers respond to sounds of increasing amplitudes with progressive increases in firing rate up to some point at which the response saturates and no further increases in firing rate occur. At the level of the cochlear nucleus and lower brainstem, increasing numbers of neurons exhibit a decreased response at high sound levels—that is, they have nonmonotonic rate-level functions. By the level of the IC, not only do many neurons have nonmonotonic rate-level functions, quite a few have upper thresholds at which their response is completely suppressed. Neurons with upper thresholds are tuned to a specific range of amplitudes. Because different neurons have different ranges, there is selective filtering for sound amplitude across the population (Covey and Casseday 1991; Covey 1993; Haplea, Covey and Casseday 1994; Casseday and Covey 1992). Fig. 17.4 shows examples of IC neurons tuned to amplitude. Each neuron responds only to some low or intermediate range of sound levels and ceases responding at levels to which auditory nerve fibers are still within their dynamic range or have saturated at some maximal firing rate. For an echolocating bat, amplitude-tuned neurons could be important for responding selectively to relatively faint echoes, while filtering out the bat's high-intensity vocalizations.

Fig. 17.5 shows examples of the effect of blocking inhibitory neurotransmitters on two IC neurons with closed frequency response areas. Both bicuculline and strychnine eliminated these neurons' upper thresholds and broadened their response areas, indicating that amplitude tuning was the direct result of synaptic inhibition at the IC neuron. Because both drugs were effective in eliminating upper thresholds and broadening the neurons' response areas, we conclude that both GABAergic and glycinergic inhibition are involved in shaping amplitude selectivity in IC neurons. Upper thresholds were eliminated by either bicuculline or strychnine in 86% of the amplitude-tuned IC neurons tested with blockers of inhibitory neurotransmitters (B. R. Johnson 1993). These results, together with findings from other laboratories (e.g., Fuzessery and Hall 1996), suggest that both GABAergic and glycinergic inhibition are involved in suppressing IC neurons' responses to high sound pressure levels.

The role of neural inhibition in creating amplitude

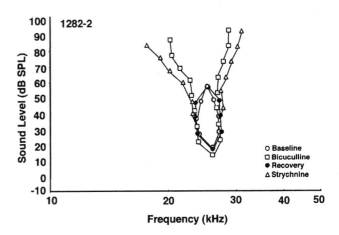

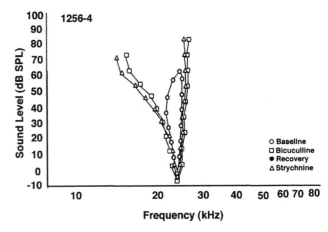

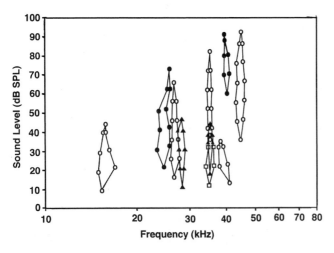

Fig. 17.4. Response areas of 10 amplitude-tuned neurons in the IC. The area inside each curve represents the combinations of frequency and amplitude that evoke a response (modified from Casseday and Covey 1992).

Fig. 17.5. Response areas of two amplitude-tuned IC neurons under baseline conditions (open circles), during iontophoretic application of bicuculline (open squares), during iontophoretic application of strychnine (open triangles), and after recovery from the effects of drugs (solid circles) (data from B. R. Johnson 1993).

tuning of IC neurons is illustrated further by intracellular recordings, such as the example shown in fig. 17.6. In this case, the IC neuron appeared to receive mainly excitatory input at low sound levels, although there was an initial inhibitory input that preceded the train of action potentials. As sound level increased, the early inhibitory input became larger, and at high levels only a net inhibitory component remained.

Typically, for IC neurons with upper thresholds or nonmonotonic rate-level functions, the progressive suppression of spike counts seen at higher sound levels was accompanied by a progressive increase in response latency. As a result, these neurons exhibited what is termed "paradoxical latency shift" (Sullivan 1982b). This term was coined because most neurons in the auditory system, starting at the auditory nerve, exhibit a progressive *decrease* in latency as sound level increases. Sullivan suggested that those neurons in the bat's auditory cortex that exhibit paradoxical latency shift could provide input to another class of cortical neurons tuned to the time interval between the bat's high-intensity emitted pulse and the low-intensity echo. These latter cells are termed delay-tuned neurons (O'Neill and Suga 1982). They are found in the auditory cortex, medial geniculate body (e.g., Olsen and Suga 1991), and inferior colliculus (e.g., bats [Mittmann and Wenstrup 1995]). If the latency difference between the response to the emitted pulse (carried by one input pathway) and the response to the echo (carried by a second input pathway) were of a magnitude that compensated for the pulse-echo delay, then coincidence of two excitatory inputs would occur, facilitating the response of the delay-tuned neuron.

RELATION OF DELAY TUNING TO EARLY INHIBITION AND PARADOXICAL LATENCY SHIFT

Several lines of evidence indicate that the IC of echolocating bats contains a systematic map of latencies, and that latency is lengthened through neural inhibition (B. R. Johnson 1993; Park and Pollak 1993; Saitoh and Suga 1995). This inhibitory input would have to arrive prior to, or simultaneously with, an onset-evoked excitatory input. In *Eptesicus,* when GABAergic inhibition was blocked by application of bicuculline, latency decreased in 84% of the neurons tested; blocking glycinergic input with strychnine caused latency to decrease in 86% of the IC neurons tested (B. R. Johnson 1993). Fig. 17.7 shows a neuron whose latency decreased after glycinergic inhibition was blocked, but not after GABAergic inhibition was blocked.

Another line of evidence indicating that the latency of IC neurons is lengthened through neural inhibition, and that the amount of lengthening is level dependent, comes from intracellular recordings of neural responses to sound at different amplitudes. Fig. 17.8 shows how the increasing size and duration of the initial inhibitory input to an IC neuron progressively truncates the early part of the response to create a paradoxical latency shift.

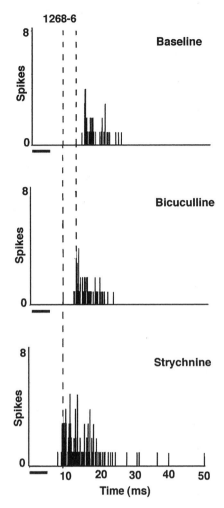

Fig. 17.6. Poststimulus time histograms showing the responses of an IC neuron under baseline conditions (top), during iontophoretic application of bicuculline (middle), and during iontophoretic application of strychnine (bottom). The leftmost dashed line indicates the response latency during application of strychnine; the rightmost dashed line indicates the response latency during application of bicuculline (data from B. R. Johnson 1993).

At sound levels near threshold, the only input to the neuron appeared to be an excitatory one that evoked a train of action potentials. At higher sound levels, an initial inhibitory input appeared. This inhibition truncated the early part of the response, increasing the latency to the first action potential by >10 ms. Further increases in sound level caused further increases in latency.

Hypothetically, if neurons with paradoxical latency shift projected onto a target neuron that functioned as a coincidence detector, then the target neuron would respond best to a pair of sounds, one at high intensity (such as the bat's vocalization) and a second at low intensity (such as the attenuated echo) separated by a delay corresponding to the amount of paradoxical latency shift. Moreover, because the amount of paradoxical latency shift is a function of sound amplitude, the target neuron would respond best to short delays at low sound

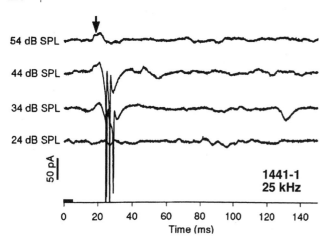

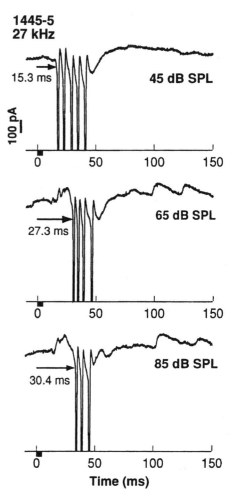

Fig. 17.7. Responses of a neuron with an upper threshold to sounds of different amplitudes recorded using the in vivo whole-cell patch clamp recording technique (Covey, Kauer, and Casseday 1996). Each trace represents one response of the neuron to a 5 ms pure-tone stimulus as indicated by the stimulus bar above the *x*-axis. Recordings were performed in voltage clamp mode so that an upward deflection represents outward current (indicative of inhibitory input) and a downward reflection represents inward current (indicative of excitatory input). Because IC neurons have extensive dendritic trees, voltage clamp was not always complete, hence the action potentials fired during large, rapid inward currents. The arrow indicates the inhibition seen at the neuron's upper threshold (modified from Covey, Kauer, and Casseday 1996).

Fig. 17.8. Responses of a neuron with paradoxical latency shift to sounds of different amplitudes recorded using the in vivo whole-cell patch clamp recording technique (Covey, Kauer, and Casseday 1996). Each trace represents one response of the neuron to a 5 ms pure-tone stimulus as indicated by the stimulus bar just above the *x*-axis. Recordings were performed in voltage clamp mode so that an upward deflection represents outward current (indicative of inhibitory input) and a downward reflection represents inward current (indicative of excitatory input). Because IC neurons have extensive dendritic trees, voltage clamp was not always complete; hence the action potentials fired during large, rapid inward currents. Arrows and numbers at the left of each trace indicate the latency of the first spike evoked by each amplitude (modified from Covey, Kauer, and Casseday 1996).

levels and long delays at high sound levels. This relationship is exactly that seen in the "amplitude shift" neurons found in the cortex of *Myotis* (Sullivan 1982a), *Pteronotus* (Taniguchi, Wong, and Suga 1986), and *Eptesicus* (Dear et al. 1993).

Although delay-tuned neurons in bats already are present at the level of the inferior colliculus, it seems likely that delay-tuning properties are further sharpened and adjusted at the thalamic and cortical levels. Furthermore, there is evidence that delay tuning in midbrain neurons is adjusted by descending cortical projections (Yan and Suga 1996a, 1996b). Nevertheless, given that paradoxical latency shift seems to be first created in the IC, it seems likely that neurons with this property represent the input stage to circuitry that includes coincidence detectors that receive the output of amplitude-tuned neurons with appropriate response delays.

Discussion

In echolocating bats, the responses of nearly all IC neurons are affected in some way by removal of neural inhibition (e.g., B. R. Johnson 1993; Park and Pollak 1993; Pollak and Park 1993; Casseday, Ehrlich, and Covey 1994; Fuzessery and Hall 1996). Both GABAergic and glycinergic inhibition interact with glutamate-mediated excitation at individual IC neurons to create emergent

properties, including duration tuning, amplitude tuning, paradoxical latency shift, and pulse-echo delay tuning. The end effect of this process is not only selectivity of neurons for biologically important sound patterns, but a fundamental change in the dynamics of neural responses.

Given the massive amount of convergence that occurs at the IC, one might suppose that the result of multiple inputs to a single neuron would be an amplification of the original signal and a sharpening of synchronization to the temporal pattern of the stimulus. Instead, the opposite generally seems to occur. Whereas the majority of neurons in the lower brainstem continue to fire action potentials throughout the duration of a stimulus, most IC neurons respond with only one or a few action potentials correlated with the sound's onset or offset, and then only if other conditions are met (e.g., the proper duration and/or amplitude). Thus, the neural circuitry that creates selectivity to biologically relevant patterns of sound severely limits the amount of activity that is transmitted from the IC to subsequent levels of the central auditory system. Another related consequence of the selection process that occurs at the IC is that at least some of the real-time patterns that arrive at the IC from lower levels are converted to spatial representations of "computed" parameters such as duration and interval between sounds. The conversion of a temporal representation to a spatial representation means that there is no longer a need for neural responses to be tightly synchronized with the temporal pattern of the stimulus. As a consequence, the neural circuitry can maintain a trace of a temporal pattern long after the actual pattern in real time is gone.

One way in which the trace of a temporal feature such as sound duration can be maintained is through activation of a population of neurons with a wide range of latencies. At the IC, the range of latencies increases manyfold, at least partly through the action of neural inhibition as discussed earlier. The result is that for an ongoing sequence of sounds—such as would occur during the course of a natural hearing task—different neurons respond to different elements in the sequence at the same time. Conversely, individual neurons within the population will time-shift the representation of the sequence to a greater or lesser extent depending on their response latencies. Such a system is ideal for performing computational operations that resemble autocorrelations or cross-correlations.

Finally, because the integrative process takes time, the selectivity of IC neurons for simple patterns or features of sound is accompanied by a slowing of the rate of neural output to subsequent stages. Casseday and Covey (1996) suggested this process helps match the rate of responses to sounds with the rate at which motor systems can generate behavior.

Acknowledgments

Our research was supported by NIH grants DC-00607 and DC-00287, and NSF grants IBN-9210299 and IBN-9511362.

{ 18 }

Sensitivity to Interaural Time Differences in the Sound Envelope in the Inferior Colliculus and Auditory Cortex of the Pallid Bat

Zoltan M. Fuzessery and *Thomas D. Lohuis*

Introduction

The highly specialized auditory systems of bats provide excellent models for elucidating the mechanisms that underlie the processing of complex sounds. Quite naturally, the computational processes that extract biosonar information have received the greatest attention. However, bats are the most behaviorally diverse group of mammals, and their auditory specializations extend beyond biosonar. The surface gleaning bats are a particularly interesting group that take prey from substrates (Neuweiler 1984; Fenton 1990). The sensory information used to detect and locate prey varies among the gleaners (Feilder 1979; Tuttle and Ryan 1981; Bell 1982; Bell and Fenton 1986; Marimuthu and Neuweiler 1987; Ryan and Tuttle 1987; Fuzessery et al. 1993). Perhaps all use passive sound localization, eavesdropping on prey-generated sounds, but some also employ echolocation and vision. All face a similar problem when using passive sound localization. Prey-generated sounds typically have peak spectral energy in the lower end of a bat's audible range. Ear directionality is broadest at the low end of their audible ranges (Guppy and Coles 1988; Coles et al. 1989; Fuzessery 1996). Consequently interaural intensity differences (IIDs), one important binaural cue, will be low and perhaps provide poor spatial information. Larger mammals (e.g., humans) solve this problem by using interaural phase differences (IPDs) at low frequencies, and IIDs at higher frequencies, a strategy formulated as the duplex theory of sound localization. Because the low end of a bat's audible range is typically a relatively high frequency (>1 kHz), IPDs are not available to probably most species. Therefore, if bats use interaural time differences (ITDs), they must obtain them from arrival-time differences in the signal envelope, not the carrier. In other mammalian groups, behavioral and physiological studies have demonstrated that at higher frequencies where phase information is not available, both IIDs and envelope ITDs influence spatial perception (Henning 1974; McFadden and Pasanen 1976; Nuetzel and Hafter 1976; Trahoitis and Bernstein 1986), but IIDs are thought to dominate perception.

This chapter describes neuronal sensitivity to envelope ITDs at two levels of the pallid bat's auditory system, the inferior colliculus (IC) and auditory cortex (AC). The pallid bat relies heavily on passive sound localization to find prey (Bell 1982; Fuzessery et al. 1993). It attends selectively to noise transients produced by prey locomotion. It is capable of accurate sound localization, perhaps 1° angular resolution, using single, short noise transients (10–20 ms) with spectral energy peaks around 5 kHz, the lower end of its audible range. The IIDs available at these frequencies are poor (Fuzessery 1996); the pallid bat may use envelope ITDs to acquire accurate spatial information.

The ITD range that the pallid bat normally experiences is only ±70 μs. If ITDs were to be used, its auditory system must have an unusually acute sensitivity to envelope ITDs. To test this, neurons in the IC and AC tuned to low frequencies and selective for noise transients used in prey localization were tested for ITD sensitivity. Results show that these neurons exhibit the sharpest envelope ITD sensitivity reported in a mammal. This sensitivity doubles from the level of the IC to the AC.

Materials and Methods

Bats were initially anesthetized with methoxyflurane (Metofane) inhalation; they were then given an intraperitoneal injection of sodium pentobarbital (Nembutal) (30 μg/g body weight) and Acepromazine (2 μg/g). To expose the IC and AC for recording, an incision was made on the midline of the scalp, and the musculature over the skull was reflected. To hold the head stable during recording, an aluminum headpin was attached to the skull using a layer of glass microbeads and dental cement. Using skull and brain surface landmarks, a small hole (0.5 mm diameter) was made in the skull over the auditory cortex with a microscalpel, and the craniotomy was filled with paraffin oil to prevent desiccation of the cortical surface. The exposed muscle tissue was covered with petroleum jelly to prevent desiccation. Bats were lightly anesthetized during recording sessions.

Recording Procedures

Recording sessions were conducted in a room lined with anechoic foam and maintained at temperatures of 35–40°C. The bat was secured in a Lucite holder and the headpin was fixed in a restraining bar. Electrodes were advanced radially. Electrode depth was remotely controlled with a David Kopf Inc. model 650 hydraulic microdrive. Single neuron activity was recorded with glass micropipette electrodes filled with 1M NaCl, with tip resistance of 3–8 M.

Stimulus Presentation and Data Collection

Stimuli were generated and neuronal responses were counted by a custom computer-controlled system (Fuzessery, Gumtow, and Lane 1991). Stimuli were delivered as closed-field dichotic sounds through Infinity Emit-K ribbon tweeters fitted with cotton-filled funnels. The tips of the funnels were inserted into the large pinnae of the bat, and the pinnae were sealed around the funnel tips with petroleum jelly to prevent acoustic crosstalk.

The choice of the stimulus used to test ITD and IID sensitivity was based on the types of sounds the pallid bat uses to locate prey. It attends to short 5–20 ms noise transients produced by a walking insect (Bell 1982; Fuzessery et al. 1993). It is capable of accurate, open-loop passive sound localization using only a single 10 ms noise transient (Fuzessery et al. 1993). IC neurons were found to respond best to short stimuli (1–5 ms) with infinitely short rise-fall times, and they were able to time-lock to trains of square-wave amplitude-modulated (AM) signals up to rates of 1400 Hz. Their responses to sinusoidally modulated signals were poor. IC neurons typically express their greatest ITD sensitivity when presented with 20–30 ms trains of square-wave AM tones or noise with modulation rates of 100–400 Hz, a modulation depth of 100%, and short AM duty cycles of 10–30%. These fast trains of AM pulses maximize the number of events/unit time that can be binaurally compared. These results parallel human psychoacoustic studies reporting that spatial resolution increases with the number of events/unit time (Hafter and Dye 1983).

While the cortical neurons recorded in the present study did not time-lock to AM sounds, we assumed that the IC neurons that indirectly provide their input would express maximal ITD sensitivity in response to trains of AM pulses and convey this information to the cortex. Therefore, the great majority of cortical neurons were tested with 200–400 Hz square-wave AM signals of 20–30 ms duration, 10–30% duty cycles, and 100% modulation depth. The carrier was a tone at the neuron's best frequency.

Binaural sensitivity was tested with static IIDs and ITDs. Sounds were presented at repetition rates of 400 ms in the IC and 1000–1200 ms in the AC. Response magnitudes were expressed as the total number of impulses obtained over 30 stimulus presentations. Binaural interactions were quantified by comparing thresholds and response magnitudes elicited by stimuli presented at the contralateral ear alone, the ipsilateral ear alone, and to both ears simultaneously. A neuron was said to be ITD sensitive if its firing rate changed by >50% in response to a ± 1000 μs delay at either ear. This large ITD range was used as a criterion for ITD sensitivity to prevent biasing our estimates of ITD sensitivity to only the most sensitive neurons, and to allow a better comparison with similar studies in other species. ITDs were varied in 50–100 μs steps while holding the IID constant. Then, IIDs were tested at an ITD of 0 μs and were created by holding the contralateral intensity level constant while varying the ipsilateral intensity level. IID sensitivity was tested at several contralateral intensity levels, 10–30 dB above the response threshold. Time-intensity trading ratios were measured by testing ITD sensitivity over a range of fixed IIDs, and by testing IID sensitivity for a range of fixed ITDs.

Results

ITD SENSITIVITY IN THE INFERIOR COLLICULUS AND AUDITORY CORTEX

Inferior colliculus

The pallid bat IC has been functionally divided into three regions: lateral IC, dorsal IC, and ventral IC (Fuzessery 1994; Fuzessery and Wenstrup 1995; Fuzessery and Hall 1996). The ventral IC represents biosonar frequencies (30–80 kHz) and contains an unusually large percentage of neurons that respond selectively to the downward frequency sweep of the biosonar pulse. The lateral IC appears to serve passive sound localization. Of 52 neurons tested, 75% had similar response thresholds for tones and noise presented at the same total sound pressure levels, and 23% either had lower thresholds for noise or responded exclusively to noise transients used in sound localization. In contrast, of 72 neurons recorded in the dorsal IC, also tuned to lower frequencies, 87% did not respond to noise. The search for evidence of a sensitivity to the available envelope ITD range of ± 70 μs therefore focused on the lateral IC.

In 27 lateral IC neurons tested, four types of ITD function configurations were seen (fig. 18.1). Cyclical neu-

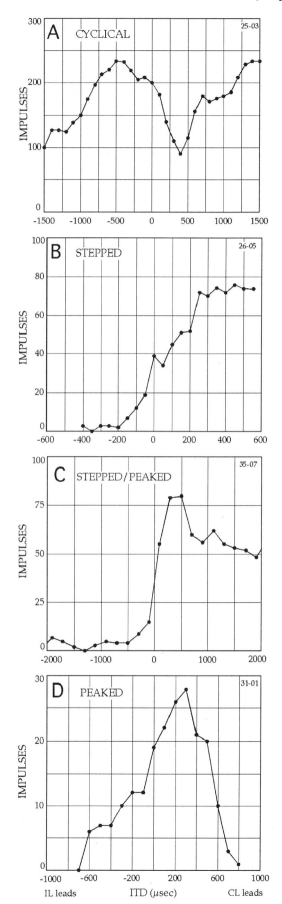

Fig. 18.1. Examples of the four basic types of ITD functions observed in the lateral IC of the pallid bat. *A:* Cyclical neurons exhibited either a peak or trough over ± 1500 μs. *B:* Neurons with stepped functions responded at maximum plateau value at positive ITDs, but they were totally inhibited as ITDs favored the inhibitory ear and remained inhibited up to -1500 μs. *C:* Neurons with stepped/peaked functions fired at a submaximal level at large positive ITDs, but they increased their firing rate prior to inhibition at ITDs favoring the inhibitory ear. *D:* One neuron exhibited a peaked function, showing it was totally inhibited at both positive and negative ITDs.

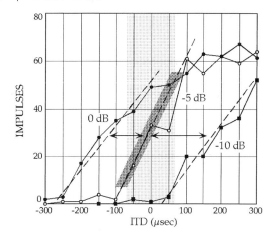

Fig. 18.2. An illustration of the calculation of dynamic ITD range and time-intensity trading ratios in an IC neuron. The three ITD functions were measured at three fixed IID values (0, −5, and −10 dB) created with the CL intensity fixed at 40 dB. The shaded area indicates the relevant ITD range of ±70 μs, the dark shaded band shows the dynamic ITD range from 10% to 90% of maximum response, and the arrows indicate where time/intensity trading ratios were measured after smoothing (dashed lines) the dynamic ITD ranges.

rons (6/27) had one response peak and trough over an ITD range of ±1500 μs. The peak could occur at ITDs favoring either the contralateral or ipsilateral ear. Neurons with stepped functions (10/27) all responded maximally to ITDs favoring the contralateral ear (positive values), and they were inhibited in ITDs favoring the ipsilateral ear (negative values). Neurons with step-peaked functions (10/27) gave a submaximal response (75% to 25%) to large positive ITDs; the response rose near 0 μs ITD, and then dropped at negative ITDs. Only one peaked ITD function was observed. The responses of neurons with peaked ITD functions dropped >25% of maximum response at those ITD values more positive and negative to their preferred ITD range.

Two measures of ITD sensitivity were obtained; the dynamic ITD range and the time-intensity trading ratio. The dynamic range is the range of ITDs over which the response drops 80%, from 90% maximum response to 10% maximum response. The dynamic range was measured on an ITD function that passed through the relevant ITD range. For example, the neuron in fig. 18.2 had a maximum response of 61 spikes when its ITD function was measured at a fixed IID of −5 dB. At 90% and 10% of maximum response, the neuron responded with 55 and 6 spikes respectively. These responses occurred at +75 and −75 μs, respectively. The dynamic range was therefore 150 μs. Overall, from 0 to 100% maximum response, the neuron's response changed 74% over the relevant ±70 μs (140 μs) ITD range. For neurons with noncyclic ITD functions, dynamic ranges extended over 125–625 μs, with an average of 304 μs. There was an average change in maximum response of 37% over the relevant ITD range. Neurons with cyclic ITD functions

were less sensitive. Their dynamic ranges spanned 250–575 μs, with an average of 483 μs. The average change in maximum response was only 23% over the relevant ITD range.

Binaural time-intensity trading ratios derive from human psychoacoustic studies, but they have been applied to neuronal responses as well. The trading ratio measures the extent to which a change in IIDs shifts a neuron's ITD function along the ITD axis. The smaller the value, the more a neuron can respond to ITDs independently of the IID. Fig. 18.2 shows a typical example and illustrates the measure. ITD functions were obtained at three different fixed IIDs. The slopes of the functions shift toward positive ITDs as the IIDs increasingly favor the ipsilateral ear. In other words, a shift in IID to favor the ipsilateral ear can be offset by a shift in ITD toward favoring the contralateral ear, hence the term time-intensity trading. The trading ratio was measured at 50% maximum response between two ITD functions with dynamic ranges centered near the relevant ITD range, in this case between the slopes of ITD functions obtained at IIDs of 0 and −5 dB. The shift in the slope was 100 μs (from −25 to −125 μs). The trading ratio was 100 μs/5 dB, or 20 μs/dB. Within the population, the trading ratios of neurons with noncyclic ITD functions ranged from 7.5 to 30 μs/dB, with an average of 17.9 μs/dB.

Auditory cortex

Data were obtained from 44 binaurally sensitive AC neurons with best frequencies ranging from 6 to 64 kHz. All of these neurons were sensitive to IIDs, and 42 (95%) were also sensitive to ITDs. The same four ITD function configurations observed in the IC (fig. 18.1) were also seen in the AC. Cyclic neurons constituted 10%, stepped neurons 48%, step-peaked neurons 17%, and peaked neurons 26%. The main difference between IC and AC was the increase in the neurons of peaked ITD functions.

The dynamic ITD ranges of AC neurons ranged from 80 to 370 μs, with an average of 175 μs. There was little difference of the average dynamic ranges of neurons with cyclic and noncyclic ITD functions (183 versus 174 μs). Thus there were significant differences in dynamic ITD ranges that occur from the midbrain to cortical level. Average ITD sensitivity approximately doubled in neurons with noncyclic ITD functions, in that dynamic ITD ranges in AC (175 μs) were about half that of IC (304 μs). IC neurons averaged a 38% change in maximum response over the relevant ±70 μs ITD range, while AC neurons averaged a 67% change. The change in sensitivity in neurons with cyclic ITD functions was greater; IC neurons had an average dynamic range of 483 μs, compared to 183 μs in the AC.

In contrast to changes in dynamic ITD ranges, the average time-intensity trading ratios of neurons with noncyclic ITD functions were remarkably similar. The values were 17.9 μs/dB in the IC and 16.7 μs/dB in the AC. This comparison indicates that dynamic ITD ranges and trading ratios are not necessarily linked. ITD sensitivity can sharpen dramatically without a change in trading ratios.

UNDERLYING BINAURAL MECHANISMS

In this section, we examine the binaural mechanisms that create ITD sensitivity and suggest that ITD sensitivity derives from the same binaural mechanisms that create IID sensitivity. We show that binaural interactions, whether inhibitory or facilitatory, are typically the same for ITDs and IIDs, and that ITD and IID functions are typically similar in shape, suggesting that these interaural disparities may have similar effects on the same binaural comparator mechanism.

Binaural inhibition and facilitation are defined relative to the response level evoked by monaural stimulation. In the IC, all recorded neurons exhibited only binaural inhibition. ITDs could reduce their response relative to monaural contralateral stimulation, but not increase it. At the cortical level, neurons could exhibit binaural inhibition alone (figs. 18.3A, 18.4A), both inhibition and facilitation (figs. 18.3B, 18.4C, E, G), or facilitation alone, where monaural stimulation evoked no response (fig. 18.3C). Almost half of the neurons (43%, 18/42) recorded in the AC required binaural stimulation to respond maximally. Five neurons did not respond at all to monaural stimulation. There was little relationship between the ITD function configurations of AC neurons and the binaural interactions they expressed (figs. 18.3 and 18.4), except in the case of those with peaked ITD functions. Four of the five neurons that expressed only binaural facilitation had peaked functions. One explanation for this change in binaural interactions from the IC to AC is that monaural stimulation has a stronger excitatory drive in the IC, and that this is diminished at the cortical level. More cortical neurons require binaural stimulation to respond maximally.

A second feature is the similarity in the shapes of IID and ITD functions in 27 of 38 (71%) cortical neurons examined (figs. 18.4A–B and 18.4C–D). In most cases where they differed, neurons would express a stepped function for one binaural cue, and a stepped-peaked or peaked function for the other (fig. 18.4E–F). The exception was neurons with cyclic ITD functions (fig. 18.4G–H). No neurons were observed in either the IC or AC that had cyclic IID functions.

Another similarity in the effects of ITDs and IIDs was the binaural inhibition and facilitation they evoked. Of 38 neurons tested, 37 showed similar binaural interactions. The neuron in figure 18.4A–B exhibited binaural inhibition in response to both ITDs and IIDs, while the neurons in figs. 18.4C–D and 18.4E–F exhibited binaural facilitation and inhibition, with monaural stimulation evoking submaximal responses. An exception is shown in fig. 18.4G–H, where ITDs evoked binaural fa-

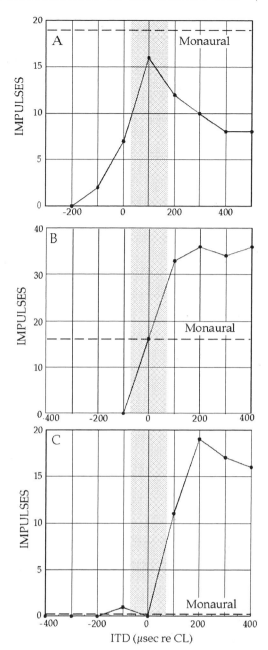

Fig. 18.3. The range of binaural interactions observed in the auditory cortex in response to changing ITDs, showing the response to monaural contralateral stimulation (dashed line, monaural) and binaural stimulation (solid line). The intensity level of the monaural stimulus and the contralateral intensity level of the IID at which the ITD was tested were the same. Gray rectangles indicate the relevant ITD range of ±70 μs.

cilitation and inhibition, while IIDs evoked only binaural inhibition.

Discussion

Sensitivity to envelope ITDs in the pallid bat IC is the sharpest reported in any mammal. The fact that this sensitivity approximately doubles at the cortical level

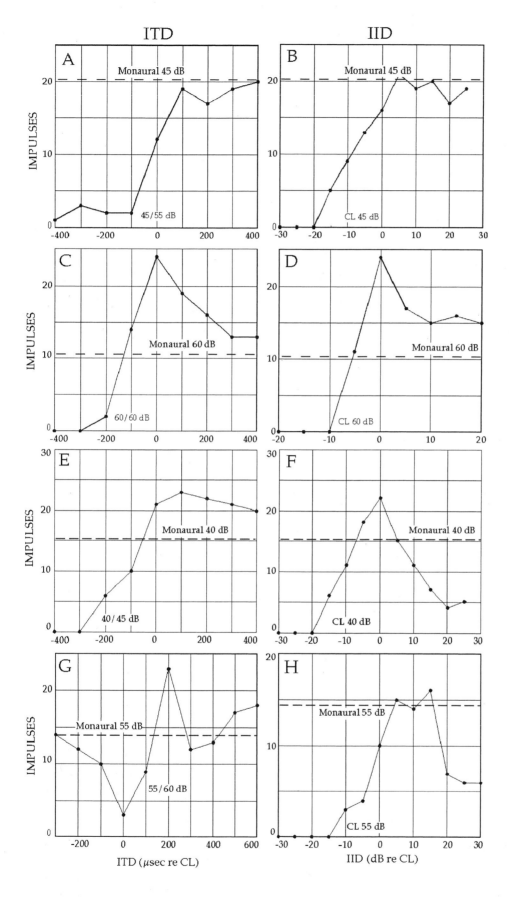

Fig. 18.4. Comparisons of the shapes of the ITD and IID functions of four neurons, and the relative responses to monaural (dashed lines) and binaural (solid lines) stimulation. All responses were obtained with the contralateral intensity fixed at the same intensity level. For example, in *A* and *B*, the monaural response was obtained at 45 dB (monaural 45 dB); the ITD function was obtained with a fixed IID created with contralateral 45 dB, ipsilateral 55 dB (45/55 dB); and the IID function was obtained with the contralateral intensity level fixed at 45 dB (CL 45 dB). Dashed lines indicate the level of response to contralateral monaural stimulation, and solid lines indicate the responses to ITDs or IIDs.

suggests that this interaural disparity may provide the pallid bat with spatial information. This sharpening of envelope ITD sensitivity through the ascending auditory system may be functionally analogous to the sharpening of low-frequency IPD sensitivity reported in the rabbit (Stanford, Kuwada, and Batra 1992; Fitzpatrick et al. 1997). The only other mammal in which a similar comparison is available is the white rat (Kidd and Kelly 1996; Kelly and Phillips 1991). No sharpening of sensitivity was observed; the average dynamic ITD range in the rat IC was 570 μs, and 590 μs in the AC. This translates into approximately a 30% change in maximum response over the rat's relevant ITD range of ± 130 μs.

Differences in envelope ITD sensitivity could reflect the behavioral needs of the species. ITD sensitivity in other bat species is much weaker. In the Mexican free-tailed bat, neurons in the lateral superior olive (LSO) and IC exhibit ITD functions that are similar to those reported here, but their sensitivity is not sufficient to allow them to be affected by ITDs within the small ± 30 μs ITD range available to this species (Pollak 1988; Grothe and Park 1995; Park et al. 1996). A similar result was obtained in another bat, *Molossus ater* (Harnischfeger, Neuweiler, and Schlegel 1985). This difference may derive from the fact that these other bat species obtain prey through echolocation at high frequencies. IIDs alone could provide sufficient spatial information. The need for accurate spatial information at the low end of its audible range, where IIDs are small, may account for the appearance of ITD sensitivity in the pallid bat.

In contrast to the change in dynamic ITD range from the IC to the AC in the pallid bat, there was essentially no change in the average time-intensity trading ratio. In the Mexican free-tailed bat, there was also little change in trading ratios from the LSO to the IC (Pollak 1988; Grothe and Park 1995). This finding suggests that the trading ratio is established in the lower auditory brainstem and reflected at higher levels of the system without modification. Present results suggest that this stability may persist all the way to the cortical level.

The average time-intensity trading ratios in the pallid bat (17–18 ms/dB) are the lowest reported for envelope ITD-sensitive neurons. Trading ratio values average about 30 μs/dB in the rat. In the domestic cat, they were estimated to be 85 μs/dB in the superior colliculus (Yin, Hirsch, and Chan 1985) and 120 μs/dB in the IC (Caird and Klinke 1987). In the bat *M. ater,* the average value is approximately 30 μs/dB (Harnischfeger, Neuweiler, and Schlegel 1985), and 41–47 μs/dB in the Mexican free-tailed bat (Pollak 1988; Grothe and Park 1995). But while the average 17–18 μs/dB trading ratios in the pallid bat are low, they are still considerably higher than those reported in low-frequency, IPD-sensitivity neurons—which, in the cat IC and cortex, average around 5 μs/dB and range from 0 to 13 μs/dB (Yin and Kuwada 1983; Yin, Chan, and Irvine 1986; Reale and Brugge

1990). The functional significance is that these IPD-sensitive neurons can encode ITDs more independently of IIDs than can the envelope ITD-sensitive neurons of the pallid bat.

UNDERLYING MECHANISMS

IIDs and envelope ITDs in our study had remarkably similar effects on the responses of the majority of cortical neurons in the pallid bat, both in terms of the configurations of binaural functions, and the expressions of binaural facilitation and inhibition. Moreover, the transformation in responses to envelope ITDs that occurred from midbrain to cortex was very similar to changes in IID sensitivity reported in all species studied (e.g., Benson and Teas 1976; Phillips and Irvine 1981; Reale and Kettner 1986; Kelly and Sally 1988; Irvine and Gago 1990; Semple and Kitzes 1993a, 1993b; Irvine, Rajan, and Aitkin 1996). These changes in responses to IIDs included an increase in the number of neurons with mixed facilitatory and inhibitory binaural interactions, and an increase in the number of neurons with peaked IID functions. In the pallid bat IC, less than 5% of the neurons recorded had peaked ITD functions. In the auditory cortex, 26% had peaked ITD functions.

The similarities in the effects of IIDs and envelope ITD on pallid bat cortical neurons are most likely the result of the two interaural disparities acting upon a single binaural processing mechanism. It has been often postulated that the encoding of IIDs in the high-frequency pathway is, in part, a time-domain process in which the relative arrival times of inputs from the two ears influence the output of the binaural comparator (Jeffress 1948; Kitzes, Wrege, and Cassady 1980; Harnischfeger, Neuweiler, and Schlegel 1985; Yin, Hirsch, and Chan 1985; Irvine 1986; Pollak 1988; Joris and Yin 1995, 1996; Park et al. 1996). At least two factors can influence these arrival times. The first is the ITD range normally experienced by a species. The second derives from changes in intensity levels at the two ears. The response latencies of neurons typically decrease with increasing intensity level. Consequently, IIDs can be encoded through what has been termed an "interaural latency difference" (Kitzes, Wrege, and Cassady 1980; Pollak 1998; Kelly and Phillips 1991; Joris and Yin 1996). The functional significance of ITDs in sound localization depends upon the extent to which the binaural comparator is sensitive to available ITDs.

FUNCTIONAL RELEVANCE

Does the pallid bat use envelope ITDs to acquire spatial information? In humans—the only mammal in which this question has been addressed behaviorally—envelope ITDs do influence high-frequency spatial perception, but the consensus is that IIDs dominate perception (see Hafter 1984). Extrapolating these results to the pallid bat is, however, not entirely appropriate. Humans use IPDs to acquire spatial information at low frequencies, and they have adequate IIDs to do so at high frequencies. Pallid bats have only envelope ITDs and IIDs available at low frequencies. Functional conclusions from physiological data are mixed. In the domestic rabbit, neuronal sensitivity to envelope ITDs was concluded to be adequate for use in sound localization (Batra, Kuwada, and Stanford 1993). It is considered marginal in the cat (Yin, Hirsch, and Chan 1985; Caird and Klinke 1987; Joris and Yin 1995) and rat (Kelly and Phillips 1991) and useless in the Mexican free-tailed bat (Pollak 1988; Park et al. 1996). In the pallid bat, neurons are sufficiently sensitive to be influenced by either interaural disparity, and both ITDs and IIDs probably act upon the same binaural comparator system to provide spatial information. Moreover, a sound emanating from a given point in space will create a unique ITD and IID combination. If neurons are tuned to these combinations, as is the case in the barn owl auditory system (Knudsen and Konishi 1984), then this dual sensitivity could enhance spatial processing. Whether this is indeed the case is not yet entirely clear, but we found (Fuzes-

sery 1997) that, in some neurons, covarying IIDs and ITDs to simulate sound moving along the azimuth sharpens their predicted azimuthal sensitivity. This, however, is only part of the picture that will emerge from the information that provides gleaning bats with their remarkable spatial resolution. The frequency-dependent directionality of their external ears generates a rich array of monaural and binaural spectral cues (Guppy and Coles 1988; Fuzessery 1996) that most likely enhance their spatial acuity in both the horizontal and vertical planes. Therefore, future studies must determine how the binaural comparator system processes information across frequency channels before our understanding of the mechanisms underlying passive sound localization in gleaners is complete.

Acknowledgments
We thank Khaleel Abdulzarak and Jeff Wenstrup for their comments on this chapter. Research was supported by funds from NIH grant DC00054, NSF grant IBN-9828599, and funds from the University of Wyoming.

{ 19 }
Temporal Dynamics of Amplitude-Tuning in the Inferior Colliculus of the Little Brown Bat

Alexander V. Galazyuk, Kenneth R. White, and *Albert S. Feng*

Introduction

Echolocating bats actively increase their sonar emission rate when they approach a target or when facing a perceptually difficult task—for example, dodging thin wires or detecting a target in the presence of noise or clutter (Griffin, Webster, and Michael 1960; Simmons, Howell, and Suga 1975; Simmons et al. 1996). For instance, big brown bats emit sonar signals at 5–20 pulses per second (pps) while searching for prey. When a potential prey is detected, this rate is increased to 20–50 pps (and concomitantly the duration is reduced) during the approach, and to 100–200 pps during the "feeding buzz" (Griffin 1958). The increase in emission rate or the target sampling rate is considered important for tracking the flying prey—that is, for determining the target range and azimuth that at short distance can vary markedly (Simmons et al. 1996). Prior physiological studies

showed that an increase in rate sharpens the auditory receptive fields (Wu and Jen 1996) and enhances the response selectivities to pulse-echo delays representing the target range (Wong, Maekawa, and Tanaka 1992) of central auditory neurons.

Rate of stimulation also influences other response properties of central auditory neurons. Phillips, Hall, and Hollett (1989) observed that when the rate of stimulation is increased, some neurons in the cat's auditory cortex showed a shift in the unit's best pulse amplitude and the response latency lengthened. Jen and his colleagues (Pinheiro, Wu, and Jen 1991; Moriyama et al. 1994) observed that a change in stimulus repetition rate can modify the rate-level functions (RLFs) of some inferior colliculus (IC) units, but the change was not easily predictable. Finally, several studies showed that while some IC neurons respond preferentially to sound pulses at low repetition rates, others prefer higher pulse rates

(Jen and Schlegel 1982; Pinheiro, Wu, and Jen 1991; Condon, White, and Feng 1994; Wu and Jen 1995), indicating that the response thresholds to acoustic stimuli are rate dependent. Taken together, these studies suggest that the basic response properties of central auditory neurons to a sound pulse likely depend on the rate of stimulation.

We have recently found that a population of IC neurons of little brown bats exhibit greater selectivity to amplitude of sound pulses when they are components of "natural" amplitude-modulated (AM) pulse trains than when they are presented in isolation (Galazyuk and Feng 1997a; see "Results," below). This AM pulse train consists of tone pulses whose repetition rate increases progressively from 20 to 100 pps, resembling the temporal sequence of sonar emission during a target pursuit. We hypothesized that this response hyperacuity is due to the higher pulse repetition rate in the pulse train, and not due to the dynamics of pulse rate, nor to the temporal ordering of sound pulses within the train. The goal of this study was to test this hypothesis directly by investigating the responses of IC neurons to tone pulses embedded in random-amplitude pulse trains presented at *fixed* repetition rates (from 1 to 100 pps) and comparing the above responses to the unit's responses to tone pulses in natural AM pulse trains.

Materials and Methods

Experimental methods are described in Galazyuk and Feng (1997b). Extracellular recordings were made from awake little brown bats (*Myotis lucifugus*) weighing 7–10 g. During a recording session, the animal was placed in a flexible plastic holder inside a sound-attenuating chamber whose walls were coated with 7 cm anaechoic foam. A rod glued to the bat's head was secured to a small holder. A sharpened tungsten wire made a small hole (approximately 50 μm) in the skull overlying the recording area in the IC. A glass micropipette (2–5 μm tip) filled with horseradish peroxidase (5% in 0.2 M Tris buffer) was inserted through the hole in the skull. After a recording session of 6–8 hours, the bat was returned to its holding cage. Such experiments proceeded every 2–3 days for a maximum of three weeks.

Tone pulses of 2 ms with a rise-fall time of 0.5 ms, ranging from 0 dB to 90 dB SPL in 2 dB increments, were presented in three different temporal configurations (see "Results" for rationales). First, tone pulses were presented at a rate of 1 pps—this shall be referred to as isolated tone pulses (fig. 19.1A). This is a conventional method for collecting the rate-level function (RLF) data. The second configuration used natural AM pulse trains—that is, short (625 ms) trains of tone pulses (fig. 19.1B); within the train the pulse rate increased progressively from 20 to 100 pps, resembling the temporal sequence of sonar emission during a prey pursuit.

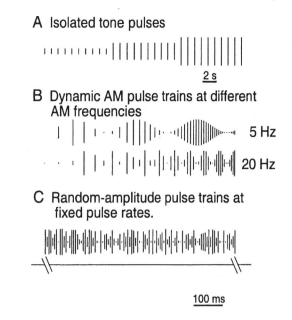

Fig. 19.1. Schematic diagram of the three types of stimuli used in the present study. *A:* Constant-amplitude tone pulses in blocks of 10–20 presented at a fixed pulse rate (ranging from 1 to 100 pps). The amplitude of tone pulses in each block is increased progressively with increments of 2 dB. *B:* Natural amplitude-modulated pulse trains with a fixed modulation frequency (ranging from 5 to 110 Hz). Each train has a fixed duration of 625 ms. Pulse rate in the train increases progressively from 20 to 100 pps. This train (at each modulation frequency) is repeated 10–20 times. *C:* A long random-amplitude pulse train with a fixed pulse rate (ranging from 20 to 100 pps), characterized by a random temporal order of tone pulses within the train.

The amplitude of tone pulses was sinusoidally, amplitude-modulated with a fixed AM frequency over the range of wing-beat frequencies of the bat's natural prey, 5–110 Hz. Each neuron was tested with four or more natural AM pulse trains, each at a different AM frequency. Note that a change in AM frequency produced a change in temporal orders of sound pulses with specific amplitudes. The third comprised random-amplitude pulse trains—that is, continuous trains of tone pulses presented at a fixed pulse rate (ranging from 20 to 100 pps), and whose amplitudes were uniformly distributed over a 90 dB range in 2 dB steps but were randomly ordered (fig. 19.1C).

The tone pulses were presented at the unit's characteristic frequency (CF). Sounds were delivered to the bat via a free-field ultrasonic loudspeaker located 60 cm in front of the bat. The absolute peak sound pressure level was measured with a 6 mm microphone situated near the concha of the ear opposite to the recording side. The parameters of acoustic stimuli were controlled by D/A hardware and software from Tucker-Davis Technologies.

For all three types of stimuli, the unit's response to

each of the tone pulses was used to construct its RLF. The average spike count to 10–20 epochs was used as the response metric. For isolated tone pulses, the stimuli were presented orderly, beginning from low sound levels and progressing to higher sound levels in incremental steps. The unit's response to specific pulse amplitude could be determined unambiguously. In the case of AM or random-amplitude pulse trains, to identify the tone pulse in the train to which the unit responded, we compensated for the delay in response latency by an amount equivalent to the average latency to isolated tone pulses. Action potentials falling within a 7 ms window relative to the onset time of a tone pulse (i.e., average latency ±3.5 ms) were defined as the response to that pulse. We found empirically that the maximum fluctuation in response latency due to a change in pulse rate was on average 1.12 ms and always <2.1 ms.

To evaluate how rate of stimulation influenced the amplitude selectivity, we measured the amplitude response widths at different rates. Amplitude response width was defined as the range of amplitudes over which the response was equal to 50% below the maximum spike count.

Results

Responses from 46 IC neurons to natural AM pulse trains were obtained. For 21 neurons (46%), their RLFs were sharp nonmonotonic. In contrast, the RLFs of these neurons to isolated tone pulses at 1 pps were either broad nonmonotonic (15 units) or monotonic (6 units). The average amplitude response width for isolated tone pulses was 39 ± 14.0 dB, whereas that for natural AM pulse trains was much narrower: 4.5 ± 2.6 dB. A representative example is shown in fig. 19.2. This unit responded well to tone pulses at 1 pps over a wide range of amplitudes (fig. 19.2A), giving rise to a monotonic RLF (fig. 19.2C). In contrast, its response to tone pulses in natural AM pulse trains was highly selective (fig. 19.2B), responding to tone pulses only when they fell narrowly within 36–40 dB SPL (fig. 19.2D). As shown in fig. 19.2B, the unit showed a preference for the stronger tone pulses in the train that had a low maximum amplitude (shown by arrows in the upper trace). The preference shifted to the weaker tone pulses in a train that had a higher maximum amplitude (shown by arrows in lower traces). The best amplitude (i.e., BA, defined as the amplitude to which the unit responded maximally) essentially was unchanged and remained at 39 dB SPL, and so was the response selectivity or tuning width. In general, these cells displayed sharp amplitude selectivity in response to stimuli whose temporal patterns resembled what bats encounter naturally, and this selectivity was independent of the peak amplitude (fig. 19.2B) and the modulation frequency (not shown) of the AM pulse train.

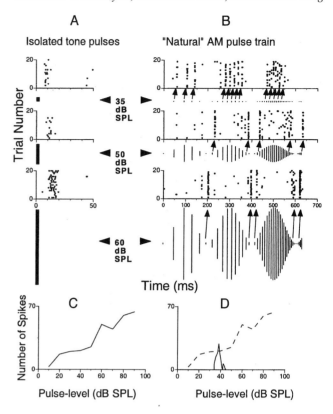

Fig. 19.2. Responses of a monotonic rate-level function (RLF) IC neuron exhibiting marked amplitude selectivity to tone pulses embedded in natural AM pulse trains. Panels *A* and *B* show the dot histograms of responses to isolated tone pulses and natural AM pulse trains having a modulation frequency of 5 Hz and different peak amplitudes (35, 50, and 60 dB SPL, respectively). For the dot histograms in *A* and *B*, the occurrence of each spike is shown as a dot at the time of its occurrence, for each of the twenty trials. *C:* The unit's RLF to isolated tone pulses. *D:* The unit's RLFs to natural AM trains with peak amplitude of 50 dB SPL (solid curve) and to isolated tone pulses (dashed curve). Tone pulses are shown as vertical bars below each histogram (in *A* and *B*); the height of the vertical bar represents the relative amplitude of the tone pulse within the natural AM pulse train.

For IC neurons showing nonmonotonic RLFs (in response to isolated tone pulses), the configuration of neurons' RLF remained nonmonotonic when tone pulses were presented in natural AM pulse trains. However, the amplitude-tuning width became narrower (fig. 19.3).

The amplitude response width to isolated tone pulses was 44 dB, which was reduced to 1.4 dB when presented with natural AM pulse trains. Reduction of response width indicated a sharpening of the neuron response selectivity to sound amplitude.

For the remaining 25 units, there was either no distinct difference in the RLFs for isolated tone pulses and natural AM pulse trains (21 units; fig. 19.4), or there was a marked drop in spike count in response to tone pulses embedded in AM pulse trains (4 units). The unit in

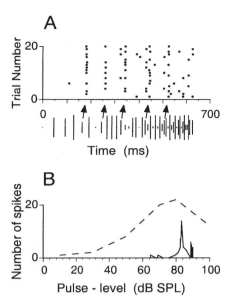

Fig. 19.3. Nonmonotonic rate-level-function (RLF) IC neuron exhibits a sharpening of unit's amplitude selectivity in response to natural AM pulse trains. *A:* The dot histograms of responses to natural AM pulse trains having a modulation frequency of 50 Hz. *B:* The unit's RLFs to natural AM train (solid curve) and to isolated tone pulses (dashed curve).

fig. 19.4 showed a monotonic RLF in response to isolated tone pulses (dashed curve in fig. 19.4B). The unit's RLF was unchanged when natural AM pulse trains had different maximum pulse amplitudes. In both cases, the unit showed a monotonic increase in the response to tone pulses at higher sound levels.

The experiments above revealed that about one half

of IC neurons showed greater amplitude selectivity to tone pulses when they were embedded in natural AM pulse train than to tone pulses presented in isolation. The two types of stimuli differ in a number of respects. One is the temporal ordering of tone pulses with different amplitudes. The other is the absolute rate of simulation, as well as the dynamic change in pulse rate. To determine the relative importance of these attributes, we created a train of tone pulses whose amplitudes were randomly ordered but had fixed pulse rates, ranging from 20 to 100 pps. The constant pulse rate and the random ordering of tone pulses of the pulse train allowed evaluation of the significance of the dynamic change in pulse rate, as well as that of the temporal order of tone pulses.

For 24 IC neurons, we investigated their responses to random-amplitude pulse trains and to natural AM pulse trains. The two pulse trains produced generally similar results, as exemplified by the data from one neuron (fig. 19.5). This unit gave a monotonic RLF in response to isolated tone pulses (dashed curve in figs. 19.5A and 19.5B). The RLFs for tone pulses embedded in two different pulse trains were remarkably similar. Both pulse trains produced highly nonmonotonic RLFs (solid curves) with a response preference for tone pulses of 51–53 dB SPL. RLFs derived from these two types of stimuli showed high statistical similarity.

Since an increase in response selectivity was evidenced for AM pulse trains, whose temporal order was random and whose pulse rate was constant, these results suggest that this increase in selectivity was not dependent on the temporal ordering of tone pulses or on the dynamic change in pulse rate. The two types of AM

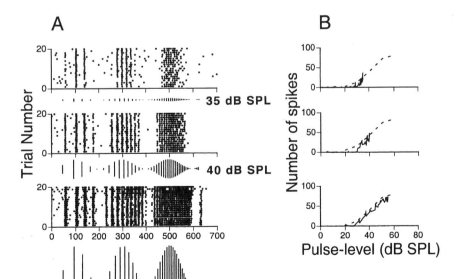

Fig. 19.4. Response patterns from an IC neuron representative of a population for which the rate-level functions (RLFs) to isolated tone pulses and to tone pulses in natural AM pulse trains showed small difference. The dot histograms represent the unit's responses to AM pulse trains having a modulation frequency of 5 Hz at various different peak amplitudes (shown to the right of each of the stimuli). The corresponding RLFs to tone pulses in AM pulse trains are shown as a solid curve in *B*, whereas the RLFs to isolated tone pulses is shown by a dashed curve in *B*.

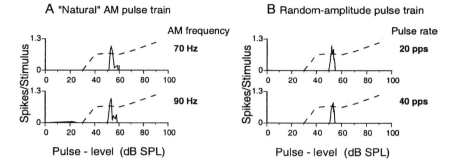

Fig. 19.5. Responses of an IC neuron to all three types of acoustic stimuli. *A:* Unit's rate-level functions (RLFs) to tone pulses in natural AM pulse trains having AM frequencies of 70 and 90 Hz are shown by solid curves in the two panels, along with the unit's RLF to isolated tone pulses (dashed curve). *B:* Unit's RLFs to tone pulses in random-amplitude pulse trains at a repetition rate of 20 and 40 pps. The unit's RLFs to tone pulses in two types of pulse trains are similar (in terms of the best amplitude and the amplitude tuning width), but drastically different from the unit's RLF to isolated tone pulses.

pulse trains shared one parameter that was different from the isolated tone pulses—that is, a higher absolute rate of stimulation. These results therefore suggest that it is the absolute rate that is most responsible for the increase in amplitude selectivity.

This interpretation has been directly substantiated by experimental work just completed in the laboratory. Briefly, we studied the RLFs of IC neurons to single-tone pulses presented at different pulse rates, from 1 to 100 pps. We found that for about one half of IC neurons, an increase in pulse rate altered a unit's RLF systematically from monotonic to nonmonotonic, or from broad nonmonotonic to increasingly sharper nonmonotonic. (Results of this study are described in detail in Galazyuk, Llano, and Feng et al. 2000.) Taken together, these studies showed that an increase in rate employed by bats during target approach and prey capture confers amplitude selectivities that are significantly higher than predicted by the unit's response to isolated tone pulses at 1 pps. This response hyperacuity presumably enhances the bat's perceptual resolution.

In response to natural AM pulse trains, different IC neurons responded best to different BAs. The BAs ranged from 5 to 85 dB SPL. The wide distribution of BAs provides high resolutions across the broad range of sound levels encountered by bats during echolocation. The amplitude tuning width, measured at 50% below the peak response, ranged from 1.2 dB to 17 dB. As described earlier, these bandwidths were much narrower than the response bandwidths of the same units to tone pulses presented at 1 pps.

Discussion

Our investigation showed that IC neurons often are more selective to sound amplitude when sound pulses are presented in a behavioral context than when the stimulus is presented in isolation. The experimental results described herein showed that this response hyperacuity is attributed to the rate of stimulation. Data from random-amplitude pulse trains provided evidence that the change in amplitude selectivity is independent of the temporal ordering of tone pulses in the train. In other words, it is the higher rate in the natural AM pulse train that confers the increase in amplitude selectivity. This interpretation is supported by results obtained from studying the unit's RLFs to isolated tone pulses presented at different pulse rates (Galazyuk, Llano, and Feng 2000), showing that an increase in rate produces a systematic increase in amplitude selectivity.

In nature, when a bat pursues a flying insect, its sonar emission is actively increased from 5–20 pps during the searching phase to 100–200 pps just prior to prey capture (Griffin, Webster, and Michael 1960; Simmons, Howell, and Suga 1975). Results of our physiological studies suggest that an increase in rate likely confers greater amplitude resolution, which could improve echolocation performance (see below). Echo amplitude is correlated with target size and distance (Griffin 1958; Kick and Simmons 1984). Interestingly, the change in emission rate is accompanied by a reduction in sonar intensity such that the perceived echo amplitude is stable and independent of the target distance (Kick and Simmons 1984). This manipulation consequently makes the echo amplitude a reliable measure of target size; thus, an active increase in rate can confer a higher resolution of target size. This enhancement in sensory capacity is useful for animals that rely on the auditory system to analyze complex auditory scenes. We found that at 100 pps, the amplitude tuning width of IC neurons could be as low as 1.2 dB. The resolution of the system, as a whole, is conceivably higher when a population of neurons is involved in the analysis. Amplitude resolution in the range of a fraction of 1 dB would provide a powerful mechanism for precise estimation of size of targets (large and small) in the real world.

Rate of stimulation was previously shown to contribute to sharpening of delay-tuned responses of neurons in the bat's auditory cortex (Wong, Maekawa, and Tanaka 1992) and of auditory receptive fields of IC neurons (Wu and Jen 1996). Taken together, an active increase of sonar emission rate can generate sharper "images" of objects in the real world that include the target range, direction, and size. At this time, it is unclear whether the rate dependence effects are applicable to lower centers in the auditory brainstem and what the underlying cellular mechanism is. Future studies are necessary to pin these down.

Acknowledgments

This study was supported by grants from the National Institute on Deafness and Other Communication Disorders of the NIH (R01-DC01951).

{ 20 }

Neural Processing of Target Distance:
Transformation of Combination-Sensitive Responses

Christine V. Portfors and *Jeffrey J. Wenstrup*

Introduction

To determine the distance to a target, an echolocating bat uses the delay between its emitted sonar pulse and a returning echo (Simmons 1971). In the central auditory systems of several species of bats (*Myotis lucifugus*, Sullivan 1982; *Eptesicus fuscus*, Feng, Simmons, and Kick 1978; *Rhinolophus rouxi*, Schuller, O'Neill, and Radtke-Schuller 1991; *Pteronotus parnellii*, O'Neill and Suga 1982), a special class of neurons respond selectively to a limited range of pulse-echo delays. These delay-tuned neurons respond to delays between the frequency-modulated (FM) sweep in the emitted pulse and returning FM sweeps in echoes, presumably encoding target distance. In the mustached bat, delay-tuned neurons respond to different FM harmonics in the emitted pulse and returning echoes. These so-called FM-FM neurons respond best to the combination of the first harmonic FM sweep (FM_1, 29–24 kHz) in the emitted pulse, and a higher harmonic FM component (FM_2, 58–48 kHz; FM_3, 89–72 kHz; FM_4, 119–96) in returning echoes (O'Neill and Suga 1982). FM-FM neurons are one type of the broader class of neurons described in vertebrates as combination-sensitive, song-selective, or having nonlinear summation in frequency or time (Fuzessery and Feng 1983; Margoliash and Fortune 1992; Suga, O'Neill, and Manabe 1978; Rauschecker, Tian, and Hauser 1995). All of these complex response properties are charac-

terized by their preference for particular spectral and temporal combinations of acoustic elements in species-specific vocalizations.

In the auditory cortex of the mustached bat, FM-FM neurons predominate in two areas. In each area, neurons sensitive to FM_2, FM_3, and FM_4 harmonics are clustered in adjacent slabs; within each echo FM harmonic slab, best delay is mapped (O'Neill and Suga 1982). This cortical organization of best delay is thought to provide the basis for topographic representations of distance. FM-FM neurons are also common in the medial geniculate body (MGB) (Olsen and Suga 1991; Wenstrup 1999) and the central nucleus of the inferior colliculus (ICC) (Mittmann and Wenstrup 1995; Portfors and Wenstrup 1999). Recent evidence suggests that nearly all FM-FM responses originate in the ICC (see Wenstrup et al., chapter 21, this volume).

In this study, we examined whether the basic response properties of FM-FM neurons related to coding of pulse-echo delay undergo transformations in the colliculo-thalamo-cortical pathway. First, we found that basic response properties of FM-FM neurons are similar between the ICC and MGB. Second, we determined that delay-tuned FM-FM neurons in the ICC are less well organized than in the auditory cortex. Our main conclusion is that substantial transformations in the organization of response properties occur between the ICC and auditory cortex. This organization results in the segregation of FM-FM neurons into distinct areas and creates a topographical representation of delay sensitivity in these areas.

Materials and Methods

Data presented in this paper are part of two larger studies examining complex response properties and their topographical organization in the ICC and MGB of the mustached bat, *Pteronotus parnellii* (Portfors and Wenstrup 1999; Wenstrup 1999). Neural responses were obtained from the ICC of 13 bats and the MGB of 21 bats.

Surgical procedures, acoustic stimulation, and recording procedures were the same as those described elsewhere (Portfors and Wenstrup 1999). Briefly, bats were anesthetized with methoxyflurane (Metofane, Pitman-Moore, Inc., Mundelein, Ill.) in combination with sodium pentobarbital (Nembutal 5 mg/kg, i.p.; Nembutal, Abbott Laboratories, North Chicago, Ill.) and acepromazine (2 mg/kg, i.p.; Med-Tech, Inc., Buffalo, N.Y.) to attach a metal pin to the skull to secure the bat's head during physiological recordings.

We delivered tone-burst stimuli to the awake bat through a speaker (Technics leaf tweeter) placed 10 cm away and 25° into the sound field contralateral to the IC or MGB under study. Tone bursts were 3–30 ms with a rise-fall time of 0.5 ms and were presented at 4/s. To record single neurons, a micropipette electrode, filled with one of several tracers, was inserted into the ICC or MGB through a hole made in the skull. The electrode was directed into the medial division of the ICC where higher frequencies of the mustached bat's hearing (72–120 kHz) are represented. For MGB recordings, the electrode was directed into all divisions to provide adequate sampling of the MGB. However, in this chapter we are interested mainly in the response properties of FM-FM neurons, which occurred mainly in the dorsal division of the MGB. A detailed discussion of response properties and their topographic distribution in all subdivisions of the MGB can be found in Wenstrup 1999.

When a single neuron was isolated, we obtained its best frequency (BF), threshold, and frequency-tuning curves. For neurons that responded to two separate frequency bands, we documented the best high-frequency response and the best low-frequency response. We then tested the neuron's sensitivity to combinations of frequencies by setting one tone at BF and at 10 dB above threshold and presenting a second tone that varied in frequency, intensity, and relative timing to the first tone. Responses to two-tone stimuli were identified as facilitatory or inhibitory if the neuron's response to the combination of tones was 20% greater (facilitation) or 20% less (inhibition) than the sum of the responses to the two tones presented individually. The neuron's frequency-tuning curves to the single stimuli and the combination stimuli were documented. Sensitivity to delay between the two tones was assessed by maintaining the onset time of the first tone and varying the onset time of the second tone in steps of 1 or 2 ms and collecting neural responses to 32 presentations at each delay. The delay that elicited the greatest response (or least response in an inhibitory interaction) was defined as the neuron's "best delay." Sharpness of delay tuning was quantified as the width of the delay response curve at 50% of the maximal facilitated response. The strength of the facilitated response, at the neuron's best delay, was quantified as a facilitation index, defined as $(Rc - Rl - Rh)/(Rc + Rl + Rh)$, where Rc, Rl, and Rh are, respectively, the neuron's responses to the combination of the high- and low-frequency signals, low-frequency signal alone, and high-frequency signal alone (Dear and Suga 1995). A neuron was classified as facilitated if the facilitation index was ≥ 0.09, corresponding to an increase in response of at least 20% above the summed responses to the low- and high-frequency sounds. A facilitation index of 1.0 indicated the strongest possible facilitation.

At the end of a penetration, we deposited the tracer to enable histological reconstruction of recording sites. Details of deposits, tissue processing, deposit visualization, and electrode reconstruction are available elsewhere (Wenstrup and Grose 1995).

Results

This study was based on 173 and 463 anatomically localized responses in the ICC and MGB, respectively. Responses in the ICC were recorded from the medial division and had best frequencies in the 72–89 kHz range (FM_3 of the bat's sonar signal). The 463 MGB responses were recorded from all divisions of the MGB. In making our physiological comparisons of basic response properties of delay-tuned neurons, we compared the response properties of all the ICC neurons with 80 single unit responses in the MGB that also had best frequencies corresponding to the higher harmonics of the bat's sonar signal. These neurons were in the dorsal division of the MGB where FM-FM responses are common (Wenstrup 1999). Topographical comparisons were made using all responses in the ICC and MGB. In describing the transformations that occur in the colliculo-thalamic pathway, we first examined changes in the physiological response properties of FM-FM neurons and then changes in the organization of these response properties.

PHYSIOLOGICAL RESPONSES OF FM-FM NEURONS

We tested the neurons for sensitivity to combinations of signals and found that in the ICC only 29 of 173 neurons were tuned to a single frequency band. The majority of the neurons showed some combination-sensitive interaction between signals in two separate frequency bands, with 51% ($n = 88$) facilitated by the combination of the two frequencies (table 20.1). These facilitatory neurons responded better to the combination of signals in the 24–29 kHz and 72–89 kHz range than to

TABLE 20.1. Physiological response properties of neurons recorded in the FM₃ representation of the ICC and in the dorsal division of the MGB.

	Inferior colliculus *n* (%)	Medial geniculate body *n* (%)
Single tuning	29 (17)	6 (7.5)
Facilitatory FM-FM	88 (51)	68 (85)
Inhibitory FM-FM	43 (25)	5 (6.25)
Multiple tuning	13 (7)	1 (1.25)
Total responses	173	80

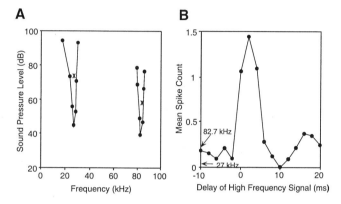

Fig. 20.1. Frequency and delay tuning of a single neuron recorded in the ICC. *A:* Facilitatory frequency-tuning curves obtained by fixing the frequency and intensity of one tone burst (denoted by the *X*) while varying the frequency and intensity of the other tone burst to obtain threshold facilitatory responses. *B:* Delay-tuning curve obtained by varying the relative timing between the two facilitating tones. The arrows indicate the neural response to each tone presented separately. This FM-FM neuron had a best delay of 2 ms.

the individual signals. They were delay-tuned, in that facilitation was strongest at a particular delay between the two sounds. In the dorsal division of the MGB facilitated FM-FM neurons accounted for 85% of responses (table 20.1).

In fig. 20.1 the frequency (fig. 20.1A) and temporal (fig. 20.1B) response characteristics of an ICC facilitatory FM-FM neuron are shown. This neuron responded poorly to individual tones, but it responded

well to the combination of a 27 kHz (FM₁) and an 82.7 kHz (FM₃) tone, when the delay between the low- and high-frequency tones was 2 ms. A distinguishing characteristic of FM-FM neurons is their sensitivity to the timing between the low- and high-frequency signals. Fig. 20.1B shows that the neuron responded maximally when the high-frequency signal was delayed 2 ms from the onset of the low-frequency signal. Best delays among ICC and MGB neurons were similar (fig. 20.2A), ranging from 1 to 20 ms in the ICC and from 1 to 24 ms in the MGB.

The sharpness of delay tuning, an indicator of the precision in encoding target distance, was variable among FM-FM neurons, both in the ICC and MGB. However, sharpness of delay tuning was similar between the ICC and MGB neurons (fig. 20.2B). Sharpness of delay tuning was quantified as the width of the delay curve at response rates evoking 50% of the maximally facilitated response ("50% delay width"). Delay widths ranged from 2 to 13 ms (average 6.6 ± 3.4 ms) in the ICC, and between 1.8 and 14.5 ms (average 6.2 + 3.3) in the MGB.

The strength of the facilitated response indicated the degree of selectivity for preferred delays. The neuron in fig. 20.1 demonstrates a strongly facilitated response. At a delay of 2 ms the facilitated index was 0.73. The average facilitation index among the ICC delay-tuned neurons was 0.38 + 3.4, corresponding to an average response increase of 122%. The strength of facilitation among MGB neurons was similar to ICC neurons (fig. 20.2C). The average facilitation index for FM-FM neurons in the MGB was 0.40 + 0.34.

A second type of FM-FM response we recorded was inhibitory. With these neurons, the excitatory response to the high-frequency signal was inhibited by the simultaneous presentation of a low-frequency (FM₁) signal. In the ICC, 25% (*n* = 43) of the recorded neurons were inhibitory FM-FM, in constrast to only 6% in the dorsal division of the MGB (table 20.1).

ORGANIZATION OF FM-FM RESPONSES

The tonotopic organization of the mustached bat ICC has been well documented (Zook et al. 1985) and is illustrated in fig. 20.3A. Frequencies in the third (72–89 kHz) harmonic of the sonar call of the mustached bat

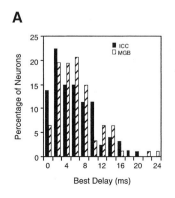

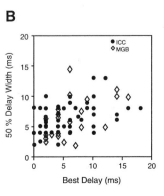

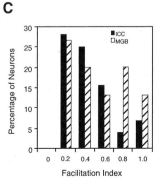

Fig. 20.2. Comparison of response properties of FM-FM neurons in the ICC and MGB. *A:* Distribution of best delays. *B:* Sharpness of delay tuning quantified as the width of the delay curve at 50% maximal response. *C:* Strength of facilitation.

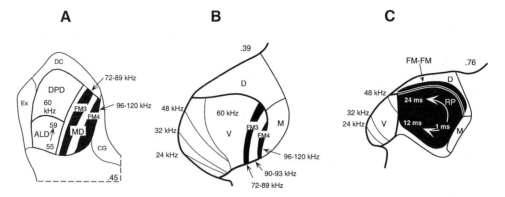

Fig. 20.3. Schematic summary of the tonotopic organization of the ICC and MGB. *A:* Schematic illustration of the organization of the ICC in a transverse section. The number at the lower right indicates the location of the section along the caudal-rostral axis of the ICC. In the medial division (MD), FM-FM neurons occur throughout the FM₃ (72–89 kHz) and FM₄ (96–119 kHz) representations (in black). *B:* Schematic illustration of the tonotopic representation in the ventral division of the MGB. The FM₃ and FM₄ representations are blackened to indicate their underrepresentation in the ventral division. *C:* Schematic illustration of the dorsal division of the MGB. FM-FM responses predominated in this region. The blackened area indicates FM-FM responses, with the numbers representing the best delays. Comparing the topography of the ICC and MGB shows that in the MGB, FM-FM responses occur outside the tonotopic representation and are organized in terms of the best delay. In the ICC, FM-FM responses occur within the tonotopic representation, and there is no apparent organization of their best delay. DC: dorsal cortex of the IC; Ex: external nucleus of the IC; DPD: dorsoposterior division of the ICC; ALD: anterolateral division of the ICC; CG: central gray; D: dorsal division of the MGB; M: medial division of the MGB; RP: rostral pole of the MGB.

are represented in the medial division (MD). Within the MD, FM-FM responses constituted about 75% of all acoustic responses. Because the MD is tonotopically organized, the delay-tuned responses occurred within the tonotopic representations. In other words, FM-FM neurons were not grouped together in a separate area.

In contrast, FM-FM neurons in the MGB occurred in distinct regions, outside the tonotopically organized regions. The MGB of the mustached bat is divided into ventral, medial, and dorsal divisions (fig. 20.3B–C). The ventral division contains the only clear tonotopic organization of frequency, as illustrated in fig. 20.3B (Wenstrup 1999). Three features of this frequency organization are apparent: there are relatively few neurons tuned to the 24–29 kHz (FM₁) range; the 60–63 kHz frequency representation is hypertrophied; and the representations of the third (72–89 kHz) and fourth (96–119 kHz) FM harmonics are very small. Furthermore, very few delay-tuned neurons occur in the tonotopically organized ventral division.

In contrast, the dorsal MGB is not tonotopically organized. In the dorsal MGB, FM-FM responses constitute at least 70% of all responses. Thus, FM-FM neurons group together outside the tonotopic representation in the MGB (fig. 20.3C). This is a major transformation from the ICC, where FM-FM responses occur within the tonotopic frequency representations.

A second transformation of response properties between the ICC and MGB appears to be the organization of delay sensitivity. A gradient of delay sensitivity could exist along either the dorsomedial-to-ventrolateral or caudo-rostral dimensions of the ICC. To identify any gradient of best delay, we plotted the location of each FM-FM neuron within its ICC or MGB and computed correlation coefficients of location and best delay. In the ICC, there was no correlation between best delay and dorsomedial-ventrolateral ($r = 0.22$, $df = 87$, $p > 0.05$) or caudo-rostral ($r = 0.21$, $df = 87$, $p > 0.05$) location. Thus, delay sensitivity is not topographically organized in the high-frequency representations of the ICC. In the MGB, the best delay of the population of FM-FM neurons was significantly correlated with medial-lateral location ($r = 0.44$, $df = 23$, $p < 0.05$).

Discussion

In this study, we describe the response properties of delay-tuned neurons and their topographical organization in the ICC and MGB to examine what modifications and reorganizations occur with ascending auditory processing. In the high-frequency representations of the ICC, facilitatory delay-tuned neurons constituted over 50% of all responses. In the regions of the MGB where FM-FM neurons were found, facilitatory responses accounted for at least 70% of all responses. The abundance of FM-FM neurons in the ICC and MGB indicates their importance in processing species-specific vocalizations. In comparing response properties in the ICC and MGB, we found that basic response properties related to coding pulse-echo delay were similar, but modifications of their topographical organization were apparent. Thus, a major role of the ascending auditory pathway is to organize neuronal projections so that auditory cortex neurons with similar response features occur together in functionally specialized areas. Further, certain response properties, such as pulse-echo delay sensitivity, become mapped across the functional area. The reorganization of responses begins in the tecto-thalamic projection, but there is further reorganization from the MGB to the auditory cortex.

PHYSIOLOGICAL RESPONSE PROPERTIES

Basic response properties of FM-FM neurons related to coding of pulse-echo delay include best delay, sharpness of delay tuning, and strength of facilitation. The distribution of best delays among the population of delay-tuned neurons indicates the range of distances over which the neurons are responsive. Best delays among our sample of ICC and MGB neurons were similar, ranging between 1 ms and 24 ms. These delays correspond to target distances up to about 4 m. The sensitivity to delay of FM-FM neurons indicates their importance in encoding target distance information. The broad range of best delays of FM-FM neurons indicates that a large expanse of space can be analyzed by the population of neurons. In both the ICC and MGB, the majority of neurons had best delays of less than 10 ms, indicating that distances up to 1.7 m may be most functionally relevant to the mustached bat. Short target distances are also emphasized in the auditory cortex of the mustached bat (O'Neill and Suga 1982). There may be more short best-delay neurons in the mustached bat because it forages for insects in cluttered environments where space is limited and the operational range for echolocation is short.

An important response property of delay-tuned neurons related to encoding target distance information is the sharpness of delay tuning; the sharper the tuning, the better the precision of delay coding. A sharply tuned neuron will only respond over a narrow range of distances, whereas a more broadly tuned neuron will respond over a much greater range of distances. The sharpness of delay tuning (50% delay width) was similar between ICC and MGB FM-FM neurons. Furthermore, delay widths were variable with 50% delay widths as low as 2 ms and as high as 14 ms. It is apparent from these values that the delay tuning of an individual neuron is not sharp enough to encode the distance of a particular target precisely. However, precisely encoding the distance to a target may not be the only functional role of delay-tuned neu-

rons. Instead, they could function as filters to encode features of the target when it is within a particular extent of space. In other words, each neuron is responsible for analyzing features of targets within a block of space. In this way, a population of delay-tuned neurons would sample different regions of space and function together to analyze the entire scene. It is unknown what other features of a target delay-tuned neurons respond to, but they could encode several different aspects of the target—such as horizontal and vertical location, texture, or wing-beat frequency. Evidence from the auditory cortex of a different bat, the little brown bat, suggests that delay-tuned neurons are involved in multidimensional analyses of targets (Paschal and Wong 1994b).

The final physiological comparison of response properties that we make between the ICC and MGB is the strength of facilitation. This provides a measure of how well the neuron responds to the combination of the two sounds, compared to the responses to the two sounds individually. Neurons in the ICC and MGB displayed strong facilitation. In the ICC the average facilitation index was 0.38, which corresponds to an increase in response of 122%. The average value in the MGB was 0.40. Although the average values were similar between the ICC and MGB neurons (fig. 20.2C), the MGB had more neurons with facilitation index values of 1.0, indicating that these neurons did not respond to either the low- or high-frequency signals alone—they only responded to the combination of the sounds.

The similarities in the range of best delays, sharpness of delay tuning, and strength of facilitation between FM-FM neurons in the ICC and MGB (fig. 20.2) suggest that basic response properties related to coding pulse-echo delay are not modified in this part of the colliculo-thalamo-cortical pathway.

ORGANIZATION OF FM-FM RESPONSES

FM-FM neurons occurred throughout the high-frequency tonotopic representations of the ICC; they were not grouped in a distinct area. This was not the case in the MGB. In the ventral division of the MGB, where there is a tonotopic organization of frequencies, very few FM-FM neurons were found. In contrast, most FM-FM neurons were in the dorsal division. The grouping of FM-FM responses in the MGB indicates that projections to the MGB from the ICC project outside the regular tonotopic arrangement. Anatomical evidence shows that the ICC outputs from different frequency bands project to different regions of the MGB. ICC neurons tuned to the second, third, and fourth FM sonar harmonics project to the rostral one third of the MGB in the dorsal division (Frisina, O'Neill, and Zettel 1989; Wenstrup, Larue, and Winer 1994), presumably leading to the distinct grouping of FM-FM neurons in the dorsal division and not in the ventral division. Furthermore, the ICC representations of the FM_3 and FM_4 harmonic project only weakly to the ventral division (Wenstrup, Larue, and Winer 1994). This causes the minimal higher harmonic frequency representation in the ventral division and the grouping of FM-FM responses in the dorsal division of the MGB. Modifications in the ICC projections produce an organization of response properties in the MGB similar to that in the auditory cortex (O'Neill and Suga 1982). However, the organization in the auditory cortex appears to be even more distinct than in the MGB, indicating that further reorganization of FM-FM responses occurs in the thalamo-cortical pathway.

Another transformation of response properties that occurs in the ascending auditory pathway is the emergence of a highly systematic organization of best delay in the auditory cortex. In the FM-FM and dorsal fringe areas of the auditory cortex, delay-tuned neurons are organized by echo harmonic sensitivity and best delay (O'Neill and Suga 1982). In each area, best delay increases from rostral to caudal within each echo FM harmonic slab. In the dorsal division of the MGB, FM-FM neurons are also organized by best delay, although the organization is less clear than in the auditory cortex. While short best delays occur medial-ventral and long best delays occur more lateral-dorsal, there is not a clear map of best delay as is apparent in the auditory cortex. In contrast, in the ICC there appears to be no clear organization of best delay. Thus, in the colliculo-thalamo-cortical pathway, outputs from the ICC are reorganized so there is some organization of delay sensitivity in the MGB, and then further reorganization occurs to produce the highly organized mapping of delay sensitivity in the auditory cortex.

This study shows that the physiological response properties related to coding pulse-echo delay in the mustached bat are not substantially modified in the ascending auditory pathway between the ICC and MGB. We suggest that FM-FM responses are created in the ICC, and that the major modifications that occur within the pathway are reorganizations of outputs to create the distinct functionally specialized areas in the auditory cortex and to generate the maps of response properties in these specialized regions. Although dramatic reorganization occurs between the ICC and MGB, further reorganizations occur between the MGB and the auditory cortex to create the highly organized representation of target distance in the auditory cortex.

Acknowledgments

We thank C. Grose for technical assistance; B. Grothe and an anonymous reviewer for helpful comments on the chapter; and J. Thomas, C. Moss, and M. Vater for editorial comments. We are grateful to the Natural Resources Conservation Authority of Jamaica and the Wildlife section of the Ministry of Agriculture, Land, and Marine Resources of Trinidad and Tobago for permission to export the bats. This work was supported by research grant 5 R01 DC00937 to Jeffrey Wenstrup from the National Institute on Deafness and Other Communication Disorders.

{ 21 }

Neural Mechanisms Underlying the Analysis of Target Distance

Jeffrey J. Wenstrup, Scott A. Leroy, Christine V. Portfors, and *Carol D. Grose*

Introduction

Echolocating bats use delays between their emitted sonar pulse and returning echoes (Simmons 1971) to evaluate the distance of objects. These delays are encoded within the central auditory system of bats by a special class of neurons that respond selectively to a limited range of pulse-echo delays (Feng, Simmons, and Kick 1978; O'Neill and Suga 1982). Populations of these delay-tuned neurons (tuned to different delays)

are believed to contribute to the bat's perception of the distance of sonar objects (Feng, Simmons, and Kick 1978; O'Neill and Suga 1982). This chapter describes mechanisms underlying delay-tuned response properties in the auditory system of the mustached bat (*Pteronotus parnellii*).

Most bats, including the mustached bat, use frequency-modulated (FM) sweeps in their sonar signals to measure pulse-echo delay. In the mustached bat, however, delay-tuned neurons respond to different FM

A

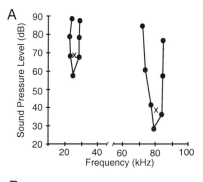

B

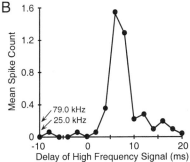

Fig. 21.1. Frequency and delay tuning for a facilitatory FM-FM neuron in the inferior colliculus (IC). *A:* Facilitatory frequency-tuning curves, obtained by fixing the frequency and intensity of one 4 ms tone burst (marked by the *X*) while varying the frequency and intensity of another to obtain threshold facilitatory responses. Facilitating inputs were tuned to 25.0 kHz and 79.0 kHz, within frequency ranges of first and third harmonic FM sweeps of the sonar call. *B:* Delay-tuning curve, obtained by presenting two-tone bursts at best facilitating frequencies, varying relative timing of the two signals. Arrows indicate the response of the neuron to each tone burst presented separately. Neuron showed strong facilitation that was best when higher frequency stimulus was delayed by 6 ms.

harmonics in the emitted pulse and returning echoes. These so-called FM-FM neurons are facilitated by a combination of the first harmonic FM (FM_1, 29–24 kHz) in the emitted pulse and a higher harmonic FM component (FM_2, 59–48 kHz; FM_3, 89–72 kHz; FM_4, 119–96 kHz) in delayed echoes (O'Neill and Suga 1982). FM-FM neurons are members of a functional class of vertebrate auditory neurons, called combination-sensitive, that analyzes information within two or more spectrally or temporally distinct elements of vocalizations.

Fig. 21.1 displays characteristic response properties of FM-FM neurons in the mustached bat's inferior colliculus (IC) (Mittmann and Wenstrup 1995; Portfors and Wenstrup 1999). First, these neurons show sensitivity to auditory inputs in two distinct frequency bands (fig. 21.1A). The higher frequency response is, by definition, tuned to frequencies within the second, third, or fourth harmonic of the FM sonar component, while the lower frequency response is tuned to frequencies within the first FM harmonic. A second characteristic property

is facilitation; the combination of the lower and higher frequency stimuli evokes a response significantly greater than responses to the individual signals (fig. 21.1B). The third characteristic property is delay tuning; the low-frequency signal must precede the high-frequency signal by a particular interval to obtain the maximum facilitated response. In fig. 21.1B, peak facilitation is obtained when the high-frequency stimulus occurs 6 ms after the low-frequency stimulus. The delay that evokes the maximum response is called the "best delay" and is thought to underlie the neuron's coding of distance of sonar targets. Among delay-tuned neurons in the IC, the range of best delays is 1–20 ms (Portfors and Wenstrup 1999), corresponding to distances of 0.2–3.4 m.

Although previously thought to occur only in the forebrain of the mustached bat, delay-tuned FM-FM neurons are common in the central nucleus of the inferior colliculus (ICC) (Mittmann and Wenstrup 1995; Yan and Suga 1996b; Portfors and Wenstrup 1999). In the ICC, they occur in tonotopic representations of the higher of their frequency sensitivities. Thus, an FM_1-FM_3 neuron in the ICC occurs within the tonotopic representation of frequencies in the third harmonic FM sweep (FM_3: 72–89 kHz), but not in the ICC representation of the first sonar harmonic (FM_1: 24–29 kHz) (Mittmann and Wenstrup 1995; Portfors and Wenstrup 1999).

A full explanation of the origin of delay-tuned, FM-FM neurons considers mechanisms underlying all three response features—dual frequency sensitivity, delay sensitivity, and the facilitated response—as well as the brain regions in which these mechanisms operate. We describe three experiments addressing these issues. The first examines whether FM-FM response properties occur in a major auditory center below the IC, the nucleus of the lateral lemniscus. In the second, retrograde transport methods examine possible sources of the dual frequency sensitivity. Third are microiontophoresis experiments, examining whether inhibitory neurotransmitters play a role in the response properties of collicular FM-FM neurons.

Materials and Methods

General Experimental Procedures

Neurophysiological, anatomical, and pharmacological experiments were conducted on greater mustached bats (*P. parnellii*) from Jamaica or Trinidad. Surgical preparation has been described in detail elsewhere (Wenstrup and Grose 1995). One or more days before experiments began, bats were surgically prepared to expose the dorsal surface of the IC and to attach a metal pin onto the skull for use in a stereotaxic device. Before surgery, animals were anesthetized with methoxyflurane (Metofane, Pitman-Moore, Inc., Mundelein, Ill.; to effect) and sodium pentobarbital (Nembutal, Abbott Laboratories, North Chicago, Ill.; 5 mg/kg, intraperitoneal injection). A small hole (approximately 0.25 mm diameter) was placed in the skull over the IC. After application of a local anesthetic (lidocaine) and a topical antibiotic, animals were placed in a holding cage to recover from the surgery.

In the mustached bat's IC, neurons responding to frequencies in FM sweeps also respond well to tone bursts (O'Neill 1985). We used tone-burst stimuli to test delay-tuned responses in order to assess frequency sensitivity of the neurons. Acoustic stimulation and data acquisition were computer controlled (Wenstrup and Grose 1995). Two different tone-burst stimuli (4 ms, including 0.5 ms rise-fall times, 4/sec) were separately generated, switched, and attenuated. Signals from the two channels were added, amplified, and broadcast to the bat by a speaker placed 10 cm away from the bat and 25° into the sound field contralateral to the IC under study. Distortion components in the speaker output were not detectable 60 dB below the signal level.

Single-unit responses were obtained using micropipette electrodes (resistances of 10–20 MΩ). Multiunit responses were obtained with broken-tipped micropipettes (resistances of 5–15 MΩ). Action potentials were amplified, filtered, and displayed audiovisually using conventional techniques. Pulses indicating the timing of action potentials were sent to the computer for quantitative analyses of the number and temporal distribution of spikes.

Units were recorded from the IC or nuclei of the lateral lemniscus. When a unit was isolated, best frequency and threshold were obtained audiovisually. Responses were assessed to sound throughout the audible range, usually at more than one intensity. Next, a two-tone paradigm examined sensitivity to combinations of signals; the neuron's best frequency was presented 10 dB above threshold, and the response monitored as a second tone was presented at frequencies between 10 kHz and 110 kHz. Intensity and timing of this second tone burst were varied. Quantitative data were obtained to document a neuron's multiple-frequency tuning and sensitivity to delays between the different spectral elements in sounds.

The strength of combination-sensitive interactions (facilitation or inhibition) was quantified as the index of interaction (I), where $I = (R_{comb.} - R_1 - R_2)/(R_{comb.} + R_1 + R_2)$. R_1 is the response to one signal, R_2 is the response to the other signal, and $R_{comb.}$ is the response to the two signals presented together. An interaction index value of 0.09 corresponds to 20% facilitation, our criterion for combination-sensitive facilitation. An index value of +1 indicates maximum facilitation, in which the unit responds only when presented with combination stimuli. Negative numbers indicate combination-sensitive inhibition, in which the excitatory response to one signal is suppressed by the other signal. A value of −1 indicates maximum inhibition, in which the excita-

tory response to one signal is completely suppressed by presentation of the other signal.

DEPOSITS OF RETROGRADE TRACERS

After several FM-FM single units were recorded in a single penetration, the high-resistance electrode was replaced by a tracer-filled, lower-resistance electrode (5–15 MΩ) at the same location on the dorsal surface of the IC. Electrodes were filled with wheat germ agglutinin conjugated to horseradish peroxidase (WGA-HRP, Sigma Chemical Co., St. Louis, Mo.) or cholera toxin B-subunit (CTb, List Biologicals, Campbell, Calif.). The lower-resistance electrode was lowered to a depth where, in the previous penetration, a combination-sensitive single unit was recorded. At that or a nearby depth, a single-unit or multiunit FM-FM response was obtained; then the tracer was deposited iontophoretically. Deposits were placed to include at least one documented FM-FM recording site and all were in the ICC.

After an appropriate survival period, animals were perfused under deep anesthesia with Nembutal (60 mg/kg, i.p.). When nociceptive reflexes were eliminated, the chest cavity was opened and the animal perfused through the heart with phosphate-buffered saline and aldehyde fixative. The brain case was opened and the brain was blocked in a consistent plane inclined about 15% from dorsal and caudal to ventral and rostral (Wenstrup and Grose 1995). Brains were refrigerated overnight in 30% sucrose-phosphate buffer solution before transverse sectioning on a freezing microtome at a thickness of 30–40 μm. All sections from the cochlear nucleus through the auditory cortex were collected. Alternating series were processed by different protocols; one series was stained with cresyl violet. Animals were perfused 24–48 hours after WGA-HRP deposits, using a mixed aldehyde fixative (1.25% glutaraldehyde and 1% paraformaldehyde) followed by 10% sucrose in phosphate buffer (4°C). Tissue was reacted with tetramethylbenzidine to visualize the tracer. After CTb deposits, animals were perfused 3–5 days later with a 4% paraformaldehyde fixative. CTb was visualized using immunohistochemistry and an avidin-biotin-peroxidase procedure (Vector Laboratories). Free-floating sections were incubated in normal rabbit serum to block nonspecific background staining, then in goat anti-CTb (1:40,000, List Biologicals) for 60–65 hours at 4°C. (For details of histological processing, see Wenstrup, Mittman, and Grose et al. 1999.)

MICROIONTOPHORESIS METHODS

We used standard techniques to record from single units during local application of drugs to block inhibitory neurotransmission by glycine, with strychnine, and by gamma-amino butyric acid (GABA), with bicucul-line. A recording electrode was glued at a 20° angle to a multibarreled pipette used for drug delivery (Havey and Caspary 1980). The recording electrode was filled with 0.9 % NaCl, with resistances of 15–25 MΩ. A five-barreled pipette (World Precision Instruments, Inc.) was pulled and broken to a total tip diameter of 10–30 μm. The recording electrode extended 10–25 μm beyond the multibarreled pipette. Solutions in the multibarreled pipette included strychnine-hydrochloride (10 mM in deionized water, 3.5 pH, Fluka, Milwaukee, Wisc.), bicuculline methiodide (10 mM in 0.165 M NaCl solution, 3.5 pH, Sigma, St. Louis, Mo.), and 0.9 % NaCl (sum and control barrels). Each barrel was connected to a channel of a microiontophoresis current generator (Dagan, model 6400). A sum channel balanced currents across all channels.

Responses of FM-FM neurons were recorded before, during, and sometimes after application of strychnine or bicuculline. Prior to drug application, tests of facilitation and delay sensitivity were completed. We then iontophoresed strychnine for 8–12 min or bicuculline for 3–8 min, and repeated all tests while the drug was applied. Strychnine was applied with 20–40 nA positive current and retained with −15 nA. Bicuculline was applied with 10–30 nA positive current and retained with −15 nA. To determine whether the applied current affected the neuronal response, we applied current through a barrel containing a 0.9% NaCl drug delivery solution at the same levels used to iontophorese drugs. This control test produced no noticeable changes in the neuronal response.

Results

Physiological Studies of Lateral Lemniscal Nuclei

Our working hypothesis is that delay-tuned, FM-FM response properties originate in the central nucleus of the IC. One test of this hypothesis was to examine whether neurons in brainstem auditory nuclei show similar response properties. In seven bats, we recorded 101 histologically localized single units distributed throughout the four nuclei of the lateral lemniscus (Portfors and Wenstrup 2001). Of these units, three displayed facilitated combination-sensitive response properties. An additional nine displayed combination-sensitive inhibitory response properties (fig. 21.2), in which a neuron's response to a higher frequency signal is inhibited by a simultaneously presented low-frequency sound. These numbers differ sharply from the IC, where results across two studies found that 48% of neurons showed facilitatory combination-sensitive interactions, and 26% showed inhibition (Portfors and Wenstrup 1999; Leroy and Wenstrup 2000). These findings support the hypothesis that most FM-FM responses are created in the IC.

Ascending Inputs to Collicular FM-FM Neurons

In the IC, delay-tuned FM-FM neurons only occur in tonotopic representations of frequencies corresponding to the second, third, or fourth FM harmonics. If the dual frequency sensitivity of FM-FM neurons is created in

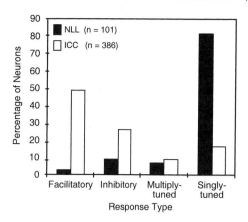

Fig. 21.2. Occurrence of combination-sensitive and other spectrally complex response properties in nuclei of lateral lemniscus and IC. "Facilitatory" and "Inhibitory" indicate combination-sensitive facilitation and inhibition, respectively, as defined in "Materials and Methods." "Multiply-tuned" neurons have more than one peak of sensitivity in their frequency-tuning curve, but no facilitatory or inhibitory interactions between these different spectral elements. Data are from single units. Locations of units within lateral lemniscal nuclei were histologically verified. Combination-sensitive response properties are abundant in the IC but are uncommon in lateral lemniscal nuclei.

the IC, as our hypothesis proposes, neurons in these higher frequency representations of the ICC must receive one or more low-frequency inputs in addition to the expected high-frequency input. We placed deposits of retrograde tracer at collicular recording sites of FM-FM neurons to examine possible sources of this low-frequency input (Wenstrup, Mittman, and Grose 1999). In six animals, we examined topographic and quantitative features of labeling in the brainstem and IC after tracer deposits in collicular representations of the third (72–89 kHz) or fourth (96–119 kHz) harmonic FM signals.

Topographic features of labeling after a deposit at an FM-FM recording site are illustrated in fig. 21.3A–D. In most brainstem nuclei and in the IC, labeling occurred within a single restricted area of the nucleus. This labeling pattern occurs in dorsal and posteroventral cochlear nuclei, lateral and medial superior olivary nuclei, the dorsal nucleus of the lateral lemniscus, and the columnar division of the ventral nucleus of the lateral lemniscus. In the midbrain as well, significant retrograde labeling occurs only in one part of the contralateral IC. All of this labeling occurs in regions of these nuclei that, on the basis of physiological studies (see "Discussion," below), are believed to encode higher frequencies within the bat's audible range—for example, the frequencies of the third harmonic of the sonar call. In contrast, more complex labeling was observed in three regions. In the anteroventral cochlear nucleus, labeling was observed in anterior and marginal subdivisions that are thought to respond to frequencies of the first sonar harmonic

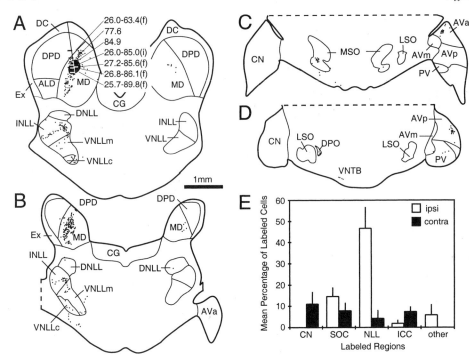

Fig. 21.3. Inputs to FM-FM neurons of the IC, shown by retrograde transport. *A–D:* Retrogradely labeled neurons (dots) in the brainstem after a deposit of tracer (CTb) at a facilitatory FM-FM recording site in the central nucleus of the IC (*A*). Transverse sections are arranged from rostral (*A*) to caudal (*D*). Blackened area in (*A*) shows estimate of effective size of the tracer deposit. Frequency tuning and combination-sensitive interactions (*f,* facilitation; *i,* inhibition) are indicated in the electrode track. *E:* Distribution of brainstem and midbrain labeling after tracer deposits at FM-FM recording sites in the IC of six animals. Lines above bars indicate standard deviation. ALD: anterolateral division of the central nucleus of the IC; AVa: anteroventral cochlear nucleus, anterior part; AVm: anteroventral cochlear nucleus, marginal cell division; AVp: anteroventral cochlear nucleus, posterior part; CG: central gray; CN: cochlear nuclei; DC: dorsal cortex of the IC; DNLL: dorsal nucleus of the lateral lemniscus; DPD: dorsoposterior division of the central nucleus of the IC; DPO: dorsal periolivary nucleus; Ex: external nucleus of the IC; ICC: central nucleus of the IC; INLL: intermediate nucleus of the lateral lemniscus; LSO: lateral superior olive; MD: medial division of the central nucleus of the IC; MSO: medial superior olive; NLL: nuclei of the lateral lemniscus; PV: posteroventral cochlear nucleus; SOC: superior olivary complex; VNLLc: ventral nucleus of the lateral lemniscus, columnar division; VNLLm: ventral nucleus of the lateral lemniscus, magnocellular division; VNTB, ventral nucleus of the trapezoid body.

(fig. 21.3C–D). In the magnocellular part of the ventral nucleus of the lateral lemniscus and the intermediate nucleus of the lateral lemniscus (fig. 21.3A–B), complex patterns of labeling could not be assigned to a single tonotopic representation. The more complex labeling in these nuclei may occur as the result of projections from two or more frequency representations to the FM-FM recording site in the IC.

The strongest projections to delay-tuned neurons in the IC are from, in order, the ipsilateral nuclei of the lateral lemniscus, the ipsilateral superior olivary complex, and the contralateral cochlear nucleus (fig. 21.3E). The regions with complex labeling patterns, including the

anteroventral cochlear nucleus (fig. 21.3C–D), the magnocellular part of the ventral nucleus of the lateral lemniscus (fig. 21.3A–B), and the intermediate nucleus of the lateral lemniscus (fig. 21.3A–B), together constitute 49.8% of the labeling. The labeling patterns in these nuclei may possibly contain both low-frequency-tuned neurons and high-frequency-tuned neurons that project to FM-FM neurons in the ICC. Moreover, this label is sufficiently strong that low-frequency input could be substantial.

Very few labeled cells (<0.15% on average) were in the rostral part of the IC, where frequencies of 24–29 kHz (frequency band of the first FM harmonic) are

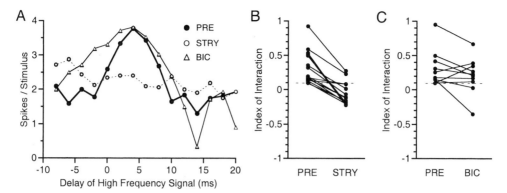

Fig. 21.4. Effects of local strychnine and bicuculline application on delay-tuned facilitation in FM-FM neurons of the inferior colliculus. *A:* In predrug testing (PRE), this delay-tuned neuron showed maximum facilitation when the 85.7 kHz signal was delayed 4 ms after onset of the 26.9 kHz signal. Strychnine (STRY) application eliminated delay-tuned facilitation in this FM-FM neuron. Strychnine was iontophoresed for 7 min at +32 nA. Application of bicuculline (BIC, +30 nA, 5 min) did not eliminate delay-tuned facilitation but altered some aspects of the delay-tuning curve. *B:* Compared to predrug conditions, strychnine decreased or eliminated delay-tuned facilitation in all 15 facilitated FM-FM neurons tested. Once facilitation was eliminated by strychnine, many responses to combination stimuli at the former best delay had interaction indices more negative than −0.11, meeting our criterion for combination-sensitive inhibition. This suggests the presence of an additional underlying inhibitory input normally masked by glycine-dependent facilitation. *C:* Bicuculline had no consistent effect on delay-tuned facilitation, eliminating facilitation in only two of nine neurons. In *B* and *C,* the dashed line indicates our criterion for facilitation (interaction index value of 0.09 or 20% facilitation).

represented (fig. 21.3E). This indicated that delay-tuned neurons in high-frequency representations of the IC do not receive substantial input from collicular neurons tuned to frequencies within the first sonar harmonic.

GLYCINERGIC INPUT TO ICC REQUIRED FOR DELAY-TUNED FACILITATION

Evidence from other studies (Olsen and Suga 1991; Saitoh and Suga 1995) implicated inhibitory processes in delay-tuned, facilitatory responses. We recorded facilitated FM-FM responses in the IC before and after local application of strychnine (to block glycine receptors) and bicuculline (to block GABAa receptors) in 11 bats (Wenstrup and Leroy 2001). The major result, that local strychnine application eliminated delay-tuned facilitation, is illustrated in fig. 21.4A. The neuron showed a facilitated response tuned to 26.9 kHz and 85.7 kHz, with peak facilitation occurring at 4 ms delay. After strychnine application, delay-tuned facilitation was eliminated. Strychnine eliminated facilitated responses in 13 of 15 neurons and greatly reduced the facilitation in the other two neurons (fig. 21.4B).

Two other results of strychnine application are important. First, strychnine application had no consistent effect on temporal aspects of the neuron's response. Among the 15 neurons, there was no significant effect on

the latencies of response to the high-frequency stimulus alone ($p > 0.1$, paired t-test) or to the combination stimulus ($p > 0.05$, paired t-test). Moreover, strychnine did not shift the delay at which facilitation occurred; rather, it eliminated the facilitated response altogether. Second, strychnine had no consistent effect on the magnitude of response to combination stimuli. For some units (fig. 21.4A), the response magnitude at best delay decreased with strychnine, even when response magnitude at other delays increased. Among other units, response at best delay was unchanged with strychnine, but responses at other delays were elevated; among the sample, there was no significant change ($p > 0.1$, paired t-test). There was no significant correlation between the change in the response magnitude at best delay and the change in facilitation as measured by the interaction index ($r = 0.18$, $df = 13$, $p > 0.1$).

Local application of bicuculline had less effect on delay-tuned facilitation than did strychnine, as shown for the unit in fig. 21.4A. In seven of nine neurons, delay-tuned facilitation persisted after bicuculline application, even when the drug elevated the overall response rate of neurons (fig. 21.4C). This preliminary result raises doubt that GABAa receptors play a significant role in the creation of delay-tuned, facilitated responses of collicular FM-FM neurons.

Discussion

In the mustached bat's IC, many neurons tuned to frequency ranges of the FM sweep in the bat's sonar call display a special set of complex response features. These so-called FM-FM neurons are characterized by sensitivity to distinct frequency bands, in which neurons show tuned sensitivity to two different harmonics of sonar sounds; interaction, in which the two spectral components either show facilitatory or inhibitory interactions; and delay tuning, in which the neuron shows selectivity for particular delays between the two spectral elements. Our research seeks the site(s) of construction for these response features and an understanding of underlying mechanisms. The present results support several conclusions. First, nearly all facilitatory FM-FM response properties originate in the IC. Second, FM-FM neurons, located in high-frequency representations of the IC, could receive low-frequency (24–29 kHz) input from nuclei of the lateral lemniscus or subdivisions of the anteroventral cochlear nucleus. Third, facilitation in the IC depends on glycinergic input, which must arrive from one or more of the auditory brainstem nuclei.

Origin of Most Facilitated FM-FM Responses in the IC

In single-unit recordings from lateral lemniscal nuclei, few neurons (<3%) showed combination-sensitive facilitation. Thus, nuclei that send the largest projection to collicular FM-FM neurons do not show FM-FM response properties. Moreover, because lateral lemniscal neurons receive their main input from many of the same lower brainstem nuclei that also project to the IC (Covey and Casseday 1995), we infer that these lower nuclei also show very few or no combination-sensitive facilitatory responses. These results favor a collicular origin for FM-FM responses. Additional evidence comes from microiontophoresis experiments. Local application of strychnine onto collicular FM-FM neurons eliminated delay-tuned facilitation in 87% of FM-FM neurons without altering other response properties. This result suggests that facilitated responses depend on glycinergic input to collicular neurons. Together, evidence from these experiments strongly supports the conclusion that facilitated, delay-tuned responses of FM-FM neurons originate in the IC.

Brainstem Sources of Low-Frequency Input

What are the sources of low-frequency input to high-frequency tonotopic representations of the IC? Our retrograde transport studies suggested several possibilities. One is the anteroventral cochlear nucleus. Rostral labeling in both anterior (AVa) and marginal (AVm) parts of this nucleus were in areas thought to respond primarily to low-frequency sounds (Kössl and Vater 1990). Other possibilities include intermediate nucleus and magno-

cellular parts of the ventral nucleus of the lateral lemniscus (INLL and VNLLm, respectively). Because frequency organization of these two nuclei has not been described for mustached bats, it is unknown whether the retrograde labeling is in regions containing low-frequency responses. However, topographic patterns of labeling in these nuclei are quite complex, allowing the possibility that the complex patterns are due to projection from both low- and high-frequency-sensitive neurons. These complex topographic patterns of labeling contrast with labeling in dorsal and posteroventral cochlear nuclei, lateral and medial superior olivary nuclei, and the dorsal nucleus of the lateral lemniscus, each of which show a single focus of labeling consistent with high-frequency sensitivity (Kössl and Vater 1990; Covey and Casseday 1995). In addition, the lack of retrograde labeling in low-frequency representations of the IC indicates that this nucleus does not provide low-frequency input to the high-frequency representations. These results suggest that INLL, VNLLm, AVa, and AVm are the most likely sources of low-frequency input to FM-FM neurons of the IC and warrant further study. However, these results do not conclusively identify any source of low-frequency input to FM-FM neurons in the IC.

Glycinergic Input as a Requirement of Facilitated FM-FM Responses in the IC

A major finding here is that locally applied strychnine eliminates delay-tuned facilitation in nearly all FM-FM neurons. The effect of strychnine is highly specific for delay-tuned facilitation. It produced no consistent change in the magnitude or latency of responses to combination stimuli. The source of this glycinergic input must be in the auditory brainstem, since no glycinergic cell bodies occur in the auditory midbrain, thalamus, or cortex (Winer, Larue, and Pollak 1995). It is noteworthy that glycinergic neurons are found in each brainstem auditory nucleus that could provide low-frequency input to FM-FM neurons of the IC (Kemmer and Vater 1997; Winer, Larue, and Pollak 1995).

In contrast to the effect of strychnine, delay-tuned facilitation usually remained after bicuculline application. This occurred even though response rates were elevated, indicating that the drug was having some effect on the neuron. Bicuculline had no significant effect on the latency of responses to individual or combination stimuli. These data suggest that GABAergic inputs to IC neurons, as mediated by GABAa receptors, play less of a role in construction of the delay-tuned facilitation. However, further study is required for a definitive conclusion.

These results also bear on mechanisms underlying response facilitation. At issue is the fact that glycinergic synapses are regarded as inhibitory, yet the main effect on these responses of blocking glycinergic input was to eliminate facilitation. It is conceivable that glycinergic input may evoke an apparent facilitated response by inhibiting a local collicular inhibitory interneuron, which in turn projects to the delay-tuned neuron under study. That inhibitory neuron would most likely be GABAergic. For this scenario, bicuculline application should eliminate delay-tuned facilitation as effectively as strychnine. But it does not, suggesting that this scenario is less likely. A second scenario is that glycinergic inhibitory input activates a postinhibitory rebound excitation. If rebound excitation occurs at the right time, the delay-tuned response would be facilitated. Other investigators proposed similar mechanisms to underlie duration selectivity in the big brown bat (Casseday, Ehrlich, and Covey 1994), and glycinergic postinhibitory rebound has been reported in spinal neurons in rats (Bertrand and Cazalets 1998). In the spinal cord, however, rebound excitation occurs tens to hundreds of milliseconds after glycinergic application, whereas these auditory effects occur within a few milliseconds. Thus, the cellular mechanisms that play a role in delay-tuned facilitation are probably different from those operating in the spinal cord.

ORIGIN OF DELAY TUNING

Delay-tuned facilitation depends on the coincidence of low-frequency-evoked excitation (presumably from the emitted pulse) and high-frequency-evoked excitation (presumably from the delayed echo) (Olsen and Suga 1991; Portfors and Wenstrup 1999). Since the low-frequency signal occurs earlier than the high-frequency signal, low-frequency excitation must be delayed within the central nervous system (i.e., have a longer latency) in order to achieve the coincidence. This clearly occurs in the IC; for FM-FM neurons that respond to both the low- and high-frequency components presented separately, latencies for the low-frequency response are always longer than the latencies for the high-frequency response. Moreover, the difference in latencies correlates strongly with a neuron's best delay, as predicted by coincidence detection models (Portfors and Wenstrup 1999). The glycinergic postinhibitory rebound mechanism that we believe underlies facilitation could also determine the latency of the low-frequency excitation, similar to what may occur in duration-tuned neurons (Casseday, Ehrlich, and Covey 1994). A major goal of our future work will be to identify mechanisms underlying delay tuning of collicular FM-FM neurons.

Acknowledgments

We thank the Natural Resources Conservation Authority of Jamaica and the Wildlife Section of the Ministry of Agriculture, Lands, and Food Production of Trinidad and Tobago for permission to collect bats. Our study was supported by research grant 5 R01 DC00937 from the National Institute on Deafness and Other Communication Disorders. The Institutional Animal Care and Use Committee approved experimental procedures.

{ 22 }

Temporal Processing of Rapidly Following Sounds in Dolphins: Evoked-Potential Study

Alexander Ya. Supin and *Vladimir V. Popov*

Introduction

Many natural sounds contain rapid amplitude fluctuations. Therefore, it is of interest to know how well the auditory system responds to these fluctuations—particularly in the case of dolphins, which use rapid successive locating sound pulses of very short duration (Au 1993).

In a number of behavioral studies, a very high temporal resolution of the dolphin's hearing was found. Analysis of echolocation data showed an integration time of between 200 and 300 μs (Au, Moore, and Pawloski 1988;

Au 1990). Experiments with discrimination between pulse pairs also have suggested that pulses merge into an "acoustic whole" when separated by no longer than 200–300 μs (Dubrovskiy 1990). Backward masking in dolphins also is possible at intervals of up to 200–300 μs (P. Moore et al. 1984; Dubrovskiy 1990).

Preceding studies showed that recordings of evoked potentials can be advantageous in determining many features of the dolphin's auditory system. In particular, the auditory brainstem response (ABR) can be used for measuring a number of auditory characteristics in dolphins (Ridgway et al. 1981; Popov and Supin 1990a, 1990b; Supin, Popov, and Klishin 1993). ABR recordings have confirmed high temporal resolution of the dolphin's hearing. In conditions of double-click stimulation, a just-detectable auditory brainstem evoked response (ABR) to the second click appeared at intervals of 200–300 μs (Supin and Popov 1985, 1995a). In conditions of stimulation by noise with a short gap, a just-detectable ABR was evoked by a gap as short as 100 μs (Supin and Popov 1995b; Popov and Supin 1997). Computation of an integration transfer function based on evoked-potential data resulted in an integration time of around 300 μs, which corroborates behavioral data.

The present study focuses on a few topics related to temporal processing of auditory stimuli that have called for more detailed investigations. During echolocation, dolphins emit high-rate trains of pulses and perceive corresponding trains of echoes. Therefore, responses to high-rate pulse trains are of specific interest. However, because complex interactions between successive stimuli are possible, integration times cannot help to predict how the auditory system responds to a train of high-rate stimuli. It should be noted that different results have been obtained in studies with different types of rhythmic stimulation. With the use of rhythmic clicks, a noticeable ABR amplitude decrease was observed at stimulation rates greater than 100 Hz; small evoked potentials were detectable at rates of up to a few hundred Hz (Ridgway et al. 1981; Popov and Supin 1990a, 1990b). However, the ability to follow rhythmic stimulation was much better in experiments with a sinusoidally amplitude-modulated sound. The evoked-potential sequence (the envelope following response, or EFR) had maximal amplitude at rates of 600–1000 Hz, and EFR of a significant amplitude was detectable at rates up to 1700 Hz (Dolphin 1995; Dolphin, Au, and Nachtigall 1995; Supin and Popov 1995c). Thus, because estimates of reproducible rates differed, the source of this discrepancy and the ability of the dolphin's auditory system to transfer rapid sound fluctuations may be significant.

A source of disagreement between data obtained with rhythmic sound pulses and sinusoidal amplitude modulation may be a long-term adaptation. In the above-mentioned experiments, ABRs were collected from long, steady-state pulse sequences. With the pulse amplitude constant, the mean power of the pulse sequence is proportional to the pulse rate; hence the long-term adaptation could increase with rate, thus resulting in the ABR amplitude decrease. Not speculating about the nature of adaptation, one can expect that at a higher mean stimulus level, the adaptation is stronger. This effect is impossible in sinusoidally amplitude-modulated stimuli, since the mean level is independent of the modulation rate; thus any adaptation effect, if it exists, cannot be rate specific.

Contrary to the steady-state experimental stimula-

tion, dolphins in natural echolocation conditions use click trains of a restricted duration. Here, the effect of a long-term adaptation could be much less pronounced. Therefore, in this study we measured the ability of ABR in dolphins to follow rhythmic sound pulses (clicks and pips) presented as short trains separated by longer intervals (to more closely simulate natural conditions).

We also investigated combined frequency-temporal interactions in the dolphin's auditory system. Frequency tuning of the dolphin's auditory system has been measured in a number of studies using both behavioral (C. S. Johnson 1968, 1971; C. S. Johnson, McManus, and Skaar 1989; Au and Moore 1990) and evoked-potential (Supin, Popov, and Klishin 1993; Popov, Supin, and Klishin 1995, 1997) methods. The latter studies showed rather high frequency tuning, several times more acute than in terrestrial mammals. In view of very high temporal resolution and frequency tuning of the dolphin's auditory system, its processing of complex acoustic signals—in particular, those consisting of two or more carrier frequencies modulated at high rates—could be significant. This could explain interactions between different frequency channels in stimulation conditions mimicking echolocation, especially with stimulation by high-rate sound pulses.

Materials and Methods

SUBJECTS

The studies reported herein were carried out during the summers of 1995–1997 at the Utrish Sea Station of the Russian Academy of Sciences (Black Sea coast). The experimental animals were bottlenose dolphins, *Tursiops truncatus,* males and females, captured 1–2 months prior to each study season. The animals were kept in seawater pools measuring $9 \times 4 \times 1.5$ m.

EXPERIMENTAL CONDITIONS

During experimentation, a dolphin was placed in a small pool ($4 \times 0.6 \times 0.6$ m) filled with seawater. The animal was supported by a stretcher so that the dorsal part of the body and blowhole were above the water surface. The stretcher was made of a sound-transparent material (thin net). The animal was neither anesthetized nor curarized. The daily experimental session lasted for 3–4 hours, after which the animal was returned to its home pool.

EVOKED POTENTIAL COLLECTION

Evoked potentials were recorded using 1 cm disk electrodes secured at the body surface with a drop of adhesive electric-conductive gel. The active electrode was placed at the head vertex, 5–7 cm behind the blowhole, and the reference electrode at the dorsal fin, both above the water surface. The recorded potentials were amplified within a passband of 200–5000 Hz, digitized using an A/D converter, and averaged using a standard personal computer and custom-made software. The record window was 10–30 ms long; 500–1000 sweeps were averaged to collect one evoked-response record.

Stimuli

Short tone clicks and pips, rhythmic sequences of clicks or short pips, amplitude-modulated and non-modulated tone bursts, and continuous tones were used as stimuli. Clicks were generated by activation of a spherical piezoceramic transducer (3 cm in diameter) by rectangular pulses of 10 μs duration. Thus, the click-frequency spectrum was determined mainly by the transducer frequency response. The spectrum peaked at 50 kHz and had a bandwidth up to 92 kHz at a level of −30 dB. Pips and amplitude-modulated and nonmodulated tone bursts were generated digitally at a 500 kHz sampling rate and played through a D/A converter, amplifier, attenuator, and the same transducer. Their carrier frequency varied from 32 kHz to 90 kHz. Pip envelopes were a single cycle, cosine function. Modulated tone-burst envelopes comprised bursts of cosine functions with various numbers of cycles.

Trains of sound pulses (clicks or pips) were presented in series of 20 ms; amplitude-modulated bursts were also 20 ms long. Pulse or modulation rate within a train (burst) varied from 50 to 4000 Hz, maintaining a whole number of modulation cycles during the 20 ms window. The last pulse or the last modulation cycle started at the twentieth ms; thus, the number of pulses or modulation cycles in a burst was $0.02f + 1$, where f is the pulse rate, in Hz. Trains or bursts were repeated at a rate of 10 Hz; thus, the duration of silence intervals was 80 ms. Single pulses also were presented at the rate of 10 Hz.

The transducer was immersed in water at a depth of 30 cm, 0.5–1 m in front of the animal's head. Intensity and duration of stimuli were monitored through a probe hydrophone with a passband of 150 kHz, also located near the animal's head.

Results and Discussion

ABR Waveform and Spectrum

ABR waveform and frequency spectrum were significant in the analysis of rhythmic evoked-potential sequences. The waveform of ABR provoked by short clicks is shown in fig. 22.1, which presents evoked potentials to rhythmic click series of various frequencies, including low rates when each stimulus evokes a separate response (10 Hz). Each ABR consisted of a few waves occurring mainly within the first 5 ms after the click; late waves (positivity at 3.7 ms, negativity at 4.2 ms) were of the highest amplitude. This corresponded to ABR properties described in earlier studies (Ridgway et al. 1981; Popov and Supin 1985, 1990a,

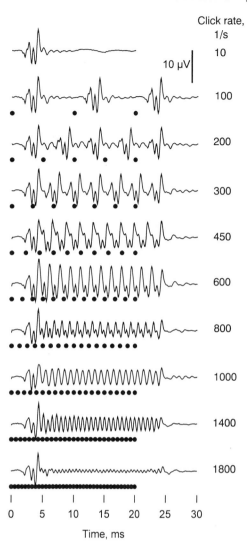

Fig. 22.1. RFR to rhythmic clicks at rates from 10 to 1800 Hz. Stimulus level is 50 dB above the ABR threshold; dots show instances of click presentation; the double-headed arrow at the time scale shows the time window for analysis of quasi-sustained RFR.

1990b). At high-intensity clicks, response amplitudes exceeded 10 μV peak-to-peak.

To interpret further results of rhythmic stimulation, it is important to know the proper ABR frequency spectrum. Fig. 22.2 presents frequency spectra of ABRs obtained at several stimulus intensities, from 80 to 120 dB re 1 μPa peak-to-peak sound pressure (these levels were 20–60 dB above ABR threshold, respectively). The spectrum was similar at all stimulus intensities. It contained a few peaks and valleys: the most prominent peaks were at 200–250, 500–600, 1000–1200, and above 2000 Hz; valleys were at 400–450, 700–800, and above 2000 Hz. Peaks at 500–600 and 1000–1200 Hz were the most prominent. This rippled spectrum pattern is a direct consequence of a few successive waves in the

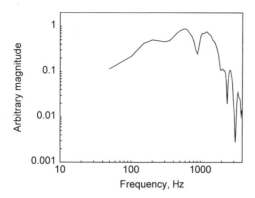

Fig. 22.2. Frequency spectrum of single ABR to a click (click level 60 dB above the ABR threshold)

ABR waveform. A significant spectrum magnitude (more than 0.1 re maximum) was observed in a wide frequency range, from 50 to 2500 Hz.

RESPONSES TO RHYTHMIC CLICK TRAINS

At rhythmic stimulation with rates up to 200–300 Hz, successive clicks evoked distinct ABRs, similar to those evoked by single clicks (fig. 22.1). At higher rates, ABRs merged into a complex quasi-sustained waveform. Only small but noticeable changes in the response waveform and amplitude were observed. By analogy with the envelope-following response (EFR) to amplitude-modulated sounds, we designated this response type as a rate-following response (RFR). At click rates up to 800 Hz, the RFR waveform was complex and nonsinusoidal. At higher rates, the RFR waveform became simple and close to a sinusoid. At rates higher than 600 Hz, the response contained an initial transient response (ABR) to the train beginning and the subsequent quasi-sustained RFR of a smaller amplitude.

At stimulation rates up to 450–600 Hz, continuous transition could be traced from RFR waves closely resembling the single ABR waveform to smooth waveforms at higher rates. This indicates that RFR is a rhythmic sequence of ABRs. This is also substantiated by the 4–4.5 ms end lag of RFR (the ongoing RFR after the last pulse of a series), which corresponds to latencies of the highest ABR waves.

To evaluate the ability of the auditory system to reproduce high-rate sound modulation, RFR amplitude was used. This amplitude remained comparable with that of single ABRs, at least at rates up to 1000 Hz. Significant amplitude decrease was observed at rates above 1400 Hz (fig. 22.1). RFR amplitude as a function of stimulation rate is shown in more detail in fig. 22.3. To evaluate RFR amplitude, peak-to-peak amplitudes of all RFR cycles (taking a one-cycle amplitude as a difference between the highest positive and negative peaks during the cycle) were averaged within a 20 ms time window, from 5.5 to 25.5 ms. As fig. 22.3 shows, at all stimu-

lus intensities, RFR amplitude was almost constant at stimulation rates up to 200 Hz. Further rate increase resulted in a few small peaks and valleys of the function: peaks at 500–600, 1000–1200, and around 2400 Hz; valleys at 400–450, 700–800, and around 2000 Hz. These peaks and valleys corresponded fairly to those of the ABR spectrum (see fig. 22.2). Superimposed on these peaks and valleys, a general trend of the function was observed with the amplitude decrease at rates above 1000 Hz, so that a 10-fold amplitude decrease occurred at a rate of 1700 Hz. However, small responses appeared at much higher rates (100-fold amplitude decrease at 2800–3200 Hz). A stimulus level shift resulted in a function shift along the ordinate axis (the higher level, the higher amplitude), but its form was little changed.

Although the function resembled the ABR spectrum in respect to the presence of characteristic peaks and valleys, it did not reproduce the ABR spectrum exactly. At low rates (below 200 Hz), the amplitude versus rate function was constant, whereas the ABR spectrum magnitude decreased at low frequencies. At high rates (above 1700 Hz), the amplitude versus rate function passed below the ABR spectrum.

To characterize RFR properties in terms of a frequency transfer function, frequency spectra of the rhythmic responses were calculated. For this purpose, a 20 ms part of the record (from 5.5 to 25.5 ms after the stimulus beginning) was Fourier transformed. Examples of spectra obtained at a few stimulation rates are presented in fig. 22.4. All the spectra had clearly visible peaks corresponding to the fundamental and harmonics of the stimulation rate. The RFR spectral composition resembled to a large extent the spectrum of a single ABR. At low rates (100–200 Hz), the magnitude of the fundamental was much less than those of higher harmonics, and the highest harmonic magnitudes were at 500–600 and 1000–1200 Hz—that is, at frequencies corresponding to the main peaks of the single ABR spectrum. At higher rates, magnitudes of the fundamentals increased. Fig. 22.4 shows the highest fundamental mag-

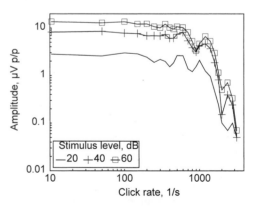

Fig. 22.3. RFR peak-to-peak amplitude dependence on click rate. Click levels are 20–60 dB above the ABR threshold.

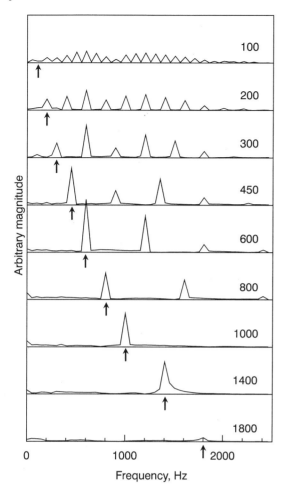

Fig. 22.4. Frequency spectra of RFR to click rates from 100 to 1800 Hz. Time and magnitude scales are linear; stimulus level is 50 dB above the ABR threshold; arrows show fundamentals of the stimulation rates.

nitudes at 600 and 1000 Hz and rather high magnitudes at rates from 450 Hz to 1400 Hz. These were the same frequencies that had high magnitudes in the ABR spectrum. At stimulation rates above 1400 Hz, the magnitude of fundamentals decreased markedly. At a rate of 1800 Hz, the fundamental peak was very small. The similarity of RFR and ABR frequency spectra also confirms an origin of RFR as a rhythmic sequence of ABRs, and a few peaks and valleys of the RFR amplitude versus stimulation rate function obviously reflect corresponding features of the ABR spectrum.

The latter conclusion is important to evaluate the ability of the dolphin's auditory system to reproduce stimulation rates. Although the function describing RFR amplitude dependence on frequency has several peaks and valleys, it does not indicate that rates of 500–600, 1000–1200 Hz, and so on are transferred more efficiently in the auditory system than rates of 450–500, 700–800 Hz, and so on. The ABR waveform consists of a few waves with different delays. Overlapping of potentials pro-

duced by several generators with different delays results in peaks and valleys in both the ABR spectrum and RFR transfer function. This spectral pattern has little to do with the real temporal resolution of the auditory system.

Neglecting these alternating peaks and valleys, the RFR transfer function looks like a low-pass filter. Regardless of the evoked potential waveform, appearance of potentials at a certain rate indicates a capability of neuronal structures to follow this rate. Thus, the temporal resolution of the dolphin's auditory periphery is not expected to be less than the obtained RFR transfer function, which has a cutoff frequency of about 1700 Hz at a 0.1 amplitude level.

On the other hand, the real ability of the auditory system to transfer modulation rates is hardly higher than the cutoff frequency of the RFR transfer function. Indeed, the upper frequency limit of the RFR transfer function was slightly but significantly less than the ABR frequency bandwidth, as comparison of the ABR spectrum (fig. 22.2) with the RFR transfer function (fig. 22.3) shows. It is noteworthy also that at a high stimulation rate (1800 Hz), the 1800 Hz fundamental magnitude was less than the 1800 Hz harmonics of lower stimulation rates (200–600 Hz) (fig. 22.4). Thus, it was not the ABR waveform that limited reproduction of frequencies higher than 1700 Hz. Hence, the RFR frequency limit of around 1700 Hz reflects the real ability of the auditory pathways to transfer rapid temporal modulations. We stress, however, that this cutoff frequency estimate was defined at an arbitrarily chosen criterion (10-fold amplitude decrease). A smaller but still noticeable RFR was observed at stimulation rates up to 3200 Hz (fig. 22.3).

The presented estimate of the cutoff frequency of RFR transfer function differs markedly from that obtained in preceding studies with the use of steady-state rhythmic clicks (Ridgway et al. 1981; Popov and Supin 1990a, 1990b). This difference may be explained by a long-term adaptation in conditions of steady-state stimulation: when the click rate increases, the mean sound level and hence the adaptation are bound to increase. In short-train stimulation conditions, the adaptation was insignificant. The lack of adaptation is indicated by the fact that transient on-responses to the burst onsets were almost independent of stimulation rate (fig. 22.1). On the other hand, the RFR transfer function obtained with the use of short pulse trains was similar to that obtained with amplitude-modulated sounds (Dolphin 1995; Supin and Popov 1995c): in both cases the cutoff frequency exceeded 1000 Hz.

Thus, data presented herein confirmed the highest estimations of the capability of the dolphin's auditory system to follow rhythmic stimulation. These results correspond to the integration time of the dolphins' hearing of around 300 μs: the low-pass cutoff frequency of 1700 Hz corresponds to an integration time of 300 μs. Because integration times shorter than a few milliseconds have never been found in humans nor in a number of animals (Fay 1988), dolphins clearly show a remarkable difference in terms of the temporal resolution of the auditory system.

Lateral Suppression of Responses to Rhythmic Sound Stimuli

EFR evoked by the amplitude-modulated probe could be modified markedly by simultaneous presentation of a tone that differed in frequency from the probe. We designated this as the "extra tone." The result depended, both quantitatively and qualitatively, on the frequency difference between the probe and the extra tone. Typical modes of interactions are exemplified in figs. 22.5 and 22.6 for EFR evoked by a sinusoidally modulated tone (modulation rate of 1000 Hz). In these experiments, the probe stimulus (*Pr* record) did not have a shallow rise ramp, so the evoked-potential waveform consisted of both a transient on-response and a quasi-sustained rhythmic response (EFR). The on-response had a latency of around 1.5 ms. (In the pre-

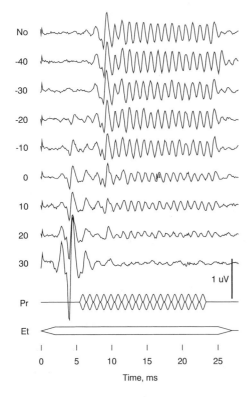

Fig. 22.5. ABR and EFR suppression by extra tones. Probe: sinusoidally amplitude-modulated tone burst at 76 kHz, 100 dB re 1 μPa (40 dB above response threshold), burst duration 18 ms, modulation rate 1000 Hz. Extra tone: nonmodulated 25 ms tone burst with 2 ms rise-and-fall ramps. Extra-tone frequency 78 kHz; extra-tone levels are shown next to records in dB re probe level. *No:* no extra tone; *Pr* and *Et:* envelopes of the probe and extra tone.

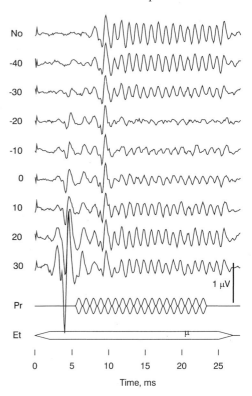

Fig. 22.6. The same as fig. 22.5, extra-tone frequency 85 kHz.

sented cases, the stimulus onset occurred 5.5 ms after the beginning of the record; thus, the on-response occupied a part of the time scale from approximately 7 to 11 ms.) A subsequent sustained rhythmic response reproduced the modulation rate of 1000 Hz during the full stimulation time.

When the probe and extra tone were similar in frequency (76 kHz in fig. 22.5), low-intensity extra tones (20 dB below the probe level and lower) influenced neither transient ABR nor sustained EFR. As the intensity of the extra tone increased, both ABR and EFR were suppressed. Note that at high levels, the extra tone itself evoked an ABR in spite of the prolonged rise-ramp. This response occupied the time scale from 2 to 5 ms and was not considered a probe response. The probe response subjected to suppression appeared after 7 ms of the record. The observed suppression had typical features of regular tone-tone masking as described earlier in dolphins (Supin, Popov, and Klishin 1993; Popov, Supin, and Klishin 1995).

A very different effect was observed when the extra-tone frequency was 5–20 kHz higher than that of the probe (76 kHz probe and 85 kHz extra tone in fig. 22.6). In this case, suppression of the probe EFR appeared at very low extra-tone levels. A small but noticeable decrease in the EFR amplitude occurred at an extra-tone level as low as 40 dB below the probe level. The extra-tone level increase resulted in a deeper EFR suppression, but the suppression still appeared at rather low

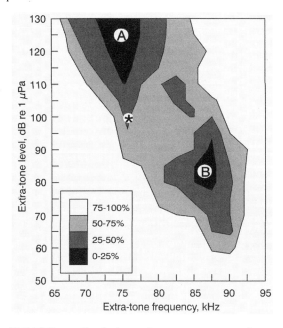

Fig. 22.7. EFR amplitude dependence on extra-tone frequency and level. Probe stimulus is the same as in fig. 22.6. Asterisk shows the probe frequency and level (76 kHz, 100 dB re 1 μPa). EFR amplitude is shown by shadowing according to the legend in % re no-extra-tone response. *A* and *B:* areas of masking and lateral suppression, respectively.

extra-tone levels. The maximum suppression was observed when the extra-tone level was 20 dB below the probe level: EFR disappeared almost completely. Further increases in the extra-tone level paradoxically resulted not in further response suppression but instead in releasing the probe response. At +20 dB extra-tone level, EFR amplitude recovered to almost the same level as seen in no-extra-tone conditions. At very high extra-tone levels (+30 dB in fig. 22.6), the EFR was slightly suppressed again. This suppression can be attributed to a masking effect that occurs at high masker levels—even when probe and masker frequencies are different (Supin, Popov, and Klishin 1993; Popov, Supin, and Klishin 1995). An important feature of this effect was that low-level extra tones suppressed only the sustained EFR while the transient ABR remained unchanged.

To obtain a complete pattern of the suppression effects, a wide variety of extra-tone frequencies and intensities were tested at each of the probe frequencies. A representative experiment is presented in fig. 22.7, which shows EFR amplitude as a function of both extra-tone frequency and level. Two distinctly separate areas of EFR suppression are seen in the plot. The first, arbitrarily designated as area *A*, presents suppression appearing at extra-tone intensities higher than that of the probe. This suppression was most effective when the extra-tone frequency coincided with that of the probe. Obviously this is the regular masking effect, and isolevel lines in this area of the plot represent frequency-tuning

curves that reveal the frequency selectivity of the masking. Another area of suppression, area *B*, appeared at higher frequencies and lower intensities than those of the probe. This suppression was mostly effective at the extra-tone frequencies 10–12 kHz above the probe frequency and levels 15–20 dB below the probe level. At these parameters, the suppression can be assessed as a kind of lateral suppression, since it appeared when frequencies of the probe and extra tone differed markedly. Thus, we refer to this phenomenon as "lateral suppression."

The lateral suppression was only observed at extra-tone frequencies higher than that of the probe. No suppression was produced by low-level extra tones of frequencies lower than that of the probe. The two areas of suppression, *A* (masking) and *B* (lateral suppression), were separated by a gap—an area of less effective suppression, demonstrating that the two suppression areas reflect two different mechanisms: one (area *A*) obviously is a regular masking, while the other (lateral suppression, area *B*) reflects another kind of neuronal interaction.

In principle, lateral suppression is a well-known effect in the auditory system (Evans 1992). However, the phenomenon described above differs from the known cases of lateral suppression in two paradoxical features: (1) the suppression was observed when the extra-tone (suppresser) level was much lower than the probe level; and (2) the extra-tone level increase over a certain level led not to the further suppression increase but rather to its decrease. There is, however, a third feature that can explain these paradoxical properties: only the sustained rhythmic response (EFR) was subject to this kind of suppression—the transient on-response (ABR) was not. Thus extra tones of low levels do not reduce the overall excitability in the auditory structures; rather, they reduce their ability to reproduce high stimulation rates.

Further studies have shown that the lateral suppression depends on stimulation rate. To investigate the lateral suppression at a wide variety of probe rates, we used stimuli consisting of a train of short tone pips of constant duration (1 ms), but of variable rates (fig. 22.8). Contrary to sinusoidally modulated sounds, the rise-fall time of the pips was independent of the presentation rate, so this kind of stimulus effectively provoked evoked potentials at both low and high presentation rates.

Fig. 22.8 presents responses to pip trains at rates from 200/s to 1000 Hz. In the absence of the extra tone (*A*), all the rates produced robust responses. The 200 Hz sequence evoked separated ABRs; at 1000 Hz, the 1 ms pips fused into a sinusoidally modulated sound, and evoked potentials fused into a sustained EFR. In the presence of the extra tone (*B*), responses to low pip rates were slightly influenced, but suppression increased with rate, and at a rate of 1000 Hz the suppression became the most effective. Thus, the lateral suppression

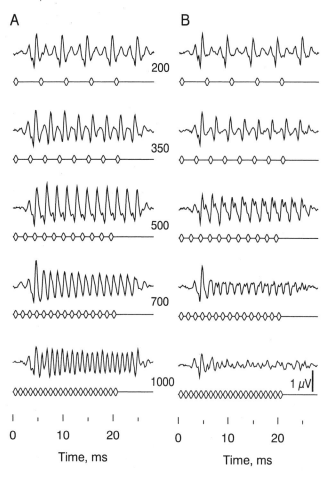

Fig. 22.8. Lateral suppression at various rates of probe pips. Probes: trains of cosine-enveloped pips of 64 kHz carrier, 1 ms duration, 100 dB re 1 μPa. Pip rates (s^{-1}) are indicated near records. In each pair of records, the upper one is the evoked-response record, and the lower one is the stimulus envelope. *A:* no extra tone; *B:* extra tone of 75 kHz, 80 dB re 1 μPa.

demonstrated an obvious rate dependence: the higher pip rate, the deeper the suppression. These data confirm that low-level extra tones inhibit the reproduction of high stimulation rates. This explains why a weak sound can suppress a response to a much stronger one: responses are actually suppressed by the preceding pulses of the probe itself, while the presence or absence of the extra tone enables or disables this suppression.

The experiments described above presented interactions between two sounds when one (the probe) was amplitude-modulated and capable of evoking rhythmic response, whereas the other (the extra tone) was nonmodulated. Thus, only the influence of a nonmodulated sound on the response to a modulated one was studied—not vice versa. It was also of interest to study the interaction between two carriers of different frequencies and levels when both of them were capable of evoking rhythmic response. For this purpose, the two carriers were modulated by different rates—that is, each was "la-

beled" with its modulation rate, so the response frequency showed which of them evoked the response.

Experiments with two modulated carriers showed that lateral suppression appeared in these conditions as well. Fig. 22.9A exemplifies the interaction between two carriers of different frequencies and levels: *St1* of 76 kHz 100 dB, and *St2* of 85 kHz 80 dB re 1 μPa. The carrier *St1* was modulated by a rate of 1000 Hz; the carrier *St2* by a rate of 600/s. Presented alone, carrier *St1* evoked a robust EFR at the rate of 1000 Hz (fig. 22.9A-1) while the carrier *St2*—although its level was 20 dB lower—also evoked a significant EFR at the rate of 600 Hz (fig. 22.9A-2). Fig. 22.9A-3 shows a result of the two carriers' overlapping. During the action of *St1* alone (the initial part of the record), the 1000 Hz EFR was observed. As soon as *St2* was added (the later part of the

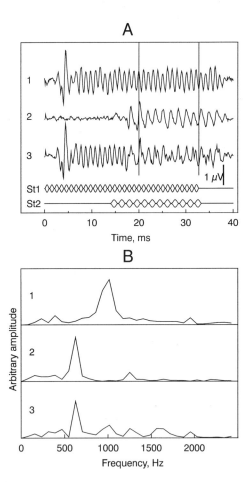

Fig. 22.9. EFR evoked by two carriers and their combination. *A*(1): EFR to carrier 1 of 76 kHz, 100 dB re 1 μPa, modulation rate 1000 Hz. *A*(2): EFR to carrier 2 of 85 kHz, 80 dB re 1 μPa, modulation rate 600 Hz. *A*(3): EFR to both carriers superimposed. *St1* and *St2*: envelopes of carriers 1 and 2, respectively. (Note: *St1* and *St2* records show only envelope waveforms, not their amplitude relations.) Vertical lines delimit the time window for Fourier transform. *B* (1–3): Fourier transforms of the record fragments *A* (1–3) respectively.

record), the EFR of 1000/s was markedly suppressed, and the 600 Hz EFR appeared and prevailed over the 1000/s EFR—despite the lower level of the carrier modulated by 600 Hz.

To evaluate the effect quantitatively, the 12.8 ms fragments of all the presented records were Fourier transformed to obtain their frequency spectra. The results are shown in fig. 22.9B. The EFR to *St1* alone had a prominent spectral peak at 1000 Hz (fig. 22.9B-1). Accordingly, EFR to *St2* had a peak at 600 Hz (fig. 22.9B-2). The EFR to the combination of the both carriers (fig. 22.9B-3) had a prominent peak at 600 Hz—almost the same magnitude as to *St2* alone. A peak at 1000 Hz was several times less than that of response to *St1* alone. Consequently, the contribution of the weaker sound to the combined response was much more substantial than that of the stronger one.

Suppression, therefore, makes the dolphin's auditory system more sensitive to pulsations of weak sounds of higher frequency rather than to pulsations of stronger sounds of lower frequency. It may be supposed that the effect is connected with the biosonar function. The dolphin's hearing demonstrates increased sensitivity to high sound frequencies in frontal directions, where a target is usually located (Au and Moore 1984; Supin and Popov 1993). Thus, the echo can be heard with a greater proportion of higher frequencies than the emitted pulses. Weaker intensity and higher frequency of echoes compared to the emitted sounds are just the conditions allowing the suppression of the auditory response to the emitted train of sonar pulses and advantageous echo perception. Moreover, better sensitivity to high sound frequencies in frontal directions can increase the proportion of higher frequencies in echoes as compared to reverberating sounds. This could result in suppression of responses to reverberation, enhancing the echo perception.

Acknowledgments

We followed the guidelines established by the Russian Ministry of Higher Education on the use of animals in biomedical research. The study was supported by the Russian Foundation for Basic Research (RFBR), grant 97-04-49024. Alexander Ya. Supin greatly appreciates the RFBR travel grant and the financial support of the Office of Naval Research that enabled him to participate in the Biological Biosonar Conference held in Portugal.

{ 23 }

Chemical Neuroanatomy of the Inferior Colliculus in Brains of Echolocating and Nonecholocating Mammals: Immunocytochemical Study

Ilya L. Glezer, Patrick Hof, Peter J. Morgane, Anna Fridman, Tamara Isakova, Donald Joseph, Arun Nair, Preety Parhar, Abraham Thengampallil, Sanjay Thomas, Ravi Venugopal, and *Gariel H. Jung*

Introduction

Morphological studies of the cetacean brain have concentrated on the cerebral cortex. Data of many studies, including ours, on cytoarchitecture and Golgi organization of the cetacean neocortex (particularly, in *Tursiops truncatus*) showed many features that are similar to those in the archicortex and paleocortex of terrestrial mammals (Morgane and Jacobs 1972; Kesarev 1977; Morgane et al. 1985, 1986a, 1986b; Morgane and Glezer 1990; Glezer et al. 1988). These features include a very narrow cortical plate, homogeneous cytoarchitecture, extensive morphological development of layers I (in thickness) and II (in cellularity) compared to other cortical layers, and overall agranularity of the neocortex, which is particularly noticeable in the absence or very weak development of layer IV in all regions of the neocortex (Morgane et al. 1985, 1986a, 1986b, 1988, 1990; Glezer et al. 1988). As we previously showed in Golgi materials, layer II of odontocete cetaceans is enriched in so-called extraverted neurons, spreading their dendrites widely into layer I (Morgane et al. 1986a, 1986b).

In our study (Glezer et al. 1998), we showed that these proisocortical features are present in both visual and auditory cortices of odontocete cetaceans. The presence of these features of the cetacean neocortex coincides with the enrichment of both cortical and subcortical structures of the visual and auditory systems with populations of calretinin (CR) and calbindin D-28k (CB) immunoreactive neurons, whereas parvalbumin (PV) immunoreactive neurons are seen in relatively small numbers. These three calcium-binding (CaBP) proteins are characteristic for major groups of inhibitory cortical and subcortical neurons in the mammalian brain. CaBP proteins are widely distributed in the central nervous system of different mammals and in most brain structures are co-localized with the major inhibitory neurotransmitter, aminobutyric acid (GABA) (Jones 1986; Rogers 1987; Demeulemeester et al. 1991; Hendry et al. 1989; Blümcke et al. 1990, 1991; Hof and Nimchinsky 1992; Hof et al. 1992, 1995, 1996, 1999; Glezer et al. 1993, 1995, 1998). Various studies established that 90–95% of GABAergic neurons are co-localized with one of the three calcium-binding proteins that is, calbindin, parvalbumin, or calretinin (Andressen et al. 1993; DeFelipe and Jones 1985; Jones 1986; Kosaka et al. 1987; Celio 1986, 1990; Hendry et al. 1989; Hendry and Jones 1991; Blümcke et al. 1990, 1991; Demeulemeester et al. 1991; Hof et al. 1992, 1999; Glezer et al. 1990, 1992, 1993a, 1993b, 1995, 1998). Thus, use of calcium-binding proteins as markers for local circuit neurons facilitates the study of chemically and morphologically distinct subsets of GABAergic neurons. In other words, by studying the CaBP-immunoreactive neurons one can reveal the distribution of the subpopulations of the inhibitory neurons and thus indirectly characterize organization of the microcircuits in the particular regions of the central nervous system.

The major goal of the present study was to find possible specific features in the inferior colliculus (IC) of echolocating mammals. To date, biosonar abilities were documented only in odontocete cetaceans and insectivorous bats (Sokolov et al. 1972; Supin et al. 1978; Zook et al. 1988; Zettel et al. 1991).

Based on physiological and behavioral data we anticipated the presence of some convergent specific features of central processing units of the auditory system in echolocating unrelated evolutionary species of aquatic and terrestrial mammals. The most probable site of the convergence of features we can expect is in auditory structures related to the time measurement of the arriving auditory stimuli. It is possible that processing of acoustic information used in echolocation is performed by chemically identified subpopulations of neurons. We hypothesized that evolutionary changes in behavioral traits that resulted in acquisition of biosonar capabilities profoundly changed the morphology and physiology of auditory structures. Consequently, this would be reflected not only in specific but also in general morphology and physiology of the CNS in echolocating mammals.

As is well known in terrestrial mammals, the inferior colliculus (IC) is one of the most important processing centers in the acoustic sensory system (Aitkin 1986; Morest and Oliver 1984; Faingold et al. 1991). In the physiology of the auditory sensory system, IC plays a major role as a relay center conveying acoustic information transformed into nerve impulses from the cochlear nuclei of the pontine tegmentum and then relayed to higher centers of the diencephalon (medial geniculate nucleus) (Rose et al. 1963; Kuwada et al. 1984; Oliver and Shneiderman 1989). The major afferent input to the IC is reached via the lateral lemniscus, consisting of the ipsi- and contralateral axons originating in the cochlear nuclei, the superior olivary complex, and nuclei of the lateral lemniscus (Coleman and Clerici 1987). Also, IC receives many inputs from the contralateral IC through commissural fibers (González-Hernández et al. 1986; Herrera et al. 1988; Salada and Merchán 1992). IC also is controlled by the inputs from the primary auditory cortex (Andersen et al. 1980; Games and Winer 1988; Faye-Lund 1988; Herbert, Aschoff, and Ostwald 1991). Other systems have substantial inputs into IC (somatosensory, visual, pyramidal, and extrapyramidal motor) (Itaya and Hoesen 1982). The multiple inputs into the IC coincide with the complex patterns of responses of the IC neurons to acoustic stimuli. It was shown that these responses can be both excitatory and inhibitory depending on laterality, intensity, delay, and frequency of the stimuli (Aitkin 1986; Pollak 1988; Fuzessery 1994; Park and Pollak 1994; Ehrlich, Casseday, and Covey 1997). Based on immunocytochemical and physiological studies in the last 20 years it was firmly established that GABAergic neurons play a crucial role in functions of IC (Thompson et al. 1985; Roberts and Ribak 1987; Moore and Moore 1987; Bristow and Martin 1988; Faingold et al. 1991; Vater, Lenoir, and Pujol 1992; Ribak et al. 1993; Oliver et al. 1994; Park and Pollak 1994). In terrestrial mammals, a substantial part of the neuronal population of the IC is GABA-immunoreactive (Roberts and Ribak 1987; Ribak et al. 1993; Oliver et al. 1994). Not only neuronal perikarya but also numerous pre- and postsynaptic structures of IC are GABA-immunoreactive (Ribak et al. 1993; Vater and Brown 1994). Some of these terminals are intrinsic, but others originate in different structures of the acoustic system (Ribak et al. 1993). Alterations in the GABAergic neurons in IC play an important role in changing of acoustic functions with age and in genetic epilepsy (Faingold et al. 1991; Ribak et al. 1993). The leading role of GABAergic neuronal populations in functions of the IC (Roberts and Ribak 1987) correlates with immunocytochemical data showing the presence of large populations of the IC neurons immunoreactive to the three calcium-binding proteins (CaBP): calbindin (CB), calretinin (CR), and parvalbumin (PV). Dorsal and lateral nuclei of IC in rats and bats are enriched with CB- and CR-immunoreactive perikarya, whereas the central nucleus contains mostly PV-immunoreactive neurons (Zettel et al. 1991).

One of the key findings in our recent comparative study of auditory and visual systems of the bottlenose dolphin (*T. truncatus*) pertains to the inferior colliculus (Glezer et al. 1998). We found that the IC in the bottlenose dolphin displays a vertical columnar pattern (Glezer et al. 1998) in addition to the typical horizontal lamination of the central nucleus, as found in terrestrial mammals (Geniec and Morest 1971; Malmierca et al. 1993; Oliver et al. 1994). These vertical columns of neurons correspond to irregular, densely stained domains in tangential sections and contain mostly CB-immunoreactive neurons and smaller numbers of CR-immunoreactive neurons. The interdomain spaces contain mostly large CR-immunoreactive neurons. Since the auditory system in odontocete whales, including the bottlenose dolphin, plays a major role as a source of information essential for their specialized behavior (Popper 1980a, 1980b; Nachtigall 1986; Herman 1990; Schusterman 1988, 1990), these morphologically and cytochemically defined domains and interdomains in the IC of *T. truncatus* can be interpreted as highly specialized adaptations for processing of auditory signals. It is possible that this unusual organization, combining both columnar and laminar patterns of the IC in the bottlenose dolphin, relates to echolocation abilities in odontocete cetaceans. In the present study, we tested this hypothesis by comparing the morphochemical features of IC of several odontocetes (with documented echolocating abilities) with the features of the IC in nonecholocating mysticete cetaceans and pinnipeds. The terrestrial nonecholocating (primates, ungulates) and echolocating (chiropterans) were also examined for comparison.

Material and Methods

SPECIES

In this study, inferior colliculi (IC) of the tectum opticum of the midbrain from the following species were investigated: *T. truncatus; Delphinapterus leucas; Lagenorhynchus obliquidens; Inia geoffrensis; Sotalia pallida; Stenella coeruleoalba; Globicephala melaena; Megaptera novaengliae; Balaenoptera physalus; Arctocephalus pusillus; Zalophus californianus; Myotis myotis; Macaca fascicularis; Ovis aries;* and *Homo sapiens* (see table 23.1).

FIXATION

In cases of *T. truncatus, S. coeruleoalba, D. leucas, M. myotis,* and *M. fascicularis* the brains were perfused in situ with 2.5% glutaraldehyde and 4% paraformaldehyde solution in 0.1 M phosphate buffer. After removal, brains were postfixed in a mixture of 2.5% glutaraldehyde and 4% of paraformaldehyde in phosphate or cacodylate 0.1 M buffers for 2–4 hours; and they were kept in 0.5% glutaraldehyde/4% paraformaldehyde solution in 0.1 M phosphate buffer with 10% sucrose for at least 12 hours before collecting the brain samples (see below). In all other cases, brains were fixed by immersion in 10% neutral formalin and then postfixed in a mixture of 2.5% glutaraldehyde and 4% paraformaldehyde in 0.1 M phosphate buffer.

IMMUNOCYTOCHEMICAL PROCEDURES

The following primary antibodies were used: monoclonal antiparvalbumin (dilution 1:5,000–10,000), monoclonal anticalbindin-28 (dilution 1:5,000–10,000), and polyclonal anticalretinin (dilution 1:500–800). (All three monoclonal antibodies were obtained from Sigma, St. Louis, Mo.) We also used polyclonal primary antibodies against calbindin and parvalbumin (dilution 1:1000, SWANT, Bellinzona, Switzerland). We did not find

TABLE 23.1. Species in which the inferior colliculi (IC) of the tectum opticum of the midbrain were investigated.

Species	*n*	Order, suborder, or family	Common name
Tursiops truncatus	5	Cetacea, Odontoceti	Bottlenose dolphin
Delphinapterus leucas	2	Cetacea, Odontoceti	Beluga or white whale
Lagenorhynchus obliquidens	1	Cetacea, Odontoceti	Pacific white-sided dolphin
Inia geoffrensis	1	Cetacea, Odontoceti	Amazon River dolphin
Sotalia pallida	2	Cetacea, Odontoceti	Bouto or boto
Stenella coeruleoalba	3	Cetacea, Odontoceti	Spotted dolphin
Globicephala melaena	1	Cetacea, Odontoceti	Pilot whale
Megaptera novaengliae	1	Cetacea, Mysticeti	Humpback whale
Balaenoptera physalus	1	Cetacea, Mysticeti	Fin whale
Arctocephalus pusillus	1	Pinnipedia, Otariidae	Fur seal
Zalophus californianus	2	Pinnipedia, Otariidae	Californian sea lion
Myotis myotis	2	Chiroptera, Vespertilionidae	Little brown bat
Macaca fascicularis	3	Primates, Cercopithecidae	Long-tailed macaque
Ovis aries	3	Arziodactyla, Bovidae	Sheep
Homo sapiens	2	Primates, Hominidae	Human

significant differences between monoclonal and polyclonal antibodies in terms of specificity or intensity of the immunostaining. However, in most cases we used polyclonal antibodies, since theoretically they reveal more immunoreactive profiles—especially in poorly studied species, such as whales (Larsson 1993). As secondary antibodies, antirabbit and antimouse biotinylated immunoglobulins were used at a dilution of 1: 250 (Sigma, St. Louis, Mo.). A pre-embedding immunostaining procedure modified from Leranth and Feher 1983; Leranth and Frotscher 1986; and Leranth et al. 1988 was used. Serial sections (45–50 μm) from all specimens were prepared using a Vibratome-2000 (Ted Pella Instruments, Redding, Calif.). Sections were incubated in primary serum for 48–72 hours at 4° C. This was followed by the incubation in a secondary antibody. Vectastain ABC reagent staining kits (Vector Laboratories, Burlingame, Calif.), as well as 3,5-diaminobenzidine (DAB) with 0.3% hydrogen peroxide, were used for the visualization of the immunostaining followed by osmication in 0.005% of OsO_4. Two controls of the specificity of the immunoreaction were performed in all specimens and for all antibodies used in this study. For the first control, the primary antibody was omitted from the reaction, while for the second control the secondary antibody was omitted.

OTHER HISTOLOGICAL METHODS

To clarify the anatomical organization of the regions of interest we used the following methods: the Nissl method (cresyl violet and toluidine blue) for cytoarchitecture; the Weigert-Pal method (iron hematoxylin) for myeloarchitecture; and the Nissl and Bielschowsky-Plien methods (cresyl violet and silver impregnation) for combined analysis of cytoarchitecture, angioarchitecture, and myeloarchitecture.

COMPUTERIZED IMAGE ANALYSIS AND STATISTICS

Quantitative analyses were performed using an image analyzing system (Microcomp SMI, Southern Micro Instruments, Atlanta, Ga.) and Planomorphometry software. For measurements of immunoreactive perikarya, 25–30 neurons were examined in each species in each cortical layer or subcortical nucleus. Thirty random samples were used on each of 12 serial sections to calculate the numerical density of the immunoreactive perikarya in IC. The areas for cell counting and cell measuring were sampled randomly and projected from the microscope to the screen of the computer monitor. These areas were marked using a digitizing tablet and mouse. Within these areas the immunoreactive perikarya were labeled and their x/y coordinates recorded. The computer program then automatically calculated the numerical density of the perikarya per mm^2. Knowing the average thickness of the individual sections (45–50 μm), the numerical density per mm^2 was transformed into the numerical density in mm^3. In the case of Nissl

sections, only cells with a clearly defined nucleus and nucleolus were measured and counted. Sections adjacent to the sections used for immunoreactive staining were counterstained with cresyl violet, and the same image analysis was performed on these sections for computing the density of the total population of neuronal perikarya per mm^3. The thickness of the sections was measured using the readings on the calibrated microfocusing knob of the microscope (Olympus, Vanox). After dehydration and mounting, the thickness of 45–50 μm sections was diminished to 23–25 μm. To check the accuracy of measurements, the optical disector method was used on randomly selected sections (Sterio 1984). These data were used to calculate percentages of immunoreactive perikarya relative to the total neuron density in each sample. The Microcomp SMI Image Analysis statistics software was used to compute the major statistical parameters for each structure (mean, standard deviation, standard error). The following parameters were measured in all tissue sections: (1) numerical density of the immunoreactive perikarya, as well as total numerical density of neurons in counterstained Nissl sections per mm^3; (2) percentage of immunoreactive perikarya of the total neuronal population; (3) ratios between different calcium-binding protein-immunoreactive neuronal populations; (4) frequency of myelinated fibers per 1 mm; and (5) ratio between density of neurons and frequency of myelinated fibers.

Results

MACROSCOPY

On sagittal and horizontal sections through the whole brain the evident difference in size and shape of IC between two odontocete cetaceans (*T. truncatus, S. coeruleoalba*) and primate (*H. sapiens*) is shown (figs. 23.1 and 23.2). The difference in size is especially prominent in view of relatively comparable brain sizes. Overall, the volume of IC in bottlenose dolphins is approximately 40 times larger than in humans. On the other hand, the size of the IC in a mysticete whale (*B. physalus*) is only 2.2 times larger than in humans and 18 times smaller than in the bottlenose dolphin. In pinnipeds, the size of the IC relative to the whole brain is larger than in humans but much smaller than in odontocete cetaceans—comparable in relative size with mysticete whales.

CYTOARCHITECTONICS

In all species of odontocete cetaceans studied, the cytoarchitectonic preparations revealed the presence of convoluted domains with highest packing density of neuronal perikarya, separating these domains by interdomains with much lower packing density of neuronal perikarya (fig. 23.3). In different species of odontocete cetaceans, the diameters of the domains and interdomains varied: they were larger in *T. truncatus* and *D. leucas* and relatively smaller in *S. pallida, I. geoffrensis,* and *L. obliquidens*. In all studied species, domains were

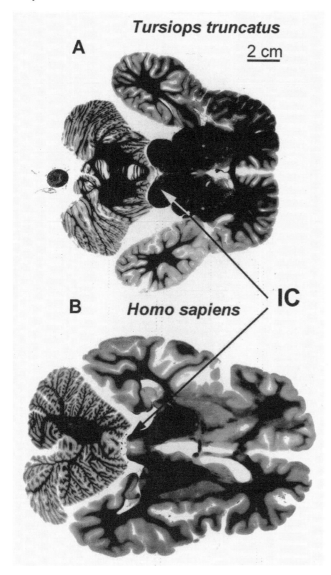

Fig. 23.1. Digital photographs of horizontal sections through the whole brain. *A:* Bottlenose dolphin (*Tursiops truncatus*). *B:* Human (*Homo sapiens*). Reduction: 20%. Celloidin, Bielschowsky-Plien. Note the extreme size of the inferior colliculus (IC) in the bottlenose dolphin brain in comparison to the human brain.

characterized not only by higher concentrations of neurons but also by darker intercellular matrix. In mysticete whales (*B. physalus* and *M. novaeanglia*), as well as in pinnipeds (*Z. californianus* and *A. pusillus*), the domain/interdomain arrangement of the distribution of the neuronal perikarya was present in small regions of IC central nucleus, whereas in most of the IC volume the distribution of neuronal perikarya was more or less homogenous (fig. 23.4). Among terrestrial mammals, we found domain/interdomain arrangement of cells in the IC only in insectivorous bats (*M. myotis*). On the other hand, in primates (*H. sapiens* and *M. fascicularis*) the distribution of neuronal perikarya in cytoarchitectonic preparations was more homogenous.

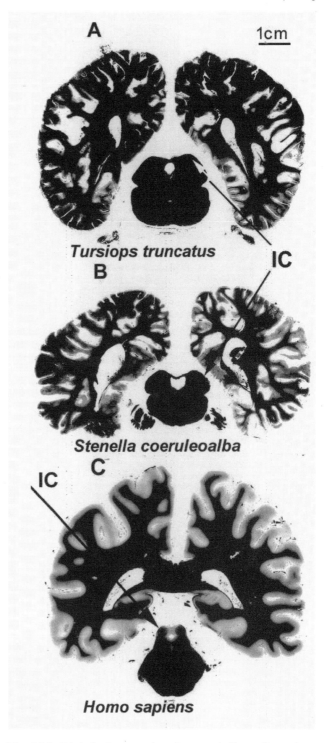

Fig. 23.2. Digital photographs of coronal (frontal) sections through the whole brain. *A:* Bottlenose dolphin (*Tursiops truncatus*). *B:* Spotted dolphin (*Stenella coeruleoalba*). *C:* Human (*Homo sapiens*). Reduction 30%. Celloidin, Bielschowsky-Plien. Note the extreme size of the inferior colliculus (IC) in odontocete brains in comparison to the human brain.

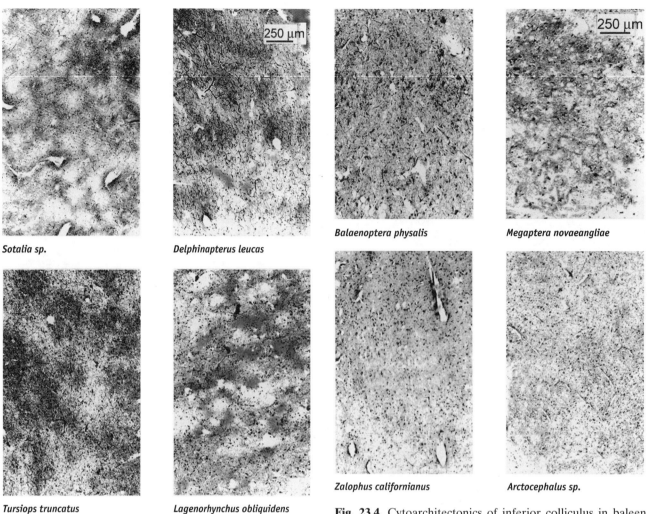

Sotalia sp.

Delphinapterus leucas

Balaenoptera physalis

Megaptera novaeangliae

Tursiops truncatus

Lagenorhynchus obliquidens

Zalophus californianus

Arctocephalus sp.

Fig. 23.3. Cytoarchitectonics of inferior colliculus in odontocete whales. Magnification ×40, Nissl stain. Note the presence of patches of the more densely and less densely neuronal groups (domain/interdomain arrangement) in all four species of odontocete whales.

Fig. 23.4. Cytoarchitectonics of inferior colliculus in baleen whales and seals. Magnification ×40, Nissl stain. Note the more or less homogenous distribution of neurons of IC. Only in *Megaptera* are there arrangements similar to the domain/interdomain network.

DISTRIBUTION OF CaBP-IMMUNOREACTIVE NEURONS

In odontocete cetaceans the distribution of CaBP-immunoreactive neurons roughly corresponds to domain/interdomain arrangements seen in cytoarchitectonics. Calbindin-immunoreactive (CB) neurons had small-to-medium perikarya and slender dendrites with weak immunostaining. The calbindin-immunoreactive perikarya were concentrated mostly in the domain areas (fig. 23.5), whereas CR-immunoreactive perikarya were found in domain and interdomain regions. In mysticete whales and pinnipeds, the distribution of CB perikarya had some domainlike clustering, but it was not as regular as in odontocete whales. Among terrestrial mammals (primates, ungulates) the distribution of CB-immunoreactive neurons was more or less homogenous (fig. 23.6). However, in chiropterans (*Myotis myotis*) there was domain clustering of CB neurons. In all studied species of odontocete whales, CR-immunoreactive

neurons were characterized by medium-to-large perikarya of angular shape with well-developed immunoreactive dendrites (fig. 23.7A, B). These neurons were located mostly in interdomains, but they were also found in domain areas. The roughly triangular shape of CR perikarya and organization of their dendrites were similar in odontocetes, mysticetes, and pinnipeds (fig. 23.8A, B, C, E). On the other hand, in primates CR-immunoreactive perikarya were oval- or pear-shaped with thin dendrites (fig. 23.8F). However, in ungulates the shape and distribution of dendrites of CR neurons were similar to those found in odontocete and mysticete whales (fig. 23.8D).

ANGIOARCHITECTONICS

Distribution of the capillaries in the inferior colliculi of the studied species of mammals in general reflected the distribution of neurons in cytoarchitectonic prepa-

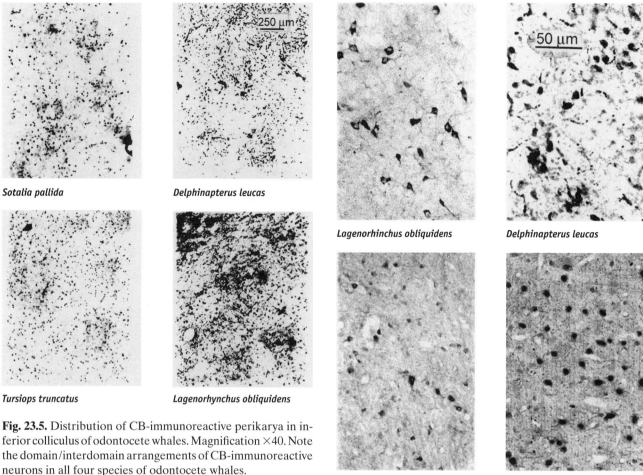

Sotalia pallida *Delphinapterus leucas*

Tursiops truncatus *Lagenorhynchus obliquidens*

Lagenorhinchus obliquidens *Delphinapterus leucas*

Zalophus californianus *Macaca fascicularis*

Fig. 23.5. Distribution of CB-immunoreactive perikarya in inferior colliculus of odontocete whales. Magnification ×40. Note the domain/interdomain arrangements of CB-immunoreactive neurons in all four species of odontocete whales.

Fig. 23.6. Distribution and structure of CB-immunoreactive perikarya in inferior colliculus of aquatic, semiaquatic, and terrestrial mammals. Magnification ×320. Note that in odontocete whales, CB-immunoreactive neurons have angular or oval shapes and their distribution is patchy. In seals, CB-immunoreactive neurons are smaller than in odontocete whales, but their distribution is more or less homogeneous. In the primate, most CB-immunoreactive neurons are oval in shape and their distribution is more homogenous.

rations. Thus, in odontocete whales the high density of clusters of capillaries corresponds to domains with a loose network of capillaries corresponded to the interdomains (fig. 23.9A, C). In mammalian species with absence of the domain/interdomain cytoarchitectonics of the IC (primates) the distribution of the vessels is also homogenous, reflecting the distribution of the neuronal perikarya (fig. 23.9B, D).

MYELOARCHTECTONICS

In all species of odontocete and baleen whales, the inferior colliculi are characterized by a high density of myelinated fibers (fig. 23.10A, B). In contrast, in primates the density of the myelinated axons is significantly lower (fig. 23.10C).

QUANTITATIVE ANALYSIS

Comparison of percentages of different subpopulations of CaBP-immunoreactive neurons in the IC showed that in both odontocete and baleen whales, PV-immunoreactive neurons are totally absent, whereas CB- and CR-immunoreactive neurons in sum constitute 80–100% of total neurons (fig. 23.11). In primates, each of the CaBP subpopulations constitute only 10–15% of the total neurons and in sum approximately 40% of to-

tal neurons (fig. 23.11). The ratio of densities of CaBP-immunoreactive neurons of the IC relative to the CaBP neurons of the primary auditory cortex was highest in odontocete whales and bats (2.7) and lowest in primates (0.8–1), whereas mysticete whales showed intermediate ratios (1.5–2) (fig. 23.12).

The frequency of myelinated fibers in the IC was highest in insectivorous bats and lowest in primates, whereas odontocete and baleen whales are characterized by intermediate frequencies (fig. 23.13). These differences were statistically significant for primate/odontocete comparison and cetacean/chiropteran comparison ($p < 0.001$). However, for odontocete/mysticete comparison, the difference was not statistically significant ($p < 0.09$).

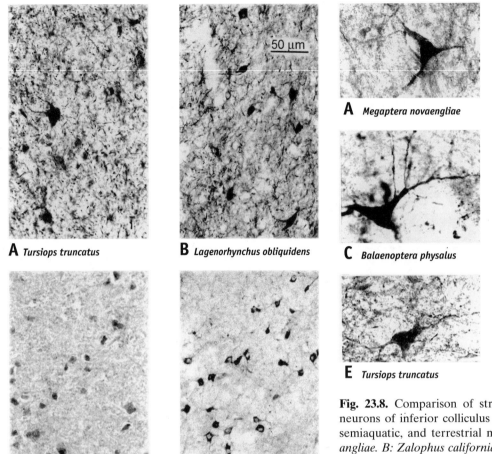

A *Tursiops truncatus*   **B** *Lagenorhynchus obliquidens*

C *Tursiops truncatus*   **D** *Lagenorhynchus obliquidens*

Fig. 23.7. Comparison of CB- and CR-immunoreactive neurons of inferior colliculus in two species of odontocete whales. *A: Tursiops truncatus* CR-immunoreactive neurons. *B: Lagenorhynchus obliquidens* CR-immunoreactive neurons. *C: Tursiops truncatus* CB-immunoreactive neurons. *D: Lagenorhynchus obliquidens* CB-immunoreactive neurons. Magnification ×320. Note that in both species of odontocetes, CR–immunoreactive neurons are approximately twice as large as CB-immunoreactive neurons and their packing density is significantly lower.

The ratio of the myelinated fibers to the number of the neurons of the inferior colliculus was highest in odontocete and baleen whales (70–75 fibers per neuron) and lowest in primates (8–10 fibers per neuron). In insectivorous bats the ratio was 30 fibers per neuron. All differences between the groups were statistically significant ($p < 0.001$), except the difference between odontocete and baleen whales ($p < 0.1$) (fig. 23.14).

SUMMARY OF THE MAJOR FINDINGS
 1. The IC in echolocating mammals is enlarged absolutely and relatively to the whole brain.
 2. The IC in echolocating odontocete cetaceans, as

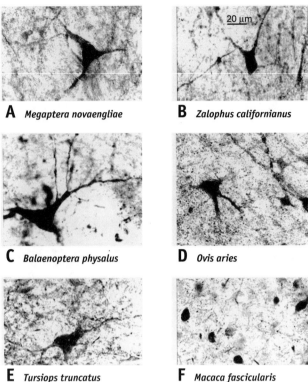

A *Megaptera novaengliae*   **B** *Zalophus californianus*

C *Balaenoptera physalus*   **D** *Ovis aries*

E *Tursiops truncatus*   **F** *Macaca fascicularis*

Fig. 23.8. Comparison of structure of CR-immunoreactive neurons of inferior colliculus in different species of aquatic, semiaquatic, and terrestrial mammals. *A: Megaptera novaeangliae. B: Zalophus californianus. C: Balaenoptera physalus. D: Ovis aries. E: Tursiops truncatus. F: Macaca fascicularis.* Note the similarity in shape of CR-immunoreactive perikarya in aquatic, semiaquatic, and ungulate species (A–E), where CR-immunoreactive neurons are medium to large and mostly triangular in shape. CR-immunoreactive neurons in the primate (*F*) are significantly smaller and have oval to round shapes.

well as in insectivorous bats, shows modular (domain/interdomain) arrangements of neuronal and nonneuronal (glia, vessels) components, whereas in nonecholocating terrestrial mammals (primates) the distribution of neuronal and nonneuronal structures of IC is more homogenous.
 3. In mysticetes and pinnipeds, the cytoarchitectonic structure of the IC has intermediate features between terrestrial mammals and odontocete whales—that is, the domain/interdomain arrangement of neuronal and nonneuronal components is present but is less regular and exists only in some regions of IC.
 4. In both Cetacea and Pinnipedia, calretinin and calbindin-immunoreactive subpopulations in cortical and subcortical structures are significantly higher than parvalbumin subpopulations. In terrestrial mammals (primates, chiropterans) CB, CR, and PV subpopulations were almost equal in numbers.
 5. The difference between echolocating odontocete

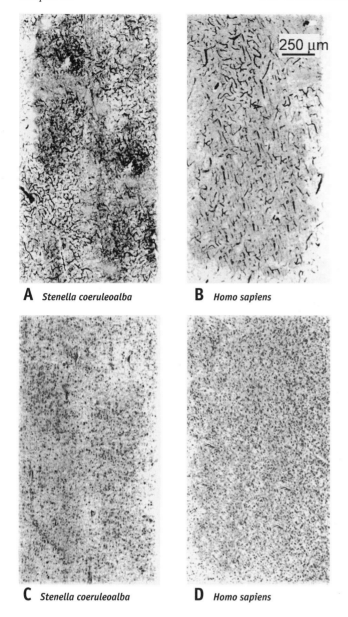

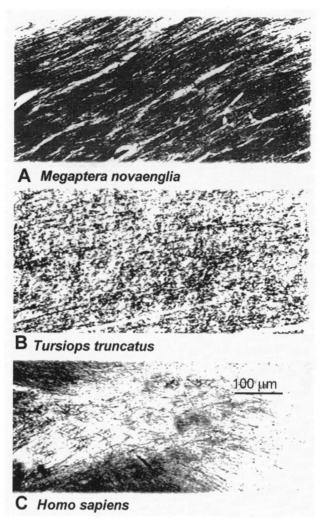

Fig. 23.10. Comparison of myeloarchitectonics in an odontocete whale with that in a mysticete whale and a human. *A: Megaptera novaeanglia. B: Tursiops truncatus. C: Homo sapiens.* Note that the highest density of myelinated fibers is in the mysticete whale IC, while the lowest is in the human IC.

Fig. 23.9. Angioarchitectonics (Bielschowsky-Plien) and cytoarchitectonics (Nissl) of IC in an odontocete whale and a primate. Magnification ×40. *A: Stenella coeruleoalba,* angioarchitectonics. *B: Homo sapiens,* angioarchitectonics. *C: Stenella coeruleoalba,* cytoarchitectonics. *D: Homo sapiens,* cytoarchitectonics. Note the correspondence of the distribution of neurons and vessels in both species. In the odontocete whale, the domain/interdomain arrangements of the neuronal perikarya correspond to the patchy distribution of vessels, with domains having a higher concentration of vessels than interdomains. In *Homo sapiens,* the homogenous distribution of neurons corresponds to the homogenous distribution of vessels.

cetaceans and supposedly nonecholocating mysticetes is expressed mostly at the subcortical level (IC), whereas the cerebral cortex of both families is quite similar in cytoarchitectonics and chemical neurohistology.

6. Some features of CaBP-immunoreactive neuronal subpopulations (especially CR-immunoreactive subpopulation) of the inferior colliculi and primary auditory cortex showed close similarity between cetaceans and ungulates.

Discussion

The combination of peculiarities of cetacean major sensory systems reflects, in our view, the specifics of evolutionary adaptations to the aquatic environment of the ancestral terrestrial species. The adaptation of the terrestrial mammals to the aquatic environment involved practically all bodily functions—but especially the sensory and motor abilities of the evolving organism. The factors of the adaptive evolutionary changes of the peripheral and central structures of the sensory systems in cetaceans are likely to be numerous and multifactorial.

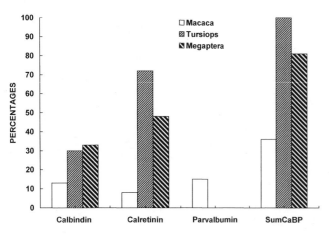

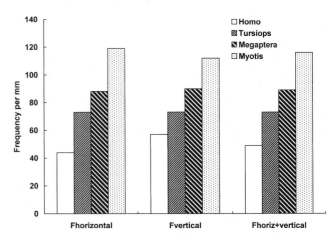

Fig. 23.11. Graph comparing percentages of CaBP-immuno-reactive neurons of the total neuronal population in IC of primate (*Macaca fascicularis*), odontocete whale (*Tursiops truncatus*), and mysticete whale (*Megaptera novaengliae*). Note that in both odontocete and mysticete whales, the percentages of CaBP are significantly higher than in primate ($p < 0.001$). Also, percentages of CB-immunoreactive neurons are similar in mysticete and odontocete whales, whereas the percentages of CR neurons are significantly higher in the odontocete whale ($p < 0.001$). In both families of whales, PV-immunoreactive neurons were not found in IC, whereas in primate these neurons constitute a prominent part of the neuronal populations of IC.

Fig. 23.13. Graph comparing the frequency of myelinated fibers in IC of four mammalian species: primate (*Homo sapiens*), odontocete whale (*Tursiops truncatus*), mysticete whale (*Megaptera novoaengliae*), and insectivorous bat (*Myotis myotis*). Note that the highest frequency of the fibers is in the IC of the insectivorous bat and lowest in the human IC. Also, the mysticete whale has a significantly higher frequency of fibers than the odontocete whale ($p < 0.02$). The frequency of fibers in IC of the insectivorous bat is significantly higher than in both mysticete and odontocete whales ($p < 0.001$).

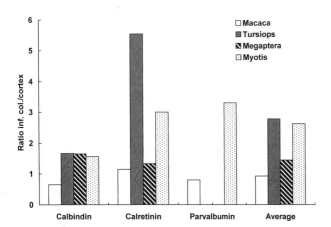

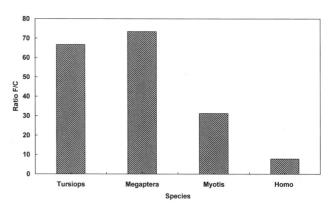

Fig. 23.12. Graph showing ratios of CaBP-immunoreactive neurons in the IC to those neurons in the primary auditory cortex (A1). Note that in general this ratio is significantly higher in bats and whales than in primates ($p < 0.001$). Ratios for CB-immunoreactive neurons are practically the same in mysticete and odontocete whales, whereas for CR-immunoreactive neurons ratios are significantly higher in odontocete whales ($p < 0.0001$).

Fig. 23.14. Graph showing ratios of number of myelinated fibers to the number of neuronal perikarya per 1 mm³ in IC of four mammalian species. Note that the highest ratio of fibers to the number of neurons is in odontocete and mysticete whales. There is no significant difference between their respective ratios. The lowest ratio is found in primates. Also, this ratio is significantly higher in cetaceans than in insectivorous bats ($p < 0.0001$).

The selective pressures of the new environment often presented contradictory demands from different systems, thus resulting in drastic reduction or hypertrophy of different structures. For example, hydrodynamics of swimming and full separation of the respiratory and digestive canals caused drastic changes in anatomy of the visceral and cephalic cranium in cetaceans. In turn, these changes constitute quite possibly major factors in the evolutionary reductive changes in the peripheral and central structures of the olfactory sensory system. Similar reductive changes occurred due to the transformation of the extremities into flippers and fins. These transformations, combined with the hydrodynamic reconstructions of the skull, in turn, resulted in important changes in the functional map and size of the motor zone and reduction of whole frontal lobe of the cerebral cortex in cetaceans. On the other hand, transfer into an aquatic milieu with absolutely different physical visual and acoustic characteristics resulted in extensive changes in visual and, especially, the acoustic sensory systems. The latter became the major instrument for signaling, orientation, and communication in cetaceans.

In our previous studies, we have suggested that in cetaceans a paradoxical combination of the highly convoluted and large neocortex with an archi-/paleocortical type of its architecture might result from the retention of an ancestral type of cortical modules with an extreme multiplication of these modules (Glezer et al. 1988; Morgane, Glezer, and Jacobs 1990; Glezer et al. 1993). On the other hand, certain features of the cetacean brain also may be result of dedifferentiation of an initially neocortical modular structure; thus, the overall brain organization in extant toothed whales would represent a derived stage (Innocenti 1984; Ridgway and Wood 1988; Deacon 1990). However, it is even more probable that during the complicated evolutionary changes involved in the transition from a terrestrial ecological niche to an aquatic one, both retention of ancestral traits and acquisition of the new features took place in all systems of cetaceans, including the brain. According to the paleontological record and molecular data, the most probable archetypal groups for the origin of cetaceans were early ungulates (Miyamoto and Goodman 1986; Gingerich et al. 1990; Millinkovitch et al. 1993). One of the possible reasons why this group and not, for example, carnivores are the ancestors of cetaceans could be related to an early physical maturity of the whale calf after birth. This factor is of a prime importance for the survival of the species confined to aquatic environment. Contemporary theories of evolution recognize that one of the most important initial evolutionary changes occurs in the early embryonic period (Waddington 1975; Løvtrup 1974, 1984). These initial epigenetic changes that occur in early periods of development are then incorporated into hereditary mechanisms through mutations and natural selection (Reid and Masters 1985). In this context, heterochrony in ontogenetic development could be one of the mechanisms of evolutionary transition of the terrestrial ancestors to the new aquatic environment. For successful survival of the newborn delivered in an aquatic environment, the animal has to be considerably more mature than in the case for terrestrial environment. The sensory and motor systems of the newborn whale have to be ready for immediate adaptation to the environment to prevent drowning. In the case of the wild ungulates, calves also have to adapt quickly to ecological pressures to avoid the high probability of predation. This can be achieved through the process proposed by Severtsov (1939). In this process, pedomorphosis, the characteristics that are initially present in ancestral juvenile forms are shifted to the adult stage (Severtsov 1939; Gould 1977; Schoch 1986; Schad 1993; Roth et al. 1993; Marcus et al. 1995; Godfrey and Sutherland 1996). The pedomorphosis is a "genetically easy response to an environmental pressure" (Raff and Kaufman 1983). It is well known that in all cetaceans, especially in toothed whales, the newborn calves are large relative to their mother's body and are able to follow the adults immediately (Kellog 1928). Some features of the possible pedomorphic changes in brain evolution of cetaceans are seen, for example, in the presence of the mesencephalic, pontine, and cephalic flexures in the adult brain of the *T. truncatus,* whereas these flexures in primates are seen only in embryos (Morgane and Jacobs 1972). The pedomorphic tendencies in ontogenesis of the extant toothed whales have been well documented in an extensive study on Pacific *Stenella* species (Miyazaki et al. 1981). In two species of *Stenella* the relative brain weight achieves its maximum at the end of the fetal period and then sharply decreases after birth. After six months of postnatal life, the percentage of brain weight of the body weight remains practically unchanged for the rest of ontogenesis. In the case of primate development, the relative weight of the brain increases over a much longer period of time. In chimpanzees, it becomes constant after 4 years and in humans after 14–15 years of postnatal life (Blinkov and Glezer 1968). We speculate that many other features of the cetacean brain also evolved using this pedomorphic evolutionary model—in particular, the proisocortical characteristics of neocortex in cetaceans (Morgane et al. 1986a, 1986b, 1990; Glezer et al. 1993). In different classes of vertebrates, the "cost" for the fast pedomorphic evolutionary changes is relative simplification of the particular structure or system, especially such as the skull and central nervous system (Schad 1993; Roth et al. 1993). It is suggestive that extreme multiplication of the relatively simplified cortical modules (Glezer et al. 1988) and, as a consequence, development of the large cortical surface in cetaceans are compensatory evolutionary features helping to overcome the pedomorphic ontogenetic shift. The relative prevalence of the CR- and CB-immunoreactive over PV-immunoreactive neurons in major sensory systems of the cetacean brain also could serve as an additional indication of pedomorphic evolutionary traits of these species. This is supported by the fact that in ontogenesis of the mammalian (rat, cat, monkey) central nervous system the most precocious neurons are CR-immunoreactive, followed by CB-immunoreactive neurons, while PV-immunoreactive neurons appear last (Enderlin et al. 1987; Solbach and Celio 1991; Anstrom et al. 1997; Hof et al. 1999). It is possible that a putative pedomorphic shift in the cetacean brain development resulted in the embryo-like deficit of PV-immunoreactive neurons in the adult cetacean brain and its enrichment with the CR- and CB-immunoreactive neurons.

The enrichment of the IC of cetaceans in CB- and CR-immunoreactive neurons also has other intriguing aspects in view of data from terrestrial echolocating mammals (bats) (Zettel et al. 1991; Vater and Braun 1994). It was shown that in bats CB- and CR-immunoreactive neurons in MGN and IC participate in preservation of the temporal information during echolocation (Zettel et al. 1991). Thus, a pedomorphic shift resulting in enrichment of the adult brain in CR- and CB-immunoreactive neurons of cetaceans also was under selective pressure to develop effective echolocating devices in an aquatic environment.

To date, echolocating abilities are experimentally documented only for two groups of mammals (toothed whales and insectivorous bats) and one group of birds (owls). There are some scattered data on ultrasound frequencies in seals, some Mysticeti, and insectivores.

In the present study, we found two features of the IC that are common for echolocating terrestrial and aquatic mammals. First is the presence of domain/interdomain arrangement of neurons of IC in odontocete cetaceans and bats. It was found that irregular, densely stained domains contain mostly CB-immunoreactive neurons and smaller numbers of CR-immunoreactive neurons, whereas interdomains contain mostly large CR-immunoreactive neurons. Since the auditory system in odontocete whales and bats plays a major role as a source of information essential for their specialized behavior (Popper 1980; Nachtigall 1986; Ralston and Herman 1989; Herman 1990; Schusterman 1990), these morphologically and cytochemically defined domains and interdomains in the IC of odontocete species and chiropterans can be interpreted as highly specialized adaptations for processing of auditory signals. Since this neuronal arrangement of IC is present only in echolocating mammals, both terrestrial and, especially, totally aquatic (Chiroptera, Odontoceti) and absent in nonecholocating mammals (primates, Pinnipedia, Mysticeti), we may tentatively suggest that domain/interdomain organization of the IC somehow relates to echolocation abilities in aquatic and terrestrial species. The hypertrophied "columnization" of the lower brainstem nuclei in the auditory system in mammals with echolocation abilities was demonstrated previously for the ventral nucleus of lateral lemniscus in bats and dolphins (Zook et al. 1988). Thus, the observed dark domains (termed tectosomes in the present study) and the light (matrixlike) interdomains could provide an anatomical substrate for isofrequency coding of frequency bands during echolocation, submarine sound analysis, and orientation.

The second feature of the echolocating mammals is the high ratio of CaBP-immunoreactive neurons in IC to the neurons with the corresponding immunoreactivity in the primary auditory cortex. It could relate to the necessity to analyze significantly higher amounts of information in the primary auditory cortex of echolocating mammals than in nonecholocating mammals. However, there is one inconsistency with our hypothesis: we found both higher concentrations of myelinated fibers and neurons in IC also in Mysticeti, which according to the contemporary data have no developed biosonars. Higher concentrations of the fibers in IC of these species could relate to the overall ability to orient with sounds and not necessarily with high-frequency sounds.

It should be noted that some features of the CaBP-immunoreactive neuronal subpopulations in central auditory structures, such as the cerebral cortex and inferior colliculi, could reflect either a common ancestry or convergent traits of evolution. Thus, the general morphology of the CR-immunoreactive neurons in odontocetes, mysticetes, pinnipeds, and ungulates are very similar both in the auditory cortex and in the inferior colliculi. On the other hand, morphology of the same subpopulation of neurons in primates differs significantly from the mammalian taxa mentioned above. According to the paleontological record and molecular data, the most probable archetypal groups for origin of cetaceans were early ungulates (Miyamoto and Goodman 1986; Gingerich et al. 1990; Millinkovitch et al. 1993). We tentatively speculate that the observed similarities in morphology of the CR-immunoreactive neurons in auditory structures of mammalian species (which differ so much by their niche, as do aquatic cetaceans and terrestrial ungulates) may reflect the ancestral relationships between these mammalian groups (Hof et al. 1999). On the other hand, the similar morphology of these neurons in pinnipeds, which are derived from the archetypal carnivores (McKenna 1975; Novacek et al. 1983; Shoshani and McKenna 1998), could reflect convergent evolution in these aquatic mammals.

Acknowledgments

This study was supported by the PSC-CUNY grant 662232, MBRS/CRS grants of NIH, NIH grant HD 25539, and by the Mount Sinai School of Medicine. We thank T. F. Ladygina (Moscow State University, Moscow, Russia) and A. Ya. Supin, A. M. Mass, and A. V. Revishchin (Severtsov Institute of Evolutionary Morphology and Ecology of Animals, Moscow, Russia) for providing *T. truncatus* brain material.

{ 24 }

Feature Extraction and Neural Activity:
Advances and Perspectives

Nobuo Suga

Introduction

INFORMATION-BEARING ELEMENTS AND PARAMETERS

The concepts of acoustic features and feature extraction were discussed by about 100 participants in a full session of a conference held at Dahlem in 1976. Most conference participants agreed with the notion that acoustic signals have no features but have physical properties characterized by particular values of parameters, and that acoustic features are determined by the auditory system of animals. Therefore, to study feature extraction, one must understand first the auditory behaviors of the species and the information-bearing elements (IBEs) and parameters (IBPs) of acoustic signals eliciting the behavior—that is, acoustic features. Then one must systematically study responses of auditory neurons as a function of individual IBPs (e.g., Suga 1992).

Since an animal commonly hears a large variety of sounds, all sounds heard by the animal potentially are behaviorally relevant and there is an infinitely large number of acoustic features. Therefore, it is unlikely that the auditory system processes acoustic signals with neurons tuned to IBPs or IBEs, but rather with spatial and temporal patterns of electrical activity of neurons that simply are tuned to a single frequency. This may be true for processing potentially relevant acoustic signals, but may not be true for processing narrowly defined behaviorally relevant sounds (see below), because a large amount of neuroethological data indicates that the central auditory system contains many neurons tuned to IBPs. The auditory system evolved together with the vocal system, so that the acoustic signals produced by a given species must be particularly relevant to its behavior. Also, detection and analysis of the sounds produced by both prey and predators of a species must be very important for its survival. Therefore, "behaviorally relevant sounds" could be defined narrowly as species-specific communication sounds, orientation sounds in bats, and sounds produced by both prey and predators.

In bats, IBEs and IBPs for echolocation have been described in numerous studies (e.g., Sales and Pye 1974; Simmons, Howell, and Suga 1975). Neuroethologists have used this information to investigate the neural mechanisms of processing biosonar information (e.g., Suga 1984, 1990a, 1990b; Popper and Fay 1995). The

large amount of data accumulated over the last 40 years indicates that the central auditory system creates neurons tuned to IBPs and that each species is specialized in its own way. Therefore, the auditory cortex physiologically differs from species to species, according to the properties of species-specific biosonar signals and auditory behavior (see below). It also can differ among individuals (in particular, among males and females) of the same species, according to the properties of their own biosonar signals (Suga et al. 1987).

NEURONS TUNED TO IBPS AND NEURAL MAPS

At the auditory periphery, frequency is represented by the location along the basilar membrane and phase- and stimulus-locked discharges. There are no anatomical representations for stimulus amplitude and time (i.e., duration and interval between stimuli). In the central auditory system, however, divergent and convergent neural projections repeatedly take place, and multiple cochleotopic (frequency) and computational maps and neuronal response properties differ from those at the periphery emerge.

Many central neurons apparently are tuned to IBPs or combinations of IBPs in the frequency, amplitude, and/or time domains. (The IBPs in the time domain include echo delays, rates of frequency or amplitude modulations, durations, interaural time differences, etc.) Neural mechanisms for creating response properties consist of excitation, inhibition, facilitation, and short- or long-delay lines (e.g., Suga 1990b; Covey and Casseday 1999). Combinations of facilitation and delay lines in the frequency, amplitude, and/or time domains are particularly interesting and important, because facilitation means amplification based on coincidence and because both facilitation and delay lines are the neural mechanisms for creating different types of neurons tuned to IBPs—such as echo delays, temporal pattern of sounds (including repetition rate and rate of amplitude modulation), durations, depth and rates of frequency modulations, and interaural time differences (e.g., Suga 1990b; Covey and Casseday 1999). Short-delay lines are created by conduction delays of action potentials (Konishi 1992) and synaptic delays at excitatory synapses, whereas long-delay lines are created by inhibition of different durations that evoke a rebound off-response (Sullivan

1982a, 1982b; Suga 1990b; Saitoh and Suga 1995; Covey and Casseday 1999). Facilitation (coincidence detection and amplification)—for example, by NMDA receptors (Butman 1992)—is particularly important for the adjustment of signal processing based on associative learning (Weinberger 1998; Gao and Suga 1998, 2000).

Recent progress in the neurophysiology of bats indicates that the corticofugal system, which forms positive and negative feedback loops, plays an essential role in shaping the response properties of subcortical neurons and, thereby, in shaping cortical neurons' own input. (See the chapters by Jen et al. and Suga et al. in this volume.) There is evidence that the corticofugal system continuously adjusts and improves auditory signal processing according to auditory experience (Gao and Suga 1998, 2000).

Many types of neurons showing different response properties are created in the central auditory system. They separately cluster in the auditory cortex and form so-called functional subdivisions. (The functional significance of the subdivisions has been physiologically identified, but not yet behaviorally explored.) A cluster may form a band parallel to the frequency axis within the primary auditory cortex (e.g., binaural I-E and E-E bands) or an independent area that has no clear frequency axis when studied with single-tone bursts (e.g., the FM-FM area). Different values of an IBP are mapped systematically in certain clusters (e.g., the FM-FM area), but not in other clusters (e.g., I-E or E-E bands). The formation of a map in a cluster appears to depend on whether analysis of small variations in the value of an IBP is particularly important for behavior and whether one or more mechanisms to create an array of neurons tuned to different values of the IBP systematically operate according to an anatomical arrangement. The map is the emergent organization to create the array of IBP-tuned neurons and has an advantage in both incorporation of lateral inhibition circuit and interfacing with the motor system.

The frequency (cochleotopic) map in the auditory cortex was first found in the cat by Woolsey and Walzl (1942). Since then, not only the frequency map but also other types of maps (computational maps) have been found. Certain maps, such as a time-interval (echo-delay) map in the FM-FM area, can be easily justified as a map because of a systematic topographic representation of an IBP (Suga and O'Neill 1979; O'Neill and Suga 1982). However, other maps, such as those representing sharpness of frequency tuning along the iso-best-frequency line (frequency versus sharpness map), show not a systematic representation of an IBP but rather a gradient or tendency for a topographic representation (Schreiner and Mendelson 1990). Table 24.1 lists the maps or functional organizations found thus far in the auditory cortex of different species of animals. Some maps or organizations found only in nonchiropterans may likely be found in the future in chiropterans, and vice versa. However, an identical map may have different behavioral significance among species.

In biosonar signals, acoustic features for biosonar

TABLE 24.1. Information space in the auditory cortex. In certain maps, values of an acoustic parameter to which neurons are tuned do not systematically vary but tend to vary along the cortical surface. Note that identical maps can be described in different ways (see 1, 5, 6, and 13). Freq.: frequency; CF: constant-frequency components (similar to formants); FM: frequency-modulated components (similar to transitions). Only the first report of a given map or organization is listed.

Map or organization (animal)	Citation
1. Frequency map; cochleotopic map; tonotopic map (cat)	Woolsey and Walzl 1942
2. Freq.-vs.-threshold map (dog)	Tunturi 1952
3. Freq.-vs.-amplitude map (mustached bat)	Suga 1977
4. Freq.-vs.-binaural bands (cat)	Imig and Adrian 1977
(mustached bat)	Manabe, Suga, and Ostwald 1978
5. Time-interval map; echo-delay (distance) map; FM-vs.-FM map (mustached bat)	Suga and O'Neill 1979
6. Freq.-vs.-freq. map; Doppler-shift (velocity) map; CF-vs.-CF map (mustached bat)	Suga et al. 1983
7. Freq. = azimuth axis (mustached bat)	Kujirai and Suga 1983
8. Freq. = AM rate axis (cat)	Schreiner and Urbas 1986
9. Freq.-vs.-AM rate map (myna bird)	Hose, Langner, and Scheich 1987
10. Freq.-vs.-sharpness map (cat)	Schreiner and Mendelson 1990
11. Freq.-vs.-FM rate map (myna bird)	Scheich 1991
12. Freq.-vs.-spectral gradient map (ferret)	Shamma et al. 1993
13. Center freq.-vs.-bandwidth map; noise burst map (macaque monkey)	Rauschecker, Tian, and Hauser 1995
14. Freq.-vs.-duration map (little brown bat)	Galazyuk and Feng 1997b

have been known, so that certain cortical maps in the mustached bat (e.g., the FM-FM area) can be related directly to feature extraction. However, without further studies, we cannot say the same about some other maps or organizations (e.g., binaural bands). Although our understanding of the functional organization of the auditory cortex is limited, it is clear that the central auditory system creates neurons tuned to particular values of an IBP or combinations of particular values of IBPs, and that it is organized with these single neurons in a species-specific way, regardless of whether such neurons are theoretically necessary for feature extraction.

FEATURE EXTRACTION AND SPATIOTEMPORAL PATTERN OF NEURAL ACTIVITY

In the auditory cortex of the mustached bat (*Pteronotus parnellii*), a cortical minicolumn (20–25 μm radius, approximately 900 μm thick) contains 45–50 neurons. Single-unit studies indicate that so-called very sharp tuning curves of neurons to an IBP are not so sharp that single neurons respond only to a single value of the IBP. Accordingly, any acoustic stimulus would excite neurons in several minicolumns, representing slightly different values of the IBP. Therefore, the acoustic signal is expressed by a spatial and temporal pattern of neural activity. A number of synchronized action potentials evoked by the stimulus would be largest in the optimally excited minicolumn. In general, feature extraction is based on the spatiotemporal pattern of neural activity. However, it should be noted that cortical neurons possibly involved in feature extraction are quite different in response properties from peripheral neurons.

In the auditory cortex of the mustached bat, echo-delay (i.e., target-range) tuned neurons are clustered in the FM-FM, DF, and VF areas. Suzuki and Suga (1991) performed a computer simulation experiment of these areas and obtained data indicating that a just-noticeable change in echo delay at the range of up to 60 cm is 30–85 μs in terms of the locations of maximally responding cortical minicolumns along the echo-delay axes in the three delay-tuned areas, 0.4–1.7 μs in terms of the weighted sum of responses, and 0.8–1.8 μs in terms of the center of mass activity.

According to behavioral studies on the big brown bat (*Eptesicus fuscus*), the just-noticeable difference in target range was approximately 13 mm (corresponding to 75 μs echo delay) for triangular targets placed on the left or right side of the bat (Simmons 1971). However, the just-noticeable jitter in echo delay was 0.5 μs (corresponding to 87 μm distance) for a computer-generated jittering or nonjittering phantom target placed on the left or right side while the bat emitted several pulses aimed in one direction (Simmons 1979). The just-noticeable jitter was approximately 0.01 μs (corresponding to ap-

proximately 1.7 μm distance) when the same jitter experiment as above was repeated with a computer capable of introducing much finer jitter than 1 μs. The large just-noticeable target-range difference of approximately 13 mm would likely be due to head movements on each trial of the discrimination experiment. Therefore, the above three values for a just-noticeable target-range (or echo-delay) difference are simply due to a different method. The bat can detect an echo-delay difference as small as approximately 0.01 μs (corresponding to approximately 1.7 μm distance) in an ideal situation (Simmons 1989). Another experiment with computer-generated jittering phantom targets (Simmons, Ferragamo, and Moss 1998), designed to estimate two-point resolution, showed that just-noticeable separation of two echoes was approximately 2 μs in delay (corresponding to approximately 300 μm distance).

Since there are no critical behavioral data to correlate cortical neural activity with feature extraction or acuity, it may be speculated that an animal uses neural activities in a given cortical area in different ways to perform similar tasks in different situations—that is, tuning of single neurons to an IBP, weighted sum of responses, center of mass activity, or fine-temporal pattern of nerve impulses. Feature extraction and acuity based on the location of an optimally excited minicolumn may not be fine, but it could be tolerant to noise. The center of mass activity may show hyperacuity but be vulnerable to noise. The fine temporal pattern of discharges may be related to hyperacuity, but they might also be vulnerable to noise.

Since animal behavior may rely on neural activity in different ways (depending on tasks to be performed and the situation), neurophysiological studies on feature extraction may be performed in different ways: single-unit recording, multiple single-unit recording, multiunit recording, and evoked potential recording, all of which have advantages and disadvantages. The advantages of multiunit recording include ease in long-term recording with implanted electrodes, in characterizing responses of a population of neurons, and in exploring the general organization of the auditory cortex and nuclei. Multiunit recording is much more efficient for mapping the cortex or a nucleus than single-unit recording, and it is also useful for studying synchronized discharges of several neurons. A synchronized portion of neural responses may be behaviorally more relevant than an unsynchronized portion. However, multiunit recording is not adequate for exploring differences between individual neurons, even when a computer program is used to sort nerve impulses. Various forms of data should be collected, as they are generally complementary. It is essential to first find something interesting and important, regardless of whether it is single-unit, multiunit, or evoked potential recording.

{ 25 }

Feature Extraction in the Mustached Bat Auditory Cortex

William E. O'Neill

Introduction

The auditory cortex of the mustached bat (*Pteronotus parnellii*) is distinguished by a number of unusual (some might say remarkable) neurophysiological and anatomical features. Many of these features have been extensively reviewed in past publications (see O'Neill 1995). In this brief review, I highlight the most prominent aspects of cortical processing in this species. By comparison with the other species reviewed by authors in other chapters within this section, the level of cortical complexity that has evolved in the mustached bat is remarkable. The novel features of cortical organization include (1) the high proportion of neurons requiring complex stimulus combinations resembling acoustic features found in mustached bat vocalizations, (2) the segregation of neurons with distinct response properties into separate cortical areas devoted to specific processing tasks, (3) the existence of multiple processing streams linking functionally related cortical areas, and (4) hemispheric specialization for echolocation and communication sound processing.

Echolocation Signals

The mustached bat is a selective predator of flying insects (Goldman and Henson 1977). Like horseshoe bats, the mustached bat uses CF-FM sonar signals and employs Doppler-shift compensation to stabilize echo frequency (Schnitzler 1970). These behaviors are adaptations for capturing flying insect prey in cluttered habitats, such as the understory of neotropical forests. Each call consists of a long constant-frequency (CF) component followed by a short, downward frequency-modulated (FM) sweep. Emitted pulses vary in duration from as long as 30 ms (during search and early approach phase) to as short as 5–7 ms (during terminal phase of pursuit) (Novick and Vaisnys 1964). Emission rates increase correspondingly from about 10/s to about 80/s. Each sonar pulse contains four prominent harmonics, designated H_1-H_4 (each CF or FM component is also addressed by a subscripted number, e.g., FM_1, CF_3, etc.). Although the bat can vary the energy distribution in the harmonics, the second harmonic is usually dominant (Gooler and O'Neill 1987), its

exact frequency ranging from about 58 kHz in the largest subspecies (*P. p. rubiginosus*) to over 62 kHz in the smallest (*P. p. portoricensis*).

Combination Sensitivity

Neurons in the mustached bat's auditory cortex are classified according to the CF or FM component of the sonar signal to which they are most sensitive. Table 25.1 summarizes the properties of these classes. It is striking that the majority of cortical fields contain so-called combination-sensitive neurons. Most of these neurons respond *exclusively* to combinations of at least two signal elements. With regard to echolocation signals, one of the essential signal elements always derives from the first harmonic of the emitted pulse, while the other derives from higher harmonics in the echo. In communication sounds, the two essential elements may be contained within a single utterance.

Combination-sensitive neurons fall into three classes: FM-FM cells, CF/CF cells, and FM_1-CF_2 cells. FM-FM cells are facilitated by a combination of the FM_1 component in the pulse, and the FM_2, FM_3, and/or FM_4 component in the echo (Suga et al. 1983). They are typically unresponsive to either of these FM sweeps presented alone, but they are strongly facilitated when the two components are paired with a specific delay. All FM-FM neurons exhibit some degree of delay tuning and are thought to play a role in echo-range coding.

CF/CF cells are facilitated by combinations of CF_1 in the pulse and either CF_2 or CF_3 components in the echo (Suga, O'Neill, and Manabe 1979). Again, they are typically unresponsive to these tones presented in isolation. However, unlike FM-FM neurons, CF/CF cells show very broad delay sensitivity. Facilitation occurs whenever there is any degree of temporal overlap between the CF_1 and CF_2 or CF_3 components in the stimulus. However, CF/CF neurons are very selective for the frequency of the CF_2 or CF_3 components. These neurons have been linked to both echo-flutter detection and Doppler-shift compensation behavior (Gaioni, Riquimaroux, and Suga 1990; Suga et al. 1983).

The third type of combination sensitivity, called FM_1-CF_2, involves the FM_1 component in the emitted pulse

TABLE 25.1. Cortical units in *Pteronotus parnellii*.

Signal type	Frequency tuning	Delay dependency	Key parameters	Field(s)	Organization	Information encoded
CF$_2$	extremely sharp	none	frequency and amplitude modulation	DSCF	bicoordinate: frequency vs. level	target flutter; velocity
FM$_1$-FM$_2$ FM$_1$-FM$_3$ FM$_1$-FM$_4$	broad	sharp	pulse-echo delay	FM-FM DF VF	• clustered by type • chronotopic (delay axis)	target range
CF$_1$/CF$_2$ CF$_1$/CF$_3$	CF$_1$: broad CF$_2$, CF$_3$: extremely sharp	broad	frequency of CF components, modulation of echo CF	CF/CF	• clustered by type • bicoordinate: tonotopic	relative target velocity (Doppler-shift magnitude), flutter
FM$_1$-CF$_2$ (H$_1$-H$_2$)	FM$_1$: broad CF$_2$: sharp	broad	pulse-echo delay; echo frequency and level	DSCF VA	see DSCF above; VA indeterminate	target detection at distance
multipeaked	very broad	unknown	unknown	DM, VP	nontonotopic	unknown

and a time-delayed CF$_2$ component in the echo (Fitzpatrick et al. 1993; Tsuzuki and Suga 1988). These cells express a hybrid of properties associated with FM-FM and CF/CF neurons: they are both delay sensitive and sharply tuned for the echo CF$_2$ component.

Cortical Processing Streams

Based on global mapping of response properties and interconnections of defined cortical fields, Fitzpatrick, Olsen, and Suga (1998a) have provided evidence of at least two, and possibly three, processing "streams." Processing streams consist of strongly interconnected, functionally similar cortical fields. Each stream is incompletely segregated, making weak connections with other streams composed of functionally dissimilar fields. Specifically, the CF stream (fig. 25.1B) comprises both non-combination-sensitive neurons responsive to echo CF$_2$ or CF$_3$ components, and combination-sensitive FM$_1$-CF$_2$ neurons sensitive to echo delay. Neurons in this stream are very responsive to echo flutter (i.e., amplitude and frequency modulations), particularly for distant targets, and are hypothesized to direct transitions between search and approach stages of target pursuit (Fitzpatrick, Olsen, and Suga 1998b). The CF/CF stream (fig. 25.1C) links cortical fields containing CF/CF neurons, and it is therefore highly sensitive to Doppler shifts and minute frequency and amplitude modulations in Doppler-shifted echoes. This stream presumably is involved in Doppler-shift compensation, and detection/categorization of fluttering targets (Gaioni, Riquimaroux, and Suga 1990). The FM-FM stream (fig. 25.1D) links cortical fields with FM-FM neurons selective to pulse-echo time delays, and it is therefore presumably involved in echo ranging. Highlights of the feature extraction properties

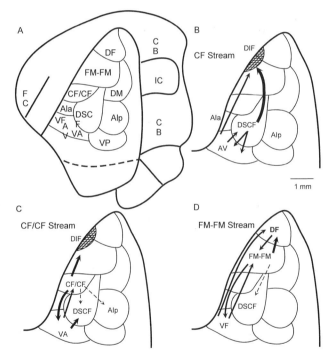

Fig. 25.1. Functional organization of the mustached bat's auditory cortex. *A:* Left hemisphere showing known auditory cortical fields in relation to the frontal cortex (*FC*), cerebellum (*CB*), and inferior colliculus (*IC*). *B:* CF processing stream, including DSCF, AV, and dorsal intrafossal zone (DIF, *crosshatched area*). Arrow thickness indicates strength of connections. *C:* CF/CF processing stream, including CF/CF, DIF, and VA. VA projects back to CF/CF and DSCF. *D:* FM-FM processing stream, including FM-FM, DF, VF. Adapted from Fitzpatrick, Olsen, and Suga 1998a, © Wiley-Liss, Inc.

expressed in each of these three streams are presented in the following sections.

THE CF PROCESSING STREAM:
DSCF, AIa, AIp, AND AV AREAS

The core of the mustached bat's auditory cortex consists of three fields that together are homologous to the primary auditory cortex (area AI) in other mammals: the *DSCF, AIa,* and *AIp* areas (fig. 25.1B, Asanuma, Wong, and Suga 1983; Fitzpatrick, Olsen, and Suga 1998a). All three areas are organized tonotopically, but each over-represents a different part of the spectrum.

AIa is by far the smallest of the three tonotopic areas. It represents CF_3 frequencies from 88 to 95 kHz (Asanuma, Wong, and Suga 1983). AIp represents frequencies between approximately 10 kHz and 55 kHz (Fitzpatrick, Olsen, and Suga 1998b). In both areas, the tonotopic axes are aligned rostrocaudally and isofrequency contours are oriented dorsoventrally, resembling AI in many nonecholocating mammals. DSCF represents a very narrow frequency band within 2–3 kHz of the CF_2 component, and it has concentrically organized isofrequency contours (fig. 25.2A, Suga and Jen 1976; Suga and Manabe 1982). Taken together, all three areas represent frequencies from 10 to 95 kHz. However, certain frequencies found in the sonar signal are notably *under*represented. H_1 and FM_2 echolocation frequencies are underrepresented in AIp. AIp represents instead the nonecholocation frequencies below 22 kHz and between 32 and 45 kHz (Fitzpatrick, Olsen, and Suga 1998b). Similarly, the FM_3 frequency band between 75 and 88 kHz is nearly absent. As will be seen, these FM frequency bands are "displaced" into the FM-FM stream of the auditory cortex.

Area AV lies ventral to AIa (fig. 25.1B) and is not tonotopically organized. It contains FM_1-CF_2 neurons and therefore might appear as an extension of the DSCF area. However, experiments have revealed a reversal of best CF_2 frequencies indicative of a border between the AV and DSCF areas (Asanuma, Wong, and Suga 1983). It will be treated as a distinct field in this review.

Sandwiched between AIa and AIp, the DSCF area is by far the largest of the CF processing areas and shows the most pronounced overrepresentation of frequency (fig. 25.2A). DSCF neurons respond well to tones, and they have best excitatory frequencies (BEF) in the narrow band between 59 kHz and 62 kHz, around the dominant CF_2 component of the sonar signal. Spectral response areas (fig. 25.2B) are extremely sharp in DSCF cells (Q10 dB from 15 to 300, Suga and Jen 1976; Suga and Manabe 1982). Inhibition plays a major role in sharpening these tuning curves. Inhibitory areas can be quite extensive, covering more than an octave above and below BEF in some cells. In the extreme, some DSCF cells, dubbed "level tolerant," have very narrow spindle-shaped excitatory response areas completely circum-scribed by inhibition (fig. 25.2B, *right panel*). The vast majority of DSCF neurons have nonmonotonic rate-level functions, giving rise to a "best amplitude" (BA) at the maximum discharge rate (Suga 1977; Suga and Manabe 1982). Therefore, DSCF neurons are tuned in both the frequency and intensity domains.

The DSCF area is unique in that its tonotopic axis is radial rather than linear, with isofrequency contours organized into concentric rings (fig. 25.2A). At the center of the DSCF area, BEFs match the individual bat's CF_2 "resting frequency" (Suga and Jen 1976; Suga and Manabe 1982; Suga et al. 1987). At the periphery, BEFs may be only 1.5–2.0 kHz higher. Moreover, Suga (1977) showed that not only frequency but also tone level was represented systematically within the DSCF area. Suga and Manabe (1982) coined the term "amplitopic representation" to describe this organization. The iso-BA contours are arrayed orthogonal to the iso-BF contours, much like spokes in a wheel (fig. 25.2A). Frequency and sound pressure level are thereby organized in a bicoordinate map in the DSCF area.

Facilitation and Delay Tuning
in DSCF and AV Neurons

AI neurons often have rather linear spectrotemporal response properties. However, in the mustached bat DSCF and AV areas, Fitzpatrick et al. (1993) and Fitzpatrick, Olsen, and Suga (1998b) found widespread non-linearity, in the form of facilitation by spectral and temporal stimulus combinations. About 75% of DSCF neurons show FM_1-CF_2 facilitation and are delay sensitive (fig. 25.2D). The delay at which facilitation is maximal, or "best delay" (BD), is relatively long, from 5 to 30 ms, with a mean at about 21 ms, corresponding to a target distance of 3.6 m. In essence, FM_1-CF_2 neurons express a time-dependent amplification that enhances sensitivity to weak echoes reflected from distant objects —for example, during the search phase of target pursuit. Such amplification presumably improves the target-to-background echo level ratio. Interestingly, Fitzpatrick et al. (1993) point out that the echo delay at which facilitation first occurs in DSCF neurons corresponds to the distance at which mustached bats typically initiate the approach phase of target pursuit.

Binaural Properties of DSCF, AIa, and AIp Neurons

Two studies have investigated binaural properties in the DSCF, AIa, and AIp areas (Liu and Suga 1997; Manabe, Suga, and Ostwald 1978). Three classes of binaural neurons were found: binaurally excited (EE), binaurally suppressed (EI), and (rarely) monaural (E0). EE neurons, which summate binaural input, are found in the ventral half of the DSCF area in the region with low BAs (fig. 25.2C, filled circles). EI neurons, which respond to interaural intensity differences, are found in the dorsal half of the DSCF area, in the region with

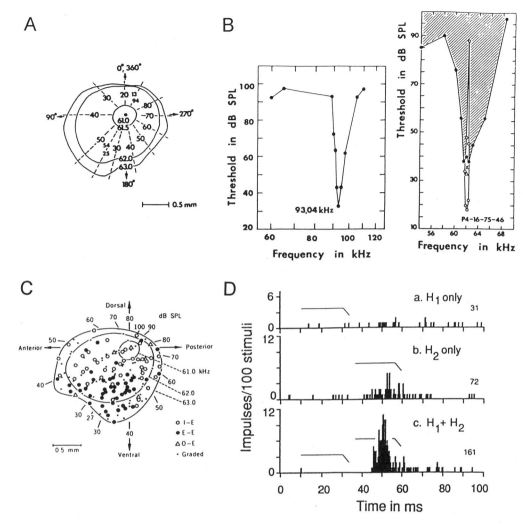

Fig. 25.2. CF processing stream. *A:* Bicoordinate organization of DSCF area. Isofrequency contours (in kHz) are concentric; isolevel contours (in dB SPL) are radial. (Adapted from Suga 1977, © Amer. Assoc. Adv. Science.) *B:* Frequency-tuning curves from AIa (*left*) and DSCF (*right*) neurons. For the AIa unit, the curve depicts excitatory response area. For the DSCF unit, *open circles* delineate excitatory area, *filled circles/shaded* region denote inhibitory area. (Adapted from Asanuma, Wong, and Suga 1983; and Suga and Manabe 1982, © American Physiological Society.) *C:* Binaural representation in DSCF area. EE units (E-E, *filled circles*) are located in the ventral half of the DSCF in the region of lower best amplitudes. EI units (I-E, *open circles*) are found dorsally in the region of higher best amplitudes. Contralateral monaural EO units (O-E, *open triangles*) are few and scattered. (Adapted from Manabe, Suga, and Ostwald 1978, © Amer. Assoc. Adv. Science.) *D:* Facilitation by FM_1-CF_2 combinations in DSCF area. Peristimulus time histograms showing responses to H_1 (CF_1/FM_1) alone (*top*), H_2 (CF_2/FM_2) alone (*center*), and paired H_1-H_2 with a 26 ms delay (*bottom*). (Adapted from Fitzpatrick et al. 1993, © Soc. for Neuroscience.)

higher BAs (fig. 25.2C, open circles), as well as in AIa and AIp. Manabe, Suga, and Ostwald (1978) asserted that EE neurons are best suited for target detection, because they have low BAs and, presumably, large spatial receptive fields. EI neurons, on the other hand, are better suited to target localization, because they have higher BAs and smaller receptive fields than EE cells. As in the cat auditory cortex (Imig and Reale 1980), Liu

and Suga (1997) found that the EE bands in the DSCF area of the two hemispheres are interconnected, but EI regions are not.

THE CF/CF STREAM: CF/CF, VA, AND DIF AREAS

The CF/CF stream involves three cortical fields (fig. 25.1C). The CF/CF area is the largest, and it is interposed between AIa and the FM-FM area, anterodorsal to DSCF. The VA area is quite small, and typically borders areas AV and DSCF. The recently identified DIF (dorsal intrafosal) field receives a heavy projection from the CF/CF area and a more modest projection from VA (Fitzpatrick, Olsen, and Suga 1998a). DIF has not been physiologically characterized as yet.

In the CF/CF stream, neurons are subdivided into two types, CF_1/CF_2 and CF_1/CF_3. In the CF/CF area, these cell types are segregated into two narrow strips running rostrocaudally (fig. 25.3A; Suga, O'Neill, and Manabe 1979; Suga et al. 1983). The facilitation tuning curves for CF/CF cells (fig. 25.3B) resemble a plot combining broadly tuned CF_1 cells with sharply tuned CF_2 (cf. fig. 25.2B, right panel) or CF_3 cells (cf. fig. 25.2B, left panel) from the DSCF and AIa fields (Suga et al. 1983b; Suga, O'Neill, and Manabe 1979; Suga and Tsuzuki 1985). Suga and Tsuzuki (1985) showed that inhibitory sidebands affect amplitude tuning (fig. 25.3B) and sharpen the excitatory area of these cells, so that the spectral tuning is level tolerant even at very high stimulus levels.

By and large, CF/CF neurons are not delay tuned (fig. 25.3C). Although some cells have a preferred echo delay, strong facilitation occurs whenever there is temporal overlap. The temporal overlap can be as short as 1 ms and still produce noticeable facilitation. Such broad delay tolerance enables CF/CF cells to encode relative velocity over a wide range of target distances.

Suga, Niwa, and Taniguchi (1983a) found that CF/CF cells show exquisite phase-locking to sinusoidal FM (SFM) of the echo CF component (fig. 25.3D). SFM crudely approximates the frequency modulations of an echo caused by the Doppler effect from the wing movements of a fluttering insect target. This suggests a possible role for the CF/CF area in local target motion (flutter) analysis.

Functional Organization of CF/CF Area: Velocity Map

Because tuning to the CF_2 or CF_3 components is so sharp, CF/CF neurons are particularly suited to encoding the magnitude of echo Doppler shifts (Suga et al. 1983b). This is because the majority of CF/CF cells are "mistuned"—that is, the best facilitation frequency (BFF) for the CF_2 or CF_3 component is not harmonically related to the BFF of the CF_1 component. Because by definition all components of either a pulse or an echo are harmonically related, these cells cannot respond to pulses or echoes alone. Instead, they respond to two overlapping signals with different fundamental frequencies, such as an emitted pulse and a Doppler-shifted echo. Because of their extraordinary tuning sharpness, each cell in effect represents a single Doppler-shift magnitude. Since echo Doppler shifts are caused by relative velocity differences between the bat and its surroundings, each CF/CF cell could represent a particular relative velocity.

CF/CF cells recorded in single cortical columns es-

sentially have identical best facilitation frequencies and are therefore tuned to the same Doppler-shift magnitude. Suga et al. (1983b) showed that there is a bicoordinate map of BFFs, with iso-BFF contours for the CF_1 component oriented more or less orthogonal to those for the CF_2 or CF_3 components (fig. 25.3A, top). The map encodes Doppler shifts ranging from 0 ventrocaudally to about 3 kHz rostrodorsally. This translates into velocities ranging from about 0 to 9 m/s, with an overrepresentation of the range from 0 to 5 m/s (fig. 25.3A, bottom).

Therefore, the CF/CF area could represent both the flight velocity of the bat and flutter from insect prey. Trappe and Schnitzler (1982) showed that foraging horseshoe bats Doppler-shift compensate their flight velocity relative to the surroundings, not that of pursued insect prey. If this holds for mustached bats, echoes from the surroundings would excite a specific "isovelocity" contour in the CF/CF area encoding the bat's flight velocity. Echoes from fluttering insects moving at different relative velocities would excite other parts of the CF/CF area in a less predictable pattern. Cells activated in those regions would then encode the echo modulation patterns with phase-locked discharges encoding wing-beat frequency (e.g., fig. 25.3D).

FM-FM S TREAM : FM-FM, DF, AND VF A REAS

Three interconnected rostral areas bordering the fundus of the Sylvian fossa constitute the FM-FM stream (fig. 25.1D, Fitzpatrick, Olsen, and Suga 1998a; Fritz et al. 1981). The FM-FM area lies dorsal to the CF/CF and DSCF areas. The dorsal fringe (DF) area lies dorsal to the FM-FM area, whereas the ventral fringe (VF) area lies either ventral to AIa or dorsal to AIa but below CF/CF. Each area is dominated by delay-tuned FM-FM neurons (Edamatsu, Kawasaki, and Suga 1989; Fitzpatrick, Olsen, and Suga 1998b; Suga and Horikawa 1986; Suga, O'Neill, and Manabe 1978). Neurons are classified as FM_1-FM_2, FM_1-FM_3, or FM_1-FM_4, according to which echo FM harmonic they are tuned. Each cell type is segregated into one of three rostrocaudal bands in both the FM-FM and DF areas (fig. 25.4A; O'Neill and Suga 1979; O'Neill and Suga 1982; Suga and Horikawa 1986). At the borders between the bands, one finds "hybrid" units (e.g., FM_1-$FM_{2,3}$) facilitated by more than one FM component in the delayed sweep (O'Neill and Suga 1982).

Many FM-FM neurons are able to respond to single tones, but they are typically much more sensitive to FM sweeps (fig. 25.4B). Unlike CF/CF neurons, FM-FM units are tuned rather broadly in the frequency domain. This "frequency tolerance" enables these cells to respond to the broadband FM components of pulses and echoes, regardless of the magnitude of Doppler-shift compensation or echo Doppler shift, without impairing the ability to encode target distance.

By contrast, FM-FM cells often are tuned sharply in

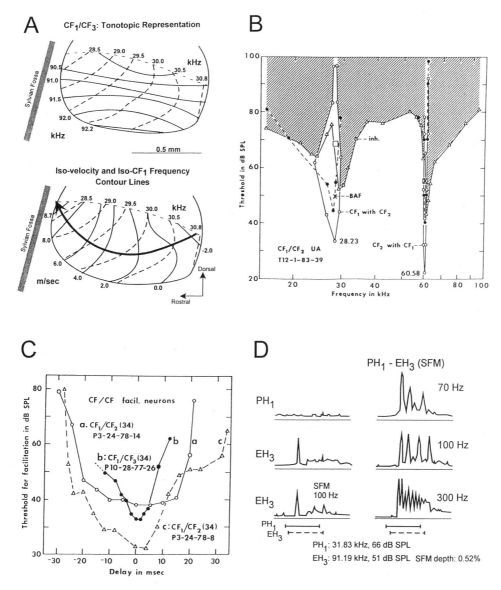

Fig. 25.3. CF/CF processing stream. *A:* Functional organization of CF_1/CF_3 area. *Top:* tonotopic organization with CF_1 (*dashed lines*) and CF_3 (*solid lines*) isofrequency contours. *Bottom:* isorelative velocity contours (*solid lines*) calculated from Doppler-shift magnitudes in upper figure. The *arrow* depicts the target-velocity axis. (Adapted from Suga et al. 1983, © Amer. Physiol. Soc.) *B:* Facilitation areas of a CF_1/CF_2 neuron. The *shaded* region shows the inhibitory area surrounding the excitatory areas (*unshaded*). Lowest thresholds were obtained with paired CF_1/CF_2 stimuli. Excitatory response areas for the CF_1 and CF_2 stimuli presented alone are indicated by *filled circles/dashed lines*. (Adapted from Suga and Tsuzuki 1985, © Amer. Physiol. Soc.) *C:* Delay tuning in CF/CF neurons. Facilitation occurs only with temporal overlap of the essential CF components. (Adapted from Suga et al. 1983b, © Amer. Physiol. Soc.) *D:* Responses of a CF/CF neuron to sinusoidal frequency modulations (*SFM*) simulating a fluttering target. *Left:* poor response to unmodulated pulse H_1 alone (PH_1), unmodulated echo H_3 alone (EH_3), and EH_3 alone modulated at 100 Hz. *Right:* facilitation by paired PH_1/EH_3 stimuli phase-locked to the modulated CF component of the EH_3 stimulus at rates of 70 Hz and higher. (Adapted from Suga, Niwa, and Taniguchi 1983, © Plenum Press.)

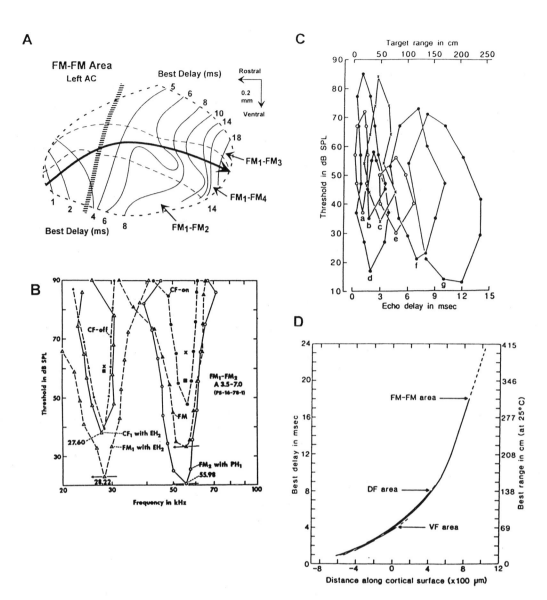

Fig. 25.4. FM-FM processing stream. *A:* Functional organization of FM-FM area, showing rostrocaudally oriented best delay axis (*thick arrow*). Iso-BD contours are oriented dorsoventrally across bands of FM_1-FM_2, FM_1-FM_4, and FM_1-FM_3 neurons. BDs between 5 and 8 ms are overrepresented. (Adapted from O'Neill and Suga 1982, © Soc. for Neuroscience.) *B:* Spectral facilitation areas of an FM_1-FM_2 neuron. Response areas for tones alone ("CF-on," "CF-off") are shown by *filled circles/ dashed lines.* Response area for FM_2 sweeps alone ("FM") is shown by *filled triangles/ dashed lines.* Curves with lowest thresholds are facilitation-tuning curves for paired FM-FM stimuli. *Length of arrows* at minima of FM response areas indicates bandwidth of FM stimulus; *arrowhead* indicates sweep direction (downward). (Adapted from Suga et al. 1983b, © Amer. Physiol. Soc.) *C:* Delay-tuning curves of FM-FM area neurons. Tuning curves are sharper for neurons with lower BD. (Adapted from O'Neill and Suga 1982, © Soc. for Neuroscience.) *D:* Range of best delays encoded in the FM-FM, DF, and VF areas. Best delays extend out approximately to 18 ms in the FM-FM area, but only to 9 and 5 ms in the DF and VF areas, respectively. (Adapted from Suga, Olsen, and Butman 1990, © Pergamon Press, plc.)

the intensity domain. Both target size and target range affect echo sound pressure level. The facilitation of FM-FM cells typically is related nonmonotonically to the level of the echo FM component, and most cells have a "best facilitation amplitude" (Suga et al. 1983b).

Delay Tuning

The most striking attribute of FM-FM neurons is that they are delay sensitive (O'Neill and Suga 1979). The vast majority of cells are delay tuned (fig. 25.4C), which means that their delay-tuning curves are stable over a wide range of pulse emission rates. By contrast, a small number of FM-FM neurons, called "tracking" neurons, have labile best delays that become shorter with increasing emission rate. Delay-tuned neurons only fire when a target is located in a specific range window and thereby provide a stable representation of echo range. Tracking neurons apparently "lock on" to targets throughout the attack (O'Neill and Suga 1979, 1982).

FM-FM neurons show evidence for suppressive interactions that serve to sharpen delay tuning. Edamatsu and Suga (1993) found that when two echoes are presented after a pulse, one inside the cell's delay-tuning curve and the other outside at shorter or longer delays, the delay-tuning curve becomes sharper. The function of these temporal inhibitory areas is analogous to that of lateral (sideband) inhibition in the spectral domain, but in this instance it improves range acuity and echo-amplitude tolerance.

It has been proposed that delay tuning is created by a system of neuronal delay lines (for review, see Suga 1990b). Olsen and Suga (1991) found FM-FM neurons in the medial geniculate body of the mustached bat. Unlike cortical FM-FM neurons, thalamic FM-FM neurons often have detectable responses to the pulse FM_1 and echo FM signal presented alone. Olsen and Suga also found that the response latency for the FM_1 component was longer than it was for the echo FM component. Best delay was directly related to the *difference* in latency between the pulse and echo FM responses. This implies that FM-FM neurons are facilitated when the inputs from the pathways carrying pulse and echo FM information are temporally coincident. For a given neuron, temporal coincidence only occurs when the pulse FM_1 stimulus leads the echo FM by an amount that offsets the latency difference for response to the two components presented alone.

Multiple Maps of Target Range in the FM-FM Stream

Neurons within a cortical column have similar BDs, but BD differs among columns. The BDs represented in the FM-FM area range from approximately 0.4 to 18 ms, but the majority of cells have BDs between 3 and 8 ms (Suga and O'Neill 1979; O'Neill and Suga 1982). In terms of target range, this distribution of BDs spans the biologically relevant echo delays encoding distances

between a few cm and about 3 m. Suga and O'Neill (1979) discovered an orderly "chronotopic" representation of BD along the rostrocaudal axis of the FM-FM area, with an overrepresentation of echo delays of 3–8 ms (fig. 25.4A, D). Iso-BD contours traverse the boundaries of the FM_1-FM_2, FM_1-FM_3, and FM_1-FM_4 cortical strips. Thus, the FM-FM area contains a single map of echo delay, and thereby target range, that is congruent across the three subdivisions of different FM-FM cell types. Because targets emphasize different echo frequencies depending upon size, the three bands of delay-tuned cells tuned to the three different higher harmonics provide a substrate for encoding distance over a wide range of target sizes and textures.

The DF area is strongly interconnected with the FM-FM area (fig. 25.1D; Fitzpatrick, Olsen, and Suga 1998a) and is similarly organized, with three rostrocaudally elongated bands tuned to the different echo harmonics and a chronotopic representation of BD (fig. 25.4A; Suga and Horikawa 1986). However, the range of BDs found along the rostrocaudal axis only extends out to 9 ms (fig. 25.4D). Consequently, DF neurons may only respond to nearby targets.

VF is another even smaller field containing FM-FM neurons (Edamatsu, Kawasaki, and Suga 1989). VF lies beneath the rostral end of the tonotopic AI region and receives a projection from both FM-FM and DF areas (fig. 25.1A, D; Fitzpatrick, Olsen, and Suga 1998a, 1998b; Fritz et al. 1981). The distribution of BDs in VF is even more truncated than that in the DF area, extending only to 5–6 ms (fig. 25.4D). The small size of VF has made it impossible to determine whether there are discrete bands, clusters, or randomly intermingled FM-FM neurons tuned to different echo harmonics.

Differences in neuronal response properties exist between the three areas of the FM-FM stream, but the differences are small and do not indicate a clearly different function for each area. Based on earlier evidence of a one-way interconnection of the FM-FM → DF → VF pathway (Fritz et al. 1981), Edamatsu and Suga (1993) hypothesized that neurons in the three areas show increasing specialization. Instead, they found that by certain criteria, only subtle differences distinguish the FM-FM and VF fields, an observation more in agreement with reciprocal interconnections (Fitzpatrick, Olsen, and Suga 1998a). For example, in both areas 80% of the cells show temporal inhibitory sidebands. However, using other criteria, distinct differences became clear. For example, most FM-FM area cells can respond to pulse-echo pairs presented at rates up to at least 100 stimuli/s (simulating terminal phase) with little change in either BD or delay-tuning curve shape. However, VF cells are not able to respond at such high repetition rates and thus cannot respond during the late approach and terminal phase—when, paradoxically, echo delays are short enough to facilitate response. This suggests that different cortical fields take part in target range processing only during particular phases of echolocation.

Spatial Receptive Fields in the FM-FM Area

Delay-sensitive FM-FM neurons encode spatial information along the dimension related to target distance. But might they also encode the location of a target in all three dimensions? Suga, Kawasaki, and Burkard (1990) measured the free-field spatial receptive fields of range-tuned neurons in the FM-FM area and found that they were very large, >70° in both azimuth and elevation. Best azimuths were typically confined between midline and 25° contralateral to the recording site. In some cells, receptive field boundaries ended at the midline, suggesting that they were EI units suppressed by signals in the ipsilateral hemifield. In other cells, activity extended across both hemifields, suggesting binaural activation (EE). Best delays remained more or less constant within a receptive field. Based on the large size of the spatial receptive fields, Suga, Kawasaki, and Burkard (1990) concluded that delay-tuned cortical cells are unsuited for horizontal or vertical target localization but can provide reliable target range estimates for objects anywhere in contralateral frontal auditory space.

However, Middlebrooks et al. (1994) argue that localization is encoded by distributed ensembles of cortical neurons with broad spatial tuning. According to this model, three-dimensional spatial information processing could be carried out in any of the delay-tuned areas of the mustached bat cortex, despite the large spatial receptive fields of units in these areas.

Responses to Communication Sounds

Kanwal et al. (1994) described an impressive repertoire of over 20 discrete mustached bat communication sounds. Many of these utterances incorporate CF and/or FM components that resemble echolocation signals. Ohlemiller, Kanwal, and Suga (1996) reported that combination-sensitive neurons possibly play a role in both echolocation and communication. They found that about one third of FM-FM cells responded strongly to both echolocation sounds and to paired syllables of multisyllabic communication calls. Contrary to expectation, when stimulated with the latter signals, the intersyllable delays eliciting maximal facilitation were much longer (e.g., 20–30 ms) than the best delays for pulse/echo FM-FM stimuli resembling sonar signals. Moreover, combinations of spectral elements in communication sounds causing facilitation did not necessarily match the biosonar FM components causing facilitation. These surprising findings suggest that combination sensitivity could be expressed very differently for classes of biologically relevant sounds.

When combined into complex sequences for communication, mustached bat vocalizations have a primitive syntactic structure. That is, different syllables can be combined to produce a wide variety of different utterances. FM-FM neurons may actually be selective for syntax (Esser et al. 1997). Facilitation is strongest when "natural" (species-typical) temporal sequences of two or more syllables are presented, as opposed to the same syllables presented singly or in temporally scrambled "unnatural" sequences. This study is the first evidence at the single-unit level of a form of higher-order syntactical processing in mammalian auditory cortex.

Kanwal and Suga (1995) have also provided preliminary evidence for hemispheric differences in the responsiveness of CF_2 and FM_1-CF_2 neurons to echolocation and communication sounds. DSCF neurons in the left hemisphere are more strongly facilitated by social vocalizations than are DSCF neurons in the right hemisphere. By contrast, the right hemisphere appears to be more responsive to echolocation sounds than to communication calls.

Role of Cortical Fields in Mustached Bat Echolocation

Reversible lesion experiments examined the possible involvement of the DSCF and FM-FM areas in frequency and target range discrimination. Riquimaroux, Gaioni, and Suga (1992) applied the selective GABA receptor agonist muscimol to either the DSCF or the FM-FM area, and then determined the bat's acuity for frequency and distance discriminations using a conditioned avoidance procedure. When muscimol was applied to the DSCF area, bats could discriminate large, but not small, frequency differences. By contrast, target distance discrimination remained unimpaired. When muscimol was applied to the FM-FM area, then fine, but not coarse, target range discrimination was impaired, and frequency discrimination remained normal. Thus, as predicted from the response characteristics of cells and their functional organization, the FM-FM area is involved in the perception of distance but has apparently little to do with frequency discrimination. The opposite is true for the DSCF area. Individual cortical fields appear to play pivotal roles in specific aspects of perception related to particular features of acoustic signals.

Concluding Remarks

Based on both anatomical and functional criteria, the mustached bat's auditory cortex is highly differentiated into discrete subdivisions. Three processing streams are evident. One stream (FM-FM) encodes target range information carried by the FM components of the sonar signal. A second stream (CF/CF) encodes Doppler-shift magnitude and target motion information. A third stream (CF) is devoted to target motion analysis, particularly at the earliest stages of target detection and pursuit. All three streams contain combination-sensitive neurons that are critical to the functional differentiation of information processing. At the level of the auditory cortex, there is little evidence that these cortical streams segregate the analysis of echolocation and communication sounds, since most cortical cells respond to both types of sounds. However, the apparent hemispheric differences in responsiveness to these two types of signals is perhaps the first clue that there are separate processing streams mediating echolocation and acoustic communication. This important finding awaits comprehensive verification beyond the preliminary evidence currently available.

At this moment, functional streams have not been described in studies of the auditory cortex in other bats. Rauschecker et al. (1997) have suggested that there are separate processing streams in the squirrel monkey auditory cortex subserving spatial and spectral processing. Current studies on bats extend our understanding of these functional streams into the frontal cortex (Esser and Kanwal 1996; Esser and Eiermann 1999; Kanwal

et al. 2000), which receives moderately strong projections from the auditory cortex (Esser and Eiermann 1999; Fitzpatrick, Olsen, and Suga 1998a; Kobler, Isbey, and Casseday 1987). Descending projections of the auditory cortex to the inferior colliculus recently were shown to mediate recurrent adaptive plasticity in the ascending pathways, which provide input to the cortex itself (Gao and Suga 1998; Yan and Suga 1996a, 1996b; Yan and Suga 1998; Zhang and Suga 1997; Zhang, Suga, and Yan 1997). Whether the different cortical streams mediate distinct aspects of auditory behavior remains to be investigated more fully.

Acknowledgments

Many thanks are given to D. C. Fitzpatrick, J. F. Olsen, and N. Suga for permission to reproduce their figures. J. Housel and M. L. Zettel provided technical support. This study was supported by NIH/NIDCD grant R01-DC02179.

{ 26 }

The Auditory Cortex of the Little Brown Bat, *Myotis lucifugus*

Donald Wong

Introduction

In echolocation, FM bats employ brief broadband signals over a wide emission rate that permits relatively good target resolution at the expense of short operating distances. In the little brown bat, *Myotis lucifugus,* the biosonar signal is a frequency-modulated (FM) call, primarily comprising one harmonic sweep downward from about 100 kHz to 40 kHz in 1–4 ms. Bats actively change their pattern of sonar emissions during the different stages of echolocation, presumably to meet the perceptual demands of target detection, characterization, and classification. For example, *Myotis* dramatically increases pulse repetition rate (PRR) from 10/s in the search phase to 150–200/s in the terminal phase (fig. 26.1) where its pulse duration is <1 ms and higher harmonics often are added (Sales and Pye 1974). The biosonar pulse of this FM species is a relatively simple acoustic signal suited for foraging in an open environment. In

contrast, CF-FM bats employ a complex, multiharmonic signal for foraging in heavy foliage, an acoustically cluttered environment where target-reflected echoes are embedded in background noise. The neural mechanisms for processing biosonar signals are not only species specific but also likely exploited to optimized target perception.

Functional Organization of the *Myotis* Auditory Cortex

The *Myotis* auditory cortex comprises two functional regions: a tonotopic and a rostrodorsal region (fig. 26.2). In electrophysiological mapping experiments using pure-tone stimulation, Suga (1965) described a tonotopic area with a frequency axis running from a rostral to caudal direction. Neurons with high best frequencies (BF) (approximately 100 kHz) were located more rostrally, and those with low BF were located caudally. Sound frequencies represented in this tonotopic area spanned the

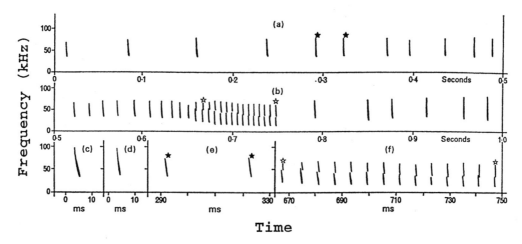

Fig. 26.1. Sonagrams of frequency-modulated (FM) biosonar signal of *Myotis lucifugus*. *a, b:* Pattern of sonar emission changes over a 1 s echolocation cycle. Note that the pulse repetition rate increases and duration shortens from over the search, approach, and terminal phases. *c, d:* Broadband FM sweeps downward from about 100 to 40 kHz in 4 ms during the search phase. *e:* FM sweeps during the approach phase (shaded star in *a* on an expanded time scale [ms]). *f:* Series of FM sweeps during the terminal phase (open star in *b* on an expanded time scale [ms]). A higher harmonic is added (from Sales and Pye 1974).

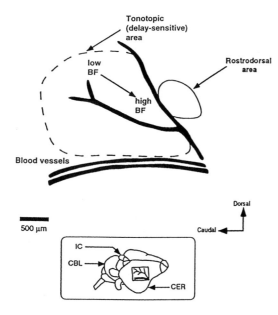

Fig. 26.2. Functional organization of the *Myotis* auditory cortex. The auditory cortex comprises a relatively large tonotopic area and a small rostrodorsal area. Both functional regions are shown with respect to a characteristic surface pattern of major blood vessels. BF: best frequency, CBL: cerebellum, CER: cerebrum, IC: inferior colliculus.

species' frequency range of hearing. This single tonotopic area, which constitutes most of the *Myotis* auditory cortex, exhibits a striking feature: its cortical response properties are altered by changes in the stimulus conditions that mimic the different phases of echoloca-

tion. In other FM and CF-FM bats, multiple tonotopic regions exhibiting differing specializations have been mapped in the auditory cortex (O'Neill 1995). The rostrodorsal area of *Myotis* contains neurons whose BFs are restricted to a relatively small part of the frequency representation of the tonotopic area.

TONOTOPIC AREA

In the tonotopic area, combination sensitivity to stimulus pairs emerges when stimulus repetition rate increases to at least 10/s (Wong, Maekawa, and Tanaka 1992). Thus, this tonotopic area could also be defined as a combinative-sensitive region when mapped with FM stimulus pairs presented at PRRs ranging from about 10 to 200/s. The FM-FM sensitivity is a delay sensitivity characterized by facilitative responses evoked by stimulus pairs at specific echo delays between the pulse FM and echo FM. To map multiple, functional cortical areas analogous to those in CF-FM bats (Suga 1984; Schuller, O'Neill, and Radtke-Schuller 1991), earlier studies of *Myotis* revealed tonotopic and delay-sensitive areas (e.g., Sullivan 1982; Berkowitz and Suga 1989) that largely overlapped (Wong and Shannon 1988). When mapping was conducted with pure tones or stimulus pairs presented systematically at different PRRs, the single tonotopic area defined using single-sound (pure-tone) stimulation was transformed into a delay-sensitive area defined using sound-pair (FM-FM) stimulation at PRRs >10/s (Paschal and Wong 1994b). In the auditory cortex of CF-FM bats, combination-sensitive areas are separate from or largely outside the primary tonotopic area.

Representation of best echo delay in the tonotopic area

A species-specific difference governing the functional organization of FM bats is the apparent lack of a neural map for best echo delay (BD). In a mapping experiment of the *Myotis* auditory cortex, both BF and BD were measured in the same neurons using pure-tone (single-sound) stimulation and FM stimulus pair, respectively (Paschal and Wong 1994b). No topographic organization according to BD was found for neurons, despite their clear frequency organization. In fact, several BDs were associated with a particular cortical locus, since neurons sampled at different cortical depths in a orthogonal penetration (similar BFs) exhibited BD values that differed by as much as 5 ms. In characterizing delay sensitivity, BD measurements were obtained at relatively low PRRs (approximately 10–20/s) that evoked clear and consistent facilitative responses. Although earlier studies demonstrated that stimulus repetition rates influence delay-tuning properties (Wong, Maekawa, and Tanaka 1992), the BD disparity within a cortical column or the apparent lack of topographic organization according to BD could not be completely accounted for by BD measurements with somewhat differing PRRs. In *Eptesicus fuscus,* a BD map was not evident in any of the multiple tonotopic regions that constitute the auditory cortex of this FM bat (Dear, Simmons, and Fritz 1993). In contrast, a BD map was demonstrated in an FM-FM area, separate from the tonotopically organized and, presumably, primary auditory cortex of CF-FM bats (O'Neill and Suga 1982). In the mustached bat, virtually all delay-sensitive neurons exhibit a type of delay sensitivity, or delay tuning, in which the BD remains constant as the PRR is increased over a range that mimics the sonar emission rate used by this species in the different phases of echolocation (O'Neill and Suga 1982). Thus, the differences in cortical representation of BD by these major classes of bats probably reflect not only different organizational principles for processing target-range information but also different cortical strategies for target perception.

Tracking neurons underlying population encoding of BD

A distinctive delay-tuning property of neurons in the tonotopic part of the *Myotis* cortex is the BD shift with changes in PRR (Wong, Maekawa, and Tanaka 1992). This tracking type of delay sensitivity was initially reported in a few FM-FM cortical neurons of the mustached bat (O'Neill and Suga 1982). Tracking neurons not only constitute a majority of *Myotis* auditory cortical neurons, but more recent studies suggest that a large majority of delay-sensitive neurons exhibit some degree of tracking (Chen and Wong 1995). Given a "dynamic range" of PRRs at which individual neurons encode BD, it is perhaps not surprising that no clear topographic distribution of BD was mapped. The BD among neurons

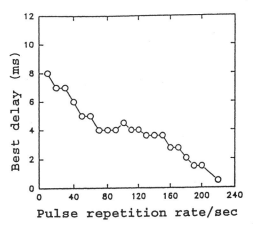

Fig. 26.3. Tracking pattern of a representative delay-sensitive neuron in which the best delay shortens with increase in the pulse repetition rate

differs most at low PRRs, stimulus rates that often were not uniform when mapping BD. However, it is noteworthy that as PRR is increased to values indicative of the emission rates at later stages of echolocation (>50/s), the BDs of most tracking neurons not only become shorter (fig. 26.3) but converge to similar values. It remains to be demonstrated whether a more orderly spatial distribution of short BDs emerges at the higher PRRs. Interestingly, the same cortical region is increasingly devoted to shorter BDs under stimulus conditions that the bat encounters at later stages of echolocation. From a computational perspective, the temporal acuity probably increases as the tonotopic cortical area becomes more dedicated to processing a narrow BD range and exhibits an overrepresentation of short BDs.

Representation of pulse duration

The pattern of sonar emission is shaped closely in the time domain. FM bats shorten the duration of their emitted pulses as their emission rate increases. This gating of the call is largely to avoid overlap of the biosonar signal with the returning echo as the bat approaches a target (Griffin 1958). This interplay between two stimulus temporal parameters (PRR and duration) is consistent with the fact that the BD of neurons shortens as stimulus duration decreases (Tanaka, Wong, and Taniguchi 1992) or PRR increases (Wong, Maekawa, and Tanaka 1992).

Duration selectivity was studied in the *Myotis* cortex, using tone bursts presented at different durations (1–400 ms) (Galazyuk and Feng 1997b). Duration-selective neurons were short pass (36%), all pass (31%), long pass (18%), and band pass (15%). Only the short- and long-pass neurons appear to be clustered along nonoverlapping slabs running rostrodorsally, with the short-pass slab located more ventral than the long-pass slab. Moreover, 65% of these neurons progressively decreased their best duration as stimulus intensity was increased (intensity-dependent duration). Given that response threshold is a function of stimulus duration, a threshold decrease induced by increase in duration would shift the optimal responses to a shorter duration, thereby shifting a neuron's duration selectivity from long to short pass. As the overall echo intensity becomes greater at shorter target ranges, the relative population of short-pass neurons is increased, and this overrepresentation of short delays emerges cortically to optimize perception of near targets. The closer target ranges dictate higher PRRs, and thus the selectivity to short durations necessitates a narrow "time window" for target-echo analysis.

ROSTRODORSAL AREA

The rostrodorsal area is much smaller than the tonotopic area of the *Myotis* auditory cortex (Paschal and Wong 1994a). In cortical mapping studies, a sharp frequency discontinuity is evident between the high frequency associated with the tonotopic area and the relatively low frequencies in the rostrodorsal area. The rostrodorsal neurons were tuned more broadly to frequency than those in the tonotopic, delay-sensitive area. The rostrodorsal area has a re-representation of sound frequencies between 20 and 40 kHz. Both single- and dual-peaked neurons (harmonically related) were found with relatively high minimum thresholds. Of the relatively small sample studied, no delay sensitivity or clear tonotopy was found.

Given the frequency range represented in the rostrodorsal area, Paschal and Wong (1994a) speculated that this area could be involved in processing echolocation or communication sounds. Most of the energy of FM signals is concentrated in the frequency range represented by this area. Moreover, the behavioral audiogram of *Myotis* showed a high sensitivity at frequencies ranging from 20 kHz to 50 kHz with a maximal hearing sensitivity at 40 kHz (Dalland 1965). This area is similar to the ultrasonic field in the mouse auditory cortex, which was speculated to process species-specific communication sounds (Hoffstetter and Ehret 1992).

Functional Properties of Neurons Underlying Target-Feature Extraction

DELAY-SENSITIVE NEURONS:
ENCODING TARGET VELOCITY

Bats perceive relative target velocity either by computing the rate at which target range (echo delay) changes in successive echoes (range-rate information) or by evaluating the Doppler shift in target-reflected echoes (Schnitzler 1984). Extracting range-rate information from the FM signal was demonstrated behaviorally in the short CF-FM bat, *Noctilio leporinus* (Wenstrup and Suthers 1984). Since FM bats do not exhibit

the Doppler-shift-compensating behavior of long CF-FM bats, target velocity was hypothesized to be encoded using range-rate information. Delay-sensitive neurons in the *Myotis* cortex were presented with sequences of pulse-echo pairs with the echo-delay step fixed at specific values (Tanaka and Wong 1993). Echo-delay step is a measure of the rate of change of echo delay, and thus changing the echo-delay step in this study is presumably related to encoding of different target velocities. While a majority of neurons were sensitive to echo-delay step, 40% of the neurons showed clear preference to echo-delay steps between 1 and 4 ms, which correspond to target velocities between 1.8 and 7.0 m/s. These values are similar to target velocities between 1 and 5 m/s reported in behavioral studies of other FM bats (for review, see Simmons and Grinnell 1988).

AMPLITUDE SENSITIVITY:
ROLE IN TARGET CHARACTERIZATION

During target-directed flight, echolocating bats reduce the amplitude of their emitted pulse to compensate for the increase in echo amplitude returning at shorter echo delays (closer target) (Griffin, Webster, and Michael 1960). Studies of amplitude tuning suggest that the amplitude at which a bat emits echolocation pulses provides a measure of target distance (Teng and Wong 1992a, 1992b). The pulse facilitation amplitude showed a significant positive correlation with both pulse facilitation latency and BD. Thus, neurons that showed a preference for higher pulse amplitudes were tuned to longer echo delays, and those that preferred a low pulse amplitude were tuned to shorter echo delays. In contrast, the *echo* amplitude of a target does not provide a simple measure of target distance. Echo amplitude is not an unambiguous cue of target range, because other target attributes (e.g., size) also influence echo amplitude in a complex manner. This is supported by two neurophysiological findings in amplitude tuning. First, the echo best facilitation amplitude has shown a weak correlation with either the echo best facilitation latency or BD. Second, the echo facilitation latency also has shown no correlation with BD. Thus, there is no neural basis by which echo amplitude could encode useful target-distance information. Yet, it is conceivable that target echo amplitude in conjunction with echo-delay information could provide some measure of the target subtended angle and target size (Suga 1990b).

SENSITIVITY TO MODULATION FREQUENCY:
ENCODING OF TARGET WING BEAT

Echoes reflected from the wings of fluttering insects are amplitude modulated (AM) with the modulation frequencies determined by specific wing-beat frequencies of different insect preys. Bats that exhibit Doppler-shift behavior are well suited for using their relatively long CF signals to encode AM derived from multiple cycles of wing beats (Schnitzler 1987). In contrast, FM bats use short biosonar signals (<5 ms), and a single echo is too short in duration to register the wing-beat frequency of most insects (the period between 10 and 100 ms). Yet, behavioral studies (Sum and Menne 1988) have demonstrated that FM bats are capable of discriminating wing-beat frequencies at the same resolution as long CF-FM bats. A key strategy by which FM bats perceive wing-beat frequency is to integrate the AM information conveyed across a sequence of echoes to compute neurally the modulation frequency.

Neurophysiological studies of the *Myotis* auditory cortex have also provided evidence for cortical encoding of specific wing-beat frequencies (Condon et al. 1997). By imposing an amplitude modulation across a train of sounds at different fixed PRRs, cortical neurons (shown separately to be delay sensitive) were found to exhibit ,sensitivity to specific modulation frequencies. First, the greater responses to modulated versus unmodulated trains suggest that animated targets play an important role in attracting the attention of bats and shaping the responsiveness of cortical neurons. Second, the PRR plays a critical role in determining not only the filter properties (spike-count response function to modulated pulse train at different repetition rates) but, more importantly, in shifting the relative proportion of neurons with different filter properties. As the PRR was increased, the number of neurons exhibiting a best modulation frequency not only increased (over half at PRR of 50/s, and over two thirds at PRRs of 100 and 200/s), but the AM frequency shifted to higher best modulation. These best modulation frequencies encompassed the entire range of wing beat frequencies encountered by bats when hunting. Moreover, when the response magnitude of modulated versus unmodulated pulse trains were compared at each PRR, most units showed a facilitation that depended on PRR, especially at high PRRs (100 and 200/s) and in a restricted range of modulation frequencies (50–70 Hz) (fig. 26.4). Such stimulus conditions reflect the sonar emission at later stages of echolocation, when bats identify their prey and use wing-beat frequencies for selecting or rejecting insect prey.

Summary: Multidimensional Analysis of Target

Although the *Myotis* auditory cortex consists primarily of a single tonotopic region, its species-specific,

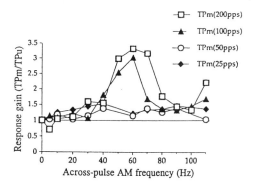

Fig. 26.4. Response gain function of a neuron at different modulation frequencies. The response gain is a spike-gain function of the modulated and unmodulated pulse trains (TPm/TPu). At TPm of 100 and 200 pulses per second (pps), this neuron exhibits a strong facilitated response to specific modulation frequencies (50–70 Hz) (from Condon et al. 1997).

functional organization can dynamically change to meet the behavioral demands of echolocation: *a frequency-organized region responsive to single pure tones becomes transformed into a delay-sensitive region comprising neurons engaged in target perception.* Individual delay-sensitive neurons have the capability to track echo delays, and thus encode target distance throughout the different stages of echolocation. These neurons also are endowed with the capability for multiple encoding of other target attributes. By integrating information across multiple target echoes, such target features as target velocity and wing-beat information can be extracted further. The perception of target structure (size, subtended angle) could entail extraction of multiple target cues (amplitude, echo delay). At stimulus conditions similar to those encountered at later stages of echolocation where target selection is critical, the disproportionate cortical representation of short echo delays provides a neural basis for the processing at higher temporal resolution, possibly using a population code (Palakal and Wong 1999). A future challenge is to elucidate how encoding of multiple target dimensions contributes to the generation of a coherent acoustic image of the target.

Acknowledgments

This research was supported by grants from NIDCD R01 00600 and NSF BCS-9307650.

{ **27** }

Processing of Frequency-Modulated Sounds in the *Carollia* Auditory and Frontal Cortex

Karl-Heinz Esser and *Arne Eiermann*

Introduction

Microchiropteran bats are well known for their ability to echolocate for short-distance orientation, obstacle avoidance, and acquisition of food. Orientation calls of CF-FM bats (i.e., rhinolophids, hipposiderids, and the mormoopid *Pteronotus parnellii*) are dominated by a long constant-frequency (CF) component preceding the call's frequency-modulated (FM) end. In contrast, FM bats (essentially all other species) rely on short frequency-modulated signals (for a more differentiated view, see Neuweiler and Fenton 1988; Fenton 1995). Apart from the mustached bat, *P. parnellii* (Suga 1990b), the auditory cortices of only two other species of CF-FM bats (*Rhinolophus ferrumequinum:* Ostwald 1980; *Rhinolophus rouxi:* Schuller, O'Neill, and Radtke-Schuller 1991; Radtke-Schuller and Schuller 1995) and of three species of FM bats (*Myotis lucifugus:* Wong and Shannon 1988; Shannon-Hartman, Wong, and Maekawa 1992; *Eptesicus fuscus:* Jen, Sun, and Lin 1989; Dear et al. 1993; and *Carollia perspicillata:* Esser and Eiermann 1999) have been mapped electrophysiologically at the single- or multiunit level. In the purely insectivorous species mentioned above (i.e., all except *Carollia*), a number of features are recognized as typical of "bat" auditory cortical organization (O'Neill 1995): (1) at least one tonotopically organized cortical field, (2) overrepresentation of sonar frequencies, and (3) specialized delay-sensitive neurons (so-called FM-FM neurons) predominantly or exclusively responsive to paired FM-sweep stimuli (FM-FM combinations). In both *Eptesicus* and *Myotis*, such delay-tuned neurons are embedded in clusters of unspecialized (i.e., tone-responsive) cells within the tonotopically organized area; whereas in *Rhinolophus* and perhaps most elaborately in *Pteronotus*, FM-FM neurons are segregated into distinct cortical fields (O'Neill 1995). The functional parcellation of the *Carollia* auditory cortex (AC) was established only recently (Esser and Eiermann 1999; fig. 27.1). Here, both the relative position of four of the auditory fields and the representational principles of sound parameters within these areas seem to reflect a "general plan" of mammalian auditory cortical organization. As an obvious adaptation for echolocation, however, a coherent dorsally

displaced high-frequency representation (HF = high-frequency field) was found, covering about 40% of the total auditory cortical surface. Although consistent differences in minimum thresholds distinguish the two subfields (HFI and HFII, see fig. 27.1), response properties of HF neurons are unknown. This chapter reports recent unpublished single-unit data from the high-frequency fields of the *Carollia* auditory cortex.

Apart from the "classic" auditory pathway, which reaches the auditory cortex via the midbrain inferior colliculus and the ventral division of the thalamic medial geniculate body, an alternative pathway exists (Kobler, Isbey, and Casseday 1987; Casseday et al. 1989). This pathway comprises the nucleus of the central acoustic tract and the suprageniculate nucleus of the auditory thalamus, finally reaching the auditory cortex and a circumscribed region in the frontal cortex. According to Casseday et al. (1989), the existence of such an extra-

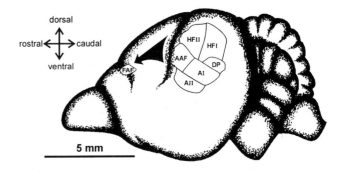

Fig. 27.1. Cortical organization in *Carollia perspicillata*. Physiologically, the *Carollia* auditory cortex is subdivided into six fields: AI, primary auditory cortex; AAF, anterior auditory field; AII, secondary auditory cortex; DP, dorsoposterior field; HFI, high-frequency field I; HFII, high-frequency field II (Esser and Eiermann 1999). In contrast to the central core of the auditory cortex (i.e., AI and AAF) and the adjacent AII, tonotopic gradients are absent in the dorsally displaced high-frequency fields (HFI and HFII). Generally, HF harbors units with characteristic frequencies of above 60 kHz. Lowest response thresholds (often below 0 dB SPL) are represented in the center of HFI, whereas substantially higher average thresholds are characteristic of HFII units. As indicated by the arrow, the auditory cortex projects to a frontal cortical region (FAF, frontal auditory field).

collicular auditory pathway (i.e., the "central acoustic tract") is a general mammalian feature but particularly conspicuous in the bat. Physiological studies in the CF-FM bat *P. parnellii* indicate that the frontal cortical target of this pathway is involved in the processing of biosonar information (e.g., Kanwal et al. 2000; Kanwal, Peng, and Esser, chapter 29, this volume). By using both metabolic brain mapping and retrograde tracing techniques (Eiermann and Esser, unpublished), we identified a corresponding frontal cortical region (hereafter frontal auditory field, FAF) in the FM bat *Carollia perspicillata* (fig. 27.1). Herein, we describe the basic stimulus preferences and response properties of neurons from the *Carollia* FAF (Eiermann and Esser 2000).

Materials and Methods

Adult short-tailed fruit bats (*C. perspicillata*) of both sexes were taken from a breeding colony at the University of Ulm. Experiments were carried out in a heated (28°C), double-walled soundproof chamber lined with convoluted, anechoic polyurethane foam. For immobilization of the bat's head, a small metal bolt was affixed to the dorsal surface of the skull and attached to a stereotaxic device, while the animal's trunk was restrained in a padded holder suspended with an elastic band (for details of the surgical procedure see Esser and Eiermann 1999). Through a tiny hole (30–50 μm) in the skull, a lacquer-coated tungsten microelectrode (impedance: 2–4 MΩ) was inserted into the cortex to record action potentials from single units and neuron clusters (multiunits, 2–4 neurons) extracellularly. During the 6–8 hour recording session, the awake animal (recovery time after surgery $\geq$ 24 h) was monitored continuously via a videocamera for signs of discomfort or distress. Neuronal responses to constant-frequency tone bursts (5–140 kHz), frequency-modulated sweeps (downward or upward), and sweep combinations all presented under free-field conditions were quantified by spike-count and PSTH (peristimulus-time histogram) analysis. In the FAF, band-limited noise and clicks also were employed as acoustic stimuli. For each stimulus type and neuron, sound parameters (e.g., sound pressure level [SPL in dB re 20 μPa], frequency, initial and terminal sweep frequency, repetition rate) were adjusted to be most effective for eliciting neuronal responses. Except for low-frequency tones, the stimulus duration was constant at 1.2 ms, corresponding to the duration of typical search- or early approach-phase echolocation calls of *C. perspicillata* (Thies, Kalko, and Schnitzler 1998). Since repetition rates of <5 Hz in the auditory cortex were found to be most effective, FAF stimuli were presented at a fixed rate of 2.5–3.5/s. Units responding to downward FM sweeps (and/or FM-FM combinations) were further tested systematically for their delay sensitivity. In this paradigm, the time delay between two successively presented FM stimuli mimicking components (e.g., harmonics) of the emitted biosonar signal and the returning echo (i.e., the pulse-echo delay) and intensities of both pulse and echo were optimized for a particular neuronal response. Subsequent testing at various shorter and longer delays was carried out using the best pulse and the best echo amplitude. A facilitatory interaction (criteria adopted from Fitzpatrick et al. 1993) between the pulse and the echo FM sweeps was regarded as a prerequisite for a unit's delay sensitivity. In tone-responsive units, neuronal best frequencies were determined as the frequency that elicited the highest spike rate during tones presented at best SPL (H. Thomas et al. 1993).

Results

Auditory Cortex

Recordings were obtained from 31 single- and 6 well-isolated multiunits from the high-frequency fields of the *Carollia* auditory cortex (n = 5 bats). Except for absolute spike numbers, no differences in response properties between single units and the few well-isolated multiunits (both denoted as units or neurons below) were found. The majority of HF units studied (47%, n = 15) preferred downward FM sweeps (fig. 27.2A). In other cases (37%), pure-tone stimuli were most effective for response generation. The remaining 6 neurons (16%) responded equally well to FM sweeps or tones (8%) or to FM sweeps independent of their sweep direction (8%). Neurons responding preferentially to upward FM sweeps were never encountered.

For 23 HF neurons, a preferred FM sweep was determined (fig. 27.2B). As a trend, in HFII, FM sweeps in the frequency range of the biosonar third and fourth harmonic were most effective stimuli, whereas in HFI, the second harmonic also was well represented.

Eight of 22 units responding to downward FM sweeps exhibited pulse-echo delay sensitivity with best delays ranging from 2 to 25 ms. In delay-tuned HF neurons (fig. 27.2C), the frequency ranges of the best pulse and best echo FM sweep differed by no more than 3 kHz.

Rate-level functions were obtained for the preferred downward FM stimulus in 20 single units (examples in fig. 27.2D). In 13 units, these functions were monotonic (top graph), whereas a nonmonotonic course (middle and bottom) was obvious in 7 cases. With respect to bat echolocation, multipeaked rate-level functions (fig. 27.2D, bottom) are of particular interest (see "Discussion" below).

Excitatory frequency-tuning curves of tone-responsive HF single units were obtained by pseudorandom presentation of different frequency-level combinations. Most tuning curves were either V-shaped or W-shaped (fig. 27.2E). In a few cases, however, tuning was more complex, as reflected by the occurrence of at least two spectrally well-separated response areas or almost closed

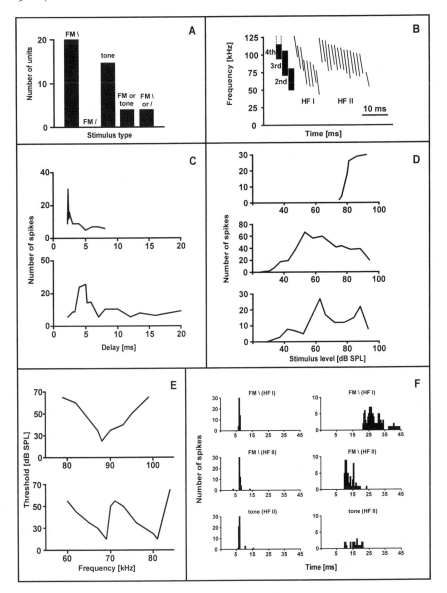

Fig. 27.2. Responses from the *Carollia* auditory cortex. *A:* Stimulus preferences of HF neurons (*n* = 32; FM \ = downward FM sweep; FM / = upward FM sweep). *B:* Preferred FM stimuli of HF neurons (*n* = 23) compared to the mean bandwidths of the second to fourth harmonic of the species' echolocation call (Eiermann and Esser, unpublished). Since initial and terminal sweep frequencies were varied at a fixed duration of the FM stimulus (i.e., 1.2 ms), minor differences in FM rate were unavoidable. *C:* Delay-tuning curves of units A13 (*top*) and A9 (*bottom;* stimulus repetitions per delay: 100; window for spike counting: 30 ms). *D:* Rate-level functions of units A16, A40, and A9 (other parameters as in C). *E:* Excitatory frequency-tuning curves of units A6 and A19. *F:* Temporal response properties of six selected HF units (A19, A9, and A17, *left;* A33, A22, and A18, *right*); for stimulus and neuronal subtype, see headings; stimulus repetitions per PSTH: 200; binwidth: 0.5 ms).

tuning curves (not shown). In 35 of 36 HF units studied, neuronal best frequencies were above 60 kHz (compare fig. 27.1).

As revealed by PSTH analysis, first-spike latencies of HF neurons ranged from 4–4.5 to 21.5 ms. Latencies ($\bar{x} \pm$ SD) of neurons responding solely or preferentially to pure-tone stimulation were shorter than those of neurons preferring FM sweeps (HFI: 7.0 ± 3.8 versus 14.1 ± 4.8 ms; HFII: 6.3 ± 2.3 versus 9.3 ± 2.1 ms). Both HF subfields contained neurons with phasic (fig. 27.2F, left) and sustained responses (fig. 27.2F, right).

FRONTAL AUDITORY FIELD

Single-unit responses from the *Carollia* frontal auditory field were more difficult to record because of a low peak firing rate and irregular response patterns of neurons. Hence, in this forebrain area, it seemed advantageous to record from well-isolated multiunits (2–4 neu-

rons) instead of single units. In contrast to auditory cortical HF neurons (fig. 27.2A), a substantial fraction of FAF-neuron clusters was not adequately driven by bursts of single downward FM sweeps or tones. Consequently, for characterization of neuronal response selectivities, FM-FM combinations, band-limited noise, and click stimuli were also employed (fig. 27.3A). The majority of units studied (47%, *n* = 34 of 73; 8 bats) preferred FM-FM combinations. Pure-tone stimuli or downward FM sweeps were most effective for response generation in 25% and 18% of units, respectively. Units responding preferentially to upward FM sweeps (1%) or click and/ or noise stimuli (10%) were rarely encountered.

For 43 FAF units, a preferred (downward) FM sweep or FM-FM combination was determined (fig. 27.3B). More than half of these FM sweeps (*n* = 40 of 73) were characterized by start frequencies either in the range of 80 kHz to <100 kHz or ≥130 kHz (compare fig. 27.2B).

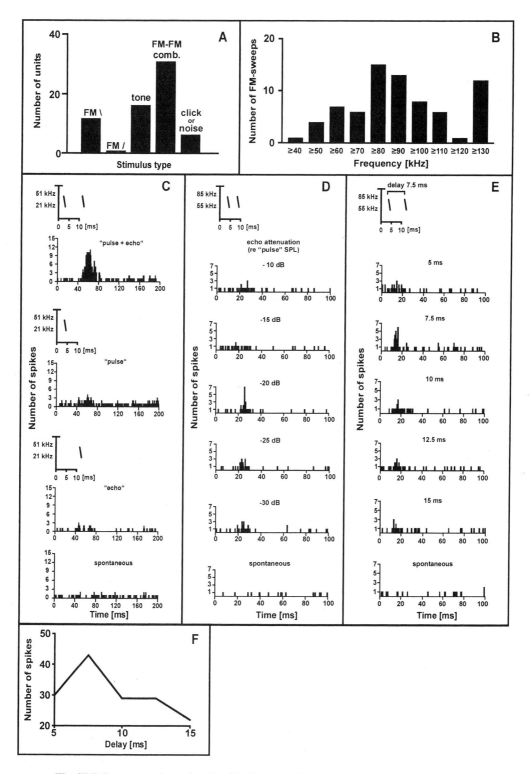

Fig. 27.3. Responses from the *Carollia* FAF. *A:* Stimulus preferences of FAF neurons (*n* = 73; FM-FM comb. = FM-FM combination; click or noise = units responding preferentially to click or/and noise stimuli). *B:* Frequency distribution of start frequencies of neuronally preferred FM stimuli (*n* = 73 downward FM sweeps; both pulse and echo stimuli considered). *C:* Pulse-echo facilitation in a well-isolated FAF unit (F37; stimulus repetitions per PSTH: 240; binwidth: 0.5 ms; onset of pulse stimulus at 0 ms). For stimulus design see insets (note different time scales; echo SPL re pulse SPL: −10 dB). *D:* Determination of best echo intensity (found at −20 dB re pulse SPL) in unit F2 (other parameters as in *C*). *E:* Determination of best pulse-echo delay (found at 7.5 ms) in unit F23 (echo SPL re pulse SPL: −11 dB; other parameters as in *C*). *F:* Delay-tuning curve of unit F23 (stimulus repetitions per delay: 240; window for spike counting: 200 ms; see *E* for comparison).

FM-FM responses were facilitative without any exception ($n = 34$ units; PSTH example in fig. 27.3C) and, apart from 2 units, the initial frequencies of the best pulse and best echo FM sweep always differed by less than 3 kHz. Generally, the magnitude of the facilitative response was less dependent on the FM-sweep width (not shown) than the echo SPL (re pulse SPL; example in fig. 27.3D). Typically, the former was set to 30 kHz, which roughly corresponds to the FM-sweep widths of the dominant biosonar second and third harmonic (Eiermann and Esser 2000).

All units facilitated by FM-FM combinations ($n = 34$) were further tested for delay sensitivity. Three of these units exhibited pulse-echo delay sensitivity (best delays: 7.5, 12.5, and 25 ms). An example (unit F23) is shown in figure 27.3E (PSTHs) and 27.3F (delay-tuning curve).

Best frequencies, obtained from 15 units preferring tone stimuli, ranged from 40 to 140 kHz (99.27 ± 32.64 kHz).

Due to the low peak firing rates and irregular response patterns of FAF units, latencies of the peak responses in the PSTHs were measured (15–60 ms) instead of first-spike latencies.

Rate-level functions and excitatory frequency-tuning curves (see above, "Auditory Cortex") are not available for FAF units for similar reasons.

Discussion

AUDITORY CORTEX

In the high-frequency fields (HFI and HFII) of the *Carollia* auditory cortex, the majority of units studied responded preferentially to FM stimuli (55%) as opposed to tones (37%). An even larger number of auditory cortical neurons preferring FM sweeps was found in *Myotis lucifugus* (96%, Shannon-Hartman, Wong, and Maekawa 1992). Comparable to the auditory cortices of other FM bats and to the FM-processing areas of CF-FM bats (for review, see O'Neill 1995), FM sweeps in the frequency ranges of the dominant harmonics of the species' biosonar signal were the most effective stimuli. Delay-dependent responses occurred in 54% of the units in the AC of *M. lucifugus* (Sullivan 1982), in 15% of *E. fuscus* (Dear et al. 1993), and in 36% of *C. perspicillata* (this chapter). A striking difference between CF-FM and FM bat auditory cortical organization is that in the former, facilitative pulse-echo interactions (e.g., in FM-FM neurons) were mediated by different harmonic components ("heteroharmonic" facilitation; Suga, Niwa, and Taniguchi 1983), whereas in the latter, neurons relied on the same harmonic in the pulse and the echo (Berkowitz and Suga 1989; O'Neill 1995). Accordingly, responses of all delay-sensitive units from the *Carollia* HF (and frontal cortex) were facilitated by pairs of virtually identical FM sweeps. By contrast to CF-FM bats, in which the FM components of the biosonar signal are

represented in separate, nontonotopic fields dorsal to the tonotopically mapped area (O'Neill and Suga 1979; Suga 1990b; Ostwald 1980; Schuller, O'Neill, and Radtke-Schuller 1991; Radtke-Schuller and Schuller 1995), in both *Myotis* and *Eptesicus,* delay-tuned cells and cells not sensitive to delay are intermingled within the tonotopically organized "primary field" (O'Neill 1995; Dear et al. 1993). Interestingly, in *Carollia,* delay-sensitive neurons were found colocated with units with simple response properties (i.e., a preference for tones) in the dorsally displaced nontonotopic high-frequency fields of the auditory cortex. Although FM signals have not been employed as stimuli in the core fields of the *Carollia* auditory cortex (AI and AAF; fig. 27.1), this situation closely parallels findings in the FM-FM area analogue of the rufous horseshoe bat (*R. rouxi;* Radtke-Schuller and Schuller 1995).

As reviewed by Clarey, Barone, and Imig (1992) for the auditory cortex, the proportions of neurons with monotonic and nonmonotonic rate-level functions vary substantially both between different cortical fields and across species. In bats, auditory cortical units exhibiting nonmonotonic rate-level functions were abundant in both CF-FM and FM species (e.g., *P. parnellii,* Suga and Manabe 1982; *M. lucifugus,* Suga 1965; *E. fuscus,* Jen, Sun, and Lin 1989; *C. perspicillata,* this study). Complex (i.e., multipeaked) nonmonotonic rate-level functions, such as those described here, indicate that neurons can be simultaneously tuned to the SPLs of the emitted pulse and the returning echo.

V- or W-shaped excitatory frequency-tuning curves were also found in the auditory cortex of *Myotis* (Suga 1965) and *Eptesicus* (Dear et al. 1993). Neurons with two threshold minima (multipeaked cells; Dear et al. 1993) are likely involved in the processing of multi-wave-front echoes with spectral peaks and notches as originating from insonified complex targets (Dear et al. 1993).

In our sample of HF neurons, both purely phasic and more sustained responses were frequently found. In the auditory cortices of other species of FM bats (*M. lucifugus,* Suga 1965; *E. fuscus,* Jen, Sun, and Lin 1989), however, phasic responses strongly prevail. In contrast, Sullivan (1982) noted that in delay-sensitive units from the *Myotis* auditory cortex, duration of a response could be much longer than the duration of the stimulus (e.g., 10–20 versus 1.5 ms). The possibility of response latencies of auditory cortical neurons as short as 4–4.5 ms (present study) is supported by measurements of synaptic delays in a variety of species (e.g., Wang, Cleland, and Burke 1985), but also might be suggestive of a routing of auditory information through the central acoustic tract.

In contrast to bats studied at the auditory cortical level to date (see "Introduction," above), the short-tailed fruit bat is easy to breed in the laboratory. With respect to its feeding habits, *Carollia* is a facultative insecti- and nectarivorous species that mainly feeds on fruits (e.g., Fleming 1988). Since the functional parcellation of the species' AC was recently established (Esser and Eiermann 1999) and since the data presented here are consistent with previous findings in purely insectivorous microchiropteran bats, *Carollia* has the obvious potential to become a common FM bat for modeling the auditory cortex.

Frontal Auditory Field

There are only a few reports on the physiology of bat FAF (Eiermann and Esser 2000; Kanwal et al. 2000; Kanwal, Peng, and Esser, chapter 29, this volume). In *P. parnellii,* the majority of FAF units responded to tones, and frequency tuning of neurons was particularly sharp around the frequencies of the CF components (CF and/or CF_2 and/or CF_3) of the echolocation signal (Kanwal et al. 2000). Some FAF neurons also exhibited combination sensitivity in that their responses were facilitated by presenting combinations of either CF_1-CF_2 and/or CF_1-CF_3 components (Kanwal et al. 2000). Delay-dependent FM-FM facilitation of cortical neurons outside the auditory cortex, as seen in the present study of *Carollia* FAF, has not previously been described.

In accordance with our findings on the *Carollia* FAF, many mustached bat frontal cortical neurons show irregular firing patterns and often the duration of a response was much longer than the duration of the stimulus (e.g., >100 versus 30 ms, Kanwal et al. 2000; compare to fig. 27.3C). Although differences in magnitude and temporal pattern of the neuronal response clearly exist between bat auditory cortical and FAF units (Eiermann and Esser 2000; Kanwal et al. 2000; present study) several parallels are obvious as well: (1) a neuronal preference for downward FM sweeps, (2) FM-FM facilitation, (3) tuning to low echo amplitudes (re pulse SPL), and (4) delay tuning. The fact that responses of FM-FM neurons in the *Carollia* FAF were strongly facilitated but rarely (re the species' auditory cortex) exhibited delay tuning, together with the pronounced tolerance for FM-sweep width, is suggestive for a certain functional role. Since these neurons maximally respond to a variety of combinations of an emitted biosonar pulse and a returning echo instead of being tuned to a particular delay or range of delays (fig. 27.2C), or respond best to specific portions of the FM sweeps (Maekawa, Wong, and Paschal 1992), they likely function as "novelty detectors"—for example, indicating the presence of an insonified object irrespective of distance. However, responses of FAF

units under such behaviorally relevant conditions have yet to be studied. For long-range target detection, low-frequency components are particularly suited (Neuweiler and Fenton 1988). A minor fraction of neuronally preferred downward FM sweeps started around 50 kHz (fig. 27.3B). Interestingly, sonagrams of echolocation calls of adult *C. perspicillata* either show a distinct fundamental in the corresponding frequency range (Gould 1977; Esser and Eiermann, unpublished) or not (Hartley and Suthers 1987; Thies, Kalko, and Schnitzler 1998). These differences point to a voluntary control of the fundamental-reject filter in the species' vocal tract (Hartley and Suthers 1987).

A second probable function of the *Carollia* FAF is indicated by the FM-sweep preferences and best frequencies of neurons. About one sixth of FAF units studied preferred FM sweeps starting around 130 kHz (fig. 27.3B). Since it is generally assumed that the emission of higher harmonics (and the central-nervous processing of the corresponding echo components) improves the resolution of the echolocation system, neurons tuned to these sound components are likely involved in object recognition. Consistent with this idea, best frequencies of FAF units averaged about 100 kHz with maximum values at 140 kHz.

Conclusions

Data presented here indicated that the response properties of auditory cortical neurons in a bat mainly feeding on fruits (i.e., *C. perspicillata*) are similar to those reported in purely insectivorous microchiropteran species.

Response selectivities of units from the *Carollia* frontal auditory field suggest that FAF neurons are likely involved in both novelty detection and object recognition. Hence, with respect to bat echolocation, it seems reasonable to assume that the representations of biologically important sound parameters in this forebrain area might be important for planning actions based on auditory information.

Acknowledgments

This study was supported by the DFG (ES 124/2-3 to Karl-Heinz Esser). P. Hackel provided valuable assistance in our data collection, and G. Ehret kindly helped in refining the manuscript. The care and use of animals in these experiments were approved by the RP Tübingen.

{ 28 }

Corticofugal Modulation of Midbrain Auditory Sensitivity in the Bat

Philip H.-S. Jen, Xinde Sun, Qi Cai Chen, Jiping Zhang, and *Zhou Xiaoming*

Introduction

During sensory signal processing, the brain has the built-in ability to edit and adjust the flow of information. This feedback control, which appears to be a common principle across all sensory systems, is executed through pathways leading from the central brain centers to the periphery. Previous studies have shown that the descending auditory pathways exert either inhibitory or excitatory influences on activity in the ascending auditory pathways. For example, electrical stimulation in the auditory cortex (AC) synchronized with acoustic stimulation briefly decreases auditory responses and sharpens spatial sensitivity and frequency tuning of corticofugally inhibited neurons in the central nucleus of the inferior colliculus (ICc) but produces the opposite effects on corticofugally facilitated collicular neurons (Sun, Chen, and Jen 1996; Zhang, Suga, and Yan 1997; Jen, Chen, and Sun 1998). This brief corticofugal modulation disappears upon cessation of cortical electrical stimulation. In addition, repetitive acoustic stimulation synchronized with cortical electrical stimulation for 6–30 min shifts the best frequencies (BFs) of ICc neurons not only toward the BFs of electrically stimulated cortical neurons but also toward the frequency of a repetitively delivered acoustic stimulus. This short-term corticofugal modulation increases the representation of the repetitive stimulus frequency in the IC and reorganizes the collicular frequency map for as long as 3 h (Gao and Suga 1998, 2000; Suga, Yan, and Zhang 1997; Yan and Suga 1998).

In this chapter, we report the studies of brief corticofugal modulation of auditory sensitivity of ICc neurons in the intensity, spatial, and frequency domains. The study of short-term corticofugal modulation of subcortical auditory sensitivity is presented in chapter 31 of this volume.

Materials and Methods

Procedures for surgery and recording have been described previously (Jen et al. 1987). Briefly, one or two days before a recording session, the flat head of a 1.8 cm long nail was glued onto the exposed skull of Nembutal-anesthetized (45–50 mg/kg b.w.) big brown bats, *Eptesi-cus fuscus,* with acrylic glue and dental cement. During recording, each bat was administered neuroleptanalgesic Innovar-Vet (0.08 mg/kg b.w. of fentanyl, 4 mg/kg b.w. of droperidol) and was tied onto an aluminum plate with a plastic band. The head was immobilized and oriented toward 0° in azimuth and 0° in elevation of the bat's frontal auditory space. Small holes were bored in the skull above the primary AC and the ICc to allow electrical stimulation and/or to record acoustically evoked neural responses.

Two "piggybacked" multibarrel electrodes described previously (Sun, Chen, and Jen 1996; Jen, Chen, and Sun 1998) were used in this study. The first multibarrel electrode consisted of two tungsten-in-glass electrodes (tip: 15μm), which were piggybacked to a single-barrel glass electrode (tip: 50 μm). The two tungsten-in-glass electrodes were used to electrically stimulate the AC (4 ms train consisting of 4 monophasic pulses of 0.1 ms delivered at 2 trains/s with electrical current between 5.0 and 70 μA) or to record acoustically evoked single- or multiunit responses from the AC. The single-barrel glass electrode was used to pressure-inject 0.3–0.5 μl of a 1% solution of lidocaine in physiological saline into the cortical stimulation/recording site. The second multibarrel electrode, which comprised a three-barrel electrode (tip: 10–15 μm) piggybacked to a 3M KCl single-barrel electrode (tip: 1–m; impedance: 5–10 MΩ), was used to record acoustically evoked responses and to eject biculline methiodide (10 mM, pH 3.0; Sigma) or GABA (500 mM, pH 3.5, Sigma) iontophoretically into the recording site of ICc neurons.

A sound stimulus (4 ms duration, 0.5 ms rise-fall times at 2 pulse/s) was used to isolate acoustically evoked AC neurons and corticofugally affected ICc neurons. The interval between the electrical and sound stimuli was adjusted to produce the maximal corticofugal effect. Then, the rate-intensity function, 5 dB auditory spatial response area, and frequency-tuning curve of corticofugally affected ICc neurons were determined both before and during cortical electrical stimulation. In addition, these auditory response properties were determined 3 min after each of the following procedures: (1) pressure-injection of lidocaine solution into the cortical stimulation site; (2) cortical electrical stimulation after lidocaine injection; and (3) iontophoretic applica-

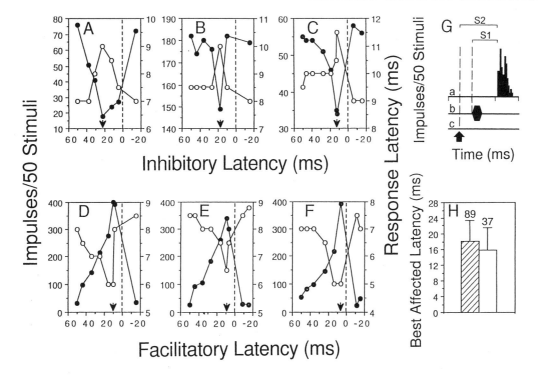

Fig. 28.1. Variation in the number of impulses (solid circles refer to left ordinates) and acoustic response latency (open circles refer to right ordinate, S1 in *G*) of corticofugally inhibited (*A, B, C*) or facilitated (*D, E, F*) ICc neurons due to variations in interval between electrical and acoustic stimuli. For convenience, the inhibitory or facilitatory latency was expressed as a positive or negative value relative to the onset of acoustically evoked responses. The arrows indicate the best inhibitory (*A, B, C*) and the best facilitatory (*D, E, F*) latencies. *G*: PST histogram (*a*), the acoustic response latency (S1), and the corticofugal inhibitory latency (S2) of an ICc neuron during acoustic (*b*) and cortical electrical stimulation (*c,* arrow). *H*: Average best inhibitory (shaded bar) or facilitatory (open bar) latency of ICc neurons. The number of ICc neurons and positive standard deviation are shown atop each bar. The BF (kHz), MT (dB SPL) and depth (μm) of these 7 ICc neurons were 41.7, 38, 834 (*A*); 24.8, 8, 230 (*B*); 21, 57, 667 (*C*); 30.7, 48, 797 (*D*); 33.0, 48,920 (*E*); 44.9, 49, 1079 (*F*); 24.6, 45, 695 (*G*).

tion of bicuculline methiodide or GABA into the collicular recording site.

Recordings of neural activities were conducted inside a double-wall, soundproof room (temperature 28°–30°C). Recorded action potentials were amplified, bandpass filtered, and fed through a window discriminator before being sent to a computer (Gateway 2000, 486) for acquisition of peri-stimulus-time (PST) histograms to 50 sound presentations. PST histograms quantitatively describe the discharge pattern of each neuron under different stimulation conditions. The number of impulses in each PST histogram was used to quantify a neuron's response under each stimulation condition.

Results

CORTICOFUGAL MODULATION OF RESPONSES OF ICc NEURONS VERSUS INTERSTIMULUS INTERVAL

Cortical electrical stimulation synchronized with acoustic stimulation decreased the number of impulses and lengthened the acoustic response latency of corticofugally inhibited ICc neurons, but it produced opposite effects for corticofugally facilitated ICc neurons. When stimulated at a constant electrical current, the degree of corticofugal inhibition or facilitation varied with the interval between acoustic and electrical stimuli. Fig. 28.1 shows variations in the number of impulses (solid circles refer to left ordinates) and acoustic response latencies (open circles refer to right ordinates, i.e., fig. 28.1G, S1) of corticofugally inhibited (fig. 28.1.A–C) and facilitated (fig. 28.1D–F) ICc neurons at different interstimulus intervals.

The number of impulses and acoustic response latency varied in opposite ways when the electrical stimulus was delivered before, simultaneously, or after the acoustic stimulus (fig. 28.1.A–F, solid circles versus open circles). The number of impulses of corticofugally inhibited ICc neurons was the smallest while the acoustic response latency was the longest at a specific interstimulus interval (fig. 28.1.A–C). In contrast, the number of

impulses of corticofugally facilitated ICc neurons was
the largest while the acoustic response latency was the
shortest at a specific interstimulus interval (fig. 28.1D–
F). For convenience, we define the lag time between the
onset of the electrical stimulus and the neuron's responses as the corticofugal-affected (inhibitory or facilitatory) latency (fig. 28.1.G, S2). We also define the corticofugal inhibitory or facilitatory latency that produced
the longest or the shortest acoustic response latency as
the best inhibitory (arrows in fig. 28.1.A–C, abscissa)
or facilitatory latency (arrows in fig. 1.D–F, abscissa).
The average best corticofugal inhibitory latencies were
18.1 ± 5.3 ms (*n* = 89) (fig. 28.1H, shaded bar) and the
average best facilitatory latencies (F) were 15.8 ± 5.8 ms
(*n* = 37) (fig. 28.1H, open bar). The average optimal interstimulus interval was 3.68 ± 5.64 ms during corticofugal inhibition and 2.5 ± 1.0 ms during corticofugal
facilitation.

CORTICOFUGAL MODULATION OF THE RATE-INTENSITY FUNCTION, SPATIAL RESPONSE AREA, AND EXCITATORY FREQUENCY-TUNING CURVE

Fig. 28.2 shows the rate-intensity functions, auditory
spatial response areas, and excitatory frequency-tuning
curves of corticofugally inhibited (fig. 28.2.A–C) and
corticofugally facilitated (fig. 28.2.D–F) ICc neurons
obtained under different experimental conditions. Cortical electrical stimulation lowered the rate-intensity
function (fig. 28.2.Aa versus Ab), reduced the auditory
spatial response area (fig. 28.2.Ba versus Bb), and narrowed the excitatory frequency-tuning curve (fig. 28.2.Ca
versus Cb) of the corticofugally inhibited ICc neuron.
Injection of lidocaine to block neural activity at the cortical stimulation site raised the rate-intensity function
(fig. 28.2.Aa versus Ac), expanded the auditory spatial
response area (fig. 28.2.Ba versus Bc), and broadened
the excitatory frequency-tuning curve (fig. 28.2.Ca ver-

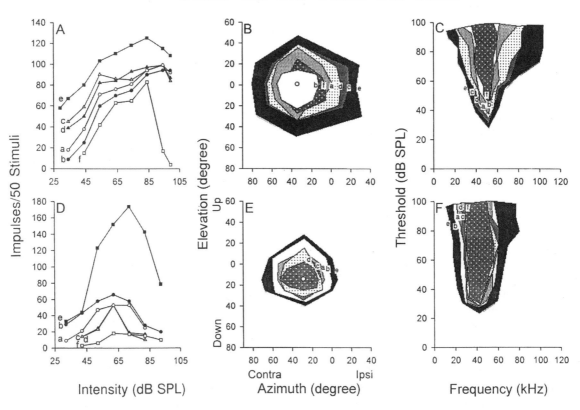

Fig. 28.2. Rate-intensity functions (*A, D*), auditory spatial response areas (*B, E*), and excitatory frequency-tuning curves (*C, F*) of corticofugally inhibited (*A, B, C*) and facilitated (*D, E, F*) ICc neurons determined under different stimulus conditions (shown at the top). *A:* Control: acoustic stimulation only. *B:* Es: cortical electrical stimulation (36.7 μA) plus acoustic stimulation. *C:* Lid, acoustic stimulation after injection of lidocaine (0.5 μl, 1%) into the cortical stimulation site. *D:* Lid + Es, acoustic stimulation plus cortical electrical stimulation after lidocaine injection into cortical stimulation site. *E:* Bic, acoustic stimulation after iontophoretic application of bicuculline (30 nA, 3 min) into collicular recording site. *F:* GABA: acoustic stimulation after iontophoretic application of GABA (30 nA, 3 min) into collicular recording site. The BF (kHz), MT (dB SPL), and depth (μm) of these two neurons were 51.2, 34, 976 (*A, B, C*); 42.9, 32, 1879 (*D, E, F*).

sus Cc) of the ICc neuron. The rate-intensity function, auditory spatial response area, and frequency-tuning curve were affected very little when the AC was electrically stimulated after lidocaine injection to the cortical stimulation site (fig. 28.2.Ac, Bc, Cc versus Ad, Bd, Cd). However, bicuculline application to the ICc recording site greatly raised the rate-intensity function (fig. 28.2.Aa versus Ae), expanded the auditory spatial response area (fig. 28.2.Ba versus Be), and broadened the excitatory frequency-tuning curve (fig. 28.2.Ca versus Ce) of the neuron. GABA application to ICc recording sites produced the opposite effects (fig. 28.2.Aa, Ba, Ca versus Af, Bf, Cf).

For the corticofugally facilitated ICc neuron, cortical electrical stimulation raised the rate-intensity function (fig. 28.2.Da versus Db), expanded the auditory spatial response area (fig. 28.2.Ea versus Eb), and broadened the excitatory frequency-tuning curve (fig. 28.2.Fa versus Fb). Injection of lidocaine to the cortical stimulation site lowered the rate-intensity function (Fig 2Da versus Dc), decreased the auditory spatial response area (fig. 28.2.Ea versus Ec), and narrowed the frequency-tuning curve (fig. 28.2Fa versus Fc) of this neuron. These response properties were affected very little when the AC was electrically stimulated after lidocaine injection (fig. 28.2.Dc, Ec, Fc versus Dd, Ed, Fd). However, bicuculline application to the ICc recording site greatly raised the rate-intensity function (fig. 28.2.Da versus De), expanded the auditory spatial response area (fig. 28.2.Ea versus Ee), and broadened the excitatory frequency-tuning curve of this neuron (fig. 28.2.Fa versus Fe). GABA application to the ICc recording sites produced the opposite effects (fig. 28.2.Da, Ea, Fa versus Df, Ef, Ff).

Discussion

CORTICOFUGAL MODULATION OF
FREQUENCY TUNING OF ICc NEURONS

Previous studies showed that surround or lateral inhibition determines the size of the auditory receptive field of midbrain neurons in the barn owl (Knudsen and Konishi 1978) and sharpens the frequency-tuning curves of central auditory neurons (Suga 1995a). In frequency tuning, the lateral inhibition provides a means to reduce ambiguity in encoding frequency at high stimulus intensities. Our recent study showed that corticofugal pathways modulate the excitatory and inhibitory frequency-tuning curves (plotted with two-tone inhibition paradigm) of ICc neurons in opposite ways (Jen and Zhang 1999). Cortical electrical stimulation sharpens the excitatory frequency-tuning curves and broadens the lateral inhibitory frequency-tuning curves of corticofugally inhibited ICc neurons (fig. 28.3A-1 versus A-2). In contrast, cortical electrical stimulation broadens the excitatory frequency-tuning curves and decreases the lateral inhibitory frequency-tuning curves of corticofu-

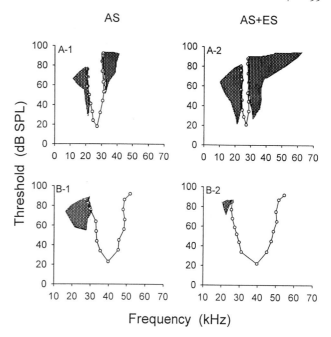

Fig. 28.3. Excitatory (open circles) and inhibitory (shaded) frequency-tuning curves of corticofugally inhibited (A-1, A-2) and facilitated (B-1, B-2) ICc neurons obtained with sound stimulation (AS) alone (A-1, B-1) and with sound stimulation plus cortical electrical stimulation (AS + ES)(A-2, B-2). Inhibitory frequency-tuning curves were obtained with a two-tone inhibition paradigm with the excitatory tone set at the BF and 10 dB above the MT of the excitatory frequency tuning. Cortical electrical stimulation (25 μA) narrowed the excitatory frequency-tuning curve but broadened both inhibitory frequency-tuning curves of the corticofugally inhibited neuron (A-1 versus A-2). In contrast, cortical electrical stimulation (10 μA) broadened the excitatory frequency-tuning curve but decreased the inhibitory frequency-tuning curve of the corticofugally facilitated neuron (*B-1* versus *B-2*). The BF (kHz), MT (dB SPL), and recording depth (μm) of these two neurons were 27.1, 18, 813 (*A*); 39.7, 23, 1434 (*B*).

gally facilitated ICc neurons (fig. 28.3B-1 versus B-2). Whether or not corticofugal modulation of frequency tuning of ICc neurons is mediated through lateral inhibition remains to be determined.

We found that cortical electrical stimulation and iontophoretic application of GABA to ICc recording sites produced the same inhibitory effect on responses of ICc neurons (fig. 28.2.Ab, Bb, Cb versus Af, Bf, Cf). This finding suggests that corticofugal modulation of ICc auditory responses could be mediated through GABAergic inhibition.

Possible corticofugal modulation pathways

Previous anatomical studies have shown that parallel auditory corticofugal fibers project separately to the pericentral nucleus, the external nucleus of the IC, and the dorsomedial division of the ICc (Andersen, Snyder, and Merzenich 1980; Games and Winer 1988; Herbert, Aschoff, and Ostwald 1991; Saldana, Feliciano, Mu-

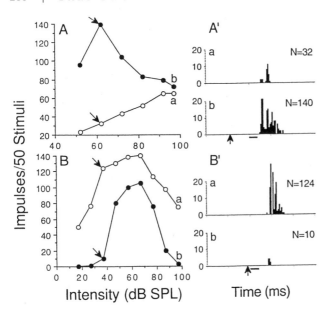

Fig. 28.4. Rate-intensity functions of neurons in the external nucleus of the IC (*A*, BF: 26.5 kHz; MT: 48 dB SPL) and the central nucleus of the IC (*B*, BF: 30.1 kHz; MT: 38 dB SPL) obtained with BF sound alone (*Aa, Ba*) or when the AC (*Ab*) or the external nucleus of the IC (*Bb*) was electrically stimulated during sound stimulation. PST histograms of each neuron obtained at two indicated stimulation conditions (arrows in *A, B*) are shown in *A′a,b* and *B′a,b*. *N* = the total number of impulses within each histogram. The interstimulus interval between electrical (arrows, 87 μA in *A′b*; 29 μA in *B′b*) and sound (horizontal bars) stimuli was set at optimal interval for each neuron. Cortical electrical stimulation facilitated responses and raised the rate-intensity function of the neuron in the external nucleus of the IC to varying degrees (*Aa, Aa′* versus *Ab, A′b*), In contrast, electrical stimulation of the external nucleus of the IC inhibited responses and lowered the rate-intensity function of the neuron in the central nucleus of the IC to varying degrees (*Ba, Ba′* versus *Bb, B′b*).

gnaini 1996). These parallel auditory corticofugal projections represent possible corticocollicular pathways underlying the corticofugal modulation of auditory sensitivity in ICc.

The fact that cortical electrical stimulation produced a longer best inhibitory latency than facilitatory latency of ICc neurons (fig. 28.1.H) suggests that corticofugal facilitation could act through a shorter neural pathway than corticofugal inhibition. We (Jen et al. 2001) recently showed that cortical electrical stimulation facilitated responses of all neurons studied in the external nucleus of the IC (fig. 28.4.Aa, A′a versus Ab, Ab′). This corticofugal facilitation is mediated at least in part through NMDA receptors, since application of APV, an antagonist for NMDA, decreases these response properties of neurons in the ICx. Electrical stimulation in the external nucleus of the IC reduced responses of neurons in the ICc (fig. 28.4.Ba, B′a versus Bb, Bb′). This inhibi-

tion is mediated at least in part through GABAa receptors, since application of bicuculline, an antagonist for GABA, increases these response properties of neurons in the ICc.

Taken together, all these data suggest that corticofugal facilitation could be mediated by activation of excitatory pathways to the ICc neurons (for example, through an interneuron in the collicular dorsal cortex or pericentral nucleus) or by disinhibition of GABAergic inputs to the ICc neurons. On the other hand, corticofugal inhibition could be mediated through excitatory projections from the AC to the external nucleus of the IC, which then sends inhibitory inputs to the ICc. Alternatively, corticofugal inhibition could be mediated through inhibition within the AC to reduce responses of neurons in the ICc. Future anatomical work correlated with electrophysiological studies are needed to confirm these possibilities.

Biological relevance of corticofugal modulation

In auditory signal processing, auditory information carried by complex sounds traditionally has been explained by neural interactions of divergent and convergent projections within the ascending auditory system (Suga 1997). We have shown that lidocaine application and cortical electrical stimulation respectively blocks and activates corticofugal modulation of auditory responses of ICc neurons (fig. 28.2). These findings offer evidence that the brain has the built-in ability to edit and adjust the ascending auditory information. Corticofugal modulation of subcortical auditory signal processing represents a neural mechanism in which an animal can actively improve ascending auditory signal analysis. For example, during hunting, bats could use corticofugal facilitation (i.e., lowering MT, expanding spatial response area and frequency-tuning curve; fig. 28.2.D–F) to enhance target detection during early phases of hunting. They could use corticofugal inhibition (i.e., increasing MT, decreasing auditory response area and frequency-tuning curve; fig. 28.2.A–C) to attenuate the echo intensity and improve fine analysis of target features as they approach the target.

Acknowledgments

We thank Cynthia Moss, Jeanette Thomas, Nobuo Suga, Gerry Summers, and Marianne Vater for reading an earlier version of this manuscript. This work was supported by a research grant from the National Science Foundation (NSF IBN 9604238) and a grant from the Research Board of the University of Missouri (RB 98–014). The experiments were conducted in compliance with NIH publication No. 85-23, "Principles of Laboratory Animal Care," and with the approval of the Institutional Animal Care and Use Committee (#1438) of the University of Missouri–Columbia.

Auditory Communication and Echolocation in the Mustached Bat: Computing for Dual Functions within Single Neurons

Jagmet S. Kanwal, J. P. Peng, and *Karl-Heinz Esser*

Introduction

Recent studies in the mustached bat indicate that in addition to echolocation, communication among conspecifics is another important auditory function in this highly social species (Kanwal et al. 1994). In contrast to the large body of literature available on echolocation in mustached bats, studies of communication sound processing were initiated in this species only recently (Ohlemiller et al. 1994; Ohlemiller, Kanwal, and Suga 1996; Esser et al. 1997; Kanwal, Ohlemiller, and Suga 1992; Kanwal 1997, 1999). To address whether two disparate auditory functions are organized within the same cortical areas—and if so, how they are organized—the response properties of auditory cortical neurons were studied in detail. That is, the responses of single neurons were obtained to systematic presentation of tones and echolocation signals, as well as to communication sounds.

This chapter focuses on the response properties and the role of cortical neurons for processing communication sounds or simply "calls" in view of what is known about neural processing of echolocation signals. We first describe auditory responses to both echolocation signals and calls within single neurons in the auditory cortex. This is followed by a description of the auditory projections to the frontal cortex, including recently obtained responses from this area in the mustached bat's cortex (Kanwal et al. 2000). Finally, the structure of species-specific calls is discussed in parallel with that of echolocation signals to demonstrate which features or groups of "information-bearing-parameters" in calls may be extracted at different levels of the auditory system.

Responses of Neurons within the Auditory Cortex

During echolocation, mustached bats emit sounds consisting of four harmonics (H_1–H_4), each with a constant-frequency component (CF_1–CF_4) and a downward-sweeping frequency-modulated component (FM_1–FM_4). The higher CF components are harmonics of an approximately 30 kHz fundamental. The flying bat hears both its own pulse and the returning echo. The auditory periphery is exquisitely sensitive to the echo CF_2 (near 61 kHz), but it is relatively insensitive to the bat's emitted CF_2 pulse component (near 59 kHz). The Doppler-shifted constant-frequency (DSCF) area is the largest portion of the primary auditory cortex (AI) and contains frequency-versus-amplitude coordinates for mapping the CF_2 component of the echo from a target (Suga and Manabe 1982). The large FM-FM area processes combinations of FMs separated by a specific delay (O'Neill and Suga 1982; Fitzpatrick, Olsen, and Suga 1998b). Neurons here are tuned both in the time and frequency domains to combinations of the pulse FM_1 with the echo CF_2 in the echolocation signal (Fitzpatrick et al. 1993; Kanwal, Fitzpatrick, and Suga 1999). Cortical maps allow yet higher levels of features to be extracted from the computed parameters by the creation of neurons that are combination-sensitive for additional parameters and/or dimensions—for example, FM_1-$FM_{2,3}$ neurons (Suga 1990a, 1990b). Yet, echolocation pulses emitted by bats are just one of the over thirty different types of sounds that must be processed largely by the same cortex.

A considerable effort was made over the last few years to understand the neural mechanisms for call processing in the auditory cortex of mustached bats (Ohlemiller et al. 1994; Ohlemiller, Kanwal, and Suga 1996; Esser et al. 1997; Kanwal 1997, 1999). First, it was shown that DSCF and FM-FM neurons respond equivalently to a simple syllable and to the relevant components of a pulse-echo pair when each is presented at its best amplitude (fig. 29.1; Kanwal, Ohlemiller, and Suga 1992; Ohlemiller, Kanwal, and Suga 1996). Second, in the FM-FM area, a neuron that is sharply tuned to a pulse-echo delay of 1 ms also shows tuning, albeit broad, to an intersyllable interval of about 25 ms to a pair of bent downward FM syllables. Yet another range of tuning to about 40 ms is observed for the descending ripple and FM syllable pair in the same neuron (Ohlemiller, Kanwal, and Suga 1996). These studies clearly indicated that within the auditory cortex, both echolocation signals and calls can be processed by the same neurons.

Finally, these call processing neurons exhibit combination sensitivity or a nonlinear increase in the mag-

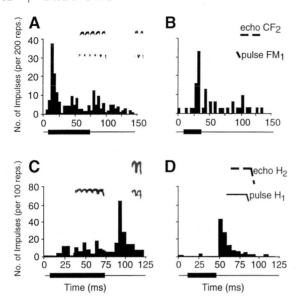

Fig. 29.1. Peristimulus-time-histograms (PSTHs) showing the processing of (*A*) excitatory components in the descending rippled FM (dRFM) call and (*B*) excitatory components in an echolocation signal within the same neuron in the primary auditory cortex (DSCF area). This is a typical response pattern observed in well-resolved single-unit recordings from over 100 DSCF neurons. *C* and *D:* PSTHs showing simultaneous processing of echolocation signals and calls within a FM₁–FM₂ neuron in the secondary auditory cortex (FM-FM area; adapted from Ohlemiller, Kanwal, and Suga 1996). Both neurons respond better to communication call components compared to essential elements in a pulse-echo pair. Spectrograms of stimuli used are shown above each PSTH. The rectangular box at the bottom of each PSTH indicates the onset and duration of the stimulus. Each stimulus is presented at its best amplitude. Binwidth = 5 ms. Stimuli were presented 200 times (*A* and *B*) and 100 times (*C* and *D*).

nitude of their response when a combination of two or more components of a syllable are presented together. This response is significantly larger than the sum of the responses to each component presented alone (Suga, O'Neill, and Manabe 1978; Suga, Niwa, and Taniguchi 1983). Combination-sensitive responses occur to combinations in both the frequency and time domains. For example, for FM-FM neurons, the FM₁ of the pulse and the FM₂ of the echo when presented together produce a "facilitated" response that is much greater (by 1.2–5 times) than the sum of the responses to the pulse and echo components presented alone. Some of the same neurons also produce a facilitated response to a combination of two syllables within a syllable train when presented with an intersyllable silence interval equivalent to the naturally present silent interval between two syllables in a train. This is a surprising result, leading to the suggestion that many neurons within the auditory cortex process both echolocation signals and calls with an equivalent level of specialization—that is, show combi-

nation sensitivity to two distinct components within each of two types of complex sounds. Facilitation of the neural response was demonstrated both for trains of simple syllables (with relatively fixed silence intervals), as well as composites (combination of two simple syllables without a silent interval) (Ohlemiller, Kanwal, and Suga 1996; Esser et al. 1997).

Auditory Pathways to the Frontal Cortex

Both echolocation signals and calls are processed within and also beyond the auditory cortex, such as in a frontal auditory (FA) field. In the mustached bat, *Pteronotus parnellii,* there are two thalamocortical pathways along which auditory information ascends to the FA field. One is via the medial geniculate body (MGB) and the auditory cortex; the other is via the suprageniculate nucleus in the thalamus (Kobler, Isbey, and Casseday 1987; Casseday et al. 1989). The suprageniculate nucleus projects directly to the frontal cortex—that is, to the FA field (fig. 29.2A). The pathway via the MGB constitutes the main lemniscal pathway, whereas that via the suprageniculate is an extension of a less well studied extralemniscal pathway known as the central acoustic tract (CAT). Thus auditory information can reach the frontal cortex via the latter pathway in as few as four synapses. The CAT also exists in cats and mice, as well as humans (Papez 1929).

Very little is known about the nature of auditory processing in the frontal cortex, which could be different from that in the auditory cortex (Eiermann and Esser 2000; Kanwal et al. 2000). For example, single-unit recordings from the ventral extent of the external nucleus of the inferior colliculus (ICXv), which projects to the FA field (fig. 29.2A), revealed a population of cells where the majority shows a preference for upward frequency modulations (Gordon 1998). Coincidentally, upward frequency modulations (UFMs) are common components of social calls produced by this species (Kanwal et al. 1994). ICXv neurons show the greatest selectivity to the direction of a FM sweep. Thus, among other functions, the CAT transmits neural activity triggered by rapid UFMs to the frontal cortex. Furthermore, the frontal cortex has a direct ascending projection to the anterior cingulate (Gooler and O'Neill 1987). In the anterior cingulate (ACg), microstimulation at specific loci triggers either echolocation or social vocalizations suggesting a segregated processing/output scheme in contrast to the largely multiplexed processing within single neurons in the auditory cortex (Gooler and O'Neill 1987). Where and how does this transition from multiplexed to separate processing at two hierarchically connected cortical areas take place? As described above, the FA field is interposed between the auditory cortex and the ACg. Therefore, to address these questions, it is important to first study the response properties of neurons within the FA field.

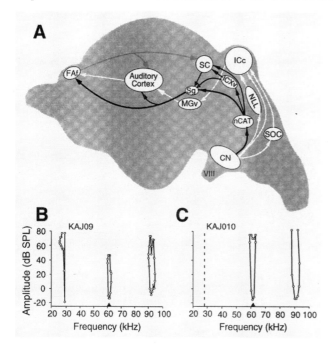

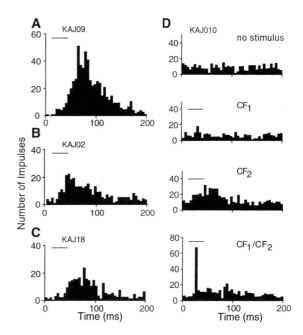

Fig. 29.2. *A:* Schematic showing the lemniscal pathway (white arrows) that reaches the cortex by way of the inferior colliculus (IC) and the ventral nucleus of the medial geniculate body (MGv). A parallel, extralemniscal pathway—an extension of the central acoustic tract—bypasses the IC and terminates in the auditory cortex as well as the FA field. Frontofugal projections from the FA field are also shown (adapted from Casseday et al. 1989). *B* and *C* show facilitatory response areas of two neurons with tuning in the CF_1, CF_2, or CF_3 region. *B:* This neuron exhibited response areas with lowest thresholds at about 30, 60, and 90 kHz. *C:* This neuron did not show a clear excitatory response to a near 30 kHz frequency tone. However, the CF_1 (shown by the dashed line) when paired with a second tone resulted in two facilitatory response areas at near 60 and 90 kHz. The mean of the bat's resting CF_2 frequency is indicated by small triangles. CN: cochlear nucleus; FAf: frontal auditory field; SOC: superior olivary complex; SC: superior colliculus; (Sg)ICc: suprageniculate nucleus, central nucleus of the inferior colliculus; ICXv: external capsule of the IC; VIII: auditory nerve.

Fig. 29.3. *A–C:* Peristimulus-time-histograms (PSTHs) computed from single neuron (same as in figure 29.2B) responses to 30 ms tone bursts generated at the neurons' best frequency (binwidth = 5 ms; SPL = 35 dB SPL above threshold; stimulus repetition rate = 2/s). *D:* PSTHs for a combination-sensitive (CF/CF) neuron (same as in figure 29.2C). *Bottom to top:* facilitated CF_1/CF_2 response, response to CF_1 (25.77 kHz, 67 dB SPL), response to CF_2 (61.81 kHz, 7 dB SPL), and spontaneous activity (binwidth = 5 ms; stimulus duration = 30 ms; stimulus repetition rate = 3/s; stimulus repetitions = 1000). Stimulus onset and duration are indicated by horizontal lines; acoustic delay between speaker and the bat's ear is considered (adapted from Kanwal et al. 2000).

RESPONSE PROPERTIES OF FRONTAL AUDITORY NEURONS

The majority (75%, $n = 60$) of neurons in the FA field are tuned to frequencies near 60 kHz, as examined with 30 ms tone bursts (Kanwal and Esser 1996; Kanwal et al. 2000). Other neurons have two, or even three, excitatory/facilitatory response areas (fig. 29.2B–C). The best frequencies for these response areas are close to the fundamental frequency of the pulse (just under 30 kHz) and at the resting CF_2 (approximately 60 kHz), and CF_3 (approximately 90 kHz) frequencies in an echo. CF-responsive neurons exhibit sharp tuning with Q30 dB values of over 180 (fig 29.2B–C). Most of the sharply tuned units studied show extremely low thresholds in the CF_2 range—that is, an average ($n = 26$) of -10.8 dB

SPL re 20 μPa at their best frequencies, whereas a few others ($n = 4$) show relatively high thresholds in the 50 dB SPL range. Although summed response patterns obtained in the frontal cortex appear to be stable and may be reproducible, up to 1000 stimulus repetitions are necessary to observe a clear response in the peristimulus-time histogram because of a high level of spontaneous activity (fig. 29.3B). About 20% of the neurons exhibit combination sensitivity in that their responses are facilitated by presenting combinations of either CF_1-CF_3 and/or CF_1-CF_2 components of the mustached bat's echolocation signal (fig. 29.3D).

Call processing was tested at eight recording sites located in the rostral part of the FA field. Single- and/or multiunits responded to playback of digitally stored simple-syllabic calls, but not very well to pure tones in the CF_1, CF_2, or CF_3 range of the mustached bat's echolocation signal. In most of these call-responsive neurons, each of 14 different syllable types were tested at sound pressure levels ranging between 98 and 45 dB SPL. Seven syllable variants—shifted in frequency to match the ±1, ±2, and ±3 sigma levels of the natural range of variation of the fundamental frequency in calls—were also presented for each syllable type. Combinations of CF stimuli were also occasionally tested and not found to be an effective stimulus for these neurons. Nearly all call-responsive FA-field neurons responded selectively to call syllables: no more than two calls elicited a response within 50% of the peak response to the most effective call syllable (normalized to 100). This most-effective or best-syllable variant was determined after testing all of the frequency-shifted variants for each of the 14 syllables. All variants were presented 10 times at the same three intensities (roughly 95, 70, and 45 dB SPL; the corrected amplitude based on root-mean-square value varied by a few dB SPL for each syllable). Best syllable variants were calculated from the sum of responses for all intensities of a particular syllable and ranged from CF type to FM and broadband type of simple syllables (fig. 29.4, left panel). Call-responsive neurons often preferred noise-band stimuli ranging from about 5 to 10 kHz, and an incremental increase in the bandwidth beyond these limits typically resulted in a decline of the response magnitude (fig. 29.4, right panel). Noise bands were generated with the same center frequency as the pure tone corresponding to the best frequency of the neuron.

As reported, the response properties of FA neurons are intriguing (Kanwal et al. 2000). Unlike neurons in the auditory cortex, those in the FA field did not appear to be dually specialized for processing echolocation and call stimuli. However, FA neurons exhibit the following response properties: (1) They have highly variable response-onset latencies (4–50 ms and occasionally longer) depending on the neuron; a response latency of 4 ms is consistent with the relatively direct magnocellular pathway from nCAT to the frontal cortex (Kobler, Isbey, and Casseday 1987). (2) They have a peak of excitatory response at long latencies (up to 100 ms) relative to stimulus (pure tone) onset. (3) They have long- as well as short-duration responses to pure-tone stimuli (typically about 20–200 ms). These data suggest that the response of FA neurons can be modulated over relatively long periods. FA neurons can therefore participate in the short-term integration of spatiotemporal patterns of cortical activity arriving from different brain regions prior to or during the period of its response. This is reflected in the inhibitory, as well as excitatory, responses observed in these neurons. Some of these differences between the FA neurons and those in the auditory cortex are best seen in raster plots and peristimulus-time histograms of responses (fig. 29.5).

Earlier, we established the convergent processing of communication and echolocation sounds within and possibly at levels below the cortex. If a significant portion of the information resides in the response rate, peak response magnitude, and the total number of spikes gen-

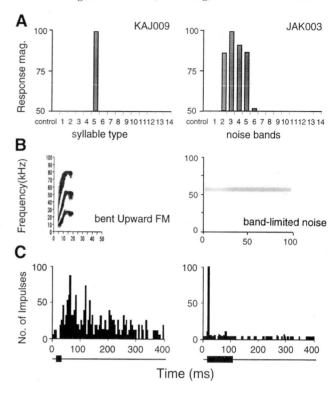

Fig. 29.4. *A:* Representative bar graphs showing relative preference (responses) of two specialized neurons in the FA field to simple monosyllabic calls (*left panel*) and band-limited noise (*right panel*). For calls, the best syllable variant for each syllable type was used. For noise bands, the pure tone was presented first, followed by noise bands of increasing (by 2 kHz) bandwidth numbered sequentially on the *x*-axis. All stimuli were presented in a sequence that was repeated 10 times for five intensities ranging from 95 to 45 dB SPL. The peak responses (100 ms binwidth) to syllables and noise bands 1 to 14 were normalized as a percentage of the maximal response to the best stimulus. *B:* Spectrograms of the "best" syllable (bUFM; *left panel*) and most effective noise band (bandwidth of 4 kHz; *right panel*) for each recording. *C:* Peristimulus-time histograms (PSTHs) resulting from 200 presentations of the "best" syllable variant and noise band for each neuron. Each stimulus was presented at its best amplitude (stimulus repetition rate =1/s; binwidth = 5 ms).

erated by a neuron (as measured in our studies), then it is virtually impossible to spatially segregate echolocation from call processing once it has converged within dually specialized neurons in the auditory cortex. A lack of segregated processing also makes it difficult to imagine how specific movement patterns associated with echolocation versus communication can be independently controlled. One possibility is that at the level of the frontal cortex a significant amount of information may be encoded in the temporal domain. This possibility can be tested using a time-series type of analysis of spike trains. If tenable, this would mean that FA neurons may become dually specialized by encoding echolocation as well as call stimuli in the time domain.

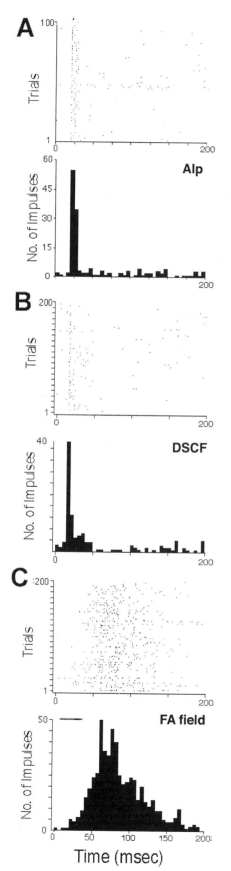

Discussion

Responses from neurons in the auditory cortex show that their level of specialization for processing calls is comparable to that for processing echolocation signals. What can we learn about call processing from what we already know about neural processing for echolocation? Because of the dual specialization of cortical neurons, we can speculate about feature extraction for processing calls based on our knowledge of neural processing of and feature extraction from echolocation signals. In the interest of future experiments, it is informative to take a comparative approach to define the structure of complex (echolocation versus communication) sounds and specify which features of calls may be extracted at lower levels of the auditory system and regrouped at higher levels of processing.

ACOUSTIC FEATURE EXTRACTION AND AUDITORY GROUPING

Basic acoustic patterns within complex sounds consist of either a tone burst or constant frequency (CF), and/or a frequency-modulated (FM) sweep, and/or a noise burst (NB). The specific parameters of a basic acoustic pattern that carry behaviorally relevant information about a sound source (e.g., its identity, location, etc.), have been referred to as information-bearing-parameters (IBPs) (Suga 1995b). The basic acoustic patterns together with their IBPs are referred to as the information-bearing elements or IBEs in a complex sound (Suga 1972). Previous studies have shown that within different regions of the auditory cortex of bats and primates, neurons are specialized to respond solely or preferentially to only one type of acoustic pattern (CF, FM, or NB) within species-specific sounds (Suga, Niwa, and Taniguchi 1983; Suga et al. 1987; Rauschecker, Tian, and Hauser 1995). The particular set of parameters of an acoustic pattern to which a neuron responds best varies with the perceptual and cognitive functions of the cortical area in question. Thus, in the absence of behavioral studies, these parameters may be estimated on the basis of neurophysiological studies where information is rep-

Fig. 29.5. Raster plots and PSTHs for responses of three different cortical neurons to each of 200 repetitions of a single 30 ms tone burst (stimulus onset and duration are indicated by horizontal line). *A:* A typical response pattern to 100 repetitions of a pure tone (24.4 kHz at 85 dB SPL) for a neuron in the unspecialized primary auditory cortex (AI-p). *B:* A typical response pattern to 200 repetitions of a pure tone (60.2 kHz at 40 dB SPL) for a neuron in the DSCF area. *C:* One of the common types of response patterns to 200 repetitions of a pure tone (61.6 kHz at 35 dB SPL) observed in the frontal cortex. Note the lack of time locking of the dual-peaked response for the frontal auditory field neuron. Binwidth for all PSTHs = 5 ms.

resented within neural responses and specializations of neurons. Because behavioral studies are relatively difficult and time-consuming, the neurophysiological data frequently provide the only clue or "best" estimate of these parameters. These data may consist of the shape of a response area for different intensities of a particular IBP or the presence of specializations—for example, combination sensitivity.

The above approach for understanding feature extraction has proven successful for understanding how a bat's auditory system, especially the cortex, is organized to process echolocation signals (Suga 1973, 1982; Suga, Niwa, and Taniguchi 1983; Suga et al. 1987). Studies on neural processing of bat communication signals, however, are still in their infancy; but they have already yielded some very interesting results (Ohlemiller, Kanwal, and Suga 1996; Esser et al. 1997; Kanwal 1999). It is clear that sounds used for communication are far more complex than those used for echolocation, and little is known about the behavioral significance of the different sounds. Nevertheless, a hierarchy in the acoustic structure of calls can be delineated on the basis of information about IBPs as indicated by responses of neurons in the auditory cortex and at other levels of the auditory system (Kanwal et al. 1994; Ohlemiller et al. 1994; Ohlemiller, Kanwal, and Suga 1996; Esser et al. 1997; Kanwal 1999). Herein, we use the available acoustic and neurophysiological data to decompose complex calls into their constituent elements or groups of acoustic parameters that may correspond to a set of IBPs (fig. 29.6).

At the periphery, nerve fibers respond to the different frequencies present within the basic acoustic patterns in complex sounds—for example, CFs and FMs within an echolocation signal (bottom part of fig. 29.6A) and calls (bottom part of fig. 29.6B). Neighboring frequencies within a narrow band of the sound spectrum are transduced and transmitted by frequency-tuned hair cells and nerve fibers, respectively. At lower levels of the ascending auditory pathway, such as the olivary nuclei, neurons are not only tuned to binaural parameters, but also compute simple frequency and amplitude modulations within sounds (Grothe, Park, and Schuller 1997). Some of this processing continues in the inferior colliculus; here, neurons may even encode acoustic motion (O'Neill, Frisina, and Gooler 1989; Kleiser and Schuller 1995). In addition, neurons in the midbrain are tuned to the duration of tones (Casseday, Ehrlich, and Covey 1994). These duration-tuned neurons provide a secondary level of information extraction that can be used to segregate complex sounds. Interestingly, the duration of a simple syllable in mustached bat calls is not closely correlated with other parameters such as fundamental frequency and depth of FM, etc., as suggested by multidimensional scaling of syllable parameters. Duration, therefore, is a useful independent parameter to encode for discriminating between and classifying calls (Kanwal

et al. 1994). After the basic parameters associated with echolocation signals and/or calls are extracted, combinations of specific parameters can be encoded within combination-sensitive neurons. These combination-sensitive neurons also exist within the inferior colliculus (Mittman and Wenstrup 1995). Combination sensitivity allows neurons to be further tuned to new parameters not present within a simple sound, such as temporal relationships between a pure tone and an FM sweep and/or their relative amplitude levels. When combination-sensitive neurons are first created, they are broadly tuned to the combination of parameters (Kawasaki, Margoliash, and Suga 1988; Olsen and Suga 1991; Suga, Olsen, and Butman 1990; Taniguchi, Wong, and Suga 1986; Suga 1995a; Portfors and Wenstrup 1999). At the level of the auditory cortex, many of these neurons become sharply tuned. Finally, at levels of the frontal cortex, complex pulse-echo sequences of over a hundred milliseconds could influence the pattern of rhythmically bursting or irregularly firing neurons (Kanwal et al. 2000). These neurons probably encode perceptual information contained within multiple echoes of pulse-echo pairs or sequences as a "gestalt" that, together with information stored from past experience, could guide target tracking and prey capture behavior.

Whereas much is known about the neural mechanisms that create neural specializations and tuning for echolocation, we know virtually nothing about the mechanisms that are important for call processing at subcortical levels. Even for the cortical level (level above dashed line in fig. 29.6), we have little information about call processing compared to the processing of echolocation signals. In general, neurons within the auditory cortex encode combinations of CFs, FMs and NBs, as well as the temporal relationships among these parameters. Thus, neurons at the level of the primary and secondary auditory cortices in the mustached bat have been shown to process isosyllabic pairs and composites (Kanwal et al. 1994; Ohlemiller, Kanwal, and Suga 1996; Kanwal 1999). At frontal cortical levels, neurons could uniquely encode a group of syllables—for example, a train of simple syllables and/or a phrase that together communicate emotions or trigger vocalizations and/or social behaviors by the cingulate and motor cortex. Alternatively, modifications of their temporal response pattern could encode an appropriate behavioral response, for example, when an echolocation call is used as a communication signal (Balcombe and Fenton 1986). It is foreseeable that certain simple syllables and composites are extracted already within the brainstem so that they can trigger reflexive behaviors (e.g., those to alarm calls), but no information exists to date in support of or against this part of the scheme presented here. We hope that this scheme will function as a working hypothesis and stimulate further investigations of call processing in a bat's auditory system, especially at subcortical levels.

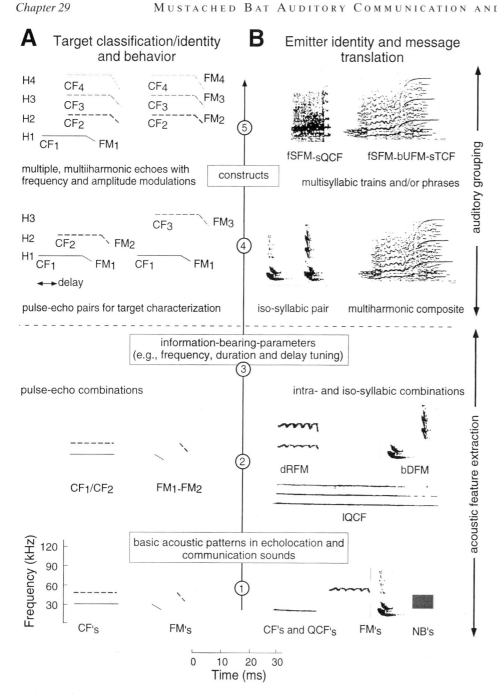

Fig. 29.6. Schematic of parallel-hierarchical processing and reconstruction of echolocation and communication sounds at different levels of the auditory system in the mustached bat. *A:* Echolocation sounds are resolved into CF and FM types of elements, from which information-bearing parameters are extracted and processed separately and in parallel. At medullary, midbrain, and thalamic levels (represented by circles numbered 1, 2, and 3 below the dashed line), neurons are tuned to and become specialized to group these parameters within combination-sensitive neurons. At cortical levels (represented by circles numbered 4 and 5 above the dashed level), combination-sensitive neurons show an increasingly sharp tuning to these combinations and become organized in the form of maps. *B:* Schematic of an analogous process of segmentation of complex calls—for example, bent downward FM (bDFM), descending rippled FM (dRFM), and long, quasi-CF (lQCF)—during early auditory processing. According to this scheme, call segments would be regrouped as higher order constructs within neurons at cortical levels for purposes of cognition.

MULTIMODAL PROCESSING AND DUAL SPECIALIZATIONS

Recent studies on the mustached bat show that the same cortical networks, down to the level of a single neuron, process both echolocation signals and calls. These neurons exhibit equivalent levels of specializations for processing the two kinds of sounds. Furthermore, single cortical neurons can switch rapidly (within a second) between the two modes of processing, as evident from our routine recordings of neural responses (i.e., a neuron can respond well to an echolocation stimulus followed by a call stimulus and vice versa in awake mustached bats). The dynamics of this switching and the

structure of the cortical networks involved remains to be systematically examined. This type of "dual specialization" or "multifunctional sensory processing" was also recently demonstrated within single cortical cells of a pallid bat (Razak, Fuzessery, and Lohuis 1999). These data substantiate the theory that single cortical neurons can have multiple functions or domains for processing auditory information (Suga 1994). Among other nervous systems, output neurons in the stomatogastric ganglion of a crab also can switch their function and participate within multiple pattern generating circuits (Weimann and Marder 1994). Furthermore, intracellular recordings from the same laryngeal motor neuron in decerebrate paralyzed cats show complex combinations of excitation and inhibition that vary between vocalization versus upper airway defensive reflex patterns (Shiba et al. 1999). This switch in the firing patterns of laryngeal motoneurons is based on changes in dynamic connections between the respiratory and nonrespiratory neuronal networks. In the FA field, a dual mode of auditory processing within single neurons is less obvious, and neurons are less likely to be able to quickly switch between the two modes because of the long response duration (up to a few hundred milliseconds) often observed in these neurons. However, a dual processing mode could still be embedded in the temporal pattern of impulses within single FA neurons. We hope future studies of auditory processing that examine these questions in greater detail will lead to new ideas about the functional organization of the cortex and an understanding of how neural computations accomplish multiple functions within single neurons.

Acknowledgments

This work was supported in part by NIH grant DC02054 to Jagmet S. Kanwal and by a grant from the University of Ulm to Karl-Heinz Esser (Anfangsförderung von Forschungsvorhaben). We thank Nobuo Suga for providing the facilities and equipment at Washington University in St. Louis, where this research was initiated. We also thank Michael Gordon, Nick Fuzessery, and the editors for their useful comments on earlier versions of the manuscript; and the Natural Resource Conservation Authority of Jamaica and the Department for Wildlife in Trinidad, who gave permission to export bats, without which this research would not be possible.

{ 30 }

The Auditory Cortex of the Horseshoe Bat:
Neuroarchitecture and Thalamic Connections
of Physiologically Characterized Fields

Susanne Radtke-Schuller

Introduction

The auditory cortex (AC) of bats has been the subject of extensive physiological investigations, and its functional organization in several species has been characterized in detail (e.g., Suga 1984; Wong and Shannon 1988; Dear et al. 1993; Esser and Eiermann 1999). Neither its neural architecture nor its afferent connections with the thalamus have been adequately characterized. A review in 1970 even stated that "there is little evidence ... of an extensive auditory representation in either insectivorous Microchiroptera or Megachiroptera" (O. W. Henson 1970, p. 34). More recently, Fitzpatrick and Henson (1994) investigated the morphological characteristics of neurons and the cyto- and myeloarchitecture, as well as the topographical distributions of cytochrome oxidase and GABAergic neurons, in the AC of the mustached bat, *Pteronotus p. parnellii*. These investigators could not identify morphological correlates of the functional subdivisions so elegantly depicted by Suga and coworkers (see Suga 1984, 1990b for reviews). Further anatomical studies are necessary to elucidate the relationship between structural organization and functional properties of bat AC.

The present report reviews the results of combined investigations of the physiology (Radtke-Schuller and Schuller 1995; Schuller, O'Neill, and Radtke-Schuller 1991), neuroarchitecture (Radtke-Schuller 2001), and connectivity pattern (Radtke-Schuller 1997) of the horseshoe bat AC. The rufous horseshoe bat, *Rhinolophus rouxi* (Old World family Rhinolophidae) uses a similar type of echolocation call to the mustached bat, *P. parnellii*. These two species are not closely related phylogenetically, but they show strong convergence in their acoustical behavior, and neurons in the AC have similar functional properties (Schuller, O'Neill, and Radtke-Schuller 1991).

OUTLINES AND STRUCTURE OF THE NEOCORTEX

The forebrain hemispheres of *Rhinolophus* are lissencephalic. Because a fissura rhinalis is missing, there are no external features to delineate the neocortex. The

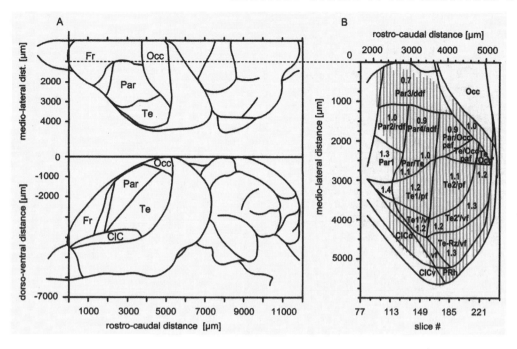

Fig. 30.1. *A:* Neocortical regions in *Rhinolophus rouxi.* The topographic position of the neocortical regions is represented on a standardized brain in a dorsal (upper) and lateral view (lower). The regions are labeled as frontal (Fr), parietal (Par), temporal (Te), and occipital (Occ). The dorsal part of the claustrocortex (ClC) is included in the neocortex. *B:* The neocortical areas constituting the AC are shown in a flattened surface projection starting at 1000 μm lateral from the midline. The gross regions are divided further into neuroarchitectural fields (Par 1 . . . 4, Par/Occ, Te 1 . . . 2, Te/Occ, Te1' . . . 2', Par/Te, Te-Rz). Perirhinal cortex (PRh) and claustrocortex (ClCd: dorsal ClC; ClCv: ventral ClC) border the temporal region. The numbers give the ratio of the widths of layers III–IV and layer V and illustrate the predominance of either layers III–IV (>1) or layer V (=1) in the respective fields. In addition to the neuroarchitectural field names, the names of the neurophysiologically outlined auditory cortical areas are indicated (following the /). The primary field (PF) and the ventrally adjacent ventral field (vf) correspond to the central and ventral part of the temporal region. The dorsal fields (rdf, ddf, adf) constitute most of the parietal region. The posterior auditory field (paf) corresponds to the most caudal parts of the temporal and parietal regions, already transitional to the occipital region. The target area of auditory thalamic projections is delineated by vertical lines and matches mostly the physiologically derived borders.

neocortex is relatively small (Stephan and Pirlot 1970) and forms a dorsal cap that does not reach the temporal and occipital poles of the hemispheres (fig. 30.1A). It shows the typical mammalian six-layered structure, in a frontal section stained for neurofibrils by the Bodian-silver impregnation (fig. 30.2). Layer I in *Rhinolophus* is thick relative to that of more advanced mammals, despite large regional variations. Layer II is relatively thin and is characterized by a high packing density of neuronal somata. This layer contains primarily "extraverted" neurons, a type of phylogenetically old pyramidal neurons with prominent apical dendrites that extend far into layer I, often reaching the cortical surface. The above features are hallmarks of a "primitive" neocortex as found typically in the isocortex of basal insectivores, other groups of bats, and cetaceans (e.g., Rehkämper 1981; Sanides and Sanides 1974; Ferrer 1987; Morgane,

Glezer, and Jacobs 1990). Layer III is occupied by medium-sized pyramidal cells having a high packing density, which decreases toward the deeper layer. There is no sharp border separating layer III from layer IV. In layer IV, the cell density is equal to or lower than that in layer III. Granular cells are not the dominant cell types in this layer throughout all neocortical areas, including the primary AC. Thus, the cytoarchitecture of layer IV of the neocortex of *Rhinolophus* differs substantially from that of mammals with a more developed neocortex. In the latter, layer IV of the sensory cortices has a high cell density, especially layer IV of the koniocortex (the area of neocortex structurally corresponding to a primary sensory representation; Sanides 1970). It comprises predominantly granular elements—that is, small stellate cells (spiny or nonspiny) that function as interneurons. In contrast, the middle layers of the neocortex

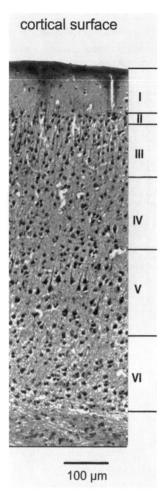

cortical surface

I
II
III
IV
V
VI

100 µm

Fig. 30.2. Frontal section (10 µm) from the primary auditory cortex (AC, Te1) of *Rhinolophus*. The neocortex shows the typical mammalian six-layered structure. Remarkable are some conservative features, like a wide layer I, extraverted neurons in a sharply accentuated layer II, and the generally poor granularization. (Staining: Bodian-silver impregnation for neurofibrils.)

of microchiropterans, insectivores, and dolphins contain giant nonspiny stellate cells (see above references). It is believed that these cells are precursors of inhibitory cortical interneurons found in advanced mammals, while pyramidal cells are precursors of the excitatory (spiny) stellate cells (e.g., Nieuwenhuys 1994). Layer V has a low cell density and contains primarily pyramidal cells of various size. In the middle of this layer, large loosely packed pyramidal cells form a prominent band. Deeper in layer V, a rather sparse cell stripe can be observed in most regions of the neocortex. Layer VI shows a higher cell packing density than layer V. It contains a variety of mostly middle- or small-sized cells of various types.

OUTLINES OF THE AUDITORY CORTEX

The AC has been delineated by neurophysiological recordings (Radtke-Schuller and Schuller 1995) and by anterograde transport studies following tracer in-

jections into the auditory thalamus (Fromm 1990). Neurophysiological recordings and cytoarchitectural and connectional data have been analyzed using a common brain atlas (Radtke-Schuller, unpublished) as a reference, so that the positional definitions and correspondences among the different experimental approaches could be reliably determined. Fig. 30.1B summarizes the projection patterns of the different regions of the medial geniculate body (MGB; compare legend). The cortical areas receiving thalamic projections closely match the borders of the AC as defined by physiological recordings. The primary field (pf) and the adjacent ventral field (vf) correspond to the central and ventral part of the temporal cortex as distinguished anatomically. The dorsal fields (rdf, ddf, adf) occupy most of the parietal cortex. The posterior auditory field (paf) corresponds to the most caudal parts of the temporal and parietal cortices, encroaching the transitional zone with the occipital cortex.

CHARACTERISTICS OF THE TEMPORAL
AND PARIETAL REGIONS

The temporal and the parietal regions of the neocortex can be visualized clearly by the pattern of cytochrome oxidase activity in frontal sections (fig. 30.3A–B). In the parietal region, large pyramidal cells with an intense somatic cytochrome oxidase reactivity form a prominent band in the middle of layer V, also seen as dark band in Nissl stained materials. Layer V is thicker than in other regions and the myelin stain is intense. In the temporal cortex, the size as well as the density of pyramidal cells in the middle of layer V diminish progressively along the dorsoventral axis. At the same time, the cytochrome oxidase reactivity in layer IV and lower layer III increases progressively and emerges as a dark stripe. Layer IV also shows an intense myelin stain. In an intermediate position these features overlap (see ventral Par4 in fig. 30.3A). Thus, a characteristic feature of the temporal cortex is the dominance of layers III and IV, which are markedly thicker than layer V. The characteristic dominance of layers III and IV in the temporal region and layer V in the parietal region is most evident by measurements of the thickness (th) ratio, th(III + IV)/th(V). In the temporal region, this ratio is greater than 1 (fig. 30.1B). In contrast, this value equals 1 for all parietal fields, with the exception of Par1. Interestingly, Par1 is the only parietal area that does not receive auditory thalamic input.

The thickness of the neocortical regions shows marked rostrocaudal variations, especially in the temporal cortex, which is not even half as thick at its caudal end than at its rostral beginning. Along the same gradient, the intensity of myelin stain, Nissl stain, and cytochrome oxidase reactivity decreases progressively (see fig. 30.3A–B). In addition to distinct cytoarchitectural features, temporal and parietal parts of the AC

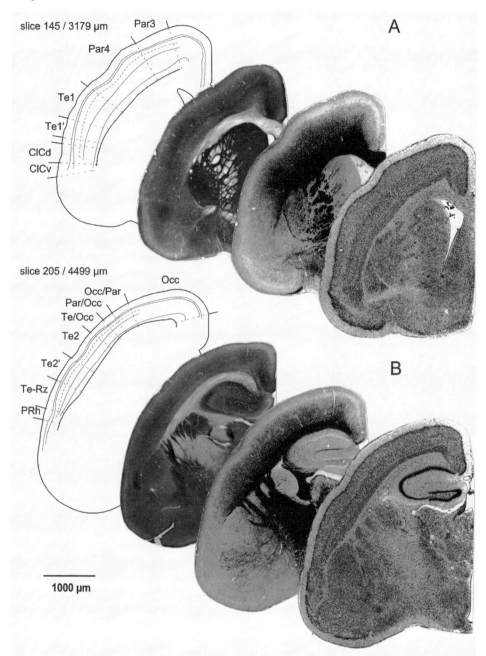

slice 145 / 3179 µm   Par3
  Par4
Te1
Te1'
ClCd
ClCv

slice 205 / 4499 µm
             Occ
      Occ/Par
    Par/Occ
   Te/Occ
  Te2
 Te2'
Te-Rz
PRh

1000 µm

A

B

Fig. 30.3. Frontal sections at a central (*A*) and more caudal (*B*) location of AC, illustrating the neuroarchitectural differences of the auditory cortical fields. Note the dramatic changes in cortical thickness and staining intensities along the rostrocaudal position. (*From left to right:* schematic drawing, cytochrome oxidase stain (42 µm), fiber stain (Gallyas, myelin silver impregnation, 22 µm), cell stain (Nissl, 22 µm). (For rostrocaudal location within auditory cortical areas, compare fig. 30.1.)

show differences with respect to physiology and connectivity patterns. Neurons with multiple-frequency tuning and neurons responding exclusively during the bat's vocalizations are found throughout the parietal AC, but they are almost lacking in the temporal AC. The temporal AC receives its characteristic afferents from he "main sensory nucleus of the MGB—that is, the ventral MGB (MGBv). In contrast, the parietal AC has strong afferent projections from the associated nuclei of the auditory thalamus, especially from the nuclei of the anterior dorsal MGB (AD), the medial MGB (MGBm), and the transitional zone between MGBm and the intralaminar part of the posterior nucleus (M/PO). The ef-

ferent projection patterns of the two regions also differ. Whereas the dorsal (parietal) AC projects to nonsensory brain areas—for example, to the pontine nuclei (Schuller, Covey, and Casseday 1991), the paralemniscal tegmental area (Metzner 1996), and the deep layers of the superior colliculus (Reimer 1989)—such projections are weak from the temporal AC.

There is a good correspondence between the cytoarchitecturally recognized regions of the neocortex in *Rhinolophus* and those distinguished in the neocortex in other Chiroptera and some other lissencephalic mammals (e.g., *Myotis:* Rose 1912; *Pteropus:* Brodmann 1909; *Rattus:* Zilles 1985). In those species (e.g., flying fox, rat) in which a functional mapping of the neocortex was done, the somatosensory cortex was found to dominate postcentral and parietal regions, whereas the auditory cortex was mostly confined to the temporal region. Responses to acoustic stimuli in the parietal region have been reported for several mammalian species, but such neurons are typically restricted to the multisensory or auditory-somatosensory fields (e.g., *Pteropus, Sciurus:* Krubitzer, Calford, and Schmid 1993; Krubitzer, Sesma, and Kaas 1986; *Mus:* Carvell and Simons 1987; *Felis:* Clemo and Stein 1983). These fields receive somatosensory thalamic afferents from the posterior nucleus and/or the ventroposterior nucleus, and also auditory projections from the MGBm.

In *Rhinolophus,* some of the thalamic input to the parietal fields of the AC (e.g., projections from the ventral nuclei of the lateral thalamus and/or posterior nucleus) is most probably comparable to that of the auditory-somatosensory fields in other mammals. However, the projections from the MGB in bats (perhaps uniquely) are dominant. These projections do not only originate from the medial nucleus and other multisensory nuclei of the MGB, but also prominently from the anterior dorsal MGB and even from the MGBv. The connectivity patterns of the parietal fields of the AC suggest that these represent auditory-somatosensory fields with strong accentuation of the auditory component. Taking into account the strong efferents to nonsensory structures, the parietal fields even seem to play an important role in sensory-motor integration in respect to acoustically guided behavior, like orientation in space and guided attention.

The two neocortical regions described so far can be divided further on the basis of more subtle differences in neuroanatomical characteristics. Only the most outstanding ones are mentioned below. These subdivisions closely correspond to those found physiologically (fig. 30.1B). However, in some cases the so-far physiologically defined fields comprise more than one cytoarchitectural subdivision.

FIELDS OF AC–FUNCTIONAL CORRESPONDENCE
TO NEUROANATOMICAL SUBDIVISIONS

Primary field

The pf comprises the neuroarchitectural subdivisions Te1 and Te2 of the temporal region. It shows clear tonotopic organization with increasing best frequencies in the caudorostral direction. A notable feature in *Rhinolophus* is the strong overrepresentation of frequencies at, and a few kHz above, the bat's resting frequency. In the auditory cortex, this narrow frequency band occupies 6–12 times more neural space per kHz compared to representations of both lower and higher frequencies. Moreover, neurons tuned to these frequencies have very narrow tuning curves. The area of overrepresentation is localized in Te1 (fig. 30.3A), a subdivision having (1) the

greatest overall cortical thickness, (2) the thinnest layer I, (3) the greatest relative thickness for its layers III–IV, and (4) the strongest cytochrome oxidase reactivity and the most intense myelin staining in layer IV. It would be interesting to know how the AC of bats without an overrepresentation of a narrow frequency band (e.g., *Myotis;* Wong and Shannon 1988) is organized, and whether mammals for which low frequencies are of biological relevance (e.g., *Spalax*), have a similar structural-functional relationship.

Caudal to Te1 lies Te2 (fig. 30.3B), where lower frequencies are represented. The total cortical thickness in Te2 decreases progressively. This decrease in cortical thickness is accompanied by a decrease in cytochrome oxidase reactivity, and even more so in the intensity of myelin stain. Layers I–II are thicker than those in Te1. Layers III–IV are prominent compared to layer V, but not quite to the degree of Te1.

In spite of the great relative thickness and the intensity of myelin and cytochrome oxidase stain, the cell density in the middle layers of the primary AC is relatively low, and these layers exhibit a generally weak granularization (fig. 30.2). Thus, the population of typical interneurons is smaller than in mammals with a more developed neocortex. The functional significance of this difference is unclear.

The primary AC receives characteristic topographically ordered afferents from the central MGBv. Its rostral part (Te1) additionally receives projections from the deep dorsal nucleus of the MGB (DD).

Ventral field

The neuroanatomical features of the vf of the temporal region (Te1′, Te2′ and edge zones of the temporal cortex) are similar to those of the central temporal fields, but in general less pronounced (fig. 30.3A–B). Toward the ventral and caudal borders, characteristic allocortical features become more prominent. Specifically, the thickness of layers I–II increases, while the cell density of layer IV, and the intensity of myelin stain, and cytochrome oxidase reactivity decrease. The ventral fields receive afferent projections mainly from border regions of the MGBv, the caudal part of the suprageniculate nucleus (SGN) and the MGBm. Unlike the primary AC, neurons in these fields mostly show inconsistent responses to auditory stimuli.

Anterior dorsal field

The adf is located dorsally from the high-frequency region of the primary AC in the parietal cortex (Par4; fig. 30.3A). This field is tonotopically organized, but it is less orderly than the primary field; and the distribution of neuronal best frequencies also indicates an overrepresentation of frequencies around the bat's resting frequency (Radtke-Schuller and Schuller 1995). The high-frequency representation is continuous with that in the primary AC, leading to a reversal of the tonotopic gradient between primary AC and adf. A notable feature is the high concentration of neurons responding consistently during spontaneous vocalizations or showing modified activity to acoustic stimuli when concurrent vocalization occurs.

In terms of the neuroarchitecture and afferent connections, the adf exhibits intermediate properties between the rostral primary and the dorsal dorsal field (Radtke-Schuller 1997). It has the most complex pattern of afferent projections among all fields of AC. Characteristic main afferents are from the dorsomedial MGBv; parts of the dorsal MGB, the dorsal nucleus of MGB (D), DD, AD, the anterior deep dorsal nucleus of MGB (ADD); SGN; MGBm; M/PO; the intralaminar nuclei of the thalamus (ILN); and the nucleus of the brachium of the colliculus inferior (BICN). Thus the adf is comparable with the anterior auditory field (AAF) of the cat (e.g., Andersen, Knight, and Merzenich 1987; Mitani, Itoh, and Mizuno 1987).

Rostral dorsal field

The rdf corresponds to Par2 of the parietal cortex and exhibits no tonotopic organization. This field is characterized by neurons that show (1) strong preference for pure tones and no response to any frequency-modulated stimulus or noise band, and (2) facilitated responses to combinations of pure tones in separate frequency bands (Schuller, O'Neill, and Radtke-Schuller 1991).

Par2 is characterized by a thick layer V with an intense myelin stain and a high concentration of darkly Nissl stained pyramidal cells that show intense cytochrome oxidase reactivity. Afferent projections of Par2 are mainly from the rostral MGBd (AD, ADD), SGN, MGBm, M/PO, and ILN.

Dorsal dorsal field

The ddf corresponds to anatomical subdivision Par3 in the parietal cortex (fig. 30.3A). Many neurons in ddf show a preference for linear frequency-modulated stimuli and are sensitive to pairs of frequency-modulated stimuli represented at characteristic delays (Schuller, O'Neill, and Radtke-Schuller 1991). The neurons' characteristic delay is topographically organized in stimulus pairs ("chronotopy" map), with an overrepresentation of the biologically most important delays of 2–4 ms. Within the ddf, structural features—for example, maximum cortical thickness and most intense staining of myelin—are obvious in the region of overrepresentation. Among the parietal fields, the ddf shows the most pronounced structural differences to the primary AC—for example, layer V is thickest and layer IV is thinnest. A relatively high cell density, which seems to be evenly distributed over layers III to V, gives the impression of a less distinct layered organization. Layer VI is strongly accentuated by a high cell density.

Afferents to ddf/Par3 are mainly from AD, MGBm, M/PO, and ILN, and from the intralaminar part of the posterior nucleus (PO/ILN).

The connectivity patterns of the rdf and ddf with the thalamus closely resemble that of the primate AC. In *Macaca,* there is an ascending pathway (via the anterodorsal MGB to the posteromedial/centromedial field) paralleling the well-established ascending projection from the MGBv to the primary AC (Molinari et al. 1995; Rauschecker et al. 1997).

Posterior auditory field

The paf occupies two anatomical subdivisions, the parietal-occipital and the temporal-occipital fields (fig. 30.3). There is no clear-cut tonotopic organization, but a trend for low frequencies to be represented caudoventrally (i.e., adjacent to the representation of these frequencies in the primary AC) and middle frequencies dorsorostrally. A small number of neurons are tuned to the bat's resting frequency range, and these cells are located at the dorsorostral edge of the paf. In the dorsal part of the paf (i.e., Par/Occ), neurons are sharply tuned (Q10 dB values > 20) to frequencies that fall into the FM range of the bat's sonar signal. Interestingly, these neurons are not encountered in the ventral part of the paf (i.e., Te/Occ), or in the primary AC.

The cytoarchitecture of Par/Occ is similar to that of the occipital region. In particular, layers III through V are not readily differentiable, but layer V in Par/Occ is thicker than layers III–IV together. Main afferents are from the rostral MGBv and the AD. Te/Occ is small and has a less distinct layered appearance in Nissl stained materials. Also, it stains weakly for myelin, and the cytochrome oxidase reactivity is relatively low. However, the characteristic features of the temporal cortex (i.e., thicker layers III–IV compared to V, and a contrasting layer IV in the cytochrome oxidase stain) can be observed. Afferent projections are mainly those from the MGBv and the caudal SGN.

The thalamic connectivity pattern of *Rhinolophus* AC fits well into the general concepts of parallel ascending auditory thalamocortical pathways in mammals. A very comprehensive scheme proposed by Rouiller et al. (1991) has been modified here to include computational representations.

The ascending thalamocortical pathways comprise three parallel functionally distinct subsystems with differential representation in the auditory cortical fields.

One subsystem shows topographical organizations: the cochleotopic representation of a stimulus parameter that is already established in the periphery (central ventral MGB–AC fields with tonotopic organization) and the computational representation of one or several parameters of the auditory stimulus (anterodorsal MGB–dorsal fields of AC).

The second subsystem is not or only weakly topographically organized. It is located mainly in edge zones of the MGBv and projects to the AC ventral to the primary field. The posterior part of the MGBd, as well as the dorsal rim of the anterior dorsal MGB, also belong to the nontopographic system. Each field of the AC has at least minor input from this system.

The third subsystem, the multisensory subsystem, comprises those nuclei of the MGB that are classifiable as multisensory due to the sources of their afferents (M, M/PO, SGN, BICN, caudal part of lateral posterior nucleus [LPc, ILN]). These nuclei are thought to be important for acoustically guided behavior, like orientation in space and guided attention. Except for the primary field, every field of the AC has strong afferents from at least one nucleus of the multisensory subsystem.

Each auditory cortical field, as defined physiologically, receives a mixture of afferent projections from the three subsystems, but to a different degree. The primary field, with its tonotopic organization, shows the least convergent input (only main source of input MGBv) and corresponds most closely with the cochleotopic part of the topographic subsystem. The ventral field mostly represents the nontopographic subsystem and also to some extent the multisensory one. The posterior auditory field is dominated by the topographic subsystem (main afferents from MGBv and AD, to some extent tonotopically organized), but it also has a strong component of the multisensory subsystem. The tonotopically organized anterodorsal field shows the largest convergence. All three subsystems are strongly represented. The dorsal dorsal field with the chronotopy of the FM-FM neurons closely relates to the computational part of the topographic subsystem (main afferents from AD) and has additionally very strong input from the multisensory system. The multisensory system is most dominantly represented in the rostral dorsal field (CF-CF field), which also receives afferents from the computational part of the topographic subsystem (AD).

Compared to the concept of Winer and Morest (1983), the tonotopic component of the topographic subsystem corresponds to the main sensory nucleus—that is, the ventral MGB—whereas the computational component of this system (and possibly also the nontopographic subsystem) is represented by the aligned nuclei of the associated nuclei of the MGB. The multisensory system thus can be referred to as the unaligned nuclei of the associated nuclei.

Bats, with their unique ability to fly and to echolocate, often have been considered as "exotic" representatives of the mammalian order—compared, for example, to cats or rats. However, these investigations demonstrate the strong similarities between the bat's auditory cortex and that of other mammals, despite its specializations.

Acknowledgments

This study was supported by Deutsche Forschungsgemeinschaft, SFB 204 ("Gehör," TP10).

{ 31 }

Modulation of Frequency Tuning of Thalamic and Midbrain Neurons and Cochlear Hair Cells by the Descending Auditory System in the Mustached Bat

Nobuo Suga, Yunfeng Zhang, John F. Olsen, and *Jun Yan*

Introduction

Neurons in the deep layers of the auditory cortex (AC) project to the medial geniculate body (MGB) of the thalamus, the inferior colliculus (IC) in the midbrain, or the subcollicular auditory nuclei (Saldana, Feliciano, and Mugnaini 1996; Huffman and Henson 1990). These corticofugal projections are organized tonotopically (An-

dersen, Knight, and Merzenich 1980; Herbert, Aschoff, and Ostwald 1991; Malmierca et al. 1995). Corticothalamic fibers project only ipsilaterally to the MGB and the thalamic reticular nucleus (Rouiller and De Ribaupierre 1990; Ojima 1994; Bajo et al. 1995). Corticocollicular fibers bilaterally project to the IC. The ipsilateral projection is much heavier in density, much more extensive in area, and much more topographically organized than the contralateral projection in cats (Andersen, Knight, and Merzenich 1980; Rouiller, Hornung, and de Ribaupierre 1989), in rats (Saldana, Feliciano, and Mugnaini 1996; Herrera et al. 1994), and in guinea pigs (Feliciano and Potashner 1995). Therefore, ipsilateral corticofugal modulation would be much larger than contralateral corticofugal modulation in the IC and MGB, and would be frequency dependent.

Some corticofugal fibers project further down beyond the IC, to the superior olivary complex and cochlear nucleus on both sides (Feliciano, Saldana, and Mugnaini 1995). Therefore, corticofugal modulation would take place even in the cochlea via olivocochlear (OC) neurons in the superior olivary complex. The central nucleus of the IC projects not only to the MGB and the superior colliculus, but also to medial OC neurons. OC neurons bilaterally project to the cochlea, but medial OC neurons mostly project to contralateral cochlear outer hair cells (Warr 1992; Bishop and Henson 1987). Contralateral ear stimulation excites all IC, MGB, and AC neurons, while ipsilateral ear stimulation either inhibits, excites, or has no effect. Therefore, corticofugal and colliculofugal modulations are expected to be larger for the contralateral cochlea than for the ipsilateral cochlea.

Past physiological data of corticofugal effects on MGB and IC neurons has been controversial; the effects have been described as inhibitory (Massopust and Ordy 1962; Watanabe et al. 1966; Amato, La Grutta, and Enia 1969; Sun, Chen, and Jen 1996), excitatory (Andersen, Junge, and Sveen 1972; Villa et al. 1991), or both (Ryugo and Weinberger 1976; Syka and Popelar 1984; Sun et al. 1989). However, this controversy has probably been settled by the finding that the excitatory or inhibitory corticofugal effect depends on topographic and physiological relationships between cortical and subcortical neurons studied in a pair, as explained below.

In the mustached bat, *Pteronotus parnellii parnellii*, cortical neurons, via the corticofugal system, mediate a highly focused positive feedback to augment the auditory responses of "matched" subcortical neurons. This feedback is incorporated with widespread lateral inhibition to suppress the auditory responses of "unmatched" subcortical neurons. Corticofugal feedback adjusts and improves auditory information processing in the subcortical auditory nuclei not only in the time domain (Suga et al. 1995; Yan and Suga 1996a), but also in the frequency domain (Zhang, Suga, and Yan 1997). These cortical functions, named "egocentric selection," are also found in the big brown bat (Yan and Suga 1998; Gao and Suga 1998) and the cat (He 1997), so that egocentric selection could be a general function of the corticofugal system. As a matter of fact, corticofugal positive feedback associated with inhibition has also been found in the visual system (Tsumoto, Creutzfeldt, and Legendy 1978).

Muscimol is a potent agonist to GABAa receptors, which mediate synaptic inhibition (de Feudis 1978; Lester and Peck 1979). The Doppler-shifted constant-frequency (DSCF) processing area of the primary auditory cortex of the mustached bat is highly specialized for representing echoes from flying insects by spatiotemporal patterns of neural activity. Muscimol applied to the DSCF area inactivates the corticofugal fibers originating from this area and evokes a prominent decrease in excitatory response (number of impulses/stimulus) of ipsilateral subcortical DSCF neurons to single tones: 34% in the IC and 60% in the MGB on the average (Zhang and Suga 1997). The same experiment as above with the big brown bat (*Eptesicus fuscus*) evoked a 38% decrease in responses of ipsilateral IC neurons to single tones on the average (Gao and Suga 1998). The first aim of this chapter is to show data on the corticofugal modulation of the auditory responses of MGB and IC neurons, primarily responding to single tones.

Corticofugal signals potentially influence the activity of cochlear hair cells through the OC neurons, which are inhibitory (Huffman and Henson 1990; Feliciano, Saldana, and Mugnaini 1995; Warr 1992). In the guinea pig, electrical stimulation of the IC reduces the temporal threshold shift of the summated auditory nerve response (N1), perhaps because of the excitation of OC neurons evoked by the collicular stimulation (Rajan 1990). The functional significance of the corticofugal system in signal processing in the frequency domain in the subcollicular nuclei and the cochlea remains unknown. To explore the function of OC neurons physiologically, we have to activate or inactivate an OC neuron with a particular BF and study its effect on individual hair cells or primary auditory neurons with different BFs. Our present experiment on the cochlea of the mustached bat is a preliminary study.

In the mustached bat, cochlear microphonic response (CM) recorded from the cochlear aqueduct (Pollak, Henson, and Novick 1972) or the round window (Suga, Simmons, and Jen 1975) is tuned sharply to approximately 61 kHz: the frequency of the second harmonic of the constant-frequency component of the species-specific biosonar pulse. Because of this sharp tuning, the CM evoked by a tone burst at approximately 61 kHz shows a prominent after-response: CM-aft—that is, damped oscillation, which occurs at a fixed frequency irrespective of the frequencies of tone bursts (Suga and Jen 1977). The corticofugal modulation of the hair cells tuned to approximately 61 kHz can be studied by electrically stimulating the AC or IC. Since the IC is anatomically closer than the AC to the hair cells, we stimulated the IC and found that the frequency tuning of CM recorded from the contralateral cochlea was modulated by the stimulation. The second aim of this chapter is to report these intriguing preliminary data.

Materials and Methods

ANIMALS AND SURGERY

Materials, surgery, recording of neural activity, acoustic stimulation, and data acquisition and processing in this study were mostly the same as described elsewhere (Suga et al. 1983; Zhang, Suga, and Yan 1997). The essential portions of these are summarized below.

Seven adult mustached bats, *P. parnellii*, from Jamaica were used for the present experiments: five for corticothalamic and corticocollicular modulations and two for colliculocochlear modulation. Under neuroleptanalgesia (Innovar 4.08 mg/kg b.w.), a 1.8 cm long metal post was glued onto the exposed dorsal surface of the bat's skull. Two to four days after the surgery, the unanesthetized and untranquilized bat was placed in a polyethylene-foam restraint suspended by an elastic band at the center of a soundproof, echo-attenuated room maintained at 30–32°C. The head was immobilized by fixing the post to a metal rod with set-screws and adjusted to face directly at a condenser loudspeaker located 74 cm away. The surgical wound was treated with local anesthetic (Xylocaine) and antibiotic ointment (Furacin). Experiments were performed as described below. Water was given occasionally to the bat during the experiments. After the experiments, the bat was returned to the animal room. Each bat could be used for 3–14 days.

EXPERIMENTS ON CORTICOFUGAL MODULATIONS

For the experiments, multiple neurons were recorded first with a tungsten-wire electrode with a 6–8 μm tip diameter at 4–5 loci in the DSCF area of the primary auditory cortex, which represented 60.47–62.30 kHz (Suga and Manabe 1982; Suga et al. 1987), and their best frequencies (BFs) were measured. Then, the responses of a single DSCF neuron to tone bursts were recorded in the ventral division of the MGB or the dorsoposterior division (DPD) of the IC with a tungsten-wire electrode; its BF and frequency-tuning curve were then measured, as in the experiments performed by Zhang, Suga, and Yan (1997). To inactivate one of the cortical loci where BFs were measured, 90 nl of 1.0% lidocaine (local anesthetic) was injected into the approximately 900 μm thick cortex at a depth of 600–700 μm with a mechanical microinjection unit. To activate the locus, a 100 nA, 0.2 ms monophasic electric pulse was delivered at the 600–700 μm depth at the onset of each 23 ms tone burst. The frequency of the tone burst was set at the best frequency of the electrically stimulated neurons. The amplitude of the tone burst was set at approximately 60 dB SPL re 20 μPa. The responses of the subcortical DSCF neuron to 23 ms tone bursts of different frequencies and amplitudes were collected with a computer before, during, and after the focal cortical inactivation or activation. The *t*-test was used to examine whether the difference in auditory response was statistically significant before and after lidocaine inactivation or electrical activation or between thalamic and collicular neurons (see Zhang, Suga, and Yan 1997).

EXPERIMENTS ON COLLICULOFUGAL MODULATION

For the experiments, responses of neurons in the DPD of the IC to tone bursts were recorded with a pair of tungsten-wire electrodes, and their best frequencies were measured. Then, these electrodes were implanted and used to electrically stimulate a given site in the DPD. Cochlear microphonic responses (CMs) to tone bursts were recorded with a tungsten-wire electrode (approximately 30 μm tip diameter) implanted in or at the cochlear perilymphatic sac, as described by Henson and Pollak (1972).

Experiments were performed without anesthesia. Acoustic stimuli comprised 2.0 ms tone bursts, with a 0.01 ms rise-decay time. Their frequency was set at or near the resonance frequency of a given cochlea (approximately 62 kHz). The tone bursts were delivered at approximately 75 dB SPL. The electrical stimulus delivered to the IC was a 10 μA, 0.1 ms monophasic electric pulse. Three electric pulses with a 2.0 ms time interval were delivered to the DPD through the bipolar electrodes as a 4.0 ms train, followed by the 2.0 ms tone burst after a 1.0 ms time interval. The train was delivered at a rate of 20/s for 20 min.

CM responses to tone bursts were recorded continuously before, during, and after the electric stimulation with a Racal tape recorder. The tape-recorded CMs were played back, digitized, and analyzed using the SIGNAL program (Engineering Design Inc.) running on an 80486 CPU (Intel Corp.). The sampling rate of the A/D board was 250 kHz. Fast-Fourier transforms (FFT) were computed with a resolution of 30.5 Hz. The CM response consisted of CM-on (CM during the tone-burst stimulation) and CM-aft (CM after the cessation of the tone burst) (Suga and Jen 1977). The CM-aft was sampled over a 5.0 ms period between the moment of the cessation of the tone burst and 5.0 ms after the cessation. For digital FFT analysis, the CM-aft obtained from the consecutive five CM responses to identical tone bursts were concatenated and zero-padded front and back by 10 ms.

Results and Discussion

CORTICOFUGAL MODULATIONS OF FREQUENCY TUNING OF THALAMIC AND COLLICULAR DSCF NEURONS

Thalamic and collicular DSCF neurons normally showed large variations in the shape of frequency-tuning curves (fig. 31.1, A–D). Focal cortical inactivation with lidocaine did not shift the frequency-tuning curves of "matched" subcortical neurons, but it shifted those of "unmatched" subcortical neurons, together with their BFs, along the frequency axis. (Here, matched and

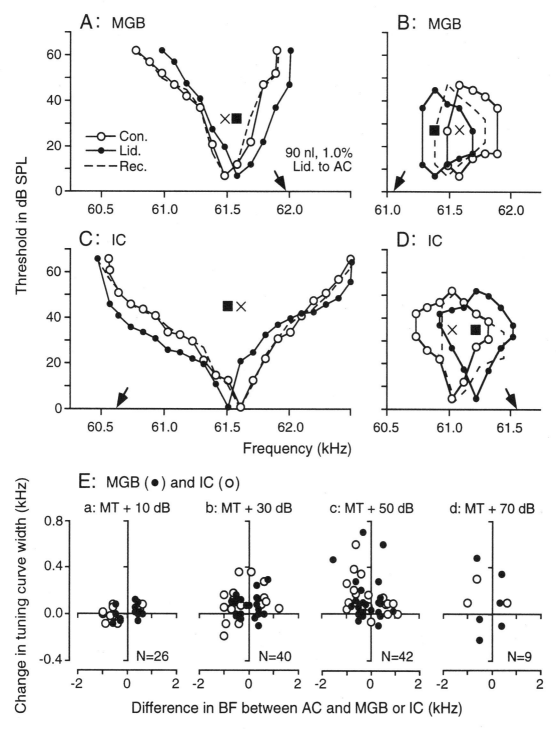

Fig. 31.1. Changes in the frequency-tuning curves of thalamic (*A* and *B*) and collicular neurons (*C* and *D*) evoked by a focal inactivation of ipsilateral cortical neurons. The best frequencies of the inactivated cortical neurons are indicated by the arrows. The curves were measured before (control; open circles), during (filled circles), and after (recovery; dashed lines) the cortical inactivation. The tuning curves shift toward the best frequencies of the inactivated cortical neurons. *X*s and squares indicate the best amplitudes measured before and during the cortical inactivation, respectively. *E* shows the amounts of changes in the widths of frequency-tuning curves at 10 (*a*), 30 (*b*), 50 (*c*), and 70 dB (*d*) above minimum threshold (MT) evoked by a focal cortical inactivation. The abscissae represent the differences in BF between cortical (AC) and thalamic (MGB, filled circles) or collicular (IC, open circles) neurons in the control condition (Zhang, Suga, and Yan 1997).

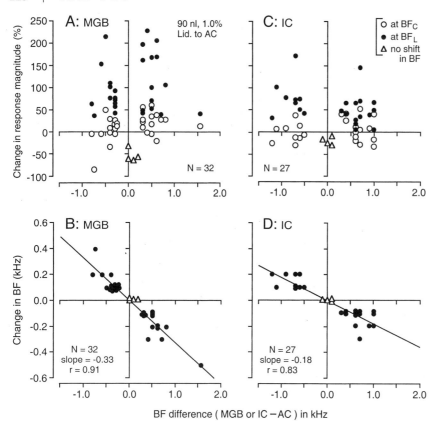

Fig. 31.2. Changes in the response magnitude (*A* and *C*) and the best frequency (*B* and *D*) of thalamic (*A* and *B*) and collicular neurons (*C* and *D*) evoked by a focal inactivation of ipsilateral cortical neurons. The abscissae represent the differences in BF between cortical (AC) and thalamic (MGB) or collicular (IC) neurons in the control condition. The triangles and circles represent the data obtained from matched and unmatched subcortical neurons, respectively. The filled circles in *A* and *C* represent the amount of increase in response magnitude at the BFs shifted by cortical inactivation (BFL), whereas the open circles represent changes in response magnitude at the BFs in the control condition (BFc). The data were based on the frequency-response curves of single neurons measured at the best amplitude of a given neuron, usually approximately 30 dB above the minimum threshold of a neuron. In *B* and *D*, the regression line, its slope, and the correlation coefficient (*r*) are shown. Triangles are "matched" neurons; circles are "unmatched." Lid.: lidocaine (Zhang, Suga, and Yan 1997).

unmatched respectively mean that the BF of a subcortical neuron was the same as or different from the BF of cortical neurons paired for the experiments.) The inactivation evoked little change in both minimum threshold and best amplitude to excite them. The shifted curves returned to the curves in the control condition in 1.5–3.2 hours. Fig. 31.1 shows this shift and recovery in the frequency-tuning curves of two thalamic (A and B) and two collicular (C and D) neurons that had BFs lower (A and D) or higher (B and C) than those of inactivated cortical neurons. The shift was accompanied by a small amount of broadening in tuning curve in 25% of the subcortical neurons studied (e.g., at 40–60 dB SPL in fig. 31.1C). The amount of broadening varied with stimulus amplitude: the higher the stimulus level, the larger the amount of broadening (fig. 31.1E).

After focal cortical inactivation, BF did not shift for matched subcortical neurons (fig. 31.2, B and D, open triangles), but shifted for unmatched subcortical neurons: the larger the difference in BF, the larger the shift (fig. 31.2, B and D, filled circles). The rate of BF shift per BF difference was −0.33 for the thalamic neurons and −0.18 for the collicular neurons. Therefore, the magnitude of BF shift was 1.8 times larger for the thalamic neurons than the collicular neurons (*p* < 0.05).

Changes in frequency tuning evoked by focal cortical inactivation were always associated with changes in response magnitude: a decrease in response of matched neurons and an increase in response of unmatched neurons. These changes in response magnitude were larger for thalamic neurons than for collicular neurons. The decrease in the response of matched neurons was 2.6 times larger for the thalamic neurons (54.2% ± 14.5, mean ± SD; *n* = 4) than for the collicular neurons (20.8% ± 8.90, *n* = 4) (fig. 31.2, A and C, open triangles, *p* < 0.05). The increase in the response of unmatched neurons measured at the BFs shifted by the focal cortical inactivation was 1.8 times larger for the thalamic neurons (104% ± 59.5, *n* = 28) than for the collicular neurons (59.0% ± 39.5, *n* = 23) (fig. 31.2, A and C, filled circles; *p* < 0.01). These data indicated that the decrease or increase in response and the shift in frequency tuning of the subcortical neurons take place in the colliculus through the corticocollicular projection, and that additional changes take place in the thalamus through the corticothalamic projection.

The decreases in the responses of matched thalamic and collicular neurons were also evoked by inactivation of the entire DSCF area with muscimol: 59.8% ± 15.8 (*n* = 5) for thalamic DSCF neurons and 33.5% ± 7.8 (*n* = 6) for collicular DSCF neurons. These data indicated that, in the normal condition, the corticofugal system amplifies collicular single-tone responses by 1.5 times and thalamic single-tone responses by 2.5 times on the average. Inactivation of cortical areas outside the DSCF area had no effect on the auditory responses and

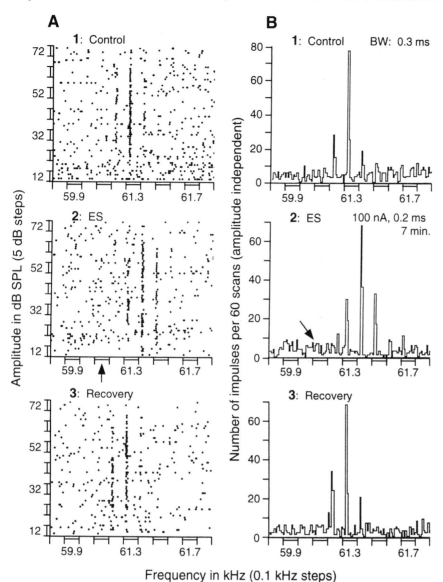

Fig. 31.3. Shift in a frequency-tuning curve of a thalamic DSCF neuron, evoked by focal electrical stimulation (ES) of the auditory cortex. *A:* Responses to frequency-amplitude (F-A) scans were recorded (*1*) prior to, (*2*) immediately after, and (*3*) 3 hours after the ES of cortical neurons tuned to 61.03 kHz and 62 dB SPL. Each dot indicates an action potential. In the F-A scan, the frequency of a tone burst was varied in 0.1 kHz steps. Every five identical frequency scans, the amplitude of the tone bursts was varied in 5.0 dB steps. The best frequency (BF) of the thalamic (MGB) neuron shifted from 61.32 to approximately 61.40 kHz. *B:* "Amplitude-independent" frequency-response curves, corresponding to those in *A*. ES was a 100 nA, 0.2 ms electric pulse delivered at a rate of 5/s for 7 min. The arrows indicate the BF of the electrically stimulated cortical neurons (Zhang and Suga 2000).

the frequency-tuning curves of subcortical DSCF neurons (Zhang and Suga 1997).

Focal activation with electric stimulation of the DSCF area in the cortex produced effects on subcortical DSCF neurons opposite to those produced by focal cortical inactivation with lidocaine (Zhang and Suga 2000). Fig. 31.3 shows the responses of a single thalamic neuron to the frequency-amplitude scan recorded before, immediately after, and 3 hours after a focal cortical electrical stimulation. This neuron was sharply tuned to a 61.3 kHz tone burst in the control condition 3 hours after the electrical stimulation (fig. 31.3A-1). The response to the 61.3 kHz tone burst was strong over a wide range of amplitudes (22–62 dB SPL). However, the neuron showed no response when the frequency of a tone burst was increased or decreased from 61.3 kHz by more than 0.1 kHz. The width of the tuning curve was less

than 0.33% of 61.3 kHz (best frequency of the neuron), regardless of stimulus amplitudes. This is one of the typical "level-tolerant" neurons recorded. When cortical DSCF neurons tuned to 61.03 kHz was electrically stimulated, the frequency-tuning curve of this thalamic neuron shifted higher by 0.1 kHz—that is, away from the BF of the activated cortical neurons (fig. 31.3A-2). The tuning curve returned to its control condition 3 hours after the electrical stimulation (fig. 31.3A-3). To substantiate the shift in tuning curve, the numbers of impulses discharged at different stimulus amplitudes at a given frequency were summed; these are shown as the frequency-response curves in fig. 31.3B. The rate of BF shift per BF difference between paired cortical and subcortical DSCF neurons was 0.30 for thalamic neurons and 0.20 for collicular neurons. (Note that the rate of BF shift evoked by cortical inactivation was −0.33 for thalamic neurons and

−0.18 for collicular neurons.) The effects of cortical electrical stimulation lasted up to 3.0 hours for both collicular and cortical neurons.

It is clear that for frequency-domain processing, cortical neurons mediate a highly focused positive feedback, incorporated with widespread lateral inhibition, via corticofugal projections, and adjust and improve auditory signal processing in the frequency domain. "Egocentric selection," first found for processing echoes in the time domain (Yan and Suga 1996a), also operates for the adjustment and improvement of processing echoes in the frequency domain.

Egocentric selection is the corticofugal function for augmentation of the auditory responses of matched subcortical neurons, suppression of the auditory responses of unmatched subcortical neurons, and shifting away the tuning curves of unmatched subcortical neurons from those of activated cortical neurons. Yan and Suga (1996a) hypothesized the following: Under natural conditions, acoustic signals received vary with time, so that all cortical neurons tuned to particular acoustic parameters are probably in a semisteady state. When an identical signal is frequently received by the animal, egocentric selection will enhance the neural representation of that signal in the IC, MGB, and AC. Egocentric selection is probably involved in the adjustment to long-term changes in the overall functional organization of the IC, MGB, and AC (Gao and Suga 1998, 2000).

In the mustached bat, frequency-tuning curves of peripheral neurons tuned to approximately 61 kHz are roughly symmetrical and very sharp (Suga and Jen 1977), so that the effect of egocentric selection on collicular and thalamic DSCF neurons is also symmetrical (fig. 31.2; Zhang, Suga, and Yan 1997). In the little brown bat and, presumably, the big brown bat, frequency-tuning curves of peripheral neurons are asymmetrical and much wider than those of 61 kHz-tuned neurons of the mustached bat, so that any single tone can excite many neurons tuned to different frequencies. Accordingly, it was expected that for the big brown bat, egocentric selection in the frequency domain is less discrete than that at around 61 kHz-tuned neurons of the mustached bat, and evokes asymmetrical adjustment in BF for overrepresentation of the BFs of activated cortical neurons. The effect of egocentric selection is thus expected to be different between different species of animals and, perhaps, between different portions of a frequency map of a given species, reflecting the shape and sharpness of frequency-tuning curves. As a matter of fact, the experiments with the big brown bat by Yan and Suga (1998) indicated that focal electrical stimulation of the AC lowers the BFs of collicular neurons toward the BF of stimulated cortical neurons. This results in the overrepresentation of the frequency at and near the BF of the stimulated cortical neurons in the colliculus. Moreover, in the big brown bat, changes similar to those evoked by cortical electric

stimulation can be induced by repetitive acoustic stimuli delivered at 50 dB SPL over 30 min (Yan and Suga 1998; Gao and Suga 1998).

We, therefore, hypothesized that egocentric selection adjusts the frequency map of the IC according to auditory experience and that this adjustment becomes larger when auditory experience is behaviorally relevant. To test this hypothesis, Gao and Suga (1998, 2000) delivered to the bat a 1.0 s train of acoustic stimuli paired with electric leg-stimulation every 30 s over 30 min, because such paired stimuli allowed the animal to learn that the acoustic stimulus was behaviorally important and to make behavioral and neural adjustments based on the acquired importance of the acoustic stimulus (Weinberger 1995, 1998). They found that the 1.0 s train of acoustic stimuli alone evoked only a subthreshold change in the frequency map of the IC; that this change in the IC became greater when the acoustic stimulation was made behaviorally relevant by pairing it with electrical stimulation; that the collicular change was mediated by the corticofugal system; and that the IC itself could sustain the change evoked by the corticofugal system for 3.0 hours.

Extensive research on the corticofugal modulation of collicular neurons in the big brown bat has been performed by P. Jen and colleagues (chapter 28, this volume).

COLLICULOFUGAL MODULATION OF FREQUENCY TUNING OF COCHLEAR HAIR CELLS

The corticofugal system is expected to modulate auditory responses of subcortical neurons and cochlear hair cells. Therefore, preliminary experiments were performed to examine the effects of electrical stimulation (ES) of the IC on hair cells. In one mustached bat, a CM response to a 62.38 kHz tone burst showed the most prominent CM-aft, which lasted 5.3 ms (fig. 31.4A-1), so that the tone burst was set at this frequency and a moderate amplitude (74 dB SPL). Then, the contralateral IC neurons tuned to 62.32 kHz was electrically stimulated. Within one minute, the waveforms of the CM-on and the CM-aft started to change in shape and amplitude. The CM-aft became longer in duration (12 ms at 15 min after ES and 16 ms at 130 min after ES) and showed beats. The duration of each beat was 3.37 ms on the average (fig. 31.4A-2 and A-3), so that the beat frequency was 297 Hz. Since the longer the damped oscillation, the sharper the frequency tuning, the lengthening of the CM-aft indicates that ES evoked sharpening in frequency tuning. Beats indicate that two resonators showed damped oscillations after the tone-burst stimulus. ES alone did not evoke any CM.

Fig. 31.4B shows the amplitude spectra of the CM-afts obtained in the control, stimulation, and recovery conditions. The frequency of the CM-aft at the peak (FP) was $62,515 \pm 15$ Hz in the control condition. It

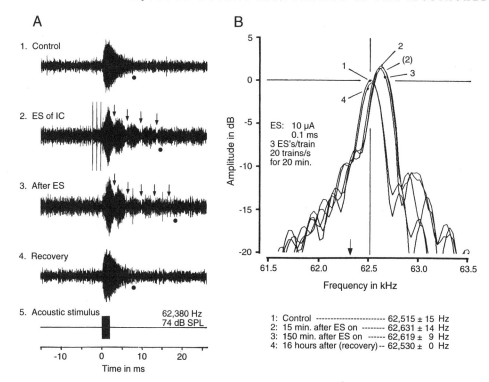

Fig. 31.4. Sharpening and shifting of frequency tuning of cochlear hair cells. Cochlear microphonic responses (CMs) recorded from the perilymphatic sac of the cochlea of the mustached bat before, during, and after the electrical stimulation (ES) of the contralateral collicular neurons tuned to approximately 62.32 kHz (arrow in *B*). *A:* Oscillograms of CMs. *B:* Amplitude spectra of the after-potentials of CMs (CM-aft) shown in *A;* 1–4 were respectively recorded prior to ES (control condition), during ES (15 min after the beginning of ES), 150 min after the beginning of ES, and 16 hours after the ES (recovery condition). In *A,* each dot and arrow indicates the end of CM-aft and a node of beat, respectively. The lengthening of CM-aft indicates sharpening of frequency tuning of hair cells. In *B,* the peaks of the amplitude spectra indicate the resonance frequencies (FP), listed at the bottom. The FP shift from 62,515 Hz to 62,631 Hz probably indicates an increase in resonance frequency of hair cells. The acoustic stimulus was a 2.0 ms, 62,380 Hz tone pulse at 74 dB SPL. The parameters of the ES are listed in *B*.

shifted by 116 Hz—that is, to 62,631 ± 14 Hz—10 min after the onset of the ES. It was the same when remeasured 15 min after the ES (62,631 ± 14 Hz). When the FP was remeasured, for example, 150 min after the beginning of the ES, it was 62,619 ± 9 Hz, showing some sign of recovery (fig. 31.4B-3). It stayed nearly the same in the shifted value for several hours. The FP measured 16 hours after the ES (i.e., the following day) was 62,530 ± 0.0 Hz (fig. 31.4B-4). The FP almost returned to that in the control condition. In fig. 31.4B, there are many sidebands, which are separated by approximately 200 Hz. These are related to analyzing CM-aft sampled by a 5.0 ms period. When the IC was electrically stimu-

lated on the second day, the CM changed in the same way that it had on the first day. The FP shifted from 62,530 to 62,628 ± 13 Hz. These intriguing preliminary data indicate that the colliculofugal system modulates the frequency tuning of hair cells.

Acknowledgments

This work has been supported by a research grant from the National Institute on Deafness and Other Communicative Disorders (DC00175). The protocol for this research was approved by the animal studies committee of Washington University.

{ 32 }

Relationship between Frequency and Delay Tuning in the FM Bat *Myotis lucifugus*

Donald Wong and *Kejian Chen*

Introduction

The auditory cortex of echolocating bats exhibits a functional organization that is species-specific for processing of their biosonar pulses. In the FM bat *Myotis lucifugus* the auditory cortex largely comprises a single tonotopic region, where most of the delay-sensitive neurons are found (Paschal and Wong 1994b). Although the cortical neurons are organized topographically according to their best frequency (BF), these neurons do not show any organization according to best delay (BD). Rather than being delay tuned to a specific BD, individual neurons exhibit BDs that depend on the pulse emission rate (Wong, Maekawa, and Tanaka 1992) and duration (Tanaka, Wong, and Taniguchi 1992). The lack of a BD organization is also found in the FM bat *Eptesicus fuscus* (Dear et al. 1993). Thus, the tonotopic population dynamically encodes target distance over the echolocation cycle.

Spectral and temporal cues conveyed in the target echoes underlie the perception of multiple target cues (Simmons, Moss and, Ferragamo 1990). As a step toward clarifying possible cortical mechanisms underlying target perception, the relationship between frequency- and temporal-tuning was examined in individual neurons in the tonotopic area of *Myotis*. Classical studies of neurons in the auditory cortex of this FM bat have described the frequency-tuning characteristics and response pattern evoked by FM stimuli sweeping through their excitatory and inhibitory areas (Suga 1965). Yet, subsequent studies that focused on delay tuning used FM stimuli without regard to the extent in which the excitatory and/or inhibitory areas are swept by the stimulus frequencies (e.g., Sullivan 1982). Selection of FM stimuli often was determined by the sweep frequencies of the stimulus pair evoking the maximal facilitative response. Thus, this approach possibly could introduce variability in measuring the delay-tuning characteristics (e.g., BD, range of pulse repetition rates [PRRs] evoking facilitative responses). In the present study, the frequency tuning of individual neurons initially was characterized. The identification of the excitatory area in the frequency-tuning curve then served as the basis for selecting the sweep frequencies of the FM stimuli. The BD was characterized at PRRs naturally employed by *Myotis* at different echolocation phases (Sales and Pye 1974). With such an approach, the delay-tuning responses can be studied under dynamic stimulus conditions, where the spectral stimuli are at least known to be restricted to a neuron's excitatory areas.

Materials and Methods

ELECTROPHYSIOLOGICAL RECORDINGS AND ACOUSTIC STIMULATION

Eight little brown bats, *M. lucifugus,* were used in this study. Extracellular recordings of single units were obtained in the tonotopic auditory cortex from the *awake* bat (for details of experimental methods, see Wong, Maekawa, and Tanaka 1992).

A condenser loudspeaker delivered sounds 75 cm directly in front of the bat's head. FM stimuli were generated electronically and shaped to sweep linearly downward through 15 kHz in 1 ms, with rise-fall times of 0.1 ms. These FM stimuli were presented either singly or in pairs to simulate the loud biosonar pulses and soft echo used by *M. lucifugus* (Griffin 1958). An FM stimulus with a 15 kHz bandwidth was chosen to simulate the terminal phase of echolocation (see spectrogram in Sales and Pye 1974). Delay tuning could then be compared over a range of PRRs corresponding to emission rates from the search to terminal phases. Selecting this FM bandwidth was also based on the neurophysiological findings that only a small part of the FM stimulus was essential for evoking the delay-sensitive responses at low PRRs corresponding to the search phase (Paschal and Wong 1994b). Compared to the measured Q values, this stimulus bandwidth permitted the acoustic energy to be restricted in the excitatory area of the neuron's frequency-threshold curve (FTC).

Pulse-echo pairs were presented in trains with the echo delay (time interval between pulse and echo onsets) of each successive sound pair decreased by 1 ms. An unpaired pulse and an unpaired echo also were presented at the end of each trial to permit direct comparison between sound-pair and single-sound responses. Stimulus amplitudes were expressed in dB (re 20 μPa).

222

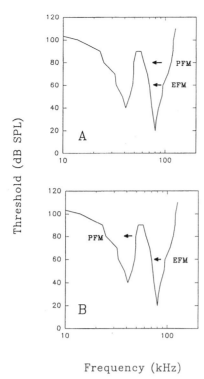

Threshold (dB SPL)

Frequency (kHz)

Fig. 32.1. Selection of spectral stimulus combinations with reference to the frequency-threshold curve of a dual-peaked neuron. Two spectral combinations were used to evoke facilitative responses: spectrally identical (*A*) or spectrally distinct (*B*). The FM stimuli of both the pulse (PFM) and echo (EFM) swept 15 kHz downward in 1 ms. Arrows denote direction of sweep.

DATA ACQUISITION

Excitatory FTCs were measured initially in all neurons using pure-tone stimuli. A stimulation rate of 1/s typically elicited maximal single-sound responses and facilitated accurate threshold measurements. Threshold response was determined audiovisually using a criterion of 0.1 spike/stimulus. Cortical neurons fire phasically with little to no spontaneous activity.

The inhibitory area was measured in some neurons using two pure-tone stimuli (each 1 ms in duration) presented without time delay. A *fixed* tone was typically presented in the excitatory FTC at the neuron's BF and 10 dB above BF threshold. The *test* tone was typically presented at stimulus amplitudes that changed in steps of 1–10 dB over a broad frequency range both outside and within the excitatory FTC.

The excitatory FTC of a unit served as a reference for selecting the stimulus parameters used in examining delay-tuning characteristics (fig. 32.1). The FM sweep of the pulse and echo stimuli had starting frequencies typically 5 kHz above the unit's BF. To test for delay tuning in *all* neurons, the pulse and echo stimuli were identical in sweep frequencies (termed spectrally identical) and traversed the unit's BF. In dual-peaked neurons, the FM

stimuli swept through BF_2 (excitatory peak at the *higher* best frequency and typically the unit's lowest threshold). Dual-peaked neurons were examined further for delay tuning with pulse and echo stimuli that differed in sweep frequencies (termed spectrally distinct): the echo FM stimulus swept through BF_2, and the pulse FM stimulus swept through BF_1 (excitatory peak at the *lower* best frequency). When single- and dual-peaked neurons were examined with spectrally identical combinations, the pulse and echo amplitudes were measured and typically fell within 40–50 dB and 20–30 dB above BF threshold, respectively. In dual-peaked neurons examined with spectrally distinct combinations, the pulse amplitude was measured and typically fell within 10–30 dB above BF_1 threshold, and the measured echo amplitude was 20 dB lower than the pulse amplitude.

A computer stored spike-arrival times (binwidth of 0.5 ms) and represented sound-evoked neural responses as peristimulus-time (PST) histograms from 50 trials at different PRRs. Delay-dependent facilitation was examined with sound-pair stimulation at PRRs as high as 300/s. A response was considered facilitative when pulse-echo stimuli at a particular echo delay evoked a response greater than the sum of responses evoked by the pulse alone and echo alone (O'Neill and Suga 1982). The BD is defined as the echo delay of pulse-echo stimuli that evoke the maximal facilitative response (Suga and Horikawa 1986). Statistical analysis of data was performed using a student's *t*-test.

Results

FREQUENCY TUNING IN DELAY-
SENSITIVE AUDITORY CORTEX

Neurons in the tonotopic area exhibit either single- or dual-peaked FTCs. Single-peaked frequency tuning was found in 46% (*n* = 39) of the 85 neurons (fig. 32.2A–B). The FTC was typically V-shaped. A few neurons also exhibited a second peak, but these neurons were still classified as single-peaked if the minor peak in the FTC was a shallow dip <10 dB.

Dual-peaked frequency tuning was characterized in 54% (*n* = 46) of the neurons. The FTC is characterized by a major peak at the lowest threshold, and a second peak > 10 dB (fig. 32.2C–D). In 83% (*n* = 38) of these neurons, there was a *single* FTC in which the two excitatory domains were adjoining at the higher stimulus amplitudes (fig. 32.2C). In the remaining 17% (*n* = 8) of these neurons, there were two, unconnected FTCs— that is, both excitatory domains were completely separated at all excitatory amplitudes (fig. 32.2D). The higher-tuned frequency (BF_2) typically had a lower threshold than the lower-tuned frequency (BF_1).

The BF_2/BF_1 ratio averaged 2.03 ± 0.26 (mean ± SD; *n* = 46). In most neurons (43/46), BF_2 had the lower threshold of the two peaks, and this BF_2 threshold was

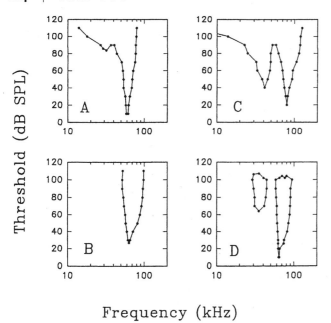

Fig. 32.2. Excitatory frequency-tuning curves of representative single-peaked (*A, B*) and dual-peaked (*C, D*) neurons

between 10 and 50 dB SPL. BF_1 thresholds ranged from 40 to 80 dB SPL.

In single-peaked neurons (*n* = 39), the Q_{10dB} and Q_{30dB} were 6.75 ± 4.44 and 2.78 ± 1.76, respectively. In dual-peaked neurons (*n* = 43), the Q_{10dB} and Q_{30dB} for the major peak were 10.55 ± 6.91 and 3.50 ± 1.39 respectively, and the Q_{10dB} for the minor peak was 3.62 ± 1.83. The Q values were not measured in three dual-peaked neurons with very high BF_2 thresholds.

INHIBITORY SIDEBAND

Inhibitory areas also were mapped with respect to the excitatory FTCs, largely to provide further evidence that the excitatory peaks of dual-peaked neurons were in fact separated by a nonexcitatory area, whether inhibitory or nonresponsive. In single-peaked neurons with relatively sharp tuning ($Q_{10dB} > 10$), an inhibitory sideband typically flanked the excitatory FTC on the lower-frequency side (6/6) (fig. 32.3A). Two single-peaked neurons with sharp frequency tuning had an inhibitory sideband flanking the FTC on both low- and high-frequency sides (fig. 32.3B). No inhibitory sideband was measured in broadly tuned, single-peaked neurons (0/4).

In both types of dual-peaked neurons, an inhibitory area was interposed between the two excitatory areas (6/6) (fig. 32.3C–D). In some dual-peaked neurons, an inhibitory area was also measured at the low-frequency side of the BF_1 peak (3/6) (fig. 32.3D). No inhibitory band was measured at the high-frequency side of the BF_2 peak (0/6).

DELAY-TUNING PROPERTIES OF NEURONS WITH DIFFERENT FTC TYPES

Both types of dual-peaked neurons consistently exhibited delay-tuning properties within a range of PRRs that depended on whether spectrally identical or distinct stimulus combinations were used. Most neurons showed facilitative responses to spectrally identical combinations at PRRs ranging as wide as from 10/s to 220/s. In contrast, these neurons showed a narrower range of PRRs evoking facilitation (10/s to 120/s) for spectrally distinct combinations. When the facilitative PRRs in dual-peaked neurons (*n* = 12) was compared at the two spectral stimulus combinations, the *maximal* facilitative PRR was significantly larger (*p* < .0005) to stimulation with spectrally identically (≥200/s) than with spectrally distinct (≤120/s) combinations.

Single-peaked neurons with relatively sharp FTCs also exhibited delay tuning within a range of PRRs at which spectrally identical combinations were used. In the few neurons with relatively broad single-peaked FTCs, the facilitative responses were not clear at most PRRs used to examine delay tuning. This may not be due simply to the FM bandwidth employed, since larger FM bandwidths were used further in an attempt to elicit delay-sensitive responses.

The best echo delay of a neuron progressively decreased as the PRR was increased—that is, from a maximal BD at a low PRR (e.g., 10/s) to a BD near 0 ms at a high PRR (e.g., 220/s). Such a pattern of BD shift pro-

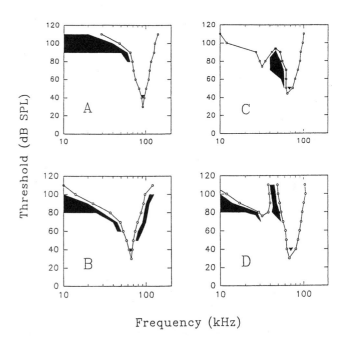

Fig. 32.3. Excitatory (○) and inhibitory areas (*darkened area*) of dual- and single-peaked neurons. The inhibitory area was measured using two, 1 ms tones presented simultaneously. ▼: fixed tone.

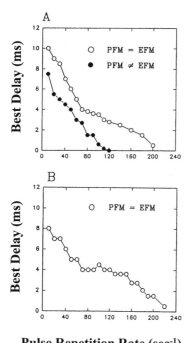

Pulse Repetition Rate (sec⁻¹)

Fig. 32.4. Patterns of BD shift with increase in PRR for a dual-peaked (*A*) and single-peaked (*B*) neuron. *A:* Tracking pattern of delay tuning with either spectrally identical (○: pulse FM = echo FM = 67 → 52 kHz) or spectrally distinct stimulus combinations (●: pulse FM = 37 → 22 kHz; echo FM = 67 → 52 kHz). The maximal facilitative PRR and maximal BD were both greater for the spectrally identical than for the spectrally distinct stimulus combinations. *B:* Tracking pattern of a single-peaked neuron with identical FM stimulus combinations (pulse FM = echo FM = 70 → 55 kHz).

duced with increase in PRR has previously been described as a "tracking" type of delay-tuning property (O'Neill and Suga 1982). In dual-peaked neurons, the maximal BD was significantly larger when stimulated with spectrally identical (≥10 ms) than with spectrally distinct (≤8 ms) combinations ($p < .005$, $n = 12$).

Fig. 32.4 depicts the pattern of BD shift across a range of PRRs in a dual-peaked (*A*) and single-peaked (*B*) neuron stimulated with specific spectral combinations. In dual-peaked neurons, spectrally distinct combinations produced a BD shift over a narrower range of PRRs than that produced by spectrally identical combinations (fig. 32.4A). The maximal BDs obtained with spectrally distinct combinations were 8 to 2 ms. Of the neurons with relatively small maximal BD, the BD shift would necessarily be minimal (<2 ms) with changes in PRR, a behavior resembling the "delay-tuned" type of delay sensitivity commonly found in CF-FM bats (O'Neill and Suga 1982). In single-peaked neurons (fig. 32.4B), the pattern of BD shift is similar to that found in dual-peaked neurons.

Discussion

Multipeaked, frequency-tuned neurons have provided the neural basis for combination-sensitive responses of cortical neurons in CF-FM bats. The present study examined delay-tuning characteristics (BDs, facilitative PRRs) in neurons with FM stimulus frequencies that sweep excitatory regions identified from their tonal FTCs. The frequency-tuning characteristics in the *Myotis* tonotopic auditory cortex are consistent with and extend those previously described (Suga 1965; Paschal and Wong 1994b; Condon et al. 1997). About half of the neurons in the *Myotis* tonotopic auditory cortex exhibit harmonically related (2:1) multipeaked, frequency-tuned neurons. Delay-sensitive responses are evoked by activating one (higher-frequency peak) or both excitatory domains with FM stimuli. About half of the neurons exhibit single-peak frequency tuning, and these neurons are also delay-sensitive. Thus, delay-sensitivity is not unique to multipeaked frequency-tuned neurons in FM as in CF-FM bats.

The present study of this species revealed that the delay-tuning characteristics could vary depending on the selection of the stimulus FM frequencies. This is particularly evident in the BD measurement of over half of the neurons stimulated at low PRRs (10–20/s): the BD determined in dual-peaked neurons by spectrally distinct versus spectrally identical stimulus combinations could differ by as much as 8 ms. Yet, the PRRs were typically fixed at relatively low values in the earlier *Myotis* studies of delay tuning (e.g., Wong and Shannon 1988). Furthermore, FM stimuli with relatively large FM bandwidths (30–60 kHz) often were used to mimic the natural biosonar signals emitted by *Myotis* in the search phase. Whether the starting sweep frequencies used for searching delay-sensitive responses were either fixed (e.g., 100 kHz) or varied (e.g., Maekawa, Wong, and Paschal 1992), it was likely that the FM segment that effectively traversed through the excitatory region would vary depending on the neuron's BF and frequency-tuning characteristics. A search stimulus *fixed* at a high starting frequency would influence the number of delay-sensitive neurons actually detected or optimally driven. This is evident from the present findings, since the terminal frequency components of such an FM sweep might not even traverse the excitatory region of low-BF neurons. Despite such methodological-related variability, a consistent pattern of BD shift with increasing PRR is found for virtually all neurons.

The neural mechanism underlying delay tuning is currently not well understood. A few studies have demonstrated that the initial pulse stimulus induces an early excitation followed by a later inhibition, which varies in duration (Berkowitz and Suga 1989; Casseday, Ehrlich, and Covey 1994). A facilitative response occurs when an echo stimulus arrives as the neuron recovers from this inhibitory period. The current study agrees with the fact that excitatory stimulus frequencies trigger inhibition. Furthermore, spectrally identical or distinct stimulus combinations could evoke facilitative responses in dual-peaked neurons. Yet, facilitation to spectrally identical stimulus combinations rarely resulted when reversing the presentation of the soft (echo) and loud (pulse) stimuli. Thus, the high amplitude required in the initial stimulus (i.e., pulse) of the stimulus pair lends further support to the critical role of the amplitude (rather than spectral) dimension in delay tuning as originally reported in this species (Sullivan 1982b).

A unique feature of this FM bat is the large majority of neurons that exhibit the capability to track target distance over the operational range of echolocation. Such a neural property also provides a basis for encoding other target features that require integrating information conveyed over target-echo sequences. These neurons could be further exploited in target perception by the nature in which the spectral content (harmonic or identical combination) is represented using a place code in the tonotopic region. The spatiotemporal firing pattern of these cortical neurons could thus provide a population code for representing target features.

Acknowledgments

This research was supported by NSF grant BCS-9307650.

Part Two / Literature Cited

ALLEN, J. B., and P. F. FAHEY. 1993. A second cochlear frequency map that correlates distortion product and neural tuning measurements. *Journal of the Acoustical Society of America* 94:809–816.

ALTRINGHAM, J. D. 1999. *Bats: Biology and behaviour.* Oxford: Oxford University Press.

AMATO, G., V. LA GRUTTA, and F. ENIA. 1969. The control exerted by the auditory cortex on the activity of the medial geniculate body and inferior colliculus. *Archivio di Scienze Biologiche* 53:291–313.

ANDERSEN, P., K. JUNGE, and O. SVEEN. 1972. Corticofugal facilitation of thalamic transmission. *Brain Behavior Evolution* 6:170–184.

ANDERSEN, R. A., P. L. KNIGHT, and M. M. MERZENICH. 1980. The thalamocortical and corticothalamic connections of AI, AII, and the anterior auditory field (AAF) in the cat: Evidence for two largely segregated systems of connections. *Journal of Comparative Neurology* 194:663–701.

ANDERSEN, R. A., R. L. SNYDER, and M. M. MERZENICH. 1980. The topographic organization of corticocollicular projections from physiologically defined loci in the AI, AII and anterior auditory cortical fields of the cat. *Journal of Comparative Neurology* 191:479–494.

ASANUMA, A., D. WONG, and N. SUGA. 1983. Frequency and amplitude representations in anterior primary auditory cortex of the mustached bat. *Journal of Neurophysiology* 50:1182–1196.

ASHMORE, J. F. 1987. A fast motile response in guinea pig outer hair cells: The cellular basis of the cochlear amplifier. *Journal of Physiology* 288:323–347.

AU, W. W. L. 1990. Target detection in noise by echolocating dolphins. Pp. 203–216 in *Sensory abilities of cetaceans: Laboratory and field evidence,* ed. J. Thomas and R. Kastlelein. New York: Plenum Press.

———. 1993. *The sonar of dolphins.* New York: Springer-Verlag.

AU, W. W. L., and P. W. B. MOORE. 1984. Receiving beam patterns and directivity indices of the Atlantic bottlenose dolphin *Tursiops truncatus. Journal of the Acoustical Society of America* 75:255–262.

———. 1990. Critical ratio and critical band width for the Atlantic bottlenose dolphin. *Journal of the Acoustical Society of America* 88:1635–1638.

AU, W. W. L., P. W. B. MOORE, and D. A. PAWLOSKI. 1988. Detection of complex echoes in noise by an echolocating dolphin. *Journal of the Acoustical Society of America* 83:662–668.

BAJO, V. M., E. M. ROUILLER, E. WELKER, S. CLARKE, A. E. VILLA, Y. DE RIBAUPIERRE, and F. DE RIBAUPIERRE. 1995. Morphology and spatial distribution of corticothalamic terminals originating from the cat auditory cortex. *Hearing Research* 83:161–174.

BALCOMBE, J. P., and M. B. FENTON. 1986. The communication role of echolocation calls in vespertilionid bats.

Pp. 625–629 in *Animal sonar: Processes and performance,* ed. P. E. Nachtigall and P. W. B. Moore.

BATRA, R., S. KUWADA, and T. R. STANFORD. 1993. High-frequency neurons in the inferior colliculus that are sensitive to interaural delays of amplitude-modulated tones: Evidence for dual binaural interactions. *Journal of Neurophysiology* 70:64–80.

BÉKÉSY, G. VON. 1960. *Experiments in hearing.* New York: McGraw-Hill.

BEL'KOVICH, V. M., and SOLNTSEVA, G. N. 1970. Anatomy and function of the ear in dolphins. *Zoologicheskiy Zhurnal* 2:275–282.

BELL, G. P. 1982. Behavioral and ecological aspects of gleaning by a desert insectivorous bat, *Antrozous pallidus* (Chiroptera: Vespertilionidae). *Behavioral Ecology and Sociobiolology* 10:217–223.

BELL, G. P., and M. B. FENTON. 1986. Visual acuity, sensitivity, and binocularity in a gleaning insectivorous bat, *Macrotus californicus* (Chiroptera: Phyllostomidae). *Animal Behavior* 34:409–414.

BENSON, D. A., and D. C. TEAS. 1976. Single unit study of binaural interaction in the auditory cortex of the chinchilla. *Brain Research* 103:313–338.

BERKOWITZ, A., and N. SUGA. 1989. Neural mechanisms of ranging are different in two species of bats. *Hearing Research* 41:255–264.

BERTRAND, B., and J.-R. CAZALETS. 1998. Postinhibitory rebound during locomotor-like activity in neonatal rat motoneurons in vitro. *Journal of Neurophysiology* 79:342–351.

BISHOP, A., and M. M. HENSON. 1988. The efferent auditory system in Dopplershift-compensating bats. Pp. 307–311 in *Animal sonar systems,* ed. R. G. Busnel and J. F. Fish. New York: Plenum Press.

BISHOP, A. L., and O. W. HENSON JR. 1987. The efferent cochlear projections of the superior olivary complex in the mustached bat. *Hearing Research* 31:175–182.

BLÜMCKE, I., P. R. HOF, J. H. MORRISON, and M. R. CELIO. 1990. The distribution of parvalbumin in the visual cortex of Old World monkeys and humans. *Journal of Comparative Neurology* 301:417–432.

BLÜMCKE, I., P. R. HOF, J. H. MORRISON, and M. R. CELIO. 1991. Parvalbumin in the monkey striate cortex: A quantitative immunoelectron-microscopy study. *Brain Research* 554:237–243.

BRILL, R. L., and HARDER, P. J. 1991. The effects of attenuating returning echolocation signals at the lower jaw of a dolphin (*Tursiops truncatus*). *Journal of the Acoustical Society of America* 89(6): 2851–2857.

BRILL, R. L., M. L. SEVENICH, T. J. SULLIVAN, J. D. SUSTMAN, and R. E. WITT. 1988. Behavioral evidence for hearing through the lower jaw by an echolocating dolphin, *Tursiops truncatus. Marine Mammal Science* 4:223–230.

BRODMANN, K. 1909. Vergleichende Lokalisationslehre der Großhirnrinde. Leipzig: Barth.

BROWN, A. M., and D. T. KEMP. 1984. Suppressibility of the 2f1-f2 stimulated acoustic emission in gerbil and man. *Hearing Research* 13:29–37.

BROWN, A. M., GASKILL, S. A., and D. M. WILLIAMS. 1992. Mechanical filtering of sound in the inner ear. *Proceedings of the Royal Society of London B* 250:29–34.

BROWNELL, W. E., C. F. BADER, D. BERTRAND, and Y. DE RIBEUAPIERRE. 1985. Evoked mechanical responses of isolated cochlear outer hair cells. *Science* 227:194–196.

BRUNS, V. 1976a. Peripheral auditory tuning for fine frequency analysis by the CF-FM bat, *Rhinolophus ferrumequinum.* I. Mechanical specializations of the cochlea. *Journal of Comparative Physiology A* 106:77–86.

———. 1976b. Peripheral auditory tuning in the Doppler shift compensating bat, *Rhinolophus ferrumequinum:* II. Frequency mapping in the cochlea. *Journal of Comparative Physiology A* 106:86–96.

———. 1980. Basilar membrane and its anchoring system in the cochlea of the greater horseshoe bat. *Anatomy and Embryology* 161:29–51.

BRUNS, V., and M. GOLDBACH. 1980. Hair cells and tectorial membrane in the cochlea of the greater horseshoe bat. *Anatomy and Embryology* 161:65–83.

BRUNS, V., and E. T. SCHMIESZEK. 1980, Cochlear innervation in the greater horseshoe bat: Demonstration of an acoustic fovea. *Hearing Research* 3:27–43.

BRUNS, V., H. BURDA, and M. J. RYAN. 1988. Ear morphology of the frog-eating bat (*Trachops cirrhosus,* family Phyllostomidae): Apparent specializations for low frequency hearing. *Journal of Morphology* 199:103–118.

BULLOCK, T. H., A. D. GRINNELL, E. IKEZONO, K. KAMEDA, Y. KATSUKI, M. NOMOTO, O. SATO, N. SUGA, and K. YANAGISAWA. 1968. Electrophysiological studies of central auditory mechanisms in cetaceans. *Zeitschrift für Vergleichende Physiologie* 59:117–156.

BURDA, H., J. FIEDLER, and V. BRUNS. 1988. The receptor and neuron distribution in the cochlea of the bat, *Taphozous kachensis. Hearing Research* 32:131–136.

BURGER, R. M., and G. D. POLLAK. 1998. Analysis of the role of inhibition in shaping responses to sinusoidally amplitude-modulated signals in the inferior colliculus. *Journal of Neurophysiology* 80:1686–1701.

BUTMAN, J. A. 1992. Synaptic mechanisms for target ranging in the mustached bat, *Pteronotus parnellii.* Ph.D. dissertation, Washington University (St. Louis, Mo.).

CAIRD, D., and R. KLINKE. 1987. Processing of interaural time and intensity differences in the cat inferior colliculus. *Experimental Brain Research* 68:379–392.

CARVELL, G. E., and D. J. SIMONS. 1987. Thalamic and corticocortical connections of the somatic sensory area of the mouse. *Journal of Comparative Neurology* 265:409–427.

CASSEDAY, J. H., and E. COVEY. 1992. Frequency tuning properties of neurons in the inferior colliculus of an FM bat. *Journal of Comparative Neurology* 319:34–50.

———. 1996. A neuroethological theory of the operation of the inferior colliculus. *Brain Behavior and Evolution* 47:311–336.

CASSEDAY, J. H., E. COVEY, and B. GROTHE. 1997. Neural selectivity and tuning for sinusoidal frequency modulations in the inferior colliculus of the big brown bat, *Eptesicus fuscus. Journal of Neurophysiology* 77:1595–1605.

CASSEDAY, J. H., D. EHLRICH, and E. COVEY. 1994. Neural tuning for sound duration: Role of inhibitory mechanisms in the inferior colliculus. *Science* 264:847–850.

———. 2000. Neural calculations of sound duration: Control by excitatory-inhibitory interactions in the inferior colliculus. *Journal of Neurophysiology* 84:1475–1487.

CASSEDAY, J. H., J. B. KOBLER, S. F. ISBEY, and E. COVEY. 1989. Central acoustic tract in an echolocating bat: An extralemniscal auditory pathway to the thalamus. *Journal of Comparative Neurology* 287:247–259.

CELIO, M. R. 1986. Parvalbumin in most gamma-aminobutyric acid-containing neurons of the rat cerebral cortex. *Science* 213:995–997.

CHEN, K., and D. WONG. 1995. Influence of frequency-tuning on the adaptive properties of cortical neurons of *Myotis lucifugus. Society for Neuroscience Abstract* 21:667.

CLAREY, J. C., P. BARONE, and T. J. IMIG. 1992. Physiology of thalamus and cortex. Pp. 232–334 in *The mammalian auditory pathway: Neurophysiology,* ed. A. N. Popper and R. R. Fay. New York: Springer-Verlag.

CLEMO, H. R., and B. E. STEIN. 1983. Organization of a fourth somatosensory area of cortex in cat. *Journal of Neurophysiology* 50:910–925.

COLES, R. B., A. GUPPY, M. E. ANDERSON, and P. SCHLEGEL. 1989. Frequency sensitivity and directional hearing in the gleaning bat, *Plecotus auritus* (Linneaus). *Journal of Comparative Physiology A* 165:269–280.

CONDON, C. J., K. R. WHITE, and A. S. FENG. 1994. Processing of amplitude-modulated signals that mimic echoes from fluttering targets in the inferior colliculus of the little brown bat, *Myotis lucifugus. Journal of Neurophysiology* 71:768–784.

CONDON, C. J., A. GALAZYUK, K. R. WHITE, and A. S. FENG. 1997. Neurons in the auditory cortex of the little brown bat exhibit selectivity for complex amplitude-modulated signals that mimic echoes from fluttering insects. *Auditory Neuroscience* 3:269–287.

COVEY, E. 1993. Response properties of single units in the dorsal nucleus of the lateral lemniscus and paralemniscal zone of an echolocating bat. *Journal of Neurophysiology* 69:842–859.

COVEY, E., and J. H. CASSEDAY. 1991. The ventral lateral lemniscus in an echolocating bat: Parallel pathways for analyzing temporal features of sound. *Journal of Neuroscience* 11:3456–3470.

———. 1995. The lower brainstem auditory pathways. Pp. 235–295 in *Hearing by bats,* ed. A. N. Popper and R. R. Fay. New York: Springer-Verlag.

———. 1999. Timing in the auditory system of the bat. *Annual Review of Physiology* 61:457–476.

COVEY, E., J. A. KAUER, and J. H. CASSEDAY. 1996. Whole-cell patch-clamp recording reveals subthreshold sound-evoked postsynaptic currents in the inferior colliculus of awake bats. *Journal of Neuroscience* 16:3009–3018.

COVEY, E., VATER, M., and J. H. CASSEDAY. 1991. Binaural properties of single units in the superior olivary complex of the mustache bat. *Journal of Neurophysiology* 66:1080–1094.

DALLAND, J. I. 1965. Hearing sensitivity in bats. *Science* 150:1185–1186.

DALLOS, P. 1992. The active cochlea. *Journal of Neuroscience* 12:4575–4585.

DALLOS, P., and B. N. EVANS. 1995. High-frequency motility of outer hair cells and the cochlear amplifier. *Science* 267:2006–2009.

DANNHOF, B. J., and V. BRUNS. 1991. The organ of Corti in the bat *Hipposideros bicolor. Hearing Research* 53:253–268.

DANNHOF, B. J., B. ROTH, and V. BRUNS. 1991. Length of hair cells as a measure of frequency representation in the mammalian cochlea? *Naturwissenschaften* 78:570–573.

DEAR, S. P., and N. SUGA. 1995. Delay-tuned neurons in the midbrain of the big brown bat. *Journal of Neurophysiology* 73:1084–1100.

DEAR, S. P., J. A. SIMMONS, and J. FRITZ. 1993. A possible neuronal basis for representation of acoustic scenes in auditory cortex of the big brown bat. *Nature* 364: 620–623.

DEAR, S. P., J. FRITZ, T. HARESIGN, M. FERRAGAMO, and J. A. SIMMONS. 1993. Tonotopic and functional organization in the auditory cortex of the big brown bat, *Eptesicus fuscus*. *Journal of Neurophysiology* 70: 1988–2009.

DE BOER, E. 1984. Auditory time constants: A paradox? Pp. 141–158 in *Time resolution in auditory system,* ed. A. Michelsen.

DEFELIPE, J., and E. G. JONES. 1985. Vertical organization of gamma-aminobutyric acid-accumulating intrinsic neuronal systems in monkey cerebral cortex. *Journal of Neuroscience* 5:3246–3260.

DE FEUDIS, F. V. 1978. Can binding of GABA, glycine and beta-alanine to synaptic receptors be determined in presence of physiological concentration of Na$^+$? *Experientia* 34:1314–1315.

DEMEULMEESTER, H., L. ARCKENS, F. VANDESANDE, C. HEIZMANN, and R. POCHET. 1991. Calcium-binding proteins and neuropeptides as molecular markers of GABAergic interneurons in the cat visual cortex. *Experimental Brain Research* 84:538–544.

DOLPHIN, W. F. 1995. Steady-state auditory-evoked potentials in three cetacean species elicited using amplitude-modulated stimuli. Pp. 25–47 in *Sensory systems of aquatic mammals,* ed. R. A. Kastelein, J. A. Thomas, and P. E. Nachtigall. Woerden, The Netherlands: De Spil Publishers.

DOLPHIN, W. F., W. W. L. AU, and P. NACHTIGALL. 1995. Modulation transfer function to low-frequency carriers in three species of cetaceans. *Journal of Comparative Physiology A* 177:235–245.

DUBROVSKIY, N. A. 1990. On the two auditory systems in dolphins. Pp. 233–254 in *Sensory abilities of cetaceans: Laboratory and field evidence,* ed. J. Thomas and R. Kastlelein. New York: Plenum Press.

ECHTELER, S. M., R. R. FAY, and A. N. POPPER. 1994. Structure of the mammalian cochlea. Pp. 134–172 in *Comparative hearing: Mammals,* ed. R. R. Fay and A. N. Popper. New York: Springer-Verlag.

EDAMATSU, H., and N. SUGA. 1993. Differences in response properties of neurons between two delay-tuned areas in the auditory cortex of the mustached bat. *Journal of Neurophysiology* 69:1700–1712.

EDAMATSU, H., M. KAWASAKI, and N. SUGA. 1989. Distribution of combination-sensitive neurons in the ventral fringe area of the auditory cortex of the mustached bat. *Journal of Neurophysiology* 61:202–207.

EDDINS, D. 1993. Amplitude modulation detection of narrow-band noise: Effect of absolute bandwidth and frequency region. *Journal of the Acoustical Society of America* 93:470–479.

EGAN, J. P., G. Z. GREENBERG, and A. I. SCHULMAN. 1961. Operating characteristics, signal detectability, and the method of free response. *Journal of the Acoustical Society of America* 33:993–1007.

EHRLICH, D., J. H. CASSEDAY, and E. COVEY. 1997. Neural tuning to sound duration in the inferior colliculus of the big brown bat, *Eptesicus fuscus*. *Journal of Neurophysiology* 77:2360–2372.

———. 2000. Auditory responses from the frontal cortex in the short-tailed fruit bat *Carollia perspicillata*. *NeuroReport* 11:421–425.

ESSER, K.-H., and A. EIERMANN. 1999. Tonotopic organization and parcellation of auditory cortex in the FM-bat *Carollia perspicillata*. *European Journal of Neuroscience* 11:3669–3682.

ESSER, K.-H., and J. S. KANWAL. 1996. Auditory responses from neurons in the frontal cortex of the mustached bat. *Association for Research in Otolaryngology Abstracts* 19:447.

ESSER, K.-H., C. J. CONDON, N. SUGA and J. S. KANWAL. 1997. Syntax processing by auditory cortical neurons in the FM-FM area of the mustached bat *Pteronotus parnellii*. *Proceedings of the National Academy of Sciences* 94:14019–14024.

FELICIANO, M., and S. J. POTASHNER. 1995. Evidence for a glutamatergic pathway from the guinea pig auditory cortex to the inferior colliculus. *Journal of Neurochemistry* 65:1348–1357.

FELICIANO, M., E. SALDANA, and E. MUGNAINI. 1995. Direct projections from the rat primary auditory cortex to nucleus sagulum, paralemmiscal region, superior olivary complex and cochlear nuclei. *Auditory Neuroscience* 1:287–308.

FENG, A. S., J. A. SIMMONS, and S. A. KICK. 1978. Echo detection and target ranging neurons in the auditory system of the bat *Eptesicus fuscus*. *Science* 202: 645–648.

FENTON, M. B. 1990. The foraging behavior and ecology of animal-eating bats. *Canadian Journal of Zoology* 68:411–422.

———. 1995. Natural history and biosonar signals. Pp. 37–86 in *Hearing by bats,* ed. A. N. Popper and R. R. Fay. New York: Springer-Verlag.

FENTON, M. B., and G. P. BELL. 1981. Recognition of species of insectivorous bats by their echolocation calls. *Journal of Mammology* 62:233–243.

FERRER, I. 1987. The basic structure of the neocortex in insectivorous bats (*Miniopterus schreibersi* and *Pipistrellus pipistrellus*): A Golgi study. *Journal für Hirnforschung* 28:237–243.

FIEDLER, J. 1979. Prey catching with and without echolocation in the Indian false vampire bat (*Megaderma lyra*). *Behavioral Ecology and Sociobiology* 6:155–160.

———. 1983. Vergleichende Cochlea-Morphologie der Fledermausarten *Molossus ater, Taphozous nudiventris kachhensis* und *Megaderma lyra*. Ph.D. dissertation, University of Frankfurt.

FINNERAN, J. J., C. E. SCHLUNDT, D. A. CARDER, J. A. CLARK, J. A. YOUNG, J. B. GASPIN, and S. H. RIDGWAY. 2000. Auditory and behavioral responses of bottlenose dolphins (*Tursiops truncatus*) and a beluga whale (*Delphinapterus leucas*) to impulsive sounds resembling distant signatures of underwater explosions. *Journal of the Acoustical Society of America* 108(1): 417–431.

FITZPATRICK, D. C., and O. W. HENSON. 1994. Cell types in the mustached bat auditory cortex. *Brain, Behavior, and Evolution* 43:79–91.

FITZPATRICK, D. C., J. S. KANWAL, J. A. BUTMAN, and N. SUGA. 1993. Combination-sensitive neurons in the primary auditory cortex of the mustached bat. *Journal of Neuroscience* 13:931–940.

FITZPATRICK, D. C., J. F. OLSEN, and N. SUGA. 1998a. Connections among functional areas in the mustached bat auditory cortex. *Journal of Comparative Neurology* 391:366–396.

———. 1998b. Distribution of response types across entire hemispheres of the mustached bat's auditory cortex. *Journal of Comparative Neurology* 16:353–365.

FITZPATRICK, D. C., R. BATRA, T. R. STANFORD, and S. KUWADA. 1997. A neuronal population code for sound localization. *Nature* 388:871–874.

FLEISCHER, G. 1978. Evolutionary principles of the mammalian middle ear. *Advances in Anatomy, Embryology, and Cell Biology* 55:1–70.

FLEMING, T. H. 1988. *The short-tailed fruit bat.* Chicago: University of Chicago Press.

FRANCIS, C. M., and J. HABERSETZER. 1998. Interspecific and intraspecific variations in echolocation call frequency and morphology of horseshoe bats, *Rhinolophus* and *Hipposideros*. Pp. 169–181 in *Bats, phylogeny, morphology, echolocation, and conservation biology*, ed. T. H. Kunz and P. A. Racey. Washington, D.C.: Smithsonian Institution Press.

FRANK, G., W. HEMMERT, and A. W. GUMMER. 1999. Limiting dynamics of high-frequency electromechanical transduction of outer hair cells. *Proceedings of the National Academy of Sciences* 96:4420–4425.

FRISINA, R. D., W. E. O'NEILL, and M. L. ZETTEL. 1989. Functional organization of mustached bat inferior colliculus: II. Connections to the FM2 region. *Journal of Comparative Neurology* 284:85–107.

FRITZ, J. B., J. OLSEN, N. SUGA, and E. G. JONES. 1981. Connectional differences between auditory fields in a CF-FM bat. *Society for Neuroscience Abstracts* 12:12.

FROMM, S. 1990. Anatomische und physiologische Charakterisierung des medialen Geniculatums der Hufeisennase, *Rhinolophus rouxi*. Diplomarbeit, Ludwig-Maximilians-University, Munich.

FUZESSERY, Z. M. 1994. Response selectivity for multiple dimensions of frequency sweeps in the pallid bat inferior colliculus. *Journal of Neurophysiology* 72:1061–1079.

———. 1996. Monaural and binaural spectral cues created by the external ears of the pallid bat. *Hearing Research* 95:1–17.

———. 1997. Acute sensitivity to interaural time differences in the inferior colliculus of a bat that relies on passive sound localization. *Hearing Research* 109: 46–62.

FUZESSERY, Z. M., and A. S. FENG. 1983. Mating call selectivity in the thalamus and midbrain of the leopard frog (*Rana p. pipens*): Single and multiunit analyses. *Journal of Comparative Physiology A* 150:333–344.

FUZESSERY, Z. M., and J. C. HALL. 1996. Role of GABA in shaping frequency tuning and creating FM sweep selectivity in the inferior colliculus. *Journal of Neurophysiology* 76:1059–1073.

FUZESSERY, Z. M., and J. J. WENSTRUP. 1995. Functional organization of the inferior colliculus of a gleaning bat: Rewiring for parallel processing of active and passive hearing. *Fourth International Congress for Neuroethology.*

FUZESSERY, Z. M., R. R. GUMTOW, and R. LANE. 1991. A microcomputer-controlled system for use in auditory physiology. *Journal of Neuroscience Methods* 36:45–52.

FUZESSERY, Z. M., P. BUTTENHOFF, B. ANDREWS, and J. M. KENNEDY. 1993. Passive sound localization of prey by the pallid bat (*Antrozous p. pallidus*). *Journal of Comparative Physiology A* 171:767–777.

GAIONI, S. J., H. RIQUIMAROUX, and N. SUGA. 1990. Biosonar behavior of mustached bats swung on a pendulum prior to cortical ablation. *Journal of Neurophysiology* 64:1801–1817.

GALAZYUK, A. V., and A. S. FENG. 1997a. Natural sound sequence creates hyperacuity to sound amplitude. *Abstract of the 27th Annual Meeting of Society for Neuroscience* 288.3.

———. 1997b. Encoding of sound duration by neurons in the auditory cortex of the little brown bat, *Myotis lucifugus. Journal of Comparative Physiology A* 180:301–311.

GALAZYUK, A. V., D. LLANO, and A. S. FENG. 2000. Temporal dynamics of acoustic stimuli enhance amplitude tuning of inferior colliculus neurons. *Journal of Neurophysiology* 83:128–138.

GALE, J. E., and J. F. ASHMORE. 1997. An intrinsic frequency limit to the cochlear amplifier. *Nature* 389:63–66.

GAMES, K. D., and J. A. WINER. 1988. Layer V in rat auditory cortex: projections to the inferior colliculus and contralateral cortex. *Hearing Research* 34:1–25.

GAO, E., and N. SUGA. 1998. Experience-dependent corticofugal adjustment of midbrain frequency map in bat auditory system. *Proceedings of the National Academy of Sciences* 95:12663–12670.

———. 2000. Experience-dependent plasticity in the auditory cortex and the inferior colliculus of bats: role of the corticofugal system. *Proceedings of the National Academy of Sciences* 97:8081–8086.

GOLDMAN, L. J., and O. W. HENSON. 1977. Prey recognition and selection by the constant frequency bat, *Pteronotus p. parnellii. Behavioral Ecology and Sociobiology* 2:411–419.

GOOLER, D. M., and A. S. FENG. 1992. Temporal coding in the frog auditory midbrain: The influence of duration and rise-fall time on the processing of complex amplitude-modulated stimuli. *Journal of Neurophysiology* 67:1–22.

GOOLER, D. M., and W. E. O'NEILL. 1987. Topographic representation of vocal frequency demonstrated by microstimulation of anterior cingulate cortex in the echolocating bat, *Pteronotus parnelli parnelli. Journal of Comparative Physiology A* 161:283–294.

GORDON, M. 1998. Temporally dependent neural mechanisms underlying selectivity for rapid frequency modulations in the external nucleus of the inferior colliculus of the mustached bat, *Pternotus parnellii.* Ph.D. dissertation, University of Rochester, Rochester, N.Y.

GORDON, M., and W. E. O'NEILL. 1998. Temporal processing across frequency channels by FM selective auditory neurons can account for FM rate selectivity. *Hearing Research* 122:97–108.

GOULD, E. 1977. Echolocation and communication. Pp. 247–279 in *Biology of bats of the New World family Phyllostomatidae,* part 2, ed. R. J. Baker, J. K. Jones, and D. C. Carter. Lubbock: Texas Tech Press.

GREEN, D. M. 1984. Temporal factors in psychoacoustics. Pp. 123–140 in *Time resolution in auditory system,* ed. A. Michelsen. New York: Springer-Verlag.

GREEN, D. M., H. A. DEFERRARI, D. MCFADDEN, J. S. PEARSE, A. N. POPPER, W. J. RICHARDSON, S. H. RIDGWAY, and P. L. TYACK. 1994. *Low-frequency sound and marine mammals: Current knowledge and research needs.* Washington, D.C.: National Academy Press.

GREENE, C. R., JR. 1995. Ambient noise. Pp. 87–100 in *Marine mammals and noise,* ed. W. J. Richardson, C. R. Greene Jr., C. I. Malme, and D. H. Thomson. San Diego: Academic Press.

GRIFFIN, D. R. 1958. *Listening in the dark.* New Haven, Conn.: Yale University Press.

GRIFFIN, D. R., F. A. WEBSTER, and C. R. MICHAEL. 1960. The echolocation of flying insects by bats. *Animal Behavior* 8:141–154.

GROTHE, B. 1994. Interaction of excitation and inhibition in processing of pure tone and amplitude-modulated stimuli in the medial superior olive of the mustached bat. *Journal of Neurophysiology* 71:706–721.

GROTHE, B., and PARK, T. S. 1995. Time can be traded for intensity in the lower auditory brainstem. *Naturewissenschaften* 82:521–523.

GROTHE, B., T. J. PARK, and G. SCHULLER. 1997. Medial superior olive in the free-tailed bat: Response to pure tones and amplitude-modulated tones. *Journal of Neurophysiology* 77:1553–65.

GUPPY, A., and R. B. COLES. 1988. Acoustical and neural aspects of hearing in the Australian gleaning bat, *Macroderma gigas* and *Nyctophilus gouldi. Journal of Comparative Physiology A* 162:653–668.

HABERSETZER, J., and G. STORCH. 1992. Cochlea size in extant chiroptera and middle eocene microchiroptera from Messel. *Naturwissenschaften* 79:462–466.

HACKETT, J. T., H. JACKSON, and E. W. RUBEL. 1982. Synaptic excitation of the second and third order auditory neurons in the avian brain stem. *Neuroscience* 7:1455–1469.

HAFTER, E. R. 1984. Spatial hearing and the duplex theory: How viable is the model? Pp. 425–448 in *Dy-*

namic aspects of neocortical function, ed. G. M. Edelman, W. M. Cowan, and W. E. Gall. New York: Wiley.

HAFTER, E. R., and R. H. DYE. 1983. Detection of interaural differences of time in trains of high-frequency clicks as a function of interclick interval and number. *Journal of the Acoustical Society of America* 73: 644–651.

HALL, J. C., and FENG, A. S. 1986. Neural analysis of temporally patterned sound in the frog's thalamus: Processing of pulse duration and pulse repetition rate. *Neuroscience Letters* 63:215–220.

HAPLEA, S., E. COVEY, and J. H. CASSEDAY. 1994. Frequency tuning and response latencies at three levels in the brainstem of the echolocating bat, *Eptesicus fuscus. Journal of Comparative Physiology A* 174: 671–683.

HARNISCHFEGER, G., G. NEUWEILER, and P. SCHLEGEL. 1985. Interaural time and intensity coding in superior olivary complex and inferior colliculus of the echolocating bat *Molossus ater. Journal of Neurophysiology* 53:89–109.

HARTLEY, D. J., and R. A. SUTHERS. 1987. The sound emission pattern and the acoustical role of the noseleaf in the echolocating bat, *Carollia perspicillata. Journal of the Acoustical Society of America* 82:1892–1900.

HASKO, J. A., and G. P. RICHARDSON. 1988. The ultrastructural organization and properties of the mouse tectorial membrane matrix. *Hearing Research* 35: 21–38.

HAVEY, D. C., and D. M. CASPARY. 1980. A simple technique for constructing "piggy-back" multibarrel microelectrodes. *Electroencephalography & Clinical Neurophysiology* 48:249–251.

HE, J. 1997. Modulatory effects of regional cortical activation on the onset response of the cat medical geniculate neurons. *Journal of Neurophysiology* 77: 896–908.

HEITHAUS, E. R., T. H. FLEMING, and P. A. OPLER. 1975. Foraging patterns and resource utilization in seven species of bats in a seasonal tropical forest. *Ecology* 56:841–885.

HEMILÄ, S., S. NUMMELA, and T. REUTER. 1999. A model of the odontocete middle ear. *Hearing Research* 133: 82–97.

HENDRY, S. H. C., E. G. JONES, P. C. EMSON, D. E. M. LOWSON, C. W. HEIZMANN, and P. STREIT. 1989. Two classes of cortical GABA neurons defined by differential calcium binding protein reactivities. *Experimental Brain Reseach* 76:467–472.

HENNING, G. B. 1974. Detectability of interaural delay in high-frequency complex waveforms. *Journal of the Acoustical Society of America* 55:84–90.

HENSON, M. M. 1973. Unusual nerve-fiber distribution in the cochlea of the bat *Pteronotus p. parnellii* (Gray). *Journal of the Acoustical Society of America* 53:1739–1740.

HENSON, M. M., and O. W. HENSON JR. 1988. Tension fibroblasts and the connective tissue matrix of the spiral ligament. *Hearing Research* 35:237–58.

———. 1991. Specializations for sharp tuning in the mustached bat: The tectorial membrane and spiral limbus. *Hearing Research* 56:122–132.

HENSON, M. M., O. W. HENSON JR., and L. J. GOLDMAN. 1977. The perilymphatic spaces in the cochlea of the bat, *Pteronotus p. parnellii* (Gray). *Anatomical Records* 187:767.

HENSON, O. W., JR. 1970. The central nervous system. Pp. 57–152 in *Biology of bats,* vol. 2. New York: Academic Press.

HENSON, O. W., JR., and G. D. POLLAK. 1972. A technique for chronic implantation of electrodes in the cochleae of bats. *Physiological Behavior* 8:1185–1187.

HENSON, O. W., JR., G. SCHULLER, and M. VATER. 1985. A comparative study of the physiological properties of the inner ear in Doppler shift compensating bats (*Rhinolophus rouxi, Pteronotus parnellii*). *Journal of Comparative Physiology A* 157:587–597.

HENSON, O. W., JR., D. H. XIE, A. W. KEATING, and M. M. HENSON. 1995. The effect of contralateral stimulation on cochlear resonance and damping in the mustached bat: The role of the medial efferent system. *Hearing Research* 86:111–124.

HENSON, O. W., A. BISHOP, A. KEATING, J. KOBLER, M. HENSON, B. WILSON, and R. HANSEN. 1987. Biosonar imaging of insects by *Pteronotus p. parnellii,* the mustached bat. *National Geographic Research* 3:82–101.

HERBERT, H., A. ASCHOFF, and J. OSTWALD. 1991. Topography of projections from the auditory cortex to the inferior colliculus in the rat. *Journal of Comparative Neurology* 304:103–122.

HERMAN, L. M. 1990. Cognitive performance of dolphins in visually-guided tasks. Pp. 455–462 in *Sensory abilities of cetaceans: Laboratory and field evidence.* New York: Plenum Press.

HERRERA, M., J. F. HURTADO-GARACIA, F. COLLIA, and J. LANCIEGO. 1994. Projections from the primary auditory cortex onto the dorsal cortex of the inferior colliculus in albino rats. *Archives Italiennes de Biologie* 132:147–164.

HOFFSTETTER, K. M., and G. EHRET. 1992. The auditory cortex of the mouse: connections of the ultrasonic field. *Journal of Comparative Neurology* 323:370–386.

HOLLEY, M. 1996. Outer hair cell motility. Pp. 386–435 in *The cochlea* (Springer handbook of auditory research vol. 8), ed. P. Dallos, A. N. Popper, and R. R. Fay. New York: Springer-Verlag.

HOSE, B., G. LANGNER, and H. SCHEICH. 1987. Topographic representation of periodicities in the forebrain of the mynah bird: One map for pitch and rhythm? *Brain Research* 422:367–373.

HUFFMAN, R. F., and O. W. HENSON, JR. 1990. The descending auditory pathway and acousticomotor systems: connections with the inferior colliculus. *Brain Research* 15:295–232.

HUFFMAN, R. F., P. C. ARGELES, and E. COVEY. 1998. Processing of sinusoidally frequency modulated signals in the nuclei of the lateral lemniscus of the big brown bat, *Eptesicus fuscus*. *Hearing Research* 126:161–180.

IMIG, T. J., and H. O. ADRIAN. 1977. Binaural columns in the primary field (AI) of cat auditory cortex. *Brain Research* 138:241–257.

IMIG, T. J., and R. A. REALE. 1980. Patterns of corticocortical connections related to tonotopic maps in cat auditory cortex. *Journal of Comparative Neurology* 192:293–332.

IRVINE, D. R. F. 1986. The auditory brainstem: A review of the structure and function of auditory brainstem processing mechanisms. Pp. 1–290 in *Progress in sensory physiology,* ed. D. Ottoson. Berlin: Springer-Verlag.

IRVINE, D. R. F., and G. GAGO. 1990. Binaural interactions in high frequency neurons in the cat: Effects of sound pressure level on sensitivity to interaural intensity differences. *Journal of Neurophysiology* 63:570–591.

IRVINE, D. R. F., R. RAJAN, and L. M. AITKIN. 1996. Sensitivity to interaural intensity differences of neurons in primary auditory cortex of the cat. I. Types of sensitivity and effects of variations in sound pressure level. *Journal of Neurophysiology* 75:75–96.

JEFFRESS, L. A. 1948. A place theory of sound localization. *Journal of Comparative Physiology and Psychology* 41:35–39.

JEN, P. H.-S., and R. B. FENG. 1999. Bicuculline application affects discharge pattern and pulse-duration tuning characteristics of the bat inferior colliculus. *Journal of Comparative Physiology* 184:185–194.

JEN, P. H.-S., and P. A. SCHLEGEL. 1982. Auditory physiological properties of neurons in the inferior colliculus of the big brown bat *(Eptesicus fuscus)*. *Journal of Comparative Physiology A* 147:351–362.

JEN, P. H.-S., and J. P. ZHANG. 1999. Corticofugal regulation of excitatory and inhibitory frequency tuning curves of bat inferior collicular neurons. *Brain Research* 841:184–188.

JEN, P. H.-S., Q. C. CHEN, and X. D. SUN. 1998. Corticofugal regulation of auditory sensitivity in the bat inferior colliculus. *Journal of Comparative Physiology A* 183:683–697.

JEN, P. H.-S., X. D. SUN, and Q. C. CHEN. 2001. An electrophysiological study of neural pathways for corticofugally inhibited neurons in the central nucleus of the inferior colliculus of the big brown bat, *Eptesicus fuscus*. *Experimental Brain Research* 137:292–302.

JEN, P. H.-S., X. SUN, and P. J. J. LIN. 1989. Frequency and space representation in the primary auditory cortex of the frequency modulating bat *Eptesicus fuscus*. *Journal of Comparative Physiology A* 165:1–14.

JEN, P. H.-S., X. D. SUN, D. M. CHEN, and H. B. TENG. 1987. Auditory space representation in the inferior colliculus of the FM bat, *Eptesicus fuscus*. *Brain Research* 419:7–18.

JOHNSON, B. R., 1993. GABAergic and glycinergic inhibition in the central nucleus of the inferior colliculus of the big brown bat, *Eptesicus fuscus*. Ph.D. dissertation, Duke University, Durham, N.C.

JOHNSON, C. S. 1967. Sound detection thresholds in marine mammals. Pp. 247–260 in *Marine bio-acoustics,* vol. 2, ed. W. N. Tavolga. New York: Pergamon Press.

———. 1968a. Masked tonal thresholds in the bottlenose porpoise. *Journal of the Acoustical Society of America* 44(4): 965–967.

———. 1968b. Relation between absolute thresholds and duration-of-tone pulses in the bottlenose porpoise. *Journal of the Acoustical Society of America* 43:757–763.

———. 1991. Hearing thresholds for periodic 60-kHz tone pulses in the beluga whale. *Journal of the Acoustical Society of America* 89:2996–3001.

JOHNSON, C. S., M. W. MCMANUS, and D. SKAAR. 1989. Masked tonal hearing thresholds in the beluga whale. *Journal of the Acoustical Society of America* 85(6): 2651–2654.

JONES, E. G. 1986. Neurotransmitters in the cerebral cortex. *Journal of Neurosurgery* 65:135–153.

JORIS, P. X., and T. C. T. YIN. 1995. Envelope coding in the lateral superior olive. I. Sensitivity to interaural time differences. *Journal of Neurophysiology* 73:1043–1062.

———. 1996. Envelope coding in the lateral superior olive. III. Comparison with afferent pathways. *Journal of Neurophysiology* 79:253–269.

KANWAL J. S. 1997. A multidimensional code for processing social calls in the auditory cortex of the mustached bat. 33rd IUPS:L081.05.

———. 1999. Processing species-specific calls by combination-sensitive neurons in an echolocating bat. In *Causal mechanisms of animal communication: Essays in honor of Peter Marler,* ed. M. D. Hauser and M. Konishi.

KANWAL, J. S., and K.-H. ESSER. 1996. Auditory-responsive units in the frontal cortex of the mustached bat, *Pteronotus parnellii.* Pp. 236 in *Göttingen neurobiology report, proceedings of the 24th Göttingen Neurobiology Conference,* vol. 2, ed. N. Elsner and H.-U. Schnitzler. New York: Thieme Verlag.

KANWAL, J., and N. SUGA. 1995. Hemispheric asymmetry in the processing of calls in the auditory cortex of the mustached bat. *Abstracts of the Association for Research in Otolaryngology* 18:104.

KANWAL, J. S., D. C. FITZPATRICK, and N. SUGA. 1999. Facilitatory and inhibitory frequency tuning for combination-sensitive neurons in the primary auditory cortex of mustached bats. *Journal of Neurophysiology* 82:2327–2345.

KANWAL, J. S., K. K. OHLEMILLER, and N. SUGA. 1992. Selective representation of call-syllables in the DSCF area in the primary auditory cortex of the mustached bat. *Society for Neuroscience Abstracts* 18:884.

KANWAL, J. S., M. GORDON, J. P. PENG, and K.-H. ESSER. 2000. Auditory responses from the frontal cortex in the mustached bat, *Pteronotus parnellii. NeuroReport* 11:367–372.

KANWAL, J. S., S. MATSUMURA, K. OHLEMILLER, and N. SUGA. 1994. Analysis of acoustic elements and syntax in communication sounds emitted by mustached bats. *Journal of the Acoustical Society of America* 96:1229–1254.

KAWASAKI, M., D. MARGOLIASH, and N. SUGA. 1988. Delay-tuned combination-sensitive neurons in the auditory cortex of the vocalizing mustached bat. *Journal of Neurophysiology* 59:623–635.

KEATING, A. W., O. W. HENSON JR., M. M. HENSON, W. C. LANCASTER, and D. H. XIE. 1994. Doppler-shift compensation by the mustached bat: Quantitative data. *Journal of Experimental Biology* 188:115–29.

KELLOGG, W. N., and R. KOHLER. 1952. Responses of the porpoise to ultrasonic frequencies. *Science* 116:250–252.

KELLY, J. B., and D. P. PHILLIPS. 1991. Coding of interaural time differences of transients in auditory cortex of *Rattus norvegicus:* Implications for the evolution of mammalian sound localization. *Hearing Research* 55:39–44.

KELLY, J. B., and S. L. SALLY. 1988. Organization of the auditory cortex in the albino rat: Binaural response properties. *Journal of Neurophysiology* 59:1756–1769.

KEMMER, M., and M. VATER. 1997. The distribution of GABA and glycine immunostaining in the cochlear nucleus of the mustached bat (*Pteronotus parnellii*). *Cell and Tissue Research* 287:487–506.

KEMP, D. T. 1978. Stimulated acoustic emissions from within the human auditory system. *Journal of the Acoustical Society of America* 64:1386–1391.

KESAREV, V. S., L. I. MALOFEYEVA, and O. V. TRYKOVA. 1977. Structural organization of the cerebral cortex in cetaceans. *Archiv Anat. Gistol. Embriol* 73:23–30.

KETTEN, D. R. 1984. Correlations of morphology with frequency for odontocete cochlea: Systematics and topology. Ph.D. dissertation, Johns Hopkins University, Baltimore.

———. 1992. The marine mammal ear: Specializations for aquatic audition and echolocation. Pp. 717–750 in *The evolutionary biology of hearing,* ed. D. B. Webster, R. R. Fay, and A. N. Popper. New York: Springer-Verlag.

———. 1994. Functional analyses of whale ears: Adaptations for underwater hearing. *IEEE Proceedings in Underwater Acoustics* 1:264–270.

———. 1997. Structure and function in whale ears. *Bioacoustics* 8:103–135.

———. 1998. Marine mammal auditory systems: A summary of audiometric and anatomical data and its implications for underwater acoustic impacts. NOAA Technical memorandum, U.S. Department of Commerce.

———. 2000. Cetacean ears. Pp. 43–109 in *Hearing by whales and dolphins,* ed. W. W. L. Au, A. N. Popper, and R. R. Fay. New York: Springer-Verlag.

KIANG, N. Y. S., T. WATANABE, E. C. THOMAS, and L. F. CLARK. 1965. Discharge patterns of single fibers in the cat's auditory nerve. *Research Monograph No. 35.* Cambridge: MIT Press.

KICK, S. A., and J. A. SIMMONS. 1984. Automatic gain control in the bat's sonar receiver and the neuroethology of echolocation. *Journal of Neuroscience* 4:2725–2737.

KIDD, S. A., and J. B. KELLY. 1996. Contribution of the dorsal nucleus of the lateral lemniscus to binaural responses in the inferior colliculus: Interaural time delays. *Journal of Neuroscience* 16:7390–7397.

KITZES, L. M., K. S. WREGE, and J. M. CASSADY. 1980. Patterns of responses of cortical cells to binaural sound. *Journal of Comparative Neurology* 192:455–472.

KLEISER A., and G. SCHULLER. 1995. Responses of collicular neurons to acoustic motion in the horseshoe bat, *Rhinolophus rouxi*. *Naturwissenchaften* 82:337–340.

KNUDSEN, E. I., and M. KONISHI. 1978. Center-surround organization of auditory receptive field in the owl. *Science* 202:778–780.

KOBLER, J. B., S. F. ISBEY, and J. H. CASSEDAY. 1987. Auditory pathways to the frontal cortex of the mustache bat, *Pteronotus parnellii*. *Science* 236:824–826.

KOCH, U., and B. GROTHE. 1998. GABAergic and glycinergic inhibition sharpens tuning for frequency modulations in the inferior colliculus of the big brown bat. *Journal of Neurophysiology* 80:71–82.

KONISHI, M. 1992. The neural algorithm for sound localization in the owl. *Harvey Lectures* 86:47–64.

KÖPPL, C., G. A. MANLEY, and O. GLEICH. 1993. An auditory fovea in the barn owl cochlea. *Journal of Comparative Physiology A* 171:695–704.

KOSAKA, K., K. HAMA, and I. NAGATSU. 1987. Tyrosine hydroxylase-immunoreactive intrinsic neurons in the rat cerebral cortex. *Experimental Brain Research* 68: 393–405.

KÖSSL, M. 1992. High frequency distortion products from the ears of two bat species, *Megaderma lyra* and *Carollia perspicillata*. *Hearing Research* 60:156–164.

———. 1994a. Otoacoustic emissions from the cochlea of the "constant frequency" bats, *Pteronotus parnellii* and *Rhinolophus rouxi*. *Hearing Research* 72:59–72.

———. 1994b. Evidence for a mechanical filter in the cochlea of the "constant frequency" *Rhinolophus rouxi* and *Pteronotus parnellii*. *Hearing Research* 72:73–80.

———. 1997. Sound emission from cochlear filters and foveae: Does the auditory sense organ make sense? *Naturwissenchaften* 84:9–16.

KÖSSL, M., and I. J. RUSSELL. 1995. Basilar membrane resonance in the cochlea of the mustached bat. *Proceedings of the National Academy of Sciences* 92: 276–279.

KÖSSL, M., and M. VATER. 1985a. The cochlear frequency map of the mustache bat, *Pteronotus parnellii*. *Journal of Comparative Physiology A* 157:687–697.

———. 1985b. Evoked acoustic emissions and cochlear microphonics in the mustache bat, *Pteronotus parnellii*. *Hearing Research* 19:157–170.

———. 1990. Tonotopic organization of the cochlear nucleus of the mustache bat, *Pteronotus parnelli*. *Journal of Comparative Physiology A* 166:695–709.

———. 1995. Cochlear structure and function in bats. Pp. 191–235 in *Hearing by bats*, ed. A. N. Popper and R. R. Fay. New York: Springer-Verlag.

———. 1996a. A tectorial membrane fovea in the cochlea of the mustached bat. *Naturwissenchaften* 2:89–92.

———. 1996b. Further studies on the mechanics of the cochlear partition in the mustached bat. II. A second cochlear frequency map derived from acoustic distortion products. *Hearing Research* 94:78–87.

KRUBITZER, L. A., M. B. CALFORD, and L. M. SCHMID. 1993. Connections of somatosensory cortex in megachiropteran bats: The evolution of cortical fields in mammals. *Journal of Comparative Neurology* 327: 473–506.

KRUBITZER, L. A., SESMA, M. A., and J. H. KAAS. 1986. Microelectrode maps, myeloarchitecture, and cortical connections of three somatotopically organized representations of the body surface in the parietal cortex of squirrels. *Journal of Comparative Neurology* 250:403–430.

KUJIRAI, K., and N. SUGA. 1983. Tonotopic representation and space map in the non-primary auditory cortex of the mustached bat. *Auris Nasus Larynx* (Tokyo) 10:9–24.

KUWADA, S., R. BATRA, and V. MAHER. 1986. Scalp potentials of normal and hearing impaired subjects in response to sinusoidally amplitude modulated tones. *Hearing Research* 21:179–192.

LEROY, S. A., and J. J. WENSTRUP. 2000. Spectral integration in the inferior colliculus of the mustached bat. *Journal of Neuroscience* 20:8533–8541.

LESTER, B. R., and E. J. PECK. 1979. Kinetic and pharmacologic characterization of gamma-aminobutyric acid receptive sites from mammalian brain. *Brain Research* 161:79–97.

LIM, D. J. 1986. Functional structure of the organ of Corti: A review. *Hearing Research* 22:117–146.

LIU, W., and N. SUGA. 1997. Binaural and commissural organization of the primary auditory cortex of the mustached bat. *Journal of Comparative Physiology A* 181:599–605.

LLANO, D. A., and A. S. FENG. 1999. Response characteristics of neurons in the medial geniculate body of the little brown bat to simple and temporally patterned sounds. *Journal of Comparative Physiology A* 184:371–385.

MACKAY, R. S., and H. M. LIAW. 1981. Dolphin vocalization mechanisms. *Science* 212:676–678.

MAEKAWA, M., D. WONG, and W. G. PASCHAL. 1992. Spectral selectivity of FM-FM neurons in the auditory cortex of the echolocating bat, *Myotis lucifugus*. *Journal of Comparative Physiology A* 171:513–522.

MALMIERCA, M. S., A. REES, F. E. LE BEAU, and J. G. BJAALIE. 1995. Laminar organization of frequency-defined local axons within and between the inferior colliculi of the guinea pig. *Journal of Comparative Neurology* 357:124–44.

MANABE, T., N. SUGA, and J. OSTWALD. 1978. Aural representation in the Doppler-shifted-CF processing area of the primary auditory cortex of the mustached bat. *Science* 200:339–342.

MANLEY, G. A, L. GALLO, and C. KÖPPL. 1996. Spontaneous otoacoustic emissions in two gecko species, *Gekko gecko* and *Eublepharis macularius*. *Journal of the Acoustical Society of America* 99:1588–1603.

MANLEY, G. A., D. R. F. IRVINE, and B. M. JOHNSTONE. 1972. Frequency response of bat tympanic membrane. *Nature* 237:112–113.

MARGOLIASH, D., and E. S. FORTUNE. 1992. Temporal and harmonic combination-sensitive neurons in the zebra finch's Hvc. *Journal of Neuroscience* 12:4309–4326.

MARIMUTHU, G., and G. NEUWEILER, G. 1987. The use of acoustical cues for prey detection by the Indian false vampire bat, *Megaderma lyra*. *Journal of Comparative Physiology A* 160:509–515.

MASSOPUST, L. J., and J. ORDY. 1962. Auditory organization of the inferior colliculi in the cat. *Experimental Neurology* 66:465–477.

MCBRIDE, A. F. 1956. Evidence for echolocation by cetaceans. *Deep Sea Research* 3:153–154.

MCCORMICK, J. G., E. G. WEVER, G. PALIN, and S. H. RIDGWAY. 1970. Sound conduction in the dolphin ear. *Journal of the Acoustical Society of America* 48:1418–1428.

MCCORMICK, J. G., E. G. WEVER, S. H. RIDGWAY, and J. PALIN. 1980. Pp. 449–467 in *Sound reception in the porpoise as it relates to echolocation in animal sonar systems,* ed. R. G. Busnel and J. F. Fish. New York: Plenum Press.

MCFADDEN, D., and E. G. PASANEN. 1976. Lateralization at high frequencies based on interaural time difference. *Journal of the Acoustical Society of America* 59:634–639.

METZNER, W. 1996. Anatomical basis for audio-vocal integration in echolocating horseshoe bats. *Journal of Comparative Neurology* 368:252–269.

MIDDLEBROOKS, J. C., A. CLOCK-EDDINS, X. LI, and D. M. GREEN. 1994. A panoramic code for sound location by cortical neurons. *Science* 264:842–844.

MILLINKOVITCH, M. C., G. ORTÍ, and A. MEYER. 1993. Revised phylogeny of whales suggested by mito-chondrial and ribosomal DNA-sequences. *Nature* 361:346–348.

MILLS, D. M., and E. W. RUBEL. 1997. Development of distortion product emissions in the gerbil: "Filter" response and signal delay. *Journal of the Acoustical Society of America* 101:395–411.

MITANI, A., K. ITOH, and N. MIZUNO. 1987. Distribution and size of thalamic neurons projecting to layer I of the auditory cortical fields of the cat compared to those projecting to layer IV. *Journal of Comparative Neurology* 257:105–121.

MITTMANN, D. H., and J. J. WENSTRUP. 1995. Combination-sensitive neurons in the inferior colliculus. *Hearing Research* 90:185–191.

MOLINARI, M., M. E. DELL'ANNA, M. G. RAUSELL, T. LEGGIO, T. HASHIKAWA, and E. G. JONES. 1995. Auditory thalamocortical pathways defined in monkeys by calcium binding protein immunoreactivity. *Journal of Comparative Neurology* 362:20–43.

MOORE, B. C. J., B. R. GLASBERG, C. J. PLACK, and A. K. BISWAS. 1988. The shape of the ear's temporal window. *Journal of the Acoustical Society of America* 83:1102–1116.

MOORE, P. W. B., R. W. HALL, W. A. FRIEDL, and P. E. NACHTIGALL. 1984. The critical interval in dolphin echolocation: What is it? *Journal of the Acoustical Society of America* 76:314–317.

MORGANE, P. J., I. GLEZER, and M. S. JACOBS. 1990. Comparative and evolutionary anatomy of the visual cortex of the dolphin. Pp. 215–262 in *Cerebral Cortex,* vol. 8B, *Comparative structure and evolution of cerebral cortex,* part 2, ed. E. G. Jones and A. Peters. New York: Plenum Press.

MORIYAMA, T., T. HOU, M. WU, and P. H.-S. JEN. 1994. Responses of inferior collicular neurons of the FM bat (*Eptesicus fuscus*) to pulse trains with varied pulse amplitudes. *Hearing Research* 79:105–114.

MOULTON, J. M. 1960. Swimming sounds and the schooling of fishes. *Biological Bulletin* 119:210–223.

MÜLLER, M., B. LAUBE, H. BURDA, and V. BRUNS. 1992. Structure and function of the cochlea in the African mole rat (*Cryptomys hottentottus*), evidence for a low frequency acoustic fovea. *Journal of Comparative Physiology A* 171:469–476.

NACHTIGALL, P. E. 1986. Vision, audition, and chemo-reception in dolphins and other marine mammals. Pp. 79–113 in *Dolphin cognition and behavior: A comparative approach,* ed. R. J. Schusterman, J. A. Thomas, and F. G. Wood. Hillsdale, N.J.: Lawrence Erlbaum Associates.

NARINS, P. M., and R. R. CAPRANICA. 1980. Neural adaptations for processing the two-note call of the Puerto Rican treefrog, *Eleuthereodactylus coqui. Brain and Behavioral Evolution* 18:48–66.

NEUWEILER, G. 1984. Foraging, echolocation, and audition in bats. *Naturwissenschaften* 71:446.

———. 1990. Auditory adaptations for prey capture in echolocating bats. *Physiological Reviews* 70:615–641.

NEUWEILER, G., and M. B. FENTON. 1988. Behaviour and foraging ecology of echolocating bats. Pp. 535–549 in *Animal sonar: Processes and performance,* ed. P. E. Nachtigall and P. W. B. Moore. New York: Plenum Press.

NEUWEILER, G., and M. VATER. 1977. Response patterns to pure tones of cochlear nucleus units in the CF-FM bat, *Rhinolophus ferrumequinum. Journal of Comparative Physiology A* 115:119–133.

NIEUWENHUYS, R. 1994. The neocortex. An overview of its evolutionary development, structural organization and synaptology. *Anatomy and Embryology* 190:307–337.

NORRIS, K. S. 1964. Some problems of echolocation in cetaceans. Pp. 317–336 in *Marine bio-acoustics,* vol. 2, ed. W. N. Tavolga. New York: Pergamon Press.

———. 1968. The evolution of acoustic mechanisms in odontocete cetaceans. Pp. 298–323 in *Evolution and environment,* ed. E. T. Drake. New Haven, Conn.: Yale University Press.

———. 1969. The echolocation of marine mammals. Pp. 391–423 in *The biology of marine mammals,* ed. H. T. Andersen. New York: Academic Press.

———. 1980. Peripheral sound processing in odontocetes. Pp. 495–509 in *Animal sonar systems,* ed. R. G. Busnel and J. F. Fish. New York: Plenum Press.

NORRIS, K. S., and G. W. HARVEY. 1974. Sound transmission in the porpoise head. *Journal of the Acoustical Society of America* 56:659–664.

NOVACEK, M. J. 1985. Evidence for echolocation in the oldest known bat. *Nature* 315:140–141.

NUETZEL, J. M., and E. R. HAFTER. 1976. Lateralization of complex waveforms: Effects of fine structure, amplitude and duration. *Journal of the Acoustical Society of America* 60:1339–1346.

NUMMELA, S., T. REUTER, S. HEMILÄ, P. HOMBERG, and P. PAUKKU. 1999a. The anatomy of the killer whale middle ear (*Orcinus orca*). *Hearing Research* 133:61–70.

NUMMELA, S., T. WÄGAR, S. HEMILÄ, and T. REUTER. 1999b. Scaling of the cetacean middle ear. *Hearing Research* 133:71–81.

NOVACEK, M. J. 1985. Evidence for echolocation in the oldest known bats. *Nature* 315:140–141.

NOVICK, A., and J. R. VAISNYS. 1964. Echolocation of flying insects by the bat, *Chilonycteris parnellii. Biological Bulletin* 127:478–488.

OBRIST, M., M. B. FENTON, J. L. EGER, and P. A. SCHLEGEL. 1993. What ears do for bats: A comparative study of pinna sound pressure transformation in chiroptera. *Journal of Experimental Biology* 180:119–152.

OELSCHLÄGER, H. A. 1990. Evolutionary morphology and acoustics in the dolphin skull. In *Sensory abilities of cetaceans: Laboratory and field evidence,* ed. J. Thomas and R. Kastlelein. New York: Plenum Press.

OHLEMILLER, K. K., J. S. KANWAL, and N. SUGA. 1996. Facilitative responses to species-specific calls in cortical FM-FM neurons of the mustached bat. *Neuroreport* 7:1749–1755.

OHLEMILLER, K. K., J. S. KANWAL, J. BUTMAN, and N. SUGA. 1994. Stimulus design for auditory neuroethology: Synthesis and manipulation of complex communication sounds. *Auditory Neuroscience* 1:19–37.

OJIMA, H. 1994. Terminal morphology and distribution of corticothalamic fibers originating from layers 5 and 6 of cat primary auditory cortex. *Cerebral Cortex* 6:646–663.

OLSEN, J. F., and N. SUGA. 1991. Combination-sensitive neurons in the medial geniculate body of the mustached bat: Encoding of target range information. *Journal of Neurophysiology* 65:1275–1296.

O'NEILL, W. E. 1985. Responses to pure tones and linear FM components of the CF-FM biosonar signal by single units in the inferior colliculus of the mustached bat. *Journal of Comparative Physiology A* 157:797–815.

———. 1995. The bat auditory cortex. Pp. 416–480 in *Hearing by bats,* ed. A. N. Popper and R. R. Fay. New York: Springer-Verlag.

O'NEILL, W. E., and N. SUGA. 1979. Target-range sensitive neurons in the auditory cortex of the mustache bat. *Science* 203:69–73.

———. 1982. Encoding of target range and its representation in the auditory cortex of the mustached bat. *Journal of Neuroscience* 2:17–31.

O'NEILL, W. E., R. D. FRISINA, and D. M. GOOLER. 1989. Functional organization of mustached bat inferior colliculus: I. Representation of FM frequency bands important for target ranging revealed by 14C-2-deoxyglucose autoradiography and single unit mapping. *Journal of Comparative Neurology* 284:60–84.

O RDY , J. 1962. Auditory organization of the inferior colliculi in the cat. *Experimental Neurology* 66:465–477.

O STWALD , J. 1980. The functional organization of the auditory cortex in the CF-FM bat *Rhinolophus ferrumequinum.* Pp. 953–956 in *Animal sonar systems,* ed. R. G. Busnel and J. F. Fish. New York: Plenum Press.

P ALAKAL , M. J., and D. W ONG . 1999. Cortical representation of spatiotemporal pattern of firing evoked by echolocation signals: Population encoding of target features in real time. *Journal of the Acoustical Society of America* 106(1): 479–490.

P APEZ , J. W. 1929. Central acoustic tract in cat and man. *Anatomical Record* 42:60.

P ARK , T. J., and G. D. P OLLAK . 1993. GABA shapes a topographic organization of response latency in the mustached bat's inferior colliculus. *Journal of Neurophysiology* 13:5172–5187.

P ARK , T. J., B. G ROTHE , G. D. P OLLAK , G. S CHULLER , and U. K OCH . 1996. Neural delays shape selectivity for interaural intensity differences in the lateral superior olive. *Journal of Neuroscience* 16:6554–6566.

P ASCHAL , W. G., and D. W ONG . 1994a. A cortical region specialized for processing echolocation pulses of *Myotis lucifugus. Association for Research in Otolaryngology Abstract* 17:93.

———. 1994b. Frequency organization of delay-sensitive neurons in the auditory cortex of the FM bat *Myotis lucifugus. Journal of Neurophysiology* 72:366–379.

P HILLIPS , D. P., and D. R. F. I RVINE . 1981. Responses of single neurons in the physiologically defined area A1 of the cat cerebral cortex: Sensitivity to interaural intensity differences. *Hearing Research* 4:299–307.

P HILLIPS , D. P., S. E. H ALL , and J. L. H OLLETT . 1989. Repetition rate and signal level effects on neuronal responses to brief tone pulses in cat auditory cortex. *Journal of the Acoustical Society of America* 85:2537–2549.

P IERCE , G. W., and D. R. G RIFFIN . 1938. Experimental determination of supersonic notes emitted by bats. *Journal of Mammology* 19:454–455.

P INHEIRO , A. D., M. W U , and P. H.-S. J EN . 1991. Encoding repetition rate and duration in the inferior colliculus of the big brown bat, *Eptesicus fuscus. Journal of Comparative Physiology A* 169:69–85.

P LACK , C. J., and B. C. J. M OORE . 1990. Temporal window shape as a function of frequency and level. *Journal of the Acoustical Society of America* 87:2178–2187.

P OLLAK , G. D. 1988. Time is traded for intensity in the bat's auditory system. *Hearing Research* 36:107–124.

P OLLAK , G. D., and T. J. P ARK . 1993. The effects of GABAergic inhibition on monaural response properties of neurons in the mustache bat's inferior colliculus. *Hearing Research* 65:99–117.

P OLLAK , G. D., O. W. H ENSON J R ., and A. N OVICK . 1972. Cochlear microphonic audiograms in the "pure tone" bat, *Chilonycteris parnellii parnellii. Science* 176:66–68.

P OPOV , V. V., and A. Y A . S UPIN . 1985. Determining the hearing characteristics of dolphins according to brainstem evoked potentials. *Doklady Biological Sciences* 283:524–527.

———. 1990a. Auditory brain stem responses in characterization of dolphin hearing. *Journal of Comparative Physiology A* 166:385–393.

———. 1990b. Electrophysiological studies of hearing in some cetaceans and a manatee. Pp. 405–415 in *Sensory abilities of cetaceans: Laboratory and field evidence,* ed. J. Thomas and R. Kastlelein. New York: Plenum Press.

———. 1997. Detection of temporal gaps in noise in dolphins: Evoked-potential study. *Journal of the Acoustical Society of America* 102:1169–1176.

P OPOV , V. V., A. Y A . S UPIN , and V. O. K LISHIN . 1995. Frequency tuning curves of the dolphin's hearing: Envelope-following response study. *Journal of Comparative Physiology A* 178:571–578.

———. 1997. Paradoxical lateral suppression in the dolphin's auditory system: Weak sounds suppress response to strong sounds. *Neuroscience Letters* 234(1): 51–54.

P OPPER , A. N., and R. R. F AY , EDS . 1995. *Hearing by bats.* New York: Springer-Verlag.

P OPPER , A. N., H. L. H AWKINS , R. C. G ISINER . 1997. Questions in cetacean bioacoustics: Some suggestions for future research. *Bioacoustics* 8:163–182.

P ORTFORS , C. V., and J. J. W ENSTRUP . 1999. Delay-tuned neurons in the inferior colliculus of the mustached bat: Implications for analyses of target distance. *Journal of Neurophysiology* 82:1326–1338.

———. 2001. Responses to combinations of tones in the nuclei of the lateral lemniscus. *Journal of the Association for Research in Otolaryngology* 2:104–117.

P OTTER , H. D. 1965. Patterns of acoustically-evoked discharges of neurons in the mesencephalon of the bullfrog. *Journal of Neurophysiology* 28:1155–1184.

P ROBST , R., B. L. L ONSBURY -M ARTIN , and G. K. M ARTIN . 1991. A review of otoacoustic emissions. *Journal of the Acoustical Society of America* 89:2027–2067.

PUJOL, R., M. LENOIR, S. LADRECH, F. TRIBILLAC, and G. REBILLARD. 1992. Correlation between the length of outer hair cells and the frequency coding of the cochlea. Pp. 45–52 in *Auditory physiology and perception, advances in biosciences,* ed. Y. Cazals, L. Demany, and K. Horner. New York: Pergamon Press.

PYE, J. D. 1980. Adaptiveness of echolocation signals in bats: Flexibility in behaviour and in evolution. *Trends in Neuroscience* 232–235.

RADTKE-SCHULLER, S. 1997. Struktur und Verschaltung des Hörcortex der Hufeisennasenfledermaus *Rhinolophus rouxi*. Ph.D. dissertation, Ludwig-Maximilians-University, Munich.

———. 2001. Neuroarchitecture of the auditory cortex in the rufous horseshoe bat (*Rhinolophus rouxi*). *Anatomy and Embryology* 204:81–100.

RADTKE-SCHULLER, S., and G. SCHULLER. 1995. Auditory cortex of the rufous horseshoe bat: I. Physiological response properties to acoustic stimuli and vocalizations and the topographical distribution of neurons. *European Journal of Neuroscience* 7:570–591.

RAJAN, R. 1990. Electrical stimulation of the inferior colliculus at low rates protects the cochlea from auditory desensitization. *Brain Research* 506:192–204.

RALL, W. A. 1964. Theoretical significance of dendritic trees for neuronal input-output relations. Pp. 73–97 in *Neural theory and modelling,* ed. R. F. Reiss. Palo Alto, Calif.: Stanford University Press.

RAMPRASHAD, F., K. E. MONEY, J. P. LANDOLT, and J. LAUFER. 1978. A neuroanatomical study of the cochlea of the little brown bat (*Myotis lucifugus*). *Journal of Morphology* 160:345–358.

RAUSCHECKER, J., B. TIAN, and M. HAUSER. 1995. Processing of complex sounds in the macaque nonprimary auditory cortex. *Science* 268:111–114.

RAUSCHECKER, J. P., B. TIAN, T. PONS, and M. MISHKIN. 1997. Serial and parallel processing in rhesus monkey auditory cortex. *Journal of Comparative Neurology* 382:89–103.

RAZAK, K. A., Z. M. FUZESSERY, and T. D. LOHUIS. 1999. Single cortical neurons serve both echolocation and passive sound localization. *Journal of Neurophysiology* 81:1438–1442.

REALE, R. A., and J. F. BRUGGE. 1990. Auditory cortical neurons are sensitive to static and continuously changing phase cues. *Journal of Neurophysiology* 64:1247–1260.

REALE, R. A., and R. KETTNER. 1986. Topography of binaural organization in primary auditory cortex (A1) of the cat: Effects of changing interaural intensity. *Journal of Neurophysiology* 56:663–682.

REES, A., G. GREEN, and R. H. KAY. 1986. Steady-state evoked responses to sinusoidally amplitude-modulated sounds recorded in man. *Hearing Research* 23:123–133.

REHKÄMPER, G. 1981. Vergleichende Architektonik des Neocortex der Insectivora. *Zeitschrift für zoologische Systematik und Evolutionsforschung* 19:233–263.

REIMER, K. 1989. Bedeutung des Colliculus superior bei der Echoortung der Fledermaus, *Rhinolophus rouxi.* Ph.D. dissertation, Ludwig-Maximilians-University of Munich.

RENAUD, D. L., and A. N. POPPER. 1975. Sound localization by the bottlenose porpoise *Tursiops truncatus. Journal of Experimental Biology* 63:569–585.

REYSENBACH DE HAAN, F. W. 1956. Hearing in whales. *Acta Otolaryngologica Supplement* 134:1–114.

RICHARDSON, W. J. 1995. Marine mammal hearing. Pp. 205–240 in *Marine mammals and noise,* ed. W. J. Richardson, C. R. Greene Jr., C. I. Malme, and D. H. Thomson. San Diego: Academic Press.

RICHARDSON, W. J., C. R. GREENE JR., C. I. MALME, and D. H. THOMSON, EDS. 1995. *Marine mammals and noise.* San Diego: Academic Press.

RIDGWAY, S. H. 1972. Homeostasis in the aquatic environment. Pp. 590–747 in *Mammals of the sea,* ed. S. H. Ridgway. Springfield, Ill.: Charles A. Thomas.

———. 1988. The cetacean central nervous system. Pp. 20–25 in *Comparative neuroscience and neurobiology,* ed. L. N. Irwin. Boston: Birkhäuser.

RIDGWAY, S. H., and D. A. CARDER. 1988. Nasal pressure and sound production in an echolocating white whale, *Delphinapterus leucas.* Pp. 53–60 in *Animal sonar: Processes and performance,* ed. P. E. Nachtigall and P. W. B. Moore. New York: Plenum Press.

———. 1993. High-frequency hearing loss in old (25+ years old) male dolphins. *Journal of the Acoustical Society of America* 94:1830 (abstract).

———. 1997. Hearing deficits measured in some *Tursiops truncatus,* and discovery of a deaf/mute dolphin. *Journal of the Acoustical Society of America* 101:590–593.

RIDGWAY, S. H., T. H. BULLOCK, D. A. CARDER, R. L. SEELEY, D. WOODS, and H. GALAMBOS. 1981. Auditory brainstem response in dolphin. *Proceedings of the National Academy of Sciences* 78:1943–1947.

RIDGWAY, S. H., D. A. CARDER, P. L. KAMOLNICK, and D. J. SKAAR. 1991. Acoustic response times (RTs) for

Tursiops truncatus. Journal of the Acoustical Society of America 89(4): 1967–1968 (abstract).

RIDGWAY, S. H., D. A. CARDER, T. KAMOLNICK, R. R. SMITH, C. E. SCHLUNDT, and W. R. ELSBERRY. 2001. Hearing and whistling in the deep sea: Depth influences whistle spectra but does not attenuate hearing by white whales (*Delphinapterus leucas*) (Odontoceti, Cetacea). *Journal of Experimental Biology* 204:3829–3841.

RIDGWAY, S. H., D. A. CARDER, R. R. SMITH, T. KAMOLNICK, C. E. SCHLUNDT, and W. R. ELSBERRY. 1997. Behavioral responses and temporary shift in masked hearing threshold of bottlenose dolphins, *Tursiops truncatus,* to 1-second tones of 141 to 201 dB re 1 μPa. Tech. Rep. 1751. Report from Naval Command, Control and Ocean Surveillance Center, RDT&E Division, San Diego, Calif., for U.S. Office of Naval Research.

RIQUIMAROUX, H., S. J. GAIONI, and N. SUGA. 1992. Inactivation of DSCF area of the auditory cortex with muscimol disrupts frequency discrimination in the mustached bat. *Journal of Neurophysiology* 68:1613–1623.

ROSE, M. 1912. Histologische Lokalisation der Großhirnrinde bei kleinen Säugetieren (Rodentia, Insectivora, Chiroptera). *Journal für Psychologie und Neurologie* (suppl. 2), 19:391–479.

ROSOWSKI, J. J. 1992. Hearing in transitorial mammals: Predictions from the middle-ear anatomy and hearing capabilities of extant mammals. Pp. 615–632 in *The evolutionary biology of hearing,* ed. D. B. Webster, R. R. Fay, and A. N. Popper. New York: Springer-Verlag.

———. 1994. Outer and middle ears. Pp. 172–248 in *Comparative hearing: Mammals,* ed. R. R. Fay and A. N. Popper. New York: Springer-Verlag.

ROUILLER, E. M., and F. DE RIBAUPIERRE. 1990. Arborization of corticothalamic axons in the auditory thalamus of the cat: A PHA-L tracing study. *Neuroscience Letters* 108:29–35.

ROUILLER, E. M., J. P. HORNUNG, and F. DE RIBAUPIERRE. 1989. Extrathalamic ascending projections to physiologically identified fields of the cat auditory cortex. *Hearing Research* 40:233–246.

ROUILLER, E. M., G. SIMM, A. VILLA, Y. DE RIBEAUPIERRE, and F. DE RIBEAUPIERRE. 1991. Auditory corticocortical interconnections in the cat: Evidence for parallel and hierarchical arrangement of the auditory cortical areas. *Experimental Brain Research* 86:483–505.

RUSSELL, I. J., and M. KÖSSL. 1999. Micromechanical responses to echolocation signals in the cochlear fovea of the mustached bat. *Journal of Neurophysiology* 82:676–686.

RYAN, M. J., and M. D. TUTTLE. 1987. The role of prey-generated sounds, vision and echolocation in prey localization by the African bat *Cardioderma cor* (Megadermatidae). *Journal of Comparative Physiology A* 161:59–66.

RYUGO, D. K., and N. M. WEINBERGER. 1976. Corticofugal modulation of the medial geniculate body. *Experimental Neurology* 51:377–391.

SAITOH, I., and SUGA N. 1995. Long delay lines for ranging are created by inhibition in the inferior colliculus of the mustached bat. *Journal of Neurophysiology* 74:1–11.

SALDANA, E., M. FELICIANO, and E. MUGNAINI. 1996. Distribution of descending projections from primary auditory neocortex to inferior colliculus mimics the topography of intracollicular projections. *Journal of Comparative Neurology* 371:15–40.

SALES, G., and D. PYE. 1974. *Ultrasonic communication by animals.* New York: John Wiley and Sons.

SANIDES, D., and F. SANIDES. 1974. A comparative Golgi study of the neocortex in insectivores and rodents. *Zeitschrift für Mikroskopie und anatomische Forschung* 88:957–977.

SANIDES, F. 1970. Functional architecture of motor and sensory cortices in primates in the light of a new concept of neocortex evolution. In *The primate brain: Advances in primatology,* vol. 1, ed. Ch. R. Noback and W. Montagna. New York: Appleton-Century-Crofts.

SCHEICH, H. 1991. Auditory cortex: Comparative aspects of maps and plasticity. *Current Opinion in Neurobiology* 1:236–247.

SCHLUNDT, C. E., J. J. FINNERAN, D. A. CARDER, and S. H. RIDGWAY. 2000. Temporary shift in masked hearing thresholds (MTTS) of bottlenose dolphins, *Tursiops truncatus,* and white whales, *Delphinapterus leucas,* after exposure to intense tones. *Journal of the Acoustical Society of America* 107(6): 3496–3508.

SCHNITZLER, H.-U. 1970. Echoortung bei der Fledermaus *Chilonycteris rubiginosa. Zeitshrift fur Vergleichende Physiologie* 68:25–38.

———. 1984. The performance of bat sonar systems. Pp. 211–224 in *Localization and orientation in biology and engineering,* ed. D. Varjú and H.-U. Schnitzler. Berlin: Springer-Verlag.

———. 1987. Echoes of fluttering insects: Information for echolocating bats. Pp. 223–243 in *Recent advances in the study of bats,* ed. M. B. Fenton, P. Racey, and J. M. V. Rayner. Cambridge: Cambridge University Press.

SCHNITZLER, H.-U., and O. W. HENSON JR. 1980. Performance of airborne animal sonar systems, I. Microchiroptera. Pp. 109–81 *Animal sonar systems,* ed. R. G. Busnel and J. F. Fish. New York: Plenum Press.

SCHNITZLER, H.-U., and E. KALKO. 1998. How echolocating bats search and find food. Pp. 183–197 in *Bats, phylogeny, morphology, echolocation, and conservation biology,* ed. T. H. Kunz and P. A. Racey. Washington, D.C.: Smithonian Institution Press.

SCHREINER, C. E., and J. R. MENDELSON. 1990. Functional topography of cat primary auditory cortex: Distribution of integrated excitation. *Journal of Neurophysiology* 64:1442–1459.

SCHREINER, C. E., and J. V. URBAS. 1986. Representation of amplitude modulation in the auditory cortex of the cat. I. The anterior auditory field (AAF). *Hearing Research* 21:227–241.

SCHULLER, G., and G. D. POLLAK. 1979. Disproportionate frequency representation in the inferior colliculus of Doppler-compensating greater horseshoe bats: Evidence for an acoustic fovea. *Journal of Comparative Physiology A* 132:47–54.

SCHULLER, G., E. COVEY, and J. H. CASSEDAY. 1991. Auditory pontine grey: Connections and response properties in the horseshoe bat. *European Journal of Neuroscience* 3:648–662.

SCHULLER, G., W. E. O'NEILL, and S. RADTKE-SCHULLER. 1991. Facilitation and delay sensitivity of auditory cortex neurons in CF-FM bats, *Rhinolophus rouxi* and *Pteronotus p. parnellii. European Journal of Neuroscience* 3:1165–1181.

SCHUSTERMAN, R. J. 1980. Behavioral methodology in echolocation by marine mammals. Pp. 11–41 in *Animal sonar systems,* ed. R. G. Busnel and J. F. Fish. New York: Plenum Press.

SEMPLE, M. N., and L. M. KITZES. 1993a. Binaural processing of sound pressure level in cat primary auditory cortex: Evidence for representation based on absolute levels rather than interaural level differences. *Journal of Neurophysiology* 69:449–461.

———. 1993b. Focal selectivity for binaural sound pressure level in cat auditory cortex: Two-way intensity network tuning. *Journal of Neurophysiology* 69:462–473.

SEVERTSOV, A. N. 1939. *Morphological principles of evolution.* Nauka, Moskva.

SHAMMA, S. A., J. W. FLESHMAN, P. R. WISER, and H. VERSNEL. 1993. Organization of response areas in ferret primary auditory cortex. *Journal of Neurophysiology* 69:367–383.

SHANNON-HARTMAN, S., D. WONG, and M. MAEKAWA. 1992. Processing of pure-tone and FM stimuli in the auditory cortex of the FM bat, *Myotis lucifugus. Hearing Research* 61:179–188.

SHIBA, K., I. SATOH, N. KOBAYASHI, and F. HAYASHI. 1999. Multifunctional laryngeal motoneurons: An intracellular study in the cat. *Journal of Neuroscience* 19:2717–2727.

SIMMONS, J. A. 1971. Echolocation in bats: Signal processing of echoes for target range. *Science* 171:925–928.

———. 1979. Perception of echo phase information in bat sonar. *Science* 204:1336–1338.

———. 1989. A view of the world through the bat's ear: The formation of acoustic image in echolocation. *Cognition* 33:155–199.

SIMMONS, J. A., and A. D. GRINNELL. 1988. The performance of echolocation: The acoustic images perceived by echolocating bats. Pp. 353–385 in *Animal sonar: Processes and performance,* ed. P. E. Nachtigall and P. W. B. Moore. New York: Plenum.

SIMMONS, J. A., M. J. FERRAGAMO, and C. F. MOSS. 1998. Echo-delay resolution in sonar images of the big brown bat, *Eptesicus fuscus. Proceedings of the National Academy of Sciences* 95:12647–12652.

SIMMONS, J. A., D. J. HOWELL, and N. SUGA. 1975. The information content of bat sonar echoes. *American Scientist* 63:204–215.

SIMMONS, J. A., C. F. MOSS, and M. FERRAGAMO. 1990. Convergence of temporal and spectral information into acoustic images of complex sonar targets perceived by the echolocating bat, *Eptesicus fuscus. Journal of Comparative Physiology A* 166:449–470.

SIMMONS, J. A., S. P. DEAR, M. J. FERRAGAMO, T. HARESIGN, and J. FRITZ. 1996. Representation of perceptual dimensions of insect prey during terminal pursuit by echolocating bats. *Biological Bulletin* 191:109–121.

SKINNER, B. F. 1961. *Cumulative record.* New York: Appleton-Century-Crofts.

SLEPECKY, N. B. 1996. Structure of the mammalian cochlea. Pp. 44–129 in *The cochlea,* ed. P. Dallos, A. N. Popper, and R. R. Fay. New York: Springer-Verlag.

STANFORD, T. R., S. KUWADA, and R. BATRA. 1992. A comparison of interaural time sensitivity of neurons in the inferior colliculus and thalamus of the unanesthetized rabbit. *Journal of Neuroscience* 12:3200–3216.

STEELE, C. R. 1997. Three-dimensional mechanical modeling of the cochlea. Pp. 455–460 in *Diversity in auditory mechanics,* ed. E. R. Lewis, G. R. Long, R. F. Lyon, P. M. Narins, C. R. Steele, and E. Hecht-Poinar. Singapore: World Scientific.

STEPHAN, H., and P. PIRLOT. 1970. Volumetric comparisons of brain structures in bats. *Zeitschrift für zoolo-*

logische Systematik und Evolutionsforschung 8:200–236.

SUGA, N. 1965. Functional properties of auditory neurones in the cortex of echolocating bats. *Journal of Physiology* (London) 181:671–700.

———. 1969. Classification of inferior collicular neurones of bats in terms of responses to pure tones, FM sounds and noise bursts. *Journal of Physiology* (London) 200:555–574.

———. 1972. Analysis of information bearing elements in complex sounds by auditory neurons of bats. *Audiology* 11:58–72.

———. 1973. Feature extraction in the auditory system of bats. Pp. 675–754 in *Basic mechanisms in hearing,* ed. A. R. Möller. New York: Academic Press.

———. 1977. Amplitude-spectrum representation in the Doppler-shifted-CF processing area of the auditory cortex of the mustache bat. *Science* 196:64–67.

———. 1982. Functional organization of the auditory cortex: Representation beyond tonotopy in the bat. Pp. 157–218 in *Cortical sensory organization: Multiple auditory areas,* ed. C. N. Woolsey. Clifton, N.J.: Humana Press.

———. 1984. The extent to which biosonar information is represented in the bat auditory cortex. Pp. 315–373 in *Dynamic Aspects of Neocortical Function,* ed. G. M. Edelman, W. E. Gall, and W. M. Cowan. New York: John Wiley & Sons.

———. 1988. What does single-unit analysis in the auditory cortex tell us about information processing in the auditory system? Pp. 331–349 in *Neurobiology of the neocortex,* ed. P. Rakic and W. Singer. New York: John Wiley & Sons.

———. 1990a. Biosonar and neural computation in bats. *Scientific American* 262:60–66.

———. 1990b. Cortical computational maps for auditory imaging. *Neural Networks* 3:3–21.

———. 1994. Multi-function theory for cortical processing of auditory information: implications of single-unit and lesion data for future research. *Journal of Comparative Physiology A* 17:135–44.

———. 1995a. Sharpening of frequency tuning by inhibition in the central auditory system: Tribute to Yasuji Katsuki. *Neuroscience Research* 21:287–299.

———. 1995b. Processing of auditory information carried by species-specific complex sounds. Pp. 295–313 in *The cognitive neurosciences,* ed. M. Gazzaniga. Cambridge: MIT Press.

———. 1997. Parallel-hierarchical processing of complex sounds for specialized auditory function.

Pp. 1409–1418 in *Encyclopedia of acoustics,* ed. M. J. Crocker. New York: John Wiley & Sons.

SUGA, N., and J. HORIKAWA. 1986. Multiple time axes for representation of echo delay in the auditory cortex of the mustached bat. *Journal of Neurophysiology* 55:776–805.

SUGA, N., and P. H.-S. JEN. 1976. Disproportionate tonotopic representation for processing CF-FM sonar signals in the mustache bat auditory cortex. *Science* 194:542–544.

———. 1977. Further studies on the peripheral auditory system of "CF-FM" bats specialized for fine frequency analysis of Doppler-shifted echoes. *Journal of Experimental Biology* 69:207–232.

SUGA, N., and T. MANABE. 1982. Neural basis of amplitude-spectrum representation in auditory cortex of the mustached bat. *Journal of Neurophysiology* 47:225–255.

SUGA, N., and W. E. O'NEILL. 1979. Neural axis representing target range in the auditory cortex of the mustached bat. *Science* 206:351–353.

SUGA, N., and K. TSUZUKI. 1985. Inhibition and level-tolerant frequency tuning in the auditory cortex of the mustached bat. *Journal of Neurophysiology* 53:1109–1145.

SUGA, N., M. KAWASAKI, and R. F. BURKARD. 1990. Delay-tuned neurons in auditory cortex of mustached bat are not suited for processing directional information. *Journal of Neurophysiology* 64:225–235.

SUGA, N., H. NIWA, and I. TANIGUCHI. 1983. Representation of biosonar information in the auditory cortex of the mustached bat, with emphasis on representation of target velocity information. Pp. 829–867 in *Advances in vertebrate neuroethology,* ed. J.-P. Ewert, R. R. Capranica, and D. J. Ingle. New York: Plenum Press.

SUGA, N., J. F. OLSEN, and J. A. BUTMAN. 1990. Specialized subsystems for processing biologically important complex sounds: Cross-correlation analysis for ranging in the bat's brain. *Cold Spring Harbor Symposium on Quantitative Biology* 55:585–597.

SUGA, N., W. E. O'NEILL, and T. MANABE. 1978. Cortical neurons sensitive to combinations of information-bearing elements of biosonar signals in the mustache bat. *Science* 200:778–781.

———. 1979. Harmonic-sensitive neurons in the auditory cortex of the mustache bat. *Science* 203:270–274.

SUGA, N., J. A. SIMMONS, and P. H.-S. JEN. 1975. Peripheral specialization for fine analysis of Doppler-shifted echoes in the auditory system of the "CF-FM" bat,

Pteronotus parnellii. Journal of Experimental Biology 63:161–192.

SUGA, N., J. YAN, and Y. F. ZHANG. 1997. Cortical maps for hearing and egocentric selection for self-organization. *Trends in Cognitive Sciences* 1:13–20.

SUGA, N., H. NIWA, I. TANIGUCHI, and D. MARGOLIASH. 1987. The personalized auditory cortex of the mustached bat: Adaptation for echolocation. *Journal of Neurophysiology* 58:643–654.

SUGA, N., W. E. O'NEILL, K. KUJIRAI, and T. MANABE. 1983. Specificity of combination-sensitive neurons for processing of complex biosonar signals in the auditory cortex of the mustached bat. *Journal of Neurophysiology* 49:1573–1626.

SUGA, N., J. A. BUTMAN, H. B. TENG, J. YAN, and J. F. OLSEN. 1995. Neural processing of target-distance information in the mustache bat. Pp. 13–30 in *Active hearing,* ed. A. Flock, D. Ottoson, and M. Ulfendahl. England: Elsevier. Sci. Ltd., Pergamon Press.

SULLIVAN, W. E. 1982. Neural representation of target distance in auditory cortex of the echolocating bat *Myotis lucifugus. Journal of Neurophysiology* 48:1011–1032.

SUM, Y. W., and D. MENNE. 1988. Discrimination of fluttering targets by the FM-bat *Pipistrellus stenopterus? Journal of Comparative Physiology A* 163:349–354.

SUN, X., Q. CHEN, and P. H.-S. JEN. 1996. Corticofugal control of central auditory sensitivity in the big brown bat, *Eptesicus fuscus. Neuroscience Letters* 212:131–134.

SUN, X., P. H.-S. JEN, D. SUN, and S. ZHANG. 1989. Corticofugal influences on the responses of bat inferior collicular neurons to sound stimulation. *Brain Research* 495:1–8.

SUPIN A. YA., and V. V. POPOV. 1985. Recovery cycles of the dolphin's brain stem responses to paired acoustic stimuli [in Russian]. *Doklady Akademii Nauk SSSR* [Proceedings of the Academy of Sciences of USSR] 283:740–743.

———. 1995a. Temporal resolution in the dolphin's auditory system revealed by double-click evoked potential study. *Journal of the Acoustical Society of America* 97:2586–2593.

———. 1995b. Frequency tuning and temporal resolution in dolphins. Pp. 95–110 in *Sensory systems of aquatic mammals,* ed. R. A. Kastelein, J. A. Thomas, and P. E. Nachtigall. Woerden, The Netherlands: De Spil Publishers.

———. 1995c. Envelope-following response and modulation transfer function in the dolphin's auditory system. *Hearing Research* 92:38–46.

SUPIN, A. YA., V. V. POPOV, and V. O. KLISHIN. 1993. ABR frequency tuning curves in dolphins. *Journal of Comparative Physiology A* 173:649–656.

SUZUKI, K., and N. SUGA. 1991. Acuity in ranging based upon neural responses in the FM-FM area of the mustache bat. *Society of Neuroscience Abstracts* (1991) 445:181–188.

SYKA, J., and J. POPELAR. 1984. Inferior colliculus in the rat: neuronal responses to stimulation of the auditory cortex. *Neuroscience Letters* 51:235–240.

TANAKA, H., and D. WONG. 1993. The influence of temporal pattern of stimulation on delay tuning of neurons in the auditory cortex of the FM bat, *Myotis lucifugus. Hearing Research* 66:58–66.

TANAKA, H., D. WONG, and I. TANIGUCHI. 1992. The influence of stimulus duration on the delay tuning of cortical neurons in the FM bat, *Myotis lucifugus. Journal of Comparative Physiology A* 171:29–40.

TANIGUCHI, I., H. NIWA, D. WONG, and N. SUGA. 1986. Response properties of FM-FM combination-sensitive neurons in the auditory cortex of the mustached bat. *Journal of Comparative Physiology A* 159:331–337.

TENG, H., and D. WONG. 1992a. Dual amplitude sensitivity of delay-sensitive neurons in the auditory cortex of *Myotis lucifugus. Association for Research in Otolaryngology Abstract* 15:141.

———. 1992b. Neural mechanisms of delay shift with amplitude variation in delay-sensitive cortical neurons of *Myotis lucifugus. Society for Neuroscience Abstract* 18:152.

THIES, W., E. K. V. KALKO, and H.-U. SCHNITZLER. 1998. The roles of echolocation and olfaction in two Neotropical fruit-eating bats, *Carollia perspicillata* and *C. castanea,* feeding on *Piper. Behavioral Ecology and Sociobiology* 42:397–409.

THOMAS, H., J. TILLEIN, P. HEIL, and H. SCHEICH. 1993. Functional organization of auditory cortex in the Mongolian gerbil (*Meriones unguiculatus*). I. Electrophysiological mapping of frequency representation and distinction of fields. *European Journal of Neuroscience* 5:882–897.

TRAHIOTIS, C., and L. R. BERNSTEIN. 1986. Lateralization of bands of noise and sinusoidally amplitude-modulated tones: Effects of spectral locus and bandwidth. *Journal of the Acoustical Society of America* 79:1950–1957.

TRAPPE, M., and H.-U. SCHNITZLER. 1982. Doppler-shift compensation in insect-catching horseshoe bats. *Naturwissenschaften* 69:193–194.

TSUMOTO, T., O. D. CREUTZFELDT, and C. R. LEGENDY. 1978. Functional organization of the corticofugal system from visual cortex to lateral geniculate nucleus in the cat (with an appendix on geniculo-cortical monosynaptic connections). *Experimental Brain Research* 32:345–364.

TSUZUKI, K., and N. SUGA. 1988. Combination-sensitive neurons in the ventroanterior area of the auditory cortex of the mustached bat. *Journal of Neurophysiology* 60:1908–1923.

TUNTURI, A. R. 1952. A difference in the representation of auditory signals for the left and right ears in the isofrequency contours of the right middle ectosylvian auditory cortex of the dog. *American Journal of Physiology* 168:712–727.

TUTTLE, M. D., and M. J. RYAN. 1981. Bat predation and the evolution of frog vocalizations in the Neotropics. *Science* 214:677–678.

VARNASSI, U., and D. G. MALINS. 1971. Unique lipids of the porpoise (*Tursiops gilli*): Differences in triacyl glycerols and wax esters of acoustic (mandibular canal and melon) and blubber tissues. *Biochemica et Biophysica Acta* 231:415–418.

VATER, M. 1982. Single unit responses in cochlear nucleus of horseshoe bats to sinusoidal frequency and amplitude modulated signals. *Journal of Comparative Physiology A* 149:369–388.

———. 1997. Evolutionary plasticity of cochlear design in echolocating bats. Pp. 49–55 in *Diversity in auditory mechanics,* ed. E. R. Lewis, G. R. Long, R. F. Lyon, P. M. Narins, C. R. Steele, and E. Hecht-Poinar. Singapore: World Scientific.

———. 1998. Adaptations of the auditory periphery of bats for echolocation. Pp. 231–247 in *Bats, phylogeny, morphology, echolocation, and conservation biology,* ed. T. H. Kunz and P. A. Racey. Washington, D.C.: Smithonian Institution Press.

VATER, M., and K. BRAUN. 1994. Parvalbumin, Calbindin D-28K, and Calretinin immunoreactivity in the ascending auditory pathway of horseshoe bats. *J. Comp. Neurol.* 341:534–558.

VATER, M., and D. DUIFHUIS. 1986. Ultra-high frequency selectivity in the horseshoe bat: Does the bat use an acoustic interference filter? Pp. 23–31 in *Auditory frequency selectivity* (NATO ASI Series), ed. B. C. J. Moore and R. D. Patterson. New York: Plenum Press.

VATER, M., and M. KÖSSL. 1996. Further studies on the mechanics of the cochlear partition in the mustached bat. I. Ultrastructural observations on the tectorial membrane and its attachments. *Hearing Research* 94: 63–78.

VATER, M., and M. LENOIR. 1992. Ultrastructure of the horseshoe bat's organ of Corti. I. Scanning electron

microscopy. *Journal of Comparative Neurology* 318: 367–379.

VATER, M., and P. SCHLEGEL. 1979. Comparative auditory neurophysiology of the inferior colliculus of two molossid bats, *Mollosus ater* and *Mollosus mollosus.* II. Single unit responses to frequency-modulated signals and signal and noise combinations. *Journal of Comparative Physiology A* 131:147–160.

VATER, M., and W. SIEFER. 1995. The cochlea of *Tadarida brasiliensis,* specialized functional organization in a generalized bat. *Hearing Research* 91:178–195.

VATER, M., A. S. FENG, and M. BETZ. 1985. An HRP-study of the frequency-place map of the horseshoe bat cochlea: Morphological correlates of the sharp tuning to a narrow frequency band. *Journal of Comparative Physiology A* 157:671–686.

VATER, M., M. LENOIR, and R. PUJOL. 1992. Ultrastructure of the horseshoe bat's organ of Corti. II. Transmission electron microscopy. *Journal of Comparative Neurology* 318:380–391.

VEL'MIN V. A., and N. A. DUBROVSKIY. 1976. On the analysis of pulsed sounds by dolphins [in Russian]. *Doklady Akademii Nauk SSSR* [Proceedings of the Academy of Sciences of USSR] 225:470–473.

VIEMEISTER, N. F. 1979. Temporal modulation transfer functions based upon modulation thresholds. *Journal of the Acoustical Society of America* 66:1364–1380.

VILLA, A. E., E. M. ROUILLER, G. M. SIMM, P. ZURITA, Y. DE RIBAUPIERRE, and F. DE RIBAUPIERRE. 1991. Corticofugal modulation of the information processing in the auditory thalamus of the cat. *Experimental Brain Research* 86:506–517.

WANG, C., B. G. CLELAND, and W. BURKE. 1985. Synaptic delay in the lateral geniculate nucleus of the cat. *Brain Research* 343:236–245.

WARR, W. B. 1992. Organization of olivocochlear efferent systems in mammals. Pp. 410–448 in *The mammalian auditory pathway: Neurophysiology,* ed. A. N. Popper and R. R. Fay. New York: Springer-Verlag.

WARTZOK, D., and D. KETTEN. 1999. Marine mammal sensory systems. Pp. 117–176 in *Biology of marine mammals,* ed. J. E. Reynolds and S. A. Rommel. Washington, D.C.: Smithsonian Press.

WATANABE, T., K. YANAGISAWA, J. KANZAKI, and Y. KATSUKI. 1966. Cortical efferent flow influencing unit responses of medial geniculate body to sound stimulation. *Experimental Brain Research* 2:302–317.

WEIMANN, J. M., and E. MARDER. 1994. Switching neurons are integral members of multiple oscillatory networks. *Current Biology* 4:896–902.

WEINBERGER, N. M. 1995. Dynamic regulation of receptive fields and maps in the adult sensory cortex. *Annual Review of Neuroscience* 18:129–158.

———. 1998. Physiological memory in primary auditory cortex: Characteristics and mechanisms. *Neurobiology of Learning and Memory* 70:226–251.

WENSTRUP, J. J. 1999. Frequency organization and responses to complex sounds in the medial geniculate body of the mustached bat. *Journal of Neurophysiology* 82:2528–2544.

WENSTRUP, J. J., and C. D. GROSE. 1995. Inputs to combination-sensitive neurons in the medial geniculate body of the mustached bat: The missing fundamental. *Journal of Neuroscience* 15:4693–4711.

WENSTRUP, J. J., and S. A. LEROY. 2001. Spectral integration in the inferior colliculus: Role of glycinergic inhibition in response facilitation. *Journal of Neuroscience* 21:RC124 (1–6).

WENSTRUP, J., and R. A. SUTHERS. 1984. Echolocation of moving targets by the fish-catching bat, *Noctilio leporius. Journal of Comparative Physiology A* 155:75–89.

WENSTRUP, J. J., D. T. LARUE, and J. A. WINER. 1994. Projections of physiologically defined subdivisions of the inferior colliculus in the mustached bat: Targets in the medial geniculate body and extrathalamic nuclei. *Journal of Neuroscience* 346:207–236.

WENSTRUP, J. J., D. H. MITTMANN, and C. D. GROSE. 1999. Inputs to combination-sensitive neurons of the inferior colliculus. *Journal of Comparative Neurology* 409:509–528.

WEVER, E. G., J. G. McCORMICK, J. PALIN, and S. H. RIDGWAY. 1971a. The cochlea of the dolphin, *Tursiops truncatus:* General morphology. *Proceedings of the National Academy of Sciences* 68:2381–2385.

———. 1971b. The cochlea of the dolphin *Tursiops truncatus:* The basilar membrane. *Proceedings of the National Academy of Sciences* 68:2708–2711.

———. 1971c. The cochlea of the dolphin *Tursiops truncatus:* Hair cells and ganglion cells. *Proceedings of the National Academy of Sciences* 68:2908–2912.

———. 1972. Cochlear structure in the dolphin, *Lagenorhynchus obliquidens. Proceedings of the National Academy of Sciences* 69:657–661.

WILSON, J. P., and V. BRUNS. 1983. Middle-ear mechanics in the CF-bat *Rhinolophus ferrumequinum. Hearing Research* 10:1–13.

WINER, J. A., and D. K. MOREST. 1983. The medial division of the medial geniculate body of the cat: Implications for thalamic organization. *Journal of Neuroscience* 3:2629–2651.

WINER, J. A., D. T. LARUE, and G. D. POLLAK. 1995. GABA and glycine in the central auditory system of the mustache bat: Structural substrates for inhibitory neuronal organization. *Journal of Comparative Neurology* 355:317–353.

WONG, D., and S. L. SHANNON. 1988. Functional zones in the auditory cortex of the echolocating bat *Myotis lucifugus. Brain Research* 453:349–352.

WONG, D., M. MAEKAWA, and H. TANAKA. 1992. The effect of pulse repetition rate on the delay sensitivity of neurons in the auditory cortex of the FM bat (*Myotis lucifugus*). *Journal of Comparative Physiology A* 170:393–402.

WOOLSEY, C. N., and E. M. WALZL. 1942. Topical projection of nerve fibers from local regions of the cochlea to the cerebral cortex of the cat. *Johns Hopkins Hospital Bulletin* 71:315–344.

WU, M. I., and P. H.-S. JEN. 1995. Responses of pontine neurons of the big brown bat, *Eptesicus fuscus,* to temporally patterned sound pulses. *Hearing Research* 85:155–168.

———. 1996. Temporally patterned pulse trains affect directional sensitivity of inferior collicular neurons of the big brown bat, *Eptesicus fuscus. Journal of Comparative Physiology A* 179:385–393.

XIE, D. H., and O. W. HENSON JR. 1998. Tonic efferent-induced cochlear damping in roosting and echolocating mustached bats. *Hearing Research* 124:60–68.

XIE, D. H., M. M. HENSON, and A. L. BISHOP. 1993. Efferent terminals in the cochlea of the mustached bat, quantitative data. *Hearing Research* 66:81–90.

YAN, J., and N. SUGA. 1996a. Corticofugal modulation of time-domain processing of biosonar information in bats. *Science* 273:1100–1103.

———. 1996b. The midbrain creates and the thalamus sharpens echo-delay tuning for the cortical representation of target-distance information in the mustached bat. *Hearing Research* 93:102–110.

———. 1998. Corticofugal modulation of the midbrain frequency map in the bat auditory system. *Nature Neuroscience* 1:54–58.

YANG, L. C., and G. D. POLLAK. 1997. Differential response properties to amplitude modulated signals in the dorsal nucleus of the lateral lemniscus of the mustache bat and the roles of GABAergic inhibition. *Journal of Neurophysiology* 77:324–340.

YIN, T. C. T., and S. KUWADA. 1983. Binaural interaction in low-frequency neurons in inferior colliculus of the cat: II. Effects of changing rate and direction of inter-aural phase. *Journal of Neurophysiology* 50:1000–1019.

YIN, T. C. T., J. C. K. CHAN, and D. R. F. IRVINE. 1986. Effects of interaural time delays of noise stimuli on low-frequency cells in the cat's inferior colliculus. I. Responses to wide-band noise. *Journal of Neurophysiology* 55:280–300.

YIN, T. C. T, J. A. HIRSCH, and J. C. K. CHAN. 1985. Responses in the cat's superior colliculus to acoustic stimuli. II. A model of interaural intensity sensitivity. *Journal of Neurophysiology* 53:746–758.

ZHANG, Y., and N. SUGA. 1997. Corticofugal amplification of subcortical responses to single tone stimuli in the mustached bat. *Journal of Neurophysiology* 78:3489–3492.

———. 2000. Modulation of responses and frequency tuning of thalamic and collicular neurons by cortical activation in mustached bats. *Journal of Neurophysiology* 84:325–333.

ZHANG, Y., N. SUGA, and J. YAN. 1997. Corticofugal modulation of frequency processing in bat auditory system. *Nature* 387:900–903.

ZILLES, K. 1985. *The cortex of the rat: A stereotaxic atlas.* Berlin: Springer-Verlag.

ZOOK, J. M., and P. A. LEAKE. 1989. Connections and frequency representation in the auditory brainstem of the mustache bat, *Pteronotus parnellii. Journal of Comparative Neurology* 290:243–261.

ZOOK, J. M., J. A. WINER, G. D. POLLAK, and R. D. BODENHAMER. 1985. Topology of the central nucleus of the mustache bat's inferior colliculus: Correlation of single unit response properties and neuronal architecture. *Journal of Comparative Neurology* 231:530–546.

ZWICKER, E. 1952. Die Grenzen der Horbakeit der Amplituden-modulation und der Frequenzmodulation eines Tones. *Acustica* 2:125–133.

ZWISLOCKI, J. J., and L. K. CEFARATTI. 1989. Tectorial membrane II. Stiffness measurements *in vivo. Hearing Research* 42:211–228.

PART THREE

Performance and Cognition in Echolocating Mammals

Performance and Cognition in Echolocating Mammals

W. Mitchell Masters and *Heidi E. Harley*

Introduction

Microchiropteran bats and odontocete cetaceans (toothed whales and dolphins) are the two mammalian groups in which echolocation has been most highly developed over a long evolutionary period (Novacek 1985; N. B. Simmons and Geisler 1998; Berta and Sumich 1999). Because only one species of megachiropteran, the tongue-clicking bat *Rousettus,* is known to echolocate (Henson and Schnitzler 1979), our discussion is restricted to the microchiropteran bats. In both bats and dolphins, individuals can, when necessary, use sonar as a nearly complete replacement for vision, a statement that is, as far as we know, not true for any species of any other group of vertebrates (e.g., *Rousettus,* oilbirds, cave swiftlets, shrews, or rodents [Henson and Schnitzler 1979]). The introductory chapter by Au (this volume) and chapter 11 in Au 1993 compare a number of aspects of echolocation in bats and dolphins. After a brief consideration of some important similarities and differences between bats and dolphins, we discuss here the performance and cognitive abilities of the two groups.

Probably the most striking difference between a bat and dolphin is size. This translates, by the usual allometric relationships, into dolphins having much larger brains than bats. All other things being equal, a larger brain should be capable of more complex processing, and certainly some experiments conducted with dolphins may be impossible to duplicate with bats (e.g., chapter 40 in this volume on cross-modal matching). We might therefore expect dolphins to have a more highly developed cognitive capability. Also, their generally more complex social lives (Pryor and Norris 1991; Norris et al. 1994) should contribute to dolphins' cognitive sophistication. Nevertheless, one lesson of cognitive ethology is that animals are adapted to solve the types of problems they must solve to survive, so, despite the typical microchiropteran having a (literally) pea-sized brain, bats are quite adept at many complex sonar-related tasks, such as target detection, identification, tracking, and interception (e.g., Webster 1967b).

Size also affects an animal's ability to direct its sonar beam in a particular direction. The energy needs to be concentrated in the most important direction (usually straight ahead) or on a particular target (e.g., prey or obstacle). The frequency range used by bats and dolphins is roughly comparable (approximately 20–150 kHz, depending on species and situation), but because the speed of sound in water is about five times greater than in air, the wavelength of a given frequency is five times longer. In order to project a narrow beam, the size of the sound-radiating structure should be comparable to, or larger than, the wavelength of the sound being projected. Bats emit sounds through their mouth or nose; dolphins through their forehead (melon). The longer wavelength in water makes forming a narrow beam more difficult, but the much larger head size of dolphins more than offsets this difficulty, and hence the beamwidth of their sounds is narrower than that for bats. The 3 dB beamwidth measured for seven odontocete species ranges from about 6° to 16°, depending mainly on the size of the animal (Au 1993; Au et al. 1999). For bats, beamwidth usually exceeds 30°, but width varies with frequency. As expected, high frequencies (short wavelengths) show more directionality than low frequencies (for references and discussion, see Au 1993; Hartley and Suthers 1989, 1990). Directionality of reception (hearing) is also size dependent in a way analogous to sound transmission and shows similar trends (Schnitzler and Henson 1979; Coles et al. 1989; Au 1993).

The interaction of directional sound transmission, directional reception at the ear, and differences in the waveform (time of arrival and spectra) received by the two ears all affect an animal's ability to determine the angular direction of an echo. Ear spacing and head size have a strong influence on the time and frequency cues contained in the acoustic waveform at the two ears, with larger animals generally able to localize sounds more precisely (Heffner and Heffner 1992). The higher speed of sound in water reduces the magnitude of timing differences between the ears, and the longer wavelength of sounds in water makes sound shadowing by the head less effective; but, again, because of their much larger size, the relationships favor dolphins over bats. For the big brown bat, *Eptesicus fuscus,* the accuracy of angle resolution is about 1.5° in the horizontal plane (Simmons et al. 1983; Masters, Moffat, and Simmons 1985) and about 3° in the vertical plane (Lawrence and Simmons 1982a). The corresponding echolocation measurements have not been made with dolphins (see Au, the introduction to this volume), but for passive sound localization of sonar-type clicks, bottlenose dolphins, *Tursiops truncatus,* perform slightly better than this, with a resolution of about 0.8° in the horizontal plane and 0.9° in the vertical (Renaud and Popper 1975). Using sonar echoes, their resolution might be even better due to the directionality of transmission. For big brown bats, and perhaps others, angular resolution of echoes is apparently much better than for exogenous sounds arising in the environment, suggesting that active and passive localization of sounds may be subserved by different neural mechanisms (Koay et al. 1998; Razak, Fuzessery, and Lohuis 1999).

The ability of dolphins to resolve time intervals, as measured by range-discrimination experiments and two-wave-front experiments, appears to be slightly, although not markedly, better than for bats (Au, the introduction to this volume; but see Simmons, Ferragamo, and Moss 1998 for more recent results with bats). Since the timing of impulses at the level of the individual neuron is not extremely precise, the ability of the whole system (bat or dolphin brain) to make accurate time measurements probably depends on averaging across a population of many neurons. If so, dolphins' better time resolution could again be due to their size, inasmuch as their larger brain should allow access to a greater population of neurons for averaging.

Besides size, another obvious difference between bats and dolphins is their sonar signals. Odontocete signals are typically clicklike. Some species use very short clicks (<70 µs) whose waveforms contain only a few cycles, and some use longer clicks (100 to about 400 µs) containing more cycles (Au 1993). In a stationary setting, dolphins produce several spectrally distinct click types (Houser, Helweg, and Moore 1999). The proportion of each type appears to depend both on the individual dolphin and on the task, but a functional significance has not yet been attached to the different types. For short-click odontocetes, the frequency of peak energy increases as the click intensity increases (Au, the introduction to this volume). Dolphins also change the interclick interval as target range changes (Au 1993). However, in comparison to bats, current data from dolphins positioned in stationary hoops suggest that they do not modify their echolocation signals greatly. This view may change as clicks from freely moving dolphins tracking moving targets are analyzed.

The sonar signals used by bats are extremely diverse compared to dolphins, not only across species but within individuals of a given species (Simmons and Stein 1980; Neuweiler and Fenton 1988; Fenton 1995). Bats commonly alter the time-frequency structure of their sonar emissions as they approach and intercept a target, when they are close to "clutter" such as the ground or vegetation, or when there are other competing sounds in the environment, such as sonar from other bats (Griffin,

Webster, and Michael 1960; Simmons et al. 1978; Simmons and Kick 1983; Miller and Degn 1981; Schnitzler et al. 1987; Rydell 1993; Obrist 1995; Kalko and Schnitzler 1998; Jensen and Miller 1999; Siemers and Schnitzler 2000). The particular signal a bat uses seems to be determined largely by the information needed at the time, so signal modification can be viewed as sensory prefiltering to enhance certain aspects of the environment while suppressing others (Webster 1967a; Simmons and Stein 1980; Simmons and Kick 1983; Schnitzler and Kalko 1998; Wadsworth and Moss 2000). This approach to echolocation may be necessitated by the more limited cognitive capabilities of bats compared to dolphins, or because the quick reactions needed by bats to intercept prey and avoid obstacles puts a premium on rapid extraction of pertinent information, which can be optimized by proper signal design (Webster 1967a).

The foraging signals of bats customarily are divided into two categories: CF (constant-frequency) and FM (frequency-modulated) sounds. CF signals are also called "high duty cycle" signals, because the signal is on for more than 80% of the cycle and off for less than 20%. Conversely, FM signals are low duty cycle, because the signal is on for less than 20% of a cycle (Fenton 1995). CF signals are usually long and contain a constant-frequency (or narrowband) portion longer than 10 ms (Simmons and Stein 1980). A downward frequency sweep normally terminates the CF portion of the signal, so these signals may be designated CF-FM. FM signals are characterized by wide bandwidth, which is achieved by using a broad frequency sweep, or by using multiple harmonics, or both. Sometimes, especially in the search phase of echolocation, a narrowband, nearly CF tail is added to the end of the frequency-swept portion of the signal while the frequency sweep itself is much reduced, so these signals may be designated short FM-CF. The CF versus FM distinction is overly simplistic and fails to recognize important properties of the signals actually used by bats, particularly the extent to which they can be varied (Simmons and Stein 1980; Neuweiler and Fenton 1988; Fenton 1995), but it is embedded in the literature as a convenient label distinguishing two fundamentally different signal types. The longest dolphin signals barely overlap the shortest bat signals.

CF bats specialize in detecting moving prey, using either the Doppler shift of the constant frequency, which arises from the difference in velocity between bat and insect, or by detecting the changes in echoes of a flying insect caused by its wing beat, or both (Schnitzler et al. 1987; Ostwald 1988). These bats are also capable of Doppler compensation to remove Doppler shifts arising from the echoes of background objects (Schnitzler and Henson 1979; Gaioni, Riquimaroux, and Suga 1990). Dolphins are not known to detect targets by a Doppler shift, and in any event Doppler shifts from natural targets in water are quite small due to their low velocity compared with the velocity of sound in water. Instead, dolphins appear to operate more like the FM bats, which are specialized for detecting the position of an object rather than its velocity. In terms of range-Doppler ambiguity (Woodward 1953), the signals of FM bats and dolphins are strongly biased toward the "range" side; hence, a single emission has good ability to determine target distance but poor ability to determine its velocity. A CF signal has the opposite properties. Presumably the poor range resolution of CF signals is one reason why CF bats include an FM component in their signal, since knowing the range of an object is important for either interception or avoidance. The prominence of the FM component in the signals of both FM and CF bats becomes larger as a bat approaches a prey or a landing spot. Besides providing more precise timing information, which is useful for determining range, broadening signal bandwidth also gives more information about target shape, which is useful for identifying a target.

Although signals of dolphins and FM bats are probably not useful for detecting Doppler shift, this does not mean they cannot sense target motion, for instance by observing the change in target range. For dolphins, Zaitseva and Korolev (1996) reported that the efficiency of target detection is related to its acceleration. Bats, too, are sensitive to target motion, as revealed by "jitter" experiments in which the apparent range of an electronically synthesized echo from a phantom target is made to jump back and forth between two ranges on successive emissions. Jitter sensitivity, in terms of the difference in time of arrival of echoes from the two target ranges, is certainly under 1 μs (Menne et al. 1989; Moss and Schnitzler 1989; Masters et al. 1997) and may be as small as 10 ns (Simmons et al. 1990a). Similar experiments have not yet been done using dolphins.

It might seem that the much longer signals used by FM bats, as compared to the short clicks of dolphins, would require a significantly different method of signal analysis, but from a signal-processing point of view the important parameters are bandwidth and energy. Bandwidth is roughly comparable in the sonar signals of bats and dolphins, although, as just mentioned, the bandwidth of a bat's signal depends strongly on the behavioral situation (e.g., searching for versus intercepting prey). The energy content of the sonar signal of a large dolphin such as *T. truncatus* is greater than that of a bat such as *E. fuscus*, but for some of the smaller dolphins the reverse is true (Au 1993, table 11.1). Pulse compression (dispersive filtering) can convert a long batlike signal into a short dolphinlike signal without changing either energy or bandwidth. In other words, although the sonar waveforms used by bats and dolphins are different, and no doubt bats and dolphins process their signals differently (if nothing else, millions of years of separate evolution virtually guarantees this), it is nonetheless true that from a fundamental point of view the differ-

ences between an FM bat cry and a dolphin click are not profound. It could be that the principal reason for the different signals used in the two groups is that bats have a more limited peak-power output than dolphins and consequently must use a longer signal to produce enough energy in their emission (McCue 1966). On the other hand, the difference could just as easily be due to the different methods of signal generation. In bats, the signal is produced by vibration of the vocal folds in the larynx (Suthers 1988). In dolphins, clicks are generated below the blowhole in a region where air is forced through two ridged, interlocking arcs with attached fatty bursae. After air forces the arcs apart, the bursae reunite with a slap to produce the signal, which is radiated through the dolphin's melon (Cranford, Amundin, and Norris 1996; Cranford and Amundin, chapter 4, this volume).

Both bats and dolphins use sonar to navigate in their highly three-dimensional environments, but they exhibit differences in their use of sonar, some of which can be traced to the effectiveness of sonar in air and water, and some to the behavior of their typical prey. In air, the effective range of sonar is much more limited than in water, largely due to the higher absorption of sound in air, especially at high frequencies. As a result, a bat can detect an insect-sized object at a distance of about five meters at most (Kick 1982), and in nature probably much less (Griffin, Webster, and Michael 1960; Schnitzler and Henson 1979; Simmons and Kick 1983). A dolphin, conversely, can detect small targets more than a hundred meters away (Au and Snyder 1980). Since a bat can fly at 5 m/sec or faster, the time between detection and interception of a small target is often extremely short, necessitating quick reactions and high maneuverability. In effect, targets tend to "pop up" at close range on the bat's "display screen," whereas this is probably not the usual case for a dolphin, whose sonar can illuminate a much larger volume. As mentioned above, the difference in the time available to evaluate signals might account at least in part for the relative invariability of dolphin signals. Time constraints may also account for the highly stereotypic changes in a bat's signal as it goes through a capture sequence (Griffin, Webster, and Michael 1960; Webster 1967a; Simmons and Kick 1983).

Bats encounter another problem related to prey capture that dolphins meet less often, namely, that their prey may hear them coming. Many insects, particularly those that frequent bat-populated places, have organs of hearing that allow them to detect a bat's approach (Fullard 1987; Rydell, chapter 43, this volume; Faure and Hoy, chapter 51, this volume). In contrast, most fish are deaf in the ultrasonic range used by odontocetes, although fish in the family Clupeidae detect ultrasound, and thus may avoid predation by dolphins (Mann, Lu, and Popper 1997; Mann et al. 1998). An interesting case involves pods of killer whales (*Orcinus orca*) specializing on either sea lions or fish. Pods pursuing sea lions (mammals with good hearing) use echolocation much more sparingly than those seeking fish (Barrett-Lennard, Ford, and Heise 1996). Insects that hear an approaching bat may simply fly away and never be detected by the bat, but if a bat is very close, the insect is likely to go into evasive maneuvers (power dives, loops, etc.), thus taxing the prey-capture ability of bats to the utmost. Yet a further development in the ongoing "arms race" between bats and their insect prey is that some insects reply to the approach of a bat with ultrasonic signals of their own. Whether these signals serve to startle the bat, jam its sonar, or warn of the insect's noxiousness, or all three to some degree, is an area of active investigation (see below, and Tougaard, Miller, and Simmons, chapter 50, this volume; Miller, Futtrup, and Dunning, chapter 52, this volume). As far as is known, "talking back" by prey is not a behavior dolphins encounter.

One ability of dolphins unknown in bats is that they can probe beneath surfaces with their sonar. For instance, dolphins have been observed hunting over a sandy bottom for buried fish (Rossbach and Herzing 1997; Herzing, chapter 56, this volume), and dolphins have been trained for military purposes to detect buried mines (P. Moore 1997). They are able to perform such feats because sound penetrates through the bottom to some depth. Likewise, dolphins are able to discriminate between metal cylinders with different wall thicknesses because sound penetrates the metal, providing an echo from both the front and back surfaces (and potentially other surfaces) (Au and Pawloski 1992). Bats are not known to use sonar in this way, although they are known to evaluate echoes coming from the same direction, as if one target were behind another (Masters et al. 1997). Fishing bats, such as *Noctilio leporinus,* cannot detect fish below the water surface but can detect a fish's fin if it breaks the surface by as little as a millimeter (Suthers 1965) or when the fish jumps (Schnitzler et al. 1994). Presumably the tremendous difference in acoustic impedance between air and water (or between air and almost any solid or liquid) prevents significant penetration by the incident sound. On the other hand, the lesser difference in impedance between water and metal, or between water and water-saturated ocean bottom, allows dolphins to detect and use subsurface echoes.

One final caution is perhaps in order. Despite our frequent reference to bats and dolphins as homogenous groups, there really is no such thing as "the" bat or "the" dolphin sonar signal. There are over 700 different species of microchiropterans, and over 60 species of odontocetaceans, each with its own peculiar mode of behavior and lifestyle. While the similarities among species may be responsible for the general success of the two groups, the differences are undoubtedly critical in allowing species to coexist in ecological space, and just as certainly such differences will find expression in how each species' sonar system operates and is used by the animal. Unfortunately, at the moment our knowledge of comparative echolocation is rudimentary. Most of what we know comes from just a few species of bats and perhaps a dozen species of dolphins (with the bottlenose dolphin, *T. truncatus,* by far the most intensively studied). This situation will improve in time, but we emphasize that a search for generalities—necessary and informative as it is—will not reveal the whole story. Doubtless, many fascinating chapters are yet to be written about the differences between what we now think of as very similar species. For a telling example from neurophysiology, we can compare the differences between the FM bats (*Eptesicus fuscus* and *Myotis lucifugus*), or between the CF bats (*Pteronotus parnellii* and *Rhinolophus ferrumequinum*), in the organization of their auditory cortex (see Grinnell 1995).

Performance and Cognition in Bats

With this introduction to their different ecological milieus, we now turn to the performance and cognition of bats and dolphins. More is known about performance-related aspects of the sonar, such as range resolution, angle discrimination, and less about more cognitive aspects, such as target identification, cross-modal matching, and scene analysis. Some aspects of performance have already been covered in the introductory chapter by Au (this volume) or in the preceding section of this chapter. Because sonar research in bats and in dolphins has historically been disjunct, with different investigators studying each taxon and concentrating on different aspects of the sonar systems, we discuss research on these two groups separately for the most part. Our principal goal is to provide context for the contributions following this chapter for readers not familiar with the literature. In the last section of the chapter, we discuss the need for closer integration of sonar research in bats and dolphins.

PREY DETECTION BY BATS

The majority of microchiropterans, including nearly all bats in temperate regions, make their living by capturing flying insects. Before a prey item can be caught, it must be detected, so one might expect bats to be very good at detecting faint echoes. The optimal detector of a faint echo—that is, one near the background noise level (which might be of either environmental or neural origin)—is a matched-filter receiver (Woodward 1953). (As discussed below in the section on neural processing and learning, a matched filter is also the optimal receiver for determining pulse-echo delay.) One way to characterize a matched-filter receiver is as a filter designed to respond best to a signal having the particular time-frequency structure of the expected echo and to reject other signals. Because an echo would have, to a good first approximation, the same time-frequency structure as the outgoing pulse, we would expect a bat's matched filter, if it exists, to be good at detecting both the outgoing emission and the echo. Møhl (1986) and Masters and Jacobs (1989) used this idea to test whether bats use matched-filter detection of echoes. In their experiments, every time a bat emitted a call, it received a playback of an electronically synthesized model echo resembling its sonar emission. The synthesized echo was either played forward (i.e., with the typical downward frequency sweep) or backward (with upward sweep), and the bat's task was simply to detect the echo. If a matched filter is involved in echo detection, there should have been a noticeable difference in the detection threshold for normal and reversed signals, since only the normal sweep would have the time-frequency structure expected by the matched filter. However, in both experiments thresholds were about equal for normal and reversed signals, suggesting that bats do not detect echoes by matched filtering and may instead use echo energy or envelope for detection. One caveat concerning these experiments, however, is that even when the synthesized echo was played in the forward direction and therefore had the nominally correct time-frequency structure, it probably only rarely matched the exact time-frequency structure of the bat's emission. Mismatch is likely because the synthesized echo had an unvarying time-frequency structure while a bat's emissions are normally variable (Masters and Raver 2000). If the bat constructs its matched filter anew on each outgoing emission (e.g., Simmons et al. 1996), then the synthesized echo might have failed to match the emission closely enough for the hypothetical matched filter to operate properly.

Whether dolphins detect targets as efficiently as a matched filter was investigated by Au and Pawloski (1989), who found that dolphins performed about 7–8 dB worse than an ideal receiver in a noise-limited situation. Although their performance was therefore not ideal, it is substantially better than obtained for the serotine bat, *Eptesicus serotinus,* by Troest and Møhl (1986). Given the task of detecting a target in noise, the bats proved inferior to an ideal detector by 40–50 dB.

Even if bats do not use matched filtering for echo detection, it does not necessarily follow that the time-frequency structure of the signal is of no consequence for target detection. The search-phase calls of many FM, low-duty-cycle bats are relatively long, having extended "tails" of nearly constant frequency. Concentrating call energy in a narrow frequency band might be a design feature allowing bats to improve detection of faint echoes by integrating the energy received in the narrowband region over time (Simmons and Stein 1980). Indeed, neurophysiological evidence suggests that the best auditory thresholds for big brown bats are in this region (Casseday and Covey 1992). Unfortunately, the only behavioral data to date (Surlykke 1995) suggest that the length of the tail is not related to how easily a bat detects the signal, a result that raises questions about the purpose of an extended narrowband tail in search calls.

In addition to the problem of detecting faint echoes, bats must avoid self-deafening by their own loud emissions. Humans find it harder to detect a sound that immediately follows another sound, a phenomenon termed forward masking, and the louder the masking sound, or the nearer in time, the more difficult it becomes to detect the second sound (B. Moore 1997). A bat's emission is very loud compared with most echoes, so it might well act as a masker. In fact, echo detection becomes more difficult when the echo closely follows the emission (Kick and Simmons 1984; Hartley 1992; Simmons, Moffat, and Masters 1992), as would occur when a bat approaches a target. Previous experiments have not, however, differentiated between two possible causes of increasing threshold with decreasing emission-echo delay: (1) forward masking or (2) the middle-ear reflex, in which a loud sound causes small muscles of the middle ear to contract and reduce sound transmission from tympanum to cochlea (Henson 1965). In fact, both effects could operate. The experiments of Siewert, Schilinger, and Schmidt (chapter 36, this volume) separate the effect of forward masking from the middle-ear reflex. They found forward masking persists for pulse-echo separations up to 30–40 ms, corresponding approximately to the likely maximum range of target detection of about 5 m (Kick 1982). Siewert et al. found a change in threshold of about 6 dB per doubling of target distance, but Kick and Simmons (1984) and Simmons, Moffat, and Masters (1992) obtained about double this slope. The discrepancy in slope might be due to the additional effect of the middle-ear reflex, but other factors discussed by Siewert et al. may contribute (see also Hartley 1992 for a discussion of threshold versus target distance, especially the possible contribution of backward masking to these results).

If a sonar receiver has multiple chances to detect an echo, it can combine information from several echoes to detect a fainter target, or detect it more reliably, than a receiver that looks at only a single echo. Bats apparently integrate information across several emissions (Roverud and Grinnell 1985), but how they do so is not known. The experiments by Surlykke (chapter 37, this volume) address this question. Using electronically synthesized echoes, Surlykke measured bats' detection threshold as the number of echoes returned to the bat per trial was varied. When bats received fewer than 10 echoes per trial, their threshold was elevated compared to when they received an echo after every emission. In fact, when the number of echoes was limited to 3 or fewer, the detection threshold was about 25 dB worse than when the number of echoes was unlimited. The elevation in threshold is substantially more than predicted by either energy integration or "multiple looks" models, leaving open the question of how bats integrate information across echoes. How bats incorporate information across echoes to build up a perceptual image of their surroundings, a more complicated task that simply detecting the presence of a target, is also a subject for future investigation.

PREY IDENTIFICATION BY BATS

After detecting potential prey, bats must evaluate its echoes to decide whether or not to intercept it. In a laboratory setting, bats display great ability to distinguish targets to be captured from those to be ignored or rejected (Griffin 1967; Simmons and Chen 1989), although in nature bats may be less discriminating (Barclay and Brigham 1994). Signal design determines the information available to a bat about the target. As discussed earlier, high-duty-cycle CF bats use the long-CF portion of their signal to separate stationary objects from moving prey. They can also derive additional information from prey-induced modifications of the CF portion to identify particular targets or classes of targets. For instance, changes in echo amplitude or frequency occurring at the wing-beat rate of a flying insect could be used to identify the insect (von der Emde and Schnitzler 1990; Kober and Schnitzler 1990). Sum and Menne (1988), working with the FM bat *Pipistrellus stenopterus,* found that these bats, like CF bats, can detect differences in the "flutter rate" of a whirling blade, and, in fact, are not markedly inferior to CF bats in this regard. The lack of a long-CF component in the sonar signals of *P. stenopterus* suggested to Sum and Menne that FM bats might use a different mechanism for flutter discrimination than CF bats. They proposed that the bats perceived both Doppler-shifted echoes from the moving blade and non-Doppler-shifted echoes from the stationary holder as a two-wave-front target. Further experimental and theoretical work by Grossetête and Moss (1998) supports this idea.

As already mentioned, bats using short-duty-cycle FM calls emit longer calls with a nearly constant frequency tail when searching for prey. As with CF bats, prey-induced changes in the tail region might provide information about target identity. To investigate this idea, von Stebut and Schmidt (chapter 35, this volume) tested frequency discrimination by big brown bats at frequencies within the spectral region of the tail and found no obvious specialization. Stebut and Schmidt also tested bats' ability to discriminate 50 Hz frequency modulation (a rate matching the wing-beat frequency of some insects) from frequency modulation at other rates. They found that big brown bats are comparable to CF bats in discriminating small differences in modulation frequency, suggesting that modulation of the echo tail could help FM bats categorize targets. Because these tests employed electronically synthesized targets rather than real targets, and the test stimuli were long compared with the short FM sweeps of bats in the experiments of Sum and Menne (1988) and Grossetête and Moss (1998), the two-wave-front mechanism of flutter detection suggested by these latter investigators would not apply to Stebut and Schmidt's results.

BAT-INSECT INTERACTION

Bats and insects have shared the night skies for at least 50 million years, so it is not surprising that many nocturnal insects developed strategies for avoiding bat predation, nor that bats evolved countermeasures. For instance, many night-flying insects, including many moths, crickets, lacewings, and praying mantids, are able to hear the ultrasonic sounds of bats and take evasive action (Roeder 1965; Moiseff, Pollack, and Hoy 1978; Yager et al. 1990; May 1991). Some moths have gone beyond mere detection of bats and have acquired the ability to produce clicklike ultrasonic sounds of their own, which they emit as a bat approaches. A long-standing debate exists about the mode of action of moth clicks (Surlykke 1988; Bates and Fenton 1990; Miller 1991; Fullard, Simmons, and Saillant 1994). The three main hypotheses—not mutually exclusive—are (1) the startle hypothesis: that because a bat does not expect them, clicks distract the bat and increase the chance it will miss the insect; (2) the aposematism hypothesis: because click-producing insects often are protected, chemically or otherwise, clicks warn of the insect's noxiousness, leading a bat to reject the prey; and (3) the jamming hypothesis: because clicks overlap the frequency range of the sonar signal and often are produced toward the end of the capture sequence, they interfere in some way with bats' sonar and make interception difficult. This last possibility is particularly interesting, because, if true, we might learn something about how bats process sonar signals by studying the jamming effect of moth clicks. Tougaard, Casseday, and Covey (1998) and Tougaard, Miller, and Simmons (chapter 50, this volume) found that moth clicks may suppress or disrupt neural discharges in the lateral lemniscus, thus disrupting the temporal processing of echoes.

The experiments of Miller, Futtrup, and Dunning (chapter 52, this volume) focused on bats' response to extraneous sounds. These investigators attempted to mimic, as naturalistically as possible, the situation in which a bat in the process of intercepting an insect receives distracting sounds (either clicks from moths or echolocation pulses from other bats) during the last part of its approach. Surprisingly, they found that capture success was not decreased by either moth clicks or bat pulses—if anything success was higher on trials with extraneous sounds. These results contrast with those of an earlier experiment by Dunning and Roeder (1965) in which bats avoided capturing targets associated with moth clicks but captured those associated with sounds from another bat. Why did bats behave differently in the two experiments? One possibility is simply that the earlier study mimicked the natural conditions during prey capture with less fidelity than the present one.

SENSORY INTEGRATION IN BATS

Although their remarkable sonar system is an important reason for the scientific attention bats have received, it is not their only sensory system. Moreover, we can be sure that bats, like other animals, function as an integrated whole, making use of all their capabilities, including vision, olfaction, and kinesthetic sense. Bats are not blind (despite public confusion on this point) so an obvious question is how their sonar and visual systems interact, since both are largely concerned with creating a three-dimensional representation of the environment. We know, for instance, that bats are attracted to light when seeking to escape confinement (Mistry and McCracken 1990). But whether in light or in dark, bats ordinarily emit echolocation calls whenever they are moving, suggesting that sonar is an essential component of their sensorimotor system. This contrasts with dolphins, which may forgo echolocation when visibility is good, echoic sea conditions are poor, they are resting, and so on (e.g., Goodson and Mayo 1995; Norris et al. 1994; Dos Santos et al., chapter 53, this volume). In bats, some attention has been focused recently on the superior colliculus in the midbrain as a likely location for sensorimotor integration, both for directing the bat's movement and for controlling the sonar system (e.g., head and pinnae positions, production of vocalizations, etc.) to acquire needed information (Valentine and Moss 1997, 1998).

While bats obviously acquire a great deal of information via sonar, they also show remarkable spatial memory—memory that sometimes overrides sonar information. Griffin (1974) described cases in which bats flying familiar pathways crashed headlong into unexpected barriers. Conversely, bats accustomed to flying through a small opening between two rooms continued to fly through the former position of the opening even after the opening itself had been greatly enlarged. Observations like these raise questions about how bats integrate idiothetic information (that is, positional information derived from proprioceptive cues) and allothetic information (that is, positional information derived from monitoring the environment via sonar and vision). The chapter by Schmidt, Schlangen, Krasemann, and Höller (chapter 34, this volume) takes a behavioral approach to the salience of external and internal cues in an orientation task and concludes that both play a role in bats' spatial orientation.

NEURAL PROCESSING AND LEARNING

As discussed in the introduction, a matched filter is the optimal receiver for detecting a known signal in noise, but experiments have not supported the idea that bats use matched-filter detection (Møhl 1986; Masters and Jacobs 1989). A matched filter also happens to be the ideal receiver for determining the time delay of an echo relative to the emission. Bats use this delay to determine the distance (range) to a target (Simmons 1973). Masters and Jacobs (1989) used normal and time-reversed echoes to test whether echo time-frequency structure is important in target ranging (as opposed to target detection) and found it was. With time-forward (normal) echoes bats could discriminate a target at 80 cm from another at 81–82 cm, but with time-reversed echoes

a bat typically could not distinguish a target at 80 cm from one at 100 cm. This experiment did not prove that bats use a matched filter, such as an engineer might design for radar or sonar, but did strongly suggest that the echo's time-frequency structure is relevant to bats. A more recent set of experiments in which echo structure was altered in various ways also supported the idea that proper time-frequency structure is critical for echo processing (Masters and Raver 2000).

Masters and Jacobs (1989) found that a bat given the signal of another bat as an electronically simulated echo is impaired in its ability to determine target range. On the other hand, Miller (1991) and Surlykke (1992a, 1992b) found bats could use a signal from another bat quite well for range discrimination. The discrepancy between these results may be explained by the amount of prior practice bats received with the foreign signal. Masters and Raver (chapter 38, this volume) tested whether practice with a novel echo improves a bat's ability to determine target range. They found that practice with an echo that had previously caused difficulty did help, suggesting that learning has a role in echo processing.

Masters and Raver's experiment, as well as others that have used playback of a standard, unvarying signal stored in computer memory to synthesize a "phantom" target (Miller 1991; Surlykke 1992a, 1992b), raises an important question about the neurophysiological mechanisms bats use to analyze echoes. Most current theories of range determination require a close match between the time-frequency structure of emission and echo (Suga 1990; Park and Pollak 1993; Simmons et al. 1996). Of course, in nature a good match is achieved automatically because the echo from an insect or other target has nearly the same time-frequency structure as the incident sound (the bat's emission). However, when a phantom target is synthesized from a stored, unvarying echo, some mismatch between emission and echo is inevitable—because while the emissions vary from one to the next, the echoes do not. Despite such mismatch, bats discriminate range quite well. How they do so, and whether they use the same echo-processing method for normal echoes and for unvarying echoes, remain unanswered questions.

Performance and Cognition in Dolphins

Although researchers studying bats and dolphins share general questions about echolocation performance and mechanisms, the topics in this section are quite different from those just discussed for bats. Reasons for this difference include considerations of the goals of the major funding agency, ethical and legal constraints, expectations about large-brained mammals, and difficulty accessing wild and captive dolphins. In general, studies of bat echolocation are more ecologically focused, whereas most studies of dolphin echolocation are conducted with trained captive dolphins in which both dolphins and targets are stationary (although cf., e.g., Zaitseva and

Korolev 1996). In this situation a dolphin's echolocation emissions tend to be fairly stable across targets; in general, they are short (20–50 μs), loud (up to 200 dB SPL), broadband signals with peak frequencies from 40 to 130 kHz (Au 1993). In parallel, the dolphin's sound receiving system processes a wide range of frequencies (from 75 Hz to 150 kHz, Johnson 1967) and is designed for fine temporal resolution (Supin and Popov 1995, chapter 22, this volume). Forward- and backward-masking studies suggest that dolphins process echoes as an energy detector with an integration time of 264 μs—that is, multiple echoes received within 264 μs are analyzed as one (P. Moore et al. 1984; Vel'min and Dubrovskiy 1976). (Part 2 of this volume has more on the auditory system.) Studies of target detection and discrimination reveal that dolphins can determine an object's material, size, shape, and wall thickness (see Au 1993 and Nachtigall 1980 for reviews).

In the last decade, most investigations of dolphin cognition during echolocation tasks have targeted the dolphin's recognition of objects across disparate sensory experiences. Essentially, the question is one of representation: what does the dolphin represent when it is echolocating? After hearing an object's echo, the dolphin may identify and represent the object itself or may merely represent relevant features of the echo without object recognition (Roitblat, Helweg, and Harley 1995). Humans typically represent objects. For example, if one hears keys jangling, one does not merely represent one's proximal experience of the frequency, amplitude, and duration of the sound. Rather, one represents keys. Recent work in dolphin echolocation probes this ability in the bottlenose dolphin, *Tursiops truncatus*. Data come mainly from three different sources: eavesdropping studies in which a silent dolphin must process echoes generated by an echolocating neighbor; studies using aspect-dependent objects that produce different echoic signals when presented at different angles; and multimodal studies in which a dolphin must use vision and echolocation to solve a matching task.

EAVESDROPPING

Although bats appear to use emissions of other bats to identify fruitful foraging locations (e.g., Leonard and Fenton 1984), dolphins' use of echolocation signals from conspecifics is mostly speculative (Jerison 1986). However, recent work by Xitco and Roitblat (1996) suggests that it is possible for a nonecholocating dolphin to identify an object by eavesdropping on the echoes from that object when it is ensonified by an echolocating neighbor.

In the Xitco and Roitblat study, two male dolphins were trained separately to perform an echoic matching task. A subject was presented with a sample object placed behind a visually opaque but echoically transparent black polyethylene screen and rewarded for choosing an object identical to the sample that was presented

with two other objects behind a screen in another area of the tank. After learning this task, one dolphin (the listener) was trained to remain at the surface with its click-emitting structures (Cranford, Amundin, and Norris 1996) in air and its echo-receiving lower jaw (Brill et al. 1988) under water. The second dolphin (the inspector) echolocated from a hoop 30 cm below the water's surface. The dolphins' rostrums were within 5° of each other relative to the sample. After the inspector echolocated the sample, each dolphin swam to its own array of alternatives located on either side of a cement "coral head" in a 22 million-liter tank.

When objects were familiar to both dolphins, the performance of the listener was 50%—that is, significantly above the 33% chance rate. With objects familiar only to one dolphin, the listener's performance was only significantly above chance when the inspector was correct. However, the inspector's errors were not predictive of the listener's choices, suggesting that the listener was not cued by the inspector's choices. Apparently, echo quality was not good enough for successful matching by either dolphin when the inspector was incorrect. Overall, the results suggest that a nonecholocating dolphin can gain information about objects by eavesdropping on an echolocating neighbor.

The implications of these findings are important for several reasons. First, echolocation may play a role in social or communicative behaviors in dolphins (Harley, Xitco, and Roitblat et al. 1995), as it may also do in bats (e.g., Masters, Raver, and Kazial 1995). Second, the neural mechanisms underlying bat and dolphin echolocation may be quite different. In bats, production of a sonar emission appears to be critical to processing a returning echo (Suga 1990). Although similar physiological work in dolphins is not possible in the United States, the fact that a dolphin can analyze the echoes generated by a neighbor's clicks suggests that echo processing can occur without click production in the dolphin (Xitco and Roitblat 1996). Finally, a dolphin may recognize an object as the same even when the echoes it receives from it are quite variable. Echoes are highly directional. When an echo is received only 10° off-axis, it can be much quieter (−10 dB) and lower (−114 kHz peak frequency) than when received on-axis (Au 1993). It is probable that a dolphin hears very different echoes when eavesdropping versus actively echolocating identical objects.

ASPECT-DEPENDENT OBJECTS

Studies of matching aspect-dependent objects presented at different angles also suggest that dolphins can recognize an object as the same even when its echoes vary. When a cube or pyramid rotates, echo spectra and the timing of the amplitude peaks (highlight structure) in the waveforms change. Hence, if a dolphin represents only the characteristics of an object's echo rather than the object itself, it should not recognize that echoes from a cube presented face-on versus edge-on represent the same object.

In the earliest study of recognition of objects intentionally presented at different angles, a blindfolded male dolphin echolocated a foam cube and a cylinder (three sizes of each) presented pairwise (Nachtigall, Murchison, and Au 1980). The dolphin was always rewarded for choosing the cylinder. The cube was typically presented with one face toward the dolphin; the cylinder was presented vertically. Baseline performance accuracy with these objects in these positions was 91%. Performance varied in probe trials in which objects were presented in new orientations: performance with vertical cylinder versus cube edge was 93%, horizontal cylinder versus cube edge was 71%, and flat end of the cylinder versus face of the cube was 57%. With the first probe pair, the dolphin probably recognized the rewarded cylinder in its standard orientation. With the third probe pair, the dolphin's performance was at chance—perhaps because the greater variability in echo amplitudes from flat faces (of either cylinder or cube) led to greater overlap between the echoes of the cylinder and cube. However, the dolphin's performance with the second pair suggested it could recognize the cylinder (at least some of the time) whether horizontal or vertical. Hence, these results suggest some tolerance for variability in the echoes returning from the same object.

A later study confirmed the ability of a dolphin to recognize that a cylinder, whether horizontal or vertical, was the same object (Au and Turl 1991). An echolocating dolphin was trained to discriminate between a hollow aluminum cylinder and a hollow steel cylinder presented at angles of 0°, 10°, 45°, 80°, and 90°. (From the dolphin's point of view, the cylinder at 0° was vertical and at 90° was horizontal.) The dolphin could discriminate between the cylinders at all angles. Then the dolphin was trained to discriminate between the hollow aluminum cylinder and a hollow coral rock cylinder at angles of 0°, 45°, and 90°. It was tested in probe fashion (i.e., rewarded for all responses) at new angles of 15°, 30°, 60°, and 75°. The dolphin's performance at novel angles was nearly perfect, even though the waveforms and frequency spectra of the echoes varied with object angle. Again, these data suggest that a dolphin can identify an object as the same even when it produces very different echoes.

In a third study, a blindfolded dolphin performed a matching task with aspect-dependent objects (Helweg et al. 1996a, 1996b). The stimuli included three aspect-dependent objects (a polyurethane foam rectangular prism, pyramid, and cube) and one aspect-independent object (a water-filled stainless-steel sphere). All objects were free to rotate and echoes varied with aspect (except for the sphere). The dolphin produced more clicks when echolocating aspect-dependent objects, but it could discriminate among all objects. Recognition of

the sphere was best. Statistical and network models suggest that the dolphin used amplitude and frequency information integrated across successive echoes. Again, echo variability did not preclude object recognition.

Of course, given that dolphins swim at varying depths and encounter objects from varying angles, life without this ability would be difficult. Perceptual constancy may arise from sensitivity to some invariant feature experienced by the dolphin that we have not yet identified. From a Gibsonian perspective (Gibson 1960, 1979), echoes produced from different aspects of an object, though variable, are still generated by, and hence organized by, that object. A moving dolphin producing click trains directed at a single object may gain a great deal of redundant, overlapping information helping it to perceive this organization. If so, recognition should be better if the dolphin integrates information over successive echoes, just as recent models of dolphin echolocation suggest (e.g., Roitblat et al. 1991). Recognition would require the more distal object-based representational system that experiments indicate dolphins indeed possess.

Multimodal Matching

Research on integration of visual and echoic information suggests that an echolocating dolphin does not rely simply on a proximal sound-based representational system. If a dolphin performs echolocation-matching tasks merely by sound matching, then vision should not improve performance. However, vision helps.

In one study, a dolphin performed a three-alternative matching task in which it either echolocated or saw objects (a gray metal cylinder, a gray PVC cylinder, or a white, thicker-walled PVC cylinder) that required the use of both visual and echoic object features for good discrimination (Harley, Roitblat, and Nachtigall 1996). The two gray cylinders were easy to discriminate visually from the white cylinder but difficult to discriminate visually from each other. The two PVC cylinders were easy to discriminate echoically from the metal cylinder but difficult to discriminate echoically from each other. The dolphin performed at about 70% accuracy in the single modality conditions, but its performance accuracy jumped to over 90% when it could use both vision and echolocation. Apparently, it could integrate visual and echoic information to make discriminations.

In a second experiment (Harley, Roitblat, and Nachtigall 1996), three familiar three-alternative object sets were presented in various modality combinations, including two cross-modal conditions: visual-only sample to echoic-only alternatives, and echoic-only sample to visual-only alternatives. Matching successfully from one modality to another required that different sensory experiences of the same object be linked through object identity. The dolphin's performance accuracy was above chance levels with most objects.

In another cross-modal study, a dolphin matched familiar objects across vision and echolocation (Pack and Herman 1995; Pack, Herman, and Hoffmann-Kuhnt, chapter 41, this volume). The dolphin previously had never simultaneously both seen and echolocated the objects; hence, it could not have learned to associate her visual and echoic experiences of those objects. Overall and first-trial performance accuracy was above chance across eight pairs of PVC objects presented in reciprocal cross-modal conditions (visual-to-echoic, echoic-to-visual). The dolphin maintained good accuracy even when samples were presented via television. With completely novel objects, the dolphin's performance was above chance level with two object pairs and at chance level with another two object pairs in both cross-modal conditions (Herman, Pack, and Hoffmann-Kuhnt 1998).

In general, these data suggest that dolphins integrate visual and echoic information and that they associate their visual and echoic experiences with a single object. This implies that dolphins have a representational system in which an object's characteristics as perceived by different systems are integrated into a single representation. This means that if a dolphin can see, hear, and/or taste a mullet, then it will associate all those characteristics with the mullet itself, as would a human. We do not know whether bats have a similar representational system.

METHODOLOGICAL SUGGESTIONS

Although cognitive work on dolphin echolocation is off to a productive start, we need more studies in which simultaneous acoustical and behavioral data are integrated. Of the studies mentioned, only those on the perception of aspect-dependent objects include acoustical analyses of the stimuli; and, of these, only Helweg et al.'s (1996a, 1996b) work includes echoes generated by an echolocating dolphin. To gain additional insight into the echo features dolphins use, we need to analyze the echoes produced during a discrimination task on a trial-by-trial basis to compare the echoes when the dolphin succeeds or fails. In addition, because dolphins appear to do better with familiar objects, we should track how their investigation of an object changes as it becomes better known.

Another improvement would be to expand the stimulus sets. As Roitblat (chapter 39, this volume) points out, perception does not occur through the processing of sensory information alone; expectations also direct behavior. Studying dolphin echolocation in situations in which a few objects are presented over and over allows the dolphin to create a situation-specific strategy to discriminate among the objects in this restricted world; this may not reflect its strategies for object recognition in an ever-changing environment.

Conclusions and Future Work

As the preceding two sections illustrate, researchers studying bats and dolphins have often taken rather different tacks. In some cases, this probably makes good sense. For example, the eyes of microchiropterans are small and probably not suited for detailed image formation (Suthers 1966). Furthermore, bats typically hunt at night, when small eyes are at a further disadvantage. Therefore, we do not expect bats to be capable of cross-modal matching of vision and echolocation for small prey. The situation is, of course, different for dolphins, with their large, well-developed eyes. On the other hand, we might expect both bats and dolphins to be capable of aspect-independent recognition of prey by echolocation. This idea has been investigated with dolphins but has received less attention in bats. Using the CF bat *Rhinolophus ferrumequinum,* von der Emde and Schnitzler (1990) found that bats could discriminate different prey having the same wing-beat frequency, despite changes in angle of the prey relative to the bat. It is not known whether aspect-independent recognition of nonfluttering insects is possible, although the ability of FM bats to capture mealworms and avoid similar-sized disks suggests that it might be (Griffin 1967; Simmons and Chen 1989). Simmons, Moss, and Ferragamo (1990b) and Simmons, Ferragamo, and Moss (1998) suggest that the FM bat *E. fuscus* can construct a "range profile" of targets from temporal and spectral information contained in the echo. If so, bats might use this information to associate the echoes of similar prey viewed from different angles. Aspect-independent recognition in bats is an area calling for further investigation.

Just as researchers studying bats would benefit by taking a page from the research notebook of their dolphin-studying colleagues, researchers studying dolphins could benefit from the more ecological perspective common in bat work. Bat researchers have devoted extensive research to understanding how bats solve the problems posed by their environment. Prey detection, prey identification, prey capture, responses to the avoidance techniques of insects, navigation, and so on are all significant areas of bat research, and there is an interplay between field and laboratory work. Such interplay is rarer with dolphins, although it exists (e.g., Rossbach and Herzing's 1997 discovery that dolphins find buried fish occasioned an experiment by Nachtigall et al. [2000] on the discrimination of objects in sediment). At present, we have little detailed information on when and how dolphins actually use echolocation in the wild. For example, although echolocation signals have been recorded from feeding dolphins (e.g., Marten et al. 1988; Norris et al. 1994), they do not always echolocate (e.g., Dos Santos and Almada, chapter 55, this volume). In fact, Evans reported that a captive dolphin wearing eyecups and trained in an echolocation task did not echolocate when it repeatedly retrieved a live fish introduced into its tank (Wood and Evans 1980). As Roitblat (chapter 39, this volume) emphasizes, animal senses, including echolocation, have evolved to allow animals to survive in their ecological niche by providing them with useful information about the world. Further studies on wild dolphins, like those in part 4 of this volume, will likely advance studies of dolphins in the laboratory.

Another area where more information is needed is the response of fish to dolphin signals. A fair amount is known about how insects respond to bat echolocation (both on a proximal, physiological level, and on an ultimate, evolutionary level), but we lack similar information for the prey of dolphins. Some researchers have hypothesized that odontocetes can debilitate their prey by producing loud "bangs" with energy between 200 and 6000 Hz, possibly generated either through the apparatus for echolocation emissions, by the tail's impact with the water, by the jaws clapping together, or by some other means (Marten et al. 1988; Norris and Møhl 1983; Smolker and Richards 1988). However, 13 species of fish exposed to electronic clicks and to dolphins trained to echolocate loudly (around 215 dB at 1 m re 1 μPa) were apparently unaffected by the clicks (Marten et al. 1988). (Note: high-intensity dolphin clicks always have high center frequencies, usually above 100 kHz; e.g., P. Moore and Pawloski 1990. A bang is probably not an echolocation click.) On the other hand, yellowfin tuna may be attracted by the dolphin's low-frequency sounds, some of which they may detect up to 1000 m away (Finneran et al. 2000). Tuna may join groups of spotted, spinner, and common dolphins for protection and to find prey (Norris et al. 1994). As mentioned earlier, however, recent work suggests that clupeid fish (herrings and shads) can detect ultrasound and are sensitive to simulated dolphin clicks (Mann, Lu, and Popper 1997; Mann et al. 1998). These fish may well have a strategy for avoiding dolphins, but at present we do not know.

Advances in technology will continue to drive work on both bats and dolphins. Many researchers studying bats (see Moss and Schnitzler 1995 for a review) and some studying dolphins (Au, Moore, and Martin 1987) have used phantom targets—that is, electronically synthesized echoes—to probe mechanisms of echo processing. In dolphins, however, the synthesized echoes have not been well matched to the animal's outgoing signal.

The system recently developed by Aubauer et al. (see Aubauer, Au, and Nachtigall, chapter 68, this volume) apparently removes this restriction. Future work with both dolphins and bats will undoubtedly benefit from our increasing control over echoic stimuli, thereby permitting more refined study of the sonar systems of these mammals.

We also expect continued use of computer-based models to yield better understanding of, and insights into, the echolocation systems of bats and dolphins, thus informing future work. For example, data from neural networks suggest that dolphins probably integrate information across successive echoes. Dolphins tend to emit many clicks toward a single stimulus in an echolocation scan, and they consistently emit more clicks to some stimuli than others in discrimination tasks (Helweg et al. 1996b; Roitblat, Penner, and Nachtigall 1990). When information was integrated across echoes, a network model classified targets almost as well as a dolphin, whereas a standard back-propagation model without the integrator function performed much more poorly (P. Moore et al. 1991). Computer models also predicted that dolphins could discriminate among objects in sediment (Gaunaurd et al. 1998; Roitblat et al. 1995), which later empirical study confirmed (Nachtigall, Roitblat, and Pawloski 2000). The spectrogram correlation and transformation (SCAT) model of echolocation, based on FM bats, has received much attention (Simmons et al. 1996; Peremans and Hallam 1998). Recent neurophysiologically based models of echolocation also appear promising (Palakal and Wong 1999; Inoue et al. 2000).

A final question of interest to everyone studying echolocation is how echolocators construct their three-dimensional world using sonar (e.g., Simmons 1989; Altes 1992; Roitblat, Helweg, and Harley 1995; Popper, Hawkins, and Gisiner 1997; Pack, Herman, and Hoffmann-Kuhnt, chapter 41, this volume). Humans, for example, create spatial-analog representations (Shepard and Metzler 1971). Do bats or dolphins? If so, how? Perhaps some of the preceding suggestions and approaches will help answer such questions.

Acknowledgments

We thank Jeanette A. Thomas for her advice and many helpful comments.

{ 34 }

The Significance of Multimodal Orientation in Phyllostomid Bats

Uwe Schmidt, Miriam Schlangen, Philipp Krasemann, and *Patrick Höller*

Introduction

Bats possess highly developed orientation systems like echolocation, vision, and olfaction. Despite the obvious importance of echolocation in Microchiroptera, the other sensory modalities are used to a great extent (Chase 1983; Hessel and Schmidt 1994; Höller 1995; Laska 1990). In addition to this multimodal orientation capabilities, bats are known for their extraordinary spatial memory (first postulated in Möhres and Oettingen-Spielberg 1949 for *Rhinolophus ferrumequinum* and *Myotis myotis*). Neuweiler and Möhres (1967) trained *Megaderma lyra* to fly through a wire grating. The bats learned to fold their wings to pass the grating without any wing contact. After the obstacle was removed, the bats kept on folding their wings while passing the location where the grating had been. *Rhinolophus ferrumequinum* have been shown to use spatial memory to approach a familiar landing site (Heblich 1993). When the landing was moved aside, the bats often failed the new target, approaching the former location instead.

It has been discussed that this spatial memory is based partially on idiothetic orientation (Höller and Schmidt 1996). The term idiothesis was introduced by Mittelstaedt and Mittelstaedt (1979) to specify an orientation mechanism, where all spatial memory is gathered through the subject's monitoring of its own movements. The combined use of different orientation systems may provide spatial information that is robust against disturbances. In our investigations, we demonstrate the capability of two phyllostomid species (*Desmodus rotundus* and *Phyllostomus discolor*) to make simultaneous use of exogenous and intrinsic information when flying in a familiar space.

Materials and Methods

For the experiments, three lesser spearnosed bats (*Phyllostomus discolor*) and three vampire bats (*Desmodus rotundus*) were trained to proceed inside a V-shaped flight arena (length 1.8 m, height 1 m; fig. 34.1) directly from a defined starting point to a defined landing. They took off from a circular starting box, installed at the center angle of the arena. The starting box could be rotated

horizontally, altering the alignment of the takeoff. The target was near the center of the arc-shaped backwall of the arena. The target location was specified by two equidistant landmarks (yellow LED, plastic block for echoacoustic detection, 25 × 4.5 × 4.5 cm; landmarks 35 cm apart). The food reward (pupae of *Tenebrio molitor* for *P. discolor;* blood for *D. rotundus*) was concealed at the landing in a groove that extended along the whole length of the backwall. All flights were recorded on videotape under infrared light; the echolocation sounds were audiotaped (for details of the experimental design, see Schlangen 2000).

During standard experiments, the start was aligned toward the center of the backwall. To investigate the combined or exclusive use of landmark and/or idiothetic orientation we set up three critical experiments: (1) starting box rotated 15° to the left; (2) landmarks shifted 30 cm to the left; and (3) both starting box and landmarks changed simultaneously. Each animal ran 20 trials per day. Critical experiments were included once in 10 flights, to prevent bats from adapting their orienta-

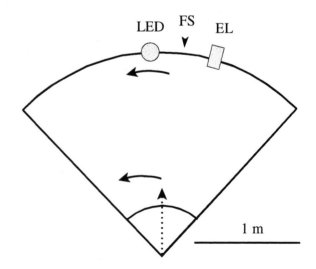

Fig. 34.1. Diagram of the arena and the experimental setup. The dotted arrow indicates the alignment of the starting box in standard flights; black arrows symbolize the changes of the setup during the critical flights (visual landmark, *LED;* echoacoustic landmark, *EL;* feeding site, *FS*).

tion strategy to changes in the setup. Food reward was presented at both prospective feeding sites.

Results

INITIAL TRAINING AND STANDARD FLIGHTS

After an initial training phase, all bats landed precisely at the standard feeding site between the two landmarks (fig. 34.2a and b). The flight routes displayed individual characteristics, but they were more or less constant for each bat. The flight corridors never deviated from the "beeline" course between the starting point and target by more than 19 cm. Counts of wingbeats and measurements of the head-trunk angles revealed significant differences of motor action during the individual flights.

CRITICAL EXPERIMENTS

In both species, the altered alignment of the starting box did not affect the position of the landing—the trajectories perfectly ended at the rewarded standard landing (fig. 34.2a and b). When the landmarks were shifted,

the bats altered their flight direction toward the new landmark position within the first 30–40 cm of flight. Hence, the flight routes and the landing sites of the bats differed markedly from standard (fig. 34.2a–c). Nonetheless, they failed to hit the new rewarded target but instead landed at an intermediate site between old and new targets. If, however, landmarks and starting alignment were shifted simultaneously, the bats landed at the new target (fig. 34.2c and d). In both species, the new landing locations differed significantly from standard and the two other critical situations (Mann-Whitney U-test, $p < 0.01$).

The number of echolocation sounds per flight did not change significantly in the different setups (16–18 calls/flight in the *Desmodus;* 23–27 in the *Phyllostomus*).

Discussion

Idiothetic orientation has been demonstrated in many animal species, ranging from arthropods (Burger 1972; Görner 1973; Mittelstaedt and Mittelstaedt 1979) to nonhuman mammals (Etienne, Maurer, and Seguinot 1996;

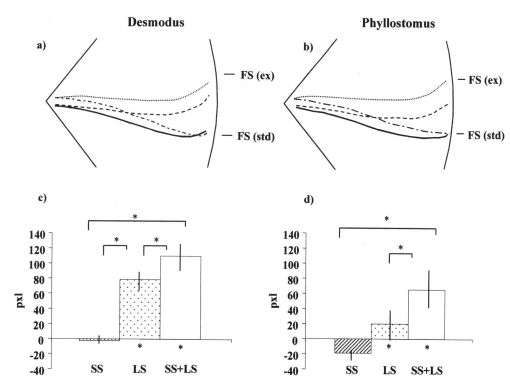

Fig. 34.2 (*a* and *b*) Average flight paths (*n* = 20 flights per experiment) of one *Desmodus* and one *Phyllostomus* under different experimental conditions: *solid line,* standard condition; *dashed and dotted line,* starting alignment shifted; *dashed line,* landmarks shifted; *dotted line,* both parameters shifted simultaneously; *FS (std),* standard feeding site; *FS (ex),* prospective feeding site in the critical trials. (*c* and *d*) Landing position in the different experimental conditions (relative distance from standard feeding site): *hatched,* starting alignment shifted, *SS; dotted,* landmarks shifted, *LS; blank,* both parameters shifted simultaneously (*SS + LS*). Data was pooled from three bats per species (mean SD; 20 flights per experiment and individual; asterisk, U-test $p < 0.01$).

Mittelstaedt and Mittelstaedt 1980) and humans (Israel et al. 1996; Mittelstaedt and Glasauer 1991). There are hints that bats also use idiothesis (Höller 1995), but experimental proof is lacking. It is quite hard to deprive bats of their allothetic contact to the environment. Their echolocation system cannot simply be shut off by the experimenter without running the risk of affecting the whole orientation behavior.

We approached the problem indirectly, checking for the effects of shifted landmarks, of altering the starting alignment, and of a combination of both. When only the starting alignment was altered, the bats landed at the standard feeding site as if nothing had changed. When the landmark position shifted, all bats clearly responded, but they did not use this information for an exact target localization. Only when both changes were presented simultaneously did the bats land at the new feeding site.

Considering the obvious effects of the landmark shift, we conclude that the bats predominantly rely on information gathered from their exteroceptive sensory systems. Navigational errors might be due to external landmarks (e.g., side walls of the arena or extra-arena cues) or to idiothetic orientation. The use of intrinsic spatial information became evident when both experimental landmarks and starting alignment were shifted, and the bats accurately approached the new feeder.

In the standard situation, each bat used rather constant flight routes. This might lead to the conclusion that they reiterated motor patterns of preceding flights. Analyses of the dynamics of the head-trunk angles during flight and the rhythm of wing beats revealed that each flight was characterized by an individual motor pattern. We conclude that kinesthesis does not play any role for this orientation behavior; the idiothetic influence most likely is achieved by acceleration and/or velocity measurements (inertial idiothesis, Mittelstaedt and Glasauer 1991).

These experiments demonstrate that bats use multimodal orientation when flying in a familiar space. The combined use of different sensory modalities includes not only the well-known exteroceptive orientation systems, but endogenous information as well.

{ 35 }

Frequency Processing at Search Call Frequencies in *Eptesicus fuscus:* Adaptations for Long-Distance Target Classification?

Boris von Stebut and *Sabine Schmidt*

Introduction

Information that bats obtain by echolocation largely depends on the structure of the emitted signal (Simmons and Stein 1980). When searching for prey, many species enhance the range of their sonar by concentrating the call energy in a narrow low-frequency band (Lawrence and Simmons 1982b; Obrist 1995). The search calls emitted by the big brown bat (*Eptesicus fuscus*) consist of narrowband sweeps with durations of up to 20 ms. The first, most intense harmonic of these calls is modulated from about 28–22 kHz (Simmons 1987). In the inferior colliculus (IC) of *E. fuscus,* this frequency range is covered by a large population of neurons that are characterized by the lowest detection thresholds. Each of these neurons is narrowly tuned to a 1–2 kHz band, with Q10 dB values above 20 (Casseday and Covey 1992; Haplea, Casseday, and Covey 1994; Ferragamo, Haresign, and Simmons 1998). Compared to the sharply tuned neurons found in CF-FM bats (Neuweiler 1990), these neurons in *E. fuscus* have been termed "filter neurons" (Casseday and Covey 1992). Casseday and Covey (1992) suggested that the narrow tuning of IC neurons is related to the detection of biologically significant features contained in the bats' search call echoes.

We investigated whether these specializations of IC neurons in the 28–22 kHz range are reflected in an improved performance in psychoacoustic experiments. First we determined the frequency difference limen (FDL) of *E. fuscus* at search call frequencies. Then we investigated the ability of this species to discriminate between different modulation frequencies.

Materials and Methods

Experiments were run in a sound-attenuated chamber (3.4 m × 2.2 m × 2.2 m). A Y-shaped platform was mounted in the middle of this chamber 1 m above the floor. Bats were trained in a two-alternative, forced-choice paradigm to enter a decision platform from a small box. According to the stimuli presented, the bat had to crawl onto one of two response platforms (Y-arms) to indicate its choice. Correct responses were rewarded with a bridgetone (whistle) and a mealworm. After each trial, the bat had to crawl back into the small box. False responses were followed by a ten-second timeout. To start a new trial, the box was turned by 180°, so that the bat's head was again directed toward the loudspeaker.

Bats were trained to discriminate between a probe and a reference signal. The probe was approximated to the reference in steps until the bat's performance dropped to chance level. At least 20 trials were run per day. In experiment 1, each probe signal was presented in not less than 15 trials (resulting in a minimum of 30 trials for each absolute deviation from the reference frequency). In experiment 2, a total of at least 30 trials were run at each condition. A sigmoidal curve was fitted to the data. Threshold was defined at 75% correct responses.

Stimulus presentation and data collection were controlled by a PC. The signals were generated with a DSP32C PC system board (Loughborough Sound Images Ltd.), D/A converted, high-pass filtered at 20 kHz using a Wavetek Dual Hi/Lo filter model 442, and fed to a Harman Kardon HK 6150 integrated amplifier. The stimuli were attenuated by 40 dB, monitored with an oscilloscope, and broadcast via a Technics A400 loudspeaker from the opposite side of the chamber (distance between platform and loudspeaker was 2.0 m). The presentation level of the stimuli at the bat's position on the decision platform was 65 dB SPL in experiment 1 and 70 dB SPL in experiment 2. Presentation level between trials was randomized by ±2 dB. Signals were checked with a B&K measuring amplifier type 2610 and a quarter-inch condenser microphone type 4135. Signal spectra were checked with a Stanford Research System model SR 770 FFT Network Analyzer.

To determine the FDL of *E. fuscus*, three bats were trained to compare a pure tone with a previously presented reference pure tone of 25 kHz. Bats had to classify the frequency of the probe as higher or lower than that of the reference. Both the duration of the reference and of the probe signal were 50 ms, separated by a 50 ms silent interval. Signal pairs were presented at a repetition rate of 1 Hz.

In the second experiment, just-noticeable differences (JND) in modulation frequency were determined in two bats. The animals were trained to discriminate a probe modulation frequency from a reference modulation frequency of 50 Hz. Either the probe or the reference were presented during a trial. The stimulus presentation was triggered by the bat's vocalizations. The stimuli were played back with a total delay of 11 ms. Stimuli consisted of a sinusoidally frequency-modulated signal centered at 25 kHz. Duration was set to 20 ms, respectively 50 ms. The modulation depth Δf was 2.5 kHz. The reference stimulus was rewarded at the left response platform, the probe stimulus at the right.

Results

In the first experiment, bat 1 was able to discriminate mean frequency differences of 481 Hz; bat 2, mean differences of 388 Hz; and bat 3, mean differences of 387 Hz (fig 35.1a). The mean FDL in this frequency range was 418 Hz, with a standard deviation of 44 Hz. The corresponding Weber ratios amounted to 0.019 in *Ef1*, 0.016 in *Ef2*, and 0.015 in *Ef3*, with a mean value of 0.017.

At a signal duration of 20 ms, the JND in modulation frequency for bats 4 and 5 amounted to 6.8 Hz and 7.3 Hz, corresponding to a mean value of 7.1 Hz (fig. 35.2a). For the 50 ms signal, the JNDs were 6.7 Hz and 8.5 Hz, with a mean of 7.6 Hz.

Discussion

FREQUENCY DISCRIMINATION
AT SEARCH CALL FREQUENCIES

E. fuscus is able to resolve differences in frequency of about 420 Hz at a reference frequency of 25 kHz. In the 20–30 kHz range, comparable FDLs of about 200–400 Hz are also common in nonecholocating mammals (cf. fig. 35.1b). In *E. fuscus*, a Weber ratio of 0.009 was determined at a carrier frequency of 50 kHz (Roverud and Rabitory 1994). This value is even better than the Weber ratio of 0.017 we obtained. Thus, the FDL at 25 kHz does not reflect the neuronal specializations found in the IC of *E. fuscus* at search call frequencies.

In contrast, the greater horseshoe bat (*Rhinolophus ferrumequinum*) is able to keep constant the resting frequency of its CF component at about 82 kHz within a range of less than 100 Hz (Schuller, Beuter, and Schnitzler 1974). With this Weber ratio of about 0.001 the frequency discrimination ability of this species in the range of its CF component is far better than that of *E. fuscus*.

It is very unlikely that we missed a similarly sensitive region in the frequency range comprised by the search call of *E. fuscus* by using a 25 kHz reference, for the following reasons. "Filter neurons" with similar tuning characteristics are found in the whole search call frequency range. In addition, *E. fuscus* does not show any cochlear specializations comparable to CF-FM bats.

The differences in FDL at the frequency of the narrowband call component may reflect the specific use of these components in FM and CF-FM bats. In CF-FM

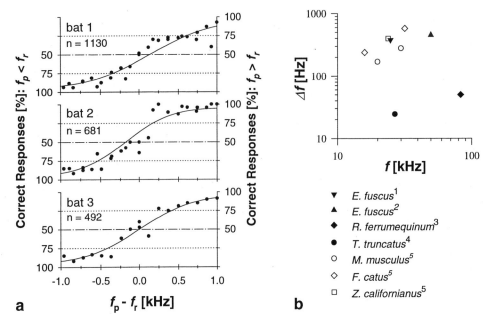

a   b

Fig. 35.1. Frequency discrimination. (*a*) Psychophysical frequency discrimination ability of three bats. Dotted lines indicate 75% correct responses both to higher and to lower probe frequencies relative to the reference. Chance level is marked by the dashed-dotted line. The abscissa is normalized according to the reference frequency at 25 kHz. A Gaussian-shaped psychometric function was fitted to the data. Each data point results from at least 15 trials (f_p, frequency of the probe; f_r, reference frequency). (*b*) Frequency difference limen Δf (Hz) of six species of mammals at the reference frequency f (kHz). Filled symbols represent echolocating species; open symbols, nonecholocating species. [1]this study; [2]Roverud and Rabitory 1994; [3]Schuller, Beuter, and Schnitzler 1974; [4]Herman and Arbeit 1972; [5]Fay 1988.

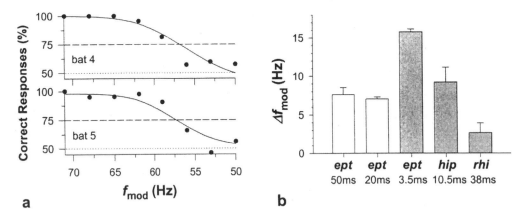

a   b

Fig. 35.2. Discrimination of modulation frequencies. (*a*) Psychometric functions of two bats at a signal duration of 20 ms and a reference modulation of 50 Hz. The correct responses (%) are plotted as a function of the probe modulation frequency f_{mod} (Hz). Each data point results from at least 30 trials. (*b*) Modulation frequency difference limen Δf_{mod} (Hz) of three bat species. On the abscissa, the species and the durations of the signals used are specified *ept, E. fuscus; hip, Hipposideros lankadiva; rhi, Rhinolophus rouxi.*

bats, the CF component remains dominant throughout prey pursuit and capture (Grinnell 1995). These bats exploit the Doppler shift of the echo due to their own flight velocity to spectrally separate the CF components of calls and echoes (Neuweiler 1990). This mechanism al-

lows these bats to exploit the CF component as a carrier of glints produced by insects throughout target approach.

In contrast, *E. fuscus* emits long and narrowband calls only during the search phase. When a bat approaches an insect, it gradually switches to FM signals

(Kick and Simmons 1984) and adapts call duration to target distance to avoid a call-echo overlap. Whereas the Doppler-shift-compensation behavior requires an exact frequency determination of both the emitted CF component and its echo, a similar precision is not essential for the analysis of glints per se. Thus, *E. fuscus* may classify its prey by evaluating the amplitude and frequency glints of the search call echoes reflected from fluttering insects.

DISCRIMINATION OF MODULATION FREQUENCIES IN FM SIGNALS

We used sinusoidally frequency-modulated stimuli to determine the ability of *E. fuscus* to discriminate between different modulation rates. These stimuli differ from echo glints by their lack of amplitude modulations and a smooth change in frequency (as opposed to abrupt spectral broadenings). However, they are similar to echo glints with respect to the occurrence of cyclical modulations, while the bats were forced to base their decision solely on modulation rate.

Roverud, Nitsche, and Neuweiler (1991) correlated the ability of bats to discriminate between different wing-beat rates with the duration of the narrowband component of the emitted sonar calls. They used a rotating bar to modulate the bats' emissions and found that *Rhinolophus rouxi,* a species emitting calls with a rather long CF component, perceived differences in modulation frequency as small as 2.7 Hz at a reference modulation frequency of 60 Hz. *Hipposideros lankadiva,* with its 10.5 ms CF-FM emissions, detected differences in modulation rate of 9.2 Hz. *E. fuscus,* using exclusively short FM sweeps, achieved a discrimination threshold of only 15.8 Hz (fig. 35.2b).

Our study showed that the ability of *E. fuscus* to detect modulation frequencies is considerably improved by the use of long, narrowband signals. The JNDs of 7.1 Hz for the 20 ms signals, and 7.6 Hz for the 50 ms signals, were even better than that of *H. lankadiva* (Roverud, Nitsche, and Neuweiler 1991). Thus, *E. fuscus* perceived differences in modulation frequencies carried by a signal of typical search call duration with an acuity similar to that of short CF-FM bats. Interestingly, increasing the stimulus duration to that of CF-FM bats using long CF components did not further enhance the discrimination of modulation frequencies in *E. fuscus* (fig 35.2b). The hypothesis of Roverud, Nitsche, and Neuweiler (1991), therefore, is only partly confirmed. We suggest instead that the duration of a narrowband stimulus affects the resolution of modulation information only below a species-specific critical duration of temporal summation. The time constant of temporal summation has not yet been determined in *E. fuscus*. However, in the FM bat *Tadarida brasiliensis,* this time constant has corre-

sponded to the duration of the search calls (Schmidt and Thaller 1994).

Several attempts have been made to account for the ability of FM bats to discriminate between modulation frequencies produced by fluttering targets. Sum and Menne (1988) suggested that the discrimination is based on the perception of two-wave-front echoes, produced by moving and nonmoving parts of a reflecting object. Feng, Condon, and K. R. White (1994) assumed that FM bats can determine the wing-beat frequency of insects by a synchronization with their own call emission (stroboscopic hearing). This hypothesis, however, is problematic. Due to the short duration and the low duty cycle of the calls, only very few echoes may contain glints. Further, FM bats often emit less than 10 calls between the detection of prey and the final buzz (Simmons 1987). It remains an open question whether FM bats are able to classify targets by measuring the glint period during approach (cf. Kober and Schnitzler 1990).

The prey selectivity observed in *E. fuscus* (Hamilton and Barclay 1998; Whitaker 1995) may be due to active selection, that is, a decision process after target detection—or it may be a simple consequence of the call structure. The low-frequency calls used during the search phase enable bats to detect prey over a distance of at least several meters. Because small insects only reflect faint echoes, prey size is a limiting factor for target detection. Moreover, the resolution of low-frequency calls may not be sufficient to detect small insects even at close range Barclay and Brigham 1994. Accordingly, insects below a minimum size would escape detection by bats that use low-frequency search calls. However, the echo strength of insects is virtually independent of the emitted frequency (Waters, Rydell, and Jones 1995). Thus it can be assumed that *E. fuscus* actively selects prey.

The high sensitivity for differences in modulation frequency reported here suggests a special importance of the search call for evaluating prey specific information. Casseday and Covey (1996) postulated that auditory midbrain neurons are filters for biologically important sounds requiring immediate action—that is, sounds made by prey or predators. In view of the neuronal emphasis of the search call frequency range in the IC, we suggest that prey selectivity is also determined by a classification of prey on the basis of frequency modulations in the search call echoes.

Acknowledgments

We thank Gerhard Neuweiler, Volker Nitsche, and Lutz Wiegrebe. The present noninvasive experiments did not require an animal experimentation approval. Research was supported by the DFG/SFB 204-TP20.

{ 36 }

Forward Masking and the Consequences on Echo Perception in the Gleaning Bat, *Megaderma lyra*

Ilonka Siewert, Tanjamaria Schillinger, and *Sabine Schmidt*

Introduction

Forward masking is a phenomenon in which hearing thresholds increase when acoustic stimuli are presented shortly after a sound of similar spectral composition. This phenomenon must affect echo perception in bats, since echolocating bats have to analyze weak echoes returning shortly after the emission of their own comparatively intense sonar calls. Yet this effect has not been systematically studied in any bat species.

In the big brown bat, *Eptesicus fuscus,* detection thresholds measured as a function of the time delay between the bats' own vocalizations and the returning echo deteriorated with decreasing time delay (Kick and Simmons 1984; Simmons, Moffat, and Masters 1992; Hartley 1992). In addition to mechanisms to protect the ear from the full strength of the bats' own emissions (middle-ear muscle reflex) and to stabilize echo amplitudes by adjusting the sound pressure level of the calls (automatic gain control), forward masking could account for this threshold increase.

In the present study, we investigated forward masking in the gleaning bat, *Megaderma lyra.* The species has a flexible strategy for preying on insects and small vertebrates, which includes hunting from perches and search flights close to the ground (Audet et al. 1991). The bats use broadband sonar for orientation and target tracking. The short (<2 ms), multiharmonic calls are downward frequency modulated with a fundamental sweeping from 23 kHz to 19 kHz (Möhres and Neuweiler 1966; Marimuthu and Neuweiler 1987). Although the bats are well adapted to listen to faint prey-generated sounds (Fiedler 1979), they use and emit sonar calls during all stages of prey hunting (Schmidt, Hanke, and Pillat 2000). Especially at short distances between a bat and the target, considerable forward-masking effects could arise.

Materials and Methods

Four *M. lyra* were trained in a two-alternative, forced-choice task to fly toward a feeding dish associated with the speaker from which the test signal was presented. The side of signal presentation was chosen according to a pseudorandom schedule. After four trials, the stimulus was attenuated in steps of 5 dB or 2.5 dB (close to the threshold), until the bats' performance dropped to chance level. At least 30 trials per stimulus level, time delay, and signal duration were conducted. A Gaussian-shaped psychometric function was fitted to the data points and its 75% correct value was taken as a measure for the threshold. The standard error for the threshold value was calculated from the quadratic deviation of the data points from the fit curve (cf. Schmidt 1995).

The digitally generated test signals resembled a typical approach phase call of *M. lyra.* All signals used in this study consisted of four harmonics and were downward frequency modulated, with a fundamental sweeping from 21 kHz to 19 kHz. The third harmonic was most prominent; the second, fourth, and first harmonic were attenuated by 6 dB, 12 dB, and 24 dB, respectively.

All maskers used in this study consisted of broadband noise (8–100 kHz). The noise spectrum decreased linearly by 10 dB between 10 kHz and 100 kHz. The masker level L^* is specified as sound pressure level of the continuous broadband noise from which the rectangular masking pulses were cut out (Zwicker 1984).

EXPERIMENT 1: ECHOLOCATION THRESHOLDS VERSUS PASSIVE ACOUSTIC DETECTION THRESHOLDS IN *M. LYRA*

In this experiment, *M. lyra* had to detect test signals of variable duration. The durations were set to $t_p = 0.5 - 512$ ms, corresponding to equivalent rectangular durations (Gerken, Bhat, and Hutchinson-Clutter 1990) of $t_r = 0.186 - 511.471$ ms. All test signals were played back in the presence of a continuous broadband noise with a constant level of $L = 21$ dB SPL.

In a first experimental series, the test signals were presented free-running. The interpulse intervals were set to 500 ms for signals <160 ms, 1000 ms for signal durations of 160–320 ms, and 1500 ms for signals >320 ms. In a second series, the signal presentation was coupled to the bats' call emission. The test signals were played back with a constant time delay of 8.5 ms after vocalization.

EXPERIMENT 2: PSYCHOACOUSTICAL FORWARD MASKING IN *M. LYRA*

In the second experiment, we determined detection thresholds in *M. lyra* as a function of the time delay between the end of a broadband masker and the onset of the test signals. The durations of the signals were set to $t_{p1} = 1$ ms and $t_{p2} = 2$ ms, corresponding to equivalent rectangular durations of $t_{r1} = 0.471$ ms and $t_{r2} = 1.471$ ms. The masker consisted of broadband noise pulses, with a constant level of $L_m^* = 45$ dB SPL, and a constant duration of $T_m = 40$ ms. The time delay (Δt) between the end of the masking pulses and the beginning of the test signals varied between 24 ms and −6 ms. The masking pulses and the related test signals were presented free-running with a repetition rate of 1 Hz.

Results

In experiment 1, the detection thresholds for test durations above 32 ms were identical both for the echolocation and the passive acoustic paradigm (fig. 36.1). For shorter signals, the detection thresholds were up to 8.5 dB higher in the echolocation paradigm compared to the free-running condition.

In experiment 2, the initially measured median thresholds in quiet (fig. 36.2) amounted to 12.6 dB SPL and 6.7 dB SPL for the 1 ms and 2 ms test signals, respectively. Masked thresholds deteriorated by about 30 dB with decreasing time delay (Δt) from 24 ms to −6 ms, compared to the thresholds in quiet. The amount of masking fell to half its maximum value after about 6 ms. At the 8.5 ms delay, it was 11 dB for both signals. Thresholds in quiet are reached at time delays of about 30–40 ms, as can be estimated from fitting a regression line to the masked threshold values as a function of the logarithm of the time delay.

Discussion

The results of this study indicate that forward masking has a considerable effect on echo perception in *M. lyra*. During gleaning close to the ground, echo delays of only 3–6 ms are common. In our second experiment, forward masking amounted to more than 15 dB at these time delays. Masking effects were found for time delays up to 40 ms, corresponding to a target range of about 6 m. Thus, forward masking affects echo perception over the whole range relevant for target tracking. In behavioral experiments in the bat *E. fuscus*, detection thresholds did not improve beyond pulse-echo delays corresponding to ranges of 3–5 m (Kick and Simmons 1984). Despite the differences in species and method, these results fit well to the present data.

In our experiment, the detection thresholds increased by about 6 dB per delay half reduction. Detection thresholds of *E. fuscus* increased by about 11–12 dB for each twofold reduction in target range (Kick and Simmons 1984; Simmons, Moffat, and Masters 1992). The 5–6 dB discrepancy between the two studies is likely a result of the high-intensity emissions by *E. fuscus* com-

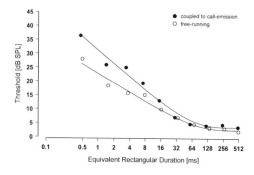

Fig. 36.1. Average thresholds (dB SPL) of four *M. lyra* as a function of the equivalent rectangular duration (ms). Thresholds were measured by presenting the test signals either free-running (open symbols) or coupled to the bats' own call emissions (solid symbols) with a constant time delay of 8.5 ms. Note that the detection thresholds for test tone durations below 32 ms are higher in the echolocation paradigm, with significant threshold differences for durations below 8 ms.

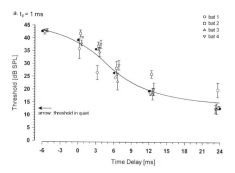

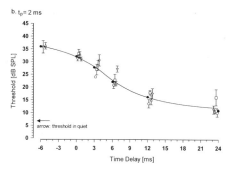

Fig. 36.2. Median masked thresholds (dB SPL) for (*a*) 1 ms and (*b*) 2 ms test signals (solid symbols) as a function of the time delay (ms) between the end of the maskers and the test tone onset. Open symbols represent the individual thresholds of the four bats, and the error bars give the standard error of the thresholds. The horizontal arrow indicates the median threshold in quiet. A sigmoid exponential function was fitted to the data.

pared to the low intensity of the masking pulses in our experiment. We would not expect the middle-ear reflex to have a considerable effect on the thresholds determined in the present study. Moreover, part of the threshold increase found by Kick and Simmons could be due to a reduction in call amplitude by *E. fuscus*, an effect that cannot be accounted for in our passive acoustic measuring paradigm.

The extent to which forward masking influenced the detection thresholds in the active echolocation situation can be assessed by comparing the results of the present experiments at the time delay of 8.5 ms: detection thresholds for short signals in the echolocation context were up to 8.5 dB higher than in the passive acoustic paradigm, whereas the forward masking amounted to 11 dB. The difference of about 3 dB measured in the two experiments could be caused by the longer duration of the masking pulses compared to the duration of the bats' own sonar calls.

Alternatively, neuronal mechanisms for the suppression of forward masking during active sonar could account for this finding. Neurophysiological and behavioral studies established the existence of a gated time window in the auditory pathway of different bat species (e.g., Roverud and Grinnell 1985; Suga 1970; Sullivan 1982). Within this time window, neuronal responses to the second of two tone pulses are facilitated. Although a slight improvement of detection thresholds due to facilitation cannot be excluded on the basis of the present data, our results showed that this mechanism cannot fully compensate for forward masking during echolocation.

For a 2 kHz tone, forward masking in broadband noise amounts to 17–25 dB in humans, depending on the duration of the probe and the duration and level of the masker; thresholds in quiet are reached at time delays of about 50–100 ms (Zwicker and Fastl 1972; B. Moore and Glasberg 1983; Carlyon 1988). Thus, both the time course and increase of the forward-masked thresholds for short signals within the most sensitive frequency range of *M. lyra* are comparable to those in humans.

We conclude that possible mechanisms for the reduction of forward-masking effects in *M. lyra* are more likely to be found in the sonar calls than in adaptive mechanisms of the auditory pathway. In fact, call durations of *M. lyra* are considerably shorter than needed to avoid a mere physical overlap between call and echo during all stages of target approach. Rather, they may constitute an adaptation to reduce the physiological masking effect.

Acknowledgments

This study was supported by the DFG/SFB 204-TP 20. We thank Gerhard Neuweiler Luta Wiegrebe, and Juergen Pillat. The noninvasive experiments described in this study did not require animal experimentation approval.

{ 37 }

The Relationship of Detection Thresholds to the Number of Echoes in the Big Brown Bat, *Eptesicus fuscus*

Annemarie Surlykke

Introduction

In most situations where animals react behaviorally to sensory input, they can exploit the redundancy of the signals—for example, when an animal is listening to a calling conspecific. Echolocating bats may use the information from several echoes. Vespertilionid bats emit sonar signals with a repetition rate of 5–20 Hz in the search phase and up to 200 Hz in the terminal buzz phase. Thus, in a typical pursuit sequence lasting around one second, the bat receives at least 20–25 echoes. The information content of the echoes and the task for the bat change during the pursuit, but there is probably no

instant in which the bat has to make decisions based on only a single echo. The same applies to behavioral experiments with trained bats in the laboratory. So far, all psychophysical experiments have been designed to let the bat receive as many echoes as wanted before responding. Generally, untrained bats emit many signals. In our own setup (Surlykke 1992a), new bats often emit more than 200 signals in a single trial, whereas more experienced bats emit as few as 5 to 10 signals when the task is easy. When the task gets more difficult, they usually increase the number of signals. The purpose of the experiments described below was to determine how the detection threshold in actively echolocating bats depends upon the number of echoes the bats receive. Hence big brown bats, *Eptesicus fuscus,* were trained to detect simulated echoes from a loudspeaker in a target simulator setup.

Several studies have shown that detection thresholds decrease with increased repetition rate of long or continuous pulse trains, both in insects (see, e.g., Fullard 1984), in passively listening dolphins (Dubrovskiy 1990), and in other mammals, (e.g., Turnbull and Terhune 1993). In echolocating dolphins, the threshold has been shown to decrease when each simulated echo consisted of several clicks instead of a single click (Au, Moore, and Pawloski 1988). However, none of these experiments addressed the question of how the total number of echoes affects the detection threshold in an echolocating animal.

Theoretically, the combined energy content of two identical signals is +3 dB relative to the energy of a single signal; therefore, the detection threshold intensity should decrease by 3 dB for two signals compared to one. However, in bats such a threshold reduction is only expected if the two signals are very close together in time, since the energy integration time for actively echolocating bats is only a few milliseconds (Surlykke and Bojesen 1996). Furthermore, power integration is not the only factor influencing detection thresholds. Simple statistical considerations also imply that multiple signals should be more detectable than a single signal (Green 1985). For instance, Viemeister and Wakefield demonstrated that in humans the detection threshold will decrease by 1.5 dB per doubling of number of signals when given the possibility of "multiple looks"—that is, more signals on which to base the decision. This improvement was not affected by the noise level between the pulses, suggesting that the information is somehow stored and can be accessed and processed selectively (Viemeister and Wakefield 1991).

Materials and Methods

Two *E. fuscus* were trained to detect echoes in a one-channel phantom target simulator (Simmons 1973; Surlykke 1992a). They echolocated toward a ⅛″ Brüel & Kjær 4138 microphone at 20 cm. A small (15 mm diameter) custom-built loudspeaker was positioned immediately above the microphone (for details see Surlykke 1992a). In test trials, the bat's signal triggered playback of an artificial echo (fig. 37.1A) after a delay of 3000 μs (corresponding to a distance of around 50 cm to the virtual target). In control trials, no echo was played back. The echo was a 2.5 ms approach-like FM sweep recorded previously from one of the bats. The detection threshold was determined using a modified staircase procedure (Levitt 1971), in which a "hit" resulted in a 1.5 dB decrease in phantom echo intensity and a "miss" in a 4.5 dB increase in echo intensity. A positive midrun was defined as the average echo intensity between the first

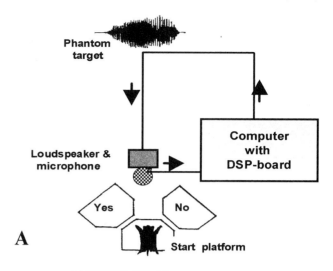

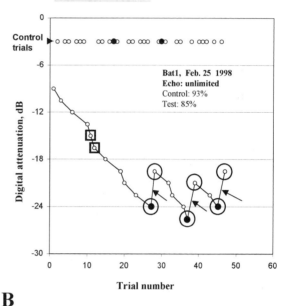

Fig. 37.1. *A:* The single channel target simulator. The microphone and loudspeaker were approximately 20 cm in front of the bat's head. The phantom target was a stored FM sweep at a virtual distance of 50 cm (3000 μs delay). *B:* A session with unlimited echo number. Open circles are correct responses; filled circles are failures. There were three positive midruns (arrows) in this session. The bats' acoustic behavior was analyzed in four types of trial: "easy test trials" (big open squares), "difficult test trials" around the midruns (big open circles), and control trials immediately following the easy or difficult test trials, respectively.

miss and the following hit of a reversal (fig. 37.1B). The session was stopped after 2–5 reversals to avoid frustrating the bat with too many impossible decisions. Test and control trials had the same probability (50%) and were mixed following a pseudorandom schedule. The bats' success in control trials did not influence the echo intensity, but sessions were only included in the database if the success (correct rejections) in the control trials was at least 80%. The threshold was defined as the average of all positive midruns of all sessions with the same echo number. Thresholds are given in dB peSPL (peak equivalent, rms).

First, the detection thresholds of both bats were determined without limiting the number of phantom echoes. These thresholds served as reference thresholds. Next, the number of phantom echoes was reduced systematically to 20, 10, 7, 5, 4, 3, 2, and 1. For each echo number, the threshold was determined before further reducing the number of echoes. Unfortunately, bat 2 died before completing the whole scheme, and hence no threshold was determined for this bat at 2 and 1 echoes.

In each trial, the playback of phantom echoes started from the bat's first sonar emission after trial onset and continued for as many consecutive emissions as the set number of echoes in that experiment. For example, in test trials where the limit was set to 5 echoes, the first 5 sonar cries emitted by the bat would each trigger the playback of a phantom echo, while no playback would be triggered by all the subsequent bat cries in that trial. The number of midruns contributing to a threshold was 30–50 in most cases (table 37.1). Each midrun corresponded to approximately eight trials around the threshold, four test trials and four control trials (see fig. 37.1B).

The acoustic behavior of the bats was monitored during all experiments using specially developed software (S. Boel Pedersen, Center for Sound Communication, Odense University). Briefly, the system was based on a DSP (digital signal processing) board. The signals recorded by the microphone were continuously A/D (analog-to-digital) converted (sample rate 250 kHz) and stored in a ring buffer (FIFO: first in, first out). Up- and down-trigger levels were set to detect the beginning and end of all signals. Thus, only the signals were stored, together with the time for exceeding the up-trigger level. The time between signals was not stored, which not only saved around 90% data-storage space but also enabled us to scroll quickly from signal to signal within a trial during analysis. The software displayed all experimental paradigms (e.g., number of echoes, test/control trial, echo attenuation, bat's success, etc.); when scrolling from cry to cry, the signal energy, duration, and spectrum were displayed. To perform a more detailed analysis of the bats' acoustical behavior, signals emitted by the bats were divided into four groups, depending on the category of trials in which the signals were emitted: group 1, the "easy test trials," or trials well above threshold defined as the two trials at 5 and 6 steps (7.5 and 9 dB, respectively) above the first miss of the first midrun (see fig. 37.1B); group 2, the "difficult test trials," or trials in the midruns—that is, at echo intensities around the threshold; group 3, the "easy control trials"; and group 4, the "difficult control trials," defined as control trials that followed immediately after the test trials in the two first groups.

Results

Detection thresholds for both bats were highly dependent on the number of echoes. Without limits on echo number, bat 1 had a reference threshold of 35.0 ± 0.8 dB peSPL and Bat 2 of 34.5 ± 1.3 dB peSPL (fig. 37.2). Reducing the echo number to 20 and 10 echoes did not affect the threshold, but at ≤7 echoes the thresholds of both bats increased steeply. At 3 echoes, the threshold had increased by 25 dB in bat 1 and 26 dB in bat 2 compared to the reference threshold. Bat 1 worked for a long period with 3 echoes (seven sessions distributed over 20 days), but there was no improvement in threshold with experience and time. Bat 1, continuing with 2 and 1 echoes, showed a further increase in threshold with reduced echo number, but now at a slower rate. However, with results from only one bat (bat 1) it is not possible to say whether this plateau on the curve is real or merely coincidence.

Both bats adjusted the number of sonar signals emitted to the number of phantom echoes received (fig. 37.3, table 37.1). In spite of the decrease, both bats—but especially bat 2—kept the signal number well above the echo number. Duration of trials was a measure of how much time the bats used to make a decision. Duration followed the number of emitted signals closely (table 37.1), such that the pulse repetition rate remained approximately the same, irrespective of the number of phantom echoes. The detailed analysis revealed individual differences in the acoustic behavior of the two bats. Bat 1 always emitted fewer signals in easy test trials than in difficult test trials (one-tailed t-test, $t = 1.81, p < 0.05$).

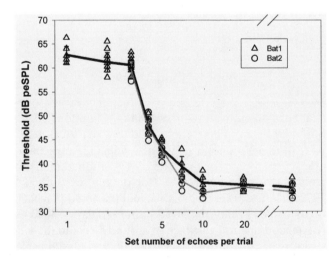

Fig. 37.2. Detection thresholds for both bats increased when the echo number was reduced ≤7 echoes. Individual positive midruns are depicted as open symbols, a triangle for bat 1 and a circle for bat 2. Threshold means (closed symbols) are connected with lines and standard deviations are given as vertical bars. Bat 1: triangles and black lines; bat 2: circles and gray lines.

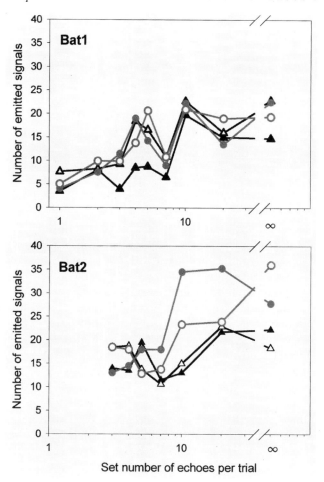

Fig. 37.3. Number of signals emitted by the bats. Black lines are test trials; gray lines are control trials; closed symbols are values from easy trials well above the threshold; and open symbols are values from difficult trials around the threshold.

The number of emitted signals in both groups of control trials was approximately the same as the number emitted in difficult test trials. Hence, bat 1 emitted most signals in trials with no or low-intensity phantom echoes. Bat 2 emitted more signals than bat 1 under all conditions. Bat 2 did not show the same pronounced difference between number of emitted signals in easy and difficult test trials, but it emitted more signals in control trials than in test trials of ≥7 echoes.

Acoustic analysis also included the duration and sound pressure level of the emitted signals. The number of echoes did not affect the signal duration in either of the bats. Bat 1 reduced the average output level by around 10 dB when the echo number was reduced from no limit to 1, whereas bat 2 did not change the output level in any systematic way (table 37.1).

Although we did not monitor the bats' behavior on the platform systematically, we noted changes in behavior that were correlated with echo number. Reducing echo number from unlimited to 20 and 10 echoes did not result in obvious changes in behavior. At 7, 5, or 4 echoes, the bats seemed slightly more motivated. However, when the echo number was reduced to 3 (on different days for the two bats) the behavior of both bats was remarkable: they refused to work and seemed frustrated. It took a couple of days before they cooperated and performed down to threshold. After that, bat 1 accepted working with 2 and 1 echoes without delay.

Discussion

The most important result of this study was the large (up to 25 dB) increase in detection threshold when the number of echoes was reduced. Data from more bats are

TABLE 37.1. Acoustic behavior of the bats depending on the number of echoes. At selected echo numbers from unlimited down to 1, the table shows average values + standard deviations for number of signals the bats emitted, duration of trials, duration and energy (in relative dB) of the emitted signals, and number of midruns used for threshold calculations. The first column ("Emitted signals all trials") lists the number of emitted signals pooled from all types of trials. All other values are from difficult test trials around threshold.

	Emitted signals all trials	Emitted signals	Trial duration (s)	Signal duration (ms)	Signal energy (dB)	Number of midruns
Bat 1						
unlimited	20.5 ± 8.3	22.8 ± 8.7	2.7 ± 1.9	2.4 ± 0.8	48.4 ± 5.1	33
10	21.6 ± 11.1	22.7 ± 12.7	5.8 ± 3.3	2.1 ± 0.5	50.4 ± 12.9	51
5	16.3 ± 9.7	16.6 ± 9.8	1.6 ± 1.5	2.1 ± 0.7	49.4 ± 12.8	54
3	9.0 ± 6.2	9.3 ± 5.9	1.8 ± 1.1	2.2 ± 0.6	44.7 ± 4.2	76
1	5.5 ± 3.9	7.6 ± 4.5	0.7 ± 0.8	2.1 ± 0.5	40.5 ± 16.8	39
Bat 2						
unlimited	25.8 ± 17.8	18.3 ± 9.7	8.8 ± 6.7	2.5 ± 0.4	53.1 ± 2.5	7
10	20.5 ± 11.5	15.0 ± 3.6	6.2 ± 5.1	2.4 ± 0.8	52.4 ± 15.1	34
5	13.8 ± 5.3	13.8 ± 6.3	3.4 ± 1.9	2.4 ± 0.4	53.3 ± 4.1	60
3	16.0 ± 2.9	18.5 ± 2.1	8.4 ± 8.3	3.4 ± 2.1	54.1 ± 4.6	16
1	—	—	—	—	—	—

needed, especially at low echo numbers, but the effect in our study was pronounced. The data also showed that the bats' acoustic behavior was affected, reducing the number of emitted signals when the number of phantom echoes was reduced. Dolphins echolocating in target simulators emit more clicks during test than control trials (Au, Penner, and Kadane 1982; Au and Turl 1983). Interestingly, both our bats did the opposite, emitting more signals in control than in test trials.

The observed threshold increase was much larger than predicted on the basis of the two simplistic models. The expected increase in threshold for a power integrator would be 10 dB for a 10-fold reduction in echo number. However, the energy integration time for echolocating bats is only a few milliseconds (Surlykke and Bojesen 1996), and the bats' repetition rates in the setup never exceeded around 20 Hz. Thus, the echoes arrived at intervals of at least 50 ms, and improvement of the detection threshold by long-term power integration seems impossible. Furthermore, the duty cycle of the sonar signals is low (<10%); thus the noise energy increased more than the signal energy when integrating the total energy over the time of several echoes. Therefore, the signal-to-noise ratio would not improve by 10 dB per 10-fold increase of echo number.

The "multiple-look" model based on results for humans (Viemeister and Wakefield 1991) would only predict a 5 dB increase for a 10-fold decrease in echo number, which is even further from our results here for the bats than the prediction by the power integration model. However, the 1.5 dB per doubling predicted by the "multiple-look" model is based upon the assumption that all echoes are equally detectable. Experimental evidence from humans indicates that all pulses in a series are almost equally detectable (Buus 1999). However, this may not be the case for an echolocating bat. If the detectability of the first echoes in a trial increases such that echo number two is more detectable than echo number one, and echo number three is more detectable than echo number two, the curve would be more steep than 1.5 dB/doubling. For bats echolocating in a phantom target simulator, the situation may actually support this assumption, because the bats seem to need a few warm-up echoes in each trial before they start seriously solving the problem. This would explain (partly) the very steep performance curves. However, instead of further speculation, the present results call for experiments with echoes placed at other positions than after the first emission in a trial. It would also be interesting to test whether thresholds depend on echoes coming after con-

secutive emissions or after randomly chosen emissions in a trial.

Another explanation for the large increase in threshold could be that compared to nonecholocators (e.g., human observers), the situation for a bat is normally very favorable, since the bat itself decides when to emit a pulse and thus when to expect an echo. If this expectation is not fulfilled, the effect could be more dramatic than for a passive listener. A passive listener simply experiences a cessation, but a bat that continues to emit sonar pulses after the playback has stopped will get contradicting information, since now the trial has shifted from a test to a control trial.

Finally, these bats were trained in the setup for a considerable time without limits on the echo number. Trained bats are known to be rather conservative (see Tougaard, Casseday, and Covey 1998), and it cannot be ruled out that the thresholds would have decreased after further training. However, two facts speak against this: First, none of the bats improved their thresholds at conditions where they were tested over a period of several days. Second, the bats' behavior did not indicate frustration at some echo numbers where there was a clear increase in threshold (e.g., 5 and 4 echoes).

The reference thresholds were close to values obtained earlier in the same setup (Surlykke and Bojesen 1996) and also to thresholds from a closed-loop target simulator, where the echo was a delayed and attenuated version of its own emitted signal (Møhl and Surlykke 1989). The resemblance of thresholds in open-loop (as in the present study) and closed-loop phantom-target simulators indicate that bats do accept a prerecorded signal as an echo. This suggests that a reduction in echo number will result in a threshold increase in all kinds of setups, which should be considered when reporting thresholds. So far, behavioral thresholds have been determined without limiting the echo number, but have been reported as the energy of a single signal (Møhl and Surlykke 1989).

Although this present study was done on echolocating bats, the results also should be relevant to the general problem of how animals integrate and exploit redundant information.

Acknowledgments

The work was supported by the Danish Natural Research Foundation. Jakob Tougaard and Ole Næsbye Larsen gave valuable suggestions to the manuscript.

{ 38 }

Bats Learning to Use Echoes with Unfamiliar Time-Frequency Structures

W. Mitchell Masters and *Kelley A. S. Raver*

Introduction

Electronically synthesized "phantom" targets are of great value in studies of echolocation because of the control they allow in testing bats or dolphins. There are two main techniques for generating phantom echoes for bats. The first picks up a sonar emission using a microphone near the bat and plays it back via a loudspeaker (i.e., a rebroadcast echo). The second detects the emission and uses it to trigger playback of a signal contained in digital memory (i.e., a stored echo). In both approaches, the distance to the phantom target (its range) is determined by the time delay between emission and echo. Time delay can be set either by the propagation delay from bat to microphone and speaker to bat, or by electronic delay added to the propagation delay. An important difference between the two approaches is that a rebroadcast echo closely matches the bat's emission in duration and time-frequency structure, even if these characteristics change from emission to emission. An unvarying stored signal cannot track such changes.

Masters and Jacobs (1989) were the first to use a stored echo in a range-discrimination task. In that experiment the stored signal closely matched the bat's typical emission. "Typical" was defined by the average of about 60 emissions recorded while the bat performed range discrimination using rebroadcast echoes. Because the signals used by bats are individually distinct (Masters and Jacobs 1989; Masters, Jacobs, and Simmons 1991; Masters, Raver, and Kazial 1995), the stored signal was different for each bat. We assumed, based on theories of bat ranging (Simmons 1973, 1979; O'Neill and Suga 1982; Sullivan 1982; Beuter 1980; Pollak and Casseday 1989), that bats need a good match between emission and echo to determine range accurately. We did not expect bats to be as successful with a stored echo as with a rebroadcast signal, because bat emissions vary from call to call (Masters and Raver 2000), whereas stored signals do not. Consequently, the mismatch between stored signal and emission should degrade a bat's ability to make emission-to-echo comparisons. Surprisingly, however, the range-discrimination thresholds obtained using a stored echo were as good as those obtained using a rebroadcast echo (Masters and Jacobs 1989). This

fact suggests that current theories of neural processing based on emission-echo comparison are incomplete, or possibly invalid (see "Discussion," below).

At the time, we assumed that matching the stored signal to the characteristics of the individual bat was critical to success. In fact, two bats given each other's signals did very poorly in range discrimination (Masters and Jacobs 1989). Shortly thereafter, however, Miller (1991) and Surlykke (1992a, 1992b) reported range-discrimination experiments with stored echoes *not* matched to the individual bat. The thresholds they obtained were in the range of 0.6–2.0 cm, thresholds as good as those obtained using rebroadcast echoes or real targets (see Moss and Schnitzler 1995 for a summary of range-discrimination thresholds for various species). When we compared our procedures with those of Miller and Surlykke, we felt the critical difference was most likely the amount of practice the bat received with the stored signal. Bats given each other's signals by Masters and Jacobs had little chance to become familiar with the new echo before threshold measurement, whereas the bats used by Miller and Surlykke had received extensive practice with the stored echo prior to testing. If practice is important, then evidently learning is involved, with implications for theories of sonar ranging by bats.

In the present study, we investigated the effect of practice on a bat's ability to use a signal known to cause initial difficulty. A previous range-discrimination experiment (Masters and Raver 2000), in which selected aspects of a model echo were systematically changed, showed that bats had difficulty using an echo whose frequency sweep was either more curved or less curved than its typical call. On the other hand, changes not involving the signal's time-frequency structure produced no decrease in range-discrimination performance. We report here on the effect of practice with model echoes whose frequency-sweep curvature differ from the bat's normal curvature.

Materials and Methods

We analyzed two adult female big brown bats, *Eptesicus fuscus* (Giggles and Guppy), 22 months old at the start of the experiment, who had just completed a range-discrimination experiment in which parameters of the electronically synthesized echo were systematically modified to determine the importance of each parameter to the bat (Masters and Raver 2000). Bats were born in the laboratory to mothers who were pregnant when captured. The results of Miller (1991), Masters et al. (1997), and Masters and Raver (2000) suggest that the range-discrimination ability of captive-born bats is as good as wild-caught bats. Bats were housed separately and given *ad libitum* access to vitamin- and mineral-enriched water. When not being tested, bats were provided with mealworms (*Tenebrio molitor*) raised on an enriched medium following the recommendations of Barnard (1995).

The setup used to generate artificial echoes consisted of an elevated platform 15 cm in front of an ultrasonic microphone (Brüel & Kjær type 4135, 6 mm diameter). A condenser loudspeaker (13 mm diameter) was mounted directly above the microphone. Detection of a sonar emission via the microphone triggered playback of a digitally stored signal (the model echo) via the loudspeaker. Digital-to-analog conversion of the model echo took place at 500 kHz. Zeroes were inserted before the signal to adjust the delay—and hence the apparent range—of the phantom target. The frequency response of the system measured at the position of the bat on the platform was flat ±3 dB from 25 to 110 kHz.

To begin a trial, the bat crawled from the experimenter's hand onto the platform and began echolocating toward the phantom target. The intensity of the echo at the bat's location was approximately 39 dB peSPL (peak-equivalent SPL = sound pressure relative to 20 μPa rms of a continuous sinusoid whose peak amplitude equals that of the echo). The bat's task was to move to the right or left, depending on the target's apparent range. When the target was at its near position, a (nominal) distance of 80 cm, the correct response was to turn right; when the target was at a greater distance, the correct response was to turn left. For each correct response, the bat received a piece of mealworm; for an incorrect response, the experimenter made a soft "shush" sound and gave the bat a short time-out. Trials with the target at 80 cm (near) alternated with trials at greater than 80 cm (far) following a pseudorandom schedule. Bats received 20 trials (10 near, 10 far) per day, with 2 to 4 warm-up trials preceding the day's test. The range difference was constant on a given day, and tests were typically run five days a week. The bats' weights were maintained between 15 and 22 g during the experiment. Because both bats had just completed a range-discrimination experiment, they were very experienced with the procedure.

Bats received three kinds of model echoes. The "normal" model echo corresponded to the average parameters of the bat's emissions. To obtain the average, we recorded approximately 60 calls (500 kHz digitization rate, 12-bit resolution) as the bat performed range discrimination using rebroadcast echoes. These calls were analyzed by a customized software program (Masters

and Jacobs 1989; Masters, Jacobs, and Simmons 1991; Masters, Raver, and Kazial 1995) that determined the values of the critical call parameters, including starting and ending frequency of the first harmonic (fundamental), amplitude envelopes of the first three harmonics, and the decay constant of the exponentially falling frequency sweep—which was the function best matching the calls of these two bats (sweep function given by eq. 1 in Masters, Jacobs, and Simmons 1991). The average values of parameters were used to create the normal model for each bat, a signal corresponding closely to that bat's typical emission in terms of duration, starting and ending frequency, sweep shape, relative energy in the first three harmonics, and amplitude envelopes of each harmonic.

The two other model echoes were altered versions of the normal model, in which all parameters were the same except for the decay constant of the frequency sweep. In model d_1, the decay constant was increased by 3 dB (41%) from the bat's average value; in model d_2, the decay constant was decreased by 3 dB (29%) from the average value. Model d_1, therefore, was less curved than the normal, and model d_2 was more curved than the normal. These two alterations were chosen because they were the only 2 of 12 alterations tested in the preceding range-discrimination experiment (Masters and Raver 2000) that had a negative effect on range discrimination.

To begin, we determined the before-practice range-discrimination threshold of the two bats using one of the two altered model echoes. Giggles was tested with model d_2 and Guppy with model d_1. A bat received 40 trials (two days of testing) at each range separation of 3, 2, 1.5, 1, and 0.5 cm (i.e., near target at 80 cm, far at 83 cm; near target at 80 cm, far at 82 cm; etc.). Probit analysis (Masters and Raver 1996) was used to determine the range separation at which the bat's performance fell to 75% correct (i.e., threshold). Each bat was then given 80 trials (four days) of additional practice with the altered model (Giggles continuing with d_2, Guppy with d_1) at a range separation of 5 cm. The range separation was reduced to 4 cm for a further 80 trials. This procedure was continued through range separations of 3, 2, 1.5, 1, 0.5, and 0.25 cm, with 80 trials at each separation. From the psychometric curve obtained, the after-practice threshold was again determined using probit.

At the end of the period of practice with the altered model, bats were retested with all three models using an interleaved design intended to equalize the effect of further practice with each model. Bats received 12 days of tests (240 trials) at a range separation of 1 cm. On each day, the electronic target the bat received was produced using one of the three model echoes, in the following order: d_1, d_2, normal, d_2, normal, d_1, normal, d_2, d_1, d_2, d_1, normal. At the end of this series, the bats received a further 12 days of tests at a range separation of 2 cm, with the order of presentation of models as before. Data were analyzed using three-way G-tests (Sokal and Rohlf 1981).

Results

The dotted curves in the two panels of fig. 38.1 show the before-practice range-discrimination performance of the two bats. The 75% correct threshold with the altered model was 22.6 ± 3.9 mm (mean ± SE) for Giggles (using model d_2) and 23.6 ± 4.1 mm for Guppy (using model d_1). For comparison, the dashed curves in fig. 38.1 show each bat's performance using its normal model. This curve is the average of two determinations (data not shown) made 10 months apart at the beginning and end of the previous experiment (Masters and Raver 2000). During this time, the threshold did not change for either bat. Giggles's thresholds using the normal model were 6.7 ± 1.5 mm and 6.6 ± 1.4 mm for the first and second determinations, while Guppy's were 5.4 ± 1.9 mm and 6.0 ± 1.1 mm. The threshold using the altered model was therefore about three times worse than with the normal model.

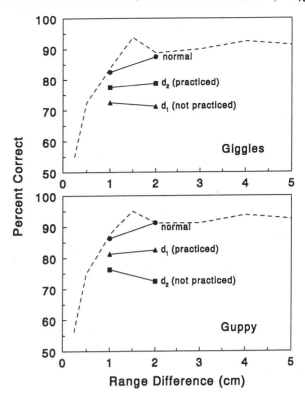

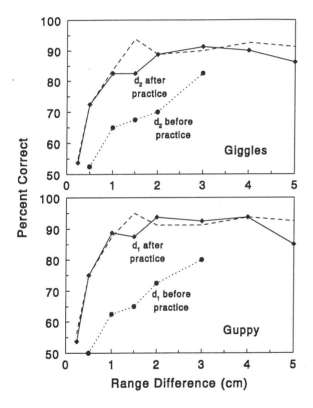

Fig. 38.1. Psychometric curves of range discrimination for two bats (Giggles and Guppy). Each panel shows the bat's percentage of correct decisions versus the range difference between the near and far phantom targets. Zero range difference corresponds to both near and far targets at a nominal distance of 80 cm from the bat. The dotted curve shows the bat's performance with one model's time-frequency structure altered before practicing with that model; each point represents 40 trials (two days of testing). The solid curve shows the bat's performance after more practice using the altered model; each point represents 80 trials (four days of testing). The dashed curve shows the bat's performance with the normal model as determined prior to the experiment.

Fig. 38.2. The performance of the two bats using the three models presented in an interleaved fashion (a different model on successive days) at a range separation of 1 and 2 cm (i.e., 80 versus 81 cm; and 80 versus 82 cm). Each bat received practice with one of the altered models but not the other. The dashed curve is the psychometric function for the bat using its normal model, as determined prior to the experiment (same as the dashed curve in fig. 38.1). Each point represents 80 trials (four days of testing).

After additional practice with the altered models, the performance of both Giggles and Guppy improved substantially (fig. 38.1, solid curves). The after-practice thresholds calculated from these curves were 6.8 ± 2.4 mm for Giggles and 5.0 ± 0.5 mm for Guppy; these were not significantly different from those obtained earlier with the normal model echoes (maximum $z = 0.08$ and 0.80, two-tailed $p \leq 0.94$ and 0.42, for Giggles and Guppy, respectively). These thresholds were, however, significantly better than the bats' before-practice thresholds ($z \geq 3.8$, two-tailed $p \leq 0.001$ in both cases).

The results of the interleaved test of all three models are shown in fig. 38.2. Bats did best using the normal model and worst using the unpracticed altered model (d_1 for Giggles, d_2 for Guppy). They showed intermediate performance using the practiced altered model (d_2 for Giggles, d_1 for Guppy). G-tests showed an interaction between type of model and performance (number right and wrong) for both bats ($G_2 = 8.26$, $p \leq 0.016$ for Giggles; $G_2 = 11.25$, $p \leq 0.004$ for Guppy), suggesting that the type of model is indeed important in predicting performance. For both bats, the results with the normal model were different from the results with the unpracticed model ($G_1 = 8.26$, $p \le 0.004$ for Giggles; $G_1 = 11.24$, $p \le 0.0008$ for Guppy). However, comparison of the normal versus the practiced models did not show a significant difference ($G_1 = 2.53$, $p \le 0.11$ for Giggles; $G_1 = 3.05$, $p \le 0.08$ for Guppy). Likewise, comparison of the practiced versus the unpracticed models showed no significant difference ($G_1 = 1.67$, $p \le 0.19$ for Giggles; $G_1 = 2.64$, $p \le 0.10$ for Guppy).

Because the behavior of the two bats appeared to be consistent—in that performance with the normal model tended to be better than with the practiced model, which in turn tended to be better than with the unpracticed model—we combined data for the two bats in a single three-way G-test in which the first factor was bat, the second was type of model (normal, practiced, or unpracticed), and the third was performance (number right and wrong combined over the two range differences). The interaction between bat and performance was not significant ($G_1 = 1.70$, $p \le 0.20$), supporting the impression that the bats behaved similarly. As before, there was a significant effect of type of model ($G_2 = 19.26$, $p \le 0.0001$). Using the combined data, the effect of normal versus practiced was significant ($G_1 = 5.52$, $p \le 0.02$), as was the effect of practiced versus unpracticed ($G_1 = 4.23$, $p \le 0.04$).

Discussion

The marked improvement between the first range-discrimination threshold obtained using the altered model (fig. 38.1, dotted curves) and the threshold obtained after the bats were given more exposure to the altered model (fig 38.1, solid curves) indicates that experience with a signal can lead to better processing of that signal. Indeed, after practice, the bats' thresholds with the altered model were as low as reported for any bat (Moss and Schnitzler 1995). The fact that a period of learning was involved before the bats were able to make effective use of the signal for range discrimination explains why Masters and Jacob (1989) found that bats receiving a novel signal did poorly in range discrimination, whereas the bats used by Miller (1991) and Surlykke (1992a, 1992b) did well. Other differences in procedure, such as comparatively shorter signals (about 1 ms, as compared to 2 and 3 ms for the model echoes in the present study) and louder signals (about 68–75 dB peSPL as compared to about 39 dB peSPL in the present study), probably are not of great significance.

The limits on the signal a bat can learn to use for range discrimination are unknown. In the present experiment, the altered model that a bat learned to process differed in only one major respect from its typical call, namely in the time-frequency trajectory of the signal from starting to ending frequency. The signals learned by the bats in the experiments of Miller (1991) and Surlykke (1992a, 1992b) probably differed in a number of parameters from

the signal the bat would normally emit, yet the range-discrimination thresholds were normal. The experiment of Surlykke (1992a) showed that, although their threshold was poor (7–8 cm), bats could obtain some range information even with noise bursts that were either stable (a stored noise burst) or variable (one generated anew on each emission). It would be interesting to know whether a bat could improve performance using a stored noise burst if given practice.

A suggestion that the limits on an acceptable signal could be quite broad comes from Miller's (1991) experiment. He tested the effect of the warning clicks of an arctiid moth (brief, broad-bandwidth pulses) on range discrimination. A stored click burst was played to the bat just before the arrival of the artificial echo from a phantom target. The click burst was strongly disruptive if it arrived within a time window beginning about 1.5 ms before the echo. However, Miller was surprised when one bat, after extended experience with the interfering clicks, suddenly ceased being affected and behaved as if there were no interference. Miller believed that the bat had learned to range using the click burst rather than the echo—a strategy that was successful because the time interval between burst and echo was constant in a given block of trials; thus the "range" of the click burst corresponded (with an offset) to the range of the target. If this interpretation is correct, then the minimum requirements for an echolocation signal may be only broad bandwidth and reasonable brevity.

The finding that the match between emission and echo need not be particularly exact poses a problem for current theories of sonar ranging in bats (e.g., Suga 1990; Park and Pollak 1993; Saillant et al. 1993; Simmons et al. 1996). These theories assume that ranging requires comparison of a stored version of the emission with the returning echo to determine emission-to-echo delay. Specifically, the theories propose that a bat determines echo delay by delay-tuned neurons that measure the interval between the occurrence of energy at a particular frequency in the emission and in the echo (not necessarily the same frequency). Such theories require a close match between the time-frequency structure of the emission and the echo. Otherwise, different delay-tuned elements would produce inconsistent delay estimates, and hence the range of the target would be ambiguous. This is exactly the situation with stored echoes. We found that in range-discrimination tests, a bat's emissions vary from one to the next in every measurable parameter (Masters and Raver 2000). A stored echo, on the other hand, does not vary at all from echo to echo, thus introducing mismatch between emission and echo. It is not known whether a naive bat, given a stored echo closely matching its typical emission (i.e., a normal model), would accomplish range discrimination without a period of familiarization to the stored echo. However, the fact that bats can use a stored echo of any sort, and especially that they can learn to use the signal of another bat (Miller 1991; Surlykke 1992a), or perhaps even the clicks of an arctiid moth (Miller 1991), suggests that the current theories of ranging are incomplete.

Our results with bats learning to use new echoes parallel in many ways those of Hofman, Van Riswick, and Van Opstal (1998) with humans learning to use new "ears." In their experiment, humans wore molds that changed the shape of their external ears. Because the folds and ridges of the molds changed the path of sounds entering the ear, and thus changed the head-related transfer function of sounds arriving from different directions, the ability of wearers to determine the elevation of a sound source was strongly disrupted. However, wearers steadily regained their ability to judge elevation over the course of three to six weeks. The processes of localizing the elevation of a sound source by a human (or other mammal), and of determining the range to a target by a bat, are similar in that both require substantial neural processing of the sensory input. Simple one-to-one receptor-to-brain mapping, as illustrated by maps of visual space derived from the pattern of light falling on the retinas, is not possible in the case of sound localization or echo ranging. Therefore, the neural changes involved in bats and humans in solving the new stimulus situation could occur at any level of the brain, not just the most peripheral, and at present their precise nature is unknown. It is possible to say, however, that unlike the situation in the barn owl, where rewiring of the auditory spatial map to align with the visual spatial map appears to be possible only during a critical period in a juvenile's life (Knudsen 1985), the neural plasticity required for rewiring to use altered ears (humans) or altered echoes (bats) persists into adulthood.

One difference between our results and those of Hofman, Van Riswick, and Van Opstal (1998) is that, after the period of adaptation to the modified ears, humans localize equally well with and without the modified ears. On the other hand, when we tested bats with their normal model and the two altered models presented on successive days, bats did best with their normal model. This is despite the fact that, prior to the test, bats achieved thresholds using the practiced altered model identical to those achieved using the normal model, and despite the fact that they had not heard the normal model for 83 days. However, this difference in persistence of the new processing pathways could be one of degree only, since Hofman, Van Riswick, and Van Opstal (1998) reported that subjects' ability to localize accurately while wearing ear molds decreased slowly after they ceased wearing the molds.

So far we have made the tacit assumption that the normal model, being based on the average parameters of a bat's call, ought to be the one the bat is able to use most easily. If true, this might explain why a bat does best with its normal model when challenged to use all three models more or less simultaneously (i.e., on successive days). But perhaps bats in this study did best with their normal model merely because they received more exposure to it. Prior to the present study, Giggles and Guppy had just completed a range-discrimination experiment, lasting about 11 months with approximately 1200 trials using the normal model. For comparison, prior to the challenge of using all three models simultaneously, the bats had received only about 920 trials with the practiced altered model. Although a 25% difference does not seem overwhelming, the fact that the trials with the normal model were spread over more time, or possibly occurred during a time when the bat's auditory system was relatively more plastic, could have promoted better learning of the normal model relative to the altered model.

In summary, it appears that bats can modify their neural echo processing to accommodate echo signals not matched to their emission, if given practice with the novel echo. Explaining this behavior presents a challenge for theories of sonar ranging in bats.

Acknowledgments

We thank S. C. Burnett, K. A. Kazial, A. J. M. Moffat, and A. Surlykke for their comments on the manuscript. The study was supported by NIH grant R01-DC001251 to W. Mitchell Masters.

{ 39 }

Object Recognition by Dolphins

Herbert L. Roitblat

Introduction

Dolphin echolocation is one of the most sophisticated cognitive processes ever studied. When a dolphin uses its biological sonar to recognize objects, its brain performs the equivalent of some extraordinarily complex computations. As remarkable as these computations are, however, they are not much different from the computations performed by our own brains. There is much that can be learned from comparing the kinds of cognitive processes in which the two species engage.

There are three levels at which one can couch a cognitive theory (Marr 1982). At the lowest level, descriptions concern the basic elements (e.g., neurons, transistors) by which the cognitive or computational process is performed. In hearing, for example, such a theory might describe how the hair cells in the cochlea convert sound energy into a neural signal or how tonotopic fields in the brain are organized.

The second level is the theory of the algorithm. What algorithm does the cognitive system use? In hearing, an algorithmic-level theory might involve a description of the successive forms of representation that make up the transformation of a signal into a neural pattern that can be processed by the rest of the brain.

Finally, the third level is the theory of computation—what work does the cognitive system do? For what "purpose" was it "designed?" The main job of a perceptual system is to provide veridical information about features and objects in the environment. The senses are ecological systems that have evolved to allow the animal to make its way in the world. Any sensory system that systematically gave misleading information would be evolutionarily disadvantageous and would quickly be lost from the population. Similarly, object recognition evolved in an ecological context, enabling the animal to obtain information about specific classes of biologically pertinent objects.

The three levels of cognitive theorizing are only fuzzily distinct and only partially independent of one another. They represent a continuum of explanation, rather than a set of clearly demarcated classes. Although there are many kinds of basic elements that could be used to compute the same algorithm, the choice of algorithm depends in part on the kind of basic element available. Similarly, in evolution, the nature of the basic ele-

ments can change in response to the kind of algorithms computed. Analogously, any given class of computation can usually be performed by a large number of algorithms. Consider, for example, the number of algorithms available for computing a Fourier transform.

Formal analyses of the functions that an animal performs tell us the conditions under which a certain capacity can be accomplished. This result is not the product of a detailed study of how the animal implements the function, but rather is based on observations of the existence of the capacity and an analysis of the function's properties. The computational theory tells us about the intrinsic properties of the task. It tells us about the requirements that the organism must meet somehow, using some algorithm, but it does not specify the algorithm.

Ullman's (1979) analysis of the kinetic depth effect is an example of such a theory. Ullman showed that the recovery of three-dimensional shape from the motion of depth features on the retina can be performed provided that certain conditions are met—and only if certain assumptions are made about the object projecting those points. The object's shape can be recovered only if there are enough distinct views of the object and only if enough distinct features are presented. Further, the system must be constrained in the possible interpretations it considers (e.g., it must limit the possibilities to rigid shapes under rotation). Without these conditions and assumptions, a unique mapping of retinal motion to object structure is not possible, because the same proximal feature movements (i.e., the same movement of points on the retina) can originate from an arbitrarily large number of distal object configurations. The rigidity assumption (that the object projecting the features is rigid) allows a unique interpretation of the features. Humans tend to correctly recognize objects that meet these requirements; they fail to interpret projections of objects that do not meet them. The theory of computation says what conditions and assumptions are necessary to solve the problem of recovering three-dimensional shape from motion, but says almost nothing about the algorithms used to compute the relevant relationships or the mechanisms by which this computation is performed. Empirical research was necessary to test the validity of this theory.

How an organism or a model of the organism solves these problems is not specified directly by a computational theory and is a subject of empirical investigation. Knowing some of the properties of the function that the visual system computes helps us understand why perception is generally veridical. The conditions under which the inverse mapping—for example, of the kinetic depth effect—happen to be appropriate are those that happen to be frequently met. The rigidity assumption, for example, happens generally to be valid.

Most approaches to biosonar have focused on the lower levels of Marr's hierarchy. By focusing on the levels of basic elements and algorithms, we have mimicked the fragments of the capacity without coming to grips with its algorithms, goals, or principles.

Some of the problems solved by the echolocation system include recognizing objects, dealing with occlusion and clutter, aspect dependence, source separation, identifying material composition, and detecting barriers. We do not have fully articulated theories of these phenomena in the same way that Ullman has an articulated theory of recovering shape from motion. Instead, most of the work in dolphin and bat echolocation has attempted to reconstruct the algorithm from observations of the physical signals that are available to the animal and our metaphorical intuitions for how those signals might be used. We mistake the proximal stimulus—for example, the sound that arrives at the ear—for the distal stimulus, the object that returns the echo. Yet, the problem that the animal solves is to recognize characteristics of the object, not the echo.

Echolocation is foreign to people; few have any real experience with echolocation and so we grasp at analogies to help us understand what is going on. For example, a number of investigators have tried to pursue an analogy between echolocation and vision—arguing, for instance, that dolphins might use echolocation to "paint pictures" of the objects they scan. Humanmade sonar systems, such as side-scan sonars, tend to translate echoic information into a visual display so that it can be processed by people, exploiting the exquisite pattern recognition capacity of the human visual system. The fact that humanmade systems employ such perceptual conveniences, however, does not imply that dolphins do the same. Other investigators have attempted to exploit the analogy between echolocation and hearing. They conceive of dolphin echolocation as basically a listening problem. This leads to the prediction that a dolphin recognizes an object when it detects sounds like the one it is seeking.

As algorithmic descriptions, these analogies are probably both wrong. Neither approach has proven particularly effective. Perhaps they can be translated into something more useful by taking a step backward and considering them as clues to a theory of computation for dolphin echolocation. A complete account of dolphin echolocation will require descriptions at all three of Marr's levels.

How Dolphins Represent Objects

The two echolocation theories suggested by intuition provide at best only a partial account of echolocation. There is not enough information in a small number of echoes to "paint" the equivalent of a visual scene (not if each echo returns one or a few pixels of that scene), and echolocation is inherently three-dimensional (whereas vision is inherently two-dimensional). The particular waveforms that return in an echo are so sensitive to the orientation of the object that searching for specific sounds is also an unlikely mechanism for dolphin echolocation.

A third hypothesis, based on a theory-of-computation approach to echolocation, views it as an object-recognition process, according to which dolphins use echo information to construct object representations. It rests on an understanding of the kinds of information that would be required to recover object characteristics from the complex echo signals and their changes under varying perspectives. In developing such a theory, it may be helpful to examine other sensory systems—not for a direct analogy, but for guidance.

According to almost all theories of object recognition, the process of recognizing objects involves both bottom-up (sensory driven) and top-down (conceptually driven) processes. Part of what humans perceive is driven by the sensory qualities of the stimulation; and part, perhaps the greater part, comes from our own minds.

There is ample evidence from various areas of cognitive science that the physical pattern striking the sensory surface—for example, the retina—is not sufficient to determine the object that is being perceived. The same perceptual pattern, for example, can often result in the perception of different objects. The most famous example of this phenomenon is the so-called Rubin's vase or goblet. The very same retinal pattern can constitute either two silhouetted faces looking at each other or a white goblet on a black background. What the viewer sees as figure (the object) and what he or she sees as ground (the undifferentiated background) depends on the pattern of stimulation and on the momentary interests of the viewer.

Similar phenomena occur in auditory object perception. In Bach's two-part inventions, for example, a given voice (musical line) can sometimes be followed as the melody and at other times be perceived as the equivalent of the background.

A cocktail party provides another auditory example of separating a sound object from its background. It is relatively easy to attend to the voice of a single person, even while many people around you are simultaneously speaking. Attention extracts the speech sounds of your conversational partner from the background of the other voices, but if someone nearby mentions your name, or

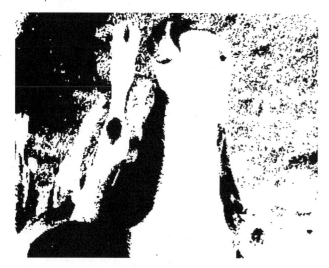

Fig. 39.1. The important role of top-down cognitive processing in recognizing objects

one of your special interests, your attention is drawn to that voice and it becomes figure, at least momentarily (Cherry 1953). These attentional mechanisms may not be much different from those used by dolphins to extract target object echoes from clutter.

Fig. 39.1 shows a very difficult-to-recognize object. Look at it for a while and see whether you can determine what it depicts. Although some people can recognize this object easily, most people have a difficult time figuring it out. Two points need to be made here: first, one has to *figure out* what the picture represents; second, the stimulus by itself is insufficient to indicate what object is being perceived.

If you have not yet recognized the object in fig. 39.1, try turning it upside down to a more typical orientation. Objects have typical and atypical orientations. Recognition is usually easier when the object is in its typical orientation, which implies that this orientation information is part of what we represent about an object. The orientation specificity also illustrates that object recognition is an interaction between stimulus features and the perceiving organism.

In addition to resolving ambiguity, perceptual systems also combine multiple sources of information. The McGurk effect (McGurk and McDonald 1976) is one example of such an integrative phenomenon. A combination visual display of a face and an acoustic display, a syllable, is presented to a human observer. The face moves as if it were articulating the syllable /ga/ and the sound plays the syllable /ba/. The observer perceives the syllable /da/. The back consonant /ga/ sound is articulated by closing the back of the throat; the front consonant in /ba/ is articulated by closing the vocal tract at the lips; and the sound that is heard, /da/, is ordinarily articulated as a middle consonant with the closure behind the alveolar ridge (behind the teeth).

The same effect occurs with more complex speech signals. A videotape with the audio nonsense sentence "My bab pop me poo brive," combined with the video nonsense sentence "My gag kok me koo grive," produces the percept, "My dad taught me to drive." Experimental participants generally cannot interpret the video by itself, and they interpret the audio alone as the nonsense sentence (Massaro 1998).

Human listeners also show evidence of integrating information within the auditory modality. Human speech sounds contain a large number of harmonics, but most of the information is carried by the first two harmonics, which are called formants, of the fundamental. In the demonstration depicted in fig. 39.2, the same first formant was combined with one of thirteen second formants, whose starting pitch varied in 12 steps of about 135 Hz each. These sounds, when heard as a combination of the first formant and one of these second formants, are heard as progressing from the syllable /ba/ to the syllable /da/ and then eventually to the syllable /ga/ as the starting frequency increases from the first to the thirteenth example. When the second formant is played by itself, however, it seems to bear no resemblance to a speech sound, sounding instead like a bleat or chirp, which appears to rise in frequency over the same thirteen examples. Only by integrating the first and second formant does the sound come to resemble speech. Neither part by itself can be understood.

Furthermore, there is substantial ambiguity in the raw spectrum of the speech sound. The same sound pattern is heard as /p/ in /pi/ and as /k/ in /ka/ (Schatz 1954). People never mistake one for the other because we use the context (in this case, whether the vowel after the sound is /i/ or /a/) to select one phoneme or the other. As another example of using context to identify a speech sound, the words "writer" and "rider" and the words "latter" and "ladder" in many parts of the United States differ only in the duration of the initial vowel. The vowel

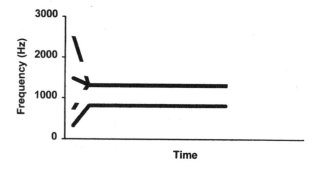

Fig. 39.2. A simplified speech spectrogram showing three phonemes made from the same first formant and different second formants. The first formant is the lowest line in the graph, corresponding to a low-frequency upsweep followed by a steady tone. The second formant is one of the higher frequency sweeps followed by a steady tone.

Fig. 39.3. Another depauperate stimulus

duration determines whether /d/ or /t/ is heard (Fodor, Bever, and Garrett 1974).

The McGurk effect and these other speech effects remind us that perceivers tend to interpret events to make them maximally consistent with all of the available sensory information and their expectations. In ordinary speech perception, the visual and auditory cues are consistent—people find it easier to recognize speech when they can both see the face of the speaker and hear what is being said (Binnie, Montgomery, and Jackson 1974; Sumby and Pollack 1954).

As these examples show, the stimulus by itself is insufficient to determine what is perceived. The stimulus is "depauperate," a biological term meaning poorly or imperfectly developed. Another example of a stimulus that illustrates the depauperate nature of perception is shown in fig. 39.3. If you have never seen this picture before, it may take you some time to recognize what it depicts. Once you have recognized it, however, it is very difficult ever again to "recapture your ignorance"—the depiction will usually be obvious when you look at it again.

All of the visual figures described so far can be "seen" in more than one way. Obviously, the pattern of light and dark on the page does not change when you perceive them in these different ways, but the object that you perceive in the picture does. Rubin's goblet has two stable states: one in which the white goblet is the figure and the black area is background, and one in which the black is seen as two faces on a white background. The other two figures have one state that is stable and another state, the disorganized perception of the figure, that is only temporarily stable. Once the object in these figures has been perceived, it becomes a new, much more stable, state. Once the organization is perceived, it is difficult ever to see it in its unorganized unstable state. A change occurs in the observer's brain that transforms how the pattern is perceived afterward.

Perception can, thus, be conceived not as a low-level process that copies the distal stimulus in the environment into a proximal stimulus that is processed by the brain.

Perception is not a passive process of information pickup, but a constructive dynamic process. You may have noticed that it took some time inspecting these figures before they became organized (if they are still not organized for you, then please see the end of the chapter for some hints). These examples suggest that it may be more fruitful to conceive of perception as a dynamic constraint-satisfaction process, rather than as a simple process of picking up or describing the patterns that impinge on the sense organs.

On this view, the organized percept can be called an attractor. Any disturbance or change in the pattern causes your perceptual system to be attracted once again to one of these stable states. In the case of Rubin's goblet, there are two about equally stable attractors. In the other figures, there seems to be only one. The dynamic aspects of this point of view involve the convergence of the percept onto one of the stable states. A stable state is one where most of the constraints are satisfied. Constraints include those imposed by the physical stimulus pattern, as well as those imposed by such top-down influences as expectations and intentions. A configuration is stable when no nearby states are as effective at satisfying the constraints as is the stable configuration, when small disturbances in the state of the system are followed by a return to the same configuration.

Attractor perception is not limited to visual stimuli. The same kind of multistability and attractor dynamics can also occur with auditory stimuli. Bach's two-part inventions (e.g., "Two-Part Inventions, No. 10") show how an ambiguous set of melody lines can be constructed in a musical score. An invention is a short two- or three-voice piece of music that is intended to demonstrate the composer's ability to write polyphonic music (music that contains more than one independent voice).

Listening to one of these inventions is also an active process. The two voices compete with each other for perception as the melody. Our perception switches from one voice to another and back again. Sometimes one voice carries the main melody of the piece, sometimes the other voice carries it—even in polyphonic pieces in which the voices do not switch.

Biosonar Recognition of Complex Objects

Two points of the preceding discussion are critical. First, the stimulus by itself is often inadequate to specify perception. Second, organisms often have to integrate disparate sources of information to specify perception. It follows that perception consists not of the mere registration of sensory information, but of interpretation and integration of that information to yield a percept of the object that is responsible. When people hear speech sounds, they do not perceive the frequency bands as independent objects; rather they perceive the phoneme that corresponds to the received sound in the context in

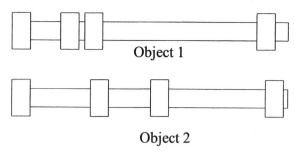

Object 1

Object 2

Fig. 39.4. Two complex objects constructed from ⅝ inch (1.6 cm) hex screws and three nuts. The spacing of the nuts on the screw differentiate the two objects.

which it was emitted. In the case of recognizing objects on the basis of their reflected sonar signal, the organism has to extract from the complex sonar signal the characteristics of the object that returned that signal. On this argument, the dolphin perceives the characteristics of the objects rather than the sound elements that convey its characteristics.

A major barrier to such object recognition using biosonar is the strong dependence of the echo on the specific aspect from which an object is viewed. Viewing a cylinder from the side (i.e., perpendicular to its longitudinal axis) produces an echo that varies substantially as the angle of the cylinder changes (longitudinally relative to the sonar beam axis). Much of the work on humanmade object recognition has been dedicated to finding echo properties that are invariant over changes in aspect and whose presence would allow a cylinder to be recognized from any angle. So far, there do not seem to be any such invariant features.

Vision is also aspect dependent. Although there have been some attempts to discover invariant visual properties, it is generally assumed that the ability of humans and other organisms to recognize visual targets across aspect angles is due to invariant properties that are derived in the brain, rather than being imminent in the pattern projected on the retina.

To illustrate the problem and the benefits of aspect dependence for complex object recognition, fig. 39.4 shows two complex objects that were built from ⅝ inch (1.6 cm) hex screws and three nuts. These objects were hung in a test pool and ensonified with an artificial dolphin biosonar signal. The echoes from these two objects, oriented normal to the sonar beam, are shown in fig. 39.5, along with the corresponding spectra in fig. 39.6. The echoes from these objects are obviously very similar to one another, despite differences in their structure. Echoes from the same objects rotated 3° relative to the sonar beam are shown in fig. 39.7, and the corresponding spectra in fig. 39.8. This very small rotation produced substantial differences relative to the echoes at 0° and also resulted in substantial differences between the echoes produced by the two objects.

These data suggest that aspect dependence can some-

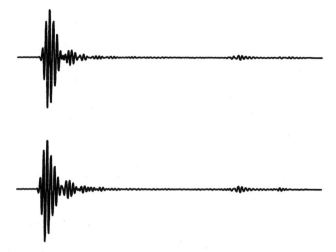

Fig. 39.5. Echoes from object 1 (upper) and object 2 (lower) both oriented at 0° (perpendicular) to the sonar beam

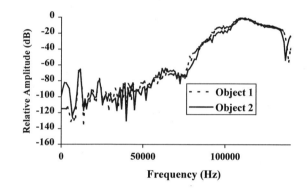

Fig. 39.6. The spectra for the two echoes shown in fig. 39.5

Fig. 39.7. The waveforms from the same two objects, rotated 3° from perpendicular

times be an aid to object recognition rather than a hindrance. By rotating the objects slightly, the task is made from a difficult one (Euclidean distance of 0.72 between the echoes) to an easier one (distance of 1.38). By combining the information from both aspects, the distance between objects increases even further (to 1.56).

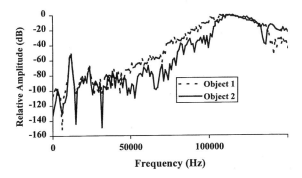

Fig. 39.8. The spectra from two objects, rotated 3°

Conclusions

Whatever the sensory modality, object recognition is not a simple process of responding directly to the pattern of stimulation on the sensory organ. The stimulus itself is inadequate to specify the object that is the source of the stimulation. Most accounts of perceptual systems have focused on the proximal stimulus (the pattern of stimulation as received at the sensory surface) rather than the distal stimulus (the object that is the source of that stimulation). These accounts fail to note that the proximal stimulus varies substantially over minor transformations of position and other factors (e.g., in vision and lighting), whereas the distal stimulus remains constant. As a partial solution, investigators have sought invariant properties of the proximal stimulus that could be used to identify the object across variations. If the invariant feature is present, then the object can be identified. After much effort, however, such properties turn out to be very difficult to identify and to exploit.

The problem with trying to identify the distal stimulus directly from the proximal sensory information is that the proximal stimulus is depauperate. There simply is not enough information in the proximal stimulus to determine what will be perceived. Stated another way, there is not enough information in the proximal stimulus to uniquely identify the object. Context can provide this needed information, but there is no obvious way to include context in computational models of perception. If context consisted of simple invariant features that could be recognized in an all-or-none manner, then the search for such invariants would already have been successful.

Rather, context consists of a set of constraints that help to limit the range of possible objects that could be the source of the proximal stimulus as received. No solitary contextual feature is sufficient either; but when taken together, the pattern of stimulation that is received and the set of contextual features that are simultaneously available help to limit the range of possible objects. The use of context in this way demands a different style of computation that is more like a dynamic system than like a standard computational algorithm. No single piece of information is either necessary or sufficient for recognition, and no linear sum of such bits is adequate either. Instead, perception is determined as a complex nonlinear function of all sources of information in interaction with the perceptual system itself. Perception is not just a way for the brain to pick up information; it is a way for the brain to change itself as a result of its experience. To be successful, computational models must employ similar kinds of mechanisms along with multiple sources of information.

Note: Fig. 39.1 depicts the face of a cow; fig. 39.3 depicts a Dalmatian under a tree.

{ 40 }

Identity versus Conditional Cross-Modal Matching by the Bottlenose Dolphin

Heidi E. Harley

Introduction

Most studies of dolphin sonar focus on the characteristics of sounds, and how they are produced and received (e.g., over half of the chapters on dolphins in this book).

Clearly, this knowledge is important to understand echolocation. However, it is also important to know what information dolphins extract from these sounds. The central focus of this chapter is to present a method that is useful for investigating dolphins' representations of object features detected through echolocation. Some recent data that suggest the method will be successful also are presented.

The method outlined here employs a version of the more general cross-modal identity matching task used with many species, including dolphins (e.g., Harley, Roitblat, and Nachtigall 1996). In a cross-modal matching task, a subject is presented with a sample object in one modality (e.g., vision) and is required to identify a match for that object from among some number of alternatives using a second modality (e.g., touch or echolocation). Humans, including infants, are typically good at cross-modal tasks (e.g., Garbin 1988; Rose and Ruff 1987).

The performance of such tasks requires a subject to associate different sensory experiences of a stimulus: the first experience with the object is with one sense and then it must be identified again using another sense. The mechanisms used to associate these different sensory experiences have been debated since the fifth century B.C. (Marks 1978). There are at least two possible mechanisms through which these associations occur.

The first mechanism, *association learning,* is necessary for associating object features that cannot be detected through the same sensory systems—for example, an object's color and its taste. Detection of color requires eyes; detection of taste requires the tongue. However, one can learn to associate certain colors with certain flavors. For instance, candy in the United States is typically artificially colored and flavored. Over time, many Americans have learned that purple is correlated with grape flavor, yellow is correlated with lemon, bright red is correlated with cherry flavor, and so on. Given this consistent correlation, many Americans could successfully perform a cross-modal matching task in which they tasted a sour apple candy without seeing it, and then picked the candy that they had tasted from a visual array of differently colored candies. However, their ability to perform this task has occurred purely through their learning to associate the candies' colors and flavors. There is nothing about a candy's color that dictates its flavor. If candy makers had chosen to match light green with grape flavor, people could have learned that association just as easily. People are good at this sort of arbitrary, conditional matching.

Dolphins are also good at learning to associate arbitrarily related sensory experiences. For example, dolphins can learn to associate artificial sounds with objects and locations (Herman, Richards, and Wolz 1984) and to produce arbitrary sound labels to objects presented visually (Richards, Wolz, and Herman 1984). As with humans learning about color and flavor, the dolphin could learn to associate a Frisbee with a down-sweep sound or a pure tone with equal effort. This flexibility occurs because the sensory experiences are not intrinsically related.

However, when the *same* object feature can be detected through two different sensory systems, then these different sensory experiences are intrinsically related. Their pairing is not an arbitrary one. For example, humans perceive information about texture through both visual and tactile systems; humans can both see that an object is smooth and feel that an object is smooth. The tactile experience and the visual experience are tied to each other through the object's texture, its smoothness. Humans can recognize an equivalence between their different sensory experiences of the same object feature. This *recognition of equivalence* is the second mechanism that humans use to associate different sensory experiences.

The capacity of humans to recognize an equivalence between their different sensory experiences of the same object feature is evidenced in their ability to immediately perform cross-modal matching tasks with unfamiliar objects—that is, without having a chance to learn about the new objects specifically. In a matching task human subjects felt but could not see unfamiliar oddly shaped clay objects (Garbin 1988). When they were required to determine whether or not they had felt an object identical to the sample by viewing (without touching) it, they could often do so immediately without the opportunity for learning.

In summary, humans use two mechanisms to associate different sensory experiences with a single object: (1) association learning to pair arbitrarily related sensory experiences of a single object (such as a piece of candy's color and flavor) and (2) recognition of equivalence to pair intrinsically related but different sensory experiences of the same object feature (such as a clay object's shape detected through touch and sight). Data from dolphins suggest that they can use the first mechanism, association learning. Can dolphins use the second, recognition of equivalence?

The current data suggesting that dolphins can recognize an equivalence between different sensory experiences of the same object are somewhat equivocal. One dolphin was successful in immediate cross-modal matching of oddly shaped familiar PVC objects that had never been simultaneously seen and echolocated, thereby suggesting that the dolphin recognized an equivalence between its echoic and visual experiences of at least some object features (Pack and Herman 1995). However, when the same dolphin was presented with four pairs of novel objects, first-session performance was above chance with only two pairs (Herman, Pack, and Hoffmann-Kuhnt 1998). This finding is problematic, because immediate transfer is necessary to assert that the dolphin recognizes an equivalence between its echoic and visual experiences using this method. After a few trials, the dolphin has a reinforcement history that can affect its association of its visual and echoic experiences—that is, the dolphin has had an opportunity to learn to associate its visual and echoic experiences of that object. But there are other reasons for dolphins to fail with new stimuli that have nothing to do with recognition of visual-echoic equivalence. For example, dolphin matching performance in general often declines with new stimuli and then improves with experience (e.g., Xitco and Roitblat 1996). Hence, the use of a method that does not require immediate transfer should lead to a better understanding of the mechanisms the dolphin is using to perform cross-modal matching tasks. Such a method has been used with rhesus monkeys (Cowey and Weiskrantz 1975; Weiskrantz and Cowey 1975) and can be adapted for use with dolphins.

Because we already know that dolphins can use association learning, this new method must allow us to determine whether the dolphin can also use recognition of equivalence. If we assume that the dolphin cannot use this second mechanism, then the dolphin must always use association learning to pair its visual and echoic experiences of an object. In this case, when it performs a cross-modal matching task, it must learn which visual experience goes with which echoic experience by the presence or absence of reward. That is, the dolphin learns that its visual experience of object A goes with its echoic experience of object A when the dolphin receives a fish for matching one to the other.

Typically, we do reward the dolphin for matching its different sensory experiences of object A together—that is, we reward the dolphin for performing an identity-matching task. However, we could just as easily reward the dolphin for matching its visual experience of object A with its echoic experience of object B—that is, for performing a conditional-matching task. If the dolphin merely learns to associate these experiences and does not recognize an equivalence between the two experiences, then it should not be confused by this procedure. However, if the dolphin has been trained to perform an identity matching task (i.e., choose alternative object A after experiencing sample object A), and if it recognizes an equivalence between its visual and echoic experiences of the same object features, then the dolphin should make mistakes when it is suddenly expected to choose object B after experiencing object A. In this case, the dolphin should choose, at least initially, object A after experiencing object A no matter how we reinforce the dolphin.

The method suggested here takes advantage of the fact that the use of the different mechanisms leads to different predictions. This method rewards the dolphin for making both identity and conditional (nonidentity) matches. That is, the dolphin is rewarded for making conditional matches with two object pairs (visual object A and echoic object B; visual object B and echoic object C) and an identity match with one object pair (visual object D and echoic object D). If the dolphin only uses association learning to perform the cross-modal task, then the dolphin should learn all three matches with equal ease, and its errors should be equally distributed. However, if the dolphin can also use recognition of equivalence, then it should learn the identity match (D-D) most quickly, and it should make identity-based errors (B-B) with the other object pairs.

Materials and Methods

Animal Subject

The subject, Toby, was an 18-year-old male dolphin (*Tursiops truncatus*) housed at the Living Seas, Epcot, Walt Disney World, Orlando, Florida. Toby had participated in many behavioral research activities, including a passive echolocation study (Xitco and Roitblat 1996). His work in this study gave him a great deal of practice performing echoic identity matching. Therefore, it was expected that if the dolphin was presented with the opportunity to perform an identity match, he would.

Procedure and Materials

Toby was previously trained to perform a three-alternative identity matching-to-sample task for Xitco and Roitblat's (1996) passive echolocation study. Before current testing began, we reacquainted Toby with the matching-to-sample task by presenting some of the same objects from the original study in an apparatus essentially identical to the one previously used. Sessions were conducted in the main tank (67 m in diameter and 9 m deep) of the Living Seas.

At the beginning of each cross-modal trial, the dolphin was positioned in front of a black polyethylene screen (0.59 m width by 0.46 m height) situated 0.25 m above the water's surface. A trainer presented the sample object in air (thus it could not be investigated using echolocation due to density differences between water and air) in front of the screen. The dolphin looked at the sample object as long as desired and then swam to the middle of the tank where three choice objects were centered within three equal-sized square sections (0.90 m × 0.90 m) of a PVC frame (front: 2.76 m width × 0.90 m height; sides: 0.98 m width × 0.90 m height) covered with tight 0.10 mm thick, black polyethylene sheeting. The polyethylene was visually opaque but echoically transparent, thereby giving the dolphin echoic but not visual access to the choice objects. The top of the screen was 1 m below the surface of the water.

At the screen, the dolphin indicated his choice by stationing in front of an object for a minimum of 3 s. A research intern naive to the identity of the sample object indicated the dolphin's choice. If the dolphin was correct, the trainer blew a whistle and fed the dolphin two small fish. If the dolphin was incorrect, he was recalled by the trainer to await the next trial. Intertrial intervals averaged 43 s.

Each sample object occurred as a sample an equal number of times, and each alternative object occurred as an alternative an equal number of times in each position (left, center, right) within each 18-trial session. The order of the trials was randomized.

Two object sets were used. The first consisted of three familiar objects (Xitco and Roitblat 1996): a 10.2 cm × 10.2 cm × 10.2 cm solid wooden cube, a 26 cm × 0.6 cm aluminum plate, and a 15.5 cm × 11.5 cm plastic pipe with 0.9 cm thick walls. This object set was reintroduced in an echoic identity task, and then transferred to the visual-echoic task described above. The dolphin performed five visual-echoic identity-matching sessions with this object set before going on to the next set.

The second object set consisted of four unfamiliar stimuli. They were a 10 oz. glass soda bottle with a 5.6 cm diameter base that tapered into a 2.5 cm diameter top (*Glass Bottle*); the same bottle wrapped in plastic bubble packaging containing 236 air bubbles (0.89 cm diameter) per meter^2 (*Bubble Bottle*); a cluster of three of the glass soda bottles, two wrapped in plastic bubble packaging, connected at the top with clear monofilament (*Three Bottles*); and a 5.6 cm diameter tan, hard plastic ball normally used as a cue ball for playing billiards (*Sphere*). The dolphin performed 21 visual-echoic sessions with these objects. The dolphin always viewed Three Bottles, Sphere, and Bubble Bottle as samples after which he was reinforced for echoically choosing Sphere, Glass Bottle, and Bubble Bottle, respectively. Therefore, the dolphin was reinforced for making two conditional matches (Three Bottles-Sphere; Sphere-Glass Bottle) and one identity match (Bubble Bottle-Bubble Bottle). The dolphin never simultaneously saw and echolocated any of these stimuli. The dolphin had no previous experience with these objects of any kind, including intramodally (visual and visual; echoic and echoic).

Results

The dolphin's performance accuracy averaged 92% over the first 5 sessions (89%, 100%, 94%, 83%, 94%) of the visual-echoic matching task using familiar objects, at which point the new object set was introduced in the visual-echoic format.

The data set was divided into three parts (early, middle, and late) based on the dolphin's tendency to choose a single object or distribute choices across the set. Table 40.1 presents the distribution of choices during each session. During the first 7 sessions, when he was presumably learning about the object set, the dolphin defaulted to a single object over 50% of the time (to Sphere in sessions 1, 2, 6, and 7, and to Glass Bottle in sessions 3, 4, and 5). It is not atypical for this dolphin to require some time to learn about objects from set to set (cf. Xitco and Roitblat 1996). During the middle 4 sessions the dolphin distributed choices fairly evenly across the objects (35% to Sphere, 33% to Glass Bottle, 32% to

TABLE 40.1. The distribution of the dolphin's choices on each session.

Trials	Stimulus set		
	Sphere	Glass bottle	Bubble bottle
Early sessions			
1	16	2	0
2	17	1	0
3	3	14	1
4	3	15	0
5	4	14	0
6	10	7	1
7	11	6	1
Middle sessions			
8	6	6	6
9	9	6	3
10	5	6	7
11	5	6	7
Late sessions			
12	0	6	12
13	0	8	10
14	0	4	14
15	0	6	12
16	0	3	15
17	0	3	15
18	1	2	15
19	0	0	18
20	0	3	15
21	0	0	18

Bubble Bottle). Across the final 10 sessions the dolphin moved from choosing a single object (i.e., Bubble Bottle) more than 50% of the time to choosing it exclusively.

The dolphin's tendency to choose a single object in the early sessions gave him an opportunity to learn about the reinforcement contingencies for every stimulus pair. During the first 7 sessions, he was reinforced for 25 of 126 trials: 11 for Sphere after seeing Three Bottles, 12 for Glass Bottle after seeing Sphere, and 2 for Bubble Bottle after seeing Bubble Bottle. However, reinforcement only seemed to affect the choice of Bubble Bottle. After being reinforced for choosing Bubble Bottle once in session 6 and once in session 7, there were only 6 trials when he did not choose Bubble Bottle after seeing it over the remaining 14 sessions. By session 12 it became the default object—that is, the dolphin chose it over 50% of the time no matter what sample was presented. Bubble Bottle remained the default object throughout the final 10 sessions. In sessions 19 and 21, he chose it 100% of the time. The dolphin was never successful at matching the arbitrarily associated object pairs for which he was reinforced.

An error analysis of choices during the middle sessions (sessions 8–11) was conducted. (See the confusion matrix in table 40.2 for a presentation of choices related to each sample.) Choices in these sessions were dependent on the samples he had seen, $\chi^2_{(4)} = 138.08, p < .001$.

Table 40.2. Dolphin choices based on sample identity for sessions 8–11 (choices on the diagonal represent reinforced trials).

Samples	Choices		
	Sphere	**Glass bottle**	**Bubble bottle**
Three bottles	0	24	0
Sphere	20	0	4
Bubble bottle	5	0	19

After viewing Three Bottles, the dolphin chose the Glass Bottle 100% of the time. After viewing Sphere, the dolphin chose Sphere 83% of the time. After viewing Bubble Bottle, the dolphin chose Bubble Bottle 79% of the time. Of these matches, he was only reinforced for the Bubble Bottle–Bubble Bottle match.

Discussion

The data suggest that the dolphin recognized an equivalence between visual and echoic experiences of the same object features. When the dolphin was distributing choices, his responses suggested identity matching even when he was reinforced for performing conditional matches. In addition, the dolphin ultimately was only successful at matching the identity-matched pair (Bubble Bottle–Bubble Bottle). The dolphin associated Three Bottles with Glass Bottle and Sphere with Sphere in spite of being reinforced for associating Three Bottles with Sphere and Sphere with Glass Bottle. In fact, reinforcement apparently only affected his choice of the stimulus that formed the identity match, Bubble Bottle with Bubble Bottle. A consistent default to any object would have resulted in the dolphin's receiving reinforcement 33% of the time; however, he only invested heavily in Bubble Bottle. The dolphin's reinforcement experience during the middle sessions (8–11), when he consistently associated objects based on their similarity, could have been predicted, since he was primarily reinforced only for choosing Bubble Bottle during those sessions.

The dolphin originally was trained to perform an identity-matching task. It should not be surprising that he would continue to do so given the opportunity. However, the only way for him to perform an identity-matching task in this procedure was dependent on his recognizing that his visual and echoic experiences of the same or similar objects were equivalent in some way—that is, he had to use recognition of equivalence. Without this recognition, a cross-modal task would be a conditional matching task in any case—whether the sample-choice object pairs were identical or not. However, these data suggest that the dolphin did continue to perform an identity-matching task, because he recognized an equivalence between his visual and echoic experiences of the same or similar objects. Hence, these data are consistent with previous findings that dolphins integrate and recog-

nize equivalencies between their visual and echoic experiences of the same objects (Harley, Roitblat, and Nachtigall 1996; Herman, Pack, and Hoffmann-Kuhnt 1998; Pack and Herman 1995).

These data, along with the success of this method with other species, suggest a good procedure for studying the mechanisms underlying cross-modal matching performance. Each mechanism used for associating multimodal experiences predicts different outcomes in the dolphin's choices, and the resulting data are helpful in suggesting which mechanisms the dolphin used.

Results presented here also suggested a way to strengthen this design. In this study, reinforcement of the object pairs was designed to be A → B, B → C, and D → D. Because it was not originally clear whether the dolphin could recognize matching features across objects, several of the objects had overlapping features. The Three Bottles stimulus was created by forming a pyramid from tying two bubble-wrapped bottles together with a plain glass bottle. Therefore, this stimulus shared features with Glass Bottle and Bubble Bottle. After sessions 3, 4, and 5—in which the dolphin predominantly chose Glass Bottle, regardless of the sample's identity—he chose Glass Bottle after seeing Three Bottles very consistently from session 6 onward. That is, of the 72 times that the dolphin chose Glass Bottle in sessions 6–21, 68 (94%) of them were after seeing Three Bottles. Recall that the dolphin was never reinforced for this match. It was assumed that the dolphin saw the glass bottle as the defining feature in Three Bottles, although admittedly the evidence for this is somewhat preliminary. However, the dolphin's performance with this pair suggested the design could be improved to provide more information. Specifically, the conditionally associated object pairs could be crossed such that the reinforced object pairs would be A → B, B → A, and C → C. This would allow a more direct method of assessing errors in relation to object identity. Future research should include this design change, as well as testing echoic to visual matching.

Acknowledgments

This work was possible due to the financial and institutional support provided by Walt Disney World. I thank the Walt Disney Company in general, as well as many individuals associated with the Living Seas. In particular, I thank administrators Elizabeth Stevens, Conrad Litz, Tom Hopkins, Jane Davis; trainers Kim Odell, Mike Muraco, Cathy Goonen, Mark Barringer, Bobbie Cavanaugh, Jane Capobianco; researchers Erika Weber, Dru-Ann Clark, John Gory, Mark Xitco; and the many Seas interns. In addition, talks with Gordon Bauer, Herb Roitblat, and New College students in comparative cognition were also invaluable.

{ 41 }

Dolphin Echolocation Shape Perception:
From Sound to Object

Adam A. Pack, Louis M. Herman, and *Matthias Hoffmann-Kuhnt*

"Echolocation is not just sensing the presence of an echo. It requires the ability to interpret, evaluate, and identify that echo" (Kellogg 1961, p. 158).

Introduction

The discovery by Herman and Pack (1992) that a bottlenose dolphin (*Tursiops truncatus*) could recognize objects across the senses of echolocation and vision set into motion a new theory of dolphin echolocation perception. This theory stated that echolocation, like vision, yields a direct percept of the object being inspected. Subsequent reports (Herman, Pack, and Hoffmann-Kuhnt 1998; Pack and Herman 1995) provided extensive data to corroborate and extend the original findings. These studies revealed that object shape could be recognized spontaneously across the two senses in either direction, from vision to echolocation and from echolocation to vision. Furthermore, it was established that the reported capability was not dependent on associative learning. The implication was that the ability to spontaneously recognize shapes through echolocation must depend on mechanisms and processes beyond the raw acoustic cues (amplitude, highlight structure, spectral content) typically invoked to explain dolphin shape discrimination (see Au 1993). In the following sections, we describe the background and context for these findings on echolocation, discuss their implications for understanding echoic shape perception, and present suggestions for future studies.

THE PROBLEM AND ITS BACKGROUND

When a dolphin echolocates on an object that it cannot see, does it perceive only the "raw" acoustic characteristics present in the backscatter information (e.g., highlight structure, amplitude, spectral content), or does it perceive the actual shape of that object? This question asks in essence whether the dolphin's representations of ensonified objects are acoustic-based (i.e., have characteristics of sounds exclusively) or object-based (i.e., have characteristics of objects including their spatial structure). Although this seems a fundamental issue in the study of dolphin echolocation, it is only recently that research has been directed toward this question.

The issue may be best appreciated through a visual analogy. From a strictly mechanical perspective, light energy reflected from an object is refracted by the cornea and lens and focused onto the retina, where it stimulates a series of light-sensitive cells. What is *perceived*, however, is not the physical properties of light, but rather the appearance of the object reflecting that light. The object appears three-dimensional, and its shape remains stable over changes in viewing angle, illumination, and distance, even though these manipulations result in changes in the physical energy striking the retina and/or the retinal image of the object (Marr and Nishihara 1978). Although the particular processes responsible for visual object perception are still far from being well understood (Tarr and Bulthoff 1998), and the mechanics of dolphin echolocation are quite different from those of vision, does echolocation perception nevertheless function in a similar fashion to visual perception? In particular, are echoically interrogated objects, like visually inspected objects, perceived holistically, as "gestalts" (Spelke et al. 1993)? And are echoically derived representations subject to the same type of shape constancy as visual representations (Jolicoeur and Keith-Humphrey 1998)?

Shape discrimination using two-alternative forced-choice tasks and echoic matching tasks

Early studies of dolphin echolocation shape discrimination determined that bottlenose dolphins could learn to echoically discriminate cylinders from spheres (Au, Schusterman, and Kersting 1980), cylinders from cubes (Nachtigall, Murchison, and Au 1980), and flat circles from either flat triangles or squares (Barta 1969, cited in Au 1993). One or more acoustic cues present in the backscatter information from these objects (e.g., differences in highlight structure, amplitude, spectral content) were implicated as responsible for these discriminative abilities (Au 1993). However, the possibility for direct shape perception was not considered in these analyses. Nor could it be determined, because the particular two-alternative forced-choice tasks used in these studies made it impossible to distinguish between explanations of performance based on the dolphin's reliance on acoustic cues alone, and those based on direct perception of differences in shape. In the two-alternative successive version of the forced-choice task (Au, Schusterman, and Kersting 1980), the dolphin is blindfolded and rewarded for pressing a particular paddle in the presence of one shape, and a different paddle in the presence of the other shape. In the "simultaneous" version (Nachtigall, Murchison, and Au 1980), the dolphin is required to indicate on which side (left or right) a particular shape appears. In either case, if the dolphin is successful, it cannot be unequivocally determined whether it has learned to associate particular echoes or echo characteristics with particular responses, or whether it directly perceives the shape of the different objects. This is because no conditions allow for the demonstration of direct echoic shape perception without the possibility for associative learning. In order to distinguish between these possibilities, transfer tests must be conducted using objects that maintain shape information but alter backscatter information (see below).

Similar difficulties in interpretation of results exist for the echolocation identity matching-to-sample (MTS) task (e.g., Roitblat, Penner, and Nachtigall 1990). In this task, a dolphin ensonifies a "sample" object and then echoically examines two or more alternative objects to find a match to the sample. Even when the dolphin performs the task accurately on the first trials in which novel objects differing in shape alone are exposed (Pack and Herman 1995), it is unknown whether the dolphin is matching object shape or matching echoes of the sample and the matching alternative. In other words, is it performing shape matching or sound matching? Unless transfer tests are conducted with objects that alter backscatter information while retaining shape information, the echolocation identity matching-to-sample task, like the two-alternative forced-choice task, cannot determine unambiguously whether mental representations of objects perceived echoically are object-based or acoustic-based.

Echoic recognition of object shape at different orientations

If representations of ensonified objects are object-based, then certain changes in the object's orientation that produce changes in acoustic features of the echo should have little impact on object recognition (see Roitblat, Helweg, and Harley 1995), as long as these manipulations do not entirely hide all previously examined object features. For example, a cube whose face is perpendicular to the echolocation beam of the dolphin will yield different raw echo characteristics than one that is rotated 45° so that a corner faces the echolocation beam (Helweg et al. 1996a). In theory, if the dolphin forms object-based representations and its perceptual system allows for shape constancy, it should classify the rotated cube and the unrotated cube as the same object. Nachtigall, Murchison, and Au (1980) explored this by training a dolphin in a two-alternative forced-choice echoic discrimination task using foam cubes and cylinders pre-

sented simultaneously in standard positions (i.e., cylinder = upright, cube = flat face toward dolphin). The dolphin was blindfolded and reinforced for selecting the side on which the cylinder appeared. After successful discrimination was achieved (417 trials), objects were presented at different aspect angles relative to the dolphin. The only aspect change yielding a discrimination significantly above chance was a comparison between the cube rotated 45° along its horizontal axis and the cylinder remaining in the standard position. Nachtigall et al. implicated amplitude differences in the echoes (i.e., target strength) from the objects as likely cues allowing the dolphin to achieve this discrimination. Thus, this experiment provided little evidence for object-based representations. However, the task itself may have led the dolphin to expect a reward only if it responded to the upright cylinder.

Helweg et al. (1996b) tested a dolphin's ability to echoically match three aspect-dependent objects: a cube, a rectangular prism, and an equilateral pyramid. The objects were suspended underwater from thin monofilament lines and "allowed to rotate freely in the horizontal plane" (p. 22). The authors reasoned that if the dolphin performed this task above chance levels, it must form aspect-independent representations, inasmuch as the acoustic features in the reflections from the same object would vary with aspect angle. After extensive training, performance was above chance levels for all objects other than the cube, although the high degree to which this object was avoided as a choice by the dolphin suggested to the experimenters that it was at least discriminated echoically. Although the overall results appear to support the hypothesis that the dolphin's representations were object-based, the weakness of this study is its inability to specify precisely what aspects of an object were being revealed to the dolphin during echolocation. It is conceivable, for example, that the dolphin simply waited for a familiar aspect to be revealed from a rotating object and then made its judgments based on that aspect. The dolphin also could have learned to associate different echoes from several different aspects of the same object. Additionally, it is possible that on some trials, no rotations occurred. To provide unconfounded evidence that representations of rotated objects are aspect-independent and object-based, accurate *first-trial* performance must be reported over a significant number of *known* novel aspects with a variety of objects.

Recognition of object shape cross-modally

Herman and Pack (1992) pioneered a new approach to the study of echoic object perception by using a cross-modal task to investigate whether echolocation may directly yield a shape percept. Their approach was broadly modeled after studies of cross-modal perception of object equivalence through vision and touch (haptic sense), examined for both humans and nonhuman primates (e.g., Davenport and Rogers 1970; Rose and Orlian

1991). For example, in chimpanzees, spontaneous (first-trial) cross-modal recognition of novel objects differing in shape alone provided compelling evidence for the integration of vision and touch (Davenport and Rogers 1970). Herman and Pack (1992) investigated whether analogous integration might occur for dolphins using vision and echolocation. They examined whether a shape inspected through echolocation alone could be matched to an identical shape inspected through vision alone, and vice versa. Their rationale was that if associative learning were eliminated, by examining first-trial recognition performance with a variety of novel objects, then accurate matching must reflect the dolphin's ability to form a shape percept through echolocation (and, of course, vision). That is, inasmuch as the raw physical stimuli arising from echoes and those arising from reflected light are different and not correlated, first-trial matching across these two senses can only be based on matching representations that preserve the shape or appearance of the objects (i.e., are object-based). Spontaneous, first-trial cross-modal matching can neither be explained by the dolphin learning that a particular echo is associated with a particular response (as is possible in the two-alternative forced choice task), nor by it spontaneously matching identical acoustic characteristics of echoes (as is possible in the identity echolocation matching-to-sample task).

To prevent echolocation during visual object inspection, Herman and Pack (1992) held objects in air, a medium in which dolphin echolocation is ineffective. They also developed an anechoic box to occlude vision during active echoic interrogation (and to eliminate the need for applying or removing eyecups). The box, unlike eyecups, allowed the dolphin to transition easily between echoic and visual examination of an object. A detailed description of the anechoic box can be found in Pack and Herman (1995). Briefly, the box measures 1.1 m wide × 0.6 m deep × 1.0 m high. The sides and back consist of a series of angled redwood slats to absorb and scatter extraneous echoes. The key feature is the front panel, a 0.32 cm thick black Plexiglas sheet. Although visually opaque, in water the Plexiglas sheet transmits sound well and allows objects suspended within the box to be inspected through echolocation. Using one or more of these anechoic boxes, two cross-modal paradigms are possible, echoic-visual (E-V) MTS and visual-echoic (V-E) MTS. (In our terminology, the first letter refers to the sense to which the sample is exposed, and the second letter the sense to which the alternatives are exposed.) Additionally, within-modal echoic-echoic (E-E) MTS can be conducted. We tested all these paradigms, as well as within-modal visual-visual (V-V) MTS.

For E-V MTS, a single box was suspended inside the dolphin's habitat from the tank wall. Seawater entered the box through a small opening in the bottom panel. A "sample" object was hung in the water column inside the

box, using thin monofilament lines. The dolphin approached the front panel and interrogated the object inside through echolocation. Two alternative objects were then displayed to the dolphin's visual sense by holding them in air. The dolphin's task was to choose the visual alternative (by stationing in front of it) that was identical to the sample inspected through echolocation.

For V-E MTS, the sample object was displayed in air, and the alternatives were presented within two anechoic boxes. Following visual inspection of the sample, the dolphin pressed a paddle attached to the anechoic box containing the matching alternative object. Several controls were employed to guard against any confounding of variables or cueing. These included the use of masking objects to protect against any passive acoustic cues arising from the splashing sound of objects being placed within the anechoic boxes. Two masks were immersed simultaneously with a test object within a box and then the masks were immediately withdrawn, leaving only the test object inside the box. Also, "blind" procedures were used throughout. For example, the dolphin's choices among alternatives were judged using a strict criterion by an observer having no knowledge of what sample had been displayed (see Pack and Herman 1995 for additional controls). Importantly, objects were never exposed to echolocation and vision simultaneously, precluding any direct learned association between echolocation and vision.

Herman and Pack (1992) trained a dolphin in E-V, V-E, and E-E MTS. (The dolphin was already familiar with V-V MTS.) Four sets of six objects that differed from one another in material composition, size, and shape were used. Most of these objects were "real-world" objects (e.g., a metal bowl, a terra-cotta flowerpot, etc.). After being taught to use echolocation to interrogate the contents of an anechoic box, the dolphin readily acquired each cross-modal task as well as E-E MTS (for exact numbers of trials in each task see Herman and Pack 1992).

Following these procedures, Pack and Herman (1995) tested the dolphin's ability to recognize cross-modally eight pairs of novel complexly shaped objects (16 objects in total), all constructed of sand-filled PVC pipe and fittings. The same combinations of thicknesses of PVC were used across each pair member, and frontal surface areas of pair members were equated to within 4%. Fig. 41.1 shows computer-scanned images of the objects (labeled *A* through *P*) composing these eight pairs, both from a frontal view (the orientation in which they were presented) and at 45° from the frontal plane to better reveal their three-dimensional aspects.

Using these objects, both E-V and V-E MTS were tested. However, before beginning the cross-modal tests, each of the eight pairs was tested separately for 24 trials in E-E and 24 trials in V-V MTS. The rationale behind this within-modal pretesting was that if the dolphin could not discriminate between pair members within each sense, it was unlikely that it would have been able to recognize these objects cross-modally (see, e.g., Bryant 1968). The pretests determined that each of the eight pairs were discriminable by echolocation alone and by vision alone. During this testing, no object was ever exposed to vision and echolocation simultaneously, and tests of E-E and V-V with the same pair were never contiguous. Following pretests, each of the eight pairs was tested for 48 trials in E-V and 48 trials in V-E MTS. (For the eight pairs tested cross-modally, there were respectively 113, 166, 179, 84, 104, 79, 153, and 98 days between the final day of within-modal testing and the initial day of cross-modal testing.) Sessions consisted of 24 trials, within which 12 trials with a particular test pair were interleaved among 12 "baseline" trials using six familiar objects from Herman and Pack (1992). Tests of E-V and V-E MTS were counterbalanced to control for order effects. Fig. 41.2A shows for each pair (labeled 1 through 8) the percentage of correct matches. Performance accuracy on all pairs was significantly above chance in each cross-modal test ($p < 0.05$) using the cumulative binomial test (CBT).

The critical measure of the dolphin's ability to directly perceive object shape in the cross-modal task was its performance on the first trial with each member of a pair acting as sample. Anything beyond first-trial performance was subject to interpretations of associative learning between the visual appearance of an object and the echo from that object. First-trial data are indicated within each bar of fig. 41.2A. A plus sign indicates a correct first trial, and a zero indicates an incorrect first trial. The dolphin was correct on 13 of 16 (81.3%) first trials in E-V MTS, and on 14 of 16 (87.5%) in V-E MTS, ($p = 0.011$ and $p = 0.002$, respectively, CBT). These results clearly demonstrated spontaneity of object recognition across echolocation and vision, and provided the first strong evidence for direct perception of shape through echolocation. Furthermore, performance accuracy in E-V versus V-E was not significantly different, supporting the idea that the representations derived through echolocation and vision are *functionally* equivalent. That is, both provide enough similar information on the spatial structure of the object for the dolphin to categorize them as comparable.

Pack and Herman (1995) also tested the dolphin in V-E MTS with the sample object displayed on a small television monitor placed behind an underwater window. Pairs 1, 2, 5, and 9 (fig. 41.1) were tested. The first three pairs had been tested previously by Pack and Herman (1995), but pair 9 had never been tested previously in any cross-modal task. The televised images of the samples were 40–49% smaller than the real objects. The alternatives suspended in the anechoic boxes were unaltered. Despite these differences in size between sample and alternatives, object constancy was maintained by the dol-

phin who spontaneously matched pairs 1, 2, and 5 in the novel television condition, as well as pair 9 (see fig. 41.2A). These findings provided further evidence that the dolphin's representation of a sample object was based on the object's shape. The results also suggest that an important characteristic of the dolphin's representation of an object is the relative relation of object features to one another and not the size of the object.

In light of the results by Pack and Herman (1995), Herman, Pack, and Hoffmann-Kuhnt (1998) examined whether within-modal object experience was required for spontaneous cross-modal recognition. They first tested whether the dolphin could spontaneously match new *pairings* of objects cross-modally without any prior within-modal matching experience of these pairings. All 25 new two-way combinations of objects E, F, H, K, L, N, O, and P (fig. 41.1) from Pack and Herman (1995) were tested for 24 trials in E-V and 24 trials in V-E MTS. (Pairings E-F, K-L, and O-P, tested previously by Pack and Herman [1995], were not tested again.) The dolphin spontaneously matched both objects of a novel pairing for 24 of 25 pairings under E-V and 20 of 25 pairings under V-E MTS ($p < .005$ for both tests, CBT), indicating that cross-modal recognition was not dependent on having experienced particular pairings previously in E-E and V-V MTS. Furthermore, a subsequent videotape analysis of V-E trials revealed that echoic decision time was remarkably brief, averaging only 1.9 s, indicating that even in the absence of within-modal experiences, shape percepts can develop easily.

Herman, Pack, and Hoffmann-Kuhnt (1998) next tested whether spontaneous cross-modal recognition could proceed in the absence of *any* prior experience with objects. Four completely novel pairs of complexly shaped objects (pairs 10 through 13 in fig. 41.1) were tested directly in E-V and V-E MTS without first testing them in E-E and V-V MTS. The objects were constructed of PVC, under the same constraints as those of Pack and Herman (1995), with the additional requirement that each object subtend the same 43 cm^2 area. Herman, Pack, and Hoffmann-Kuhnt reasoned that if at least some of the novel pairs could be spontaneously matched across the senses, it would constitute evidence that prior within-modal experience with given objects was not necessarily required for spontaneous cross-modal recognition of the objects. Two of the four pairs, 11 and 13, were matched successfully ($p < .05$, CBT) in both E-V and V-E MTS (fig. 41.2B). Moreover, with these two pairs, the dolphin was correct on 7 of 8 first trials in E-V and V-E ($p = .035$, CBT). Neither overall performance nor first trial performance (5 of 8 first trials correct) were significantly above chance on the remaining two pairs. These results show that spontaneous cross-modal recognition can be achieved without any prior object experience, but that not all pairings yield this result.

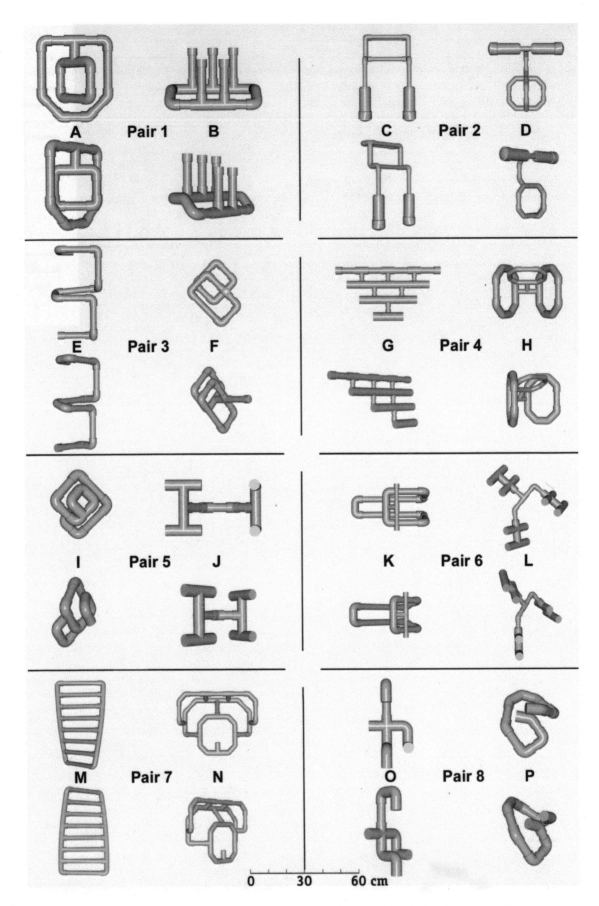

Fig. 41.1. Computer-generated images of 13 pairs of objects (26 objects in total) that were tested cross-modally by Pack and Herman (1995) and Herman, Pack, and

(continued)

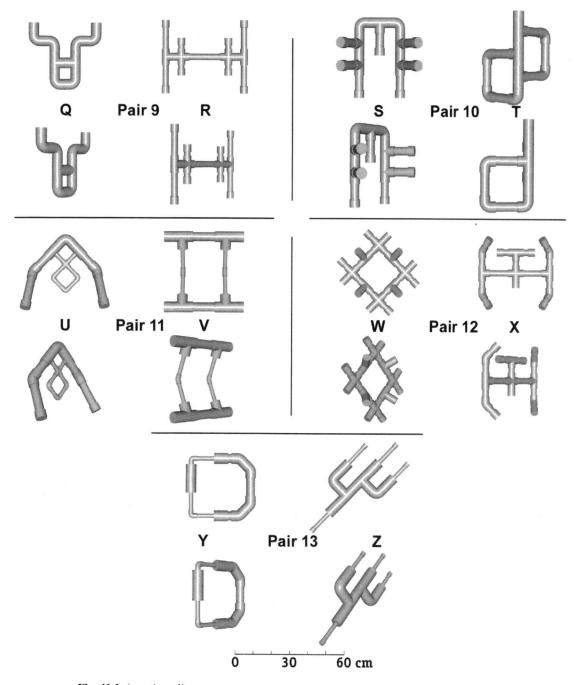

Fig. 41.1. (*continued*)
Hoffmann-Kuhnt (1998). Pair 9 was tested only in visual-echoic (V-E) matching with the sample object displayed on a television screen placed behind an underwater tank window. Two views are shown of each object, a frontal view (to which the dolphin was actually exposed when it faced the object), and a 45° view to show the object's extent in the Z dimension. All objects were constructed of PVC pipe and fittings. All pair members, other than pair 9, were composed of the same diameter PVC and were equated for surface area to within 4%. Additionally, members of pairs 10–13 all subtended the same square area.

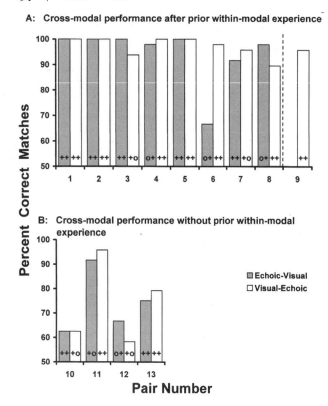

Fig. 41.2. The dolphin's performance accuracy in echoic-visual (E-V) and visual-echoic (V-E) cross-modal matching with 13 pairs of objects (labeled pairs 1–13 as shown in fig. 41.1) that had not been tested previously cross-modally. The dolphin's performance on its first exposure to the two unique trials in which each object of a particular pair appeared as sample in each condition is shown within each bar (plus = correct, zero = incorrect). Panel A shows the percentage of correct matches (48 trials per bar) and first-trial data reported by Pack and Herman (1995) for pairs 1–8 in E-V and V-E tests with sample objects presented "live." Also shown is the percentage of correct matches (*n* = 24) and first trial data for pair 9 in V-E tests with the sample object presented on a television screen. Objects in pairs 1–9 were pretested separately in echoic-echoic (E-E) and visual-visual (V-V) matching. Panel B shows the percentage of correct matches (24 trials per bar) and first-trial data during initial E-V and V-E tests reported by Herman, Pack, and Hoffmann-Kuhnt (1998, experiment 2) with four novel pairs of objects (pairs 10–13) that had not been pretested in E-E and V-V.

Cross-modal studies using blindfold methods

Other researchers have attempted to examine cross-modal recognition in the dolphin using different methods from those of Herman and Pack (1992). Azzali, Manzini, and Buracchi (1995) and Harley, Roitblat, and Nachtigall (1996) placed latex suction cups over their dolphin's eyes to occlude vision in their cross-modal tests. Azzali, Manzini, and Buracchi (1995) used a three-alternative forced-choice task to examine for echoic recognition of a single object (the target) that had been previously inspected through vision alone. The target and five other nonmatching objects were all constructed

from air-filled copper cylinders and differed from one another only in shape. Azzali, Manzini, and Buracchi first trained the dolphin for visual recognition of the target, an inverted "U" shape. On each trial, the U was pitted against two other objects selected from the set of five. All objects were shown in air. The dolphin first viewed the U alone and then selected an identical U shape from the array placed 5 m away. Over several months the alternatives were exposed in different viewing conditions. The dolphin was most accurate when alternatives were affixed to the brown background (95% accuracy over 22 trials).

Approximately eight weeks after the termination of visual tests, the dolphin was tested in a single session of 10 trials to determine if it could select the inverted U among the alternatives through echolocation alone. As before, the trainer first showed the dolphin the U in air. The trainer then placed eyecups over the dolphin's eyes and signaled it to make a selection from among three alternatives now suspended underwater 13 m away. Overall, depending on what criterion one accepts for a correct response, the dolphin was correct on either 7 or 9 of 10 trials (on two trials the dolphin reportedly briefly contacted an incorrect alternative and switched to the correct alternative). In either case, performance was significantly above chance ($p < .05$, CBT, chance = .33). Most importantly, the dolphin was correct on the first five test trials, indicating spontaneous cross-modal recognition and corroborating the findings by Pack and Herman (1995) of the dolphin's ability to directly perceive shape through echolocation.

Harley et al. (1996, experiment 3) also tested a dolphin on cross-modal recognition using a three-alternative procedure in which eyecups were applied to occlude vision during echoic inspection. Unlike the Azzali, Manzini, and Buracchi study, however, Harley et al.'s procedure involved true matching-to-sample in that the sample identity changed across trials and thus required selection of different "matching" alternatives. In E-V MTS, eyecups were applied to the dolphin prior to its echoic inspection of the sample and then were removed prior to the dolphin's choice among the alternative objects, which were held in air. In V-E MTS, after visual inspection of the sample, eyecups were applied to the dolphin prior to the presentation of the alternatives underwater. After some success using objects that had been previously exposed to both echolocation and vision simultaneously, Harley et al. tested their dolphin with six sets of novel objects (three objects per set) in V-E and E-V MTS. Objects in each of these sets varied along one or more dimensions (e.g., size only, shape only, brightness only, material + brightness + texture, etc.). Overall, performance accuracy on the cross-modal transfer tests remained at chance levels (approximately 39%), a result that contrasts markedly with the findings of Pack and Herman (1995). However, Harley et al.'s results might have been expected given their dolphin's chance performance level on E-E and V-V MTS tests with the same novel objects used in the cross-modal tests. Additionally, possible interference in memory for novel objects could have occurred during the delays (up to 30 s) that were an inevitable consequence of eyecup application or removal when transitioning between echolocation and vision.

DISCUSSION AND THEORETICAL IMPLICATIONS

The findings from Pack and Herman (1995), Azzali, Manzini, and Buracchi (1995), and Herman, Pack, and Hoffmann-Kuhnt (1998) support the hypothesis that shape is directly perceived and represented by the dolphin through echolocation. The robustness of the dolphin's capability for direct echoic shape perception becomes clear when one considers the wide variety and large number of complex shapes that were spontaneously recognized cross-modally. For example, a total of 26 differently shaped objects (see fig. 41.1) were tested cross-modally by Pack and Herman (1995) and Herman, Pack, and Hoffmann-Kuhnt (1998)—24 of the 26 in E-V and all 26 in V-E (objects Q and R were tested in the V-E television condition only). The dolphin was correct on 19 of 24 first trials in E-V and 21 of 26 first trials in V-E ($p < .005$ for both tests, CBT). These results strongly suggest that the dolphin's representations of ensonified objects (as well as those inspected visually) were object-based. The results of Helweg et al. (1996b) are consistent with this theory, inasmuch as identical objects echoically inspected from different aspects were judged to be equivalent.

Furthermore, the findings of spontaneity of matching novel pairings of objects by Herman, Pack, and Hoffmann-Kuhnt (1998, experiment 2) argues against any imputation that spontaneous cross-modal performance across the pairs previously tested by Pack and Herman (1995) was the result of the dolphin generalizing some hypothetical characteristic common to all pairs. If such were the case, poor performance accuracy would have resulted when the objects of these earlier pairs were recombined into the 25 new pairings, because many of these pairings would have pit objects having common characteristics against each other.

Does the dolphin form echoic images? For bats, echoic imagery has been conceived of as a form of spatial imagery in which representations derived from sounds (echoes) having acoustic dimensions preserve the spatial dimensions of an object (Simmons 1989; Simmons and Grinnell 1988). Theoretically, this could also hold for dolphins—the ensonified object being represented as an echoic image that retains analog information about that object's spatial structure. The data reviewed earlier indicating spontaneous cross-modal recognition by the dolphin provide the strongest evidence that its representations of ensonified objects are object-based and preserve information about the spatial structure of those objects. Together, these findings support Pack and Herman's (1995) characterization of the

mental representations of objects perceived through echolocation as spatial "images." The dolphin's ability to spontaneously match the image of a televised sample object that is reduced in size by as much as 49% to the unaltered "live" object presented in an anechoic box (Pack and Herman 1995) suggests that size constancy is an important characteristic of the dolphin's echoic image.

Echoic imagery should also result in object constancy during within-modal echolocation matching tests when the echoically perceived sample object and its matching alternative are constructed of different materials or are of different sizes. In each manipulation, "raw" acoustic information preserved in the echoes will be altered, but shape will remain constant. Nevertheless, because the dolphin's representations are object-based, not acoustic-based, it should recognize the equivalency or constancy of the shapes of these objects. On the other hand, echoically inspected objects that are rotated in depth (e.g., Helweg et al. 1996a) may or may not be spontaneously recognized as equivalent to their unrotated counterparts. For humans and pigeons using vision, recognition of rotated stimuli can be impaired, depending on the complexity of objects and whether rotations reveal previously unexamined features or hide previously examined features (e.g., Biederman and Gerhardstein 1993; Wasserman et al. 1996).

Whether echoic imagery in the dolphin also allows for object representations to undergo the same types of transformations observed in visual "mental rotation" tasks (e.g., Shepard and Metzler 1971) is an open question.

How might the dolphin construct a representation of object shape from returning echoes? Amplitude, frequency, and temporal information are all preserved as "raw" acoustic features in the echoes returning from an ensonified target. These features contribute to measures of an object's target strength, spectral composition, and highlight structure. Traditionally, as discussed earlier, these parameters have been inferred as the acoustic cues on which dolphins rely to make accurate discriminations among objects (Au 1993). However, Popper, Hawkins, and Gisiner (1997) stressed that these cues are insufficient to account for the findings of spontaneous cross-modal recognition by Pack and Herman (1995), because they do not translate directly into the appropriate spatial dimensions of the objects. Additionally, simple associative mechanisms—such as echo intensity correlating with object size—cannot explain the cross-modal results, inasmuch as novel pair members often shared the same surface area and in some instances subtended the same square area (Herman, Pack, and Hoffmann-Kuhnt 1998). What processes and mechanisms might therefore be responsible for the dolphin's ability to determine the spatial structure of an ensonified object?

Herman, Pack, and Hoffmann-Kuhnt (1998) speculated that echoic shape perception is probably a result of specialized peripheral and central processing, resulting in a representation of the contours of an object. Through peripheral processing, the dolphin can take advantage of both spatial and temporal information in the echoes returning from an ensonified object to create a spatial-temporal representation that is then interpreted by central mechanisms as a three-dimensional shape. Currently, the central processing mechanisms involved in echoic shape perception are unknown. More information is available on potentially important peripheral processes.

Pack and Herman (1995) suggested that fine range-difference resolution capabilities and cross-range resolution capabilities in the dolphin are important peripheral processes contributing to shape perception. The ability of the echolocating big brown bat (*Eptesicus fuscus*) to perceive the spatial structure of objects appears to derive from its fine range-difference resolution capabilities (as small as 1–2 cm at distances from 30 to 240 cm) (Simmons 1973). Range-difference resolution is a function of the ability to perceive temporal differences in echo arrival. Shape perception has been theorized as a process involving the translation of these temporal differences in echoes into spatial dimensions of that target (Simmons 1989; Simmons, Moss, and Ferragamo 1990). According to the model of acoustic imaging proposed by Simmons (1989), when the bat ensonifies a target, echoes reflecting from different points along the target (called "glints" by Altes [1976]) will arrive at the bat at slightly different times. The resulting temporal array represents the array of different range values along that target. The bat interprets this "range profile" as a spatial dimension of the object: its shape.

Range-difference resolution capabilities in the dolphin have been examined through discrimination tasks. Murchison (1980) trained a dolphin to respond to the closer of two identical lead-filled 7.62 cm polyurethane foam spheres separated in azimuth by 40°. The smallest resolvable distance difference was 0.9 cm at a range of 1 m. To achieve this level of resolution, it is theoretically necessary for the dolphin to detect echo-delay differences of approximately 12 μsec, applying the same model as Simmons (1989). This value is in close agreement with the time-resolution constants for *Tursiops* reported by Au (1993, table 10.1), which varied between 12 and 15 μsec, and translates to resolvable ranges of 1.0 to 1.1 cm between point targets. (Au [1993, p. 206] considered the study by Murchison [1980] to be better characterized as a range-difference-discrimination experiment rather than a range-resolution experiment, because "true" range resolution involves resolving targets at different distances along the same line of acoustic propagation.)

Cross-range resolution in the bat has been investigated through vertical and horizontal angular resolution experiments. Lawrence and Simmons (1982a) measured angular resolution capabilities of bats in the vertical plane. The bats were required to indicate which of two arrays of bars had smaller angular separation between the rods. At threshold the vertical angular resolution was 3–3.5 °. Simmons et al. (1983) conducted similar experiments in the horizontal plane and determined an angular separation resolution capability of 1.5°.

Branstetter et al. (in press) conducted echolocation horizontal angular resolution experiments with a bottlenosed dolphin. The blindfolded dolphin was required to echoically discriminate horizontal angular differences between two simultaneously presented arrays of vertically oriented PVC pipes arranged in an arc. Branstetter et al. (in press) obtained a threshold of 1.6°, which is very similar to the threshold obtained for echolocating bats (Simmons et al. 1983). It is also in close agreement with the results from passive sound localization experiments with dolphins that indicated, in the horizontal plane, a minimum audible angle threshold of 0.9° for click sounds and 1.2° for pure tones (Renaud and Popper 1975). Together, the evidence suggests that the dolphin's range-difference resolution and cross-range resolution capabilities are fine enough to provide significant spatial information for the direct perception of the types of relatively large complex shapes used in our cross-modal studies (see fig. 41.1).

In conjunction with fine range- and azimuthal-resolution capabilities, the dolphin's movement during echoic inspection of targets in the cross-modal task also may contribute to its ability to form shape percepts. The dolphin swims freely while echoically interrogating the contents of an anechoic box in our E-V and V-E MTS tasks (as well as in the E-E task). Theoretically, movement of the dolphin relative to the target yields different "views" of it, providing aspect information that may enable the three-dimensional representation of that object. Azzali (1992) stated that an echolocating dolphin "is required to move to establish the 3-D image of a stationary object and its motion must not be in the line of sight direction, but for example, along a polygonal-like curve" (p. 581). A better understanding of where and when the dolphin echoically interrogates a target along its path in the cross-modal task could provide valuable insights into the role that movement plays in the development of a shape percept.

If, as theorized above, dolphin echo imaging is accomplished through a combination of range and azimuthal profiling integrated with aspect information obtained through target ensonification at different angles, artificial sonar devices may benefit by using this same process. Altes (1995) designed theoretical sonar models of target recognition to emulate the dolphin's "echoic-imaging" capabilities as demonstrated by Pack and Herman (1995). These models integrated azimuth, elevation, and range information obtained from different "viewpoints" of a target that allowed localization of points along that target. Altes showed that a "vision-like" acoustic image could be developed with a rotated wavelet transform that "permits recursive delay-and-sum beamforming with a sparsely sampled synthetic aperture constructed with a moving multibeam sonar system" (1995, p. 1275). Characterizing the dolphin's movements relative to the target in an anechoic box may reveal whether dolphin echolocation functions in some ways like a synthetic aperture sonar, constructing an image of an object by integrating information from different reflecting points sampled over slices in time from different aspects.

Conclusions

Although we are still in the early stages of examining the characteristics of the representations formed by dolphins through echolocation, there is now sufficient evidence to conclude that dolphins can directly perceive and represent object shape through echolocation. This was clearly demonstrated through the dolphin's spontaneous recognition of numerous complex shapes across echolocation and vision (Herman, Pack, and Hoffmann-Kuhnt 1998; Pack and Herman 1995). These findings imply that the dolphin's representations of objects are object-based. That is, even though the raw material in the echo return from an object is sound, the dolphin's *perception* appears to be of the object itself and not of the acoustic qualities of the echoes.

The short inspection time used by the dolphin when echolocating on an object in an anechoic box (on the order of 2 s or less) indicates that shape percepts are developed easily through echolocation. Moreover, cross-modal matching appeared to be a relatively easy task for the dolphin, with decisions among alternative objects made with little or no vacillation. The implication is that a substantial amount of information about an object's spatial structure is readily available to the dolphin through echolocation. Although representations of objects formed through echolocation may differ in many ways from those formed through vision, enough information about the spatial structure must be available to each sense for complex shapes to be recognized easily from one sense to the other. Thus, the representations formed through echolocation and vision may be characterized as functionally equivalent. Exactly how faithful the representation of the ensonified object is to the spatial structure of the "live" object requires further study. A deeper understanding of this issue may benefit from new cross-modal experiments that use trials in which none of the alternatives match the sample, or use alternatives that differ only in the particular spatial arrangement of features (see Pack et al. in press).

The "freeform" nature of the cross-modal procedures developed by Herman and Pack (1992) represents a new generation of echolocation studies that more closely emulate those functional aspects of echolocation

observed from dolphins in the wild (see also Azzali, Manzini, and Buracchi 1995). Dolphins in their natural habitat echolocate while physically unconstrained and often while moving (e.g., Rossbach and Herzing 1997). As we noted earlier, the ability to move while echolocating allows the freedom to examine objects from different viewpoints, and Azzali (1992) considered movement essential for the construction of a three-dimensional image of objects. If this is true, then movement relative to the target could be a major contributor to the dolphin's success in direct echoic shape recognition in any cross-modal tasks. This hypothesis could be tested by examining cross-modal performance when the dolphin's movement during echoic inspection is constrained or its ability to derive aspect information is limited.

The spontaneity and apparent ease with which objects were recognized cross-modally suggests that the ability for direct echoic shape perception is a fundamental cognitive characteristic of the dolphin that offers distinct advantages in the dolphin's natural world. One of the most impressive aspects of the dolphin's cross-modal performance in the studies of Pack and Herman (1995) and Herman, Pack, and Hoffmann-Kuhnt (1998) is the symmetry in matching accuracy between E-V and V-E. This is unlike the results of cross-modal studies in other species using the senses of vision and touch in which asymmetries are common. For example, in human infants (Bushnell and Weinberger 1987) and rhesus monkeys (DiMattia, Posley, and Fuster 1990), generally greater accuracy is evidenced in visual-haptic matching than in haptic-visual matching, although in a few studies the opposite asymmetry was found (e.g., Rose and Orlian 1991). Jones (1981), reviewing some early work on cross-modal abilities in humans, noted that asymmetries emerged predominantly in successive cross-modal matching tasks where the sample was removed from inspection prior to the presentation of the alternatives. The successive matching procedure allows for delays between the removal of the sample and the presentation of the alternatives. Asymmetries in performance therefore could reflect differences in the ability to maintain the representation of a haptic versus a visual sample. Jones (1981) suggested that visual exploration of an object more easily yields a holistic "image" for storage. Haptic exploration is less likely to do this because of its sequential nature. Thus far, no comparable echolocation-vision cross-modal delay studies have been conducted with the dolphin.

Finally, although our understanding of dolphin echolocation (summarized in Au 1993) has increased dramatically since Kellogg's book *Porpoises and Sonar* was first published, even 40 years later his cautionary words (quoted at the beginning of this chapter) still ring true. If we are to understand how dolphins perceive their world, explanations of dolphin echolocation performance must extend beyond the identification of the possible acoustic cues that may be responsible for discriminative abilities. Obviously, there is more to echoic object perception than the simple use of these cues. Our work strongly suggests that dolphins can and do perceive the shape of objects that they inspect through echolocation. The cross-modal task (Herman and Pack 1992) has been instrumental in this discovery, providing new data to challenge long-standing conceptions of echolocation. The result is an emerging view of dolphin echolocation as a form of spatial perception.

Acknowledgments

The work reported in this chapter was funded by grants from the National Science Foundation (IBN-9121331), Earthwatch and its Research Corps, and the Office of Naval Research. It was also supported by The Dolphin Institute, and by donations from Apple Computer, Inc. We thank all the staff, students, interns, volunteers, and short-term participants at the Kewalo Basin Marine Mammal Laboratory for their assistance. We also thank Brian Branstetter, Alison Craig, Bill Friedl, Denise Herzing, and an anonymous reviewer for helpful comments on earlier drafts of this chapter.

Part Three / Literature Cited

ALTES, R. A. 1976. Sonar for generalized target description and its similarity to animal echolocation systems. *Journal of the Acoustical Society of America* 59:97–105.

———. 1992. The line segment transform and sequential hypothesis testing in dolphin echolocation. Pp. 317–355 in *Marine mammal sensory systems,* ed. J. A. Thomas, R. A. Kastelein, and A. Supin. New York: Plenum Press.

———. 1995. Signal processing for target recognition in biosonar. *Neural Networks* 8:1275–1295.

AU, W. W. L. 1993. *The sonar of dolphins.* New York: Springer-Verlag.

AU, W. W. L., and D. A. PAWLOSKI. 1989. A comparison of signal detection between an echolocating dolphin and an optimal receiver. *Journal of Comparative Physiology A* 164:451–458.

———. 1992. Cylinder wall thickness discrimination by an echolocating dolphin. *Journal of Comparative Physiology A* 172:41–47.

AU, W. W. L., and K. J. SNYDER. 1980. Long-range target detection in open waters by an echolocating Atlantic bottlenose dolphin (*Tursiops truncatus*). *Journal of the Acoustical Society of America* 68:1077–1084.

AU, W. W. L., and C. H. TURL. 1983. Target detection in reverberation by an echolocating Atlantic bottlenose dolphin (*Tursiops truncatus*). *Journal of the Acoustical Society of America* 73:1676–1681.

———. 1991. Material composition discrimination of cylinders at different aspect angles by an echolocating dolphin. *Journal of the Acoustical Society of America* 89:2448–2451.

AU, W. W. L., P. W. B. MOORE, and S. W. MARTIN. 1987. Phantom electronic target for dolphin sonar research. *Journal of the Acoustical Society of America* 82:711–713.

AU, W. W. L., R. J. SCHUSTERMAN, and D. A. KERSTING. 1980. Sphere-cylinder discrimination via echolocation by *Tursiops truncatus.* Pp. 859–862 in *Animal sonar systems,* ed. R.-G. Busnel and J. F. Fish. New York: Plenum Press.

AU, W. W. L., D. A. CARDER, R. H. PENNER, and B. L. SCRONCE. 1985. Demonstration of adaptation in beluga whale echolocation signals. *Journal of the Acoustical Society of America* 77:726–730.

AU, W. W. L., R. A. KASTELEIN, T. RIPPE, and N. M. SCHOONEMAN. 1999. Transmission beam pattern and echolocation signals of a harbor porpoise (*Phocoena phocoena*). *Journal of the Acoustical Society of America* 106:3699–3705.

AU, W. W. L., P. W. B. MOORE, and D. A. PAWLOSKI. 1988. Detection of complex echoes in noise by an echolocating dolphin. *Journal of the Acoustical Society of America* 83:662–688.

AU, W. W. L., R. H. PENNER, and J. KADANE. 1982. Acoustic behavior of echolocating Atlantic bottlenose dolphins. *Journal of the Acoustical Society of America* 71:1269–1275.

AUDET, D., D. KRULL, G. MARIMUTHU, S. SUMITHRA, and J. BALA-SINGH. 1991. Foraging behavior of the Indian false vampire bat, *Megaderma lyra* (Chiroptera: Megadermatidae). *Biotropica* 23:63–67.

AZZALI, M. 1992. New optical and acoustic system to study perception and motor-control of a *Tursiops truncatus.* Pp. 575–600 in *Marine mammal sensory systems,* ed. J. A. Thomas, R. A. Kastelein, and A. Supin, New York: Plenum Press.

AZZALI, M., A. MANZINI, and G. BURACCHI. 1995. Acoustic recognition by a dolphin of shapes. Pp. 137–156 in *Sensory systems of aquatic mammals,* ed. R. A. Kastelein, J. A. Thomas, and P. E. Nachtigall. Woerden, The Netherlands: De Spil Publishers.

BARCLAY, R. M. R., and R. M. BRIGHAM. 1994. Constraints on optimal foraging: a field test of prey discrimination by echolocating insectivorous bats. *Animal Behaviour* 48:1013–1021.

BARNARD, S. M. 1995. *Bats in captivity.* Springville, Calif.: Wild Ones Animal Books.

BARRETT-LENNARD, L. G., J. K. B. FORD, and K. A. HEISE. 1996. The mixed blessing of echolocation: Differences in sonar use by fish-eating and mammal-eating killer whales. *Animal Behaviour* 51:553–565.

BARTA, R. E. 1969. Acoustic pattern discrimination by an Atlantic bottle-nosed dolphin. Unpublished manuscript, Naval Undersea Center, San Diego, Calif.

BATES, D. L., and M. B. FENTON. 1990. Aposematism or startle? Predators learn their response to the defense of prey. *Canadian Journal of Zoology* 68:49–52.

BERTA, A., and J. L. SUMICH. 1999. *Marine mammals: Evolutionary biology.* New York: Academic Press.

BEUTER, K. J. 1980. A new concept of echo evaluation in the auditory system of bats. Pp. 747–761 in *Animal sonar systems,* ed. R.-G. Busnel and J. F. Fish. New York: Plenum Press.

BIEDERMAN, I., and P. C. GERHARDSTEIN. 1993. Recognizing depth-rotated objects: Evidence and conditions for three-dimensional viewpoint invariance. *Journal of Experimental Psychology: Human Perception and Performance* 19:1162–1182.

BINNIE, C. A., A. A. MONTGOMERY, and P. L. JACKSON. 1974. Auditory and visual contributions to the perception of consonants. *Journal of Speech and Hearing Research* 17:619–630.

BRANSTETTER, B. K., S. J. MEVISSEN, A. A. PACK, L. M. HERMAN, and S. P. ROBERTS. In press. Horizontal angular discrimination by an echolocating bottlenose dolphin (*Tursiops truncatus*). *Bioacoustics.*

BRILL, R. L., M. L. SEVENICH, T. J. SULLIVAN, J. D. SUSTMAN, and R. E. WITT. 1988. Behavioral evidence for hearing through the lower jaw by an echolocating dolphin (*Tursiops truncatus*). *Marine Mammal Science* 4:223–230.

BRYANT, P. E. 1968. Comments on the design of developmental studies of cross-modal matching and cross-modal transfer. *Cortex* 4:127–137.

BURGER, M.-L. 1972. Der Anteil propriozeptiver Erregung an der Kurskontrolle bei Arthropoden (Diplopoden und Insekten). *Verhandlungen der Deutschen Zoologischen Gesellschaft* 65:220–225.

BUSHNELL, E. W., and N. WEINBERGER. 1987. Infants' detection of visual-tactual discrepancies: Asymmetries that indicate a directive role of visual information. *Journal of Experimental Psychology: Human Perception and Performance* 13:601–608.

BUUS, S. 1999. Temporal integration and multiple looks, revisited: Weights as a function of time. *Journal of the Acoustical Society of America* 105:2466–2475.

CARLYON, R. P. 1988. The development and decline of forward masking. *Hearing Research* 32:65–80.

CASSEDAY, J. H., and E. COVEY. 1992. Frequency tuning properties of neurons in the inferior colliculus of an FM bat. *Journal of Comparative Neurology* 319:34–50.

———. 1996. A neuroethological theory of the operation of the inferior colliculus. *Brain, Behaviour and Evolution* 47:311–336.

CHASE. J. 1981. Visually guided escape responses of microchiropteran bats. *Animal Behaviour* 29:708–713.

———. 1983. Differential responses to visual and acoustic cues during escape in the bat *Anoura geoffroyi:* Cue preferences and behaviour. *Animal Behaviour* 31:526–531.

CHERRY, C. 1953. Some experiments on the recognition of speech with one and two ears. *Journal of the Acoustical Society of America* 25:975–979.

COLES R. B., A. GUPPY, M. E. ANDERSON, and P. SCHLEGEL. 1989. Frequency sensitivity and directional hearing in the gleaning bat, *Plecotus auritus* (Linnaeus 1758). *Journal of Comparative Physiology A* 165:269–280.

COWEY, A., and L. WEISKRANTZ. 1975. Demonstration of cross-modal matching in rhesus monkeys (*Macaca mulatta*). *Neuropsychologia* 13:117–120.

CRANFORD, T. W., M. AMUNDIN, and K. S. NORRIS. 1996. Functional morphology and homology in the odontocete nasal complex: Implications for sound generation. *Journal of Morphology* 228:223–285.

DAVENPORT, R. K., and C. M. ROGERS. 1970. Intermodal equivalence of stimuli in apes. *Science* 168:279–280.

DIMATTIA, B. V., K. A. POSLEY, and J. M. FUSTER. 1990. Cross-modal short-term memory of haptic and visual information. *Neuropsychologia* 28:17–33.

DUBROVSKIY, N. A. 1990. On the two auditory subsystems in dolphins. Pp. 233–254 in *Sensory abilities of cetaceans: Laboratory and field evidence,* ed. J. A.

Thomas and R. A. Kastelein. New York: Plenum Press.

DUNNING, D. C., and K. E. ROEDER. 1965. Moth sounds and the insect-catching behavior of bats. *Science* 147:173–174.

EDELMAN, S. 1999. *Representation and recognition in vision.* Cambridge: MIT Press.

ETIENNE, A. S., R. MAURER, and V. SEGUINOT. 1996. Path integration in mammals and its interaction with visual landmarks. *Journal of Experimental Biology* 199:201–209.

EVANS, W. E. 1973. Echolocation by marine delphinids and one species of fresh-water dolphin. *Journal of the Acoustical Society of America* 54:191–199.

EVANS, W. E., and B. A. POWELL. 1967. Discrimination of different metallic plates by an echolocating delphinid. Pp. 363–382 in *Animal sonar systems,* ed. R.-G. Busnel and J. F. Fish. New York: Plenum Press.

FAY, R. R. 1988. *Hearing in vertebrates: A psychophysics databook.* Winnetka, Ill.: Hill and Fay Associates.

FENG, A. S., C. J. CONDON, and K. R. WHITE. 1994. Stroboscopic hearing as a mechanism for prey discrimination in frequency-modulated bats? *Journal of the Acoustical Society of America* 95:2736–2744.

FENTON, M. B. 1995. Natural history and biosonar signals. Pp. 37–86 in *Hearing by bats,* ed. A. N. Popper and R. R. Fay. New York: Springer-Verlag.

FERRAGAMO, M. J., T. HARESIGN, and J. A. SIMMONS. 1998. Frequency tuning, latencies, and responses to frequency-modulated sweeps in the inferior colliculus of the echolocating bat, *Eptesicus fuscus. Journal of Comparative Physiology A* 182:65–79.

FIEDLER, J. 1979. Prey catching with and without echolocation in the Indian false vampire (*Megaderma lyra*). *Behavioral Ecology and Sociobiology* 6:155–160.

FINNERAN, J. J., C. W. OLIVER, K. M. SCHAEFER, and S. H. RIDGWAY. 2000. Source levels and estimated yellowfin tuna (*Thunnus albacares*) detection ranges for dolphin jaw pops, breaches, and tail slaps. *Journal of the Acoustical Society of America* 107:649–656.

FODOR, J. A., T. G. BEVER, and M. F. GARRETT 1974. *The psychology of language.* New York: McGraw-Hill.

FULLARD, J. H. 1984. Listening for bats: Pulse repetition rate as a cue for a defensive behavior in *Cycnia tenera* (Lepidoptera: Arctiidae). *Journal of Comparative Physiology A* 154:249–252.

———. 1987. Sensory ecology and neuroethology of moths and bats: Interactions in a global perspective. Pp. 244–272 in *Recent advances in the studies of bats,* ed. M. B. Fenton, P. Racey, and J. M. Rayner. Cambridge: Cambridge University Press.

FULLARD, J. H., J. A. SIMMONS, and P. A. SAILLANT. 1994. Jamming bat echolocation: The dogbane tiger moth *Cycnia tenera* times its clicks to the terminal attack calls of the big brown bat *Eptesicus fuscus. Journal of Experimental Biology* 194:285–294.

GAIONI, S. J., H. RIQUIMAROUX, and N. SUGA. 1990. Biosonar behavior of mustached bats swung on a pendulum prior to cortical ablation. *Journal of Neurophysiology* 64:1801–1817.

GARBIN, C. P. 1988. Visual-haptic perceptual nonequivalence of shape information and its impact upon cross-modal performance. *Journal of Experimental Psychology: Human Perception and Performance* 14 (4): 547–553.

GAUNAURD, G. E., D. BRILL, H. HUANG, P. W. B. MOORE, and H. C. STRIFORS. 1998. Signal processing of the echo signatures returned by submerged shells insonified by dolphin "clicks": Active classification. *Journal of the Acoustical Society of America* 103:1547–1557.

GERKEN, G. M., V. K. H. BHAT, and M. HUTCHINSON-CLUTTER. 1990. Auditory temporal integration and the power function model. *Journal of the Acoustic Society of America* 88:767–778.

GIBSON, J. J. 1960. The concept of stimulus in psychology. *American Psychologist* 16:694–703.

———. 1979. *The ecological approach to visual perception.* Boston: Houghton-Mifflin.

GOODSON, A. D., and R. H. MAYO. 1995. Interactions between free-ranging dolphins (*Tursiops truncatus*) and passive acoustic gill-net deterrent devices. Pp. 365–379 in *Sensory systems of aquatic mammals,* ed. R. A. Kastelein, J. A. Thomas, and P. E. Nachtigall. Woerden, The Netherlands: De Spil Publishers.

GÖRNER, P. 1973. Beispiele einer Orientierung ohne richtende Außenreize. *Fortschritte der Zoologie* 21: 20–45.

GREEN, D. M. 1985. Temporal factors in psychoacoustics. Pp. 122–138 in *Time resolution in auditory systems,* ed. A. Michelsen. Berlin: Springer-Verlag.

GRIFFIN, D. R. 1967. Discriminative echolocation by bats. Pp. 273–300 in *Animal sonar systems: Biology and bionics,* ed. R.-G. Busnel. Jouy-en-Josas: Laboratoire de Physiologie Acoustique.

———. *Listening in the dark: The acoustic orientation of bats and men.* New York: Dover.

GRIFFIN, D. W., F. A. WEBSTER, and C. R. MICHAEL. 1960. The echolocation of flying insects by bats. *Animal Behaviour* 8:141–154.

GRINNELL, A. D. 1995. Hearing in bats: An overview. Pp. 1–36 in *Hearing by bats,* ed. A. N. Popper and R. R. Fay. New York: Springer-Verlag.

GROSSETÊTE, A., and C. F. MOSS. 1998. Target flutter rate discrimination by bats using frequency-modulated sonar sounds: Behavior and signal processing models. *Journal of the Acoustical Society of America* 103: 2167–2176.

HAMILTON, I. M., and R. M. R. BARCLAY. 1998. Diets of juvenile, yearling, and adult big brown bats (*Eptesicus fuscus*) in southeastern Alberta. *Journal of Mammalogy* 79:764–771.

HAPLEA, S., J. H. CASSEDAY, and E. COVEY. 1994. Frequency tuning and response latencies at three levels in the brainstem of the echolocating bat, *Eptesicus fuscus. Journal of Comparative Physiology A* 174: 671–683.

HARLEY, H. E. 1993. Object representation in the bottlenose dolphin (*Tursiops truncatus*): Integration and association of visual and echoic information. Ph.D. dissertation, University of Hawaii, Honolulu.

HARLEY, H. E., H. L. ROITBLAT, and P. E. NACHTIGALL. 1996. Object representation in the bottlenose dolphin (*Tursiops truncatus*): Integration of visual and echoic information. *Journal of Experimental Psychology: Animal Behavior Processes* 22(2): 164–174.

HARLEY, H. E., M. J. XITCO, and H. L. ROITBLAT. 1995. Echolocation, cognition, and the dolphin's world. Pp. 515–528 in *Sensory systems of aquatic mammals,* ed. R. A. Kastelein, J. A. Thomas, and P. E. Nachtigall. Woerden, The Netherlands: De Spil Publishers.

HARTLEY, D. J. 1992. Stabilization of perceived echo amplitudes in echolocating bats. I. Echo detection and automatic gain control in the big brown bat, *Eptesicus fuscus,* and the fishing bat, *Noctilio leporinus. Journal of the Acoustical Society of America* 91:1120–1132.

HARTLEY, D. J., and R. A. SUTHERS. 1989. The sound emission pattern of the echolocating bat, *Eptesicus fuscus. Journal of the Acoustical Society of America* 85:1348–1351.

———. 1990. Sonar pulse radiation and filtering in the mustached bat *Pteronotus parnellii rubiginosus. Journal of the Acoustical Society of America* 87:2756–2772.

HEBLICH, K. 1993. Ortsgedächtnis und Echoortung bei der Großen Hufeisennase (*Rhinolophus ferrumequinum*). Ph.D. dissertation, University of Tübingen, Germany.

HEFFNER, R. S., and HEFFNER, H. E. 1992. Evolution of sound localization in mammals. Pp. 691–715 in *The evolutionary biology of hearing,* ed. D. B. Webster, R. R. Fay, and A. N. Popper. New York: Springer-Verlag.

HELWEG, D. A., W. W. L. AU, H. L. ROITBLAT, and P. E. NACHTIGALL. 1996a. Acoustic basis for recognition of aspect-dependent three-dimensional targets by an echolocating bottlenose dolphin. *Journal of the Acoustical Society of America* 99:2409–2420.

HELWEG, D. A., H. L. ROITBLAT, P. E. NACHTIGALL, and M. J. HAUTUS. 1996b. Recognition of aspect-dependent three-dimensional objects by an echolocating Atlantic bottlenose dolphin. *Journal of Experimental Psychology: Animal Behavior Processes* 22:19–31.

HENSON, O. W., JR. 1965. The activity and function of the middle ear muscles in echolocating bats. *Journal of Physiology* (London) 180:871–887.

HENSON, O. W., JR., and H.-U. SCHNITZLER. 1979. Performance of animal airborne sonar systems: II. Vertebrates other than Microchiroptera. Pp. 183–195 in *Animal sonar systems,* ed. R.-G. Busnel and J. F. Fish. New York: Plenum Press.

HERMAN, L. M., and W. R. ARBEIT. 1972. Frequency difference limens in the bottlenose dolphin: 1–70 kc/s. *Journal of Auditory Research* 12:109–120.

HERMAN, L. M., and A. A. PACK. 1992. Echoic-visual cross-modal recognition by a dolphin. Pp. 709–726 in *Marine mammal sensory systems,* ed. J. A. Thomas, R. A. Kastelein, and A. Supin. New York: Plenum Press.

HERMAN, L. M., A. A. PACK, and M. HOFFMANN-KUHNT. 1998. Seeing through sound: Dolphins (*Tursiops truncatus*) perceive the spatial structure of objects through echolocation. *Journal of Comparative Psychology* 112 (3): 292–305.

HERMAN, L. M., D. G. RICHARDS, and J. P. WOLZ. 1984. Comprehension of sentences by bottlenosed dolphins. *Cognition* 16:1–90.

HESSEL, K., and U. SCHMIDT. 1994. Multimodal orientation in *Carollia perspicillata* (Phyllostomidae). *Folia Zoologica* 43:339–346.

HOFMAN, P. M., J. G. A. VAN RISWICK, and A. J. VAN OPSTAL. 1998. Relearning sound localization with new ears. *Nature Neuroscience* 1:415–421.

HÖLLER, P. 1995. Orientation by the bat *Phyllostomus discolor* (Phyllostomidae) on the return flight to its resting place. *Ethology* 100:72–83.

HÖLLER, P., and U. SCHMIDT. 1996. The orientation behaviour of the lesser spearnosed bat, *Phyllostomus discolor* (Chiroptera) in a model roost: Concurrence of visual, echoacoustical and endogenous spatial information. *Journal of Comparative Physiology A* 179:245–254.

HOUSER, D. S., D. A. HELWEG, and P. W. MOORE. 1999. Classification of dolphin echolocation clicks by energy and frequency distribution. *Journal of the American Acoustical Society* 106:1579–1585.

INOUE, S., M. KIMYOU. Y. KASHIMORI, O. HOSHINO, and T. KAMBARA. 2000. A neural model of medial geniculate body and auditory cortex detecting target distance independently of target velocity in echolocation. *Neurocomputing* 32–33:833–841.

ISRAEL, I., A. M. BRONSTEIN, R. KANAYAMA, M. FALDON, and M. A. GRESTY. 1996. Visual and vestibular factors influencing vestibular "navigation." *Experimental Brain Research* 112:411–419.

JENSEN, M. E., and L. A. MILLER. 1999. Echolocation signals of the bat *Eptesicus serotinus* recorded using a vertical microphone array: Effect of flight altitude on searching signals. *Behavioral Ecology and Sociobiology* 47:60–69.

JERISON, H. J. 1986. The perceptual worlds of dolphins. Pp. 141–166 in *Dolphin cognition and behavior: A comparative approach,* ed. R. J. Schusterman, J. A. Thomas, and F. G. Wood. Hillsdale, N.J.: Erlbaum.

JOHNSON, S. C. 1967. Sound detection thresholds in marine mammals. Pp. 247–260 in *Marine bioacoustics,* ed. W. Tavolga. New York: Pergamon Press.

JOLICOEUR, P., and G. KEITH-HUMPHREY. 1998. Perception of rotated two-dimensional and three-dimensional objects and visual shapes. Pp. 69–123 in *Perceptual constancy: Why things look as they do,* ed. V. Walsh and J. Kulikowski. Cambridge: Cambridge University Press.

JONES, B. 1981. The developmental significance of cross-modal matching. Pp. 109–136 in *Intersensory perception and sensory integration,* ed. R. D. Walk and H. L. Pick Jr. New York: Plenum Press.

KALKO, E. K. V., and H.-U. SCHNITZLER. 1998. How echolocating bats approach and acquire food. Pp. 197–204 in *Bat biology and conservation,* ed. T. H. Kunz and P. A. Racey. Washington, D.C.: Smithsonian Institution Press.

KELLOGG, W. N. 1961. *Porpoises and sonar.* Chicago: University of Chicago Press.

KICK, S. A. 1982. Target detection by the echolocating bat, *Eptesicus fuscus. Journal of Comparative Physiology* 145:431–435.

KICK, S. A., and J. A. SIMMONS. 1984. Automatic gain control in the bat's sonar receiver and the neuroethology of echolocation. *Journal of Neuroscience* 4:2725–2737.

KNUDSEN, E. I. 1985. Experience alters the spatial tuning of auditory units in the optic tectum during a sensitive period in the barn owl. *Journal of Neuroscience* 5:3094–3109.

KOAY, G., D. KEARNS, H. E. HEFFNER, and R. S. HEFFNER. 1998. Passive sound localization ability of the big brown bat (*Eptesicus fuscus*). *Hearing Research* 119:37–48.

KOBER, R., and H.-U. SCHNITZLER. 1990. Information in sonar echoes of fluttering insects available for echolocating bats. *Journal of the Acoustical Society of America* 87:882–896.

LASKA, M. 1990. Olfactory discrimination ability in short-tailed fruit bat, *Carollia perspicillata* (Chiroptera: Phyllostomidae). *Journal of Chemical Ecology* 16:3292–3299.

LAWRENCE, B. D., and J. A. SIMMONS. 1982a. Echolocation in bats: The external ear and perception of the vertical positions of targets. *Science* 218:481–483.

———. 1982b. Measurement of atmospheric attenuation at ultrasonic frequencies and the significance for echolocation by bats. *Journal of the Acoustical Society of America* 71:585–590.

LEONARD, M. L., and M. B. FENTON. 1984. Echolocation calls of *Euderma maculum* (Vespertilionidae): Use in orientation and communication. *Journal of Mammalogy* 65:122–126.

LEVITT, H. 1971. Transformed up-down methods in psychoacoustics. *Journal of the Acoustical Society of America* 49:467–477.

MANN, D. A., Z. LU, and A. N. POPPER. 1997. A clupeid fish can detect ultrasound. *Nature* 389:341.

MANN, D. A., Z. LU, M. C. HASTINGS, and A. N. POPPER. 1998. Detection of ultrasonic tones and simulated dolphin echolocation clicks by a teleost fish, the American shad (*Alosa sapidissima*). *Journal of the Acoustical Society of America* 104:562–568.

MARIMUTHU, G., and G. NEUWEILER. 1987. The use of acoustical cues for prey detection by the Indian False Vampire Bat, *Megaderma lyra. Journal of Comparative Physiology A* 160:509–515.

MARKS, L. E. 1978. *The unity of the senses: Interrelations among the modalities.* New York: Academic Press.

———. 1987. On cross-modal similarity: Perceiving temporal patterns by hearing, touch, and vision. *Perception and Psychophysics* 42 (3): 250–256.

MARR, D. 1982. *Vision: A computational investigation into the human representation and processing of visual information.* New York: W. H. Freeman.

MARR, D., and H. K. NISHIHARA. 1978. Representation and recognition of the spatial organization of three-dimensional shapes. *Proceedings of the Royal Society of London B* 200:269–294.

MARTEN, K., K. S. NORRIS, P. W. B. MOORE, and K. A. ENGLUND. 1988. Loud impulse sounds in odontocete predation and social behavior. Pp. 567–579 in *Animal sonar: Processes and performance,* ed. P. E. Nachtigall and P. W. Moore. New York: Plenum Press.

MASSARO, D. W., and D. G. STORK. 1998. Sensory integration and speech reading by humans and machines. *American Scientist* 86:236–244.

MASTERS, W. M., and S. C. JACOBS. 1989. Target detection and range resolution by the big brown bat (*Eptesicus fuscus*) using normal and time-reversed model echoes. *Journal of Comparative Physiology A* 166:65–73.

MASTERS, W. M., and K. A. S. RAVER. 1996. The degradation of distance discrimination in big brown bats (*Eptesicus fuscus*) caused by different interference signals. *Journal of Comparative Physiology A* 179:703–713.

————. 2000. Range discrimination by big brown bats (*Eptesicus fuscus*) using altered model echoes: Implications for signal processing. *Journal of the Acoustical Society of America* 107:625–637.

MASTERS, W. M., S. C. JACOBS, and J. A. SIMMONS. 1991. The structure of echolocation sounds used by the big brown bat, *Eptesicus fuscus:* Some consequences for echo processing. *Journal of the Acoustical Society of America* 89:1402–1413.

MASTERS, W. M., A. J. M. MOFFAT, and J. A. SIMMONS. 1985. Sonar tracking of horizontally moving targets by the big brown bat *Eptesicus fuscus. Science* 228:1331–1333.

MASTERS, W. M., K. A. S. RAVER, and K. A. KAZIAL. 1995. Sonar signals of big brown bats, *Eptesicus fuscus,* contain information about individual identity, age and family affiliation. *Animal Behaviour* 50:1243–1260.

MASTERS, W. M., K. A. S. RAVER, K. KORNACKER, and S. C. BURNETT. 1997. Detection of jitter in intertarget spacing by the big brown bat *Eptesicus fuscus. Journal of Comparative Physiology A* 181:279–290.

MAY, M. 1991. Aerial defense tactics of flying insects. *American Scientist* 79:316–328.

MCCUE, J. J. G. 1966. Aural pulse compression in bats and humans. *Journal of the Acoustical Society of America* 40:545–548.

MCGURK, H., and J. MACDONALD. 1976. Hearing lips and seeing voices. *Nature* 264:746–748.

MENNE D, I. KAIPF. I. WAGNER, J. OSTWALD, and H.-U. SCHNITZLER. 1989. Range estimation by echolocation in the bat *Eptesicus fuscus:* Trading of phase versus time cues. *Journal of the Acoustical Society of America* 85:2642–2650.

MILLER, L. A. 1991. Arctiid moth clicks can degrade the accuracy of range difference discrimination in echolocating big brown bats, *Eptesicus fuscus. Journal of Comparative Physiology A* 168:571–579.

MILLER, L. A., and H. J. DEGN. 1981. The acoustic behavior of four species of vespertilionid bats studied in the field. *Journal of Comparative Physiology A* 142:67–74.

MISTRY, S., and G. F. MCCRACKEN. 1990. Behavioural response of the Mexican free-tailed bat, *Tadarida brasiliensis mexicana,* to visible and infra-red light. *Animal Behaviour* 39:598–599.

MITTELSTAEDT, M.-L., and S. GLASAUER. 1991. Idiothetic navigation in gerbils and humans. *Zoologisches Jahrbuch der Physiologie* 95:427–436.

MITTELSTAEDT M.-L., and H. MITTELSTAEDT. 1979. Interaction of gravity and idiothetic course control in millipedes. *Journal of Comparative Physiology* 133:267–281.

————. 1980. Homing by path integration in a mammal. *Naturwissenschaften* 67:566–567.

MØHL, B. 1986. Detection by a pipistrelle bat of normal and reversed replica of its sonar pulses. *Acustica* 61:75–82.

MØHL, B., and A. SURLYKKE. 1989. Detection of sonar signals in the presence of pulses of masking noise by the echolocating bat, *Eptesicus fuscus. Journal of Comparative Physiology A* 165:119–124.

MÖHRES, F. P., and G. NEUWEILER. 1966. Die Ultraschallorientierung der Großblattfledermäuse (Chiroptera-Megadermatidae). *Zeitschrift für vergleichende Physiologie* 53:195–227.

MÖHRES, F. P., and T. OETTINGEN-SPIELBERG. 1949. Versuche über die Nahorientierung und das Heimfindevermögen der Fledermäuse. *Verhandlungen der Deutschen Zoologischen Gesellschaft* 42:248–25.

MOISEFF, A., G. S. POLLACK, and R. R. HOY. 1978. Steering responses of flying crickets to sound and ultrasound: Mate attracting and predator avoidance. *Proceedings of the National Academy of Sciences* (USA, Biological Science) 75:4052–4056.

MOORE, B. C. J. 1997. *An introduction to the psychology of hearing.* San Diego: Academic Press.

MOORE, B. C. J., and B. R. GLASBERG. 1983. Growth of forward masking for sinusoidal and noise maskers as a function of signal delay; implications for suppression in noise. *Journal of the Acoustic Society of America* 73:1249–1259.

MOORE, P. W. B. 1997. Mine hunting dolphins of the Navy. *Society of Photo-Optical Instrumentation Engineers Proceedings,* series 3079 (0277-786X/97): 2–6.

MOORE, P. W. B., and D. A. PAWLOSKI. 1990. Investigations on the control of echolocation pulses in the dolphin (*Tursiops truncatus*). Pp. 305–316 in *Sensory abilities of cetaceans: Laboratory and field evidence,* ed. J. A. Thomas and R. A. Kastelein. New York: Plenum Press.

MOORE, P. W. B., R. W. HALL, W. A. FRIEDL, and P. E. NACHTIGALL. 1984. The critical interval in dolphin echolocation: What is it? *Journal of the Acoustical Society of America* 76:314–317.

MOORE, P. W. B., H. L. ROITBLAT, R. H. PENNER, and P. E. NACHTIGALL. 1991. Recognizing successive dolphin echoes with an integrator gateway network. *Neural Networks* 4:701–709.

MOSS, C. F., and H.-U. SCHNITZLER. 1989. Accuracy of target ranging in echolocating bats: Acoustic information processing. *Journal of Comparative Physiology A* 165:383–393.

———. 1995. Behavioral studies of auditory information processing. Pp. 87–145 in *Hearing by bats,* ed. A. N. Popper and R. R. Fay. New York: Springer-Verlag.

MURCHISON, A. E. 1980. Detection range and range resolution of echolocating bottlenose porpoise (*Tursiops truncatus*). Pp. 43–70 in *Animal sonar systems,* ed. R.-G. Busnel and J. F. Fish. New York: Plenum Press.

NACHTIGALL, P. E. 1980. Odontocete echolocation performance on object size, shape and material. Pp. 71–95 in *Animal sonar systems,* ed. R.-G. Busnel and J. F. Fish. New York: Plenum Press.

NACHTIGALL, P. E., and P. W. B. MOORE, eds. 1988. *Animal sonar: Processes and performance.* New York: Plenum Press.

NACHTIGALL, P. E., A. E. MURCHISON, and W. W. L. AU. 1980. Cylinder and cube discrimination by an echolocating bottlenose dolphin. Pp. 945–947 in *Animal sonar systems,* ed. R.-G. Busnel and J. F. Fish. New York: Plenum Press.

NACHTIGALL, P. E., W. W. L. AU, H. L. ROITBLAT, and J. L. PAWLOSKI. 2000. Dolphin biosonar: A model for biomimetic sonars. Pp. 115–121 in *Proceedings of the First International Symposium on Aqua Bio-Mechanisms.*

NEUWEILER, G. 1990. Auditory adaptations for prey capture in echolocating bats. *Physiological Reviews* 30: 615–641.

NEUWEILER, G., and M. B. FENTON. 1988. Behavior and foraging ecology of echolocating bats. Pp. 535–549 in *Animal sonar: Processes and performance,* ed. P. E. Nachtigall and P. W. B. Moore. New York: Plenum Press.

NEUWEILER, G., and F. P. MÖHRES. 1967. Die Rolle des Ortsgedächtnisses bei der Orientierung der Großblatt-Fledermaus *Megaderma lyra. Zeitschrift für vergleichende Physiologie* 57:147–171.A.

NORRIS, K. S., and B. MØHL. 1983. Can odontocetes debilitate prey with sound? *American Naturalist* 122:85–104.

NORRIS, K. S., B. WÜRSIG, R. S. WELLS, and M. WURSIG. 1994. *The Hawaiian spinner dolphin.* Berkeley and Los Angeles: University of California Press.

NOVACEK, M. J. 1985. Evidence for echolocation in the oldest known bat. *Nature* 315:140–41.

OBRIST, M. 1995. Flexible bat echolocation: The influence of individual, habitat, and conspecifics on sonar signal design. *Behavioral Ecology and Sociobiology* 36:207–219.

O'NEILL, W. E., and N. SUGA. 1982. Encoding of target range and its representation in the auditory cortex of the mustached bat. *Journal of Neuroscience* 2:17–31.

OSTWALD, J., H.-U. SCHNITZLER, and G. SCHULLER. 1988. Target discrimination and target classification in echolocating bats. Pp. 413–434 in *Animal sonar: Processes and performance,* ed. P. E. Nachtigall and P. W. B. Moore. New York: Plenum Press.

PACK, A. A., and L. M. HERMAN. 1995. Sensory integration in the bottlenosed dolphin: Immediate recognition of complex shapes across the senses of echolocation and vision. *Journal of the Acoustical Society of America* 98:722–733.

PACK, A. A., L. M. HERMAN, M. HOFFMANN-KUHNT, and B. K. BRANSTETTER. 2002. The object behind the echo: Dolphins (*Tursiops truncatus*) perceive object shape globally through echolocation. *Behavioural Processes* 58:1–26.

PALAKAL, M. J., and D. WONG. 1999. Cortical representation of spatiotemporal pattern of firing evoked by echolocating signals: Population encoding of target feature is real time. *Journal of the Acoustical Society of America* 106:479–490.

PARK, T. J., and G. D. POLLAK. 1993. GABA shapes a topographic organization of response latency in the mustache bat's inferior colliculus. *Journal of Neuroscience* 13:5172–5187.

POLLAK, G. D., and J. H. CASSEDAY. 1989. *The neural basis of echolocation in bats.* New York: Springer-Verlag.

POPPER, A. N., H. L. HAWKINS, and R. C. GISINER. 1997. Questions in cetacean bioacoustics: Some suggestions for future research. *International Journal of Animal Sound and Its Recording* 8:163–182.

PRYOR, K., and K. S. NORRIS, eds. *Dolphin societies: Discoveries and puzzles.* Berkeley and Los Angeles: University of California Press.

RAZAK, K. A., Z. M. FUZESSERY, and T. D. LOHUIS. 1999. Single cortical neurons serve both echolocation and passive sound localization. *Journal of Neurophysiology* 81:1438–1442.

RENAUD, D. L, and A. N. POPPER. 1975. Sound localization by the bottlenose porpoise *Tursiops truncatus. Journal of Experimental Biology* 63:569–585.

RICHARDS, D. G., J. P. WOLZ, and L. M. HERMAN. 1984. Vocal mimicry of computer-generated sounds and vocal labeling of objects by bottlenosed dolphins, *Tursiops truncatus. Journal of Comparative Psychology* 98:10–28.

ROEDER, K. D. 1965. Moths and ultrasound. *Scientific American* 212:94–102.

ROITBLAT, H. L., D. A. HELWEG, and H. E. HARLEY. 1995. Echolocation and imagery. Pp. 171–181 in *Sensory systems of aquatic mammals,* ed. R. A. Kastelein, J. A. Thomas, and P. E. Nachtigall. Woerden, The Netherlands: De Spil Publishers.

ROITBLAT, H. L., R. H. PENNER, and P. E. NACHTIGALL. 1990. Matching-to-sample by an echolocating dolphin. *Journal of Experimental Psychology: Animal Behavior Processes* 16:85–95.

ROITBLAT, H. L., W. W. L. AU, P. E. NACHTIGALL, R. SHIZUMURA, and G. MOONS. 1995. Sonar recognition of targets embedded in sediment. *Neural Networks* 8:1263–1273.

ROITBLAT, H. L., P. W. B. MOORE, P. E. NACHTIGALL, and R. H. PENNER. 1991. Natural dolphin echo recognition using an integrator gateway network. Pp. 273–281 in *Advances in neural information processing systems 3,* ed. D. S. Touretsky and R. Lippman. San Mateo, Calif.: Morgan Kaufmann.

ROSE, S. A., and E. K. ORLIAN. 1991. Asymmetries in infant cross-modal transfer. *Child Development* 62:706–718.

ROSE, S. A., and H. A. RUFF. 1987. Cross-modal abilities in human infants. Pp. 318–362 in *Handbook of infant development,* 2nd edition, ed. J. Osofsky. New York: Wiley.

ROSSBACH, K. A., and D. L. HERZING. 1997. Underwater observations of benthic-feeding bottlenose dolphins (*Tursiops truncatus*) near Grand Bahama Island, Bahamas. *Marine Mammal Science* 13:498–504.

ROVERUD, R. C., and A. D. GRINNELL. 1985. Echolocation sound features processed to provide distance information in the CF/FM bat, *Noctilio albiventris:* Evidence for a gated time window utilizing both CF and FM components. *Journal of Comparative Physiology A* 156:457–469.

ROVERUD, R. C., and E. R. RABITORY. 1994. Complex sound analysis in the FM-bat *Eptesicus fuscus,* correlated with structural parameters of frequency modulated signals. *Journal of Comparative Physiology A* 174:567–573.

ROVERUD, R. C., V. NITSCHE, and G. NEUWEILER. 1991. Discrimination of wingbeat motion by bats, correlated with echolocation sound pattern. *Journal of Comparative Physiology A* 168:259–263.

RYDELL, J. 1993. Variation in the sonar of an aerial-hawking bat (*Eptesicus nilssonii*). *Ethology* 93:275–284.

SAILLANT, P. A., J. A. SIMMONS, S. P. DEAR, and T. A. McMULLEN. 1993. A computational model of echo processing and acoustic imaging in frequency-modulated echolocating bats: The spectrogram correlation and transformation receiver. *Journal of the Acoustical Society of America* 94:2691–2712.

SCHATZ, C. D. 1954. The context in the perception of stops. *Language* 30:47–56.

SCHLANGEN, M. 2000. Untersuchungen zur multimodalen Orientierung bei der Gemeinen Vampirfledermaus, *Desmodus rotundus* (fam. Phyllostomidae). Ph.D. dissertation, University of Bonn, Germany.

SCHMIDT, S. 1995. Psychoacoustic studies in bats. Pp. 123–134 in *Methods in comparative psychoacoustics 6,* ed. G. M. Klump, R. J. Dooling, R. R. Fay, and W. C. Stebbins. Zuerich: Birhaeuser Verlag.

SCHMIDT, S., and J. THALLER. 1994. Temporal auditory summation in the echolocating bat, *Tardarida brasiliensis. Hearing Research* 77:125–134.

SCHMIDT, S., S. HANKE, and J. PILLAT. 2000. The role of echolocation in the hunting of terrestrial prey—new evidence for an underestimated strategy in the gleaning bat, *Megaderma lyra. Journal of Comparative Physiology A* 186:975–988.

SCHNITZLER, H.-U., and O. W. HENSON. 1979. Performance of animal airborne sonar systems: I. Microchiroptera. Pp. 109–181 in *Animal sonar systems,* ed. R.-G. Busnel and J. F. Fish. New York: Plenum Press.

SCHNITZLER, H.-U., and E. K. V. KALKO. 1998. How echolocating bats search and find food. Pp. 183–196 in *Bat biology and conservation,* ed. T. H. Kunz and P. A. Racey. Washington, D.C.: Smithsonian Institution Press.

Schnitzler, H.-U., E. Kalko, I. Kaipf, and A. D. Grinnell. 1994. Fishing and echolocation behavior of the greater bulldog bat, *Noctilio leporinus,* in the field. *Behavioral Ecology and Sociobiology* 35:327–345.

Schnitzler, H.-U., E. Kalko, L. Miller, and A. Surlykke. 1987. The echolocation and hunting behavior of the bat, *Pipistrellus kuhli. Journal of Comparative Physiology A* 161:267–274.

Schuller, G., K. Beuter, and H.-U. Schnitzler. 1974. Response to frequency shifted artificial echoes in the bat *Rhinolophus ferrumequinum. Journal of Comparative Physiology* 89:275–286.

Shaw, M. 1990. Visual matching by a language-naive dolphin (*Tursiops truncatus*). Master's thesis, University of Hawaii, Honolulu.

Shepard, R. N., and L. A. Cooper. 1982. *Mental images and their transformations.* Cambridge: MIT Press.

Shepard, R. N., and J. Metzler. 1971. Mental rotation of three-dimensional objects. *Science* 171:701–703.

Siemers, B. M., and H.-U. Schnitzler. 2000. Natterer's bat (*Myotis nattereri* Kuhl, 1818) hawks for prey close to vegetation using echolocation signals of very broad bandwidth. *Behavioral Ecology and Sociobiology* 47:400–412.

Simmons, J. A. 1973. The resolution of target range by echolocating bats. *Journal of the Acoustical Society of America* 54:157–173.

———. 1979. Perception of echo phase information in bat sonar. *Science* 204:1336–1338.

———. 1987. Acoustic images of target range in the sonar of bats. *Naval Research Review* 39:11–26.

———. 1989. A view of the world through the bat's ear: The formation of acoustic images in echolocation. *Cognition* 33:155–199.

Simmons, J. A., and L. Chen. 1989. The acoustical basis for target discrimination by FM echolocating bats. *Journal of the Acoustical Society of America* 86:1333–1350.

Simmons, J. A., and A. D. Grinnell. 1988. The performance of echolocation: Acoustic images perceived by echolocating bats. Pp. 353–385 in *Animal sonar: Processes and performance,* ed. P. E. Nachtigall and P. W. B. Moore. New York: Plenum Press.

Simmons, J. A., and S. A. Kick. 1983. Interception of flying insects by bats. Pp. 267–279 in *Neuroethology and behavioral physiology,* ed. F. Huber and H. Markl. Berlin: Springer-Verlag.

Simmons, J. A., and R. A. Stein. 1980. Acoustic imaging in bats: Echolocation signals and the evolution of echolocation. *Journal of Comparative Physiology A* 135:61–84.

Simmons, J. A., M. B. Fenton, and M. J. O'Farrell. 1978. Echolocation and pursuit of prey by bats. *Science* 203:16–21.

Simmons, J. A., M. J. Ferragamo, and C. F. Moss. 1998. Echo-delay resolution in sonar images of the big brown bat, *Eptesicus fuscus. Proceedings of the National Academy of Sciences* 95:12647–12652.

Simmons, J. A., J. M. Moffat, and W. M. Masters. 1992. Sonar gain control and echo detection thresholds in the echolocating bat, *Eptesicus fuscus. Journal of the Acoustical Society of America* 91:1150–1163.

Simmons, J. A., C. F. Moss, and M. Ferragamo. 1990. Convergence of temporal and spectral information into acoustic images of complex sonar targets perceived by the echolocating bat, *Eptesicus fuscus. Journal of Comparative Physiology A* 166:449–470.

Simmons, J. A., M. Ferragamo, C. F. Moss, S. B. Stevenson, and R. A. Altes. 1990. Discrimination of jittered sonar echoes by the echolocating bat, *Eptesicus fuscus:* The shape of target images in echolocation. *Journal of Comparative Physiology A* 167:589–616.

Simmons, J. A., S. A. Kick, B. D. Lawrence, C. Hale, C. Bard, and B. Escudié. 1983. Acuity of horizontal angle discrimination by the echolocating bat, *Eptesicus fuscus. Journal of Comparative Physiology A* 153:321–330.

Simmons, J. A., P. A. Saillant, M. J. Ferragamo, T. Haresign, S. P. Dear, J. Fritz, and T. A. McMullen. 1996. Auditory computations for biosonar target imaging in bats. Pp. 401–468 in *Auditory computation,* ed. H. L. Hawkins, T. A. McMullen, A. N. Popper, and R. R. Fay. New York: Springer-Verlag.

Simmons, N. B., and J. H. Geisler. 1998. Phylogenetic relationships of *Icaronycteris, Archaeonycteris, Hassianycteris,* and *Palaeochiropteryx* to extant bat lineages, with comments on the evolution of echolocation and foraging strategies in Microchiroptera. *Bulletin of the American Museum of Natural History* 235:4–182.

Slater, A. 1998. Visual organization and perceptual constancies in early infancy. Pp. 6–30 in *Perceptual constancy: Why things look as they do,* ed. V. Walshand and J. Kulikowski. Cambridge: Cambridge University Press.

Smolker, R., and A. Richards. 1988. Loud sounds during feeding in Indian ocean bottlenose dolphins. Pp. 703–706 in *Animal sonar: Processes and performance,* ed. P. E. Nachtigall and P. W. B. Moore. New York: Plenum Press.

SOKAL, R. R., and F. J. ROHLF. 1981. *Biometry.* New York: Freeman.

SPELKE, E. S., K. BREINLINGER, K. JACOBSON, and A. PHILLIPS. 1993. Gestalt relations and object perception: A developmental study. *Perception* 22:1483–1500.

SUGA, N. 1970. Echo ranging neurons in the inferior colliculus of bats. *Science* 170:449–452.

———. 1990. Cortical computational maps for auditory imaging. *Neural Networks* 3:3–21.

SULLIVAN, W. E. 1982. Neural representation of target distance in the auditory cortex of the echolocating bat, *Myotis lucifugus. Journal of Neurophysiology* 58:1011–1032.

SUM, Y. W., and D. MENNE. 1988. Discrimination of fluttering targets by the FM-bat *Pipistrellus stenopterus? Journal of Comparative Physiology A* 163:349–354.

SUMBY, W. H., and I. POLLACK 1954. Visual contribution to speech intelligibility in noise. *Journal of the Acoustical Society of America* 26:212–215.

SUPIN, A. Y., and V. V. POPOV. 1995. Frequency tuning and temporal resolution in dolphins. Pp. 95–110 in *Sensory systems of aquatic mammals,* ed. R. A. Kastelein, J. A. Thomas, and P. E. Nachtigall. Woerden, The Netherlands: De Spil Publishers.

SURLYKKE, A. 1988. Interaction between echolocating bats and their prey. Pp. 551–566 in *Animal sonar: Processes and performance,* ed. P. E. Nachtigall and P. W. B. Moore. New York: Plenum Press.

———. 1992a. Target ranging and the role of time-frequency structure of synthetic echoes in big brown bats, *Eptesicus fuscus. Journal of Comparative Physiology A* 170:83–92.

———. 1992b. Ranging in the FM bat, *Eptesicus fuscus:* Accuracy with artificial FM, FM-CF, and CF signals. Third International Congress of Neuroethology, Montreal, Canada.

———. 1995. Detection of short FM signals and long almost-CF signals in the FM bat, *Eptesicus fuscus. Assoc. for Research in Otolaryngology* 18:68.

SURLYKKE, A., and O. BOJESEN. 1996. Integration time for short broad band clicks in echolocating FM-bats (*Eptesicus fuscus*). *Journal of Comparative Physiology A* 178:235–241.

SUTHERS, R. A. 1965. Acoustic orientation by fish-catching bats. *Journal of Experimental Zoology* 158:319–348.

———. 1966. Optomotor responses by echolocating bats. *Science* 152:1102–1104.

———. The production of echolocation signals by bats and birds. Pp. 23–45 in *Animal sonar: Processes and performance,* ed. P. E. Nachtigall and P. W. B. Moore. New York: Plenum Press.

TARR, M. J., and H. H. BULTHOFF. 1998. Image-based object recognition in man, monkey and machine. *Cognition* 67:1–20.

TOUGAARD, J., J. H. CASSEDAY, and E. COVEY. 1998. Arctiid moths and bat echolocation: Broad-band clicks interfere with neural responses to auditory stimuli in the nuclei of the lateral lemniscus of the big brown bat. *Journal of Comparative Physiology A* 182:203–215.

TROEST, N., and B. MØHL. 1986. The detection of phantom targets in noise by serotine bats; negative evidence for the coherent receiver. *Journal of Comparative Physiology A* 159:559–567.

TURNBULL, S. D., and J. M. TERHUNE. 1993. Repetition enhances hearing detection thresholds in a harbour seal (*Phoca vitulina*). *Canadian Journal of Zoology* 71:926–932.

ULLMAN, S. 1979. *The interpretation of visual motion.* Cambridge: MIT Press.

VALENTINE, D. E., and C. F. MOSS. 1997. Spatially selective auditory responses in the superior colliculus of the echolocating bat. *Journal of Neuroscience* 17:1720–1733.

———. 1998. Sensorimotor integration in bat sonar. Pp. 220–230 in *Bat biology and conservation,* ed. T. H. Kunz and P. A. Racey. Washington, D.C.: Smithsonian Institution Press.

VEL'MIN, V. A., and N. A. DUBROVSKIY. 1976. The critical interval of active hearing in dolphins. *Soviet Physics of Acoustics* 2:351–352.

VIEMEISTER, N. F., and G. H. WAKEFIELD. 1991. Temporal integration and multiple looks. *Journal of the Acoustical Society of America* 90:858–865.

VON DER EMDE, G., and H.-U. SCHNITZLER. 1990. Classification of insects by echolocating greater horseshoe bats. *Journal of Comparative Physiology A* 167:423–430.

WADSWORTH, J., and C. F. MOSS. 2000. Vocal control of acoustic information for sonar discriminations by the echolocating bat, *Eptesicus fuscus. Journal of the Acoustical Society of America* 107:2265–2271.

WASSERMAN, E. A., J. L. GAGLIARDI, B. R. COOK, K. KIRKPATRICK-STEGER, S. L. ASTLEY, and I. BIEDERMAN. 1996. The pigeon's recognition of drawings of depth-rotated objects. *Journal of Experimental Psychology: Animal Behavior Processes* 22:205–221.

WATERS, D. A., J. RYDELL, and G. JONES. 1995. Echolocation call design and limits on prey size: A case study using the aerial hawking bat *Nyctalus leisleri. Behavioural Ecology and Sociobiology* 37:321–328.

WEBSTER, F. 1967a. Discussion. Pp. 626–671 in *Animal sonar systems: Biology and bionics,* ed. R.-G. Busnel. Jouy-en-Josas: Laboratoire de Physiologie Acoustique.

———. 1967b. Interception performance of echolocating bats in the presence of interference. Pp. 673–613 in *Animal sonar systems: Biology and bionics,* ed. R.-G. Busnel. Jouy-en-Josas: Laboratoire de Physiologie Acoustique.

WEISKRANTZ, L., and A. COWEY. 1975. Cross-modal matching in the rhesus monkey using a single pair of stimuli. *Neuropsychologia* 13:257–261.

WHITAKER, J. O. 1995. Food of the big brown bat *Eptesicus fuscus* from maternity colonies in Indiana and Illinois. *American Naturalist* 134:346–360.

WOOD, F. G., and W. E. EVANS. 1980. Adaptiveness and ecology of echolocation in toothed whales. Pp. 381–426 in *Animal sonar systems,* ed. R.-G. Busnel and J. F. Fish. New York: Plenum Press.

WOODWARD, P. M. 1953. *Probability and information theory, with applications to radar.* London: Pergamon Press.

XITCO, M. J., JR., and H. L. ROITBLAT. 1996. Object recognition through eavesdropping: Passive echolocation in bottlenose dolphins. *Animal Learning and Behavior* 24(4): 355–365.

YAGER, D. D., M. L. MAY, and M. B. FENTON. 1990. Ultrasound-triggered, flight-gated evasive maneuvers in the flying praying mantis, *Parasphendale agrionina.* I: Free flight. *Journal of Experimental Biology* 152:17–39.

ZAITSEVA, K. A., and V. I. KOROLEV. 1996. Studies of the ability of dolphin sonar to detect the acceleration of a moving target. *Sensory Systems* 10:192–195.

ZWICKER, E. 1984. Dependence of post-masking on masker duration and its relation to temporal effects in loudness. *Journal of the Acoustic Society of America* 75:219–223.

ZWICKER, E., and H. FASTL. 1972. Zur Abhaengigkeit der Nachverdeckung von der Stoerimpulsdauer. *Acustica* 26:78–82.

PART FOUR

Ecological and Evolutionary Aspects of Echolocating Mammals

Ecological and Evolutionary Aspects
of Echolocation in Bats

Annette Denzinger, Elisabeth K. V. Kalko, and *Gareth Jones*

Introduction

Bats are ecologically one of the most diverse groups of mammals. Numerous morphological, physiological, behavioral, and sensory adaptations permit bats access to a wide range of habitats and resources at night. Based on results of comparative studies, several authors recently refined the conceptual frameworks about the adaptive value of echolocation call structure and call patterns, notably in the context of foraging mode, habitat type, and diet. This has inspired a number of propositions about the evolution of echolocation and flight in bats. A related topic that has drawn further attention within the last years concerns interactions between echolocating bats and insects. In this introductory chapter, we first give an overview regarding different scenarios of how echolocation and flight evolved in bats, and discuss the possible impact of echolocation on bat evolution. Next, we describe recent findings on the relationships between signal structure and ecology. Finally, we highlight new discoveries of interactions between echolocating bats and insects.

The Evolution of Flight and Echolocation in Microchiropterans

The key characters that distinguish microchiropteran bats from other mammals are the use of echolocation as an active orientation system and the capacity for powered flight. The development of echolocation and flight allowed bats to radiate into a wide range of nocturnal niches. Unfortunately, we lack fossil records that give sufficient information about the origin of flight and echolocation. The oldest fossil Microchiroptera records extend back to the early Eocene (approximately 50–55 million years ago), when at least eight genera of bats were present. A variety of characters including postcranial osteology, cochlear size, wing morphology, and stomach content demonstrate convincingly that, at that time, flight and echolocation already were highly evolved in bats (Habersetzer, Richter, and Storch 1992). In the Eocene, a burst of diversification took place. Currently, there are 24 genera recognized from Eocene deposits, and many extant microchiropteran bat lineages can be traced back to fossils from Middle to Late Eocene (Simmons and Geisler 1998).

There is an ongoing debate in chiropteran systematics concerning bat monophyly (for review see Simmons 1994). The oldest fossil record of flying foxes (Megachiroptera) dates back to the Late Eocene (Ducrocq, Jaeger, and Sigé 1993). Pettigrew (1995) proposed a diphyletic origin of bats and placed the Megachiroptera closer to primates than to the Microchiroptera, implying that flight evolved twice. This contrasts with a wide range of studies supporting bat monophyly on the basis of morphological features of the wings, the skeleton, and inner organs (Thewissen and Babcock 1991), as well as molecular data including DNA hybridization and sequencing of mitochondrial and nuclear genes (Miyamoto 1996). Recently Hutcheon, Kirsch, and Pettigrew (1998), Teeling et al. (2000, 2002), and Springer et al. (2001) proposed a paraphyletic origin of microchiropterans. DNA hybridization and sequence data from mitochondrial and nuclear genes has revealed that microbat families in the superfamily Rhinolophoidea are more closely related to megabats than they are to other microbats. This implies that echolocation either evolved twice in microchiropteran bats or was lost in Pteropodidae.

SCENARIOS FOR THE EVOLUTION OF FLIGHT AND ECHOLOCATION

Partly due to the discussion on bat monophyly or diphyly, the question about the timing of the origin of flight and echolocation in microchiropteran bats has received increasing attention. Several scenarios have been developed that address possible successive evolutionary steps. Since we lack fossils of pre-bats, any scenario concerning the origin and evolution of bats must remain speculative and is developed on the basis of plausibility.

Overall, three theories have been discussed: the "echolocation-first" hypothesis (Fenton et al. 1995), the "flight-first" hypothesis (Norberg 1989, 1994; Simmons and Geisler 1998), and the "tandem evolution" hypothesis (Speakman 1993). All theories agree that the ancestors of echolocating bats were most likely small quadrupedal, arboreal insectivores. The proposed scenarios differ considerably, however, in the assumptions about the sequence in which echolocation and flight evolved.

The echolocation-first hypothesis

The echolocation-first hypothesis (Fenton et al. 1995) states that the ability to use echolocation to detect, track, and assess flying insects evolved before powered flight. The authors postulated that the ancestor of bats was a small nocturnal glider that lived in forests. This pre-bat used short, low-intensity, and broadband clicks for general orientation, allowing only a short operational range of echolocation. Because clicks are the signal type prevailing in the primitive echolocation systems of other mammals, the authors concluded that clicks were also ancestral in the echolocation systems of modern bats. The next evolutionary step assumes that the pre-bats made a transition from short, rather faint clicks to most likely longer, tonal signals of higher intensity, thus effectively increasing their range of echolocation. The authors suggested that the increased operational range gave the pre-bats more time between detection and direct contact with the prey, and thus was crucial to adapt echolocation for hunting airborne insects. Since faint incoming echoes are masked during signal emission, the authors stated, the transition from clicks to stronger tonal echolocation signals must have been accompanied by adaptations that minimize "self-deafening" or forward masking.

Fenton et al. (1995) presumed that the pre-bats were still gliders at this evolutionary stage. They detected and tracked airborne prey from a perch in relatively open areas in the subcanopy and determined the interception course. Then pre-bats took off, glided toward the prey, intercepted it, landed at a somewhat lower spot, and climbed back to the perch. Thus, echolocation already was used for the detection and capture of airborne prey before controlled flapping flight evolved. The subsequent development of flapping flight allowed the first bats to hunt on the wing with strong, low duty-cycle, broadband, tonal signals in the subcanopy, where the first echo returning to the bat was from the prey.

In recent bats, Fenton et al. (1995) distinguished between "high-duty-cycle" and "low-duty-cycle" echolocation. High-duty-cycle bats produce constant-frequency echolocation calls with a terminal, frequency-modulated component (CF-FM). When in flight, they compensate for the Doppler shifts caused by their own flight movement and thus separate pulse and echo in frequency. Low-duty-cycle bats produce short echolocation signals with relatively long pulse intervals. Due to the short signals, they separate pulse and echo in time. In the proposed scenario, gleaning and high-duty-cycle echolocation are specialized conditions derived from the low-duty-cycle echolocators. The authors suggested that high-duty-cycle echolocation evolved most likely in relatively open areas. Low-duty-cycle gleaners detect prey by using cues other than echolocation. When closing in on prey, these bats continue to emit short broadband tonal signals of low intensity to assess the background. According to Fenton et al. (1995), echolocation signals of gleaners should be short, broadband clicks if gleaning were an ancestral foraging strategy.

Speakman (1999) pointed out that the echolocation-first hypothesis involves an intermediate hypothetical period during evolution where the pre-bats hunted for insects that flew toward their perch. Prey was intercepted by reaching outward with the forelimbs from perches to snatch passing insects, a behavior Speakman (1999) termed "reach hunting." With this hunting strategy, selection would have favored the extension of the forearm and the digits, as well as the webbing between them to increase the surface area. This specialization could have improved capture success and at the same time would also have represented a pre-adaptation for flight. Considering the energetic feasibility of the reach hunting strategy, Speakman (1999) concluded that, even under the most favorable conditions, daily energy demands could not be met with this foraging strategy. His modeling suggests that the evolution of bats did not include a period of reach hunting behavior and provides evidence against the echolocation-first hypothesis.

The echolocation-first hypothesis does not explicitly address the question of bat monophyly or diphyly. According to Arita and Fenton (1997), it could be adjusted to either of these theories. In their opinion, flight evolved twice in the case mega- and microchiropterans are diphyletic, and echolocation appeared independently in both suborders. In the case of mega- and microchiropteran monophyly, echolocation was lost early in the megachiropteran lineage and only reappeared in some species of the genus *Rousettus*. Simmons and Geisler (1998), however, questioned whether the echolocation-first hypothesis is compatible with bat monophyly. To them, it is highly unlikely that megachiropterans lost a very sophisticated and advantageous echolocation system and redeveloped a rather rudimentary form as seen at least in the "clicking" behavior of some flying foxes (*Rousettus*). The broad distribution of *Rousettus*, however, underlines the advantage of echolocation and makes it highly unlikely that it was given up during evolution (see also Speakman 1993, 1999).

The flight-first hypothesis

The flight-first hypothesis (Norberg 1994; Simmons and Geisler 1998) states that flapping flight evolved before echolocation was used to detect, track, and assess flying insects. In this scenario, proposed by Norberg (1994), pre-bats also were gliding, nocturnal insectivores that used a primitive form of echolocation, possibly clicks, for orientation in space and probably also for communication. Norberg (1994) further assumed that gliding from tree to tree reduced the energy and time required for foraging and, at the same time, maximized the net energy gain. Maximizing foraging efficiency could have initiated strong selection for better gliding performance. Presumably, a decrease in gliding speed, due to the development of an increased glide surface, permitted safer landings and more control over the gliding path. Consequently, due to the probably high selection pressure on better gliding performance, the pre-bats could have improved stability and movement control before flapping flight evolved, and were able to change the direction of a glide to reach a particular destination.

Finally, the ability to flex the wings and coordinate wing movements led to the evolution of horizontal flapping flight. In contrast to Fenton et al. (1995), Norberg (1994) assumed that gliding was used exclusively for transportation, but not for catching insects. She argued that aerial hawking for insects requires high maneuverability, which could only be guaranteed after bats attained true flight. A sophisticated echolocation system that allows detection and tracking of airborne prey was most likely established only after bats were able to maneuver well enough for aerial hawking with powered flight (Norberg 1989, 1994).

Simmons and Geisler (1998) proposed a somewhat different scenario. On the basis of a cladistic analysis of the phylogeny of bats, using the total evidence approach, the authors concluded that flight evolved before echolocation. Fundamental to this scenario is the assumption of bat monophyly, stating that powered flight evolved only once among Chiroptera. According to Simmons and Geisler (1998), early bats lived in an arboreal habitat and used short flights from a perch to glean food, such as insects and perhaps fruits, from the vegetation and possibly the ground. Bats probably used vision for orientation. Potential food items were localized by passive acoustic cues generated by the prey and/or by vision. At that point, bats already had developed modern flapping flight.

The authors assumed that in a subsequent step, a primitive laryngeal echolocation system with short signals and relatively long pulse intervals evolved that improved obstacle avoidance and orientation in space. Most likely, the echolocation calls originated from communication signals. Additionally, vision was used for orientation. Stationary prey were detected from a perch, localized, and classified by vision and by listening for prey-generated sounds. In the next step, early bats started to detect and track flying prey with echolocation, whereas stationary prey continued to be detected by passive means. The transition from the use of echolocation for orientation alone to prey detection occurred most likely in bats that foraged along the edges of rivers or lakes and in gaps where insect echoes were adjacent to relatively clutter-free open spaces. Due to the low energy costs of echolocation when flight is mechanically coupled to ventilation (Speakman and Racey 1991), the substantial benefits of aerial insectivory favored the development of a sophisticated echolocation system capable of detecting, tracking, and assessing airborne prey. The improved performance of the echolocation system increased the foraging options of the bats, allowing them to exploit nocturnal, flying insects as a food source. In a further step, echolocation capabilities allowed the bats to combine perch hunting and aerial hawking in continuous flight. In the scenario proposed by Simmons and Geisler (1998), bats continued to use a combination of vision and echolocation for spatial orientation. Finally, it is assumed that the bats left the perches and hunted continuously on the wing. Thereby, echolocation was used for prey detection and tracking; and vision, as well as echolocation, for spatial orientation and obstacle avoidance. This foraging strategy was most likely primitive to the microchiropteran crown group. Perch hunting and gleaning using passive cues, which is also known in extant bats, are thus secondarily derived foraging strategies that evolved several times, rather than a retention of primitive habits.

A critical step in this scenario is the transition from an entirely vision-based orientation to echolocation. The authors speculated that this transition was a multistep process. Most likely, the early bats relied, at first, on both vision and echolocation. As the importance of echolocation for orientation and food detection increased, a complex auditory system developed in parallel that competed with the visual system for the available resources. The authors concluded that this process finally resulted in a reduction of the visual system.

The tandem evolution hypothesis

The hypothesis that flight and echolocation evolved simultaneously, as proposed by Speakman (1993), also assumed bat monophyly, but with a nonflying common ancestor for both Mega- and Microchiroptera. Further, a tight link between the cost of echolocation and flight is regarded as a key factor in the assumed parallel evolution of flight and echolocation. Speakman, Anderson, and Racey (1989) calculated the energetic costs for the production of echolocation pulses from respirometry measurements of resting bats, and Speakman and Racey (1991) compared the costs for echolocation with the energy costs for flight. They found a linear relationship between the logarithm of the costs of flight and the logarithm of body mass for echolocating bats, as well as for nonecholocating pteropodids and birds. Speakman and Racey (1991) concluded that bats that echolocate in flight do not spend more energy than nonecholocating flying foxes and birds in flight. Thus, the high energy costs of echolocation call production, as measured in resting bats, are somehow covered by the total costs of flight. The high air flow necessary to produce high-intensity echolocation sounds is generated by the contraction of the pectoralis and scapularis muscles during expiration. Furthermore, contraction of the same muscles provides the power for flight. By coupling sound emission with wing beat, the costs for echolocation are assimilated by the costs for flight (see also Lancaster, Henson, and Keating 1995).

In his scenario, Speakman (1993) suggested that Microchiroptera evolved from small nocturnal insectivores that were predominantly specialized for audition. Due to the energetic constraints, echolocation was only affordable in combination with flight. Hence, both echolocation and flight were closely linked during evolution. For these insectivores, the auditory system was already the most important sensory system, thus favoring the development of a sophisticated echolocation system, because it did not require a reorganization of the brain. The model, however, lacks a detailed description of successive evolutionary steps of how flight and echolocation coevolved. In chapter 49 of the present volume, Speakman et al. present new data on the energy costs of echolocation in *Eptesicus serotinus* and *Rhinolophus ferrumequinum* and discuss their significance in the context of the evolution of echolocation.

In contrast to Speakman (1993), Arita and Fenton (1997), and Simmons and Geisler (1998) noted that the energetic coupling between echolocation and flight supports the flight-first hypothesis. To their opinion, flight definitely can develop without echolocation as it is evident in birds and Megachiroptera. They argued further that the observed tight coupling of signal emission and wing beat implies that echolocation was only affordable after flapping flight was established.

The discussion on the evolution of echolocation and flight in Microchiropterans is enriched by a further scenario proposed by Schnitzler, Kalko, and Denzinger (chapter 44, this volume). Based on results of comparative studies of echolocation and foraging behavior in extant bats, the authors outlined a plausible sequence of evolutionary steps that led to the diversity of contemporary echolocation systems and foraging behaviors. In this scenario, gliding and finally fluttering ancestors used echolocation for landing control, obstacle avoidance, and, with increasing maneuverability, also for spatial orientation. They gleaned prey from substrates, whose presence was advertised by prey-generated acoustic signals. The transition to using echolocation to detect, localize, and assess airborne insects was the last step in evolution and occurred after flapping flight was established.

THE POSSIBLE IMPACT OF ECHOLOCATION ON BAT EVOLUTION

Flight and the reliance on an echolocation system for prey acquisition could have brought with it several evolutionary constraints, especially limitations on body size. Compared with other flying vertebrates, bats, especially aerial-feeding insectivorous species, are small, with most species weighing less than 30 g. Barclay and Brigham (1991) discussed several factors that could limit body size in microchiropteran aerial insectivores. It has been proposed that bats feed on small insects that are too small to be energetically adequate for a larger predator. Further, the smaller size of aerial insectivorous bats could be imposed by the mechanics of flight. To capture prey, they rely on agility and maneuverability, which both

decrease with increasing body size (Norberg and Rayner 1987). Comparisons of body mass and wing shape of aerial insectivorous bats with that of insectivorous birds, however, revealed that neither aerial feeding on insects nor aerodynamic constraints necessitate small size in bats (Barclay and Brigham 1991). Even within the microchiropterans, aerial-feeding bats are smaller than gleaners and noninsectivorous bats. Barclay and Brigham (1991) concluded that it is the short range of echolocation and the concomitant need to be maneuverable to pursue and catch insects at short range that limits the body size of aerial insectivorous bats. Due to the atmospheric and geometric attenuation of sound traveling in air, the sound pressure level (SPL) of echoes decreases rapidly with increasing target distance. Thereby, the maximum detection distance depends on signal frequency, relative humidity, temperature, and prey size (Kober and Schnitzler 1990). Overall, compared with other sensory systems, such as vision, echolocation only works over short distances. The short operational range of echolocation necessitates high agility and maneuverability and thus could indirectly limit body size. Since maneuverability and agility decrease with body size, bats can compensate for the disadvantage by emitting low-frequency signals, which attenuate less and thus increase the detection range. To a certain extent, there is a trade-off between body size and signal frequency reflected in the tendency for bigger bats to use lower signal frequencies (Barclay and Brigham 1991). However, low-frequency signals have the disadvantage that they are reflected only weakly by small targets. Barclay and Brigham (1991) predicted that larger species cannot detect small prey early enough to maneuver for capture and thus have to rely on large, relatively rare insects. They concluded that aerial insectivores are small because their prey detection system limits the availability of insects and makes the existence of large bats energetically difficult or even impossible. However, recent data on the echolocation frequencies of large bats that eat aerial insects, such as *Cheiromeles torquatus* (Heller 1995), showed that body size might only be a weak constraint on call frequency.

Another factor frequently discussed in relation to limitations of body size is the coupling of respiration, wing beat, and echolocation due to the high energy costs of echolocation. By comparing wing-beat cycle, body mass, and pulse repetition rate (the latter mainly obtained from recordings in the field), Jones (1994) found that in aerial insectivores, both wing-beat frequency and pulse repetition rate decreased with increasing body mass. He concluded that the larger the bat is, the lower is its wing-beat rate and the coupled pulse rate. This reaches a limit if wing-beat rates of large-bodied bats are so low that the resulting low pulse repetition rate does not transmit an information flow sufficient to encounter small prey rapidly. The low encounter rate becomes critical for large bats because it does not meet their high energy demands and thus could constrain upper body size. This argument implies that big bats have a very low wing-beat rate and that aerial insectivores are unable to produce more than one pulse per wing beat in search flight. The role of pulse interval in echolocating bats is also discussed by Schnitzler, Kalko, and Denzinger and Fenton (chapters 44 and 47, both in this volume).

Perspectives

Until fossils of pre-bats are found, any scenario about the evolution of flight and echolocation must remain speculative and can only be developed on the basis of plausibility. Nevertheless, a better understanding of constraints acting on the echolocation systems of extant bats and new data on phylogenetic relationships of Chiroptera are providing exciting new insights on the evolution of flight and echolocation.

Relationships between Signal Structure and Ecology

Structure and pattern of echolocation calls in microchiropteran bats vary substantially within and across species and taxonomic groups. Over the years, a large number of echolocation calls from a wide range of bats have been described (for summaries see Fenton 1995; Neuweiler 1990; Schnitzler and Kalko 1998). Generally, three functional components of echolocation calls can be discriminated: (1) narrowband, shallow frequency-modulated (FM); (2) wideband, steep frequency-modulated (FM); and (3) constant-frequency (CF) elements in hipposiderids, rhinolophids, and *Pteronotus parnellii* (Mormoopidae). Signals consist of either one or combinations of these three elements. Narrowband, shallow frequency modulated components are also called quasi-constant frequency (QCF). Call energy in echolocation signals is either concentrated mostly in one (e.g., Hipposideridae, Molossidae, Mormoopidae, Rhinolophidae, Vespertilionidae) or in several (e.g., Emballonuridae, Megadermatidae, Nycteridae, Phyllostomidae, Rhinopomatidae) harmonics.

Signal patterns also differ within and among species. The parameters most commonly used to characterize differences in echolocation behavior include sound duration, pulse interval, duty cycle (percentage of time filled with sound), starting and terminal frequency, bandwidth (measured for the strongest harmonic; or, in multiharmonic signals, also for the whole signal), best or peak frequency (measured as the frequency containing most signal energy), and number of harmonics.

Signal variability depends largely on the basic perceptual tasks performed by all bats when foraging, commuting, or inspecting new environments. These tasks include landing control, obstacle avoidance, spatial orientation, detection, localization, identification of food, and probably also recognition of conspecifics (e.g., Fen-

ton 1995; Neuweiler 1990; Obrist 1995; Schnitzler and Kalko 1998).

Influence of Habitat on the Structure of Search Calls

Interspecific variability

Ecological conditions, particularly habitat structure, are major factors that have shaped the structure of search signals of bats during evolution. The main problem a bat has to solve is the separation of target echoes (i.e., food) from interfering signals, including internal noise, echoes from the background (clutter echoes), other bats, and its own sound emission (e.g., Fenton 1990, 1995; Obrist 1995; Schnitzler and Kalko 1998). Interfering signals that precede a target echo create "forward-masking" effects, while interfering signals, such as clutter that follow target echoes, produce "backward-masking" effects. Most bats using overlap-sensitive, narrow- and wideband FM calls avoid forward as well as backward masking by changing the duration of their calls (e.g., Kalko and Schnitzler 1998; Rydell 1993). For instance, bats flying closer to or within vegetation produce shorter signals than bats searching for insects away from obstacles. Prey detection occurs mainly in the "overlap-free window." One exception is *Myotis nattereri*, which hunts very close to vegetation and produces extremely broadband signals (Siemers and Schnitzler 2000), obviously tolerating at least some overlap between prey and clutter echoes.

Bats that emit CF-FM calls tolerate pulse-echo overlap of the CF element. They compensate for the Doppler shifts caused by their own flight movement by adjusting their emission frequency. Hence, returning echoes of the CF components are separated from the outgoing signal in the frequency domain. Furthermore, the characteristic echo patterns (amplitude and frequency modulations) created by the wing beats of fluttering insects (acoustic glints) and a highly specialized hearing system ("acoustic fovea") adapts these bats to detect minute modulations on the target echoes amid clutter echoes (e.g., Moss and Schnitzler 1995; Neuweiler 1990; Schnitzler 1987).

Comparative field studies of bats have demonstrated that a series of collective intra- and interspecific trends of signal structure and pattern are closely associated with a set of ecological conditions that include habitat type, foraging mode, and diet (e.g., Aldridge and Rautenbach 1987; Crome and Richards 1988; Fenton 1990, 1995; Neuweiler 1989, 1990; Schnitzler and Kalko 1998). Based on the amount of clutter, several distinct habitat types are proposed for bats (Fenton 1990, 1995; Neuweiler 1989; Schnitzler and Kalko 1998) (fig. 42.1). Uncluttered or open spaces refer to areas with prey far away from the background, encompassing open areas above the ground or the canopy. Here, clutter echoes are undetectable for a bat or they are far apart from the emitted signal and from the echoes of potential prey. Background-cluttered

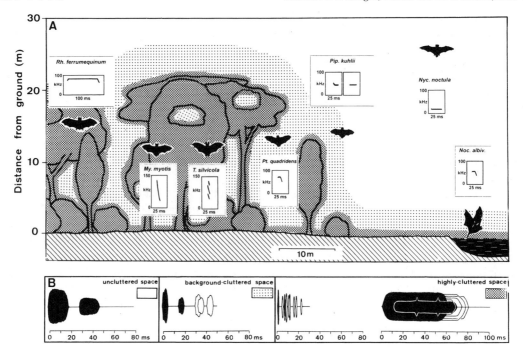

Fig. 42.1. *A:* Schematic display of three habitat types with flight silhouettes and search signal sonograms of representative species of open space (uncluttered), edge and gap (background cluttered), narrow-space FM and narrow-space CF (highly cluttered) bats (from Schnitzler and Kalko 1998). *B:* Clutter situations in the three habitats: the pulse emitted by the bat and the returning echo are depicted in black, clutter echoes are given as open symbols.

space refers to areas with prey at forest edges, in large gaps or in openings in the forest between canopy, sub-canopy, or understory level. Here, prey echoes are followed closely by but do not overlap with clutter echoes. Bats in highly cluttered or narrow (closed) spaces forage low over ground, in close proximity to or within vegetation where target echoes are in part or fully buried in background clutter.

The tight association of bat behavior and ecological constraints allows the characterization of subgroups of bats where members of each subgroup face similar sets of constraints and echolocation tasks imposed on their sensory and motor systems (Schnitzler and Kalko 1998, 2001). Bats foraging in open space emit intense, narrow-band, shallow FM search signals of rather long duration, low to medium frequency, and large pulse intervals, thereby facilitating long-range detection of prey (open-space foragers). Emission of calls is tightly coupled with wing beat. Typically, open-space foragers often make one or more wing beats without sound emission. This leads to rather low duty cycles (<10%) despite the rather long search calls.

Bats hunting for flying insects at edges and gaps typically emit rather intense signals of medium duration and medium frequency (edge-and-gap foragers), often consisting of a short wideband, steep FM element and a longer narrowband, shallow FM component (also called

QCF). The narrowband component facilitates medium-range detection of prey. The short, wideband component improves classification and localization of targets and allows a more refined description of extended background targets (e.g., vegetation, landmarks). Generally, edge and gap foragers produce one search signal per wing beat. Duty cycle is low (<10%).

Bats that forage in narrow space (narrow-space foragers) face the problem that the prey echoes overlap with clutter echoes, thus creating masking effects. Narrow-space foragers have evolved two strategies to cope with the clutter problem. Bats of the families Hipposideridae, Rhinolophidae, and *Pteronotus parnellii* (Mormoopidae) mainly catch insects in the aerial mode, either in sally flights from a perch or on the wing. They typically produce long CF signals ending with a downward-modulated, wideband FM-component (narrow-space flutter-detecting CF foragers). CF components are either long in rhinolophids and *Pteronotus parnellii* or of medium duration in hipposiderids. The CF component in combination with a specialized hearing system is used to recognize acoustic glints in the echoes from fluttering prey amid unmodulated clutter echoes (e.g., Schnitzler 1987). The overlap-sensitive FM component facilitates target localization. Its bandwidth is increased and its duration is shortened when the bats approach a target (e.g., Neuweiler et al. 1987; Schnitzler, Hackbarth, and

Heilman 1985; Tian and Schnitzler 1997). The rather long signals emitted in the rhythm of the wing beat lead to a high duty cycle.

Bats foraging in a narrow space in the gleaning mode emit overlap-sensitive, short and wideband, uni- or multiharmonic FM calls at low intensity (narrow-space gleaning FM foragers). Since echoes of food are often buried in echoes from the substrate, these bats rely mainly on sensory cues originating from the food source (e.g., advertisement calls of insects and frogs, rustling and flight noises of insects, scent of fruits and flowers) for detection, classification, and rough localization. Echolocation is used mainly for orientation in space. There are, however, a few examples where echoes originating from potential foods in highly cluttered space are so characteristic that bats can detect them with echolocation alone. Examples are dangling fruit and exposed flowers with acoustic "reflectors" (von Helversen and von Helversen 1999; Kalko and Condon 1998). Narrow-space gleaning FM foragers emit one and sometimes two or more orientation or search calls per wing beat. Because of the short signal duration, duty cycle is low.

Alternatively, Fenton (1990, 1995, 1999) suggested that all Microchiroptera be divided into high-duty-cycle and low-duty-cycle bats. This proposition is less fine-grained than the scenario outlined above (Schnitzler and Kalko 1998, 2001), as it lumps together all bats that forage in open space, at edges, and in gaps and bats that glean food from surfaces in narrow space (FM strategy).

An important step forward in the past years was the classification of bats into functional groups or guilds (see Fenton 1990; Neuweiler 1990; Schnitzler and Kalko 1998, 2001). Following the original guild concept by Root (1967), each guild or subgroup is characterized by similarities in habitat type, diet, and foraging mode. Habitat is defined with respect to the proximity of a bat to the clutter-producing background and ranges from open spaces with no interfering background clutter echoes to narrow spaces where echoes from food are buried in clutter echoes (fig. 42.1). With regard to feeding modes, bats that catch flying insects in the air (aerial mode) are distinguished from bats that take mostly stationary food from surfaces (gleaning mode). Bats that hunt from perches are seen as "aerial insectivores" when they catch their prey in the air, as is the case with many rhinolophids and hipposiderids, and as "gleaners" when they take food from surfaces, as is known, for example, for New World leaf-nosed bats (Phyllostomidae). Food refers to the main diet of a species.

Intraspecific variability

Field studies revealed a high degree of plasticity in signal structure of bats. Many aerial insectivorous bats are highly flexible and frequently switch among foraging habitats (e.g., Fenton 1990). As with between-species variability, signal structure within a species changes from

longer and shallow-modulated calls emitted in open habitats to shorter calls consisting of a narrowband, shallow-modulated element associated with a wideband, steep FM component in more cluttered habitats (e.g., Fenton 1995; Kalko and Schnitzler 1993; Obrist 1995; Rydell 1990, 1993; Schnitzler and Kalko 1998). Changes in signal structure depend largely on the horizontal distance of a bat to obstacles and probably to a lesser degree on its flight height (Jensen and Miller 1999; Kalko and Schnitzler 1993). In individual search sequences of vespertilionids, the terminal and/or peak frequencies of calls are shifted upward with increasing bandwidth and simultaneous reduction in sound duration (Kalko 1995a; Rydell 1993).

Studies on flight and echolocation behavior of bats also point to limitations concerning the flexibility of habitat use. Individual bats readily access habitats that are less cluttered than the space to which they are mostly adapted; however, successful exploitation of spaces that are more cluttered is difficult to impossible (Fenton 1990). Although clutter poses a challenge for the sensory abilities of bats, laboratory and field studies have shown that all bats produce the short FM calls necessary to cope with cluttered situations (Kalko and Schnitzler 1998; Schnitzler and Kalko 1998). However, due to constraints imposed by wing shape and flight performance, bats are limited to certain habitat types and foraging modes. For instance, bats that forage in open space are characterized by long, pointed wings for fast flight but reduced maneuverability, which is necessary to cope successfully with a more cluttered environment. Conversely, bats that forage in more cluttered habitats possess short, broad wings that give more maneuverability but reduce their ability for fast and energetically inexpensive flight over long distances needed to successfully forage in open space (e.g., Norberg and Rayner 1987).

INFLUENCE OF FORAGING MODE
ON SIGNAL STRUCTURE AND PATTERN

During evolution, echolocation behavior of bats has been adapted to the specific perceptual tasks of the bats living under different ecological conditions (see also Schnitzler, Kalko, and Denzinger, chapter 44, this volume). Aerial insectivores rely almost exclusively on echolocation to track mobile prey in the air. As shown first by Griffin, Webster, and Michael (1960), aerial insectivores switch upon detection of a target from the search to the approach phase by gradually reducing pulse interval and duration, and by increasing the bandwidth of their calls (e.g., Britton et al. 1997; Jones and Rayner 1988, 1991; Kalko and Schnitzler 1989; Kalko 1995a; Surlykke et al. 1993) (fig. 42.2). The functional significance of short, wideband FM components lies in the improved localization of targets. Shortly before reaching the target, all aerial insectivores emit the terminal phase or buzz with many short signals at a high repetition rate

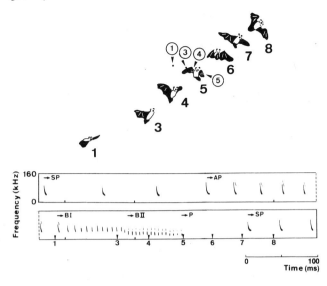

Fig. 42.2. Approach flight and capture of an insect with corresponding echolocation sequence by a free-flying *Pipistrellus pipistrellus* (Vespertilionidae) in the field (redrawn from a multiflash sequence) (from Kalko and Schnitzler 1998). The number of flashes is given in consecutive numbers on the capture sequence and on the echolocation sequence. Small arrows with circled numbers indicate the position of the insect. SP = search phase; AP = approach phase; BI + BII: terminal phase; P = pause in echolocation.

(e.g., Griffin, Webster, and Michael 1960) (fig. 42.2). Bats with overlap-sensitive signals omit the shallow FM components of their calls during approach and terminal phase (e.g., Kalko and Schnitzler 1998). Bats with CF signals maintain a small CF component through the end of the terminal phase (e.g., Schnitzler et al. 1987). All bats reduce or avoid echo overlap of the FM components of their calls until the last signals of the terminal phase (e.g., Britton et al. 1997; Kalko et al. 1998; Wilson and Moss, chapter 3, this volume).

Duration of the terminal phase is partly species-specific (see Wilson and Moss, chapter 3, this volume) and partly influenced by prey behavior. Evasive maneuvers of insects, for instance, can elicit long pursuits accompanied by long terminal phases (Acharya and Fenton 1992; Kalko 1995a). After the terminal phase, a silent period or pause of varying duration occurs that is associated with prey handling (fig. 42.2). The assumption (proposed in earlier works) that the duration of the silent period can be taken as an indicator of capture success could not be verified in the field (Schnitzler et al. 1987; Kalko 1995a; Acharya and Fenton 1992), although pulse repetition rate is lower after successful prey capture than following unsuccessful capture attempts (Britton and Jones 1999).

Gleaning bats mostly take stationary food and rely on cues such as scent and prey-produced acoustical signals (passive listening) for detection, classification, and rough localization of food. Echolocation is mostly used to orient in space and to guide the approach to the site with food. The reaction toward the background is indicated by a reduction in sound duration and pulse interval and an increase in duty cycle and bandwidth. Gleaning bats do not produce a terminal phase when approaching a site with food (e.g., Faure, Fullard, and Dawson 1993; Neuweiler 1989; Schnitzler and Kalko 1998; Thies, Kalko, and Schnitzler 1998). Some gleaning bats may stop echolocating before reaching the food site and rely on spatial memory and other sensory cues. However, since the calls of gleaning bats are particularly weak, some of these observations could reflect recording difficulties.

Field and laboratory studies revealed a high degree of plasticity in some of the insectivorous bats that switch between aerial captures and gleaning. Changes in foraging mode are accompanied by differences in echolocation behavior, particularly the absence of a terminal phase while gleaning and the presence of a terminal phase during aerial captures (e.g., Faure and Barclay 1994; Schumm, Krull, and Neuweiler 1991). Search calls range from rather long, very shallow-modulated signals, used to detect and capture prey on the wing in open spaces, to short and broadband FM signals while foraging for insects close to the vegetation.

INFLUENCE OF BODY MASS ON SIGNAL STRUCTURE

Interspecific variability

Among bats living under similar ecological conditions, larger species tend to emit longer calls with lower peak frequencies at longer pulse intervals than smaller species with shorter calls at higher peak frequencies and with shorter pulse intervals (e.g., Barclay and Brigham 1991; Bogdanowicz, Fenton, and Daleszczyk 1999; Fenton et al. 1999; Jones 1999; Waters, Rydell, and Jones 1995). Although call frequency scales with body mass, hipposiderids tend to emit higher frequencies for their body mass than other families (Francis and Habersetzer 1998; Heller and von Helversen 1989).

Intraspecific variability

In some bats, such as *Myotis adversus* (Vespertilionidae), larger individuals, as measured by forearm length, produce lower frequencies than smaller individuals of the same species (Jones and Rayner 1991). In other species, no clear relationships have been found between forearm length or body size and calling frequency, including *Eptesicus nilssoni* (Rydell 1993), two phonic types (cryptic species) of *Pipistrellus pipistrellus* (Jones and Van Parijs 1993), and several species of *Rhinolophus* (Neuweiler et al. 1987; Jones, Gordon, and Nightingale 1992; Jones and Ransome 1993).

INFLUENCE OF GENDER AND AGE ON SIGNAL STRUCTURE

For a number of bats, gender and age influence signal frequency. With one exception (*Hipposideros speoris*;

Jones et al. 1994), all females of the species studied so far emitted higher frequencies than males (Neuweiler et al. 1987; Jones, Gordon, and Nightingale 1992; Jones et al. 1993), with or without sexual dimorphism. Further, first-year bats emit calls of lower frequency than older bats (Jones and Kokurewicz 1994; Jones, Gordon, and Nightingale 1992; Jones et al. 1993). Call frequency can change over the lifetime of individuals, with lower frequencies in older bats (Jones and Kokurewicz 1994; Jones 1995a). Although young CF bats that were raised deaf produced adultlike calls, fine-tuning of frequencies within populations could include a learning component (Jones and Ransome 1993; Rübsamen and Schäfer 1990). Overall, effects created by differences in gender and age is small and thus may not affect the overall performance of the animals. The information contained in the frequency differences, however, could serve for intraspecific communication.

INFLUENCE OF GEOGRAPHIC DISTRIBUTION ON SIGNAL STRUCTURE

Several studies suggest regional and geographic variation in echolocation calls (Francis and Habersetzer 1998; Jones, Gordon, and Nightingale 1992; Jones et al. 1993; Heller and von Helversen 1989; Parsons, Thorpe, and Dawson 1997). There is evidence for distinct differences in calling frequencies of syntopic hipposiderids and rhinolophids (Jones, Gordon, and Nightingale 1992; Jones et al. 1993). Although the distribution of call frequencies is variable geographically in the five European rhinolophids, it is considerably narrower within smaller geographic areas (Heller and von Helversen 1989). Francis and Habersetzer (1998) found interpopulation variations in the best frequency of the CF component in hipposiderids in addition to size differences in morphological characters, suggesting at least subspecific status within some species.

In some studies, particularly of non-CF bats, the significance of local and regional differences in signal structure, especially of the best frequency, is difficult to judge because recording situations might not be fully comparable. Thus, it cannot be discounted that some of the observed differences are caused by ecological factors such as distance to clutter-producing objects or body size, rather than by geographic differences in call structure. For instance, although Rydell (1993) found regional differences in the search calls of *Eptesicus nilssonii*, the regional differences disappeared when flight altitude was taken into account.

INFLUENCE OF CONSPECIFICS ON SIGNAL STRUCTURE

Several field studies address the possible influence of conspecifics on signal structure. Under the assumption that bat echolocation calls originate from communication calls, it is postulated that they still carry individual-specific information. Further, bats that fly or forage within hearing distance of one another need to distinguish their own calls from conspecifics to avoid jamming. Obrist (1995) investigated signal structure of search calls from several species of bats (Vespertilionidae) flying in similar habitats. After statistical analysis, differences in calling behavior among sites were attributed to the presence or absence of conspecifics. His results suggest that some bats have individual-typical sonogram shapes. Jamming avoidance by shifting frequencies when flying in groups still lacks convincing data (Jones et al. 1994). Rydell (1993) found that marked, individual *E. nilssoni* covered a broad range of frequencies and did not use exclusive frequency bands. Moreover, the bats emitted slightly higher calls when conspecifics were around. This, however, also can be interpreted as a reaction of the bats to their conspecifics as potential obstacles. In vespertilionids, an increase in signal bandwidth leads to higher best frequencies of the call (Kalko 1995a; Rydell 1993).

PERFORMANCE OF ECHOLOCATING BATS IN THE FIELD

The importance of clutter for prey detection was tested experimentally in several field studies. These studies demonstrated convincingly that physical clutter negatively affects bat foraging behavior. Mackey and Barclay (1989) and Rydell, Miller, and Jensen (1999) added structural clutter to the surface of ponds, which led to a decline in foraging activity of little brown bats (*Myotis lucifugus*) and Daubenton's bats (*Myotis daubentonii*). Likewise, Boonman et al. (1998) showed that *M. daubentonii* avoids foraging over water surfaces that are covered with duckweed. Schnitzler et al. (1994) noted that the fisherman bat, *Noctilio leporinus*, ignored foraging in areas with prominent wind-induced ripples on the water surface. Probably, clutter-producing objects on the water surface make it more difficult for bats to detect and discriminate prey. In another study, Brigham et al. (1997) artificially introduced three-dimensional clutter at clear-cut forest edges, which negatively affected the foraging activity of small *Myotis* species. However, larger bats (*E. fuscus*, *Lasiurus cinereus*, and *Lasionycteris noctivagans*) remained unaffected by the clutter treatment. The larger bats likely flew outside the artificial clutter zone. Another factor that may have played a role in all these experiments are covariable changes in insect abundance, which are difficult to control for in the field.

Although laboratory experiments demonstrated that FM bats make fine discriminations among target shapes and textures (e.g., Moss and Schnitzler 1995; Schmidt 1988b; Simmons, Moss, and Ferragamo 1990), many anecdotal observations suggest that bats may not make full use of these abilities to classify food (see also Fenton, chapter 47, this volume). Barclay and Brigham (1994) showed low discrimination performance in two species of FM bats (*Myotis* sp.) under natural conditions. They let free-living bats choose in a cafeteria-style presentation over water among a range of edible and inedible targets of different sizes and shapes. The bats attacked almost any target of appropriate size. Moving targets were preferred over stationary ones. The authors suggest that rapid flight and short detection distances of the bats preclude fine-grained prey discrimination. However, Houston, Boonman, and Jones (chapter 45, this volume) demonstrate that another species of *Myotis* (*M. daubentonii*), when foraging over water in a two-alternative forced-choice test in the laboratory and in the field, performs a nonrandom search and chooses larger over smaller targets (here, different-sized mealworms). Probably, in addition to echolocation information, learning effects contribute to the increased performance of the bats.

ROLE OF ECHOLOCATION IN PREY SELECTION AND RESOURCE PARTITIONING

The role of echolocation behavior and its possible influence on prey selection received considerable attention in the past decade. Sensory constraints imposed by call structure, size, prey type, and its movement likely play important roles in resource partitioning. In aerial insectivorous bats with overlap-intolerant search signals, undisturbed detection without masking effects is only possible in the overlap-free window where the returning echo from the target (i.e., food) does not overlap with the outgoing signal (signal-overlap zone) or with clutter echoes (clutter-overlap zone) (Schnitzler and Kalko 1998). Sound duration sets a minimum detection distance beyond which the target echo (i.e., prey) does not overlap with the emitted signal, and it determines the minimum distance of prey toward clutter-producing background necessary for clutter-free target detection.

Maximum detection distance is influenced by the intensity and the frequency of the call, transmission loss (i.e., atmospheric attenuation), target strength (i.e., size of prey), detection threshold of the bat, and, potentially, also pulse interval (Fenton et al. 1999; Kober and Schnitzler 1990; this volume: Fenton, chapter 47; Houston, Boonman, and Jones, chapter 45; Schnitzler, Kalko, and Denzinger, chapter 44). Generally, longer signals with lower frequencies allow larger search volumes because of their lower directionality and lower atmospheric attenuation. However, they are less suited for detection of small insects due to their long wavelength. Conversely, shorter calls with higher frequencies lead to smaller search volumes. Due to their shorter wavelength, they are best suited for the detection of small insects.

Based on the detection limits set by signal duration and frequency, a size-filtering hypothesis was proposed for bats searching for insects with overlap-sensitive signals (Schnitzler and Kalko 1998). Bats with long signals and low frequencies should hunt for larger insects than bats with shorter signals of higher frequencies. The general trend that larger bats tend to forage for larger insects at larger detection distances, whereas smaller bats

forage for smaller insects at shorter detection distances, was confirmed by a number of field studies (e.g., Barclay 1985, 1986; Kalko 1995a; Houston, Boonman, and Jones, chapter 45, this volume). Nevertheless, contrary to expectation, the diet of the lesser noctule bat (Vespertilionidae: *Nyctalus leisleri*), a medium-sized, open-space bat that emits rather low frequency calls, includes many small insects (Waters, Rydell, and Jones 1995). Two explanations account for this observation. First, *N. leisleri* is highly flexible in its call pattern and switches occasionally to shorter and more wideband signals better suited for detection of smaller prey. Second, in a modeling approach, the echo strength of real insects was very similar for frequencies in the range of 20–100 kHz (Waters, Rydell, and Jones 1995). Potentially, the acoustic glints created by the beating wings of insects and low frequencies of the echolocation calls increase the effective range of echolocation, allowing species of large and medium-sized aerial insectivores to hunt for relatively small insects (see Houston, Boonman, and Jones, chapter 45, this volume).

For narrow-space gleaning CF foragers, the acoustic information contained in the echoes from fluttering insects can be used to classify and hence select prey. The fish-eating, greater bulldog bat, *Noctilio leporinus* (Noctilionidae), which emits CF and CF-FM search calls, attacked the water surface near an underwater pump that produced water splashes mimicking the glint patterns created by jumping, small fish (Schnitzler et al. 1994). Behavioral studies in the laboratory demonstrated the ability to discriminate among wing-beat rates of insects of a range of species from three families (Hipposideridae, Rhinolophidae, Mormoopidae: *Pteronotus parnellii*) (Roverud, Nitsche, and Neuweiler 1991; von der Emde and Schnitzler 1986, 1990; von der Emde and Menne 1989). In a generalization experiment, *R. ferrumequinum* even recognized phantom echoes of a particular insect species at novel aspect angles, suggesting cognitive abilities in pattern recognition (von der Emde and Schnitzler 1990). Dietary studies and fecal analysis of free-living greater horseshoe bats (*Rhinolophus ferrumequinum*) confirmed prey selection for this species. The bats fed on larger proportions of Lepidoptera and Coleoptera than were available in the insect samples collected simultaneously. The bats only ate small unprofitable prey when large insects were scarce (Jones 1990). In contrast, prey selection was not evident in *Hipposideros ruber* (Bell and Fenton 1984)

The ability to detect glints in echoes also was shown for two vespertilionid FM bats under laboratory conditions (Sum and Menne 1988; Roverud, Nitsche, and Neuweiler 1991). It remains to be seen to which degree these bats use this information for classification and prey selection in the field (see Bogdanowicz, Fenton, and Daliszczyk 1999a; Houston, Boonman, and Jones, chapter 45, this volume). Many FM signals are too short to carry enough echo information of an insect's wing-beat cycle (Moss and Zagaeski 1994). However, wing-beat cycle information could perhaps be extracted from several successive echoes, as suggested by Kober and Schnitzler (1990) and Moss and Zagaeski (1994). Furthermore, small insects deliver more glint information because of their high wing-beat rates. Frequency distribution of echolocation signals among syntopic species was implied as a factor in structuring local communities. Heller and von Helversen (1989) showed that for 12 syntopic species of *Rhinolophus* and *Hipposideros* in Malaysia, the frequency distribution of the species over the full frequency range (40–200 kHz) was significantly more even than would be expected from chance. The authors concluded that this was a consequence of resource partitioning with respect to prey selection based on differences in echolocation frequencies. Interestingly, a recent study by Kingston et al. (2000), including more species of rhinolophids and hipposiderids in the same area over a longer time period, could not reject the null hypothesis that call frequencies were distributed randomly. They suggest that differences in wing morphology are more informative in determining resource partitioning than call frequencies. Alternatively, recognition of conspecifics by call frequencies could also play a role in creating the observed patterns. Overall, it remains to be tested experimentally what small to larger differences in the CF frequencies cause in terms of prey detection, discrimination, and classification and, hence, in resource partitioning.

SIGNAL STRUCTURE AND SPECIES IDENTIFICATION

Because echolocation calls, particularly search signals of aerial insectivorous bats, are mostly species-specific, biologists use this information to identify bats in the field and use this information to conduct field surveys of habitat use and community structure (e.g., Aldridge and Rautenbach 1987; Carmel and Safiel 1998; Crome and Richards 1988; Fenton et al. 1998b; Kalko 1995b; McKenzie and Rolfe 1986; Mills et al. 1996; O'Farrell and Miller 1999). The literature on species identification and its applications is accumulating rapidly and is beyond the scope of this introduction (see Obrist et al., chapter 64; Gannon et al., chapter 63, both this volume). It is clear, however, that signal design can be very plastic within species both among and within individuals (e.g., Barclay 1999; Kalko 1995a; Obrist 1995; Schnitzler and Kalko 1998). Hence, a detailed knowledge of factors affecting signal design such as body size, foraging mode, and distance of a bat toward clutter is necessary to allow reliable identification. The most promising advances in species identification were made by using multivariate models that include several frequency and time measurements from echolocation calls (e.g., Zingg 1990; Vaughan, Jones, and Harris 1997a; Betts 1998) and by applying discriminant function analysis and artificial networks (Parsons and Jones 2000). Species identification is difficult for many bats, yet absolute for others, but at least the degree of confidence in identification can be quantified by comparing calls with a library of calls from known species.

CRYPTIC SPECIES

One spin-off from research on inter- and intraspecific differences in signal structure has been the discovery that certain bat species that are difficult to distinguish from morphological characteristics may differ in echolocation calls. Indeed, Europe's most widespread bat, the common pipistrelle (*Pipistrellus pipistrellus:* Vespertilionidae), recently was found to comprise two cryptic species that differ in the peak frequency of their echolocation calls by up to 10 kHz. Jones and Barlow review the diversity and evolution of cryptic species of echolocating bats (chapter 46, this volume).

PERSPECTIVES

A wide range of field studies supports the assumption that environmental constraints have largely shaped the structure of the echolocation signals and the auditory system during evolution. With a functional guild concept in place, more quantitative data are crucial to assessing just how fine-grained resource partitioning is based on perceptual and morphological niche differentiation. Moreover, the growing database of signal types across a wide range of taxonomic groups of bats offers promising perspectives for future research to analyze the possible influence of phylogenetic relationships on call structure. Further, a better understanding of factors that cause signal variation and the application of neural networks for the analysis of complex call patterns are likely to further enhance the resolution of species identification in the field, particularly of aerial insectivores.

Interactions between Echolocating Bats and Prey

Hearing in insects has polyphyletic origins, and is used both in communication and for the detection of predators, especially echolocating bats. In many cases, auditory and sound-producing adaptations that originally evolved for defense against echolocating bats subsequently fulfilled roles in the detection and production of courtship signals (W. E. Conner 1999). Insect ears typically use tympanic membranes (eardrums) for sound detection, and here we concentrate on those taxa of tympanate insects that have evolved sensitivity to ultrasound. Auditory sensitivity to ultrasound evolved in the Orthoptera, Lepidoptera, Neuroptera, Dictyoptera, Coleoptera, and Diptera, and is especially associated with winged, nocturnal taxa (Hoy, Nolen, and Brodfuehrer 1989).

Here, we describe some recent discoveries on the hearing of ultrasound by insects. Obviously listening for conspecifics and listening for predatory bats present different challenges to tympanate insects. We discuss how insects separate acoustic signals from potential mates from those of predators. We then describe new findings on mechanisms that tympanate insects use to avoid being captured by echolocating bats, and focus on some hypotheses about the function of "clicking" in arctiid moths. We next consider adaptations other than auditory ones that reduce the chances of insects being eaten by bats. Such adaptations are especially important for insect taxa that lack ears. Biologists recently became very interested in several ultimate or "why" questions about the evolution of insect hearing. Are there costs associated with the evolution of hearing in insects? Are bats that call outside of the frequency range of best hearing in insects better able to prey on tympanates? Did bats evolve certain call designs specifically to facilitate the capture of tympanate prey? Some of these questions are central to answering whether there is an "arms race" between tympanate insects and echolocating bats—that is, does the scenario fulfill the criteria needed for coevolution? Finally, we suggest some fruitful avenues for future research in the area of bat-insect interactions.

RECENT DISCOVERIES OF AUDITORY SENSITIVITY TO ULTRASOUND IN INSECTS

Ears occur in a range of locations on the insect body and are found in at least seven orders (Hoy and Robert 1996). Many tympanal organs are sensitive to ultrasound and evolved as bat detectors. Fullard and Yack (1993) and Fullard (1998) reviewed recent advances in the study of insect hearing from an evolutionary perspective. Some more recent discoveries of ultrasound sensitivity are highlighted below.

Diptera

Parasitic flies (*Ormia ochracea:* Diptera, Tachinidae) possess tympanal hearing organs that seemingly evolved so that the flies can detect their hosts, field crickets (*Gryllus* spp.: Orthoptera). The hearing organs are quite different from those previously described in Diptera. The females' ears are most sensitive between 4 and 6 kHz, matching the dominant frequencies in the crickets' songs. Male and female flies are also sensitive to ultrasound between 15 and 50 kHz, however, with auditory thresholds similar to those of mantids and lacewings (Robert, Amoroso, and Hoy 1992). Although these ears probably evolved primarily to detect hosts for parasitism, they will at least in theory detect the calls of some echolocating bats, many of which fly at similar times. Whether these flies show evasive responses to the calls of echolocating bats remains an interesting avenue for future research.

The bat fly (*Mystacinobia zealandica:* Diptera, Mystacinobidae) lives in roosts of the New Zealand short-tailed bat (*Mystacina tuberculata*). It possesses long sen-

silla trichodea that Dey (1995) suggested are used for the detection of ultrasound by nontympanate means. Although this suggestion is highly speculative, it remains unknown how many parasites that rely on bats either for feeding or for dispersal actually find their hosts. Sensitivity to ultrasound would facilitate the location of hosts.

Coleoptera

Tympanal hearing organs were first found in the Coleoptera in tiger beetles (Cicindellidae) (Spangler 1988a). The ears of tiger beetles (*Cicindela marutha*) have relatively broad tuning curves, and could be used in intraspecific communication, as well as in defense against echolocating bats. The ears have best frequencies close to 30 kHz, with sensitivities of 50–55 dB (Yager and Spangler 1995). In addition, *C. marutha* produces trains of ultrasonic clicks after exposure to ultrasound. The clicks have energy peaks at 30–40 kHz and could function in ways similar to the sounds produced by arctiid moths (see below). Like many arctiid moths, tiger beetles produce distasteful secretions when disturbed (Yager and Spangler 1997).

Free-flying scarab beetles (*Euetheola humilis:* Scarabaeidae) can be trapped under speakers that broadcast calls (40 kHz), resembling the echolocation calls of bats. The beetles are sensitive to frequencies between 20 and 70 kHz with thresholds of 60–70 dB, and produce a startle response to ultrasound that seemingly functions to evade potential approaches by echolocating bats (Forrest, Farris, and Hoy 1995). Similar frequency tuning was found in ears of two other genera of scarab beetles in the subfamily Dynastinae (Forrest et al. 1997).

Lepidoptera

Tympanal organs are known in at least six families of Lepidoptera (Hoy and Robert 1996), including nocturnal hedylids, the closest ancestors of extant butterflies (Yack and Fullard 2000). Little was known about hearing in moths in the family Geometridae until recently. Geometrids typically have four sensory cells used in hearing. Seven relatively large geometrids (wingspans 35–47 mm), studied by Surlykke and Filskov (1997), had best frequencies between 20 and 30 kHz and dynamic ranges of 45–50 dB (the dynamic range of the two-celled noctuid ear is about 35–45 dB). Because the audiograms of geometrids were similar to those of sympatric noctuids, it seems that bat predation was a common selection pressure acting on hearing in both moth families. Sensitivity to ultrasound is also found in two clades of hawkmoths (Sphingidae), and could have evolved independently in two subfamilies. Best frequencies are typically between 30 and 70kHz, depending on species, and hearing organs are located in the mouthparts (Göpfert and Wasserthal 1999a, 1999b). Some hawkmoths show startle responses to ultrasound, and may even produce acoustic signals in response (Göpfert and Wasserthal 1999a).

Conflicts between Hearing Conspecific Song and Predatory Bats

Although many insect ears are sensitive to ultrasound, they often need to interpret signals that convey very different messages. For example, katydids must decide if a certain signal is attractive (e.g., the song of a male), or if it poses a threat to survival (the call of an echolocating bat). The katydid can be attracted to the former, but elicit an evasive response to the latter. How can the insect discriminate the two signals? Katydids (*Neoconocephalus ensinger*) possess an auditory interneuron (the T-cell) that responds best to ultrasound above the frequency range of conspecific song (Faure and Hoy, chapter 51, this volume). The T-cell may therefore facilitate the discrimination of a potential mate from a possible predator, hence initiating a positive phonotactic response rather than an evasive maneuver.

How Insects Evade the Approach of Echolocating Bats

Considerable effort has been recently devoted to describing how insects avoid bats. Many insects initiate "acoustic startle responses" (ASRs) soon after hearing ultrasonic pulses (Hoy, Nolen, and Brodfuehrer 1989). ASRs include active phonotactic and evasive flight maneuvers in tympanate moths (Roeder 1967), passive wing-folding and dropping to the ground in lacewings (Miller 1975), and negative phonotactic flight in crickets (Nolen and Hoy 1986). The neural mechanisms underlying ASRs can be complex, and are beyond the scope of this review (see Hoy, Nolen, and Brodfuehrer 1989). Recent studies identified kinematic and aerodynamic responses of ASRs in crickets (May, Brodfuehrer, and Hoy 1988), katydids (Libersat and Hoy 1991) and locusts (Dawson et al. 1997). Male praying mantises (*Parasphendale agrionina*) grade their evasive responses according to the intensity of the ultrasound stimulus (Yager, May, and Fenton 1990). At low sound intensities (probably corresponding to calls from distant bats), the mantids steer away from the sound source. At high intensities, they make steep diving turns or spirals. The graded response is similar to that proposed for moths by Roeder (1967), and evasion seems to reduce the mantises' chances of being captured by wild bats (Yager, May, and Fenton 1990). Undeafened mantises appear to be better at avoiding echolocating bats than do deafened individuals (Cumming 1996). Although stationary mantises can hear ultrasound, they do not respond. Gleaning bats are especially sensitive to the sounds made by the movement of prey on substrates (Yager, May, and Fenton 1990).

Moths that are sensitive to ultrasound also modify their mating behavior to perceived threats from echolocating bats. Males abort flights toward female pheromones, and females cease pheromone release when subjected to ultrasound resembling the calls of aerial-hunting echolocating bats. Thus, tympanate insects must titrate how much effort they invest in mating relative to the level of predation risk (Acharya and McNeil 1998).

The Functions of Clicking in Arctiid Moths

One specialized response to the calls of bats is to produce sounds that resemble echolocation calls back at the bat. This process was described above for tiger beetles, but the clicks produced by arctiid moths have received considerably more attention. Bats tend to eat fewer arctiid moths than are available to them, and moths that have been muted experimentally are more likely to be captured (Acharya and Fenton 1992; Dunning et al. 1992). Thus, clicking protects arctiid moths against bat predation. Three hypotheses, which are not mutually exclusive, have been advanced: that the sounds startle an approaching bat (Bates and Fenton 1990), clicks interfere with the bats' ability to process echoes (e.g., Fullard, Fenton, and Simmons 1979), and/or signal to the bat that the moth is distasteful (the aposematism hypothesis, e.g., Dunning 1968). There is recent support for all three hypotheses.

Startle

Bates and Fenton (1990) startled big brown bats (*Eptesicus fuscus*) trained to fly to a platform by broadcasting arctiid clicks at them. There is support, therefore, for the startle hypothesis, though it could be of limited function if it only deters naive bats.

Interference with echo processing

Fullard, Fenton, and Simmons (1979) suggested that arctiid clicks produce neural responses in a bat's brain that create illusions of multiple targets, with the clicks acting as phantom echoes. Fullard, Simmons, and Saillant (1994) claimed that moths (*Cycnia tenera*) time their clicks to best jam a bat's echolocation because they wait until they hear terminal phase echolocation calls before clicking. Perhaps moths should emit clicks earlier if clicking functions to warn the bats about the distasteful nature of the prey. Rapid clicking by *C. tenera* could result in the clicks merging during auditory processing, interfering with the bats' ability to process sound (Fullard, Simmons, and Saillant 1994).

Miller (1991) reduced performance in range discrimination in *Eptesicus fuscus* when clicks were presented in a window beginning 1.5 ms before a phantom echo. Hence clicks can mask echoes and reduce ranging performance. Clicks could disrupt the neural mechanisms associated with temporal processing of echoes, thus making it difficult for the bat to judge the range of the moth. This hypothesis of temporal interference is different from Fullard, Fenton, and Simmons's (1979) "illusion" hypothesis.

Some fascinating experiments recently examined neural correlates of the perception of clicks similar to those produced by arctiid moths (Tougaard, Casseday, and Covey 1998). These experiments suggest a neural mechanism for Miller's (1991) findings that clicks reduce ranging ability. The nuclei of the lateral lemniscus provide time markers for the occurrence of a bat's call, and for the reception of returning echoes. Several units in the nuclei of the lateral lemniscus respond differently to clicks and FM sweeps, suggesting that bats have neural mechanisms to discriminate the two stimuli, as required by the aposematism hypothesis. Fullard, Fenton, and Simmons's (1979) illusion theory required that clicks should not be discriminable from calls or echoes, and this was not generally supported by Tougaard, Casseday, and Covey's experiments. Because the timing of responses to echolocation calls are altered by clicks in 77% of units in the lateral lemniscus, there is support for the theory that temporal processing is affected by clicks. Tougaard, Casseday, and Covey's experiments suggested that clicks must be emitted in a time window <2 ms before the echo (or extending partly into it) for temporal disruption to occur. The best way for a moth to achieve this would be to produce batches of clicks for each echolocation call emitted by the bat. This was the strategy shown by *Cycnia tenera* (Fullard, Simmons, and Saillant 1994), though other arctiids produce only a few clicks, and these may function by aposematism.

Aposematism

Clicking arctiid species from South Africa are less palatable to bats than are nonclicking species from other moth families matched for body size (Dunning and Krüger 1995). Clicking arctiids are more likely to survive presentations to little free-tailed bats (*Tadarida pumila*) than are nonclicking species. Similar results were found for arctiids and control species fed to *Eptesicus fuscus* in Ontario (Dunning et al. 1992). Dunning and Krüger (1995) argued that their findings support the aposematic theory for the evolution of clicking in arctiids, though the moths could gain additional protection if clicks interfere with echolocation, or if they startle bats. Arctiid moths capable of clicking were rarely eaten by free-living Sundevall's leaf-nosed bats (*Hipposideros caffer*) in South Africa, although the bats' echolocation calls (128–153 kHz peak frequency, Jones et al. 1993) are probably above the frequency range of hearing in the moths (Dunning and Krüger 1996). Perhaps the bats rejected these unpalatable moths after capture, and clicking played no role in the arctiids' defense against bats that emit high-frequency calls.

NONAUDITORY DEFENSES AGAINST ECHOLOCATING
BATS AND AUDITORY DEGENERATION

Although many insects evolved sensitivity to ultrasound and ASRs, many taxa lack ears. Nonauditory defenses of insects against echolocating bats include temporal isolation and flying close to clutter, adaptations reviewed by Rydell (chapter 43, this volume). Tympanate ears can be large, taking up space that could be devoted to other functions in an insect's body. It is therefore conceivable that possession and maintenance of hearing is costly. If so, then moths would be expected to lose hearing when predation risk from echolocating bats is low or absent. Several recent studies investigated whether moths that are temporally or spatially isolated from bats show poorer hearing responses. In general, there is little evidence of auditory generation when moths and bats are isolated temporally or geographically, unless females become flightless (Cardone and Fullard 1988; Fullard et al. 1997; Rydell et al. 1997; Surlykke and Treat 1995). Even limited temporal overlap between echolocating bats and tympanate moths may still exert considerable selection pressure for the maintenance of hearing that is sensitive to ultrasound (Fullard and Dawson 1999; Rydell et al. 1997). There is some evidence for degeneration of hearing in moths from French Polynesia, where bat predation is absent and sensitivity to high frequencies is poor (Fullard 1994). A geometrid moth (*Archiearis parthenias*) is diurnal in Scandinavia, and thus temporally isolated from bat predation (Surlykke et al. 1998). Interestingly, this moth has poor hearing above 25 kHz, and a best frequency of 12 kHz. It therefore shows degenerative hearing at frequencies emitted by most echolocating bats, though the function of its sensitivity to low frequencies remains unclear (Surlykke et al. 1998).

In many insect species, females emit pheromones to attract males. Males are therefore airborne more than females, since they search for mates (Acharya 1995; Jones 1990). Male moths fly more than females, and fly at times in the night when bats are more abundant (Acharya 1995). Males are therefore more vulnerable to predation by echolocating bats, and therefore can be expected to show greater sensitivity to ultrasound. Although females often showed degenerate hearing in taxa where females have vestigial wings (Cardone and Fullard 1988; Rydell et al. 1997; Yager 1990), no studies have shown degenerative hearing in females for taxa where both sexes fly (Acharya 1995).

IS THERE AN EVOLUTIONARY "ARMS RACE" BETWEEN
TYMPANATE INSECTS AND ECHOLOCATING BATS?

The evolution of ultrasound sensitivity in nocturnal insects evolved in response to predation pressures exerted by echolocating bats. Have bats modified their biosonar signals in response to the evolution of insect hearing? Such a response is necessary if the interactions between bats and moths are examples of coevolution—that is, if they involve reciprocal influences that result in lineages evolving together (Janzen 1980), or as an evolutionary "arms race" (Dawkins and Krebs 1979). Although often cited as such, the interaction between tympanate insects and echolocating bats does not always fulfill the criteria needed to show coevolution. Fullard (1998) argued that the case for bat-moth coevolution was clear, but this can only be verified if the call structures used by bats that eat large numbers of eared moths are not solely advantageous for other reasons.

One argument sometimes used in a coevolutionary context is that bats that call outside the frequency range of greatest sensitivity in eared moths (typically 20–50 kHz) are best able to prey upon eared moths (the allotonic frequency hypothesis, Fullard 1987). Indeed, bats that call at allotonic frequencies often do eat large quantities of moths (Bogdanowicz et al. 1999b; Jones 1992) (fig. 42.3), and bat species that call at frequencies closer to those at which eared moths show greatest sensitivity may sometimes be restricted to eating noneared taxa (Pavey and Burwell 1998). The apparency of different bat signals to moth ears was elucidated recently by playback experiments of calls to tympanate preparations of moths (Fullard and Dawson 1997; Waters and Jones 1996). Studies relating what bats eat to whether prey can hear ultrasound also should consider that studies around streetlights could represent an artificial situation. Not only are moths attracted to streetlights, the light might also inhibit escape behavior (Acharya and Fenton 1999; Svensson and Rydell 1998).

For coevolution to occur, bats must specifically change their call frequencies to better catch moths. It is in fact difficult to reject alternative explanations of the allotonic frequency hypothesis. For example, bats could use allotonic frequencies for reasons unrelated to the hearing sensitivities of their prey. The calls of lesser horseshoe bats (*Rhinolophus hipposideros*) are over 21 dB more audible to large yellow underwing moths (*Noctua pronuba*) at the frequency of the first harmonic (55 kHz), which is usually suppressed (Jones and Waters 2000). Why do many horseshoe bats call by emphasizing the second harmonic? Although horseshoe bats that emit high frequencies eat most moths (Jones 1992), high frequencies could have evolved for greater directionality, for reduced range, or for the detection of small prey, rather than to facilitate the capture of tympanate prey. Large changes in frequency and time have relatively small changes on the audibility of short, broadband calls to moths. Thus the potential for coevolution in bats that emit broadband calls may be less than that in species that emit narrowband or CF signals (Jones and Waters 2000).

Bats that use exceptionally low frequencies in biosonar exploit apparently tympanate insects as prey—for example, the spotted bat, *Euderma maculatum* (peak frequency 9–12 kHz, Fullard and Dawson 1997) and the European free-tailed bat, *Tadarida teniotis* (peak frequency 11–12 kHz, Rydell and Arlettaz 1994). These bats could have evolved the use of low frequencies for

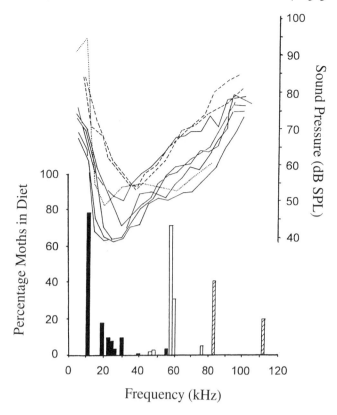

Fig. 42.3. Audiograms of eight European species of moths, and the incorporation of moths in the diets of European bats in relation to the dominant frequencies of their echolocation calls. Note how bat species that call at frequencies between 20 and 50 kHz eat few moths, and moth sensitivity to bat calls is greatest at these frequencies. Bats that hunt by gleaning, or which call at frequencies <15 or >60 kHz often eat large numbers of moths. Audiograms (upper part of figure) are from *Agrotis segetum, Diarsia mendica, Cerapteryx graminis, Apamea crenata,* and *A. maillardi* (Noctuidae—continuous lines), *Phalera bucephala* and *Phoesia tremula* (Notodontidae—dashed lines), and *Ephestia kuehniella* (Pyralidae—dotted line). The lower part of the figure shows the importance (% volume or % frequency) of moths in the diets of European bats in relation to the frequency of most energy in echolocation calls normally emitted in search flight. Black bars are aerial-hawking species (from left to right: *Tadarida teniotis, Nyctalus noctula, N. leisleri, Vespertilio murinus, Eptesicus serotinus, E. nilssonii, Myotis daubentonii* and *Pipistrellus pipistrellus*), white bars are gleaners (*M. myotis, M. nattereri, Plecotus austriacus, P. auritus* and *M. emarginatus*), and hatched bars are flutter detectors (*Rhinolophus ferrumequinum* and *R. hipposideros*). Sources are detailed in Rydell, Jones, and Waters (1995) from where the figure is taken, permission of the Nordic Ecological Society.

the detection of distant or large targets, and could feed on tympanate prey as a by-product of the original adaptation. However, the long wavelengths emitted by these bats reduce the detection distances of typical noctuid moths by Rayleigh scattering, so there appears to be a cost associated with their use. The hearing sensitivity of moths typical falls sharply at frequencies <15 kHz, however, and the benefits of reduced detection by moths appear to outweigh the costs of reduced detection range in echolocation (Norman and Jones 2000). Low-frequency allotonic bats therefore provide strong evidence for bats evolving calls to exploit the limited hearing capacities of moths, perhaps an example of coevolution after all (Fullard 1998; Norman and Jones 2000; Jones and Waters 2000).

Do bats use lower intensity calls or even switch off echolocation as an adaptation to facilitate capture of tympanate prey (Anderson and Racey 1991; Faure, Fullard, and Barclay 1990; Faure and Barclay 1992; Faure, Fullard, and Dawson 1993)? Echolocation could be of limited use in cluttered situations, where echoes from background objects interfere with those from prey (Schnitzler and Kalko 1998). In cluttered situations, bats therefore could use other sensory cues for prey detection, and switch off echolocation because of its limited use, rather than to reduce apparency to tympanate prey.

Overall, the selection pressures that drive the evolution of defenses often are stronger than those that shape predation strategies, because the costs of making a mistake are often greater for prey (death) than they are for predators (a missed meal—the "life-dinner" principle, Dawkins and Krebs 1979). Moreover, moths have higher reproductive rates than bats and have a faster potential for evolutionary change. In Europe, moths are rarely eaten by aerial-hawking bats that call between 20 and 50 kHz (fig. 42.3). Indeed, Rydell, Jones, and Waters (1995) argued that hearing moths won the evolutionary contest against large, aerial hawking bats that probably posed the greatest threat to pretympanate and early tympanate moths in their evolutionary past. Thus, overall, the evidence for coevolution interactions between echolocating bats and tympanate insects needs careful scrutiny. We need more convincing evidence to show that echolocation call strategies of bats alter in direct response to ultrasound sensitivity in insects to demonstrate coevolution. Fruitful avenues for future research could involve determining whether the echolocation calls of bat species differ according to the presence of tympanate prey, and studies of hearing organs in fossil insects that could clarify the timescale over which tympanal hearing evolved.

PERSPECTIVES

Despite the major advances in studies of the interactions between echolocating bats and tympanate prey in the last 10 years, several areas remain underresearched. In particular, most studies on insect hearing remain laboratory-based. Few studies have investigated real-life interactions in the field, yet this is where natural selection operates. Although advances have been made in quantifying call intensities of bats in the laboratory (Waters and Jones 1995), we need to know more about the intensities of calls emitted by bats in the field, and at what distances tympanate prey react to natural calls. We also need better estimates of target strengths of prey taxa, so we can better estimate the distances at which bats detect tympanate insects. Recent studies suggest that bigger moths are better able to detect bat calls. Although bigger moths reflect stronger echoes (Norman and Jones 2000; Surlykke et al. 1999), this disadvantage is offset by possession of more sensitive ears (Surlykke et al. 1999). Field studies linking ultrasound recording with images of flying bats and insects also could clarify the function of clicking in arctiid moths and better document whether ASRs are unpredictable in nature. More experiments on the capture success of deafened versus intact moths by bats (Acharya and Fenton 1999) would clarify the survival value of tympanate hearing. Finally, we feel that many of the principles developed in studies of bat-insect interactions could be applied to echolocating cetaceans and their prey, especially in light of discoveries of ultrasound sensitivity in fish (e.g., Mann, Lee, and Popper 1997).

Acknowledgments

We are most grateful to Nancy Simmons (NSF grant DEB-9873663), Hans-Ulrich Schnitzler, Brock Fenton, Jeannette Thomas, and an anonymous referee for their constructive comments on previous drafts of the manuscript.

{ 43 }

Evolution of Bat Defense in Lepidoptera: Alternatives and Complements to Ultrasonic Hearing

Jens Rydell

Introduction

The appearance of echolocating bats in the early Tertiary (Habersetzer, Richter, and Storch 1994) changed the situation drastically and permanently for contemporary insects, because eluding predators that detect their prey by sound, rather than by vision, required a different set of antipredation adaptations. Moths and butterflies today are the products of 50–60 million years of natural selection imposed by bats. Many physiological and morphological traits of extant moths probably evolved because they served in predator defense, although some of them now could have other functions. Lepidoptera are partially designed by bat predation, as shown by the fact that ultrasonic hearing organs have evolved several times and are now possessed by members of at least six superfamilies (Yack and Fullard 2000).

A selective agent such as bat predation likely affected the prey in many ways, not only through the evolution of a single defensive adaptive system. Nevertheless, the research on this subject has been dominated by studies of ultrasonic hearing and associated evasive responses. Many other interesting aspects of moth biology with obvious or possible relevance to bat defense have received little attention. The structure and function of tympanic organs in Lepidoptera has been reviewed elsewhere (e.g., Roeder 1967; Fullard 1987; Spangler 1988b; Hoy 1992; Fullard and Yack 1993). I attempt to broaden the subject of bat defense in Lepidoptera by reviewing recent research on alternatives and complements to ultrasonic hearing. Hopefully, this will also extend the knowledge about the evolution of hearing in Lepidoptera and put this trait in a slightly different evolutionary perspective. Although moth larvae sometimes are eaten by gleaning bats (e.g., Bauerová 1986), I only consider predation on and antipredation adaptations of adult insects—that is, the flying stage.

Occurrence of Nontympanate Lepidoptera

A survey of the larger Scandinavian species of nocturnal moths suggests that 94% (*n* = 1093) belong to families or subfamilies that possess ultrasonic hearing organs in one form or another (Rydell and Lancaster 2000). The survey excluded the Microlepidoptera except the Pyralidae and also families of Macrolepidoptera that consist of exclusively diurnal species (e.g., "butterflies," Sesiidae and Zygaenidae).

Although most of the larger nocturnal moths are tympanate, the fact that there are also species that probably are deaf raises the interesting question of how these can persist. Deaf moths presumably are not defenseless against bats, but use other methods to avoid them (Roeder 1974). It also seems likely that tympanate moths do not rely entirely on their hearing for defense against bats, but also possess complementary systems. Predator defense systems usually are complex, and traits that result in lower predation are favored by natural selection, regardless of why they evolved in the first place. Most Lepidoptera species belong to the Microlepidoptera, the majority of which are small and probably earless. They are not dealt with extensively herein, primarily because there is almost no information on how they avoid bats.

Alternative and Complementary Defenses

Body Size in Lepidoptera

Roeder (1974) hypothesized that some of the largest moths, such as Sphingidae and Saturniidae, escape bat predation by their size alone, being too large to catch and handle for most bats. Indeed, the biggest moths weigh several grams, and they are thus of sizes comparable to small insectivorous bats. However, some large moths (such as sphingids, for example) also possess a sensitive ultrasonic hearing system, which probably serves as defense against bats when the moths hover and feed in front of flowers (Roeder, Treat, and Vande Berg 1968), and this suggests that large size alone may not always provide enough protection against bats.

It is also possible that some of the smallest Lepidoptera, such as Nepticulidae and Elachistidae, automatically achieve protection by their size, because they may be too small (2–5 mm wingspan) to be detected efficiently by many echolocating bats (Waters, Rydell, and Jones 1995). Nevertheless, the great majority of the Lepidoptera are without question large enough to be detectable by bats. Although it seems likely that the extreme sizes of some moths make them less available to echolocating bats, there is little evidence that bat predation has directly favored increased or decreased body sizes in moths.

TEMPORAL AVOIDANCE OF PREDATORS BY LEPIDOPTERA

The foraging activity of bats is influenced by predation from diurnal raptors, and this is an important reason why bats avoid flying during the day (Speakman 1991). This constraint on bat activity is potentially exploited by insects that have become diurnal (Fullard et al. 1998) or crepuscular (Andersson, Rydell, and Svensson 1998). Indeed, diurnal species occur in most Lepidoptera families. The selection pressure behind the evolution of this trait is usually unknown, however, although Yack and Fullard (2000) suggested that bats "invented" the butterflies, by preying on their nocturnal ancestors.

Insectivorous bats are more active during the reproductive season (June–August in the Northern Hemisphere) than at other times of the year (Rydell 1991). The predictable variability in bat activity perhaps also is exploited by insects that emerge and fly when bats are less active, such as very early or late in the season (Fullard 1977). Yack (1988) suggested that members of the generally earless bombycoid moth families, including the Lasiocampidae and Saturniidae, emerge as adults either before or after the summer peak in bat reproduction, thereby presumably facing less predation from bats. However, there are also other reasons why non-feeding moths, such as the lasiocampids, emerge outside the main growing season. For example, allocating mating and egg laying to late autumn or early spring, before the vegetation turns green, could ensure that larvae feed on young leaves relatively free of defensive compounds. It is thus by no means certain that the off-season emergence of many lasiocampids is a bat defense trait.

Relaxed selection pressure from bats presumably would lead to degenerate hearing in insects, provided that hearing is used mainly for defense against bats (Cardone and Fullard 1988). Tests of this hypothesis have provided ambiguous results, however. Among spring- and autumn-emerging species of Lepidoptera, some of the fast-flying cuculline noctuids show slightly reduced hearing sensitivity (Surlykke and Treat 1995), whereas this is not the case in slow-flying geometrids (Rydell et al. 1997). This suggests that the flight season itself only provides efficient protection against bats in some Lepidoptera groups, but not in all. In Cucullinae, fast flight may in itself provide protection against bats and other predators. Hence, the fast flight could be an indirect reason for the reduction in hearing sensitivity in the Cucullinae. In contrast, for the geometrids, which lack this defense system, acute ultrasonic hearing could be more essential.

The assumption that some moths are seasonally isolated from bats was recently examined by Svensson, Rydell, and Brown (1999). They used the "winter moth" *Operophtera brumata,* which flies in late October and November in Scandinavia. Its flight season is terminated by the onset of winter. Nevertheless, the flight activity of *O. brumata* frequently coincides with that of bats (*Eptesicus nilssonii*) even late in the season (Svensson, Rydell, and Brown 1999), and its responses to artificial bat calls are highly adaptive and without any sign of degeneration (Svensson and Rydell 1998). So, at least in northern Europe, avoidance of bats on a seasonal basis may not be effective enough to result in substantially decreased selection pressure on any moth.

In contrast, moths seem to escape bat predation efficiently by becoming diurnal or crepuscular, although they may then need adaptations to protect them from visual predators such as birds. The hearing sensitivity at ultrasonic frequencies of some diurnal moths, such as the Neotropical Dioptinae (Notodontidae), are changed drastically compared to their nocturnal relatives, and may, in some cases, no longer be functional against bats (Fullard et al. 1998).

ACOUSTIC CONCEALMENT BY LEPIDOPTERA

Roeder (1974) suggested that earless moths of the family Lasiocampidae reduce the risk of detection by bats by flying close to the ground, thus taking advantage of the fact that prey detection by echolocation is complicated by clutter. Lewis, Fullard, and Morrill (1993) tested this acoustic concealment hypothesis, and found that two species of Lasiocampidae (*Malacosoma americanum* and *M. disstria*) fly significantly closer to the ground than sympatric tympanate species. Hence, flying near the vegetation or the ground could be an adaptation that reduces predation from bats. Acoustic concealment from bats also seems to explain why the males of the lekking ghost moth (*Hepialus humuli*) stay <0.5 m of vegetation when displaying (fig. 43.1). Within this zone, they are relatively safe from attacks by patrolling northern bats (*Eptesicus nilssonii*), but staying too close to vegetation is probably suboptimal with respect to the visual signaling that attracts females (Rydell 1998). Thus mating success probably is compromised by predation risk in this species, as in some other moths (Acharya and McNeil 1998).

Acharya (1995) showed that males are more exposed to predation by aerial-hawing bats hunting near streetlights than females, presumably because they spend more time in the air. Hence, moths probably are concealed to such aerial-hawking bats as long as they do not fly. Accordingly, the heavy and flightless female gypsy moth (*Lymantria dispar*) shows reduced hearing sensitivity at ultrasonic frequencies compared to the normally flying male, suggesting that females may not be subject to predation by aerial-hawking bats (Cardone and Fullard 1988). Acoustic concealment in moths can also be enhanced if the wing area (i.e., the target strength) is reduced. This is the case in some geometrid "winter moths," which have only vestigial wings. As expected, these females also have drastically reduced ultrasonic

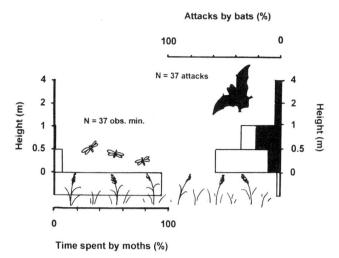

Fig. 43.1. Flight height of displaying male ghost swifts (*Hepialus humuli*) over a hay field in relation to the top of the panicle-bearing grass stems (left bars) and the height where the displaying moths were attacked (right bars) and caught (black sections) by northern bats *Eptesicus nilssonii*. Attacks on moths occurred at significantly higher elevations than expected based on the moths' display positions ($\chi^2 = 59.0$, $p < 0.0001$), and successful attacks occurred significantly higher than unsuccessful ones ($\chi^2 = 7.3$, $p < 0.01$). From Rydell (1998), with permission from the Royal Society of London.

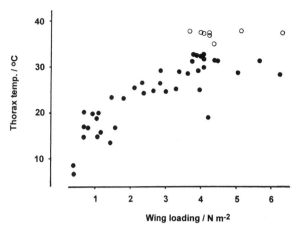

Fig. 43.2. Thorax temperature in flight in relation to wing loading of earless (open dots) and tympanate (black dots) moths from Scandinavia. Across all species wing loading was closely correlated with thoracic temperature in flight ($r = 0.826$, $p < 0.001$). There were no earless species with low wing loading. At high wing loading (>3.6 N m^{-2}), the thoracic temperature was significantly higher in the earless species (ANCOVA; $p < 0.001$). From Rydell and Lancaster (2000), with permission from the Nordic Ecological Society.

hearing capacity and are practically deaf to bat echolocation calls, in sharp contrast to the males (Rydell et al. 1997). However, the wing reduction in these moths is not primarily an effect of bat predation. By abandoning flight the females get rid of an important energetic constraint, and can therefore produce more eggs (Marden 1995).

Avoiding flight apparently is an efficient way to escape bat predation, and moths might therefore minimize time in flight (Morrill and Fullard 1992). Males of the ghost swift (*Hepialus humuli*), an earless moth, display on leks for less than 30 min each evening, and since they do not fly much otherwise, they probably use a time minimization strategy (Andersson, Rydell, and Svensson 1998). How the search strategy of male moths is influenced by bat predation remains an interesting subject for future studies.

Significance of Endothermy in Lepidoptera

Some moths characteristically have very high thoracic temperatures (near 40°C) while flying (Casey and Joos 1983). Others (e.g., *Operophtera* spp. and many other geometrids) fly with body temperatures that approach ambient temperature even when the latter is near freezing (Heinrich and Mommsen 1985). High body temperatures are found in moths that have high wing loading (Bartholomew and Heinrich 1973), and together these traits favor fast and erratic flight at the expense of higher energy consumption (Chai and Srygley 1990). In contrast, the low body temperature and wing loading of

some moths presumably minimize the energy consumption but results in slow flight. Fast and erratic flight could be an alternative anti-bat adaptation in earless moths (Lewis, Fullard, and Morrill 1993), and perhaps a complement to the hearing-based bat defense in some tympanate forms as well.

Indeed, Rydell and Lancaster (2000) showed that Scandinavian members of nontympanate superfamilies of "Macrolepidoptera" (Hepialoidea, Cossoidea, Zygaenoidea, Bombycoidea, and Sphingoidea) do have significantly higher wing loadings than members of typically tympanate superfamilies (Pyraloidea, Drepanoidea, Geometroidea and Noctuoidea) (4.3 ± 2.4 as compared to 2.1 ± 1.3 N m^{-2}), and they also consistently fly with higher thorax temperature (36–40°C as compared to 8–35°C) (fig. 43.2).

Low wing loading (i.e., around 1 N m^{-2} or lower), implying low body temperature and slow flight, is typical among the Geometroidea and Pyraloidea, and such characteristics are also typical for some families and genera of Drepanoidea and Noctuoidea, including many species of Arctiidae, Lymantriidae, and Noctuidae. In contrast, all species of earless moths have relatively high wing loading (2 N m^{-2} or higher), regardless of systematic affinity. In particular, the combination of earlessness and slow flight does not exist among the Macrolepidoptera, with the exception of butterflies. Slow flying, deaf, nocturnal species presumably would have been virtually defenseless against bats. The fact that many tympanate moths, particularly many noctuids and notodontids, also have high wing loading and elevated body temperature, although not as extreme as in the deaf forms, and therefore fly fast, suggests that these adaptations may not necessarily have evolved as anti-bat adaptations in the first place. Indeed, it seems more likely that endothermy first evolved in moths as they became nocturnal and no longer could warm up by basking in the sun. It could then have evolved toward higher sophistication by selection pressure from bats.

APOSEMATISM AND MIMICRY BY SOUND

Considering the widespread use of visual aposematic signals among diurnal Lepidoptera, it is not surprising that aposematic sounds seem to have evolved in some nocturnal and tympanate moths as a defense against bats (Dunning and Roeder 1965). When exposed to bat echolocation calls, some arctiids emit trains of ultrasonic clicks, produced by tymbal organs situated on the metathorax (Blest, Collett, and Pye 1963), and it is clear that the clicks protect the moths against bats (Dunning et al. 1992). The mechanism by which this system works is a matter of some controversy, however. The clicks may not be purely aposematic, but also may include a startling effect (Miller 1991), or they may jam the bat's echolocation calls and provide the moth with acoustic camouflage (Fullard, Fenton, and Simmons 1979). In any case, the arctiids seem particularly well equipped to deal with echolocating bats.

Another example of sound-producing Lepidoptera is the nymphalid butterflies, which often share hibernaculum in winter with bats. Hibernating peacocks (*Inachis io*) open their wings in a characteristic manner and produce intense ultrasonic clicks from a special area of the wing membrane (Møhl and Miller 1976). Since many nymphalids are known to be distasteful, the clicks hypothetically warn bats of noxious prey, just as in the Arctiidae. Nevertheless, nymphalid wings sometimes pile up in large numbers in bat hibernacula (Rydell unpublished), suggesting that this hypothetical defense system may not always be efficient against bats. Quite surprisingly, there are no documented examples of acoustic mimicry in Lepidoptera. The defensive clicking of arctiids would otherwise seem to be vulnerable to such exploitation by other tympanate but palatable moths.

PREDATOR SWAMPING BY LEPIDOPTERA

The highly synchronous mass emergence of a small lasiocampid (*Malacosoma americanum*) may provide protection from predation (Lewis, Fullard, and Morrill 1993). Although such predator swamping is common in insects, such as mayflies and chironomids, it is probably less important in Lepidoptera, because this group generally possesses other sophisticated means of predator defense. In fact, the synchronous emergence in *Malacosoma* could be an evolutionary effect of the extremely high energy expenditure of these earless moths (see above) necessary to accommodate mate finding within their short adult life span. If so, predator swamping could be a by-product of this adaptation.

THE THREAT OF GLEANING BATS

It seems as if most evasive responses of Lepidoptera are induced in flight, and presumably evolved through predation pressure from aerial-hawking bats (Rydell, Jones, and Waters 1995). However, there are also gleaning bats, some of which feed extensively on moths. Moths respond to ultrasound even when not in flight, either by freezing (Werner 1981), stopping the release of pheromones (Acharya and McNeil 1998), or releasing their grip and falling to the ground (Rydell et al. 1997). However, gleaning bats normally use low-intensity echolocation calls (Faure, Fullard, and Barclay 1990) or rely on passive listening (Anderson and Racey 1991) or vision (Bell 1985), and they do not seem to be detected at sufficient distance by moths in most situations (Faure and Barclay 1992). This discrepancy is puzzling, and, although gleaning bats are abundant components in some bat faunas (e.g., in northern Europe) and often eat many moths (Rydell, Jones, and Waters 1995), at present it seems as if moths generally lack efficient defense against them.

The warming-up process may be an additional cost for warm-blooded moths. A shivering individual may be vulnerable to predation, particularly if it cannot detect approaching gleaning or flutter-detecting bats. Gleaning perhaps represents an example of counteradaptations in bats, coevolved in response to the defense systems in moths. However, gleaning techniques could just as well have evolved because detection of prey in clutter was facilitated, and increased availability of moths and other tympanate insects was again just a fortunate side effect.

A Speculative Evolutionary Scenario

Lepidoptera appeared as a distinct group in the fossil record in the Triassic or Jurassic periods and several families of "Microlepidoptera" evolved by the end of the Cretaceous period (Whalley 1986). However, these early forms belong to families that are largely diurnal. It seems possible that nocturnal behavior first evolved in moths as a response to increased predation during the day. The earliest convincing specimens of tympanate moth families, which presumably were nocturnal, are in the Baltic Amber from the late Eocene or early Oligocene (beginning of the Tertiary period; Ross and Jarzembowski 1993). This followed the appearance and radiation of bats in the Paleocene and early Eocene (Habersetzer, Richter, and Storch 1994).

The early bats presumably imposed a strong selective pressure favoring moths that could elude them either by detecting biosonar signals and taking evasive action, by fast and erratic flight, or by becoming diurnal or crepuscular. Tympanic organs in moths most likely evolved in this context (Spangler 1988b). Endothermy had perhaps evolved in moths earlier, as they became nocturnal, but particularly in the earless taxa it seems to have been refined by bat predation, since high body temperature improves the flight performance. Hearing organs, as well as adaptations that permit fast flight, are polyphyletic adaptations in moths, and both provide spectacular examples of convergent evolution. Hence, bat predation seems to have exerted a profound effect on the morphology (size and wing loading), physiology (flight body temperature), and behavior (activity patterns, flight style, defensive reactions) of nocturnal moths.

Because earless moths lack early warning systems, their evasive behavior must operate more or less continuously while in flight, and since this is energetically costly, a higher energy expenditure is the price they pay for being earless. The increased energy expenditure may in turn have ecological and evolutionary consequences, including a shortened adult life span or a need to feed regularly. The high energy demand of the nontympanate moths could underlie their limited evolutionary success, as currently reflected in their relatively low species diversity (6% of the nocturnal species of Macrolepidoptera).

Evolution of tympanic organs made moths less vulnerable to bat predation but probably had other important effects. For example, ultrasonic hearing may have indirectly facilitated cold adaptation in moths, because it solved the problem of efficient bat defense at slow flight speed and low body temperature (Svensson, Rydell, and Brown 1999). Therefore, moth ears are "key innovations" (Heard and Hauser 1995) in the sense that they released their bearers from the energetic constraints imposed by a high flight metabolism necessary to avoid bats, and thus set the stage for adaptive radiation of slow-flying energy-conserving forms.

Conclusions

Like any perceptual system, ultrasonic bat echolocation has limitations and shortcomings, which potentially can be exploited by nocturnal insect prey. The most obvious example is that ultrasonic hearing has evolved in many insect groups, which thus take advantage of the fact that echolocating bats always reveal their presence. However, there are also other constraints of bat echolocation that are exploited by insects. For example, bats that use high-intensity, low-duty-cycle echolocation seem to have problems detecting and identifying small targets near clutter (e.g., Barclay and Brigham 1994). Accordingly, some earless moths, and probably many other insects, make use of this constraint by preferentially flying near vegetation and thus avoiding free air space, where bats are most likely to hunt (Lewis, Fullard, and Morrill 1993). Likewise, bats generally fly slowly because of constraints imposed by the short range of echolocation (Barclay and Brigham 1991), and are therefore largely restricted to the dark hours due to predation pressure from birds (Speakman 1991). This constraint probably is exploited by some moths that have become diurnal or crepuscular (Yack and Fullard 2000). Despite these constraints, echolocating bats, without question, had an important impact on the evolution of the Lepidoptera, not only because they triggered the evolution of ultrasonic hearing, but also because they influenced many other aspects of their morphology, physiology, and behavior.

Acknowledgments

I thank Lalita Acharya, Paul Faure, Bengt Gunnarsson, and Winston C. Lancaster for help and/or comments on the manuscript, the Swedish Research Council VR for financing my work on bats and moths, and SAAB Dynamics AB and Lunds Djurskyddsfond for providing equipment and additional grants.

{ 44 }

Evolution of Echolocation and Foraging Behavior in Bats

Hans-Ulrich Schnitzler, Elisabeth K. V. Kalko, and *Annette Denzinger*

Introduction

Numerous morphological, physiological, and behavioral adaptations of sensory and motor systems permit bats nighttime access to a wide range of habitats and resources. Echolocation and flight are two of the adaptations that make bats so successful (e.g., Neuweiler 1989; Fenton 1990; Schnitzler and Kalko 1998). The evolution of both echolocation and flight is of particular interest, because of ongoing discussions as to whether one trait preceded the other and whether they evolved independently or in tandem.

Echolocating bats emit ultrasonic signals and analyze the returning echoes to perform a multitude of tasks including landing, obstacle avoidance, spatial orientation, and, in many species, recognition and acquisition of food. Landing and avoiding obstacles in the flight path require rather similar echolocation abilities. However, recognition and acquisition of food (i.e., the ability to detect, localize, identify, track, and successfully catch prey) pose different tasks depending on ecological conditions set by habitat type, foraging mode, and diet. Bats that forage in the open encounter different conditions than bats searching for insects near the edges of vegetation, in gaps, within vegetation, or near the ground. The tasks also differ depending on whether bats catch insects in flight (aerial mode) or collect food from surfaces (gleaning mode).

To date, echolocation has mostly been discussed under the assumption that its main role lies in detection, localization, and identification of single targets such as flying insects in more or less obstacle-rich environments. This view of echolocation also is reflected in the design of most behavioral experiments where specific echolocation tasks are set up to measure the performance of bats. In such experiments, the echolocation scene often consists of real or simulated point targets that differ in range, angle, and/or sound pressure level (SPL). The strong emphasis on the primary role of echolocation for recognition and assessment of food is also reflected in discussions of whether echolocation or flight evolved first. In these discussions, the role of echolocation is seen mostly in the context of detection and localization of flying insects (Fenton et al. 1995; Norberg 1994). To our understanding, this approach covers only some of the conditions that molded the echolocation systems of bats during evolution. If we want to understand the adaptive value of echolocation systems, we must consider the complete set of tasks performed by echolocating bats.

The echolocation abilities of bats evolved step by step, increasing the bats' fitness and leading to the echolocation systems and foraging behaviors found today. Several authors have presented their views on the evolution of echolocation and flight in bats, in particular Arita and Fenton (1997), Fenton et al. (1995), Norberg (1994), Pettigrew (1986, 1991), Simmons and Geisler (1998), and Speakman (1993, 1999). We present an alternative scenario where, in accordance with Simmons and Geisler (1998), detection, localization, and identification of flying insects represent a late step in the evolution of echolocation. We propose a series of evolutionary steps in the foraging behavior of microchiropteran bats and link these to plausible steps in the evolution of echolocation performance (fig. 44.1). In the context of foraging, we consider motor behavior, foraging habitat, search mode, mode of prey perception, and foraging mode.

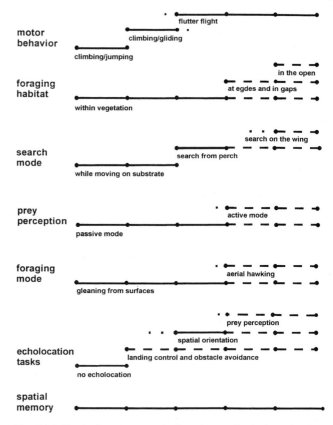

Fig. 44.1. Evolutionary steps in foraging and echolocation behavior in microchiropteran bats.

Proposed Scenario for the Evolution of Foraging and Echolocation Behavior in Bats

We hypothesize that microchiropteran bats arose from nocturnal arboreal insectivorous mammals that climbed bushes and trees to find and capture prey that sat on vegetation (J. D. Smith 1976). These ancestors presumably relied strongly on their ears to detect, localize, and classify distant prey that revealed itself through self-generated sounds (passive mode), such as the advertisement calls and rustling noises of insects. In these animals, the approach to distant prey was mainly guided by acoustic information produced by the prey (indirect approach). After the pre-bats reached an insect, they may have used olfaction to probe the prey before gleaning it from the substrate. For spatial orientation in the foraging habitat, these ancestors may have used a well-developed spatial memory and probably also vision, particularly for long-range orientation.

The evolution of echolocation may have started when these ancestors began to jump and finally to glide from branch to branch and from tree to tree to maximize net energy gain during foraging (Norberg 1994) and to reduce their risk of predation. In the dark, vision alone does not guarantee a safe landing on a neighboring branch or tree within the forest. Therefore, in tandem with the evolution of jumping and finally gliding, the pre-bats may have evolved the ability to perceive echoes of their high-pitched communication signals returning from distant targets. They may have used this initial form of echolocation to localize distant landing sites. In the beginning, the pre-bats probably emitted the localization signals just before jumping to probe the distance to a landing site as is still seen in other mammals with primitive sonar systems (see Henson and Schnitzler 1980; Thomas and Jalili, chapter 72, this volume). With increasing glide distances, the pre-bats may have continued signal emission while in air for better landing control.

Due to the limited operational range of echolocation, the pre-bats had to rely on their spatial memory and possibly also on vision to determine the correct direction for takeoff for longer gliding flights to feeding sites on neighboring trees. However, we hypothesize that the pre-bats used echolocation during the gliding flights to control landing and perhaps also to avoid obstacles in their path as much as their limited maneuverability allowed. We further hypothesize that the pre-bats climbed through the trees to inspect for feeding sites after they landed. We propose that they detected and located prey in the passive mode, first approaching the spot where the acoustic cues originated and then gleaning food from the surface of the vegetation. Fundamental to this scenario is our assumption that gliding from tree to tree saved much more energy than the costs expended for the production of echolocation signals.

We agree with Norberg (1994) that gliding evolved primarily for transportation to maximize net energy gain and not to catch aerial insects. However, the limited maneuverability during glides prevented pre-bats from directly reaching all sites where they had detected and localized potential prey. This disadvantage could have resulted in a strong selection pressure on gliding performance, eventually leading to the development of controlled flapping flight.

The development of flight increased the maneuverability of the pre-bats and ultimately led to the evolution of gleaning microchiropteran bats that were able to fly, guided by echolocation, directly to sites where an insect revealed its position acoustically. As seen in extant bats (e.g., *Tonatia* and *Trachops*, Phyllostomidae; Schnitzler, Denzinger, and Kalko 1995; Servatius 1997), insects crashing into vegetation could have been especially attractive to evoke such flights. After landing, the early bats then gleaned the prey from the substrate. Consequently, they would not have to climb on trees to inspect their feeding sites, but could now listen for prey from perches, then approach the food site directly, catch the prey, and return to the perch to process the food. Furthermore, this new step also diminished the risk of predation, particularly when hunting for prey on the ground. We hypothesize that like their gliding ancestors, early perch-hunting bats continued to use echolocation to con-

trol their landing at the site with prey and at their perch and to avoid obstacles in their flight paths.

With increasing maneuverability, bats needed more spatial information about their foraging area. This demand could have fostered the use of echolocation in bats to characterize landmarks in the darkness and to integrate this information into the cognitive map of the bats' home range. Therefore, we propose that increasing specialization in the echolocation system allowed bats to use echolocation to orient in the foraging area, to find perches and roosts, and to use caves as roosts.

Thus far, we have postulated that the early bats used echolocation only to describe stationary extended targets such as landing sites, sites with prey, obstacles, and landmarks but not to detect, classify, and localize flying insects. We hypothesize that the use of echolocation for recognition and acquisition of prey (active mode) represents a late step in the evolution of echolocation systems. Simmons and Geisler (1998) also hypothesized that at first a simple form of echolocation with short laryngeal pulses evolved, permitting obstacle avoidance and orientation. The detection and tracking of airborne prey is thus a derived behavior. However, in contrast to our scenario, Simmons and Geisler (1998) assumed that flight evolved before echolocation, and that the first bats used vision for orientation in their arboreal/aerial environment.

The foraging behavior of extant bats can give clues about how the transition from passive to active mode was achieved. Extant gleaners waiting for prey at perches not only initiate foraging flights in response to acoustic cues from insects that are sitting on surfaces or crashing into vegetation, but they can also be alerted by the flight tone of large flying insects. For instance, when hearing a flying katydid at close range, *Trachops cirrhosus* (Phyllostomidae), hanging at a perch, immediately emits echolocation signals and begins pursuit. It often catches the insect in the air or overwhelms it by pushing it down onto the vegetation or ground (Schnitzler, Denzinger, and Kalko 1995). The early perch-hunting gleaners also may have encountered situations where the acoustical cues indicating a landing insect or an insect crashing into the vegetation were preceded by the flight tone of the insect. Bats that associated flight tones of large insects with prey and reacted toward them with foraging flights may have had a selective advantage. They most likely emitted echolocation signals while approaching this moving sound source in a way similar to their approach of stationary sound sources indicating prey on surfaces. We propose that with the evolving ability in those bats to perceive echoes of flying insects and to distinguish them from background echoes, bats could now approach prey directly under the guidance of echolocation. Upon reaching the insect, the early bats likely used their wings and body to push the insect down onto the vegetation or the ground to overwhelm it, as is still seen in extant bats

Hans-Ulrich Schnitzler, Elisabeth K. V. Kalko, and Annette Denzinger

(Servatius 1997). Finally, we postulate that these bats eventually evolved to catch insects in the air, without landing on vegetation or the ground, and to return with the prey to their perch to eat it. However, we hypothesize that these bats only pursued flying insects after being alerted by the flight tone. Presumably, they did not scan the surroundings with echolocation signals to detect prey in the active mode. Such behavior still is seen in some extant gleaning bats hunting at least partly from perches (e.g., *Trachops cirrhosus*). These bats occasionally go for flying insects and emit echolocation signals only after being alerted. Exceptions are the hipposiderids and rhinolophids, which have a highly specialized echolocation system that, in our opinion, evolved later, after the bats' transition from the passive to active mode of prey detection (see below).

In a further step of our scenario, bats began to search for flying insects upon leaving their perches for short foraging flights along vegetation and in gaps while using echolocation for initial detection of prey. This innovation completed the transition from the passive to the active mode of prey detection in bats and from the gleaning of stationary prey to the aerial hawking of flying insects. Subsequently, echolocation was not only used for the evaluation of stationary, extended targets such as landing sites, obstacles, and landmarks but also for the perception of small moving targets like flying insects. Finally, some bats evolved out of the perch-hunting mode and foraged for flying insects in continuous search flight ever more in the open, a behavior typical for extant "edge-and-gap bats" and "open-space bats."

Each additional echolocation task performed by bats resulted in new search-and-foraging modes that allowed the exploitation of new niches. However, the phylogenetically older behaviors were conserved in some lineages and are still seen today. For instance, while all microchiropteran bats use echolocation for landing control, obstacle avoidance, and spatial orientation, echolocation is additionally used in many, but not all, bats for prey perception.

Multiple behaviors presumably indicate a mix of older and newer styles as well as considerable evolutionary plasticity. For example, habitat types can range from uncluttered, to background-cluttered, to highly cluttered space (Schnitzler and Kalko 1998). Although some bats mostly use one foraging habitat—for example, *Tadarida* (Molossidae), which hunts for insects in uncluttered space, or *Tonatia* (Phyllostomidae), which gleans insects from surfaces in highly cluttered space—many other bats are very flexible in their foraging behavior and switch among foraging habitats (e.g., Fenton 1990). For instance, some species that mainly capture insects in the gleaning mode in highly cluttered space also forage for prey in the aerial mode in background-cluttered space. Many bats that mostly forage in background-cluttered space also search for insects in uncluttered space. For

instance, *Trachops cirrhosus* (Phyllostomidae), which mostly gleans insects and small vertebrates from the ground in the gleaning mode, also catches flying insects in the aerial mode. Other bats search for insects from perches, as well as in continuous search flights (e.g., *Rhinolophus ferrumequinum;* Rhinolophidae).

The evolution of aerial hawking in continuous search flight improved the ability of bats to radiate to new niches. For instance, we hypothesize that open-space bats with their good dispersal ability could have colonized large areas, even continents, subsequently radiating into a wide range of niches. A good example is the radiation of the Noctilionoidea, which are restricted to the New World. Like J. D. Smith (1976), we assume that they probably originated from an Old World aerial insectivore that was the common ancestor of all bats of the families Mormoopidae, Noctilionidae, and Phyllostomidae. Aerial hawking of insects is found in all three groups. Additionally, most phyllostomids are gleaners. They readapted to what we describe here as the phylogenetically older foraging behavior by detecting prey in the passive mode and employing the gleaning mode to take insects, arthropods, and small vertebrates from surfaces. Some of these phyllostomid gleaners evolved the use of olfaction—and, if suitable, also the use of echolocation—to find *distant* fruits and flowers (Kalko and Condon 1998; von Helversen and von Helversen 1999; Thies, Kalko, and Schnitzler 1998).

To our understanding, gleaning and perch hunting in phyllostomatid bats and possibly also in most other gleaners (maybe with the exception of Nycteridae and Megadermatidae) represents a secondarily derived trait. This corresponds well with the opinion of Simmons and Geisler (1998), who argue, based on the examination of fossil forms basal to extant microchiropterans, that the common ancestor of living microbats used continuous aerial hawking. We additionally suggest that these early aerial hawking bats descend from perch-hunting gleaners as described above.

Proposed Scenario of the Evolution of Signal Structure in Bats

Comparative studies in extant bats reveal that in evolution echolocation has been adapted to ecological constraints set by habitat type, foraging mode, and prey (e.g., Aldridge and Rautenbach 1987; Fenton 1990; Neuweiler 1989; Schnitzler and Kalko 1998). Similarities in echolocation tasks correlate with similarities in signal structure. Not only are the structures and patterns of search signals intimately linked to habitat type and foraging mode, but so are motor behaviors. The intimate linkage between bat behavior and ecological constraints makes it possible to use the preferred foraging habitat and strategy to characterize subgroups of bats. The members of each subgroup must confront and solve a common set of constraints and tasks imposed on their sensory and motor systems (Schnitzler and Kalko 1998, 2001).

Open-space bats mainly forage for insects in the open and emit rather long and narrowband signals of low frequency that are adapted for long-range detection of insects. *Edge-and-gap bats* search for flying insects mainly near edges, in gaps, or near the ground and often emit mixed search signals of medium duration consisting of a narrowband component of medium frequency and a frequency-modulated component of broad bandwidth. These signals are adapted for medium-range detection of insects as well as for orientation in relation to background targets. *Narrow-space bats* either glean their food from the substrate or capture prey close to it. In both situations, the prey echo overlaps with the clutter echoes from the substrate. This overlap is likely to mask important information. To solve this problem, two behavioral strategies have evolved. We categorize both of them according to the signal types that are associated with these specific behavioral strategies. (We put the abbreviations of these signal types between quotation marks to avoid misinterpretations [Fenton 1999] and to make clear that they indicate a behavior strategy.) *Narrow-space "CF" bats* search for fluttering insects mostly in narrow spaces often very close to vegetation or to the ground. They use signals consisting of a long component of constant frequency (CF) followed by a terminal frequency-modulated (FM) component. With their long CF-FM signals and their specialized hearing system, these bats are able to discriminate prey echoes from unmodulated background echoes by evaluating flutter information. *Narrow-space "FM" bats* are gleaners that take food from surfaces and mainly use prey-generated acoustic cues (e.g., animal-eating bats) or olfactory cues (e.g., frugivorous and nectarivorous bats) to detect, localize, and classify their prey. While flying in narrow space near vegetation or the ground, these bats emit broadband uni- or multiharmonic FM signals of short duration and low sound pressure level (SPL), which are used mainly for spatial orientation. Therefore, we propose to name these signals "orientation signals" instead of search signals. Only if a food target delivers a specific echo (e.g., flowers that act as acoustic reflectors) can echolocation also be used for the recognition of food (von Helversen and von Helversen 1999). The structure of the search and/or orientation signals is generally species-specific and also reflects phylogenetic relationships. Additionally, signal structure varies in each species, depending on the echolocation tasks confronting the bat. For example, signal structure and pattern change when a bat approaches a target (Kalko and Schnitzler 1998, 2001). For any given bat species, the high flexibility of signal generation is demonstrated by its ability to drastically change signal structure according to the problems that have to be solved.

We hypothesize that during evolution, high flexibility in signal generation led to high variation in signal structure. This variability likely resulted in the selection of signal types and patterns that are appropriate and well adapted to a variety of new tasks. This gives us the possibility of speculating about the type of echolocation signals used by the ancestors of extant bats. Our fundamental assumption is that echolocation started in nocturnal arboreal insectivores that relied on their ears to find distant prey in the passive mode. Near the prey, olfaction may have been used for further discrimination. These animals presumably had a well-developed hearing system and a flexible laryngeal system for the production of variable high-frequency communication signals, both prerequisites for the development of echolocation. We hypothesize that the pre-bats adapted their laryngeal signals to specific tasks so that the available hearing system could extract the relevant information from the returning echoes.

The evolution of echolocation was additionally favored by the increase of auditory threshold during vocalization, resulting of a contraction of the middle-ear muscles and central attenuation mechanisms (Suga and Shimozawa 1974). This is a characteristic that bats have in common with nonecholocating mammals (Carmel and Starr 1963); it reduces forward masking of subsequent acoustic signals and thus allows bats to perceive the echoes shortly after the emission of a louder echolocation signal. Following our proposition that echolocation primarily evolved to perceive landing sites, obstacles, and later also landmarks, it is most likely that the echolocation signals of the pre- and early bats were similar to those used by extant gleaners—such as nycterids, megadermatids, and phyllostomids—for orientation in narrow space. We therefore hypothesize that the pre-bats used some kind of tonal, low-intensity, short, multiharmonic, and broadband signals well suited to detect, localize, and later on to characterize large, extended targets at short ranges. These tonal signals were produced by the larynx and thus differ fundamentally from the low-intensity clicks produced by other echolocating mammals that click with their teeth or tongue.

When the gleaning ancestors progressed from the passive to the active mode and started to hunt with echolocation for flying insects at edges and in gaps and finally in the open, the echolocation signals were adapted to the new habitat-specific echolocation tasks. The result of this adaptation process is reflected in the structure of search signals in extant bats—for example, the rather long narrowband signals of open-space bats that are optimized for long-range detection.

Comparative studies on the signal inventory of various species of extant bats, including studies of the ontogeny of echolocation signals, indicate how changes in signal structure could have occurred during evolutionary times. All signal types found within a single species of bat, as well as within related species, genera, and even families, can be arranged in such a way that a continuous transformation from one signal type to another becomes apparent. Furthermore, it was demonstrated that during ontogeny, echolocation signals develop from communication signals by a continuous change of signal parameters (Moss 1988; Moss et al. 1997). We therefore conclude that during evolution bats developed a new signal type by continuously changing signal parameters such as duration, frequency structure, SPL, modulation rate, and harmonic content of a basic, phylogenetically older signal in such a way that new echolocation tasks could be mastered. We compare the "older" signal to a moldable structure that can be transformed into a signal type best suited to deliver the information needed for new echolocation tasks. According to this *vocal-plasticity hypothesis,* echolocation signals reflect a phylogenetically determined basic call structure shaped by specific ecological conditions.

The evolution of the echolocation systems of narrow-space "CF" bats is a good example to illustrate our proposed vocal-plasticity hypothesis. The ability to use long CF-FM signals in combination with Doppler-shift compensation and an auditory fovea to evaluate flutter information from insects flying in cluttered environments evolved independently in both Old World bats (Rhinolophidae, Hipposideridae) and in one New World bat (Mormoopidae: *Pteronotus parnellii*). Rhinolophids and hipposiderids, which hunt in the aerial and gleaning mode for fluttering insects in narrow spaces, are together with the megadermatids and nycterids part of the superfamily Rhinolophoidea. Nycterids and megadermatids are narrow-space "FM" bats that partly hunt from perches and mainly glean their prey in the passive mode. Addition of a long CF component to the upper end of the second harmonic of the presumed ancestral, short, steep FM signals could have enabled a nycterid or megadermatid-like ancestor, which initially foraged in the passive mode from perches, to find fluttering insects in the active mode in a highly cluttered environment. Many rhinolophids and hipposiderids sometimes also wait for prey at perches, which supports the hypothesis that they descended from a perch-hunting gleaner. While waiting for prey at perches, they scan the surroundings with their long CF-FM signals to detect fluttering prey in a highly cluttered environment, a behavior that is not found in gleaning narrow-space FM bats. In contrast, the New World mormoopid *P. parnellii* could have evolved from an aerial insectivorous ancestor foraging in background-cluttered space at forest edges and in gaps similar to all other extant species of this genus. We propose that this ancestor acquired the ability to forage for fluttering insects also in highly cluttered space close to or within vegetation by lengthening the short CF component of a basic CF-FM signal, similar to those that are found in the smaller species of this genus. Another argument for the descent of *P. parnellii* from an aerial in-

sectivore instead of a gleaning ancestor is the observation that the perch-hunting strategy frequently used by rhinolophids and hipposiderids has not been observed in *P. parnellii,* suggesting that it continuously hunts on the wing. These scenarios for the evolution of narrow-space "CF" bats, with their ability to evaluate flutter information, are consistent with the recent phylogeny of the respective families (Conway and Simmons 1999; Simmons and Geisler 1998).

Comparison with Other Proposed Scenarios

Other published scenarios for the evolution of echolocation and flight in bats partly overlap with our view, but they also contain substantial deviations, particularly in the proposed sequence in which flight and echolocation may have evolved (see Arita and Fenton 1997; Denzinger, Kalko, and Jones, chapter 42, this volume).

The *echolocation-first hypothesis* (Fenton et al. 1995) states that aerial hawking with laryngeal echolocation signals evolved before flapping flight. This scenario implies that bats arose from gliding, nocturnal insectivores that used short, broadband clicks of low intensity for spatial orientation and that echolocation with stronger, tonal signals produced in the larynx evolved primarily for prey acquisition in perch-hunting pre-bats that attacked their prey while gliding. To our opinion, it is unlikely that the evolution of echolocation started with clicks and that the switch to tonal, laryngeal signals occurred in the proposed way. Echolocation based on clicks, rather than tonal signals, would have led to specific adaptations in the sound production and hearing system, making it difficult to switch later to an echolocation system operating with tonal signals. A click system has the disadvantage that its short signals are uniform and of low energy and cannot be adapted to deliver the information necessary to perform sophisticated echolocation tasks. The echolocation system of *Rousettus* is a good example for this limitation (Henson and Schnitzler 1980). The great success of microchiropteran echolocation systems was only made possible by a flexible laryngeal system producing tonal signals that could be adapted to specific tasks by changing their duration, frequency structure, and sound pressure level. This advantage also was recognized by Fenton et al. (1995). In contrast to Fenton et al. (1995) and in accordance with Simmons and Geisler (1998), we hypothesize that the evolution of echolocation already started with tonal signals and not with clicks. To reach a high performance with clicks, high-energy signals and a specialized hearing system are needed, as found in dolphins. Their loud-click signals are produced in the nasal sac system and not in the larynx, and their auditory system is specialized to evaluate echo information mainly in the time domain (Au 1993).

In their scenario, Fenton et al. (1995) argued that stronger signals and adaptations to minimize "self-deafening" (or forward masking in technical terms) were central to the perfection of echolocation for the perception of flying prey. In contrast, we hypothesize that echolocation was never range-limited due to low energy signals, as the pre-bats presumably already started with laryngeal signals that could be varied in SPL according to which task had to be performed. We further believe that the vocalization-induced threshold increase was already present before bats started to hunt for flying insects, because this attenuation also is found in non-echolocating mammals (Carmel and Starr 1963). In extant bats, the maximal attenuation of the perceived emitted signal reaches 35–40 dB (as described for *Myotis grisescens* by Suga and Shimozawa 1974) and reduces masking of subsequent echoes.

Speakman (1999) pointed out that the echolocation-first hypothesis includes a hypothetical phase where the pre-bat hunted by intercepting insects as they flew past a perch. He called this behavior "reach hunting" and discussed the energy gains an animal could achieve when using this foraging strategy. Even under the most favorable conditions his model suggests that the evolution of bats is unlikely to have included a period of reach-hunting behavior.

In contrast to the echolocation-first hypothesis proposed by Fenton et al. (1995), the *flight-first hypothesis* of Norberg (1994) states that flapping flight preceded the evolution of bats' sophisticated echolocation system to catch flying insects. Concerning the evolution of flight, we agree with this scenario of Norberg (1994). It implies that gliding evolved for transportation to maximize net energy gain, that lack of maneuverability precluded the capture of airborne prey, and that aerial hawking was only possible after active flapping flight was attained. Concerning the evolution of echolocation, we disagree with Norberg (1994), as well as with Fenton et al. (1995). Like Fenton et al. (1995), Norberg (1994) suggested a primitive echolocation system for pre-bats that used low-energy clicks for short-range orientation in space and an advanced echolocation system that used louder tonal signals for aerial hawking of flying insects over longer distances. In contrast to these propositions, we hypothesize that the use of echolocation to detect, localize, classify, and catch airborne prey is a late step in this continuous evolutionary process that started when the pre-bats made their first jumps and glides to distant landing sites. We believe that bats possessed a highly sophisticated echolocation system for spatial orientation long before they started to use echolocation to find and assess flying insects. Thus it is misleading to discuss the evolution of echolocation and flight in terms of simple alternatives of whether echolocation for prey acquisition or flight evolved first.

Simmons and Geisler (1998) proposed that after the evolution of flight, bats still used vision for spatial orientation and obstacle detection. According to their scenario, stationary prey was detected by passive means, including vision and/or listening for prey-generated sounds, and then approached in short flights from a perch. Only in the next step did a simple echolocation system evolve, using laryngeal signals that permitted the transition from spatial orientation guided by vision to orientation guided by echolocation. We find it difficult to understand which evolutionary pressure should have forced these early bats that were relying on a fully functional vision-based system for spatial orientation to develop a second orientation system (i.e., echolocation) to perform the same tasks. We find it more likely that the evolution of echolocation started when pre-bats began to jump and later on to glide toward distant landing sites. In the next step of their scenario Simmons and Geisler (1998) assumed perch hunting, where echolocation and vision is used for orientation. Flying prey was detected and tracked by echolocation, whereas stationary prey was found by passive means. However, it remains unclear how the bats could have made the important step from the passive to the active mode of prey perception. Overall, this step and their subsequent steps leading to perch hunting combined with continuous aerial hawking using echolocation, and finally to exclusive reliance on continuous aerial hawking, corresponds well with our scenario.

Simmons and Geisler (1998) argued that flight evolved before echolocation because powered flight evolved only once among Chiroptera. They assume that the many complex postcranial morphological synapomorphies found in Micro- and Megachiroptera indicate their origin from a common flying ancestor. Further, they hypothesize that this common ancestor used vision and not echolocation for orientation. We see the possibility that the two lineages stem from a nocturnal, arboreal, predominantly insectivorous ancestor splitting before the development of flight in a predominantly insectivorous lineage that relied mainly on its ears to find noisy prey and a lineage that used its eyes and nose to find smelling food. These pre-megabats probably foraged for fruits and flowers that advertised their presence by distinct optical and olfactory cues. Therefore, they were not exposed to a strong selection pressure to develop a sophisticated passive hearing system, as was the case for the pre-microbats that were foraging for insects. However, because both forms were under similar selection pressures concerning the reduction of predation risk and the saving of energy, both forms started to jump, glide, and finally fly in flapping flight to distant landing sites. The pre-microbats with their more sophisticated hearing system were preadapted for echolocation and started to localize distant targets in their environment with acoustic signals derived from communication sounds. They continued to use their ears for the passive localization of acoustic cues from hidden prey. The pre-megabats relying on vision and olfaction did not manage to make this step and instead improved their visual system for an efficient orientation at twilight and in the dark. They proceeded to use their eyes and nose to find the advertised food. This presumed split is in accordance with Speakman (1993), who also concluded that Megachiroptera and Microchiroptera divided before the evolution of flight. According to his scenario, flight evolved independently in the Mega- and Microchiroptera, with simultaneous development of echolocation in the Microchiroptera, but the commitment to the visual system in the Megachiroptera precluded the evolution of echolocation in that group.

We are aware that Simmons and Geisler (1998) find it rather unlikely that flight evolved twice within the monophyletic Chiroptera. If we apply their position to our scenario a common ancestor of mega- and microbats would have evolved flight and echolocation in parallel, and later on echolocation was lost in the megabat lineage. We find it unlikely that megabats switched from echolocation, which allows an efficient orientation at night, to a sophisticated night vision system and even redeveloping a rather poor echolocation system operating with tongue clicks as seen in *Roussettus*. New molecular data show that the issue of the phylogenetic origin of megabats is still unclear. Teeling et al. (2000, 2002) and Springer et al. (2001) even suggest that microbat families in the superfamily of Rhinolophoidea are more closely related to megabats than they are to other microbats.

The scenario of Fenton et al. (1995) asserts that aerial hawking started in bats that waited for prey at perches and detected flying insects while scanning the surroundings with echolocation signals. In extant bats, this behavior is only observed in the highly specialized rhinolophids and hipposiderids. All of them belong to the group of narrow-space "CF" bats. They emit long CF-FM signals suited to detecting the glint patterns in echoes from fluttering insects. All other extant perch-hunting bats belong to the group of narrow-space "FM" bats, in which prey detection by scanning with echolocation signals has not been observed so far. We have difficulties in understanding how these early perch hunters should have managed to make the sudden transition to the detection of flying insects by echolocation as proposed by Fenton et al. (1995). We feel that important intermediate steps that enabled these bats to recognize echoes from flying insects remain unexplained. We find it more likely that the early perch hunters were gleaners that listened for prey-generated signals and approached the sites of the sound's origin under the guidance of echolocation. By relying on such signals, the bats may have finally acquired the ability to recognize flying insects by their flight tone. They then presumably approached the sound source while emitting echolocation signals. Thereby, it is likely that these bats made the transition from the passive to the active mode of prey detection. This finally allowed them to recognize insect echoes and to discriminate these echoes from those of background

targets. Extant perch hunters of the group of narrow-space "FM" bats, which sometimes also pursue flying insects, still need the prey-advertising flight tone to get alerted. Only after this stimulus do they start to emit echolocation signals, take flight, and pursue the insect. Hence, we conclude that with the exception of narrow-space "CF" bats, all other bats probably never managed to detect flying insects by echolocation alone while hanging at a perch.

According to the scenario of Fenton et al. (1995) aerial hawking, using echolocation to detect, track, and evaluate prey, represents a primitive foraging strategy. This implies further that gleaning from perches in the passive mode, as found in extant bats, is a derived behavior rather than a retention of primitive habits. We propose instead that gleaning from perches with the proposed transition from passive to active mode is a more primitive strategy and was an important intermediate step in the continuous evolutionary process that led to the echolocation systems of extant bats. However, it is consistent with our scenario that most, if not all, of the modern gleaners actually had aerial hawking ancestors and reverted to gleaning. Simmons and Geisler (1998) also stated that aerial hawking using echolocation to find prey was apparently the primitive foraging strategy for the microchiropteran crown group and that gleaning and perch hunting in extant bats are secondarily derived behaviors. However, they further assumed that perch hunting both for stationary and flying prey preceded aerial hawking in the microchiropteran lineage as a whole.

An argument that plays an important role in the discussion of how echolocation could have evolved is the question whether a reduction of energy costs, gained by the coupling of sound production, ventilation, and wing beat (Speakman, Anderson, and Racey 1989; Speakman and Racey 1991), was a necessary prerequisite for the evolution of echolocation. Speakman (1993) suggested that because of the high costs for the production of signals in nonflying bats, echolocation was only affordable in combination with flight and therefore must have evolved in parallel. Arita and Fenton (1997) and Simmons and Geisler (1998) further hypothesized that because of the energetically favorable coupling, flight should have evolved even before laryngeal echolocation signals were produced. We think that only a well-founded, comprehensive energy balance, including all presumed behaviors of pre-bats, will permit justifiable assumptions as to whether it was more costly to search for prey while moving on the substrate without echolocation, or to jump and glide to distant targets with echolocation. We assert that pre-bats saved much more energy by jumping and gliding than was needed to produce echolocation signals.

The steps to powered flight and to aerial hawking with echolocation resulted in constraints limiting the further evolution of bats. We agree with Barclay and Brigham (1991) that to catch insects successfully, the rather short detection range of echolocation for insect prey requires high maneuverability and relatively small bodies of bats. This constraint could limit the body size of aerial insectivorous bats. Jones (1994) agrees with Speakman and Racey (1991) that echolocation is only energetically cheap and affordable if signal production is linked to wing beat. He concludes further that larger bats that operate with lower wing-beat frequencies should have lower pulse rates of echolocation calls because of tight coupling with wing-beat rate. He then speculates that this coupling could constrain maximal body size in aerial insectivorous bats, because very large bats would be unable to echolocate at a sufficiently high repetition rate to catch enough insects to meet the high energetic demands associated with their larger body size. We disagree with this argument, because all bats hunting for insects in the open have lower average pulse rates compared to average wing-beat frequencies. The underlying reason for this discrepancy is that these bats regularly skip calling during one, and often even during several, successive wing beats. From field data, we conclude that bats keep their pulse intervals long enough that the detection zones of succeeding pulses just overlap, and that no prey is missed. This results in repetition rates that are clearly below the wing-beat rate. To us, this indicates that coupling of echolocation sound production and wing beat does not represent a major constraint on body size.

Conclusions

We are aware that all proposed scenarios are speculative, since we lack fossils of pre-bats that would allow statements of how echolocation, foraging behaviors, and flight evolved. However, we consider it a valid approach to look at the behaviors and adaptations of extant bats and related groups to generate ideas of how echolocation and foraging behaviors evolved in a plausible continuous process. In contrast to all other proposed scenarios, we suggest that echolocation started with tonal signals produced in the larynx when pre-bats evolved to jump and then to glide to distant targets. Only this highly flexible signal generation mechanism could allow the variation in signal structure, which for each evolutionary step was the prerequisite for the selection of signal types suited for new tasks. In our opinion, the evolution of echolocation in pre-bats represented a key innovation that enabled collision-free gliding to distant landing sites and sites with prey in the dark. This innovation also allowed spatial orientation with landmarks. Subsequently, the transition to the active mode of prey detection represented another key innovation that permitted the early bats to leave perches, and to search for aerial prey in continuous search flights at vegetation edges, in gaps, and finally in the open, thereby promoting the strong species radiation that occurred during the Eocene.

Acknowledgments

We are most grateful to Bob Stallard for reading this manuscript and offering valuable suggestions for its improvement. Further we thank Nancy Simmons (NSF grant DEB-9873663), Gareth Jones, Jeanette Thomas, and Brock Fenton for their critical and constructive review of the manuscript. Our research has been supported by grants of the Deutsche Forschungsgemeinschaft and by trust funds of the U.S. Smithsonian Institutions.

{ 45 }

Do Echolocation Signal Parameters Restrict Bats' Choice of Prey?

Robert D. Houston, Arjan M. Boonman, and *Gareth Jones*

Introduction

The hypothesis that aerial-feeding, echolocating bats are limited in the size range of available prey by the wavelength of their sonar signals has been a key issue in bat sonar research for many years (Griffin 1958). A bat is predicted to have difficulty in detecting the echo of an insect smaller than the wavelength of its echolocation signal as the echo would become vanishingly weak, due to Rayleigh scattering (see Pye 1993). If a proportion of the insect assemblage in a bat's hunting habitat is smaller than its sonar signal wavelength (see insect size distributions in, e.g., Rydell, Entwistle, and Racey 1996), the bat's diet should be restricted to those insects large enough to reflect its sonar effectively. The restriction—that is, the minimum detectable prey size—should be proportional to the wavelength of the bat's sonar signal. Such a constraint on available prey sizes should be evident in the diets of bats. Therefore, we predict that when the diets of bats are examined, the diets of those species using long-wavelength signals for prey detection should contain only large prey.

An insect has such a complex shape that we cannot estimate exactly at what point and how steeply their reflectivity drops when the wavelength of ultrasound striking it increases (Pye 1993). In this study, we aim to determine whether bats experience constraints due to the Rayleigh scattering effect by measuring the ultrasound reflectivity of small insects in the frequency range used by aerial-feeding bats.

Waters, Rydell, and Jones (1995) suggested that signal duration could be an important echolocation parameter that determines the size range of available prey to echolocating bats. If non-Doppler-compensating bats (all bats treated in our study) cannot detect echoes while emitting the outgoing pulse (e.g., Suga and Jen 1976), and behave in the wild so as to avoid pulse-echo overlap (e.g., Schnitzler et al. 1987), signal duration sets a minimum operating range for a particular echolocation signal. Hence, the longer the signal duration, the greater the minimum detection distance for prey and the greater the echo attenuation. Small prey are less likely to be detected from a long distance, hence signal duration could influence prey size available to bats. Therefore, we expect the diets of bats that use long duration signals to lack very small prey, but that those small prey will be present in the diets of bats using short-duration signals.

High frequencies are more strongly attenuated than low frequencies (Bazley 1976), so we expect some interaction between signal duration and wavelength in determining prey detectability. Furthermore, among species of aerial-feeding bats, long duration signals also tend to be long in wavelength (Jones 1999). Therefore, we expect that prey detectability will only be predictable for a signal given both its wavelength and its duration.

There is little quantitative evidence of the sizes of prey taken by aerial-feeding bats. The diets of bats that emit search-phase signals of extremely long wavelength (>30 mm; *Euderma maculatum*, Leonard and Fenton 1984; *Tadarida teniotis*, Rydell and Arlettaz 1994; *Otomops martiensseni*, Rydell and Yalden 1997) are devoid of "small-bodied" taxa like Diptera. This evidence remains qualitative. Some quantitative prey size studies have concentrated on a single common component of the diet that is recoverable whole in the feces (e.g., the eyes of Chironomidae, Barlow 1997; Waters, Rydell, and Jones 1995) to provide a reliable index of the entire body sizes of a single prey type.

A bat's body and wing morphology, and the frequency and duration of its sonar signals, are in fact all related (Jones 1999) and are factors that limit prey detection or availability in echolocating aerial insectivores. Using data presented in this chapter, we quantify the effects of sonar signal duration and wavelength on an aerial-hawking bat's ability to detect prey of different sizes. We then predict the extent to which bats are limited in their diets by the design of their echolocation signals. Accurate data on sizes of prey consumed by bats is then presented as evidence of whether the diets of bats are restricted in the way we predict.

Materials and Methods

TARGET STRENGTH MEASUREMENTS

We ensonified insects with tone pulses at various frequencies and measured the amplitude of the echoes. We produced the signals using the method of Waters, Rydell, and Jones (1995). We adjusted the duration (hold time) of the pulse within the range 0.35–0.55 ms to keep the bandwidth constant at about 10 kHz (−60 dB), independent of absolute frequency. We broadcast the tone pulses from an Ultrasound Advice (23 Aberdeen Road, London N5 2UG, UK) ultrasound speaker (membrane diameter = 45 mm) via an Ultrasound Advice amplifier. We supported a thawed insect target on the intersection of two 0.06 mm diameter nylon wires, with its wings perpendicular to the direction of the sound emitted from the speaker. We mounted a solid dielectric microphone (QMC Instruments model PSM–3; Ultrasound Advice, 23 Aberdeen Road, London N5 2UG, UK) next to the speaker and directed both speaker and microphone toward the insect by eye, by centering the insect's reflections in the transducer membranes. The distance between the speaker-microphone assembly and the target was 400 mm, and the centers of the transducer membranes were 70 mm apart. The microphone sensitivity was 5–10 dB re 1 V Pa^{-1}. We frequently calibrated the components of this system against a Larson Davies (1681 West 820 North Provo, Utah 84601) model 2520 6.35 mm air-dielectric microphone (frequency response ±2 dB 1.0–100 kHz). The signal was high-pass filtered (−3 dB cutoff point: 13 kHz). Thirty-two echoes were acquired for every measurement, and averaged by a Gould model 465 digital storage oscilloscope. This reduced the noise floor of the measurements to approximately 24 dB rms SPL. We measured the averaged peak-to-peak voltage of the echo, and subtracted the averaged peak-to-peak voltage when only the target's supporting wires were present. We recalculated the resulting figures as target strength ($20 \times \log_{10}$[incident SPL/echo SPL at 1 m]) (Waters, Rydell, and Jones 1995), correcting for atmospheric attenuation according to Bazley (1976).

We measured the target strength of individuals of three insect types: small chironomids (Diptera, length of one wing 2.6–3.1 mm, *n* = 5), large chironomids (wing length 4.0–5.0 mm, *n* = 7), and caddis flies (Trichoptera, wing length 8.0–9.0 mm, *n* = 5) at 19 frequencies between 20 and 85 kHz.

Fecal Analysis

We measured the eye diameter and the length of one wing of a sample of 100 chironomids collected by sweep net from a site in southern England used by bats for foraging. The correlation between eye diameter and wing length was tight (Pearson r^2 = .893). The straight line of best fit was computed.

We analyzed the diets of a range of European aerial-feeding bats that use narrowband, FM-CF or QCF (quasi CF, Kalko and Schnitzler 1993) search-phase signals at different frequencies. Samples of bat feces were collected from maternity roosts in Sweden (*Eptesicus nilssoni*), Ireland (*Pipistrellus nathusii*), France (*Vespertilio murinus*), and England (*Nyctalus noctula, N. leisleri, E. serotinus, P. pipistrellus,* and *P. pygmaeus*). We examined a sample of approximately 50 pellets from each species, and measured the diameter of every chironomid eye found in each pellet. We estimated the wing length of the intact insect from the eye diameter, using the equation of the line of best fit calculated from the whole chironomid specimens. The "minimum" prey size eaten by the bat was estimated by quoting the lower tenth percentile of the distribution of estimated chironomid wing lengths.

Calculations

From the target strength data we calculated the echo SPL received by a bat at various distances. Absolute dB SPL calculations used 57 μPa as a reference so that our peak-to-peak amplitude measurements were comparable to peak-equivalent decibels (dB peSPL). The calculation incorporated estimates of atmospheric attenuation (Bazley 1976). Using an assumed bat echolocation signal source level of 110 dB peSPL (Surlykke et al. 1993) and echo detection threshold of 0 dB ppSPL (Kick 1982), we estimated maximum detection distances of the insects.

We collected search-phase signal parameters (frequency, *f*, and duration, *d*) from the literature, typical of each bat species considered in the fecal analyses (table 45.1). We calculated a minimum detectable insect size (wing length) for each bat species according to that species' value of *d* and *f* (table 45.1) as follows. We divided the maximum detection distance data into eight subsets corresponding to each bat's signal frequency (*f*) (for example, 20 kHz in the case of *N. noctula*). Within each subset, we computed the line of best fit for the relationship between the maximum detection distance and insect wing length. Assuming the bats concerned cannot detect echoes while emitting pulses, the *minimum* detection distance is ($c \times d$)/2, where *c* is the speed of

TABLE 45.1. Peak frequency and signal duration for eight bat species emitting QCF (quasi-CF) pulses in their natural habitats. Values are averages of echolocation parameters recorded in open environments for each species, collected by the following sources: (1) Ahlén 1981; (2) Kalko and Schnitzler 1993; (3) Rydell 1993; (4) Vaughan, Jones, and Harris 1997; (5) Waters, Rydell, and Jones 1995; (6) Zbinden 1989; (7) Zingg 1990.

Species	Peak frequency (kHz)	Signal duration (ms)	Sources
Nyctalus noctula	20	19	1, 4, 6, 7
Vespertilio murinus	25	14	1, 7
Nyctalus leisleri	25	10	4, 5, 7
Eptesicus serotinus	27	11	1, 4, 7
Eptesicus nilssoni	29	12	1, 3, 7
Pipistrellus nathusii	39	8	2, 4, 7
Pipistrellus pipistrellus	46	6	2, 4, 7
Pipistrellus pygmaeus	54	5	1, 4, 7

sound (ms^{-1}). The point at which the line of best fit exceeded this minimum detection distance was taken to be wing length of the smallest insect detectable.

Results

The reflectivity of the insects decreased when the wavelength of the emitted signal exceeded the length of the insect's wing (fig. 45.1). Low-frequency ultrasound (20–30 kHz) reflected poorly from the smaller insects (wing length 2.5–5.0 mm). The *average* slope of the relationship between target strength and signal frequency was measured as about 9 dB per octave. However, in the case of the large midges (chironomids) and the caddis flies, there is evidence of a wavelength-insensitive region (wing length : wavelength >1.3 in caddis flies and >0.8 in large midges).

When recalculated as the sound pressure level (SPL) of echoes at various distances from a hypothetical bat, the 9 dB per octave high-pass filtering effect is opposed by the low-pass filtering effect of excess atmospheric attenuation. The gradient of the low-pass filtering is proportional to the distance from the bat. This information is summarized in fig. 45.2, which describes estimated maximum detection distances of all the insects measured as a function of echolocation signal frequency. The net result of the high-pass filtering is seen in fig. 45.2, mixed with different amounts of low-pass filtering at different distances from the bat. The bat can be imagined as positioned anywhere on the *x*-axis, directing the axis of its sonar system parallel with the *y*-axis. The area of the graph below the dotted line of pulse-echo overlap is inaccessible to the bat's sonar system. The small and the large midges drop below this line (and become undetectable) at 36 and 25 kHz, respectively—partly due to Rayleigh scattering and partly due to the longer duration of low-frequency echolocation signals. In general,

Fig. 45.1. Insect target strength when ensonified by tone pulses at frequencies between 20 and 85 kHz. The *x*-axis shows the ratio between an arbitrary linear dimension of the target (wing length) and the wavelength of ultrasound. "Small midges" (chironomids): wing length 2.6–3.1 mm; "large midges" (chironomids): 4.0–5.0 mm; caddis flies: 8.0–9.0 mm.

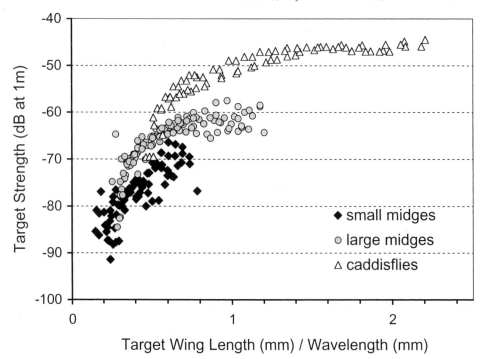

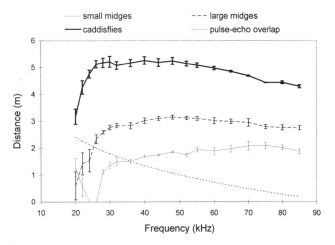

Fig. 45.2. Estimated maximum target detection distance by a hypothetical bat as a function of its signal frequency. The estimates are made assuming a bat source level of 110 dB peSPL (Surlykke et al. 1993), an echo detection threshold of 0 dB ppSPL (Kick 1982), and atmospheric attenuation figures at 12°C and 80% relative humidity (Bazley 1976). The pulse-echo overlap line (dashed) depicts the modeled relationship between frequency and duration in aerial-feeding vespertilionid search signals; $d = 40.71 - 8.91 \cdot \log_e(f)$, where d = duration (ms) and f = frequency (kHz) (Waters, Rydell, and Jones 1995). Standard error bars describe only the variation introduced by the target strength measurements ($n = 5$ small midges, $n = 7$ large midges, $n = 5$ caddis flies).

the effect of Rayleigh scattering leads to a steep decrease in maximum detection distance as frequency decreases below 30 kHz. Above this frequency, Rayleigh scattering by large midges appears to approximately balance filtering by atmospheric attenuation at a little less than 3 m. The optimum frequency for detecting these

large midges is about 48 kHz, although any frequency between 32 and 85 kHz is only slightly suboptimal. At less than 2 m, Rayleigh scattering by small midges exceeds atmospheric attenuation, as shown by the positive slope of the small midge detection distances. The optimum frequency here is about 70 kHz, although nothing greater than 40 kHz is very suboptimal. Those insects predicted to be detectable at a greater distance (the caddis flies) show a greater influence of atmospheric attenuation (i.e., low-pass filtering) in their curves, such that insects are more detectable at lower frequencies (30 kHz). However, frequencies up to 50 kHz are almost equally suitable for detection of caddis flies. These optimal frequencies are sensitive to changes in the estimates of bat source level and echo detection threshold.

The size of the smaller insects found in the diet (the tenth percentile of the size distribution) was not significantly correlated with signal wavelength (Spearman Rank Correlation Coefficient (df = 7) $\rho = 0.711, 0.10 > p > 0.05$). However, it was correlated with signal duration ($\rho = 0.778, 0.05 > p > 0.025$), and bat forearm length ($\rho = 0.892, 0.025 > p > 0.01$). All three predictors are correlated with one another.

Predicted wing lengths of the smallest insect detectable by each bat species (given a particular species-specific signal wavelength and duration) is compared to the observed minimum (tenth percentile) wing length in the diet of each species (fig. 45.3). The values of predicted wing lengths rely on the combined effect of both signal wavelength and duration, plus atmospheric attenuation, bat source level, and echo detection threshold. A rank correlation between predicted and observed prey sizes is significant ($\rho = 0.851$, and $p < 0.025$), and the correlation is closer than that between observed prey

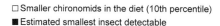
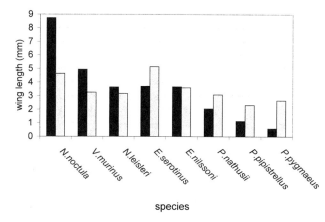

Fig. 45.3. Comparison of the estimated minimum insect size detectable and the minimum insect size found in the diet for each of the bat species whose diet we analyzed. Species are arranged from those with the longest wavelength signals (on the left) to those with the shortest wavelengths (on the right). Predicted minimum insect sizes are calculated according to the signal parameters in table 45.1.

size and either duration or wavelength alone. There is absolute correspondence between the predicted and actual prey sizes (fig. 45.3), but it is not constant across all the bat species studied. The species in figure 45.3 are arranged from those with longest wavelength signals on the left, to the shortest on the right. The correspondence is good in the middle of this range, but the prey distributions of the "short-wavelength" bats do not extend to insects as small as those predicted, while those of the "long-wavelength" bats extend well beyond the smallest insects predicted to be detectable.

Discussion

SIGNAL DESIGN AND PREY DETECTION

The insect reflectivity measurements and our predictions about signal design and target detection seem largely to confirm previous theoretical work. We confirm that sounds at the wavelengths used by bats result in Rayleigh scattering from insects (fig. 45.1), leading to high-pass filtering of insect echoes. While the shorter wavelengths tested (4–8 mm, 40–85 kHz) reflected relatively well from the insects, long wavelengths (14–17 mm, 20–25 kHz) reflected poorly.

From our measurements, we predict that high-pass filtering would be evident in echo amplitudes at close range (e.g., <2.5 m), but atmospheric attenuation would become a more important filtering effect at greater distances (fig. 45.2; Hartley 1989; Kober and Schnitzler 1990). We found that the combination of high-pass and low-pass filtering would lead to a band-pass effect. A target would be detected at maximum range by a signal of a particular optimum frequency specific to the target size. This optimum frequency should decrease as target

size increases; but as measured here, the effect is weak. The optimum frequency is not distinct, and a wide range of frequencies is appropriate for detecting each size of insect tested.

A long-wavelength signal is well suited to detecting large targets at long range; it is not suited to detecting small targets. A short-wavelength signal is suited to detecting a wider range of target sizes, but is limited to shorter ranges by atmospheric attenuation. As a minimum operating distance is imposed by pulse-echo overlap, long signal duration introduces another limit on the detectability of small targets. Among aerial-feeding bats, signal duration increases as wavelength increases (Jones 1999), so the effect of long signal wavelength and long signal duration are combined in many bat species. Barclay's characterization of "long-range" and "short-range" echolocation strategies (Barclay 1986) is therefore a useful one, at least as a starting point when studying the ecological adaptation of echolocation signals for target detection. The signals used by bat species considered in this study (table 45.1) could be described as representing long-range strategy (*Nyctalus noctula*), short-range strategy (*Pipistrellus pygmaeus*), or a number of intermediate states.

RESTRICTION OF PREY AVAILABILITY AND DIET

Is there evidence that these predicted effects of signal design restrict the diets of bats? We must disentangle numerous interrelated factors: bat body size and maneuverability, echolocation signal duration and wavelength, and prey detectability and diet. That so many factors are correlated is a pitfall of correlational studies, but we aim to identify which relationships are functional and which are not—not only by correlating but also by making quantitative predictions.

At the first stage of our analysis, the closest correlation is that between prey size and bat forearm length ($\rho = 0.892$). Among aerial-feeding bats, forearm length scales positively with body mass and wing loading, which are related negatively to maneuverability (Norberg and Rayner 1987). Because body size, signal wavelength, and signal duration also are correlated, bats with long-wavelength and long-duration echolocation signals are also less maneuverable. Maneuverability of bats with long-wavelength, long-duration signals could limit availability of small prey (Barclay and Brigham 1991).

However, on further analysis, illustrated by figure 45.3, we amalgamate signal wavelength and duration with some physical calculations, to produce a single predictor: "minimum detectable prey size." This predictor is highly correlated with the observed pattern of minimum prey sizes ($\rho = 0.851$, $p < 0.025$). Using minimum detectable prey size, we can not only correlate predicted values with observed data, we can also compare the absolute values. Indeed, there is close agreement between the absolute predicted and observed values of minimum prey size. This suggests that the wavelength and duration of signals used by bats for prey detection have a tangible and quantifiable effect on the prey available to them.

Both echolocation signal parameters and body size (hence maneuverability) are correlated with observed minimum prey size. However, we have predicted the quantitative effect of echolocation signal parameters on prey size, and we have found both correlation and absolute agreement with observed values of minimum prey size. The sensory constraint on prey detectability would remain, no matter what the body size of the bat. To add further evidence for this, it would be interesting to study species of bat that possess unusual combinations of traits (e.g., small bats with long wavelength signals).

There is some deviation between the observed and predicted prey sizes (fig. 45.3). The bats with the longest wavelength signals (*Nyctalus noctula, Vespertilio murinus*) defy predictions and manage to eat prey smaller than we predict they can detect. On the other hand, the bats with shortest wavelengths (*Pipistrellus pipistrellus, P. pygmaeus*) do not take insects as small as the minimum predicted size available. When a bat's signal performs well at detecting small targets, the observed minimum prey size may be limited not by detection constraints, but by the size distribution of insects in the bat's habitat, or by the response latency or maneuverability of the bat. Note that the zone of pulse-echo overlap extends to only 0.85 m for a 5 = ms *Pipistrellus* signal. Furthermore, bats might actively reject very small and unrewarding prey items (Houston and Jones, chapter 48, this volume). The departure from the predictions of the bats with long-wavelength signals is perhaps more remarkable, and we will return to this problem at the conclusion of our discussion.

INSECT REFLECTIVITY

The present results suggest that the reflectivity begins to decrease when wavelength exceeds the length of the insect's wing (fig. 45.1), although the precise point differs between midges and caddis flies. The target strength of insects measured in our study increased by an average of 9 dB per octave, compared to the 12 dB per octave predicted for spheres. The absolute values of target strengths are comparable to those measured by Waters, Rydell, and Jones (1995). For reference, a sphere of radius 2.0 mm should have a target strength of −60 dB ($TS = 20 \cdot \log_{10}[radius/2]$; see Møhl 1988). The 10° separation of the microphone from the loudspeaker leads to a slight underestimation of target strength. On the other hand, the insect was positioned with its wings—its largest reflecting surfaces—perpendicular to the impinging sound waves, so reflecting the most intense echo possible (a "glint" in the sense of Schnitzler et al. 1983).

DETECTION MODEL

When using the target strength data to estimate the range of bat sonar for various targets, it was necessary to make some assumptions. Our estimation of minimum

detectable prey sizes for bat species (fig. 45.3) rests on supposedly characteristic species-specific signal parameters. Many field-workers since Griffin (1958) have noted behavioral variation in signal design of individual bats, even in their search signals. Below we justify further assumptions of our model and make suggestions for future improvements. Some refinements to or adaptations of the model could lead to greater agreement with the field data and stronger evidence that signal design determines prey availability and diet.

Echo detection threshold

The threshold chosen for inclusion in the model was 0 dB ppSPL, arrived at by estimating the echo SPL of real targets (spheres) (Kick 1982) at the bat's detection threshold. Kick's detection threshold is the lowest figure for echoes available in the literature, and is perhaps the study least limited by clutter. This latter feature is desirable, because the aerial-feeding bats considered in our chapter generally are not limited by clutter during foraging. However, the low noise conditions in Kick's laboratory, and the bat's facility to integrate echoes of a stationary target over several pulses (Surlykke, chapter 37, this volume) might have lowered the bat's threshold in Kick's study. A wild bat in flight would have a higher threshold if it were unable to integrate echo energy in this way. The threshold is also modified at echo delays close to pulse-echo overlap, due to contraction of the middle ear muscles (Suga and Jen 1976).

Bats seem to integrate echo energy within a time constant, when measured by the two-click method (e.g., Surlykke and Bojesen 1996). Energy integration leads to a 3 dB improvement in detection threshold per doubling of signal duration. Bats using longer duration signals might gain an advantage in echo detection if they adapt their energy integration time to the length of each outgoing pulse.

Directionality

We assume that bats in detection experiments aim the axis of their sonar system (emission and reception) in the exact direction of the target to optimize detection. The detection estimates presented here assume targets are always encountered on-axis in the wild, which of course they are not. Due to directionality of emission and reception of their signals, bats would have a higher threshold for echoes received off-axis. Long-wavelength signals are less directional and less prone to lower off-axis detection probabilities.

Source level

Recent field studies have added to knowledge of the variation in source levels used by bats. Among "high-intensity" aerial-feeding bats, typical values range from 100 (*Myotis siligorensis*; Surlykke et al. 1993) to 121–125 dB peSPL at 10 cm (*Eptesicus serotinus*; Jensen and Miller 1999). Variation in source level could be another factor that increases maximum detection distances (fig. 45.2) and decreases minimum detectable prey size (fig. 45.3) for long-wavelength, long-duration signals.

Conclusions

Although the bats with signals of long wavelength and duration ate larger prey than the bats with short-wavelength, short-duration signals, the former group of bats nevertheless ate much smaller prey than we predicted (fig. 45.3). Their signal design restricts the availability of small prey relative to that of bats with shorter wavelength and duration signals, but the restriction is not what we predicted in absolute terms. We might find a closer agreement after some or all of the refinements to the detection model suggested above. However, the conclusions we make about detection performance apply to a given signal only, of given wavelength and duration. Most of the bats in question in fact show a great deal of behavioral flexibility in search-signal design (e.g., Jones 1995b; Kalko and Schnitzler 1993; Rydell 1993). The search signals of *N. noctula,* in particular, can range from those of peak frequency 18 kHz and 25 ms in duration, to those peaking at 27 kHz with a duration of only 5 ms (Zbinden 1989). Using the latter signal, a noctule could detect prey of 2 mm wing length, as opposed to the 9 mm minimum wing length predicted using the signal parameters in table 45.1. We propose that such plasticity in signal design can account for a great deal of variation in the detection performance of bat sonar.

Despite the departure of the "longest-wavelength" bats from the model, overall there is evidence of an influence of signal wavelength and duration on the sizes of prey taken by bats. If the detectability of small prey is affected by signal design, then there exists a potential mechanism for resource partitioning by echolocation (Jones 1995a), possibly important in the structuring of bat communities (Heller and von Helversen 1989).

Many traits in aerial-feeding bats are linked, whether simply by physiological constraint or because they are co-adapted: signal duration and wavelength scale positively with body size (Barclay and Brigham 1991), and echolocation pulse repetition rate scales negatively (Jones 1999). Barclay and Brigham (1991) proposed that the link between body size and echolocation wavelength ultimately limits body size, due to the poor reflectivity of prey under long wavelengths. The reflectivity data in our study support this idea. However, the idea requires that signal wavelength be physiologically constrained by body size. Such constraints are clearly not narrow, as *P. pipistrellus* (mass 3–8 g) produces intense social calls at low frequency (20 kHz, Barlow and Jones 1997a), while *Cheiromeles torquatus* (mass 160 g) uses surprisingly high frequency echolocation signals (28 kHz, Heller 1995). This suggests that an aerial insectivorous bat of a given body size is free to echolocate within a relatively wide range of wavelengths. If the bat uses QCF signals for hunting aerial prey in the open, the echolocation wavelength and duration it uses probably reflects its requirements for prey detection and other echolocation functions, rather than its body size.

Acknowledgments

We thank the following field-workers who kindly supplied bat feces for the dietary analyses: Raphaël Arlettaz, Kate Barlow, Jean-Daniel Blant, Laurent Duvergé, Jon Russ, and Jens Rydell. We thank Dean Waters and David Pye for their helpful comments on an earlier draft.

{ 46 }
Cryptic Species of Echolocating Bats

Gareth Jones and *Kate E. Barlow*

Introduction

Mayr (1977) defined a cryptic species as "a species the diagnostic features of which are not easily perceived." Cryptic species are usually defined according to human visual perception. It is therefore likely that taxa whose senses are different from those of humans also may comprise many cryptic species. Audition is a dominant sense in microchiropteran bats, and bats often emit ultrasonic frequencies that humans cannot hear without the use of specialized equipment. Humans are predominantly visual animals, and although human audition is proficient between about 500 Hz and 3 kHz, we are deaf to the calls of most bats. It therefore follows that bat species may remain unidentified if they vocalize at frequencies that humans have little studied.

In this chapter, we show that cryptic species may be widespread in echolocating bats, and how some cryptic bat species look similar and emit similar acoustic signals. We also review examples of cryptic bat species that differ in echolocation calls. We concentrate on recent studies of pipistrelles (*Pipistrellus* spp.) centered at the University of Bristol. Finally, we explore why bat species sometimes show acoustic divergence without differing in morphology. We examine whether acoustic divergence could be favored by sexual or natural selection. If natural selection is important, does it favor call divergence because different call frequencies allow exploitation of different prey sizes (acoustic resource-partitioning hypothesis), or to facilitate acoustic communication? We test the acoustic resource-partitioning hypothesis by measuring echo strengths from insect prey at frequencies emitted by the two cryptic pipistrelle species.

CRYPTIC SPECIES OF ECHOLOCATING BATS

Baker (1984) described the cryptic species *Rhogeessa genowaysi* by karyotyping. In Britain, where field studies on bats have been relatively detailed, 6 of the 16 breeding species (38%) are cryptic (*Plecotus auritus* and *P. austriacus*, *Myotis mystacinus* and *M. brandtii*, and two cryptic *Pipistrellus* species). At least four cryptic *Plecotus* species have been described in continental Europe (Mayer and von Helversen 2001a; Spitzenberger, Piálek, and Haring 2001). Cryptic species seem to be widespread in the genus *Myotis*: *Myotis myotis* and *M. blythii* in Europe (Arlettaz 1996; Arlettaz, Perrin, and Hausser 1997); *M. evotis*, *M. septentrionalis*, and *M. keenii* in North America (Van Zyll de Jong and Nagorsen 1994); *M. lucifugus* and *M. yumanensis* in Canada (Herd and Fenton 1983); and *Myotis nigricans*, which may be a composite of cryptic species (LaVal 1973). A new cryptic species in the *M. mystacinus* group was discovered in Greece by genetic analysis (Nemeth and von Helversen 1994). This recently was described as *M. alcathoe* (van Helverson et al. 2001). Cryptic species may also be widespread in rhinolophid bats (see below).

CRYPTIC SPECIES WITH RELATIVELY SMALL DIFFERENCES IN ECHOLOCATION CALLS

Some cryptic species have only small differences in echolocation calls. These include *Myotis mystacinus/M. brandtii* (Vaughan, Jones, and Harris 1997a), and *M. myotis/M. blythii/M. alcathoe* (von Helversen et al. 2001; R. Arlettaz and G. Jones, unpublished) (fig. 46.1). A recent molecular phylogeny (Ruedi and Mayer 2001) showed that *M. mystacinus* and *M. brandtii* were diphyletic and have come to resemble one another by remarkable convergent evolution. Sedlock (2001) identified two cryptic *Pipistrellus* species in the Philippines with small (<2 kHz) differences in terminal frequency. Although these species pairs may often be discriminated statistically by detailed analysis of calls, the differences

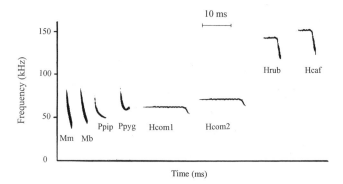

Fig. 46.1. Sonograms of echolocation calls of cryptic bat species. Species illustrated are *Myotis mystacinus* (Mm) and *M. brandtii* (Mb); *Pipistrellus pipistrellus* (i.e., 45 kHz phonic type, Ppip) and *P. pygmaeus* (i.e., 55 kHz phonic type, Ppyg); two echolocating types of *Hipposideros commersoni* (Hcom1 and Hcom2) (Hcom1, G. Jones, unpubl. recording; call of Hcom2 reconstructed from data in Pye 1972); *Hipposideros ruber* (Hrub) and *H. caffer* (Hcaf).

are so small that they are unlikely to be associated with differences in target detection abilities (Houston et al., chapter 45, this volume). Acoustic divergence may be unimportant in *M. myotis/M. blythii*, since these bats detect many prey by listening for prey-generated sounds rather than by echolocation (Arlettaz 1996).

CRYPTIC SPECIES WITH LARGE DIFFERENCES IN ECHOLOCATION CALLS

Bats in the families Hipposideridae and Rhinolophidae emit calls with a dominant CF component. Pye (1972) discovered bimodal echolocation in two groups of putative *Hipposideros commersoni* living in a cave in Kenya. The groups had CF echolocation calls averaging 56 and 66 kHz, and are probably cryptic species (fig. 46.1). Although *Hipposideros caffer* and *H. ruber* are difficult to separate on morphology (Fenton 1986), they show little overlap in the frequencies of their echolocation calls (Jones et al. 1993; see fig. 46.1). A new cryptic species similar to *Hipposideros bicolor*, but with different echolocation frequency characteristics (CFs averaging 131 and 142 kHz), was discovered in Malaysia (Kingston et al. 2001). A new cryptic species, *Hipposideros orbiculus*, was described in southeast Asia by Francis, Kock, and Habersetzer (1999). *H. orbiculus* differed by 19 kHz in echolocation call frequency from the similar *H. ridleyi*. Recent analyses of echolocation calls and genetics of *Rhinolophus rouxi* suggest that the taxon may comprise several cryptic species (N. Thomas and S. Schmidt, pers. comm.). Overall, then, cryptic species may be widespread in rhinolophoid bats, and pairs of cryptic species may differ in call frequency by >10 kHz.

CASE STUDY OF PIPISTRELLES

Pipistrelles are the most common bats in Britain, with an estimated prebreeding population size of two million

(Harris et al. 1995). They are probably the most widespread bats in Europe (Stebbings and Griffith 1986). Pipistrelles produce a diverse range of echolocation calls. Miller and Degn (1981) thought much of the variation was created when bats flew close together, varying their frequencies to avoid jamming. Zingg (1990) reported pipistrelles in Switzerland calling with end frequencies at about 45 kHz or 57 kHz, but he interpreted this difference as reflecting intraspecific variability caused by the bats emitting certain calls in particular habitats. In Britain, pipistrelle calls fall into two frequency bands (fig. 46.1), one with most energy close to 46 kHz (the 45 kHz phonic type), and the other close to 55 kHz (the 55 kHz phonic type, Jones and Van Parijs 1993). The two phonic types never mixed at maternity colonies and were sympatric over much of Britain. Jones and Van Parijs (1993) rejected the hypotheses that sex, age, or habitat caused the bimodality in call frequency. In Europe, the lower frequency bats are often found at middle latitudes, and 55 kHz pipistrelles appear more often on the edge of the pipistrelle's range—that is, in Britain, Ireland, the Mediterranean countries, and Scandinavia (Jones and Van Parijs 1993; Jones 1997; Barratt et al. 1997; Mayer and von Helversen 2001b).

Evidence amassed quickly to confirm that the two echolocating types were in fact cryptic species. Morphologically, the species are very similar, though 45 kHz pipistrelles had a statistically larger skull with longer dentary bones, longer upper canines, and a larger gape (Barlow, Jones, and Barratt 1997). The 45 kHz bats also had slightly larger forearms and wings (Jones and Van Parijs 1993; Barlow and Jones 1999).

At foraging areas, pipistrelles often emit social calls that are lower in frequency, longer in duration, and more complex in structure than echolocation calls. The social calls of 45 kHz pipistrelles are lower in frequency than those of 55 kHz bats, and most often contain four, rather than three, components (Barlow and Jones 1997a). Social calls are emitted when insects are scarce, and playback experiments caused neighboring bats of the same phonic type to leave the area (Barlow and Jones 1997b). Because social calls are agonistic, though only evoke responses in bats of the same phonic type, resource partitioning could occur between phonic types feeding in the same area (Barlow and Jones 1997b).

During autumn, male pipistrelles perform songflights, probably to attract females to roosts that they defend (Lundberg and Gerell 1986). Male songflight calls are similar to the social calls described above, except they are emitted at higher repetition rates (Barlow and Jones 1997a). The songflight calls of 45 kHz bats are lower than those of 55 kHz bats and, like their social calls, typically contain four, rather than three, components (Barlow and Jones 1997a). The function of these calls seems to change from repulsion to attraction, perhaps as a result of different repetition rates. Because bats associate as-

sortatively according to phonic type at mating roosts (Park, Altringham, and Jones 1996), female pipistrelles may use the songflight calls for species recognition (which leads to reproductive isolation) during the mating period.

Phylogenetic relationships were estimated by sequencing the cytochrome-*b* gene of mitochondrial DNA (Barratt et al. 1997). The 45 and 55 kHz pipistrelles showed large (11%) and consistent genetic differences of a magnitude similar to those described for noncryptic, congeneric bat species (reviewed in Jones 1997). The genetic differences confirm that the phonic types are separate species (Barratt et al. 1997) and further genetic differences in mitochondrial and nuclear DNA have been described by Mayer and von Helversen (2001b).

Thus, there is strong evidence from differences in echolocation call frequency, social call features, and molecular genetics that the two phonic types of pipistrelle are cryptic species. They are reproductively isolated and fulfill the criteria for separate species under both the biological and phylogenetic species concepts (Freeman and Herron 1998). Neotypes have been taken for *Pipistrellus pipistrellus* (45 kHz phonic type, or genetic Clade II in Barratt et al. 1997), with the next available synonym (*P. pygmaeus*) proposed for the 55 kHz (Clade I) bats (Jones and Barratt 1999). The proposed names are used for the remainder of this chapter, and in the phylogenetic tree showing how they differ in cytochrome-*b* sequences (fig. 46.2).

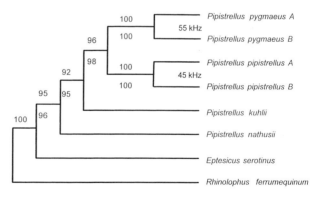

Fig. 46.2. Phylogeny of two cryptic species of pipistrelles based on 630 base pairs of the cytochrome b gene of mitochondrial DNA. Numbers indicate bootstrap values; above the branches from a neighbor-joining analysis, below from parsimony bootstrap analysis (1000 replicates each). The branches for *Pipistrellus pygmaeus* and *P. pipistrellus* A and B haplotypes showed sequence divergence <1%, while the divergence between *P. pygmaeus* and *P. pipistrellus* was >11%. *P. pygmaeus* and *P. pipistrellus* were never found in the same maternity or mating roosts, while individuals with haplotypes A and B for both species sometimes were mixed within roosts. Redrawn from Barratt et al. (1997) with permission from Macmillan Magazines Ltd.

CRYPTIC SPECIES CAN SHOW EXTENSIVE
DIFFERENCES IN RESOURCE USE

P. pygmaeus forms larger maternity roosts than *P. pipistrellus* (Barlow and Jones 1999). It feeds mainly in aquatic and riparian habitats (Vaughan, Jones, and Harris 1997b), and its maternity roosts are more likely to occur in areas with water nearby (Oakeley and Jones 1998). Although *P. pipistrellus* feeds mainly in riparian habitats, it occupies other habitats (e.g., unimproved grassland, amenity grassland, cattle pasture, conifer plantations, and mixed woodlands) more than *P. pygmaeus*. Therefore *P. pipistrellus* appears to be more of a generalist. Differences in habitat use are reflected in dietary differences between the species (Barlow 1997). *P. pygmaeus* eats mainly insects in the families Chironomidae and Ceratopogonidae, taxa associated with aquatic habitats. *P. pipistrellus* eats mainly flies in the families Psychodidae, Anisopodidae, and Muscidae. Although there was no clear difference in the eye diameters of the Diptera eaten by the two species, *P. pipistrellus* ate significantly more yellow dung flies, *Scatophaga stercoraria* — relatively large insects.

The cryptic species *Myotis myotis* and *M. blythii* also show clear partitioning of trophic resources despite having similar echolocation calls. The bats hunt mainly by listening for prey-generated sounds (Arlettaz 1996). *M. myotis* feeds mainly in meadows, orchards, and forests that lack undergrowth (where it eats terrestrial prey, mainly carabid beetles). *M. blythii* feeds more in grassland, where it eats large numbers of bushcrickets (Arlettaz 1996; Arlettaz, Perrin, and Hausser 1997). *Myotis lucifugus* and *M. yumanensis* partition resources by eating different prey in different habitats (Herd and Fenton 1983), probably as a consequence of the species having slightly different wing shapes and maneuverability (Aldridge 1986). Thus cryptic bat species partition resources, though do not necessarily emit very different echolocation calls.

HYPOTHESES FOR THE EVOLUTION
OF ACOUSTIC DIVERGENCE

It has been argued elsewhere that divergence in echolocation call frequency is unlikely through sympatric speciation (Jones 1997). Here, we assume that allopatric mechanisms result in speciation of these cryptic taxa. Why do cryptic species of echolocating bats, such as the pipistrelles and some hipposiderids, evolve different echolocation call frequencies? Also, why do some cryptic species evolve much smaller, apparently nonfunctional differences in call frequency? It is now timely to develop and explore some of the hypotheses that seek to explain why differences in call frequency occur.

Nonadaptive hypotheses

Isolated populations could evolve different call frequencies through genetic drift. Populations could diverge in call frequency during isolation, even if there is no advantage (and no cost) in doing so. Differences in frequencies could be maintained once isolated populations retain contact, and could even continue to diverge after secondary contact.

Hypotheses involving sexual selection

Many cryptic species of orthopterans, anurans, and birds have different songs. In these cases, female choice in intersexual selection can drive acoustic divergence (Jones 1997). Because bat echolocation calls are unlikely to be used in mate choice, there is no reason to expect sexual selection as an important force promoting differences in call frequency, unless the frequency of echolocation calls is correlated genetically with traits favored by sexual selection (e.g., body size, social call frequency; Jones 1997).

Acoustic benefits of using different frequency channels

Echolocation calls are often used in communication by eavesdropping (Fenton 1985), although they may not have evolved for that purpose. Selection may therefore promote acoustic divergence to facilitate communication. A likely scenario is where acoustic divergence occurs after previously isolated populations establish secondary contact, in ways similar to those proposed for character displacement (Freeman and Herron 1998).

Bats could reduce interference in their echolocation by operating on a new frequency band not shared by other species. For example, if there were large numbers of 45 kHz pipistrelles feeding in an area, a 53 kHz bat may have greater reproductive success if her offspring emitted calls at 55 kHz rather than at 51 kHz, because the lower frequency calls would overlap more with those of the 45 kHz bats, and echoes from the calls would be more susceptible to jamming or masking from the other species.

Bats could reduce interspecific competition if they partition different size categories of prey by using different call frequencies. This theory was used to explain the coexistence of many bat species in a Malaysian rainforest (Heller and von Helversen 1989). Only recently has it been confirmed that Rayleigh scattering is a potent force affecting the target strengths of insects subjected to ultrasound that resembles the echolocation calls of bats (Houston et al., chapter 45, this volume). The question therefore arises—are the frequency differences seen between cryptic species of echolocating bats sufficiently large to influence the target strengths of insects in the size range eaten, therefore allowing the evolution of a novel call frequency to enable the exploitation of different-sized prey? We test this theory for pipistrelles below.

Testing the Acoustic Resource-Partitioning Hypothesis

To test whether differences in call frequency allow resource partitioning between cryptic species, we monitored echo strengths from different sizes of prey within the frequency differences shown between cryptic species, and within the size range of their prey.

MATERIALS AND METHODS

We mimicked the situation experienced by the two cryptic species of pipistrelle. We ensonified midges (Chironomidae; wingspans 2.6–3.0 mm) and dung flies (wingspans 7.5–9.5 mm) with CF pulses of ultrasound at 44 and 55 kHz. Methods used for this experiment are detailed in Houston et al., chapter 45, this volume). The sizes of insects that were ensonified are at the lower and upper extremes of those eaten by these bats (Barlow 1997). In nature, *P. pygmaeus* (55 kHz—wavelength 6.2 mm) eats more midges and fewer dung flies than does *P. pipistrellus* (45 kHz—wavelength 7.6 mm) (Barlow 1997). Assuming that the species have similar auditory sensitivities, for acoustic resource partitioning to work we predicted 55 kHz pulses would return significantly stronger echoes from midges (which are in the Rayleigh region of the backscatter curve). The 45 kHz pulses may return the stronger echoes from dung flies. However, because the wing lengths of dung flies are close to or longer than the wavelengths of both 44 and 55 kHz pulses, there may be no difference in target strength at these frequencies.

RESULTS

Target strengths (adjusted for atmospheric attenuation, Bazley 1976) were in fact scarcely different for either frequency at both prey sizes. The 55 kHz echoes were significantly stronger from both prey types, but the differences were very small in magnitude (3.2 dB for midges, 1.9 dB for dung flies—fig. 46.3). We conclude that resource partitioning of prey sizes by the use of different echolocation call frequencies is unlikely to be important for these cryptic pipistrelle species.

Discussion

Cryptic species are fairly common among echolocating bats. More bat species will almost certainly be described as bioacoustic techniques for monitoring ultrasound in the field become more widespread. Some cryptic species differ in the frequencies of their echolocation calls. Differences in call frequency have been described for species that use echolocation for prey detection—such species often emit long relatively long CF or narrowband components in their calls. Even differences in call frequency of 10 kHz between cryptic species could contribute little to differences in target strengths for prey of the sizes eaten, at least by pipistrelle species. The most parsimonious hypotheses for acoustic divergence are that call

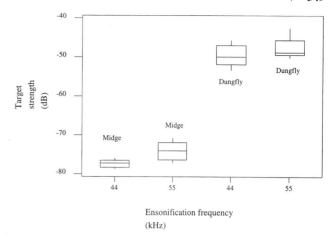

Fig. 46.3. Target strengths of midges (*n* = 5) and dung flies (*n* = 6) ensonified by 4 ms duration 44 and 55 kHz pulses similar in frequency to the narrowband tails of *P. pipistrellus* and *P. pygmaeus*. Measurements (measured at 45 cm, standardized to 1 m) include atmospheric attenuation. Categories for target strengths are 44 M is 44 kHz, midges; 55 M is 55 kHz, midges; 44 DF is 44 kHz, dung flies; 55 DF is 55 kHz, dung flies. The graphs show box plots. The line across the box is the median, the bottom of the box is the first quartile (Q1), and the top is the third quartile (Q3). The upper and lower limits of the whiskers are defined by Q1 + 1.5 (Q3-Q1). Echoes of 55 kHz pulses were slightly but significantly stronger both from small midges (2.6–3.0 mm wing length) (one-tailed Wilcoxon test, *p* = 0.02) and dung flies (wing length 7.7–9.5 mm wing length) (one-tailed Wilcoxon test, *p* = 0.05), these insects representing the extremes of prey sizes normally eaten by the bats.

frequency differences evolved either by genetic drift, or to facilitate intraspecific communication or autocommunication (Bradbury and Vehrencamp 1998), in this context processing echoes from the bat's own calls. Alternatively the differences in frequency may be adapted to fine-scale differences in habitat use that are yet to be elucidated. *P. pygmaeus* may, for example, feed more regularly along stretches of riparian woodland, and may use high-frequency calls to maintain regular contact with habitat edge features. *P. pipistrellus* may feed more in open habitats where lower frequency calls travel over greater ranges.

Acknowledgments

Collaborators included Elizabeth Barratt, Sara Oakeley, Sofie Van Parijs, Paul Racey, and Nancy Vaughan. We thank Rob Houston and Arjan Boonman for help with target strength experiments, and Nancy Vaughan for providing recordings of whiskered and Brandt's bats. David Pye, Raphael Arlettaz, and Donald Griffin provided valuable comments on an earlier draft.

{ 47 }

Aerial-Feeding Bats:

Getting the Most Out of Echolocation

M. Brock Fenton

Introduction

Microchiropteran bats that hunt airborne prey, usually flying insects, provide some of the most interesting details about the role that echolocation can play in the lives of mammals. In these species, which represent a number of families (table 47.1), echolocation allows the bat to detect, track, and assess airborne targets (Griffin 1958; Altringham 1996). Echolocation calls change dramatically in pulse design and rate of production throughout the sequence of searching, detecting, and tracking a flying insect (Obrist 1995). Variations in call design are influenced by the physical and social setting in which an individual hunts (Obrist 1995). In spite of the flexibility associated with the different phases of foraging, the echolocation calls of aerial-feeding bats are largely species-specific (Betts 1998; Kuenzi and Morrison 1998; Obrist et al., chapter 64, this volume). One excellent illustration of this is the use of echolocation calls to distinguish two phonotypes (species) previously considered to be one species, namely *Pipistrellus pipistrellus* (Barratt et al. 1997).

In one sense, echolocating aerial-feeding bats face a relatively easy task, because most often the target is an insect: a hard (reflective) object against the air, a soft (nonreflective) background. There are many variations in the situation, depending on whether the insect is flying close to other reflective objects (areas with clutter), such as other insects, the vegetation, or the ground. Although echolocation has not been "proven" as the means of orientation for the vast majority of these bats, the changes in pulse repetition rate and call design are taken as evidence that it is of primary importance (Obrist 1995). Echolocation may not be as important in food location for species that take objects from surfaces, whether they are gleaning insects or fruit.

Two basic approaches (table 47.1) to echolocation in aerial-feeding bats reflect two ways that bats avoid deafening themselves, and both are supported by observations of bats foraging in the field (Fenton et al. 1995). Species in one group, the majority of Microchiroptera, separate pulse and echo in time and avoid deafening or jamming themselves by not broadcasting and receiving at the same time. These are the " low-duty-cycle bats," which, when searching for a target, separate each outgoing signal by a much longer (than the duration of the signal) period of silence. Species in the second group, hipposiderids, rhinolophids, and the mormoopid *Pteronotus parnellii,* separate pulse and echo in frequency so that they can broadcast and receive simultaneously. These "high-duty-cycle bats" use Doppler-shift compensation and an "acoustic fovea" (Schuller and Pollak 1979) to avoid deafening themselves. High-duty-cycle bats typically produce echolocation signals almost continuously with interpulse intervals being shorter than call durations when the bat is searching for a target. This dichotomy in echolocation behavior does not coincide with the use of echolocation calls dominated by a single frequency–constant frequency or CF calls (Fenton et al. 1999a; Fenton 1999).

The purpose of this chapter is to use data from field studies to explore variety in foraging situations and to demonstrate that aerial-feeding bats are well served by their echolocation. These issues are addressed from four standpoints: (1) details of targets available to the bats, (2) their ability to operate in clutter (echoes returning from the background rather than the target of interest), (3) the effective range of echolocation, and (4) the impact of prey with hearing-based defenses. Here I focus on the low-duty-cycle species (table 47.1).

HOW MUCH DETAIL DO FORAGING BATS USE?
The time between detection and contact with a target is crucial to the successful attack by a bat. Some aerial-feeding bats ignore detailed information, using size and motion to identify suitable targets (Barclay and Brigham 1994). Their experiments demonstrated that, under natural conditions, two species of foraging, aerial-feeding bats (*Myotis lucifugus* and *M. yumanensis*) attacked moving noninsect targets such as pieces of leaf of an appropriate size as well as flying insects in the same size range. From the bats' standpoint, this behavior is adaptive because repeated past experience would have indicated that any moving object in that size range and setting was edible. Other observations show that a bat's experience affects its response to airborne targets while

TABLE 47.1. Echolocation behavior, call features, foraging behavior, and diet of bats. Most bats echolocate, some producing calls at low duty cycle (LDC), others at high duty cycle (HDC). High-intensity calls appear to be about 110 dB SPL at 10 cm, low-intensity calls closer to 60 dB SPL at 10 cm. Aerial-feeding bats hunt airborne prey, usually flying insects, while trawlers use their hind feet to take prey, usually from the water's surface. Gleaners take prey from surfaces, often directly in their mouths. Some bats eat fruit, nectar, and pollen, or blood, while most species eat other animals, mainly insects. With the exception of according Hipposideridae family status, this classification generally follows Simmons and Geisler (1998).

			Foraging behavior			Diet other than animals		
	Echolocate	Intensity of calls	Aerial-feeders	Trawlers	Gleaners	Fruit	Nectar and pollen	Blood
Suborder Megachiroptera								
Pteropodidae	No[1]	N/A	No	No	No	Yes	Yes	No
Suborder Microchiroptera								
Superfamily Rhinopomatoidea								
Rhinopomatidae	Yes—LDC	High	Yes	No	No	No	No	No
Craseonycteridae	Yes—LDC	High	Yes	No	No	No	No	No
Superfamily Emballonuroidea								
Emballonuridae	Yes—LDC	High	Yes	No	No	No	No	No
Superfamily Rhinolophoidea								
Rhinolophidae	Yes—HDC	High	Yes	No	Yes	No	No	No
Hipposideridae	Yes—HDC	High	Yes	No	Yes	No	No	No
Nycteridae	Yes—LDC	Low	Sometimes	No	Yes	No	No	No
Megadermatidae	Yes—LDC	Low	Sometimes	No	Yes	No	No	No
Superfamily Noctilionoidea								
Noctilionidae	Yes—LDC	High	Sometimes	Yes	No	No	No	No
Mormoopidae	Yes—both	High	Yes	No	No	No	No	No
Phyllostomidae	Yes—LDC	Low	No	No	Yes	Yes	Yes	Yes
Superfamily Nataloideaa								
Furipteridae	Yes—LDC	Low	?					
Thyropteridae	Yes—LDC	Low						
Natalidae	Yes—LDC							
Myzopodidae	Yes—LDC							
Superfamily Vespertilionoidea								
Vespertilionidae	Yes—LDC	High and low	Yes	Some	Some	Rare	Rare	No
Superfamily Molossoidea								
Antrozoidae	Yes—LDC	High	No	No	Yes	Yes	Yes	No
Molossidae	Yes—LDC	High	Yes	No	No	No	No	No
Not assigned to superfamily								
Mystacinidae	Yes—LDC							

[1] Except for *Rousettus egyptiacus* and perhaps some other species in the genus *Rousettus*.

hunting. For example, at Pinery Provincial Park in southwestern Ontario, *Lasiurus borealis* and *L. cinereus* often feed on insects attracted to lights (Obrist 1995; Hickey, Acharya, and Pennington 1996). Normally, these bats predictably chase and often make contact with 1 cm diameter stones lobbed into the air. After a few nights, the same individual bats (Hickey and Fenton 1991, 1996) then abort their attacks on stones, presumably learning to distinguish stones from flying insects (unpublished observations and M. K. Obrist, pers. comm.). This experiment and the observations indicate that foraging bats may not use the full potential of their echolocation when it comes to assessing the details of their airborne prey, but will increase their perception of detail when necessary.

CLUTTER

In echolocation, as in radar, "clutter" is defined as echoes returning from other than the target of interest.

The presence and degree of background clutter affects echolocation performance in aerial-feeding bats (Neuweiler 1989; Altringham 1996). In studies of echolocating bats, "clutter" remains a factor that is often invoked without any precise definition. I suspect that an operational definition of clutter must consider time factors, such as arrival of echoes and call duration. Definitions of clutter will probably have to be relatively species- and situation-specific. The degree of clutter will be influenced by the bat's flight path relative to the background clutter. For example, the angle of approach and perception of clutter by a *Noctilio leporinus* hunting over the surface of water (Schnitzler et al. 1994) differs from that confronting a *Lasiurus cinereus* attacking a moth flying directly in front of the canopy of a tree (Obrist 1995).

In either case, the timing of the arrival of echoes from the target of interest relative to those from other objects is critical. A barrage of echoes, all reaching the bat at virtually the same time, presents a different perceptual challenge than echoes from clutter arriving somewhat later than those from a potential prey. Quite simply an echolocating bat has no time to react to echoes from clutter that arrive at the same time as those from prey. The same bat has more time, but often less than 1 s, to react to echoes that arrive later. The actual details will depend upon the bat's flight speed and the proximity of the clutter to the target.

Call duration also affects a low-duty-cycle bat's perception of clutter, because while the bat is transmitting, it is very insensitive to echoes. This also has been referred to as "forward masking." For example, a 5 ms echolocation call renders a bat almost deaf to echoes returning from objects closer than 1.7 m; for a 10 ms call the deaf zone is 3.4 m. Flight speed influences the magnitude of the problem. For a bat flying 5 m/s, the 1.7 m deaf zone is covered in 340 ms; and at 10 m/s, in 170 ms. These data illustrate the potential selective relationship between flight speed and reaction time in aerial-feeding bats. The durations of search-phase echolocation calls of aerial-feeding bats range from about 4 ms to about 20 ms (Fenton et al. 1998a).

Definitions of clutter should be species-specific, reflecting differences in wing features, flight speed, and operational range, as well as interpulse interval and call duration. It is no wonder that fast-flying, aerial-feeding bats use long, narrowband search-phase echolocation calls produced at relatively long interpulse intervals when foraging in open areas. Exploitation of Doppler shifts could make the high-duty-cycle bats more clutter-resistant than the low-duty-cycle aerial foragers, whose search-phase echolocation calls are dominated by narrowband components. For some insects, flying closer to clutter is an effective defensive behavior against low-duty-cycle aerial-feeding bats (Andersson, Rydell, and Svensson 1998; Rydell 1998), but it should not work against the high-duty-cycle echolocators.

RANGE OF OPERATION

I think the effective range for detecting insect prey is a primary factor influencing the morphology and behavior of aerial-feeding bats. In aerial-feeding bats, echolocation call design is closely related to wing features that influence maneuverability and agility (Norberg and Rayner 1987). The operational range of a bat's echolocation will reflect its echolocation calls and its auditory system, as well as its flight speed, maneuverability, and agility. Two features of the echolocation calls are crucial to the range of operation: (1) The *intensity* of the call directly affects the operational range of echolocation, because a 110 dB call has more energy to lose than a 90 dB or a 60 dB call. High intensities are typical of the echolocation calls of aerial-feeding bats (>110 dB; Griffin 1958; Waters and Jones 1995), while bats with other foraging behavior use lower intensities (60–80 dB). (2) The *frequencies* of the sounds also affect range, because atmospheric attenuation is higher for frequencies >20 kHz

(Griffin 1971; Lawrence and Simmons 1982). For aerial-feeding bats using echolocation calls between 20 and 60 kHz, attenuation and spreading loss mean that the effective range of echolocation is accurately reflected by the distance sound travels between consecutive search-phase pulses (Fenton et al. 1998a). The rate at which cruising aerial-feeding bats produce echolocation calls reflects the synchronization of pulse production with the wing-beat cycle (Speakman and Racey 1991).

Samples from different zoogeographic areas (Nearctic, Neotropics, Ethiopian tropics, Australian topics) demonstrate that aerial-feeding bats most often use echolocation calls dominated by sounds between 20 and 60 kHz (Fenton et al. 1998a), although some bats achieve broader bandwidths by the addition of harmonics (Fenton et al. 1999a). Sounds in this bandwidth offer most aerial-feeding bats an effective compromise between details about targets and effective range, reflecting the combined impact of wavelength and attenuation. In many areas, however, a few aerial-feeding species use lower (<20 kHz) or higher (>60 kHz) frequencies (Fenton et al. 1998a). The situation closely resembles the morphological one, with many similar species and few distinct ones (Findley 1993). Statistically, bat size appears to be the best predictor of the frequencies dominating echolocation calls in aerial-feeding species—larger species of emballonurids, vespertilionids, and molossids use echolocation calls dominated by lower frequency sounds compared to smaller taxa (Barclay and Brigham 1991; Bogdanowicz, Fenton, and Daleszczyk 1999).

The facial features of bats can influence signal strength and directionality. Specifically, nose leaves contribute to the operational range of echolocation. Hartley and Suthers (1987) demonstrated how the nose leaf affected the pattern of sound radiation from *Carollia perspicillata*, albeit not an aerial-feeding bat. The flaps of skin around the mouths in mormoopid bats serve to direct sounds, and photographs of echolocating bats show an arrangement of the lips that may increase the directionality and range achieved by these sounds.

The structure of the external ear (pinna) also affects the overall sensitivity of the auditory system to returning echoes. The pinnae of some bats are mechanically tuned to the frequencies dominating the echolocation calls (Obrist et al. 1993). This relationship is evident among some aerial-feeding bats, such as molossids, but absent in others, such as *Lasiurus borealis* (Obrist et al. 1993), where increased auditory sensitivity involves more than pinna structure (Obrist and Wenstrup 1998). In *Megaderma lyra*, which is not usually an aerial-feeder (Audet et al. 1991), pinna structure contributes to a low hearing threshold (−25 dB, Neuweiler 1989).

The large ears of gleaning bats are mechanically tuned to the lower frequency sounds associated with movement, rather than to the sounds dominating the bats' echolocation calls (Obrist et al. 1993). But some aerial-feeding bats have remarkably large pinnae that could increase hearing sensitivity and extend the range they detect airborne prey (Fenton et al. 1998a). Included here are some molossids (e.g., *Otomops* spp., *Eumops* spp., some *Tadarida* spp.) and vespertilionids (e.g., *Euderma maculatum, Histiotus* spp.). Longer effective range should occur with echolocation calls of lower frequency (<20 kHz) because of decreased atmospheric attenuation at lower frequencies (Lawrence and Simmons 1982). In *Tadarida midas,* a big (40 g), large-eared, African molossid, echolocation calls are dominated by sounds <15 kHz, produced at interpulse intervals of almost 1000 ms. When searching for prey, these bats show almost no frequency overlap between adjacent pulses. The low frequency of the sounds reduces attenuation, albeit at the cost of being "blind" to targets whose largest axis is <3 cm. It further raises the possibility that the bat can detect and process the returning echoes of one call even after another has been produced. In theory, this could extend the range of operation, an important factor for such a large, aerial-feeding bat (Fenton et al. 1998a).

CALL DESIGN

There are several general patterns of frequency change over time in the search-phase echolocation calls of low-duty-cycle aerial-feeding bats (fig. 47.1). These calls usually sweep from high to low frequencies, although there can be an initial upward sweep. The degree of frequency modulation in the echolocation calls varies considerably within and between species. In general, species hunting in more open areas, where most echoes are from potential prey or other bats (i.e., low clutter), use calls dominated by longer, narrowband components. Broadband calls and a tendency toward more even distribution of sound across the bandwidth are more typical of the search-phase calls of species hunting in areas of higher clutter. The picture can be complicated by variations in call intensity, the presence (or absence) of harmonics, and conspecifics (e.g., Obrist 1995). Bats that are agile and maneuverable usually produce shorter echolocation calls. These bats hunt "at the end of their nose" (Brosset 1966), detecting and reacting to targets at very short range (approximately 1 m), often in situations with high clutter.

Low-duty-cycle, aerial-feeding bats use echolocation to detect the flutter associated with the wing beats of potential prey. Changes in the frequency (Doppler shift) and amplitude are generated by the insect's beating wings. Sum and Menne (1988) demonstrated this for *Pipistrellus stenopterus.* Von Stebut and Schmidt (chapter 35, this volume) have reported similar data for *Eptesicus fuscus.* Call duration appears to be crucial to flutter detection, because the longer returning echo encodes changes in frequency or amplitude associated with the wing beats of prey. The long, narrowband calls of low-duty-cycle species foraging in the open are well suited for flutter detection, improving target detection within the range of their echolocation.

Few studies of echolocation performance in the field by high-duty-cycle echolocating bats are available, leaving uncertainty about effective range in these bats. Vaughan's (1977) observations of *Hipposideros commersoni* attacking dung beetles suggested these bats detect flying prey at 5–10 m. The dichotomy between high- and low-duty-cycle echolocating bats is an ancient one (the Eocene, Simmons and Geisler 1998), yet both groups of bats are well served by echolocation as they hunt flying insects.

PREY DEFENSE

The high intensity echolocation calls of aerial-feeding bats are conspicuous to listeners with appropriate acoustic sensitivity. The roster of potential eavesdroppers is huge, including potential prey, potential predators, and biologists with bat detectors (Betts 1998). I believe that eavesdropping by prey is another critical factor affecting the echolocation behavior of aerial-feeding bats because insects such as moths with bat-detecting ears hear most approaching bats long before the bats would detect an echo from a moth (40 m versus 5 m; Fenton and Fullard 1979). Virtually all these data come from the interactions between bats and their insect prey, leaving unknown possible hearing-based defensive behavior in prey such as other arthropods or vertebrates.

Some of the best evidence of the effectiveness of echolocation as practiced by aerial-feeding bats comes directly from the hearing-based defenses of insects, immortalized both in prose (Roeder 1967) and in poetry (Pye 1968). Additional information is presented elsewhere in this volume (chapters 43, 50, and 51). Does an insect's perception of the risks presented by an echolocating bat affect its behavior? The answer can be

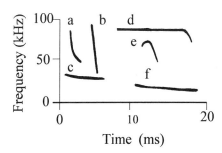

Fig. 47.1. Bats use a variety of basic echolocation call designs. Some aerial-feeding species use a combination of steep frequency-modulated sweeps and a longer component of narrower bandwidth (*a*) when searching for prey; while those hunting in more open areas use longer calls of narrower bandwidth (*c* and *f*). Some emballonurid and noctilionid bats use calls that combine upward and downward frequency-modulated sweeps (*e*), while high-duty-cycle bats use longer calls dominated by one frequency (*d*). When taking insect prey from surfaces, some *Myotis* species use very short, steep, frequency-modulated calls (*b*).

"yes." Acharya and McNeil (1998) demonstrated how, in the presence of simulated bat echolocation calls, male moths of two species (a pyralid and a noctuid) altered their flight responses to plumes of pheromones released by conspecific females. The male moths apparently considered the risk posed by the bats to be more important than the immediate opportunity to mate. Many insects have bat-detecting ears that confer a statistically significant advantage on moths when it comes to detecting and avoiding aerial-feeding bats (e.g., Rydell 1992; Acharya and Fenton 1999).

The bat-detecting ears of a variety of insects are most sensitive to sounds in the 20–60 kHz range, precisely the one that dominates the echolocation calls of low-duty-cycle aerial-feeding bats in many parts of the world (Fenton et al. 1998a). Considerable research is conducted on the diets of aerial-feeding bats whose echolocation calls are dominated by sounds outside the 20–60 kHz range, the "allotonic" species of Fullard (1987). Bogdanowicz et al. (1999) demonstrated that the low-duty-cycle echolocating bats whose calls were dominated by sounds in the 20–40 kHz range fed on beetles more than on moths. In both high- and low-duty-cycle bats there also was a relationship between the size of the bats' jaws and teeth, and the frequencies dominating the echolocation call. Within any family, compared to smaller bats, larger species have larger teeth and jaws, produce echolocation calls dominated by lower frequencies, and eat larger prey (Bogdanowicz, Fenton, and Daleszczyk 1999).

Aerial-feeding bats using echolocation calls dominated by sounds <20 kHz often feed on moths, perhaps on species with bat-detecting ears (Rydell, Jones, and Waters 1995). The best documented examples are *Euderma maculatum* (Fullard and Dawson 1997) and *Tadarida teniotis* (Rydell and Arlettaz 1994), although other possible examples include *Otomops martiensseni* and *Taphozous perforatus* (Rydell and Yalden 1997). At the other end of the spectrum, *Furipterus horrens* is a low-duty-cycle species that uses echolocation calls dominated by sounds >100 kHz and feeds heavily on moths (Fenton et al. 1999b). But moths in a bat's diet may reveal little about echolocation and hearing-based defenses, because some moths lack ears. The picture can be complicated as illustrated by Pavey and Burwell (1998), who reported two species of bats eating mainly moths, one taking eared species, the other earless (deaf) ones.

The picture differs for high-duty-cycle species, which tend to use echolocation calls dominated by frequencies >60 kHz. Here again there is a statistically significant relationship between bat size and frequency dominating the echolocation calls (Bogdanowicz et al. 1999). Jones (1992) reported that the data from high-duty-cycle bats supported the allotonic hypothesis, with species whose echolocation calls were dominated by sounds >60 kHz taking significantly more moths than species with lower frequency echolocation calls. Between 60 and 100 kHz,

the frequency dominating the echolocation calls is an accurate predictor of the incidence of moths in a bat's diet, but this is not so at >100 kHz (Bogdanowicz et al. 1999).

Two species of bats are of particular interest because they use echolocation calls dominated by sounds >100 kHz and feed mainly on moths. The low-duty-cycle *Furipterus horrens* and the high-duty-cycle *Cloeotis percivali* are both tiny (<5 g adult mass) and feed mainly on moths (Fenton et al. 1999b; Whitaker and Black 1976). At present there are no published studies of their foraging behavior. The frequencies dominating the calls, 150 and 212 kHz, respectively, suggest very short effective range for echolocation. The somewhat larger *Hipposideros ater,* whose echolocation calls are dominated by sounds around 160 kHz, take mainly eared moths (Pavey and Burwell 1998), directly supporting the allotonic hypothesis.

Species Deserving More Attention

The diversity of aerial-feeding bats and their insect prey offers many opportunities for further research into echolocation and foraging behavior. There remains the issue of just how to characterize the diet of an aerial-feeding bat and the matters of inter- and intraspecific variability in the prey taken (see Bogdanowicz et al. 1999 for discussion). But the available information hints at the complexity that we can expect to emerge from further work.

Nyctalus leisleri represents the "typical" condition because its echolocation calls should be conspicuous to moths that do not form a large portion of its diet (Waters, Rydell, and Jones 1995), but some other species of low-duty-cycle, aerial-feeding bats feed heavily on moths that should have detected their echolocation calls. *Lasiurus borealis* and *L. cinereus* are two examples (Hickey, Acharya, and Pennington 1996; Hickey and Fenton 1996). Both sometimes eat mainly moths even though sympatric moths show evasive behavior in response to attacks by these bats (Acharya and Fenton 1999). Most of these moths should first detect the calls of either species of bats at distances of >30 m, while the bats' reactions suggest they detect the moths at approximately 10 m (Acharya and Fenton 1999). At lights, however, moths are subject to conflicting sensory inputs from their eyes and their ears, so they could be more vulnerable to marauding bats.

Black (1972) called *Lasiurus cinereus* a "moth specialist," and around lights it feeds almost exclusively on moths (Hickey, Acharya, and Pennington 1996; Acharya and Fenton 1999). At lights, *L. borealis* feeds more on moths than when foraging elsewhere (Ross 1967). Lights are important feeding situations that offer significant energetic benefits to both *L. borealis* and *L. cinereus,* mainly because of shorter interattack intervals reflecting high prey density (de la Cueva et al. 1995). In southern Manitoba, *L. cinereus* often eat moths, but include a greater variety of insects in their diets (Barclay 1985). At lights in Hawaii, moths again dominate the diet of this species (Belwood and Fullard 1984).

We do not know which aspect(s) of the behavior of these two species of *Lasiurus* make them so effective at thwarting the hearing-based defenses of moths. Obrist and Wenstrup (1998) concluded that coevolution with hearing prey has meant that *L. borealis* is more effective in using echolocation for localizing and tracking prey, rather than improving detection. At lights, the second bat attacking a moth (within a minute) of the first has a higher success rate than the first (Acharya and Fenton 1999), suggesting that foraging together could be an element of success. This also explains why "chases" are commonly observed in *Lasiurus borealis* (Hickey and Fenton 1991). Other species use echolocation calls in the 20–60 kHz range and feed heavily on moths (Bogdanowicz et al. 1999), although perhaps on deaf or earless taxa. The African *Chalinolobus variegatus, Pipistrellus rueppellii,* and *Eptesicus zuluensis* deserve further attention in this regard, because they sometimes eat moths to whom their echolocation calls should be conspicuous (Fenton et al. 1998b).

Other Echolocating Microchiroptera

It is more difficult to determine whether gleaning bats (those taking prey from surfaces), fruit-eaters, nectar-feeders, and blood-feeders are served as well by echolocation, because we are quite uninformed about the role that echolocation plays in their lives. There is evidence that at least three species of phyllostomids use echolocation to find fruits (*Phyllostomus hastatus,* Kalko and Condon 1998; *Carollia* spp., Thies, Kalko, and Schnitzler 1998), but there seems little doubt that olfaction is important in the actual selection of food. In any event, the short (usually <1 ms), broadband, low-intensity calls are dominated by frequencies >30 kHz, a combination of characteristics that should mean a very short range of operation. The recent description of an ultrasonic nectar guide in some flowers pollinated by nectarivorous phyllostomids (von Helversen and von Helversen 1999) adds a new dimension to this area of echolocation research.

A large volume of evidence demonstrates that gleaning bats detect and track their targets by prey-generated sounds such as those associated with movement or advertisement (Faure, Fullard, and Barclay 1990; Tuttle and Ryan 1981). At the same time, echolocation calls of gleaning bats are relatively inconspicuous to insects such as moths (Faure, Fullard, and Barclay 1990; Surlykke and Filskov 1997). Gleaning bats can use their echolocation calls to assess fine differences in detail (Schmidt 1988b), but it is unknown how often this actually happens in the wild.

The same set of call features associated with a short range of operation (call duration, call strength, call frequencies) makes it difficult to monitor the echolocation

behavior, so it is hard to know just when these bats are echolocating. Although some species (e.g., *Myotis evotis, M. septentrionalis;* Faure, Fullard, and Barclay 1990) seem to continue call production regardless of the situation, others do not appear to produce echolocation calls all the time when approaching prey (e.g., *Megaderma lyra,* Fiedler 1979; *Macrotus californicus,* Bell 1985; *Antrozous pallidus,* Bell 1982).

Aerial-Feeding Bats Are Well Served by Echolocation

The high-intensity echolocation calls of aerial-feeding, low-duty-cycle Microchiroptera appear to provide more details about insect targets than they necessarily use. The echolocation behavior of these bats allows them access to areas of clutter, apparently scaled according to issues of timing relating to call duration and intercall interval. Echolocation by bats hunting airborne prey appears to be a relatively short range, allowing these bats to detect insect-sized targets at distances of 1 to 10 m, which often means <1 s from detection to contact with a target. The echolocation calls of these bats usually include most energy in the 20–60 kHz range. Although there are various patterns of call structure, frequency, bandwidth, and duration, longer, narrowband calls are usually produced by species hunting in open (low to no clutter) areas. Within any family of aerial-feeding bats, size is the best predictor of the frequency dominating the echolocation calls. There is greater consistency in call patterns in high-duty-cycle bats where bat size is the best predictor of call frequency. In general, species broadcasting in the 20–60 kHz bandwidth are conspicuous to moths and feed on eared moths infrequently compared to bats using calls dominated by frequencies <20 kHz. Many of the smaller, high-duty-cycle species produce calls dominated by frequencies >60 kHz, and between 60 and 100 kHz call frequency is a good predictor of the incidence of moths in the bats' diets. Some species of bats feed heavily on moths even though their echolocation calls should be conspicuous to these prey. It remains to be determined how these bats adjust their foraging strategy to counter the insects' defensive behavior. Although echolocation offers aerial-feeding bats a relatively short-range picture of their surroundings, it could provide a better range of operation than vision, given the size of the targets and the uncertain lighting conditions (Fenton et al. 1998a).

Acknowledgments

I thank Martin Obrist, Lorraine Standing, and two anonymous referees for reading earlier drafts of this manuscript and offering suggestions for its improvement. My research on bats was supported by Research and Equipment grants from the Natural Sciences and Engineering Research Council of Canada.

{ 48 }

Discrimination of Prey during Trawling by the Insectivorous Bat, *Myotis daubentonii*

Robert D. Houston and *Gareth Jones*

Introduction

Free-living, insectivorous, microchiropteran bats typically detect and locate prey by using echolocation. To forage efficiently, these bats must reject the echoes of inedible objects, and respond to the echoes of potential prey. The role of other senses in the evaluation of potential prey is uncertain, especially in aerial-feeding bats (Faure and Barclay 1992). However, the opportunity to use other senses (e.g., vision, listening for prey-generated sounds) is often restricted. Some authors have recently tested the ability of echolocating bats to identify prey and to reject dummy objects when foraging in the field. Little ability was found in *Myotis lucifugus* and *Myotis yumanensis* to discriminate dead moths from leaves when both stimuli were made to "flutter" above the surface of a lake (Barclay and Brigham 1994). However, *Myotis daubentonii* largely ignored dummy objects (dead mealworms, twigs, etc.) and attacked live mealworms, while trawling on the surface of a freshwater pool (Boonman et al. 1998). Furthermore, Griffin, Friend, and Webster (1965) trained *M. lucifugus* in the laboratory to discriminate mealworms from one or more discs of a similar size, when both were launched across the flight path of the bat.

It would benefit a bat if it could apply fine discrimination during hunting, not only to reject inedible objects but also to reject less profitable prey. This would give bats the opportunity to feed selectively, which could optimize foraging efficiency (Stephens and Krebs 1986). To discriminate consistently among prey types and to prey selectively, bats must not only discriminate but also evaluate and *classify* prey echoes. Acoustic, behavioral, and ecological work on horseshoe bats (Rhinolophidae) suggests these bats classify flying insects and prey selectively (von der Emde and Schnitzler 1990; Jones 1990; Moss and Zagaeski 1994). Horseshoe bats use a high-duty-cycle, Doppler-compensating echolocation system that yields many sources of information suitable for distinguishing different insects (Kober and Schnitzler 1990). "FM" bats have no Doppler compensation, and use short, low-duty-cycle signals that are less suitable for preserving dynamic features of fluttering insect targets

(Roverud, Nitsche, and Neuweiler 1991). However, two types of field study produce indirect or direct evidence of active selection even by FM bats. The first type involves a comparison of the assemblage of potential prey sampled from a bat's hunting habitat with the assemblage of insects found in its stomach or feces. Results of these studies often suggest selective foraging (e.g., Anthony and Kunz 1977; Hamilton and Barclay 1998). These studies are open to interpretation due to an ignorance of which insects are strictly available (Faure and Barclay 1992), which is in turn due to the sampling biases of insect collection methods and to an ignorance of the precise hunting habitats used by the bats prior to the taking of a diet sample (Arlettaz and Perrin 1995).

The second type of study is direct observation of wild, hunting bats. During such observations, bats have been reported to ignore or to avoid some types of prey in favor of other types (e.g., Acharya and Fenton 1992). Some authors compare the insects found in fecal samples with insect abundance samples taken in situ during the observed foraging (e.g., Belwood and Fullard 1984). Such data support the conclusions that the bats were foraging selectively from the "available" prey, although in these cases, artificial lights made visual cues available to an unusual degree. It is therefore difficult to draw conclusions about the use of echolocation alone in selecting prey.

Although there is no conclusive evidence of active prey selection by echolocation (Arlettaz and Perrin 1995), bats using FM signals are not necessarily incapable of discriminating insect echoes. Their ability to resolve spatial positions of different reflecting surfaces is between 1.0 mm (*Myotis myotis*, Habersetzer and Vogler 1983) and 0.2 mm (*Megaderma lyra*, Schmidt 1988b). They can also discriminate target shape and size accurately (e.g., 1.5–3.0 dB of echo intensity, *Eptesicus fuscus*, Simmons and Vernon 1971; discrimination of similar shapes, *Vampyrum spectrum*, Bradbury 1970; *E. fuscus*, Simmons and Vernon 1971). Furthermore, there is now evidence that FM bats respond to dynamic cues in echoes (e.g., Feng, Condon, and White 1994). Note that the limits of echolocation performance are often measured under ideal conditions, with no relative movement between bat and target. To discriminate targets in flight, a bat must deal with Doppler distortion of echoes, and must make decisions within tight time constraints.

The subject of this study is Daubenton's bat (*Myotis daubentonii*). This species feeds mainly by trawling over water surfaces for insect prey (Jones and Rayner 1988). Its sensory environment approximates that of an aerial-feeding bat more closely than that of a gleaning bat, because a calm, clean water surface represents very little clutter (Boonman et al. 1998). To identify prey, *M. daubentonii* does not require the target to move, although movement does encourage attack (Boonman et al. 1998). It is simple to present, in the wild, a choice between two artificial prey to which the bat will respond. It is an opportunity to test not only the target discrimination ability of the animal, but also the behavioral tendency, during natural foraging, to select some prey and reject others. We present preliminary results of an experiment with Daubenton's bat. By repeating our experiment in the field and in the laboratory, we aim to contribute data to the discussion of the difference between discrimination performance of echolocating bats in the laboratory (reviewed in Moss and Schnitzler 1995; Simmons et al. 1995) and in the field (Arlettaz and Perrin 1995; Barclay and Brigham 1994).

Materials and Methods

The study site was a freshwater pool in southern England (51°27′ N, 2°39′ W) and is a known foraging site of Daubenton's bats (*M. daubentonii*). Daubenton's bat is the only species resident in Britain that uses the trawling technique. Its echolocation signals are purely frequency-modulated, with no constant-frequency tail. During search phase, the signals are about 5 ms long and the duty cycle is less than 10% (see Jones and Rayner 1988; Kalko and Schnitzler 1989).

Field Experiment

Two-alternative prey discrimination trials were conducted while the bats were foraging naturally. A submerged apparatus was used to present instantly at the water surface a choice of two live "prey" items. The large prey was a mealworm (*Tenebrio molitor*, a flour beetle larva) of mass 40–70 mg, and the small prey was a buffalo worm (*Alphitobius diaperinus*, a buffalo beetle larva) of mass 10–25 mg. These prey differed in size but not substantially in morphology. A prey item was impaled on the end of each of the two pieces of fine wire, which were positioned 40 mm apart. The bats were free to approach the pair of prey from any angle, but the feeding station was at a constant position 1.5 m from the bank of the pool. The bats usually approached on a path parallel with the bank and the prey were aligned perpendicular to this path.

A trial began when the submerged prey were raised to the water surface by the experimenter sitting on the bank of the pool. The trial ended when a bat took a prey from one wire. The remaining prey was immediately lowered below the water surface to make it unavailable to bats. This imposed a two-alternative choice on the bats. The apparatus was then reloaded with prey for the next trial. The relative position of the large prey was randomized in successive trials, although the position of the entire feeding station was always constant.

During the field experiments, the bats were exposed to starlight, moonlight, and the light of a red-filtered head torch deflected away from the feeding station. At the beginning of each night, water surface conditions were classified as "cluttered" (debris on the surface of the pool) or "clear" (negligible debris visible). The debris consisted of leaves, mainly small pine needles, seeds, and other vegetation. Most particles were small (<5 mm), and there was generally <1 cm spacing between clutter items on cluttered nights.

Field Experiment with Clutter Manipulation

A portion of the pool surface centered on the feeding station was enclosed, using a square arrangement of floating wooden beams, 2.36 m long and protruding from the water surface by approximately 1 cm. Within this enclosure, the density of surface debris was manipulated using pine needles from the adjacent forest floor. Further prey discrimination trials were conducted while manipulating the mass of pine needles covering the water in the enclosure. During these trials, the prey were not positioned at the water surface, but raised 1–2 mm clear of the meniscus, to ensure that all the available reflecting surface of a target was exposed to the bats. Five levels of clutter density between 0.0 and 9.0 g/m^2 were tested. Two treatments were tested on a single night, and the order of the treatments within a night was random.

Laboratory Experiment

The field procedure was repeated in the laboratory with four individuals. The male Daubenton's bats were caught by mist net under license from English Nature. The bats were held captive for six weeks and trained to hunt over an artificial pool (2.38 m × 0.78 m) in a room of length 6.0 m, width 1.8 m, and height 1.9 m. After the study, they were released at the site of capture.

Prey were presented in pairs at the water surface, 1–2 mm from the meniscus (as in the second field experiment), impaled on fine wire, and placed about 40 mm apart. A trial was prepared by the light of a head torch, during which time the prey were concealed from the bat. There was no light available during the trial. Buffalo worms varied between 10 and 34 mg, while mealworms varied between 12 and 74 mg. One prey presentation consisted of (1) a large mealworm and a small buffalo worm, (2) a large and a small mealworm, or (3) a large and a small buffalo worm.

Only one bat was allowed to fly and hunt over the pool at a time. After an attack (including unsuccessful attacks), the prey item not chosen was immediately removed, and the position of the larger prey item in successive trials was randomized. The difference in mass between prey items in every trial was recorded and used as a predictor of the outcome of trials through binary logistic regression.

Trials continued for about 20 min, after which time the bat rested, and another bat flew. When not taking part in the trials, bats were housed in a room of dimensions 2 m × 2 m × 3 m, and allowed access to water at all times.

PROFITABILITY OF PREY

During each of the above trials, the period between prey capture and the end of mastication was recorded as "handling time." To assess quantitatively the bats' possible motivation to prefer one of the stimuli, the profitability of the prey in each lab and field trial was calculated in terms of rate of energy intake, using the measured handling time and other parameters gleaned from the literature.

Results

FIELD EXPERIMENT

The bats quickly learned the position of the feeding station, indicated by attacks on the position when no prey was available. They also perceived two targets during the first pass of each trial, indicated by an immediate return for the second prey if the first was captured but mishandled and dropped. The second prey was always submerged by the experimenter by this stage, but this

did not stop the bat from attacking the former position of the second prey item.

Overall, the bats in the field caught 64% large prey during 229 trials (table 48.1). This is a significant bias ($\chi^2 = 17.3, p < .001$). The bias was evident during nights classified as clear of water surface clutter (75% large prey, $n = 120$, $\chi^2 = 18.3$, $p < .001$), while on cluttered nights, there was no significant bias (51% large prey, $n = 109$, $\chi^2 = 0.09, p > .50$).

FIELD EXPERIMENT WITH CLUTTER MANIPULATION

At the two highest levels of clutter density, no trials were recorded (table 48.2) because the bats did not succeed in capturing any prey. The bias in the prey size caught by the bats was constantly high at the other three levels, at around 75% large prey taken. That is, the bias was not affected until the surface debris was so dense that the bats made no successful captures. At high clutter densities, bats made many passes over the position of the feeder without attacking either prey. Occasionally, bats attacked the introduced clutter items elsewhere in the enclosure.

LABORATORY EXPERIMENT

All four bats learned to catch and eat prey from the surface of the artificial pool within a few days. The prey taken by each bat was significantly biased toward the larger prey (table 48.3), and the degree of bias was similar to the proportions of larger prey taken in the wild (this study). The probability of the bats taking the larger prey was significantly greater than 0.5, regardless of whether the choice offered was (1) two mealworms differing in size, (2) two buffalo worms differing in size, or (3) a large mealworm and a small buffalo worm (table 48.3). There was variation in the probability of

TABLE 48.1. Probability of wild Daubenton's bats taking the larger of two prey from the surface of a freshwater pool during a field experiment. "Cluttered conditions" means that a high density of small debris particles covered the water surface (>1 particle/cm²). "Clear conditions" means the water appeared clear of surface debris.

Water surface conditions	Number of trials	Probability of large prey taken	Upper 95% confidence limit	Lower 95% confidence limit
Clear	120	0.75	0.82	0.68
Cluttered	109	0.51	0.59	0.42
Total	229	0.64	0.73	0.54

TABLE 48.2. Probability of wild Daubenton's bats taking the larger of two prey from the water surface during the clutter manipulation experiment. The "clutter" consisted of dry pine needles from the adjacent forest floor.

Clutter density (g/m²)	Number of trials	Probability of large prey taken	Upper 95% confidence limit	Lower 95% confidence limit
0.0	82	0.76	0.84	0.63
2.5	73	0.74	0.83	0.61
3.8	22	0.77	0.88	0.56
5.4	—	—	—	—
9.0	—	—	—	—

TABLE 48.3. Proportion of large prey taken by the four captive bats during discrimination trials. Figures in parentheses indicate sample sizes (number of trials). The 95% binomial confidence intervals of the proportions do not include 0.5 in any case, meaning that all proportions are significantly different to the null expectation of 0.5. The binary logistic regression model was [probability of bat taking large mealworm] = [difference in mass between mealworms], and was performed only on data from "mealworm versus mealworm" trials. With the *p*-value we evaluate the null hypothesis that the odds ratio is 1.00 (that is, no change) with respect to an increase of 1 in the predictor.

| | | Probability of bat taking large prey (*n*) | | | Binary logistic regression results | |
Bat	Total	Mealworm versus mealworm	Buffalo worm versus buffalo worm	Mealworm versus buffalo worm	Z	*p*-value
1	0.72 (340)	0.70 (196)	0.83 (12)	0.77 (193)	2.83	0.005
2	0.65 (451)	0.59 (191)	0.73 (63)	0.68 (193)	1.10	0.272
3	0.64 (246)	0.64 (246)	—	—	2.89	0.004
4	0.64 (178)	0.64 (178)	—	—	1.30	0.193

taking large prey that correlated with the combination of prey species offered in the trials. In the case of both bats 1 and 2, the probability of taking the large prey was greatest during buffalo worm versus buffalo worm trials, and least (although still significantly greater than 0.5) during mealworm versus mealworm trials. There was an intermediate probability during the dual-species trials (table 48.3).

The probability of the bat taking the large prey increased with the difference in mass between the two mealworms (fig. 48.1). This occurred in the case of each of the four bats. However, according to a binary logistic regression model, the trends apparent in the data were statistically significant only in the case of bats 1 and 3 (see last two columns of table 48.1). In the terms of logistic regression, there is no evidence in the case of the

bats 2 and 4 that the odds ratio was greater than unity with respect to an increase in the value of the predictor (difference in mass between mealworms).

PROFITABILITY

Rate of net energy gain (*R*, in Js^{-1}) resulting from taking a prey of mass *M* was estimated as:

$$\text{Energy gain} = M \cdot E \cdot D$$
$$\text{Cost}_{handling} = \text{Handling time} \times F$$
$$\text{Cost}_{search} = (\text{Search time} \times F) + \text{Cost}_{call}$$
$$R = \text{Energy gain} - (\text{Cost}_{handling} + \text{Cost}_{search})$$
$$\text{Handling time} + \text{Search time}$$

where the terms in the numerator are measured in joules, and those in the denominator in seconds; where *E* is the

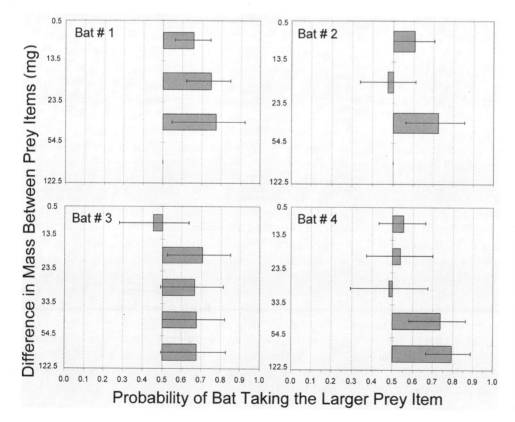

Fig. 48.1. Probability of four captive Daubenton's bats taking the larger of two mealworms (12–74 mg) presented on the surface of a pool in a flight room, as a function of difference in size between mealworms. Trials involving buffalo worms are not considered here. Error bars are 95% binomial confidence intervals.

Probability of Bat Taking the Larger Prey Item

energy density of mealworms (10.6 Jg^{-1} live mass, calculated from data in Speakman, Anderson, and Racey 1989), D is the digestive efficiency when eating mealworms (90%, Barclay, Dolan, and Dyck 1991), F is the cost of flying (1.85 Js^{-1}, calculated from data in Speakman and Racey 1991), *Search time* is the minimum required for search, approach, and capture (0.85 s, estimated from Britton 1996 and unpublished data), and *Cost$_{call}$* is an estimate of the cost of producing the extra vocalizations during the approach and feeding buzz involved in each capture (1.80 J, calculated from data in Speakman, Anderson, and Racey 1989 and Britton 1996).

Handling time measured during the laboratory trials increased linearly with mass of prey in all bats (bat 1: handling time (y, seconds) = 0.589 × mass (x, grams), $p < 0.0005$, $r^2 = 0.67$; bat 2: $y = 1.05x$, $p < 0.0005$, $r^2 = 0.25$, bat 3: $y = 0.712x$, $p < 0.0005$, $r^2 = 0.55$, bat 4: $y = 0.881x$, $p < 0.0005$, $r^2 = 0.48$. The *rate* of energy gain during handling is therefore unaffected by prey size. In other words, *Handling time* and mass (M) are proportional in the above equation. With any positive, constant cost of searching for and capturing each prey item, the total rate of energy gain, including search and capture times, is increased by taking larger prey. Therefore, the larger of the two prey items presented in the discrimination trials was always the more profitable.

Discussion

The results of the behavioral experiments show a consistent tendency in Daubenton's bats to take the larger of two prey items, both in the field and in the laboratory. If Daubenton's bat encounters more than one prey item at once, we predict that it will not take prey from its foraging environment according to abundance, but will tend to capture the larger prey items within the size range tested. Small differences in target size apparently discriminated in this study (fig. 48.1) contrast with the crude discrimination shown by *M. lucifugus* in Barclay and Brigham's study but are consistent with the discrimination ability demonstrated in prey identification/clutter rejection studies (*M. daubentonii,* Boonman et al. 1998; *M. lucifugus,* Griffin, Friend, and Webster 1965).

Most target discrimination experiments in the laboratory condition bats to fly or crawl toward one of two targets to receive a food reward (e.g., Bradbury 1970; Simmons and Vernon 1971). In this study, bats used their natural foraging technique to carry out the task: the stimuli themselves were the food reward, and the bats reported their decision by attacking and eating one stimulus. The bat's decision, if active, might be aimed at maximizing its net rate of energy intake. Using profitability calculations, we predicted that if the bats optimized prey profitability in this way, they should always take the larger of the two prey, within the size range tested.

We found that the bats were indeed more likely to take the larger prey.

Did the bats in this study really show active prey selection, based on evaluation of prey, via a sonar representation of each? Is the smaller prey rejected in favor of the larger prey? The larger prey might be taken only because it generally reflects a more intense echo, and is therefore more detectable. Our simple experimental design, lacking conditioning of the subjects, allowed us to replicate the experiment in the laboratory and in the field, but leaves us uncertain about the bats' motivation. At least two arguments, however, favor the active selection hypothesis. First, the two prey were presented in exactly the same positions (although relative positions of large and small prey were randomized), which the bats learned over hundreds of trials. That the bats learned there were always two prey items is suggested by immediate turns and attacks on the position of the (submerged) second prey item, after the first was taken by the bat and then fumbled. Second, the targets were close together (40 mm), and probably detected first as a single large target, given the bats' likely angular resolution of more than 1.5° (Simmons et al. 1983), and Daubenton's bat's average estimated detection distance of 1.3 m (Kalko and Schnitzler 1989).

There are conflicting data on the bats' response to clutter. In the first field experiment, the proportion of large prey taken was lower in the presence of naturally occurring clutter. This suggests some deleterious effect of clutter on active prey size discrimination. Conversely, the densities of clutter used in the subsequent clutter manipulation experiment did not affect the proportion of large prey taken, although high clutter densities appeared to prevent prey capture. We propose that the dense clutter either masked the prey echoes or presented an insoluble clutter rejection task (Boonman et al. 1998), which prevented prey detection. Pine needles, although a common component of the naturally occurring water surface clutter, were perhaps not the component that affected bat performance in the first experiment.

In the field, we had no way of knowing which individual bats were responsible for how many of the trials. However, we found similar results using four known individual bats in essentially the same experiment, in the laboratory. The bats would have experienced clutter from the walls and ceiling of the flight room, and from the sides of the artificial pool, which was only 70 cm wide. Their sonar signals were probably much less intense (Waters and Jones 1995), and detection of the targets probably occurred at a shorter distance, allowing less time for a decision (Britton and Jones 1999). Despite these differences between the laboratory and field setups, the overall probabilities of known bats taking large prey in the laboratory (0.72, 0.65, 0.65, and 0.64, respectively) were similar to those of unknown wild bats in the field (0.64).

In the laboratory, the dissimilarity in mass (proportional to the area of reflecting surface) between the two mealworms was taken as a measure of the "ease" of the discrimination. We expected that the probability of the bat taking the large mealworm would rise as the mass dissimilarity increased, whether the bats were actively selecting or not. Fig. 48.1 shows such a trend in all four bats, although the trend is not significant in two cases. We did not expect a model psychometric function (the classic sigmoidal curve of psychophysical experiments), because the stimuli presented in these experiments were too imprecise and variable. The mealworms were free to wriggle, and each would present a variety of echo spectra and amplitudes during each trial. If the bats were selecting prey actively, any particular cue they used to discriminate target size would have been inconsistently presented within and across trials, leading to a noisy and smeared psychometric function.

Our study tested target discrimination by echolocation in an environment that also allowed us to assess prey selection behavior. Both targets were edible and provided a food reward. Both targets were identified as prey, but one was preferred. The target that was preferred was more profitable, but also more detectable, so active selection could not be demonstrated. It is hoped that further work with prey choice experiments will distinguish between active selection and passive detection by providing bats with a stimulus that is more attractive despite being less detectable. If prey that are too large to handle are not attacked in the wild (Barclay and Brigham 1994; Belwood and Fullard 1984), we should be able to use preference for smaller, less detectable prey. By using an appropriate range of prey sizes, we could further examine prey discrimination ability and behavior.

Acknowledgments

We thank Mark Brigham for his helpful comments on an earlier draft of this manuscript.

{ 49 }

Energy Cost of Echolocation in Stationary Insectivorous Bats

John R. Speakman, Winston C. Lancaster, Sally Ward, Gareth Jones, and *Kate C. Cole*

Introduction

Measurements of the energy costs of vocalization have been made for relatively few species. However, the costs of vocalization in at least some appear to be very high. For example, vocalizing anuran amphibians and orthopteran insects have energy demands during calling that range from 20 to 100 times their resting rates of metabolism (Ryan 1988; Forrest 1991; Prestwich and Breuer 1987). Among singing birds estimates of the energy costs of vocalizing are more disparate, with some studies indicating high costs (Brackenbury 1977; Eberhardt 1994; Jurisevic, Sanderson, and Baudinette 1999), but other studies suggesting costs are much lower (Chappell et al. 1995; Horn, Leonard, and Weary 1995; McCarty 1996).

Echolocation calls produced by both the Chiroptera and odontocetes are very intense (Griffin 1958). Moreover, these animals can produce echolocation calls continuously for periods of several hours each day. There-fore, the question arises of the contribution that such intense and protracted calling makes to daily energy requirements. During the 1980s, we made the first measurements of the energy demands connected with echolocation in stationary pipistrelle bats (*Pipistrellus pipistrellus* −55 kHz phonic type = *Pipistrellus pygmaeus:* Jones and Van Parijs 1993). This study revealed that each echolocation call costs a pipistrelle bat about 0.067 joules to produce (Speakman, Anderson, and Racey 1989). Consequently, these animals are predicted to be expending approximately 7–12 times their basal metabolic rate (BMR) on vocalization when echolocating at the typical rate during searching flight (approximately 10 calls per second). This discovery led to the question of what the energy costs of flight are in echolocating bats, since flight itself is known to be energetically expensive (12–15 times BMR). Would the costs of echolocation be additive to those of flight alone, making total flight costs in these bats around 23 times BMR?

Using the doubly labeled water method to measure the energy demands of flight in pipistrelle bats (Speakman and Racey 1991), we demonstrated that the costs of flight in this species were not elevated above the expected levels for nonecholocating animals. This economy was possible because during flight bats use their flight muscles to simultaneously perform three tasks. First, they flap the wings, generating lift and thrust to keep the animal flying. Second, they coordinate with respirations to facilitate ventilation of the lungs, enabling the elevated oxygen consumption required to fuel the flight. Third, they facilitate pressurization of the thoracoabdominal cavity (Lancaster, Henson, and Keating 1995), generating a forceful expiration that is modulated by the larynx to produce the echolocation call. Hence there is a link between wing beating and echolocation pulse production (Suthers, Thomas, and Suthers 1972; Kalko 1994). When bats fly they do not have to do extra work to also echolocate. However, when bats are stationary they must contract muscles specifically to generate the expiratory burst. Thus, stationary echolocation is costly, but echolocation coupled to wing beating is effectively free (or at least has no detectable costs).

There are, however, several species of bats that echolocate frequently when they are stationary during perch hunting (Jones and Rayner 1989; Neuweiler et al. 1987; Schnitzler, Hackbarth, and Heilman 1985). The benefits of perch hunting are generally presumed to include a reduction in hunting costs, because the animals do not fly during the search phase. However, this interpretation ignores the fact that remaining stationary and echolocating may be energetically expensive. In the present study, we aimed to measure the costs of echolocation when stationary in two species of bats: the greater horseshoe bat (*Rhinolophus ferrumequinum*) and the serotine (*Eptesicus serotinus*). Both of these bats have similar body masses, weighing 25–30 g, thus controlling for body mass effects, but the serotine is a conventional aerial-hawking bat while the greater horseshoe bat is a typical perch-hunting rhinolophid.

Materials and Methods

Study Animals

Measurements were made on two representatives of each species. This is a small sample of animals, but we had problems collecting more specimens from the wild because of their protected status in the UK.

Respirometry

Measurements of energy expenditure were made using standard indirect calorimetry following similar protocols to those used by Speakman, Anderson, and Racey (1989). Bats were placed in specially modified respirometry chambers into which a small number of mealworm beetles were also introduced as a stimulus to encourage the animals to echolocate. Excurrent gases were dried using silica gel, and a subsample of the airstream was directed through a dual-channel oxygen analyzer (Applied Electrochemistry SA3; Amtek Ltd., California). Oxygen concentrations in the airstream were compared with a subsample of the incurrent gas monitored in the second channel of the analyzer. The effective system volume was 870 ml, which was only marginally greater than the actual volume (800 ml), indicating there was good gas mixing in the system (that is, there were no dead spaces that would elevate the effective volume well above the actual volume). We collected data at 5 s intervals throughout actual measurement runs. These data were corrected to instantaneous levels of oxygen consumption using the known effective volume of the system and the half-life of the washout.

Simultaneous to the measurements of oxygen consumption, we monitored the vocalization behavior of the animals. Two systems recorded bat vocalizations. In early recordings, vocalizations were monitored with a bat detector (Ultrasound Advice S25) and the output of the high-frequency channel was recorded. In later sessions, we used a model 2630 (6.25 mm) Larson Davis condenser microphone with model 2200C Larson Davis amplifier and a 15 kHz high-pass filter. In both cases, the microphone was mounted in the base of the respirometry chamber, approximately 12 cm from the bat's mouth. All signals were recorded on a four-channel Racal instrumentation tape recorder at 38 cm/s. Bats were observed continuously, and records of their activity made at 5 s intervals, coincident with the records of oxygen consumption. In each 5 s period, we counted the number of echolocation calls produced by the animal.

To compare the energy expenditure derived from the oxygen consumption estimates with the behavior, we made several manipulations of the data. We removed the first 20 min of data inside the chamber. Previous measurements indicated that this period is associated with handling stress, and although the animals make their greatest number of echolocation calls during this interval, their metabolism is elevated for reasons unconnected to calling. Pooling these data with those collected later when the animals were calm, but echolocating less, would have generated a spurious link between energy cost and echolocation. Second, we rejected any time period when the bats were physically active while echolocating. This avoided any confusion of the costs of activity with the costs of vocalization. Finally, we stepped the records of echolocation (calls per 5 s period) relative to the records of oxygen consumption, so that we could account for the lag between the chamber and the analyzer. We took the relationship at the point of greatest correlation as indicative of the link between energy expenditure and echolocation calling behavior (after Speakman, Anderson, and Racey 1989).

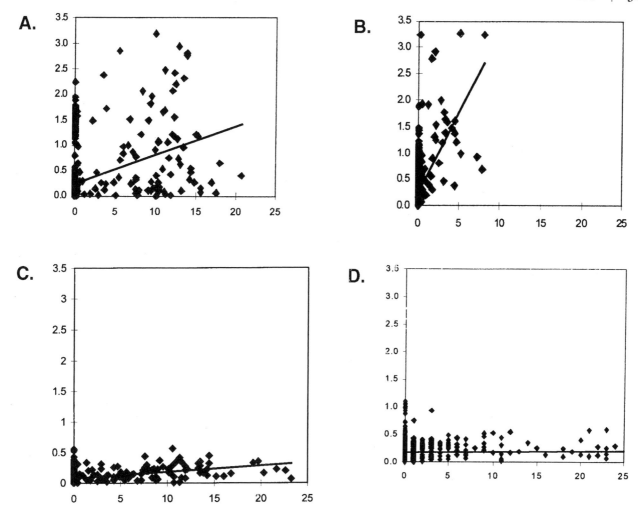

Fig. 49.1. Relationships between energy expenditure (metabolic power in watts) and rate of echolocation (calls per second) for two stationary serotine (*A* and *B*) and greater horseshoe (*C* and *D*) bats

Results

The relationships between energy expenditure and the number of calls per second over each 5 s period for the serotines (*A* and *B*) and the greater horseshoe bats (*C* and *D*) are illustrated in fig. 49.1. Two points are immediately apparent. First, the total calling rate of the greater horseshoe bats was much higher than that of the serotines. The greater horseshoe bats inside the respirometer continuously called at rates up to 20 calls per second. In contrast, the serotines seldom called, and often restricted calling to periods when we rejected the data—that is, during the first 20 min of a respirometry run or when they were active. The second very obvious feature is that the gradients linking the costs of echolocation to calling rate were different for the two species. In fact, across the two greater horseshoe individuals, the gradient of this relationship averaged 0.008 joules/call, while across the two serotines the cost was 0.187 joules/call (approximately 20 times higher in the serotine). The dif-

ference in gradients was significant ($t = 21.4$, $p < .03$, df = 2).

The resting metabolic rate (RMR) (in thermoneutral but not necessarily post-absorptive) of the serotines averaged 0.25 watts; that of the greater horseshoe bats was 0.27 watts. Assuming a rate of pulse production of 10 calls/s, the costs of stationary echolocation in the serotine bat would amount to approximately 7.5 times RMR. For the greater horseshoe bat, however, the cost was much lower, only 1.29 times RMR.

The serotine bats typically produced their calls singly. However, the greater horseshoe bats often batched their calls closely together, producing two or more calls in rapid succession, followed by a greater time interval without calls. On average, the greater horseshoe bats produced 1.7 calls in each group. For those groups consisting of two calls, we compared the first and second calls for duration, intensity, and sound energy content (after Waters and Jones 1995) and compared values with those for single calls (table 49.1).

TABLE 49.1. Durations, amplitude, and energy contents of echolocation calls produced by greater horseshoe bats for single calls, double calls, and the sum of the two calls in a double call (combined).

		Double		
	Single	**First**	**Second**	**Combined**
Duration (mean)	54.8	49.8	40.4	90.2
(se)	3.4	5.2	2.9	
n	20	10	10	
Amplitude (rms volts)				
Mean	0.629	0.481	0.387	
(se)	0.076	0.163	0.162	
n	20	10	10	
Energy content ($rms^2 \times$ duration)				
Mean	1894.9	1240	637.4	1877.4
(se)	324.1	553	293.1	
n	20	10	10	

On average, when greater horseshoe bats made two calls together, they were both shorter than single calls. Their intensity also was reduced relative to a single call, and their sound energy content was also lower. However, the summed energy across the two calls in a group matched almost exactly the sound energy of a single call.

Discussion

The extrapolated high cost of echolocation in the serotine bat calling at 10 calls/s, at around 7.5 times RMR, confirms our previous measurements for the pipistrelle bat, that stationary echolocation in a typically aerial-hawking bat is energetically demanding. Given the potentially high cost of this activity, it is unsurprising that the bats were reluctant to perform it often when in the respirometry chamber. The same was true of the pipistrelles we studied previously (Speakman, Anderson, and Racey 1989). In complete contrast, the greater horseshoe bat, which routinely uses echolocation from a stationary position (Jones and Rayner 1989), had a much lower energy demand. Presumably because the behavior was less taxing, the bats showed no hesitation in performing the behavior throughout the respirometry measurements.

Part of the reason why echolocation is cheaper in greater horseshoe bats relates to their echolocation behavior. Unlike the serotines, which typically made single calls, the greater horseshoe bats grouped calls together. The rates at which these groups were generated are consistent with the stationary ventilation rates of insectivorous bats (Hays et al. 1991). Thus greater horseshoe bats appeared to be generating multiple echolocation pulses from each expiration. In doing so, they were unable to make the same duration or intensity calls as when they generated single calls per breath. Indeed, it appeared that there was a fixed amount of energy available

from each expiration and the bats could divide this energy into one, two, or more calls. Hence, the sound energy content of two calls in a group of two was equal to the sound energy of single calls.

This division of sound energy from a single expiration probably also occurs during the terminal-buzz phase of echolocation calling in aerial-hawking bats. This suggestion is supported by analysis of terminal-buzz phase calls, which invariably show reduced sound intensities relative to search-phase calls (Kobler et al. 1985). We suggest this is not only because the bat is much closer to the target, and therefore requires less energy in the call (Kobler et al. 1985), but is probably an inevitable energetic constraint that attends the mechanism of dramatically increasing pulse repetition rate—that is, dividing the available expiration energy into several calls. If this hypothesis proves correct, then the breakdown of the coupling of wing-beat frequency and echolocation call rate during the terminal-buzz phase need not necessarily be associated with elevated echolocation costs, as has frequently been surmised—for example, by Simmons and Geisler (1998).

In the greater horseshoe bat, this effect of dividing the energy into multiple calls explains approximately one tenth of the lower energy costs per call in this species compared with the serotine. This is because on average only 1.7 calls were produced per group in the greater horseshoe bat, while the costs per call were 20 times lower. Therefore, greater horseshoe bats must have some other mechanism(s) for reducing the costs of echolocation. We suggest that these mechanisms include anatomical differences that improve the calling efficiency when stationary.

The power required for the generation of vocalizations derives from the primary expiratory musculature of the abdominal wall. Another microchiropteran, Parnell's mustached bat (*Pteronotus parnellii*), is known to produce copious vocalizations at rest (Lancaster, Henson, and Keating 1995), although it is not a perch-hunting species (Henson et al. 1987). Studies of its abdominal wall revealed a muscular and aponeurotic structure (aponeurosa ventralis of Kovaleva 1989) with a high density of elastin fibers that is considerably more derived than in closely related bats that use low-intensity echolocation (Lancaster and Henson 1995). A similarly derived abdominal aponeurosis also is seen in the horseshoe bats, in addition to similarities in the rib cage that contribute to a greater degree of stiffness in comparison with other bats (Lancaster, Henson, and Keating 1995). This elastic membrane could function as an energy storage mechanism. Set into tension during inspiration, the energy could be recovered during vocalization by facilitating rapid repressurization of the thoracoabdominal cavity following a call, hence reducing the costs of stationary vocalization (Lancaster 1994). This suggests that stationary calling is an economical alternative only in conjunction with highly derived behavioral and morphological adaptations. This interpretation therefore does not compromise our suggestion that the costs of high-intensity echolocation were incompatible with perch hunting as the ancestral foraging strategy of bats (Speakman 1999, 2001); early bats likely did not have the morphological adaptations that reduce the costs of calling.

Acknowledgments

This work was supported by BBSRC grant J/0987845 awarded to John R. Speakman. Kate C. Cole was supported by a BBSRC studentship. We are grateful to Colin Morris for helping to catch the serotine bats.

{ 50 }

The Role of Arctiid Moth Clicks in Defense against Echolocating Bats: Interference with Temporal Processing

Jakob Tougaard, Lee A. Miller, and *James A. Simmons*

Introduction

Moths of the family Arctiidae have ears specialized for the detection of ultrasound. This ear is anatomically and functionally similar to the ear of noctuids (Roeder and Treat 1957). It is located laterally on the metathorax and consists of a thin tympanal membrane onto which a chordotonal organ with two receptor cells is attached (see Ghiradella 1971 for details). The ears of arctiids and many other moths are believed to have evolved in response to selective pressure from echolocating microchiropteran bats (see, e.g., Hoy 1992; Miller and Surlykke 2001).

In addition to hearing ultrasound, many arctiids produce sound by means of two tymbal organs located on each side of the thorax at the base of the hind legs (Blest, Collett, and Pye 1963). Each tymbal organ (fig. 50.1A) consists of a modified sclerite backed by an air-filled cavity. By contracting muscles attached to the sclerite (fig. 50.1B), it can be made to buckle, which produces clicking sounds (Blest, Collett, and Pye 1963). In some species, like the garden tiger moth (*Arctia caja*), the surface of the tymbal is smooth and emits two sharp clicks when buckled; one when the tymbal snaps inward and another when it returns (Blest, Collett, and Pye 1963; Surlykke and Miller 1985). Other species, like the dogbane tiger moth (*Cycnia tenera*) and the ruby tiger moth (*Phragmatobia fuliginosa*), have a striated band on the tymbals (fig. 50.1A, C), termed microtymbals (Blest, Collett, and Pye 1963). When the tymbal buckles, each of the microtymbals in turn deforms (fig. 50.1C), which gives rise to a train of clicks—one click from each microtymbal. A similar train of clicks is emitted when the tymbal returns

(see Blest, Collett, and Pye 1963; Fullard and Heller 1990 for details). Clicks are rather variable, even from the same individual, but the overall characteristic is their broad frequency spectrum. Examples of clicks from *P. fuliginosa* are shown in fig. 50.1D. The sound intensity also varies, and values from 55 to 95 dB re 20 μPa at distances of a few centimeters have been measured (Fullard and Fenton 1977; Surlykke and Miller 1985).

The neuroethology of sound production has been studied primarily in *C. tenera*. The vigorous sound production of this species occurs in response to both tactile and acoustic stimulation (Fullard 1992). When *C. tenera* is stimulated with intense ultrasound (80–100 dB SPL; comparable to a nearby bat), an alternating sequence of clicking occurs from the two tymbal organs. The modulation cycle, which is the interval between successive click trains, is affected by stimulus intensity, but it is not synchronized to the modulation periods of a pulsed stimulus (Fullard 1992). Experiments in which recordings of bat echolocation sounds were used as stimuli showed that *C. tenera* appears to time the onset of its clicking behavior to the terminal phase of the attack of the bat (Fullard, Simmons, and Saillant 1994). However, the propensity to click varies greatly among species and among individuals of the same species. Some species with tymbal organs apparently rarely emit clicks upon tactile or acoustic stimulation (Dunning and Krüger 1995). The reason for this variability remains unclear, but some may arise as an artifact of manipulation.

Data on clicking behavior under field conditions are not available, and little is known about what actually occurs in live interactions between flying arctiids and bats. Field studies have demonstrated, however, that arctiids

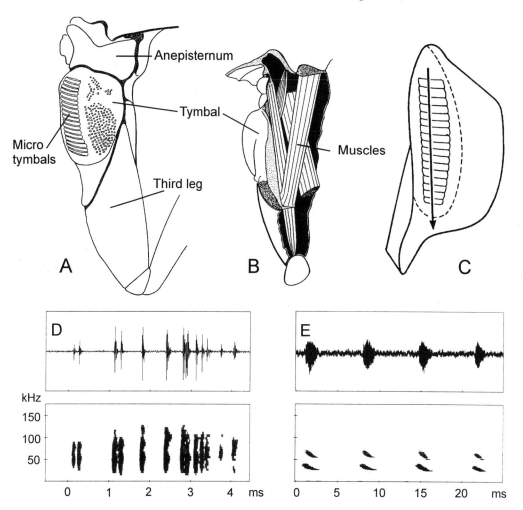

Fig. 50.1. Tymbal organs of arctiid moths. *A:* External anatomy of a tymbal organ with microtymbals. *B:* Tymbal muscles. *C:* Direction of successive buckling of microtymbals. *D:* Examples of time series and spectrograms of arctiid clicks (*Phragmatobia fuliginosa,* FFT size 64, Hann window). *E:* Cries from the terminal phase of a bat, *Eptesicus serotinus,* catching prey in the field (FFT size 512, Hann window). *A–C* modified from Blest, Collett, and Pye (1963).

suffer proportionally less predation from bats than do other moths of comparable size (Acharya and Fenton 1992), and muted arctiid moths were caught more often by bats than clicking ones in a laboratory experiment (Dunning et al. 1992). Thus, it seems without doubt that the clicks play a role in defending arctiid moths against echolocating, insectivorous bats. In addition, some species also use clicking sounds in intraspecific communication (e.g., Sanderford and Conner 1995).

Blest, Collett, and Pye (1963) first conducted a systematic study of click emission in arctiids and proposed that clicks could somehow serve as a defense against bats. They suggested two possible mechanisms: (1) clicks could interfere with the sonar of the bats: "the addition of [clicks] to an echo from the moth's body may serve to conceal its position from an approaching bat" (205), or (2) clicks could serve as aposematic signals, indicating to the bat that the moth is distasteful (see also Rothschild 1965; Dunning 1968). The first idea was taken up again

by Fullard, Fenton, and Simmons (1979), who suggested that clicks confuse bats by mimicking real echoes. Another variant of the perceptual interference idea was proposed by Miller (1991), who suggested that clicks interfere with the temporal processing of echoes in the auditory system of the bat.

A third possibility is that clicks evoke a startle reaction by the bat that may allow the moth to escape (Surlykke and Miller 1985; Stoneman and Fenton 1988; Miller 1991).

We present data from behavioral experiments on bats using an artificial arctiid click as an interfering signal and interpret the results in light of recent electrophysiological findings from the brainstem of bats.

Materials and Methods

The behavioral experiments were designed to test the ability of an echolocating bat to resolve differences in target range when an artificial arctiid click occurred within a sensitive temporal window 1 ms before a relevant echo (Miller 1991). The experiments were performed at the Department of Psychology, Brown University, on two bats.

TARGET SIMULATOR

The simulator is previously described in Simmons, Moss, and Ferragamo (1990). Fig. 50.2 shows how sounds were presented to the bats. Each bat rested on the starting platform and echolocated to the right or left. The bat's echolocation signals were picked up by two Brüel & Kjær ⅛ inch microphones placed at the end of the two response platforms. Signals were band-pass filtered (15 to 100 kHz), delayed in a digital delay line, and played back to the bat from small loudspeakers (RCA 112343) located next to the microphones. This attenuated replica of the bat's emission (artificial echo) simulates the echo of a real target in front of the bat. The amplitude of each single replica varied with the intensity of the bat's emissions, but the gain was adjusted so an *average* replica was 10–15 dB above the individual bat's detection threshold, which was about 65 dB SPL.

On one side of the platform a B-echo was played back with a constant delay (2165 μs), which corresponded to

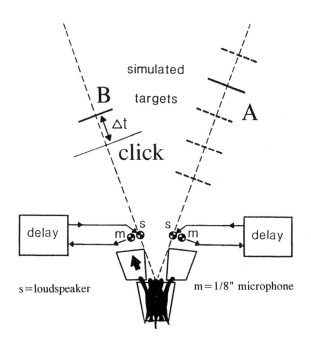

Fig. 50.2. The electronic target simulator. The bat's sonar signals were picked up at microphones, digitally delayed, and returned from loudspeakers as artificial echoes. The bats had to choose the click and B-echo pair (*arrow*). The maximum distance from the bat's head to the microphone/loudspeaker was about 150 mm. The B-echo always occurred at a delay of 2165 −s, representing a simulated target at a distance of about 520 mm. The artificial "arctiid" click appeared 1 ms before the B-echo (Δt) with an amplitude of +6 dB relative to the average B-echo. The delay of the A-echo was constant for each session and varied by 10–200 μs between sessions.

a virtual distance of the simulated (phantom) target of about 520 mm from the bat's starting position. An artificial arctiid click was added to the B-echo and occurred before the B-replica (click and B in fig. 50.2). The click was triggered by the 33 kHz portion of the bat's emission by means of a narrow band-pass filter (Q = 10). The delay of the click was adjusted for the click to occur 1000 μs before the 33 kHz part of the bat's signal. However, the actual delay depended on the frequency/time characteristics of the signal, and the actual delay was measured to 1000 ± 76 μs (n = 49 tape-recorded signals) for one bat (bat 3) and 970 ± 55 μs (n = 40 tape-recorded signals) for the other bat (bat 7).

The 50 μs click, produced from a 15 μs electronic pulse, had a bandwidth from 15 kHz to 60 kHz (−10 dB) and maximum amplitude at 30 kHz. The amplitude of the click was fixed at 6 dB above the average B-echo.

A single replica (A-echo) was presented to the bat from the other side of the platform. This signal was presented with a delay that was constant for each experimental session, but varied from session to session. The bats could receive signals from both loudspeakers simultaneously, but the strongest echo came from the side to which they directed their signals.

Bat signals from the trials of some sessions were recorded on magnetic tape (Racal Store 4, 60 ips) for later digitizing and analyzing.

TRAINING PROCEDURE

Naive bats were first trained in a two-alternative forced-choice procedure to detect a Styrofoam ball in front of either the right or left arm of a Y-shaped platform and to walk to the correct side to receive a reward (a piece of mealworm). When a bat had learned the task, it was transferred to the target simulator. On the simulator, the bat was trained to crawl to the platform where a single B-echo was presented. Later a second B-echo, 100 μs before the first, was introduced, followed by addition of an A-echo on the opposite side (A in fig. 50.2). The A-echo delay was initially about 6000 μs, and the amplitude was gradually increased until the bat accepted A and B replicas of equal amplitude. In the last stage of training, one B-echo was replaced with the artificial arctiid click (click and B in fig. 50.2). Introducing the click–B-echo pair did not affect the performance of the bats. Of seven big brown bats, *Eptesicus fuscus* (Vespertilionidae), two females (bat 3 and bat 7) completed the training regimes rather quickly and were used for experimentation.

EXPERIMENTAL PROCEDURE

The placement of the A and B + click replicas was varied from left/right to right/left in a pseudorandom way. Each bat usually ran one experimental session per day. Each session consisted of 5–10 blocks of trials, 10 trials to a block, and usually one delay per session of the A-echo. A session always started with a block of control trials without the A-echo. Additional control blocks were added when deemed necessary. Each data point represents a percentage of at least one block of trials. If two blocks of trials were run at a particular delay and the difference between them was more than 10%, at least two more blocks of trials were run and the highest and lowest block percentages discarded before averaging. The A-echo delays for sessions were either changed in progressive steps of 10–200 μs, depending of the situation, or randomized.

Results

SIGNALS USED BY THE BATS

Because the bats were working close to their threshold of detection, it seems reasonable to assume that the most intense signals are more important for the bat when deciding which side to choose. Thus, the 3 most intense signals from the first 10 signals of a trial were considered representative and were analyzed. The signals had terminal frequencies of the first harmonic from 30 to 35 kHz (bat 3, $n = 64$) and 26–32 kHz (bat 7, $n = 64$) The duration of the signals of bat 3 were mostly between 1 and 3 ms ($n = 280$, 16 signals <1 ms) and between 1 and 2 ms for bat 7 ($n = 140$, 14 signals <1 ms). Intervals between signals were from 25 to 200 ms (bat 3) and 30–300 ms (bat 7). The amplitude of the bats' signals were up to +3.8 dB relative to the average signal intensity ($n = 52$ trials × 3 signals). The average relative amplitude of the signals used by bat 3 was 5 dB greater when the click preceded the B-echo compared to when the click was absent or substituted by a second B-echo. Thus, a few of the B-echoes had greater peak-to-peak (pp) levels than those of the click.

RANGING PERFORMANCE

Poor performance was seen in the bats when the A-echo was presented between −200 and +1200 μs relative to the B-echo (fig. 50.3). The poorest performance was seen when the delay of the A-echo and the B-echo was similar, where the performance fell near to chance level (50% correct). A second region of poor performance of bat 3 appeared when the A-echo was 300–500 μs after the click (fig. 50.3A).

Discussion

BEHAVIOR

The results clearly showed that clicks interfere substantially with the bats' temporal processing. In the absence of any interfering signals, the range-difference discrimination threshold of *E. fuscus* lies between 35 and 110 μs (Moss and Schnitzler 1995). In this experiment, with clicks presented 1 ms *before* the echo, discrimination between A-echo and B-echo deteriorated when A-echoes occurred at delays approximately −200 μs to

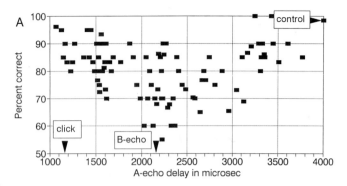

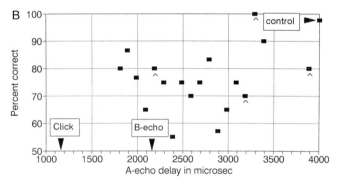

Fig. 50.3. The performance of bat 3 (*A*) and bat 7 (*B*) to various delays of the A-echo (replica) when the click occurred 1 ms before the B-echo (replica). The bats received artificial echoes from both sides, but the strongest echoes came from the side to which the bat directed its sonar signals. Each data point represents the average results of from 20 to 60 trials, except those data points indicated by carets (10 trials). The control in *A* represents 15 mistakes of 928 trials and in *B* 11 mistakes in 471 trials with the A-echo absent.

+1000 μs relative to the B-echo (fig. 50.3). In conjunction with the experiments described above, the click was placed 1 ms *after* the B-echo. In this case, the bats' performance deteriorated by only 20–30% when the A-echo followed the B-echo or the click by 200–400 μs.

The results presented here agree with those reported earlier (Miller 1991), even though the experimental designs differed substantially. Miller (1991) used a single-channel simulator, an actual arctiid click burst, and a standard *E. fuscus* signal—not the bat's own signal. This gave a constant ratio between the amplitude of clicks and echo, as well as a better controlled delay between click burst and artificial echo. The greater magnitude of interference reported earlier (a 40-fold increase in range-difference threshold, Miller 1991), as opposed to about a 7-fold increase here, may be a consequence of the differences in experimental design. Also, in Miller (1991), the largest effect of the click was seen when the click preceded the onset of the echo by about 1 ms. In this experiment, the click occurred 1000 μs before the 33 kHz portion of the echo, toward the end of the signal, resulting in a smaller delay between click and echo onset. The clicks may thus not have been at the position causing maximal interference. In any case, the consequence of the demonstrated interference to a free-flying bat would be difficulties for the bat in determining the difference in distance between prey and clutter or perhaps even in determining accurate distance to the prey should clicks appear about 1–2 ms before the prey echo.

The task of the bats in the present experiment can be viewed as either potentially very easy or potentially posing difficulties, depending on the cues used by the bat. If the task is understood by the bat as a ranging problem, the objective of the bat is to select the side containing the echo at the standard range (click + B-echo) over the A-echo at variable range. This will be difficult for the bat if the range of the A-echo is close to that of the standard. Alternatively, the bat could solve the task by selecting the standard target (click + B-echo) over the comparison (A-echo), independent of their ranges. Since there are several differences other than delay between the click + B echo and the A-echo (e.g., spectral properties and overall energy), this alternative task should not challenge the bats. The fact that the bats were unable to solve the task at some settings of A-echo delay is thus somewhat surprising. This suggests that the bats in these cases understood the task as a range-discrimination problem despite being trained using a detection paradigm and using the click + B-echo alone as a control during experimentation. At other A-echo delays, or times, the bats seem to interpret the task as a detection problem and made fewer mistakes by choosing the click + B-echo pair. If the interpretation that the bats sometimes made range discriminations and at other times relied on detection is valid, this would explain the variation in data points seen in fig. 50.3. We will not consider this further in the following, but only take the decrease in performance as evidence that the bats did in fact rely predominantly on range information in solving the task and thus worked mainly in the time domain. The important conclusion from the experiments is that the click *can* interfere with ranging under certain circumstances.

Electrophysiological Correlates

Electrophysiological recordings from the brainstem (the nuclei of the lateral lemniscus) and the midbrain (inferior colliculus) of *E. fuscus* (Tougaard, Casseday, and Covey 1998) revealed that responses from about 75% of the neurons in these areas are affected if a click is presented immediately preceding or simultaneously with a tone or a downward FM sweep. In the majority of the neurons tested, the latency of the response to tones or FM sweeps was disturbed. For neurons in the columnar division of the ventral nucleus of the lateral lemniscus (VNLLc), clicks presented immediately before pure tones or FM sweeps interfere with the latency of the responses. Examples of this are shown in figs. 50.4 and 50.5. The neurons in the VNLLc are characterized by being

strictly phasic in their response, and their latencies are little affected by both stimulus frequency and intensity (Covey and Casseday 1991). These "constant latency" neurons are believed to play a central role in coding echo arrival time and pulse-echo delay (Covey and Casseday 1991, 1995). Thus, disturbance of the latency in these neurons is expected to interfere with the bat's ability to perform range determination. These neurons show "latency ambiguity," since a click presented at the right time relative to an FM sweep will give rise to a response on its own, while at the same time suppressing the response to the sweep, as seen in fig. 50.4 and fig. 50.5A–B. This might shift the bat's perception of the arrival time of the sweep by up to 1 ms, corresponding to about 170 mm difference in range, which could interfere with the bat's ability to locate its prey. This effect is a likely candidate for a neural correlate of the results of Miller (1991).

Another effect of the clicks on neural responses was observed by Tougaard, Casseday, and Covey (1998). In the inferior colliculus and all areas of the lateral lemniscus, with the exception of the VNLLc, a suppressive effect of the clicks was seen in about 25% of the neurons. These neurons responded readily to downward FM sweeps, but a click presented before or during the first part of the sweep was able to suppress, partially or completely, the response to the sweep. An example is shown in fig. 50.5C. How this suppression affects the performance of the bat's sonar is unknown.

Role of Clicks in Defense against Bats

As mentioned in the introduction, three functions of the moth's clicks in relation to bat predation have been suggested: interference with echolocation (jamming), aposematism, and startle.

The major problem with evaluating the different possibilities is the lack of experiments in which clicks are presented to the bats in a realistic way. The studies by Acharya and Fenton (1992) and Dunning et al. (1992) are exceptions, since the clicks were produced by the moths. These experiments, on the other hand, suffered from the inability to correlate the observed behavior of moths and bats with accurate sound measurements.

Other studies presented clicks (recordings of real clicks or artificial ones) from a loudspeaker in connection to either feeding behavior of the bats (Dunning and Roeder 1965; Stoneman and Fenton 1988; Bates and Fenton 1988; and Miller, Futtrup, and Dunning, chapter 52, this volume) or in range-discrimination tasks (Surlykke and Miller 1985; Miller 1991; and present experiment).

Interference with ranging

The range-discrimination experiments (Surlykke and Miller 1985; Miller 1991, present study) were clear in their conclusions. Only for clicks presented within a window of 0–2 ms before onset of the returning echo

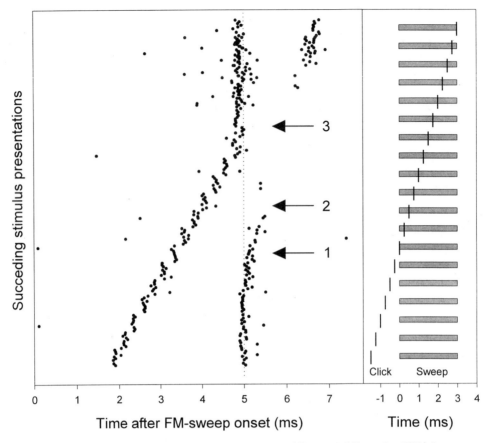

Fig. 50.4. Responses of a "constant latency" neuron (Covey and Casseday 1991) from the columnar division of the ventral nuclei of the lateral lemniscus (VNLLc) in *E. fuscus*, to different combinations of a downward FM sweep (38–18 kHz) and a broadband click. The sweep was centered around the best frequency of the neuron and had a duration of 3 ms. The intensity of the click was +15 dB peak to peak relative to the sweep. Each dot represents one spike, and 10 successive stimulations for each click-sweep pair are plotted above one another. Right column indicates the temporal relation between click (*vertical line*) and sweep (*horizontal bar*). Two groups of dots are seen. One group has a more or less constant latency of 5 ms and is the response to the sweep (more accurately, the center of the sweep, which is the best frequency of the neuron). The other is the response to the click, and the latency accurately follows the change in click position. Note that the timing of the response of the neuron to the sweep (artificial echo) begins to lag at arrow 1 and disappears between arrows 2 and 3. For experimental details see Tougaard, Casseday, and Covey (1998).

has an effect on distance-discrimination ability been demonstrated for *E. fuscus*. In this situation, the range-difference resolution of the bat deteriorated dramatically from better than 6 mm to about 25 cm (Miller 1991), and it was even shown that there need not be a one-to-one relationship between clicks and echoes for clicks to interfere with ranging. At ratios as low as 1:5 (one click burst in six echoes), the threshold was still increased by more than 400%. It is, however, not clear how clicks from real, flying arctiids affect ranging in a flying bat. Species like *C. tenera*, which possess microtymbals, could potentially rely on temporal interference. *C. tenera* emits a great number of clicks and will do so very late—within the last few hundred milliseconds before contact with the bat (Fullard, Simmons, and Sail-

lant 1994). It is conceivable that clicks are so numerous that a sufficient number of them arrive at the bat in the right temporal position relative to the echoes, causing interference with ranging. However, it is not clear how a deterioration in ranging ability at this late stage of the chase affects the ability of the bat to catch the moth. Although several experiments addressed this possibility, no results can reasonably be understood as effects of temporal interference (see Miller, Futtrup, and Dunning, chapter 52, this volume).

Multiple target confusion

The multiple target hypothesis, proposed by Fullard, Fenton, and Simmons (1979), suggested that clicks are perceived by the bat as extra "phantom echoes" and

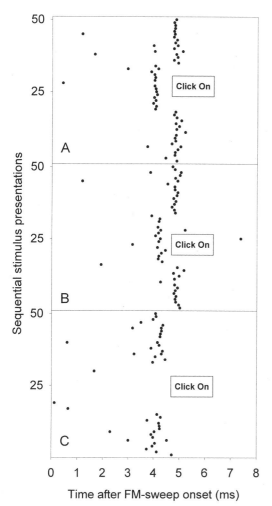

Fig. 50.5. Responses of two different single neurons from the brainstem of *E. fuscus* to downward FM sweeps and broadband clicks. *A* and *B* show the response of a "constant latency" neuron (VNLLc) to a 3 ms sweep 38–18 kHz and clicks at two different intensities (+15 dB pp and +5 dB pp re sweep, *A* and *B*, respectively), positioned 750 μs after onset of the sweep. Clicks were only presented during the middle third of the stimulus presentations. Note that the response from the neuron is time shifted when the click is on. *C* shows the suppression by a click on the response of a single neuron from the intermediate nuclei of the lateral lemniscus (INLL). The stimulus was a downward FM sweep (34–14 kHz, 2 ms, +15 dB pp re sweep). The click was presented at sweep onset and was only presented during the middle third of the stimulus presentations. Note that this neuron fails to respond when the click is on. For experimental details, see Tougaard, Casseday, and Covey (1998).

serve to disguise the moth and confuse the bat. There is little direct experimental evidence to support this theory unequivocally. The original proposal (Fullard, Fenton, and Simmons 1979) was partially based on unpublished observations that bats treat moth clicks as if they are real echoes. This claim was refuted by Surlykke and Miller (1985), who concluded that moth clicks have the same effect on the performance of the bats as broadband

noise stimuli. All ranging experiments found no effect on performance for clicks outside the critical 1–2 ms window in front of the echo. This means that either bats ignore moth clicks or accurately keep track of both click and echo, thus focusing on the echo important for the ranging task. Bats have no difficulties in detecting broadband clicks (e.g., Surlykke and Bojesen 1996) and can accurately determine the range when clicks are substituted for their own signals (Miller 1991). The difference between the ranging interference and the multiple echo theory, however, is not clear and is perhaps a matter of semantics. It seems appropriate to separate the two on the basis of whether the bat perceives the click as a real echo, which is odd due to its short duration and lack of sweep (see fig. 50.1D), or as an entirely extraneous sound unrelated to the bat's own sonar emissions.

Clicks as aposematic signals

Much evidence seems to support the aposematic theory, at least for arctiids like *A. caja*, which emit few clicks when the bat is several meters away (Surlykke and Miller 1985). It seems unlikely, for statistical reasons, that moth clicks emitted long before the final buzz will interfere with the ranging abilities of the bat. It is not possible for the moth to emit a click 1–2 ms prior to the return of the echo to the bat. Thus, the moth must rely on chance in the temporal placing of the clicks to interfere with the bat's ranging ability. Since the repetition rate in the approach phase is still relatively low, this appears to be a futile task for moths like *A. caja*.

There seems to be little, if any, simple relationship between the different arctiid species' ability and willingness to click and their palatability to bats, which would otherwise be predicted from an aposematic theory. This could suggest the presence of Batesian mimicry (some arctiids produce clicks, but are not toxic and thus rely on the nonpalatability of other clicking species). This, however, does not explain the presence of noxious but rarely clicking species. These could rely on the smell of toxins together with the fact that bats catch moths not directly with the mouth but instead in the tail pouch or wing membrane. Thus, the moth has a chance of escaping more or less unhurt even if caught, since the bat should "think twice" when smelling the moth.

Dunning and Roeder (1965) demonstrated that *Myotis lucifugus* trained to catch mealworms tossed into the air would veer away if clicks were played simultaneously with the catapulting of the worm. The authors interpreted the result in favor of the aposematic theory. However, the intensity of the clicks was fairly high (about 100 dB SPL at the location of the bat) and Miller, Futtrup, and Dunning in a recent experiment (chapter 52, this volume) were unable to show effects of more moderately intense clicks (74 dB SPL) on catch success in *Pipistrellus pipistrellus*. It is thus possible that the response observed by Dunning and Roeder (1965) was in fact a startle response.

Startle

A startle reaction should not be excluded as a possibility. The latency of the acoustic startle reflex in rats is <15 ms (Horlington 1968) and may be even less in bats, due to their small size. This may be sufficiently short to be of advantage to the moth. Møhl and Miller (1976) demonstrated that ultrasonic clicks generated by the peacock butterfly (*Inarchis io*), upon tactile stimulation from a crawling bat, elicit a dramatic startle reaction in the bat. A startle function also could explain why some arctiids apparently do not react to acoustic stimulation, but will click readily upon tactile stimulation (Dunning and Krüger 1995). This raises the possibility that these species rely on evoking a startle reaction from the bat after they are caught (i.e., while in the tail pouch).

Conclusions

During the last two decades we have learned much about the interactions between echolocating bats and arctiid moths. More than anything, we have learned that it is unlikely that there is one, simple explanation for the role of arctiid clicks as a defense against predatory bats. No theory can be completely excluded, and, because they are not mutually exclusive, it is likely that different species of arctiids use different strategies for defense. Perhaps the same species, or even the same individual, uses clicks differently, according to the situation.

Whatever the mechanisms, clicks seem to be an excellent "evolutionary choice" of arctiids in the defense against bats. Arctiids might not be alone in this "choice" since certain tiger beetles also emit clicks when confronted with bats (Yager and Spangler 1997; Yager 1999).

Additional experiments clearly are needed in order to understand the role of arctiid clicks. It seems impor-tant to determine whether effects on prey-catching ability of free-flying bats when hunting in free flight can be linked to the interference with ranging performance seen under experimental conditions. It also appears crucial to obtain high-quality sound recordings of the clicking behavior of free-flying arctiids when attacked by bats. Most knowledge about when and how arctiids emit clicks is extrapolated from experiments using unnatural sounds as stimuli (Fullard and Fenton 1977; Dunning 1968; Surlykke and Miller 1985) or carried out on tethered moths (Fullard, Simmons, and Saillant 1994). The temporal relationship between calls of the bat, the returning echoes, and the emitted clicks during interactions between arctiid moths and bats are central to understand. It is vital to record sounds at the location of either the moth or the bat during free field (or semi–free field) interactions. This could be done by accurate high-speed video tracking, perhaps together with equipping the bat with a microphone and a radio transmitter, as has been done in another context by Lancaster, Keating, and Henson (1992). In this way, there might be a good chance to answer some central questions on this fascinating system of predator-prey interactions.

Acknowledgments

The electrophysiological experiments were conducted at Duke University Medical Center together with John H. Casseday and Ellen Covey, whom we thank for their support and hospitality. We appreciated the constructive comments given by the editors and two referees. We also thank the Danish National Research Foundation, the Danish Research Council for Natural Sciences, Odense University, and Aarhus University for substantial financial support.

{ 51 }

Auditory Adaptations of Katydids for the Detection and Evasion of Bats

Paul A. Faure and *Ronald R. Hoy*

Introduction

The evolution of echolocation was a major sensory adaptation that undoubtedly played a huge role in the diversification and radiation of bats. This is because the ability to echolocate allowed bats to tap into an ecological niche and food source that previously was unavailable to other mammalian and avian predators—that food source, of course, being nocturnal flying insects. Clearly, the ability to echolocate is a distinct advantage in detecting nocturnal aerial insect prey.

However, like any biological interaction, the contest

between predator and prey is never simple. Bats evolved the ability to echolocate insects in total darkness; but over time, and given the considerable selection pressure that echolocation placed on aerial insects, many insects evolved ultrasound avoidance and escape behaviors. In this chapter we more generally refer to such behaviors as ultrasound-induced acoustic startle responses (ASRs).

Perhaps the best example of an ASR in insects is the evasive flight maneuvers of tympanate moths elicited in response to the echolocation cries of bats (Roeder and Treat 1957, 1961). Many moths possess tympanal hearing organs whose sole function, with few exceptions, is the detection of bat echolocation calls (Fenton and Fullard 1981). Upon hearing bat biosonar, flying moths initiate defensive behaviors that typically result in the insect flying away from the sound source so as to avoid being detected, or if the bat is in close proximity (and thus its call is of high amplitude), a series of last-ditch evasive maneuvers designed to thwart capture and predation (e.g., cessation of flight and dropping to the ground, Roeder 1962). The entire anti-bat response of the moth is initiated based on input from one to four primary auditory afferent neurons (Roeder 1967). Interestingly, the idea that moths may use their ears to detect "the shrill squeaking of bats" was suggested even before the discovery of echolocation itself (White 1877).

More than forty years have passed since Kenneth Roeder and Asher Treat first described this fascinating predator-prey interaction, and we now know that many species of nocturnal insects—including lacewings, crickets, locusts, katydids, mantids, beetles, and butterflies—have independently evolved ultrasound-sensitive ears and ultrasound-induced ASRs (see Hoy, Nolen, and Brodfuehrer 1989; Hoy 1992; Hoy and Robert 1996; Yager 1999 for reviews). A consistent feature of bat-detecting insect ears is their broad tuning and sensitivity to ultrasonic frequencies between 20 and 50 kHz; this best frequency (BF) range encompasses the peak spectral frequencies (PFs) commonly emitted by species of sympatrically foraging, aerially hawking bats (Fullard 1987). The fact that ASRs are so widespread among nocturnal flying tympanate insects reinforces the notion that bat echolocation was and continues to be an important driving force in the evolution of insect hearing. (Readers who are interested in alternative types of ASRs and defenses of insects to bad predation are referred to the chapters by Rydell [43]; Tougaard, Miller, and Simmons [50]; and Miller, Futtrup, and Dunning [52] in this volume.)

By virtue of their simpler nervous systems and behavior patterns, which in turn allow for the identification of specific neurons and cellular mechanisms controlling behavior, insect startle and escape responses have been particularly amenable to neuroethological investigation (e.g., Westin, Langberg, and Camhi 1977; Camhi and Tom 1978; Camhi, Tom, and Volman 1978). A startle re-

sponse can be defined as a short-latency, abrupt, fast-movement stereotypic motor act reliably elicited by stimuli with a swift and/or unexpected onset (Friedel 1999). Note that this definition lacks an absolute time reference, because the only time scale that matters is that of a prey's natural enemies (Bullock 1984). By their nature, startle behaviors are adaptive: they rapidly bring prey into a state of alertness, thereby warning or priming them for escape. Startle responses have been the focus of a number of investigations trying to identify specific neurons or networks controlling behavior (e.g., Eaton 1984); the rapidity of startle allows the neuroethologist to determine the exact timing of the response (i.e., its onset and termination), whereas the quality of motor stereotypy allows for a reliable determination and quantification of startle occurrences.

ACOUSTIC ORTHOPTERA

Given the diversity of signals produced and the varied roles that sounds plays in their lives, considerable effort has been devoted to understanding peripheral and central auditory processing in orthopteran insects (locusts, haglids, crickets, katydids, and their allies). To date, the majority of this research has focused on adaptations for the recognition and localization of mate-calling song. In most orthopterans the male stridulates to produce a species-specific song to which sexually receptive females may be attracted—a response known as positive phonotaxis. Nevertheless, orthopterans also listen for and respond to heterospecific acoustic signals, particularly the sounds of predators, whether they be active vocalizations (e.g., bat biosonar) or the incidental noises produced by predators moving through the environment, and natural selection should favor individuals that detect and respond to threatening acoustic stimuli by moving away from the sound source—a behavior called negative phonotaxis. Within this context, it should be noted that visual and vibrational signals are also important releasers of startle responses in all orthopteran insects (e.g., Pearson and O'Shea 1984; Friedel 1999).

The task of recognizing and distinguishing conspecific from predatory signals is simplified in many orthopterans (e.g., crickets, locusts, and haglids) by the fact that conspecifics emit sounds with low carrier frequencies, whereas predators often produce short-duration (transient) signals containing high-frequency ultrasound (Sales and Pye 1974). However, for many insects in the family Tettigoniidae (i.e., katydids, bushcrickets, or long-horn grasshoppers), discriminating conspecifics and predators is not so straightforward. This is because katydid songs are normally broadband and contain both audio (<20 kHz) and ultrasonic (≥20 kHz) frequencies (e.g., Keuper et al. 1988). Indeed, the songs of some katydids are purely ultrasonic (e.g., Morris et al. 1994). Hence, distinguishing conspecific from predatory sources of ultrasound is a nontrivial task for many tettigoniids. Moreover, fewer studies have devoted themselves to understanding behavioral and physiological adaptations for processing predatory sounds in the acoustic Orthoptera (but see Moiseff, Pollack, and Hoy 1978; Nolen and Hoy 1984, 1987; Belwood and Morris 1987).

ACOUSTIC STARTLE RESPONSES OF *NEOCONOCEPHALUS ENSIGER*

The eastern sword-bearer conehead katydid, *Neoconocephalus ensiger* Harris (fig. 51.1A), is a common inhabitant of fields, meadows, and roadside ditches. Its mate-calling song is one of the more conspicuous nocturnal sounds that a human listener is likely to encounter from mid-July through October in the Midwest and northeastern regions of the United States and southern Ontario, Canada. During the day, *N. ensiger* are silent and seek refuge in the ground vegetation. At dusk, however, males climb up on grass stalks where they sing to females, and continue calling, uninterruptedly for minutes at a time, well into the night. To the human ear, *N. ensiger*'s song sounds like a train of raspy or lisping syllables ("tsip-tsip-tsip," Dethier 1992).

Katydids are normally considered ground-dwelling insects and as such they are exquisitely sensitive to sounds and vibrations in their natural habitat (see Kalmring and Elsner 1985). Nonetheless, individuals will fly during dispersal events, to avoid predators, and while searching for mates. Virtually any insect that flies at night is at risk from predation by aerially hawking bats, and *N. ensiger* is no exception. Indeed, *N. ensiger* possesses a robust in-flight ASR: when stimulated with loud, short-duration pulses of batlike ultrasound, flying *N. ensiger* extend their front and middle legs posteriorly along their body and rapidly close all four wings, causing them to drop toward the ground (Libersat and Hoy 1991). Presumably this ASR confers a survival advantage by allowing katydids that perform the behavior to escape bat predation.

We have recently discovered that *N. ensiger* possess a second, previously unknown type of ASR: cessation of singing (Faure and Hoy 2000a). When stimulated with pulsed ultrasound, stridulating males cease mate-calling and remain motionless (and cryptic), drop from their perch and/or jump to the ground where they hide by burrowing into the vegetation, or take to the wing (i.e., evasive flight). As with cessation of flight, cessation of song (i.e., silencing oneself) presumably evolved as a way of avoiding predators—in this case acoustically orienting terrestrial predators such as mice, cats, shrews, and substrate-gleaning bats (Walker 1964; Sales and Pye 1974; Belwood and Morris 1987; Belwood 1990).

Threshold-tuning curves for the two ASRs are virtually identical, suggesting that a common neural mechanism underlies both forms of escape (Faure and Hoy 2000a). However, for such a mechanism to function adequately, the katydid must perform a nontrivial dis-

crimination. Because *N. ensiger*'s song is broadband and contains both sonic and ultrasonic frequencies (fig. 51.1), this immediately raises an interesting question: how does the central nervous system (CNS) of *N. ensiger* categorize and distinguish conspecific song, which contains ultrasound, from predatory ultrasound so that mate attraction and predator avoidance behaviors are reliably and appropriately performed? Despite overlap in the spectral and temporal characteristics of the two signal types, *N. ensiger* rarely startle when listening to conspecific song but reliably do so when stimulated with batlike ultrasound (Libersat and Hoy 1991; Faure and Hoy 2000a). Thus, our goal was to discover neurophysiological correlates underlying behavioral perception and sound categorization in katydids (Wyttenbach, May, and Hoy 1996). In this chapter, we present evidence that the physiology of a prothoracic auditory interneuron, the T-cell, is well suited for this purpose, and suggest that the T-cell plays a significant role in coordinating and/or mediating the ASR of both flying and singing katydids.

Materials and Methods

Adult *Neoconocephalus ensiger* were collected from fields near Ithaca, New York. Males were localized by listening to their calling song. Female *N. ensiger* do not sing; thus they were found by scanning the tops of grass seed heads with a flashlight. Katydids were housed in individual cages containing apple, cat chow, and water. Most animals were maintained in a cold room (12°C) to slow their rate of senescence, but were transferred to a rearing room (25°C) at least 1 day prior to their use in electrophysiological experiments. (For details on animal care, the methods and apparatus used in physiological recording, acoustic stimulation and calibration, and the experiments and data described below, see Faure and Hoy 2000a, 2000b, 2000c, 2000d.)

NEUROPHYSIOLOGY

Electrophysiological experiments and sound recordings were performed on adult katydids inside a 1.10 × 0.65 × 0.65 m chamber lined with acoustic foam. Standard extracellular techniques were employed. Briefly, katydids were waxed ventral-side-up to an elevated platform of a custom-made insect holder, and the CNS was exposed by removing the cervical membrane overlying the neck connectives. Katydids were minimally dissected and special care was taken not to cut or damage the internal acoustic trachea, which are important for providing acoustic gains at ultrasonic frequencies (Michelsen et al. 1994). An electrolytically sharpened tungsten hook electrode, referenced to a second tungsten electrode stabbed into the abdomen, was used to record T-cell responses. Action potentials were amplified (A-M Systems

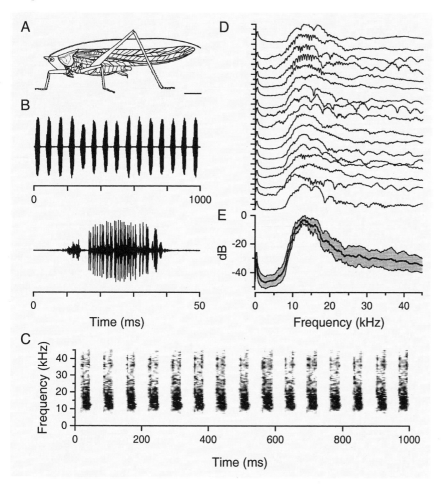

Fig. 51.1. Temporal and spectral characteristics of the mate-calling song of the eastern sword-bearer conehead. *A:* Male *Neoconocephalus ensiger* (drawing by Martha B. Lackey). Scale bar = 5 mm. *B:* Time domain display of a 1 s train of calling song syllables (temperature = 24.1 ± 0.8°C) (*upper trace*); time expanded view of one syllable (*lower trace*). *C:* Spectrogram of a 1 s train of calling song syllables. *D:* Frequency domain displays of the average power spectrum of 30 syllables from 17 katydids. *E:* Mean ± SD power spectrum of the 17 males shown in *D.* Reprinted from Faure and Hoy 2000a, with permission from Springer-Verlag.

model 1700) and were either stored on VHS tape for offline analysis (Vetter model 400 PCM) or digitized directly using a Macintosh Centris 650 computer and GW Instruments MacAdiosII data acquisition board (stimulus envelope and spike digitization rate = 10 kHz). T-cell spikes were window discriminated and analyzed using Igor Pro (Wavemetrics, Inc.) and custom software written by Robert A. Wyttenbach.

ACOUSTICS

Pure-tone pulses (2–100 kHz) with a 1 ms linear or raised cosine rise/fall time, produced by a computer with an array processor and A-to-D/D-to-A interface purchased from Tucker Davis Technologies (TDT: Apos II), were used to determine T-cell auditory thresholds and elicit suprathreshold responses. The spectral and temporal characteristics of *N. ensiger*'s song were analyzed using Canary bioacoustics software (Charif, Mitchell, and Clark 1995). For playbacks of conspecific song, a 3 s segment of male song was digitized (TDT: AD1, sampling rate = 198 kHz), edited, and stored in a buffer that was looped to provide a user-defined playback duration. Digitized song was converted to analog (TDT: DA3-2) and low-pass filtered (f_c = 50 kHz, Krohn-Hite model

3550) prior to amplification. Digitally synthesized bat-like biosonar signals consisting of linear FM sweeps (FM depth: 80 → 30 kHz; duration: 10 ms or 30 ms; rise/fall time: 1 ms raised cosine) were generated with custom software written by Timothy G. Forrest. During acoustic playback, the presentation rate of conspecific song (Katydid signal) and the 10 ms (Bat 10 signal) and 30 ms (Bat 30 signal) bat FM sweeps was held constant at 14.25 Hz, which is the natural syllable repetition rate of *N. ensiger* singing at 25°C (Faure and Hoy 2000a).

Stimulus amplitudes, measured with a Brüel and Kjær (B&K) type 2209 Impulse Precision Sound Level Meter fitted with either a type 4135 or type 4138 microphone (without protecting grid), were calibrated with a B&K type 4220 Pistonphone. Amplitudes are expressed in decibels sound pressure level (dB SPL re 20 μPa) and were adjusted using a programmable attenuator (TDT: PA4) and stereo amplifier (Nikko NA-790). All signals were broadcast with a Panasonic EAS-10TH400B leaf tweeter. The position of the loudspeaker is given by its angle relative to the katydid's rostral-caudal body axis (anterior = 0°, lateral = 90°). Ipsilateral refers to the ear and auditory spiracle closest to the loudspeaker.

Results

BIOACOUSTICS

The mate-calling song of *N. ensiger* consists of a continuous train of loud, short-duration, broadband syllables. The syllable repetition rate varies from 5 to 15 pulses/s, depending on the ambient temperature (Frings and Frings 1957). Syllables averaged 30.8 ms in duration and were composed of a series of clicks (fig. 51.1B), with each click representing the passage of the scraper (located on the dorsal surface of the right forewing) over a single tooth of the file (located on the ventral surface of the left forewing). The amplitude of the song, measured 10 cm from stridulating males, ranged from 80 to 100 dB SPL. As previously mentioned, *N. ensiger*'s song contains both audio and ultrasonic frequencies. The spectrograph in fig. 51.1C, which shows a 1 s train of syllables from a single male, reinforces the broadband structure of the song. Fig. 51.1D shows the average power spectra of 30 syllables from 17 individual males. The PF of the song ranged from 10.5 to 16.3 kHz, although significant acoustic energy was present at higher ultrasonic frequencies. The bandwidth of the song at −20 dB (re PF) ranged from 7.9 to 37 kHz (fig. 51.1E). (For additional details on *N. ensiger*'s song, see Faure and Hoy 2000a.)

NEUROPHYSIOLOGY

T-cell identification

The presence of a large amplitude, phasic-spiking prothoracic auditory interneuron known as the T-cell (T large fiber or TN1) has been recognized for four decades in katydids (Suga and Katsuki 1961a, 1961b). We used a pair of hook electrodes to monitor spike discharge patterns recorded simultaneously in the anterior and posterior connectives of the ventral nerve cord (VNC) emanating from the prothoracic ganglion ipsilateral to the loudspeaker, to confirm that the large amplitude spikes in the VNC of *N. ensiger* were produced by a T-cell. By counting the number of spikes and measuring spike latencies recorded with each electrode, we were able to verify a characteristic physiological fingerprint of the T-cell: that stimulus-evoked spikes traveling rostrally and caudally within each connective of the VNC were from one and the same cell (Suga and Katsuki 1961a). (For additional evidence confirming the identity and homology of the T-cell with previous reports in the literature, see Faure and Hoy 2000c.)

T-cell tuning

Fig. 51.2 shows the average T-cell threshold tuning curve from 25 *N. ensiger* males. As is evident, the T-cell is fairly sensitive and broadly tuned. Its average BF is approximately 25 kHz, although the neuron is quite sensitive from 12 kHz up to 65 kHz. Note that this best sensitivity range overlaps with the frequencies present in conspecific song (fig. 51.1), as well as with frequencies

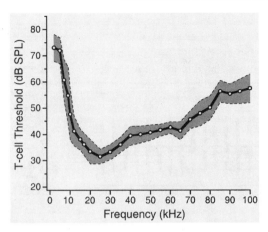

Fig. 51.2. T-cell tuning in adult male *N. ensiger*. The solid line with open circles is the mean excitatory threshold tuning curve, while the broken line with shading indicates the range of ± 1 SD (*n* = 25 male katydids; loudspeaker position = 90°).

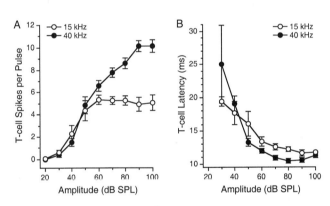

Fig. 51.3. T-cell intensity-response functions at 15 kHz (open circles) and 40 kHz (closed circles). *A:* Mean ± SE number of spikes per pulse (*n* = 4–13 katydids per point). *B:* Mean ± SE spike latency (*n* = 4–13 katydids per point). In both panels, the loudspeaker position is 90°. Data set reprinted from Faure and Hoy (2000c), with permission from The Company of Biologists Limited.

typically emitted by North American species of echolocating bats (e.g., *Eptesicus fuscus*) (Simmons 1987).

Intensity/response functions

Single, 10 ms pure-tone pulses of either 15 kHz (mimicking the PF of conspecific song) or 40 kHz (mimicking the PF of a bat cry) were used to examine T-cell suprathreshold physiological responses (fig. 51.3). With 40 kHz stimulation the T-cell showed nearly a monotonic increase in spike number from 40 to 90 dB SPL, whereas pulses of 15 kHz did not produce such a response (fig. 51.3A). At 15 kHz the number of T-cell spikes per pulse increased with increasing stimulus amplitude, but only for low to moderate SPLs; above 60 dB SPL the number of T-cell spikes often plateaued, and in some preparations a decline in response strength was apparent. As is typical for many sensory systems, the latency to the first spike decreased with increasing stimulus amplitude (fig. 51.3B). Although the difference in latency between sonic and ultrasonic stimulation is not as pronounced as the difference in spike number, T-cell latencies evoked by stimulation with 40 kHz pulses were nevertheless consistently shorter than latencies evoked with 15 kHz pulses over the majority of SPLs tested. (For suprathreshold responses at additional frequencies and loudspeaker positions, see Faure and Hoy 2000c.)

Short-term temporal pattern coding

By varying the repetition rate of 25, 10 ms pulses to mimic the syllable pattern of a calling male katydid (15 kHz carrier) or the biosonar pulses of a hunting bat (40 kHz carrier), we measured the T-cell's ability to synchronize its spiking output to stimulus trains varying in repetition rates from 1 to 100 Hz; these rates encompass the range of tempos naturally exhibited by katydids singing in the field at different ambient temperatures, as well as the range of pulse emission rates employed by aerially hawking bats during the search, approach, and terminal phases of echolocation. Fig. 51.4 summarizes the T-cell's temporal pattern copying ability at 70 dB SPL. For pulse rates ≤5 Hz, the T-cell more or less faithfully copied the temporal pattern of the stimulus at both 15 and 40 kHz. However, for repetition rates >5 Hz, the proportion of stimulus pulses eliciting one or more T-cell spikes at 15 kHz quickly declined, whereas with 40 kHz stimulation the T-cell showed proficient temporal pattern copying for repetition rates up to 30 Hz (fig. 51.4). Even at 40 Hz the T-cell was still capable of encoding more than half of the stimulus pulses when stimulating with 40 kHz ultrasound. The difference in the T-cell's temporal pattern copying ability at 15 and 40 kHz increased when the amplitude of the stimulus was increased; for 40 kHz stimulation at 90 dB SPL the T-cell was able to encode more than 50% of stimulus pulses reliably for pulse rates as fast as 40 to 100 pulses/s (Faure and Hoy 2000c).

Acoustic playback experiments

The above experiments show that the T-cell responds differentially to pure tones mimicking the spectral and temporal characteristics of katydid song and bat ultrasound. One obvious question is whether differences in T-cell responsivity are maintained when katydids are presented with playbacks of conspecific song and batlike FM sweeps. Although *N. ensiger*'s T-cell showed superior temporal copying at 40 versus 15 kHz over a broad range of pulse repetition rates (fig. 51.4), the design of the experiment that tested this used, for the most part, only short-duration stimulus trains (typically <1 s) and thus was appropriate for short-lived events such as predator-prey encounters. In nature, male *N. ensiger* stridulate continuously for minutes at a time; therefore it is equally important to examine T-cell responses when

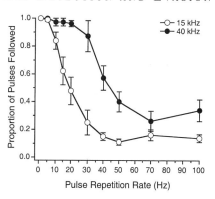

Fig. 51.4. T-cell short-term temporal pattern coding at 70 dB SPL. Shown are the mean ± SE proportion of stimulus pulses encoded by one or more T-cell spikes as a function of stimulus repetition rate with the loudspeaker position at 90°. Stimulus trains consisted of 25, 10 ms pulses (1 ms linear rise/fall time). The order of stimulus presentation was varied within and between katydids ($n = 11$ animals tested). Data set reprinted from Faure and Hoy (2000c), with permission from The Company of Biologists Limited.

presented with long-term acoustic stimulation of conspecific and predatory signals.

Example T-cell responses to playbacks of conspecific song and batlike FM sweeps presented for 3 min at 70 dB SPL are shown in fig. 51.5. Note that the data are all from the same T-cell. Within each panel, the upper graph is a dot raster diagram, while the lower graph is a poststimulus time (PST) histogram over the duration of acoustic playback. The duration of the 10 ms 80 → 30 kHz FM sweep (Bat 10 signal) was chosen to mimic the search and/or approach phase biosonar pulses of a variety of aerially hawking bats that hunt sympatrically with *N. ensiger;* however, because the average syllable duration of *N. ensiger*'s calling song is three times longer than the Bat 10 signal, we also generated a 30 ms 80 → 30 kHz FM sweep (Bat 30 signal) for comparison.

The difference in potency between the three signal types is obvious: T-cell spiking and temporal pattern copying were strongest for batlike FM sweeps. When listening to the Bat 10 signal the T-cell of this katydid responded with multiple spikes per stimulus pulse (1.6 ± 0.84 spikes/pulse), at a short latency (14.70 ± 1.67 ms), and at a high instantaneous firing frequency (227.31 ± 225.22 Hz). Visually, this is seen as a distinct band of time-locked spikes (dots) extending over the entire playback duration. Surprisingly, the longer duration Bat 30 signal elicited fewer T-cell spikes (0.40 ± 0.64 spikes/pulse), at a longer latency (26.22 ± 8.76 ms), and at a lower instantaneous spike rate (11.58 ± 53.22 Hz) than its Bat 10 counterpart. In contrast, stimulating the same T-cell with the Katydid signal resulted in even fewer spikes (0.11 ± 0.40 spikes/pulse), occurring at a relatively long latency (27.55 ± 12.60 ms), and at a lower instantaneous firing frequency (5.29 ± 41.70 Hz). In response to the Bat 10 signal at 70 dB SPL there were rel-

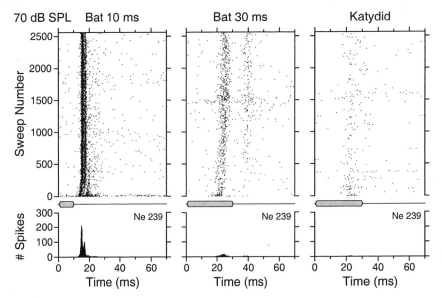

Fig. 51.5. Historaster displays of T-cell spiking in response to acoustic playback with a 10 ms 80 → 30 kHz FM sweep (*left panel*), a 30 ms 80 → 30 kHz FM sweep (*center panel*), and katydid conspecific song (*right panel*) presented at 70 dB SPL (*N. ensiger* #239, adult male). Within each panel, the upper graph shows the timing of individual T-cell spikes as dots, relative to the onset of the stimulus, with successive stimulus presentations (sweeps) stacking on top of one another. Hence, time increases with each sweep number (playback duration = 3 min, stimulus presentation rate = 14.25 Hz, ∴ the total number of stimulus pulses = 2565). The lower graph is a post-stimulus time (PST) histogram of the cumulative number of T-cell spikes over the entire playback duration (bin width = 0.1 ms). The duration of each stimulus is illustrated schematically as a gray pulse (*middle*). Reprinted from Faure and Hoy (2000d), with permission from The Company of Biologists Limited.

atively few instances of T-cell misfires (<1%)—that is, when the neuron did not respond with at least one action potential per stimulus pulse—whereas spike failures were obvious and common in response to the Bat 30 (66%) and Katydid signals (92%). In fact, assuming that the T-cell would fire at least one spike per stimulus pulse at 70 dB SPL, which is well above the neuron's average threshold (fig. 51.2), stimulation with conspecific song resulted in an average instantaneous firing frequency that was lower than expected based simply on the rate of stimulus delivery (i.e., 14.25 Hz). (For additional details on T-cell responses in male and female katydids to acoustic playback of conspecific and predatory signals presented at 50, 70, and 90 dB SPL, see Faure and Hoy 2000d.)

Discussion

Despite a long and prominent history in the insect hearing literature, the function of the T-cell in the acoustic behavior of katydids has remained unclear. Two nonmutually exclusive hypotheses have been proposed:

RECOGNITION AND LOCALIZATION OF CONSPECIFICS

Because auditory activity in the ipsilateral hearing organ has a strong inhibitory effect on responses of the con-

tralateral T-cell, T-cell spiking varies with the direction of a sound source (Suga and Katsuki 1961a; Rheinlaender and Römer 1980). A pharmacological study showed that this inhibition likely stems from an interneuron using a chemical synapse (Suga and Katsuki 1961b). It should be noted that T-cell spikes do not simply correspond to what is registered by primary tympanal afferents in the peripheral nervous system (PNS); auditory receptors fire tonically whereas T-cell responses are phasic or phasi-tonic (Suga and Katsuki 1961a; McKay 1969). The change in spike impulse pattern from the PNS to the CNS likely is determined by local (intraganglionic) inhibitory neurons, such as the pair of Ω-neurons (omega or ON1) that appear to sharpen side-to-side (binaural) contrasts in other orthopterans (Selverston, Kleindienst, and Huber 1985).

The discoverers of the katydid T-cell reported that it seemed well adapted to the reception of conspecific stridulation (Suga and Katsuki 1961a). Combined with the neuron's directional sensitivity, this resulted in a number of studies that assumed or inferred that the principal function of the katydid T-cell was in the detection and localization (lateralization) of singing conspecifics (e.g., Rheinlaender and Römer 1980; Rheinlaender, Hardt, and Robinson 1986). If true, then one would not expect T-cell responses to adapt to playbacks of conspecific song, but close inspection of Suga and Katsuki's original example response trace (see fig. 5A in Suga and Katsuki 1961a) clearly reveals that spike adaptation was manifest. Moreover, existing data on the T-cell's ability to detect (respond) and follow (encode) the syllable pattern of conspecific song are contradictory; published abilities range from little or no temporal coding (McKay 1969), to some initial coding but with adapting responses (Kalmring, Rehbein, and Kühne 1979; Zhantiev and Korsunovskaja 1983), to a one-to-one correspondence with the pulsatory stridulatory sound (Suga and Katsuki 1961a) or the doublet syllable pulse (Schul 1997).

T-cell adaptation to playbacks of conspecific song was particularly evident in *N. ensiger* males (fig. 51.5), even at the highest SPLs, whereas in females T-cell responses to the Katydid signal were somewhat improved (Faure and Hoy 2000d). This suggests that a sex difference may exist in the physiology of *N. ensiger*'s T-cell. Previously, we showed that the tuning of the T-cell differs between the sexes, with *N. ensiger* females being slightly more sensitive in the spectral band encompassing male song (Faure and Hoy 2000c). With this in mind, it may not be surprising that the T-cells of *N. ensiger* females encoded conspecific song better than in males. Perhaps females also use their T-cells to evaluate the ultrasonic components (quality?) of a male's song when they are in close proximity? Nevertheless, in both males and females, T-cell responses were clearly biased in favor of predatory batlike signals (Faure and Hoy 2000d).

DETECTION OF PREDATORS

McKay (1969, 1970) first suggested that the tettigoniid T fiber functions as a warning neuron, reporting that it responds preferentially to short-duration high-frequency sounds, that it rapidly adapts to continuous tones, and that it responds poorly or not at all to conspecific song. A decade later Kalmring, Rehbein, and Kühne (1979) published the anatomy of an auditory giant neuron (axonal diameter = 10 μm) in the VNC of the European bushcricket *Decticus verrucivorous*. Based on the cell's adaptation to conspecific song—a response characteristic previously noted by McKay (1969, 1970)—and assuming that the neuron was part of the katydid's giant fiber system, Kalmring, Rehbein, and Kühne (1979) concluded that the auditory giant neuron of *D. verrucivorous* (probably the T-cell, but the soma did not stain) was well suited to a warning or arousal function. Römer, Marquart, and Hardt (1988) reported that the onset of synaptic activity in the T fiber of the bushcricket *Tettigonia viridissima* was delayed by only 0.8–1.2 ms relative to the first spike of the afferent receptor burst recorded only microns away, implying a monosynaptic connection. Such fast throughput is consistent with the T-cell functioning in a predator detection circuit.

Physiologists have long considered giant fibers to play a significant role in the initiation and coordination of startle and escape behaviors (e.g., Wiersma 1947; Wine and Krasne 1972; Eaton, Bombardieri, and Meyer 1977; Westin, Langberg, and Camhi 1977; Zottoli 1977; Drewes, Landa, and McFall 1978; Mackie 1984). Due to their large axonal diameters, giant fibers increase action potential conduction velocities, thereby providing fast uninterrupted nervous conduction over relatively long distances (Parnas and Dagan 1971)—an indispensable feature for rapid arousal and escape. Thus, it was the T-cell's giant fiber appearance, combined with its short response latency and high sensitivity to ultrasonic frequencies, that led researchers to suggest a predator detection function. Unfortunately, few studies had examined T-cell responses specifically from a predator detection point of view.

The present results on the T-cell of *N. ensiger* strongly support the predator detection hypothesis. Although the T-cell is equally sensitive (re threshold) to pure tones centered near the PF of conspecific song and to PFs typical of North American bat biosonar (fig. 51.2), it gave stronger spiking responses at a shorter latency to short-duration pulses of ultrasound (fig. 51.3). A pronounced difference also was observed in the T-cell's ability to encode the temporal pattern of short-duration stimulus trains encompassing the range of tempos naturally emitted by singing katydids and hunting bats. The T-cell reliably copied the temporal pattern of 40 kHz pulses over a wide range of repetition rates, whereas stimulation with 15 kHz pulses resulted in significantly poorer temporal copying (fig. 51.5). Finally, acoustic playback with conspecific song and batlike FM sweeps resulted in clear differences between these two classes of behaviorally relevant stimuli. T-cell responses were distinctly biased toward predatory ultrasound; the interneuron responded with more spikes per pulse, at a shorter latency, and at a higher instantaneous firing frequency to 10 ms 80 → 30 kHz FM sweeps than to pulses of conspecific song or, surprisingly, 30 ms 80 → 30 kHz FM sweeps (fig. 51.4). The lower firing frequency and pronounced adaptation manifest in response to conspecific song—particularly in *N. ensiger* males—renders it unlikely that the excitatory responses of the T-cell play a major role in the detection and localization of singing conspecifics (Faure and Hoy 2000c, d).

Conclusion

Our goal was to identify mechanisms in the CNS of *N. ensiger* that have evolved in response to the acoustic selection pressures imposed by ultrasound-emitting predators, particularly echolocating bats. The present data show that responses from a single prothoracic auditory interneuron—the T-cell—are well suited for this purpose. Furthermore, they suggest that the T-cell is a likely candidate contributing to the ASR of both flying and nonvolant katydids (Faure and Hoy 2000a). Aerially hawking bats increase their echolocation pulse repetition rates from 5 to >20 Hz during the search phase, to 20 to >50 Hz during the approach phase, to 50 to >100 Hz during the terminal phase of insect pursuit (Kalko 1995a). The high sensitivity, short latency, and strong (nonadapting) spiking responses of the katydid T-cell to pulses of batlike ultrasound make it precisely adapted for encoding information relevant to predator detection and escape. Exactly what role the T-cell plays in the bat avoidance response of flying *N. ensiger* is still unknown and is the subject for future investigations. For instance, is T-cell activation both necessary and sufficient for eliciting the in-flight ASR, in a manner similar to the bat-detecting Interneuron-1 (Int-1) of flying field crickets (Nolen and Hoy 1984)? Nevertheless, we have demonstrated that responses from a single neuron are sufficient for discriminating behaviorally relevant acoustic stimuli pertinent to the behaviors of mate attraction (positive phonotaxis) and predator avoidance (negative phonotaxis).

Acknowledgments

We thank the numerous helpers who assisted us in collecting katydids from the field, and Margo Chiuten and Kristi Snyder for providing katy care. Past and present members of the Hoy Lab—particularly Hamilton Farris, Timothy Forrest, Andrew Mason, Daniel Robert, and Robert Wyttenbach—provided constructive criticisms of this work, as did Christopher Clark, Cole Gilbert, Carl Hopkins, and three external reviewers. We thank Cynthia Moss for the invitation to attend the Biological Sonar Conference and Jeanette Thomas for her many untold efforts in organizing the meeting and its publication. This research was conducted at Cornell University and was supported by operating and equipment grants from NIMH (NIDCD R01 DC00103) to Ronald R. Hoy. Paul A. Faure was supported by a Sir James Lougheed Award of Distinction (Alberta Heritage Scholarship Fund) and teaching and research assistantships from Cornell University and Ronald R. Hoy.

{ 52 }

How Extrinsic Sounds Interfere with Bat Biosonar

Lee A. Miller, Vibeke Futtrup, and *Dorothy C. Dunning*

Introduction

Griffin (1958) reported that big brown bats (*Eptesicus fuscus*) were attentive to false echoes and other high-frequency signals that bore no relationship to the echoes from their own calls. Since then, numerous studies confirmed Griffin's original observations. For example, the sounds of insects and the echolocation signals of neighbors attract individual foraging bats (Balcombe and Fenton 1988). Laboratory experiments also show that bats are attentive to sounds other than the echoes of their own emissions. Initially, at least, extraneous sounds like ultrasonic clicks and batlike signals attract bats that otherwise use their echolocation for ranging or orientation (Surlykke and Miller 1985; Bates and Fenton 1990). Thus, the evidence at present suggests that echolocating bats use two auditory pathways simultaneously, one for processing echoes from their own sounds (intrinsic signals) and the other for processing extrinsic signals (see Fuzessery and Lohuis, chapter 18, this volume).

Briefly, extrinsic sounds that could disturb bat echolocation fall into two classes: abiotic and biotic. Abiotic sounds in the environment, such as noises from running water, could disturb bats hunting near the water surface. Short-duration ultrasonic noise from ripples might be one reason why *Myotis daubentoni* avoids hunting insects over rippled water (Rydell, Miller, and Jensen 1999). Many bats, like *Pipistrellus pipistrellus*, are highly social and often hunt together (Kalko 1995a), so sounds from neighbors classify as extrinsic (biotic) signals—which also are interesting, as mentioned above.

Extrinsic sounds can disturb bat biosonar if these occur in strict temporal correlation with returning echoes (Miller 1991; Masters and Raver 1996). Certain insects produce ultrasonic clicks when they hear bat echolocation sounds. Individuals of different species of arctiid moths (family Arctiidae) broadcast clicks of varying numbers, temporal patterns, and intensities when confronted with bats. The short broadband clicks protect arctiid moths against bats (Acharya and Fenton 1992), but exactly how is still debated (Tougaard and Miller, chapter 50, this volume.) One idea is that moth clicks interfere with ranging. In fact, if the moth clicks fall within a short temporal window 1–2 ms *before* the expected echo, the bat's ability to determine range deteriorates by as much as 4000% (Miller 1991). Range resolution also becomes poorer by about 400% when bat sounds or modified bat sounds occur just before an echo (Masters and Raver 1996). Empirically, moth clicks appear to function best as interfering signals.

One species of arctiid, *Cycnia tenera,* actually waits until the bat is in the final phase of its pursuit before emitting copious bursts of clicks (Fullard, Simmons, and Saillant 1994). Here the bat is continuously updating the position of the prey and clicks from the arctiid moth during this critical phase of the pursuit would optimally interfere with the bat's sonar system (Fullard, Simmons, and Saillant 1994).

Here we present the results of laboratory experiments with pipistrelle bats (*Pipistrellus pipistrellus*) trained to catch catapulted prey in the presence of prerecorded sounds (a click sequence from the arctiid, *Phragmatobia fuliginiosa,* and a conspecific bat buzz sequence). We strove to present the sounds as naturally as possible in our experimental setting. The bats' echolocation signals triggered the playback of sounds, so they received the sounds during the late approach and terminal phase of their hunting signals. The intensity of the sounds was adjusted to the natural intensity of the arctiid clicks expected at the bat's position. The clicks and bat buzz sounds did not reduce the success of prey capture, and we discuss the possible reasons for this.

Materials and Methods

We trained two female bats (*Pipistrellus pipistrellus*) to fly in an oval circuit in a flight cage (7 m long × 4.8 m wide × 2.35 m high). We used overhead illumination (less than 55 Lx), but the bats caught prey as well in the dark as with low illumination. Bat 1 learned to catch and eat mealworms (larval *Tenebrio molitor*) and "miniworms" (larval *Alphitobius diaperinus*), about half the size of *T. molitor,* while bat 2 preferred moth bodies (noctuid moths without legs and wings). The apex of the trajectory was about 1 m above the floor. The trajectories varied slightly for the different prey targets and for the same prey from trial to trial. All three targets tumbled in flight, producing variable echo intensities that were not measured. Both these factors probably prevented the bats from having a priori knowledge of the exact position of the target. By the end of training, bat 1 captured

between 70% and 95% of the catapulted worms while bat 2 captured between 40% and 70% of catapulted moth bodies.

Following the training period, we ran one session per day per bat, presenting bat and moth sounds on alternate days. The bats were allowed to catch and eat prey until satiation; thus, they determined the number of trials per session. Sound trials were interspersed with nonsound trials on a pseudorandom schedule, with an attempt to keep equal numbers of sound and nonsound trials and no more than three consecutive trials under the same conditions. Trials were tallied only when both bat and prey were in the catch site. Bat 1 completed 10 sessions with a total of 591 trials and bat 2 completed 9 sessions with a total of 166 trials.

We used two types of playback sounds, both originally recorded on tape using a Racal Store 4 and a Brüel and Kjær (B&K) system (see below). One sound consisted of a single modulation cycle (two click bursts) from the arctiid moth, *Phragmatobia fuliginosa* (fig. 52.1A). The energy of the burst extended from 24 kHz to 72 kHz (−6 dB points). The duty cycle of the moth sound was 14%, which is the natural duty cycle. (The duty cycle of a moth click burst [or bat signal] was calculated by dividing the duration of a click burst [bat signal] by its duration plus the time interval to the next burst [bat signal], and expressing the quotient as a percentage.) The second playback sound consisted of five *P. pipistrellus* signals taken from the buzz I phase of a bat hunting in the field (fig. 52.1B). The energy in the bat sounds ranged from 57 to 92 kHz (−6 dB points) and the duty cycles ranged from 14% to 20%. Moths and bats of these two species are sympatric in Denmark; therefore it is probable that bats of this species naturally would encounter these moths when hunting.

The sounds were digitized at 352.8 kHz and files edited to 37.5 ms 12-bit segments with equal maximum peak-to-peak amplitudes. We used a custom-designed program to play back the digital sounds to the bats. The selected playback could be triggered by the bats' own echolocation signals when the interval between two consecutive bat signals reached ≤15.2 ms. This value was chosen because the bat would be in the late approach phase or the terminal phase by the time it reached the catch site and received the playback sounds. Moth clicks presented during the approaching bat's buzz are thought

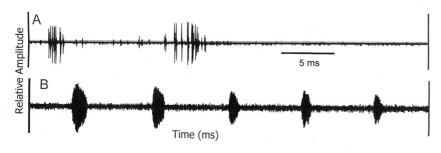

Fig. 52.1. Playback sounds. Two types of sounds were played back to bats approaching the catch site. The playback sounds consisted of a click-burst pair (*A*) recorded from an arctiid moth (*P. fuliginosa*), or a portion of a "buzz" (*B*) recorded from a bat in the field (*P. pipistrellus*). (See "Materials and Methods" for more details.)

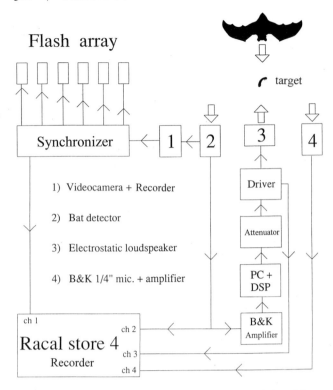

Fig. 52.2. Schematic of the setup. The nondivided signal from the bat detector (2) served as the input to trigger the playback sounds. The divide-by-ten detector signal (2) led to one audio channel of a video cassette recorder (1). (See "Materials and Methods" for more details.)

to interfere maximally with the bat's biosonar (Fullard, Simmons, and Saillant 1994).

The output of an electrostatic bat detector microphone, which was placed beneath the loudspeaker, was amplified (B&K model 2610) and served as the input to a digital signal processing (DSP) board (SPB2, Signal-Data, Copenhagen, Denmark) (fig. 52.2). On triggering, the playback sound was repeated 13 times, giving a complete playback time of 487.5 ms. The electronic playback sound from the DSP (352.8 kHz D/A rate) went through a passive conditioning filter (12–118 kHz, 8 pole), an analog attenuator, and a driver amplifier before being transduced into sound by a custom-built 60 mm diameter electrostatic loudspeaker (fig. 52.2). The output of the loudspeaker was flat to $\pm$ 3 dB from 20 to 120 kHz. The loudspeaker was placed 3 m directly in front of the center of the catch site.

The catch site was defined as a circle parallel to the plane of the loudspeaker, with the center at about the apex of the catapulted prey's trajectory. We chose the circumference to be the −6 dB points relative to the sound amplitude at the center for a 50 kHz signal. With this definition, the catch site had a radius of 29 cm. The depth of the catch site was only about 0.5 m along a 3 m axis between the target and the loudspeaker, so variations in stimulus sound pressure were much less than 6 dB in this

plane. We adjusted the maximum sound pressure for both types of playback sounds at the center of the catch site to 73 dB peak equivalent (pe) sound pressure level (SPL) re 20 μPa, which is the natural level for *P. fuliginosa* clicks at a distance of 20 cm (Surlykke and Miller 1985). Pipistrelle bats begin the terminal buzz at 20–50 cm from a target (Kalko 1995a).

All 19 sessions were video recorded using Sony DXC-101P camera (40 ms per frame) and a custom-made video/flash array (6 Nikon SB 16 flash guns), which was activated manually and produced a flash for each video frame (fig. 52.2). The video camera was located just behind and above the loudspeaker. It focused on the center of the catch site and the zoom was adjusted so the entire catch site filled the video field. A light-emitting diode (not shown in fig. 52.2) behind the catch site in the video field was illuminated for the duration of the stimulus. The video signal was recorded on a VCR (Panasonic 6200 VHS). Voice notes and bat signals from a custom-made divide-by-ten bat detector were recorded simultaneously on the audio tracks of the videotape.

We analyzed video recordings of all sessions frame-by-frame and tallied the number of prey captures. We used 2 × 2 contingency analysis for chi square to test the statistical significance of the number of prey captured versus the type of sound playback (table 52.1). To analyze for possible body reorientations after playback sounds commenced, we edited short video clips from 24 prey capture trials, 12 for each bat. Half of these trials included moth clicks triggered by the bats' signals, and the other half were no-sound trials.

Following the sessions, we devoted nine additional sessions (three for bat 1 and six for bat 2) to recording the bats' echolocation signals on a Racal Store 4 (762 mm/s) using a B&K system (microphone type 4135, 6.35 mm with cap removed, preamplifier type 2619, and measuring amplifier type 2604 with an HP filter, −3dB at 745 kHz). The microphone was placed 3 m in front of the center of the catch site and about 10 cm to one side of the loudspeaker. High-speed tape recordings of bat signals were digitized at an effective analog-to-digital sampling rate of 352.8 kHz (16-bit) using a Sona-PC hardware/software system (Waldmann EDV-Systeme, Tübingen, Germany). The same system was used for generating spectrograms, and power spectra were generated using Spectra Plus (Pioneer Hill Software, Poulsbo, Wash., USA).

Among other acoustic parameters, we also measured the number of harmonics and the relative amplitude for each of the last 15 buzz signals in nine trials without and nine trials with moth clicks chosen randomly from 24 trials. We started the analyses at about the fourth signal after the onset of playback clicks, and after an equal number of equivalent signals in trials with no sounds, giving a total of 270 (18 × 15) signals. Each power spectrum (1024 point FFT) contained only one signal at a time, and the weakest buzz II signal was at about +1 to +3 dB relative to the background noise.

We also measured the duration of the buzz phase (buzz I plus buzz II) in 12 trials with moth clicks and 20 trials without clicks for bat 2. We used a paired *t*-test to compare means at $p = 0.05$.

Results

Bat 1 was significantly more proficient at catching prey (table 52.1) in the presence of sounds ($p = 0.005$), and also when the sounds were bat buzzes ($p = 0.029$). The bat's prey-catching behavior also improved in the presence of moth click bursts, but not significantly. Under these circumstances, bat 2 was as proficient at catching prey in the presence of playback sounds as in their absence ($p = 0.1026$).

A frame-by-frame analysis of the 55 flight paths ending in successful catches showed that 76% of these occurred within the inner portion of the catch site having a radius of 14.5 cm, whether or not there were playback sounds. Though the resolution of the 12 video clips for each bat were not fine enough to show ear or head movements, we detected no differences in the bats' behavior as they closed in on the target in the presence of moth clicks.

Detailed analyses of the bats' sonar emissions revealed that more harmonics were added to the terminal signals and the relative amplitude increased when moth clicks were present. These were the only parameters analyzed that obviously changed in response to playback sounds.

Our automatic recording/playback system began registering the bats' emissions about 300–600 ms before the sounds were triggered. Most of these signals were from the approach phase (fig. 52.3). About 50 ms prior to triggering, both bats reduced the interval between signals before they decreased the signal duration. The bats received the first playback sound when they were in the early buzz or late approach phase. Prey capture occurred about 100-150 ms following the trigger.

Fig. 52.3 exemplifies spectrograms of the sonar emissions as bat 1 closed in on the target in the presence of arctiid clicks. The sequence starts with the last "search"

TABLE 52.1. Prey-capture performance by two pipistrelle bats, *P. pipistrellus,* as affected by playback sounds. The bats' behavior was tallied only when they were at the catch site. Bat 1 caught significantly more prey than expected when she triggered sounds (bat buzz and moth click trials pooled) ($\chi^2 = 7.76$, $n = 591$, $p = 0.0053$), as well as when she triggered bat sounds ($\chi^2 = 4.72$, $n = 362$, $p = 0.0299$). The number of trials in the respective category is indicated by (*n*).

Categories	Bat 1		Bat 2	
	Sessions	Catch success % (*n*)	Sessions	Catch success % (*n*)
Sounds/no sound	10	93/86 (264/327)	9	64/51 (87/79)
Bat buzz/no sound	6	92/83 (157/205)	6	65/52 (54/54)
Moth clicks/no sound	4	95/89 (107/122)	3	64/48 (33/25)

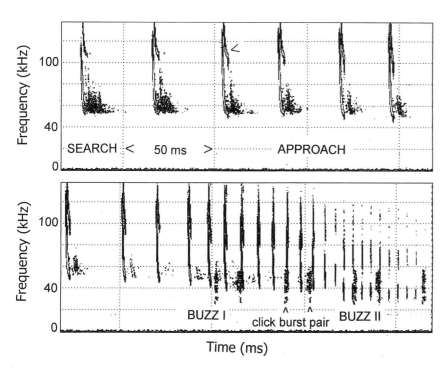

Fig. 52.3. Spectrograms of echolocation signals from bat 1 (*P. pipistrellus*) during a trial with clicks as playback sounds. The figure shows the last search signal, the approach phase, and the terminal or "buzz" phase. Buzz II started when the frequency of signals shifted downward almost 10 kHz. (The labels BUZZ I and BUZZ II signify the start of these phases.) The playback sounds (arctiid clicks) in this sequence began at about the start of buzz I and continued throughout the buzz phase. (Seven click bursts are shown in the lower panel. The arrows indicate the second click-burst pair.) The trigger occurred at the bat signal prior to the first click burst (lower panel). (To obtain the correct temporal relationship between bat signals and click bursts, the former must be shifted about 18 ms to the left.) Only the lower frequencies of the clicks are shown here, since the microphone was beside and parallel to the loudspeaker transmitting the clicks.

signal. We define the beginning of the "approach" phase as the first sonar signal having a flattened S-shape that is most easily seen in the second harmonic (arrow in fig. 52.3, upper panel). The approach phase began at about 300 ms before buzz II. We define the transition from the approach phase to buzz I as the point where the interval between emissions decreased abruptly and continued decreasing. The buzz II phase began with the sudden drop in frequency (from about 40 kHz to about 30 kHz). In fig. 52.3 the first click-burst pair began during buzz I, and the trigger occurred at the bat signal prior to the first click burst. The playback continued for 487.5 ms, which is considerably longer than the buzz phase. Buzz I signals have durations of about 1.5–0.5 ms, with intervals of 20–8 ms. Buzz II contained signals as short as 0.2 ms, with intervals of about 5 ms. The duty cycle increased by as much as 20% during the buzz I phase, but fell during the buzz II phase to below 5%. The mean duration of the terminal phase (buzz I and buzz II) was 90.4 ± 20.0 ms with moth clicks present, and 85.0 ± 19.0 ms without clicks for bat 2. Durations were not significantly different ($t = 0.76$, $n = 12, 20$). Buzz II signals can be reliably counted and the bat used essentially

the same number ($\bar{x} = 7.1 \pm 1.2$, $n = 9$, contra $\bar{x} = 7.0 \pm 1.7$, $n = 9$; $t = 0.244$; $p = 0.813$) irrespective of the situation. (Note: $\bar{x}$ is the mean.)

We analyzed the harmonic content (fig. 52.4) and relative amplitude of the last 15 echolocation signals from bat 2 in trials with and without moth clicks. When moth clicks were present, there were two to three harmonics in buzz I signals and three to four harmonics in buzz II signals ($\bar{x} = 3.20 \pm 0.11$, $n = 9$). There were mostly two harmonics in all buzz signals analyzed when clicks were absent ($\bar{x} = 2.32 \pm 0.33$, $n = 9$); the difference was significant ($t = 7.57$, $p < 0.001$). In the presence of moth clicks the bat's buzz signals were more intense ($\bar{x} = 12.11 \pm 0.95$ rel. dB, $n = 9$) as compared to those in the absence of clicks ($\bar{x} = 13.22 \pm 0.96$ rel. dB, $n = 9$). The intensity difference, when converted to a linear scale, was also significant ($t = 2.952$, $p = 0.0184$). Bat 2 added more harmonics to its terminal signals when buzz signals from another bat were presented during prey capture, but we had too few data to test for significance.

Discussion

The purpose of this study was to play back natural sounds that potentially could interfere with the echolocation of pipistrelle bats as they caught prey. The sounds we chose, moth clicks and bat buzz sounds, are known to affect bat behavior (Dunning and Roeder 1965; Balcombe and Fenton 1988). We presented the arctiid click sounds at levels that can be expected when the bat starts its terminal phase at 20–50 cm from the prey. Based on the target strength of an echo from a moth (about −48 dB at 1 m), the playback sounds were about 25 dB louder than echoes from catapulted insects. Moth clicks should optimally confuse bats when these occur in the terminal phase of prey capture (Fullard, Simmons, and Saillant 1994). Presenting arctiid clicks during the terminal phase as the bats approached an aerial target did not interfere with their abilities to capture prey. Presenting bat buzz sounds in the same manner likewise did not confuse the bats. In fact, the buzz sequence from another bat improved the success rate of prey capture by bat 1 (table 52.1). The feeding buzzes of a foraging red bat (*Lasiurus borealis*) attract conspecifics, suggesting the bats "eavesdrop" while processing their own sonar signals (Balcombe and Fenton 1988). It may be that bat 1's motivation for catching prey was heightened in the presence of conspecific buzz signals, and it was therefore more successful.

Even though the catching behavior of bat 2 was not affected by playback sounds when these were triggered by the bat's signals, the bat clearly heard the moth sounds and responded by adding harmonics and increasing the intensity of its buzz signals. Increasing the number of harmonics and the intensity of sonar signals would be advantageous if bats interpreted moth clicks as noise.

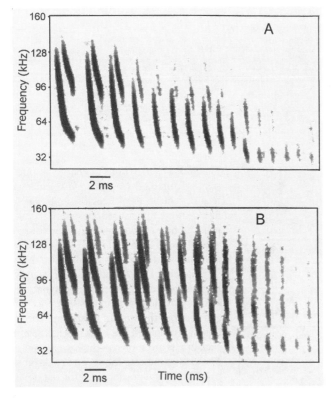

Fig. 52.4. Spectrograms of the last 15 echolocation signals bat 2 (*P. pipistrellus*) during prey capture without playback sounds (*A*) and during the playback of arctiid clicks (*P. fuliginosa*) (*B*). Echolocation signals occurring after the onset of clicks contain one to two additional harmonics (*B*). Only the spectrograms of signals are shown; thus the x-axis does not represent continuous time.

The increase in harmonics would sharpen the range resolution, and would shift more energy to higher frequencies, thus reducing spectral interference from clicks. In addition, increasing the intensity improves the signal-to-noise ratio.

Why did *P. fuliginosa* clicks and bat buzzes not interfere with prey capture? There could be several reasons why our playback signals did not interfere with the bats' abilities to capture prey. First, assuming that *P. pipistrellus* performs ranging as does *E. fuscus,* clicks falling within a 1–2 ms window preceding an echo can drastically reduce the range resolution (Miller 1991). Apparently, the low duty cycle of *P. fuliginosa* sounds means that the bats could receive enough echoes separated from moth clicks to use for determining the position of the prey. Copious clicking should increase the chances for some clicks to fall within the critical window before an echo. Arctiid moths producing clicks with high duty cycles should have a better chance to acoustically confuse their bat predators.

Secondly, the sounds in our experiments obviously were not coming from the prey. The loudspeaker was placed 3 m from the apex of the catapulted prey's trajectory. Although most prey captures occurred near the center of the catch site, the bats were not necessarily on a direct line between the prey and the loudspeaker. We are reluctant to speculate on how directionality influenced the bats' behavior. However, clicks arriving within the short temporal window before the echo and that are heard by the bat should interfere with ranging (Miller 1991). Recent neurophysiological studies have shown that arctiid-like clicks presented with specific temporal relationships to simulated echoes can interfere with the timing of discharges from brain neurons (Tougaard and Miller, chapter 50, this volume). Our results show that the bats heard the sounds, but their catching performance was unaffected when they received the sounds in the late approach or the buzz of the catch sequence.

Thirdly, if the moth clicks functioned as startle sounds, our bats habituated rapidly. Or, if our bats knew from previous experience that clicks signified noxious prey (aposematism), they evidently learned that this was not the case now (Tougaard and Miller this volume). Although we did not test the aposematic hypothesis, we know from earlier studies that both *P. pipistrellus* and *E. fuscus* quickly learn to avoid noxious prey associated with clicks (Surlykke and Miller 1985), even when clicks do not come from the prey itself (Bates and Fenton 1990). They also habituate to clicks, after an initial startle, when there are no noxious associations (Bates and Fenton 1990; Miller 1991).

Why do our results differ from those of a similar study by Dunning and Roeder (1965)? They reported that three trained little brown bats, *Myotis lucifugus,* at least touched 77% of the catapulted mealworms when exposed to a series of echolocation pulses (approach through buzz) from a conspecific bat as they flew in to catch the target. However, the bats touched only about 13% of the prey and dodged 85% when recorded clicks of the arctiid, *Halysidota tessellaris,* were played back through the same loudspeaker at the same intensity. Our bats were as successful at catching prey with moth clicks as with conspecific bat sounds. We never saw them dodge targets, under any conditions. The two bat species, *M. lucifugus* and *P. pipistrellus,* detect and capture mealworms using comparable sonar signals and behavior.

Differences in the intensities and/or duty cycles of playback sounds could account for the discrepancies between our results and those of the previous study. Dunning and Roeder (1965) adjusted the level of their playback sounds to between 100 and 110 dB SPL in the catch site. This is about 30 dB louder than what we used, and 50–60 dB louder than the calculated amplitude of wild *H. tessellaris* at 20 cm, when a pursuing bat would be in the buzz phase. The trained *M. lucifugus* surely were accustomed to hearing loud conspecific signals; thus the bat playback sounds were probably of no surprise. The duty cycles of the moth clicks and the bat signals in the 1965 experiments were similar and about 20% overall, but those of the clicks varied considerably during each stimulus sequence (Dunning, personal observations). The arctiid clicks might have been so loud and novel that the bats mostly avoided the catch site, or they could have interfered with the bats' sonar during high-duty-cycle intervals. We cannot differentiate between these two possibilities.

Acknowledgments

We thank Simon Boel Pedersen for providing software. We appreciate the constructive comments on the manuscript from Torben Dabelsteen, University of Copenhagen, Jakob Tougaard, Niels Skals, two reviewers, and the editor. We acknowledge financial support from the Danish National Research Foundation, the Danish Research Council for Natural Sciences, and Odense University. Appropriate permits were obtained.

{ 53 }

Functional Aspects of Echolocation in Dolphins

Denise L. Herzing and *Manuel E. dos Santos*

"Our perception of how dolphins utilize their sonar in the wild is based on extrapolation of knowledge obtained in 'laboratory' experiments—we do not have the foggiest idea of how dolphins utilize their sonar in a natural environment" (Au 1993, p. 271).

Introduction

This chapter surveys some key aspects of the existing knowledge on the functions of echolocation signals emitted by dolphins and other odontocetes, comparing these advances in understanding of cetacean biosonar to those obtained by similar research on bat echolocation. An attempt is made to present the issues requiring clarification, and the topics where fresh empirical research is needed.

Echolocation Signal Characteristics

The echolocation abilities and the signal characteristics of a few odontocete species have been studied in detail. Au (1993, 1997) provided recent reviews especially concerning the bottlenose dolphin (*Tursiops truncatus*), the beluga (*Delphinapterus leucas*), the false killer whale (*Pseudorca crassidens*), and the harbor porpoise (*Phocoena phocoena*). Not all odontocetes have been specifically demonstrated to echolocate, although echolocation-type signals have been recorded from most species.

Richardson et al. (1995) listed 13 species with "echolocation demonstrated" and 17 species with "echolocation-type" clicks. The sperm whale (*Physeter macrocephalus*) was not included in this list, and only indirect evidence exists for the largest of the odontocetes to be considered an echolocator (Mullins, Whitehead, and Weilgart 1988; Watkins and Daher, chapter 57, this volume).

For some species such as the Clymene dolphin, *Stenella clymene,* little is known about their signal production (Perrin and Mead 1994), but it is more than probable that their acoustic characteristics will be similar to those of related species. In other groups, like the ziphiids, information is very scarce. Lynn and Reiss (1992) described click trains recorded in a pool with two young Hubb's beaked whales (*Mesoplodon carlhubbsi*), but

the equipment was limited in frequency response. Thus the signals may not be representative of the species, and their function was not clear.

Au (1997) divided odontocete species according to their echolocation signal characteristics. Two main categories of signals were summarized: broadband and short (<100 μs), often with the energy peak, or one of the energy peaks, below 70 kHz; and narrowband, long (>125 μs), with the energy peak at about 110 kHz. Interestingly, Ketten (1997) found good basis for this gross classification in the anatomy of the cochlea, especially in terms of the basal ratios (thickness-to-width ratio of the cochlear basilar membrane), and even on an ecological basis. In her typology, the first category of echolocation signals presented above is produced by animals with basal ratios between 0.5 and 0.7, functional hearing limits below 160 kHz, and highly social lives with low-frequency communication signals. The second category is more typical of inshore phocoenid and riverine platanistid species, adapted to the needs of fine-detail discrimination in turbid waters. Au (1997) also included smaller delphinids, such as Commerson's or Hector's dolphins (*Cephalorhynchus commersonii* and *C. hectori,* respectively), in this group.

The higher intensity signals of the bottlenose dolphin, the false killer whale, the narwhal (*Monodon monoceros*), and the Atlantic spotted dolphin (*Stenella frontalis*) were measured in open water, reaching source levels of 210–230 dB re 1 μPa (Au, Herzing, and Aubauer 1998). In contrast, porpoise ultrasonic pulses usually fall below 170 dB (Au 1997), and sperm whale clicks were measured at 180 dB, at frequencies below 16 kHz (Watkins 1980).

In some species, but not all, echolocation clicks were found to be highly directional, with the strongest beam in front of the head, showing a 5° upward tilt. The 3 dB beamwidth (the cone-shaped area with a 3 dB loss compared to the center of the beam) in the bottlenose dolphin was about 10–12° and even narrower (6.5°) in the beluga (Au, Moore, and Pawloski 1986; Au, Penner, and Turl 1987). As these authors noted, the level increment due to directionality can reach 30 dB. This promotes sonar performance for the animal, but makes it harder to obtain the complete spectrum and the highest levels in recordings made at sea, with free-ranging animals.

The high-frequency echolocation components suffer a greater attenuation outside of the strongest beam than the less directional low-frequency components. In species that produce broadband signals, this low-frequency portion of clicks may be detected even when the animal's head is directed at a greater angle from the target. In the power spectra of four bottlenose dolphin clicks, representative of the variability found by Au (1997), one may notice that at 20 kHz (the useful range of many recording systems) the relative amplitudes were at about 70%, 50%, 50%, and 20% compared to the peak.

Distances at which a delphinid uses echolocation signals to discriminate small objects are in the 100 m range in experimental settings (Richardson et al. 1995). However, the disturbing number of accidental entrapments of these animals in fishing nets (Kraus et al. 1997) should alert us to the fact that the abilities demonstrated by trained and focused animals in experimental conditions do not necessarily reflect the spontaneous use of this sensory system in natural, routine situations.

In all species studied, clicks are emitted in trains that vary in duration, number of pulses, and repetition rate (or inversely, interclick intervals). Bottlenose dolphins in stationary discrimination tasks emit each click after the echo of the preceding one is received and processed. Thus, if the distance to the target is incremented by the experimenter, the interclick intervals increase, allowing for the longer two-way transit (TWT) time. There is also a lag time corresponding to the neural processing time. Au (1993) suggested that this processing time should range between 19 and 45 ms.

In observations of dolphins echolocating on objects greater than 1 m away, dolphins tend to adjust the interclick interval so that the echo is returned before the next signal is sent. This ensures that the analysis of the previous click occurs before another outgoing click is sent, perhaps avoiding any masking effect on the weaker echo. In experiments with stationary and moving animals, the interclick intervals of dolphins increased with the distance to the target, allowing for a TWT time (see Au 1980, p. 115, for summary). However, variations do occur, indicating that click rate may have other features currently not understood.

In some species, and within target ranges of less than about 0.4 m, interclick intervals are far too short for this echo-by-echo processing. Trains of 500 clicks per second (interclick interval of 2 ms), so common in some circumstances, could be instances of echo-bulk processing, or could be reserved for other functions, not strictly sensory. One beluga studied by Turl and Penner (1989) performed discrimination tasks using click trains that in some cases did not respect TWT times. Interestingly, the beluga strongly preferred interclick intervals of about 45 ms. It was unclear whether this was an individual peculiarity or a species adaptation. Very high-repetition-rate trains (2 kHz) also were recorded in social situations

with free-ranging Atlantic spotted dolphins (Herzing 1996; Herzing, chapter 56, this volume), orcas, *Orcinus orca* (Ford 1991), Hector's dolphins (Dawson 1991), and, in Sigurdson's (1998) study of moving, echolocating bottlenose dolphins.

This suggests not only that odontocetes use "shorter than two-way transit time" intervals, but that packets of clicks (burst-pulsed sounds) can also be manipulated in an "echolocation way" to provide information to the dolphins. It is unclear whether for odontocetes in their natural environment outgoing signals really mask returning echoes, thus making the very short interclick interval trains less useful for echolocation. Another interesting feature of echolocation, whose function is unknown, were double clicks recorded from pilot whales, *Globicephala* sp. (Purves 1967); harbor porpoises, *Phocoena phocoena* (Verboom and Kastelein 1997); and spinner dolphins (Lammers et al., chapter 58, this volume). Because of the paucity of information from moving dolphins in social contexts, we could have overlooked other processing possibilities of echolocation clicks.

Issues of Signal Design and Discrimination

Echolocation signals of odontocetes operate in a medium where sound travels approximately 4.4 times faster than in air. This property of seawater makes it more difficult to use time-separation cues from echoes and impossible to benefit from Doppler-shift information. Being in air gives bats advantages in this respect. On the other hand, the impedance mismatch between the medium and the targets is lower in water than in air. This enables dolphins to obtain a great deal of information about the inside of targets, while for bats most of a signal's energy is reflected by the surface of the target (Au 1993, 1997).

The research on the discrimination abilities of the bottlenose dolphin's sonar is now in its fourth decade. These studies have revealed unexpected powers of echolocation and a fine adaptation to the properties of their medium. Major reviews were presented by Nachtigall (1980) and by Au (1993), detailing the discrimination of target shapes, such as cylinder length, diameter, sphere diameter, target range, material composition, solid versus hollow targets, and wall thickness and texture. Most studies were conducted with the targets in the water, but a few also tested discrimination abilities for targets buried in sediment (Roitblat et al. 1995 for the bottlenose dolphin; Kastelein et al. 1997 for the harbor porpoise). If targets were buried only a few centimeters, both species were still quite good at discrimination. In the wild, the detection and retrieval of buried prey has been observed (Herzing 1996; Rossbach and Herzing 1997). In addition, the description of foraging techniques specific to prey species and habitat (Herzing, chapter 56, this volume) could indicate that dolphins have the ability to discriminate specific prey species.

The information-bearing parameters in the echoes could be target strength, time-separation cues, spectral (frequency) differences or echo highlight structure. When the acoustic energy of a signal penetrates a target, it creates a series of echo highlights on the inner structures and surfaces. These secondary reflections follow the first echo, produced by the front surface. The very short dolphin clicks allow for detectable differences in the arrival times of the highlight echoes (Au 1993). The number of discernible highlights and the total echo duration are also relevant parameters that echolocating animals could use to discriminate targets. As to target strength differences, one should keep in mind that dolphins can resolve as little as 1 dB differences in echo pressure (Evans 1973).

Research on energetic costs of echolocation production for odontocetes is clearly needed, to advance knowledge at the physiological level and also to allow a better understanding of the ecological, social, and evolutionary aspects of this orientation behavior in these aquatic mammals. Speakman and Racey (1991) measured and monitored the coordinated movements of wing beats and echolocation production in bats, showing the rather economical coupling of the two activities. Bats exhale during the wing upstroke and emit pulses only during exhalation, thereby minimizing the energetic costs of echolocation. Longer pulses are emitted only once per upstroke; shorter pulses are produced in bursts limited to the upstroke. In either case, bats effectively use the energy of the wing beat for sound production.

Echolocation in the Wild: Natural Habitats and Prey

In contrast to the wealth of data on the physical and design characteristics of the odontocete sonar system, little is known about how dolphins manipulate and use echolocation signals in the wild. Food preferences and hunting strategies have been obtained primarily from sampling stomach content (Barros and Wells 1998) and from observations of surface behavior including fish kicking, stranding on mud flats to retrieve fish, and others (reviewed by Shane 1990). Würsig (1986) reviewed general delphinid foraging strategies and surface behavior in different environments. The use of echolocation signals in the detection and retrieval of prey is well established, and intense sound pressure levels of over 220 dB re 1 μPa (Au, Floyd, and Haun 1978; Au 1993) emitted by bottlenose dolphins could stun prey (Norris and Møhl 1981), although it has not been experimentally demonstrated. Atlantic spotted and bottlenose dolphins echolocate while scanning and digging for buried prey in sandy bottoms, increasing the repetition rate from 200 to 500 Hz as they direct their sound into the sand (Herzing 1996).

Little is known about the use of echolocation in the dolphin's natural habitats. Critical areas of inquiry for the future include (1) how a moving dolphin's strategy might vary from a stationary dolphin emitting echolocation clicks, and (2) the possible advantages of using passive audition concomitant with an active sonar system.

How should researchers begin to think about dolphin echolocation in the wild relative to what is known about bats? What parallel features can be compared? Two major ecological features that determine features of bat (Microchiroptera) echolocation are (1) physical environment (cluttered or uncluttered with vegetation or substrate), and (2) prey species—their movement, habitats, and evasive strategies. Although bats show flexibility, their foraging strategies are associated with particular forms of echolocation (Altringham 1996; Fenton 1984). Let's look at these strategies and extrapolate some possible aquatic parallels and comparative aspects with bat ecology.

ECOLOGICAL, ACOUSTICAL, AND BEHAVIORAL PARALLELS

Despite difference in sound velocity and optimum frequency use in aerial and aquatic media, echolocating animals may have encountered similar physical constraints and obstacles regarding environmental clutter and prey movement during evolution. Considering convergent evolutionary strategies and potential parallel adaptations, three main questions emerge: (1) Do dolphins deal with cluttered acoustic environments in the same way as bats? (2) Are dolphins using prey and environmental cues analogous to those used by bats? and (3) Are there parallel passive and active strategies used by both bats and dolphins for prey detection and capture?

Bat strategies: Detection and hunting

Reviews of principles of aerial sound transmission (Altringham 1996; Rydell 1993) suggests that (1) signals that are short, have a wide bandwidth, and are emitted in pulsed series are favorable for the measurement of target distance, angle (localization), and properties (shape, texture); and (2) signals that are long, with a narrow and constant frequency, are favorable for prey detection and trajectory and velocity estimation (Bradbury and Vehrencamp 1998).

Several major features impose tradeoffs in the signal design for various kinds of bat foraging strategies. Open-site foragers need high performance in all aspects: the detection, range and angle estimation, and target properties. Therefore, they send intense signals through the mouth, wide-beam varying with the capture phase: constant frequency–frequency modulated (CF-FM) passing to FM wider bandwidth, faster repetition rate. (As discussed above, with shorter ranges, repetition rate may increase without echoes overlapping with the next signal.)

For hawking (the detection, pursuit, and eating of prey on the move) and fishing bats, on the other hand, the premium is in prey detection and velocity measurement, so they use high-duty-cycle emissions combining FM portions before and after a long CF component. Bats hawking above or between vegetation (uncluttered background) use narrowband FM or CF search calls with no harmonics (15–30 kHz). This ensures early detection over long distances. Upon targeting, hawking bats switch to short, broadband FM pulses to give details of their acquired targets. Bats hawking between vegetation use slightly higher frequency calls since their prey is at relatively shorter distance. Due to echoes from the vegetation clutter, hawking bats use two strategies to reveal their prey against background clutter, both of which involve high-frequency calls (>50 kHz). Some bats use CF calls to detect clutter, others use broadband FM calls with several harmonics as movement detectors.

Gleaner bats (those that glean insects from surfaces like fruit and flowers) also have to fight clutter, and the premium for them is in determining target angle and properties. Therefore, they use short, higher bandwidth, low-intensity calls (hence the epitome "whispering bats"), with narrow beam signals emitted through the nostrils (with nose leaves) for directionality. Gleaning bats over ground/foliage clutter hover over prey and use short (<2 ms) FM echolocation pulses, of low intensity. Their echoes allow the discrimination of texture and target movement, over short distances. Many gleaning bats rely on prey-generated sounds and visual contact as alternative senses. Although bats clearly use both active and passive strategies, the details of passive listening in bats during prey acquisition and the use of other cues has not been studied.

Bat strategies: Sound use and prey evasion

Some bats add harmonics in cluttered environments to provide more detail about their surroundings. In uncluttered background situations, fundamental frequencies are emphasized (Simmons et al. 1978). In addition, the use of harmonic structures change during different capture phases (i.e., the approach versus terminal phase). Bats use FM signals in cluttered background environments or during the final approach to a target. FM pulses are good at determining fine structure of the environment. Amplitude modulation (AM) helps determine differences in size. Spectral changes in FM pulses may be used by bats to detect movement in cluttered environments (Altringham 1996). The slightest movement by a prey will change the echo spectrum. This is possibly why bats hover motionless when gleaning, because the movement of the bat itself introduces spectral changes that could complicate analysis.

Bats produce different pulse repetition rates (PRR) during different stages of the capture process (approach versus terminal phases), including a "wind-up" or increase in PRR upon target approach. Increasingly higher repetition rates are associated with insect-catching maneuvers when there is a need to appraise the changing position of prey. However, when gleaning insects from a surface, bats do not increase their PRRs (Fenton and Bell 1979).

Prey evasion strategies can alter signal use by the predator as well. Insects adjust their "evasion" strategies to intensity and PRR levels by either moving, freezing, or sometimes "jamming" the signals. Moths respond to intense calls by diving to the ground, but they react to less-intense sounds with negative phonotaxes (Roeder 1967). Although bats use different foraging strategies for different prey, they should not be categorized restrictively by the type of strategy alone. In addition, individual differences could be more flexible than previously believed (Fitzpatrick 1980). Bats also share characteristics of information transfer with dolphins, including social signals, imitative learning, eavesdropping, and intentional signaling (Wilkinson 1995).

Dolphin strategies: Detection and hunting

A review of underwater sound principles (such as high sound speed, low absorption) may explain why: (1) Dolphins use very short pulses, high bandwidth, low-duty-cycle pulses, high intensity (actually the porpoises might be called "whispering odontocetes" at 170 dB re 1 μPa). (2) Dolphins cannot use Doppler shifts. (3) Binaural localization based on different time delays must be much more difficult but are likely used for low frequencies, and intensity differences for high frequencies >20 kHz (Renaud and Popper 1975). (4) Dolphin prey have body impedances similar to water, so when acoustic energy penetrates, echolocation highlights help discriminate their internal composition. (5) There is no spectral adaptation to specific prey, proximity, or velocity, but dolphins increase click pressure by distance to targets or relative to noise in the environment. (6) Adaptations are made in pulse repetition rate with changing distance to target.

Dolphins inhabit a variety of environments, including rivers, coastal habitats, and open oceans. For animals using echolocation signals in the water with such short wavelengths, anything larger than a few millimeters will be reflected. Therefore, clutter may come in the form of vegetation, rocks, debris, or even bubble screens. Clutter is known to seriously affect echolocation performance in odontocetes, more so at a grazing angle of 90° (perpendicular) than at 68° (Turl, Skaar, and Au 1991). However, these effects have only been studied in artificial situations, and, again, not much is understood about how this factor influences the use of echolocation in the wild. A variety of sensory strategies, including passive hearing, vision, and intraspecific and interspecific behavioral cues, are likely used by dolphins (Wood and Evans 1980). Surface observations of foraging include reports of individual versus group foraging, changing group sizes in open-water versus coastal environments, and varying interanimal distances such as dispersed or tight school formations (Würsig 1986). The difficulties in obtaining complete spectra, real source levels, and simultaneous underwater observations in free-ranging situations have precluded many studies. Only recently, underwater observations, in at least one clear-water study site, were described (Herzing 1996; Herzing, chapter 56, this volume), and real-time, high-frequency echolocation measured (Au, Herzing, and Aubauer 1998).

Dolphin strategies: Sound use and prey evasion

Echolocation signals of different species of odontocetes vary in structure, intensity, frequency, and pulse repetition rates. Although it is known that dolphins use high-frequency signals, observations on modulated frequency or amplitude, during foraging strategies, are unmeasured. One basic feature, the "wind-up," or increase in PRR during approach to a target by dolphins (Au 1993), parallels the strategy used by bats.

The coevolution of prey hearing and predator signaling seems to have influenced the design of echolocation signals (Rydell, chapter 43, this volume). Bats form an interesting analogy, since they, like dolphins, can be both predator and prey. Bats and insects have both independently evolved and coevolved. Coevolving strategies include jamming activity, countermeasures, and approach versus terminal stages of hunting both from the predator-detection and prey-evasion aspects. Predators may decrease the probability of alerting their prey by (1) increasing frequencies to extend outside their prey's hearing range, (2) using cryptic strategies of encoding information within background noise, (3) listening passively and tracking prey, or (4) reducing the intensity or duty cycle of signals.

The ability of prey to detect and adopt evasion responses can influence the "encrypting" of the signal as a strategy as well. Evasion strategies of prey also can be relevant to odontocetes and pinnipeds (S. H. Andersen and Amundin 1976; Thomas, Ferm, and Kuechle 1987). An interesting example is the recent work by Barrett-Lennard, Ford, and Heise (1996) on echolocation strategies by fish versus mammal-eating orcas. This study documented that orcas use passive listening as a primary means of locating prey. They also use echolocation patterns for different hunting strategies—for example, they emit orientation clicks in cryptic patterns (isolated or in occasional doublets), thus masking them in the background noise when hunting other cetaceans (prey that can hear their high-frequency clicks). In contrast, when hunting fish, they do not mask their high-frequency signals.

However, clupeid fish have recently been reported to respond to ultrasound and to simulated dolphin echolocation (Mann et al. 1998), which may be very relevant to dolphins feeding on such species. All extant clupeids share this auditory specialization, preceding the evolution of marine mammal hearing and sound production.

This type of preadaptation is an alternative to the more active coevolution (the adjustment to sensory features and survival strategies) between predator and prey.

It is known that changes in insect "wing" aspect modify the intensity of the echo and lead to amplitude modulation of bat echolocation signals. Bats determine target distance, speed, and size with their echolocation, but echoes also contain insect-specific information of the wing-beat frequency, length, types, and structure, providing prey-species information (Schnitzler et al. 1983). In this case, information is encoded within time intervals and changes in intensity. Could the change of angle of a fish underneath the sand, or moving in the water column, contribute to both prey identification and prey retrieval? Could these cues be used by hunting dolphins? Dolphins could use echolocation to distinguish different prey or learn prey behavior and evasion strategies. Search and approach strategies described by Herzing (chapter 56, this volume) suggest that dolphins not only recognize the type of prey under the sand but also learn the prey's typical escape mode and employ appropriate strategies to retrieve their meal. Fine-discrimination abilities of dolphins, such as those described by Roitblat et al. (1995), might allow them to select prey by species and size, thus increasing optimal foraging.

Both prey and predator strategies likely will vary with the clarity of water or whether hunting occurs during the day (when vision can also be employed) or at night. Just as bat echolocation coevolved with flight and fast-moving hunting strategies (Rydell, chapter 43, this volume), dolphin echolocation could have evolved due to fast-moving prey. In opposition to their slower moving cousins, large whales, the inner ear structure of odontocetes shows structural changes necessary for high-frequency reception (Ketten 1994). Like bats, dolphins may depend on other senses, including vision and prey-generated sounds, as supplementary cues. In the coastal waters of Florida, fish that are conspicuous sound producers constitute a disproportionate percentage of bottlenose dolphins' stomach contents (Barros and Wells 1998). Cross-modal work between vision and echolocation in dolphins (Pack and Herman 1995) and bats (Simmons, Moss, and Ferragamo 1990) has illuminated the possibility of shared information between these senses. Insectivore bats may specialize on "groups" of insects (Black 1979) and learn and modify their capture strategies accordingly (Dunning 1968).

ANATOMICAL PARALLELS

Parallel anatomy is another feature we can compare between bats and odontocetes. Both bats and dolphins have low-light nocturnal vision (Bell 1985; Nachtigall 1986) and acoustic lens structures for focusing sound. Bats in the family Rhinolophidae have large, mobile ears to focus reception of signals by rotation, and complex nose leaves that act as an acoustic lens, focusing the nasally emitted echolocation pulses (Altringham 1996). This nose leaf could be analogous to the dolphin's fatty structure in the head, the melon (Ketten 1994; Cranford, Amundin, and Norris 1996).

Bats and dolphins rotate parts of their body for better sound reception. Bats have the ear/tragus complex that improves directionality to incoming signals and helps focus reception to a field of 30–40° from either side of the midline. Ketten (1994) described the segmented sound-conduction properties of the dolphin rostrum, with the anterior channel specialized for high-frequency reception and the lateral channel for lower frequency reception. This suggests that the mouth and lower jaw are analogous to a hydrophone array, where specific frequency receptors are arranged systematically. If the lower jaw functions as a frequency-specific receptor organ, then we must look at both open-mouth and scanning behavior as potentially proactive searching behavior (Herzing, chapter 56, this volume). Such jaw rotation could be comparable to ear rotation of bats, in its ability to tune and focus reception of frequency-specific acoustic information.

Strategies Using Echolocation

DEFINITIONS OF ECHOLOCATION

Although traditionally categorized separately for their function (echolocation for orientation, burst-pulsed sounds for social interactions), there is no clear demarcation between the production of echolocation and burst-pulsed sounds. Instead, these sounds form a graded series that can be treated as a single class of sounds (Herzing 1988). Perceptual features of click rates, recently reported for orcas (Szymanski et al. 1998), have not been measured for many species.

In addition to the "gray areas" described above, the definition of echolocation by clicks only is challenged by the bat's use of FM sweeps for echolocation (Altringham 1996). In addition, CF-call overlap appears to be an integral part of echo processing in some bats. Although differences between air versus water for the transmission of sound need to be considered, speculations on the function of long-distance, low-frequency, frequency-modulated signals from large balaenopterid whales for an "echo-ranging" function challenge traditional thoughts of the possible uses of FM vocalizations (Frazer and Mercado 2000; Clark and Ellison (chapter 73, this volume). Although echolocation is traditionally thought of as high-frequency sound production, high-frequency sound is not essential for echolocation, as demonstrated in cave dwelling swiftlets who use sound in the 2–10 kHz range (Fullard, Barclay, and Thomas 1993) for gross echolocation tasks. Bradbury and Vehrencamp (1998) reported that oilbirds (*Steatornis caripensis*) emit bursts of clicks with dominant frequencies between 6 and 10 kHz and durations of 1–1.5 s, noting

they probably only detect fairly large obstacles. Similarly, least shrews use audible clicks for echolocation (Thomas and Jalili, chapter 72, this volume). Echolocation aspects of California sea lions, *Zalophus californianus* (Poulter and Jennings 1969), penguins, *Spheniscus humboldti,* and harbor seals, *Phoca vitulina* (Renouf, Galway, and Gaborko 1980) have been reported. However, the predominant view is that these capabilities have not been unequivocally demonstrated by those studies (Richardson et al. 1995). Awbrey, Thomas, and Evans (chapter 70, this volume) provide some new data related to pinnipeds.

Echolocation abilities have been studied in humans, and both the blind and the sighted show unsuspected detection performance using sounds with low dominant frequencies, such as tongue clicks or hisses (Rice, Felnstein, and Schusterman 1965). Recently, Arias and Ramos (1997) showed in various performance tests that the ability to detect and discriminate obstacles does not require "privileged ears" or musical training. They used artificial stimuli composed of clicks lasting a few ms, with dominant frequencies below 2 kHz, or noise bursts, also with dominant frequencies below 2 kHz. At distances greater than 3 m, the subjects could hear the emitted signal clearly separated from the echo. At shorter ranges, both sounds fused into a single stimulus but with a perceived pitch shift. This perceived time separation pitch (or repetition pitch) of 200–500 Hz is quite functional for humans, and performance was even better with noise stimuli than with click sounds. Blind humans show greater acuity than the sighted, probably related to their permanent reliance on acoustical cues.

The parallel rules of aerodynamic and hydrodynamic evolution (e.g., moving target detection as reviewed by Altringham 1996) supports the notion that moving target detection may be more complicated than stationary detection (the traditional means of measuring echolocation clicks in dolphins). In one of the few studies of moving echolocation clicks in dolphins, Sigurdson (1998) reported that bottlenose dolphins (1) had variable interclick intervals not necessarily conforming to the TWT time rule, (2) modulated both frequency and amplitude for enhancing signal-to-noise ratios, (3) independently modulated low-frequency and high-frequency components, and (4) optimized their detection abilities through a learning process over time. Echolocation measurements of moving dolphins therefore provide new and important information about the use of this sense in the wild.

PASSIVE LISTENING VERSUS ACTIVE ECHOLOCATION

Dolphins may use a combination of passive listening and active echolocation in hunting. Like bats, dolphins are both predator and prey in their natural environment. Bats, as prey, usually emerge at night in large groups and switch sites to avoid alerting their predators (Altringham 1996). Echolocation may be a secondary or an additional proactive searching technique, after other primary signal detection systems are employed. Fenton (1984) reported that in some cases bats use echolocation to avoid obstacles, relying on other cues for prey detection. Similarly, Barrett-Lennard, Ford, and Heise (1996) suggested that some orcas use click trains only to locate distant obstacles or prey, avoiding emission during the capture approach. When do dolphins and porpoises listen and when do they actively search? These questions are relevant in discussions of mortality reduction in nets (Kraus et al. 1997), and in recent results of the significance of "silence" in wild bottlenose (dos Santos and Almada, chapter 55, this volume) and Atlantic spotted dolphins (Herzing, chapter 56, this volume). The costs of vigilance (usually by active scanning) have been calculated for other species (Illius and Fitzgibbon 1994), but passive listening has not been addressed. Eavesdropping, by individuals or by a group, potentially would facilitate passive listening abilities of wild dolphins, as has been experimentally demonstrated (Xitco and Roitblat 1996). In bats, the need to avoid intragroup mutual jamming may have led to strategies that help to separate calls, but precluded eavesdropping (Obrist 1995).

Social Uses of Echolocation

In addition to enhancing foraging abilities and predator detection, echolocation could have a social function, and several authors have found support for this idea (Wood and Evans 1980; dos Santos and Almada, chapter 55, this volume; Herzing, chapter 56, this volume). There remains the additional possibility that the receiver of the buzz is experiencing the tactile effects of sound. Given the graded nature of echolocation clicks and burst-pulsed sounds, it should become apparent that all these sounds may produce a tactile as well as auditory effect.

The genital and mammary regions are the area of richest somatic innervation, followed by the upper rostrum, lower jaw, forehead, flukes, and pectoral and dorsal fins (Ridgway and Carder 1990). Cutaneous mechanoreceptors and their cortical responses were reported by Bullock et al. (1968). Combined with the fact that the trigeminal nerve has the greatest number of axons of any dolphin cranial nerve (Jansen and Jansen 1969), this suggests that acoustic and tactile information are intimately related in dolphins.

What sound-intensity level would be required to surpass tactile thresholds on the dolphin's body? Kolchin and Bel'kovich (1973) presented threshold levels of tactile sensitivity for the common dolphin, *Delphinus delphis*. These measurements were described and recalculated for sound pressure estimates (Herzing, chapter 56, this volume). It is clear that SPL of echolocation signals for several species (false killer whale at 225 dB, Thomas and Turl 1990; beluga at 222 dB, Au et al. 1985; bottlenose dolphin at 228 dB, Au et al. 1974; Atlantic spotted dolphin at 210 dB, Au, Herzing, and Aubauer 1998; spinner dolphin at 222 and −220 dB, Schotten et al., chapter 54, this volume) are well above the estimated SPL needed for tactile reception by dolphins.

So dolphins not only may receive social information about the receiver during high-intensity and repetition rate use of echolocation signals, they also may cause tactile sensations via sound pressure. Researchers have speculated that use of such intense sound in close proximity could provide both auditory and tactile "comfort or discomfort." Concomitantly, burst-pulsed packets of clicks were observed in both conspecific aggression (Overstrom 1983), intraspecific and interspecific aggression (Herzing 1996), and during herding behavior of conspecifics (Connor and Smolker 1996) and fish (Norris and Møhl 1981).

Levels of Environmental Information

In their recent review, Bradbury and Vehrencamp (1998) showed that echolocating animals often extract high levels of environmental information from their sonar signals, close to what would be physically and theoretically possible. These authors argued that the obvious lack of comparable sophistication in the information content of social signals cannot be explained by design or system limitations. Instead, this asymmetry could only be understood with a "game economical" reasoning—that is, considering that in traditional communication there are conflicts of interest between senders and receivers. Only if senders have substantial benefit from providing conspecifics with high levels of environmental information would there be a functional and evolutionary justified selection pressure for the development of richer signaling. The cases where such systems exist, or might exist, and where genetic economics appear to favor such development, remain hot topics for future research.

In fact, if bats avoid eavesdropping on echoes by conspecifics (Obrist 1995), the picture in the case of dolphins is not so clear. The multitude of communication modes used by these animals may provide conspecifics with details of important features of the environment, and eavesdropping on conspecific echolocation may be the norm rather than the exception. This could explain the null correlation between group size and number of click trains emitted found in bottlenose dolphins (dos Santos and Almada, chapter 55, this volume). Johnson and Norris (1994), having observed the echolocation behavior of spinner dolphins, also suggested that these animals rotate sonar duties in their groups, allowing each animal to rest its emission system regularly.

Future Areas of Inquiry

Although the physical and structural aspects of dolphin echolocation clicks are well researched, knowledge of how dolphins use this sense in the wild is at its infancy.

Critical areas for future research described in other chapters in this volume include (1) measurement of echolocation signals, especially high-frequency clicks, in free-ranging dolphins (Schotten et al., chapter 54; Lammers et al., chapter 58); (2) social and nonsocial uses of echolocation both in captivity (Blomquist and Amundin, chapter 60; Moreno, Kamminga, and Stuart, chapter 59)

and in the wild (Herzing, chapter 56); (3) signal propagation (Watkins and Daher, chapter 57); passive versus active use (dos Santos and Almada, chapter 55); and (4) cross-modal studies of echolocation (Pack, Herman, and Hoffman-Kuhnt, chapter 41). Continued research is needed in these areas to ensure the future understanding of the function of dolphin echolocation in the wild.

{ 54 }

Echolocation Recordings and Localization of Wild Spinner Dolphins (*Stenella longirostris*) and Pantropical Spotted Dolphins (*S. attenuata*) Using a Four-Hydrophone Array

Michiel Schotten, Whitlow W. L. Au, Marc O. Lammers, and *Roland Aubauer*

Introduction

Despite the large amount of data, derived from captive odontocetes, on the capabilities of the active dolphin echolocation system (see Au 1993 for an overview), virtually nothing is known about the actual use of echolocation in the wild and its ecological significance. The most important questions needing answers are from which distances dolphins usually echolocate, to what extent the use of echolocation is dependent on the type of environment and time of the day (e.g., the light-dark cycle), whether members of a dolphin school echolocate simultaneously or eavesdrop on the echolocation of one animal, and how often echolocation is used (Au 1993, 271). However, before such questions can be addressed, it is first necessary to describe the characteristics of echolocation clicks emitted by free-ranging odontocetes.

Odontocetes can be divided into two acoustic categories (Au, introduction to this volume). The first comprises all species that can produce both long-duration, frequency-modulated tonal sounds (known as whistles) as well as pulsed sounds (echolocation clicks and burst-pulses). Clicks can extend to frequencies >150 kHz, are broadband, and have a duration of 50–100 μs; while whistles are frequency-modulated tones up to 20 kHz with harmonics up to around 70 kHz (Lammers et al. 1997), lasting 0.1 to several seconds. The odontocetes in the second acoustic category are known to produce only pulsed sounds. These pulsed sounds are narrowband,

generally around a high peak frequency of up to 140 kHz, with durations in the order of 100–200 μs.

Because the proposed division of odontocetes into two acoustic categories might have implications concerning the different uses of clicks, it would be worthwhile to determine whether the division holds for all odontocete species, and to which category each species belongs. For this purpose, it is necessary to record and analyze echolocation clicks from all odontocete species using similar, high-frequency (up to 200 kHz) broadband equipment. No such click descriptions were found in the literature for either spinner dolphins (*Stenella longirostris*) or pantropical spotted dolphins (*S. attenuata*). Both species, like all species from the genus *Stenella*, are known to produce whistles (Norris et al. 1994) and therefore are expected to belong to the first acoustic category.

When recording echolocation clicks from wild dolphins at sea, there are a number of problems: (1) it is generally unknown which dolphin is producing the recorded clicks and how many animals are echolocating; (2) the peak-to-peak source level (SL) of clicks cannot be estimated with accuracy because the distance from the dolphin to the hydrophone is unknown; (3) terminations of clicks are often lost in reverberation and reflections from the water surface; and (4) the orientation of the dolphin's head with respect to the hydrophone is generally unknown, so that it cannot be ascertained whether clicks are from the main axis of the echolocation beam (Au 1993).

An array of hydrophones can be used to determine

the distance of an echolocating dolphin and whether the measured signals propagated along the animal's beam axis. By using a line array of three or more hydrophones spaced equal distances apart, such as in the study of Møhl, Surlykke, and Miller (1990), it is possible to determine the distance to the sound source but not the direction. However, with four hydrophones arranged in a configuration other than a line, it is possible to determine the exact position of the sound source to one of two points. W. A. Watkins and Schevill (1974) used an array of four hydrophones spaced 30 m apart at the vortices of a tetrahedron to localize spinner dolphins (*S. longirostris*). Due to the large size of the array, however, the directional echolocation clicks were seldom recorded at all four hydrophones. To localize dolphins by their echolocation clicks, an array would need to be small, rigid, and portable. Furthermore, by attaching an underwater camera to the array, connected to a VCR synchronized with the click recording device, the orientation of echolocating dolphins can be ascertained. In the present study, an array of four hydrophones arranged in a symmetrical star configuration, with one center hydrophone and three extending arms spaced 120° apart (adopted from Aubauer 1995), was used to measure the echolocation signals of wild spinner dolphins (*S. longirostris*) and pantropical spotted dolphins (*S. attenuata*).

Materials and Methods

Let the plane of the four-hydrophone array be the *y-z* plane of a Cartesian coordinate system with the center hydrophone (H_0) at the origin. The coordinates of an echolocating dolphin can be expressed as a distance from H_0 to the dolphin (range R), a horizontal angle φ, and a vertical angle θ, as follows (see fig. 54.1):

$$x = R \cdot \cos\varphi \cdot \cos\theta \tag{54.1}$$

$$y = R \cdot \sin\varphi \cdot \cos\theta \tag{54.2}$$

$$z = R \cdot \sin\theta \tag{54.3}$$

To localize the dolphin it is sufficient to know R, φ, and θ. If the coordinate system is defined as in fig. 54.1, these values can be derived using the above expressions and Pythagoras's theorem to be (Aubauer 1995):

$$R = \frac{c^2(\tau_{01}^2 + \tau_{02}^2 + \tau_{03}^2) - 3a^2}{2c(\tau_{01} + \tau_{02} + \tau_{03})} \tag{54.4}$$

$$\varphi = 90° \\ \pm \arccos\left(\frac{2cR(\tau_{02} - \tau_{01}) + c^2(\tau_{01}^2 - \tau_{02}^2)}{2\sqrt{3a^2R^2 - 0.75(2Rc\tau_{03} - c^2\tau_{03}^2 + a^2)^2}}\right) \tag{54.5}$$

$$\theta = -\arcsin\left(\frac{2Rc\tau_{03} - c^2\tau_{03}^2 + a^2}{2AR}\right) \tag{54.6}$$

where

$-180° < \varphi < 180°$
$-90° < \theta < 90°$
c = speed of sound in water $\approx$ 1500 m/s
a = distance between center hydrophone (H_0) and outer hydrophones (H_1, H_2, and H_3) = 0.61 m
τ_{01} = time of click arrival at H_0 − time of click arrival at H_1 (expressed in s)
τ_{02} = time of click arrival at H_0 − time of click arrival at H_2
τ_{03} = time of click arrival at H_0 − time of click arrival at H_3

The $\pm$ sign in eq. 54.5 represents the ambiguity in localization, and translates in either a positive or negative

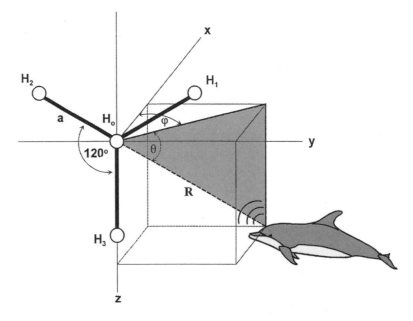

Fig. 54.1. In a three-dimensional Cartesian coordinate system, the position of a dolphin echolocating on a four-hydrophone symmetrical star array in one plane can be expressed as a range R to the center hydrophone H_0, a horizontal angle φ, and a vertical angle θ. Distance a between H_0 and each of the outer hydrophones H_1, H_2, and H_3 is 0.61 m. In this coordinate system, the echolocating dolphin has a positive *x*-coordinate, but negative *y*- and *z*-coordinates. Therefore, both φ and θ have negative values as well.

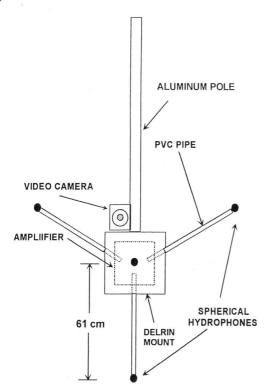

Fig. 54.2. The hydrophone array that was used for data acquisition

x-coordinate in fig. 54.1. Furthermore, eq. 54.4 shows that as the sum of time-of-arrival differences ($\tau_{01} + \tau_{02} + \tau_{03}$) approaches 0 μs, range *R* (as well as the range estimation error ΔR) increases to infinity. Therefore, only ranges up to an arbitrary value of 30 m ($\tau_{01} + \tau_{02} + \tau_{03} \leq -11$ μs) were reliable, and calculated positions with $R > 30$ m were rejected beforehand.

The hydrophone array consisted of four omnidirectional ITC 1094 A elements. Hydrophones, with a flat frequency response up to 160 kHz, were attached to a rectangular block of delrin mounted via PVC pipes as shown in fig. 54.2. The four hydrophones were connected to a rechargeable battery-driven, multichannel preamplifier/linedriver with an 18 dB gain, housed in a watertight box attached to the delrin block. The preamplifier was connected via cables feeding back to the boat to a rechargeable battery-driven, multichannel amplifier with an adjustable gain for each channel. An aluminum pole with a small video camera in a watertight transparent container was attached to the array to stick it into the water. The camera was connected to a VCR on board, synchronized with the click recording device.

The hydrophone outputs were amplified by either 36 or 42 dB and fed into a four-channel, 12-bit simultaneous analog-to-digital (A/D) converter system sampling at 500 kHz. The A/D cards were housed in a transportable "lunch-box" type personal computer. The data acquisition program was written in Qbasic 4.5. Data acquisition was triggered by the input of H_0, which caused

the transfer of 200 pretrigger points and 200 posttrigger points (800 μs) per channel to the board's memory. A maximum of 80 consecutive clicks, with the accompanying interclick intervals and times of recording, could be stored in one file each time.

The array was calibrated by transmitting trains of simulated *Tursiops* clicks under water and recording them with the hydrophone array at different distances from the transmitter. The array was held so that H_0 was at the same depth as the transmitter (thus, $\theta \approx 0°$), and the plane of the array was parallel to the plane of the transmitter (thus, $\varphi \approx 0°$). Calculations of *R* best resembled the actual ranges when the point of the maximum amplitude of the recorded click was taken as the arrival time on each channel, under the restriction that the same excursion within the click was used on each of the four channels (for that purpose, excursions could be selected manually by means of a built-in cursor option). Additionally, the best results were obtained when a three-point parabolic curve was fitted through the point of maximum amplitude and the points preceding and succeeding that point, for an exact estimate of the time of click arrival on a channel. The calculated mean ranges were plotted against the actual ranges, expressed in units of the center/outer hydrophone distance "a" (which was 0.61 m in this case). Localization was highly accurate for ranges smaller than 15 m, and sufficiently accurate for ranges up to 25 m (fig. 54.3). Standard deviations increased with range, but remained very small (<0.7 a).

Echolocation recordings from wild spinner dolphins and pantropical spotted dolphins were obtained at the Waianae coast of Oahu, Hawaii, aboard a 5.2 m Boston Whaler during four days from February to April 1997. While spinner dolphins frequently visit two sandy bottom areas of this coast, spotted dolphins are only encountered on rare occasions, and only on one occasion could their clicks be recorded. The measured water depth was 40 m, while depth varied from 6 to 21 m for the spinner dolphin click recordings. The subsequent analysis of each click was performed on the channel with the highest recorded amplitude, to increase the chance that the analyzed click was recorded from the center of the echolocation transmission beam. First, the click was manually selected on that channel by using the built-in cursor option, to separate the actual click from reverberation and from its reflection from the water surface, which often overlapped with the click itself. Because of this overlap, a subjective decision was made in differentiating the actual click from surface reflection. This was facilitated by comparing the four channels: the elapsed time (Δt) between the click and its surface reflection should be different on each channel, with the largest Δt on the channel of the deepest hydrophone H_3. To get a rough estimate of Δt, equations were derived for Δt on each channel, as specified in the appendix. After manual selection of a recorded click on the channel with the highest amplitude,

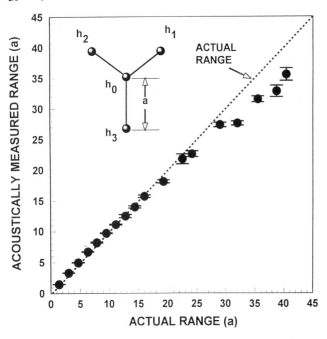

Fig. 54.3. Array calibration for calculations of R (means and standard deviations), expressed in units of a ($a = 0.61$ m)

the following click characteristics were calculated (defined as in Au 1993, 137, 216–24): normalized energy (E_N), peak frequency (f_p), center frequency (f_0), 3 dB bandwidth (BW), root mean square (rms) bandwidth (β), signal duration (τ), rms signal duration (τ_d), time bandwidth product ($\tau_d\beta$), centroid of the time waveform (t_0), Woodward time resolution constant ($\Delta\tau$), and intrinsic range resolution ($\Delta r = \frac{1}{2} \cdot c \cdot \Delta\tau$). The click characteristics were fed into a spreadsheet program, together with the coordinates of that click, the peak-to-peak source level SL (level referenced to 1 m from the source with units of dB re 1 μPa), and the source energy flux density SE (referenced to 1 m from the source with units of dB re 1 μPa²s).

Results

A total of 851 spinner dolphin clicks and 340 spotted dolphin clicks were recorded and analyzed. Of these clicks, only 131 spinner dolphin clicks and 196 spotted dolphin clicks were recorded on all four channels. The remaining click recordings suffered from a loose connection between the preamplifier and amplifier, causing a loss of one or more channels for those recordings, which therefore could not be localized. Also, the video equipment malfunctioned, so that the orientation of echolocating dolphins could not be ascertained. However, if the center hydrophone recorded the highest signal, then in all probability the animal was directing its beam toward the array.

A typical spinner dolphin click is shown in fig. 54.4A. It was recorded from a distance of 13 m, with the highest amplitude recorded by the center hydrophone, H_0. The

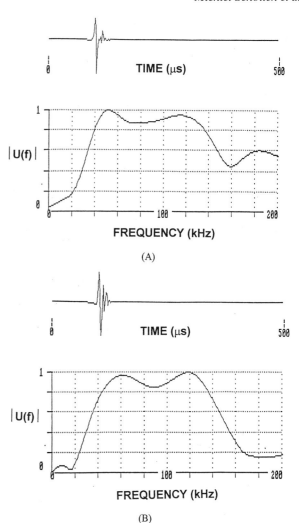

Fig. 54.4. Normalized time domain waveform $s(t)$ and frequency spectrum $S(f)$ of a typical spinner dolphin click (*A*) and spotted dolphin click (*B*)

waveform of the recorded click was a 36 μs transient signal with two main excursions and some minor excursions; most of the energy of its bimodal broadband frequency spectrum fell between 40 and 140 kHz. The calculated peak-to-peak source level was 214 dB. A typical spotted dolphin click, recorded from a distance of 12 m with the highest amplitude on the center channel, is shown in fig. 54.4B. For this click, SL = 218 dB. Generally, clicks recorded from spinner dolphins and from spotted dolphins were similar, although the waveforms of spotted dolphin clicks had minor excursions that were larger in amplitude than those of the spinner dolphin clicks. Also, there was more variation in spotted dolphin clicks. For both species, medium- to high-amplitude clicks had predominantly bimodal frequency spectra, with a low-frequency peak at 40–60 kHz and a high-frequency peak at 120–140 kHz. Clicks that were among the highest in amplitude had only a single peak in frequency, either at the low- or high-frequency peak.

Means and standard deviations of the calculated click

TABLE 54.1. Calculated click characteristics ($\bar{x} \pm$ SD), defined as in Au (1993, 137, 216–24), for all the recorded spinner dolphin (*Stenella longirostris*) and spotted dolphin (*S. attenuata*) clicks. For the spinner dolphin clicks, sample size (n) = 851, except for SL and SE (n = 131), for E_N (n = 831), and for fp (n = 836). For the spotted dolphin clicks, n = 340, except for SL and SE (n = 195), and for f_p (n = 338). For SL, dB is re 1 μPa; for SE dB is re 1 μPa^{2}s; whereas E_N is unitless.

	SL (dB)	SE (dB)	E_N (dB)	f_p (kHz)	f_0 (kHz)
S. longirostris	208 ± 5	148 ± 5	−57.5 ± 2.4	69.7 ± 23.1	80.4 ± 12.1
S. attenuata	212 ± 5	150 ± 4	−56.9 ± 1.7	69.4 ± 31.3	83.4 ± 16.8
	BW (kHz)	β (kHz)	τ (μs)	τ_d (μs)	$\tau_d\beta$
S. longirostris	76.4 ± 23.4	34.1 ± 4.9	31 ± 12	4.6 ± 1.5	0.16 ± 0.06
S. attenuata	79.8 ± 35.9	38.7 ± 6.7	43 ± 15	5.3 ± 1.9	0.21 ± 0.10
	t_0 (μs)	$\Delta\tau$ (μs)	Δr (cm)		
S. longirostris	11.6 ± 6.2	9.4 ± 2.7	0.70 ± 0.20		
S. attenuata	15.8 ± 8.2	8.9 ± 3.0	0.67 ± 0.23		

characteristics are presented in table 54.1. Note that, compared to echolocation clicks of captive *Tursiops* (Au 1993, 217), the recorded clicks had high peak-to-peak source levels (with maximum source levels of 222 and 220 dB for the spinner and spotted dolphin clicks, respectively), large 3 dB and rms bandwidths, short durations, and small values for intrinsic range resolution (with minimum values of 0.4 cm for both species). The variance of each click characteristic, except for SL, SE, and E_N, was significantly higher for the spotted dolphin clicks than for the spinner dolphin clicks ($p < 0.0001$, variance ratio test). For this reason, the nonparametric two-tailed Mann-Whitney test, rather than Student's *t*-test, was applied to test for differences between mean click characteristics. The spotted dolphin clicks had higher values for SL, SE, E_N, f_0, β, τ, τ_d, $\tau_d\beta$, and t_0 ($p < 0.0001$), while the spinner dolphin clicks had higher values for f_p ($p < 0.05$), $\Delta\tau$, and Δr ($p < 0.0001$). No significant difference in BW was found. In summary, the recorded spotted dolphin clicks were found to be louder and longer, with better intrinsic range resolution, than the spinner dolphin clicks.

Because positions of echolocating dolphins were known, it was possible to discriminate between clicks that were supposedly emitted by different dolphins. To assign the large number of recorded clicks that had one channel missing to individual animals as well, time-of-arrival differences rather than the actual positions were used—two time-of-arrival differences were in most cases already sufficient for this purpose. Successive clicks with similar coordinates were arranged into groups, each of which was considered as a single click train emitted by one dolphin. Next, click trains that had similar coordinates but were separated in time by one or more other trains were linked and assigned to one animal, taking into account the time interval between these trains and the animal's direction of movement.

After all clicks were assigned to individual animals, a linear discriminant analysis was applied (as in Lindeman, Merenda, and Gold 1980, 183–96, 221) to test whether the division could be supported by differences in click characteristics among presumed individual dolphins.

Ten click characteristics (E_N, f_p, f_0, BW, β, τ, τ_d, $\tau_d\beta$, t_0, and $\Delta\tau$) were fed into the analysis. For the 48 presumed spinner dolphin individuals, this resulted in nine significant discriminant functions ($p < 0.05$), with which the SPSS program was able to assign 44% of all clicks to the correct (i.e., previously assigned) individuals. For the 13 presumed spotted dolphin individuals, it resulted in four significant discriminant functions ($p < 0.05$), with which 40% of the clicks could be assigned to the correct individuals. Therefore, the discriminant analysis supported the performed division of clicks. However, the discriminant functions were not consistent in the weights that they assigned to each of the 10 click characteristics, so the relative importance of each click characteristic in discriminating individuals remains unclear.

For three click trains of an individual spinner dolphin and one long click train of an individual spotted dolphin (all recorded at a distance of 10–15 m, at about the same depth as the hydrophone array), several click characteristics were plotted as a function of click number (fig. 54.5). SL, SE, f_0, f_p, and BW were generally smaller at the beginning and end of a click train than in the middle part, while β remained more or less constant. There was much more variation within the spotted dolphin click train than in the spinner dolphin click trains, and the spotted dolphin click train had larger maximum values.

Additionally, for the total data set several click characteristics were plotted as functions of one another (fig. 54.6). In this way, a linear relationship was found between interclick interval (ICI) and calculated range R of each click (fig. 54.6A, B). The so-called two-way transit time, which is defined as the time needed for an echolocation click to travel from the dolphin to the hydrophone array and back to the dolphin, is also indicated in these plots and can be expressed as two-wafy transit time (ms) = $1.33 \cdot R$ (m). Note that for all recorded clicks the ICIs were longer than the two-way transit times. Also, the slopes of the linear regression lines through the data were steeper than the slope of the equation for two-way transit time by about a factor of 2. This could be an indication that when the array was located farther

Fig. 54.5. Click trains emitted by an individual spinner dolphin (*A, C,* and *E*) and by an individual spotted dolphin (*B, D,* and *F*). SL and SE of the clicks are plotted in *A* and *B*, f_0 and β are plotted in *C* and *D*, and f_p and BW are plotted in *E* and *F*. Since the first 15 clicks of the spinner dolphin's second recorded click train were not recorded on the channel from hydrophone H_2, no ranges and therefore no values for SL and SE could be calculated for those clicks in *A*.

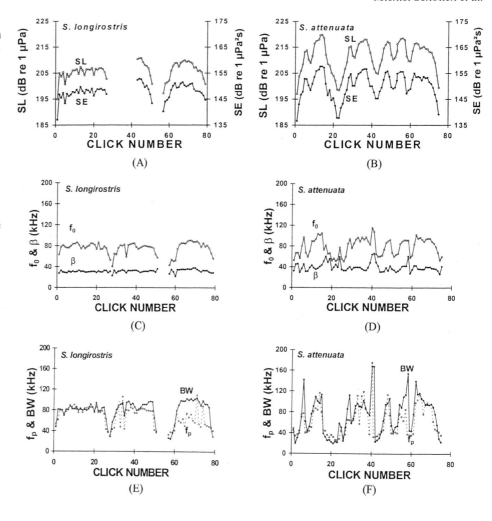

away, the dolphins needed a longer processing time between receiving an echo of one click and emitting the next click, assuming they were echolocating on the array. A second linear relationship was found between center frequency f_0 and peak-to-peak source level SL (fig. 54.6C, D). Equations of the linear regression lines through the data were similar for the spinner and spotted dolphin clicks, and also resembled the equation $[f_0 \text{ (kHz)} = 2.55 \cdot \text{SL (dB)} - 456.40]$ found by Au et al. (1995) for a false killer whale (*Pseudorca crassidens*) performing an echolocation task. Finally, a third linear relationship was found between 3 dB bandwidth BW and center frequency f_0 (fig. 54.6E, F).

Discussion

Calibration of the four-hydrophone array indicated that it was highly accurate in localizing ranges up to $25 \cdot a$ and sufficiently accurate for ranges up to $40 \cdot a$, where a is the distance between the center hydrophone H_0 and each of the three outer hydrophones. Therefore, increasing the size of the array would increase the distance at which dolphins could accurately be localized, but it would have

the disadvantage that at close ranges the directional echolocation clicks probably would not be recorded on all four channels. Another disadvantage would be that the time-of-arrival differences would increase, thus requiring more digitized points per channel to store each click.

By using the four-hydrophone array, solutions were provided to three of the four problems of recording clicks at sea. Localizing echolocating dolphins made it possible to measure peak-to-peak source levels (SL) of the clicks (with an estimation error of less than 1.5 dB at $R = 25$ m in the calibration), to discriminate recorded clicks from their surface reflections, and to assign clicks to presumed individual animals. The division of clicks was supported by a linear discriminant analysis, which indicated highly significant differences in all click characteristics among presumed individual dolphins. However, due to the uncertainty of assigning multiple click trains with similar coordinates to a single dolphin (but separated in time by one or more click trains of other dolphins), it remained unclear whether each dolphin emitted its own type of click or that all click trains emitted by a single dolphin were different from one another.

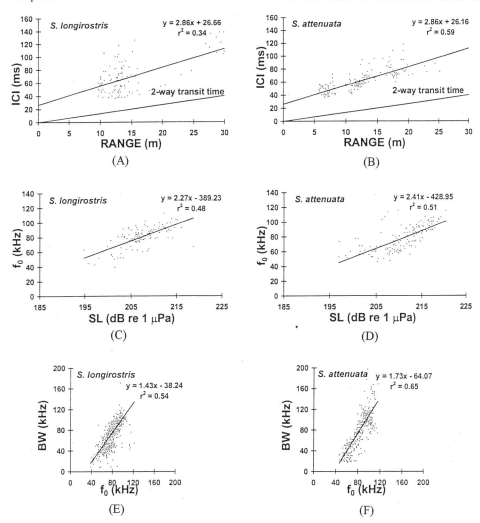

Fig. 54.6. Scatter plots of inter-click interval (ICI) on range (*A* and *B*), f_0 on SL (*C* and *D*), and BW on f_0 (*E* and *F*). *A*, *C*, and *E* show plots for all spinner dolphin clicks, *B*, *D*, and *F* show plots for all spotted dolphin clicks. The equation of the least square linear regression line through the data and its correlation coefficient are indicated in the upper right corner of each plot.

To investigate this question in future research, the array recordings should be used in combination with good video recordings of known individual dolphins, such as in the populations studied by Herzing (1996).

Concerning the fourth problem of recording clicks at sea, the orientation of echolocating dolphins could not be ascertained due to malfunctioning of the video recorder. However, the use of a four-hydrophone array (as opposed to a single hydrophone) makes it easier to discriminate clicks recorded on the axis of the echolocation beam. First, it is possible to select the channel with the highest amplitude for each click. Second, one can select only those clicks with the highest amplitude recorded by the center hydrophone H_0. In this case, it is reasonable to assume that the dolphin directed its echolocation beam at or near the center of the array. Third, the linear relationship that was found between interclick interval ICI and range *R*, and the fact that ICI was larger than the calculated two-way transit time in all cases, indicate that the majority of clicks probably were recorded from dolphins that had been echolocating directly on the array. However, variations in click characteristics within one click train (such as those in fig. 54.5B, D, and F)

should be treated with caution, since those variations also could result from scanning movements of the dolphin.

The clicks recorded from both spinner and spotted dolphins had source levels that were 30–60 dB higher than those recorded previously from wild odontocetes (e.g., Watkins 1980). Similar click source levels, however, were recorded from wild narwhals (*Monodon monoceros*) by Møhl, Surlykke, and Miller (1990), who used a three-hydrophone line array. Therefore, more studies should use hydrophone arrays to obtain reliable source levels for other species, to close the traditionally perceived "dB gap" between wild odontocetes and captive odontocetes trained in echolocation tasks. Besides high source levels, the spinner and spotted dolphin clicks also were characterized by very broad 3 dB and rms bandwidths, short durations, and very small values for intrinsic range resolution compared to a typical *Tursiops* click. However, while closely resembling the *Tursiops* click and echolocation clicks from other whistling dolphin species, the spinner and spotted dolphin clicks were different from clicks emitted by nonwhistling dolphin species. Therefore, the hypothesized division of odontocetes into two acoustic categories (Au, introduction to this volume) is further supported by these data.

Appendix

Equations of the elapsed time Δt between the click and its surface reflection on each channel were derived to obtain rough estimates of Δt. If the water surface is flat, with the pole of the array (the z-axis in fig. 54.1) exactly perpendicular, Δt can be expressed as

$$\Delta t_i = \frac{R_i - SR_i}{c}$$

where

> $i = 0, 1, 2,$ or 3, for hydrophones H_0, H_1, H_2, and H_3, respectively
> $c \approx 1500$ m/s
> R_i = direct path from the dolphin to hydrophone i
> SR_i = surface reflected path from the dolphin to hydrophone i

Now R_i and SR_i can be derived by writing range R and the three extending arms of the array as vectors (with H_0 as the origin of the coordinate system, as in fig. 54.1) and then using the cosine rule (Schotten 1998):

$$R_0 = R$$
$$R_1 = \sqrt{R^2 + a^2 + aR \cdot \sqrt{3} \cdot \sin \varphi \cdot \cos \theta - aR \cdot \sin \theta}$$
$$R_2 = \sqrt{R^2 + a^2 - aR \cdot \sqrt{3} \cdot \sin \varphi \cdot \cos \theta - aR \cdot \sin \theta}$$
$$R_3 = \sqrt{R^2 + a^2 + 2aR \cdot \sin \theta}$$

The surface reflected paths are

$$SR_0 = \sqrt{R^2 + 4D^2 - 4RD \cdot \sin \theta}$$
$$SR_1 = \\ \sqrt{R^2 + 4D^2 - 3aD + 1.5 \cdot a^2 + aR \cdot \sqrt{3} \cdot \sin \varphi \cdot \cos \theta - 2RD \cdot \sin \theta}$$
$$SR_2 = \\ \sqrt{R^2 + 4D^2 - 3aD + 1.5 \cdot a^2 - aR \cdot \sqrt{3} \cdot \sin \varphi \cdot \cos \theta - 2RD \cdot \sin \theta}$$
$$SR_3 = \sqrt{R^2 + 4D^2 + 4aD + a^2 - 2R \cdot (2D + a) \cdot \sin \theta}$$

D indicates the depth of H_0, assumed to be between 0.5 and 2 m. Due to wave action, varying values for D, and angles other than 90° between the array pole and water surface, the equations for SR_i (and therefore for Δt_i) often will be inaccurate. However, they can give a rough indication and can be used in discerning surface reflection.

Acknowledgments

The authors are grateful to Dave Lemonds for useful discussions, Jennifer Philips for support in programming in Qbasic, and Paul Nachtigall, Linda Choy, and Claudia Aubauer for general support. Michiel Schotten acknowledges Louis Herman and Adam Pack from the Kewalo Basin Marine Mammal Laboratory for providing computer hardware and software on which part of the data analysis was carried out. He also acknowledges the Royal Netherlands Academy of Arts and Sciences (K.N.A.W.), the Schuurman-Schimmel van Outeren Foundation, and the faculty of Mathematics and Natural Sciences of the University of Groningen (The Netherlands) for providing personal financial support.

{ 55 }

A Case for Passive Sonar: Analysis of Click Train Production Patterns by Bottlenose Dolphins in a Turbid Estuary

Manuel E. dos Santos and *Vítor C. Almada*

Introduction

The use of a sophisticated echolocation system to navigate and to discriminate prey and other targets has been demonstrated in many odontocetes (for reviews and comparisons with the bat's systems, see Au 1993, 1997, and the introduction to this volume). This active sonar capability is based on the emission of short, broadband clicks, usually in trains, and on the interpretation of their echoes, providing the echolocating animals, and also their companions, with detailed information about their environment.

In the best-studied species, the bottlenose dolphin (*Tursiops truncatus*), peak frequencies of clicks are at around 70 kHz (interestingly close to its frequency of best hearing). Detection of the broadband pulses of this species is somewhat facilitated by their energy levels at relatively low frequencies. In four examples of bottlenose dolphin clicks represented in fig. 3 of Au 1997, at 20 kHz the relative amplitude levels were approximately 70%, 50%, 50%, and 20% of the respective peak frequencies.

THE MIXED BLESSING QUESTION

Several researchers have reported instances in which odontocetes did not take advantage of their echolocation ability, either because it simply was not necessary, or because it might have been more adaptive to be quiet. A classical example is that described by Wood and Evans (1980) of Scylla, a bottlenose dolphin being used in the experiments of Diercks et al. (1971). Still blindfolded after the procedures, Scylla was observed following closely and capturing a live fish, several times, surprisingly without the emission of clicks.

Evans and Awbrey (1988) reported an observation of a group of beluga whales (*Delphinapterus leucas*), in turbid waters, feeding on salmon apparently without echolocation. Usually very vocal odontocetes, these belugas were silent throughout the episode. It is likely that the belugas were able to track the fish, like Scylla, using the sounds and pressure changes made by the moving prey.

Another relevant example is that reported by Barrett-Lennard, Ford, and Heise (1996) from killer whales (*Orcinus orca*); these authors compared click train production by groups feeding on fish and by groups feeding on aquatic mammals. It was observed that the latter killer whale groups emit significantly less echolocation signals, preferring apparently to detect and approach their prey with passive audition alone, perhaps because aquatic mammals can hear their clicks.

The serious fact that so many dolphins get trapped in fishing nets they could easily detect using echolocation (Au and Jones 1991) also shows the need for a better understanding of the ecological context of sonar production. In what circumstances will dolphins use the full potential of their specialized sonar system? Under what conditions may they rely on other sensory channels, or on passive audition? Borrowing the phrase used by Fenton (1980) to describe the same problem faced by bats, how do they manage "the mixed blessing of echolocation"? When will they avoid the shortcomings of active sonar and use more stealthy ways of creeping up on their prey?

Other researchers have focused on these partially unanswered questions, as other chapters in this section of the book illustrate. The purpose of this chapter is to contribute to the discussion of these bioacoustical issues by presenting data collected as part of a study of the acoustic emissions by the bottlenose dolphins resident in the Sado Estuary, Portugal.

Materials and Methods

The Sado Estuary and the adjacent coastal waters, along the western coast of Portugal, sustain a resident population of bottlenose dolphins, of the Eastern Atlantic robust form, who enter the estuary mainly to feed (dos Santos and Lacerda 1987). The population has been declining in the last decades, apparently due to high infant mortality, and now there are about 35 resident animals. Descriptions of the area have been provided by dos Santos and Lacerda (1987) and dos Santos (1998).

Although a busy harbor, a city, many industries, and agriculture generate a great deal of pollution, the dolphins find in the turbid waters of the estuary several species on which they feed. The more important prey species are probably the mullets (*Mugil cephalus, Liza ramada, Liza aurata,* and *Chelon labrosus*), cuttlefish (*Sepia officinalis*), eels (*Anguilla anguilla*), and shad (*Alosa fallax*). Other locally abundant species that have not yet been identified as dolphin prey include the Lusitanian toadfish (*Halobatrachus didactylus*).

The dolphins were followed in a small boat with an outboard engine, in trips lasting between 30 min and 6 h, during which a nonintrusive distance was maintained (usually >50 m). Behavior sampling was ad libitum, and the recording method was either continuous or with scans every 5 or every 10 min, depending on dolphin movements or position.

Underwater sound recordings were made with a B&K 8101 hydrophone and a Sony TCD-D10 Pro DAT recorder, a setup that is flat up to 22 kHz. Obviously, this equipment will only record the lower frequency component of the sonar clicks. Fortunately, we were able to verify in the field that the bottlenose dolphin clicks really are broadband and that their lower frequency component must be nearly omnidirectional. In fact, our sensitive hydrophone detected clicks up to a few hundred meters from the dolphins, even if they were facing away from it. Note that we only wanted to study the occurrence pattern of the clicks, not their acoustical properties. So we assumed for the purpose of the present analysis that, although our instrumentation was suboptimal, we recorded the lower frequency components of all sonar click trains produced by the dolphins near the boat.

This part of our larger study used samples recorded on 13 different days, in 1992, 1993, and 1995, in five different months. These tape segments were selected because they had good signal-to-noise ratios, complete commentary on the animals' positions and behavior, and samples of the different activity patterns in almost all of the habitat subareas.

Sound analysis was performed initially using the package Hypersignal-Workstation (PC-based), digitizing the signals with a 12-bit Data Translation A/D board, with a sampling frequency of 44.1 kHz. More recently, Canary 1.2.4 has been used, and the signals have been digitized at the same sampling frequency by the in-built sound board of a PowerMacintosh 7100.

The sound segments selected were arbitrarily divided in one-minute samples. In the 447 one-minute samples obtained, signals were carefully counted by two researchers, using waveforms and sonograms whenever necessary (especially when trains overlapped). Signal occurrence was quantified using the categories "click trains," "burst pulses," "bray series," "bangs," "whistles," and "total number of vocalizations" (dos Santos 1998). Analysis of these counts and the other variables (e.g., group size, activity, and zone) included Spearman rank correlations, and simulation statistics using the program ACTUS for contingency tables (Estabrook and Estabrook 1989).

Results

Recordings were made near groups of dolphins engaged in various activities. Using behavioral criteria such as division in subgroups, proximity among animals, directionality and speed of the group's movement, dive sequences, aerial behaviors, and visibility of prey, five classes of general activity patterns were identified. These were "Travel" (rapid, linear displacement), "Travel/foraging" (zigzag displacement, long dives), "Disperse foraging" (animals quite spread, variable behaviors, some feeding events), "Surface group feeding" (aroused feeding around the same area), and "Social interactions at the surface" (aroused behaviors, with physical contacts and no prey visible).

As to the acoustic production, the average number of signal units (click trains, bray series, etc.) recorded near the groups per minute was 16.6, but in 20.6% of the one-minute samples no signals were detected.

The distribution of the abundance of click trains sample was also different from the expected. Not only did 23% of the samples have zero echolocation clicks, but a considerable 44% contained less that five click trains (fig. 55.1). So we created two nominal categories of train abundance: "Low train occurrence" (to include all samples with less than five click trains in one minute) and "High train occurrence." We then compared the distribution of these two categories of click train abundance with the occurrence of the various behavioral patterns. Having the data thus organized in a contingency table, the table was tested for independence using simulation statistics. This demonstrated that lines and columns are not independent ($p < 0.001$), specifically because of the results in those cells denoted with asterisks (table 55.1). Low train occurrence was more frequent than expected in Travel and Travel/feeding. High train occurrence was only more frequent than expected in Disperse feeding, where a diversity of behaviors is usu-

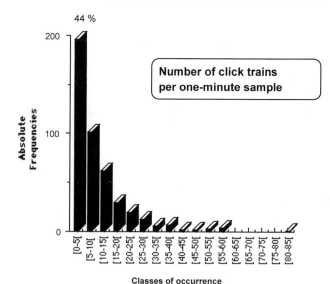

Fig. 55.1. Distribution of the abundance of click trains per sample

TABLE 55.1. Distribution of samples with less than five click trains (Low train occurrence) and more than five click trains (High train occurrence) per one-minute sample, in the various activity patterns, and results of contingency table analysis with simulation statistics (ACTUS). Asterisks denote significant cells.

Activity pattern	Low train occurrence	High train occurrence
Travel	15 ($p = 0.037$)*	5
Travel/feeding	151 ($p < 0.001$)*	110
Disperse feeding	25	107 ($p < 0.001$)*
Surface group foraging	5	23
Social interactions at the surface	0	6

ally observed in various social subunits, including prey capture at the surface and other aroused activities.

In the samples used for this study, dolphin group size ranged from 3 to 25; the class with the highest number of signal counts comprised 8–12 animals. Our data allowed us to examine the relationship between the number of click trains in a sample and the number of dolphins counted in the group whose behavior was being monitored. That is, how does sonar signal abundance correlate with group size? Also, how does the production of all signal categories correlate with group size? From these data, correlations are null and nonsignificant, both between number of trains and group size (rs = 0.04, $p = 0.43$, n = 447), and also between total number of signals in each sample (including the lower frequency pulsed signals and the whistles) and group size (rs = 0.018, $p = 0.71$, n = 447).

Discussion

These results raise some interesting questions concerning the use of echolocation signals by bottlenose dolphins in this population. It is readily admitted, however, that further studies should use broadband recording systems, or at least some kind of an ultrasound detector, and a more detailed sampling and analysis program. Even with the current limitations, a few points can be discussed.

The null correlations between signal production and group size, contrary to what would be expected, suggest that there might be some restriction mechanism that regulates acoustic production in groups. The absence of a positive correlation between number of possible emitters and number of vocalizations, in any situation and in any sound category, suggests that dolphins might follow some communication *and echolocation* "emission rule." Such a rule could help the dolphins to avoid the need to process too many signals of different origins (perhaps like in the conversations among humans).

Especially in the case of sounds with a predominant echolocation function, like click trains, there might exist emission rules possibly based on sex or age or a dominance hierarchy, or alternation routines. Johnson and Norris (1994) speculated that in spinner dolphins (*Stenella longirostris*) there might be a "trading of the duty" to allow each individual the opportunity to rest the emission tissues after intense click production. It would be relevant to point out that this sharing of sonar duties would require that dolphins analyze echoes of clicks emitted by conspecifics. The ability to use bistatic sonar was firmly established by the experiments of Xitco and Roitblat (1996) with the natural clicks of bottlenose dolphins, although it had already been strongly suggested by the experiments of Scronce and Johnson (1976) using artificial pulses.

Considering that visibility is very limited inside the Sado Estuary and that the dolphins spend a great deal of time feeding in turbid water, with no risk of predator interception, the production of sonar signals was globally much lower than expected. Table 55.1 suggests that these dolphins travel and try to find their prey mostly in silence. This means that, for orientation and prey detection, they must be using processes other than echolocation. Vision is certainly useful for tracking landmarks and underwater features in the more shallow or clear areas. Also, vision should allow dolphins to follow prey near the surface. However, visibility is very limited at the bottom of their main foraging grounds, such as in the deep south channel of the estuary. An obvious possibility is that audition, or passive listening (or passive sonar as a submariner would call it), is of more primordial importance to these animals' feeding ecology than one would predict. In that case, one wonders what reasons there could be for these animals to refrain from using their sonar, in a situation where predator interception of signals is not a risk.

One explanation might be that some preferred preys are difficult to find in the substrate through echolocation, because they hide in nests and in thick debris screens. Other prey species might show avoidance behaviors upon detection of sonar signals; in fact, this may be the case for the shad, *Alosa fallax*. Mann, Lu, and Popper (1997) showed that the related species *A. sapidissima* does detect and avoid sources of ultrasonic signals like dolphin clicks (see also Astrup and Møhl 1993 for the case of cod).

Some possible prey species, like the Lusitanian toadfish, are quite sonorous, and might be easier to find by passive sonar tracking. This toadfish's conspicuous display calls are produced by males from May to September, the nesting period of the species. The frequencies are rather low, mainly below 700 Hz (dos Santos 1998; dos Santos et al. 2000), but dolphins might be able to detect them at a useful distance. The hypothesis that bottlenose dolphins in the Sado Estuary eavesdrop on toadfish sounds requires further research. It is likely that they prey on the toadfish, considering what is known about equivalent fauna in Florida (Barros and Odell 1990). Playback experiments, at the appropriate pressure levels, could tell us whether dolphins do track these sounds, and at what distances would they be able to detect and locate their sources.

In this context, it is relevant to consider the question of the possible effects of anthropogenic underwater noise—which, in this estuary, shows pressure levels and spectra typical of industrial zones (Ferreira, Bento-Coelho, and dos Santos 1996; dos Santos 1998). If dolphins depend on audition to hear some of their prey, and if the ambient noise or the noise produced by some particular sources overlap in frequency with prey calls and masks them, then the indirect effects of underwater noise might be more stressing for the animals than the direct impacts on their auditory system.

Acknowledgments

Giorgio Caporin, António J. Ferreira, Alexandra Freitas, André Silva, other colleagues, and volunteers participated in the field recording and observation sessions, which were made possible by the supporters of Projecto Delfim–Centro Português de Estudo dos Mamíferos Marinhos. Our colleagues J. L. Bento Coelho, H. Onofre Moreira, and Rafael Serrenho at the Centro de Análise e Processamento de Sinais–Instituto Superior Técnico were of great assistance. The Biological Sonar Conference in Carvoeiro created a superb atmosphere for discussions, and Élio Vicente and Zoomarine deserve our gratitude for all the efforts to provide it. Jeanette Thomas, Kathleen Dudzinski, and Miguel Couchinho helped to improve the manuscript.

{ 56 }

Social and Nonsocial Uses of Echolocation in Free-Ranging *Stenella frontalis* and *Tursiops truncatus*

Denise L. Herzing

Introduction

Dolphins have sophisticated acoustic processing, echolocation abilities, and anatomical structures (Au 1993; Cranford, Amundin, and Norris 1996; Ketten 1997). Measuring parameters of echolocation signals primarily has occurred with physically restricted dolphins (Au 1993). Only recently have real-time recordings of echolocation signals been acquired in the wild (Au, Herzing, and Aubauer 1998). Although many aspects of sound use by dolphins have been reported (Herman and Tavolga 1980; Au 1993), most are in captivity or in field sites where behavior is often inferred through surface activity. Although interesting, the interpretation of surface behavior lacks sufficient visual confirmation to determine what the animals are actually doing under the water while signals are being recorded. It is more likely that correlating visually observable and verifiable behaviors with sound use will be an unbiased and productive approach without the problems inherent with observer inference of underwater behavior from the presence, or lack of, surface behavior. Unfortunately, there are only a few study sites in the world where dolphins can be observed on a regular basis engaging in a variety of normal behaviors. Specific behavioral and spectral descriptions of bottlenose dolphins (*Tursiops truncatus, T.t.*) foraging in the sand (Herzing 1996; Rossbach and Herzing 1997) and Atlantic spotted dolphins (*Stenella frontalis, S.f.*) using echolocation during foraging and social behavior (Herzing 1996) have been reported. Since behavioral definitions of underwater observations are less subjective than surface observations, substantiated visual observation should be required in the interpretation of the function of echolocation and other signals. This chapter offers a further detailed description of underwater visual and acoustic observations made during social and nonsocial behavior of free-ranging Atlantic spotted and bottlenose dolphins in the Bahamas from 1989 to 1998. Echolocation observations are placed in (1) behavioral contexts, by specific activity categories, and (2) ecological contexts, by describing specific habitats, depth of water, prey items, and bottom substrate. It is hoped that by reporting the use of echolocation signals during a variety of underwater behavioral contexts and specific eco-logical habitats, the functional use of these signals can be explored fully. Improved methods of capturing these signals, in a full-bandwidth capacity and in a real-life and observable field site, will be explored in the future.

Materials and Methods

The Bahama Islands are an archipelago in the tropical West Atlantic east of Florida. Although surrounded by deep water and the Florida straits, the water here is shallow (<15 m). The Bahamian banks are thick, submerged platforms of calcareous rock providing diverse habitats, including fringe and patch reefs, atolls, grassy flats, and ledges.

Since 1985, a resident community of approximately 220 spotted and 200 bottlenose dolphins have been identified, sexed, and observed in a variety of behavioral contexts in this region (Herzing 1996, 1997). During each underwater observation data were collected on group size, habitat type, environmental conditions, individual identity, age class, and behavioral context. Underwater behavioral and sound data were recorded for each encounter using underwater video cameras (Sony TR700, Yashica KV1, Sony Digital TR200) with a Labcore 76 hydrophone with frequency to 20 kHz and sensitivity of −192 dB re 1 μPa input to assure simultaneous recordings of underwater behavior and sounds. Ad libitum and behavioral event sampling was used to record foraging and social activity.

Individual click trains, or social sounds, were chosen for analysis when dolphins were alone and echolocating on bottom, playing with objects, or physically orienting/approaching another dolphin. The full-bandwidth recording of echolocation signals was not possible because (1) such mobile/remote equipment did not exist for underwater use, and (2) in the wild, the angle at which the dolphins orient during natural behavior cannot be controlled (resulting in the inability to assure the collection of head-on high-frequency signals). Unlike surface studies, where the underwater activity of echolocating dolphins is often inferred (based on the presence or absence of surface behaviors), the strength of these underwater observations include (1) our ability to observe visually underwater, and on a regular and repeated basis, the

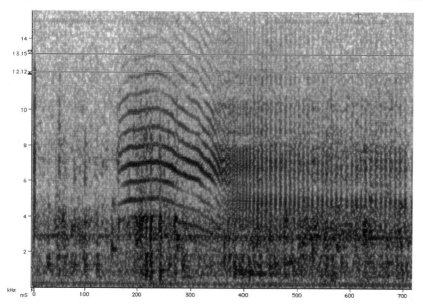

Fig. 56.1. Spectrogram of a high-repetition-rate, burst-pulsed sound grading to a click train with a slow repetition rate. High repetition rates are more easily measured by harmonic artifact intervals (vertical cursors) than individual clicks (horizontal cursors) as described by Watkins (1967).

actual activities of dolphins during echolocation, and (2) the recording click rates that can be extracted with narrowband recordings and when dolphins are recorded from a side angle. The recording and analysis of click rates (in lower frequencies and off-axis) are not restricted to full-bandwidth recording equipment and are a viable parameter to measure during echolocation.

Click repetition rates were extracted using Spectral Innovations MacDSP hardware and software. Click rates were individually counted, or, when repetition rates were high, the harmonic artifact feature representing repetition rate (as described in Watkins 1967) was used (fig. 56.1).

Results

NONSOCIAL USES OF ECHOLOCATION:
SEARCH AND APPROACH STRATEGIES

From 1989 to 1998, 306 observations of foraging and 751 of social behavior were made. The average duration of an observation was 20 min. The 159 spotted (*S.f.*) and 147 bottlenose (*T.t.*) dolphin foraging observations can be categorized as follows: horizontal and vertical scanning (32, *S.f.;* 103, *T.t.*), surface chases (83, *S.f.;* 25 *T.t.*), ledge/hole feeding (0, *S.f.;* 6, *T.t.*), percussive jaw-claps/tail-hits (4, *S.f.;* 5, *T.t.*), and rostrum hits (40, *S.f.;* 8, *T.t.*).

Table 56.1 displays the types of foraging strategies observed along with habitat information, prey types, and postural/vocal behavior. Concomitant visual behaviors are depicted in figs. 56.2–56.8. The strategy of searching was substrate and prey specific. Two phases of echolocation during foraging were observed, the search phase and the approach phase. In the search phase, dolphins actively searched using body or head movement. During this phase dolphins were either stationed in the water column (up to 15 m) scanning down, or near the bottom scanning <1 m off the sand. Echolocation click rates were

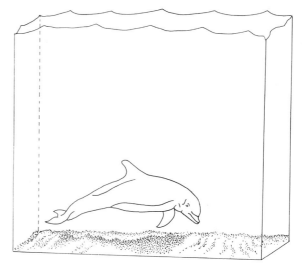

Fig. 56.2. Horizontal bottom scanning occurs while the dolphin moves along sandy bottom.

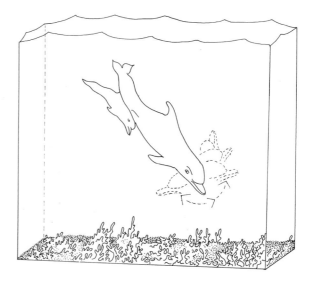

Fig. 56.3. Vertical water-column scanning occurs while the dolphin hovers in a stationary position above the bottom.

TABLE 56.1. Foraging strategies by Atlantic spotted dolphin (*Stenella frontalis: S.f.*) and bottlenose dolphin (*Tursiops truncatus: T.t.*). Species key: (A) conger eels (family Congridae); (B) snake eels (family Ophichthidae); (C) razorfish/wrasses (family Labridae); (D) flying fish (family Exocoetidae); (E) needlefish (family Belonidae); (F) ballyhoo (family Hemiramphidae); (G) flounder (family Bothidae); (H) snakefish (family Synodontidae); (I) grunts (family Pomadasyidae); (J) squirrelfish (family Holocentridae).

Strategy	Habitat	Foraging depth	Prey	Postural behavior	Echolocation observations
Horizontal bottom scanning (*T.t., S.f.*)	Dense or loose sand, no grass	<.5 m from bottom	Deeply buried fish (A, B, C)	Head bent to right and angled downward. Upon detection dolphin tips body and "digs" with rostrum.	Click rates 200 Hz, increasing to 500 Hz when digging in sand (Herzing 1996)
Vertical water-column scanning (*T.t., S.f.*)	Turtle grass beds, algae covered sand	3–15 m from bottom	Deeply buried fish (A, B, C)	Dolphin stationary, head bent to right, angled downward. Head rotation clockwise-counterclockwise.	Click rates 20–200 Hz when audible. Calf may be positioned under adult and mimic movements of head.
Surface scanning (*T.t., S.f.*)	Water surface, air surface	At surface	Schooling or moving fish (D, E, F)	Dolphin swims upright scanning; may turn inverted during chase, catch	Rapid echolocation clicks with increasing repetition rate with proximity to target
Ledge, hole scanning (*T.t.,* only)	Coral reef, rocky bottom	<.5 m from substrate	Hidden fish (I, J)	Rostrum directed under ledges or in holes. Dolphin grabs fish under ledge or hole.	Click rates 8–100 Hz, with whistles and trills
Percussive jaw claps (*T.t., S.f.*)	Scattered benthic sargassum	<.5 m from bottom	Deeply buried fish (A, B, C)	Head bent downward during intermittent jaw claps. Dolphin moves along bottom slowly	May or may not echolocate during behavior
Percussive rostrum bop (*T.t., S.f.*)	Dense or loose sand	On bottom or <.5 m	Shallow buried fish (G, H)	Dolphin stops during bottom scanning and strikes bottom multiple times with tip of rostrum until fish darts out and chase ensues.	Scanning clicks audible during bottom scanning and increased repetition rates during rostrum hits.
Percussive tail slaps and pectoral fin hits (*T.t., S.f.*)	Dense or loose sand	<.5 m from bottom	Deeply buried fish (A, B, C)	Dolphin moves along bottom, striking sand with appendage and dolphin returns to vortex for fish.	May have associated echolocation clicks during horizontal scanning on bottom and retrieval of fish.

slow (<200 Hz) until the approach phase, when, after detection, dolphins emitted increasing repetition rates of clicks (up to 500 Hz) while pointing into the sand and digging out prey. Other techniques, such as percussive sounds or the physical disturbance of the sand, were used before the approach phase and appeared to create supplemental information and access to prey.

Various surveillance strategies were observed for *Stenella* during rest and travel (table 56.2). These include underwater observation of echolocation activity and be-

havior such as age-class specific reactions to predators such as sharks. The primary mode of monitoring was listening in both clear and murky waters. During twilight hours and in deep water (>30 m), dolphins were observed underwater and did not employ active, audible (<15 kHz) echolocation until an apparent cue was detected and warranted inspection. Postdetection behavior included slow-rate echolocation clicks (8–20 Hz), close, physical contact within the dolphin group, and coordinated head scanning toward the bottom.

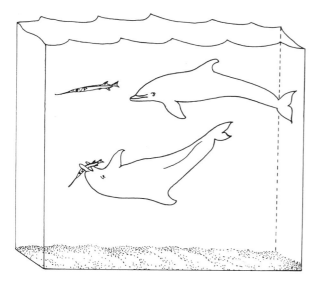

Fig. 56.4. Water surface/air interface scanning occurs as the dolphin works underneath the surface of the water.

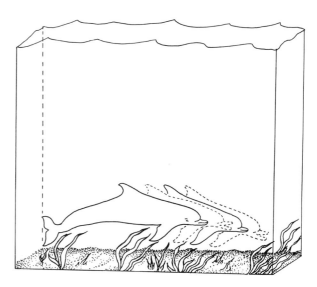

Fig. 56.6. Percussive jaw-claps occur as the dolphin is moving along bottom.

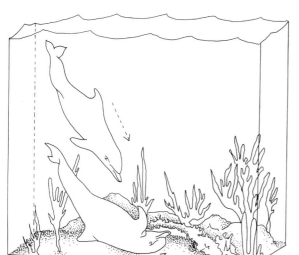

Fig. 56.5. Ledge/hole scanning occurs as the dolphin searches and explores under substrate.

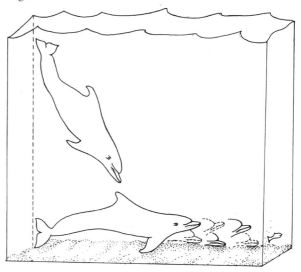

Fig. 56.7. Percussive rostrum hits occur as the dolphin is on bottom. Rostrum makes shallow and intermittent contact into sand.

Fig. 56.8. Percussive tail-slaps and pectoral fin hits occur as the dolphin moves along bottom. The dolphin may turn and seek out the vortex of sand and then dig in the center for fish.

TABLE 56.2. Surveillance strategies of Atlantic spotted dolphins (*Stenella frontalis*).

Behavioral observation	Echolocation observation	Age/class	Reaction
Daytime rest and travel	Quiet and passive vigilance	All	None
Acoustic cue of predator	Slow echolocation 8–20 Hz	All	Head-scanning and tightening of group
Predator in visual range			
• Tiger/bull sharks	Echolocation may stop while predator in sight	Mothers and calves	Sink to bottom and flee as group
• Tiger sharks	Social signals: excitement vocalization or signature whistles	Calves	Panic and erratic swimming
• Hammerheads and nurse sharks	Active echolocation and buzzing of predator	Mixed age/classes	Vigilance or active chasing of predator
• Bull sharks	Quiet	Old adults	Sink to bottom and remain motionless

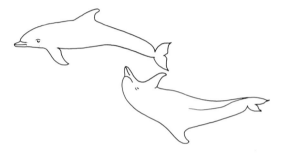

Fig. 56.9. Genital inspection and buzzing is typical during courtship and disciplinary behavior.

SOCIAL USES OF ECHOLOCATION:
CONSPECIFIC BUZZING AND SCANNING

The use of high-repetition-rate click trains during social interaction (genital buzzes of 8–2000 Hz, squawks of 200–1200 Hz—as described in Herzing 1996) were observed during social behavior including courtship (fig. 56.9), discipline and play (fig. 56.10), aggression (fig. 56.11), and during focused attention to inanimate objects (fig. 56.12). Clicks within such close proximity and with high repetition rates are not believed to have an echolocation function according to current sonar theory. To determine the possibility of tactile components to these sounds, Kolchin and Bel'kovich's (1973) pressure amplitude measurement threshold levels (10–40 mg/mm^2) for tactile sensitivity in the common dolphin (*Delphinus delphis*) were converted to sound pressure levels (SPL in dB re 1 μPa), where

$1 \mu Pa = 10^{-6} kg/ms^2$
gravity (g) = 9.8 m/s^2
mg/mm^2 = 10^{-6} kg/10^{-6}m^2 = kg/m^2

then

P1 = 10 mg/mm^2 × g = 10 kg/m^2 × 9.8 m/s^2 = 98
Pa = 98 × 10^6 μPa and SPL = 20 log P1 = *160 dB re 1 μPa*
P2 = 40 mg/mm^2 × g = 40 kg/m^2 × 9.8 m/s^2 = 392
Pa = 392 × 10^6 μPa and SPL = 20 log P2 = *172 dB re 1 μPa*

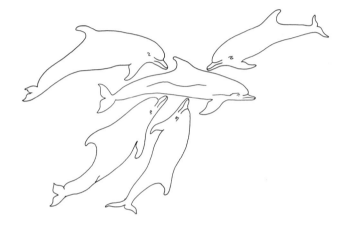

Fig. 56.10. Group buzzing occurs when one dolphin floats passively while others buzz it, usually in the midbody area.

Fig. 56.11. Head-to-head orientation and buzzing occur during aggressive behavior and juvenile play. Dolphins may be moving or stationary.

Fig. 56.12. Exploratory object buzzing is typical on objects, including floating debris and plant life.

Therefore, it would take SPL levels of 160–172 dB re 1 μPa to stimulate the dolphin's somatic pressure receptors. Peak source levels as high as 210 dB re 1 μPa were recently measured for these resident *Stenella frontalis* (Au, Herzing, and Aubauer 1998), which are above the amplitude levels required to pass the estimated levels for somatic sensation.

Discussion

NONSOCIAL USES OF ECHOLOCATION

Atlantic spotted dolphins (*Stenella frontalis*) and bottlenose dolphins (*Tursiops truncatus*) in our study used echolocation for both social and nonsocial uses. Scanning strategies during foraging included (1) horizontal bottom scanning, (2) vertical water-column scanning, (3) surface scanning, and (4) ledge/hole scanning. The presence of specific substrate or algae, depth of water, and the prey species determined the scanning techniques. Deeply buried prey were retrieved using a variety of techniques including percussive elements such as jaw-claps, tail-slaps, and pectoral fin strikes. Shallow-buried species were retrieved with direct horizontal scanning or the use of physical blows of the rostrum on the bottom to encourage the fish to move and presumably be detected moving along the sandy bottom. Dolphins scanned vertically, and from great distances, in areas of dense grass or other bottom clutter. Detection seemed to occur from these distances and did not require more horizontal scanning on the bottom. Instead, the dolphins went directly down to the bottom from the midwater column and dug out a fish. For schooling fish, dolphins arched their bodies during the chase to presumably accommodate the projection of clicks. Just before capture, dolphins often turned upside down, possibly utilizing their visual system at this stage of capture. The variety of strategies mirrored the variety of habitats and prey species, suggesting that dolphins use a multitude of behavioral and acoustic strategies in sophisticated anticipation of their prey's behavior and the effects of their sound in cluttered/uncluttered environments. Although different than bat echolocation (Au 1997), dolphin echolocation is used in habitat exploration (cluttered versus uncluttered) and for monitoring escape strategies of prey (move, dig, freeze). Both suggest that dolphins recognize specific prey and employ species-specific strategies.

Dolphins also employed various monitoring and vigilance strategies and were typically silent listeners unless cues were received that triggered active inspection. To what degree do dolphins use passive listening versus active echolocation in other locations? Barros and Wells (1998) reported that bottlenose dolphins target soniferous prey, suggesting that passive listening is employed in detection.

Clicks and pulsed sounds have similar acoustic features, with the primary difference being time intervals between pulses. A slight change in this interval can make a large difference in aural perception to both the human ear and to the visual representation on a spectrogram. The fact that click trains grade into pulsed sounds with gradual decreases in interclick intervals, and often without a change in peak frequency (Murray 1997; Herzing

1988), should caution us against making discrete functions out of purely aural differences. Although the two-way transit time interval rule (needed to explain currently understood echolocation functions of clicks) no longer holds during click trains that grade into high-repetition-rate pulsed sounds, the animals are clearly using them in a continuum of motion and activity. Perhaps it is our definition of the functions of such clicks that needs to be expanded, rather than separating the explanation based on our understanding of how biological sonar may work. The measuring of echolocation signals by moving dolphins, rather than stationary subjects, is already illuminating challenges to current sonar theories (Sigurdson 1998).

SOCIAL USES OF ECHOLOCATION

The social uses of click trains reported here included low-frequency buzzing during courtship, discipline, play, aggression, and exploration. Source levels adequate for tactile stimulation have been measured in spotted dolphins (Au, Herzing, and Aubauer 1998) and for multiple delphinid species (Au 1993). Ketten (1997) described the frequency-specific sound-conduction properties of various parts of the dolphin's rostrum. This suggests that jaw rotation, open-mouth behavior, and head rotation could be comparable to sound focusing and ear rotation in bats. Intense, close proximity sounds and percussive elements, that theoretically provide tactile sensation, have been recorded in intraspecific aggression (Overstrom 1983; Herzing 1996), interspecific aggression (Herzing and Johnson 1997), and possibly fish stunning (Norris and Møhl 1981; Marten et al. 1988; Marten et al. 2001). The use of high-repetition click trains suggests that odontocetes use "shorter than transit time" intervals, and packets of clicks (burst-pulsed sounds) could also be manipulated in an "echolocation type" way to provide both social and prey information. The proximity and angle of calves underneath scanning mothers (fig. 56.3), parameters critical for eavesdropping (Xitco and Roitblat 1996), support the idea that learning opportunities are available during foraging behavior in the wild. In the Bahamas, observations of young calves echolocating on small flounders, and following conspecifics during foraging, suggest that echolocation skills may begin very early in life. Spotted dolphins nurse as long as five years but are weaned gradually and learn to catch fish in the first year of life (Herzing 1997). The development of echolocation abilities and its directed use may be critical for survival in the wild. In captivity, the use of echolocation signals by bottlenose dolphin calves has been reported as early as six months (Reiss 1988).

Conclusion

The use of high-repetition-rate click trains (higher than defined by current sonar theory to be of echolocation value), during both social and nonsocial activity, may indicate uses of echolocation clicks not previously re-corded. High rates of clicks have been reported by Watkins (1967) in the harbor porpoise (*Phocoena phocoena*) at 700 Hz, Risso's dolphin (*Grampus griseus*) at 250–330 Hz, in bottlenose and Atlantic spotted dolphins at 2000 Hz (Herzing 1988, 1996). Murray (1997) reported click trains grading into pulsed sounds with decreasing interclick intervals (<2 ms). Gradation of click trains, to pulsed sounds to whistles, without a change in peak frequency, has been described elsewhere (Herzing 1988; Watkins 1967) and is strong evidence that decreasing interclick intervals during click trains are a graded and continuous event. Whether the function of such clicks is also continuous, as far as informational content, is not clear.

Why would dolphins use click rates so high that individual pulses are no longer distinguishable and separated by adequate transit times to discriminate one outgoing click from another? Whether such high rates of clicks are useful for return echolocation information (i.e., signals with transit times long enough to allow the return echo to be received before the next on is sent out) or have a tactile effect on a target, in their ability to escape, is unknown. The fact that dolphins play and buzz animate and inanimate objects, when echolocation information about an already visual and captive target are not needed, may suggest that there is another function for high-repetition click trains other than echolocation. Evans and Powell (1967) reported repetition rates of 200 Hz from dolphins 0.5 m from a target. Relative geometry of the dolphin's head to the sand, such as a more vertical angle, and rotation of that angle (e.g., through rostrum movement up and down, or upside-down inverting swimming), would also change the angle of the echolocation beam to the target. Our abilities to measure such subtle aspects of echolocation signals, with moving dolphins, would greatly increase our understanding of the use of echolocation during natural behavior. The development of a self-contained, mobile underwater video and full-bandwidth recording unit would ensure complete sound and visual signal collection during the activity of dolphins, thus removing the potential misinterpretation of signals recorded during surface-only studies.

Acknowledgments

Special thanks to all the staff, crew, members, and foundations that helped support the research of the Wild Dolphin Project over the years. Special thanks to Dr. Whitlow Au for help with equation conversions, Ms. Patricia Weyer for use of her illustrations, and Dr. Christine Johnson and Kelly Allman for discussion and thoughts over the years on this subject.

{ 57 }

Variable Spectra and Nondirectional Characteristics of Clicks from Near-Surface Sperm Whales (*Physeter catodon*)

William A. Watkins and *Mary Ann Daher*

Introduction

The first positive attribution of click sounds to sperm whales (*Physeter catodon*) was by Worthington and Schevill in 1957; detailed study of the acoustic behavior of these whales began in 1958 (Schevill and Watkins 1962). Results from these and subsequent studies included the following assessments of their use of sound: whales often clicked at regular rates during dives for minutes at a time, interspersed with silences of variable duration; no clicks were heard from lone whales; those near the surface produced few clicks; short click series, called codas (Watkins and Schevill 1977b), appeared to have a communicative function; young (7–8 m) whales were responsible for most aerial displays and often produced long click sequences with highly variable click rates. Clicks from near-surface whales recorded during more than 100 cruises (Woods Hole Oceanographic Institution, WHOI) consistently had variable spectra and no apparent propagation directionality. These sperm whale clicks appeared to be used mostly for communication (Watkins 1977).

Sperm whales produce broadband click sounds with frequencies from about 100 Hz to 30 kHz. They click sporadically during many activities, usually in repeated series, often at variable and temporally patterned rates (cf. Watkins and Schevill 1977b; Gordon 1987; Whitehead and Weilgart 1991).

Long-term tracking of sperm whales provided details of their behavior using radio with dive-profile telemetry, and acoustic and satellite tags that allowed correlation of sounds with the whale movements both at the surface and under water. Whale sounds were followed by three-dimensional, 100-kHz bandwidth hydrophone arrays and were related to underwater movements and dives (cf. Watkins and Schevill 1972, Watkins et al. 1993, 2002).

Adult sperm whales typically produced few sounds either when near the surface during respiration between dives or when apparently resting at the surface. During social interactions, there were occasional acoustic exchanges of short click series and codas (Watkins and Schevill 1977b). The click sounds from near-surface whales had variable levels and frequency emphases, with no discernible directionality (Watkins 1977; Watkins et al. 1999).

Juvenile, 7–8 m sperm whales characteristically had a variety of rambunctious behaviors, including repeated breaching and other aerial activity, exploring of objects, and close approaches to ships. In addition, these young whales often produced long click series with widely varying click rates, levels, and spectra. Such smaller sperm whales have been observed repeatedly (more than 10 times in our records) swimming at the surface with the head raised clear of the water while continuing to click. One such episode is analyzed here, demonstrating the variability and omnidirectional character of these click sequences from near-surface sperm whales.

Materials and Methods

Underwater acoustic studies of sperm whales in the southeast Caribbean were conducted in 1984 from the 17 m research vessel IDA-Z (W. A. Watkins, Moore, and Tyack 1985). Surface activities of the whales were monitored visually by experienced observers with 5–26 years of cetacean experience. Underwater sound sources were distinguished and tracked by sound arrival-time and phase differences at each of four hydrophones in a three-dimensional array (Watkins and Schevill 1972). The sounds illustrated in the figures were from one array hydrophone, suspended 3 m below a 25 cm, inflated rubber buoy, tied loosely to a 3 m pole extending from the bow of the ship. Sounds were received by the hydrophone (WHOI modified, Ithaco 602 and Ithaco 450 series amplifier) and recorded on tape (WHOI modified Pemtek 110 recorder) with an overall frequency response within 1 dB from 15 Hz to 100 kHz.

Analog sounds were digitized at 60 kHz and analyzed with 256 point FFT and 50% overlap for the spectrographic analysis in the figures using CSIG (Fristrup et al. 1992).

Results

Three sperm whales were within 125 m of the drifting ship on 17 March 1984 at 1515 h at N15°26' W61°32', 4 km from the shore of Dominica Island. Two of the whales were approximately 9 m in length, likely adult females, and the third was an older calf of about 8 m. The adults surfaced repeatedly with little horizontal movement, apparently resting silently, while the smaller whale lobtailed occasionally and actively swam about, producing long sequences of variable clicks, which were readily correlated with the whale's location.

At one point, the small whale approached and swam around the ship. It swam just below the surface for much of this excursion. The whale produced long sequences of clicks, and as the ship was approached, it repeatedly raised its head out of water. Fig. 57.1 depicts the whale's track during the 4.5 min visit to the ship, as it came to-

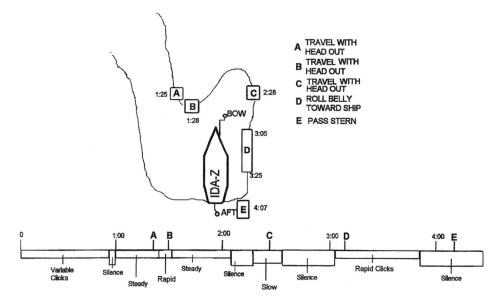

Fig. 57.1. Track of 8 m sperm whale as it approached the ship and produced a long series of clicks. Blocks A–C identify positions where the whale raised its head clear of water. After a short period of silence, the whale rolled, ventral surface toward the ship, and clicked rapidly (block D). *Below diagram:* the stream of clicks during the 4.5 min period of the whale's visit to the ship.

Fig. 57.2. Photograph of partial head-out by the 8 m sperm whale as it started to raise its head above water for the third time (full head-out not caught). Photo by P. Tyack.

ward the ship, approached to within 10 m of the hydrophone at the bow, turned from and then moved with its side toward this hydrophone, turned around, swam along the length of the far side of the ship, rolled so that its belly was toward the ship, righted itself, and passed close to and below the ship's stern, and then returned toward the other whales. The whale clicked variably throughout its visit to the ship.

The different orientations of the whale to the hydrophone during this visit included right and left approaches, turning alongside it, and moving obliquely away. In addition, the whale raised its head clear of the water three times while continuing to click (blocks A–C of fig. 57.1). This whale is shown in fig. 57.2 at the beginning of the third head-out episode.

During head-out episodes the whale's head was raised high enough out of water to expose the lower jaw, as far as the jaw hinge. In each instance, the head was above water for periods of 3–4 s. The variable stream of clicks produced during the 4.5 min period and the relative occurrence of events is diagramed at the bottom of fig. 57.1.

Sequences of clicks were analyzed in 10 and 12 s increments to show the overall variability in click spectra and level, and the relative lack of directionality in click propagation. The click sequence of fig. 57.3A was from a distance of about 20 m as the whale approached. The sequence in fig. 57.3B was from a distance of about 10 m, and included the first two head-out episodes (blocks A and B of fig. 57.1) as the whale turned directly toward and then began to move past the hydrophone. The clicks continued at the same rate and spectra as the whale swam around the hydrophone. Fig. 57.4A included the third head-out episode (block C of fig. 57.1) as the whale turned back obliquely toward the hydrophone. The whale stopped clicking, then rolled with ventral surface toward the ship, swam slowly (block D of fig. 57.1), and then righted itself while clicking rapidly and steadily. It

was silent as it rounded the ship's stern, and clicked sporadically as it headed back toward the other whales.

The repetition rates of the clicks, as well as their spectra and levels, were extremely variable. During the whale's approach (fig. 57.3A), the click rate varied from 4 to 6 clicks/s, averaging 5.25/s. It continued to click, and during the first head-out episode (fig. 57.3B), the click rate was 7.4 clicks/s. The click rate increased to 11/s as the whale submerged its head, and then decreased during the second head-out episode to 8.8/s (fig. 57.3B). The click rate decreased further to 4–5/s as the head was again submerged, and the whale swam with the hydrophone off to its side (fig. 57.1). There was a short period of silence. Then, by the third head-out episode (block C of fig. 57.1; fig. 57.4A), the click rate was 2.3/s, and clicking stopped again for a brief period after the head was submerged. As the whale turned its belly toward the ship, it resumed clicking at a much higher rate of 25/s, with increasing click level, although it was moving away from the hydrophone (block D of fig. 57.1, fig. 57.4B).

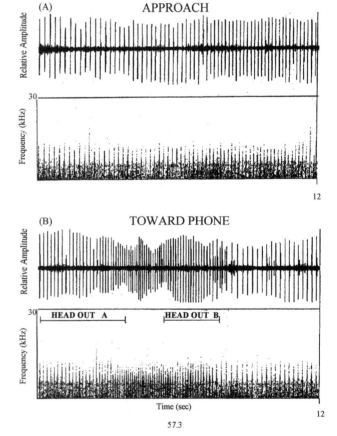

Fig. 57.3. *A:* Clicks produced at a distance of about 20 m from the hydrophone as the sperm whale approached (before the first head-out episode), analyzed in a 12 s waveform (*top*) and spectrogram (*bottom*) (scale: 200 Hz to 30 kHz). *B:* Clicks produced at a distance of about 10 m as the sperm whale turned left toward the hydrophone and then twice raised its head clear of the water (blocks A and B of fig. 57.1), analyzed in a 12 s waveform (*top*) and spectrogram (*bottom*) (scale: 200 Hz to 30 kHz).

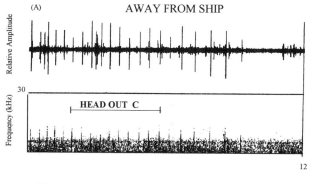

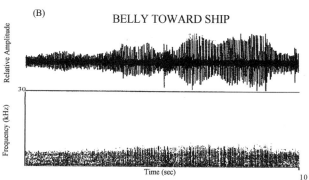

Fig. 57.4. *A:* Clicks produced from a distance of about 10 m as the sperm whale turned back to the right obliquely toward the hydrophone and raised its head clear of the water a third time (block C of fig. 57.1), analyzed in a 12 s waveform (*top*) and spectrogram (*bottom*) (scale: 200 Hz to 30 kHz). *B:* Clicks produced during the first half of the 20 s period that the sperm whale was passing with belly toward the ship (block D of fig. 57.1), analyzed in a 10 s waveform (*top*) and spectrogram (*bottom*) (scale: 200 Hz to 30 kHz).

Clicking stopped as the whale righted itself and passed under the ship's stern (block E of fig. 57.1).

The whale lifting its head out of water had no effect on either click level or frequency. There was little difference in the clicks produced with head out of water or submerged. The whale's orientation to the hydrophone also made little difference to click levels or spectra, and click repetition rates were not important. Click rates were different each time the head was lifted out of water and each time it was submerged. Click frequency emphases and levels were variable throughout the whole period. None of the click features or any changes in them could be related to the whale's orientation or proximity to objects.

Discussion

The clicks recorded from this whale were representative of those recorded generally from sperm whales (both adults and smaller whales) when they were near the surface. The variable click series for several minutes also was particularly characteristic of the sounds we have recorded from smaller 7–8 m sperm whales.

Such variability in the characteristics of click sounds from sperm whales near the surface contrasts somewhat with clicks recorded from the same whales during dives. The "usual" click sequences over a few minutes from whales during dives were relatively constant in repetition rate, frequency content, and level. Over longer periods, however, many of the features of clicks received from the same whales varied as the depth of the whale changed. Such variations, as well as the relative stability of click features over a few minutes, were demonstrated by continuous recording from tagged whales (e.g., Watkins et al. 1993, 1999, 2002).

Clicks from near-surface sperm whales, such as those analyzed here from the small whale, appear not to be produced by a sound mechanism that propagates from the forward portion of the head. Click level and spectra did not change when the head was lifted well out of water. This and the omnidirectional character of such near-surface click sounds seems to indicate a different sound-generation system than has been suggested previously (e.g., Norris and Harvey 1972; Cranford, Amundin, and Norris 1996).

We have observed neither the remarkably high levels nor the strong directionality in clicks from sperm whales reported by Møhl et al. (2000) from analysis of data recorded on an array of widely spaced hydrophones. The click sequences analyzed here from the small sperm whale exploring the ship did not appear to be used in echolocation. There was no click directionality, and none of the click features typical of echolocation in other species—such as changing repetition rates with target distance, spectra, and levels—correlated with orientation or proximity to objects. The physical behavior of the whale also did not appear to be that of a whale using clicks to explore unfamiliar objects—that is, only occasional orientation toward objects, no head scanning, no investigative delays at any part of the ship or its hydrophones with floats and cables, such as might be expected if clicks were being used for echolocation. When the whale was close to the ship, it rolled so that its belly was toward the ship, a behavior we often have associated with whales' close visual inspection of objects.

Acknowledgments

Whale observations were shared with our compeers aboard the RV IDA-Z, Karen Moore, Peter Tyack, and James Broda, as well as the ship's crew, Capt. Cecil Roper, Ian McCabe, and Laura Barr. Cruise arrangements, permitting quiet operation for good sound recording, were carefully made by Spice Island Traders, the ship's operating organization. General support was from the Office of Naval Research (current grant N00014-96-1-1130). This is contribution no. 9860 from the Woods Hole Oceanographic Institution.

{ 58 }

A Comparative Analysis of the Pulsed Emissions
of Free-Ranging Hawaiian Spinner Dolphins
(*Stenella longirostris*)

Marc O. Lammers, Whitlow W. L. Au, Roland Aubauer, and *Paul E. Nachtigall*

Introduction

Extensive research has focused on the pulsed acoustic emissions of dolphins since the discovery of their biosonar system (Kellogg 1958). While not all dolphins whistle, they all produce click trains that vary widely in click duration, interclick intervals, and spectral composition (Popper 1980). In general, two functional categories of click trains are recognized: (1) echolocation signals used in sensory tasks (Au 1993) and (2) burst-pulse signals associated with social communication (Herman and Tavolga 1980; Herzing 1988, 1996; Blomqvist and Amundin, chapter 60, this volume). While much has been experimentally learned about echolocation clicks, comparatively little is known about the characteristics and function of burst pulses. Burst pulses are often described with such qualitative terms as "yelps," "creaks," "squawks," and "blasts" (Herman and Tavolga 1980). Signals are perceived in this manner when the interval between clicks in a train drops below 5 ms, at which point humans no longer resolve individual pulses and the signal is heard as a single continuous sound (Murray, Mercado, and Roitblat 1998). Variations in the click repetition rate are perceived by the human listener as qualitatively different signals. Such labels have resulted in most analyses of burst pulses focusing on the audible spectral components of these signals (Caldwell and Caldwell 1966; Overstom 1983; Sjare and Smith 1986b; Herzing 1988, 1996).

The relatively narrowband (<20 kHz) recording equipment generally used to collect burst-pulse data has yielded peak and center frequency estimates well below those found for echolocation click trains. However, these measures may represent artifacts of the recording equipment's bandwidth limitations and be underestimates of the actual spectral energy distribution in burst pulses. Dawson (1988), using broadband equipment to examine the "cry" burst pulse produced by free-ranging Hector's dolphins (*Cephalorhynchus hectori*), a non-whistling species, reported clicks with energy centered around 120 kHz and source levels around 150 dB re

1 μPa. Au, Penner, and Turl (1987), while studying echolocation in a captive beluga (*Delphinapterus leucas*), collected burst pulses with peak-to-peak click source levels of about 206 (±6) dB re 1 μPa. Although no peak or center frequency values were reported, those source levels are comparable to echolocation click source levels. These findings suggest that burst pulses may be spectrally more similar to echolocation click trains than previously reported for many species.

An issue that remains unresolved concerns the accurate labeling of click trains as either functional echolocation signals or socially meaningful burst pulses. In other words, what specific characteristics define these classes of signals? Most efforts to date have relied primarily on qualitative aural distinctions and/or visual inspections of narrowband (<20 kHz) sonograms to classify signals. Such methods, however, make comparisons between workers difficult and can arguably lead to biologically questionable conclusions. A more quantitative and broadband approach is therefore necessary.

Experimentally, burst pulses have received little attention from workers studying echolocation. Often, the label of "burst pulse" is simply given to signals that do not conform to the types of click trains known to have an echolocation function. Target detection experiments show that bottlenose dolphins (*Tursiops truncatus*) temporally space pulses in an echolocation click train to focus their attention at a particular distance (Penner 1988). When scanning a target, clicks are spaced to account for the two-way travel time to and from a target plus an echo processing period between 19 and 45 ms long (Au 1993). Signals that do not conform to this pattern present a puzzle for researchers and raise the question: Should trains with interclick intervals less than the minimum echo processing period be treated by default as nonecholocation, social burst pulses?

To clarify the relationship between the various kinds of click trains produced by delphinids, this study examined the pulsed acoustic emissions of free-ranging Hawaiian spinner dolphins (*Stenella longirostris*). This is a species for which pulsed signals have previously only been qualitatively described (Norris et al. 1994). The objectives of this work were to (1) quantitatively characterize pulsed signals during social situations when acoustic activity is at a peak (Norris et al. 1994) and (2) look for evidence of definable classes (i.e., burst pulse versus echolocation trains) in the pulsed signaling repertoire of this species. To accomplish this, the temporal, spectral, and source level parameters of click trains were used to quantitatively relate signals to one another.

Materials and Methods

DATA COLLECTION

Spinner dolphins are an abundant and accessible species in Hawaiian waters. A population of several hundred animals resides along the leeward coast of the island of Oahu, and is readily reached by small boat. Using a 5.2 m, outboard-powered Boston Whaler, groups of spinner dolphins were approached and recordings made on 28 July, 11 August, and 13 August 1998. Group sizes ranged from 10 animals on 28 July, to 80–100 on 11 August and 60–70 on 13 August. Animals were approached as they milled about in waters between 5 and 18 m deep over a mostly flat, sandy bottom substrate. Upon encounter, the boat was either anchored or left to drift with the engine off. Behavioral observations were quantified using an unpublished ethogram based on the work of Norris et al. (1994) to establish the general behavioral state of the animals. In each case, group activity was classified as moderately to highly social. Distance to the animals at any given time ranged between 5 and 100 m. All recordings were made under Beaufort Sea State 1 or less.

Based on work with a captive false killer whale (*Pseudorca crassidens*), Murray, Mercado, and Roitblat (1998) proposed that all *Pseudorca* signals are best modeled along a graded continuum, rather than categorically. Similar ideas have been suggested in the past for the tonal signals of belugas (*Delphinapterus leucas*) (Sjare and Smith 1986b), pilot whales (*Globicephala melaena*) (Taruski 1979), and common dolphins (*Delphinus delphis*) (Moore and Ridgway 1995). With this in mind, an effort was made to keep a priori assumptions about the presence of signal classes (burst pulse versus echolocation) in the data to a minimum, so as to allow for a graded pattern, if present, to emerge. By relying on visual (a flashing LED) rather than aural cues to detect the presence of a signal (see below), bias toward collecting any one type of signal over another was minimized. The signals collected, therefore, accurately represent what free-ranging Hawaiian spinner dolphins produce in a moderate to highly social behavioral state.

RECORDING EQUIPMENT

Recordings were made using a custom-built (Au, Lammers, and Aubauer 1999), laptop computer–based, digital recording system with a bandwidth of 130 kHz. A custom-built hydrophone consisting of a 20 mm diameter spherical piezoceramic element with an omnidirectional, flat frequency response (±5 dB) from 10 kHz to 160 kHz and a sensitivity of 210 dB re 1 μPa was placed 3 m below the surface. Incoming signals were detected visually on an LED meter activated by peak-to-peak sound pressure levels >134 dB re 1 μPa. Upon detection, the operator of the recording system pressed a trigger signaling the event to the computer, which in turn stored 1 s of pretrigger data and 2 s of posttrigger data. Data were stored on the computer's hard disk drive. Water depth was recorded using an Under Sea Industries handheld personal dive sonar.

DATA ANALYSIS

Interclick interval, click center frequency, and rms bandwidth were extracted from each train using a custom-written Matlab 5.1 analysis program. Spectral measurements were obtained using either a 1024 or 512-point FFT window, depending on the interval between clicks (512 points were used when a larger window size would have overlapped two clicks). Peak-to-peak source levels of clicks (fig. 58.1) were estimated by geometrically localizing phonating animals using arrival-time differences between the direct click, the 180° phase-shifted surface reflection, and the nonphase-shifted bottom reflection (Aubauer et al. 2000). Using c as the speed of sound in seawater, the distance of the phonating animal (r) was estimated using the following equation:

$$r = \frac{\left(\frac{c \cdot \tau_{1b}}{2}\right)^2 + b \cdot (a - b) - \left(\frac{c \cdot \tau_{1s}}{2}\right)^2 \cdot \left(1 - \frac{b}{a}\right)}{\frac{c \cdot \tau_{1s}}{2} \cdot \left(1 - \frac{b}{a}\right) - \frac{c \cdot \tau_{1b}}{2}}$$

The source level (SL) of a click was established using:

$$SL = SPL_R + 20 \cdot \log r + \alpha \cdot r$$

where SPL_R is the sound pressure level relative to 1 μPa of the recorded signal and α is the absorption coefficient of the water measured in dB/m. Only click trains with very distinct surface and bottom reflections were used to estimate source level. Within these click trains only the five clicks with the highest amplitude were chosen for the calculation. No information was available on the orientation of the phonating animal, so it could not be established if the signals were measured on the beam axis. Thus, the spectral and source level values reported here are only estimates for the signals produced by spinner dolphins.

Results

A total of 133 click trains were analyzed. Fig. 58.2 shows the distribution of mean interclick intervals (ICIs) for

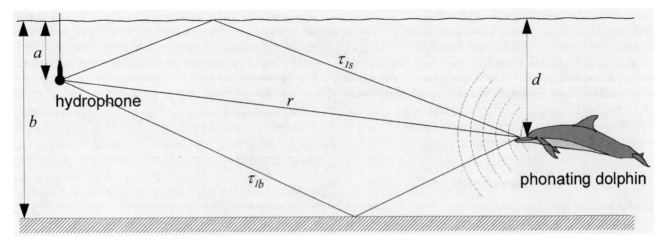

Fig. 58.1. Multipath propagation in shallow water of dolphin clicks, where *a* = hydrophone depth, *b* = water depth, *r* = distance of phonating animal to the hydroplane, *d* = depth of phonating dolphin, τ_{1b} = time delay of first order bottom reflection relative to direct click, and τ_{1s} = time delay of first order surface reflection relative to direct click.

each train. The distribution is strongly bimodal, with two peaks centered at 3.5 ms and 80.0 ms and a gap at 10 ms. To investigate how consistent ICIs remained within a train, the relationship between the first and last ICI was examined. A linear regression analysis (fig. 58.3) indicated that the two are correlated ($r = 0.87$), suggesting that within a train ICIs tended not to change drastically. Beginning ICI was a good predictor of ending ICI for trains with a mean ICI < 10 ms ($r^2 = 0.76$) and somewhat less so for trains with greater mean ICIs ($r^2 = 0.53$).

Spectrally, the mean center frequency of a train tended to increase ($r = 0.65$) with increasing mean ICI (fig. 58.4). For all trains, the mean center frequency was never below 30 kHz. The average rms bandwidth of click trains ranged between 13 and 33 kHz and did not have a strong association with mean ICI ($r = 0.4$). Thus, while trains with small ICIs generally had lower mean center

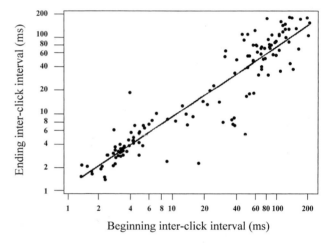

Fig. 58.3. Linear regression plot of the beginning interclick interval of a click train versus its ending interclick interval

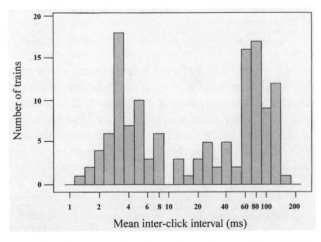

Fig. 58.2. Histogram of number of trains by mean interclick interval plotted on a semi-logarithmic scale

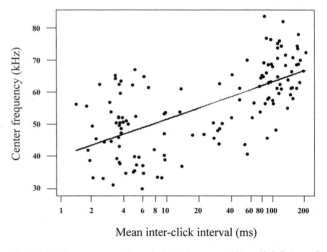

Fig. 58.4. Linear regression plot of the mean interclick interval versus the mean center frequency of each train

frequencies, their bandwidth was comparable to trains with higher ICIs.

Of the 133 trains analyzed, 33 (25%) were suitable for estimating source levels (i.e., had clear, distinguishable surface and bottom reflections). Calculated peak-to-peak source levels ranged between 191 and 216.5 dB re 1 μPa (mean = 205 dB). Linear regression revealed that ICI was a rather poor predictor of source level ($r^2 = 0.2$). A two sample t-test between the converted pressures (in μPa) of source level (SL) values for trains with ICIs greater than 10 ms (mean SL = 206 dB) and less than 10 ms (mean SL = 203 dB) did not reveal a significant difference between the two ($p = 0.093$). A somewhat better predictor of source level was the mean center frequency of the click train ($r^2 = 0.46$). As the mean center frequency of clicks increased, so did the source level.

Discussion

Although a distinction between burst-pulse sounds and echolocation trains has long been made in the literature, the characteristics that distinguish and unify these two classes of signals have seldom been quantified. Our results suggest that while certain features of click trains do in fact define two apparent classes of signals, other characteristics are also shared. Interpretations must be made with caution, however, in light of the fact that no positional information was available on phonating animals. It is impossible to say how much variation was introduced in the data from off-axis signals. The central assumption made in this study is that all click trains had an equal probability of being on- or off-axis and that relative comparisons are therefore justified. Defining the clear boundary of a functional distinction between burst-pulse and echolocation click trains will require further experimental study, but some provisional guidelines can be derived from the results obtained here.

INTERCLICK INTERVALS

The bimodal distribution of mean click intervals reveals two general modes of click train production: trains that begin and maintain a consistently short (1.5 to approximately 10 ms, mean = 3.5 ms) ICI throughout the train (fig. 58.5A) and trains with longer, more temporally variable ICIs (fig. 58.5B). Although some signals were collected that seemed to "bridge" the two modes (as predicted by the model presented by Murray, Mercado, and Roitblat [1998]), these composed only a small part (6%) of the data set.

The conspicuously low number of trains with ICIs between 10 and 20 ms suggests that this might be a functional and/or cognitive transition point in the way spinner dolphins produce and process click trains. Schotten (1998), studying the echolocation clicks of spinner dolphins using a four-hydrophone array to localize the phonating animal, calculated a minimum echo-processing delay of about 16 ms, which supports this assertion.

In the wild, therefore, target scanning appears to takes place in the same manner as under experimental conditions. A change may occur, however, when an animal approaches a target at close range. Morozov et al. (1972) showed that shortly prior to the capture of a fish (<0.5 m away) the echo-processing delay in *Tursiops* click trains could be as low as 3 ms. These results suggest that upon final approach to a target, a change takes place in the way echoes are processed, resulting in considerably diminished ICIs. This would explain the dramatic drop in ICIs observed in the few "bridging" signals mentioned above.

It appears, therefore, that the interclick interval of a train can in some cases confound the functional distinction between burst pulses and echolocation click trains. Considerable evidence, however, suggests that the behavioral contexts in which burst pulses are usually recorded tend to be social in nature (Caldwell and Caldwell 1966; Overstrom 1983; Herzing 1988, 1996). Consequently, until further evidence about the functional occurrence of burst-pulse signals suggests otherwise, it appears reasonable to presume that, in the case of spinner dolphins, those click trains with consistent ICIs between 1.5 and approximately 10 ms represent a class of signals functionally distinct from typical echolocation trains. On the other hand, those trains characterized by variable ICIs considerably greater and less than about 10–15 ms imply a type of echolocation not yet well understood, rather than a functionally separate class of signals.

SPECTRUM AND SOURCE LEVELS

The spectral and source level results reported here are likely to be underestimates because of ambiguity with respect to the orientation of animals. Nonetheless, the values obtained are similar to results reported by Schotten (1998; also see Schotten et al., chapter 54, this volume), who measured the clicks of animals echolocating directly on the hydrophone array. His mean estimated source level of 208 dB re 1 μPa ($n = 131$ clicks, maximum 222 dB) is only 3 dB higher that the one obtained here, while his mean center frequency estimate of 80.4 kHz ($n = 851$ clicks) is approximately 17 kHz higher than what was obtained in this study for trains with mean ICIs >10 ms (presumed echolocation signals). The correlation between increasing source level and center frequency agrees well with similar results reported for *Pseudorca* (Au et al. 1995). This suggests that the spectral composition of clicks is a function of the intensity with which they are produced, regardless of the interclick interval.

All the signals collected in this study contained most of their energy well above the human hearing range. This has implications with respect to the sensory and recording equipment that must be employed to successfully study these signals. Burst pulses are not always audible and therefore could be missed with some regularity if narrowband equipment is used to record them.

Fig. 58.5. *A:* Spinner dolphin burst pulse. Mean ICI = 3.4 ms, center frequency = 63.3 kHz, rms bandwidth = 22.9 kHz, and maximum peak-to-peak source level = 215.5 dB re 1 μPa. *B:* Spinner dolphin echolocation train. Mean ICI = 32 ms, center frequency = 52.1 kHz, rms bandwidth = 18.4 kHz, and maximum peak-to-peak source level = 203.5 dB re 1 μPa.

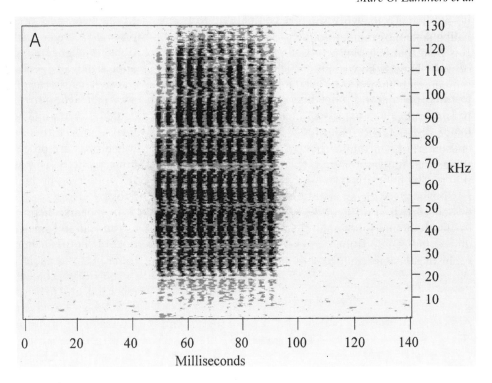

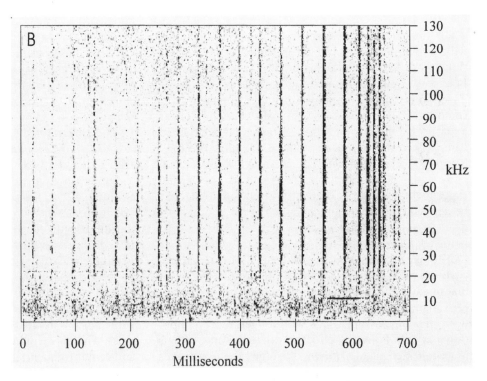

Although burst pulses tended to have lower center frequencies and slightly lower source levels, they were not as far removed from echolocation trains as reported in studies employing band-limited equipment (Caldwell and Caldwell 1966; Overstom 1983; Sjare and Smith 1986b; Herzing 1988, 1996). Diercks, Trochta, and Evans (1973) have argued that recording broadband signals with band-limited equipment results in the appearance of low-frequency artifacts not present in the original signal. Further investigation should establish whether these differences are species-specific, behaviorally related, or, in fact, undesired artifacts of methodology.

CONCLUSION

While the results presented here do not lay out a clear functional role for all the pulsed signals recorded, they do provide a framework for further studies of the signaling system of spinner dolphins and other delphinids. It is clear that dolphins produce click trains in a variety of ways. Exactly how some of them are used, however, remains ambiguous.

Important questions persist regarding the occurrence of burst pulses. For example, while the variability found in echolocation click trains is attributable to the different sonar tasks encountered by the animals, what, if anything, does the variability in burst-pulse signal production represent? Burst pulses cannot be discounted from possibly also having an echolocation function, but little evidence exists at present to suggest this. Future efforts to clearly define functional distinctions between burst pulses and echolocation signals will need to employ a sophisticated approach to closely examine associated behavioral patterns of individuals and control for the directional and broadband nature of these signals. Advances in the technology available to researchers studying these signals will help to overcome many of the limitations in bandwidth, localization, and portability encountered in the past.

Acknowledgments

The authors thank Danielle Lanyard for support and assistance with the data collection phase of this project. This work was funded partly through a grant provided by the John G. Shedd Aquarium (Chicago, Ill.) and conducted under National Marine Fisheries Service General Authorization Letter #11. This is Hawaii Institute of Marine Biology contribution no. 1113.

{ 59 }

Clicks Produced by Captive Amazon River Dolphins (*Inia geoffrensis*) in Sexual Context

Paula Moreno, Cees Kamminga, and *Avi B. Cohen Stuart*

Introduction

Recorded sounds of most odontocetes are highly diversified, even at the individual level. This variation can occur in more than one acoustic parameter, such as frequency, amplitude, duration, and rate and number of pulses (W. J. Smith 1986). Evidence suggesting the occurrence of group and individual-specific calls exists for *Orcinus orca* (J. K. B. Ford 1991) and *Tursiops truncatus.* In the latter species, these calls have been considered to be used in intraspecific recognition (Caldwell and Caldwell 1965) and group cohesion (Janik and Slater 1998) functions. Establishing the function of the suit of sounds produced by a single individual is a more difficult task. In a few species, such as *Physeter macrocephalus,* a relationship has been found between sounds and observed behavior (Whitehead and Weilgart 1991). Nevertheless, knowledge about the specific function of types of sounds is generally unknown (Richardson et al. 1995). This is largely due to the high variability in sounds and their high context-dependent nature, but also to the difficulty in identifying the sender and the subsequent behavior of the receiver (W. J. Smith 1986).

Only a limited number of studies address the communication use of pulsed sounds covering the ultrasonic range (Dawson and Thorpe 1990). This seems to be a result of the limitations of readily available equipment and also from a generalized assumption, not yet grounded, that high-frequency pulses are used only in echolocation (Dawson and Thorpe 1990). In addition, there are even fewer studies of ultrasonic pulsed sounds produced during specific behaviors (Dawson 1991), such as sexual activity. In odontocetes, sexual behaviors are not restricted to mating events, occurring often in social interactions both hetero- and homosexually (Wells, Boness, and Rathbun 1999).

The Amazon River dolphin (*Inia geoffrensis*) appears to produce only pulsed sounds, as is also known to be the case in Physeteridae and Phocoenidae (W. A. Watkins and Wartzok 1985). *I. geoffrensis* belongs to Platanistidae, a primitive family of cetaceans (Layne and Caldwell 1964) and occurs in the Amazon and Orinoco rivers and their tributaries, often found in very turbid waters (Best and da Silva 1989). To our knowledge, no studies have investigated the acoustic repertoire of *I. geoffrensis* during sexual interactions in the ultrasonic range. Caldwell and Caldwell (1966) reported *I. geoffrensis* to be less vocal than most marine delphinids and detected no sounds in the context of sexual activity in captivity. However, these studies were limited to 20 kHz; *I. geoffrensis* produces sounds up to 100 kHz (Kamminga et al. 1993). Other authors (Norris et al. 1972) analyzed ultrasounds of this species but focused on the echolocation use. Pulse repetition frequency (PRF), as opposed to the frequency of sounds in the ultrasound range, is hardly affected by directionality and can thus be recorded effectively while the sender is moving. This parameter has not been measured in *I. geoffrensis* during intraspecific interactive behavior, which is very difficult to observe in the wild. Maintenance of this species in captivity has not been successful, especially of groups, due to aggressive behavior (Caldwell, Caldwell, and Brill 1989) or disease (Tobayama and Kamiya 1989), which limits the opportunities to study interactions. The purpose of this study was to document the acoustic signals emitted during sexual behavior in *I. geoffrensis*. The objective was to ascertain whether sounds associated with five behaviors, occurring in sexual context, are distinctive using a qualitative (PRF contour) and a quantitative approach (four acoustic parameters).

Materials and Methods

The subjects were two adult male *I. geoffrensis*, caught in the Apure River, Venezuela, in 1975. They were held in Duisburg Zoological Garden, Germany, in a tank measuring 5.30 m × 6.80 m × 1.70 m deep. In this study, they were referred to as #1 (younger male) and #2 (older male). Video and acoustic recordings of five sexual behaviors were carried out 8–25 July 1995. Five behaviors from previously identified sexual sequences were selected, corresponding to an increase in the intensity of sexual interaction (Moreno 1996): Rub Object—the animal maintains the genital area in contact with an object and moves the peduncle laterally; Arching—dorsal bending of the body followed by a ventral bending, usually repeated more than twice; Rub Dolphin—the animal puts the penis in contact with any part of the other animal's body (except genital and anal areas); Rub Approach—the animal with erection swims toward the other individual and positions the genital area close to the genital or anal areas of the other animal; and Penetration—the animal inserts the penis in the genital slit or anus of the other animal. With the exception of Rub Object, which only occurs when the animal is stationary (lying on the bottom or upright), all behaviors may also be exhibited during swimming. These behaviors are relatively stereotyped and will be referred to as modal action patterns (MAPs) (Lehner 1996). However, this is a simplification, because it remains to be shown that Arching occurs in other members of this species.

To record the sounds produced during each of the five

MAPs and to select highest quality recordings, behavior was videotaped during all observations. Once the dolphin engaged in sexual activity, acoustic recording was initiated and carried out continuously until this activity ceased. Video recordings were made for 15 days, on average 1.5 h per day, taken from morning (08:30–13:30) and/or afternoon (13:30–20:30) periods. This yielded 4.2 h of acoustic recordings. To assure an adequate signal-to-noise ratio and maintain similar recording conditions, MAP samples were chosen only from video sequences where animals faced the hydrophone and that corresponded to spectrograms with an adequate signal-to-noise ratio. To allow recognition of potential patterns on click trains associated with MAPs, 10 s was established as an appropriate sample unit. To assure independence of the 10 s samples, only those that did not belong to the same MAP event were selected. An event defined as a MAP was performed without interruption. In this study it was not possible to identify which animal emitted sounds. Therefore, MAPs of both animals were recorded. However, we were limited to analyzing sounds associated with Arching, Rub Approach, and Penetration performed by individual #1 and Rub Dolphin and Rub Object performed by individual #2. This resulted from the acoustic and independent event requirements, as well as asymmetric occurrence of MAPs of the two animals. Hence, five 10 s samples of each MAP were analyzed, with the exception of Penetration (for which only four 10 s samples were available) (table 59.1). A click train was defined as a group of three or more clicks with regular or gradual change of the interclick interval. If the interval changed abruptly within subsequent intervals but was then maintained, it was considered part of the same click train.

Acoustic recordings were made with a Brüel and Kjær (B&K) 8103 hydrophone, a B&K 2635 charge amplifier, and a RACAL Store 4DS tape recorder. The system has a flat response at 30 ips recording speed from 200 Hz to 150 kHz. The hydrophone was located at 1 m depth in the center of the tank. Video recordings were made synchronously by means of a Nuriyuki time code generator/reader. The camera was located on the ceiling above the tank.

After a first aural detection, the taped clicks were digitized at a sampling rate of 1048 kHz by a 14-bit A-to-D converter in the HP 3565 signal processor, followed by an HP/Apollo 9000/425 workstation. Pulse repetition frequency (PRF) was defined as the inverse of the time interval between two subsequent clicks; it was plotted once the click train was fully processed. For each 10 s sample the PRF contours were analyzed. From these PRF plots, the acoustic parameters (maximum PRF, minimum PRF, total duration and number of clicks) of each click train were determined. Special dedicated software (SIGSYS), developed at the Information Theory Group of the Delft University of Technology, was used.

TABLE 59.1. Number of behavioral events (10 s samples) analyzed for five sexual modal action patterns (MAPs) of two Amazon River dolphins and number of click trains per event. Only samples meeting all acoustic criteria were used, which limited analyses to Rub Approach, Arching, and Penetration to dolphin #1, and Rub Dolphin and Rub Object to dolphin #2.

Dolphin	Modal action pattern	Events	Number of click trains
#1	Penetration	1	4
		2	1
		3	2
		4	2
#1	Arching	1	6
		2	3
		3	6
		4	4
		5	3
#1	Rub Approach	1	6
		2	4
		3	4
		4	5
		5	5
#2	Rub Dolphin	1	2
		2	5
		3	4
		4	2
		5	1
#2	Rub Object	1	7
		2	1
		3	5
		4	1
		5	2

To compare the acoustic parameters of Modal Action Patterns, ANOVA and post hoc multiple comparisons with Tukey test were run after $\log_{10}$ transforming the data to meet the assumptions of ANOVA (Sokal and Rohlf 1969) with SPSS 8.0 Windows. The level of significance was $\alpha = 0.05$.

Results

Only pulsed sounds were registered in all sexual MAPs (fig. 59.1). The sonograms show that Arching is accompanied by short click trains with fast increases of PRF of large amplitude (fig. 59.2). On all the other MAPs (fig. 59.2), it is difficult to detect a pattern on the PRF contour, but Rub Dolphin, which is a purring-like sound, tends to have long click trains with sinusoidal modulation of PRF of small amplitude.

Quantification of the acoustic parameters of five sexual MAPs is shown in fig. 59.3. Each acoustic parameter differed significantly with MAP, except for the number of clicks, which shows a high variation within any given MAP (PRF_{min} : $F_{4,80} = 19.6$, $p < 0.0001$; PRF_{max} : $F_{4,80} = 14.4$, $p < 0.0001$; duration : $F_{4,80} = 4.5$, $p < 0.005$; number of clicks: $F_{4,80} = 0.9$, $p < 0.5$).

Sounds produced during arching consist of high

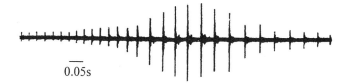

Fig. 59.1. Click train of *Inia geoffrensis* recorded July 21, 1995, in Duisburg Zoo, Germany. The clicks (vertical lines) stand out from the background noise. Note the gradual increase on time elapsed between clicks (left to right), corresponding to a decrease in pulse repetition rate.

PRF_{min} (median = 35.59 Hz, SD = 15.03) and had the highest PRF_{max} (median = 107.64 Hz, SD = 49.25) of the MAPs. PRF_{max} varied over a broader interval than PRF_{min}. Arching had the shortest duration (median = 0.46 s, SD = 0.16), ranging between 0.16 and 0.82 s.

Rub Object also shows high PRF_{min} (median = 48.61 Hz, SD = 34.31) and high PRF_{max} (median = 78.99 Hz, SD = 46.35), both with broad intervals. The duration is intermediate relative to other MAPs, averaging 0.55 s (SD = 0.81).

Rub Dolphin had the lowest PRF_{min} (median = 8.68 Hz, SD = 4.17) and PRF_{max} (median = 27.78 Hz, SD = 11.35), both with narrow intervals varying between 5.21 Hz and 17.36 Hz, and 10.42 and 52.08 Hz, respectively. It had the longest duration (median = 1.42 s, SD = 1.99), ranging between 0.28 and 3.99 s.

In comparison to the other MAPs, Rub Approach showed intermediate PRF_{min} and PRF_{max} (median = 22.57 Hz, SD = 9.53, and median = 44.27 Hz, SD = 23.13, respectively), where PRF_{max} has a broader interval than PRF_{min}. The average duration is 0.76 s (SD = 0.86) and varies between 0.11 and 3.49 s.

Likewise, Penetration showed intermediate PRF_{min} and PRF_{max} (median = 22.57 Hz, SD = 8.69, and median = 39.93 Hz, SD = 24.27, respectively). It had the second shortest duration (median = 0.52 s, SD = 0.99), varying between 0.18 and 3.26 s.

The grouping of sexual MAPs based on the three acoustic parameters that exhibited significant differences is shown in fig. 59.4. All of the acoustic parameters (PRF_{min}, PRF_{max}, and duration) combine sexual MAPs in two or three distinct subsets, which are arranged along a gradient of increasing magnitude. PRF_{min} isolates Rub Dolphin in the low category and forms intermediate and high categories. Arching appears in the latter two categories, while Rub Object belongs to the high PRF_{min}. Penetration and Rub Approach are included together in intermediate PRF_{min}. PRF_{max} groups Arching and Rub Object together, corresponding to high PRF_{max}, and all other MAPs belong to a single group. As with duration, the other acoustic parameters did not differ significantly between Rub Approach and Penetration. Duration forms only two categories. In this case, the overlapping of the low and high categories was even

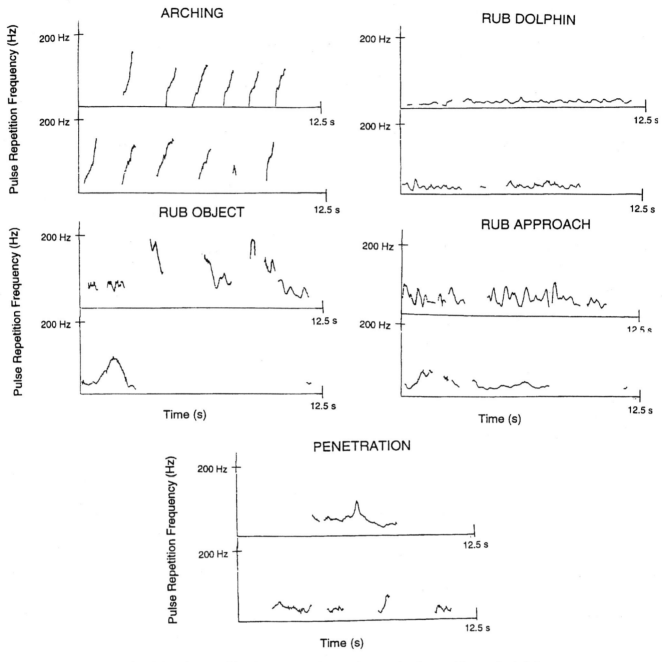

Fig. 59.2. Pulse repetition frequency contour of two series (top and bottom) of click trains recorded during five modal action patterns (Arching, Rub Dolphin, Rub Object, Rub Approach, and Penetration). Each series corresponds to a 10 s sample and was recorded from different events performed by dolphin #1 (Arching, Penetration) or dolphin #2 (Rub Dolphin, Rub Object, Rub Approach).

greater, since only Arching and Rub Dolphin belong exclusively to the low and high categories, respectively.

To summarize, through the quantitative analysis of these acoustic parameters, only Rub Dolphin can be discriminated from the other MAPs, which is accomplished through PRF_{min}. PRF_{max} led to the least overlap of MAPs, but no single MAP was distinguished. The greatest overlap among MAPs occurs in terms of number of clicks, where no differences were found, followed by

Duration. The most prevalent pairs of MAPs are Rub Approach/Penetration and Arching/Rub Object, for which PRFs were similar. However, analysis of PRF contour further distinguishes Arching from the other MAPs.

Discussion

Unlike Caldwell and Caldwell (1966), who detected no sounds of *I. geoffrensis* during sexual activity, we often

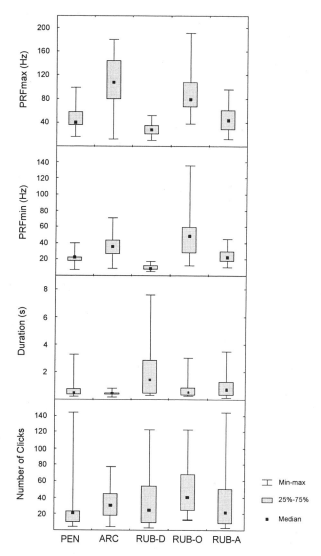

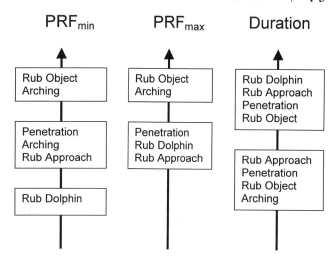

Fig. 59.4. Subsets (boxes) of sexual modal action patterns (MAPs) with nonsignificantly different (post-hoc multiple comparisons with Tukey test significant at 0.05 level) magnitude of minimum pulse repetition frequency (PRF$_{min}$), maximum pulse repetition frequency (PRF$_{max}$), and duration. Arrows indicate the direction of increase in magnitude. Note that for duration some MAPs occur in both subsets, which indicates a large variation of magnitude.

Fig. 59.3. Box plots of four acoustic parameters (PRF$_{max}$ = maximum pulse repetition frequency, PRF$_{min}$ = minimum pulse repetition frequency, duration, number of clicks) of the click trains associated with five modal action patterns recorded in two Amazon River dolphin males (#1, #2) at Duisburg Zoo, Germany, July 8–25, 1995. Values presented are median (midpoint), lower-upper quartiles (box), and minimum-maximum range (vertical lines). Codes: PEN = Penetration (#1, $n = 9$); ARC = Arching (#1, $n = 22$); RUB-D = Rub Dolphin (#2, $n = 14$); RUB-O = Rub Object (#2, $n = 16$); RUB-A = Rub Approach (#1, $n = 24$).

found more than a click train per sample. In the wild, Norris et al. (1972) also registered high acoustic activity from *I. geoffrensis* from both groups and isolated animals. Although disparity between the two studies could be due to different individual behavior of the animals analyzed, it is more likely a result of the frequency limit of the Caldwells' recording equipment (i.e., 20 kHz). We found only pulses, which agrees with the classification of *I. geoffrensis* as a nonwhistler species (Herman and Tavolga 1980) and is supported by two studies carried out

in the wild (Kamminga et al. 1993; G. Lothar, pers. comm.). However, whistles have been recorded in the wild in this species (Ding, Würsig, and Evans 1995). High-repetition-rate sounds could be misinterpreted as tonal sounds even through spectrogram analysis (W. A. Watkins 1967). On the other hand, it cannot be ruled out that those sounds were produced by other sympatric species, such as *Sotalia fluviatilis*.

It is arguable whether the recorded sounds have an echolocation or communication function. This distinction is usually based on the frequency and duration of each click or on the PRF. Thus, clicks with broadband frequency, short duration (Tavolga 1983), and medium PRF (below approximately <700 Hz) (Brownlee and Norris 1994) generally are regarded as echolocation clicks. Pulsed sounds in the low-frequency range and that have a high PRF are considered to be communication sounds. According to this PRF criterion, the signals recorded in all MAPs are in the echolocation range. However, comparisons with sounds of other species show that they are in the echolocation and communication ranges. For example, our PRF values are within the range of the "genital buzz" (8–2000 Hz) of *Stenella frontalis,* a burst-pulsed sound exhibited during courtship and disciplinary interactions in the wild (Herzing 1996). On the other hand, they are below the range of the "cry" (>200 Hz) of *Cephalorhyncus hectori;* that was found to occur more frequently during aerial and aggressive behavior than while feeding (Dawson 1991). The same study reported that high-frequency pulsed sounds were more related to social interactions than to feeding. Compared to *Orcinus orca,* our values overlap both with echolocation and social sounds (Schevill and Watkins 1966). Finally, they are above the PRF found recently for wild *I. geoffrensis* during social interactions and overlap with travel/feeding (G. Lothar, pers. comm.). These discrepancies found on PRF range of clicks used in social context, may be partially attributed to interspecific differences, but it may also result from the lack of a clear cutoff between pulsed sounds used in echolocation and communication. Brownlee and Norris (1994) cautioned about using the PRF at 700 Hz as the borderline and Tavolga (1983) referred to the possible use of echolocation-like sounds in communication. In fact, the same type of sounds have been associated with echolocation and communication in *Phocoena phocoena* (Amundin 1991). It is unlikely that the sounds registered in the present study have a predominant echolocation purpose. First, the experiment was carried out during daylight and the pool is a very restricted, familiar environment to the dolphins. Under these circumstances, vision would suffice for orientation. Second, most sounds were recorded when the animals were not swimming and apparently not involved in target classification, but instead engaged in sexual interactions. Until evidence proves otherwise, we regard high-frequency clicks and click trains with low PRF as candidates for communication, at least in close encounters, where attenuation by distance is not significant.

Presently, it is unknown whether sexual behavior between these two males is natural or induced by captivity. There are several accounts of intraspecific male-male sexual interactions in captive odontocetes, including in *I. geoffrensis* (Layne and Caldwell 1964). Accounts from wild populations of odontocetes are rare, probably due to the limitation of field observations of sexual interactions. The few cases observed were interpreted mainly as agonistic or as sexual play (Herzing and Johnson 1997). In primates, male-male mounting is common in a diversity of contexts under conditions of high arousal (Hanby 1974). It is also possible that the younger male began directing sexual behavior to the older male when it was a calf, since its mother died early in their captivity. Such behavior is often performed between mother and young to strengthen bonding (Boyd, Lockyer, and Marsh 1999).

All the modal action patterns, except Arching, have been reported widely in odontocetes both in captivity, including *I. geoffrensis* (Layne and Caldwell 1964), and in the wild in sexual and other social contexts (Herman and Tavolga 1980). Arching has not been described previously with this repetitive mode in the sexual context. Pilleri (1976) observed the only repetitive flexions that seem similar to Arching, but during defecation and rest phases in *Platanista indi*. Sylvestre (1985), who studied the same individuals, registered only a single downward flexion of the body and linked it to urination. Conversely, Nelson and Lien (1994) observed an upward flex-

ion of the body ("throwbacks") in *Lagenorhyncus acutus* without specifying a particular context. Dudzinski (1998) observed a posture similar to Arching in all age groups of *Stenella frontalis* in the wild during play and aggressive activities. She also found that in juveniles rubbing exchanges sometimes followed this posture. A visual display ("posturing") exhibited during courtship, which involves bending the back, was observed in captive *T. truncatus* (Tavolga and Essapian 1957). A similar posture was observed in *Phocoena phocoena* subadult males in connection with an erection of the penis (Amundin 1991). According to Tavolga and Essapian (1957), the "sigmoidal" posture is a reminiscent of sexual lordosis in terrestrial mammals.

Arching was found to be performed consistently with an acoustic pattern characterized by short and fast sequences of click trains and a substantial increase in the PRF. Unfortunately, few, if any, detailed studies of pulsed sound repertoire in odontocetes, covering the ultrasonic range, are available (Dawson 1991). In the low-frequency range (<21 kHz), a pattern of PRF was found in *P. phocoena* during the S-posture (Amundin 1991), which differs from the sigmoidal-shape increases that we obtained. Association of the S-posture with vocalizations was also recorded for *S. frontalis* (Dudzinski 1998), but the nature of these sounds is unclear. Sounds produced in sexual context were reported in *S. frontalis* and *T. truncatus*. The former produces the genital buzz, which occurs with the sigmoidal posture, and the latter emits the "male sex yelp" associated with courtship (Caldwell and Caldwell 1967). Other rhythmic patterns of pulsed sounds, known as "codas" and "creaks," used in communication by *P. macrocephalus* have been attributed to social and aggregation functions (Whitehead and Weilgart 1991). Non-pulsed sounds, such as whistles, have been found to be strongly associated with the exchange of rubbing between *S. frontalis* (Dudzinski 1998).

It is known from other taxa such as birds and primates that certain visual displays are coupled with acoustic displays to convey additional information about the identity or condition of the sender, like sexual receptivity, as well as intention to remain near the partner (Drickamer and Vessey 1992). Since signals can be context-dependent, combining different signaling sources could be a way of unequivocally defining the message (W. J. Smith 1986). Unfortunately, nothing is known about the sexual behavior of *I. geoffrensis* in the wild, but if courtship displays occur in turbid waters, it follows that other than visual cues are needed to communicate. It is unlikely that in the sounds associated with Arching, the PRF is a signature pattern. We had the opportunity of analyzing the signals produced when both dolphins exhibited Arching, and we obtained two distinct sets of click trains with the same described pattern of PRF.

The quantitative approach to comparing and characterizing the five sexual MAPs revealed that apart from the number of clicks, which shows a great variability, all other acoustic parameters, in particular PRFs, are useful. While the PRF contour analysis of the five MAPs clearly sets Arching apart, an analysis of PRF$_{min}$ isolates Rub Dolphin, and Arching is similar to Rub Object. Thus, it seems that the combination of the two procedures could be important to exploring the differences and similarities of click trains between modal action patterns. Despite the preliminary nature of this work and the need to increase the number of subjects and the acoustic sample size, we showed the potential of PRF analysis in understanding cetacean acoustic behavior. We suggest that future studies be carried out to test the emerging hypothesis that Arching is a visual display coupled with stereotyped acoustic signals, used in precopulatory activity of odontocetes. We also encourage other similar studies with wild populations.

Acknowledgments

Financial support was provided by Fundação para a Ciência e Tecnologia (grant no. PRAXIS XXI/BM/ 1130/94). We thank Dr. Reinhard Frese, director of the Duisburg Zoo, for supporting this study, and Dr. Manuel Garcia-Hartmann for his assistance, without which this study would not have been feasible. We also appreciate the willingness of the keepers, Ulrich Kluckner and Peter Schultz, to accommodate changes in their work routine, and thank them for their help. We thank Dr. Teresa Avelar from ISPA, Portugal, for her advice on the behavioral aspects. Finally, we express our gratitude to all those who contributed with valuable comments to the manuscript, especially to Dr. Jane Packard and Dr. Lee Fitzgerald from Texas A&M University.

{ 60 }

High-Frequency Burst-Pulse Sounds in Agonistic/Aggressive Interactions in Bottlenose Dolphins, *Tursiops truncatus*

Christer Blomqvist and *Mats Amundin*

Introduction

Most studies on dolphin communication have focused on whistles (e.g., Caldwell and Caldwell 1965). Whistles are omnidirectional (Evans, Sutherland, and Beil 1964) and convey information to all members of a dolphin school about identity, relative position, and, to some extent, emotional state of the whistler (Caldwell and Caldwell 1972). Pulsed sounds, on the other hand, have mainly been investigated in connection with echolocation (e.g., Au 1993), but a few studies suggest that pulsed sounds are also used in social contexts. Dawson (1991) found a significantly greater abundance of high-repetition-rate burst-pulse sounds, labeled "cries," during aerial and aggressive behavior situations than during feeding in Hector's dolphin (*Cephalorhynchus hectori*), suggesting that these cries were social rather than echolocation sounds. He also claimed it to be highly improbable that whistles would constitute the entire basis for intraspecific communication in odontocetes, since this would imply that nonwhistling species do not communicate acoustically at all. Amundin (1991) reported that burst-pulsed sounds in agonistic and distress situations have context-specific repetition rate patterns in the nonwhistling harbor porpoise (*Phocoena phocoena*). Connor and Smolker (1996) reported that a pulsed "pop" sound was correlated with courtship and/or dominance in the bottlenose dolphin (*Tursiops truncatus*). Overstrom (1983) reported pulsed sounds correlated with aggressive behaviors in the same species.

The objectives of this study were to investigate whether burst-pulse sounds emitted in aggressive interactions contain ultrasonic frequencies similar to the sonar sounds and to describe their repetition rate patterns and concurrent visual behavior patterns.

Materials and Methods

In a previous study (Karlsson 1997), conducted in the Kolmården captive dolphin colony, aggressive sounds in the human audible range (i.e., <20 kHz) were recorded together with concurrent behaviors of two adult female bottlenose dolphins. The females and their calves, 2 and 3 years old, were the only dolphins present in the display pool. These recordings were included (courtesy of T. Karlsson) and used to define aggressive behavior patterns to be recorded in this study (see below). The sounds in Karlsson's study were picked up by a Sonar International Hydrophone (frequency response +2 dB between 10 Hz and 20 kHz), and recorded, together with concurrent visual behaviors, on a videocassette recorder (Panasonic NV-FS 1 HQ; HiFi stereo audio frequency range 0.5–20 kHz). Spectrograms of the audible sounds emitted in selected aggressive encounters were produced using Spectra Plus (Prof. edition, v. 3.0a, Pioneer Hill Software).

New recordings were made during six consecutive days, between 16 and 21 February 1998, in the same captive colony, including all 12 dolphins in the Kolmården colony. They were kept in a 6400 m³ pool complex consisting of three pools with a surface area of 900, 800, and 185 m², respectively. Five of the dolphins were born in the facility, and the age span in the whole group ranged from 2 to 35 years. During the recordings the animals were temporarily separated into two subgroups by means of a net barrier placed in a 4.0 m long × 1.8 m wide × 2.3 m deep channel connecting the 900 m² display pool with the 185 m² holding pool (fig. 60.1). The net barrier prevented physical but not visual or acoustical contact. The channel restricted the lateral movements of dolphins engaged in social interactions across the net barrier.

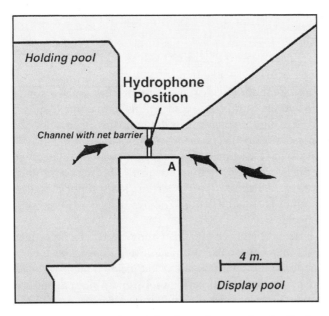

Fig. 60.1. Schematic view of the channel connecting the display pool (900 m²) with the holding pool (185 m²) at the Kolmården dolphinarium, Kolmården Wild Animal Park, Sweden. The hydrophone is attached to the net barrier (white line) at mid-depth—i.e., 1.5 m. The corner at the end of the channel (A) was used occasionally by dolphins in the display pool to hide behind during aggressive interactions over the net barrier.

rier. It thereby increased the probability, with a fixed hydrophone (fig. 60.1), of recording sounds along the axis of the rostrum where the higher frequencies (>100 kHz) would be expected if the aggressive sounds were broadband and directional similar to sonar sounds (Au 1993).

Envelope Detector Recordings

The behavior of the dolphins was recorded using a b/w Ikegami video camera, suited with a wide-angle lens and placed in an underwater housing mounted in the channel. The video images were stored on a NV FS90 HQ Panasonic VHS videocassette recorder. Underwater sounds were picked up by means of a Sonar Products HS/70 hydrophone (frequency response ±14 dB between 5 and 150 kHz), mounted on the net barrier. The hydrophone was suspended at approximately 1.5 m below the water surface.

The hydrophone was connected to an envelope detector, custom-made by Loughborough University, UK. The internal band-pass filter of the detector was supplemented with an external high-pass filter (HP-ITHACO 4302; 24 dB/octave). The frequency response of this recording system was ±2 dB between 70 and 100 kHz. Between 110 and 150 kHz it was ±3.5 dB, albeit 12 dB lower than in the 70–100 kHz range. Below 70 kHz there was a 42 dB/octave cutoff, with the sensitivity at 60 kHz in level with that at 120 kHz. The output of the envelope detector, which was in the audible frequency range, was recorded on the HiFi audio channel of the Panasonic VCR (frequency range 20 Hz to 20 kHz).

An interaction was classified as aggressive when two animals were in a face-to-face position on opposite sides of the net barrier, emitting burst-pulse sounds and showing concurrent visual aggressive behavior patterns—that is, head jerks, pectoral fin jerks, "S"-shaped body postures, and jaw claps (DeFran and Pryor 1980; Overstrom 1983). On some occasions only one of two interacting animals was in view of the underwater camera, because other dolphins, not engaged in the interaction, were playing with the camera, thus concealing the other individual. In spite of this, those encounters were included based on the sounds and the visual behavior of the individual in view.

Burst-pulse durations were measured manually from spectrograms produced by means of Spectra Plus (Prof. edition, v. 3.0a, Pioneer Hill Software).

Full-Bandwidth Recordings

In parallel with the envelope detector recordings, a selection of sounds, picked up by the HS/70 hydrophone, was also recorded using a broadband DSP card (model SPB2 from Signal-data, DK-2840 Holte, Denmark) and a Toshiba 3200 laptop. The frequency response was effectively determined by the hydrophone. The DSP card was controlled by means of custom-made

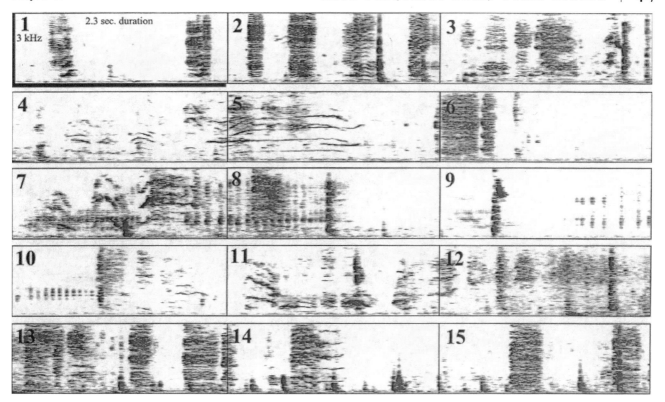

Fig. 60.2. Continuous spectrogram (FFT = 4096, overlap = 95%) of the audible sounds recorded in a typical aggressive interaction between two female bottlenose dolphins in the display pool. Each square in the figure represents approximately 2.3 s and the total duration of the sounds is 34.2 s (*y*-axis = 3 kHz). The encounter contains slow click trains (squares 7–8 and 9–10) and pulse-bursts with low pulse repetition rates (squares 1–3, 6, and 11–15), medium pulse repetition rates (squares 4–5, 11, and 14), and fast pulse repetition rates (squares 7–8, 11). "Jaw claps" (see Marten and Norris 1988) are seen in squares 3, 8, 9, 10, 12, and 15. The short and very low frequency sounds in squares 14–15 are from the hydrophone hitting the pool wall due to wave action.

software (SBP Bat Recorder v. 1.1, 11–96 CSC-OU, Odense University, Denmark). The maximum duration of each recording was 590 or 655 ms, depending on the sampling frequency used—that is, 333 and 300 kHz, respectively. The onset of each recording was manually trigged, based on the character of the sounds, which were transformed to audible range via an envelope detector and played through a speaker, as well as on the continuous sound time series displayed on the computer screen. An effort was made to get full-bandwidth samples of all the sound types shown in fig. 60.2—that is, slow-repetition-rate click trains, as well as medium- to high-repetition-rate burst pulses. Each recording was manually stored on the hard disk of the computer as a separate file.

The average power spectrum of 3–6 pulses (selected from the beginning, middle, and end of each full-bandwidth recording of burst pulses) was calculated using Waterfall (v. 3.18) software (Cambridge Electronic Design Ltd.). After corrections for the hydrophone frequency response curve, the power spectra were plotted against relative amplitude. Pulse repetition rate analysis was made using MATLAB for Windows (v. 4.2c.1, MathWorks Inc.).

Results

Analysis of the recordings made by Karlsson (1997) revealed that burst pulse sounds, with pulse repetition rates from 100 to over 900 pps, occurred frequently in aggressive interactions between the two adult female bottlenose dolphins. Fig. 60.2 shows a 34.2 s continuous spectrogram of the audible sounds emitted in a typical aggressive interaction between these two females. The frequency scale was reduced to 3 kHz in order to better display the repetition rate patterns, as revealed by the harmonic interval (Watkins 1967) and hence make them comparable to the envelope detector recordings. The interaction contained slow click trains and burst pulses with low, medium, and high pulse repetition rates.

The violent head jerks, pectoral fin jerks, S-shaped body postures, and jaw claps (see DeFran and Pryor

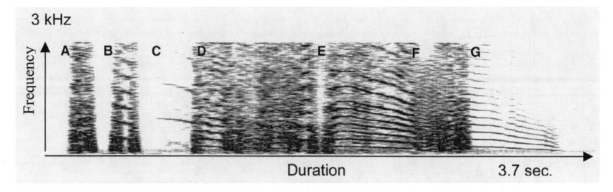

Fig. 60.3. Spectrogram of pulse-bursts recorded in an aggressive interaction between two bottlenose dolphins. The sounds were band-pass filtered between 100 and 160 kHz and recorded on a VCR using an envelope detector. Burst durations: A: 200 ms, B: 230 ms, C: 230 ms, D: 900 ms, E: 670 ms, F: 380 ms, G: 630 ms. Peak pulse repetition rates: A: 380 pps, B: 415 pps, C: 940 pps, D: 195 pps, E: 200 pps, F: 100 pps, G: 195 pps.

1980; Overstrom 1983) seen in these free-swimming aggressive encounters also occurred between dolphins interacting across the net barrier.

The distance between animals interacting across the net barrier was estimated to be between 1 and 4 m, whereas in free-swimming individuals involved in such encounters, the distance initially was in the order of 10–20 m. Often there was an escalation of the aggressive behaviors leading up to a climax of simultaneous emissions of intensive burst pulses with medium to high repetition rates and jaw claps, in concert with high-intensity aggressive behaviors. An example of such burst pulses (recorded via the envelope detector and thus representing the pulse frequency content between 60 and 150 kHz) is shown as a spectrogram in fig. 60.3.

An interaction over the net barrier most often started with two dolphins approaching the net barrier from either side. Occasionally, it seemed to be initiated by one animal closer to the net barrier, emitting click trains with a low pulse repetition rate while apparently pointing its rostrum toward animals passing by on the other side of the net barrier. The intensity and pulse repetition rate, as judged by the human ear, often increased each time another dolphin passed the net barrier. After a varying number of times ignoring this, the passing dolphin could suddenly turn and approach the net barrier, in what appeared to be a response to the other's provocation.

During long aggressive interactions, in both the free-swimming and net barrier situations, the dolphins often turned on their side or fully upside down (i.e., rotated 90–180° along their longitudinal body axis). It was obvious that both animals pointed their rostrum in the general direction of the other, and in the gate situation more than what seemed to be inevitable due to the physical restraints of the channel. In one interaction over the net barrier one of the animals made short body jerks in synchrony with intense, short burst pulses emitted by the other. On a few occasions the dolphin in the display pool was also seen to hide behind the corner at the end of the channel (see fig. 60.1A), seemingly trying to keep out of sight as well as out of the sound emission of the other animal. From time to time it exposed its head to the aggressive burst pulses of the antagonist, pointed its rostrum toward the other, and responded with similar aggressive burst pulses.

Both visual behaviors and acoustic signals were immediately interrupted if the net barrier suddenly was removed during an interaction. On such occasions, the animals in the holding pool swam silently and at high speed through the channel into the larger display pool. Continued fighting or any other aggressive behavior was never seen immediately after the animals were reunited. With the net barrier left in place an aggressive climax usually ended with slow to medium pulse rate emissions from one or both of the animals, followed by one or both of them leaving the net barrier. However, in the free-swimming encounters between the two adult females, similar aggressive climaxes sometimes resulted in both animals charging toward each other, apparently trying to bite and/or hit each other with rostrum and/or tail fin. These physical encounters were very short and did not result in any injuries. In other free-swimming encounters, one of the females fled, chased by the other, or the interaction ended with both animals just swimming away from each other, often after a final, intensive, low-repetition-rate pulse train.

ENVELOPE DETECTOR RECORDINGS

A total of 222 aggressive interactions were recorded across the net barrier with the envelope detector setup, ranging between 0.4 and 37.3 s in duration. They included 3706 burst pulses, and the presence of acoustic energy in the 60–150 kHz frequency range was confirmed in all these interactions. The average number of bursts per interaction was 16.7, ranging from 1 to 49. A total of 3435 (92.7%) burst pulses were less than 500 ms in duration,

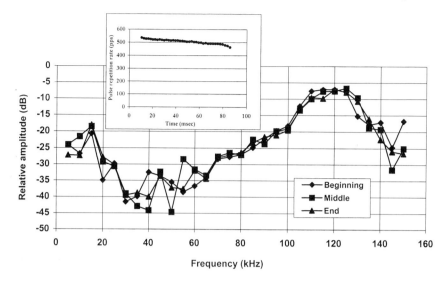

Fig. 60.4. Frequency spectrum (FFT size: 512) of three different pulses, selected from the beginning, middle, and end of an aggressive pulse-burst, 80 ms in duration. The pulse repetition rate was around 500 pps. The sound was recorded digitally with an A/D sampling rate of 333 kHz. The spectra are corrected for the frequency response curve of the hydrophone.

185 (5.0%) were between 500 and 1000 ms, and 86 (2.3%) had a duration over 1 s. The mean duration of the burst pulses within each of these classes was 130 ms, 680 ms, and 1.39 s, respectively.

FULL-BANDWIDTH RECORDINGS

A total of 80 s of full-bandwidth recordings, divided into 234 data files, were stored on the Toshiba hard disk. Twenty-six of these recordings were of occasional single pulses or other types of sounds (e.g., jaw claps). Forty-one recordings contained samples of pulse trains with a repetition rate below 100 pps. The remaining 167 recordings contained 249 burst pulses with a pulse repetition rate of more than 100 pps. Of these burst pulses, 136 (55%) had a peak pulse repetition rate between 100 and 250 pps, 64 (26%) had a repetition rate between 251 and 500 pps, and 49 (20%) had a repetition rate between 501 and 940 pps.

Four of the full-bandwidth recordings included what appeared to be two overlapping pulse trains. In each case, there was a fixed time lag between the pulses in the two trains—that is, they had identical pulse repetition rates (153–413 pps range). However, the amplitude changes were apparently independent. All individual pulses had the same phase, indicating that they were either direct sounds or direct sounds blended with a reflection from a hard surface (the pool wall or floor).

Of the 249 burst pulses, 193 were recorded without overload, and thus allowed for frequency analysis. Fig. 60.4 shows the power spectrum of three typical pulses chosen from the beginning, middle, and end of an aggressive burst. It was 80 ms in duration and had a pulse repetition rate around 500 pps. The −3 dB bandwidth was 20 kHz, centered on 120 kHz. There was a strong component (−12 dB re to the 120 kHz peak) in the audible frequency range, with a peak at 15 kHz.

Average power spectra of 3–6 pulses (1–2 selected from the start, middle, and end of each burst, respec-

tively) were calculated for all 193 pulse bursts. One hundred and thirty-one (68%) of these bursts had an average frequency peak above 100 kHz. Generally, there was a second lower peak in the 20 kHz to 85 kHz range, although spectra with a single frequency peak above 100 kHz also occurred (fig. 60.4). Sixty-two (32%) of the bursts had pulses with a peak frequency below 100 kHz. The strong frequency component in the audible range (fig. 60.4) was found in most of the burst pulses.

Discussion

This study shows that the burst pulses, which occurred in aggressive encounters between captive bottlenose dolphins, were broadband and contained strong frequency components between 60 and 150 kHz, as well as in the audible frequency range. All the characteristic pulse repetition rate patterns of these bursts were only observed in situations containing aggressive behavior elements (cf. DeFran and Pryor 1980; Overstrom 1983). Also, the increased energy content in the audible frequency range (<20 kHz) of these sounds and the synchronous escalation of concurrent aggressive behavior elements makes it likely that these sounds were social signals.

The net barrier, separating the dolphins and protecting them from the immediate consequences of their actions, may have amplified and even triggered the aggressive behaviors. However, similar encounters were frequently observed in these dolphins when swimming in the same pool (Karlsson 1997) and have also been seen in wild Atlantic spotted dolphins, *Stenella frontalis* (Dudzinski 1995). Hence, the behavior recorded in this restricted situation may still represent a sample of the normal, species-specific behavior repertoire.

The highest pulse repetition rate in the burst pulses recorded across the net barrier was 940 pps. This corresponds to a pulse interval of a little over 1 ms, that is, much shorter than that of close range sonar click "buzzes" (Evans and Powell 1967). Also the estimated distances between the dolphins involved in these interactions were 1–4 m—much longer than would be anticipated if the burst pulses were close-range sonar signals. Thus, on the basis of pulse repetition rate, it is unlikely that these very high pulse repetition rate bursts were used for echolocation.

With only one hydrophone placed between the interacting individuals, it was not possible to test whether the pulses in these social sounds were directional. However, the similarity between their frequency spectrum and that of sonar clicks (Au 1993) suggests that this may be the case. If so, the pulses with a single frequency peak in the 115–135 kHz range may be from the beam core, whereas those with dual frequency peaks may be from the beam periphery (see Au 1993).

It was also not possible to determine whether the overlapping pulse trains in the full-bandwidth recordings originated from two different animals or were a direct signal from one animal blended with a reflection from the pool wall or floor. If the latter was the case, the amplitude difference between the two overlapping pulse trains could be explained by the dolphin making small scanning movements of the head and the hydrophone, and/or the reflective surface was in the periphery of a sound beam similar to that found in the sonar clicks (Au 1993). It is also possible that one of the dolphins synchronized its pulse repetition rate to that of the other. Dolphins are adapted to match their sonar click train to the returning echoes of moving targets (Au 1993), and matching its repetition rate to that of another dolphin should not be an impossible task. If this is the case, the social significance of such a synchronization remains to be revealed. A third possibility is that the same animal operated two independent pulse sound generators (Cranford et al. 1997).

The apparently deliberate pointing of the rostrum toward the antagonist may be an indication that the proposed directional characteristic was used. This may be to ensure that maximum sound energy reaches the other individual or to address the aggressive signals to a selected individual. The more omnidirectional low-frequency components (<20 kHz) would allow the rest of the group to hear the entire interaction, but unless hit by the high-frequency beam core, they would know they were not the target for the aggression. Such a directional acoustic signaling would be a potentially powerful communication tool in a species lacking conspicuous directional visual signals, considered to be highly important in social terrestrial mammals (e.g., Altmann 1967; Goodall 1968). This aspect is currently being investigated in our dolphins.

Burst-pulse emissions and conspicuous behavior displays dominated the free-swimming aggressive interactions, and a climax including physical fighting constituted only a very small part. Thus the sounds, as well as the visual behavior patterns, may be part of a ritualized behavior sequence, with the purpose to settle rank conflicts or other disagreements between herd members with a minimum of physical fighting. Such physical fights may not only be dangerous to the combatants, but may also be potentially dangerous to the herd (Lorenz 1969). Submissive behaviors resulting in the inhibition of physical aggression have been seen in many group-living terrestrial mammals—for example, the wolf, *Canis lupus* (Schenkel 1967) and the chimpanzee, *Pan troglodytes* (Goodall 1986).

Although no accurate source levels were obtained, the aggressive burst pulses sounded loud to the human ear, relative to sonar click trains emitted by these animals. This may have been an artifact, due to the social sounds having more energy in the audible range, as compared to sonar clicks. However, the bottlenose dolphin has been demonstrated to be capable of producing very high sound pressure levels (230 dB p-p re 1 μPa at 1 m during echolocation tasks; Au 1993). It has even been suggested that dolphins may be capable of debilitating prey with intense sounds (Marten et al. 1988), a possibility originally suggested for sperm whales, *Physeter macrocephalus* (Norris and Møhl 1983). Taking this into consideration and the fact that hearing is extremely sensitive in dolphins (Au 1993), it is possible that intense burst pulses, in aggressive interactions like those reported here, can be used with the intent to cause auditory discomfort or even pain in the antagonist. The temporal summation in the ear with increasing pulse rates (Vel'min, Titov, and Yurkevich 1975 in Au 1993) may add to this effect and favor the use of high pulse repetition rates in aggressive encounters. The burst pulses, especially if they are directional, may then function as a safer alternative to physically hitting an opponent with, for example, the rostrum or the tail fin. The avoidance behaviors, such as "open mouth threat" or "tail blow," observed in response to the aggressive burst pulses, supports this hypothesis. Amundin (1991) found similar avoidance in harbor porpoises (*Phocoena phocoena*) in response to aggressive "sideward turn threats," including burst pulses with very high repetition rates (400–1000 pps). Such an acoustic "weapon" would work equally well at night, in the dark at great depths, or in murky waters, and would also allow the dolphin to keep track of where its antagonist is under these circumstances.

From this point of view, it is easy to comprehend how this aggressive use of pulse trains may have evolved from the original sonar function. The fact that some of the interactions between the two females, who were close in rank (Dolphinarium staff, pers. comm., 1998), resulted in direct, physical fights are in no conflict with this interpretation. It may be compared with, for example, the rare fights between impala (*Aepyceros melampus*) territorial males, taking place in spite of conspicuous displays of neck and horn development, in combination with a powerful roaring display (Estes 1991). These displays are usually enough to discourage weaker opponents from daring a fight with them, but may not be sufficient to intimidate equally strong males. The absence of injuries after fights between the two dolphin females in this study may be due to them not really trying to bite each other, but only performing a ritualized display fight. Such ritualized fighting is found in antelope species with potentially lethal horns—for example, the impala, *A. melampus,* and the oryx antelope, *Oryx gazella* (Estes 1991). Another example is the "bite inhibition" seen in wolves, *Canis lupus,* in connection with "passive submission," where the subordinate wolf rolls onto its back, presenting its throat and abdomen, a posture that in effect prevents a dominant wolf to kill a weaker pack mate (Mech 1970).

To study these social sounds in more detail, new methods have to be adopted where free-swimming animals can interact with each other without being restricted by a narrow channel, as in this study. At present, a sound recording unit, attached by means of suction cups to the dorsal fin of our dolphins, is being tested. It will record, in any social interaction, directional pulse sounds received by the dolphin carrying the unit.

Acknowledgments

Thanks to Ericsson Mobile Communications for funding the project. Special thanks to Lee Miller, Odense University, Denmark, for lending us the broadband PC sound card and the Toshiba laptop, and for valuable technical advice. Thanks also to Dave Goodson, Brian Woodward, Paul Lepper, Paul Connelly, and Darryl Newborough at the Underwater Acoustics Group, Loughborough University, UK, for technical support and equipment. Whitlow Au, Hawaii Institute for Marine Biology, University of Hawaii, provided prompt and helpful advice. Finally, thanks to the Kolmården Dolphinarium staff for being so tolerant and helpful and always coming up with practical solutions to problems during the recordings.

Part Four / Literature Cited

ACHARYA, L. 1995. Sex-biased predation on moths by insectivorous bats. *Animal Behaviour* 49:1461–1468.

ACHARYA, L., and M. B. FENTON. 1992. Echolocation behaviour of vespertilionid bats (*Lasiurus cinereus* and *Lasiurus borealis*) attacking airborne targets including arctiid moths. *Canadian Journal of Zoology* 70:1292–1298.

———. 1999. Bat attacks and moth defensive behaviour around street lights. *Canadian Journal of Zoology* 77:27–33.

ACHARYA, L., and J. N. MCNEIL. 1998. Predation risk and mating behavior: The responses of moths to bat-like ultrasound. *Behavioral Ecology* 9:552–558.

AHLÉN, I. 1981. Identification of Scandinavian bats by their sounds. The Swedish University of Agricultural Sciences: Department of Wildlife Ecology, Uppsala, Report 6.

ALDRIDGE, H. D. J. N. 1986. Manoeuvrability and ecological segregation in the little brown (*Myotis lucifugus*) and Yuma (*M. yumanensis*) bats (Chiroptera: Vespertilionidae). *Canadian Journal of Zoology* 64:1878–1882.

ALDRIDGE, H. D. J. N., and I. R. RAUTENBACH. 1987. Morphology, echolocation and resource partitioning in insectivorous bats. *Journal of Animal Ecology* 56:763–778.

ALTMANN, S. A. 1967. The structure of primate social communication. Pp. 325–363 in *Social communication among primates,* ed. S. A. Altmann. Chicago: University of Chicago Press.

ALTRINGHAM. J. D. 1996. *Bats: Biology and behavior.* Oxford: Oxford University Press.

AMUNDIN, M. 1991. Sound production in Odontocetes with emphasis on the harbour porpoise, *Phocoena phocoena.* Ph.D. dissertation, University of Stockholm, Sweden.

ANDERSEN, B. B., and L. A. MILLER. 1977. A portable ultrasonic detection system for recording bat cries in the field. *Journal of Mammalogy* 58:226–229.

ANDERSEN, S. H., and M. AMUNDIN. 1976. Possible predator-related adaptation of sound production and hearing in the harbour porpoise (*Phocoena phocoena*). *Aquatic Mammals* 4:56–58.

ANDERSON, M. E., and P. A. RACEY. 1991. Feeding behaviour of captive brown long-eared bats, *Plecotus auritus. Animal Behaviour* 42:489–493.

ANDERSSON, S., J. RYDELL, and M. G. E. SVENSSON. 1998. Light, predation and the lekking behaviour of the ghost swift *Hepialus humuli* (L.) (Lepidoptera: Hepialidae). *Proceedings of the Royal Society of London B* 264:1345–1351.

ANTHONY. L. P., and T. H. KUNZ. 1977. Feeding strategies of the little brown bat, *Myotis lucifugus,* in southern New Hampshire. *Ecology* 58:775–786.

ARIAS, C., and O. A. RAMOS. 1997. Psychoacoustic tests for the study of human echolocation ability. *Applied Acoustics* 51(4): 399–419.

ARITA, H. T., and M. B. FENTON. 1997. Flight and echolocation in the ecology and evolution of bats. *Trends in Ecology and Evolution* 12:53–58.

ARLETTAZ, R. 1996. Feeding behaviour and foraging strategy of free-living mouse-eared bats, *Myotis myotis* and *Myotis blythii. Animal Behaviour* 51:1–11.

ARLETTAZ, R., and N. PERRIN. 1995. The trophic niches of sympatric *Myotis myotis* and *M. blythii:* Do mouse-eared bats select prey? *Symposia of the Zoological Society of London* 67:361–376.

ARLETTAZ, R., N. PERRIN, and J. HAUSSER. 1997. Trophic resource partitioning and competition between the two sibling bat species *Myotis myotis* and *Myotis blythii. Journal of Animal Ecology* 66:897–911.

ASTRUP, J., and B. MØHL. 1993. Detection of intense ultrasound by the cod *Gadus morhua. Journal of Experimental Biology* 182:71–80.

AU, W. W. L. 1980. Echolocation signals of the Atlantic bottlenose dolphin (*Tursiops truncatus*) in open waters. Pp. 251–282 in *Animal sonar systems,* ed. R. G. Busnel and J. F. Fish. New York: Plenum Press.

———. 1986. Sonar target detection and recognition by Odontocetes. Pp. 451–465 in *Animal sonar: Processes and performance,* ed. P. E. Nachtigall and P. W. B. Moore. New York: Plenum Press.

———. 1993. *The sonar of dolphins.* New York: Springer-Verlag.

———. 1997. Echolocation in dolphins with a dolphin-bat comparison. *Bioacoustics* 8:137–162.

AU, W. W. L., and L. JONES. 1991. Acoustic reflectivity of nets: Implications concerning incidental take of dolphins. *Marine Mammal Science* 7:258–273.

AU, W. W. L., and K. J. SNYDER. 1980. Long-range target detection in open waters by an echolocating Atlantic bottlenose dolphin (*Tursiops truncatus*). *Journal of the Acoustical Society of America* 68:1077–1084.

AU, W. W. L., R. W. FLOYD, and J. E. HAUN. 1978. Propagation of Atlantic bottlenose dolphin echolocation signals. *Journal of the Acoustical Society of America* 64:411–422.

AU, W. W. L., HERZING, D. L., and AUBAUER, R. 1998. Real-time measurement of the echolocation signals of wild dolphins using a 4-hydrophone array. *World Marine Mammal Science Conference,* Monaco, 20–24 January 1998.

AU, W. W. L., M. O. LAMMERS, and R. AUBAUER. 1999. A portable broadband data acquisition system for field studies in bioacoustics. *Marine Mammal Science* 15:526–531.

AU, W. W. L., P. W. B. MOORE, and J. L. PAWLOSKI. 1986. Echolocation transmitting beam of the Atlantic bottlenose dolphin. *Journal of the Acoustical Society of America* 80(2): 688–691.

AU, W. W. L., R. H. PENNER, and C. W. TURL. 1987. Propagation of beluga echolocation signals. *Journal of the Acoustical Society of America* 82:807–813.

AU, W. W. L., D. A. CARDER, R. H. PENNER, and B. L. SCRONCE. 1985. Demonstration of adaptation in beluga whale echolocation signals. *Journal of the Acoustical Society of America* 77:726–730.

AU, W. W. L., R. W. FLOYD, R. H. PENNER, and A. E. MURCHINSON. 1974. Measurement of echolocation signals of the Atlantic bottlenose dolphin, *Tursiops truncatus* Montague, in open waters. *Journal of the Acoustical Society of America* 56:1280–1290.

AU, W. W. L., J. L. PAWLOSKI, P. E. NACHTIGALL, M. BLONZ, and R. C. GISINER. 1995. Echolocation signals and transmission beam pattern of a false killer whale (*Pseudorca crassidens*). *Journal of the Acoustical Society of America* 98:51–59.

AUBAUER, R. 1995. Korrelationsverfahren zur Flugbahnverfolgung echoortender Fledermäuse. Ph.D. dissertation, Technical University of Darmstadt, Darmstadt, Germany, and Fortschritt-Berichte VDI Reihe 17 Nr. 132., VDI-Verlag, Düsseldorf, Germany.

AUBAUER, R., M. O. LAMMERS, and W. W. L. AU. 2000. Acoustical localization of dolphins in shallow water. *Journal of the Acoustical Society of America* 107:2744–2749.

AUDET, D., D. KRULL, G. MARIMUTHU, S. SUMITHRAN, and J. B. SINGH. 1991. Foraging strategies and the use of space by the Indian false vampire, *Megaderma lyra* (Chiroptera: Megadermatidae). *Biotropica* 23:63–67.

BAILEY, W. J. 1991. Acoustic behaviour of insects: An evolutionary perspective. London: Chapman and Hall.

BAKER, R. J. 1984. A sympatric cryptic species of mammal: A new species of *Rhogeessa* (Chiroptera: Vespertilionidae). *Systematic Zoology* 33:178–183.

BALCOMBE, J. P., and M. B. FENTON. 1988. Eavesdropping by bats: The influence of echolocation call design and foraging strategy. *Ethology* 79:158–166.

BARCLAY, R. M. R. 1982. Interindividual use of echolocation calls, eavesdropping by bats. *Behavioral Ecology and Sociobiology* 10:271–275.

———. 1985. Long- versus short-range foraging strategies of hoary (*Lasiurus cinereus*) and silver-haired (*Lasionycteris noctivagans*) bats and consequences for prey selection. *Canadian Journal of Zoology* 63:2507–2515.

———. 1986. The echolocation calls of hoary (*Lasiurus cinereus*) and silver-haired (*Lasionycteris noctivagans*) bats and the consequences for prey selection. *Canadian Journal of Zoology* 64:2700–2705.

———. 1999. Bats are not birds—a cautionary note on using echolocation calls to identify bats: A comment. *Journal of Mammalogy* 80:290–296.

BARCLAY, R. M. R., and R. M. BRIGHAM. 1991. Prey detection, dietary niche breadth, and body size in bats: Why are aerial insectivorous bats so small? *American Naturalist* 137:693–703.

———. 1994. Constraints on optimal foraging: A field test of prey discrimination by echolocating insectivorous bats. *Animal Behaviour* 48:1013–1021.

BARCLAY, R. M. R., M. A. DOLAN, and A. DYCK. 1991. The digestive efficiency of insectivorous bats. *Canadian Journal of Zoology* 69:1853–1856.

BARLOW, K. E. 1997. The diets of two phonic types of *Pipistrellus pipistrellus* (Chiroptera: Vespertilionidae) in Britain. *Journal of Zoology, London* 243:597–609.

BARLOW, K. E., and G. JONES. 1997a. Differences in songflight calls and social calls between two phonic types of the vespertilionid bat *Pipistrellus pipistrellus*. *Journal of Zoology, London* 241:315–324.

———. 1997b. Function of pipistrelle social calls: field data and a playback experiment. *Animal Behaviour* 53:991–999.

———. 1999. Roosts, echolocation calls and wing morphology of two phonic types of *Pipistrellus pipistrellus*. *Zeitschrift für Säugetierkunde* 64:257–268.

BARLOW, K. E., G. JONES, and E. M. BARRATT. 1997. Can skull morphology be used to predict ecological relationships between bat species? A test using two cryptic species of pipistrelle. *Proceedings of the Royal Society of London* 264B:1695–1700.

BARRATT, E. M., R. DEAVILLE, T. M. BURLAND, M. W. BRUFORD, G. JONES, P. A. RACEY, and R. K. WAYNE. 1997. DNA answers the call of pipistrelle bat species. *Nature* 387:138–139.

BARRETT-LENNARD, L. G., J. K. B. FORD, and K. A. HEISE. 1996. The mixed blessing of echolocation: Differences in sonar use by fish-eating and mammal-eating killer whales. *Animal Behaviour* 51:553–565.

BARROS, N. B., and D. K. ODELL. 1990. Food habits of bottlenose dolphins in the southeastern United States. Pp. 309–328 in *The bottlenose dolphin,* ed. S. Leatherwood and R. R. Reeves. San Diego: Academic Press.

BARROS, N. B., and R. S. WELLS. 1998. Prey and feeding patterns of resident bottlenose dolphins (*Tursiops truncatus*) in Sarasota bay, Florida. *Journal of Mammalogy* 79:1045–1059.

BARTHOLOMEW, G. A., and B. H. HEINRICH. 1973. A field study of flight temperatures in moths in relation to body weight and wing loading. *Journal of Experimental Biology* 58:23–135.

BATES, D. L., and M. B. FENTON. 1990. Aposematism or startle? Predators learn their responses to the defenses of prey. *Canadian Journal of Zoology* 68:49–52.

BAUDINETTE, R. V., and K. SCHMIDT-NIELSEN. 1974. Energy cost of gliding flight in herring gulls. *Nature* 248:83–84.

BAUEROVÁ, Z. 1986. Contribution to the trophic bionomics of *Myotis emarginatus*. *Folia Zoologica* 35:305–310.

BAZLEY, E. N. 1976. Sound absorption in air at frequencies up to 100 kHz. *National Physical Laboratory Acoustics Report* No. Ac 74, Teddington, UK.

BELL, G. P. 1982. Behavioral and ecological aspects of gleaning by desert insectivorous bat, *Antrozous pallidus* (Chiroptera: Vespertilionidae). *Behavioral Ecology and Sociobiology* 10:217–223.

———. 1985. The sensory basis of prey location by the California leaf-nosed bat, *Macrotus californicus* (Chiroptera: Phyllostomidae). *Behavioral Ecology and Sociobiology* 16:343–347.

BELL, G. P., and M. B. FENTON. 1984. The use of Doppler-shifted echoes as a clutter rejection system: The echolocation and feeding behavior of *Hipposideros ruber* (Chiroptera: Hipposideridae). *Behavioral Ecology and Sociobiology* 15:109–114.

BELWOOD, J. J. 1990. Anti-predator defences and ecology of neotropical forest katydids, especially the Pseudophyllinae. Pp. 8–26 in *The Tettigoniidae: Biology, systematics and evolution,* ed. W. J. Bailey and D. C. F. Rentz. New York: Springer-Verlag.

BELWOOD, J. J., and M. B. FENTON. 1976. Variation in the diet of *Myotis lucifugus* (Chiroptera: Vespertilionidae). *Canadian Journal of Zoology* 54:1674–1678.

Belwood, J. J., and J. H. Fullard. 1984. Echolocation and foraging behavior in the Hawaiian hoary bat, *Lasiurus cinereus semotus*. *Canadian Journal of Zoology* 62:2113–2120.

Belwood, J. J., and G. K. Morris. 1987. Bat predation and its influence on calling behavior in neotropical katydids. *Science* 238:64–67.

Bennett, M. V. L. 1984. Escapism: Some startling revelations. Pp. 353–363 in *Neural mechanisms of startle behavior,* ed. R. C. Eaton. New York: Plenum Press.

Best, R. C., and V. M. F. da Silva. 1989. Biology, status and conservation of *Inia geoffrensis* in the Amazon and Orinoco River basins. Pp. 23–34 in *Biology and conservation of the river dolphins,* ed. W. F. Perrin, R. L. Brownell Jr., Z. Kaiya, and L. Jiankang. International Union for Conservation of Nature and Natural Resources, Species Survival Commission Occasional Paper No. 3.

Betts, B. J. 1998. Effects of interindividual variation in echolocation calls on identification of big brown and silver-haired bats. *Journal of Wildlife Management* 62:1003–1010.

Black H. L. 1972. Differential exploitation of moths by the bats *Eptesicus fuscus* and *Lasiurus cinereus*. *Journal of Mammalogy* 53:598–601.

———. 1979. Precision in prey selection by the trident-nosed bat (*Cloeotis percivali*). *Mammalia* 43:53–57.

Blest, A. D., T. S. Collett, and J. D. Pye. 1963. The generation of ultrasonic signals by a New World Arctiid moth. *Proceedings of the Royal Society of London B* 158:196–207.

Bogdanowicz, W. 1994. *Myotis daubentonii*. *Mammalian Species* 475:1–9.

Bogdanowicz, W., M. B. Fenton, and K. Daleszczyk. 1999. The relationships between echolocation calls, morphology and diet in insectivorous bats. *Journal of Zoology, London* 247:381–393.

Boonman, A. M., M. Boonman, F. Bretschneider, and W. A. van de Grind. 1998. Prey detection in trawling insectivorous bats: Duckweed affects hunting behavior in Daubenton's bat, *Myotis daubentonii*. *Behavioral Ecology and Sociobiology* 44:99–107.

Boyd, I. A., C. Lockyer, and H. D. Marsh. 1999. Reproduction in marine mammals. Pp. 218–286 in *Biology of marine mammals,* ed. J. E. Reynolds III and S. A. Rommel. Washington, D.C.: Smithsonian Institution Press.

Brackenbury, J. H. 1977. Physiological energetics of cock-crow. *Nature* 270:433–435.

Bradbury, J. W. 1970. Target discrimination by the echolocating bat, *Vampyrum spectrum*. *Journal of Experimental Zoology* 173:23–41.

Bradbury, J. W., and S. L. Vehrencamp. 1998. *Principles of animal communication*. Sunderland, Mass.: Sinauer Associates Inc.

Brigham, R. M., S. D. Grindal, M. C. Firman, and J. L. Morisette. 1997. The influence of structural clutter on activity patterns of insectivorous bats. *Canadian Journal of Zoology* 75:131–136.

Britton, A. R. C. 1996. Flight performance, echolocation and prey capture behaviour in trawling *Myotis* bats. Ph.D. dissertation, University of Bristol, Bristol.

Britton, A. R. C., and G. Jones. 1999. Echolocation behaviour and prey-capture success in foraging bats: Laboratory and field experiments on *Myotis daubentonii*. *Journal of Experimental Biology* 202:1793–1801.

Britton, A. R. C., G. Jones, J. M. V. Rayner, A. M. Boonman, and B. Verboom. 1997. Flight performance, echolocation and foraging behavior in pond bats, *Myotis dasycneme* (Chiroptera: Vespertilionidae). *Journal of Zoology, London* 241:503–522.

Brosset, A. 1966. *La biologie des chiroptères*. Paris: Masson et Cie.

Brownlee, S. M., and K. Norris. 1994. The acoustic domain. Pp. 161–185 in *The Hawaiian spinner dolphin,* ed. K. S. Norris, B. Würsig, R. S. Wells, and M. Würsig. Berkeley and Los Angeles: University of California Press.

Buchler, E. R. 1976. Prey selection by *Myotis lucifugus* (Chiroptera: Vespertiliondae). *American Naturalist* 110:619–628.

Buchler, E. R., and S. B. Childs. 1981. Orientation to distant sounds by foraging big brown bats (*Eptesicus fuscus*). *Animal Behaviour* 29:428–432.

Bullock, T. H. 1984. Comparative neuroethology of startle, rapid escape, and giant fiber-mediated responses. Pp. 1–13 in *Neural mechanisms of startle behavior,* ed. R. C. Eaton. New York: Plenum Press.

Bullock, T. H., A. D. Grinnell, E. Ikezono, K. Kameda, J. Katsuki, M. Nomota, O. Sato, N. Suga, and K. Yanagisawa. 1968. Electrophysiological studies of central auditory mechanisms in cetaceans. 1968. *Zeitschrift für Vergleichende Physiologie* 59:117–156.

Caldwell, M. C., and D. K. Caldwell. 1965. Individualized whistle contours in bottlenosed dolphins (*Tursiops truncatus*). *Nature* 207:434–435.

———. 1966. Intraspecific transfer of information via the pulsed sound in captive odontocete cetaceans.

Pp. 879–936 in *Animal sonar systems,* ed. R. G. Busnel and J. F. Fish. New York: Plenum Press.

———. 1972. Senses and communication. Pp. 466–501 in *Mammals of the sea: Biology and medicine,* ed. S. H. Ridgway. Springfield, Ill.: Charles C. Thomas.

CALDWELL, M. C., D. K. CALDWELL, and R. L. BRILL. 1989. *Inia geoffrensis* in captivity in the United States. Pp. 35–41 in *Biology and conservation of the river dolphins,* ed. W. F. Perrin, R. L. Brownell Jr., Z. Kaiya, and L. Jiankang. International Union for Conservation of Nature and Natural Resources, Species Survival Commission Occasional Paper No. 3.

CALDWELL, M. C., D. K. CALDWELL, and W. E. EVANS. 1966. Sounds and behavior of captive Amazon freshwater dolphins, *Inia geoffrensis. Los Angeles County Museum, Contributions in Science,* 108:1–24.

CAMHI, J. M., and W. TOM. 1978. The escape behavior of the cockroach *Periplaneta americana.* I. Turning response to wind puffs. *Journal of Comparative Physiology* 128:193–201.

CAMHI, J. M., W. TOM, and S. VOLMAN. 1978. The escape behavior of the cockroach *Periplaneta americana.* II. Detection of natural predators by air displacement. *Journal of Comparative Physiology* 128:203–212.

CARDONE, B., and J. H. FULLARD. 1988. Auditory characteristics and sexual dimorphism in the gypsy moth. *Physiological Entomology* 13:9–14.

CARMEL, P. W., and A. STARR. 1963. Acoustic and nonacoustic factors modifying middle-ear muscle activity in waking cats. *Journal of Neurophysiology* 26:598–616.

CARMEL, Y., and U. SAFIEL. 1998. Habitat use by bats in a Mediterranean ecosystem in Israel: Conservation implications. *Biological Conservation* 84:245–250.

CASEY, T. M., and B. A. JOOS. 1983. Morphometrics, conductance, thoracic temperature, and flight energetics of noctuid and geometrid moths. *Physiological Zoology* 56:160–173.

CHAI, P., and R. B. SRYGLEY. 1990. Predation and the flight, morphology, and temperature of neotropical rainforest butterflies. *American Naturalist* 135:748–765.

CHAPPELL, M. A., M. ZUK, T. H. KWAN, and T. S. JOHNSEN. 1995. Energy cost of an avian vocal display: Crowing in red jungle fowl. *Animal Behaviour* 49:255–257.

CHARIF, R. A., S. MITCHELL, and C. W. CLARK. 1995. Canary 1.12 user's manual. Cornell Laboratory of Ornithology, Ithaca, New York.

CLARK, C. W. 1994. Blue deep voices: Insights from the Navy's Whales '93 Program. *Whalewatcher* 28(1): 6–11.

CONNER, W. E. 1999. "Un chant d'appel amoreux": Acoustic communication in moths. *Journal of Experimental Biology* 202:1711–1723.

CONNOR, R. C., and R. A. SMOLKER. 1996. "Pop" goes the dolphin: A vocalization male bottlenose dolphins produce during consort ships. *Behaviour* 133:643–662.

CONWAY, T., and N. B. SIMMONS. 1999. Evolution of Mormoopid bats. *Bat Research News* 39:163–164.

COVEY, E., and J. H. CASSEDAY. 1991. The monaural nuclei of the lateral lemniscus in an echolocation bat: Parallel pathways for analyzing temporal features of sound. *Journal of Neuroscience* 11:3456–3470.

———. 1995. The lower brainstem auditory pathways. Pp. 235–295 in *Hearing by bats,* ed. A. N. Popper and R. R. Fay. New York: Springer-Verlag.

CRANFORD, T. W., M. AMUNDIN, and K. S. NORRIS. 1996. Functional morphology and homology in the odontocete nasal complex: Implications for sound generation. *Journal of Morphology* 228:223–285.

CRANFORD, T. W., W. G. VAN BONN, M. S. CHAPLIN, J. A. CARR, T. A. KAMOLNIK, D. A. CARDER, and S. H. RIDGWAY. 1997. Visualizing dolphin sonar signal generation using high-speed video endoscopy. *Journal of the Acoustic Society of America* 102:3123.

CROME, F. H., and G. C. RICHARDS. 1988. Bats and gaps: Microchiropteran community structure in a Queensland rain forest. *Ecology* 69:1960–1969.

CUMMING, G. S. 1996. Mantis movements by night and the interactions of sympatric bats and mantises. *Canadian Journal of Zoology* 74:1771–1774.

DAMBACH, M. 1989. Vibrational responses. Pp. 178–197 in *Cricket behavior and neurobiology,* ed. F. Huber, T. E. Moore, and W. Loher. Ithaca, N.Y.: Cornell University Press.

DAWKINS, R., and J. R. KREBS. 1979. Arms races between and within species. *Proceedings of the Royal Society of London B* 205:489–511.

DAWSON, J. W., K. DAWSON-SKULLY, D. ROBERT, and R. M. ROBERTSON. 1997. Forewing asymmetries during auditory avoidance in flying locusts. *Journal of Experimental Biology* 200:2323–2335.

DAWSON, S. M. 1988. Then high-frequency sounds of free-ranging Hector's dolphins, *Cephalorhynchus hectori. Report to the International Whaling Commission,* special issue 9:339–344.

————. 1991. Clicks and communication: The behavioural and social contexts of Hector's dolphin vocalizations. *Ethology* 88:265–276.

DAWSON, S. M., and C. W. THORPE. 1990. A quantitative analysis of the sounds of Hector's dolphin. *Ethology* 86:131–145.

DeFran, R. H., and K. Pryor. 1980. The behavior and training of cetaceans in captivity. Pp. 319–362 in *Cetacean behavior: Mechanisms and functions*, ed. L. M. Herman. New York: John Wiley & Sons.

DE LA CUEVA SALCEDO, H., M. B. FENTON, M. B. C. HICKEY, and R. W. BLAKE. 1995. Energetic consequences of flight speeds of foraging red and hoary bats (*Lasiurus borealis* and *Lasiurus cinereus;* Chiroptera: Vespertilionidae). *Journal of Experimental Biology* 198:2245–2251.

DETHIER, V. G. 1992. *Crickets and katydids, concerts and solos.* Cambridge: Harvard University Press.

DEY, S. 1995. Possible ultrasonic receptor on the bat fly *Mystacinobia zealandica. Current Science* 68:992–994.

DIERCKS, K. J., R. T. TROCHTA, and W. E. EVANS. 1973. Delphinid sonar: Measurement and analysis. *Journal of the Acoustical Society of America* 54:200–204.

DIERCKS, K. J., R. T. TROCHTA, C. F. GREENLAW, and W. E. EVANS. 1971. Recording and analysis of dolphin echolocation signals. *Journal of the Acoustical Society of America* 49:1729–1732.

DING, W., B. WÜRSIG, and W. EVANS. 1995. Comparisons of whistles among seven odontocete species. Pp. 299–323 in *Sensory systems of aquatic mammals*, ed. R. A. Kastelein, J. A. Thomas, and P. E. Nachtigall. Woerden, The Netherlands: De Spil Publishers.

DOS SANTOS, M. E. 1998. *Golfinhos-roazes do Sado—estudos de sons e comportamento.* Lisboa, Portugal: Edições ISPA.

DOS SANTOS, M. E., and M. LACERDA. 1987. Preliminary observation of the bottlenose dolphin (*Tursiops truncatus*) in the Sado estuary (Portugal). *Aquatic Mammals* 13:65–80.

DOS SANTOS, M. E., T. MODESTO, R. J. MATOS, M. S. GROBER, R. F. OLIVEIRA, and A. CANÁRIO 2000. Sound production by the Lusitanian toadfish (*Halobatrachus didactylus*). *Bioacoustics* 10:309–321.

DREWES, C. D., K. B. LANDA, and J. L. McFALL. 1978. Giant nerve fibre activity in intact, freely moving earthworms. *Journal of Experimental Biology* 72:217–227.

DRICKAMER, L. C., and S. H. VESSEY. 1992. *Animal behavior.* 3rd ed. Wm. C. Brown Publishers.

DUCROCQ, S., J.-J. JAEGER, and B. SIGÉ. 1993. Un mégachiroptère dans l'Eocène supérieur de Thaïlande: Incidence dans la discussion phylogénique du groupe. *Neues Jahrbuch für Geologie und Paläontologie, Monatshefte,* 9:561–575.

DUDZINSKI, K. 1995. Behavioral contexts: Communication and behavior in the Atlantic Spotted dolphin (*Stenella frontalis*): Relationships between vocal and behavioral activities. Ph.D. dissertation, Texas A&M University.

DUNNING, D. C. 1968. Warning sounds of moths. *Zeitschrift für Tierpsychologie* 25:129–138.

DUNNING, D. C., and M. KRÜGER. 1995. Aposematic sounds in African moths. *Biotropica* 27:227–231.

————. 1996. Predation upon moths by free-foraging *Hipposideros caffer. Journal of Mammalogy* 77:708–715.

DUNNING, D. C., and K. D. ROEDER. 1965. Moth sounds and the insect catching behavior of bats. *Science* 147:173–174.

DUNNING, D. C., L. ACHARYA, C. B. MERRIMAN, and L. DAL FERRO. 1992. Interactions between bats and arctiid moths. *Canadian Journal of Zoology* 70:2218–2223.

EATON, R. C., ed. 1984. *Neural mechanisms of startle behavior.* New York: Plenum Press.

EATON, R. C., R. A. BOMBARDIERI, and D. L. MEYER. 1977. The Mauthner-initiated startle response in teleost fish. *Journal of Experimental Biology* 66:65–81.

EBERHARDT, L. S. 1994. Oxygen consumption of Carolina wrens (*Thryothorus ludovicianus*). *Auk* 111:124–130.

ESTABROOK, C. B., and G. F. ESTABROOK. 1989. ACTUS: A solution to the problem of small samples in the analysis of two-way contingency tables. *Historical Methods* 22:5–8.

ESTES, R. D. 1991. *The behavior guide of African mammals.* Berkeley and Los Angeles: University of California Press.

EVANS, W. E. 1973. Echolocation by marine delphinids and one species of fresh-water dolphin. *Journal of the Acoustical Society of America* 54:493–503.

EVANS, W. E., and F. T. AWBREY. 1988. Natural history aspects of marine mammal echolocation: Feeding strategies and habitat. Pp. 521–534 in *Animal sonar: Processes and performance*, ed. P. E. Nachtigall and

P. W. B. Moore. NATO ASI Series. New York: Plenum Press.

EVANS, W. E., and J. BASTIAN. 1969. Marine mammal communication: Social and ecological factors. Pp. 425–475 in *The biology of marine mammals,* ed. H. T. Andersen. New York: Academic Press.

EVANS W. E., and B. A. POWELL. 1967. Discrimination of different metallic plates by an echolocating delphinids. Pp. 363–382 in *Systems sonar animaux: Biologie and bionique,* vol. 2, ed. R. G. Busnel. Jouy-en-Josas, France: Laboratoire de Physiologie Acoustique.

EVANS, W. E., W. W. SUTHERLAND, and R. G. BEIL. 1964. The directional characteristics of delphinid sounds. Pp. 1:353–372 in *Marine bioacoustics,* ed. W. N. Tavolga. New York: Pergamon Press.

FAURE, P. A., and R. M. R. BARCLAY. 1992. The sensory basis of prey detection by the long eared bat, *Myotis evotis,* and the consequences for prey selection. *Animal Behaviour* 44:31–39.

———. 1994. Substrate-gleaning versus aerial-hawking: Plasticity in the foraging and echolocation behaviour of the long-eared bat, *Myotis evotis. Journal of Comparative Physiology A* 174:651–660.

FAURE, P. A., and R. R. HOY. 2000a. The sounds of silence: Cessation of singing and song pausing are ultrasound-induced acoustic startle behaviors in the katydid *Neoconocephalus ensiger* (Orthoptera; Tettigoniidae). *Journal of Comparative Physiology A* 186:129–142.

———. 2000b. Auditory symmetry analysis. *Journal of Experimental Biology* 203:3209–3223.

———. 2000c. Neuroethology of the katydid T-cell. I. Tuning and responses to pure-tones. *Journal of Experimental Biology* 203:3225–3242.

———. 2000d. Neuroethology of the katydid T-cell. II. Responses to acoustic playback of conspecific and predatory signals. *Journal of Experimental Biology* 203:3243–3254.

FAURE, P. A., J. H. FULLARD, and R. M. R. BARCLAY. 1990. The response of tympanate moths to the echolocation calls of a substrate gleaning bat, *Myotis evotis. Journal of Comparative Physiology A* 166:843–849.

FAURE, P. A., J. H. FULLARD, and J. W. DAWSON. 1993. The gleaning attacks of the northern long-eared bat, *Myotis septentrionalis,* are relatively inaudible to moths. *Journal of Experimental Biology* 178:173–189.

FENG, A. S., C. J. CONDON, and K. R. WHITE. 1994. Stroboscopic hearing as a mechanism for prey discrimi-

nation in frequency-modulated bats. *Journal of the Acoustical Society of America* 95:2736–2744.

FELSENSTEIN, J. 1993. PHYLIP (Phylogeny Inference Package). Department of Genetics, University of Washington, Seattle.

FENTON, M. B. 1980. Adaptiveness and ecology of echolocation in terrestrial (aerial) systems. Pp. 427–446 in *Animal sonar systems,* ed. R. G. Busnel and J. F. Fish. New York: Plenum Press.

———. 1984. Echolocation: Implications for the ecology and evolution of bats. *Quarterly Review of Biology* 59:33–53.

———. 1985. *Communication in the Chiroptera.* Bloomington: Indiana University Press.

———. 1986. *Hipposideros caffer* (Chiroptera: Hipposideridae) in Zimbabwe: Morphology and echolocation calls. *Journal of Zoology, London* 210:347–353.

———. 1990. The foraging behavior and ecology of animal-eating bats. *Canadian Journal of Zoology* 68:411–422.

———. 1995. Natural history and biosonar signals. Pp. 17–36 in *Hearing by bats,* ed. A. N. Popper and R. R. Fay. New York: Springer-Verlag.

———. 1999. Describing the echolocation calls and behaviour of bats. *Acta Chiropterologica* 1:127–136.

FENTON, M. B., and G. P. BELL. 1979. Echolocation and feeding behaviour of four species of *Myotis* (Chiroptera). *Canadian Journal of Zoology* 57:1271–1277.

FENTON, M. B., and J. H. FULLARD. 1979. The influence of moth hearing on bat echolocation strategies. *Journal of Comparative Physiology* 132:77–86.

———. 1981. Moth hearing and the feeding strategies of bats. *American Scientist* 69:266–275.

FENTON, M. B., D. AUDET, M. K. OBRIST, and J. RYDELL. 1995. Signal strength, timing and self-deafening: The evolution of echolocation in bats. *Paleobiology* 21:229–242.

FENTON, M. B., C. V. PORTFORS, I. L. RAUTENBACH, and J. M. WATERMAN. 1998a. Compromises: Sound frequencies used in echolocation by aerial feeding bats. *Canadian Journal of Zoology* 76:1174–1182.

FENTON, M. B., D. H. M. CUMMING, I. L. RAUTENBACH, G. S. CUMMING, M. S. CUMMING, G. FORD, R. D. TAYLOR, J. DUNLOP, M. D. HOVORKA, D. S. JOHNSTON, C. V. PORTFORS, M. L. KALCOUNIS, and Z. MAHLANGA. 1998b. Bats and the loss of tree canopy in African woodlands. *Conservation Biology* 12:399–407.

FENTON, M. B., J. RYDELL, M. J. VON HOF, J. EKLÖF, and W. C. LANCASTER. 1999a. Constant frequency (CF) and frequency modulated (FM) components in the echolocation calls of three small bats (Emballonuridae, Thyropteridae and Vespertilionidae). *Canadian Journal of Zoology* 77:1891–1900.

———. 1999b. The diet of bats from southeastern Brazil: The relation to echolocation and foraging behavior. *Revista Brasileira de Zoologia* 16:1081–1085.

FERREIRA, A., J. L. BENTO-COELHO, and M. E. DOS SANTOS. 1996. Underwater noise in the Sado estuary. *Acustica-acta acustica* 82:S255.

FIEDLER, J. 1979. Prey catching with and without echolocation in the Indian false vampire bat (*Megaderma lyra*). *Behavioral Ecology and Sociobiology* 6:155–160.

FINDLEY, J. S. 1993. *Bats: A community perspective.* Cambridge: Cambridge University Press.

FITZPATRICK, J. W. 1980. Foraging behavior of Neotropical tyrant flycatchers. *Condor* 82:43–57.

FORD, J. K. B. 1989. Acoustic behavior of resident killer whales (*Orcinus orca*) off Vancouver Island, British Columbia. *Canadian Journal of Zoology* 67:727–745.

———. 1991. Vocal traditions among resident killer whales (*Orcinus orca*) in coastal waters of British Columbia. *Canadian Journal of Zoology* 69:1454–1483.

FORREST, T. G. 1991. Power output and efficiency of sound production by crickets. *Behavioural Ecology* 2:327–338.

FORREST, T. G., H. E. FARRIS, and R. R. HOY. 1995. Ultrasound acoustic startle response in scarab beetles. *Journal of Experimental Biology* 198:2593–2598.

FORREST, T. G., M. P. READ, H. E. FARRIS, and R. R. HOY. 1997. A tympanal hearing organ in scarab beetles. *Journal of Experimental Biology* 200:601–606.

FOWLER, J., and L. COHEN. 1990. *Practical statistics for field biology.* West Sussex: John Wiley & Sons Ltd.

FRANCIS, C. M., and J. HABERSETZER. 1998. Interspecific and intraspecific variation in echolocation call frequency and morphology of horseshoe bats, *Rhinolophus* and *Hipposideros*. Pp. 169–179 in *Bat biology and conservation,* ed. T. H. Kunz and P. A. Racey. Washington, D.C.: Smithsonian Institution Press.

FRANCIS, C. M., D. KOCK, and J. HABERSETZER. 1999. Sibling species of *Hipposideros ridleyi* (Mammalia, Chiroptera, Hipposideridae). *Senckenbergiana biologica* 79:255–270.

FRAZER, L. N., and E. MERCADO III. 2000. A sonar model for humpback whale song. *IEEE Journal of Oceanic Engineering* 25(1): 160–182.

FREEMAN, S., and J. C. HERRON. 1998. *Evolutionary analysis.* Englewood Cliffs, N.J.: Prentice-Hall.

FRIEDEL, T. 1999. The vibrational startle response of the desert locust *Schistocerca gregaria*. *Journal of Experimental Biology* 202:2151–2159.

FRINGS, H., and M. FRINGS. 1957. The effects of temperature on chirp-rate of male cone-headed grasshoppers, *Neoconocephalus ensiger*. *Journal of Experimental Zoology* 134:411–425.

FRISTRUP, K. M., M. A. DAHER, T. J. HOWALD, and W. A. WATKINS. 1992. Software Tools for Acoustic Database Management. Technical Report WHOI-92–11, Woods Hole Oceanographic Institution, Woods Hole, Mass. 02543.

FULLARD, J. H. 1977. Phenology of sound-producing arctiid moths and the activity of insectivorous bats. *Nature* 267:42–43.

———. 1987. Sensory ecology and neuroethology of moths and bats: Interactions in a global perspective. Pp. 244–272 in *Recent advances in the study of bats,* ed. M. B. Fenton, P. Racey, and J. M. V. Rayner. Cambridge: Cambridge University Press.

———. 1992. The neuroethology of sound production in tiger moths (Lepidoptera, Arctiidae). I Rhytmicity and central control. *Journal of Comparative Physiology A* 170:575–588.

———. 1994. Auditory changes in moths endemic to a bat-free habitat. *Journal of Evolutionary Biology* 7:435–445.

———. 1998. The sensory coevolution of moths and bats. Pp. 279–326 in *Comparative hearing: Insects,* ed. R. R. Hoy, A. N. Popper, and R. R. Fay. New York: Springer-Verlag.

FULLARD, J. H., and J. W. DAWSON. 1997. The echolocation calls of the spotted bat *Euderma maculatum* are relatively inaudible to moths. *Journal of Experimental Biology* 200:129–137.

———. 1999. Why do diurnal moths have ears? *Naturwissenschaften* 86:276–279.

FULLARD, J. H., and M. B. FENTON. 1977. Acoustic behavioural analyses of the sounds produced by some species of Nearctic Arctiidae (Lepidoptera). *Canadian Journal of Zoology* 55:1213–1224.

FULLARD, J. H., and B. HELLER. 1990. Functional organization of the arctiid moth tymbal (Insecta, Lepidoptera). *Journal of Morphology* 204:57–65.

FULLARD, J. H., and J. E. YACK. 1993. The evolutionary biology of insect hearing. *Trends in Ecology and Evolution* 8:248–252.

FULLARD, J. H., R. M. R. BARCLAY, and D. W. THOMAS. 1993. Echolocation in free-flying atiu swiftlets (*Aerodramus sawtelli*). *Biotropica* 25:334–339.

FULLARD, J. H., M. B. FENTON, and J. A. SIMMONS. 1979. Jamming bat echolocation: The clicks of arctiid moths. *Canadian Journal of Zoology* 57:647–649.

FULLARD, J. H., J. A. SIMMONS, and P. A. SAILLANT. 1994. Jamming bat echolocation: The dogbane tiger moth *Cycnia tenera* times its clicks to the terminal attack calls of the big brown bat *Eptesicus fuscus*. *Journal of Experimental Biology* 194:285–298.

FULLARD, J. H., J. W. DAWSON, L. D. OTERO, and A. SURLYKKE. 1998. Bat-deafness in day-flying moths (Lepidoptera, Notodontidae, Dioptinae). *Journal of Comparative Physiology A* 181:477–483.

GAUNT, A. S., T. L. BUCHER, S. L. L. GAUNT, and L. F. BAPTISTA. 1996. Is singing costly? *Auk* 113:718–721.

GHIRADELLA, H. 1971. Fine structure of the noctuid moth ear. *Journal of Morphology* 134:21–46.

GOODALL, J. 1968. Behaviour of free-living chimpanzees of the Gombe Stream area. *Animal Behavior* 1:163–311.

———. 1986. *The chimpanzees of Gombe: Patterns of behavior.* Cambridge: The Belknap Press of Harvard University Press.

GOOLD, J. C. 1996. Signal processing techniques for acoustic measurement of sperm whale body lengths. *Journal of the Acoustical Society of America* 100:3431–3441.

GÖPFERT, M. C., and L. T. WASSERTHAL. 1999a. Hearing with the mouthparts: Behavioural responses and the structural basis of ultrasound perception in acherotine hawkmoths. *Journal of Experimental Biology* 202:909–918.

———. 1999b. Auditory sensory cells in hawkmoths: Identification, physiology and structure. *Journal of Experimental Biology* 202:1579–1587.

GORDON, J. C. D. 1987. Sperm whale groups and social behaviour observed off Sri Lanka. *Report of the International Whaling Commission* 37:205–217.

———. 1991. Evaluation of a method for determining the length of sperm whales (*Physeter macrocephalus*) from their vocalizations. *Journal of Zoology (London)* 224:301–314.

———. 1996. Sperm whale acoustic behaviour. Pp. 29–33 in *European research on cetaceans*, vol. 9, ed. P. G. H. Evans and H. Nice. Proceedings of the Ninth Annual Conference of the European Cetacean Society, Lugano, Switzerland. February 9–12, 1995.

GOULD, E. 1955. The feeding efficiency of insectivorous bats. *Journal of Mammalogy* 36:399–407.

GREEN, S. 1975. Communication by a graded system in Japanese monkeys. Pp. 1–102 in *Primate behavior: Developments in field and laboratory research*, ed. L. Rosenblum. New York: Academic Press.

GREENWOOD, R. J., R. J. HARRISON, and H. W. WHITTING. 1974. Functional and pathological aspects of the skin of marine mammals. Pp. 73–110 in *Functional anatomy of marine mammals*, vol. 2, ed. R. J. Harrison. New York: Academic Press.

GRIFFIN, D. R. 1958. *Listening in the dark.* New Haven, Conn.: Yale University Press.

———. 1971. The importance of atmospheric attenuation for the echolocation of bats (Chiroptera). *Animal Behaviour* 19:55–61.

GRIFFIN, D. R., J. H. FRIEND, and F. A. WEBSTER. 1965. Target discrimination by the echolocation of bats. *Journal of Experimental Zoology* 158:155–168.

GRIFFIN, D. R., F. A. WEBSTER, and C. R. MICHAEL. 1960. The echolocation of flying insects by bats. *Animal Behaviour* 8:141–154.

GUINEE, L. N., K. CHU, and E. M. DORSEY. 1983. Changes over time in the songs of known individual humpack whales (Megaptera novaeangliae). Pp. 59–80 in *Communication and behavior of whales*, ed. R. S. Payne. Boulder, Colo.: Westview Press.

HABERSETZER, J., and B. VOGLER. 1983. Discrimination of surface-structured targets by the echolocating bat, *Myotis myotis. Journal of Comparative Physiology A* 152:275–282.

HABERSETZER, J. G., G. RICHTER, and G. STORCH. 1992. Bats: Already highly specialized insect predators. Pp. 181–191 in *Messel: An insight into the history of life and of the earth*, ed. S. Schall and W. Ziegler. Clarendon Press, Oxford.

———. 1994. Paleoecology of Early Middle Eocene bats from Messel, FRG: Aspects of flight, feeding and echolocation. *Historical Biology* 8:235–260.

HAMILTON, I. M., and R. M. R. BARCLAY. 1998. Diets of juvenile, yearling, and adult big brown bats (*Eptesicus fuscus*) in southeastern Alberta. *Journal of Mammalogy* 79:764–771.

HAMILTON III, W. J., and P. C. ARROWOOD. 1978. Copulatory vocalizations of chacma baboons (*Papio ursinus*), gibbons (*Hylobates hoolock*), and humans. *Science* 200:1405–1408.

HANBY, J. P. 1974. Male-male mounting in Japanese monkeys (*Macaca fuscata*). *Animal Behaviour* 22:836–849.

HARRIS, S., P. MORRIS, S. WRAY, and D. YALDEN. 1995. *A review of British mammals: Population estimates and conservation status of British mammals other than cetaceans.* Peterborough: Joint Nature Conservation Committee.

HARTLEY, D. J. 1989. The effect of atmospheric sound absorption on signal bandwidth and energy and some consequences for bat echolocation. *Journal of the Acoustical Society of America* 85:1338–1347.

————. 1992. Stabilization of perceived echo amplitudes in echolocating bats. I. Echo detection and automatic gain-control in the big brown bat, *Eptesicus fuscus*, and the fishing bat, *Noctilio leporinus. Journal of the Acoustical Society of America* 91:1120–1132.

HARTLEY, D. J., and R. A. SUTHERS. 1987. The sound emission pattern and the acoustical role of the nose-leaf in the echolocating bat, *Carollia perspicillata. Journal of the Acoustical Society of America* 82:1892–1900.

HATAKEYAMA, Y., and H. SOEDA. 1990. Studies of echolocation of porpoises taken in salmon net fisheries. Pp. 269–281 in *Sensory abilities of cetaceans: Laboratory and field evidence,* ed. J. Thomas and R. Kastlelein. New York: Plenum Press.

HAYS, G. C., P. I. WEBB, J. FRENCH, and J. R. SPEAKMAN. 1990. Doppler radar: A non-invasive technique for measuring ventilation rates in resting bats. *Journal of Experimental Biology* 150:443–447.

HEARD, S. B., and D. L. HAUSER. 1995. Key evolutionary innovations and their ecological mechanisms. *Historical Biology* 10:151–173.

HEINRICH, B. H., and T. P. MOMMSEN. 1985. Flight of winter moths near 0°C. *Science* 228:177–179.

HELLER, K. G. 1995. Echolocation and body size in insectivorous bats: The case of the giant naked bat *Cheiromeles torquatus* (Molossidae). *Le Rhinolophe* 11:27–38.

HELLER, K. G., and O. VON HELVERSEN. 1989. Resource partitioning of sonar frequency bands in rhinolophoid bats. *Oecologia* 80:178–186.

HENSON, O. W., JR., and H.-U. SCHNITZLER. 1980. Performance of airborne biosonar systems: II. Vertebrates other than Microchiroptera. Pp. 183–195 in *Animal sonar systems,* ed. R. G. Busnel and J. F. Fish. New York: Plenum Press.

HENSON, O. W., JR., A. BISHOP, A. W. KEATING, J. B. KOBLER, M. M. HENSON, B. WILSON, and R. HANSEN. 1987. Biosonar imaging of insects by *Pteronotus p. parnellii*, the mustached bat. *National Geographic Research* 3:82–101.

HERD, R. M., and M. B. FENTON. 1983. An electrophoretic, morphological, and ecological investigation of a putative hybrid zone between *Myotis lucifugus* and *Myotis yumanensis* (Chiroptera: Vespertilionidae). *Canadian Journal of Zoology* 61:2029–2050.

HERMAN, L. M., and TAVOLGA, W. N. 1980. Communication systems of cetaceans. Pp. 149–197 in *Cetacean behavior: Mechanisms and functions,* ed. L. M. Herman. New York: John Wiley & Sons.

HERZING, D. L. 1988. A quantitative description and behavioral associations of a burst-pulsed sound, the squawk, in captive bottlenose dolphins, *Tursiops truncatus.* Master's thesis, San Francisco State University.

————. 1996. Vocalizations and associated underwater behavior of free-ranging Atlantic spotted dolphins, *Stenella frontalis* and bottlenose dolphins, *Tursiops truncatus. Aquatic Mammals* 22:61–79.

————. 1997. The natural history of free-ranging Atlantic spotted dolphins *Stenella frontalis:* Age classes, color phases and female reproduction. *Marine Mammal Science* 13:40–59

HERZING, D. L., and C. J. JOHNSON. 1997. Interspecific interactions between Atlantic spotted dolphins *Stenella frontalis* and bottlenose dolphins *Tursiops truncatus* in the Bahamas, 1985–1995. *Aquatic Mammals* 23:85–89.

HICKEY, M. B. C., and M. B. FENTON. 1990. Foraging by red bats (*Lasiurus borealis*): Do intraspecific chases mean territoriality? *Canadian Journal of Zoology* 68:2477–2482.

————. 1996. Behavioural and thermoregulatory responses of female hoary bats, *Lasiurus cinereus* (Chiroptera: Vespertilionidae), to variations in prey availability. *Euroscience* 3:414–422.

HICKEY, M. B. C., L. ACHARYA, and S. PENNINGTON. 1996. Resource partitioning by two species of vespertilionid bats (*Lasiurus cinereus* and *Lasiurus borealis*) feeding around streetlights. *Journal of Mammalogy* 77:325–334.

HOESE, H. D. 1971. Dolphin feeding out of water in a salt marsh. *Journal of Mammalogy* 52:222–223.

HORLINGTON, M. 1968. A method for measuring acoustic startle response latency and magnitude in rats: Detection of a single stimulus effect using latency measurements. *Physiology and Behavior* 3:839–844.

HORN, A. G., M. L. LEONARD, and D. M. WEARY. 1995. Oxygen consumption during crowing by roosters: Talk is cheap. *Animal Behaviour* 50:1171–1175.

HOY, R. R. 1992. The evolution of hearing in insects as an adaptation to predation from bats. Pp. 115–129 in

The evolutionary biology of hearing, ed. D. B. Webster, R. R. Fay, and A. N. Popper. New York: Springer-Verlag.

HOY, R. R., and D. ROBERT. 1996. Tympanal hearing in insects. *Annual Review of Entomology* 41:433–450.

HOY, R. R., T. NOLEN, and P. BRODFUEHRER. 1989. The neuroethology of acoustic startle and escape in flying insects. *Journal of Experimental Biology* 146:287–306.

HULT, R. W. 1982. Another function of echolocation for bottlenosed dolphins, *Tursiops truncatus. Cetology* 47:1–7.

HUTCHEON, J. M., J. A. W. KIRSCH, and J. D. PETTIGREW. 1998. Base-compositional biases and the bat problem. III. The question of microchiropteran monophyly. *Philosophical Transactions of the Royal Society London, Series B,* 353:607–617.

ILLIUS, A. W., and C. FITZGIBBON. 1994. Costs of vigilance in foraging ungulates. *Animal Behavior* 47:481–484.

JANIK, V. M., and P. J. B. SLATER. 1998. Context-specific use suggests that bottlenose dolphin signature whistles are cohesion calls. *Animal Behaviour* 56:829–838.

JANSEN, J., and J. K. S. JANSEN. 1969. The nervous system of cetacea. Pp. 238–252 in *The biology of marine mammals,* ed. H. T. Andersen. New York: Academic Press.

JANZEN, D. 1980. When is it coevolution? *Evolution* 34:611–612.

JENSEN, M. E., and L. A. MILLER. 1999. Echolocation signals of the bat *Eptesicus serotinus* recorded using a vertical microphone array: Effect of flight altitude on search signals. *Behavioral Ecology and Sociobiology* 47:60–69.

JOHNSON, C. M., and K. S. NORRIS. 1986. Delphinid social organization and social behavior. Pp. 335–345 in *Dolphin cognition and behavior: A comparative approach,* ed. R. J. Schusterman, J. A. Thomas, and F. G. Wood. Hillsdale, N.J.: Lawrence Erlbaum Associates.

———. 1994. Social behavior. Pp. 243–286 in *The Hawaiian spinner dolphin,* ed. K. S. Norris, B. Würsig, R. S. Wells, and M. Würsig. Berkeley and Los Angeles: University of California Press.

JONES, G. 1990. Prey selection by the greater horseshoe bat (*Rhinolophus ferrumequinum*): Optimal foraging by echolocation? *Journal of Animal Ecology* 59:587–602.

———. 1992. Bats vs. moths: Studies on the diets of rhinolophid and hipposiderid bats support the allotonic frequency hypothesis. Pp. 87–92 in *Prague studies in mammalogy,* ed. I. Horáček and V. Vohralik. Prague: Charles University Press.

———. 1994. Scaling of wing-beat and echolocation pulse emission rates in bats: Why are aerial insectivorous bats so small? *Functional Ecology* 8:450–457.

———. 1995a. Variation in bat echolocation: implications for resource partitioning and communication. *Le Rhinolophe* 11:53–59.

———. 1995b. Flight performance, echolocation and foraging behaviour in noctule bats *Nyctalus noctula. Journal of Zoology* 237:303–312.

———. 1997. Acoustic signals and speciation: The roles of natural and sexual selection in the evolution of cryptic species. *Advances in the Study of Behavior* 26:317–354.

———. 1999. Scaling of echolocation call parameters in bats. *Journal of Experimental Biology* 202:3359–3367.

JONES, G., and E. M. BARRATT. 1999. *Vespertilio pipistrellus* Schreber, 1774 and *V. pygmaeus* Leach, 1825 (currently *Pipistrellus pipistrellus* and *P. pygmaeus;* Mammalia, Chiroptera): Proposed designation of neotypes. *Bulletin of Zoological Nomenclature* 56:182–186.

JONES, G., and T. KOKUREWICZ. 1994. Sex and age variation in echolocation calls and flight morphology of Daubenton's bats, *Myotis daubentonii. Mammalia* 58:41–50.

JONES, G., and R. D. RANSOME. 1993. Echolocation calls of bats are influenced by maternal effects and change over a lifetime. *Proceedings of the Royal Society of London B* 252:125–128.

JONES, G., and J. M. V. RAYNER. 1989. Flight performance, foraging tactics and echolocation in free-living Daubenton's bats (*Myotis daubentoni*) (Chiroptera: Vespertilionidae). *Journal of Zoology, London* 215:113–132.

———. 1991. Flight performance, foraging tactics and echolocation in the trawling insectivorous bat *Myotis adversus* (Chiroptera: Vespertilionidae). *Journal of Zoology, London* 225:393–412.

JONES, G., and S. M. VAN PARIJS. 1993. Bimodal echolocation in pipistrelle bats: Are cryptic species present? *Proceedings of the Royal Society of London B* 251:119–125.

JONES, G., and D. A. WATERS. 2000. Moth hearing in response to bat echolocation calls manipulated independently in time and frequency. *Proceedings of the Royal Society of London B* 267:1627–1632.

JONES, G., T. GORDON, and J. NIGHTINGALE. 1992. Sex and age differences in the echolocation calls of the lesser horseshoe bat, *Rhinolophus hipposideros.* *Mammalia* 56:189–193.

JONES, G., M. MORTON, P. M. HUGHES, and R. M. BUD-DEN. 1993. Echolocation, flight morphology and foraging strategies of some West African hipposiderid bats. *Journal of Zoology, London* 230:385–400.

JONES, G., SRIPATHI, K., WATERS, D. A., and G. MARI-MUTHU. 1994. Individual variation in the echolocation calls of three sympatric Indian hipposiderid bats, and an experimental attempt to jam bat echolocation. *Folia Zoologica* 43:347–362.

JURISEVIC, M. A., K. J. SANDERSON, and R. V. BAUDI-NETTE. 1999. Metabolic rates associated with distress and begging calls in birds. *Physiological and Biochemical Zoology* 72:38–44.

KALKO, E. K. V. 1994. Coupling of sound emission and wingbeat in naturally foraging European pipistrelle bats (Microchiroptera: Vespertilionidae). *Folia Zoologica* 43:363–376.

———. 1995a. Insect pursuit, prey capture and echolocation in pipistrelle bats (Microchiroptera). *Animal Behaviour* 50:861–880.

———. 1995b. Echolocation signal design, foraging habitats, and guild structure in six Neotropical sheath-tailed bats (Emballonuridae). *Symposia of the Zoological Society of London* 67:259–273.

KALKO, E. K. V., and M. CONDON. 1998. Echolocation, olfaction, and fruit display: How bats find fruit of flagellichorous cucurbits. *Functional Ecology* 12:364–372.

KALKO, E. K. V., and H.-U. SCHNITZLER. 1989. The echolocation and hunting behavior of Daubenton's bat, *Myotis daubentoni.* *Behavioral Ecology and Sociobiology* 24:225–238.

———. 1993. Plasticity of echolocation signals of European pipistrelle bats in search flight: Implications for habitat use and prey detection. *Behavioral Ecology and Sociobiology* 33:415–428.

———. 1998. How echolocating bats approach and acquire food. Pp. 197–204 in *Bat biology and conservation,* ed. T. H. Kunz and P. A. Racey. Washington, D.C.: Smithsonian Institution Press.

KALKO, E. K. V., H.-U. SCHNITZLER, I. KAIPF, and A. D. GRINNELL. 1998. Echolocation and foraging behavior of the lesser bulldog bat, *Noctilio albiventris:* Preadaptations for piscivory? *Behavioral Ecology and Sociobiology* 42:305–319.

KALMRING, K., and N. ELSNER. 1985. *Acoustic and vibrational communication in insects.* Berlin: Paul Parey.

KALMRING, K., H.-G. REHBEIN, and R. KÜHNE. 1979. An auditory giant neuron in the ventral cord of *Decticus verrucivorus* (Tettigoniidae). *Journal of Comparative Physiology* 132:225–234.

KAMMINGA, C., M. T. VAN HOVE, F. J. ENGELSMA, and R. P. TERRY. 1993. Investigations on cetacean sonar X: A comparative analysis of underwater echolocation clicks of *Inia spp.* and *Sotalia spp.* *Aquatic Mammals* 19(1): 31–43.

KARLSSON, T. 1997. Behaviors and some examples of pulse sounds in agonistic interactions in *Tursiops truncatus* at Kolmården Animal and Nature Park, Sweden. B.Sc. thesis, University of Linköping, Sweden.

KASTELEIN, R. A., N. M. SCHOONEMAN, W. W. L. AU, W. C. VERBOOM, and N. VAUGHAN. 1997. The ability of a harbour porpoise (*Phocoena phocoena*) to discriminate between objects buried in sand. Pp. 329–342 in *The biology of the harbour porpoise,* ed. A. J. Read, P. R. Wiepkema, and P. E. Nachtigall. Woerden, the Netherlands: De Spil Publishers.

KELLOGG, W. N. 1958. Echo-ranging in the porpoise. *Science* 128:982–988.

———. 1962. Sonar system of the blind. *Science* 137:399–404.

KETTEN, D. R. 1994. Functional analyses of whale ears: Adaptations for underwater hearing. *IEEE Proceedings in Underwater Acoustics* 1:264–270.

———. 1997. Structure and function in whale ears. *Bioacoustics* 8:103–135.

KEUPER, A., S. WEIDEMANN, K. KALMRING, and D. KA-MINSKI. 1988. Sound production and sound emission in seven species of European tettigoniids. Part I. The different parameters of the song; their relation to the morphology of the bushcricket. *Bioacoustics* 1:31–48.

KICK, S. A. 1982. Target detection by the echolocating bat, *Eptesicus fuscus.* *Journal of Comparative Physiology* 145:431–435.

KICK, S. A., and J. A. SIMMONS. 1984. Automatic gain-control in the bats sonar receiver and the neuroethology of echolocation. *Journal of Neuroscience* 4:2725–2737.

KINGSTON, T., G. JONES, A. ZUBAID, and T. H. KUNZ. 2000. Resource partitioning in rhinolophoid bats revisited. *Oecologia* 124:332–342.

KINGSTON, T., M. C. LARA, G. JONES, A. ZUBAID, T. H. KUNZ, and C. J. SCHNEIDER. 2001. Acoustic divergence in two cryptic *Hipposideros* species: A role for social selection? *Proceedings of the Royal Society, London* 268B:1825–1832.

KIRSCH, J. A. W., T. F. FLANNERY, M. S. SPRINGER, and F.-J. LAPOITE. 1995. Phylogeny of the Pteropodidae (Mammalia: Chiroptera) based on DNA hybridization, with evidence for bat monophyly. *Australian Journal of Zoology* 43:395–428.

KOBER, R., and H.-U. SCHNITZLER. 1990. Information in sonar echoes of fluttering insects available for echolocating bats. *Journal of the Acoustical Society of America* 87:882–896.

KOBLER, J. B., B. S. WILSON, O. W. HENSON JR., and A. L. BISHOP. 1985. Echo intensity compensation by echolocating bats. *Hearing Research* 20:99–108.

KOLCHIN, A., and V. M. BEL'KOVICH. 1973. Tactile sensitivity in *Delphinus delphis. Zoologichesky Zhurnal* 52:620–622.

KOVALEVA, I. M. 1989. Comparative morphology of ventral muscles in bats. Pp. 19–24 in *European bat research 1987,* ed. V. Hanák, I. Hoáček, and J. Gaisler. Prague: Charles University Press.

KRAUS, S. D., A. J. READ, A. SOLOW, K. BALDWIN, T. SPRADLIN, E. ANDERSON, and J. WILLIAMSON. 1997. Acoustic alarms reduce porpoise mortality. *Nature* 388:525.

KUENZI, A. J., and M. L. MORRISON. 1998. Detection of bats by mist-nets and ultrasonic sensors. *Wildlife Society Bulletin* 26:307–311.

KÜHNE, R., S. SILVER, and B. LEWIS. 1985. Processing of vibratory signals in the central nervous system of the cricket. Pp. 183–192 in *Acoustic and vibrational communication in insects,* ed. K. Kalmring and N. Elsner. Berlin: Paul Parey.

KUNZ, T. H. 1988. Methods of assessing the availability of prey to insectivorous bats. Pp. 191–210 in *Ecological and behavioral methods for the study of bats,* ed. T. H. Kunz and J. O. Whitaker. Washington D.C.

LAMMERS, M. O., W. W. L. AU, and R. AUBAUER. 1997. Broadband characteristics of spinner dolphin (*Stenella longirostris*) social sounds. *Journal of the Acoustical Society of America* 102:3122(A).

LAMMERS, M. O., W. W. L. AU, D. HERZING and J. OSWALD. 1999. Bandwidth characteristics of the social acoustic signals of three species of free-ranging delphinids. Abstracts of the 13th Biennial Conference on the Biology of Marine Mammals in Wailea, Maui. December, 1999.

LANCASTER, W. C. 1994. Morphological and physiological correlates of biosonar vocalization in bats. Ph.D. dissertation, University of North Carolina, Chapel Hill.

LANCASTER, W. C., and O. W HENSON. 1995. Morphology of the abdominal wall in the bat *Pteronotus par-* *nellii* (Microchiroptera, Mormoopidae): Implications for biosonar vocalization. *Journal of Morphology* 223:99–107.

LANCASTER, W. C., O. W. HENSON JR., and A. W. KEATING. 1995. Respiratory muscle activity in relation to vocalization in flying bats. *Journal of Experimental Biology* 198:175–191.

LANCASTER, W. C., A. W. KEATING and O. W. HENSON JR. 1992. Ultrasonic vocalizations of flying bats monitored by radiotelemetry. *Journal of Experimental Biology* 173:43–58.

LAVAL, R. K. 1973. A revision of the Neotropical bats of the genus *Myotis. Los Angeles County Natural History Museum Science Bulletin* 15:1–54.

LAWRENCE, B. D., and J. A. SIMMONS. 1982. Measurements of atmospheric attenuation at ultrasonic frequencies and the significance for echolocation by bats. *Journal of the Acoustical Society of America* 71:585–590.

LAYNE, J. N., and D. K. CALDWELL. 1964. Behavior of the Amazon dolphin, *Inia geoffrensis* (Blainville), in captivity. *Zoologica* 49:81–108.

LEHNER, P. N. 1996. *Handbook of ethological methods.* 2nd ed. Cambridge: Cambridge University Press.

LENDE, R. A., and B. WELKER. 1972. An unusual sensory area in the cerebral neocortex of the bottlenose dolphin (*Tursiops truncatus*). *Brain Research* 45:555–560.

LEONARD, M. L., and M. B. FENTON. 1984. Echolocation calls of *Euderma maculatum* (Vespertilionidae): Use in orientation and communication. *Journal of Mammalogy* 65:122–126.

LEVENSON, C. 1974. Source level and bistatic target strength of the sperm whale (*Physeter catodon*) measured from an oceanographic aircraft. *Journal of the Acoustical Society of America* 55:1100–1103.

LEWIS, F. P., J. H. FULLARD, and S. B. MORRILL. 1993. Auditory influences on the flight behaviour of moths in a Nearctic site. II. Flight times, heights and erraticism. *Canadian Journal of Zoology* 71:1562–1568.

LIBERSAT, F., and R. R. HOY. 1991. Ultrasonic startle behavior in bushcrickets (Orthoptera; Tettigonidae). *Journal of Comparative Physiology A* 169:507–514.

LINDEMAN, R. H., P. F. MERENDA, and R. Z. GOLD. 1980. *Introduction to bivariate and multivariate analysis.* Glenview, Ill.: Scott, Foresman.

LORENZ, K. 1969. *Das sogenannte Böse, Zur Naturgeschichte der Aggression.* Wien, Austria: Dr. G. Borotha-Schoeler Verlag.

LUNDBERG, K., and R. GERELL. 1986. Territorial advertisement and mate attraction in the bat *Pipistrellus pipistrellus*. *Ethology* 71:115–124.

LYNN, S. K., and D. REISS. 1992. Pulse sequence and whistle production by two captive beaked whales, *Mesoplodon* species. *Marine Mammal Science* 8:299–305.

MACKEY, R. L., and R. M. R. BARCLAY. 1989. The influence of physical clutter and noise on the activity of bats over water. *Canadian Journal of Zoology* 67:1167–1170.

MACKIE, G. O. 1984. Fast pathways and escape behavior in Cnidaria. Pp. 15–42 in *Neural mechanisms of startle behavior*, ed. R. C. Eaton. New York: Plenum Press.

MAGNUSSON, W. E., R. C. BEST, and V. M. F. DA SILVA. 1980. Numbers and behaviour of Amazonian dolphins, *Inia geoffrensis* and *Sotalia fluviatilis fluviatilis*, in the Rio Solimões, Brasil. *Aquatic Mammals* 8(1): 27–32.

MANN, D. A., Z. LU, and A. N. POPPER. 1997. A clupeid fish can detect ultrasound. *Nature* 389:341.

MANN, D. A., Z. LU, M. C. HASTINGS, and A. N. POPPER. 1998. Detection of ultrasonic tones and simulated dolphin echolocation clicks by a teleost fish, the American shad (*Alosa sapidissima*). *Journal of the Acoustical Society of America* 104:562–568.

MARDEN, J. 1995. Evolutionary adaptation of contractile performance in muscle of ectothermic winter-flying moths. *Journal of Experimental Biology* 198:2087–2094.

MARTEN, K., D. HERZING, M. POOLE, and K. NEWMAN-ALLMAN. 2001. The acoustic predation hypothesis: Linking underwater observations and recordings during odontocete predation and observing the effects of loud impulsive sounds on fish. *Aquatic Mammals* 27(1): 56–66.

MARTEN, K., K. S. NORRIS, P. W. B. MOORE, and K. A. ENGLUND. 1988. Loud impulse sounds in odontocete predation and social behavior. Pp. 567–579 in *Animal sonar: Processes and performance*, ed. P. E. Nachtigall and P. W. B. Moore. New York: Plenum Press.

MASTERS, W. M., and K. A. S. RAVER. 1996. The degradation of distance discrimination in big brown bats (*Eptesicus fuscus*) caused by different interference signals. *Journal of Comparative Physiology A* 179: 703–713.

MAY, M. L., P. D. BRODFUEHRER, and R. R. HOY. 1988. Kinematic and aerodynamic aspects of ultrasound-induced negative phonotaxis in flying Australian field crickets (*Teleogryllus oceanicus*). *Journal of Comparative Physiology A* 164:243–249.

MAYER, F., and O. VON HELVERSEN. 2001a. Cryptic diversity in European bats. *Proceedings of the Royal Society of London* 268B:1381–1386.

———. 2001b. Sympatric distribution of two cryptic bat species across Europe. *Biological Journal of the Linnean Society* 74:365–374.

MAYR, E. 1977. *Populations, species and evolution*. 6th ed. Cambridge: Harvard University Press.

MCCARTY, J. P. 1996. The energy cost of begging in nestling passerines. *Auk* 113:178–188.

MCKAY, J. M. 1969. The auditory system of *Homorocoryphus* (Tettigonioidea, Orthoptera). *Journal of Experimental Biology* 51:787–802.

———. 1970. Central control of an insect sensory interneurone. *Journal of Experimental Biology* 53:137–145.

MCKAY, R. S., and J. PEGG. 1988. Debilitation of prey by intense sounds. *Marine Mammal Science* 4:356–359.

MCKENZIE, N. L., and J. K. ROLFE. 1986. Structure of bat guilds in the Kimberley mangroves, Australia. *Journal of Animal Ecology* 55:401–420.

MCNALLY, R., and D. YOUNG, 1981. Song energetics of the bladder cicada *Cystosoma saundersii*. *Journal of Experimental Biology* 90:185–197.

MECH, L. D. 1970. *The wolf: The ecology and behavior of an endangered species*. Minneapolis: University of Minnesota Press.

MICHELSEN, A., K.-G. HELLER, A. STUMPNER, and K. ROHRSEITZ. 1994. A new biophysical method to determine the gain of the acoustic trachea in bushcrickets. *Journal of Comparative Physiology A* 175: 145–151.

MILLER, L. A. 1975. The behaviour of flying green lacewings, *Chrysopa carnea*, in the presence of ultrasound. *Journal of Insect Physiology* 21:205–219.

———. 1991. Arctiid moth clicks can degrade the accuracy of range difference discrimination in echolocating big brown bats, *Eptesicus fuscus*. *Journal of Comparative Physiology A* 168:571–579.

MILLER, L. A., and H. J. DEGN. 1981. The acoustic behaviour of four species of vespertilionid bats studied in the field. *Journal of Comparative Physiology* 142: 67–74.

MILLER, L. A., and A. SURLYKKE. 2001. How some insects detect and avoid being eaten by bats: The tactics and counter tactics of prey and predator. *BioScience* 51:570–581.

MILLS, D. J., T. W. NORTON, H. E. PARNABY, R. B. CUN-NINGHAM, and H. A. NIX. 1996. Designing surveys for microchiropteran bats in complex forest landscapes: A pilot study from south-east Australia. *Forest Ecology and Management* 85:149–161.

MIYAMOTO, M. M. 1996. A congruence study of molecular and morphological data for eutherian mammals. *Molecular Phylogenetics and Evolution* 6:373–390.

MOGENSEN, F., and B. MØHL. 1979. Sound radiation patterns in the frequency domain of cries from a vespertilionid bat. *Journal of Comparative Physiology* 134:165.

MØHL, B. 1988. Target detection by echolocating bats. Pp. 435–450 in *Animal sonar: Processes and performance,* ed. P. E. Nachtigall and P. W. B. Moore. New York: Plenum Press.

MØHL, B., and L. A. MILLER. 1976. Ultrasonic clicks produced by the peacock butterfly: A possible bat-repellent mechanism. *Journal of Experimental Biology* 64:639–644.

MØHL, B., A. SURLYKKE, and L. MILLER. 1990. High intensity narwhal clicks. Pp. 295–303 in *Sensory abilities of cetaceans: Laboratory and field evidence,* ed. J. Thomas and R. Kastlelein. New York: Plenum Press.

MØHL, B., M. WAHLBERG, P. T. MADSEN, L. A. MILLER, and A. SURLYKKE. 2000. Sperm whale clicks: Directionality and source level revisited. *Journal of the Acoustical Society of America* 107:638–648.

MOISEFF, A., and R. R. HOY. 1983. Sensitivity to ultrasound in an identified auditory interneuron in the cricket: A possible neural link to phonotactic behavior. *Journal of Comparative Physiology* 152:155–167.

MOISEFF, A., G. S. POLLACK, and R. R. HOY. 1978. Steering responses of flying crickets to sound and ultrasound: Mate attraction and predator avoidance. *Proceedings of the National Academy of Sciences USA* 75:4052–4056.

MOORE, K. E., W. A. WATKINS, and P. TYACK. 1993. Pattern similarity in shared codas from sperm whales (*Physeter catodon*). *Marine Mammal Science* 9:1–9.

MOORE, S. E., and S. H. RIDGWAY. 1995. Whistles produced by common dolphins from the Southern California Bight. *Aquatic Mammals* 21:55–63.

MORENO, P. 1996. Estudo preliminar do comportamento de *Inia geoffrensis* (Blainville, 1817) em cativeiro e caracterização de sinais acústicos emitidos em contexto sexual. M.Sc. thesis, Instituto Superior de Psicologia Aplicada, Lisbon, Portugal.

MOROZOV, V. P., A. I. AKOPIAN, V. I. BURDIN, K. A. ZAITSEVA, and Y. A. SOKOVYKH. 1972. Tracking frequency of the location signals of dolphins as a function of distance to the target. *Biofizika* 17:139–145.

MORRILL, S. B., and J. H. FULLARD. 1992. Auditory influences on the flight behaviour of moths in a Nearctic site. I. Flight tendency. *Canadian Journal of Zoology* 70:1097–1101.

MORRIS, G. K., A. C. MASON, P. WALL, and J. J. BELWOOD. 1994. High ultrasonic and tremulation signals in neotropical katydids (Orthoptera: Tettigoniidae). *Journal of Zoology* 233:129–163.

MORRIS, R. J. 1986. The acoustic faculty of dolphins. Pp. 369–400 in *Research on dolphins,* ed. M. M. Bryden and R. Harrison. Oxford: Clarendon Press.

Moss, C. F. 1988. Ontogeny of vocal signals in the big brown bat, *Eptesicus fuscus.* Pp. 115–120 in *Animal sonar systems,* ed. R. G. Busnel and J. F. Fish. New York: Plenum Press.

Moss, C. F., and H.-U. SCHNITZLER. 1995. Behavioral studies of auditory information processing. Pp. 87–145 in *Hearing by bats,* ed. A. N. Popper and R. R. Fay. New York: Springer-Verlag.

Moss, C. F., and M. ZAGAESKI. 1994. Acoustic information available to bats using frequency-modulated echolocation sounds for the perception of insect prey. *Journal of the Acoustical Society of America* 95:2745–2756.

Moss, C. F., D. REDISH, C. GOUNDEN, and T. H. KUNZ. 1997. Ontogeny of vocal signals in the little brown bat, *Myotis lucifugus. Animal Behavior* 54:131–141.

MULLINS, J., H. WHITEHEAD, and L. S. WEILGART. 1988. Behavior and vocalizations of two single sperm whales, *Physeter macrocephalus,* off Nova Scotia. *Canadian Journal of Zoology* 67:839–846.

MURRAY, S. O. 1997. The graded structure and neural network classification of false killer whale (*Pseudorca crassidens*) vocalizations. Thesis, University of Hawaii.

MURRAY, S. O., E. MERCADO, and H. L. ROITBLAT. 1998. Characterizing the graded structure of false killer whale (*Pseudorca crassidens*) vocalizations. *Journal of the Acoustical Society of America* 104:1679–1688.

NACHTIGALL, P. E. 1980. Odontocete echolocation performance on object size, shape and material. Pp. 71–95 in *Animal sonar systems,* ed. R. G. Busnel and J. F. Fish. New York: Plenum Press.

———. 1986. Vision, audition, and chemoreception in dolphins and other marine mammals. Pp. 79–114 in *Dolphin cognition and behavior: A comparative approach,* ed. R. J. Schusterman, J. A. Thomas, and F. G. Wood. Hillsdale, N.J.: Lawrence Erlbaum Associates.

NAKASAI, K., and A. TAKEMURA. 1975. Studies on underwater sound VI: On the underwater calls of freshwater dolphins in South America. *Nagasaki University Bulletin of the Faculty of Fisheries* 40:7–13.

NELSON, D. L., and J. LIEN. 1994. Behaviour patterns of two captive Atlantic white-sided dolphins, *Lagenorhynchus acutus. Aquatic Mammals* 20(1): 1–10.

NEMETH, A., and O. VON HELVERSEN. 1994. The phylogeny of the *Myotis mystacinus* group: A molecular approach. *Bat Research News* 35:37.

NEUWEILER, G. 1989. Foraging ecology and audition in echolocation bats. *Trends in Ecology and Evolution* 6:160–166.

———. 1990. Auditory adaptations for prey capture in echolocating bats. *Physiological Review* 70:615–641.

NEUWEILER, G., W. METZNER, U. HEILMAN, R. RÜBSAMEN, M. ECKRICH, and H. H. COSTA. 1987. Foraging behaviour and echolocation in the rufous horseshoe bat *(Rhinolophus rouxi). Behavioral Ecology and Sociobiology* 20:653–673.

NOLEN, T. G., and R. R. HOY. 1984. Initiation of behavior by single neurons: The role of behavioral context. *Science* 226:992–994.

———. 1986. Phonotaxis in flying crickets. II. Physiological mechanisms of two-tone suppression of the high frequency avoidance steering behavior by the calling song. *Journal of Comparative Physiology A* 159:441–456.

———. 1987. Postsynaptic inhibition mediates high-frequency selectivity in the cricket *Teleogryllus oceanicus:* Implications for flight phonotaxis behavior. *Journal of Neuroscience* 7:2081–2096.

NORBERG, U. M. 1989. Ecological determinants of bat wing shape and echolocation call structure with implications for some fossil bats. Pp. 197–211 in *European bat research 1987,* ed. V. Hanák, I. Horáček, and J. Gaisler. Prague: Charles University Press.

———. 1994. Wing design, flight performance and habitat use in bats. Pp. 205–239 in *Ecological morphology: Integrative organismal biology,* ed. P. C. Wainwright and S. M. Reilly. Chicago: University of Chicago Press.

NORBERG, U. M., and J. M. V. RAYNER. 1987. Ecological morphology and flight in bats (Mammalia: Chiroptera): Wing adaptations, flight performance, foraging strategy and echolocation. *Philosophical Transactions of the Royal Society London, Series B,* 316:335–427.

NORMAN, A. P., and G. JONES. 2000. Size, peripheral auditory tuning and target strength in noctuid moths. *Physiological Entomology* 25:346–353.

NORRIS, K. S., and T. P. DØHL. 1980. Behavior of the Hawaiian spinner dolphin, *Stenella longirostris. Fishery Bulletin* 77(4): 821–849.

NORRIS, K. S., and G. W. HARVEY. 1972. A theory for the function of the spermaceti organ of the sperm whale *(Physeter catodon* L.). Pp. 397–417 in *Animal orientation and navigation,* ed. S. R. Galler. National Aeronautical and Space Administration SP-262.

NORRIS, K. S., and B. MØHL. 1983. Can odontocetes stun prey with sound? *American Naturalist* 122:85–104.

NORRIS, K. S., G. W. HARVEY, L. A. BURZELL, and T. D. KRISHNA KARTHA. 1972. Sound production in the freshwater porpoises *Sotalia fluviatilis* Gervais and Deville and *Inia geoffrensis* Blainville, in the Rio Negro, Brazil. *Investigations on Cetacea* 4:251–261.

NORRIS, K. S., B. WÜRSIG, R. S. WELLS, and M. WÜRSIG, eds. 1994. *The Hawaiian spinner dolphin.* Berkeley and Los Angeles: University of California Press.

OAKELEY, S. F., and G. JONES. 1998. Habitat around maternity roosts of the 55 kHz phonic type of pipistrelle bats *(Pipistrellus pipistrellus). Journal of Zoology, London* 245:222–228.

OBRIST, M. K. 1995. Flexible bat echolocation: The influence of individual, habitat and conspecifics on sonar signal design. *Behavioral Ecology and Sociobiology* 36:207–219.

OBRIST, M. K., and J. J. WENSTRUP. 1998. Hearing and hunting in red bats *(Lasiurus borealis,* Vespertilionidae): Audiogram and ear properties. *Journal of Experimental Biology* 201:143–154.

OBRIST, M. K., M. B. FENTON, J. L. EGER, and P. SCHLEGEL. 1993. What ears do for bats: A comparative study of sound pressure transformation in Chiroptera. *Journal of Experimental Biology* 180:119–152.

O'FARRELL, M. J., and B. W. MILLER. 1999. Use of vocal signatures for the inventory of free-flying Neotropical bats. *Biotropica* 31:507–516.

OSTMAN, J. 1991. Changes in aggressive and sexual behavior between two male bottlenose dolphins *(Tursiops truncatus)* in a captive colony. Pp. 305–318 in *Dolphin societies: Discoveries and puzzles,* ed. K. Pryor and K. S. Norris. Berkeley and Los Angeles: University of California Press.

OVERSTROM, N. A. 1983. Association between burst-pulse sounds and aggressive behavior in captive Atlantic bottlenose dolphins *Tursiops truncatus. Zoological Biology* 2:93–103.

PACK, A. A., and L. M. HERMAN. 1995. Sensory integration in the bottlenose dolphin: Immediate recognition of complex shapes across the sense of echolocation and vision. *Journal of the Acoustical Society of America* 98:722–733.

PALMER, E., and G. WEDDEL. 1964. The relationship between structure, innervation and function of the skin of the bottlenose dolphin (*Tursiops truncatus*). *Proceedings of the Zoological Society of London* 143: 553–568.

PARK, K. J., J. D. ALTRINGHAM, and G. JONES. 1996. Assortative roosting in the two phonic types of *Pipistrellus pipistrellus* during the mating season. *Proceedings of the Royal Society of London* 263B:1495–1499.

PARNAS, I., and D. DAGAN. 1971. Functional organizations of giant axons in the central nervous systems of insects: New aspects. Pp. 95–144 in *Advances in insect physiology,* ed. J. W. L. Beament, J. E. Treherne, and V. B. Wigglesworth. London: Academic Press.

PARSONS, S., and G. JONES. 2000. Acoustic identification of twelve species of echolocating bat by discriminant function analysis and artificial neural networks. *Journal of Experimental Biology* 203:2641–2656.

PARSONS, S., C. W. THORPE, and S. M. DAWSON. 1997. The echolocation calls of the long-tailed bat (*Chalinolobus tuberculatus*): A quantitative description and analysis of call phase. *Journal of Mammalogy* 79:964–976.

PAVEY, C. R., and C. J. BURWELL. 1998. Bat predation on eared moths: A test of the allotonic frequency hypothesis. *Oikos* 81:143–151.

PAYNE, R. 1995. *Among whales.* New York: Scribner.

PAYNE, R., and D. WEBB. 1971. Orientation by means of long range acoustic signaling in baleen whales. *Annals of the New York Academy of Sciences* 188:110–141.

PAYNE, R. S., K. D. ROEDER, and J. WALLMAN. 1966. Directional sensitivity of the ears of noctuid moths. *Journal of Experimental Biology* 44:17–31.

PEARSON, K. G., and M. O'SHEA. 1984. Escape behavior of the locust: The jump and its initiation by visual stimuli. Pp. 163–178 in *Neural mechanisms of startle behavior,* ed. R. C. Eaton. New York: Plenum Press.

PENNER, R. H. 1988. Attention and detection in dolphin echolocation. Pp. 707–713 in *Animal sonar: Processes and performance,* ed. P. E. Nachtigall and P. W. B. Moore. New York: Plenum Press.

PERRIN, W. F., and J. G. MEAD. 1994. Clymene dolphin *Stenella clymene* (Gray, 1846). Pp. 161–171 in *Handbook of marine mammals,* vol. 5, ed. S. H. Ridgeway and R. Harrison. London: Academic Press.

PETTIGREW, J. D. 1986. Flying primates? Megabats have the advanced pathway from eye to midbrain. *Science* 231:1304–1306.

———. 1991. Wings or brain? Convergent evolution in the origins of bats. *Systematic Zoology* 40:199–216.

———. 1995. Flying Primates: Crashed, or crashed through. Pp. 3–36 in *Ecology, evolution and behavior of bats,* ed. P. A. Racey and S. M. Swift. Symposium of the Zoological Society of London.

PILLERI, G. 1976. Ethology, bioacoustics and behaviour of *Platanista indi* in captivity. *Investigations on cetacea* 6:13–69.

PILLERI, G., K. ZBINDEN, and C. KRAUS. 1979. The sonar field of *Inia geoffrensis. Investigations on cetacea* 10: 157–176.

POPPER, A. N. 1980. Sound emission and detection by delphinids. Pp. 1–52 in *Cetacean behavior: Mechanisms and functions,* ed. L. M. Herman. New York: John Wiley & Sons.

POPPER, A. N., H. L. HAWKINS, and R. C. GISINER. 1997. Questions in cetacean bioacoustics: Some suggestions for future research. *Bioacoustics* 8:163–182.

POULTER, T. C. 1969. Sonar of penguins and fur seals. *Proceedings of the California Academy of Sciences* 36:363–380.

POULTER, T. C., and R. A. JENNINGS. 1969. Sonar discrimination ability of the California sea lion, *Zalophus californianus. Proceedings of the California Academy of Sciences* 36:381–389.

PRESTWICH, K. N., and C. K. BREUER. 1987. The design of advertisement calls when energy is a limiting factor. *American Zoologist* 43A: Abstract 206.

PURVES, P. E. 1967. Anatomical and experimental observations on the cetacean sonar system. Pp. 197–270 in *Animal sonar systems,* ed. R. G. Busnel and J. F. Fish. New York: Plenum Press.

PYE, J. D. 1968. How insects hear. *Nature* 218:797.

———. 1972. Bimodal distribution of constant frequencies in some hipposiderid bats (Mammalia: Hipposideridae). *Journal of Zoology, London* 166:323–335.

———. 1993. Is fidelity futile? The "true" signal is illusory, especially with ultrasound. *Bioacoustics* 4:271–286.

RAYNER, J. M. V. 1991. Complexity and a coupled system: Flight, echolocation and evolution in bats. Pp. 173–190 in *Constructional morphology,* ed. N. Schmidt-Kittler and K. Vogel. Berlin: Springer-Verlag.

REISS, D. 1988. Observations on the development of echolocation in young bottlenose dolphins. Pp. 121–127 in *Animal sonar systems,* ed. R. G. Busnel and J. F. Fish. New York: Plenum Press.

RENJUN, L., R. J. HARRISON, and K. W. THURLEY. 1986. Characteristics of the skin *Neophocoena phocaenoides* from the Changjiang (Yangtze River), China. Pp. 23–31 in *Research on dolphins*, ed. M. M. Bryden and R. Harrison. Oxford: Clarendon Press.

RENJUN, L., W. GEWALT, B. NEUROHR, and A. WINKLER. 1994. Comparative studies on the behavior of *Inia geoffrensis* and *Lipotes vexillifer* in artificial environments. *Aquatic Mammals* 20(1): 39–45.

RENOUF, D., G. GALWAY, and L. GABORKO. 1980. Evidence for echolocation in harbour seals. *Journal of the Marine Biology Association of the U.K.* 60:1039–1042.

RHEINLAENDER, J., and H. RÖMER. 1980. Bilateral coding of sound direction in the CNS of the bushcricket *Tettigonia viridissima* L. (Orthoptera, Tettigoniidae). *Journal of Comparative Physiology* 140:101–111.

RHEINLAENDER, J., M. HARDT, and D. ROBINSON. 1986. The directional sensitivity of a bush cricket ear: A behavioural and neurophysiological study of *Leptophyes punctatissima*. *Physiological Entomology* 11:309–316.

RICE, C. E. 1967. Human echo perception. *Science* 155:656–664.

RICE, C. E., S. H. FELNSTEIN, and R. J. SCHUSTERMAN. 1965. Echo detection ability of the blind: Size and distance factors. *Journal of Experimental Psychology* 70:246–251.

RICHARDSON, W. J., C. R. GREENE, C. I. MALME, and D. H. THOMSON. 1995. *Marine mammals and noise.* San Diego: Academic Press.

RIDGWAY, S. H., and D. A. S. CARDER. 1990. Tactile sensitivity, somatosensory responses, skin vibrations, and the skin surface ridges of the bottlenose dolphin (*Tursiops truncatus*). Pp. 163–179 in *Sensory abilities of cetaceans: Laboratory and field evidence*, ed. J. Thomas and R. Kastlelein. New York: Plenum Press.

RIGLEY, L. 1983. Dolphins feeding in a South Carolina salt marsh. *Whalewatcher* 17:3–5.

ROBERT, D., J. AMOROSO, and R. R. HOY. 1992. The evolutionary convergence of hearing in a parasitoid fly and its cricket host. *Science* 258:1135–1137.

ROEDER, K. D. 1962. The behavior of free flying moths in the presence of artificial ultrasonic pulses. *Animal Behaviour* 10:300–304.

———. 1965. Moths and ultrasound. *Scientific American* 212:94–102.

———. 1967. *Nerve cells and insect behavior.* Cambridge: Harvard University Press.

———. 1974. Acoustic sensory responses and possible bat-evasion tactics of certain moths. Pp. 71–78 in *Proceedings of the Canadian Society of Zoologists Annual Meeting, June 2–5*, ed. M. D. B. Burt. University of New Brunswick, Fredericton.

ROEDER, K. D., and A. E. TREAT. 1957. Ultrasonic reception by the tympanic organs of noctuid moths. *Journal of Experimental Zoology* 134:127–158.

———. 1961. The detection and evasion of bats by moths. *American Scientist* 49:135–148.

ROEDER, K. D., A. E. TREAT, and J. S. VANDE BERG. 1968. Auditory sense in certain sphingid moths. *Science* 159:331–333.

ROITBLAT, H. L., W. W. L. AU, P. E. NACHTIGALL, R. SHIZUMURA, and G. MOONS. 1995. Sonar recognition of targets embedded in sediment. *Neural Networks* 8:1263–1273.

RÖMER, H., and M. KRUSCH. 2000. A gain-control mechanism for processing of chorus sounds in the afferent auditory pathway of the bushcricket *Tettigonia viridissima* (Orthoptera; Tettigoniidae). *Journal of Comparative Physiology A* 186:181–191.

RÖMER, H., V. MARQUART, and M. HARDT. 1988. Organization of a sensory neuropile in the auditory pathway of two groups of Orthoptera. *Journal of Comparative Neurology* 275:201–215.

ROOT, R. B. 1967. The niche exploitation pattern of the blue-gray gnatcatcher. *Ecological Monographs* 37:317–350.

ROSS, A. 1967. Ecological aspects of the food habits of insectivorous bats. *Proceedings of the Western Foundation of Vertebrate Zoology* 1:205–263.

ROSS, A. J., and E. A. JARZEMBOWSKI. 1993. Arthropoda (Hexapoda: Insecta). Pp. 363–426 in *The fossil record 2*, ed. M. J. Benton. London: Chapman and Hall.

ROSSBACH, K. A., and D. L. HERZING. 1997. Underwater observations of benthic-feeding bottlenose dolphins (*Tursiops truncatus*) near Grand Bahama Island, Bahamas. *Marine Mammal Science* 13:498–504.

ROTHSCHILD, M. 1965. The stridulation of Arctiid moths. Proceedings of the Royal Entomological Society of London (Series C, Journal of Meetings) 30:3.

ROVERUD, R. C., V. NITSCHE, and G. NEUWEILER. 1991. Discrimination of wingbeat motion by bats correlated with echolocation sound pattern. *Journal of Comparative Physiology A* 156:447–456.

RÜBSAMEN, R., and M. SCHÄFER. 1990. Audiovocal interactions during development? Vocalisation in deafened young horseshoe bats vs. audition in vocalisation impaired bats. *Journal of Comparative Physiology A* 167:771–784.

RUEDI, M., and F. MAYER. 2001. Molecular systematics of bats of the genus *Myotis* (Vespertilionidae) suggests deterministic ecomorphological convergences. *Molecular Phylogenetics and Evolution* 21:436–448.

RYAN, M. J. 1988. Energy calling and selection. *American Zoologist* 28:885–898.

RYDELL, J. 1990. Behavioral variation in echolocation pulses of the northern bat, *Eptesicus nilssoni*. *Ethology* 85:103–113.

———. 1991. Seasonal use of illuminated areas by foraging northern bats *Eptesicus nilssonii*. *Holarctic Ecology* 14:203–207.

———. 1992. The exploitation of insects around streetlamps by bats in Sweden. *Functional Ecology* 6:744–750.

———. 1993. Variation in the sonar of an aerial hawking bat (*Eptesicus nilssoni*). *Ethology* 93:275–284.

———. 1998. Bat defence in lekking ghost swifts (*Hepialus humuli*), a moth without ultrasonic hearing. *Proceedings of the Royal Society of London B* 265:1373–1376.

RYDELL, J., and R. ARLETTAZ. 1994. Low-frequency echolocation enables the bat *Tadarida teniotis* to feed on tympanate insects. *Proceedings of the Royal Society of London B* 257:175–178.

RYDELL, J., and W. C. LANCASTER. 2000. Flight and thermoregulation in moths have been shaped by predation from bats. *Oikos* 88:13–18.

RYDELL, J., and D. W. YALDEN. 1997. The diets of two high-flying bats from Africa. *Journal of Zoology, London* 242:69–76.

RYDELL, J., A. ENTWISTLE, and P. A. RACEY. 1996. Timing of foraging flights of three species of bats in relation to insect activity and predation risk. *Oikos* 76:243–252.

RYDELL, J., G. JONES, and D. A. WATERS. 1995. Echolocating bats and hearing moths: Who are the winners? *Oikos* 73:419–424.

RYDELL, J., L. A. MILLER and M. E. JENSEN. 1999. Echolocation constraints of Daubenton's bat foraging over water. *Functional Ecology* 13:247–255.

RYDELL, J., N. SKALS, A. SURLYKKE, and M. SVENSSON. 1997. Hearing and bat defence in geometrid winter moths. *Proceedings of the Royal Society of London B* 264:83–88.

SALES, G., and D. PYE. 1974. *Ultrasonic communication by animals.* London: Chapman and Hall.

SANDERFORD, M. V., and W. E. CONNER. 1995. Acoustic courtship communication in *Syntomeida epilais* Wlk. (Lepidoptera: Arctiidae, Ctenuchinae). *Journal of Insect Behavior* 8:19–31.

SCHENKEL, R. 1967. Submission: Its features and function in the wolf and dog. *American Zoology* 7:319–329.

SCHEVILL, W. E., and W. A. WATKINS. 1962. Whale and porpoise voices. A phonograph record. Woods Hole Oceanographic Institution.

———. 1966. Sound structure and directionality in *Orcinus* (killer whale). *Zoologica* 51(6): 71–76.

SCHMIDT, S. 1988a. Discrimination of target surface structure in the echolocating bat, *Megaderma lyra*. Pp. 507–512 in *Animal sonar: Processes and performance,* ed. P. E. Nachtigall and P. W. B. Moore. New York: Plenum Press.

———. 1988b. Evidence for a spectral basis of texture perception in bat sonar. *Nature* 331:617–619.

SCHNITZLER, H.-U. 1971. Fledermäuse im Windkanal. *Zeitschrift für vergleichende Physiologie* 73:209–221.

———. 1987. Echoes of fluttering insects: Information for echolocating bats. Pp. 226–243 in *Recent advances in the study of bats,* ed. M. B. Fenton, P. A. Racey, and J. M. V. Rayner. Cambridge: Cambridge University Press.

SCHNITZLER, H.-U., and E. K. V. KALKO. 1998. How echolocating bats search and find food. Pp. 183–196 in *Bat biology and conservation,* ed. T. H. Kunz and P. A. Racey. Washington, D.C.: Smithsonian Institution Press.

———. 2001. Echolocation by insect-eating bats. *BioScience* 51:557–569.

SCHNITZLER, H.-U., A. DENZINGER, and E. K. V. KALKO. 1995. Foraging and echolocation behavior of the frog-eating bat, *Trachops cirrhosus,* when catching frogs and insects. *Bat Research News* 36:107.

SCHNITZLER, H.-U., H. HACKBARTH, and U. HEILMAN. 1985. Echolocation behavior of rufous horseshoe bats hunting for insects in flycatcher-style. *Journal of Comparative Physiology A* 157:39–46.

SCHNITZLER, H.-U., E. K. V. KALKO, I. KAIPF, and A. D. GRINNELL. 1994. Fishing and echolocation behavior in the greater bulldog bat, *Noctilio leporinus. Behavioral Ecology and Sociobiology* 35:327–345.

SCHNITZLER, H.-U., E. KALKO, L. MILLER, and A. SURLYKKE. 1987. The echolocation and hunting behavior of the bat, *Pipistrellus kuhli. Journal of Comparative Physiology A* 161:267–274.

Schnitzler, H.-U., D. Menne, R. Kober, and K. Heblich. 1983. The acoustical image of fluttering insects in echolocating bats. Pp. 235–250 in *Neurophysiology and behavioral physiology,* ed. F. Huber and H. Markl. Berlin: Springer-Verlag.

Schotten, M. 1998. Echolocation recordings and localization of free-ranging spinner dolphins (*Stenella longirostris*) and pantropical spotted dolphins (*Stenella attenuata*) using a four hydrophone array. Master's thesis, University of Groningen, Groningen, The Netherlands.

Schul, J. 1997. Neuronal basis of phonotactic behaviour in *Tettigonia viridissima:* Processing of behaviourally relevant signals by auditory afferents and thoracic interneurons. *Journal of Comparative Physiology A* 180:573–583.

Schuller, G. S., and G. Pollak. 1979. Disproportionate frequency representation in the inferior colliculus of Doppler-compensating greater horseshoe bats: Evidence of an acoustic fovea. *Journal of Comparative Physiology* 132:47–54.

Schumm, A., D. Krull, and G. Neuweiler. 1991. Echolocation in the notch-eared bat, *Myotis emarginatus. Behavioral Ecology and Sociobiology* 28:255–261.

Schusterman, R. J. 1981. Behavioral capabilities of seals and sea lions: A review of their hearing, visual, learning and diving skills. *Physiological Record* 31:125 143.

Scronce, B. L., and C. S. Johnson 1976. Bistatic target detection by a bottle-nosed porpoise. *Journal of the Acoustical Society of America* 59:1001–1002.

Sedlock, J. L. 2001. Inventory of insectivorous bats on Mount Makiling, Philippines, using echolocation call signatures and a new tunnel trap. *Acta Chiropterologica* 3:163–178.

Selverston, A. I., H.-U. Kleindienst, and F. Huber. 1985. Synaptic connectivity between cricket auditory interneurons as studied by selective photoinactivation. *Journal of Neuroscience* 5:1283–1292.

Servatius, A. 1997. Das Jagd- und Echoortungsverhalten von d'Orbignys Rundohrenfledermaus *Tonatia silvicola* im Flugraum. Diploma thesis, Faculty of Biology, University Tübingen.

Shane, S. H. 1990. Behavior and ecology of the bottlenose dolphin at Sanibel Island, Florida. Pp. 245–266 in *The bottlenose dolphin,* ed. S. Leatherwood and R. R. Reeves. San Diego: Academic Press.

Shen, J.-X. 1993. Morphology and physiology of auditory interneurons in the bushcricket *Gampsocleis gratiosa. Japanese Journal of Physiology* 43:S239–S246.

Shennan, G. C., J. R. Waas, and R. J. Lavery. 1994. The warning signals of parental convict cichlids are socially facilitated. *Animal Behavior* 47:974–976.

Siemers, B. M., and H.-U. Schnitzler. 2000. Natterer's bat (*Myotis nattereri* Kuhl, 1818) hawks for prey close to vegetation using echolocation signals of very broad bandwidth. *Behavioral Ecology and Sociobiology* 47:400–412.

Sigurdson, J. E. 1998. Echolocation search and detection of bottom objects by bottlenose dolphins. Biological Sonar Conference, Carvoeiro, Portugal, May 27–June 2, 1998 (p. 40).

Simmons, J. A. 1987. Acoustic images of target range in the sonar of bats. *Naval Research News* 39:11–26.

Simmons, J. A., and R. A. Stein. 1980. Acoustic imaging in bat sonar: Echolocation signals and the evolution of echolocation. *Journal of Comparative Physiology* 135:61–84.

Simmons, J. A., and J. A. Vernon. 1971. Echolocation: Discrimination of targets by the bat, *Eptesicus fuscus. Journal of Experimental Biology* 176:315–328.

Simmons, J. A., C. F. Moss, and M. Ferragamo. 1990 Convergence of temporal and spectral information into acoustic images of complex sonar targets perceived by the echolocating bat, *Eptesicus fuscus. Journal of Comparative Physiology A* 166:449–470.

Simmons, J. A., S. A. Kick, B. D. Lawrence, C. Hale, C. Bard, and B. Escudie. 1983. Activity of horizontal angle discrimination by the echolocating bat, *Eptesicus fuscus. Journal of Comparative Physiology* 153:321–330.

Simmons, J. A., M. J. Ferragamo, P. A. Saillant, T. Haresign, J. M. Wotton, S. P. Dear, and D. N. Lee. 1995. Auditory dimensions of acoustic images in echolocation in *Hearing by bats,* ed. A. N. Popper and R. R. Fay. New York: Springer-Verlag.

Simmons, J. A., W. A. Lavender, B. A. Lavender, J. E. Childs, K. Hulebak, M. R. Rigden, J. Sherman, B. Woolmand, and M. J. O'Farrell. 1978. Echolocation by free-tailed bats (*Tadarida*). *Journal of Comparative Physiology* 125:291–299.

Simmons, N. B. 1994. The case for chiropteran monophyly. *American Museum Novitates* 3077:1–37.

Simmons, N. B., and J. H. Geisler. 1998. Phylogenetic relationships of *Icaronycteris, Archaeonycteris, Hassianycteris,* and *Palaeochiropteryx* to extant bat lineages, with comments on the evolution of echolocation and foraging strategies in Microchiroptera. *Bulletin of the American Museum of Natural History* 235:1–182.

Simpson, J. G., and M. B. Gardner. 1972. Comparative microscopic anatomy of selected marine mammals.

Pp. 298–418 in *Mammals of the sea: Biology and medicine,* ed. S. H. Ridgway. Springfield, Ill.: Charles C. Thomas.

SJARE, B. L., and T. G. SMITH. 1986a. The relationship between behavioral activity and underwater vocalizations of the white whale, *Delphinapterus leucas. Canadian Journal of Zoology* 64:2824–2831.

———. 1986b. The vocal repertoire of white whales, *Delphinapterus leucas,* summering in Cunningham Inlet, Northwest Territories. *Canadian Journal of Zoology* 64:407–415.

SMITH, J. D. 1976. Chiropteran evolution. Pp. 49–69 in *Biology in bats of the New World family Phyllostomatidae,* part 1, ed. R. J. Baker., J. K. Jones, Jr., and D. C. Carter. Special Publications of the Museum, Texas Tech University, No. 10. Lubbock: Texas Tech Press.

SMITH, W. J. 1986. Signaling behavior: Contributions of different repertoires. Pp. 315–330 in *Dolphin cognition and behavior: A comparative approach,* ed. R. J. Schusterman, J. A. Thomas, and F. G. Wood. Hillsdale, N.J.: Lawrence Erlbaum Associates.

SOKAL, R. R., and F. J. ROHLF. 1969. *Biometry.* San Francisco: W. H. Freeman and Company.

SPANGLER, H. G. 1988a. Hearing in tiger beetles (Cicindelidae). *Physiological Entomology* 13:447–452.

———. 1988b. Moth hearing, defence and communication. *Annual Review of Entomology* 33:59–81.

SPEAKMAN, J. R. 1991. Why do insectivorous bats in Britain not fly in daylight more frequently? *Functional Ecology* 5:518–524.

———. 1993. The evolution of echolocation for predation. *Symposia of the Zoological Society of London* 65:39–63.

———. 1995. Chiropteran nocturnality. *Symposia of the Zoological Society of London* 67:187–201.

———. 1999. The evolution of flight and echolocation in pre-bats: An evaluation of the energetics of reach hunting. *Acta Chiropterologica* 1:3–15.

———. 2001. The evolution of flight and echolocation in bats: Another leap in the dark. *Mammal Review* (in press).

SPEAKMAN, J. R., and P. A. RACEY. 1991. No cost of echolocation for bats in flight. *Nature* 350:421–423.

SPEAKMAN, J. R., M. E. ANDERSON, and P. A. RACEY. 1989. The energy cost of echolocation in pipistrelle bats (*Pipistrellus pipistrellus*). *Journal of Comparative Physiology A* 165:679–685.

SPITZENBERGER, F., J. PIÁLEK, and E. HARING. 2001. Systematics of the genus *Plecotus* (Mammalia, Vespertilionidae) in Austria based on morphometric and molecular investigations. *Folia Zoologica* 50:161–172.

SPOTTE, S. 1967. Intergeneric behavior between captive Amazon river dolphins *Inia* and *Sotalia. Underwater Naturalist* 4(2): 9–13.

SPRINGER, M. S., E. C. TEELING, O. MADSEN, M. J. STANHOPE, and W. W. DE JONG. 2001. Integrated fossil and molecular data reconstruct bat echolocation. *Proceedings of the National Academy of Sciences* 98:6241–6246.

STEBBINGS, R. E., and F. GRIFFITH. 1986. *Conservation and status of bats in Europe.* Institute of Terrestrial Ecology, Huntingdon.

STEPHENS, D. W., and J. R. KREBS. 1986. *Foraging theory.* Princeton: Princeton University Press.

STONEMAN, M. G., and M. B. FENTON. 1988. Disrupting foraging bats: The clicks of arctiid moths. Pp. 635–638 in *Animal sonar: Processes and performance,* ed. P. E. Nachtigall and P. W. B. Moore. New York: Plenum Press.

SUGA, N., and P. JEN. 1976. Peripheral control of acoustic signals in the auditory system of echolocating bats. *Journal of Experimental Biology* 62:277–311.

SUGA, N., and Y. KATSUKI. 1961a. Central mechanism of hearing in insects. *Journal of Experimental Biology* 38:545–558.

———. 1961b. Pharmacological studies on the auditory synapses in a grasshopper. *Journal of Experimental Biology* 38:759–770.

SUGA, N., and T. SHIMOZAWA. 1974. Site of neural attenuation of responses to self-vocalized sounds in echolocating bats. *Science* 183:1211–1213.

SUM, Y. W., and D. MENNE. 1988. Discrimination of fluttering targets by the FM bat *Pipistrellus stenopterus. Journal of Comparative Physiology A* 163:349–354.

SURLYKKE, A., and O. BOJESEN. 1996. Integration time for short broad band clicks in echolocating FM-bats (*Eptesicus fuscus*). *Journal of Comparative Physiology A* 178:235–241.

SURLYKKE, A., and M. FILSKOV. 1997. Hearing in geometrid moths. *Naturwissenschaften* 84:356–359.

SURLYKKE, A., and L. A. MILLER. 1985. The influence of arctiid moth clicks on bat echolocation; jamming or warning? *Journal of Comparative Physiology A* 156: 831–843.

SURLYKKE, A., and A. E. TREAT. 1995. Hearing in wintermoths. *Naturwissenschaften* 82:382–384.

SURLYKKE, A., N. SKALS, J. RYDELL, and M. SVENSSON. 1998. Sonic hearing in a diurnal geometrid moth,

Archiearis parthenias, temporally isolated from bats. *Naturwissenschaften* 85:36–37.

SURLYKKE, A., M. FILSKOV, J. H. FULLARD, and E. FORREST. 1999. Auditory relationships to size in noctuid moths: Bigger is better. *Naturwissenschaften* 86:238–241.

SURLYKKE, A., L. A MILLER, B. MØHL, B. B. ANDERSEN, J. CHRISTENSEN-DALSGAARD, and M. B. JØRGENSEN. 1993. Echolocation in two very small bats from Thailand: *Craseonycteris thonglongyai* and *Myotis siligorensis. Behavioral Ecology and Sociobiology* 33: 1–12.

SUTHERS, R. A., S. P. THOMAS, and B. J. SUTHERS. 1972. Respiration, wing beat and ultrasonic pulse emission in an echolocating bat. *Journal of Experimental Zoology* 56:37–48.

SVENSSON, A. M., and J. RYDELL. 1998. Mercury vapour lamps interfere with the bat defence of tympanate moths (Operophtera spp.; Geometridae). *Animal Behaviour* 55:223–226.

SVENSSON, M. G. E., J. RYDELL, and R. BROWN. 1999. Bat predation and the flight timing of winter moths, *Epirrita* and *Operophtera* (Lepidoptera, Geometridae). *Oikos* 84:193–198.

SYLVESTRE, J.-P. 1985. Some observation on behavior of two Orinoco Dolphins (*Inia geoffrensis)* humboldtiana (Pilleri and Gihr 1977), in captivity at Duisburg Zoo. *Aquatic Mammals* 11(2): 58–65.

SZYMANSKI, M. D., A. Y. SUPIN, D. E. BAIN, and D. R. HENRY. 1998. Killer whale (*Orcinus orca*) auditory evoked potentials to rhythmic clicks. *Marine Mammal Science* 14:676–691.

TARUSKI, A. G. 1979. The whistle repertoire of the North Atlantic pilot whale (*Globicephala melaena*) and its relationship to behavior and environment. Pp. 345–368 in *Behavior of marine animals: Current perspectives on research.* Vol. 3: *Cetaceans,* ed. H. E. Winn and B. L. Olla. New York: Plenum Press.

TAVOLGA, W. N. 1983. Theoretical principles for the study of communication in cetaceans. *Mammalia* 47(1): 3–26.

TAVOLGA, W. N., and F. S. ESSAPIAN. 1957. The behavior of the bottlenose dolphin, *Tursiops truncatus:* Mating, pregnancy and parturition, mother-infant behavior. *Zoologica* 42:11–31.

TEELING, E. C., O. MADSEN, R. A. VAN DEN BUSSCHE, W. W. DE JONG, M. J. STANHOPE, and M. S. SPRINGER. 2002. Microbat paraphyly and the convergent evolution of a key innovation in Old World rhinolophoid microbats. *Proceedings of the National Academy of Sciences* 99:1431–1436.

TEELING, E. C., M. SCALLY, D. J. KAO, M. L. ROMAGNOLI, M. S. SPRINGER, and M. J. STANHOPE. 2000. Molecular evidence regarding the origin of echolocation and flight in bats. *Nature* 403:188–192.

THEWISSEN, J. G. V., and S. K. BABCOCK. 1991. Distinctive cranial and cervical innervation of wing muscles: New evidence for bat monophyly. *Science* 251:934–936.

THIES, W., E. K. V. KALKO, and H.-U. SCHNITZLER. 1998. The roles of echolocation and olfaction in two Neotropical fruit-eating bats, *Carollia perspicillata* and *C. castanea,* feeding on Piper. *Behavioral Ecology and Sociobiology* 42:397–409.

THOMAS, A. L. R., G. JONES, J. M. V. RAYNER, and P. M. HUGHES. 1990. Intermittent gliding flight in the pipistrelle bat (*Pipistrellus pipistrellus*) (Chiroptera: vespertilionidae). *Journal of Experimental Biology* 149: 407–416.

THOMAS, J. A., and C. W. TURL. 1990. Echolocation characteristics and range detection threshold of a false killer whale (*Pseudorca crassidens*). Pp. 321–334 in *Sensory abilities of cetaceans: Laboratory and field evidence,* ed. J. Thomas and R. Kastlelein. New York: Plenum Press.

THOMAS, J. A., L. M. FERM, and V. B. KUECHLE. 1987. Silence as an anti-predation strategy by Weddell seals. *Antarctic Journal of the U.S.* 1987:232–234.

TIAN, B., and H.-U. SCHNITZLER. 1997. The design of echolocation signals of the greater horseshoe bat (*Rhinolophus ferrumequinum*) during transfer flight and landing. *Journal of the Acoustical Society of America* 101:2347–2364.

TOBAYAMA, T., and T. KAMIYA. 1989. Observations on *Inia geoffrensis* and *Platanista gangetica* in captivity at Kamogawa Sea World, Japan. Pp. 42–45 in *Biology and conservation of the river dolphins,* ed. W. F. Perrin, R. L. Brownell Jr., Z. Kaiya, and L. Jiankang. International Union for Conservation of Nature and Natural Resources, Species Survival Commission Occasional Paper No. 3.

TOUGAARD, J., J. H. CASSEDAY, and E. COVEY. 1998. Arctiid moths and bat echolocation: Broad-band clicks interfere with neural responses to auditory stimuli in the nuclei of the lateral lemniscus of the big brown bat. *Journal of Comparative Physiology A* 182:203–215.

TURL, C. W., and R. H. PENNER. 1989. Differences in echolocation click patterns of the beluga (*Delphinapterus leucas*) and the bottlenose dolphin (*Tursiops truncatus*). *Journal of the Acoustical Society of America* 86:497–502.

TURL, C. W., D. J. SKAAR, and W. W. L. AU. 1991. The echolocation ability of the beluga (*Delphinapterus leucas*) to detect targets in clutter. *Journal of the Acoustical Society of America* 89:896–901.

TUTTLE, M., and M. J. RYAN. 1981. Bat predation and the evolution of frog vocalizations in the Neotropics. *Science* 214:677–678.

TYACK, P. L. 1999. Communication and cognition. Pp. 287–323 in *Biology of marine mammals,* ed. J. E. Reynolds III and S. A. Rommel. Washington, D.C.: Smithsonian Institution Press.

VAN ZYLL DE JONG, C. G., and D. W. NAGORSEN. 1994. A review of the distribution and taxonomy of *Myotis keenii* and *Myotis evotis* in British Columbia and the adjacent United States. *Canadian Journal of Zoology* 72:1069–1078.

VAUGHAN, N., G. JONES, and S. HARRIS. 1997a. Identification of British bat species by multivariate analysis of echolocation call parameters. *Bioacoustics* 7:189–207.

———. 1997b. Habitat use by bats (Chiroptera) assessed by means of a broad-band acoustic method. *Journal of Applied Ecology* 34:716–730.

VAUGHAN, T. A. 1977. Foraging behaviour of the giant leaf-nosed bat (*Hipposideros commersoni*). *East African Wildlife Journal* 15:237–249.

VEL'MIN, V. A., A. A. TITOV, and L. I. YURKEVICH. 1975. Time summation of pulses in the bottlenose dolphin. Pp. 78–80 in *Morskiye mlekopitayushciye. Mater. 6-go Vses. soveshch. po izuch. morsk. mlekopitayushchikh,* part 1. Kiev: Naukova Dumka.

VERBOOM, W. C., and R. KASTELEIN. 1997. Structure of harbour porpoise (*Phocoena phocoena*) click train signals. Pp. 343–363 in *The biology of the harbour porpoise,* ed. A. J. Read, P. R. Wiepkema, and P. E. Nachtigall. Woerden, the Netherlands: De Spil Publishers.

VERFUSS, U. K., L. A. MILLER, and H.-U. SCHNITZLER. 1999. The echolocation behavior of the harbour porpoise (*Phocoena phocoena*) during prey capture. Pp. 193 in Abstracts of the 13th Biennal Conference on the Biology of Marine Mammals. Wailea, Hawaii, November 28–December 3.

VERHULST, S., and P. WIERSMA. 1997. Is begging cheap? *Auk* 114:134.

VON DER EMDE, G. 1988. Greater horseshoe bats learn to discriminate simulated echoes of insects fluttering with different wingbeat rates. Pp. 495–499 in *Animal sonar: Processes and performance,* ed. P. E. Nachtigall and P. W. B. Moore. New York: Plenum Press.

VON DER EMDE, G., and D. MENNE. 1989. Discrimination of insect wingbeat frequencies by the bat *Rhinolophus ferrumequinum. Journal of Comparative Physiology A* 164:663–671.

VON DER EMDE, G., and H.-U. SCHNITZLER. 1986. Fluttering target detection in hipposiderid bats. *Journal of Comparative Physiology A* 159:765–772.

———. 1990. Classification of insects by echolocating greater horseshoe bats. *Journal of Comparative Physiology A* 167:423–430

VON FRENCKELL, B., and R. M. R. BARCLAY. 1987. Bat activity over calm and turbulent water. *Canadian Journal of Zoology* 65:219–222.

VON HELVERSEN, D., and O. VON HELVERSEN. 1999. Acoustic guide in bat-pollinated flowers. *Nature* 398:759–760.

VON HELVERSEN, O., K.-G. HELLER, F. MAYER, A. NEMETH, M. VOLLETH, and P. GOMBKÖTÖ. 2001. Cryptic mammalian species: A new species of whiskered bat (*Myotis alcathoe* n. sp.) in Europe. *Naturwissenschaften* 88:217–223.

WALKER, T. J. 1964. Experimental demonstration of a cat locating orthopteran prey by the prey's calling song. *Florida Entomologist* 47:163–165.

WATERS, D. A., and G. JONES. 1995. Echolocation call structure and intensity in five species of insectivorous bats. *Journal of Experimental Biology* 198:475–489.

———. 1996. The peripheral auditory characteristics of noctuid moths: Responses to the search-phase echolocation calls of bats. *Journal of Experimental Biology* 199:847–856.

WATERS, D. A., J. RYDELL, and G. JONES. 1995. Echolocation call design and limits on prey size: A case study using the aerial hawking bat *Nyctalus leisleri. Behavioral Ecology and Sociobiology* 37:321–328.

WATKINS, W. A. 1967. The harmonic interval: Fact or artifact in spectral analysis of pulse trains. Pp. 15–43 in *Marine bioacoustics,* ed. W. N. Tavolga. Oxford: Pergamon Press.

———. 1976. Biological sound-source location by computer analysis of underwater array data. *Deep-Sea Research* 23:175–180.

———. 1977. Acoustic behavior of sperm whales. *Oceanus* 20:50–58.

———. 1980. Acoustics and the behavior of sperm whales. Pp. 283–290 in *Animal sonar systems,* ed. R. G. Busnel and J. F. Fish. New York: Plenum Press.

———. 1980. Click sounds from animals at sea. Pp. 291–297 in *Animal sonar systems,* ed. R. G. Busnel and J. F. Fish. New York: Plenum Press.

———. 1993. Sperm whale tracking under water and at the surface. Abstracts, Tenth Biennial Conference on the Biology of Marine Mammals, 11–15 November 1993, Galveston, Texas. The Society for Marine Mammalogy, p. 111.

WATKINS, W. A., and K. E. MOORE. 1982. An underwater acoustic survey for sperm whales (*Physeter catodon*) and other cetaceans in the southeast Caribbean. *Cetology* 46:1–7.

WATKINS, W. A., and W. E. SCHEVILL. 1972. Sound source location with a three-dimensional hydrophone array. *Deep-Sea Research* 19:691–706.

———. 1974. Listening to Hawaiian spinner porpoises, *Stenella cf. longirostris,* with a three-dimensional hydrophone array. *Journal of Mammalogy* 55:319–328.

———. 1975. Sperm whales (*Physeter catodon*) react to pingers. *Deep-Sea Research* 22:123–129.

———. 1977a. Spatial distribution of *Physeter catodon* (sperm whales) under water. *Deep-Sea Research* 24:693–699.

———. 1977b. Sperm whale codas. *Journal of the Acoustical Society of America* 62:1485–1490 and phonograph record.

WATKINS, W. A., and D. WARTZOK. 1985. Sensory biophysics of marine mammals. *Marine Mammal Science* 1(3): 219–260.

WATKINS, W. A., K. E. MOORE, and P. TYACK. 1985. Investigations of sperm whale acoustic behaviors in the southeast Caribbean. *Cetology* 49:1–15.

WATKINS, W. A., M. A. DAHER, K. M. FRISTRUP, T. J. HOWALD, and G. NOTARBARTOLO DI SCIARA. 1993. Sperm whales tagged with transponders and tracked underwater by sonar. *Marine Mammal Science* 9:55–67.

WATKINS, W. A., M. A. DAHER, N. A. DiMARZIO, A. SAMUELS, D. WARTZOK, K. M. FRISTRUP, D. P. GANNON, P. W. HOWEY, and R. R. MAIEFSKI. 2002. Sperm whale dives traced by radio and tag telemetry. *Marine Mammal Science* 18:55–68.

WATKINS, W. A., M. A. DAHER, N. A. DiMarzio, A. SAMUELS, D. WARTZOK, K. M. FRISTRUP, D. P. GANNON, P. W. HOWEY, R. R. MAIEFSKI, and T. R. SPRADLIN. 1999. Sperm whale surface activity from tracking by radio and satellite tags. *Marine Mammal Science* 15:1158–1180.

WEATHERS, W. W., P. J. HODUM, and J. J. ANDERSON. 1997. Is the energy cost of begging by nestling passerines surprisingly low? *Auk* 114:133.

WEILGART, L. S., and H. WHITEHEAD. 1990. Vocalizations of the north Atlantic pilot whale (*Globicephala melas*) as related to behavioral contexts. *Behavioral Ecology and Sociobiology* 26:399–402.

WELLS, R. S., D. J. BONESS, and G. B. RATHBUN. 1999. Behavior. Pp. 324–422 in *Biology of marine mammals,* ed. J. E. Reynolds III and S. A. Rommel. Washington, D.C.: Smithsonian Institution Press.

WERNER, T. K. 1981. Responses of non-flying moths to ultrasound: The threat of gleaning bats. *Canadian Journal of Zoology* 59:525–529.

WESTIN, J., J. J. LANGBERG, and J. M. CAMHI. 1977. Responses of giant interneurons of the cockroach *Periplaneta americana* to wind puffs of different directions and velocities. *Journal of Comparative Physiology* 121:307–324.

WHALLEY, P. 1986. A review of current fossil evidence of Lepidoptera in the Mesozoic. *Biological Journal of the Linnean Society* 28:253–271.

WHITAKER, J. O., JR., and H. L. BLACK. 1976. Food habits of cave bats from Zambia. *Journal of Mammalogy* 57:199–204.

WHITE, F. B. 1877. Scientific correspondence. *Nature* 15:293.

WHITEHEAD, H., and L. WEILGART. 1991. Patterns of visually observable behavior and vocalizations in groups of female sperm whales. *Behaviour* 118:275–296.

WIEGREBE, L., and S. SCHMIDT. 1996. Temporal integration in the echolocating bat *Megaderma lyra. Hearing Research* 102:35–42.

WIERSMA, C. A. G. 1947. Giant nerve fiber system of the crayfish: A contribution to comparative physiology of synapse. *Journal of Neurophysiology* 10:23–38.

WILKINSON, G. S. 1995. Information transfer in bats. *Symposia of the Zoological Society of London* 67:345–360.

WINE, J. J., and F. B. KRASNE. 1972. The organization of escape behaviour in the crayfish. *Journal of Experimental Biology* 56:1–18.

WOOD, F. G., and W. E. EVANS. 1980. Adaptiveness and ecology of echolocation in toothed whales. Pp. 381–425 in *Animal sonar systems,* ed. R. G. Busnel and J. F. Fish. New York: Plenum Press.

WORTHINGTON, L. V., and W. E. SCHEVILL. 1957. Underwater sounds heard from sperm whales. *Nature* 180:291.

WÜRSIG, B. 1986. Delphinid foraging strategies. Pp. 347–360 in *Dolphin cognition and behavior: A comparative approach,* ed. R. J. Schusterman, J. A. Thomas, and F. G. Wood. Hillsdale, N.J.: Lawrence Erlbaum Associates.

WYTTENBACH, R. A., M. L. MAY, and R. R. HOY. 1996. Categorical perception of sound frequency by crickets. *Science* 273:1542–1544.

XITCO, M. J., and H. L. ROITBLAT. 1996. Object recognition through eavesdropping: Passive echolocation in bottlenose dolphins. *Animal Learning and Behavior* 24:355–365.

YACK, J. E. 1988. Seasonal partitioning of atympanate moths in relation to bat activity. *Canadian Journal of Zoology* 66:753–755.

YACK, J. E., and J. H. FULLARD. 2000. Ultrasonic hearing in nocturnal butterflies. *Nature* 403:265–266.

YAGER, D. D. 1990. Sexual dimorphism of auditory function and structure in praying mantises (Mantodea: Dictyoptera). *Journal of Zoology, London* 221:517–537.

———. 1999. Structure, development, and evolution of insect auditory systems. *Microscopy Research and Technique* 47:380–400.

YAGER, D. D., and H. G. SPANGLER. 1995. Characterization of auditory afferents in the tiger beetle, *Cicindela marutha* Dow. *Journal of Comparative Physiology A* 176:587–599.

———. 1997. Behavioral response to ultrasound by the tiger beetle *Cicindela marutha* Dow combines aerodynamic changes and sound production. *Journal of Experimental Biology* 200:649–659.

YAGER, D. D., M. L. MAY, and M. B. FENTON. 1990. Ultrasound-triggered, flight-gated evasive maneuvers in the praying mantis *Parasphendale agrionina.* I. Free flight. *Journal of Experimental Biology* 152:17–39.

ZAGAESKI, M. 1987. Some observations on the prey stunning hypothesis. *Marine Mammal Science* 3:275–279.

ZBINDEN, K. 1989. Field observations on the flexibility of the acoustic behaviour of the European bat *Nyctalus noctula* (Schreber, 1774). *Revue Suisse de Zoologie* 96:335–343.

ZBINDEN, K., and P. E. ZINGG. 1986. Search and hunting signals of echolocating free-tailed bats, *Tadarida teniotis* in southern Switzerland. *Mammalia* 50:9–25.

ZHANTIEV, R. D., and O. S. KORSUNOVSKAJA. 1983. Structure and functions of two auditory neurons in the bush cricket, *Tettigonia cantans* Fuess. (Orthoptera, Tettigoniidae). *Entomologicheskoe Obozrenie* 62:462–469.

ZINGG, P. E. 1990. Akustische Artidentifikation von Fledermäusen (Mammalia: Chiroptera) in der Schweiz. *Revue Suisse de Zoologie* 97:263–294.

ZOTTOLI, S. J. 1977. Correlation of the startle reflex and Mauthner cell auditory responses in unrestrained goldfish. *Journal of Experimental Biology* 66:243–254.

PART FIVE

Echolocation Theory, Analysis Techniques, and Applications

A Biologically Plausible Framework
for Auditory Perception in FM Bats

Mathew J. Palakal and *Donald Wong*

Introduction

Bats that echolocate perceive targets as an auditory image by extracting cues from the target-reflected echoes of their emitted pulses. To optimize perception, bats actively regulate the temporal pattern and sound structure of the emitted pulses during target-directed flight (Popper and Fay 1995). The phenomenon of echolocation broadly consists of two processes: the physical process and the biological process. The *physical process* involves the transformation of the emitted pulse to a target-reflected echo at the bat's ear (Rachel and Palakal 1999). Several factors such as target characteristics (size, texture, scattering properties, etc.), the characteristics of the medium (density, humidity, velocity, viscous behavior [Rachel 1992], etc.), and the ambient acoustic field influence this process. The *biological process* involves the transformation of the target-reflected echo received at the bat's ears to the perception of the target. During this process, the acoustic signal passes through various stages in the auditory system from the ears to the brain. Deciphering the codes by which the bat processes these target-reflected sounds is essential for building auditory-system models. Neurophysiological and behavioral studies have explored strategies for echo processing with varying degree of success.

A central question in target perception is how bats derive target features from temporally spaced echoes. For accurate target ranging, bats depend critically on the orderly time-frequency structure in the broadband FM signal (Surlykke 1992). In discrimination of targets, FM bats derive temporal cues conveyed in the complex echo of real targets by the different ranges that separate multiple target surfaces (glints) (Beuter 1980; Simmons and Chen 1989). Behavioral experiments suggest that FM bats recognize real targets by the spectral notches and peaks created in the complex echo (Bradbury 1970; Habersetzer and Vogler 1983).

Echolocating bats are an excellent animal model for neurophysiological studies that examine complex-sound processing of identified stimulus features at higher centers of the auditory pathway (see Popper and Fay 1995). A major focus in understanding target perception is to characterize the functional organization of the bat's au-

ditory cortex (see O'Neill 1995). Specifically, single-unit recordings have probed the tuning of cortical neurons to different parameters of the biosonar signal and mapped the topographic organization of functional subregions specialized for processing target features. For example, in the FM bat, *Myotis lucifugus,* cortical neurons are sensitive to echo delay (Sullivan 1982; Wong and Shannon 1988); thus, it is postulated that delay-sensitive neurons in this species code for both target distance and relative velocity. Moreover, neurons in the *Myotis* cortex are sensitive to the amplitude and frequency modulations of echoes (Condon et al. 1997). Since such echo modulations mimic fluttering targets (e.g., flying insects), this neural sensitivity provides a basis for coding insect wing beat, and thus target-prey selection (see Schnitzler 1987).

The focus of this work was to develop a biologically plausible framework of the FM bat auditory system. In this framework, functional elements of the biological counterparts are incorporated into the basic framework of the auditory pathway. The proposed framework takes a modular approach to modeling the auditory system to identify the mechanisms underlying auditory perception. It emphasizes abstraction over detailed anatomical and physiological structure. This framework consists of functional units inspired by what is presently known about neurobiological elements of the bat auditory system. Analysis of this model will help identify lines of inquiry in auditory neurobiological research.

The Auditory-System Framework

The proposed auditory-system framework is largely based on key findings in FM bats. Currently, the framework consists of two stages: the auditory periphery and the central auditory system. The *auditory periphery* implements the processing in the ear and the auditory nerve. The *central auditory system* simulates a subset of the cortical neurons responsible for higher level processing.

THE AUDITORY PERIPHERY MODEL

The auditory periphery model (APM) consists of signal generation, preprocessing in the ear, and postprocessing in the auditory nerve. The signal generation module generates pulses and echoes that mimic FM bat echolocation calls. In the preprocessing module, the spike encoding stage implements processing in the basilar membrane, inner hair cells, and outer hair cells. Spike enhancement and evaluation are carried out at the postprocessing stage. Brief details about each of these processing stages are presented below. For a complete description and implementation details of APM, see Chittajallu et al. 1996 and Kohrt 1996.

Signal generation

This stage of the model generates an artificial biosonar pulse and its simulated echo. Distinguishing between the pulse and the echo is handled at the auditory cortical-level; therefore, the pulse and echo are kept separate and processed in parallel by the auditory periphery model.

Vocalizations of an FM bat can be generalized for modeling purposes as belonging to one of two classes: linear period modulated (LPM) or linear frequency modulated (LFM) (Simmons and Stein 1980). An LPM pulse changes period linearly over time, giving it a hyperbolic trace in a time-frequency domain, while an LFM pulse changes frequencies linearly over time, resulting in a linear trace in a time-frequency domain.

The LPM and LFM pulses of prescribed bandwidth and duration, as well as the CF tones, are generated based on the following equation:

$$s_i = \sin\left(\frac{2\pi}{f_s}\sum_{j=1}^{i}\left[f_0^\alpha + \left(\frac{f_1^\alpha - f_0^\alpha}{t_d}\right)\left(\frac{j-1}{f_s}\right)\right]^\alpha\right),$$
$$i = 1, 2, 3, \ldots, t_d f_s \tag{61.1}$$

where s_i is the ith sample of the desired signal vector; f_s is the sampling frequency; f_0, f_1 are the starting and ending frequencies of the chirp, respectively; t_d is the duration of the chirp; and α is a constant to determine time-frequency characteristics of the chirp. The derivation of this equation can be found in Kohrt 1996. The signal generated in eq. 61.1 sweeps from the initial frequency to the terminal frequency over the specified duration, t_d. The bandwidth of the sweep is determined by the difference between f_0 and f_1. A CF chirp is generated by assigning the same value for f_0 and f_1. The parameter α defines the time-frequency characteristics of the FM chirp but has no affect on a CF chirp. An LFM signal is produced when α is $+1$, and an LPM signal is produced when α is -1.

Preprocessing: Modeling the inner and outer hair cells

Encoding of information from each frequency channel into spike trains is the most critical stage of the auditory periphery model. This hair-cell model is based on a single reservoir model presented by Meddis (1986) and Meddis, Hewitt, and Shackleton (1990). This model is based on neurotransmitter synthesis, release, and recycling from a single hair cell.

Spikes are generated by a hair cell's afferent neurons in a random fashion, but with a probability of firing that is proportional to the amount of neurotransmitter released into the synaptic cleft by the stimulated hair cell. There is one hair-cell model for each filter channel, but each filter channel represents a sizable region of the basilar membrane. As a result, the spike train generated by a particular filter channel must reflect the activity of more than a single IHC. This increase in probability of a spike occurring in a filter channel is accomplished by lowering the CC level (as measured in percentage firing probability for one hair cell) at which 100% firing probability occurs. In essence, instead of requiring a CC level of 100 to ensure a 100% probability of firing, a CC level of 30 would be sufficient to ensure a 100% probability of firing.

The outer hair cell acts as a fine-tuning mechanism for the auditory periphery. By changing their length, OHCs either attenuate or enhance the stimulus effect on the inner hair cells. The command to change OHC length presumably comes from the central auditory system. An algorithm for implementing this feedback process is adopted from Giguere and Woodland (1994a, 1994b). The point of application of the feedback loop is a gain between the output from the filters and the input to the IHC, representing the fluid coupling between the vibrations of the basilar membrane and the stimulus on the stereo cilia of the IHCs. Details about the implementation of this algorithm are found in Kohrt 1996.

Spike encoding

The spike encoding process begins with the extraction of frequency information by the basilar membrane. The frequency-selective response of the basilar membrane to pressure fluctuations in the cochlear channels has been modeled as a bank of sharply tuned bandpass filters (Pont and Damper 1991; Giguere and Woodland 1994a). These filters are described in terms of center frequency and pass bandwidth.

The algorithm for spacing the center frequencies of the filterbank is similar to that used to determine the frequencies in an LPM sweep. The center period of the ith filter is given by

$$\tau_{c_i} = \tau_1 + \frac{i-1}{N-1}(\tau_2 - \tau_1) \tag{61.2}$$

where τ_1 and τ_2 are the center periods corresponding to first and last filters, respectively, and N is the total number of filters in the bank.

The center frequency of the ith filter of a linearly spaced filter bank is given by

$$f_{c_i} = f_1 + \frac{i-1}{N-1}(f_2 - f_1) \tag{61.3}$$

where f_1 and f_2 are the center frequencies of the first and last filters, respectively.

The bandwidth of individual filters are assigned to prescribe a constant Q value, where Q is the ratio of center frequency to filter bandwidth. Filters can also be assigned the same bandwidth throughout the filter bank, resulting in a "constant bandwidth" filter bank. Filter bandwidth is associated with the tuning curves of the auditory nerve.

Postprocessing and evaluation

In the postprocessing stage, the spike-encoded signals are enhanced by an initial processing module to eliminate noise. Spike enhancement is used to identify significant auditory events within a spike train that also contain activity due to internal (spontaneous firing) or external (background) noise. It is based on functional characteristics common to most neurons (Marmarelis 1989); details of the implementation can be found in Kohrt 1996.

Modeling the Auditory Cortex

Modeling of the cortical-level processing of information is based on three neurophysiological findings related to the functional organization of the *Myotis* auditory cortex. First, delay-sensitive neurons constitute a large majority of neurons in the auditory cortex. These cortical neurons exhibit facilitative responses to artificial pulse-echo stimulus pairs presented at particular echo and are temporally tuned to specific best delays (BDs) that evoke maximal facilitation (Wong and Shannon 1988). Second, neurons in the primary auditory cortex are tonotopically organized according to their best frequency (BF), ranging from 20 kHz to 100 kHz (Paschal and Wong 1994). Third, the delay-tuning properties of these cortical neurons are dependent on the pulse repetition rate (PRR) (Wong, Maekawa, and Tanaka 1992) and pulse duration (PD) (Tanaka, Wong, and Taniguchi 1992). Modeling at the cortical-level involves two stages: (1) response properties of delay-sensitive cortical neurons are modeled (single-neuron model); and (2) a population is created from the single-neuron model to represent the sound-evoked, cortical responses, termed *cortical response map* (CORMAP)—a multidimensional representation of time-frequency-delay-response. The CORMAP framework uses a spike-train pattern as an input arising from the peripheral auditory system described in the previous section.

The CORMAP model is based on the following assumptions: (1) targets are stationary (hence, no Doppler shift in the echoes); (2) a pulse stimulus evokes an early inhibitory response followed by a later excitatory response in delay-sensitive cortical neurons (Berkowitz and Suga 1989; Olsen and Suga 1991); (3) the pulse-evoked inhibition in cortical delay-sensitive neurons varies in duration among individual delay-sensitive neurons (Berkowitz and Suga 1989; Olsen and Suga 1991); (4) cortical delay-sensitive neurons are both temporally tuned to particular BDs and topographically organized according to their best frequency (BF) in a functional region of auditory cortex (for different bat species, see O'Neill 1995); and (5) every signal pair consists of a pulse followed by an echo with one or more glints.

Model of a delay-sensitive neuron

The delay-tuning properties of a neuron can be determined from its delay-tuning curves. A neuron's delay-tuning curve is a response property defined by the probability of the neural facilitative response for different values of echo delay. Based on these delay-tuning properties, a delay-sensitive neuron can be classified as *delay tuned* or *tracking*. Delay-tuned neurons have BDs that remain constant as PRR changes, whereas the BDs of tracking neurons become shorter as PRR increases. The delay-tuning curve of a neuron exhibits its magnitude of facilitative responses at different echo delays. Another delay-tuning characteristic of a neuron is its *delay width*, the range of echo delays at which a neuron exhibits facilitative responses. A neuron can be tuned sharply or broadly in delay (Wong, Maekawa, and Tanaka 1992). Since PRR and PD shape the delay-tuning characteristics of neurons, changes in the PRR and PD of the pulse signal would modify the delay-tuning characteristics of individual neurons.

The output function of a neuron is estimated as

$$y(\delta) = \tau \cdot e^{-i \cdot (\beta - \delta)^2 / 2 \cdot \sigma^2} \qquad (61.4)$$

where β is the best delay (BD) of the neuron; δ is the echo delay; i is the frequency channel (spike-train patterns of 0s and 1s arrive at each channel i as input to the neuron); σ is the delay width; and τ is the inhibition factor measured in time. The inhibition factor allows the model to distinguish between a pulse and an echo. This term is given by

$$\tau = \begin{cases} 1 & \text{if } \gamma = \sum_{i=0}^{PD} i + \Delta \geq PD \\ 0 & \text{otherwise} \end{cases} \qquad (61.5)$$

That is, if $\tau = 0$, then the signal is an outgoing pulse; otherwise it is an echo. In this model, the pulse duration PD is fixed at 1 ms.

Fig. 61.1 shows the input/output characteristics of the delay-sensitive neuron model. Its inputs are the pulse- and echo-evoked spike-train pattern. The neuron responds if the echo arrives at facilitative echo delays within the delay width, σ. A maximal response is evoked if the echo signal arrives at the neuron at an echo delay, δ, equal to the neuron's BD, β. Based on this mathematical model for a single neuron, an entire cortical population is constructed into a framework to represent an auditory response map. The overall framework of the auditory system comprising the periphery model and the cortical model is shown in fig. 61.2. In the periphery model, inputs are the outgoing pulses and echoes, and the output is a neural spike train evoked by the sound pair. Each echo contains n components with a constant interspike interval (δ), equal to 1/sampling frequency.

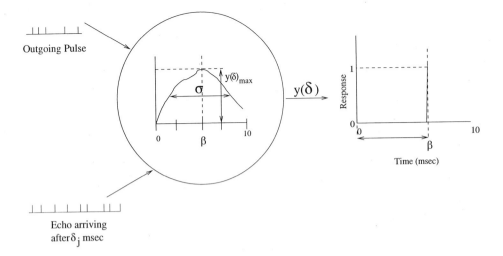

Fig. 61.1. Delay-sensitive-neuron model showing input / output characteristics. The cortical neuron responds only in the window σ, and a maximal response is evoked for an echo arriving at the neuron's BD, β. A submaximal response is evoked if the input spike arrives at a delay is greater or less than β but at values within the σ window.

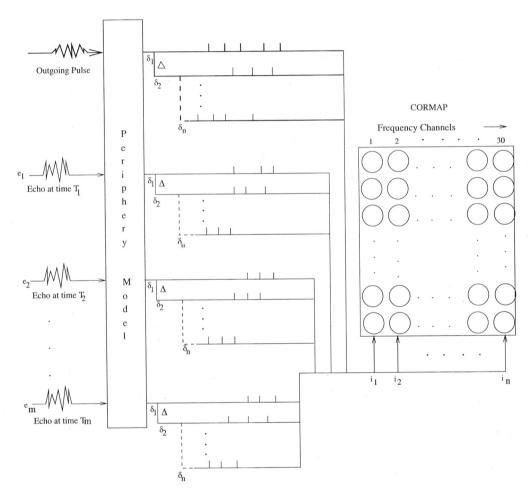

Fig. 61.2. Overall scheme of the auditory-system model consisting of the periphery and CORMAP frameworks. The acoustic inputs to the peripheral model comprise the outgoing ("emitted") pulse signal and the multiple components of the echo signal returning from a complex target (e_1, e_2, . . . , e_m is the different echo-signal components reflected from an m-glint target at T_1 . . . T_m time, respectively). The neural spike trains are generated as outputs by the peripheral model, and in turn provide the neural inputs into the cortical model. The neural spike activity evoked by each sound-signal component is represented by n time traces of spike firing, each of which is the firing pattern associated with a different sound-frequency channel. The fine delays (δ_1, δ_2, . . . , δ_n) separated by Δ of the different signal components also are represented in the different frequency channels. The CORMAP depicts the tonotopically organized cortical neuronal population. Subpopulations tuned to different best frequencies (20–100 kHz) receive only spike-firing inputs transmitted from the same frequency channel (i_1, i_2, . . . , i_n).

PULSE-ECHO PROCESSING

Inputs to the CORMAP are the neural spike-train patterns generated by the peripheral auditory system in response to the outgoing pulses and echoes. When the spike train corresponding to the pulse arrives, all neurons tuned to frequency spectrum enter an initial inhibition state followed by excitation. Neurons tuned to different frequencies are inhibited at different times. The process of obtaining spectral cues from the neural spike-train pattern is as follows:

Let t be the time at which the first pulse is emitted. When the spike train corresponding to the pulse arrives at the CORMAP, the neurons enter an inhibitory state. (The response magnitudes of the cortical population at this instance are given by eq. 61.4.)

After the pulse arrives at the CORMAP, the echo train begins to arrive. (Pulse-echo overlap is not considered here.) Let the input at channel i_n represent the highest-frequency component of the echo and δ_1 (in ms) represent echo delay of neurons tuned to this echo-frequency component. Then the response of those neurons tuned to this delay is calculated as

$$y(\delta_1) = \begin{cases} \tau \cdot e^{-i_n \cdot (\beta - \delta_1)^2/2 \cdot \sigma^2} & \text{if } (\beta - \sigma/2) \le \delta_1 \le (\beta - \sigma/2) \\ 1 & \text{if } (\beta = \delta_1) \\ 0 & \text{otherwise} \end{cases} \quad (61.6)$$

Eq. 61.6 states that a facilitative response is evoked in a neuron by a pulse-echo pair, when the echo arrives at facilitative echo delays—that is, echo delays within the delay width (σ), the facilitative response is maximal at BD $(\beta) = \delta_1$.

Assume the input at channel i_{n-1} represent the second-highest-frequency echo component and δ_2 represent its echo delay for neurons tuned to this echo frequency. This echo frequency at i_{n-1} is delayed from i_n by a time interval, Δ. For all cortical neurons frequency-tuned to echo component i_{n-1}, their facilitative response magnitude is maximal for $\beta = \delta_2$. Let this response be

$$y(\delta_2) = \begin{cases} \tau \cdot e^{-i_{n-1} \cdot (\beta - \delta_2)^2/2 \cdot \sigma^2} & \text{if } (\beta - \sigma/2) \le \delta_2 \le (\beta - \sigma/2) \\ 1 & \text{if } (\beta = \delta_2) \\ 0 & \text{otherwise} \end{cases} \quad (61.7)$$

Thus, the first echo arrived at time T_1, the i_nth echo frequency component has an echo delay of δ_1, and $y(\delta_1)$ is the response magnitude of a cortical neuron whose BD $= \beta_1$.

Therefore, the collective response magnitude, $y(\delta_1)$, $y(\delta_2), \ldots, y(\delta_n)$, is the spatiotemporal firing of the cortical population that constitutes the neural representation of the target generated at time T_1. A typical target could consist of several echo components arriving at different times. Hence, the target image construction is a process that continues for target-reflected echoes returning at times $T_1, T_2, \ldots, T_m$. The continuous image reconstruction, represented neurally by the composite cortical re-

sponses evoked by echoes impinging on the ears at times $T_1, T_2, \ldots, T_m$, can be considered as a neural image produced by the particular target under consideration.

Simulation Results

Modeling and simulation of a network of neurons provides a powerful computational tool for which large-scale reorganization can be analyzed over time in a manner not possible with isolated single-unit recording, especially in sampling a relatively small bat brain. Moreover, different manipulations of the stimulus-input characteristics could easily be made in the simulation model. The model in its current mode of implementation essentially captures the population response pattern evoked by a single pulse-echo pair fixed at a constant pulse rate and pulse duration. The auditory system framework was tested using different pulse-echo pairs of phantom targets. To mimic the phantom target, for every pulse, a known number of echoes was created at known echo delays. Two different experiments were carried out. In the first experiment, echoes were exact copies of the pulse, but delayed in time. In the second experiment, echoes were attenuated with varying dB values to mimic a real target-reflected echo. Details of each experiment are given below.

EXPERIMENT USING SIMILAR PULSE AND ECHO

A phantom target with two glints is used in this experiment. The pulse is 1.0 ms and has two echo components generated at delays of 2.0 ms and 2.5 ms. The pulse and the corresponding echoes are shown in fig. 61.3. The

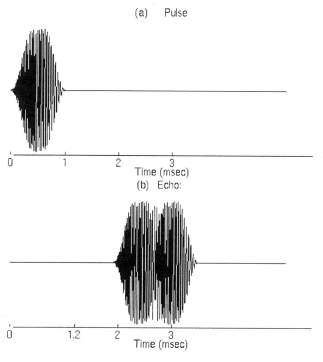

Fig. 61.3. A phantom target with echo two glints. The echoes (*b*) are exact copies of the outgoing pulse (*a*) delayed by 2.0 ms and 2.5 ms.

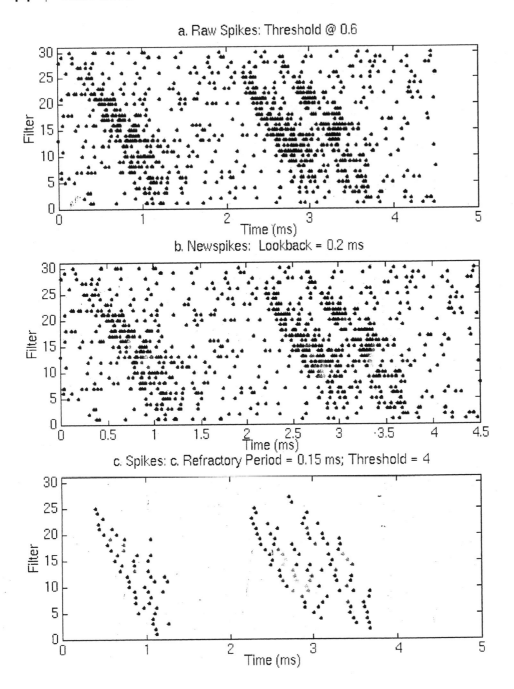

Fig. 61.4. The spike trains from the periphery model at different stages of processing (*a*) and (*b*). The *x*-axis corresponds to the filter channels ranging from 0 to 30. The spike train shown in *c* is used for further processing at the cortical level. The plots shown in *c* have a time lag of 0.5 ms; therefore, the first group of spike train, corresponding to the pulse, starts at 0.5 ms. The actual echo train thus begins approximately at 2.0 ms for the first echo.

echoes are exact copies of the pulse delayed at different times. Fig. 61.4 shows the spike trains from the periphery model at different stages of processing. The spike train generated by the periphery (fig. 61.4c) is then presented to the CORMAP model for cortical-level processing. The CORMAP begins to form neural images of the target over time. Several "snapshots" were taken during the entire processing of the echo signal, of which three are shown in fig. 61.5. In fig. 61.5, the row labels $T = 1$, $T = 4$, and $T = 6$ correspond to times at which the snapshots were taken during the process. The first column, labeled in the bottom of the figure as Output, cor-

responds to the neurons that responded to incoming echo. For example, the final snapshot at $T = 6$ under Output shows all the neurons that responded for the echo. The neurons shown in the Output column resemble the cortical organization in *Myotis*, where the *y*-axis corresponds to the frequency channels in which the neurons respond. In fig. 61.5 the snapshots in columns labeled Response and Delay are organized manually along response and best delay axes for analysis. Note that the anatomical axis is no longer represented in these figures. In Delay, the *x*-axis shows the actual arrival time of the echo—$t(\times 0.004)$ implies that the actual

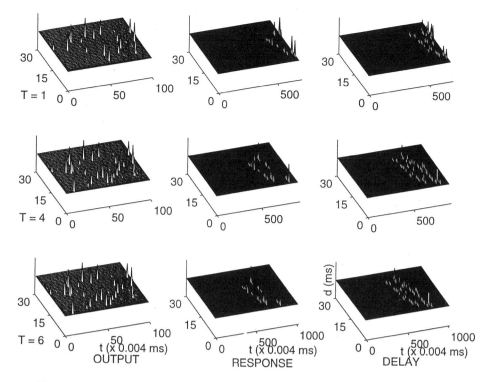

Fig. 61.5. The neural image formation of the target over time. The echo signal spans over 1000 ms. Several snapshots were taken during the entire processing of the echo signal, of which three are shown. The row labels $T = 1$, $T = 4$, and $T = 6$ correspond to times at which the snapshots were taken. The first column, labeled in the bottom of the figure as Output, corresponds to the neurons that responded to the incoming echo. The final snapshot at $T = 6$ under Output shows all the neurons that responded for the echo signal. The snapshots in columns labeled Response and Delay are organized manually along response and best delay axis for analysis purposes. In Delay, the x-axis, which is the time axis, shows the actual arrival time of the echo components along with their best frequency and best delay.

time can be obtained by multiplying the time by 0.004 — along with the best frequency and best delay. Therefore, this is a three-dimensional representation of the temporal process with $t \times f \times d$, where d is the delay and t is the actual arrival time of the echo components, which is a function of Δ, the interspike interval (t ranges from δ_1 to δ_n).

The time lag between the emitted pulse and the received echo provides an estimate of the target distance. For example, a close observation of the responses in the snapshot at $T = 6$ in the Delay column shows that neurons in the highest frequency channel fired approximately at 2.025 ms. The specified delay was 2.0 ms. Similarly, the second train of high-frequency responses arrived at approximately 2.525 ms, and the actual delay was 2.5 ms. In both cases, the error was approximately 1.25%.

EXPERIMENT USING ATTENUATED ECHO
Three glint attenuated echoes were used in this experiment. The echo delays ranged from 2.0 ms, 2.5 ms, and

3.1 ms and the attenuation factors were −3 dB, −9 dB, and −11 dB, respectively. The outgoing pulses and echoes are shown in fig. 61.6, and the spike-train output from the periphery is shown in fig. 61.7. As in the previous case, fig. 61.8 shows the snapshots of the progressive firing which took place in CORMAP. Close analysis of the snapshot at $T = 6$ in column Delay reveals that the three echoes arrived at 2.025 ms, 2.525 ms, and 3.25 ms. The third echo arrival time had an error of 4.8%.

Discussion

THE PERIPHERY MODEL
The model demonstrated adaptation to noise evidenced through several experiments (see Kohrt 1996). In the simulations of the model's adaptation capability, only 2 ms of random noise preceding the echo was used for the transient phase. The model showed a gradual decay from an onset response level to an adaptation level. The adaptation improved with an increase in the dura-

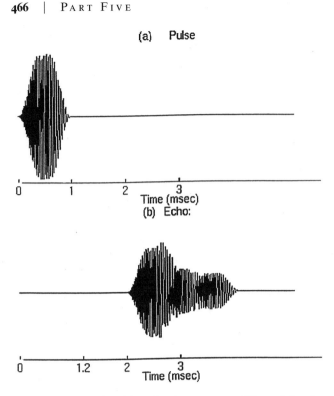

Fig. 61.6. The pulse (*a*) and echo sequence (*b*) used for the three-glint attenuated echo experiment. The echo delays of 2.0 ms, 2.5 ms, and 3.1 ms corresponded with the attenuation factors of −3 dB, −9 dB, and −11 dB.

tion preceding the arrival of the echo. The output of the model clearly registered the arrival of echo glints in the presence of noise.

The filter order influenced the signal attenuation outside the bandpass. A poorly tuned filter prompted the hair cells to respond to frequencies outside the bandpass, resulting in a depletion of the neurotransmitter reservoir and poor time encoding of signal features. However, higher order filters can minimize this deterioration. The output of the model matches very closely the half-wave rectified shape of the incoming signal, showing phase locking as observed in peripheral auditory neurons.

THE CORTICAL MODEL

The overall objective of this modeling study is to generate the population response pattern that emerges dynamically as stimulus inputs are changed. Electrophysiological evidence for dynamic cortical reorganization is inferred largely from single units whose neural-tuning properties become plastic with stimulus-input characteristics that mimic echolocation. Modeling and simulation of a network of neurons provides a powerful computational tool for which large-scale reorganization could be analyzed over time in a manner not possible with

isolated single-unit recording, especially in sampling a relatively small bat brain. Moreover, different manipulations of the stimulus-input characteristics could easily be made in the simulation model. Thus the cortical response pattern could be observed and tracked in time under different stimulus conditions: those employed electrophysiologically (changes in a few stimulus parameters), those mimicking the echolocation cycle accurately (covariation of multiple stimulus parameters), and those that are artificial (e.g., increasing echo amplitude or pulse duration with target approach). A major goal of comparing these simulation experiments using different stimulus conditions will be to determine whether specific neural ensembles are more likely to be formed and maintained with specific patterns of stimulus inputs. Such knowledge of formation of stimulus-dependent ensembles will not only confirm but also extend our understanding of self-organization derived from single-unit analysis. From a neuroethological perspective, the association of specific cortical activation patterns with stimulus-inputs characteristics for an echolocation phase may shed further insights into how dynamic reorganization is exploited for target analysis.

The CORMAP composite firing pattern of the population of delay-sensitive neurons serves as an important cue for target recognition and classification. These comprise the response magnitude of the neurons, the frequency at which the neuron is tuned, the time the echo arrived, and the best delay of the neuron that responded. Different types of target information can be derived from the combinations of these outputs. For example, the echo delay provides a measure of target distance (Simmons, Howell, and Suga 1975; Schnitzler 1984); the response magnitude of neurons provides target-size information; and the composite responses of cortical neurons tuned to different best delays convey textural information (i.e., fine structure) about the target. Thus, these three CORMAP outputs can generate a rough target signature.

The CORMAP framework was tested to validate (1) observations of the cortical response to simple and complex targets, and (2) assessments of CORMAP's ability to encode target features. Simulations for assessing various aspects of CORMAP used 100 × 30 neurons, where the 30 frequency channels span a range of 20–100 kHz and the 100 neurons provide a range of different BDs between 1.0 ms and 10.0 ms. Therefore, the resolution of CORMAP is low, although a good indication is provided of the number of neurons that respond for the different simulations. The delay information comprises the fine delays between different frequency components of the echo. As the number of glints increased from two to three, simulations revealed a larger population that becomes responsive to encode more delay values. Both fine delays and the delays between successive echoes are encoded. However, it is important to

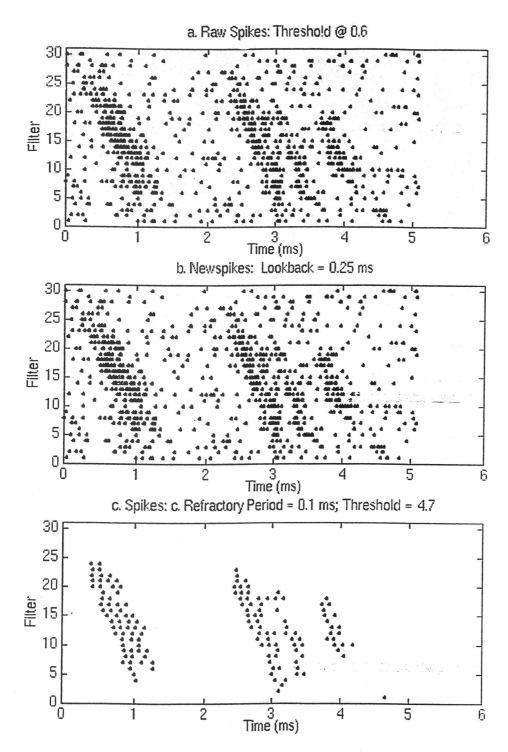

Fig. 61.7. The spike-train output from the periphery. Note that the spike-train patterns for the successive echoes significantly deteriorate due to the attenuation factor. The auditory periphery model, however, provides good firing patterns (as shown in *c*).

note that this framework cannot currently achieve fine-delay acuity. The single-unit studies that document delay tuning in the millisecond range cannot implement a CORMAP resolution in the micro- or nanosecond range as suggested in behavioral studies.

The target-distance estimation obtained from COR-MAP demonstrates that the framework is capable of performing within 2% accuracy. The distance estimation is based on the delay encoded by neurons. Once the fine-delay estimate is made possible through CORMAP, target texture can also be extracted from delays of the population of neurons that responded. In conclusion,

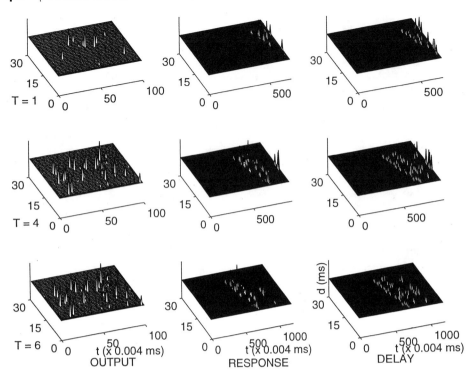

Fig. 61.8. Snapshots of the progressive firing activities for the three-glint experiment. As in the previous case, the three columns correspond to three different organizations of the activities in CORMAP. The first column, Output, contains raw firing activities where the frequency channel is the only relevant axis. (The other two columns were organized for analysis purposes.)

we presented a formulation of a biologically plausible framework to represent the spatiotemporal pattern of firing of cortical neurons evoked by echolocation signals. The current framework, based on available neurophysiological data, provides insight into the cortical firing pattern evoked by a pulse-echo stimulus pair.

Acknowledgments

The authors acknowledge the contributions by Siva Chittajallu toward the design of the auditory periphery model and extend special thanks to Kevin Kohrt and Uday Murthy for implementing the periphery and the CORMAP models. This research was supported in part by NSF grant BCS-9307650.

{ 62 }

Recent Methodological Advances in the Recording and Analysis of Chiropteran Biosonar Signals in the Field

Stuart Parsons and *Martin K. Obrist*

Introduction

Donald Griffin, in his pioneering studies of bat ultrasound, used equipment that hardly could be considered portable. Although at the cutting edge of technology at the time, the bulky equipment relied on AC power and consisted of a crystal microphone, vacuum tube amplifiers, and superheterodyne circuits. Until the 1980s, the majority of detailed studies on chiropteran echolocation had to be carried out in the laboratory. This helped to clarify the theoretical limits of chiropteran sonar systems, but little insight was gained into how bats use echolocation while foraging in the wild. Early field studies were limited to monitoring activity using microphones linked to an event recorder (Fenton and Jacobson 1973). Although bat detectors represent an invaluable tool to field-workers, the output from early leak detectors and heterodyne or frequency division models (Andersen and Miller 1977) is unsuitable for spectral analysis using Fourier transformation. High-fidelity recordings require the use of high-speed instrumentation tape recorders or portable storage oscilloscopes linked to condenser microphones (Ahlén 1981; Fenton and Bell 1981). It was (and still is) possible to use this equipment in the field, but it is easily damaged by adverse environmental conditions, very expensive, and only just portable. But eventually, researchers began to wonder if bats in "the wild" use echolocation in the same way as in the laboratory (e.g., Barclay and Brigham 1994; Kalko and Schnitzler 1989).

Recent advances in field studies of chiropteran ultrasound have been possible mainly due to advances in digital technology and the miniaturization of electronic components. High-fidelity signals now can be digitized directly to laptop computers, allowing complex spectral analysis in the field. Alternatively, signals can be time-expanded digitally by a bat detector and recorded in analog or digital form for later analysis. Techniques once used in period meters have also evolved into modern zero-crossing analysis systems. Although still constrained by laws of physics as they apply to ultrasound (Pye 1993), experiments investigating aspects of bat echolocation now can be carried out under conditions that are more natural than the laboratory. These experiments allow researchers to investigate bats' purposes for sonar, rather than simply their capabilities.

In this chapter, we discuss how the study of chiropteran sonar in the field has advanced over the past decade. Equipment and techniques for detecting, recording, visualizing, and analyzing echolocation calls of bats will be discussed. We also review some of the general pitfalls inherent in modern studies of ultrasound in the field.

Detection

Equipment for the detection of ultrasonic bat calls (i.e., beyond the range of human hearing) are either for reception of the signal or transformation to lower frequencies. The modern bat detector incorporates the receiver and transformer into one unit. Receivers detect the ultrasound and include microphones and amplifiers. Each receiver has different properties that affect their practicality and usefulness in detecting ultrasound in the field. In general, the decision of which type of receiver to use is based on cost, performance, and durability. Surveying for the presence/absence of echolocating bats does not require a receiver with a perfectly flat frequency response. However, it does require equipment that is resistant to harsh environmental conditions and

simple enough for volunteers to use. On the other hand, accurate measurement of calls requires a receiver with a linear frequency response.

Several methods are available to transform ultrasound so it is audible to humans and easier to record or analyze.

Receivers

Piezoelectric transducers

Piezoelectric transducers are widely used in low-cost brands of bat detectors. The main drawback of these devices is that their frequency response is far from flat and covers a limited range of frequencies (Pye 1992). Piezoelectric transducers are very sensitive around their resonant frequencies, but their response at other frequencies can be very variable (Waters and Walsh 1994). The limited frequency range is overcome in most bat detectors by using two transducers. For instance, the BatBox III (Stag Electronics, Steyning, UK) uses transducers with resonant frequencies of 25 kHz and 40 kHz. This means that the BatBox III is very sensitive around these frequencies but relatively insensitive above approximately 60 kHz. Such a variable response can lead to bias in basic survey work, such as activity monitoring. The uneven frequency response of piezoelectric transducers, combined with the fact that most detectors using these transducers offer only heterodyned output of signal, result in signals unsuitable for spectral analysis. However, the ruggedness and price of these transducers make them suited for use in the field.

Capacitance (condenser) microphones

Capacitance microphones fall into three broad categories: solid-dielectric, air-dielectric, and miniature electret (Pye 1993). To function, the capacitor must be charged with a constant voltage. A movement in the surface, or diaphragm, of the microphone causes a change in the capacitance. This change is a measure of signal amplitude, while the rate of change corresponds to the frequency of the signal. Solid-dielectric microphones are relatively inexpensive, are rugged, and have a relatively flat frequency response. These microphones have been incorporated into the more advanced bat detectors, such as those made by Pettersson Elektronik (Uppsala, Sweden) and Ultrasound Advice (London, UK). High-frequency, and certain transformed signals (see below for details), recorded using solid-dielectric microphones are suitable for spectral analysis. Unfortunately for field-workers, they are sensitive to changes in humidity, which can affect sensitivity by introducing excessive noise.

Air-dielectric microphones, such as those made by Brüel & Kjær (Nærum, Denmark) and Larson-Davis (Provo, Utah), are less sensitive than solid-dielectric microphones. They have a flat and stable frequency response and sensitivity, but they are very expensive and fragile. Although they have been used in field studies of bat ultrasound (e.g., Parsons, Thorpe, and Dawson 1997; Surlykke et al. 1993), their use is not recommended for those on a limited budget. To date, no air-dielectric microphones are incorporated into bat detectors, so the output signal must be made to instrumentation tape recorder, time-expansion unit, or directly to computer via analog-to-digital conversion hardware. Such recorded signals are well suited to further analysis.

Electret microphones are much smaller than other capacitance microphones. The electrical charge across the membrane is produced during manufacture, and so the microphone requires only a small power source. Electret microphones were not originally suitable for detecting ultrasound due to their limited frequency response (12–15 kHz; Pye 1992). However, this can be compensated for using an amplifier to increase the gain at higher frequencies (Andersen and Miller 1977), without degrading the signal-to-noise ratio (Pye 1993). The use of a compensating amplifier can lead to a relatively flat frequency response. High-frequency output, and certain transformed signals (see below for details), recorded using electret microphones may be suitable for spectral analysis. Compared to other microphone types, electret microphones tend to be relatively omnidirectional due to their small size.

TRANSFORMATION OF OUTPUT SIGNALS

Heterodyne

Heterodyning is probably the oldest, most successful, and most widely used method of transforming bat calls (Noyes and Pierce 1938; Pye 1992). Most detectors feature heterodyned output of signals, which allows the user to select a frequency at which the detector will "listen" through a narrow frequency window, with the selected frequency representing the center of the window. When a signal (S_{sig}) is detected, it is mixed with a signal from an input oscillator (S_{os}). The frequency of this oscillator is set using the tuning dial on the detector. The mixing of S_{sig} and S_{os} results in the production of a signal with two frequency peaks, one at $S_{os} + S_{sig}$ and another at $S_{os} - S_{sig}$. The resulting signal then is filtered so that only the low-frequency peak remains before being combined with a second signal produced by a constant-frequency oscillator. This also produces a signal with two frequency peaks, which then is filtered so that only the low-frequency peak remains. The ultimate output of this system is a signal at a much lower frequency than the original input (fig. 62.1). Unfortunately, the limited bandwidth to which the detector listens blurs the duration, absolute frequency, and the frequency-time course of the original call in the heterodyned signal, thus rendering it unacceptable for spectral analysis. However, because of the narrow frequency window through which these detectors listen, noise levels in the input signal can be very low, leading to good signal-to-noise ratios. Stronger amplification of the signal also can be achieved, as the internal oscillators use frequencies that are very different.

The relatively high sensitivity of these detectors has been demonstrated in the laboratory (Waters and Walsh 1994) and the field (Parsons 1996). However, the results can be influenced heavily by the frequency response of the receiver. The sensitivity of the heterodyne system means that these detectors can be useful in survey work where species identification is not necessary (e.g., O'Donnell and Sedgeley 1994). In particular, measures of species activity, with the use of less sensitive devices, can lead to undersampling due to differences in source levels of echolocation calls from different species. The narrow tuning range of these detectors can allow for the start and end frequencies, and frequency with most energy of calls, to be approximated. This ability may allow experienced workers some degree of species identification. The narrow frequency band is also the most limiting factor of tunable detectors, as bats calling at frequencies outside the tuned frequency window will be missed. Pye and Halls (in Pye 1992) attempted to overcome this problem by developing a heterodyne system that listened through a number of frequency windows simultaneously. The system was not successful because when a signal was detected, no indication was given as to which window (or windows) it passed through.

Time expansion

This technique, which involves the digital time expansion of a signal to lower frequency, is based on the inverse relationship that exists between time and frequency. If the duration of a finite signal is increased, the frequencies contained within the signal will decrease. Commercial time-expansion units, such as the Portable Ultrasound Processor (PUSP) produced by Ultrasound Advice, digitize the high-frequency output from a microphone or bat detector at high sampling rates. The signal is then converted back into an analog waveform using a reduced sampling rate, thus effectively increasing the signal's duration, and so time-expanding it. The resulting signal can then be recorded using a standard tape recorder or directly to a laptop computer capable of acquiring sound. The practice of slowing the replay speed of instrumentation tape recorders to lower the frequency of recorded signals has been used for many years. Unfortunately, the size, weight, and cost of these tape recorders represent only a few of the drawbacks associated with their use, particularly in the field. More recently, time expansion has been incorporated directly into bat detectors, such as those made by Pettersson Elektronik and Tranquility (Cheltenham, UK). Most time-expansion systems include an 8-bit A/D converter (40–48 dB) and lack an anti-aliasing filter, increasing the possibility of artifact recordings (see below). The great advantage of time expansion is that no informa-

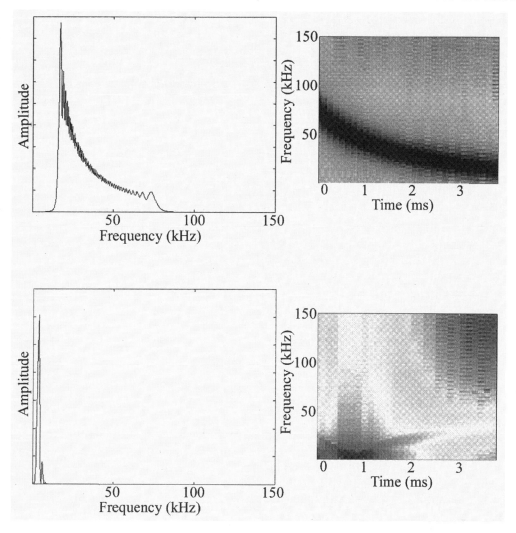

Fig. 62.1. Power spectrum and spectrogram of a simulated bat echolocation call before and after heterodyning. All were produced using 512-point Fourier transformations and Hamming windows. A 50% overlap was used in the spectrogram.

tion is lost from the incoming signal, making the output well suited to spectral analysis. When combined with a laptop computer and signal analysis software, such as Pettersson Elektronik's BatSound, the output from a time-expansion system provides field-workers with high-quality information on bat ultrasound. High sampling rates are not required, because a time-expansion factor of 10 brings signals up to 220 kHz within the range of most modern computer sound cards, including those in most newer laptops. Despite this, a number of factors can affect the quality of ultrasound recordings, necessitating care when interpreting any ultrasonic signal (Pye 1993).

Compared with heterodyne or frequency division detectors, time-expansion detectors are relatively expensive. Nevertheless, most time-expansion detectors represent excellent value when compared to the cost of traditional equipment such as air-dielectric microphones and instrumentation tape recorders. Some time-expansion detectors also are capable of outputting heterodyned, frequency-divided, and unmodified high-frequency signals, thus increasing their cost-effectiveness. It is important to note that, at present, it is not possible to sample continuously using time expansion. Most systems store between 2 and 12 s of digitized sound before outputting the expanded signal. During the output phase, the system is not sampling from the environment. For example, a time-expansion detector will take 22 s to acquire and output 2 s of ultrasound, assuming a time-expansion factor of 10. This means that, ideally, the system is only sampling 9.1% of the available time. Frequency division and heterodyne systems are not limited in this way.

Countdown or frequency division

Frequency division, or countdown, was first incorporated into bat detectors by Andersen and Miller (1977). A zero-crossing system is used to isolate the dominant or loudest harmonic of the input signal and produce a square wave of the same frequency. To minimize the amount of noise in the output signal, only input signals above a predetermined threshold are converted into square waves. The square wave then is filtered so that only every *x*th wave passes through. The value of *x* is set by the user and represents the degree of frequency division. In some systems, including that of Andersen and Miller (1977), the square wave then is converted back into a sine wave before the signal is outputted. The best systems finally multiply this output with the amplitude envelope of the original signal for optimum correspondence.

Most frequency division systems have several drawbacks. First, in most systems amplitude information from the original signal is not transferred to the square wave and, hence, is not represented in the output signal. Second, as has been mentioned earlier, the zero-crossing system tracks the harmonic with greatest amplitude, so no other harmonic information is contained in the output signal. In some species of bat, frequency overlap among the harmonics means that the harmonic with most energy may change over the course of the call. This can cause the zero-crossing system to jump between harmonics, leading to a misleading output signal, especially when analyzed spectrally. Third, the degree of division will determine the frequency range that can be heard or recorded following transformation. If a division ratio is too low, calls of bats using high frequencies may be missed. Fourth, the use of an input amplitude threshold can decrease the overall sensitivity of the detector (Parsons 1996) and lead to quieter calls (e.g., from "whispering" bats) being missed. Fifth, the frequency division method reaches its limitations in very short, high-frequency signals. When divided by 10, a 1 ms call, sweeping from 100 kHz down to 40 kHz, results in approximately eight waveform-cycles, making later spectral analysis problematic.

An obvious advantage of frequency division is that it transforms the entire bandwidth of a signal. If the amplitude envelope is preserved, the time and frequency content of calls can often be visualized in real time, making the technique useful in survey work. Detectors, such as the Anabat (Titley Electronics, Ballina, NSW, Australia), can be purchased with a variety of additional modules. This includes a zero-crossing interface module that, when combined with the Anabat software (Analook), allows for zero-crossing frequency analysis of output signals.

Storage Devices

All information contained in a signal to be conserved for later analysis or reference must be recorded to a storage device. Additionally, all noise suppression algorithms (Dolby, etc.) must be switched off and the recording set to "linear." Algorithms, such as Dolby, were de-

veloped for sounds in the human hearing range and can seriously alter the frequency composition of high- and low-frequency signals. When digitizing a recorded signal, all frequencies above the Nyquist frequency (sampling rate/2) should be filtered out to avoid potential problems with aliasing. Two different approaches to recording exist, analog and digital.

ANALOG RECORDING

All analog tape recorders can alter signals by distortions such as tape noise (hiss), changes in frequency and amplitude due to recording speed variations (wow and flutter), and other interference, such as print-through effects and dropouts. Products meant for recording music do not necessarily cope well with the rapid dynamic changes of animal vocalizations.

Reel-to-reel

Specialized instrumentation tape recorders are able to record ultrasonic frequencies, some to multiple channels simultaneously, due to their high recording speeds (up to 76 cm/s) and high-quality electronic components. Their ability to replay at slower speeds lowers the frequency and increases the duration of signals that then can be analyzed like ordinary audio-range signals. Because these devices usually use separate recording and replay heads, the signal can be monitored off-tape during recording using an oscilloscope. The dynamic range of these recorders is not generally very high (40 dB), partly due to the variable tape speeds. Low signal-to-noise ratios can be improved using bandpass filters and adequate power supplies (24 V for Racal recorders). A drawback of instrumentation tape recorders is their size, weight, and cost. Furthermore, the signals are only available after rewinding the tape; they typically contain 20–90% silence, depending on the duty cycle of the vocalizing bat. Finally, it takes some time for the tape to reach full recording speed.

Audio compact cassette tapes (CC)

The frequency characteristics of audio products are limited to the human hearing range and are not adaptable to ultrasound. However, when combined with heterodyne, frequency-division, or digital time-expansion circuitry, recording becomes feasible within the limits of the recorder's frequency response. Depending on tape quality, CCs have a dynamic range of 60–70 dB, but the sampling depths of digital time-expansion circuitry also define the effective dynamic range. Most are limited to 8-bit precision, which corresponds to 48 dB.

DIGITAL RECORDING

All digital recording devices are designed for recording sounds in the audio-frequency range, digital video, or computer data. However, digital recorders do not suffer from the same problems as analog recorders. Because these devices sample sound with an A/D converter, the dynamic range is defined by the converter's bit-depth (e.g., 16-bit sampling depth results in 96 dB dynamic range; 6 dB per bit) and the frequency range is limited by its sampling frequency (e.g., 44.1 kHz sampling results in 22.05 kHz recording bandwidth). Apart from these factors, digital devices record and reproduce signals with low noise, a flat frequency response, and no speed variations.

Video recorders

Newer models of digital video cameras can record sound at sampling rates up to 44.1 kHz, making them useful for recording calls as well as video images. Analog video recorders can record digital audio signals; however, this requires special pulse code modulation recorders (PCM, NTSC sampling at 44.056 kHz, PAL sampling at 44.1 kHz), which are not widely available. Alternatively, internal modifications to existing recorders can increase their sampling rate to approximately 96 kHz.

Digital compact cassette (DCC) and MiniDisk (MD)

Meant originally as an alternative to CC, and able to replay this media, DCC is not widely used. DCC relies on audio compression (4:1) based on psychoacoustic experiments on human hearing abilities, making the data reduction inaudible. Using magneto-optical technology, MD recorders use even more severe compression techniques (5:1) than DCC. These recorders use compression methods that change frequency and amplitude information, which may or may not be obvious. In short, neither should ever be used for recordings where the sounds will be subjected to detailed bioacoustic analysis. However, they can be extremely convenient for recording calls not intended for detailed analysis, such as those for survey work, activity monitoring, and basic species identification.

Digital audio tapes (DAT)

Rotary-head digital audiotape (R-DAT), more generally known as DAT, allows up to one hour of recording in the field. Compact and lightweight, these systems offer features such as real-time counters, date and time logs, and indexing. Most use sampling rates of 48 kHz or 44.1 kHz with 16-bit resolution or 32 kHz at 12-bit resolution with special encoding for double duration recording. On their own, these sampling rates are not sufficient for recording ultrasound. They do, however, allow very high fidelity recording of frequency-divided or time-expanded signals. Recorders often have a digital input to transfer data directly from devices, such as compact disc recorders and computers, thus avoiding the slight deterioration of the signal due to D/A–A/D conversion. DAT recorders are more sensitive to stress that typically occurs in the field (temperature, humidity, dust, physical shock) than conventional analog recorders. This is due to the narrow width of their recording tape (3.8 mm), low tape speeds, and the rapidly spinning re-

cording head. Only very expensive systems offer read-after-write control using two heads. Most consumer market products (contrasting with professional products) employ a serial code management system (SCMS), which prevents copying of previously copied digital tape. However, it does not limit the number of "first copies" from a master tape, nor does it restrict analog copies.

Compact disc (CD)

Compact audio discs can store any digital information—for example, sound sampled with 16-bit resolution at a sampling rate of 44.1 kHz. They offer 650 MB storage capacity, which translates to 18 min of ultrasound recorded at 330 kHz with 16-bit resolution. Blank CDs are U.S.$1–2 each, but recording is slow compared to hard disks, and all but a few CD recorders still depend on AC electricity. This medium is useful not only for archiving purposes, as modern portable CD recorders allow 650 Mb of data to be written in approximately 4 min (at 16x speed), even on a laptop computer.

Digital versatile disc (DVD)

Digital versatile disc allows storage capacities more than 10 times greater than a CD. Several versions of recordable DVDs presently on the market allow recording of 2.6–9.4 GB of data. However, issues of compatibility, particularly among countries, need to be resolved by manufacturers. Like CD-R, recording speed is limited. Consolidation on a single standard and the ability for high-speed recording in the field is not likely in the near future.

Random access memory (RAM) and hard disks (HD)

Computer storage media have evolved tremendously in quality, speed, and economics in the past 10 years. Digital time-expansion ultrasound detectors easily could be equipped with multimegabyte RAM storage, providing the user with minutes of sustained digital recording. Depending on requirements, these detectors could be combined with CC or DAT to produce a very versatile, low-cost field recording system. On the other hand, the versatility of computers makes them ideal tools for direct recording to RAM or hard disk. Today, portable laptop and even palmtop computers rival desktop systems in performance. With the advent of high-speed data acquisition computer cards in the PCMCIA-format, digital recording to computer in the field is feasible. The ability to display, process, store, analyze, alter, and/or replay the signals in the field promotes this method for a variety of field applications in the near future.

Hardware and Software for Monitoring of Bat Ultrasound in the Field

OSCILLOSCOPES

Pierce and Griffin (1938) first realized the unsuitability of tunable detector or tunable voltage meters for measuring bat vocalizations. While searching for a method for visualizing calls, Griffin connected the amplified output of a calibrated broadband microphone to an oscilloscope and immediately quantified the duration, intensity, and frequency content of calls (Griffin 1958). Oscilloscopes are still valuable tools for ultrasound research. Earlier models had storage screens, which kept an afterglow of a transient signal on the screen. Modern oscilloscopes hold digitized signals in memory. Unfortunately, oscilloscopes are generally bulky, mostly dependent on AC electricity or limited battery operation, and expensive.

PERIOD METER

A period meter counts the time between two crossings of an input signal through the zero volt line and displays its inverse (1/period = frequency) as a DC voltage (Simmons, Fenton, and O'Farrell 1979). When fed into a compatible oscilloscope, the trace will show the frequency structure over time of the signal. As with all zero-crossing analysis systems, the strongest harmonic will dominate the structure of the signal. A portable period meter combined with a calibrated, portable oscilloscope makes a very useful ensemble for visualizing sonar calls in the field. The major advantage of oscilloscope and period meters are their real-time performance, giving instantaneous information about temporal and dominant spectral signal content.

COMPUTER

Computer-based approaches to signal visualization rely on digitizing incoming signals and displaying the waveform and/or frequency content, after a short delay. Depending on the performance and design of hardware and software, near-real-time visualization is possible. Of the software written specifically for bat calls, Analook is used widely. The system, which monitors the output from an Anabat broadband frequency division detector, uses zero-crossing analysis (ZCA) to visualize detected signals in the frequency-time domain. In a manner similar to period-meters, zero-crossing analysis measures the number of times a waveform crosses the zero-volt point per unit time, the inverse of which gives the frequency. There are two major advantages of ZCA over Fourier analysis. First, it is not constrained by the uncertainty principle (Beecher 1988) so there is no trade-off between time and frequency resolution. Second, ZCA is fast because it does not rely on intensive computation. Unfortunately, ZCA also has some drawbacks. First, amplitude and harmonic information are lost from the input signal, thus removing potentially important analytical variables. (However, some systems do retain amplitude information in the output signal.) Second, the presence of strong harmonic components or noise can degrade the resolution of frequency measurements (Hopp, Owren, and Evans 1998). However, if this information is not required, then the system can be

extremely useful. Betts (1998) showed that identification of bats from frequency-time displays produced with Analook could be difficult for even experienced researchers. Pettersson's BatSound software (for Windows) enables the user to digitize a signal using a computer's built-in sound card (usually at 22.05 kHz or 44.1 kHz) and view its temporal and spectral content using Fourier or zero-crossing analysis. Canary (Cornell University, Ithaca, N.Y.) offers similar capabilities for Macintosh users. BatSound, in conjunction with high-speed A:D hardware, is also capable of digitizing sounds at 300–500 kHz, making real-time recording of unaltered signals possible on laptop computers in the field.

AUTOMATED MONITORING SYSTEMS

Many studies use echolocation calls to survey and monitor habitat use and foraging activity of bats (e.g., Brigham et al. 1997; Vaughan, Jones, and Harris 1997a). By recording the output from a detector, a permanent record of a night's activity can be kept. To avoid the need for trained field assistants, multiple detectors, and many nights spent in the field, O'Donnell and Sedgeley (1994) developed an automated monitoring system based around the output from a heterodyning bat detector. The system recorded the output from the detector on a voice-activated tape recorder. Incorporated into the system was a talking clock set to announce the time each hour, thus providing a reference time for the recorded activity. The system was powered by rechargeable batteries and at the time of development cost less than U.S.$270. In their analysis, O'Donnell and Sedgeley (1994) expressed activity as the number of passes (Fenton 1970) per hour. The advantages of this system are obvious: it is cheap, robust, portable, and reliable. On the other hand, tapes from monitoring boxes must be analyzed by simply listening. Bias can be introduced if different listeners are used, especially if they are not similarly experienced. Also, identification is almost impossible, even when sophisticated analysis methods are employed (Parsons 2000). Heterodyne detectors only listen to a narrow frequency range and only calls that have components within that range will be recorded. Finally, the system is triggered easily by environmental sounds with ultrasonic components (e.g., rain, insects), which can lead to tapes full of noise.

In recent years, the Anabat system has become increasingly popular. The system is based around a broadband bat detector that can be linked to a delay or time switch. This allows the system to be left in the field and activated only when a call is detected. Calls can then be downloaded to a tape recorder (digital or analog). Calls can also be digitized directly onto a computer in combination with the zero-crossing interface module for visualization and analysis. The Anabat system is extremely useful, especially in the field, and represents excellent cost value.

Many researchers improvise and use tools not specifically designed for bat work. Zbinden (1995) described a system that counted digitally the output of a heterodyne minidetector. A data logger stored counts, as well as weather conditions, all downloadable by a portable computer. The system was solar powered, making it an efficient long-term monitoring device. Unfortunately, it had a narrow spectral band and is not readily available to other researchers. Similarly, Park, Jones, and Ransome (1999) used a heterodyne bat detector to monitor *Rhinolophus ferrumequinum* activity in a hibernaculum. The output from the detector was passed through a custom-built Schmitt trigger that produced square waves in response to detected calls. The Schmitt trigger was linked to a Squirrel 1200 data logger (Grant Instruments, Shepreth, UK). The number of calls detected was downloaded each day to a laptop computer and analyzed. Thomas (1993) and Nagel and Nagel (1995) also used a bat detector linked to data loggers when monitoring activity of *R. hipposideros* and *Myotis* spp. The use of data loggers does not allow bat activity to be quantified in terms of number of passes but rather by number of calls. Therefore, interspecific results can be difficult to compare due to differences in duty cycle. Data loggers are not able to accept all data sent to them. Often, there is a time after receipt of a signal that the logger must pause to process the time and date before it is ready to receive further data (e.g., Thomas 1995). The use of data loggers is also more expensive than the technique described by O'Donnell and Sedgeley. However, it does remove the bias introduced by having to analyze an audiotape qualitatively and gives much more precise estimates of the timing of activity. Data loggers also can be left unattended in the field for extended periods of time.

SPECIES IDENTIFICATION

There are several published attempts at developing techniques for identifying bats from their echolocation calls. Zingg (1990) used discriminant function analysis (DFA) on temporal and spectral aspects of echolocation and social calls of 11 species of Swiss bats. This analysis achieved an overall correct classification rate of 86%, with rates for individual species varying from 72% for *Eptesicus fuscus* to 99% for *Pipistrellus savii* (*Hypsugo savii* in Zingg 1990). Vaughan, Jones, and Harris (1997b) also used DFA to classify the calls of 15 species of British bats. Their correct classification rates varied from 12% for *Myotis mystacinus* to 97% for *P. pipistrellus* (45 kHz phonic type).

Neefus and Krusic (1995) developed companion software for Analook that analyzes calls using DFA for species identification. The software was fast and ran on a laptop computer. No published information exists on the success rate of this software, or whether it is available to the public.

Obrist et al. (chapter 64, this volume) developed BATIT (BioAcoustic Taxa Identification Tool), a software package that analyzes high-resolution spectro-grams of acoustic signals. The system is able to digitize bat calls at up to 312 kHz and uses synergetic pattern recognition algorithms to perform species identification. The software compares incoming digitized calls to a stored reference library of prototyped calls from known species. Using the synergetic algorithm, Obrist et al. achieved an overall success rate of 80% when classifying the search-phase calls of 12 species of Swiss bats. The success rates for individual species varied from 42% for *M. daubentonii* to 98% for *Nyctalus noctula*. As with most automated recognition system, the BATIT constitutes a "black box," and its results depend on the quality of the reference library. While BATIT is an extremely promising tool for field-workers, its overall success rate does not yet warrant its release to the public.

Parsons and Jones (2000) used artificial neural networks to identify 12 species of bats from their echolocation calls. The extraction of temporal and spectral features from each call was automated and the reliability of identification reported back to the user. Using a hierarchical system whereby calls were classified first to genus level by one network, then to species level by another, three species could be identified unambiguously (*Pipistrellus pipistrellus*, *P. pygmaeus*, and *B. barbastellus*), while the lowest correct identification rate was 75% (*Myotis bechsteinii* and *M. daubentonii*). Calls from 8 of the 12 species could be identified correctly with an accuracy of more than 90%. The artificial neural networks consistently outperformed a DFA.

Many studies of bat activity and habitat use have relied on subjective analysis of echolocation calls, in combination with observation of flight behavior, to identify bats (e.g., Ahlén and Baagøe 1999; O'Farrell et al. 1999a). The equipment used in such studies is often appropriate and has been used in studies incorporating quantitative methods (e.g., Britzke et al. 1999; Murray et al. 1999). However, the calls of bats vary greatly both inter- and intraspecifically (Obrist 1995; Neuweiler 1989) and doubt has been expressed about the usefulness of subjective analysis (Parsons, Boonman, and Obrist 2000; Robbins and Britzke 1999). Subjective classifications are difficult to repeat, even when carried out by the same person (Betts 1998). The usefulness of any identification system for scientific research, including habitat surveys and species inventory work, must be questionable if other workers cannot replicate it.

Common Pitfalls When Recording and Analyzing Echolocation Calls

A variety of factors must be taken into account when recording ultrasound in the field. These factors include environmental background noise, attenuation, directionality, and Doppler effects.

Parsons, Boonman, and Obrist (2000), Pye (1993), and Pye and Langbauer (1998) provided good reviews of methodological constraints when recording and analyzing ultrasound, so we will only deal with two of the most common: aliasing and amplitude overload in recordings.

Aliasing

When digitizing analog signals, sampling rates must be used that are appropriate to the frequency range of the signal. If a signal is undersampled, aliasing can occur (fig. 62.2C–G). Frequencies above half the sampling frequency (the Nyquist frequency) simply mirror the lower frequency range and so cannot be analyzed. Increasing the sampling rate increases the Nyquist frequency and hence the upper frequency that can be safely analyzed. However, the spectral signal representation of an oversampled signal (fig. 62.2A–E) will be the same as for an appropriately sampled signal (fig. 62.2B–F). As a general rule, signals should be subjected to low-pass filtering before digitization and sampling rates set to at least twice the highest analysis frequency.

Clipping

Outdoor recordings of bats typically consist of a sequence of vocalizations emitted by a passing animal. During the bat's movement toward and away from the microphone, the spectral characteristics and perceived intensity of the signal change due to varying recording distance. If the intensity of the recorded signal surpasses the upper limit of the dynamic range of parts or all of the recording system (detector, tape, A/D converter), the signal will be clipped (fig. 62.2D–H). Clipping creates spectral components that were not present in the original signal. These components most often are harmonics of the base frequency, but lower frequency components also can appear due to undersampling (fig. 62.2). Clipping will not affect the analysis of a signal using ZCA. Clipping may not always be obvious, especially with analog recordings, where steep amplitude changes (edges) may be smoothed. VU meters or LED recording level indicators may not react quickly enough when an overloaded signal is recorded. However, some bat detectors use an LED to indicate internal circuitry overload due to intense ultrasound signals. The only effective way to monitor this phenomenon is to monitor a postrecording playback.

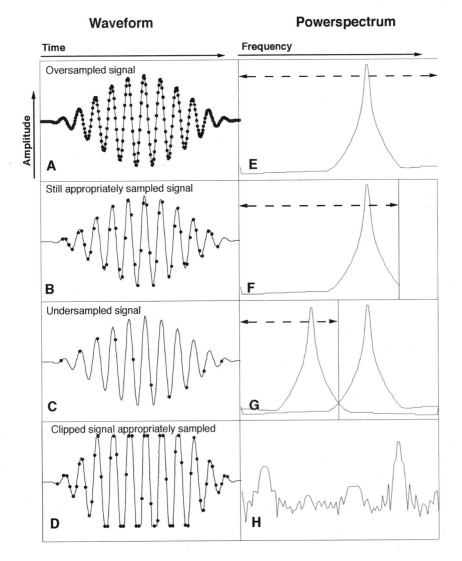

Fig. 62.2. The effect of under- and oversampling when digitizing a signal. The left-hand column shows the waveform of the signal and sampling points used in analysis. The right-hand column shows the power spectrum of each call and how sampling rate alters the representation of the call.

TABLE 62.1. Guide to the use of various biosonar equipment and analysis techniques. Fragile (•) to Extremely Rugged (••••); Cheap ($) to Very Expensive ($$$$); Not Useful/Unsuitable (−) to Very Useful/Suitable (+++).

	Durability	Cost	Usefulness/suitability			
			Detection/ observation	Automated monitoring	Species ID	Signal analysis
Microphones						
Piezoelectric transducer	••••	$	+++	+++	+	−
Electret	••	$$	+++	+++	++	++
Solid dielectric	••	$$$	+++	++	+++	+++
Air dielectric	•	$$$$	+	−	+++	+++
Transformations[1]						
Heterodyne	N/A	$	++	++	−	−
Frequency division	N/A	$$	++	+++	+	+
Time expansion	N/A	$$$$	++	+	+++	+++
HF (no transformation)	N/A	$$$$	−	−	+++	+++
Recorders—analog[1]						
Instrumentation tape recorder	•	$$$$	−	−	+	+
Audio compact cassette	••••	$	+++	+++	++[2]	++[2]
Recorders—digital[1]						
Video	••	$$$	−	−	+[2]	+[2]
Digital compact cassette	•••	$$	+++	+	−	−
MiniDisk	•••	$$	+++	+	−	−
Digital audio tape	••	$$	+++	+	+++[2]	++[2]
Compact disc (CD)	•	$$$	−	−	−	+[2]
Digital versatile disc (DVD)	•	$$$	−	−	−	+[2]
Random access memory and hard disks	•	$$$	−	−	+++	++
Visualisation[1]						
Oscilloscope	•	$$$	−	−	++	−
Period meter	•	$$$	−	−	++	+
Computer	••	$$$	+	−	+++	+++
Analysis techniques[1]						
Zero-crossing	N/A	N/A	++	++	++	++
Fourier transformation	N/A	N/A	+	+	+++	+++

[1] Assuming the input signal is appropriate.
[2] Time-expanded or frequency-divided signals only.

Conclusions

Since Donald Griffin first used a heterodyning circuit to listen to the echolocation calls produced by bats, the equipment used by bat workers to study biosonar has changed considerably. We are now able to record and analyze biosonar signals in the field, often in real time, using a variety of equipment and techniques. With the application of new methods, such as artificial neural networks and synergetic pattern recognition algorithms, the accuracy of quantitative acoustic species identification systems is also improving steadily. However, field-workers must still choose the most appropriate equipment for the task at hand. Expensive equipment can perform a wider variety of tasks, but simpler, cheaper, and more rugged equipment may be more appropriate (table 62.1). This equipment must also be used properly, and studies designed and carried out correctly, in order to obtain the most accurate results possible. The production, transmission, and receipt of sound is still subject to the laws of nature; thus, the problems associated with recording and analyzing ultrasound, such as atmospheric attenuation, remain. No technological development can relieve the observer of the necessity of having knowledge of acoustics and bat ecology. Otherwise, the interpretation of results could be erroneous.

Acknowledgments

We thank the organizers of the Biological Sonar Conference (Carvoeiro, Portugal, 27 May–2 June 1998 for inviting us to attend and for hosting an excellent and stimulating meeting. Our thanks also go to Brock Fenton, Gareth Jones, and Robert Houston for comments on earlier drafts of this manuscript. The script for simulating a hetrodyning system was written by Arjan Boonman. The New Zealand Foundation for Research, Science, and Technology funded Stuart Parsons's attendance at the conference.

{ 63 }

Call Character Lexicon and Analysis
of Field Recorded Bat Echolocation Calls

William L. Gannon, Michael J. O'Farrell, Chris Corben, and *Edward J. Bedrick*

Introduction

The role of bats as a significant North American natural resource has been emphasized in recent years. Forest management agencies now recognize the need for practical research on bats (Barclay and Brigham 1996). The U.S. Fish and Wildlife Service lists 6 species as endangered or threatened. Species classified as "of concern" number 10 in Arizona, 12 in California, and 13 in New Mexico (Jones and Schmitt 1997).

Bats are difficult to study and generally require multiple techniques simply to determine their presence (Kalko, Handley, and Handley 1996; Kapteyn 1991; Kunz 1988; O'Farrell and Gannon 1999). The use of bat detectors is one approach to determine bat activity (with or without supplemental capture), habitat use, and foraging behavior (Ahlén and Baagøe 1999; Barclay 1985, Barclay and Brigham 1996; Bell 1980; Crome and Richards 1988; Fenton 1982; Fenton, Merriam, and Holroyd 1983; Hayes 1997; MacDonald et al. 1994; Mills et al. 1996; Sherwin, Gannon, and Haymond 2000).

Within the past five years, there has been an increasing number of scientific publications using bat detectors (e.g., 81 abstracts from the annual meetings of the American Society of Mammalogists, North American Bat Research Symposium, and Wildlife Society, 75 citations in nonrefereed reports, and at least 31 peer-reviewed publications). A number of these papers cited the use of an ultrasonic bat detector with varied levels of success. We suggest that some of these studies have not used detector hardware and analytical software to its fullest potential, especially in the choice of characters to measure and the analysis techniques employed (Gannon, Sherwin, and Haymond 2002).

Our purpose is to continue a series of papers on the use and standardization of descriptive (qualitative) to quantitative approaches in identifying free-flying bats with a bat detector, specifically Anabat II (Anabat, Titley Electronics, Ballina, NSW, Australia). Recently, we described qualitative techniques for applying the Anabat detector in research and management arenas (O'Farrell and Gannon 1999; O'Farrell, Miller, and Gannon 1999; O'Farrell, Corben, and Gannon 2000). Here, we specify nomenclature to facilitate the communication of bat-call characteristics generated by field-portable bat echolocation detectors. A new lexicon is important because detectors like Anabat II allow immediate examination and analysis of the frequency-time structure of large quantities of echolocation data. Moreover, these highly portable devices (and others of this technological generation) quickly generate a large amount of data that is immediately available for analysis. Past terminology relied on the general frequency structure of a call and did not examine structural details necessary for the thorough use and understanding of this system. Previously, terms have been applied to behavioral or physiological contexts for relatively smaller numbers of calls but in greater detail than systems like Anabat provide. The primary purpose of the use of Anabat-like field detectors is to monitor and quickly identify bats free-flying in their natural habitat. Further, characters generated by these systems can be regarded as morphological characters for taxonomic goals.

To date, no paper has provided detailed comparisons among species based on call characters generated by Anabat or among classification techniques used to discriminate groups (but see Gannon et al. 2001; Hayes 1997; Vaughan, Jones, and Harris 1997a). The widespread belief that bats within the genus *Myotis* cannot be separated by call structure prompted tests using five species of *Myotis*. We show that, used in a standardized, repeatable fashion, Anabat-type detectors and analysis software are versatile and accurate for use in detection and analysis of echolocation calls, including the discrimination of *Myotis* species.

Materials and Methods

DEFINITIONS, CALL MORPHOLOGY, AND METHODS

Zero-crossings analysis

There are several ways to process ultrasonic vocalizations as a frequency-time display, one of which is zero-crossings analysis (ZCA). ZCA derives a frequency measurement from the reciprocal of the time period between two like zero-crossings of the signal. Although the ability of ZCA to determine harmonic structure or amplitude envelope is limited, it has the advantage of producing output rapidly in real time. ZCA differs from other forms of bat-call analyses (e.g., spectral analysis) in shifting emphasis from the complete analysis of a small number of calls to the rapid time-frequency examination of the dominant harmonic for entire sequences of calls.

Chris Corben designed an implementation of ZCA that produced output on the screen of an IBM-compatible computer from a countdown-type bat detector. Once developed, Titley Electronics provided Anabat as a commercial product. The Anabat II frequency-time output is displayed in real time and can be saved to a computer disk in files that average 3 KB. Part of the software evolution is the ongoing development of an auxiliary program (Analook) for managing data, viewing saved calls, and editing and measuring characteristics of a call or sequences of calls.

Clarification of traditional call terminology

Existing terminology generally does not apply to the output of all detection equipment and in many cases reflects inferences on target discrimination and habitat use, rather than specific structural features of call morphology. Anabat and other ZCA-type equipment provide information mostly for the harmonic with greatest strength (this could include the fundamental or any subsequent harmonic). Additionally, calls that are received —particularly at the higher frequencies of a given harmonic—due to attenuation, orientation, or distance of the bat to the microphone may be truncated (constant frequency [CF] and quasi-CF calls may also be affected; Kalko and Schnitzler 1993; Schnitzler, chapter 44, this volume). We designate a *call* as a single, continuous vocalization separated from all other calls by a period of silence. Bats produce a variety of other vocalizations unrelated to echolocation; these are generally referred to as social calls. A related string of calls contained within a single file makes up a *sequence*. A *pass* comprises a continuous sequence of calls, given by a single bat, from the time it was first detected until the bat travels beyond the range of the detector. A pass may be made up of one or more call sequence files.

The discovery that echolocation calls were not simply a single frequency, but rather incorporated a broad range of frequencies, gave rise to the concept of frequency-modulated calls (FM; Griffin 1958). Some bats, primarily of the family Rhinolophidae, use calls now referred to as FM-CF-FM in recognition of their profound frequency sweeps. Griffin further distinguished certain call characters that differ among species—mainly duration, intensity, and the presence or absence of harmonics—and recognized that call duration and repetition rate changed according to bat behavior (Griffin, Webster, and Michael 1960). These changes are described in three phases: *commuting calls* (similar to search-phase calls in low clutter), when calls are of greatest duration and low repetition rate; *approach phase,* when call duration de-

creases and repetition rate increases progressively; and *terminal phase,* when call duration is minimal and repetition rate has increased to the point that produces a burst of calls. Since these early studies, echolocation calls are described in terms of CF, FM, or a combination of these frequency characteristics and the implied behavior incorporated in duration and call rate (Simmons, Fenton, and O'Farrell 1979). Fenton (1999) asserted that terms such as FM and CF are inadequate for describing echolocation behavior of bats, since a free-flying bat can produce a wide range of call morphology in a single call sequence. A detailed description of characters and types of echolocation sounds was presented by Simmons and Stein (1980) and Obrist (1995): duration, harmonics, constant frequency, frequency modulation, sequence of frequency components, amplitude, and adaptive variations of the preceding six dimensions. The variety of sonar sounds includes multiple-harmonic short-CF, multiple-harmonic low-amplitude FM, short-CF/FM, FM/short-CF, short-CF or FM, and long-CF/FM. Further examination of echolocation patterns under natural conditions reveals a wider range of sound types representing more complex combinations and gradations of the basic types (Kalko and Schnitzler 1993; O'Farrell and Miller 1997).

Descriptions of call structure and variability commonly use characters measured along both frequency and temporal scales (Faure and Barclay 1994; Kalko and Schnitzler 1993; Simmons, Howell, and Suga 1975; Simmons, Fenton, and O'Farrell 1979). Variability in these characteristics are commonly reported with respect to the activity of one species or the interactions between species (Obrist 1995). Frequency, duration, and the pattern of frequency change over time were used to separate species by echolocation calls (Ahlén 1990; Fenton and Bell 1981). We propose a call character lexicon that simply describes the structural aspects of a call processed with any ZCA, without incorporating implications as to physiology or behavior of the bat, but allowing increased understanding and clarity of how these variables should be used among researchers.

Revised bat detector call character lexicon and call morphology

Bat calls exhibit a wide range of shapes over time. Figs. 63.1 and 63.2 illustrate the general variety of call shape encountered in Western Hemisphere vespertilionid and molossid bat species. However, portrayal of bat species by one call (e.g., Fenton and Bell 1981) is oversimplified, because it ignores the variation present from calls in free-flying bats. Regardless of the shape, a call comprises three useful frequency characters: maximum frequency (F_{max}), minimum frequency (F_{min}), and characteristic frequency (F_c; fig. 63.1). A call can possess three elements: initial sweep, body, and terminal sweep (fig. 63.1). Although the body is defined as the flattest portion of a call, one that is a simple down sweep of short duration is composed entirely of the body (fig. 63.1C). At times, parameter values can be redundant (e.g., when F_{min} and F_c are equal or highly correlated). In general, F_c is a more consistent measure, because F_{min} is more affected by intensity, distance, orientation to the detector, and intensity of echoes than is F_c.

Call Variables Measured by Analook

The remaining parts of call structure are best defined in relation to a standard reference point at which the char-

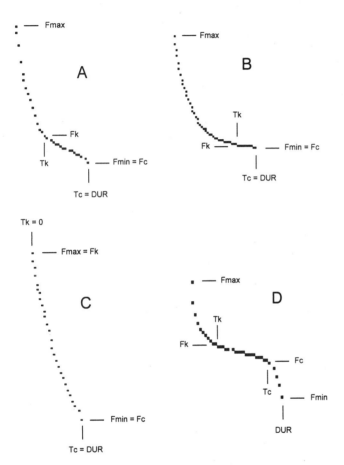

Fig. 63.1. Commonly encountered shapes of echolocation calls from North American bats. The number of calls per call sequence was recorded as *N*. The following variables were measured from each bat call: DUR = duration of a call (ms); F_{max} = maximum frequency, the highest frequency of the call; F_{min} = minimum frequency, the lowest frequency of a call; F_{mean} = mean frequency, the area under the curve divided by the duration; F_c = characteristic frequency given in kHz; T_k = time (ms) from the start of the call to the point at which F_k is measured (i.e., to the start of the body); F_k = frequency of the knee (kHz), the point at which the slope abruptly changes from the steep, initial down sweep to the flatter portion (body) of the call. T_c = time from the start of the call to the point at which F_c is measured (i.e., to the end of the body); S_1 = initial slope (decades/s; DPS), the first five points in a call; S_c = characteristic slope (decades/s; DPS), the slope of the flattest (most horizontal) part of the call.

acteristic frequency (F_c) is measured (figs. 63.1, 63.2). The body occurs prior to the F_c reference point, and the terminal sweep follows this point. If the terminal sweep forms a pronounced downward sweep, F_{min} will be measured at the end of the sweep rather than corresponding with F_c. If the slope is negative (i.e., frequency rising), F_{min} could be at the start of the call. The body is the flattest portion of the call ending at F_c. This change in slope is the *knee*, which marks the beginning of the body. Not all calls have a recognizable knee. The body is the flattest portion of the call (could be a very short segment). The knee is at the start of the body. A straight call has the knee at the very start and the body is the rest of that straight call. In a J-shaped call, the body is the flat part of the call at the bottom of the J, in which case the knee and characteristic frequency will be the same value.

In addition to the basic parameters describing a call (such as F_{max}, F_{min}, and F_c above), other measurements are generated by Analook software that may be of value in identifying species. For instance, F_k is the frequency (kHz) of the knee and T_k is the time (ms) from the start of the call to the knee (figs. 63.1, 63.2). Further, T_c is the time from the start of the call to F_c. The slope (octave/s) over the first five points of the call is the initial slope (S_1); and the slope of the body, between T_k and T_c, is the characteristic slope (S_c).

The mean frequency (F_{mean}; kHz) is calculated as $F_{mean} = (N - 1)D/2d$, where *N* is the number of points in the call counted from the call display, *D* is the division ratio, and *d* is the duration (ms) of the call. This is a weighted mean—not just the average of the frequency of points—that takes into account the fact that points constituting a call are more widely spaced at lower frequencies. The total time (ms) that a call lasts is the duration (DUR). The time between calls (TBC) is the time (ms) measured from the beginning of one call to the beginning of the next. Additional analytical features are being developed in updated versions of Analook software. For more details on analyses, additional features, and free software updates contact, Chris Corben (www .hoarybat.com).

QUANTITATIVE CALL ANALYSIS

Measurements

To demonstrate the structural characters obtained with Analook, we divided North American vespertilionids into functional groups based upon their call structure (Fenton and Bell 1981; Gannon, Sherwin, and Haymond 2002). One such group included five *Myotis* species—*M. ciliolabrum, M. volans, M. lucifugus, M. californicus,* and *M. yumanensis.* We measured commuting calls because they are the least variable type of call within a species and are less likely to be significantly influenced by proximity to clutter or foraging behavior. We collected commute-phase calls in the field from free-flying or recently captured and released bats using Anabat. Calls were edited to exclude fragmentary calls,

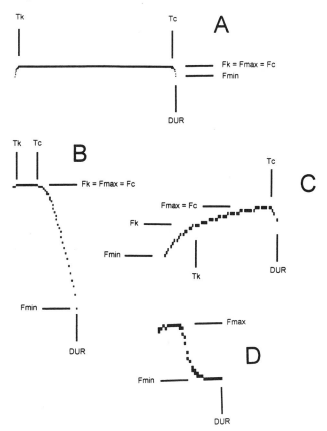

Fig. 63.2. More shapes of echolocation calls from North American bats (see fig. 63.1).

echoes, and extraneous noise prior to extraction of parameter measurements automatically calculated using Analook (figs. 63.1, 63.2).

Statistical analysis

Discriminant function analysis (DFA) is the most commonly used statistical technique for classifying calls into predetermined, a priori groupings (Sherwin, Gannon, and Haymond 2000). Logistic regression, neural networks, and classification trees are alternative statistical methods that can be used to classify bat calls (Cheng and Titterington 1994). Here, we consider the commonly applied discriminant function analysis and classification trees. A classification tree is a nonparametric classification technique (Brieman et al. 1993), in which the tree is grown using a sequential binary partitioning of the predictor variables. The algorithm is similar in spirit to forward selection of variables in multiple regression. Statistical techniques, such as pruning and shrinking, are used with cross-validation to eliminate tree branches that do not improve prediction. Unlike DFA, which is ideally suited for normally distributed data, predictor variables for a classification tree can be categorical, interval, or ordinal data. Another advantage of classification trees is that the output is easy to understand and interpret. In addition, transformations of the predictors are handled automatically, as are interactions

among variables: variables that define a "split" can be used repeatedly and split at different points in different branches of the tree. Trees can be grown using many standard statistical software packages—for example, SPSS, Splus, CART, and SYSTAT. We used Splus in this analysis.

Ten variables were measured from each bat call (figs. 63.1, 63.2): F_{max}, F_{min}, F_{mean}, F_c, F_k, T_c, S_1, S_c, DUR, and TBC. Measurements were taken on 2737 calls from *E. fuscus* (648 calls; 23 individuals), *M. californicus* (1068 calls; 21 individuals), *M. ciliolabrum* (49 calls; 2 individuals), *M. lucifugus.* (655 calls; 30 individuals), *M. volans* (73 calls; 5 individuals), and *M. yumanensis* (244 calls; 9 individuals). Data were randomly partitioned into a training set of 1369 bat calls and a test set of 1368 calls. The training set was used to construct the classification rule, which was then used to predict the species corresponding to the calls in the test set. Stepwise- and backward-discriminant analyses on the training set showed that all 10 features were important for distinguishing species. Hence, the classification rule for DFA was constructed using all variables.

Results

Almost all *E. fuscus* calls were correctly identified using DFA. Of the remaining five species, *M. lucifugus, M. ciliolabrum,* and *M. yumanensis* proved easiest to classify, with misclassification rates of only 6%, 8%, and 9%, respectively. The misclassification rates for *M. californicus* and *M. volans* were somewhat higher, 20% and 25%, respectively. Not surprisingly, most of the misclassified *M. californicus* calls were identified as *M. yumanensis,* and vice versa. The overall misclassification rate, which is the average of the species-specific error rates (assuming equal prior probabilities for all six species), was 11.6%. The raw misclassification rate, or the observed percentage of test cases misclassified, was 11.3% (155 of 1368).

Fig. 63.3 shows the classification tree grown in Splus from the 1369 bat calls in the training set. The pruned classification tree used 7 of the 10 measurements to classify calls: Dur, F_{min}, F_c, F_k, T_c, T_k, and S_1. The tree contained 12 terminal nodes that define the branches or distinct partitions of the predictor space. Each branch of the tree gives a prediction or classification for a call. For example, the terminal node labeled "mylu" corresponds to calls with $F_c < 44.02$ kHz (initial split), $F_{min} > 34.005$ kHz at the next split, and $F_c < 39.07$ kHz at the final split. The condition $F_c < 44.02$ kHz is redundant, given $F_c < 39.07$ kHz, so an equivalent characterization of this branch is $F_{min} > 34.005$ kHz and $F_c < 39.07$ kHz. Each call with these parameters was predicted to be from *M. lucifugus.* The fraction 1/333 below this terminal node was the observed misclassification rate for this branch—that is, the proportion of the 333 calls in the training data with $F_{min} > 34.005$ kHz and $F_c < 39.07$ kHz that are incorrectly classified as *M. lucifugus* in this case is 1.

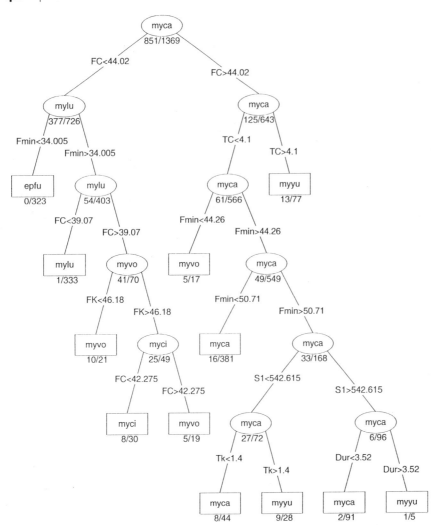

Fig. 63.3. Classification tree grown in Splus from the 1369 bat calls in call data training set. The pruned classification tree used 7 of 10 measurements to classify calls: Dur, F_{min}, F_c, F_k, T_c, T_k, and S_1. The tree contained 12 terminal nodes that define the branches or distinct partitions of the predictor space. Each branch gives a prediction or classification for a call.

The tree was far better than DFA for correctly classifying *M. californicus* calls, but it was less successful than DFA for classifying calls from *M. ciliolabrum* and *M. yumanensis* (tables 63.1, 63.2). The overall misclassification rate from the classification tree analysis was higher than DFA (20.2% versus 11.6%), in part because the tree does not work as well on species with the smallest samples sizes. However, the raw misclassification rate is lower than DFA (9.3% versus 11.3%), which suggests that classification trees might perform as well, or better, than DFA in data sets where the sample sizes for *M. ciliolabrum, M. volans,* and *M. yumanensis* are larger than our sample sizes. In general, our results suggest that classification trees provide a useful alternative to DFA for classifying bat calls.

Discussion

Classification Trees

The traditional use of discriminant function analysis may be warranted with Anabat echolocation calls to classify two species with multivariate data sets (Gannon

et al. 2001; Lance et al. 1996). However, other methods, such as classification trees, may be most appropriate with suites of species using hundreds of calls for each. A recommendation of greater sample size is not surprising, but we found that at least 300 calls are required to provide significant separation between the five *Myotis* species examined. Some authors (e.g., Barclay and Brigham 1996; Grindal 1999; Krusic et al. 1996) either cite low sample sizes or simply do not report the number of calls examined. It is imperative to state not only the number of calls measured, the number of individual bats used in the study, and the characters that were measured and analyzed, but also to list the assumptions that the investigators made in following their study design (Gannon, Sherwin, and Haymond 2002).

Although our intent was to examine classification procedures and exercise a revised call lexicon, we discovered here that we did not have adequate sample sizes for all species. Further, we note that our analyses use multiple calls from individual bats; therefore, all observations are not independent. The intraclass correlation between responses on a bat varies from 0.29 on S_1 to 0.62

TABLE 63.1. Discriminant function analysis test data (number of observations and percentage classified into species).

Species	E. fu	M. ca	M. ci	M. lu	M. vo	M. yu	Total
Eptesicus fuscus	318 98.15	0	0	6 1.85	0	0	324 100.00
Myotis californicus	0	442 80.36	4 0.73	0	7 1.27	97 17.64	550 100.00
Myotis ciliolabrum	0	0	23 92.00	2 8.00	0	0	25 100.00
Myotis lucifugus	1 0.33	1 0.33	16 5.23	287 93.79	1 0.33	0	306 100.00
Myotis volans	0	0	5 15.63	3 9.38	24 75.00	0	32 100.00
Myotis yumanensis	0	12 9.16	0	0	0	119 90.84	131 100.00
Error count estimate	0.0185	0.1964	0.0800	0.0621	0.2500	0.0916	0.1164

TABLE 63.2. Classification trees test data (number of observations and percentage classified into species).

Species	E. fu	M. ca	M. ci	M. lu	M. vo	M. yu	Total
Eptesicus fuscus	322 99.38	0	0	2 0.62	0	0	324 100.00
Myotis californicus	0	515 93.64	0	0	6 1.09	29 5.27	550 100.00
Myotis ciliolabrum	0	0	16 64.00	3 12.00	6 24.00	0	25 100.00
Myotis lucifugus	0	2 0.65	7 2.29	284 92.81	13 4.25	0	306 100.00
Myotis volans	0	0	11 34.38	0	21 65.62	0	32 100.00
Myotis yumanensis	0	48 36.64	0	0	0	83 63.36	131 100.00
Error count estimate	0.0062	0.0636	0.3600	0.0719	0.3428	0.3664	0.2020

on F_{mean}, with an average of 0.52. This indicates that the effective sample size is much smaller than 2737, and that the misclassification rates are less reliable than would have been obtained with independent samples. There are a variety of ways to deal with the dependence, of which we mention two possible approaches. We are developing a discriminant function model with a random effect for bats within species. This model provides for arbitrary numbers of correlated responses within a bat. A simple alternative method is to average the responses on each variable within bats, and then to analyze the averages. In our data, averaging the responses within bats reduces the data set from 2737 calls to 136 call averages. A discriminant function analysis on the averages, using the data both to create the classification rule and then to assess its accuracy, gives a misclassification rate of 2.3%. Although this result is promising, the sample sizes for several species were too small to unequivocally recommend this approach. We are examining these issues more carefully; nonetheless, it appears that those species with 300 calls or 20 individuals were more easily classified than those with less data.

We were able to show that, based on calls, DFA and classification trees can distinguish among species of *Myotis*. The outgroup, *Eptesicus*, separated from *Myotis* species early and consistently in the analysis because these calls are quite different morphologically. Among *Myotis*, species were grouped into 50 kHz and 40 kHz species, similar to the initial qualitative separation used by O'Farrell, Miller, and Gannon (1999). Within these two groupings, bats could be separated further to species. Some classification error (25% in one case) could have been due to our technician's inexperience in using Analook to measure call characters, which likely contributed to the overall error rate. As with many techniques used in biology (for instance, DNA extraction and sequencing), increased experience in using the Analook system and understanding the biological and natural history differences among species could increase species discrimination and classification processes. These results are important for several reasons. Critics (e.g., Barclay 1999; Corben and Fellers 2001; Fenton 2000; O'Farrell et al. 1999; Thomas and LaVal 1988) have cited the impossibility of distinguishing certain species acoustically, particularly among *Myotis*. O'Farrell, Miller, and Gannon (1999) have shown that it is possible to show qualitatively that there are distinguishable differences between species, even within the genus *Myotis*. Here, we have shown that repeatable, quantitative differences exist among six bat species based on defined structural characters of commute-phase echolocation calls.

F IELD R ECORDING OF C ALLS

Because calls were recorded in a field setting, and not all conditions affecting sound propagation and bat behavior were controlled, analysis and interpretation of call characters should not be assumed to be simple or straightforward. For instance, the transmission and reception of sound can be affected by Doppler shifts, echo, or environmental factors (e.g., temperature, humidity, vegetation structure of background). Given the high number of calls we examined, Doppler shift affected all species equally and was not likely a reason for differences found in the analysis. Some species such as *M. yumanensis* (and to some extent, *M. lucifugus*) were recorded directly over water where recordings contained strong echoes.

Although both DFA and classification trees were equivocal in discriminating groups, classification trees may work well or better with programs such as Analook to predict group or species membership through program-controlled filters bundled with analysis packages such as Splus. Lastly, programs such as Analook, and those of other systems, are being improved regularly with greater utility and analysis capability. It is the duty of the practitioner to keep up with these changes and to use the equipment in an informed manner.

Acknowledgments

We gratefully acknowledge all those students who have worked diligently to master the new generation of echolocation call detectors and have applied the methods described herein (and elsewhere) to scientific questions that produce repeatable results. We especially thank Lisa T. Arciniega and Natalie M. Gannon for allowing William L. Gannon the time to thoroughly learn Anabat and its application. The authors thank M. Scott Burt, Luis A. Ruedas, Richard E. Sherwin, and two anonymous reviewers for greatly improving the clarity of this paper. Support from agencies such as the U.S. Bureau of Land Management, the U.S. Forest Service, and the New Mexico Department of Game and Fish (Share with Wildlife program) is immensely appreciated.

{ 64 }

Who's Calling? Acoustic Bat Species Identification Revised with Synergetics

Martin K. Obrist, Ruedi Boesch, Peter F. Flückiger, and Ulrich Dieckmann

Introduction

Nocturnal activity, small size, and secretive roosting habits make species of the Microchiroptera among the most elusive of mammals. As bats constitute about 20% of mammalian species, the assessment of their occurrence and distribution should be part of any efforts to evaluate, conserve, and monitor biodiversity, but bats are often overlooked. Some species of microchiropteran bats are, however, conspicuous because they echolocate to detect, track, and assess airborne prey (usually insects).

Since the early years of echolocation research (Griffin 1958), considerable information has been collected about ecology, echolocation, and physiology of bats (Busnel and Fish 1980; Fenton, Racey, and Rayner 1987; Nachtigall and Moore 1986; Neuweiler 1999). Advancing technology has allowed characterization of echolocation calls of many species. In addition, a moderate financial investment in electrotechnical tools allows interested laypeople to identify some species of bats by their distinctive echolocation calls.

One complication is the fact that bats vary the structure of their calls (Kalko and Schnitzler 1993; Obrist 1995; Rydell 1993) and can emit signals similar to those of other species when facing comparable orientational tasks, such as flying close to clutter. The variety of echolocation calls that any one species of bat produces can lead to partial or complete overlap of the time and/or frequency structure of calls of other species—making species identification impossible despite the use of very expensive tape recorders and analytical tools. In an extensive study on the variability and flexibility of bat echolocation, Obrist (1995) repeatedly discriminated signals not discernible by standard parameter measurements (highest, lowest, loudest frequency, duration, interval, etc.) only by the "pattern" or "shape" of the frequency contour in a spectrogram.

Recent approaches to signal recognition have tried to holistically identify bioacoustic signals by pattern recognition algorithms, artificial neural networks, and methods like decision trees in different organisms like birds (Kogan and Margoliash 1998; H. Mills 1995), frogs (Taylor et al. 1996), whales (Mellinger and Clark 1997), other marine mammals (Fristrup and Watkins 1995), and in humans (Gish and Schmidt 1994). Few studies have tackled bats (Gannon et al., chapter 63, this volume; Herr, Klomp, and Atkinson 1997; O'Farrell, Miller, and Gannon 1999).

Automatic and robust acoustic species identification would be extremely helpful in a variety of research areas. A method capable of recognizing "patterns" of species-specific signals has the potential to identify environmental, behavioral, or individual differences in sonar or communicative sounds from whales to birds to crickets. Therefore, it would be a tremendous tool in scientific research and ecosystem monitoring.

With this project, we wanted to identify all Swiss bat species by their sonar calls using signal features like variations in sweep rate or shape of sonograms. Even in such acoustically similar species as the members of the *Myotis* group, we predicted that a new pattern recognition approach should allow species recognition.

The study had three goals: (1) find a suitable and economical method to record bat echolocation calls in the field, (2) evaluate a pattern recognition algorithm with the potential for fast and automated identification of bat species and optimize it for correct identification of Swiss

bats, and (3) make the whole "recognizer" operable under field conditions, preferably in an unsupervised, automated mode. This text will focus primarily on the second topic.

Materials and Methods

RECORDING

Capture of animals

All recordings were made in northern Switzerland between 28 June and 27 September 1995. From 12 species, a total of 172 recorded sequences (equaling individuals, assuming extremely unlikely rerecording of individuals), comprising 816 calls, were included in the analysis (table 64.1). Calls were recorded from either individual bats when emerging from "single-species" roosts or from bats captured at cave roosts. The latter recordings were made shortly before dawn, when releasing the bats. We tried to minimize environmental noise (wind, leaves, echo clutter) by choosing an appropriate release location. Only the more species-specific search calls or commuting calls (Griffin 1958), starting a few seconds after release, were recorded—avoiding the first few very short and rapid calls typical of bats reorienting after release. Fig. 64.1 shows waveforms and spectrograms of calls typical of the 12 species recorded.

Signal acquisition

All ultrasonic recordings were accomplished with a D140 ultrasonic bat detector (Lars Pettersson Electronics AB, Uppsala, Sweden). The detector's built-in electret microphone picks up the frequencies from 10 kHz to 120 kHz, but sensitivity drops progressively above 80 kHz. Using the available digital time-expansion mode, signals of 0.87 s were sampled at 350 kHz with 8-bit resolution. Such samples contained between 3 (*Nyctalus noctula*) and 15 (*Myotis mystacinus*) calls and are

TABLE 64.1. Recording statistics. Abbreviation of species names (ab), sample size *n* (seq = number of sequences; call = number of calls included in analysis), and call characteristics of the 12 recorded species. Mean ± standard deviation, as well as coefficients of variation (cv), are given for the following parameters: DUR = call duration, LFR = lowest frequency in call, MFR = frequency of main (highest) energy in call, and HFR = highest frequency in call. The sample sizes for species indicated with an asterisk are low.

Call parameter	*n*			DUR (ms)		LFR (kHz)		MFR (kHz)		HFR (kHz)	
Species	ab	seq	call	mean ± SD	cv(%)	mean ± SD	cv(%)	mean ± SD	cv(%)	mean ± SD	cv(%)
Eptesicus serotinus	ES	21	80	7.2 ± 1.7	24	25.5 ± 1.3	5	30.3 ± 2.1	7	57.0 ± 4.4	8
*Myotis bechsteini**	MB	1	5	4.1 ± 0.1	3	23.2 ± 0.9	4	45.2 ± 1.4	3	96.5 ± 4.5	5
*Myotis blythii**	ML	3	16	2.8 ± 0.9	32	27.7 ± 6.0	22	41.4 ± 5.1	12	72.0 ± 14.5	20
*Myotis brandti**	MR	3	19	3.4 ± 0.7	22	27.3 ± 1.5	6	43.4 ± 2.8	6	83.1 ± 11.3	14
Myotis daubentoni	MD	32	146	3.8 ± 0.8	20	27.2 ± 2.0	7	42.2 ± 2.8	7	76.1 ± 10.8	14
Myotis myotis	MM	32	155	4.0 ± 1.2	30	23.7 ± 2.6	11	37.0 ± 3.6	10	77.1 ± 13.3	17
Myotis mystacinus	MY	10	60	3.1 ± 0.9	30	27.8 ± 2.7	10	45.0 ± 4.0	9	85.2 ± 9.0	11
Myotis nattereri	MN	27	149	3.0 ± 0.8	28	17.4 ± 4.6	26	33.7 ± 7.0	21	78.0 ± 14.4	18
Nyctalus noctula	NN	12	24	13.0 ± 3.0	24	21.5 ± 1.4	6	24.2 ± 1.7	7	32.5 ± 8.6	26
Pipistrellus pipistrellus	PP	10	43	5.4 ± 0.9	16	42.5 ± 1.0	2	45.7 ± 1.4	3	84.6 ± 8.6	10
Plecotus auritus	PA	15	90	3.3 ± 0.9	27	23.0 ± 1.8	8	33.3 ± 4.4	13	53.2 ± 2.8	5
Vespertilio murinus	VM	6	29	5.5 ± 1.8	33	22.5 ± 1.0	4	28.7 ± 2.5	9	49.0 ± 5.3	11

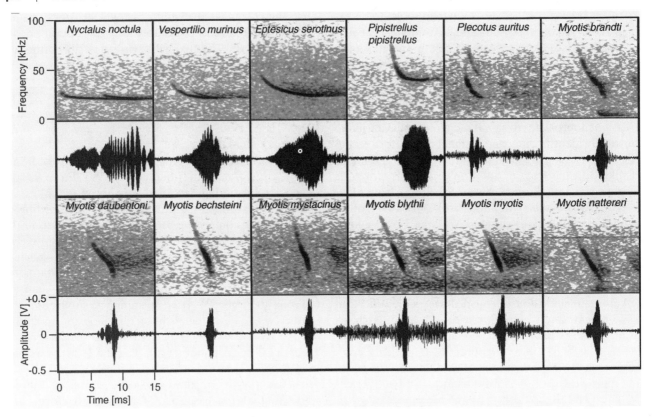

Fig. 64.1. Sonograms and waveform displays of representative calls of 12 bat species. Calls are arranged according to the apparent similarity of their spectrograms.

further referred to as a sequence. Replays performed at one tenth of the original sampling rate were recorded using a Sony WM-D6C Walkman Professional for subsequent analysis in the laboratory.

Direct recording

To allow future field use of our setup, we also implemented software on a Macintosh PowerBook computer to make direct recording of ultrasound signals using a high-speed analog-to-digital converter PCMCIA-card. Recording and pattern recognition algorithm were integrated in one software package, BATIT (BioAcoustic Taxa Identification Tool), and the provided interface to the Macintosh scripting language (AppleScript) enables implementation in an automatic acoustic "species-logger."

CONVENTIONAL SIGNAL ANALYSIS

Cassette recordings were digitized using an Apple Macintosh PowerPC 8500 with 22.05 kHz sampling rate at default 16-bit sampling depth. After extraction with the automated algorithm (see "Preprocessing," below), signals submitted to the synergetic computer (SC, see below) also were analyzed with Canary software (Charif, Mitchell, and Clark 1995) to measure parameters from the sonograms (table 64.1). Time and frequency measurements (duration, DUR; lowest frequency, LFR;

highest frequency, HFR; frequency of maximum energy, MFR) were taken from spectrographic representations of the calls. Spectrograms contained temporally overlapping windows, which possibly led to systematic but minimal error in measurements of call durations (time-smearing) when compared to published signals of the species in question. However, analyzing a spectrographic display was preferred over analyzing an amplitude display and a spectrum display, because low-intensity parts of calls (invisible in the latter) led to decreased readings for high frequency, low frequency, and duration.

SIGNAL ANALYSIS WITH A SYNERGETIC COMPUTER

Pattern recognition

For the identification of bat calls we used the classical pattern recognition system. It consists of several modules (fig. 64.2). In general, the system is able to learn a representative set of calls from each species. Using these sets, the system assigns newly recorded calls of an unknown species to one of the learned species. The output consists of a position table on which the most likely species heads a list of other possible species, in decreasing order of probability.

Preprocessing

Call sequences stored in binary format with Canary or BATIT can be input directly to preprocessing and the

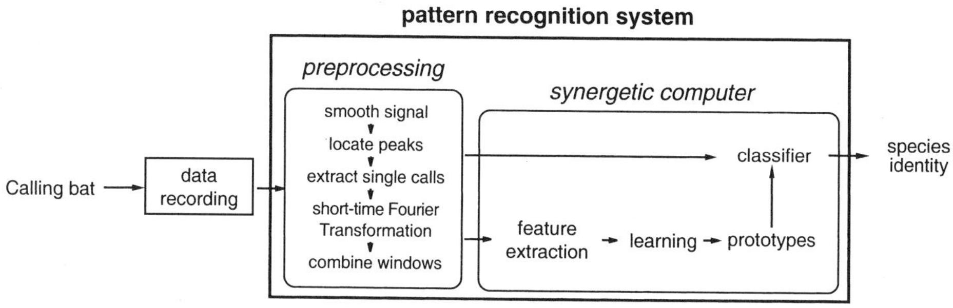

pattern recognition system

Fig. 64.2. Schematics of the data processing, using synergetic computer explained in text

classification algorithm of the SC. Recorded sequences show three main different parts: the call itself, the echo (not always present), and noise. Although the inter-call interval or repetition rate can be species-specific in some bat populations (Fenton and Bell 1981), we neglected this parameter and concentrated on analyses of single vocalizations for the following reason: preliminary tests with short-time Fourier transformations (STFT) of whole sequences showed bad recognition results, because (a) too much noise was included; (b) the time interval between two consecutive calls often varied too widely; (c) signal characteristics of echolocation calls did not resolve enough in the available spectrogram resolution (memory restrictions); and (d) long silent intervals dominated spectrograms of complete sequences, leading to overall less discernible segments. To analyze and classify, it was therefore necessary to extract each call from a sequence.

Call extraction

Peaks were identified in a signal previously smoothed with a 256-point moving average filter. To ensure a good signal-to-noise-ratio and the detection of all suitable calls, only values 16.9 dB in amplitude ($10 \cdot \log [50]$) above the average signal level were considered as peaks. This level was commonly encountered in quiet recording situations or achieved in noisy recordings by high-pass filtering above 10 kHz. After detection of a peak, 3072 samples before and 5120 samples after the peak's position were extracted, thereby creating short signals of 8192 samples.

Separating the actual call from its echo was attempted in some species (e.g., *Nyctalus noctula* in fig. 64.1). All modes of manipulation such as filling in with noise or cutting before echo introduced artifacts. This led us to leave extracted signals untouched. Therefore, the extracted calls often contained a considerable amount of interfering echoes from ground or water reflections, and ambient noise, wind, Doppler shifts, and other interferences (Pye 1993).

Fourier transformation

After detection, the STFT for each call was calculated (amplitude spectra = sqrt[real · real + imag · imag],

256 samples, Hamming window, 81% overlap). These calculated windows were concatenated to one feature vector of 20352 floating point numbers normalized to 1, which is equivalent to a spectrographic display of 159 spectra each containing 128 data points (see fig. 64.2). Phase was therefore neglected from analysis; only magnitude was considered.

Synergetic computer

Synergetics is an interdisciplinary field that deals with self-organizational phenomena in nature (Haken 1978; Kohonen 1984) characterized by many microscopic parts in an unsorted order (chaos) transforming themselves in a sorted order. The importance of each part is minor; only the properties of the whole system are relevant and can be described through synergetic differential equations.

The *synergetic computer* (SC) is not an actual computer per se, but a new set of algorithms emanating from this interdisciplinary field; it has only recently been used for classification tasks (Haken 1988; Haken 1996; Wagner et al. 1993; Wagner, Schramm, and Boebel 1995). For the classification of bat calls we used an algorithm termed SC-MELT (Dieckmann 1997). One significant advantage of this algorithm is its ability to combine several training patterns per class into one feature vector without losing any information about the training patterns. The training patterns are "melted" into one prototype, which has the same dimension as the training vectors and is normalized to length 1. This enables the SC to handle big dimensions in contrast to artificial neural networks (ANN). The computational power needed to train an ANN with input vector of 16384 features is prohibitive. Description of the synergetic computer using adjoint prototypes (SCAP) are given by Hogg and Talhami (1996) and Wagner (1993). The adjoined prototypes of the SC-MELT are achieved through simple addition of the corresponding, adjoined prototypes. One of the most interesting properties of the SC is that it emphasizes pattern contents that are unique among all others while diminishing pattern contents common to all others. The learning time of the SC is easily determined and very fast. The generation of the prototypes of the bat calls takes only a couple of minutes on a 200 MHz 603e

RISC processor (such as an Apple Macintosh Power-Book 3400). The classification is even faster, because it is simply a scalar or dot product.

Classification

Assume we let the SC train three calls of each of 10 species (classes), resulting in 10 prototype feature vectors. We now test 50 echolocation calls of an unknown species contained in the training base. The SC computes the scalar product of each test call with each class prototype, resulting in 10 values per call, varying between 1 (identical call as in training base) and 0 (no resemblance to any of the training calls). The training class with the highest scalar product identifies the species where the calls most likely came from. Some signals invariably get assigned to wrong species, resulting in a frequency distribution over the 10 classes. The peak in the frequency distribution finally identified the best-fitting species.

Dependencies of recognition success

To evaluate the effect of training call selection on recognition success, we repeated the training and classification process 115 times with our data set. Calls included in the training base were drawn from one to nine sequences. The number of calls per sequence was varied from one to five. Each combination of sequence and call selection was repeated five times, with different sequences and calls each time. These selection criteria were restricted by available computer memory, not allowing for more than 12 calls per species (144 training calls). For 2 species (*Myotis blythii* and *Myotis bechsteini*), we had to use identical training bases repeatedly due to a lack of recordings. For each of the resulting 115 repetitions, we counted for every species the number of correctly assigned calls and the number of misclassified calls, and we noted whether the correct species was the most common classification or not. These classification results were analyzed by multivariate analysis of variance (MANOVA) for the effect of number of sequences, number of calls, and number of calls per sequence on recognition success.

As described above, the higher the maximum scalar product, the clearer the recognition of a tested call. Similarly, a great difference between the highest and second-highest scalar product indicated a clear differentiation between classes. Both parameters—maximum scalar product (MSP) and difference to second-best (DSB)—were used to delimit the number of valid calls. We systematically tested all combinations of the MSP (0, 0.4, 0.5, 0.6) and the DSB (0, 0.1, 0.2, 0.3) to reject calls from the classification. Similar to a rejection criterion like signal-to-noise ratio, this eliminated a number of calls from the original base of calls, thereby changing the number of cases in subsequent analyses.

STATISTICAL ANALYSIS

Using DataDesk software (Data Description Inc., Ithaca, N.Y.), we performed a linear regression analysis to test whether the increase in delimiting parameters (MSP, DSB) improved classification. The influence of species, number of learned calls, and number of sequences (individuals) on the classification was tested using ANOVA. To compare the identification abilities of the SC, we performed multivariate analyses on the four parameters measured with Canary. We used DataDesk for a MANOVA with subsequent post hoc tests (Sheffé) to separate means of species' call characteristics. With the same software we performed a single linkage cluster analysis. SAS (SAS Institute Inc., Cary, N.C.) statistical software was used for a discriminant function analysis with reclassification for a cross-validation. A significance level of $p \leq 0.05$ was applied in all statistical tests.

Results

CLUSTER ANALYSIS

The cluster analysis clearly separated *Nyctalus noctula* from the other species due to the bandwidth and the duration of its calls (compare spectrograms in fig. 64.1). *Pipistrellus pipistrellus* also stands apart from the other species, but less clearly. *Vespertilio murinus, Eptesicus serotinus,* and *Plecotus auritus* are separated from the *Myotis* (fig. 64.3).

MULTIVARIATE ANALYSIS OF VARIANCE

MANOVA showed significant variance in all four call parameters. The Scheffé post hoc test separated most species by one or more of their mean call characteristics, most often by MFR and LFR followed by HFR and DUR, reflecting increasing variability (CV) in these parameters (table 64.1). However, *Myotis mystacinus, M. bechsteini, M. blythii,* and *M. brandti* were difficult to separate solely by the measurements taken with Canary.

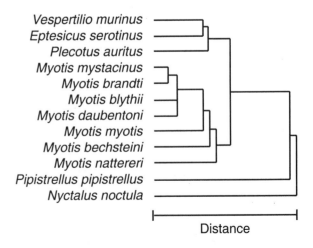

Fig. 64.3. Single-linkage cluster analysis of species according to mean call parameters DUR, LFR, MFR, and HFR

The reason remains unclear for the first species, but it might be the relatively small sample size available from the latter three species (5, 16, and 19 calls respectively; see table 64.1).

DISCRIMINANT FUNCTION ANALYSIS

Using only the parameters in table 64.1, discriminant component analysis properly identified 46% of all *Myotis* calls and 60% of all species' calls, when splitting the data pool in half and testing the second half. Considering only non-*Myotis* species, a recognition success of 82% was achieved. But after splitting the data pool in four parts and testing the remaining three quarters of the signals, recognition success deteriorated to 45% (*Myotis* group), 58% (all species), and 78% (non-*Myotis* species).

SYNERGETIC COMPUTER

Using the synergetic computer to classify patterns of spectrograms of echolocation calls was surprisingly effective. In some cases, a single call was sufficient to characterize the species, as seen in fig. 64.4. However, most species showed considerable variations in call contour (see CV in table 64.1). Accordingly, at least five optimally chosen vocalizations had to be learned from every species to cover a reasonable part of the flexibility and make classification reliable and results consistent.

We performed a variety of exploratory statistical analyses to evaluate the influences of different variables on classification. In the remaining statistics, *Myotis bechsteini, Myotis blythii,* and *Myotis brandti* were not considered, because too few recordings from these species were available for statistical analysis.

As expected, ANOVA indicated a highly significant influence of species on the classification success. Averaged over 115 trials with one to nine training calls per

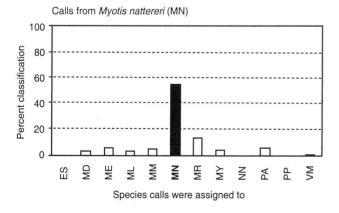

Fig. 64.4. Histogram of classification results, indicating percentage of calls assigned to various species. In this run, 82 of 148 test calls recorded from *Myotis nattereri* (55%) were assigned to the correct species. Only one call from each species was used as training base. (See table 64.1 for species abbreviations.)

TABLE 64.2. Average percentage of correctly classified calls relative to the total number of calls passing the delimiter restrictions. Results are shown for five and nine training calls with delimiter restrictions MSP > 0.5 and DSB > 0.2. The last column indicates the percentage of all calls rejected by the delimiter restrictions. The sample sizes for species indicated with an asterisk are too low to be conclusive (see also table 64.1.).

No. of training calls	% correctly classified		% rejected
	5	9	9
MSP/DSB	0/0	0.5/0.2	0.5/0.2
Species			
Eptesicus serotinus (ES)	42	60	34
Myotis bechsteini (ME)*	74	94	77
Myotis blythii (ML)*	64	100	3
Myotis brandti (MR)*	31	90	18
Myotis daubentoni (MD)	18	42	43
Myotis myotis (MM)	16	53	51
Myotis mystacinus (MY)	34	64	42
Myotis nattereri (MN)	39	93	30
Nyctalus noctula (NN)	71	98	17
Pipistrellus pipistrellus (PP)	60	93	16
Plecotus auritus (PA)	36	80	42
Vespertilio murinus (VM)	58	90	19
Avg. *Myotis*	39	77	38
Avg. non-*Myotis*	53	84	26
Average %	**45**	**80**	**33**

species, *Nyctalus noctula* scored best with 75% correct classifications, while calls of *Myotis daubentoni* had a correct classification rate of only 32%. Example results of tests with five and nine training calls are given in table 64.2.

Classification success (percentage of correctly classified calls) increased with the number of calls in the training base, except in *Nyctalus noctula,* where classification success reached a plateau at 100% with only one training call. The general impression, that the increase in success rate started to level out toward nine training calls per species, could not be further investigated due to lack of recordings and computer memory restrictions (figs. 64.5, 64.6).

In all species, the recognition success was dependent on the number of sequences included in the training base, probably due to the inclusion of more variability occurring among recordings of different individuals.

Increasing the number of training calls drawn per sequence leads to a marginal increase in recognition success, underlining high inter-individual and lower intra-individual variability in echolocation calls (Obrist 1995). Recording and hardware restrictions again prohibited further analysis; it was not known whether increasing the number of calls per sequence would saturate or even decrease recognition success at higher training call numbers.

We used the maximum scalar product (MSP) and the

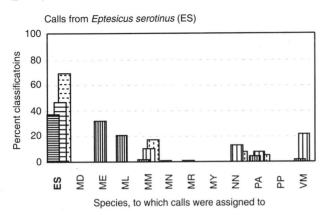

Calls from *Eptesicus serotinus* (ES)

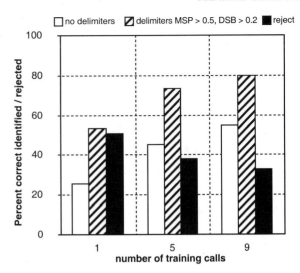

☐ no delimiters   ▨ delimiters MSP > 0.5, DSB > 0.2   ■ reject

number of training calls

Fig. 64.6. The effect of varying numbers of training calls (all taken from different sequences) and delimiter values (MSP, DSB) on classification performance and rejection of calls. Due to the limited number of recorded sequences and computer memory restrictions, *N* varies: 25, 10, and 5 runs are averaged for one, five, and nine training calls respectively.

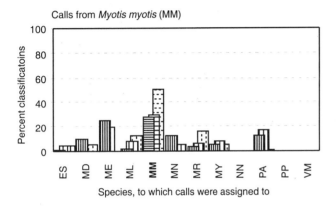

Calls from *Myotis myotis* (MM)

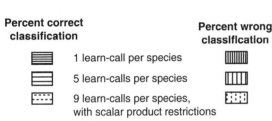

Percent correct classification		Percent wrong classification
▤	1 learn-call per species	▥
▤	5 learn-calls per species	▥
▦	9 learn-calls per species, with scalar product restrictions	▦

Fig. 64.5. Three example results of the classification of two species' calls (*Eptesicus seritonus* and *Myotis myotis*) with the synergetic computer. The algorithm was trained using one, five, and nine calls. When including five or more calls in the training base, the maximum identification score is always (also in the other 10 species not shown here) at the correct species (horizontally striped bars). Misclassified calls (vertically striped bars) most often fall into classes of species, which proved very similar in the cluster analysis (fig. 64.3). (See table 64.1 for species abbreviations.)

difference of the second-best scalar product (DSB) to evaluate a combination of both, which achieves the highest classification results with the lowest rejection of calls. For all species, any increase in MSP (0, 0.4, 0.5, 0.6) and any increase in DSB (0, 0.1, 0.2, 0.3) resulted in a significantly higher recognition success ($p \geq 0.001$; calls of *Myotis blythii* and *Myotis bechsteini* not considered), at the cost of rejecting a number of calls (figs. 64.5, 64.6). While 57% of all calls were correctly classified (0% call

rejection) when MSP and DSB were both set to 0, average recognition rate increased to 62% (with 23% rejection) for MSP > 0.4 and DSB > 0.1. Even more (71%) were correctly classified with MSP > 0.6 and DSB > 0.3, but with these settings 56% of all calls where rejected. Judging from these results, we considered a combination of MSP > 0.5 and DSB > 0.2 as adequately delimiting rejection criteria for a call, offering the best compromise between recognition success and call rejection. Using nine training calls per species and applying the above rejection criteria, we reached an average recognition rate of 80% at the cost of rejecting 33% of all detected signals. For reasons of call similarity (fig. 64.1), this result was slightly different between *Myotis* and non-*Myotis* species. Increasing scalar product criteria further led to a loss of more than half the data recorded.

Discussion

Using a synergetic computer for pattern recognition, we identified calls of 12 bat species from spectrograms of their sonar calls with variable, mostly high, degrees of success. Recognition success varied with species and with the size and variety of the training base. Applying various restrictions helped to improve classification results.

CLASSICAL RECOGNITION TASKS

Since Griffin first reviewed bat echolocation (1958), species have been known to differ in the structure of their vocalizations. Many publications have since dealt with the identification of echolocating bats by their calls

(Ahlén 1981; Fenton and Bell 1981; Vaughan, Jones, and Harris 1997b; Zingg 1990). Despite the fact that Griffin (1958) observed plasticity in call structures of individual bats, this topic only recently has received extensive attention (Betts 1998; Obrist 1995; Rydell 1993; Zbinden 1989). Variability is one problematic aspect of acoustic monitoring, the training of qualified personnel with detectors yet another. The intensity of vocalizations can vary within 30–50 dB between bat species. Therefore, some species will invariably be more conspicuous than others and are overrepresented in any acoustic monitoring study. Comparing mist-net catches and acoustically recorded bat passes can show corresponding activity patterns in temperate regions (Kunz and Brock 1975), but this seems not to be the case in southern Africa (Rautenbach, Fenton, and Whiting 1996). Nevertheless, acoustic monitoring coupled with the ability to distinguish species makes monitoring studies feasible (Fenton 1970; Hickey and Neilson 1995), at least in temperate regions and when counting all *Myotis* as a single species. Several thorough studies proved the potential of acoustic monitoring of habitat use by bats, as well as its limitations (e.g., Furlonger, Dewar, and Fenton 1987; Vaughan, Jones, and Harris 1997a). Limitations include the necessity for trained personnel to spend long nights in the field, and the inability to separate species within the genus *Myotis* or similarly calling groups (e.g., Phyllostomatidae) in the field. *Myotis* contribute significantly to biodiversity: 9 out of 25 bat species (36%) occurring in Switzerland (Hausser 1995) belong to *Myotis*. Only recently was an acoustic identification of *Myotis* species by statistical analysis achieved (Vaughan, Jones, and Harris 1997b).

PATTERN RECOGNITION APPROACHES

In the past decade there has been an increased interest in automatic recognition of acoustic patterns. Human speaker identification is most prominent and voluminous in literature (Gish and Schmidt 1994) and has mainly commercial aspects, in contrast to the identification of species (Mills 1995; Taylor et al. 1996) or behaviors (Mellinger and Clark 1997; Potter, Mellinger, and Clark 1994). Species recognition approaches generally use decision trees (Quinlan 1993) or artificial neural networks. Synergetic recognition algorithms have so far been applied to speaker recognition (acoustic and visual, Dieckmann 1997) or industrial part identification tasks (Wagner et al. 1993) with impressive results. The ability to learn test patterns unsupervised and to classify full spectrograms at high rates makes synergetics a promising technique for animal vocalization recognition.

Statistical analyses can reach high recognition rates for non-*Myotis* species vespertilionids. Zingg (1990) achieved 86% using five signal parameters; Vaughan, Jones, and Harris reached 89% (1997b, 67% for Myotis) using six parameters, including the call interval. Using only four call measures, we achieved comparable recognition (82%) with the discriminant function analysis. For the genus *Myotis,* performance dropped below 50% in our tests. However, the a priori probability of identifying the proper species by chance is only 8%.

The success rate of the SC compares favorably to these numbers. When carefully choosing the learning base and rejecting ambiguous calls, the SC reached average call recognition rates of 77% for the genus *Myotis* and 84% for other vespertilionids (average 80%). It is possible to improve these results further. First, digital recording of the ultrasound as 12-bit data will substantially improve signal quality. Second, the preparation of the spectrogram (window size and shift) could be further optimized for the SC, and, with more available RAM, we will be able to train the SC with more species and better evaluate the effect of increasing call numbers per species. Finally, the inclusion of the intercall interval in the species' characterization could further improve performance. Based on results from studies using similar algorithms in industry (99% recognition, Wagner et al. 1993) and "human" applications (93% recognition, Dieckmann 1997), and taking into consideration the substantial variability occurring in echolocation (Obrist 1995), we hope to reach an overall bat species recognition success of approximately 90%. Only then do we intend to make the SC system available to fellow scientists.

For the characterization of and quantitative comparison among species or behaviorally specific call characteristics, standard statistical analysis of measured call parameters remains the only choice. But, for monitoring purposes, the pattern recognition approach rivals this method and certainly is faster and more economical.

We estimate a total throughput (cut calls, calculate spectrograms, classify, write output) of five to six calls per second on a current Macintosh laptop. As this is roughly the rate at which an average bat emits vocalizations, the system approximates real-time performance, but it lags behind real time by the duration of the recording plus the processing time. If none or only some of the signals are kept for later reference, the system could autonomously log species-specific bat activity for a number of nights, powered by a portable 12 V battery. With falling prices of increasingly powerful laptop computers and the availability of high-speed PCMCIA data acquisition cards, direct field recording of ultrasound to computer hard disk will become economic and widespread in the very near future.

Our proposed system constitutes a so-called black box. The user inputs signals and the machine identifies the emitter of the sounds. Many field-workers will object to this approach because no control can be attained over species classification. We do not encourage inexperienced people to feed data into our system without prior knowledge of bat echolocation calls and inherent variability. Special care has to be taken when compiling a valid training base for the synergetic computer. Considering geographic variations of sonar characteristics, compilation of a single training base for a certain species will likely never be feasible. Rather, a variable approach will be more realistic, where qualified observers compile their own database for their regional coworkers. However, our system is attractive because of its ease of operation, cost-effectiveness, and the potential for auto-mated monitoring of any acoustically conspicuous species assemblage.

Acknowledgments

We are very grateful to Lars Pettersson for his generous help with recording hardware.

{ 65 }

Synthetic Aperture and Image Sharpening Models for Animal Sonar

Richard A. Altes

Introduction

Animal sonar systems typically are characterized by large bandwidths, motion of the transmitter/receiver, and small aperture (array size) relative to many human-made sonars. Such systems have cross-range (azimuth and/or elevation) resolution that is much worse than their range resolution. For bats and dolphins, the theoretical disparity between range and cross-range resolution becomes large for ranges in excess of half a meter. This disparity can be mitigated by using synthetic aperture sonar (SAS). SAS forms an acoustic image in which cross-range resolution is commensurate with range resolution. Conventional SAS, however, may be too complicated for implementation by biological systems.

Recent results indicate that synthetic aperture processing can be greatly simplified. By considering simplifications that allow for biological implementation and generalizations that can emulate animal echolocation capabilities, SAS processing has been advanced beyond the previous state-of-the-art humanmade systems. These advances involve high-resolution feature images, image-based tracking for motion compensation, and the utilization of all available knowledge for acoustic imaging. Such knowledge includes prior expectations, nonacoustic sensory information, and acoustic information not explicitly associated with imaging, such as resonances.

Doppler-Based SAS

Some synthetic aperture systems depend upon Doppler sensitivity, while others are Doppler tolerant. A Doppler-based system utilizes the angle dependence of range rate for a moving sonar and a stationary scattering point. A relatively large range rate is observed along the path of motion, and zero range rate is observed orthogonal to the path of motion. Thus, a mapping exists between azimuth angle and range rate. For side-looking systems, an object appears to rotate relative to the sonar, and this rotational motion can be used to form an image of the object.

Range rate can be measured with a single transmission if the transmitted waveform has sufficient time-bandwidth product to estimate relevant Doppler-induced time compressions of the signal (Altes 1971, 1995; Altes and Skinner 1977). If a single pulse has insufficient time-bandwidth product for Doppler-based angle measurements, then a fully coherent system can use multiple pulses for range-rate estimation. Many echoes are stored and processed as though they are all obtained from a long-duration transmitted signal that is composed of many transmitted pulses. Even if a new pulse is transmitted only after the echo from the previous pulse is received, the new pulse is part of the composite signal. This "pulse-Doppler" process uses a coherent integration time that extends over multiple transmissions and receptions. Different scattering points correspond to different range-rate versus time histories, and a coherent pulse-Doppler processor forms a multipulse matched filter for each image point (Brown and Porcello 1969; Cutrona 1975; Wehner 1987).

Doppler-Tolerant, Tomographic SAS

It is uncertain whether echolocating animals are capable of multipulse, coherent processing (Menne and Hackbarth 1986; Menne et al. 1989; Simmons, Moss, and Fer-ragamo 1990). A conservative model assumes that this capability is lacking. The following discussion shows that a wideband, Doppler-tolerant, tomographic synthetic aperture processor can be simplified to remove the requirement for fully coherent processing. In the case of dolphins, even the single-pulse matched-filter assumption (or a process that is equivalent to matched filtering) is not required.

A moving sonar transmits signals and receives echoes from a sequence of points along its path of motion. The receiving points along the sonar's path are regarded as the locations of elements that are part of a large, synthetic array. This synthetic array can focus on a particular point by delay-and-sum beam forming. A compensatory delay is inserted at the output of each element, such that all the echoes from a given scattering point occur at the same time. The resulting time-registered echoes are then added. The delay-and-sum process is equivalent to forming a spatial matched filter for echoes from the chosen scattering point.

The delay-and-sum process is also equivalent to reconstructing an image from its projections (Munson, O'Brien, and Jenkins 1983). Projections occur naturally in radar/sonar data. All scattering points that are within the physical beamwidth of the sonar and that are at the same range (i.e., that lie along the same constant-range surface) contribute to the same echo sample. The sequence of echo samples on an A-scan (matched-filter response versus range) represents a projection of the scatterer reflectivity distribution along the range axis. Different transmitter/receiver locations correspond to different propagation directions, and thus to different projections of the scatterer reflectivity distribution as shown in fig. 65.1.

Several methods can be used to reconstruct the reflectivity distribution from its projections (Macovski

1983). The back-projection algorithm is nearly identical to delay-and-sum beam forming (Altes, Moore, and Helweg 1998). Back-projection or delay-and-sum beam forming can be implemented sequentially, such that the reflectivity estimate of each image pixel is updated with each new echo. The echo sample that corresponds to a given target point or image pixel is added to the sum of previous samples (one from each previous echo) that correspond to the same pixel. At a given sonar location, the echo sample corresponding to a chosen pixel also corresponds to all the other target and clutter points at the same range. At a new location, the constant-range surface is rotated, and the echo sample for the chosen pixel corresponds to other points that are located on a *different* constant-range surface, as in fig. 65.1.

Delay-and-sum synthetic aperture processing does not require Doppler information. In fact, processing would be simplified if the sonar were to stop at each synthetic array element location, transmit a signal, receive the resulting echoes, and then move to the next transmit/receive location. To take advantage of this simplification without stopping, the system can use Doppler-tolerant signals, such that the matched-filter response is not sensitive to range rate.

Another simplification is to use noncoherent delay-and-sum beam forming, such that matched-filter envelopes are used and phase is discarded. This simplification is feasible with wideband signals, since the envelope of the matched-filter response contains comparatively few oscillatory "fine structure" peaks, which correspond to phase information. The resulting processor is semi-coherent; a matched filter is used for each echo, but different echoes are noncoherently combined by summing the envelope-detected matched-filter responses.

Wideband, short-duration pulses such as those used by dolphins are Doppler tolerant with or without semi-coherent processing. Long-duration wideband signals

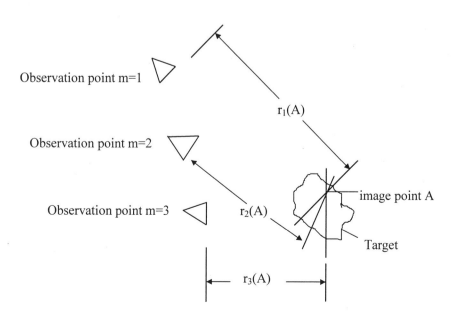

Observation point m=1

Observation point m=2

Observation point m=3

$r_1(A)$

$r_2(A)$

$r_3(A)$

image point A

Target

Fig. 65.1. As a sensor moves, it creates a synthetic array with elements $m = 1, 2, 3 \ldots$ The range of point A is different for each of these elements. To focus the synthetic array on point A, the corresponding delays are compensated by a delay-and-sum operation. Each line through point A represents part of a constant-range surface that lies within the beamwidth of the transmitter. All the scattering points that are on a line contribute to the mth echo at range $r_m(A)$.

with hyperbolic frequency modulation (linear period modulation) such as those used (or approximated) by many FM bats are Doppler tolerant when they are processed by a semicoherent receiver (Altes and Titlebaum 1970; Altes 1973, 1990; Altes and Skinner 1977). In the case of dolphins, a matched filter may be approximated by the bandpass operation of the receiver, since the signal has very small time-bandwidth product. A tomographic SAS model for dolphins can thus use noncoherent processing without a matched-filter assumption.

A tomographic SAS model for bats requires pulse compression via matched filtering, inverse filtering, or an equivalent process, together with noncoherent echo summation capability. A process that is equivalent to matched filtering or inverse filtering can be synthesized by spectrogram correlation (Altes 1980; Saillant et al. 1993) or by a time-frequency plane version of the top-down, bottom-up gradient descent process to be discussed below.

The Range, Cross-Range Ambiguity Function (SAS Point-Spread Function)

The delay-and-sum receiver response to a point scatterer is the sum of the rotated constant-range curves in fig. 65.1. This sum is shaped like an asterisk. For M different sensor positions, the center point of the asterisk is M times larger than an individual line, as shown in fig. 65.2.

The back-projection SAS image of a sampled reflectivity distribution is a superposition of weighted, shifted versions of the function in fig. 65.2, where the weights correspond to the sample point reflectivities and the shifts correspond to the locations of the sample points. This weighted sum is a discrete convolution operation.

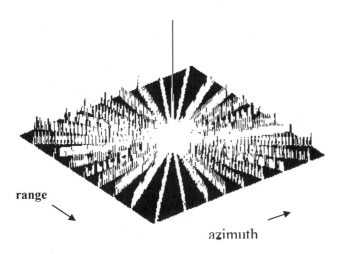

range

azimuth

Fig. 65.2. The point-spread function or SAS range, cross-range ambiguity function (RCRAF) of a tomographic SAS with 12° angle increments over 180°. Peak-to-sidelobe ratio (P/S) equals the number of echoes used to construct the image: 180/12 = 15.

The image of the reflectivity distribution is the convolution of the function in fig. 65.2 with the actual distribution. If envelope detection is used (as in a semicoherent processor), the estimated acoustic reflectivity samples are constrained to be real and nonnegative. In this case, the image is the convolution of the function in fig. 65.2 with a real, nonnegative version of the actual distribution.

For semicoherent processing, the image (i.e., the estimated reflectivity distribution) is a nonnegative version of the actual distribution convolved with (or smeared by) the function in fig. 65.2, which is known as the point-spread function. The function in fig. 65.2 is also a type of ambiguity function, representing the response of a receiver that hypothesizes a single point scatterer at the center of the asterisk, when the point scatterer is actually at various points on the image plane in the figure. The function in fig. 65.2 is the range, cross-range ambiguity function (RCRAF) of the imaging system, as well as the point-spread function (Altes 1979, 1995). This ambiguity function has a peak-to-sidelobe ratio of M, where M is the number of different sensor positions or elements in the synthetic array. When M is small (as in a monaural or binaural system with no SAS capability), a strongly reflecting point at one location can severely affect the image of a weakly reflecting point at a different location. The strength of this interaction depends upon the relative locations of the two points.

If the peak-to-sidelobe ratio (P/S) of the RCRAF is small, different points on a distributed target can interfere with one another, leading to a self-clutter effect. A target that is surrounded by other scatterers also will be difficult to detect. A psychometric procedure that measures angular accuracy or resolution between closely spaced points may depend only on the sharpness of the central peak of the binaural RCRAF, and it can be misleading with respect to detection in clutter and classification of distributed targets. For a binaural system, the RCRAF has P/S = 2 and is scissor-shaped. As a binaural sonar approaches a target, the angle between the scissor blades increases. If range resolution is sufficiently fine (if the width of the scissor blades is small), then P/S becomes larger and the image becomes less ambiguous as the target is approached. Binaural processing thus can be used with forward-looking SAS to create an acoustic image via semicoherent echo summation.

The Algebraic Reconstruction Technique (ART) and Top-Down, Bottom-Up Processing

A range sample of a pulse-compressed echo amplitude versus range plot (A-scan) corresponds to the projection (the sum of the reflectivities) of all the pixels in a constant-range surface. If the aspect changes, the measured sample value represents one of many simultaneous linear equations that theoretically can be solved to obtain the reflectivity distribution. The corresponding matrix equation can be solved by a gradient descent optimization technique. Echo samples that are generated from a model of the reflectivity distribution are compared with actual echo data samples, and the difference is used iteratively to improve the model. This method is called the algebraic reconstruction technique (ART) (Macovski 1983). ART is a special case of an iterative sharpening or deconvolution algorithm.

Top-down, bottom-up processing is a cognitive model that describes the interaction between an internal (top-down) representation and sensory input (bottom-up) data (Anderson 1980). Comparison of predicted and observed data yields a correction or modification of the internal representation. The comparison and correction can be implemented with a gradient descent algorithm in which a high-resolution, cortical model is used to synthesize input data that is observed at a peripheral neuronal processing center or at the sensor. The comparison can occur in a low-resolution representation—for example, in a cochlear time-frequency representation or in the superior colliculus (Capuano and McIlwain 1981).

For a cochlear time-frequency representation, the algorithm implements an adaptive version of a spectrogram correlator (Saillant et al. 1993; Altes 1980). The spectrogram correlation process is not an end in itself but part of a mean-square difference computation. The corresponding difference between predicted and observed spectrogram data is used to update an internal model. The final version of this model after convergence of the iterative algorithm is the desired high-resolution time-frequency distribution.

For a range-bearing (ART-SAS) image, the mean-squared error between the echo and model representations is

$$MSE(\underline{A}) = (2\pi M R_{\max})^{-1} E$$
$$\cdot \left\{ \sum_{m=1}^{M} \int_{0}^{2\pi} \int_{0}^{R\max} [echo_m(r,\theta) - model_m(\underline{A};r,\theta)]^2 \, drd\theta \right\}$$
$$(65.1)$$

where the parameter matrix A contains sample (pixel) values a_{ij} of the high-resolution representation and $E\{\bullet\}$ denotes an ensemble average over various realizations of the echo and the model. The echo varies because it is corrupted by noise, and the model may vary because it can include stochastic neuronal responses. The sum over m represents observations from M different aspect angles, as in fig. 65.1. The simplest echo model is a smeared version of the high-resolution representation:

$$model_m(\underline{A};r,\theta) = \sum_{i,j} a_{ij} smear_m(r - r_i, \theta - \theta_j) \quad (65.2)$$

The smearing function represents the loss of angular resolution associated with a wide physical beamwidth as

well as smearing introduced by the auditory system. If the signal is a short-duration pulse, the mth echo has a linelike smearing function that is orthogonal to the propagation direction, as in fig. 65.1.

The gradient descent technique iteratively solves for the high-resolution samples a_{ij} via the recursion

$$a_{kl}(n+1) = a_{kl}(n) - \mu \frac{\partial MSE(\underline{A})}{\partial a_{kl}} \quad (65.3)$$

for all values of k and l where r_k, θ_l is a point on the image plane, a_{kl} is the image amplitude at r_k, θ_l, and n is the number of times the iteration has been repeated. The LMS stochastic gradient algorithm uses only the squared error rather than the mean-squared error, yielding a simpler updated equation:

$$a_{kl}(n+1) = a_{kl}(n) - \mu \frac{\partial SE(\underline{A};r,\theta,m)}{\partial a_{kl}} \quad (65.4)$$

for each point on the image plane.

In the LMS algorithm, the iteration is computed at successive values of the variables r, θ and observations $m = 1, \ldots, M$. This iteration approximates an ensemble average over multiple observations (Widrow and Stearns 1985; Honig and Messerschmitt 1984; Rumelhart, Hinton, and Williams 1986).

The gradient in eq. 65.4 depends only upon the error at a particular r, θ value and the partial derivative of the model with respect to a_{kl}:

$$\frac{\partial SE(\underline{A};r,\theta,m)}{\partial a_{kl}} = -2[echo_m(r,\theta) - model_m(\underline{A};r,\theta)]$$
$$\cdot [\partial model_m(\underline{A};r,\theta)/\partial a_{kl}] \quad (65.5)$$

In eq. 65.5, the error between the (bottom-up) echo and the (top-down) model is measured in a low-resolution or smeared representation. For the simple convolution or smearing model in eq. 65.2,

$$\partial model_m(\underline{A};r,\theta)/\partial a_{kl} = smear_m(r - r_k, \theta - \theta_l) \quad (65.6)$$

The gradient of the squared error depends upon the error at r, θ measured in a low-resolution representation multiplied by the smearing function for a relevant part of the image. If the smearing function is broad in bearing and narrow in range as in fig. 65.1, the update equation uses the error at a given range value to update a swath of high-resolution pixels in the internal model. *The high-resolution representation is updated by using the error in the low-resolution representation along with the known smearing function.*

The iterative operations in eqs. 65.4–6 can be accelerated via parallel processing. All of the elements of the A-matrix, for example, can be updated simultaneously by a parallel implementation of eq. 65.4 for all (k, l) values, as in an ideal gradient computation. A global minimum can be pursued in spite of local minima by using the simplex method (Nash 1979) or a genetic algorithm (Fogel 1995), which is similar to the simplex technique. For a given local minimum, an associative memory can suggest other solutions (other *A*-matrices or interpretations of the same data) that may correspond to a smaller, global minimum.

Perceptual alternation (e.g., the Necker cube) and visual illusions suggest that the brain uses prior knowledge and other information to speed up the iteration process and to organize sparse image data into a picture that is commensurate with the viewer's experience. High-resolution image hypotheses are suggested by a relatively small number of observations and are introduced into the top-down data representation (Gregory 1973). Multiple, competitive hypotheses can be considered simultaneously via parallel processing. Image hypotheses can be generated from prior expectations and an associative memory that is activated by a smeared (low-resolution) image, resonance phenomena, echo time-frequency distributions, data from nonacoustic sensors, and other data that are not directly associated with imaging. A model for an associative memory is a *k*-nearest neighbor classifier in feature space (Cover and Hart 1967). Many different kinds of relevant information can be inserted into an ART-like tomographic SAS imaging algorithm in order to accelerate the iterative process and to obtain a global minimum. This process might be called "associative gradient descent."

If the predicted, top-down echo incorporates hypotheses about multiple propagation paths, then the resulting comparator is part of a RAKE or matched field receiver. A RAKE receiver correlates input data with the expected version of the data (e.g., with a predicted target echo that is passed through a multipath channel). A "matched field" receiver performs the same operation at multiple receiving sites (the locations of physical or synthetic array elements). Correlation is implemented as part of a mean-square error computation. A correlation-equivalent receiver is obtained by squaring the difference between received and predicted echo data $echo_m(r, \theta) - model_m(A; r, \theta)$ for energy normalized echoes and models, and integrating the squared error over range and cross-range coordinates.

Top-down, bottom-up gradient descent can be applied in the time-frequency plane to yield a process that is equivalent to inverse filtering (deconvolution of echoes with respect to the transmitted signal). The *A*-matrix, which represents the high-resolution, top-down internal model, is a sampled version of the hypothesized target impulse response. The top-down cochlear representation is formed by convolving the hypothesized target impulse response with the transmitted signal and passing the resulting echo through a cochlear model. This top-down version of the cochlear output is com-

pared with the cochlear response to received echoes, and the comparison is used to improve the target impulse response model. An envelope-detected version of the estimated, high-resolution target impulse response is the desired pulse-compressed echo representation for semicoherent SAS.

The cochlear time-frequency representation involves nonlinear processing, but the LMS algorithm easily handles nonlinearities. The only inconvenience is that the partial derivative in eq. 65.6 may be difficult to evaluate analytically, requiring empirical evaluation via the difference equation

$$\partial model(\underline{A}; r, \theta)/\partial a_{kl} \approx [model(\underline{A}_{kl}^+; r, \theta) - model(\underline{A}; r, \theta)]/\varepsilon \tag{65.7}$$

where A_{kl}^+ is the same as *A* except that a small increment ε has been added to element *kl*. For energy-normalized representations, the mean-square error calculation is equivalent to spectrogram correlation.

Inverse filtering, as in the foregoing discussion of cochlear deconvolution, is applicable to ART-based SAS. *Back-projection SAS uses a spatial matched filter, but ART-SAS implements a temporal-spatial inverse filter.* Higher resolution is thus expected for ART-SAS at the expense of multiple iterations.

The relation between matched and inverse filtering can be understood by considering a filter that forms a minimum mean-square error estimate of the target impulse response from an echo time series (Turin 1957; Altes 1977). If the prior estimate of the target transfer function (the Fourier transform of the unknown target impulse response) is $H_{est}(f)$, the estimating filter has transfer function

$$V(f) = \frac{E\{|H_{est}(f)|^2\}U \cdot (f)}{\Delta N(f) + E\{|H_{est}(f)|^2\}|U(f)|^2} \tag{65.8}$$

where $U(f)$ is the Fourier transform of the transmitted signal, Δ is the expected duration of the target impulse response, and $N(f)$ is the noise power spectral density. Two approximations to the right-hand side of eq. 65.8 are

$$V(f) \approx [U(f)]^{-1} \tag{65.9}$$

if signal-to-noise ratio (SNR) is large, and

$$V(f) \approx \left[\frac{E\{|H_{est}(f)|^2\}}{\Delta N(f)} \right] U \cdot (f) \tag{65.10}$$

if SNR is small.

If SNR is large, then the target impulse response (which has been modeled as a projection of the reflectivity distribution onto the range axis) is estimated with an inverse filter—that is, a filter that deconvolves the transmitted signal from the echo. If SNR is small, if the noise is white, and if there is no prior information about the target transfer function, then the target impulse response is estimated by a filter that is matched to the transmitted signal—that is, a filter with transfer function proportional to $U \cdot (f)$. The filter in eq. 65.8 performs pulse compression regardless of SNR.

Dolphin head scanning behavior suggests that SAS may be augmented by beam deconvolution. This deconvolution process can be included in an ART-type gradient descent algorithm. A high-resolution internal model is convolved or smeared with the known beam patterns and is then compared with incoming multibeam data to generate corrections to the model. Overlapping beam patterns can be generated by a binaural system that implements multiple direction-of-arrival hypotheses in parallel. Head scanning generates extra independent observations by changing the cross-range distribution of transmitted power. Binaural data can be incorporated into an ART process by predicting the echo at each ear.

Sonar platform and target motion hypotheses can be incorporated into an ART processor. The best motion hypothesis corresponds to the least error between the echo data and predictions. Updated motion hypotheses can be used to track the target, to predict its location and orientation at the next echo, and to characterize body motion in a fish or wing beats in a bat. This process is a form of image-based tracking.

A problem in neurophysiology is understanding how locations of point stimuli (e.g., in the retina) can be inferred from the collective discharge of a neuronal population (McIlwain 1986). The use of a low-resolution map to improve a high-resolution model via top-down, bottom-up gradient descent provides a hypothetical solution to this problem. The collective discharge is a consequence of divergent connections from various sensor elements, resulting in a low-resolution map. The locations of the point stimuli can be estimated with top-down, bottom-up stochastic gradient descent, using the LMS algorithm. To implement this process, the estimator must have a sufficiently accurate ensemble-average model of the divergent process (the smearing function) that maps a single point into a collective discharge. A payoff for such divergent stimulus coding is that neighboring sensors with overlapping excitation curves contribute to the representation of a point stimulus. The gradient descent deconvolution process can use these extra observations to obtain a more accurate stimulus representation than can be obtained from a single sensor.

A similar deconvolution process provides a mechanism for hyperacuity (Altes 1989a). Hyperacuity involves sensitivity to a very small difference between a reference stimulus and another stimulus—for example, two end-to-end line segments on a vernier scale, which may be colinear or slightly displaced (——— versus ——┴——). For sensitivity to small differences, the mean-squared error in eq. 65.1 can be changed to mean absolute error:

$$MAE(\underline{A}) = (2\pi M R_{\max})^{-1} E$$
$$\cdot \left\{ \sum_{m=1}^{M} \int_{0}^{2\pi} \int_{0}^{R\max} |echo_m(r, \theta) - model_m(\underline{A}; r, \theta)| \, dr d\theta \right\}$$

$$(65.11)$$

The corresponding LMS update equation is

$$a_{kl}(n + 1) = a_{kl}(n) - \mu \frac{\partial AE(\underline{A}; r, \theta, m)}{\partial a_{kl}} \qquad (65.12)$$

at the image pixel with coordinates r_k, θ_l, where

$$\frac{\partial AE(\underline{A}; r, \theta, m)}{\partial a_{kl}} = -\text{sgn}[echo_m(r, \theta) - model_m(\underline{A}; r, \theta)]$$
$$\cdot [\partial model_m(\underline{A}; r, \theta)/\partial a_{kl}] \qquad (65.13)$$

In eq. 65.13, sgn(error) equals 1 if error > 0 and -1 if error < 0. The gradient changes its sign but not its magnitude when the error becomes positive rather than negative, even for extremely small absolute error values. This behavior is analogous to computing the difference between the responses of two tuned neurons with slightly displaced and extremely steep tuning curves, where the actual stimulus value is midway between the best stimulus values for the two neurons.

Feature Images

A simplified, conservative model for biological SAS forms an image by noncoherent summation of all echo samples corresponding to each target scattering point (image pixel) as the sonar moves. This simplification allows the formation of images that represent features other than reflectivity as a function of location. A volume clutter feature image, for example, is bright when a pixel contains one or more Rayleigh scatterers—for example, bubbles or particles that are small relative to a wavelength. If the small scatterers are displaced by a comparatively large target, the volume clutter feature image becomes dark.

Another feature image can represent smooth or rough surfaces. A smooth surface is specular, with large reflectivity variation when the aspect changes. A rough surface consists of scatterers that are small relative to a wavelength and that have comparatively small aspect-induced reflectivity variation.

Other feature images can be sensitive to motion. Semicoherent processing of HFM/LPM (hyperbolic frequency modulated, linear period modulated) batlike signals, for example, can be used to estimate acceleration (Altes 1990), and these acceleration estimates can be represented by an image.

Volume clutter feature images and motion feature images are obtained by replacing the echo amplitude at each range sample by a feature other than amplitude.

The noncoherent delay-and-sum operation for SAS imaging is then applied to the resulting echoes. For smooth and rough feature images, the sum in the delay-and-sum SAS beam former is replaced by a quantity that depends upon the aspect-dependent variation of the echo samples that contribute to the sum.

Feasibility of a Biological Version of SAS

A Doppler-tolerant, semicoherent, tomographic SAS model is comparatively simple to implement and thus appears to be feasible for animal sonar. For dolphins, the model can be completely noncoherent; *echoes from a given scattering point are noncoherently summed as the animal moves.* Massively parallel processing can be used to obtain similar noncoherent sums from many neighboring points, thus forming an image. For bats, a pulse compression operation is necessary—for example, an operation as in eq. 65.8 that may be equivalent to matched filtering or inverse filtering. A deconvolution process that is equivalent to inverse filtering can be implemented by applying top-down, bottom-up gradient descent to cochlear echo representations. Pulse compression is followed by envelope detection for semicoherent SAS. After pulse compression, the bat model is the same as for dolphins.

Existing evidence indicates that echolocating animals can form a noncoherent sum of echoes from a given scattering point, although more experiments are needed. Echo summation or integration capability can be inferred from dolphin target recognition experiments (Moore et al. 1991). Summation can also be inferred from neurophysiological experiments on bats (Grinnell 1963) in which the excitation threshold of a neuron is decreased by repeated stimulation of the neuron. If the neuron is modeled as a sequential likelihood ratio test for a particular stimulus in Gaussian noise (Altes 1989b), then the prior probability of the stimulus increases monotonically with the sum of preceding stimuli. For a Bayesian hypothesis test, an increase in the prior probability of the stimulus is associated with a decrease of the excitation threshold (Van Trees 1968). Echo summation is thus encoded as a decrease in neuronal threshold. The ability to concentrate on the same point in space as the animal moves is implied by range-tracking neurons in bats (Suga and O'Neill 1980; Wong, Maekawa, and Tanaka 1992).

As the animal moves, the ability to average multiple echo samples from each target point seems to depend upon a topographic neuronal map, as in the superior colliculus (Drager and Hubel 1975). The relatively poor resolution of the superior colliculus map can be improved by a top-down, bottom-up gradient descent sharpening process, as discussed above. This improvement is manifested not in the map itself, but in an interpretation of the map by a higher processing center.

An important counterargument for the existence of biological SAS is that high-resolution topographic neuronal maps of reflectivity as a function of range and direction have yet to be discovered at higher processing centers in echolocating animals. There are several possible explanations for this deficit:

1. Amplitopic representations have been found in the bat auditory cortex. A neuron's best amplitude varies monotonically with its physical position relative to other amplitude-specific neurons (Suga and O'Neill 1980). If an amplitopic representation is used for reflectivity in an acoustic image, then a map of reflectivity as a function of range, azimuth, and elevation requires four dimensions. Such a map cannot be constructed topographically with a three-dimensional neuronal array. The brain is forced to use either a nontopographic representation or multiple maps that are different projections of a higher dimensional representation. Range, azimuth, and elevation, for example, may be coded nontopographically as additional constraints to neuronal excitations within a population of amplitopically organized neurons.

2. An advantage of a topographic map is that stimulus representations can be sharpened by localized lateral inhibition. Range/angle sharpening, however, also can be implemented via top-down, bottom-up gradient descent.

3. Images may be dynamically coded, such that a neuronal topographic map represents changes in an acoustic image rather than the image itself. Pulse-to-pulse jitter in the range and cross-range coordinates of a point scatterer, for example, may be represented topographically, while a point that does not move or scintillate between transmissions may not be represented. Range rate and rate of angular change may be represented by an ordered neuronal map.

4. A relevant map may be associated with (or projected onto) a different sensory modality such as vision or somatosensation. This type of projection or association is suggested by facial sensations that are experienced by blind people (Rice 1966). A visual or somatosensory map with comparatively high resolution is the top-down part of a top-down, bottom-up gradient descent sharpening process. Comparisons between predictions and echo data occur in a low resolution representation using registered spatial maps from different sensors, as in the superior colliculus.

Image-Based Tracking

A realistic acoustic imaging model for a biological sonar system must allow for freedom of sensor and target (prey) motion. One way to achieve this goal is to use the images themselves to compensate for deviations between actual motion and predicted motion (Altes 1998).

Image-based tracking has been introduced in the context of ART processing, but it also can be used for delay-and-sum (back-projection) SAS imaging. A delay-and-sum SAS processor sequentially constructs an image of a given target point by adding the appropriate time sample from the latest echo to a cumulative sum of sample values from previous echoes (one sample from each echo). Hypothesized motion is represented by delay corrections that are inserted into the delay-and-sum SAS beam former. For each new echo, different motion hypotheses result in various "test images." The best test image corresponds to the best motion hypothesis.

A block diagram of an image-based tracker is shown in fig. 65.3. Delay corrections are utilized by a dynamic model that predicts the next-delay hypotheses. This model can include sensor motion and translation/rotation of various parts of a moving target (Altes 1998).

Coherent SAS

Present evidence is insufficient to conclude that bats or dolphins can perform coherent pulse-to-pulse integration (necessary for conventional, coherent SAS), but the following observation may be significant. The coherent *SAS* RCRAF is a coherent sum of rotated versions of the *physical* sonar RCRAF. The physical RCRAF can be designed such that the coherent, broadband SAS RCRAF has very high peak-to-sidelobe ratio (P/S), even

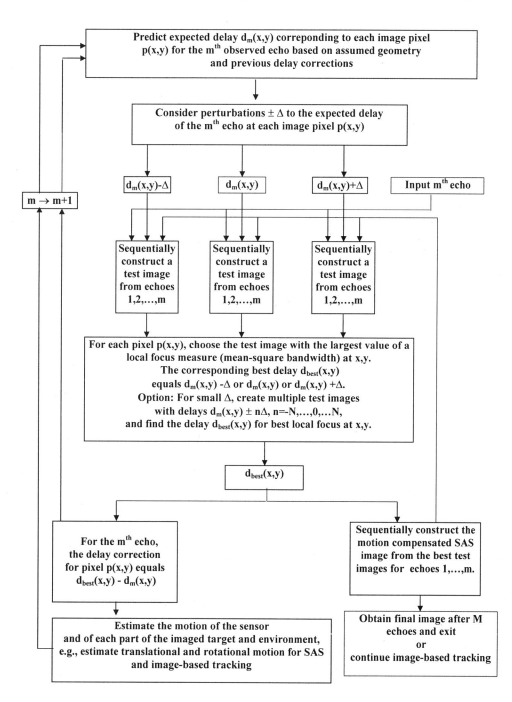

Fig. 65.3. Flow diagram showing delay-and-sum SAS adaptive focusing operations for target/sensor motion compensation and image-based tracking

with large angle increments (very few aspects). The required physical RCRAF for high coherent SAS P/S corresponds to particular signals observed around the sonar transmitter. These signals resemble waveforms measured around an echolocating dolphin (Altes 1992, 1995).

Results with Semicoherent Back-Projection SAS

Experimental results are from targets suspended in lake water. The targets were completely contained within the transmitter/receiver sonar beamwidth, and echoes were obtained as the targets were rotated. The transmitted signal was a dolphinlike pulse with 10 dB bandwidth between 50 and 150 kHz. The images were constructed from echoes measured over 360° of rotation with 3.7° increments between echoes. The echoes were recorded at the Applied Research Laboratory, University of Texas at Austin, and were organized and reformatted for a PC by P. Moore, D. Helweg, and J. Sigurdson, Code D351, SPAWAR Systems Center, San Diego.

Fig. 65.4 shows a SAS reconstruction of a target (top left) that resembles a large, trapezoidally shaped clam shell. This target was deliberately designed to have very small reflectivity. The target is surrounded by volume clutter consisting of small bubbles and particulate matter. The volume clutter appears as a hazy, foglike image surrounding the target. The top right part of fig. 65.4 shows a volume clutter feature image constructed from

the same data. In this image, the target is suppressed and the volume clutter is accentuated. Since the image has high resolution, the target appears as a hole or cavity with a distinctive shape. The bottom left part of fig. 65.4 shows an enhanced target image obtained by multiplying the clutter feature image by a constant and subtracting it from the reflectivity image. The bottom right part of fig. 65.4 shows an enhanced clutter image obtained by multiplying the reflectivity image by a constant and subtracting it from the clutter feature image.

If the volume clutter were to become more reflective relative to the target, the image at the top left in fig. 65.4 would become nearly uniform, and detection/classification of the target with a reflectivity image would become very difficult. The clutter feature image on the top right of fig. 65.4, however, would become even better at revealing the presence and shape of the target. Target detection with a clutter feature image is not predicted by the sonar equation, although this type of detection is familiar in medical ultrasound, where some objects appear as distinctively shaped dark areas, surrounded by "spackle" (volume clutter). The sonar equation is a logarithmic version of the signal-to-interference ratio at the receiver output, where the "signal" is associated with target reflectivity and the "interference" is associated with clutter echoes and noise (Urick 1975). Detection and classification with a volume clutter feature image as in fig. 65.4 may explain how dolphins find buried fish (Herzing, chapter 56, this volume).

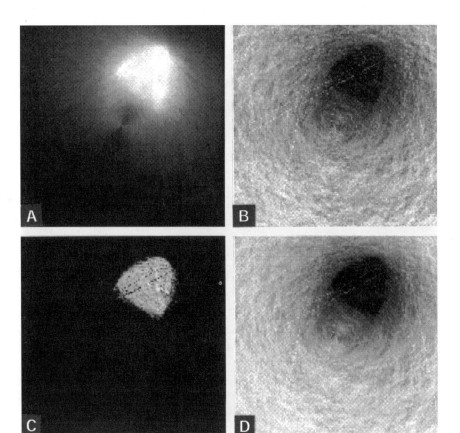

Fig. 65.4. *A–B:* A SAS reflectivity image and a volume clutter feature image of the same low-reflectivity target. *C–D:* Images constructed from weighted sums of the reflectivity and feature images in order to enhance or suppress the target relative to the volume clutter.

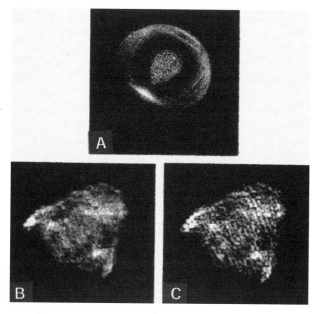

Fig. 65.5. *A:* A composite rough/smooth SAS feature image for a target that resembles a truncated cone. Pixel values in the composite image represent the larger of two feature images at each sampling point. *B:* The left tomographic SAS image is deliberately degraded by artificially decreasing the target range by 1 cm halfway through the imaging process. In the right-hand version (*C*), the image-based tracker in fig. 65.3 is used to compensate for the delay perturbation.

Fig. 65.5 (top) shows a composite of rough and smooth SAS feature images for a target that resembles a truncated cone. A SAS reflectivity image shows the target's outline; the interior appears to be hollow. A smooth-target feature image also shows only the outline. A rough-target feature image, however, shows only the interior part of the image. The smooth-target feature image and the rough-target feature image compete

for representation of each pixel in the composite image shown in fig. 65.5 (top).

The bottom of fig. 65.5 shows two contrast-enhanced reflectivity images of the trapezoidal shell-shaped target in fig. 65.4. The left-hand image is deliberately degraded by artificially decreasing the target range by 1 cm halfway through the imaging process. The right-hand image is obtained with the same data, using the image-based tracker in fig. 65.3 to compensate for the delay perturbation.

Discussion and Conclusion

A biological version of synthetic aperture imaging appears to be feasible, and it is advantageous for detection and classification of prey (e.g., finding buried fish). The simplifications and generalizations that are used to obtain a biological model can lead to new insights and capabilities for humanmade SAS systems. Biologically inspired SAS image sharpening utilizes a mathematical correspondence between sharpening with top-down, bottom-up cognitive models, gradient descent optimization, iterative deconvolution, algebraic reconstruction (ART), and spectrogram correlation. This correspondence provides insight into the utilization of a low-resolution neuronal map to obtain a high-resolution image—for example, estimation of point stimulus locations in the retina from the collective discharge of a neuronal population.

Acknowledgments

This work was supported by the Office of Naval Research. Sonar data were supplied by P. W. B. Moore and D. A. Helweg (SPAWAR Systems Center, San Diego, Calif.).

{ 66 }
Biomimetic Sonar Objects from Echoes

Roman Kuc

Introduction

Bats and dolphins employ echolocation for prey identification and tracking by emitting a series of acoustic pulses and processing the echoes (Simmons, Fenton, and O'Farrell 1979; Au 1993). In this chapter, I combine techniques employed by biological systems to im-

plement a mobile airborne sonar system that recognized objects directly from their echo waveforms. The sonar system emitted wide-bandwidth clicklike pulses and had a pair of receivers to detect echoes. Humanmade sonars have had limited success for target recognition, because wide-bandwidth echo waveforms vary significantly within the frequency-dependent transmitter and receiver beams. Previous attempts to overcome this echo variation included careful target placement in a stationary sonar field and using relatively large targets (compared to the wavelength) in shapes (spheres, cubes, pyramids and cones) that exhibit very different scattering properties (Roitblat et al. 1989). In contrast, the system described in this chapter recognized targets that were nearly identical and that were comparable in size to the emission wavelength. The system solved the target localization problem by moving the sonar, guided by pulse-echo time delays, so that it always examined a given target from a reproducible location. The echo was processed to extract the envelope, which was then described by a set of parameters forming a feature vector to classify the target.

Materials and Methods

Our airborne sonar, shown in fig. 66.1, consisted of three Polaroid electrostatic transducers (Biber et al. 1990): a center transmitter (Tx) flanked by two rotatable receivers, Rx-R on the right and Rx-L on the left. Each transducer had a 3.75 cm diameter circular aperture and a nominal resonance at 60 kHz. A protective mesh, normally covering the transducer, was removed to eliminate internal reflections and broaden the bandwidth. The transmitter was excited with an electrical impulse, which caused the emitted pulse to have a frequency bandwidth extending from 20 to 120 kHz. The center-to-center distance between the transmitter and receivers was $D - 4.8$ cm, and the three transducer axes lay in the same plane. We also investigated a batlike configuration with the ears lying above the mouth (Kuc 1994) and a four-eared configuration (Kuc 1993). The entire assembly was mounted on the end of a robot arm whose tip had three translational (x, y, z) and two rotational (azimuth and elevation) degrees of freedom. The sonar then could translate and rotate to view a target from almost any aspect.

In addition to its biological morphology, this sonar exploited the following biological principles:

Adaptation

The sonar adapted its configuration using echo information. The sonar design mimicked bat ears that react by rotating to the direction of the echo source (Mogdans, Ostwald, and Schnitzler 1988). This adaptation maximized the detected echo amplitude and bandwidth, achieved by placing the target's first echo-producing feature along all three axes. Maximizing the amplitude is important for small-feature detection in the presence of noise, while maximizing the detected echo bandwidth is important for classifying similar objects (Kuc 1997a, 1997b).

Mobility

The sonar was located at the end of a robot arm that positioned the target at a reproducible location within the beam patterns. The target range r was set in the transmitter far field to insonify the target with plane waves. I defined the far field as $r > a^2/\lambda$, where a is the aperture radius ($a = 1.875$ cm) and wavelength $\lambda = c/f$,

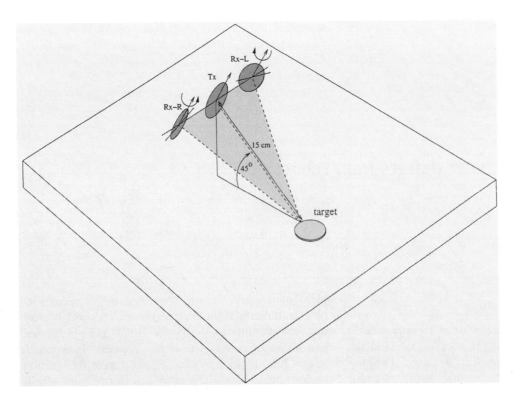

Fig. 66.1. Diagram of biomimetic sonar positioned at the end of a robot arm. Each transducer had a 3.75 cm diameter. The center transducer (Tx) was the transmitter and the two side transducers (Rx-R and Rx-L) were receivers for binaural reception. A stepper motor rotated the receivers for focusing.

with sound speed c ($c = 343$ m/s) and f the highest usable frequency in the emitted pulse (100 kHz). The smallest wavelength is then 3.4 mm, and the far field starts at 12 cm. Since the attenuation of sound increases with range, a short operating range was desirable and set at 15 cm. Sonar mobility reduced the echo complexity, since a given target then produced a unique echo waveform determined only by the target's orientation. This complexity reduction greatly simplified the object recognition task.

Feature extraction

The signal processing and feature extraction employed to recognize targets was an initial attempt to mimic primitive cochlear models of hearing (Dallos 1992). The sonar moved on a horizontal plane located 11 cm above a smooth surface. The sonar elevation initially was set to $-45°$ relative to the horizontal, allowing only the information-bearing echoes to be detected by the receivers. The main beam reflected from the surface was directed away from the receivers.

To demonstrate the feasibility of recognizing targets directly from their echoes, simple targets verified the echo-production process. Targets consisted of ball bearings, steel machine washers, and rubber O-rings, which varied in size from 6 to 25.4 mm. All the targets were small enough to fit within the three transducer beam patterns, allowing echoes from significant target features to be detected without scanning. These initial targets were symmetric to avoid issues related to target orientation. As my experience increased, I included more complex targets. Coins represented asymmetric targets, which required careful placement so that the same orientation was viewed in the learning and recognition stages. To recognize an asymmetric target independent of orientation, the target had to be scanned at a set of different orientations and the observed features stored in the database. The echo-waveform change with orientation has been investigated in the references (Kuc 1997a, 1997b). In searching for targets on the work surface, the transmitter and receiver axes were parallel, to maximize the echo-producing region. The sonar rotated about the vertical axis to examine the surface with minimal effort. Echoes were acquired from each receiver using one of the two channels of an analog-to-digital converter (Gage CS1012) sampling at 2 MHz per channel with 12-bit resolution. The echoes drove the sonar to position the target along the transmitter axis at a range of 15 cm. A new interrogation pulse was transmitted as soon as the echoes were detected and processed, a feature also observed in bats (Simmons, Fenton, and O'Farrell 1979). Controlled by a Pentium 120 processor, pulses were emitted as often as every 100 ms. Exciting the transmitter with a 6 μs duration electrical impulse emitted an interrogation pulse. This duration produced a large amplitude pulse with a large bandwidth. The

echoes were processed to extract their time-of-flights (TOFs) for adjusting the sonar azimuth and the receiver vergence angles. The TOF was determined from the time that the echo envelope first exceeded a threshold. The echo envelope was also used for target characterization. The acquired waveform consisted of 512 samples, corresponding to a data window lasting approximately 0.25 ms. The data window began 32 samples prior to the time the threshold was exceeded. Such pretriggering included low-level sections of the waveform at the beginning of the echo packet in the analysis. This 0.25 ms data window was long enough to contain all the echo components produced by the largest object in the target collection, a 2.54 cm diameter rubber O-ring.

The TOF from each receiver defined an ellipsoid of possible target locations, with the receiver and transmitter centers being the foci. The target location in the plane containing the transducer axes (in terms of range r and bearing angle θ with respect to the transmitter) was determined from the intersection of two ellipses. For the transmitter located at the origin of a Cartesian coordinate system and the right receiver located at $x = D$, the locus of points x, y at which a small target produced the TOF, T_r, was given by

$$(x^2 + y^2)^{1/2} + [(x - D)^2 + y^2]^{1/2} = cT_r \qquad (66.1)$$

A similar equation holds for the transmitter and left receiver. If T_l and T_r are the TOF values detected by the left and right receivers, the range (r) and bearing angle (θ) of the target are given by (Kuc and Barshan 1992):

$$r = \frac{(cT_l)^2 + (cT_r)^2 - 2D^2}{2c(T_l + T_r)}$$

and

$$\theta = \arcsin\left[\frac{(c^2 T_l T_r + D^2)(cT_l - cT_r)}{D(c^2 T_l^2 + c^2 T_r^2 - 2D^2)}\right] \qquad (66.2)$$

To position the target along the transmitter axis, the sonar azimuth adjusts to drive θ to zero to within the resolution of the robot arm wrist joint ($\pm 0.225°$). While the sonar positioned the target toward the transmitter axis, the receivers rotated to point toward the target using the vergence angle:

$$\varphi = \arctan\left(\frac{r}{D}\right) \qquad (66.3)$$

Results

ECHO PROCESSING

Changing φ formed a type of focusing, which had two main advantages. First, it maximized the echo amplitude, thus enhancing echo detection in the presence of noise. The waveforms of echoes reflected from a Ping-Pong ball suspended in air are shown in fig. 66.2 with and without focusing. When the receivers focused, the echo had larger amplitude. The second advantage of focusing was maximizing the echo bandwidth, which facilitated target identification. The power spectra of echoes reflected from the Ping-Pong ball are shown in fig. 66.3, with and without focusing. The range of frequencies above the noise level was larger when focusing occurred. The echo spectrum in the focused state was within 20 dB of the maximum from 20 kHz to 100 kHz. The target information was extracted from echoes by performing envelope detection and feature extraction, as shown in fig. 66.4. The rectified data underwent a logarithmic compression to reduce the dynamic range. The compression emphasized the important small-amplitude structure present in the echo packet. The envelope was computed by low-pass filtering the compressed rectified waveform. Since the envelope typically exhibited a slow variation in time, its 512 samples were reduced to a smaller number of data points without losing information. To do this, the envelope was divided into a set of 16 contiguous segments, each containing 32 samples. Each segment was represented by its average envelope value. The 32-point segment duration corresponded to approximately one period at the resonant frequency of the transducer (60 kHz). This data reduction resulted in a feature vector containing 32 elements (16 from each receiver). These vectors formed a database indexed by the target identity.

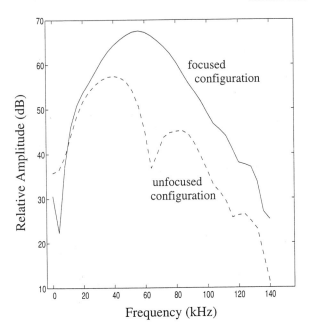

Fig. 66.3. Power spectra of waveforms shown in fig. 66.2. *Solid line:* spectrum of echo in focused configuration; *dashed line:* spectrum of echo in unfocused configuration.

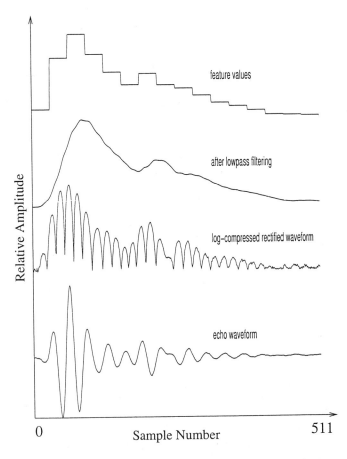

Fig. 66.4. Steps for extracting features from echo waveforms. The echo waveform (*bottom trace*) contained 512 samples; the resulting feature vector from each receiver (*top trace*) contained 16 values. Right and left feature vectors combined to form a 32-element database entry.

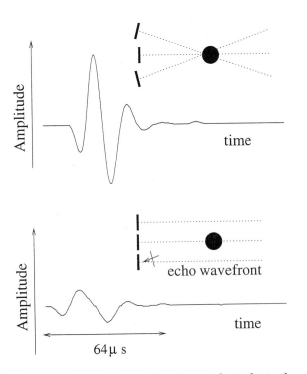

Fig. 66.2. Echo waveforms. *Top:* the receivers focused on the echo-producing sphere (Ping-Pong ball) suspended in air; *bottom:* the receiver axes parallel to the transmitter axis.

LEARNING PHASE

The system learned by detecting a known target, positioning it at a reproducible location determined by the echoes' TOFs, and then extracting a feature vector from the echoes. Each feature vector was then associated with the target identity in a database. The sonar moved to position the nearest echo-producing feature on the target along the transmitter axis at a range of 15 cm. Once at this location, a scan in elevation of $\pm 3°$ about the nominal $-45°$ elevation was performed in $0.225°$ steps. This scan was necessary to compensate for errors in elevation that could occur in the recognition phase due to finite robot accuracy. While the binaural TOFs controlled the bearing of the target, the elevation was not controlled, but set by the tolerances of the robot stepper motors and gears. As a result, elevation errors of up to $+1.5°$ could occur, depending on the robot configuration. Performing the scan in elevation in the learning phase assured that the sonar would encounter similar echo waveforms from a target in the recognition stage.

An additional step was required in the learning phase to compensate for random fluctuations that can occur in the echo waveforms caused by thermal gradients in the transmission medium, which any sonar system encounters (Sabatini 1997). In constructing the database, 15 interrogation pulses were emitted at each elevation angle and each echo pair was processed to extract the 32-element feature vector. The observed feature vector was then compared with existing database entries by computing the sum of the squared difference between each vector element pair. If the sum was larger than a perceptual threshold, the observed feature vector was included in the database; otherwise, it was discarded, because a representative database entry already existed. This procedure produced clusters of points in feature space for each target that compensated for the random fluctuations. These fluctuations caused deviations in feature values that were typically $+2\%$, but occasionally as large as $+20\%$. These fluctuations produced errors in the envelope and time registration, since the latter was determined from the time the envelope amplitude exceeded the threshold. Storing multiple feature vectors increased the likelihood that an observed feature vector matched one in the database and improved the resolution of the recognition system. At the end of the learning phase, each target was represented by approximately 20 entries in the database.

RECOGNITION PHASE

In the recognition phase, a detection and localization procedure identical to that used in the learning phase positioned a target at a reproducible location. The feature vector was extracted from the observed echoes and compared to entries in the database to find the nearest neighbor. This was done by computing the sum of the squared differences between each of the 32 elements in the observed feature vector and the corresponding ele-

ments in each database entry. The database entry that produced the minimum sum indicated the target identity. The system was surprisingly sensitive: ball bearings differing <1 mm in diameter were distinguished. This is remarkable because the wavelength at 60 kHz is about 6 mm. Targets of different types were also easily differentiated—even targets approximately the size of a wavelength, such as 6.3 mm diameter ball bearings and washers. Washers also were easily differentiated from O-rings. (For more information, see Kuc 1997a, 1997b.)

Symmetric targets produced identical echoes at each ear, independent of target orientation. For asymmetric targets, such as coins, the echoes detected at each ear were different, a consequence of the target lying in the sonar near field ($r < D^2/\lambda = 76$ cm). Asymmetric targets also produced echoes that were target-orientation dependent: the echo waveforms relate to the particular view of the target. Consider the task of recognizing the head and tail side of a U.S. penny, a 1.4 mm thick coin measuring 19 mm in diameter. In the initial learning experiment, the head side was placed on top in a standard (upright) orientation and the database was formed. Several recognition experiments were performed. First, the head side was observed in the upright orientation without moving the sonar. Then the coin was rotated to determine the effect of orientation. Finally, the coin was flipped so the tail side was observed in its upright orientation. The right and left receiver echo waveforms from the upright head and tail sides are shown in fig. 66.5. The large echo component starting at sample 30 was from the near edge of the coin and from the corner formed by the coin and work surface. The echoes from the far edge of the coin occurred around sample number 200. Later-occurring echoes were multiple reflections from the coin surface, typically from one edge to the other and then back to the receiver. Echo differences were caused by the coin relief patterns as viewed by each receiver.

The following results were observed when the feature vectors were processed:

When the recognition phase followed immediately after the database formation in the learning phase, there was no movement of the sensor between the two phases. Then the sum of squared errors between the observed feature vector and the best database match had a mean value of 200 (SD 20) arbitrary units. This value represented the error produced only by noise and random fluctuations in the echoes, indicating the limiting factors in the system.

When the robot was forced to detect and localize the upright head in the work surface in the recognition phase, the mean squared error sum increased to 400 (SD 80). This value represented the tolerances in the robot joints, which may not exactly reproduce the sensor position in the learning phase.

When the robot was forced to detect and localize the upright tail in the work surface, the mean squared error sum was 16,000 (SD 80). That is, the distance in feature

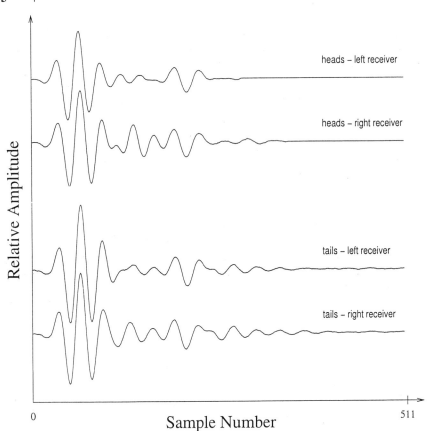

Fig. 66.5. Echo waveform pairs from the two sides of a coin (U.S. penny)

Relative Amplitude

heads – left receiver

heads – right receiver

tails – left receiver

tails – right receiver

0

Sample Number

511

space between the upright head feature vectors in the database and the feature vectors extracted from tail waveforms was much greater than the separation in the clusters due to random fluctuations and robot positioning errors. This increase was due primarily to the difference in the head and tail relief patterns. This result was a measure of the system resolution.

When the robot detected the head whose orientation differed from the upright, the error sum increased to approximately 1,000 (SD 80) as the deviation in orientation from the upright increased to 5°. This indicated that deviations of 5° could still yield reliable recognition. If more accurate orientation determination is required, feature vectors must be stored in the database as the coin is rotated in small angular increments.

Discussion

This study described an adaptive sonar mounted on the end of a robot arm to detect, localize, and identify targets using echolocation, thus demonstrating that a relatively simple sensory system could make effective discriminations. For reliable operation, the sonar standardized the echo waveform and maximized the signal-to-noise ratio by being mobile and adaptively adjusting its configuration. The learning phase formed a database from detecting and scanning each target in a target collection, and processing the observed echoes to extract a feature vector. Storing multiple feature vectors at each viewing angle treated random fluctuations in the echoes caused by medium heterogeneity. The recognition task compared an observed feature vector with database entries using a nearest neighbor criterion. Small targets simplified the recognition task, because they could fit within the transducer beams. Echoes from the significant target features could then be detected without additional scanning. Initial targets were symmetric to avoid orientation problems. Targets differing by <1 mm in size were reliably distinguished. Coins represented asymmetric targets, which require the same orientation to be viewed in the learning and recognition phases. The head and tail sides of coins were reliably recognized. To identify asymmetric targets having an arbitrary orientation, the sensor had to move around the target to obtain views from a set of aspects. The database was then enlarged to include these additional views indexed by the orientation angle. The rotation angle must be small enough so that the difference in the feature vectors is less than the error produced by the nearest neighbor (Kuc 1997a, 1997b).

Acknowledgments

The research described in this paper was supported by the National Science Foundation under grant IRI-9504079.

{ 67 }

An Investigation of Active Reception Mechanisms for Echolocators

Ashley Walker, Herbert Peremans, and *John Hallam*

Introduction

In the microchiropteran literature, active emission mechanisms, such as Doppler-shift compensation (Schnitzler 1968) and interpulse interval control (Schnitzler and Henson 1980), enable species to modify echo content to meet their informational needs. Similarly, characteristics of the receiving apparati can be altered during sensing to create additional cues. Regarding the latter, two basic questions remain unanswered. What sort of active adjustments are made during echo reception? What information is generated as a result? Here we discuss three different mechanisms that echolocators could use to alter receiver directionality and thereby gain information for target localization.

In the most familiar dynamic reception mechanism, the directional sensitivity of the auditory system is adjusted vocally. For example, by altering call frequency, many broadband-emitting microchiropteran species adjust the directional sensitivity of their pinnae. This is a *morphological* adaptation, as opposed to a *motion* approach, because it exploits the passive filtering properties of stationary pinnae. To understand the underlying acoustics, one can consider the pinnae as acoustic magnifying lenses that, when held in front of the ear canal(s), focus the auditory attention in a particular direction. Species with broadband calls and relatively fixed pinnae appear to localize targets by integrating the measurements of each ear's magnifying lens—perhaps in the form of isofrequency interaural intensity differences (IIDs)—to form a single auditory image (Grinnell and Grinnell 1965; Fuzessery, Hartley, and Wenstrup 1992).

This approach is appropriate for echolocators with broadband calls; however, narrowband echolocators cannot "call up" multiple acoustic magnifying lenses or multiple isofrequency IIDs. Nevertheless, they could create additional IIDs via dynamic reorientation of the single acoustic magnifying lens created at their central call frequency (CF) (Walker, Peremans, and Hallam 1998; Schnitzler and Henson 1980). What sort of motion would this require? One example is provided by the microchiropteran species that employ stereotyped pinnae movements during target localization: in Rhinolophidae, *Rhinolophus fumigatus, Rhinolophus ferrumequinum,*

Rhinolophus alcyone, Rhinolophus euryale, Rhinolophus hipposideros, Rhinolophus landeri; in Hipposideridae, *Hipposiderus commersoni, Triaenops afer, Asellia tridens, Hipposiderus caffe* (Pye, Flinn, and Pye 1962); and phyllostomid bat *Trachops cirrhosus* (Denzinger and Schnitzler 1998, personal communication). In this behavior, the pinnae move independently, through opposing vertical scanning motions, principally involving a rotation of one ear forward while the other rotates backward. One such movement appears to be made for each CF pulse/echo—with the right and left pinnae moving along equal but opposite arcs during one pulse/echo and reversing during the next pulse/echo, and so on (Pye, Flinn, and Pye 1962; Schneider and Möhres 1960; Griffin et al. 1962; Pye and Roberts 1970).

It is interesting to note that when deprived of the use of the muscles that move the pinnae relative to the head, *R. ferrumequinum* become disoriented and lose localization acuity—until they learn to compensate by moving the head vigorously (Schneider and Möhres 1960). *Pteronotus parnellii* have relatively immobile pinnae, but have been observed to move their heads in much this same way when emitting CF pulses (Pye and Roberts 1970).

Materials and Methods

The active reception mechanisms described in the previous section were analyzed and tested both (1) aboard a physical model (i.e., a robotic sonarhead), and (2) in a computer simulation model (i.e., in a three-dimensional Echolocation Simulator written in the C language and run on a Sparc workstation) (Walker 1997). The former provides a testbed with realistic environmental features such as noise, acoustic clutter, and natural target and background movement; while the latter affords greater exploratory capabilities (e.g., a wider range of sensor motions) and anatomical modeling fidelity (i.e., the sensor can be made arbitrarily small). The three components of each model are discussed below.

TRANSDUCERS AND MOUNTING MODULES

The robotic sonarhead consisted of a centrally mounted transmitter flanked by two receivers. As in-

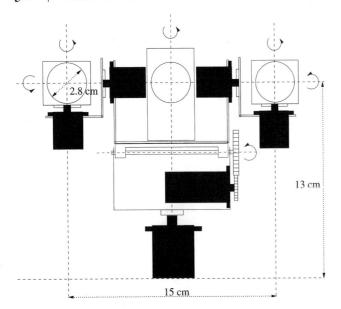

Fig. 67.1. Sonarhead architecture. Dimensions of simulated sonarhead are 10% of those shown for the robotic sonarhead, and the smaller receivers are given an azimuth offset from the vertical midline.

dicated in fig. 67.1, it had six degrees of freedom (DOF), allowing panning and tilting of the neck, and independent panning and tilting of each of the two receivers. The motors driving the different axes are standard radio-control servo motors, controlled via pulse-width modulated signals generated by a transputer.

The simulated sonarhead also consisted of three transducers mounted in a headlike configuration. Here, the acoustic response of the artificial ears is modeled as circular pistons mounted in an infinite baffle. This is a good approximation to the response of the Polaroid Series 7000 transducers used aboard the robotic sonarhead (Peremans 1997). The simulated sonarhead, scaled to more realistically represent the dimensions of micro-

chiropteran heads, had an extra degree of freedom—facilitating lateral head rolling (i.e., movement of the receivers toward and away from the notional shoulders).

TRANSMISSION AND RECEPTION MODULES

The transmitter module aboard the robotic sonarhead was split in two parts: the amplitude and frequency modulator (AM/FM) plus the power amplifier; and the actual transmitter consisting of a transformer and a Polaroid transducer. The transmitter module had two inputs for AM and FM modulations to be imposed on a carrier wave. The two modulation signals were generated by a two-channel, transputer controlled, 16-bit D/A converter. The FM signal used was a 5 ms linear sweep from 90 to 30 kHz, while the CF signal was a 100 ms tone centered at 50 kHz. Detection and amplification of the reflected echoes was performed by the receiver modules mounted, together with the transducers, on the pan/tilt servos. The output signals from the receivers were sampled by a 16-bit A/D converter at 200 kHz.

In the simulator, the frequency and amplitude characteristics of a digital call were stored in an array and modified at each sampling period while the acoustic beam was traced through the three-dimensional virtual environment. This process takes into account environmental phenomena (beam spreading and frequency-dependent atmospheric absorption) and factors introduced at the reflecting and/or receiving surfaces (scattering, Doppler shift, and directionally dependent filtering).

SIGNAL PROCESSING

The signal processing operations performed on the received signals were based upon a simple model (fig. 67.2) of the processing performed by the mammalian cochlea (Shamma et al. 1986; Saillant et al. 1993): the basilar membrane was modeled by a set of parallel bandpass filters; the transduction of the movement of the membrane into neural activity by hair cells was modeled by a

Fig. 67.2. Model of cochlear filter bank in two earlike receivers

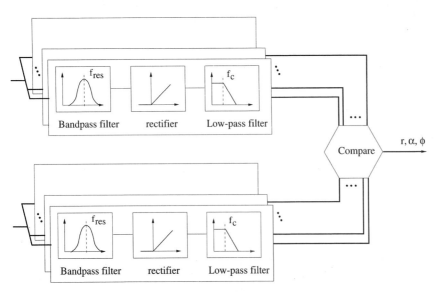

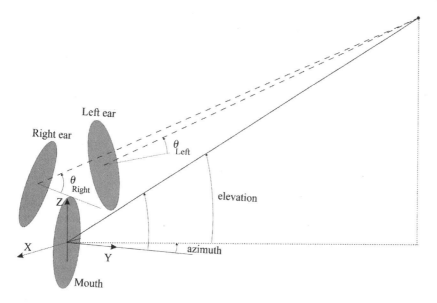

Fig. 67.3. Location of a reflecting object (fixed distance r = 0.5 m) with respect to the sonarhead

halfwave rectifier and a low-pass filter. First, the signals were processed by a filter bank consisting of 11 bandpass filters. The central frequencies and bandwidths of those filters depend on the frequency structure of the call. For FM-swept pulses, they were distributed logarithmically along the frequency range of the FM-sweep —that is, 30 to 90 kHz—and had 4 kHz bandwidths. For the CF call, corresponding echoes were processed in a narrow (Q > 200) bandpass filter centered at 1.5 kHz above the call frequency to intercept Doppler-shifted echoes from a moving target. The filters were implemented by eleventh-order Butterworth bandpass filters. Next, the outputs of the filter bank were halfwave rectified and smoothed. A second-order Butterworth lowpass filter was used with a cutoff frequency at 2 kHz.

In the robotic experiments, the reflector for the narrowband signal was a small computer cooling fan with a combined blade effective rotation frequency of approximately 500 Hz. The target for the broadband signal was a small acoustically opaque sphere; a computer mouse roller ball (radius approximately 1 cm). In the simulator, reflecting targets were modeled as a cluster of one or more pointlike reflectors—each producing one echo if within the forward hemisphere defined with respect to the transmitter and receivers. In all experiments, simulated and real targets were held at the constant range of 0.5 m.

Results

THEORETICAL

To explore mechanisms for actively reconfiguring the sonarheads' directionality, I quantify which variables affect its directionality characteristics. In this work, I was concerned with directional cues derived from IIDs (interaural intensity differences). In a binaural listening system the IID value corresponding to a particular echo

can be determined from the directionally dependent sensitivity (D_{piston}) of the receivers only, as other factors influencing intensity—like transmitter directionality, spreading losses, absorption in air, and reflection losses—can be assumed to be identical for both receivers. Modeling this directionally dependent sensitivity by that of a circular piston, the left and the right receiver respectively were

$$D_{piston}(\theta_{left}) = 20 \cdot \log\left(\frac{|J_1[ka \cdot \sin(\theta_{left})]|^2}{|ka \cdot \sin(\theta_{left})|^2}\right),$$

$$D_{piston}(\theta_{right}) = 20 \cdot \log\left(\frac{|J_1[ka \cdot \sin(\theta_{right})]|^2}{|ka \cdot \sin(\theta_{right})|^2}\right) \quad (67.1)$$

where a is the radius of the circular receivers, $k = 2\pi f/c$, f is the call frequency, c is the speed of sound, and θ_{left}, θ_{right} are the angles between the maximal sensitivity axes of the receivers and the respective reflecting object's lines of sight. Fig. 67.3 clarifies the definition of these two angles θ_{left} and θ_{right}; it also shows their relationship with the azimuth and elevation angles used in the remainder of this text to describe the location of a reflecting object (fixed distance r = 0.5 m) with respect to the sonarhead. Fig. 67.4(a–c) plots the resulting IID —that is,

$$IID(azimuth, elevation) = D_{piston}(\theta_{left}(azimuth, elevation)) \\ - D_{piston}(\theta_{right}(azimuth, elevation)) \quad (67.2)$$

for the smaller sonarhead listening at two different frequencies. Notice that, although each isofrequency IID map has ambiguities (i.e., iso-IID contours that run through numerous bearings), the ambiguous regions in different frequency maps corresponding with particular IID values usually do not coincide. Ambiguities of this

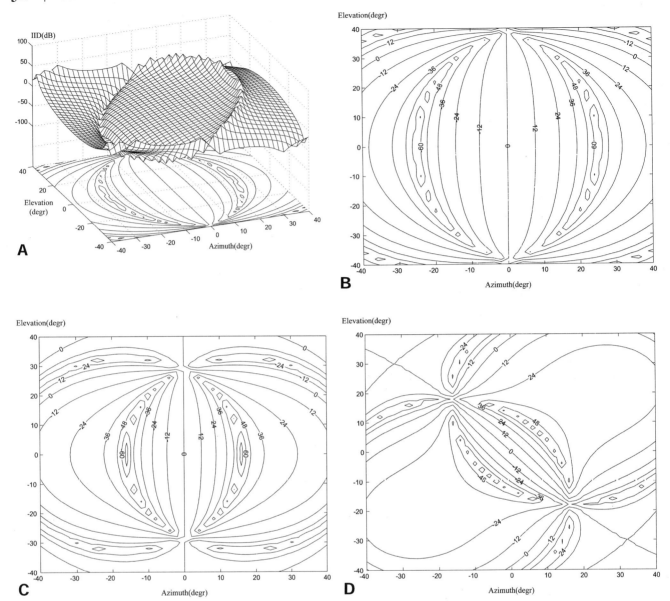

sort exist in the receptive fields of all binaural hearing systems, although the exact placement of iso-IID contours varies between mechanical systems and animal species. Thus, the combination of a set of IID maps—each corresponding with a different echolocation call frequency—could reduce target positional ambiguity. The observant reader will notice that this sonarhead has a persistent ambiguity: a 0 dB iso-IID line runs along the vertical midline in all maps. I suggest a method of overcoming this ambiguity in the experimental part of this section. I now turn to the more extreme case of an echolocator employing only one frequency and, thus, only one IID map.

As suggested earlier, by actively moving its receivers, a narrowband echolocator can collect IIDs from different regions of space and integrate these measurements in much the same way as the broadband echolo-

cator. Fig. 67.4d–e shows how an isofrequency IID map changes when the receivers are held at two vertically opposing orientations. Fig. 67.4f–g shows a similar effect achieved by rolling the head to two laterally opposing orientations. Both movements produced IID maps that break the 0 dB vertical midline symmetry. Moreover, movement of the acoustic axes swings the *IID maximum gradient line* (the region in space along which IID values and, therefore, measurement resolution vary most sharply) through the frontal sound field.

In the case of head rolling, the resulting IID maximum gradient line can be defined more strictly as a plane containing the acoustic axes. If the echolocator uses receiver, rather than head, movement to alter the orientation of the IID maximum gradient line, the acoustic axes no longer lie in the same plane, and the IID maximum gradient line is no longer strictly a planar sur-

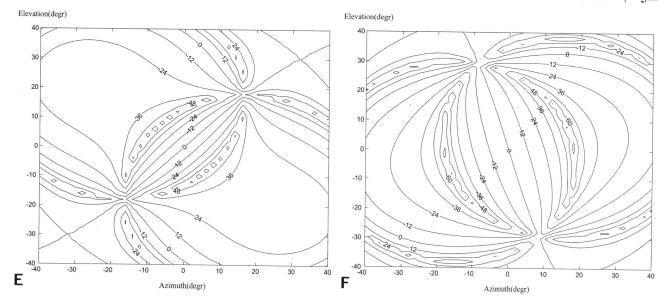

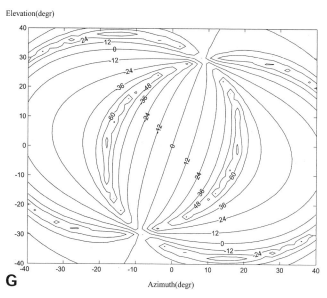

Fig. 67.4. The relationship between the IID surface and the corresponding contour plot (*A*) (*f* = 70 kHz). Directional sensitivity (iso-IID contours) are plotted in 12 dB increments (*B*): *f* = 70 kHz and (*C*) *f* = 85 kHz. Iso-IID contours for *f* = 85 kHz and receiver are offset vertically by (*D*) −15°/+15° (left/right ear) and (*E*) +15°/−15° (left/right ear). Iso-IID contours represent two lateral head roll positions (*F*) +15° and (*G*) −15°. All contours were predicted using a small simulated head. (Note: the sensitivity dips depicted are a consequence of a zero in the contralateral receiver's sensitivity pattern. The presence of noise on the actual measurements would limit the dip to the signal-to-noise ratio.)

face. However, for the same degree of rotation, employing opposing receiver movements can bring about more dramatic changes in the orientation of the IID maximum gradient line. As shown in fig. 67.4d–e, receivers with a natural azimuth orientation of 15° offset from the vertical midline can swing the IID maximum gradient line through approximately 40° when rotated through opposing vertical receiver motions of only 15°.

How many different maps (i.e., receiver sampling positions) are needed for good localization? The simple answer is that, because accuracy depends on the target's orientation relative to the orientation of the IID maximum gradient lines and ambiguity intersection points, "the more the better." To quantify this statement with respect to this system, I first defined a probabilistic evidence accumulation scheme that allowed a combination of multiple IID maps. By monitoring the entropy of the

resulting location uncertainty, we quantified the accuracy/ambiguity of the position estimate as a function of the number of IID maps integrated.

In this work, a probability map is defined as a two-dimensional array, where the value of each cell represents the probability, denoted by $P(cell_i A_m)$, that a target at the corresponding position caused a particular measured IID value A_m. Employing Bayes' rule, the posterior probability of a particular cell after n statistically independent IID values were measured, denoted by

$$P(cell_i | A_m^n, A_m^{n-1}, \varphi, A_m^1)$$

is given by

$$\frac{P(A_m^n | cell_i) \cdot P(cell_i | A_m^{n-1}, \varphi, A_m^1)}{P(A_m^n | A_m^{n-1}, \varphi, A_m^1)} \tag{67.3}$$

The term in the denominator is a normalization term, calculated by normalizing the posterior probability map. The first term in the numerator represents the measurement model. We used additive, zero mean, Gaussian noise:

$$P(A_m|cell_i) = \frac{e^{-(A_m - A_i)^2/2\sigma^2}}{\sqrt{2\pi} \cdot \sigma} \qquad (67.4)$$

where σ was determined from noise measurements with the real system and the expected value, A_i, derived from averaging over a set of measured IID maps. Changing the value of σ or employing a different distribution does not alter the results significantly.

The second term in the numerator (i.e., the prior probability map) represents the evidence accumulated through the previous $n - 1$ measurements. At startup, this prior probability map is set equal to a uniform distribution. If additional information about the target distribution is known, a maximum entropy approach could be used to select a more appropriate prior distribution. Later, the posterior probability map resulting from integrating the previous measurement is used as the prior probability map for the integration of the new measurement. The amount of uncertainty left in a particular probability map can be characterized by its entropy.

Using this Baysian map integration algorithm and integrating CF IID measurements from five pinnae orientations (6° increments between ± 12° (Walker 1997), the entropy decreases steadily—suggesting that more maps yield better accuracy. Indeed, doubling the number of maps—that is, moving through the same receiver orientations in 3° increments—drives the average error to below 1° across the frontal sound field and nearly halves the average entropy (doubles the confidence) in the angular target location measurements. Fig. 67.5 shows how the location uncertainty, quantified by the entropy of the corresponding probability map, evolves as more measurements are added (simulated results). For fixed pinnae, adding measurements results in a small decrease in uncertainty due to a noise-averaging effect, whereas using different pinnae orientations results in a much faster and much more pronounced decrease in the remaining uncertainty. Furthermore, I also conclude that although rolling the head is a less efficient strategy than flapping the ears, it is still a considerable improvement on the fixed pinnae/head approach.

AN EXPERIMENTAL EXAMPLE

It is clear that combining IID maps generated by moving receivers with less regular directionality profiles (e.g., receivers with irregular pinnae) will require fewer IID maximum gradient line orientations to resolve ambiguities in target bearing. Alternatively, the number of IID maximum gradient line orientations can be traded for call bandwidth. In this section, I present measurements resulting from the robotic sonarhead that show how combining multiple isofrequency IID maps collected at just two receiver orientations enables the robotic sonarhead to arrive at the correct three-dimensional target position—despite the highly symmetric nature of the Polaroid transducer's directionality profile.

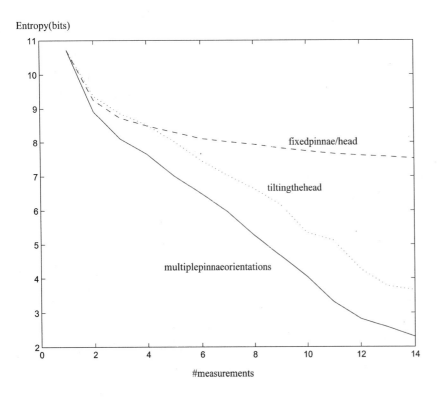

Fig. 67.5. Evolution of the location uncertainty/ambiguity, as expressed by the entropy of the corresponding probability map, as measurements are added (simulation results)

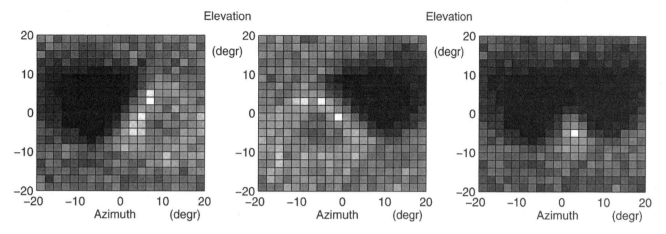

Fig. 67.6. Combining the IID maps, with localization results for receivers at vertical offset orientations of $\varphi_R = 5$, $\varphi_L = -5$; localization results for receivers at $\varphi_R = 5$, $\varphi_L = -5$; and the result of combining the previous two maps. No azimuth offset from the vertical midline was introduced.

The sonarhead was set up with the target dangling from a wire at the azimuth and elevation coordinates of $\sigma = 2°$ and $\varphi = -6°$, respectively. Fig. 67.6 shows the resulting probabilities associated with combining the IID maps for each ear configuration, as well as the subsequent combination of those two probability maps.

As can be seen from fig. 67.7, the entropy goes down smoothly while combining the IID measurements at different frequencies (lowest frequencies first) for each re-

ceiver configuration. It is also clear that the combination of corresponding maps from the two receiver configurations has much lower entropy than either alone.

From the final probability map, we derived a single object position—that is, (r, α, φ)—by collapsing the final probability map into a single, most representative, position estimate. This position estimate could be the maximum posterior probability position or the mean position—or, by introducing weights, some weighted av-

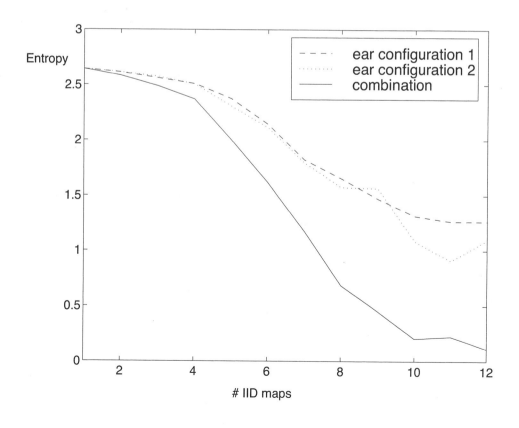

Fig. 67.7. Evolution of the location uncertainty as measurements are added (measured results)

erage. In this example, the maximum posterior probability position estimate was $\hat{\alpha} = 2°$, $\hat{\varphi} = 6°$ and the mean position estimate was $\hat{\alpha} = 1.94°$, $\hat{\varphi} = -6.07°$.

Discussion

This work tested a simple model of echolocation reception mechanisms that actively exploit either receiver *morphology* or *motion* to focus acoustic attention (i.e., directional sensitivity) during echo reception. The morphological approach—thought to be used by many broadband bat echolocators—uses call frequency sweeping to exploit the frequency-specific directionality of stationary pinnae to sample IID values from different parts of the frontal sound field. In the motion approach, active repositioning of the receivers by head or ear movements was shown to be a means by which an echolocator can achieve the same acoustic effect.

In the case of the latter, this model provides a plausible explanation for the pinna movements of rhinolophids and hipposiderids, and also could explain the head movements exhibited by these animals after pinna immobilization. Although the IID characteristic of the engineered sonarheads used in this study were vastly more simple than evolved sonarheads, the IID resolution is comparable to that measured for *R. ferrumequinum* across the frontal sound field. The model also could be used to explain the movements of other species—for example, the phyllostomid bat *Trachops cirrhosus*, which rotates its pinnae in vertical arcs when listening for the CF portion of prey calls. The model deserves further study as a means by which some species may achieve versatility in maximizing the spatial sensitivity of individual neurons during echolocation.

{ 68 }

Acoustic Simulation of Phantom Target Echoes in Dolphin Behavioral Experiments

Roland Aubauer, Whitlow W. L. Au, and *Paul E. Nachtigall*

Introduction

The biological sonar of bats and dolphins is characterized by outstanding discrimination and classification abilities. In investigations of the echolocation system of these mammals, behavioral experiments are valuable for measuring perceptual thresholds and for examining the classification of certain stimuli unknown to the animal (Au 1993). Echoes in behavioral experiments that studied the discrimination and classification capabilities of echolocating animals were in the past either echoes from real targets or simplified electronically generated echoes. Both types of stimuli have disadvantages. Because of the complex interdependence between the physical properties and the backscattering characteristics of a real target, it is difficult to control echo parameters like amplitude, duration, and phase independently. The deficiency of the echo parameter control limits the design of real target experiments, because control of both the animal and stimulus is required (Hammer and Au 1980).

Simplified electronically generated echoes allow independent parameter control of the signal, but often

were unrelated to real targets. The echoes often were "canned" or typical waveforms and did not change according to the animal's outgoing signal, or they consisted only of a few wavefronts that could not accurately simulate real reflections (Mogdans, Schnitzler, and Ostwald 1993). This contribution describes a new method of simulating real targets that combines the advantages of both echo-generation techniques and thus avoids the above mentioned problems.

The target impulse response, $h(t)$, known from system theory, can be used to describe the acoustic backscattering from a target (Aubauer and Au 1998). The target is treated here as a black box, where only the incoming (incident acoustic signal) and outgoing signals (reflected echo) are known. The back-scattered target echo, $e(t)$, is the convolution of the incidental signal, $s(t)$, with the target impulse response; it can be expressed both in the time and in the frequency domain:

$$e(t) = h(t) \cdot s(t) = \int_{-\infty}^{\infty} h(\tau) \cdot s(t - \tau)\, d\tau \qquad (68.1)$$

$$E(\omega) = H(\omega) \cdot S(\omega) \qquad (68.2)$$

With these equations, it is possible either to determine the impulse response of a real target or, if the impulse response is known, to simulate an echo of a real target for any incidental signal.

For an echo simulation, the sound of the echolocating animal has to be measured with a hydrophone and either transformed via convolution or by Fourier and inverse Fourier transform into the so-called phantom echo that is played back to the animal with a projector hydrophone. Both receiver and projector hydrophones have to be positioned away from the phantom target, since a hydrophone at the same location as the phantom target would cause unwanted overlap of the echoes from the hydrophone and the phantom target. Consequently, the propagation time delay, τ, and an attenuation, A, as a result of spherical spreading and absorption losses must be taken into account. At the same time, it is necessary to pre-equalize the phantom echo to compensate for the hydrophone characteristic, $P(\omega)$. Attenuation, time delay, and hydrophone pre-equalization are independent from the target transfer function, so that the complex phantom echo spectrum, $E_p(\omega)$, becomes:

$$E_p(\omega) = A \cdot \frac{H(\omega) \cdot S(\omega)}{P(\omega)} \cdot e^{-j\omega\tau} \quad (68.3)$$

The phantom echo generation can be done in either the time or frequency domain and executed by digital filtering techniques. The introduced target impulse response represents a target only from one aspect angle and is therefore a function of the target orientation relative to the sound transmitting and receiving location.

Materials and Methods

EXPERIMENTAL SETUP

A typical setup of a phantom echo experiment is shown in fig. 68.1. During the task, the dolphin places its head through a hoop up to the pectoral fins and echolocates onto the phantom target as soon as a screen in front of the hoop is lowered. Echolocation clicks are recorded with the receiving hydrophone and fed into the phantom echo generator for signal transformation. The resulting phantom echoes are played back to the dolphin with a projector hydrophone. The dolphin, both hydrophones, and the virtual location of the phantom target are placed in line at the same depth.

A go/no-go response paradigm was used (Schusterman 1980). A go response was associated with the presentation of the standard stimulus after which the dolphin backed out of the hoop and responded within a prescribed time by touching the response paddle. In the no-go response, when a comparison target was presented the dolphin stayed in the hoop until the trainer blew a whistle to indicate a correct no response.

PHANTOM ECHO GENERATOR

The phantom echo generator (PEG) contains a DT3809 digital signal processor (DSP) board from Data Translation plugged into a PC computer. The system is completed by various signal input and output blocks and a hand-control panel. A block diagram of the PEG is shown in fig. 68.2.

The input signal received with the Brüel & Kjær 8103 hydrophone is amplified and filtered with a 500 Hz high-pass and a 250 kHz low-pass filter before going to the analog input of the DSP system. When the analog input signal exceeds the user-set threshold, a pretrigger signal is generated. The DSP system is able to collect data shortly before and after the trigger event, so even the signal before it exceeds the trigger level is acquired. The signal is digitized with a sampling rate of 1 MHz and 12-bit resolution. The signal transformation based on the fast convolution method is programmed on the DSP that allows the generation of a phantom echo of 512 digital samples (0.512 ms) in less than 3 ms. After a time delay corresponding to the distance of the virtual position of the phantom target from the echolocating dolphin, the phantom echo is projected with the same sampling rate and resolution as when the input signal was ac-

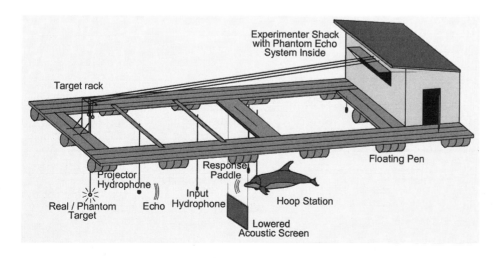

Fig. 68.1. Experimental configuration showing the dolphin's hoop station, the hydrophone and target placement, and the phantom echo generator

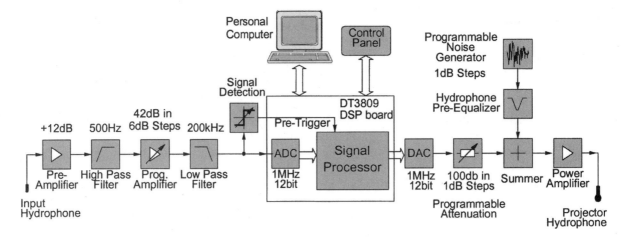

Fig. 68.2. Block diagram of the phantom echo generator

quired. Programmable input and output amplifiers allow individual control of the echo strength. Continuous white Gaussian noise for masking experiments can be added to the output signal. To compensate for the projector frequency characteristics, the masking noise is pre-equalized with a combination of sixth-order analog low- and high-pass filters, whereas the phantom echo is pre-equalized on the DSP system by inverse digital filtering technique. Therefore, the overall frequency response of the phantom echo signal path is from 20 kHz to 250 kHz ± 1 dB and the projected masking noise spectrum is from 12 kHz to 160 kHz ± 5 dB (fig. 68.3a and b). The spherical ITC1042 hydrophone from International Transducer Corporation serves as the projector hydrophone.

An example of a typical dolphin echolocation click is shown in fig. 68.3d. Fig. 68.3c shows the spectrum of the click. To verify the signal reproduction quality of the PEG, a phantom echo was generated that represents just the incidental signal measured with the B&K 8103 hydrophone. The corresponding target impulse response equals zero [$h(t \neq 0) = 0$], except for $h(t = 0) = 1$ and therefore represents a single wavefront target. The re-

sulting phantom echo is displayed in fig. 68.3e. Only minor differences from the original dolphin click were observed. The PEG is therefore suited to reproduce short clicks with very high fidelity.

TARGET IMPULSE RESPONSES

The theoretical evaluation of the acoustical target impulse response is difficult for all but the simplest targets. Transverse and longitudinal waves must be dealt with, along with a variety of target surface waves. Solutions of the problem can be found in the literature for simple geometries (Neubauer 1986), but unfortunately they are based on simplifying assumptions. The impulse responses of certain, particularly useful, targets therefore were determined experimentally. A hollow water-filled stainless-steel (HS), a solid stainless-steel (SS), and a solid brass (SB) sphere, all 7.62 cm in diameter, were selected because of their aspect-angle independent but diverse impulse responses and echoes.

The impulse responses were measured in a seawater-filled tank (Aubauer and Au 1998). Dirac impulses of one digital sample length were used as output signal. The impulses were band limited to the frequency range

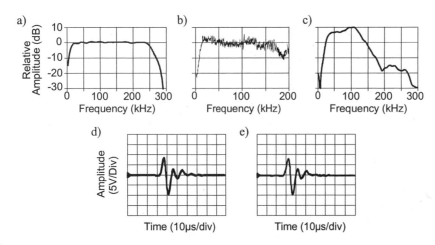

Fig. 68.3. (*a*) Overall system frequency response; (*b*) frequency spectrum of the projected masking noise; (*c*) frequency spectrum of the dolphin's echolocation click shown in (*d*); (*d*) waveform of a typical dolphin echolocation click measured with the B&K 8103 hydrophone; (*e*) waveform of a single wavefront phantom echo measured with the B&K 8103 hydrophone

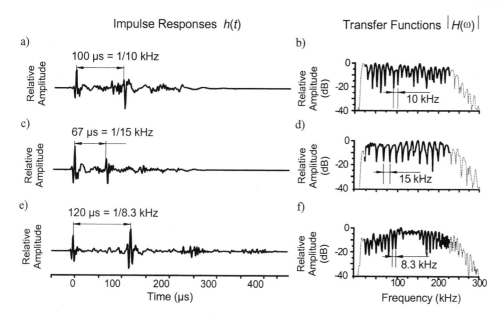

Fig. 68.4. Target impulse responses, $h(t)$, of the solid brass sphere (*a*), of the solid stainless-steel sphere (*c*), and of the hollow water-filled steel sphere (*e*); and target transfer functions, $|H(\omega)|$, of the solid brass sphere (*b*), of the solid stainless-steel sphere (*d*), and of the hollow water-filled steel sphere (*f*). Dashed lines indicate the frequency band limitation of the transfer functions due to the insufficient transmitting characteristics of the measuring system rather than the spheres themselves.

of 20 kHz to 250 kHz to avoid nonlinearity of the transmitter hydrophone. The echo signals were measured with a B&K 8103 hydrophone and averaged 2000 times to improve the signal-to-noise ratio to about 52 dB. In addition, background clutter was subtracted from the echoes. Target transfer functions $H(\omega)$ were determined by eq. 68.2 and were compensated for the time delay and attenuation of the propagation path from the target to the measuring hydrophone (right side of fig. 68.4). The impulse responses of the spheres—that is, the inverse Fourier transform of the target transfer functions—are shown on the left side of fig. 68.4.

The magnitude of the target transfer functions of all spheres is relatively flat but has a distinct periodic ripple structure. The target impulse responses show separate peaks that correspond to the wavefronts of the target echoes. The frequency difference between two ripples in the transfer function is about one over the time interval between the main peaks of the impulse response.

Both solid spheres have relatively similar impulse responses though the impulse response of the stainless-steel sphere is compressed relative to that of the brass sphere due to the higher sound velocity in steel (Mason 1958). The hollow water-filled steel sphere is clearly different; the impulse response is significantly longer and the transfer function relatively flat from 115 kHz to 155 kHz.

Phantom echoes determined with the target impulse response technique matched the real target echoes very well for several tested dolphin clicks. The cross-correlation coefficient between the phantom echo and the real echo measured in the water tank was between 0.95 and 0.98 for all three spheres (Aubauer and Au 1998). This result confirmed that the backscattering process of these spheres can be accurately simulated with the presented method.

Applications

This new PEG can be used in a wide variety of animal echolocation experiments. Very complex target echoes of the actual dolphin echolocation signals can be simulated in real time. The acoustic parameters of the simulated echoes (e.g., amplitudes, durations and delays, the distances between echo wavefronts, the frequency bandwidth, and the number of returned echoes) can be controlled separately. The target impulse response allows the echo manipulation not only in time but also in the frequency domain, where magnitude and phase information can be manipulated independently from each other. This feature is very important for future experiments, where temporal echo perception is studied. White noise can be added to investigate masking effects. The phantom echo system is suited for both echo detection and discrimination experiments.

In a first experiment, the capability of a dolphin to discriminate between acoustically simulated phantom replicas of targets and their real equivalents was tested. Phantom replicas were presented in a probe technique during a material discrimination experiment. In the probe technique, a small number of probes are presented which share some characteristics with the baseline targets, but systematically differ (Hammer and Au 1980). The animal accepted the phantom echoes and classified them almost identically to the real targets (Aubauer et al. 2000).

Results and Discussion

The PEG allows the simulation of underwater target echoes in dolphin behavioral experiments. A comparison of phantom echoes with their equivalent real target echoes showed a very good agreement for various dolphin echolocation signals. The quality of the phantom echo generation was confirmed in a first classification experiment, where a dolphin classified the phantoms like the real target models.

The PEG gives an experimenter fully programmable control over the acoustic parameters and thus allows the direct extraction of echo parameters that are relevant to the dolphin in the echolocation experiment. With all these features, the system has a tremendous potential in investigating animal echolocation.

Acknowledgments

This work was generously supported by several research grants provided by the German Academic Exchange Service (DAAD), by the German Research Foundation (DFG), and by the Office of Naval Research (ONR).

Part Five / Literature Cited

AHLÉN, I. 1981. Identification of Scandinavian bats by their sounds. The Swedish University of Agricultural Sciences, Department of Wildlife Ecology.

———. 1990. *Identification of bats in flight.* Stockholm: Swedish Society for Conservation of Nature. Item Number 0312:1–50.

AHLÉN, I., and H. J. BAAGØE. 1999. Use of ultrasound detectors for bat studies in Europe: Experiences from field identification, surveys, and monitoring. *Acta Chiropterologica* 1(2): 137–150.

ALTES, R. A. 1971. Optimum waveforms for sonar velocity discrimination. *Proceedings of the IEEE* 39: 1615–1617.

———. 1973. Some invariance properties of the wideband ambiguity function. *Journal of the Acoustical Society of America* 53:1154–1160.

———. 1977. Estimation of sonar target transfer functions in the presence of clutter and noise. *Journal of the Acoustical Society of America* 61:1371–1374.

———. 1979. Target position estimation in radar and sonar, and generalized ambiguity analysis for maximum likelihood parameter estimation. *Proceedings of the IEEE* 67:920–930.

———. 1980. Detection, estimation, and classification with spectrograms. *Journal of the Acoustical Society of America* 67:1232–1246.

———. 1988. Unconstrained minimum mean-square error parameter estimation with Hopfield networks. IEEE International Conference on Neural Networks, IEEE catalog no. 88CH2632–8, II.541–548.

———. 1989a. An interpretation of cortical maps in echolocating bats. *Journal of the Acoustical Society of America* 85:943–952

———. 1989b. Ubiquity of hyperacuity. *Journal of the Acoustical Society of America* 85:934–942.

———. 1990. Radar/sonar acceleration estimation with linear-period modulated, hyperbolic frequency modulated (LPM/HFM) waveforms. *IEEE Transactions on Aerospace and Electronic Systems* 26:914–923.

———. 1992. The line segment transform and sequential hypothesis testing in dolphin echolocation. Pp. 317–355 in *Marine mammal sensory systems,* ed. J. A. Thomas, R. A. Kastelein, and A. Ya. Supin. New York: Plenum Press.

———. 1995. Signal processing for target recognition in biosonar. *Neural Networks* 8:1275–1295.

———. 1998. Adaptive ISAR focusing of distributed time-varying targets. Pp. 261–272 in *Radar processing, technology, and applications III,* vol. 3462, ed. W. J. Miceli. Bellingham, Wash.: Society of Photo-Optical Instrumentation Engineers.

ALTES, R. A., and D. P. SKINNER. 1977. Sonar velocity estimation with a linear-period modulated pulse. *Journal of the Acoustical Society of America* 61: 1019–1030.

ALTES, R. A., and E. L. TITLEBAUM. 1970. Bat signals and optimally Doppler tolerant waveforms. *Journal of the Acoustical Society of America* 48:1014–1020.

ALTES, R. A., D. A. HELWEG, and P. W. B. MOORE. 2001. Biologically inspired synthetic aperture imaging. Technical Document 1848. SPAWAR Systems Center, San Diego, Calif., and Defense Technical Information Center, Fort Belvoir, Va.

ALTES, R. A., P. W. B. MOORE, and D. A. HELWEG. 1998. Tomographic image reconstruction of MCM targets using synthetic dolphin signals. Technical Document 2993. SPAWAR Systems Center, San Diego, Calif., and Defense Technical Information Center, Fort Belvoir, Va.

ANDERSEN, B. B., and L. A. MILLER. 1977. A portable ultrasonic detection system for recording bat cries in the field. *Journal of Mammalogy* 58:226–229.

ANDERSON, J. R. 1980. *Cognitive psychology and its implications.* San Francisco: W. H. Freeman and Company.

AU, W. W. L. 1993. *The sonar of dolphins.* New York: Springer-Verlag.

AU, W. W. L., and C. W. TURL. 1991. Material composition discrimination of cylinders at different aspect angles by an echolocating dolphin. *Journal of the Acoustical Society of America* 89:2448–2451.

AUBAUER, R., and W. W. L. AU. 1998. Phantom echo generation: A new technique for investigating dolphin echolocation. *Journal of the Acoustical Society of America* 104(3): 1165–1170.

AUBAUER, R., W. W. L. AU, P. E. NACHTIGALL, D. A. PAWLOSKI, and C. M. DELONG. 2000. Classification of electronically generated phantom targets by an Atlantic bottlenose dolphin (*Tursiops truncatus*). *Journal of the Acoustical Society of America* 107(5): 2750–2754.

BARCLAY, R. M. R. 1985. Long- versus short-range foraging strategies of hoary *(Lasiurus cinereus)* and silver-haired (*Lasionycteris noctivagans*) bats and the consequences of prey selection. *Canadian Journal of Zoology* 63:2507–2515.

———. 1999. Bats are not birds—cautionary note on using echolocation calls to identify bats: Comment. *Journal of Mammalogy* 80:290–296.

BARCLAY, R. M. R., and R. M. BRIGHAM. 1994. Constraints on optimal foraging: A field test of prey discrimination by echolocating insectivorous bats. *Animal Behavior* 48:1013–1021.

———. 1996. Bats and Forests Symposium: October 19–21, 1995, Victoria, British Columbia, Canada. Research Branch, British Columbia Ministry of Forestry, Victoria, B.C., Working Papers 23/1996.

BEECHER, M. D. 1988. Spectrographic analysis of animal vocalizations: Implications of the "Uncertainty Principle." *Bioacoustics* 1:187–208.

BELL, G. P. 1980. Habitat use and response to patches of prey by desert insectivorous bats. *Canadian Journal of Zoology* 58:1876–1883.

BERKOWITZ, A., and N. SUGA. 1989. Neural mechanisms organization are different in two species of bats. *Hearing Research* 41:255–264.

BETTS, B. J. 1998. Effects of interindividual variation in echolocation calls on identification of big brown and silver-haired bats. *Journal of Wildlife Management* 62:1003–1010.

BEUTER, K. J. 1980. A new concept on echo evaluation in the auditory systems of bats. In *Animal sonar systems,* ed. R. G. Busnel and J. F. Fish. New York: Plenum Press.

BIBER, C., S. ELLIN, E. SHECK, and J. STEMPECK. 1980. The Polaroid ultrasonic ranging system. Proceedings of the 67th Audio Engineering Society Convention, New York, 1696 (A-8).

BRADBURY, J. W. 1970. Target discrimination by the echolocating bat *Vampyrum spectrum. Journal of Experimental Zoology* 173:23–46.

BRIEMAN, L., J. H. FRIEDMAN, R. A. OLSHEN, and C. J. STONE. 1993. *Classification and regression trees.* New York: Chapman and Hall.

BRIGHAM, R. M., J. E. CEBEK, and M. B. C. HICKEY. 1989. Intraspecific variation in the echolocation calls of two species of insectivorous bats. *Journal of Mammalogy* 70:426–428.

BRIGHAM, R. M., S. D. GRINDAL, M. C. FIRMAN, and J. L. MORISETTE. 1997. The influence of structural clutter on activity patterns of insectivorous bats. *Canadian Journal of Zoology* 75:131–136.

BRITZKE, E. R., D. W. BOSSI, B. M. HADLEY, and L. W. ROBBINS. 1999. The acoustic identification of bats in Missouri. *Bat Research News* 39:158.

BROWN, W. M., and L. J. PORCELLO. 1969. An introduction to synthetic aperture radar. *IEEE Spectrum* 6(Sept.): 52–62.

BUSNEL, R. G., and J. F. FISH. 1980. *Animal sonar systems.* New York: Plenum Press.

CAPUANO, U., and J. T. MCILWAIN. 1981. Reciprocity of receptive field images and point images in the superior colliculus of the cat. *Journal of Comparative Neurology* 196:13–23.

CHARIF, R. A., S. G. MITCHELL, and C. W. CLARK. 1995. Canary 1.2 users manual. Cornell Laboratory of Ornithology, Report .

CHENG, B., and D. M. TITTERINGTON. 1994. Neural networks: A review from a statistical perspective. *Statistical Science* 9:2–54.

CHITTAJALLU, S. K., K. G. KOHRT, M. J. PALAKAL, and D. WONG. 1996. Computational model of bat auditory periphery. *Journal of Mathematical Computing and Modeling* 24(1): 67–78.

CONDON, C. J., A. GALAZYUK, K. R. WHITE, and A. S. FENG. 1997. Neurons in the auditory cortex of the little brown bat exhibit sensitivity to complex amplitude modulation signals that mimic echoes from fluttering insects. *Auditory Neuroscience* 3:269–287.

CORBEN, C., and G. M. FELLERS. 2001. Choosing the "correct" bat detector: A reply. *Acta Chiropterologica* 3:253–256.

COVER, T. M., and P. E. HART. 1967. Nearest neighbor pattern classification. *IEEE Transactions on Information Theory* 13:21–27.

CROME, F. H. J., and G. C. RICHARDS. 1988. Bats and gaps: Microchiropteran community structure in a Queensland rain forest. *Ecology* 69:1960–1969.

CUTRONA, L. J. 1975. Comparison of sonar performance achievable using synthetic aperture techniques with the performance achievable by more conventional means. *Journal of the Acoustical Society of America* 58:336–348.

DALLOS, P. 1992. The active cochlea. *Journal of Neuroscience* 12:4575–4585.

DEJONG, J. 1995. Habitat use and species richness of bats in a patchy landscape. *Acta Theriologica* 40:237–248.

DIECKMANN, U. 1997. SESAM: A biometric person identification system using sensor fusion. *Pattern Recognition Letters* 18:827–833.

DOWNES, C. H. 1982. A comparison of sensitivities of three bat detectors. *Journal of Mammalogy* 63:343–345.

DRAGER, U. C., and D. H. HUBEL. 1975. Responses to visual stimulation and relationship between visual, auditory, and somatosensory inputs in mouse superior colliculus. *Journal of Neurophysiology* 38:690–713.

DUBROVSKIY, N. A., and L. M. FADEYEVA. 1973. Discrimination of spherical targets by delphinids. Tez. dokl. 4-y Vses. bion. konf., Moscow: 29–34.

FAURE, P. A., and R. M. R. BARCLAY. 1994. Substrate-gleaning versus aerial-hawking: Plasticity in the foraging and echolocation behaviour of the long-eared bat, *Myotis evotis*. *Journal of Comparative Physiology A* 174:651–660.

FENTON, M. B. 1970. A technique for monitoring bat activity with results obtained from different environments in southern Ontario. *Canadian Journal of Zoology* 48:847–851.

———. 1982a. Echolocation, insect hearing and feeding ecology of insectivorous bats. Pp. 261–285 in *Ecology of bats,* ed. T. H. Kunz. New York: Plenum Press.

———. 1982b. Echolocation calls and patterns of hunting and habitat use of bats (Microchiroptera) from Chillagoe, North Queensland. *Australian Journal of Zoology* 30:417–425.

———. 1999. Describing the echolocation calls and behavior of bats. *Acta Chiropterologica* 1:127–136.

———. 2000. Choosing the "correct" bat detector. *Acta Chiropterologica* 2:215–224.

FENTON, M. B., and G. P. BELL. 1979. Echolocation and feeding behaviour in four species of *Myotis*. *Canadian Journal of Zoology* 57:1271–1277.

———. 1981. Recognition of species of insectivorous bats by their echolocation calls. *Journal of Mammalogy* 62:233–243.

FENTON, M. B., and S. L. JACOBSON. 1973. An automatic ultrasonic sensing system for monitoring the activity of some bats. *Canadian Journal of Zoology* 51:291–299.

FENTON, M. B., H. G. MERRIAM, and G. L. HOLROYD. 1983. Bats of Kootenay, Glacier, and Mount Revelstoke National Parks in Canada: Identification by echolocation calls, distribution, and biology. *Canadian Journal of Zoology* 61:2503–2508.

FENTON, M. B., P. RACEY, and J. M. V. RAYNER. 1987. *Recent advances in the study of bats.* Cambridge: Cambridge University Press.

FENTON, M. B., I. L. RAUTENBACK, J. RYDELL, H. T. ARITA, J. ORTEGA, S. BOUCHARD, M. D. HOVORKA, B. LIM, E. ODGREN, C. V. PORTFORS, W. M. SCULLY, D. M. SYMS, and M. J. VONHOF. 1998. Emergence, echolocation, diet and foraging behavior of *Molossus ater* (Chiroptera: Molossidae). *Biotropica* 30:314–320.

FOGEL, D. B. 1995. *Evolutionary computation.* New York: IEEE Press.

FRISTRUP, K. M., and W. A. WATKINS. 1995. Marine mammal sound classification. *Journal of the Acoustical Society of America* 97:3369.

FURLONGER, C. L., H. J. DEWAR, and M. B. FENTON. 1987. Habitat use by foraging insectivorous bats. *Canadian Journal of Zoology* 65:284–288.

FUZESSERY, Z. M., D. J. HARTLEY, and J. J. WENSTRUP. 1992. Spatial processing within the mustache bat echolocating system: possible mechanisms for optimization. *Journal of Comparative Physiology A* 170:57–71.

GANNON, W. L., R. E. SHERWIN, and S. HAYMOND. 2002. On the importance of articulating assumptions when conducting acoustic studies of habitat by bats. *Wildlife Society Bulletin* 31:45–61.

GANNON, W. L., R. E. SHERWIN, T. N. deCARVALHO, and M. J. O'FARRELL. 2001. Pinnae and echolocation call differences between *Myotis californicus* and *M. ciliolabrum* (Chiroptera: Vespertilionidae). *Acta Chiropterologica* 3:77–91.

GERSTEIN, G., D. PERKEL, and J. DAYHOFF. 1985. Cooperative firing activity of simultaneously recorded population of neurons: Detection and measurement. *Journal of Neuroscience* 5:881–889.

GIGUERE C., and P. C. WOODLAND. 1994a. A computational model of the auditory periphery for speech and hearing research. I. Ascending path. *Journal of the Acoustical Society of America* 95(1): 331–342.

———. 1994b. A computational model of the auditory periphery for speech and hearing research. II. Descending paths. *Journal of the Acoustical Society of America* 95(1): 343–349.

GISH, H., and M. SCHMIDT. 1994. Text-independent speaker identification. *IEEE Signal Processing Magazine* 10:18–32.

GORMAN, R. P., and T. J. SEJNOWSKI. 1988. Analysis of hidden units in a layered network trained to classify sonar targets. *Neural Networks* 1:75–89.

GREGORY, R. L. 1973. *Eye and brain.* New York: McGraw-Hill.

GRIFFIN, D. R. 1958. *Listening in the dark: The acoustic orientation of bats and men.* New Haven, Conn.: Yale University Press (reprint, 1986, Ithaca, N.Y.: Cornell University Press).

———. 1967. Discriminative echolocation by bats. Pp. 273–300 in *Animal sonar systems,* ed. R. G. Busnel and J. F. Fish. New York: Plenum Press.

GRIFFIN, D. R., F. A. WEBSTER, and C. R. MICHAEL. 1960. Echolocation of flying insects by bats. *Animal Behaviour* 8:141–154.

GRIFFIN, D. R., D. DUNNING, D. CAHLANDER, and F. A. WEBSTER. 1962. Correlated orientation sounds and ear movements of horseshoe bats. *Nature* 196:1185–1186.

GRINDAL, S. D. 1999. Habitat use by bats, *Myotis* spp., in western Newfoundland. *Canadian Field-Naturalist* 113:258–263.

GRINNELL, A. D. 1963. The neurophysiology of audition in bats: Resistance to interference. *Journal of Physiology* 167:114–127.

GRINNELL, A. D., and V. GRINNELL. 1965. Neural correlates of vertical localization in echolocating bat. *Journal of Physiology* 181:830–851.

HABERSETZER, J. 1981. Adaptive echolocation sounds in the bat *Rhinopoma hardwickei. Journal of Comparative Physiology A* 144:559–566.

HABERSETZER, J., and B. VOGLER. 1983. Discrimination of surface-structured targets by the echolocation bat *Myotis* during flight. *Journal of Comparative Physiology* 152:275–282.

HAKEN, H. 1978. Synergetics: An introduction. *Non-equilibrium phase transitions and self-organization in physics, chemistry and biology.* New York: Springer-Verlag.

———. 1988. Learning in synergetic systems for pattern recognition and associative action. *Zeitschrift für Physik B* 71:521–526.

———. 1996. Future trends in synergetics. Nonlinear physics of complex systems. Current status and future trends. Berlin: Springer-Verlag.

HAMMER, C. E., and W. W. L. AU. 1980. Porpoise echorecognition: An analysis of controlling target characteristics. *Journal of the Acoustical Society of America* 68:1285–1293.

HAUSSER, J. 1995. *Säugetiere der Schweiz: Verbreitung, Biologie, Ökologie.* Basel: Birkhäuser.

HAYWARD, T. J. 1997. Classification by multiple-resolution statistical analysis with application to automated recognition of marine mammal sounds. *Journal of the Acoustical Society of America* 101:1516–1526.

HAYES, J. P. 1997. Temporal variation in activity of bats and the design of echolocation-monitoring studies. *Journal of Mammalogy* 78:514–524.

HERR, A., N. I. KLOMP, and J. S. ATKINSON. 1997. Identification of bat echolocation calls using a decision tree classification system. Complexity International, 4, URL http://www.csu.edu.au/ci/vo14/herr/batcall.html, last update Jan. 1997, last verified 27 March 2002.

HICKEY, M. B. C., and A. L. NEILSON. 1995. Relative activity and occurrence of bats in southwestern Ontario as determined by monitoring with bat detectors. *Canadian Field- Naturalist* 109:413–417.

HOGG, T., and I. TALHAMI. 1996. A competitive nonlinear approach to object recognition: The generalised synergetic algorithm. Pp. 47–50 in *Proceedings of the Australian New Zealand Conference on Intelligent Information Systems.* New York: IEEE; Adelaide, South Australia.

HONIG, M. L., and D. G. MESSERCHMITT. 1984. *Adaptive filters.* Boston: Kluwer Academic Publishers.

HOPP, S. L., M. J. OWREN, and C. S. EVANS. 1998. *Animal acoustic communication.* Berlin: Springer-Verlag.

JONES, C., and C. G. SCHMITT. 1997. Mammals species of concern in New Mexico. Pp. 179–206 in *Life among the muses: Papers in honor of James S. Findley.* Special Publication, The Museum of Southwestern Biology, 3:1–290.

KALKO, E. K. V., and H.-U. SCHNITZLER. 1989. The echolocation and hunting behavior of Daubenton's bat, *Myotis daubentoni. Behavioral Ecology and Sociobiology* 24:225–238.

———. 1993. Plasticity in echolocation signals of European pipistrelle bats in search flight: Implications for habitat use and prey detection. *Behavioral Ecology and Sociobiology* 33:415–428.

KALKO, E. K. V., C. O. HANDLEY JR., and D. HANDLEY. 1996. Organization, diversity, and long-term dynamics of a Neotropical bat community. Pp. 503–553 in *Long-term studies of vertebrate communities,* ed. M. Cody and J. Smallwood. New York: Academic Press.

KAPTEYN, K. 1991. *Proceedings of the First European Bat Detector Workshop.* Netherlands Bat Research Foundation, Amsterdam, The Netherlands.

KOGAN, J. A., and D. MARGOLIASH. 1998. Automated recognition of bird song elements from continuous recordings using dynamic time warping and hidden Markov models: A comparative study. *Journal of the Acoustical Society of America* 103:2185–2196.

KOHONEN, T. 1984. Self-organization and assoziative memory. Berlin: Springer-Verlag.

KOHRT, K. 1996. A computational model of the auditory periphery. Master's thesis, Indianapolis, Indianapolis University, Purdue University.

KRUSIC, R. A., and C. D. NEEFUS. 1996. Habitat associations of bat species in the White Mountain national forest. Pp. 185–198 in *Bats and forests symposium,* ed. R. M. R. Barclay and R. M. Brigham. British Columbia Ministry of Forests, Victoria, British Columbia, Canada.

KRUSIC, R. A., M. YAMASAKI, C. D. NEEFUS, and P. J. PERKINS. 1996. Bat habitat use in White Mountain National Forest. *Journal of Wildlife Management* 60:625–631.

KUC, R. 1993. Three-dimensional tracking using qualitative bionic sonar. *Robotics and Autonomous Systems* 11:213–219.

———. 1994. Sensorimotor model of bat echolocation and prey capture. *Journal of the Acoustical Society of America* 96(4): 1965–1978.

———. 1997a. Biomimetic sonar locates and recognizes objects. *IEEE Journal of Oceanic Engineering* 22(4): 616–624.

———. 1997b. Biomimetic sonar recognizes objects using binaural information. *Journal of the Acoustical Society of America* 102(2): 689–696.

KUC, R., and B. BARSHAN. 1992. Bat-like sonar for guiding mobile robots. *IEEE Control Systems Magazine* 12(4): 4–12.

KUNZ, T. H. 1988. *Ecological and behavioral methods for the study of bats.* Washington, D.C.: Smithsonian Institution Press.

KUNZ, T. H., and C. E. BROCK. 1975. A comparison of mist nets and ultrasonic detectors for monitoring flight activity of bats. *Journal of Mammalogy* 56:907–911.

LANCE, R. F., B. BOLLICH, C. L. CALLAHAN, and P. L. LEBERG. 1996. Surveying forests—bat communities with Anabat detectors. Pp. 175–184 in *Bats and forests symposium,* ed. R. M. R. Barclay and R. M. Brigham. British Columbia Ministry of Forests, Victoria, British Columbia, Canada.

MACDONALD, K., E. MATSUI, R. STEVENS, and M. B. FENTON. 1994. Echolocation calls and field identification of the eastern pipistrelle (*Pipistrellus subflavus:* Chiroptera: Vespertilionidae), using ultrasonic bat detectors. *Journal of Mammalogy* 75:462–465.

MACOVSKI, A. 1983. *Medical imaging systems.* Englewood Cliffs, N.J.: Prentice-Hall.

MARMARELIS, V. Z. 1989. Signal transformation and coding in neural systems. *IEEE Transactions on Biomedical Engineering* 36(1): 15–24.

MASON, W. P. 1958. *Physical acoustics and the properties of solids.* Princeton, N.J.: D. van Nostrand Company Inc.

MCILWAIN, J. T. 1986. Point images in the visual system: New interest in an old idea. *Trends in Neuroscience* 9:354–358.

MEDDIS, R. 1986. Simulation of mechanical to neural transduction in the auditory receptor. *Journal of the Acoustical Society of America* 79(3): 702–711.

———. 1988. Simulation of auditory-neural transduction: Further studies. *Journal of the Acoustical Society of America* 83:1056–1063.

MEDDIS, R., M. J. HEWITT, and T. M. SHACKLETON. 1990. Implementation details of a computational model of the inner hair-cell/auditory-nerve synapse. *Journal of the Acoustical Society of America* 87(4): 1813–1816.

MELLINGER, D. K., and C. W. CLARK. 1997. Methods for automatic detection of mysticete sounds. *Marine and Freshwater Behaviour and Physiology* 29:163–181.

MENNE, D., and H. HACKBARTH. 1986. Accuracy of distance measurement in the bat *Eptesicus fuscus:* Theoretical aspects and computer simulations. *Journal of the Acoustical Society of America* 79:386–397.

MENNE, D., I. KAIPF, I. WAGNER, J. OSTWALD, and H.-U. SCHNITZLER. 1989. Range estimation by echolocation in the bat *Eptesicus fuscus:* Trading of phase versus time cues. *Journal of the Acoustical Society of America* 85:2642–2650.

MILLS, D. J., T. W. NORTON, H. E. PARNABY, R. B. CUNNINGHAM, and H. A. NIX. 1996. Designing surveys for microchiropteran bats in complex forest landscapes:

A pilot study from south-east Australia. *Forest Ecology and Management* 85:149–161.

MILLS, H. 1995. Automatic detection and classification of nocturnal migrant bird calls. *Journal of the Acoustical Society of America* 97:3370.

MOGDANS, J., and H.-U. SCHNITZLER. 1990. Range resolution and the possible use of spectral information in the echolocating bat, *Eptesicus fuscus*. *Journal of the Acoustical Society of America* 88:754–757.

MOGDANS, J., J. OSTWALD, and H.-U. SCHNITZLER. 1988. The role of pinna movement for localization of vertical and horizontal wire obstacles in the greater horseshoe bat, *Rhinolopus ferrumequinum*. *Journal of the Acoustical Society of America* 84:1676–1679.

MOGDANS, J., H. U. SCHNITZLER, and J. OSTWALD. 1993. Discrimination of two-wavefront echoes by the big brown bat, *Eptesicus fuscus:* Behavioral experiments and receiver simulations. *Journal of Comparative Physiology A* 172:309–323.

MOORE, P. W. B., H. L. ROITBLAT, R. H. PENNER, and P. E. NACHTIGALL. 1991. Recognizing successive dolphin echoes with an integrator gateway network. *Neural Networks* 4:701–709.

MUNSON, D. C., JR., J. D. O'BRIEN, and W. K. JENKINS. 1983. A tomographic formulation of spotlight- mode synthetic aperture radar. *Proceedings of the IEEE* 71:917–925.

MURRAY, K. L., E. R. BRITZKE, B. M. HADLEY, and L. W. ROBBINS. 1999. Surveying bat communities: A comparison between mist nets and the Anabat II bat detector system. *Acta Chiropterologica* 1:105–112.

NACHTIGALL, P. E. 1980. Odontocete echolocation performance on object size, shape and material. Pp. 71–95 in *Animal sonar systems,* ed. R. G. Busnel and J. F. Fish. New York: Plenum Press.

NACHTIGALL, P. E., and P. W. B. MOORE, eds. 1986. *Animal sonar: Processes and performance.* New York: Plenum Press.

NAGEL, A., and R. NAGEL. 1995. Utilisation of an underground roost by the lesser horseshoe bat (*Rhinolophus hipposideros*). Pp. 97–108 in *Zur Situation der Hufeisennasen in Europa: Arbeitskreises Fledermäuse,* ed. E. V. Sachsen-Anhalt. Berlin: IFA Verlag GmbH.

NASH, J. C. 1979. *Compact numerical methods for computers.* Bristol, UK: Adam Hilger Ltd.

NEEFUS, C. D., and R. A. KRUSIC. 1995. Computer-aided identification of bat species based on broad-band detection of echolocation calls. *Bat Research News* 36:94.

NEUBAUER, W. G. 1986. *Acoustic reflection from surfaces and shapes.* Washington, D.C.: Naval Research Lab.

NEUWEILER, G. 1984. Foraging, echolocation and audition in bats. *Naturwissenschaften* 71:446–455.

———. 1989. Foraging ecology and audition in echolocating bats. *Trends in Ecology and Evolution* 4:160–166.

———. 1999. *Biology of bats,* trans. E. Covey. New York: Oxford University Press.

NEUWEILER, G., V. BRUNS, and G. SCHULLER. 1980. Ears adapted for detection of motion, or how echolocating bats have exploited the capacities of the mammalian auditory system. *Journal of the Acoustical Society of America* 68:741–753.

NOYES, A., JR., and G. W. PIERCE. 1938. Apparatus for acoustic research in the supersonic frequency range. *Journal of the Acoustical Society of America* 9:207–211.

OBRIST, M. K. 1995. Flexible bat echolocation: The influence of individual, habitat and conspecifics on sonar signal design. *Behavioral Ecology and Sociobiology* 36:207–219.

OBRIST, M., M. B. FENTON, J. EGER, and P. SCHLEGEL. 1993. What ears do for bats: A comparative study of Pinna sound pressure transformation in Chiroptera. *Journal of Experimental Biology* 180:119–152.

O'DONNELL, C. F. J., and J. SEDGELEY. 1994. *An Automatic Monitoring System for Recording Bat Activity.* Department of Conservation Technical Series No. 5. Wellington, New Zealand: Department of Conservation.

O'FARRELL, M. J. 1997. Use of echolocation calls for the identification of free-flying bats. *Transactions of the Western Section of the Wildlife Society* 33:1–8.

O'FARRELL, M. J. and W. L. GANNON. 1999. A comparison of acoustic versus capture techniques for the inventory of bats. *Journal of Mammalogy* 80:24–30.

O'FARRELL, M. J., and B. W. MILLER. 1997. A new examination of echolocation calls of some neotropical bats (Emballonuridae and Mormoopidae). *Journal of Mammalogy* 78:954–963.

O'FARRELL, M. J., C. CORBEN, and W. L. GANNON. 2000. Geographic variation in the echolocation calls of the hoary bat (*Lasiurus cinereus*). *Acta Chiropterologica* 2:75–83.

O'FARRELL, M. J., B. W. MILLER, and W. L. GANNON. 1999. Qualitative identification of free-flying bats using the Anabat detector. *Journal of Mammalogy* 80:11–23.

O'FARRELL, M. J., C. CORBEN, W. L. GANNON, and B. W. MILLER. 1999. Confronting the dogma: A reply. *Journal of Mammalogy* 80:297–302.

OLSEN, J. F., and N. SUGA. 1991. Combination-sensitive neurons in the medial geniculate body of the mustached bat: Encoding of target range information. *Journal of Neurophysiology* 65:1275–1296.

O'NEILL, W. E. 1995. The bat auditory cortex. Pp. 416–480 in *Hearing by bats,* ed. A. N. Popper and R. R. Fay. New York: Springer-Verlag.

OWENS, A. L., T. J. DENISON, H. VERSNEL, M. REBBERT, M. PECKERAR, and S. A. SHAMMA. 1995. Multielectrode array for measuring evoked potentials from surface of ferret primary auditory cortex. *Journal of Neuroscience Methods* 58(1–2): 209–20.

PARK, K. J., G. JONES, and R. D. RANSOME. 1999. Winter activity of a population of greater horseshoe bats *Rhinolophus ferrumequinum. Journal of Zoology* 248: 419–427.

PARSONS, S. 1996. A comparison of the performance of a brand of broad-band and several brands of narrow-band bat detectors in two different habitat types. *Bioacoustics* 7:33–43.

———. 2001. Identification of New Zealand bats in flight from analysis of echolocation calls by artificial neural networks. *Journal of Zoology* (London) 253:447–456.

PARSONS, S., and G. JONES. 2000. Acoustic identification of 12 species of echolocating bat by discriminant function analysis and artificial neural networks. *Journal of Experimental Biology* 203:2641–2656.

PARSONS, S., A. M. BOONMAN, and M. K. OBRIST. 2000. Advantages and disadvantages of techniques for transforming and analyzing chiropteran echolocation calls. *Journal of Mammalogy,* in press.

PARSONS, S., C. W. THORPE, and S. DAWSON. 1997. Echolocation calls of the long-tailed bat: A quantitative analysis of types of call. *Journal of Mammalogy* 78: 964–976.

PASCHAL, W. G., and D. WONG. 1994. Frequency organization of delay-sensitive neurons in the auditory cortex of the FM bat *Myotis lucifugus. Journal of Neurophysiology* 72:366–379.

PEREMANS, H. 1997. Broad beamwidth ultrasonic transducers for tri-aural perception. *Journal of the Acoustical Society of America* 102:1567–1572.

PEREMANS, H., V. A. WALKER, and J. C. T. HALLAM. 1997. A biologically inspired sonarhead. University of Edinburgh, Department of Artificial Intelligence technical paper No. 44, September 1997.

PIERCE, G. W., and D. R. GRIFFIN. 1938. Experimental determination of supersonic notes emitted by bats. *Journal of Mammalogy* 19:454–455.

PONT, M. J., and R. I. DAMPER. 1991. A computational model of afferent neural activity from the cochlea to the dorsal acoustic stria. *Journal of the Acoustical Society of America* 89:1213–1228.

POPPER, A. N., and R. R. FAY. 1995. *Hearing by bats.* New York: Springer-Verlag.

POTTER, J. R., D. K. MELLINGER, and C. W. CLARK. 1994. Marine mammal call discrimination using artificial neural networks. *Journal of the Acoustical Society of America* 96:1255–1262.

PYE, J. D. 1992. Equipment and techniques for the study of ultrasound in air. *Bioacoustics* 4:77–88.

———. 1993. Is fidelity futile? The true signal is illusory, especially with ultrasound. *Bioacoustics* 4:271–286.

PYE, J. D., and W. R. LANGBAUER. 1998. Ultrasound and infrasound. Pp. 221–250 in *Animal acoustic communication: Sound analysis and research methods,* ed. S. L. Hopp, M. J. Owren, and C. S. Evans. Berlin: Springer-Verlag.

PYE, J. D., and L. ROBERTS. 1970. Ear movements in a hipposiderid bat. *Nature* 225:285–286.

PYE, J. D., M. FLINN, and A. PYE. 1962. Correlated orientation sounds and ear movements of horseshoe bats. *Nature* 196:1186–1188.

QUINLAN, J. R. 1993. *C4.5: Programs for machine learning.* San Mateo, Calif.: Morgan Kauffman.

RACHEL, D. R. 1992. Cetacean perception of the environment—a theoretical approach. The 124th Meeting of the Acoustical Society of America, New Orleans, USA.

RACHEL, D. R., and M. J. PALAKAL. 1999. On interfacing physics with electrophysiology in echolocation bats and cetaceans. The 137th Meeting of the Acoustical Society of America/2nd Convention of the European Acoustics Association/25th German Acoustics DAGA Convention, Berlin, Germany.

RAUTENBACH, I. L., M. B. FENTON, and M. J. WHITING. 1996. Bats in riverine forests and woodlands: A latitudinal transect in southern Africa. *Canadian Journal of Zoology* 74:312–322.

RICE, C. E. 1966. The human sonar system. Pp. 719–755 in *Animal sonar systems: Biology and bionics,* ed. R. G. Busnel. Jouy-en-Josas, France: Laboratoire de Physiologie Acoustique.

ROBBINS, L. W., and E. R. BRITZKE. 1999. Discriminating *Myotis sodalis* from *Myotis lucifugus* with Anabatcritique. *Bat Research News* 40:75–76.

ROITBLAT, H. L., P. W. B. MOORE, P. E. NACHTIGALL, R. H. PENNER, and W. W. L. AU. 1989. Natural echolocation with an artificial neural network. *International Journal on Neural Networks* 1:239–248.

ROVERUD, R. C. 1989. Harmonic and frequency structure used for echolocation sound pattern recognition and distance information processing in the rufous horseshoe bat. *Journal of Comparative Physiology A* 166:251–255.

RUMELHART, D. E., G. E. HINTON, and R. J. WILLIAMS. 1986. Learning internal representations by error propagation. Pp. 318–362 in *Parallel distributed processing,* vol. 1, ed. D. E. Rumelhart and J. L. McClelland. Cambridge: MIT Press.

RYDELL, J. 1990. Behavioural variation in echolocation pulses of the northern bat, *Eptesicus nilssonii. Ethology* 85:103–113.

———. 1993. Variation in the sonar of an aerial-hawking bat (*Eptesicus nilssonii*). *Ethology* 93:275–284.

RYDELL, J., A. BUSHBY, C. C. COSGROVE, and P. A. RACEY. 1994. Habitat use by bats along rivers in northeast Scotland. *Folia Zoologica* 43:417–424.

SABATINI, A. M. 1997. A stochastic model of the time-of-flight noise in airborne sonar ranging systems. *IEEE Transactions on Ultrasonics, Ferroelectrics, and Frequency Control* 44(3): 606–614.

SAILLANT, P. A., J. A. SIMMONS, S. P. DEAR, and T. A. McMULLEN. 1993. A computational model of echo processing and acoustic imaging in frequency–modulated echolocating bats. *Journal of the Acoustical Society of America* 94(5): 2691–2712.

SCHMIDT, S. 1988. Perception of structured phantom targets in the echolocating bat, *Megaderma lyra. Journal of the Acoustical Society of America* 91:2203–2223.

SCHNEIDER, H., and F. MÖHRES. 1960. Die Ohrbewegungen der Hufeisenfledermäuse (Chiroptera, Rhinolophidae) und der Mechanismus des Bildhörens. *Zeitung fuer vergleichende Physiologie* 44:1–40.

SCHNITZLER, H.-U. 1968. Die Ultraschall-ortungslaute der Hufeisen-fledermäuse (Chiroptera-Rhinolophidae) in verschiedenen Orientierungssituationen. *Zeitung fuer vergleichende Physiologie* 57:376–408.

———. 1973. Die Echoortung der Fledermäuse und ihre hörphysiologischen Grundlagen. *Fortschritte der Zoologie* 21:136–189.

———. 1984. The performance of bat sonar systems. Pp. 211–224 in *Localization and orientation in biology and engineering,* ed. Varju and H.-U. Schnitzler. Berlin: Springer-Verlag.

———. 1987. Echoes of fluttering insects: Information for echolocating bats. Pp. 223–243 in *Recent advances in the study of bats,* ed. M. B. Fenton, P. Racey, and J. M. V. Rayner. Cambridge: Cambridge University Press.

SCHNITZLER, H.-U., and O. W. HENSON. 1980. Performance of airborne animal sonar: I. Microchiroptera. Pp. 109–181 in *Animal sonar systems,* ed. R. G. Busnel and J. F. Fish. New York: Plenum Press.

SCHNITZLER, H.-U., E. K. V. KALKO, L. A. MILLER, and A. SURLYKKE. 1987. The echolocation and hunting behavior of the bat, *Pipistrellus kuhli. Journal of Comparative Physiology A* 161:267–274.

SCHULLER, G., K. BEUTLER, and H.-U. SCHNITZLER. 1974. Response to frequency shifted artificial echoes in the bat *Rhinolophus ferrumequinum. Journal of Comparative Physiology A* 89:275–286.

SCHUSTERMAN, R. J. 1980. Behavioral methodology in echolocation by marine mammals. Pp. 11–41 in *Animal sonar systems,* ed. R. G. Busnel and J. F. Fish. New York: Plenum Press.

SHAMMA, S. A., R. CHADWICK, J. WILBUR, J. RINZEL, and K. MOORISH. 1986. A biophysical model of cochlear processing: Intensity dependence of pure tone responses. *Journal of the Acoustical Society of America* 80:133–145.

SHERWIN, R. E., W. L. GANNON, and S. HAYMOND. 2000. The efficacy of acoustic techniques to infer differential use of habitat by bats. *Acta Chiropterologica* 2:145–153.

SHIRLEY, D. J., and K. J. DIERCKS. 1970. Analysis of the frequency response of simple geometric targets. *Journal of the Acoustical Society of America* 48: 1275–1282.

SIMMONS, J. A., and L. CHEN. 1989. The acoustic basis for target discrimination by FM echolocating bats. *Journal of the Acoustical Society of America* 86: 1333–1350.

SIMMONS, J. A., and R. A. STEIN. 1980. Acoustic imaging in bat sonar: Echolocation signals and the evolution of echolocation. *Journal of Comparative Physiology* 135:61–84.

SIMMONS, J. A., M. B. FENTON, and M. J. O'FARRELL. 1979. Echolocation and pursuit of prey by bats. *Science* 203:16–21.

SIMMONS, J. A., D. J. HOWELL, and N. SUGA. 1975. The information content of bat sonar echoes. *American Scientist* 63:204–215.

SIMMONS, J. A., C. F. MOSS, and M. FERRAGAMO. 1990. Convergence of temporal and spectral information into acoustic images of complex sonar targets perceived by the echolocating bat, *Eptesicus fuscus. Journal of Comparative Physiology A* 166:449–470.

SUGA, N., and W. E. O'NEILL. 1980. Auditory processing of echoes: Representation of acoustic information from the environment in the bat cerebral cortex. Pp. 589–611 in *Animal sonar systems,* ed. R. G. Busnel and J. F. Fish. New York: Plenum Press.

SULLIVAN, W. E. 1982. Neural representation of target distance in auditory cortex of the echolocating bat *Myotis lucifugus. Journal of Neurophysiology* 48:1011–1032.

SURLYKKE, A. 1992. Target ranging and the role of time-frequency structure of synthetic echoes in the big brown bats *Eptesicus fuscus. Journal of Comparative Physiology A* 170:83–92.

SURLYKKE, A., L. A. MILLER, B. MOHL, A. A. ANDERSON, J. CHRISTIENSEN-DALSGAARD, and M. B. JORGENSON. 1993. Echolocation in two very small bats from Thailand: *Craseonycteris thonglongyai* and *Myotis siligorensis. Behavioural Ecology and Sociobiology* 33:1–12.

TANAKA, H., D. WONG, and I. TANIGUCHI. 1992. The influence of stimulus duration on the delay tuning of cortical neurons in the FM bat, *Myotis lucifugus. Journal of Comparative Physiology A* 171:29–40.

TAYLOR, A. 1995. Bird flight call discrimination using machine learning. *Journal of the Acoustical Society of America* 97:3370.

TAYLOR, A., G. GRIGG, G. WATSON, and H. MCCALLUM. 1996. Monitoring frog communities: An application of machine learning. Pp. 1564–1569 in *Proceedings of the Thirteenth National Conference on Artificial Intelligence and the Eighth Innovative Applications of Artificial Intelligence Conference.* American Association for Artificial Intelligence, Menlo Park, Portland, Oregon.

THOMAS, D. W. 1993. Lack of evidence for a biological alarm in bats (*Myotis* spp.) hibernating under natural conditions. *Canadian Journal of Zoology* 71:1–3.

———. 1995. Hibernating bats are sensitive to non-tactile human disturbance. *Journal of Mammalogy* 76:940–946.

THOMAS, D. W., and R. K. LAVAL. 1988. Survey and census methods. Pp. 77–89 in *Ecological and behavioral methods for the study of bats,* ed. T. H. Kunz. Washington, D.C.: Smithsonian Institution Press.

TURIN, G. L. 1957. On the estimation in the presence of noise of the impulse response of a random, linear filter. *IRE Transactions on Information Theory* 3:5–10.

URICK, R. J. 1975. *Principles of underwater sound.* New York: McGraw-Hill.

VAN TREES, H. L. 1968. *Detection, estimation, and modulation theory,* part 1. New York: John Wiley & Sons.

VAUGHAN, N., G. JONES, and S. HARRIS. 1997a. Habitat use by bats (Chiroptera) assessed by means of a broad-band acoustic method. *Journal of Applied Ecology* 3:716–730.

———. 1997b. Identification of British bat species by multivariate analysis of echolocation call parameters. *Bioacoustics* 7:189–207.

WAGNER, T., U. SCHRAMM, and F. G. BOEBEL. 1995. Synergetic learning for unsupervised texture classification tasks. *Physica D* 80:140–150.

WAGNER, T., F. G. BOEBEL, U. HASSLER, H. HAKEN, and D. SEITZER. 1993. Using a synergetic computer in an industrial classification problem. Pp. 206–212 in *Proceedings of the International Conference on Artificial Neural Nets and Genetic Algorithms.* Berlin: Springer-Verlag.

WALKER, V. A. 1997. An investigation of qualitative echolocation strategies in synthetic bats and real robots. Ph.D. diss., University of Edinburgh.

WALKER, V. A., H. PEREMANS, and J. C. T. HALLAM. 1998. One tone, two ears, three dimensions: A robotic investigation of pinnae movements used by rhinolophid and hipposiderid bats. *Journal of the Acoustical Society of America* 104:569–579.

WALSH, A. L., and S. HARRIS. 1996. Factors determining the abundance of vespertilionid bats in Britain: Geographical, land class and local habitat relationships. *Journal of Applied Ecology* 33:519–529.

WATERS, D. A., and A. L. WALSH. 1994. The influence of bat detector brand on the quantitative estimation of bat activity. *Bioacoustics* 5:205–221.

WEHNER, D. R. 1987. *High resolution radar.* Norwood, Mass.: Artech House.

WELLER, T. J., V. M. SEIDMAN, and C. J. ZABEL. 1998. Assessment of foraging activity using Anabat II: A cautionary note. *Bat Research News* 39:61–65.

WIDROW, G., and S. D. STERNS. 1985. *Adaptive signal processing.* Englewood Cliffs, N.J.: Prentice-Hall.

WONG, D., and S. L. SHANNON. 1988. Functional zones in the auditory cortex of the echolocating bat *Myotis lucifugus. Brain Research* 453:349–352.

WONG, D., M. MAEKAWA, and H. TANAKA. 1992. The effect of pulse repetition rate on the delay sensitivity of neurons in the auditory cortex of the FM bat, *My-*

otis lucifugus. Journal of Comparative Physiology A 170:393–402.

ZBINDEN, K. 1989. Field observation on the flexibility of the acoustic behaviour of the European bat, *Nyctalus noctula* (Schreber, 1774). *Revue Suisse de Zoologie* 96:335–343.

———. 1994–1995. Computerized monitoring of meto data and bat echolocation activity. *Myotis* 32–33: 91–98.

ZINGG, P. E. 1990. Akustische Artidentifikation von Fledermäusen (Mammalia: Chiroptera) in der Schweiz. *Revue Suisse de Zoologie* 97:263–294.

PART SIX

Possible Echolocation Abilities in Other Mammals

Pinniped Sensory Systems and the Echolocation Issue

Ronald J. Schusterman, David Kastak, David H. Levenson,
Colleen Reichmuth Kastak, and *Brandon L. Southall*

Introduction

Each species lives in its own sensory world, or *Umwelt.* Therefore, it is surprising that even though pinnipeds inhabit dark or murky water, as do odontocete cetaceans, pinnipeds apparently have not evolved sonar. Unlike odontocete cetaceans, pinnipeds do not seem to use echolocation to obtain prey, orient themselves with regard to relocating breathing holes after lengthy underwater excursions, or navigate at sea. While dolphins produce high-frequency clicks using specialized structures, exquisitely differentiate frequencies at the high end of their auditory range, and demonstrate very acute directional hearing and temporal resolution, comparable underwater abilities in pinnipeds are mundane at best. We suggest that most pinnipeds have not evolved echolocation because, unlike dolphins that are completely aquatic, pinnipeds are truly amphibious. Presumably, there has been little selective pressure on pinnipeds to evolve echolocation. Instead, they appear well adapted to combine and synthesize a variety of sensory cues, along with spatial memory to orient under water. The preponderance of evidence shows that pinnipeds use passive cues such as light, sound, and hydrodynamic and tactile stimuli to forage, avoid predators, and navigate.

At the first Animal Sonar Systems Symposium in 1966, significant issues were raised regarding the possibility that pinnipeds (seals, sea lions, and walruses) could echolocate. This proposition followed logically from the fact that echolocation had recently been demonstrated in the odontocete cetaceans (toothed whales and dolphins), which, like pinnipeds, must forage and navigate at times in dark and murky waters. At the symposium, two papers were presented that purported to evaluate the ability of California sea lions (*Zalophus californianus*) to echolocate. These papers caused quite a stir at the symposium, first because their results were in complete contradiction, and, ironically, because both scientists came not only from the same institution (the Stanford Research Institute) but also from the same laboratory (the Diving Mammals and Biosonar Lab). Despite some evidence to the contrary (Evans and Haugen 1963), Thomas Poulter believed that he had already demonstrated sophisticated echolocation abilities in California sea lions and suggested that other pinniped species also had developed specialized sonar abilities (Poulter 1963, 1967). However, Poulter's colleague, Ronald Schusterman, presented an opposing view. Schusterman concluded that his data, collected under highly controlled conditions, demonstrated that California sea lions did not use echolocation and instead relied on other sensory modalities, particularly vision, to navigate and forage under water (Schusterman 1967).

More than three decades later, due to the lack of any convincing supporting evidence, the pinniped echolocation hypothesis remains unproven within the field of marine mammal acoustics and behavior. Since the first Animal Sonar meeting, the idea that pinnipeds use echolocation has been periodically revived (Renouf and Davis 1982; Thomas et al. 1983) and refuted (Oliver 1978; Scronce and Ridgway 1980; Wartzok, Schusterman, and Gailey-Phipps 1984). In this volume, Evans et al. (chapter 71) and Awbrey et al. (chapter 70) revisit the pinniped echolocation hypothesis in light of new recordings of Weddell seals (*Leptonychotes weddelli*) navigating under the ice, and on the basis of expanded and updated descriptions of earlier observations (Thomas et al. 1983) of a single captive leopard seal (*Hydrurga leponyx*) that emitted high-frequency signals while chasing fish under relatively darkened conditions. These authors suggest that Antarctic pinnipeds, which inhabit ice-covered areas and live in relative darkness during part of the year, are the most likely pinniped species to have evolved a specialized echolocation system for use in foraging and navigation.

The aim of this chapter is to review the echolocation hypothesis in the context of our current understanding of pinniped sensory systems. We find it necessary to first make a clear distinction between echolocation in a general sense and echolocation as a specialized ability. For instance, there is some experimental evidence that blind and sighted humans can be trained to detect, locate, and discriminate targets by listening for reflected echoes (Rice, Feinstein, and Schusterman 1965). Blinded rats, on the other hand, are capable of opportunistically using echoes of the sounds generated by their own movements to find their way through a maze (Riley and Rosenzweig

1957). While neither humans nor rats have evolved specialized echolocation abilities, these performances have been shown to emerge through experience.

In the following pages, we consider the issue of whether some pinniped species possess specialized sound production, detection, and signal-processing abilities for underwater echolocation similar to the type shown by odontocete cetaceans. We will argue that unlike the aquatic dolphins, pinnipeds have not developed active biosonar; rather, the amphibious lifestyle of pinnipeds has resulted in relatively nonspecialized underwater hearing abilities. Instead of using active biosonar, pinniped species appear to depend on alternative sensory capabilities—such as underwater hearing (passive biosonar), enhanced vision, and acute hydrodynamic reception—to orient in the marine environment. We address the issue of echolocation in pinnipeds by discussing sound production mechanisms and amphibious hearing capabilities, each in comparison to similar capabilities in odontocetes. Finally, we explore, from an experimental and ecological point of view, the use of the supplementary senses of vision and tactile/hydrodynamic sensitivity by pinnipeds to forage and navigate under water.

Sound Production

In addition to social vocalizations, bottlenose dolphins (*Tursiops truncatus*) produce broadband echolocation clicks with significant energy at frequencies above 90 kHz (see Au 1993 for a comprehensive review of cetacean echolocation performance). The clicks are extremely loud (218–228 dB re 1μPa), probably to compensate for transmission losses associated with spreading and absorption in water. Dolphins possess specialized sound production and transmitting structures not found in other mammals—sound is transmitted from the forehead and received primarily through acoustically sensitive areas of the lower jaw. Dolphins also appear to exercise fine control over the beam pattern and spectral and temporal content of their short (50–70 μs per pulse) echolocation click trains. The acoustic properties of the signal directly relate to environmental variables such as target range and ambient noise.

Pinnipeds also produce a wide variety of signals under water, including whines, grunts, roars, chirps, trills, chugs, and pulsed sounds (reviewed by Richardson et al. 1995). Many of these sounds are known to be related to social behavior and reproduction; however, the emission of high-frequency and pulsed sounds by a few species has led some researchers to speculate about the existence of echolocation in pinnipeds. Because pinnipeds produce underwater sounds, it is worthwhile to ask whether these sounds could function in a manner analogous to the echolocation pulses produced by odontocetes. The features of pinniped underwater sounds that make them appealing as potential echolocation signals are those that resemble dolphin biosonar sounds. Many pinniped sounds are pulsatile and repetitive, and at least one captive recording contained ultrasonic components (Awbrey et al., chapter 70, this volume). Further, some of these sounds have been recorded in nonsocial contexts, such as foraging or spatial orientation (Evans et al., chapter 71, this volume).

However, the sounds produced by pinnipeds differ from those produced by dolphins in some important ways. First, source levels are relatively low, ranging from only about 90–190 dB (re 1μPa) (see Richardson et al. 1995). Second, some of these sounds—for example, FM sweeps—are long in duration and emitted almost continuously. It is unlikely that this pattern of sound production allows for detection of echo returns between outgoing signals. Further, the relatively long durations of the calls increase the likelihood of contamination from reverberation, an important consideration for under-ice echolocation. Even the shorter-duration pulses produced by pinnipeds (0.3 to >10 ms) are much longer than those produced by dolphins. While potentially useful in the detection of large objects, the pulse durations typical of pinnipeds would not allow the fine target resolution seen in odontocetes. Third, in contrast to the dolphins, pinnipeds do not appear to modify the temporal patterning of click signals according to target range in laboratory tests of echolocation performance (Schusterman 1967); however, field observations of Weddell seals suggest that pulse rates increase as the seals approach ice holes (Evans et al., chapter 71, this volume). Fourth, the underwater sound detection capabilities of pinnipeds are inferior by about 20 dB at best sensitivity to those of dolphins. Finally, pinnipeds are less adept at localizing sound and detecting very brief stimuli than are the odontocetes (see Richardson et al. 1995). As a result of all of these factors, it is unlikely that the pinnipeds are consistently producing sounds with sufficient intensity to provide much environmental information by way of echo returns. All of these observations suggest an auditory/acoustic channel that is qualitatively different from that present in odontocete cetaceans, in terms of sound production, sound reception, and orienting behavior.

Attention has recently focused on the Antarctic leopard and Weddell seals, which are known to be extremely vocal under water and to forage under conditions of near darkness. These factors have led some researchers to consider the possibility of echolocation in these species (Awbrey et al., chapter 70, this volume; Evans et al., chapter 71, this volume). However, underwater observations of foraging Weddell seals made by Davis et al. (1999) contradict echolocation-based explanations of foraging. These investigators found that seals fitted with portable video cameras did not emit vocalizations while pursing prey. Rather, the seals appeared to use vision to detect and track the movements of fish. Given the generally low frequencies and source levels of Weddell seal calls, as well as the weak target strengths of typical prey items, it is unlikely that this species is able to detect prey using active biosonar. Additionally, speculation of under-ice navigation by Weddell seals via biosonar is not supported by the results of another recent study, which indicated that only half of the seals surfacing in ice holes produced any sounds at all (Evans et al., chapter 71, this volume).

What, then, are the functions of the underwater sounds produced by pinnipeds, if not for echolocation? The answers likely lie in social behavior. Phocids (true seals), which produce underwater sounds most frequently, generally breed in water, and their underwater vocalizations are almost certainly related to breeding behavior. In otariids (sea lions and fur seals), variability in signal strength, repetition rate, and other vocal parameters often is related to affect or motivational state. This observation explains the underwater sounds produced by subjects in a state of arousal while chasing fish in a tank (Schusterman 1967). The specific functions of underwater vocalizations are likely to be further elucidated when more data on the underwater behavior of these animals becomes available. Based on our current limited understanding of certain species, it remains possible that some pinnipeds are capable of specialized echolocation. However, given the extreme paucity of evidence after more than thirty years of investigation, to say that it is unlikely would be an understatement, especially when this ability has been so easily demonstrated in the odontocetes.

Pinniped Auditory Sensitivity

When examining the pinniped auditory system, it is helpful to adopt a phylogenetic perspective—all pinnipeds are carnivores, and their closest relatives are semiaquatic or terrestrial. Apart from a few anatomical modifications, the pinnipeds retain the basic, air-adapted carnivore ear, even after more than 25 million years of divergence from a common ancestor (Repenning 1976). Thus, it is likely that the pinniped auditory system functions in much the same way as that of most terrestrial mammals. In contrast, the odontocete ear has become extremely modified from its original air-adapted form and bears little resemblance to any sort of terrestrial mammalian ear. When the results of behavioral experiments on aerial and underwater hearing sensitivity in pinnipeds are expressed in terms of the detection of sound energy, they clearly support the proposition that the pinniped ear is adapted to hear in water rather than in air. However, when the metric of comparison is changed to sound pressure, a different picture emerges—the otariids are adapted to hear in air, while the phocids are generally adapted for amphibious hearing (Kastak and Schusterman 1998). The northern elephant seal

(*Mirounga angustirostris*), a deep diving phocid, shows differences in aerial and underwater sensitivity paralleling those of the bottlenose dolphin, implying aquatic specialization (Kastak and Schusterman 1998). When pairwise comparisons are made of air-water sensitivity between all pinnipeds tested to date in air and water, significant differences emerge only between phocids and otariids, and between the elephant seal and all other pinnipeds. The patterns of air and water hearing sensitivity appear to correspond to the patterns of life history of three pinniped assemblages: the otariids, the generalized phocids (*Phoca* spp.), and the northern elephant seal, a specialized phocid (Kastak and Schusterman 1998, 1999). Contrary to earlier interpretations of air-water sensitivity differences, which labeled the pinniped ear as "water-adapted," the logical conclusion based on this reinterpretation is that all pinnipeds have retained some degree of airborne hearing sensitivity while developing enhanced underwater hearing. Retention of airborne sensitivity constrained the development of the exceptional underwater auditory sensitivity necessary for echolocation. In other words, the pinniped ear has evolved through natural selection to function amphibiously, perhaps at the cost of evolving an active biosonar system.

Passive Biosonar

Laboratory experiments showing that pinnipeds can locate prey items rapidly under darkened conditions (Schusterman 1967), and reports of free-ranging, apparently blind pinnipeds surviving for long periods of time (Schusterman 1981), suggest that visually impaired seals and sea lions are able to find and capture prey, as well as orient in water and on land. Thus, one or more sensory mechanisms other than vision can apparently be used by pinnipeds when vision is of limited use. As previously discussed, the underwater hearing abilities of pinnipeds are not as acute as those of odotocetes. However, at low and intermediate frequencies, the underwater hearing sensitivity and sound localizing capabilities of pinnipeds (reviewed in Richardson et al. 1995) suggest that simply by listening, they may obtain information regarding the presence and general location of prey, predators, conspecifics, and navigational landmarks.

The practical functions of underwater hearing in free-ranging pinnipeds remain largely uninvestigated. However, experiments in the field by Wartzok et al. (1992a) showed that blindfolded ringed seals (*Phoca hispida*) and Weddell seals used experimentally produced acoustic cues to find novel holes cut through the ice. When these acoustic cues were removed, the blindfolded seals were able to return to the same ice holes, presumably using spatial memory or ambient noises rather than active biosonar, evidenced by the fact that they did not produce sounds while searching for ice holes.

Given that many pinnipeds hear well in water, what underwater sounds are of biological relevance to them? At least one experiment suggests that predator detection is an essential function of underwater hearing. Schusterman and Kastak (1996) found that a northern elephant seal, in contrast to a California sea lion and a harbor seal (*Phoca vitulina*), exhibited an exaggerated startle response and quickly became sensitized to pulsatile sounds resembling the echolocation clicks of killer whales. These observations, in addition to reports that mammal-hunting killer whales echolocate less frequently and in a more unpredictable fashion than those that hunt fish (Barrett-Lennard, Ford, and Heise 1996), indicate that predator detection is accomplished, at least in part, simply by listening.

Vision

One of the most striking features of pinniped anatomy is the large size of the eyes. Early in the study of pinniped sensory systems, it was realized that the enhanced light-gathering ability of an enlarged eye was of tremendous advantage for visual foraging in a relatively dark environment. Studies on anatomy and photorefraction have shown that the eyes of these amphibious mammals are suited primarily for vision in water rather than in air, possessing large, round lenses to compensate for the absence of corneal refraction under water (Johnson 1893; Walls 1942). The pupil of pinnipeds is stenopaic (slit-like) and probably functions as a pinhole aperture, providing relatively clear vision in air, as long as the pupil remains small (Walls 1942). Behavioral studies confirmed the pupil's role in aerial visual acuity when it was demonstrated that acuity drops off much faster in air than under water as ambient light levels decrease (Schusterman and Balliet 1971).

Like other mammals adapted for vision in dim conditions, pinnipeds possess large lenses and pupils, and densely packed, rod-dominated retinas with well-developed choroidal tapeta (Walls 1942; Landau and Dawson 1970). In circumstances where light levels change dramatically in a short period of time, it is advantageous to possess a pupil with a large dynamic range. This allows for efficient regulation of the illumination of retinas, functioning only secondarily to improve visual resolution. Substantial differences in pupillary dynamic range between shallow and deep-diving pinnipeds have been observed (Levenson and Schusterman 1997), illustrating the importance of the pupillomotor response in maintaining sufficient photoreceptor stimulation. The deepest-diving pinniped, the northern elephant seal, exhibited a greater than 400-fold increase in pupillary area, while shallow and moderately deep divers demonstrated considerably smaller ranges of only 25- to 200-fold (Levenson and Schusterman 1997). The functional significance of differences in pupillary structure was clearly shown in a comparison of the dark adaptation rates of pinnipeds (Levenson and Schusterman 1999). The time necessary to reach maximum sensitivity was substantially faster for the elephant seal than for the shallower diving California sea lion and harbor seal. From an ecological standpoint, the six-minute dark adaptation time of the elephant seal (Levenson and Schusterman 1999) matches the time it takes these seals to dive from the surface to foraging depths of 300–700 m (LeBoeuf and Laws 1994).

The elephant seal is as extreme in terms of adaptation for underwater vision as it is for underwater hearing. In addition to having an extremely rapid dark adaptation rate, the eye of the elephant seal is very light sensitive and possesses specialized rod photopigments. Consequently, the elephant seal visual system is designed not only to respond rapidly to changes in light level, but to take advantage of very low levels of the shorter wavelengths of light that predominate deep under water (Lythgoe and Dartnall 1970). The shallower diving harbor seals and California sea lions are less light sensitive than the elephant seal, and they possess rod pigments with sensitivity only slightly shifted to short wavelengths compared to those of terrestrial mammals (Lavigne and Ronald 1975; Jacobs et al. 1993; Levenson and Schusterman 1999).

Pinniped visual systems, though amphibious, are primarily water adapted. As suggested by earlier studies (reviewed in Schusterman 1981), underwater foraging and navigation likely depend to a great extent on the ability to see well in this medium. Natural selection in these animals clearly has favored the development of visual systems suited to foraging and navigation in an aquatic environment.

Tactile/Hydrodynamic Sensitivity

While benthic feeding dolphins have been observed to direct echolocation pulses into the substrate to detect prey (Rossbach and Herzing 1997), pinnipeds such as walruses that forage on the muddy sea floor tend to have modified vibrissae and facial structures to detect and extract prey from the bottom (Riedman 1990). Even pinnipeds that forage on swimming prey have vibrissae that are highly specialized with respect to size, length, vascularization, and innervation. These structures are apparently used to detect tactile and hydrodynamic cues very close to the source, and thus the tactile sense has been implicated in foraging (Davis et al. 1999) and very short-range navigation (see Riedman 1990).

Because pinniped vibrissae are exquisitely sensitive to vibration, it is possible that they function in detecting particle motion associated with the flow field of a moving fish or the acoustic near field of struggling prey (Dehnhardt, Mauck, and Bleckmann 1998). This ability would be especially useful in the final stages of prey pursuit, when vision would be of questionable use because the visual field is directed forward, above the head, and away from the mouth. One can envision the sensory systems used during prey detection, pursuit, and capture by pinnipeds as follows: The prey is detected visually (preferentially backlit against the surface water or ice) or acoustically (by eavesdropping on intraspecific signals produced by the fish or incidental sounds produced by swimming). Pursuit is guided visually and perhaps acoustically, and fine-scale orientation on the prey is mediated by detection of the flow field around the fish by the vibrissae, leading to prey capture.

Conclusions

Odontocete cetaceans possess a highly sophisticated system of active biosonar. This system involves the production of high-intensity, high-frequency clicks in a directional beam. Received echoes can provide information about target location and identity, and typical targets can be detected accurately at distances of over 100 m (Au 1993). Echolocation in this group is tied to the development of sophisticated sound production, reception, and signal-processing systems, including excellent high-frequency sensitivity, highly directional hearing, and acute temporal and frequency resolution (Au 1993). This type of echolocation system evolved only once in the marine environment, in a group of organisms that became completely tied to an aquatic existence. Because odontocetes were not evolutionarily constrained to give birth on land, the dolphin auditory system became fully adapted for underwater functioning, allowing a refinement of the biosonar system not possible in am-

phibious mammals, such as the pinnipeds. Because even the most aquatic pinnipeds must return to shore periodically and because airborne vocal communication appears to play an important role in most, if not all, pinniped social systems, selection pressures for highly sensitive, acute underwater hearing have not shaped the pinniped auditory system to as great an extent as in dolphins. Rather, the retention of in-air hearing abilities has limited the sensitivity of hearing under water, where all pinnipeds must forage. Consequently, instead of developing a primarily sound-based system of underwater orientation, pinniped visual, tactile, hydrodynamic, and acoustic sensory systems were refined and incorporated into overlapping underwater perceptual channels that permit efficient underwater foraging and navigation without the use of active biosonar.

Acknowledgments

This work was supported by grant N00014-99-1-0164 from the Office of Naval Research to Ronald J. Schusterman. The authors thank the research team at the Pinniped Cognition and Sensory Systems Lab at Long Marine Laboratory for assistance with animal care and experiments in vision and audition. Discussions with Whit Au, Bill Evans, and Bertel Møhl helped to frame our thinking regarding the adaptive significance of echolocation. The ideas in this paper were augmented by detailed descriptions of experimental work with free-ranging seals provided by Doug Wartzok. This paper is dedicated to the memories of Winthrop Niles Kellogg and Kenneth Stafford Norris.

{ 70 }

Ultrasonic Underwater Sounds from a Captive Leopard Seal (*Hydrurga leptonyx*)

Frank T. Awbrey, Jeanette A. Thomas, and *William E. Evans*

Introduction

The leopard seal (*Hydrurga leptonyx*) is a large Antarctic pack-ice phocid that feeds predominantly on krill, fish, and cephalopods, but also is known to prey on seal pups and penguins during the austral summer. In this solitary species, males and females are seen together

only during a brief mating period in November (Siniff 1991). The underwater sound repertoire of leopard seals has been studied in the wild by Stirling and Siniff (1979), Thomas and DeMaster (1982), Rogers, Cato, and Bryden (1995), and Thomas and Golladay (1995) using audio recording equipment. These investigators reported 5–12 underwater sound types, depending on the location in Antarctica. In comparison, another Antarctic phocid, the monogamous crabeater seal (*Lobodon carcinophagus*), produces only a single underwater sound (Stirling and Siniff 1979). The polygynous Weddell seal (*Leptonychotes weddellii*) has an underwater repertoire of 34 sound types (Thomas and Kuechle 1982). Thomas (1991) suggested that leopard seals have only a few underwater sounds because of their solitary social system or the need as a predator to be quiet.

Published sounds from pinnipeds are almost exclusively based on audio recordings—that is, below 20 kHz. This is largely due to the limited number of recording devices available to document ultrasonic sounds under water and the logistic problems of using a large, cumbersome reel-to-reel system in a remote field situation, especially in a pack-ice habitat. In addition, ultrasonic sounds attenuate quickly over distance; so to document very high frequency sounds a researcher needs to be close to the study animal. Thomas et al. (1983) had the unique opportunity to have access to a captive leopard seal at Hubbs–Sea World Research Institute (HSWRI) using ultrasonic recording equipment for close-range recordings. They conducted a series of experiments on a captive leopard seal to examine whether it produced high-frequency signals. In this chapter, we conduct a more detailed analysis of these data and discuss the results relative to the theory of echolocation in pinnipeds.

Echolocation abilities in bats and dolphins were discovered following the advent of ultrasonic sensing and recording equipment that identified high-frequency sounds from these species. The topic of whether pinnipeds too echolocate has been debated over the last 30 years (Airapet'yants and Konstantinov 1974; Renouf, Galway, and Gaborko 1980; Renouf and Davis 1985; Schusterman et al., chapter 69, this volume). There is speculation that echolocation would be especially adaptive for polar pinnipeds who inhabit ice-covered areas and live in total darkness during part of the year. The vocal repertoire of many polar pinnipeds is characterized by pulse sounds similar to those in echolocating odontocetes and frequency-modulated (FM) sweeps similar to echolocation sounds by microchiropterans. Documentation of ultrasonic sound production in a pinniped provides a new area of research related to documenting which families/species of pinnipeds produce ultrasounds and to understanding the function(s) of such sounds.

Materials and Methods

The leopard seal "Coogee" was originally beached as a subadult in 1978 near Coogee Beach, Australia. At the time of the study, this four-year-old female weighed 417 kg, measured 3 m in length, and was held in a 5 m diameter pool at HSWRI in San Diego, California. This seal had basic training and was fed only frozen fish.

To reduce ambient light, and thus visual detection

of fish, studies were conducted at night (approximately 2400 hours), in an unlighted "behind the scenes" pool area at HSWRI. As with the Renouf, Galway, and Gaborko (1980) study, black curtains were draped around the pool to darken the area. However, no measurements of ambient light levels were made; we just attempted to minimize all available ambient light sources in the area.

Recordings were made on four nights during April 1981: one with only live anchovies, one with only live striped bass, one with the seal and live anchovies, and one with the seal and live striped bass. We made recordings of the seal alone in this pool on several previous occasions, but not at ultrasonic ranges. Live anchovies or striped bass were introduced into the pool and the seal's sounds recorded under water until she consumed all the fish. The seal normally was only fed frozen fish by a trainer during the daytime; live fish in the pool at night was a novel situation to her. We thus presumed that under these reduced light conditions, she might use sounds to help locate the swimming fish.

Because ultrasonic sounds tend to be directional, two Brüel & Kjær hydrophones (a type 8103 with a linear frequency response to 200 kHz + 3 dB and a type 8104 with a linear frequency response to 150 kHz + 3.0 dB) were placed on opposite sides of the pool. Two 17.8 cm tapes were recorded under each of the four test situations. Recordings were made using two channels on a Racal Store-4 reel-to-reel recorder at 152 cm/s (frequency response linear to 300 kHz + 3 dB).

For analysis, playback speed was reduced eight times to bring sounds into the frequency range appropriate for the sound analyzers. Spectrograms, power spectra, and oscillograms were produced using a Uniscan II real-time sonogram system and Canary or CoolPro Edit software for a personal computer. For pulsed sounds, average peak frequency, number of pulses per series, and duration of the series were measured. For FM sounds, maximum frequency per sweep, number of sweeps, and total duration were documented.

Results

We found no sounds produced by anchovies or striped bass alone in the pool. The captive leopard seal responded to both species of fish by rapidly chasing and eating them, while producing a variety of low-amplitude sounds. Two types of leopard seal sounds were audible and designated as C1 and C2 (figs. 70.1 and 70.2; table 70.1). The majority of audible sounds (22 of 24) were recorded in the presence of striped bass. Eight of 10 C1 sounds began with a low-frequency prefix near 776 Hz, averaging 0.63 s in duration. The prefix was never seen alone. The main part of the C1 sound was a series of pulses with a low peak frequency (mean = 896 Hz), averaging 44.2 pulses per series. Two of 10 C1 sounds had two separate groups of pulses.

C1 Sonogram

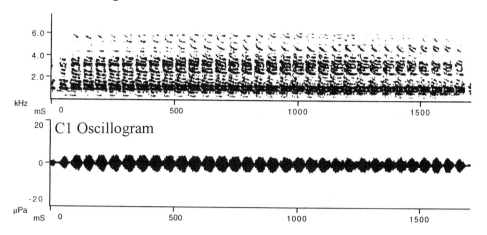

C1 Oscillogram

Fig. 70.1. Sonograms of the audible C1 underwater sound from a captive leopard seal. The upper oscillogram is a zoomed display of part of the sound. The lower oscillogram is a time-compressed waveform, showing the two-part structure and a typical duration of 27.02 s.

TABLE 70.1. Audible sound characteristics from a captive leopard seal (C) compared to sounds from wild leopard seals near the Palmer Peninsula (P) and McMurdo Sound (M), Antarctica (Thomas and Golladay 1995).

Call type	Mean peak frequency (Hz)	Mean total duration (s)
C1	896	2.8
P1	453	2.1
M1	442	2.6
C2	3176	3.1
P2	4116	2.3
M2	4662	2.0

C2 was a series of pulses, too numerous to count by listening. Six of nine C2 sounds had two separate groups of pulses (first group averaged 2.16 s and the second averaged 1.65 s in duration). The average peak frequency of C2 was 3176 Hz. The waveform of C2 was simple and consistent among pulses (fig. 70.2); whereas the waveform of C1 was complex (fig. 70.1).

Compared to audible underwater sounds from wild leopard seals at McMurdo Sound (M) and Palmer Penninsula (P), Antarctica, Coogee's sounds were most similar in frequency, duration, and having two groups of pulses to call types 1, 2, and 4 (table 70.1; Thomas and Golladay 1995). We noted that audible sounds from Coogee were not identical to any of the sounds reported by Thomas and Golladay (1995), perhaps due to geographic variations. Coogee stranded in Australia, so she presumably came from the region of Antarctica just below Australia, a site far away from Palmer Peninsula (south of Tierra del Fuego) and McMurdo Sound (south of New Zealand), Antarctica. Alternatively, something about the captive environment (e.g., reverberation or ambient noise) could have caused the seal to produce slightly different types of sounds.

At ultrasonic frequencies, FM sweeps (C3 and C4), buzzes (C5), and pulses (C6) were recorded (fig. 70.3; table 70.2). The majority of ultrasonic sounds were produced in the presence of anchovies. The most common peak frequencies were between 50 kHz and 60 kHz. Frequency variability can be attributed partially to the unknown direction of the signal relative to the hydrophone. FM signals swept as much as 60 kHz in 4 ms; each sweep lasted only 4–10 ms. C3 was a series of 4–10 rapid, descending FM sweeps. The maximum frequency of C3 sweeps ranged from 160 kHz to 64 kHz. C4 was a series of 4–5 descending FM sweeps, repeated slower than C3 and somewhat lower in frequency (maximum frequency of sweeps ranged from 110 kHz to 80 kHz). C5 was a buzz or burst pulse series, with maximum frequency of pulses descending from 132 kHz to 64 kHz. C6 sounds were heard as single, double, or trains of high-frequency pulses (fig. 70.4). Pulses from the seal while chasing the

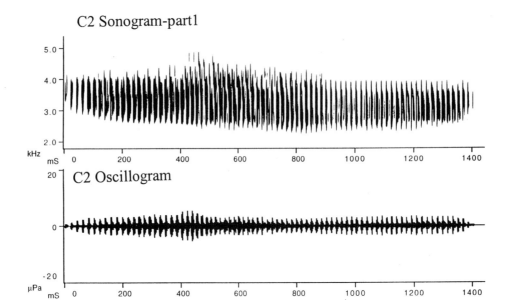

C2 Sonogram-part1

C2 Oscillogram

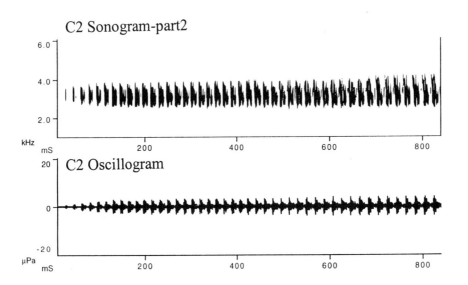

C2 Sonogram-part2

C2 Oscillogram

C2 Oscillogram

Fig. 70.2. Sonograms of two parts of the audible C2 sound from a captive leopard seal. The upper two oscillograms are a zoomed display of part of the sound. The lower oscillogram is a time-compressed waveform, showing the two-part structure and a typical duration of 30.02 s.

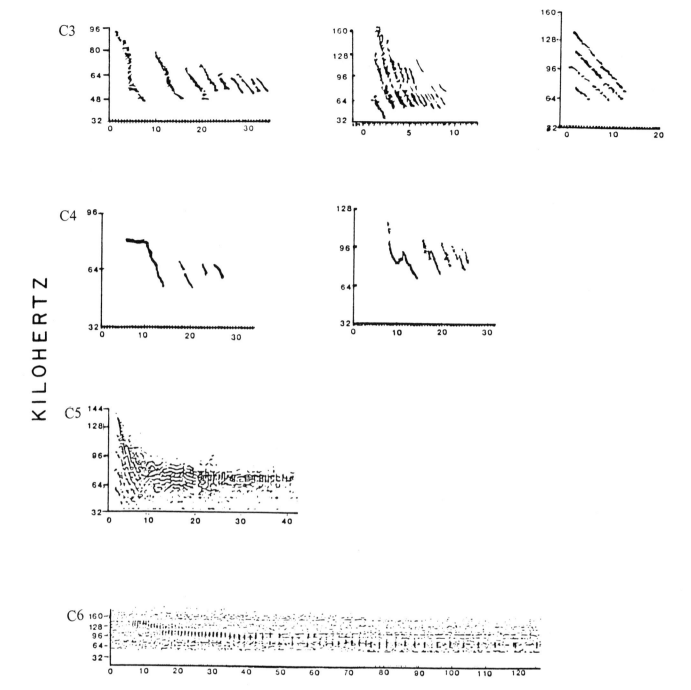

KILOHERTZ

MILLISECONDS

Fig. 70.3. Sonograms of four types of ultrasonic underwater sounds (C3–C6) from a captive leopard seal

TABLE 70.2. Summary of ultrasonic sound characteristics from a captive leopard seal.

Call type	Maximum frequency range (Hz)	Total duration range (ms)
C3	64,000–160,000	9–32.0
C4	80,000–110,000	20–25.0
C5	130,000–132,000	38–40.0
C6	128,000–132,000	0.3 (single pulse); 120.0 (train)

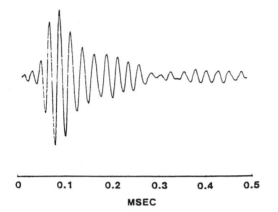

0   0.1   0.2   0.3   0.4   0.5

MSEC

Fig. 70.4. Waveform of a simple ultrasonic C6 pulse used by a captive leopard seal chasing live anchovies (frequencies [kHz] at peaks 55.6, 47.6, 43.4, 41.7, 40.0, 37.7, 43.5, 43.5, 43.5, respectively)

smaller anchovies had a simple waveform with substantial energy between 40 kHz to 50 kHz (fig. 70.4); whereas pulses while chasing the larger striped bass had highest energies at frequencies near 10 kHz.

Discussion

Renouf and Davis (1985) reported that harbor seals (*Phoca vitulina*), a temperate species, have echolocation abilities. They noted that long sequences of pulses were common in wild harbor seals, but less common in captive animals. Twice as many pulses of Miquelon Island harbor seals were recorded during nighttime compared to daylight. As a result, the authors proposed that harbor seals used echolocation during the dark night.

In another study, Renouf, Galway, and Gaborko (1980) recorded captive harbor seals given live capelin at night. During moonlight nights, there were no pulses and the seals still caught fish. In a pool draped with black neoprene, a harbor seal produced only a few pulses and not all fish were captured. The authors proposed that echolocation was used to estimate whether fish were present, then other senses—such as vision or tactile senses of the vibrissae—were used to capture individual fish. They also trained a harbor seal to fetch an air-filled and water-filled ring. When experiments were con-

ducted at night in a draped pool, a few single pulses and trains were recorded during each trial. The investigators suggested that few pulses were used because of the highly reflective walls of the pool and the close range of the ring targets.

Wartzok (1984) and Schusterman et al. (chapter 69, this volume) disputed these studies. One of the strongest arguments against an echolocation theory is the extremely good vision that seals have under water. They have large eyes in proportion to their head and a tapetum that aids in light gathering in low ambient light conditions. Efforts to demonstrate echolocation abilities in pinnipeds experimentally have been unsuccessful in California sea lions (*Zalophus californianus*) by Evans and Haugen (1963) and by Schusterman (1966, 1967) and in gray seals (*Halichoerus grypus*) by Oliver (1978) and by Scronce and Ridgway (1980). However, none of the species studied previously were examined at ultrasonic frequencies; nor were any of the species polar pinnipeds.

Sounds of such high frequencies as those in our study have never been reported in pinnipeds, perhaps because researchers rarely attempt recordings at ultrasonic frequencies. Such recordings require specialized, bulky equipment that is not practical for field situations. Sounds from this captive leopard seal share some acoustic characteristics with the ultrasonic FM sweeps of echolocating bats and the pulse trains used by dolphins to find prey or to orient in their environment.

Echolocation ability in polar pinnipeds was proposed by Poulter (1963, 1967) and Kooyman (1968), because it could help seals find food or navigate in dark oceanic depths, especially during the totally dark austral winter. Echolocation in pinnipeds (if employed as by dolphins and bats) theoretically would be most useful at short ranges to help discriminate prey types, track and capture prey, or navigate in irregular terrain. However, given advanced visual abilities in pinnipeds, prey (especially bioluminescent species) could be seen easily at dark oceanic depths, thus making echolocation unnecessary.

There is an emerging literature suggesting that marine mammals actively use sounds, other than echolocation signals, to capture prey. Norris and Møhl (1983) proposed that dolphins produce high amplitude "bangs" or pulses that acoustically debilitate or stun prey for easier capture. Evans et al. (chapter 71, this volume) reported bubble-blowing behavior by Weddell seals that cause fish to swim away from ice pockets for easier capture. An alternate idea from this study is that pinnipeds use sound at dark depths to startle prey and get them to move a bit, thus creating enough bioluminescence or motion for visual detection.

Mann, Lu, and Popper (1996) reported that most teleost fish cannot detect sounds greater than 3 kHz, but they noted a some fish that can hear or respond to ultrasound: cod (*Gadus morhua*), the American shad (*Alosa sapidissimia*), and the blueback herring. All recordings in this study had sonic sounds (< 4 kHz) known to be produced by leopard seals in the wild, so the fish probably heard these sounds. We are not sure if the anchovies or striped bass could hear the ultrasonic portion of leopard seal sounds. Because no ultrasonic sounds were heard during fish-only recordings, we can rule out that ultrasonics were produced by the fish rather than the leopard seal. It is interesting to note that leopard seal sounds were ultrasonic pulses when chasing the smaller anchovies, but audible sounds when chasing the larger striped bass. Such a change in frequency would be consistent with an echolocation function in that high-frequency sounds provide more detail and would be used with smaller targets. We cannot say whether Antarctic leopard seals produce ultrasonic sounds routinely during prey pursuit in the wild or whether the sounds recorded in this study were merely a short-term response to a novel situation. Leopard seals navigate and stalk prey under sea ice; perceptual abilities necessary to accomplish this certainly include vision but also might include active use of sounds.

Further examination of the theory of echolocation in pinnipeds requires carefully designed experiments with controls, in which seals have vision eliminated by eye cups or a blindfold and are trained to discriminate targets using only sounds. Our results indicate the importance of further investigations on ultrasonic sound production by pinnipeds.

Perhaps because high-frequency signals provide fine details for target discrimination, researchers traditionally have associated echolocation abilities only with ultrasonic sound production. However, in the following chapters by Evans et al. and by Clark and Ellison, there is some evidence that audible, even low-frequency, sounds by Weddell seals and baleen whales could provide sonar-like cues to navigating or foraging marine mammals.

Acknowledgments

Sheldon Fisher provided acoustic analysis of sounds at the Bioacoustics Laboratory of Hubbs–Sea World Research Institute. Sandra Barnett was the trainer of the leopard seal and provided invaluable support in conducting this research. We thank L. H. Cornell, D.V.M., and Bruce Stephens of Sea World, Inc., for their assistance and support in husbandry, care, and advice on training the leopard seal.

{ 71 }

Vocalizations from Weddell Seals (*Leptonychotes weddellii*) during Diving and Foraging

William E. Evans, Jeanette A. Thomas, and *Randall W. Davis*

Introduction

Marine mammals are among the most difficult groups of mammals to study in their natural habitat. Most of what is known about their behavior comes from studies of animals in captive and well-controlled environments. Fortunately, technology has advanced since the studies of the 1960s through the 1980s to provide some partial solutions to these problems. Advancements in computer hardware/software and miniaturization of video and acoustic recording systems allow not only records of sounds but also of associated behaviors (Burgess et al. 1998; Dudzinski 1995; Davis et al. 1999). Use of active and passive sounds for orientation has been well established in delphinids, but whether this capability is used by pinnipeds is still speculative. It is possible that underwater vocalizations supplement their extremely well developed visual capabilities (Kooyman 1968). Furthermore, supplemental use of vocalizations for orientation could be selectively used by only certain individuals in particular situations.

Most pinnipeds are vocal under water (King 1983; Riedman 1990). Among the most difficult to study are the ice-dwelling seals of the Arctic and Antarctic. Looking at the physical nature of their underwater habitats, there has been speculation that they might use some form of active acoustics to avoid objects under water or to detect fish during foraging (Poulter 1966). A few species of both eared seals (otariids) and true seals (phocids) have been tested (Evans and Haugen 1963), but most studies have provided negative results (Schusterman 1981; Schusterman et al., chapter 71, this volume).

William E. Evans, Jeanette A. Thomas, and Randall W. Davis

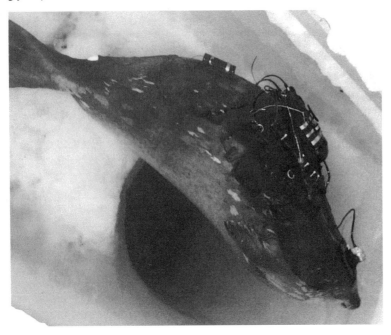

Fig. 71.1. The underwater video system and data logger positioned on the back of a Weddell seal

Past acoustic studies of the behavior of polar seals, while vocalizing during dives in the wild, have had limited success (Wartzok, Schusterman, and Gailey-Phipps 1984). Because polar seals experience periods of total darkness during the winter, Thomas (1991) and Kooyman (1968) noted that of all pinnipeds, these species should be most likely to exhibit active or passive orientation via sound cues. In fact, Antarctic Weddell seals are among the most vocal pinnipeds under water. They have an extensive vocal repertoire (Thomas and Kuechle 1982), with up to 34 types near McMurdo Sound. Geographic variation in calls has been documented around the continent (Thomas and Stirling 1983; Thomas et al. 1988). Many of these calls have a social function, including mate attraction, underwater territorial defense, and aggression, but the specific function of most of sounds is unknown (Thomas, Zinnel, and Ferm 1983). Pinnipeds actively use sounds for social functions, just as many mammals do, but pinnipeds may also use ambient sound in navigation or orientation.

Materials and Methods

In 1997, Davis et al. began a three-year study of the behavioral and energetic adaptations that enable Weddell seals to forage in the fast-ice environment of McMurdo Sound, Antarctica. One objective was to determine the hunting strategies Weddell seals use to locate and capture prey. A related objective was to document vocalizations during various underwater behaviors. To accomplish these objectives, a small video system and data logger were attached to the seal's back during voluntary dives from an isolated hole 5 km from the nearest tidal crack and other seals. The video system and data logger recorded video images of the seal's head and the envi-ronment immediately in front of the seal. A hydrophone attached to the video system recorded vocalizations made by the animal, as well as ambient sound. A data logger recorded time, depth of dive, swim speed, compass bearing, and flipper stroke rate (see Davis et al. 1999). During October to December 1997 and 1998, the video system/data recorder was attached to nine adult Weddell seals (five males, four females); see fig. 71.1.

Results

More than 100 h of underwater video and audio recordings were collected. Nineteen vocalization episodes were documented in 1997 and 73 during the 1998. Over 80% were produced by a single female in 1997 (seal 3) and a single male (seal 9) in 1998. For this reason, we present detailed analyses of only these two seals. Most of the vocalizations made by seal 3 were short, repetitive types (fig. 71.2). Only three behaviors were associated with vocalization: (1) approaching the breathing hole, (2) looking at the under-ice surface, and (3) examining a novel object (e.g., ropes or the underwater observation chamber). The vocalizations by seal 9 consisted of longer duration calls as well as shorter duration clicks and chirps as he approached the breathing hole (figs. 71.3 and 71.4). Time-depth profiles from seal 9 showed that the animal only vocalized during ascent—never during descent (fig. 71.5). Dive profile data for seal 9 (1998) with deployment, number of vocalizations, dive duration, and dive depth are presented in table 71.1. Since the hydrophone was mounted in the midsection of the animal's back facing caudally, sound pressure level was lower than reported in the literature (Thomas and Kuechle 1982). Sound pressure levels of calls significantly increased in 1998 by moving the hydrophone to a forward-

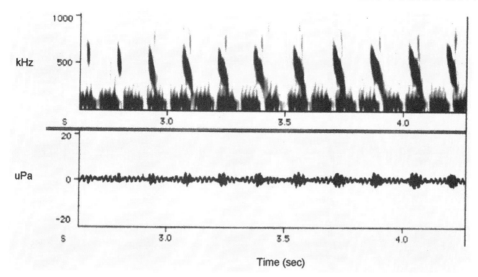

Fig. 71.2. Sonogram and oscillogram of Weddell seal chirps produced by a female while approaching breathing hole

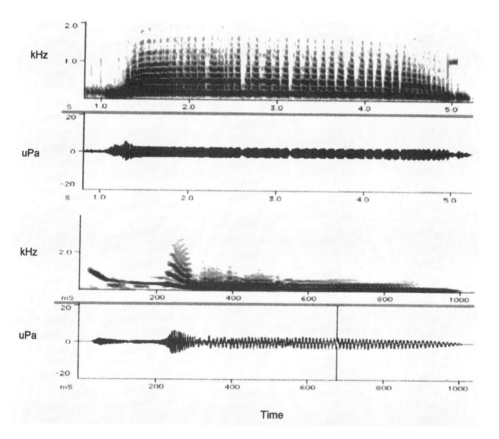

Fig. 71.3. Sonogram of long-duration trills (*T*) and gutteral-glugs (*G*) by a male Weddell seal during ascent from a shallow, 10 m dive

facing position. During the 1997 deployments, most calls of all types were below 1 kHz. During 1998, calls with energy at and above 15 kHz were recorded.

One of the more interesting initial observations was that during many calls, the seal's head bobbed up and down at about the same repetition rate as the vocalization. The male during the 1998 season produced trills accompanied by evident head and chest movements, either up and down or side to side. On occasion, head movement was observed, but it was not accompanied by any audible sound.

Suggested uses of Weddell seal vocalizations include

orientation, threat sounds, mother/pup sounds, and territorial sounds. Since most previous observations during dives were not from the seal's perspective, sounds associated with feeding have not been observed. We now have opportunity to observe this type of behavior. Interactions with three large Antarctic toothfish (*Dissostichus mawsoni*) were observed (Davis et al. 1999). There were other encounters with smaller, nototheniid fish, but none evoked vocalization. Call types and description for seal 3 and seal 9, using the method presented by Thomas and Kuechle (1982), are presented in table 71.2.

Because the hut was far from breeding colonies,

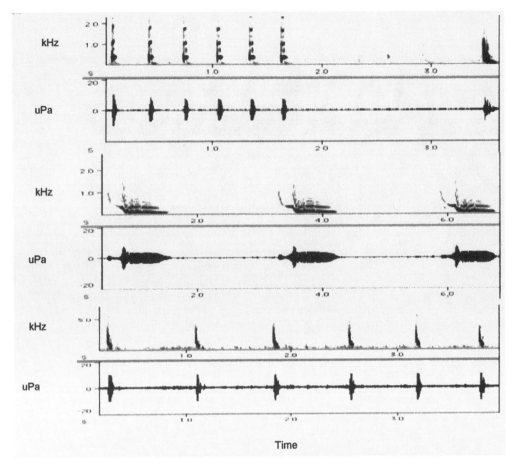

Fig. 71.4. Sonograms and oscillograms of short-duration chugs (*C*) made by a male Weddell seal while approaching a breathing hole

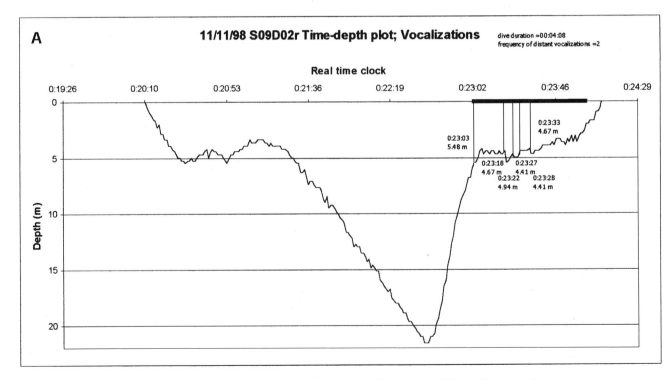

Fig. 71.5. Dive profiles for (*A*) a shallow dive (<100 m), (*B*) a mid-depth dive (>100 m), and (*C*) deep dives (>200 m) (depths and time of vocalizations are indicated). Vertical lines indicate when the experimental seal vocalized; horizontal bars indicate when the vocalizations of other seals would be detected.

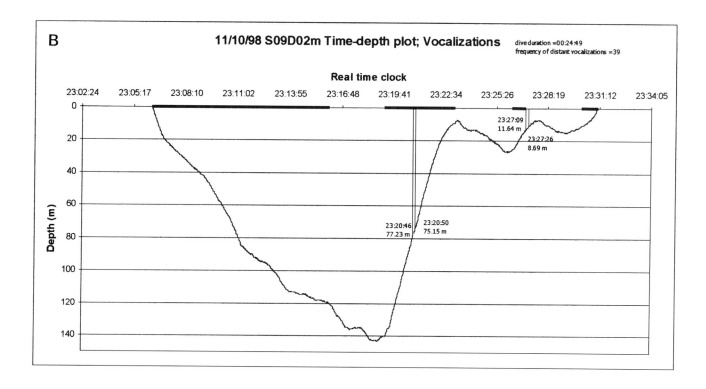

B

11/10/98 S09D02m Time-depth plot; Vocalizations
dive duration =00:24:49
frequency of distant vocalizations =39

Real time clock

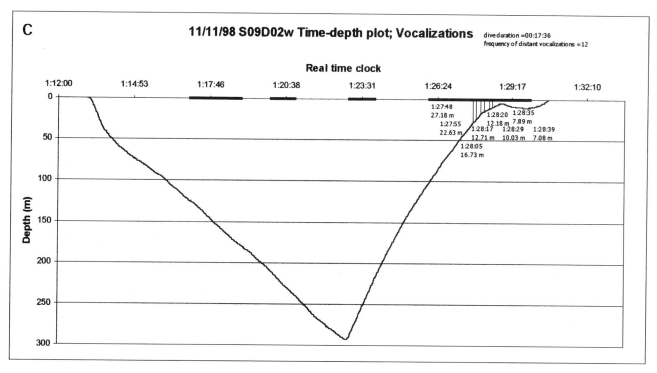

C

11/11/98 S09D02w Time-depth plot; Vocalizations
dive duration =00:17:36
frequency of distant vocalizations =12

Real time clock

Fig. 71.5. (continued)

TABLE 71.1. Dive profile data for Weddell male seal 9 (1998 season). (Dive profile data for first-year animals are not available for dives less than 25 m deep due to technical problems.)

Deployment number	Number of pulsed calls	Number of long calls	Dive duration (min)	Number of distant calls	Dive depth (m)	Number of dives
1	6	4	17–26	11–37	50–120	4
2	11	5	3–24	0–39	10–300	8
3	3	3	10–17	12–31	190–180	2
4	1	1	33	116	190	1
5	5	8	12–31	12–96	148–180	3
6	8	4	8–18	0	80–155	4
7	6	3	5–26	0	60–560	3

TABLE 71.2. Call types and description (from Thomas and Kuechle 1982) for Weddell female seal 3 during the 1997 season and male 9 during 1998. Female 3 chased a smaller fish (*P. borchgrevinski*) underneath the ice by blowing bubbles. Male 9 was the most vocal, calling during every dive.

Subject	Sex	Deployment	Call type	Description
Seal 3	female	2, 3, 5, and 6	P1–P5, C1–C4,	chirps, chugs, clicks
Seal 9	male	1–7	T, C, R, P, I	chugs, chirps, clicks, trills

sounds from these seals were most likely not related to reproduction. The 11 types of trills described by Thomas and Kuecle (1982) are thought to be a graded signal used by males to convey varying levels of warning to territorial intruders. No trills were produced by instrumented seals during 1997; however, in 1998, male seal 9 trilled frequently when approaching the breathing hole, alternating trills with pulsed chirps, clicks, and chugs and a very loud longer sound (possibly a "mew" identified by Thomas and Kuechle 1982).

In 1998, very faint vocalizations from distant Weddell seals were occasionally detected. During that season, the vocal male 9 produced trills, and there were some indications that this seal could hear colonial seals 5–10 km from the hut.

Discussion

Speculations on the evolution of echolocation exist (Pilleri 1990; Speakman 1993; Wartzok, Schusterman, and Gailey-Phipps 1984). It has been suggested that some species of mammal (bears, otters, dogs) vocalize when orienting into an unknown environment (cave, underground burrow, etc.). When navigating in caves, South American oilbirds, *Steatornis caripensis* (Griffin 1954), and Southeast Asian cave swiftlets, *Aerodramus* cf. *Collocala* (Fenton 1975), use sounds in a similar range (1–12 kHz) and repetition rate as Weddell seals (Griffin and Suthers 1970). There seems to be a mindset among researchers that signals used for echolocation must be ultrasonic. Examples from these species of bird and shrews (Thomas and Jalili, chapter 72, this volume) suggest otherwise.

The vigorous head bobbing not associated with vocalizations by Weddell seals could be a visual display.

However, elephants move their foreheads while producing inaudible, infrasonic sounds (Payne, Langbauer, and Thomas 1986), and shrews bob their heads and twitch their snouts while producing both audible and ultrasonic sound (Thomas and Jalili, chapter 72, this volume). Head bobbing by Weddell seals in the absence of audible sounds also could indicate sounds produced above the frequency range of the recorder. Evidence does exist (Awbrey et al., chapter 70, this volume) that another species of Antarctic seal, the leopard seal (*Hydrurga leptonyx*), can produce sounds both in the sonic and ultrasonic frequency ranges (30–150 kHz). Given that another Antarctic species produces ultrasonic, as well as sonic, vocalizations leaves this possibility quite open. If Weddell seals do have ultrasonic components to their vocalizations, it remains to be demonstrated whether they would supplement their well-developed visual capability when navigating. The tactile sense cannot be ruled out either. Clearly, the vibrissae extend forward when a Weddell seal approaches a potential fish prey. Until a field-portable, broadband system is available for recording ultrasonics, this question remains unanswered.

What is echolocation? It depends on how a biological sonar system is classified. Humanmade sonars are described as "active" or "passive." Further labels describe the function or desired range of the sonar—that is, "low-frequency" sonars are for long-distance or great depth probing; "high-frequency" sonars are for fine, detailed discrimination tasks. So the characteristic of an animal's signal really depends on the target and range to be detected. For finding a large breathing hole over long distances, a high priority for an air-breathing mammal, using low-frequency sonar would be desired.

Wartzok et al. (1992a, 1992b) reported that none of the ringed (*Phoca hispida*) seals or Weddell seals vocalized during an experiment on "hole finding." This contradicts our 1997 and 1998 observations of Weddell seals, in which all the seals observed to vocalize did so during ascent to their breathing hole and increased their repetition rate as distance to the hole decreased. Of the 34 types of Weddell seal sounds from McMurdo, all sounds produced by instrumented, diving seals while approaching the breathing hole were low in frequency, repeated in a series, and had an accelerating repetition rate as the seal ascended to a hole. The seals in the studies by Wartzok et al. had acoustic transmitters attached to the their backs. Although ultrasonic (60–70 kHz, source level of 159 dB re 1 μPa), the transmitters were probably audible to the seals wearing them. At 65 kHz, this source level would be 45 dB down from the ringed seal's maximum sensitivity at 32 kHz (Wartzok et al. 1992b). A signal at effectively 114 dB re 1 μPa should be audible to the seal. Furthermore, the rate of pulsing (every 1.5 s) was modulated by depth, so it changed as the seal dove. We wonder whether the changing pulsing sound could have inhibited the seal's vocal behavior, thus obscuring the researcher's ability to examine the possible use of sounds for orientation. The video-data logger system used in our study did not use a pinger and so was relatively quiet (except for low level motor noise from the camera).

Why was prey capture by the Weddell seals not always associated with sounds? Why were only two instrumented seals quite vocal? These are good questions that should be addressed. Perhaps there is a "social etiquette" dictating when echolocation-like signals should be used, such that not all seals vocalize; yet all seals hear and use the information. It makes sense that if a colony of 50–100 seals all produced sound signals at the same time, it might be difficult to evaluate the reflections. The

idea of passive listening to echolocation sounds from other animals has a foundation in Xitco and Roitblat's (1996) bottlenose dolphin studies. They demonstrated that a neighbor dolphin could perform an echolocation detection task simply by listening to the test animal's signal.

Lastly, there is the notion that echolocation signals are not restricted to only that purpose. There could be a gradient from active use of sounds for only navigation, to dual use of some sounds for navigation and communication, to the passive use of typical communication sounds for orientation cues as a supplement to vision. It is not impossible that seals vocalizing while ascending to their breathing holes are warning others ("I am coming, get out of my hole") with additional information as to how far away they are. The Weddell seal, and other polar pinnipeds, pose many unanswered questions related to the use of sound for navigation or foraging.

There is no positive evidence that seals, including Weddell seals, use echolocation. It is interesting to note, however, that some seals do have all the acoustic tools necessary for such behavior. This preliminary study demonstrates the value of studying these behaviors from instrumentation carried by the seal.

Acknowledgments

We acknowledge the support of the National Science Foundations, Polar Program, grant OPP9614857, and the National Undersea Program, grant UAF 98-0040. We also acknowledge the individuals who participated in obtaining data in the field: T. Williams, L. Fuiman, M. Horning, W. Hagey, S. Kanatous, S. Cohin, D. Calkins, S. Collier, and R. Skrovan.

{ 72 }

Echolocation in Insectivores and Rodents

Jeanette A. Thomas and *Mersedeh S. Jalili*

Introduction

In the 1950s and 1960s, development of equipment capable of sensing, recording, and analyzing ultrasonic sounds facilitated the discovery of echolocation abilities in chiropterans (bats) and odontocetes (toothed whales and dolphins). Echolocation aids navigation in visually limited environments or in capture of fast-moving prey.

For example, echolocation is particularly useful for nocturnal, insectivorous bats that need to coordinate their flight pattern relative to moving insect prey. Echo ranging is adaptive for dolphins that need to swim quickly and accurately to capture fast-moving fish at dark oceanic depths. The emphasis of echolocation studies on these two groups of mammals perhaps was driven by a special fascination for them or availability of funds for modeling animal echolocation for the improvement of humanmade sonars. Only a few investigators examined the potential of echolocation in other mammals (Airapet'yants and Konstantinov 1970; Airapet'yants 1974; Sales and Pye 1974). Clark and Ellison (chapter 73, this volume) discuss the use of acoustic cues for orientation by baleen whales. Schusterman et al. (chapter 69, this volume) summarize studies on possible echolocation in seals and sea lions. Herein, we summarize the few studies on echolocation abilities in insectivores and rodents and provide some new data for each group.

According to Wilson and Reeder (1993), the six extant families of Insectivora include Soricidae (shrews), Erinaceidae (hedgehogs), Chrysochloridae (golden moles), Solenodontidae (solenodonts), Tenrecidae (tenrecs), and Talpidae (moles and desmans). Sometimes echolocation aids in pursuit of fast-moving prey, but prey sought by insectivores are not particularly quick moving (e.g., ground insects, mollusks, worms, grubs, insect larvae) and insectivores do not stalk them over great distances. However, echolocation also is useful in environments with limited light, and many insectivores live in dark environments (table 72.1). For example, moles and golden moles are totally fossorial and feed, move, and raise young exclusively under ground. Shrews are nocturnal, nest in underground burrows, and move within dense herbaceous ground cover, runways under humus, or runways made by small rodents (George, Choat, and Genoways 1986; Whitaker 1974).

Wilson and Reeder (1993) recognized 29 families of rodents, constituting 43% of all mammalian species. Like insectivores, rodents generally do not hunt fast-moving prey; many are, in fact, herbivorous or granivorous. Nevertheless, nocturnal, semifossorial, or fossorial

TABLE 72.1. Summary of diet and geographic distribution in insectivores. (Data taken from Nowak 1991.)

Family	Subfamily/Genera	Habit	Diet	Distribution
Talpidae	15 genera, 32 species			
	Subfamily: Talpinae (Old World moles, *Talpa*)	fossorial	worms, insect larvae	Eurasia
	Subfamily: Scalopinae (New World moles, *Scalopus, Neurotrichus, Parascalopus, Scapanus*)	fossorial	worms, insect larvae	North America
	Subfamily: Condylurinae (star nose moles, *Condylura*)	fossorial	aquatic insects, small fish, crustaeans	North America
	Subfamily: Desmaninae (*Desmana* and *Galemys*)	semiaquatic	small fish, crustaceans, insects, worms	Russia; Spain and Portugal
Tenrecidae	10 genera, 23 species			
	Potamogale 1 species	semiaquatic	crabs, fish, amphibians	Central Africa
	Micropotamogale 2 species	semiaquatic	worms, insects, crabs, fish, frogs	Guinea, Liberia, Zaire
	Microgale 11 species	terrestrial and semiaquatic	insects	Madagascar
	Hemicentetes 1 species	terrestrial	earthworms	Madagascar
Chrysochloridae	7 genera, 18 species	fossorial	worms, insects, invertebrates	Africa, Cameroon/ Uganda south to Cape Good Hope
Solenodontidae	1 genus, 2 species			
	Solenodon	semifossorial	invertebrates, reptiles, vegetation	Cuba and Hispaniola
Soricidae	22 genera, 289 species			
	Subfamily: Soricinae (*Sorex, Cryptotis, Blarina, Notiosorex, Megasorex, Neomys, Nesiotites, Soriculus, Nectogale, Chimarrogale, Anourosorex*)	terrestrial	invertebrates, insects	North America, South America, Europe, Asia, Africa
	Subfamily: Crocidurinae (*Crocidura, Paracrocidura, Suncus, Sulvisorex, Ruwenzorisorex, Ferocuuls, Solisorex, Myosorex, Diplomesodon, Scutisorex*)	terrestrial	invertebrates, insects	North America, South America, Europe, Asia, Africa
Erinaceidae	9 genera, 20 species			
	Subfamily: Galericinae (gymnures, moonrats)	terrestrial	invertebrates, carrion	Africa, Europe, Asia
	Subfamily: Erinaceinae (hedgehogs)	terrestrial	invertebrates, insects	Africa, Europe, Asia

TABLE 72.2. Summary of diet and geographic distribution in fossorial rodents. (Data taken from Nowak 1991.)

Family	Genera	Habit	Diet	Distribution
Muridae	*Nannospalax* 3 species	fossorial	tubers, roots, underground vegetation	Balkan Peninsula, Asia Minor, Syria, Palestine, Egypt, Libya
Muridae	*Spalax* 5 species	fossorial	tubers, roots, bulbs	Romania, Ukrane
Muridae	*Myospalax* 6 species	fossorial	roots, grains	China
Muridae	*Tachyroryctes* 2 species	fossorial	roots, rhizomes, tubers, bulbs	Ethiopia, Somalia, Uganda, Rwanda, Kenya, Tanzania
Bathyergidae	*Georychus* 1 species	fossorial	tubers, roots, bulbs	Cape of Good Hope, South Africa
Bathyergidae	*Cryptomys* 3 species	fossorial	roots, tubers, invertebrates	western Zambia to northwestern South Africa
Bathyerigidae	*Heliophobius* 1 species	fossorial	tubers, bulbs	southern Kenya, Tanzania, southeast Zaire, eastern Zambia, and Mozambique
Bathyerigidae	*Bathyergus* 2 species	fossorial	bulbs, roots	Namibia and southern South Africa
Bathyerigidae	*Heterocephalus* 1 species	fossorial	tubers, roots	eastern Ethiopia, central Somalia, and Kenya
Ctenomyidae	*Ctenomys* 44 species	fossorial	roots, stems, grass	South America
Geomyidae	6 genera, 37 species	fossorial	roots, tubers, stems	North America, Canada south to Colombia

rodents could benefit from extrasensory inputs, such as echolocation (table 72.2).

Bats and rodents evolved directly from the most primitive placental mammal, the insectivores (Alderton 1996). We suggest that echolocation is a primitive characteristic in mammals, perhaps evolving as early as in insectivores, then further refined by bats and perhaps some rodents (see discussion in Denzinger, Kalko, and Jones, chapter 42, this volume). Clearly, more research is needed on the possible echolocation abilities in insectivores and rodents.

From a search of the literature, it appears that certain adaptations or environmental conditions are related to echolocation abilities. Below, we describe these factors for the known echolocators, such as bats and dolphins, and discuss these characteristics in some insectivores and rodents.

LIMITED AMBIENT LIGHT

Most echolocating mammals live at least part of their life in darkness or low light environments. For example, most bats are nocturnal and maneuver easily in dark caves or crevices. Ambient light for toothed whales and dolphins is limited during nighttime foraging, in deep water, turbid water, ice-covered areas, or amid a plankton bloom. Some insectivores and rodents are active mainly at night (Whitaker 1974), creating the need to orient and forage in limited environmental light. For semifossorial or fossorial mammals that nest in underground burrows, forage in underground tunnels, or move in dense herbaceous covered runways, ambient light is limited at all times.

Feldhamer et al. (1999) identified groups of mammals that are fossorial or semifossorial: insectivore families Soricidae, Talpidae, Solenodontidae, Chrysochloridae; rodent families Geomyidae (pocket gophers), Ctenomyidae (tuco tucos), Bathyergidae (blesmois), and Mu-

ridae (subfamilies Myosplacinae, East Asian mole-rats; Spalacinae, blind mole-rats; and Rhizomyinae, African mole-rats); a xenarthra family, Dasyuridae (armadillos); and a marsupial family, Notoryctidae (marsupial moles). Herein, we only discuss possible echolocation in fossorial insectivores and rodents. (Marsupial moles are not discussed; however, they also have characteristics reminiscent of echolocation—e.g., fossorial, small ear opening, vestigial eyes, no lens or pupil, reduced optic nerve, diet of earthworms and insect larvae, production of low-intensity sharp squeaks [Nowak 1991]. If echolocation occurred in this family, it would represent a separate evolution from placental mammals.)

LIMITED VISION

Bats

Mammals with poorly developed vision must rely on other senses, like olfaction, tactition, passive acoustics, or echolocation for orientation or foraging. In bats, echolocation is the predominant sensory mode. Anatomically, bat vision is not limited. In fact, "blind as a bat" is only an anecdotal description (Fenton 1983).

Odontocetes

In odontocetes, acoustics (passive listening or active echolocation) is the predominant sensory channel. Eyes of toothed whales are compromised anatomically for vision both under water and in air, but they see well over short distances in either environment (1999). River dolphins are the exception, being essentially blind from a reduced visual anatomy.

Insectivores

All insectivores have small eyes and probably poor vision (table 72.3). The optic region in the shrew brain is small and poorly developed. Churchfield (1990) described shrews as having underdeveloped eyes, only 0.7–

TABLE 72.3. Summary of sensory anatomy in insectivores. (Data taken from Nowak 1991.)

Family	Subfamily/ Genera	Eyes	Ears	Vibrissae	Sounds
Talpidae					
	Subfamily: Talpinae	minute, hidden in fur; only detect light	no external ear	keen sense of vibration; odiferous, long snout	no information given
	Subfamily: Scalopinae	minute, hidden in fur	no external ear	keen sense of vibration	*Neuortrichus* taps nose to hunt; faint audible twitter
	Subfamily: Condylurinae	minute, visible	no external ear	nose has bare tentacles that move when feeding; odiferous	no information given
	Subfamily: Desmaninae	small, poor vision	small external ear	long snout, scent glands	*Galemys* echolocates (Richard 1973)
Tenrecidae					
	Potamogale	minute	external ear present	present	no information given
	Micropotamogale	minute	external ear present	present	no information given
	Microgale	minute	prominent external ear	present	*M. dobsoni* echolocates (Gould 1965)
	Hemicentetes	minute	external ear present	present	audible "putt-putt" and "crunch"; echolocates; Gould 1965
Chrysochloridae					
	Eremitalpa	eyes fused at early age; skin thickens over eye with age; lacks temporal bullae	small, hidden in fur	long snout, leathery nose	no information given
	Cryptochloris and *Chrysochloris*	eyes vestigial, skin-covered; temporal bullae present	small, hidden in fur	long snout, leathery nose	no information given
	Other species	eyes vestigial and covered	small, hidden in fur	long snout, leathery nose, with hairy skin	no information given
Solenodontidae					
	Solenodon	small	external ear present	present	high-frequency clicks, −31 kHz, 0.1–3.6 ms; perhaps echolocation
Soricidae					
	Subfamily: Soricinae	small, poor vision	small ears hidden in fur, keen hearing	many present; keen smell, odiferous	suspected echolocation
	Subfamily: Crocidurinae	small, poor vision	small ears hidden in fur	many present	no evidence of echolocation
Erinaceidae					
	Subfamily: Galericinae	small, poor vision	no information	many present	no information
	Subfamily: Eriaceinae	small, poor vision	keen hearing	many present	hiss, scream

1.5 mm in diameter (figure 72.1), and recognized three categories of eyes in insectivores: small, but visible among the fur (in terrestrial and scansorial species); extremely small and obscured by the fur (in psammophilic and semifossorial species); and totally covered by skin (in semiaquatic species). The primary function of insectivore eyes could be in sensing shades of light and dark. George, Choat, and Genoways (1986) described ocular

structure and function in the short-tailed shrew (*Blarina brevicauda*) as "degenerate": "the optic nerve is diminutive and only slight motion is possible, ocular muscles do not rise directly from the skull, vision probably is limited to perception of light, the lachrymal gland is larger than the eyeball and secretes tears to protect and wash away dirt from the eye" (p. 4).The fossorial coast mole, *Scapanus orarius,* has small eyes, the auditory meatus is

Fig. 72.1. Representatives of the two genera of Soricidae. *Top: Blarina brevicauda* (photograph by R. Altig). *Left: Cryptotis parva* (photograph by F. A. Cervantes Reza).

concealed by fur, and the animal is primarily nocturnal (Hartman and Yates 1985). The talpid, shrew-mole (*Neurotrichus gibbsii*) is the only vertebrate to have a pigmentation layer covering the surface of the lens of the eye.

Churchfield (1990) held that the sense of touch plays an important role in shrew orientation—"judging the size of holes and crevices through which it squeezes or the nearness of objects in its path." She described shrew snouts as "long with many vibrissae" and noted that they "constantly wiggle to and fro, sniffing the air and probing the substratum" (p. 18). Moles have specialized touch-sensitive, Emier's organs on their snouts (Church-

field 1990). The talpid, shrew-mole has eight pairs of vibrissae on the snout and locates prey by tapping its nose on the ground directly in line with the prey, then 40 degrees to one side, then the other. Finally, the animal moves one step forward and continues this process until its nose touches the prey. The semiaquatic, nocturnal Iberian desman (*Galemys pyrenaicus*) has Eimer's organs dispersed among the vibrissae that are believed to play a role in orientation and detection of prey (Palmeirim and Hoffmann 1983). It actively explores the bottom of river streams, moving sediment with the snout, head, and forefeet. The Iberian desman essentially is

TABLE 72.4. Summary of sensory anatomy in fossorial rodents. (Data taken from Nowak 1991 and Rado et al. 1987.)

Family	Genera	Eyes	Ears	Vibrissae	Sounds
Muridae	*Nannospalax*	no eye opening	reduced to low ridge	facial bristles	during mating tap head against burrow
Muridae	*Spalax*	no eye opening	reduced to low ridge; keen hearing	line bristles aid navigation; poor olfaction	recognize objects by touch
Muridae	*Myospalax*	small, hidden in fur; sensitive to light	no external ear	few short whiskers	squeal when frightened
Muridae	*Tachyoryctes*	small, visible, functional	small ears	stiff hairs on face tactile role	no information given
Bathyergidae	*Georychus*	minute	small opening surrounded by skin	present	males drum hind feet during breeding
Bathyergidae	*Cryptomys*	very small cornea; sensitive to air current	very small; sensitive to sound or vibration	present	squeals, grunts, growls
Bathyerigidae	*Heliophobius*	extremely reduced eyes, unreceptive to light	reduced external ears; keen hearing	present	no information given
Bathyerigidae	*Bathyergus*	small	no external ear	present	chattering noise while burrowing
Bathyerigidae	*Heterocephalus*	minute	small external ear	nearly naked, lip vibrissae	no information; eusocial
Geomyidae	6 genera, 37 species	small	small external ear	present	squeals and hisses
Ctenomyidae	*Ctenomys*	small	external ear reduced, auditory bullae enlarged	present; tail sensory hairs	"tloc-tloc-tloc" by males during breeding, 10–20 s, accelerating rate

silent, but Richard (1973) reported aggressive sounds in captive Iberian desmans. Richard (1973) speculated desmans use sounds from the slapping of the forefeet against the water for echolocation.

The sense of smell in long-snouted insectivores also plays an important role in orientation and foraging. Many insectivores are notably odiferous. Some species have scent glands and use them in interspecific contexts (Churchfield 1990).

Rodents

In rodents, both hearing and smell are vitally important senses. Rodents active during daylight are more reliant on vision, whereas rodents active at night or living under ground depend less on sight (Alderton 1996). Nocturnal rodents, like rats and flying squirrels, generally have relatively large eyes to compensate for low light levels. Fossorial rodents have small eyes, indicating their uselessness under ground (table 72.4). In the family Bathyergidae, two species of African mole rats, *Bathyergus* spp., occur in sandy soils in hot dry regions of Africa, burrow extensively with complex tunnels nearly 300 m long, and feed on underground bulbs and tubers (1999). Several mole-rat species (eight in *Cryptomys,* two in *Georychus,* one in *Heliophobius,* and one in *Heterocephalus*) have minute eyes concealed beneath the skin and are entirely blind; they have small pinnae and their auditory acuity is reduced in favor of enhanced tactile and olfactory sensitivity. Similarly reduced or ob-

scured eyes, long vibrissae, and small ears characterize some muridae (e.g., *Spalax, Nannospalax, Myospalax,* and *Tachyoryctes;* Feldhamer et al. 1999). In these fossorial murids, animals recognize objects by touching with the nose. Tactile and auditory senses are acute, but olfaction is considered weak (Feldhamer et al. 1999).

BROADBAND HEARING

Broadband, ultrasonic hearing often (but not always) is associated with echolocation (Airapet'yants and Konstantinov 1970). Both bats and dolphins typically have ultrasonic hearing (see Vater and Koessl, chapter 12, this volume). Using a short, broadband echolocation signal (as in both bats and dolphins) allows rapid sampling of target details across a range of frequencies.

Hearing in insectivores has not been tested behaviorally; however, anatomical data by Henson (1961) suggests that shrews have ultrasonic hearing up to 50 kHz. However, he believed that moles probably are unable to detect ultrasounds. Broadband or ultrasonic hearing is expected in insectivorous mammals with small head sizes. Airapet'yants (1974) used conditioned reflex techniques in hedgehogs (family Erinaceidae) to show they hear up to 45 kHz. Shrews have a unique hearing apparatus in that the auditory bullae are completely unossified and appear as an open hole on the underside of the skull. The significance of these open bullae is unknown. The temporal bullae are present in some moles and golden moles, but not all (table 72.4).

The prominence of an external ear in shrews varies with habit and habitat. Scansorial, or ground-dwelling, species have the largest pinnae. Temperate species have quite large hairy ears, while Old World tropical species have large, nearly naked ears. Semifossorial shrews have small pinnae, often obscured by fur. Some aquatic shrews lack pinnae entirely (Churchfield 1990).

Hearing has been studied extensively in rodents (Airapet'yants and Konstantinov 1970; Fay and Popper 1994), especially in chinchillas (*Chinchilla laniger*), laboratory rats (*Rattus norvegicus*), gerbils (*Meriones unguiculatus*), house mice (*Mus musculus*), and guinea pigs (*Cavia porcellus*). Of these species, all showed ultrasonic hearing. The golden hamster (*Mesocricetus* spp.) hears up to 23 kHz. Guinea pigs and redbacked voles (*Clethrionomys* spp.) hear up to 50 kHz. Laboratory rats, garden dormice (*Eliomys* spp.), and harvest mice (*Reithrodontomys* spp.) hear up to 60 kHz. The house mouse hears up to 100 kHz. In many species there are two regions of sensitivity, one sonic and one ultrasonic. For example, laboratory mice were sensitive from 4 to 10 kHz and again at 40 to 45 kHz, and redbacked voles were sensitive at 13 kHz and 50 kHz (Airapet'yants 1974). Kahmann and Ostermann (1951) acknowledged the possibility that besides functioning in social contexts, ultrasonics in rodents could serve in echolocation.

Ultrasonic Sound Production

Echolocating mammals often produce ultrasonic sounds, either in a series of frequency-modulated (FM) sweeps or in a train of pulses. Repetition rate of sounds increases as the sender approaches a target (i.e., prey or object in the environment). The short wavelength of high-frequency signals provides good details of targets. However, mammalian production of ultrasonic vocalizations alone is not evidence for echolocation, because some ultrasonics are used for communication.

Hamilton (1940) observed that smokey shrews "when foraging for food . . . utter an almost indiscernible twitter . . . while rooting through the leaf mold or appearing on the forest litter about rotted logs, the twitching nose and vibrissae held aloft and this faint, almost inaudible twittering was kept up continually." Churchfield (1990) reported that shrews are very vocal and emit a wide range of sounds, including ultrasonics. She categorized shrew sounds into calls of alarm, defense, and aggression; calls during courtship; sounds associated with mother-young interactions; and sounds produced during exploration and foraging. High-frequency "twitters" and other sounds are produced by shrews as they explore novel situations. They emit these sounds less often as they become familiar with their environment (Churchfield 1990). Unlike rodents, shrews do not use ultrasound in the nest for mother-infant communication (Churchfield 1990). Shrews produce sounds in three ways: "puts" and "hisses" coupled with exhalation and

inhalation through the nose; "clicks" produced by the tongue; and "chirps," "twitters," and "shrieks" produced by the larynx (Churchfield 1990).

Kahmann and Ostermann (1951) noted that high-frequency sounds in small mammals could be used for acoustic location. Several rodents in the family Muridae produce a variety of ultrasonic sounds for mother-young communication or as distress calls (Airapet'yants 1974; Bell 1979; Elwood and Keeling 1981; Newman 1988; Blake 1992). In these contexts, the need to vocalize for interspecific communication must be balanced against the costs of increased detection by a predator (Sales and Pye 1974; Alderton 1996). Blumberg and Alberts (1990) suggested that the ultrasonic components of infant rodents are only a by-product of respiration; however, Blake (1992) disputed this.

Some insectivores and rodent have similar characteristics correlated with echolocation in bats and dolphins. Species that fit this profile need to be identified and studied for possible echolocation abilities.

Previous Studies on Echolocation in Insectivores

Sales and Pye (1974), Buchler (1976), Buchler and Mitz (1980), and Gould (1980) proposed a variety of echolocation abilities among insectivore groups. Most information on insectivore acoustic abilities is reported in the family Soricidae (22 genera, approximately 322 species). Species in this family are thick-bodied, robust animals able to push their way through soil and leaf litter using the long snout, teeth, and forefeet to excavate a passage. Shrews have small eyes and pinnae obscured by the fur (fig. 72.1, table 72.3). They use high-frequency sounds for intraspecific communication, orientation, and prey detection (Feldhamer et al. 1999). Shrews eat mainly earthworms and soil-dwelling insect larvae.

Studies by Gould, Negus, and Novick (1964), Gould (1969), Buchler (1976), Tomasi (1979), and Forsman and Malmquist (1988) suggested that only shrews of the subfamily Soricinae echolocate. Gould (Airapet'yants 1974, reference 523) first studied possible echolocation in shrews by having them run an obstacle course of glass bubbles. The lightest touch of a bubble would cause it to drop, signifying the shrew ran into it. Intact shrews and shrews with no vibrissae ran the course with few collisions; however, shrews with plugged auditory canals ran the course more slowly and made more errors. Because this technique was imprecise, Gould, Negus, and Novick (1964) tested possible echolocation in *Sorex palustris, S. vagrans,* and *S. cinereus* using a disk-platform apparatus. To receive a reward, the shrew entered a vinyl tube, exited onto a disk platform, and jumped down to a runway containing a mealworm and water. Shrews were observed in the dark with an infrared "snoop scope." Nonauditory cues, such as smell or touch, were eliminated. After a learning period, performance of all shrews was high. However, when the ears of shrews were plugged,

TABLE 72.5. Pulse characteristics of several shrew species.

Species	Pulse mean frequency	Short pulse mean duration	Long pulse mean duration
Sorex vagrans	37 kHz	5.3 ms	18.0 ms
Sorex palustris	36 kHz	5.1 ms	18.7 ms
Blarina brevicauda	42 kHz	2.3–7.0 ms	20.0 ms

performance dropped to chance. Initially, shrews produced low-amplitude, short pulses, but once the disk platform was recognized they produced higher amplitude, longer duration pulses. Typical pulses were 0.02–0.14 μbar, 4.9–33.0 ms, and 30–60 kHz (Sales and Pye 1974). Pulses were not repeated often, only up to 13–16 pulses in an unfamiliar situation. Table 72.5 lists the characteristics of pulses by species.

Buchler (1976) conducted studies on the performance of wandering shrews (*Sorex vagrans*) using a similar disk platform with microphones surrounding the apparatus. He noted the same high performance for normal shrews, but decreased performance in hearing-impaired shrews.

Using a Y-maze, Buchler tested the wandering shrew's ability to detect a plate blocking one arm of the maze and then choose the other, open-arm of the maze for a food reward. Wandering shrews could not detect a plate smaller than 3.5×3.5 cm. Beyond 65 cm, shrews could not detect a 15×15 cm plate. Encountering a 4×4 cm plate with a hole (to mimic a tunnel opening), shrews detected accurately out to 30 cm. Buchler felt shrews do not use echolocation to detect prey or predators, but rather in a "search-image" mode to explore their environment. Shrews often chose the correct response but did not eat the mealworm. He believed that in these cases the open-arm of the Y-maze was selected because the opportunity to explore and, thus, "know" their environment was itself rewarding. Buchler (1976) noted that echolocation could be energetically more efficient than detailed tactile exploration of the environment, a particular advantage for species with a high metabolic rate, such as shrews. He also suggested that the use of low-amplitude pulses is adaptive for shrews concerned with short distance exploration and the need to avoid detection by predators.

Forsman and Malmquist (1988) studied echolocation abilities in *Sorex araneus* using a different apparatus: a central chamber from which six tubes (200 mm long, 16 mm diameter) radiated. The shrew's task was to select the single open tube for escape. Studies were conducted in the dark and in a soundproof box. Shrews produced ultrasonic sounds from 25 kHz to 95 kHz and discriminated the open tube between 30 s and 90 s. These authors thus concluded that this species has echolocation abilities.

Gould, Negus, and Novick (1964) monitored the performance of short-tailed shrews (*Blarina brevicauda*) on the disk-platform apparatus and reported that these shrews produced clicks, puts, twitters, ultrasonic trains, chirps, and buzzes during the task (fig. 72.1). They clicked each time they extended their snout over the rim of a disk platform. He gave little data on durations or repetition rates and no data on peak frequency or bandwidth. He concluded that clicks were produced by forward movement of the tongue, puts and twitters as a consequence of sniffing, and chirps and buzzes by the larynx. Short-tailed shrew infants clicked as their mother approached to return them to the nest. Shrews frequently bobbed their head and twitched their snout while moving about in an unfamiliar environment; however, twitching of the snout was not always associated with audible sounds.

Gould (1969) noted that the sniffing rate of *Blarina* was close to the "putt" rate of another shrew, *Suncus,* and that putts could be the acoustic consequence of more intense smelling. Twitters were detected from short-tailed shrews during upward body extensions and "testing the air." According to Gould (1969), chirps, buzzes, and chirps graded into buzzes were used during agonistic encounters between short-tailed shrews. Acoustic information on the presence or absence of vegetative ground cover was suggested as an important adaptation for avoiding predation.

Tomasi (1979) conducted investigations on the echolocation abilities of short-tailed shrews trained to select open versus closed glass tubes. The open tube (imitating a tunnel) led to the shrews' home cage. No food reward was involved. Observations were conducted in the dark, under red lights, and olfactory cues were eliminated. Shrews could detect open versus closed tubes <61 cm long. Tomasi noted that glass tubes are highly reflective, so the detection distance probably was overestimated. When the diameter of a tube was varied, shrews discriminated diameters >0.63 cm. Even when the tube was bent at various angles from 0 to 90°, shrews detected the open versus the closed tube. Sound recordings made during some experiments showed the shrew produced at least one pulse before choosing which tube to enter. Pulses occurred singly, in pairs, or in triplets. Up to 20 pulses per series were given, with peak frequency between 30 kHz and 50 kHz. No waveforms or summary of acoustic parameters were published. Tomasi (1979) proposed that the primary function of echolocation in the short-tailed shrew is for exploring tunnels or avoiding obstacles, rather than for detecting prey.

In contrast, white-toothed shrews (*Crocidura russula*) in the Soricidae subfamily Crocidurinae produce clicks ranging from 10 kHz to 110 kHz, but Grunwald (1969) could not demonstrate echolocation.

Airapet'yants (1974) speculated about the usefulness of echolocation in shrews, given the low amplitude of the pulses and the cluttered vegetative environment they inhabit. He hypothesized that "in contrast to bats, which make continual use of echolocation, shrews rely on it only sporadically . . . when an open area is crossed, echolocation could be an effective means for detecting the nearest shelters (stumps, bushes) . . . use of echolocation in unfamiliar terrain is especially important to small mammals which must conceal themselves from various predators in a suitable shelter." Airapet'yants proposed that water shrews (*Neomys fodiens*) sometimes migrate over land, and echolocation would be useful in that context. He further speculated that they could use echolocation to catch fish fingerlings that are especially active at night. Sales and Pye (1974) reported a variety of sounds between 5 kHz and 55 kHz made by the water shrew. In captivity, these shrews produced ultrasonic pulses before and after jumping from a platform to the floor of their cage. But because no sounds were detected when a shrew explored on the platform, the authors suggested that echolocation is unlikely.

Novick and Gould (1964) and Gould (1965) carried out intensive studies on echolocation in three insectivore species (*Hemicentetes semispinosus, Echinops telfari,* and *Microgale dobsoni*) in the family Tenrecidae (10 genera and 25 species), using the same disk-platform apparatus as in their shrew studies. Tenrecs produced clicks as they searched the platform and successfully found the platform in darkness, when blindfolded, and when open plastic tubes were placed in their ears. However, performance dropped sharply when their ears were plugged with cotton. The authors believed tenrecs echolocate by producing a series of high-amplitude, low-frequency clicks with their tongues (0.1–2.0 ms, from 9 kHz to 17 kHz). The duration of tenrec clicks was short (0.2–1.2 ms). Peak frequency in *Hemicentes* clicks was 10–16 kHz. Clicks were produced singly, in doublets, or in triplets. When emitting clicks, *Echinops* licks its lips, moving the tongue laterally and upward above the lip, while *Hemicentetes* and *Microgale* simply open and close their mouths when clicking. Another tenrec, *Centetes,* also produce clicks this way. Amplitude of tenrec clicks varied, being very low intensity in *Centetes* and audible 7 m away in *Microgale.*

Tenrecs (fig. 72.2; table 72.3) produce a variety of sounds, including squeals, sniffs, twitters, and tongue-clicks. Kingdon (1974) and Nowak (1991) described tenrecs as having exceptionally small eyes, no auditory bullae, and an incomplete zygomatic arch. Sales and Pye (1974) reported sensitive hearing in the streaked tenrec (*Hemicentetes semispinosus*) from 2 kHz to 60 kHz and suggested that the tongue-clicks are produced at the best hearing sensitivities and stridulations sounds at the upper, less sensitive, part of their hearing range. Eisenberg (1975) reported that tenrec species, such as the streaked tenrec and the long-tailed tenrec (*Microgale longicaudata*), echolocate during foraging. The long-tailed tenrec resembles a shrew and occupies thick vegetation and ground litter in various habitats.

Novick and Gould (1964) described foraging by

groups of streaked tenrecs that have dorsal spines, which vibrate against each other to produce ultrasonics from 10 kHz to 70 kHz. The authors believed these sounds are related to communication rather than to echolocation.

Kingdon (1974) discussed two aquatic genera in the family Tenrecidae that forage along rivers in Africa and feed on aquatic invertebrates. These, the dwarf otter shrew (*Micropotamogale ruwenzorii*) and giant otter shrew (*Potamogale velox*), have several adaptations for aquatic life including long vibrissae, webbed feet, and long tails for steering. They have spectacular vibrissae that aid in underwater sensing, but also could function as a hydrofoil. The giant otter shrew feeds at night in the water and is thought to have very poor vision (Kingdon 1974). No data are available on sound production or possible echolocation abilities in the otter shrew, but its natural history is typical of mammals with echolocation abilities.

Sales and Pye (1974) reported that some hedgehogs (fig. 72.2) in the insectivore family, Erinaceidae (10 genera, 20 species), respond to ultrasonic whistles up to 84 kHz by jerking their bodies or turning their heads. Henson's (1961) studies on the anatomy of the middle ear of the European hedgehog, *Erinaceus europaeus,* indicated it hears ultrasonics. Another species of hedgehog, *Hemiechinus auritus,* responded to tones up to 40 kHz, but the most sensitive frequency was at 8 kHz (Sales and Pye 1974). Hedgehogs are terrestrial and nocturnal, with an omnivorous diet, feeding on invertebrates, small vertebrates, eggs, fruits, and carrion (Feldhamer et al. 1999). Hedgehogs have relatively small eyes and poor vision (Kingdon 1974). No studies on echolocation have been conducted on hedgehogs.

Moles (fig. 72.2; table 72.3) in the insectivore family Talpidae (17 genera, approximately 42 species) eat earthworms and soil invertebrates. Moles live in semipermanent underground tunnels and feed day and night in their tunnels. According to Gorman and Stone (1990), talpids have "poor vision, acute hearing, and exquisite sense of touch." The pinnae are reduced or absent and the eyes are minute (Feldhamer et al. 1999). Mole snouts are tipped with thousands of tactile-sensing Emier's organs. The star-nosed mole (*Condylura cristata*) exhibits a ring of tentacle-like sensing organs on the snout. Hairs on the forefeet and tail also have a sensory ability in some moles. Gorman and Stone (1990) concluded that moles rely primarily on tactile senses for navigation and foraging. Although they should be, studies designed to test echolocation abilities have not yet been conducted on true moles.

The family Talpidae also includes desmans and shrew-moles (table 72.3). Desmans live near streams or rivers, have a diet of aquatic invertebrates and fish, and are good swimmers. Richard (1973) reported that the Iberian desman echolocates and uses its long flexible snout to maneuver and locate prey under water. Again, desmans are a good candidate for studies on echolocation.

Fig. 72.2. Representative fossorial mammals in the order Insectivora. From top to bottom— *A:* hedgehog (*Ernaceus algiurus*); *B:* tenrec (*Tenrec eucaudatus*); *C:* golden mole (*Chrysochloris* spp.); *D:* eastern mole (*Scalopus aquaticus*). Drawing by Jennifer Webster.

No information is available on the hearing, sound-production, or echolocation abilities of African golden moles (fig. 72.2, table 72.4) in the insectivore family Chrysochloridae (7 genera, 19 species). Golden moles have no pinnae, large auditory bullae, and an "unusual hearing apparatus" (Kingdon 1974). Their eyes are vestigial and usually covered by skin (Kingdon 1974). Eyes of the golden mole remain closed at birth and eventually are covered with fur (Gorman and Stone 1990). Golden moles burrow deep and, despite their lack of external eyes, have an excellent sense of direction. They feed on earthworms, beetles, grubs, slugs, and snails. In the presence of a worm, they respond with movements of the head and twitching of the snout, "as if to get a sense of direction" of the prey, but they are indifferent to the smell and sounds of insect prey (Kingdon 1974). Golden moles are active during the day or night, but they remain generally silent. Golden moles are thus also a good candidate for echolocation studies.

Little is known about sound production in the insectivore family Solenodontidae (table 72.3), the Cuban and Haitian solenodonts (2 extant species). Solenodonts have long pinnae, small eyes, and a sensitive snout used in capturing prey (Feldhamer et al. 1999). They live in rocky, brushy, or forested areas and are omnivorous. Eisenberg and Gould (1966) recorded broadband clicks from *Solenodon paradoxus* that were 0.1–3.6 ms, between 9 kHz and 31 kHz, and produced in bursts of one to six clicks when the animal explored a new area or encountered a stranger. No controlled echolocation studies have been conducted in this family of insectivores.

New Data on Echolocation in Shrews

In the wild, short-tailed shrews (fig. 72.1) inhabit woodland or damp areas of central and eastern North America, with leaf litter or humus ground cover. It is a scansorial species foraging on worms, snails, slugs, and insect larvae (Rood 1958). Using an Anabat II bat detector set with a division rate of eight (i.e., monitoring up to 160 kHz), we recorded vocalizations from two adult short-tailed shrews (*Blarina brevicauda*) on a Marantz cassette recorder (fig. 72.3). Shrews were tested individually in the dark in a 20 gallon aquarium and in a closed

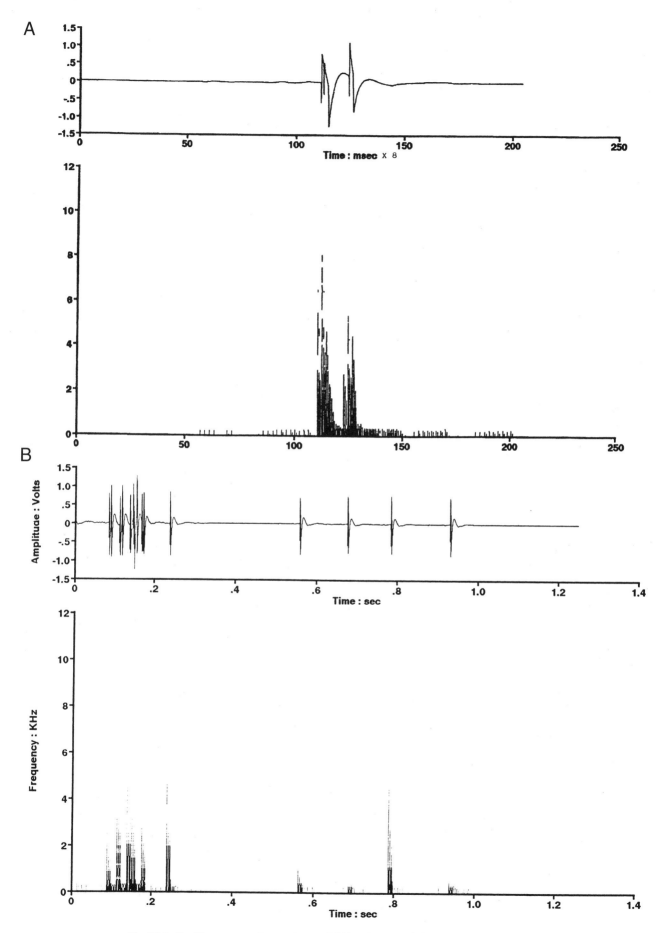

Fig. 72.3. Oscillograms and sonograms of (*A*) pulses and (*B*) trains produced by the short-tailed shrew in a dark, soundproof box

box lined with neoprene. Shrews bobbed their heads and twitched their snouts while producing ultrasonic pulses. Sonograms of clicks duplicated those reported by Gould (1969, fig. 2H). However, we did not record putts, twitters, chirps, or buzzes. Clicks were produced singly or in trains. Sometimes clicks were in pairs, but there was no phase shift to indicate the second pulse was an echo. The peak frequency (in fig. 72.3 this translates into 30 kHz to 50 kHz) was the same noted by Tomasi (1979) and by Gould (1969) for this species.

Least shrews, *Cryptotis parva* (fig. 72.1), are nocturnal and typically inhabit woodlands and open fields of the eastern North and Central America (Whitaker 1974). This shrew either uses the runways of other small mammals or requires the presence of dense herbaceous vegetation for cover. It forages on surface-dwelling invertebrates and probes among vegetation or leaf litter, but it does not actively dig for prey (Churchfield 1990).

Highly social, least shrews huddle together in the wild and in captivity and often produce clicks that have "a high localization valence" (Gould 1969). They readily produce low-amplitude clicks accompanied by head bobbing and snout twitching. *Cryptotis* produce putts and twitters, chirps, and buzzes, but mostly clicks (Springer 1937; Moore 1943; Gould 1969). One sound grades into another—that is, a chirp into a buzz and a put into twitter. Perhaps because of equipment limitations, little data

on sound duration or repetition rates are reported, and no data on peak frequency or bandwidth are available. Gould (1969) reported that clicks were contact calls used among individuals and produced by young pups at only nine days old. Both adult males and females produced clicks during solitary exploration, courtship, mouth washing, investigation of a small object, and when the snout was extended over the edge of an object. The prominence of clicks in the least shrew's repertoire compared to other soricids reflects this shrew's social nature.

We tested eight adult least shrews individually in a series of 5 min orientation and foraging tasks on an experimental board. The carpet-covered experimental board (27 × 54 cm) had an inner and outer box of equal area (fig. 72.4). Orientation tasks included open board, presence of Plexiglas walls around outer edge of board, and the presence of a Plexiglas tube in the center of the board. Foraging contexts included no mealworm present, mealworm moving freely on the board, and a mealworm pinned in center of the board. Trials were run in ambient daylight and again in a dark room. Behaviors and sounds from shrews were recorded using a video or night vision camcorder (Sony model CCD-TRV815) and a Marantz (model PMD-420) cassette recorder. The receiving beam pattern of an Anabat II bat detector (division rate of four or a monitoring range up to 80 kHz)

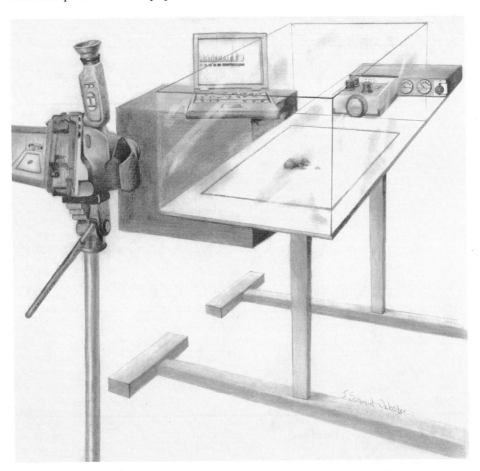

Fig. 72.4. The experimental board, bat detector, video camera, and computer with software used to examine echolocation abilities of the least shrew. Note that the area in the outer box equals the area in the inner box. Plexiglas walls surround the board with a shrew in the middle. Drawing by Jennifer Webster.

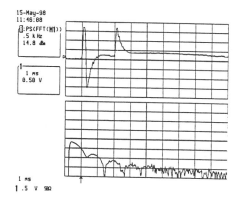

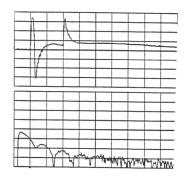

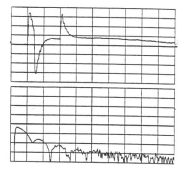

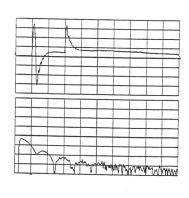

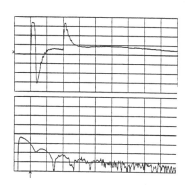

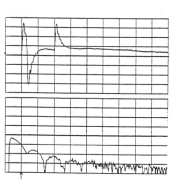

Fig. 72.5. Oscillograms and power spectra of representative pulses produced by the least shrew during various orientation and foraging tasks on the experimental board

covered the entire board. Preliminary tests indicated that a higher division rate was not necessary. Sounds were monitored using Audioscope real-time sonogram software on a laptop computer (Gateway Solo 2000). Data transcribed from video and/or cassette tapes for each pulse included whether the shrew was in the inner or outer box, the number of pulses in a train, whether the shrew searched the board, the number of times the shrew ate a worm, and total time spent eating.

The only sounds detected from least shrews during the orientation/foraging tasks were clicks, although we heard twitters and putts when the shrews were in their home cage. Least shrews showed little variation in waveform and power spectra of pulses among individuals. Typical clicks (fig. 72.5) had a doublet waveform (Buchler [1976] reported similar double pulses in *Sorex vagrans*). Pulses by least shrews were low in amplitude (ac-

tually audible when held close to the ear). We expected pulses would be ultrasonic, as in other shrews, but in fact pulses were between 200 and 12000 Hz, with the average peak frequencies between 242 and 690 Hz (table 72.6). Gould (1969) also reported the frequency range of low-intensity sounds from least shrews as between 500 and 1200 Hz and detectable 5 cm away. In retrospect, the bat detector simply amplified signals and was unnecessary for detecting ultrasonics in this species.

All shrews produced pulses at the same rate and foraged and oriented in the same manner during light and dark trials, confirming the least shrews' poor vision. The mean number of pulses was higher during orientation tasks than foraging (table 72.6). The highest number of pulses per trial (table 72.6) was produced during the wall-present trials, compared to the least number during free-moving worm trials. The mean peak frequency of

TABLE 72.6. Summary of individual pulse characteristics for least shrews (*n* = 8 shrews).

Trial condition	Task type	Mean number of pulses per trial	Mean pulse peak frequency (Hz)	Mean pulse bandwidth (Hz at +3dB)	Mean pulse duration (ms)
Open board	orient	10.0	242	271	65.8
Wall	orient	20.2	308	285	79.4
Tube	orient	16.3	564	489	42.5
Worm pinned	forage	7.3	578	509	29.8
Worm free	forage	4.0	690	702	19.2

TABLE 72.7. Summary of pulse train characteristics for least shrews (*n* = 8 shrews).

Trial condition	Task type	Mean number of pulses per train	Mean "train" duration (ms)
Open board	orient	2.7	0.3
Wall	orient	2.5	0.3
Tube	orient	3.0	0.5
Worm pinned	forage	13.8	2.8
Worm free	forage	12.9	2.6

TABLE 72.8. Summary of time spent by least shrews in different locations and tasks on the experimental board (*n* = 8 shrews).

Trial condition	Task type	Inner box (s)	Outer box (s)	Mean eating time (s)	Mean searching time (s)
Open board	orient	71.6	224.1	N/A	300.0
Wall	orient	41.3	259.5	N/A	300.0
Tube	orient	170.4	109.7	N/A	300.0
Worm pinned	forage	179.3	120.6	129.7	169.5
Worm free	forage	177.2	117.1	128.3	172.0

pulses was lowest during open-board trials, higher during orientation tasks (wall and tube trials), and highest during foraging tasks (pinned and free worm trials). Bandwidth and mean pulse duration showed a similar progression. These patterns make sense in that a shrew can increase the rate of pulses for more detailed discriminations when investigating its environment.

The largest number of pulses per train and the longest trains were produced during actual consumption of the worm (table 72.7). During orientation tasks, the pulses typically were two to three per train and 0.2–0.5 ms. On an open board, shrews spent more time in the outer box, searching the perimeter, but produced few pulses (tables 72.6 and 72.8). When walls were present, there was a significantly higher number of pulses in the outer box ($F = 64.5, p < 0.01$). When the tube was present, shrews produced a significantly higher number of pulses in the inner box containing the tube ($F = 67.7$, $p < 0.02$) and spent less time in the outer box. When no worm was present, shrews produced more pulses at all locations on the board and had a significantly longer

search time ($F = 564.5, p < 0.01$). As the number of pulses increased, so did search time ($r = 0.31$) and the number of times the shrew successfully found ($r = 0.35$) and successfully ate the worm ($r = 0.36$). Collectively, these experiments showed that least shrews produced pulse trains in both foraging and orientation tasks.

Discussion of Echolocation in Insectivores

According to Gould (1969), there is no way to distinguish between pulses used for communication and those used for echolocation. Bateson (1966) noted that these two contexts are not necessarily exclusive. A pulse produced by a shrew often has a sonic portion (a "ticklaut") associated with an ultrasonic sound (Dijkgraaf 1932). Pulses in shrews "may owe their shortness and intensity to the fact that they are a lower variant of supersonic pulses used in echolocation and would have an immediate value if used in echolocation of the source, as well as a potential value in communication, both functions may have evolved together in Insectivora" (Andrew 1964, pp. 287–288). Similarly, "calls emitted during an early period of an animal's life history may occur later in life in a different context" (Andrew 1964, p. 231). Pulses emitted by shrews about to suckle, during caravan formation, and separation from the mother occur in adults during exploration and at increased rates in courting males. Gould (1969) noted that an infant *Cryptotis* emitted twitters when a dispersed group of four was reunited with their mother in the nest. The putt (300–1000 Hz, 0.6–0.9 s, repeated 10–14 per s) was produced by both sexes of *Cryptotis* and detectable only 5 cm away Gould (1969). Putts were used when a lone shrew explored a strange situation (Gould 1969). *Cryptotis* usually emitted twitters during horizontal movement and by adults when infants were returning to the nest. Twitters probably were used at greater interanimal distances than puts. Ultrasonic trains of pulses (50–420 ms, typically 12 pulses in 162 ms) were produced by least shrews during intense exploration, and Gould (1969) felt they were used for echolocation. High-intensity buzzes were used in close encounters between shrews.

The amplitude of shrew pulses is much lower than that reported in echolocating bats. According to Churchfield (1990), this is adaptive, because "if shrews were to use high-intensity emissions in their cluttered habitats of plant stems, roots, logs, and rocks they would be inundated with echoes from every conceivable surface. So high-frequency, but low-intensity pulses are probably more useful. These low-intensity ultrasounds attenuate rapidly and are scattered easily by small objects. This means that predators, even if they can detect ultrasounds, would have difficulty locating the source. So echolocation in shrews may be a means of exploring the habitat without attracting the attention of predators."

Noting the very different tooth structure in shrews compared to other mammals (fig. 72.6), we wonder whether the first incisor functions in sound production. In red-toothed shrews (i.e., subfamily Soricinae), all teeth are tipped with a dark red coloration, an oddity among mammals. The pigment is a form of iron. Shrews have considerable tooth wear and the red coloration could be deposited to strengthen the tooth (Churchfield 1990). In addition, soricids have an exceptionally large, bicuspid first upper incisor, and a large, procumbent first lower incisor, both rarities among mammals. The upper and lower first incisors work together as tweezers to pick up insect prey (Feldehamer et al. 1999). Kingdon (1974) suggested that some shrews "crack their teeth together and make a number of squeaks and chatters." The first incisors are larger than all other teeth, and we wonder whether they provide a large surface area against which the tongue (or lower incisors) strikes to produce a click. The notable tooth wear reported in shrews could be explained if the teeth are used for both eating and pulse production. We further speculate that the double structure of the pulses could result from striking of the two cusps of the upper first incisor. Some odontocetes in the family Phocoenoidea (*Phocoenoides* and *Ponteporia*) also produce double echolocation pulses. The situations under which a mammal produces a single or double pulse for echolocation should be investigated.

Previous Studies on Echolocation in Rodents

Airapet'yants and Konstantinov (1970) first summarized studies on possible echolocation in rodents. The dormouse (*Glis glis*) found food on a slim perch at 40 cm away in utter darkness by jumping onto it, even though the location was changed many times. Kahmann and Ostermann (1951) conducted a series of experiments in the dark with golden hamsters and redbacked voles in which normal and blind animals needed to jump from a table to a platform below leading to a food reward. The vertical and horizontal position and distance of the platform relative to the table was varied. After just a few trials, performance was high. Golden hamsters had a critical distance of 6–8 cm and redbacked voles from 10 to 14 cm. Although the authors monitored up to 70 kHz, ultrasonic sounds were not detected during the trials. Removing the vibrissae from blind redbacked voles did not affect their performance. However, blind red-

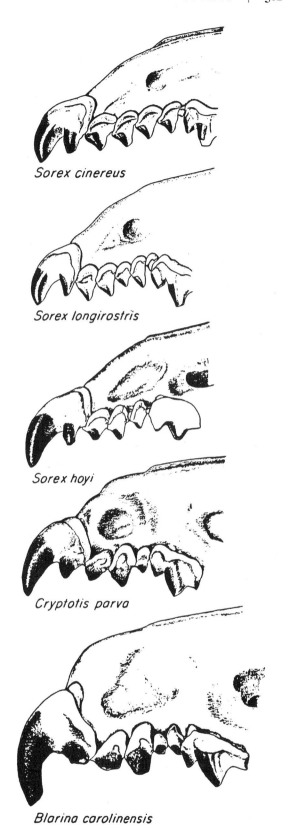

Sorex cinereus

Sorex longirostris

Sorex hoyi

Cryptotis parva

Blarina carolinensis

Fig. 72.6. Lateral view of the rostrum of five shrews. Note the prominent bicuspid first incisor and dark coloration on teeth (from Hoffmeister 1989).

backed voles with plugged ears were not able to perform and jumped randomly from the tabletop. The authors concluded that these species shifted to alternate senses when blind, but they could not conclusively prove echolocation.

In contrast, Airapet'yants (1974) tested three blind dormouse species in the dark with a similar table/platform apparatus used by Kahmann and Osterman (1951). None of the dormice could find the food except when they were close enough to use their vibrissae.

Airapet'yants (1974) conducted another study on six golden hamsters using two platforms; the lower one could be adjusted in a vertical and horizontal position to the upper platform. The hamster had to jump from the highest platform to the lower platform to receive food. After initial training, the hamsters were blinded. The hamsters ran around the edge of the upper platform, hanging their body over the side to explore below, then raising back onto the top platform. Faint "sniffling" sounds were heard when they suspended themselves, but a single attempt to monitor at 70 kHz produced no ultrasonics. By the third experiment, hamsters were performing at 70% of baseline trials. When the vibrissae were removed, hamsters could not perform the task. The authors concluded touch, rather than echolocation, was needed for the task.

Airapet'yants (1974) conducted studies on the voles *Microtus socialis* and *Microtus gregalis* with the identical setup to the hamster study; however, the results were different. Blind voles of these two species readily found the lower platform when it was <4 cm away. In contrast to hamsters, blind voles without vibrissae did not have impaired orientation to the lower task. The platform was washed between trials to eliminate olfactory cues, but performance continued to be high. In further studies, the blind voles with partially blocked hearing or perforated eardrums could no longer perform the task. The authors did not hear sounds associated with jumps and did not have equipment to record during the experiments. They believed they confirmed echolocation in these two species. However, they conducted another study with the bank vole (*Clethrionomys* spp.) using a platform from which one of eight paths radiated to a food reward. This species when blind conducted the task successfully, but they failed when vibrissae were removed.

Three sets of experiment were published on echolocation on blind laboratory rats (*Rattus norvegicus*) orienting in a Y-maze. In these experiments, one arm of the maze was obstructed, and rats were trained to find and move through the remaining open arm to find food. With training, rats were successful in 75–80% of trials. Authors concluded that hearing was involved in successful orientation, because if the ears were blocked (Rosenzweig, Riley, and Krech 1955; Chase 1980), or the hearing temporarily masked by loud white noise for seven days (Bell, Noble, and Daves 1971), success dropped to the level of chance. When the plate was moved so it was

>23 cm from the rat (Bell, Noble, and Daves 1971) or the angle of the plate was changed to 45° (Rosenzweig, Riley, and Krech 1975), success dropped to chance levels, indicating the placement of the plate was important, as it would be if rats were bouncing acoustic signals off it. These reports did not indicate whether they controlled for the odor of the food reward. Rosenzweig, Riley, and Krech (1955) noted a "variety of other auditory cues, such as foot scratching, stepping, teeth clicking, sniffing, and sneezing could not be ruled-out as possible passive auditory cues for the task."

Chase (1980), using laboratory rats in a Y-maze, reported a correlation between ultrasonic vocalizations and correct detection. Chase (1980) only published an oscillogram analysis, so characteristics of rat echolocation signals are minimal: four to eight low-amplitude pulses per series, with most energy between 40 kHz and 50 kHz. Pulses often were in pairs.

Collectively, these studies on echolocation in rodents paint a mixed picture. A more careful selection of study animals based on the ecology of the species and the aforementioned profile of correlates with echolocation could provide a better understanding of senses used in orientation in rodents.

New Data on Echolocation in Rats

We made broadband recordings of sounds from laboratory rats in daylight and darkness. Using an Anabat II detector set at a division rate of eight (i.e., monitoring up to 160 kHz), we made 5 min recordings on a Marantz cassette recorder of five adult laboratory rats in their cages in ambient light. Rats produced ultrasonic sounds, but none were pulses. When each rat was enclosed in a dark, neoprene-lined box (to prevent passive acoustic cues), several single and trains of pulses were produced with waveforms reminiscent of echolocation pulses (fig. 72.7). The fact that only isolated rats in the dark produced these pulses gives further support to echolocation abilities.

Discussion of Echolocation in Insectivores and Rodents

Mammals use many types of sensory cues to monitor their environments, navigate, and find food. In addition to toothed bats and dolphins, echolocation abilities in mammals such as insectivores and rodents merit further investigations. Touch is probably a very important sense in mammals living in limited light environments. Most have vibrissae and many have specialized tactile organs (such as the Eimer's organ in moles or the elaborate vibrissae of otter shrews), but most studies on echoloca-

Fig. 72.7. (*opposite*) Oscillograms and sonograms of (*A*) pulses and (*B*) trains produced by a laboratory rat in a dark, soundproof box

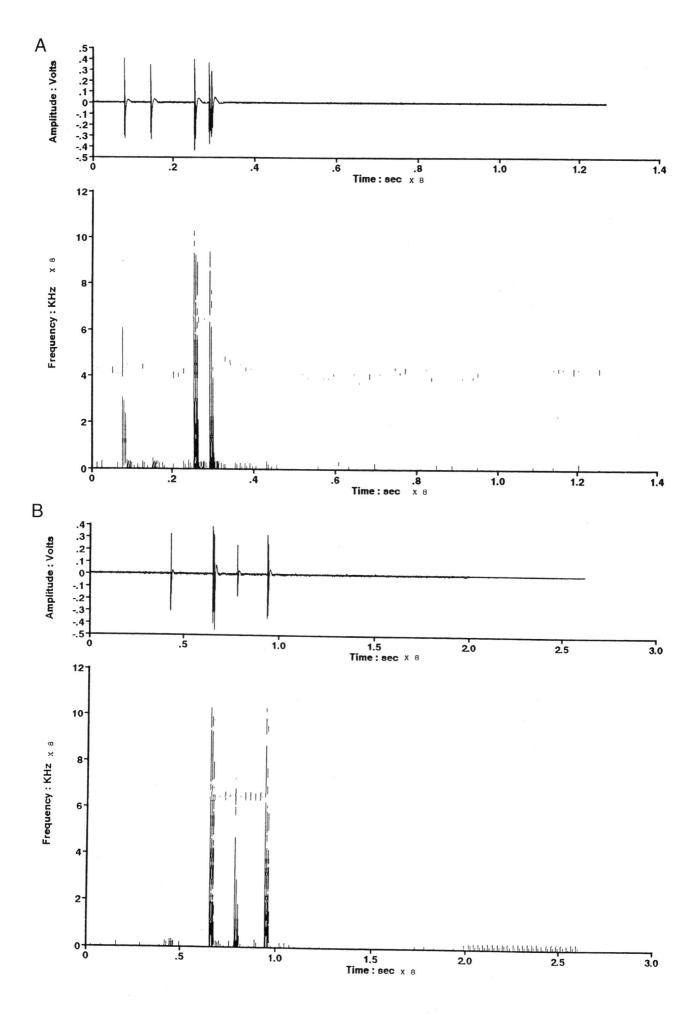

tion have not controlled adequately for possible touch or olfactory senses. Special consideration should be given to nocturnal, semifossorial, fossorial, or visually impaired mammals, and to species with broadband hearing or ultrasonic vocalizations. Echolocation seems to occur in many insectivore families examined to date. More research is needed on signal characteristics, detection, and discrimination abilities in all groups of echolocating insectivores.

One might speculate that because shrews, rats, and bats echolocate in air, rather than in water like dolphins, their signals would be similar to those of bats. Interestingly, the echolocation signal of shrews and the laboratory rat is a single or double broadband, short pulse like the signal of echolocating dolphins, rather than the longer FM, CF, CF-FM signals of bats. This could relate to the close ranges at which shrews would use echolocation. Studies on aquatic insectivores (Sergeev 1973), like water shrews (*Sorex palustris* or *Neomys fodiens*), desmans, or otter shrews, or on aquatic rodents (in the sub-family Ichtyomyinae—especially *Chibchanomys, Anatomys, Ichthyomys,* and *Rheomys*) that spend significant amounts of time foraging under water, might provide some interesting data on signal type versus medium. Fortunately, technology now is available to collect and to analyze short ultrasonic sounds often associated with echolocation in mammals.

Acknowledgments

The authors thank Bob Connour and Jaime Welling for their assistance in making recordings of shrews and rats. We thank the John G. Shedd Aquarium for use of the Bioacoustics Laboratory and the Biology Department of Western Illinois University for the acquisition of equipment. Paul Anderson, Bob Timm, and Barbara Blake provided very useful comments on this manuscript, and Jennifer Webster created the artwork for this chapter.

{ 73 }

Potential Use of Low-Frequency Sounds by Baleen Whales for Probing the Environment: Evidence from Models and Empirical Measurements

Christopher W. Clark and *William T. Ellison*

Introduction

Over the course of evolution organisms change or adapt through the processes of natural and sexual selection and genetic drift (Gould and Lewontin 1979). Natural variability within a population is the fodder for selection, and an organism with traits best adapted to its environment is most likely to transfer genes to the next generation. By this mechanism, heritable traits are selected for and become common throughout a population. The time scales over which selected traits become common in a population are dependent on a number of internal and external factors. Internal factors include such things as physiology, morphology, and phylogeny, while external factors primarily include the ecological, social, and physical environments. Generally, fixation of an adapted trait is thought to take many thousands of generations. However, recent observations of natural se-

lection in birds, fish, and insects indicate that in some cases selection, as manifested in the frequency distributions of phenotypic traits, can begin to occur within as short a period as several generations (Schluter and McPhail 1992).

Behavioral traits are subject to many of the same selective mechanisms, but not necessarily the same selective pressures as nonbehavioral (e.g., morphological) traits, and are generally more labile than morphological characteristics. In this discussion of mysticete acoustics, we take the simplistic view that characteristics shared broadly across multiple species are considered more ancestral than features shared only at the species level. We introduce this conceptual framework as the basis for gauging the degree and approximate order of the physical environment's effect on the features of sounds within a specialized group of marine mammals, the baleen whales (mysticetes).

We propose that in the marine environment the influences of physical acoustics imposed strong selective pressures on the acoustic features of baleen whale sounds. This concept extends the original idea of Morton (1975) for long-range communication in the terrestrial habitat to the marine habitat. As stated by Morton (1977), "Because these signals are broadcast through the environment, the environment may produce selection pressures favoring certain physical properties that increase their propagation" (p. 856). This concept will be demonstrated by careful consideration of acoustic features within the mysticete group in combination with the physical acoustics and properties of the ocean environment. Through further consideration of a more speculative nature, but grounded in basic principles, we propose that in some species for which selection has favored extremely long-range communication signals, a secondary function for these sounds has evolved, and that these species use the reflections of their sounds from natural boundaries as a simple form of echo ranging. That is, certain baleen whales use the echoes emanating from boundaries to navigate and orient relative to physical features of the ocean. We believe that consideration of these matters advances the understanding of the biological functions of baleen whale sounds.

In this chapter, we first describe the types and basic acoustic characteristics of sounds produced by different baleen whales and review the basic principles of the sonar equations constraining underwater sound propagation. We then compare selected acoustic features of baleen whales based on several predictions of the sonar equations. From this will emerge a distinction between the sounds from species that primarily inhabit shallow water and those that primarily inhabit deep water. This will reveal that the distinguishing acoustic characteristics for shallow-water coastal species and for deep-sea pelagic species are matched to their respective environments to optimize underwater sound transmission for long-range communication. These considerations also will show that several species, specialized for the deep-sea environment and which feed along steep bathymetric features, produce sounds that are highly redundant and stereotyped (as per Bradbury and Vehrencamp 1998). Given the importance of detection threshold on communication range, we use the comparative method to speculate on a generalized hearing threshold for baleen whales using considerations of ambient noise and critical ratios. We then extend the broadband sonar equation to include terms for signal reflection, and expand the relationship between acoustic transmission properties and sound features in support of a derived echo-ranging function.

BALEEN WHALE ACOUSTICS

There are presently 11 recognized extant species of baleen whales (Mysticeti; Ridgeway and Harrison 1985). Recent molecular work reveals further details on the evolution of the cetacea (toothed and baleen whales; Messenger and McGuire 1998). Baleen whales are monophyletic, and molecular data have identified the hippopotamus as the closest living relative to the cetacean order (Nikaido, Rooney, and Okada 1999). Within the mysticetes, the family Balaenidae, which includes the southern right whale (*Eubalaena australis*), the northern right whale (*E. glacialis*) (see Rosenbaum et al. 2000), and the bowhead whale (*Balaena mysticetus*), are the least derived (Messenger and McGuire 1998; Berta and Sumich 2000). The remaining eight species have been classified into three families. Two families, Neobalaenidae and Eschrictiidae, include only a single extant species, the pygmy right whale (*Capera marginata*) and the gray whale (*Eschrichtius robustus*), respectively. The remaining six species are in the family Balaenopteridae, which includes blue whales (*Balaenoptera musculus*), fin whales (*B. physalus*), and humpback whales (*Megaptera novaeangliae*). Phylogenetic relationships among these species remain largely unresolved.

All 11 species produce sounds, but representations of sound repertoires are still incomplete for six species. The five species for which essentially complete repertoires are available include the bowhead, gray, humpback, and southern and northern right whales (Chabot 1988; Clark 1982; Clark and Johnson 1984; Crane and Lashkari 1996; Dahlheim, Fisher, and Schempp 1984; Ljungblad, Thompson, and Moore 1982; Payne and McVay 1971; Silber 1986). There are good acoustic representations for blue and fin whales (Clark and Fristrup 1997; Stafford, Nieukirk, and Fox 1999; Watkins et al. 1987), but fewer representative recordings for Bryde's (*B. edeni*) (Cummings 1985; Edds, Odell, and Tershy 1993), minke (*B. acutorostrata*) (Edds-Walton 2000; Mellinger, Carson, and Clark 2000; Schevill and Watkins 1972; Winn and Perkins 1976), sei (*B. borealis*) (Thompson, Winn, and Perkins 1979), and pygmy right whales (Dawbin and Cato 1992). Table 73.1 lists some basic acoustic features for all species as presently available from the literature or through our own recordings (tabulated in Richardson et al. 1995).

Baleen whale sounds are often very loud. Source levels are typically reported as band (power) levels in dB re $1\mu Pa$ at 1 m/Hz (Charif et al. 2002; Thode, D'Spain, and Kuperman 2000). Maximum band levels have been reported as high as 188 dB re 1 μPa at 1 m (tabulated in Richardson et al. 1995). As Payne and Webb (1971) pointed out, increasing source level is one common mechanism for reaching a greater audience. However, under certain environmental conditions (e.g., water depth that is not well matched to the signal's frequency band, or highly reverberant environments), there is little or no advantage to increased source power level; and, as we will show, selection should favor changes in other

TABLE 73.1. Selected characteristics of some baleen whale sounds (LF, low-frequency band <100 Hz; HF, high-frequency >1000 Hz; FM, frequency modulated; AM, amplitude modulated; CW, constant frequency).

Whale species	Frequency band (Hz)	Bandwidth (Hz)	Unit duration (s)	Duration of redundant units (s)	Source band level (dB re 1 μPa)	Signal type	Stereotypy
Blue	10–100	2–9	6–40	20–100	180–188	AM, CW, FM	high
Finback	15–170	2–35	0.5–1.5	2–20	159–183	AM, FM	high
Minke	40–300	10–70	0.1–0.5	10–50	160–165	CW, FM	high
Humpback	20–800	100–400	0.5–5.0	20–1000	175	AM, FM, complex, highly variable	high for LF low for HF
Gray	20–200	70–100	0.5–1.0	20–100	111–185	AM, FM, complex	moderate?
Bowhead	35–400	50–300	0.5–2.0	20–1000	150–189	AM, FM, complex, highly variable	high for LF low for HF
Right	30–500	10–300	0.5–2.0	20–1000	172–187	AM, FM, complex, highly variable	high for LF low for HF

acoustic features to optimize communication effectiveness and range.

All species for which there are substantially complete acoustic repertoires (bowhead, gray, humpback, and southern and northern right whales) prefer coastal and/or shallow-water habitats during major portions of their lives in areas where calving occurs and where functional mating is assumed to occur. All balaenopteran species spend a large proportion of their lives in the open ocean. With few exceptions, balaenoptera mating and calving areas and behaviors are largely unknown, but they are not thought to occur in shallow water or to involve well-defined, traditional locations as with coastal species. The limited reports on balaenoptera behaviors come from studies conducted during seasonal episodes of feeding along coastal upwelling zones (Croll et al. 1998; Edds 1988; Panigada et al. 1999). Thus, the present understanding of baleen whale behavioral ecology is uneven (Tyack and Clark 2000). Most is known about coastal and/or shallow-water species; least is known about species that are pelagic.

Details of acoustic characteristics are particularly interesting for those species where fairly complete samples are available. We use the bowhead, humpback, and southern right whales as representative of coastal species; and blue and fin whales as representative of pelagic species. This division between coastal and pelagic recognizes that none of the species spends its entire life in only one oceanographic domain. It is presented to emphasize the primary environment in which the critical biological activities of breeding and feeding are known to typically occur and are assumed to have evolved.

Calls

We refer to sounds other than songs as calls, with the general assumption that calls and songs are communicative. By far the majority of information on whale calls comes from studies in which researchers reliably observed and recorded individual animals over long periods of time and inferred function through association (e.g., Chabot 1988; Clark 1982; Dahlheim, Fisher, and Schempp 1984; Silber 1986; Tyack 1983). In a few studies, researchers have used playback of natural sounds to elicit behavioral responses that support inferences of function (Clark and Clark 1980; Tyack 1983). So, for example, contact calls from right and bowhead whales are simple, frequency-modulated sweeps (Clark 1982, 1983; Clark and Johnson 1984; Würsig and Clark 1993), while feeding calls from humpbacks, which attract distant whales to the aggregation, are long, relatively constant frequency sirens (Mobley, Herman, and Frankel 1988). In right whales, Clark (1982, 1983) found a significant relationship between the complexity of a call and the social context; with the simplest, frequency-modulated (FM) calls associated with long-range communication and the most complex calls (i.e., mixtures of FM and amplitude-modulation sweeps, and broadband pulses) associated with large, surface active groups engaged in sexual activity. There have been a few observations of blue and fin whales in social groups with associated recordings of discrete sounds (Edds 1988; McDonald et al. 2001; Thode, D'Spain, and Kuperman 2000; Watkins 1981; Watkins et al. 1987). All these balaenopteran sounds were simple FM sweeps or simple AM tonals. At a subjective comparative level, the most variable transient calls from blue and fin whales are equal to the simplest transient calls from right and humpback whales. Fig. 73.1 shows spectrographic representations of calls for each of the five representative species.

Songs

Bowheads and humpbacks produce long patterned combinations of sounds referred to as songs (Payne and McVay 1971; Würsig and Clark 1993). The songs are composed of relatively complex sounds arranged hierarchically from individual units that are combined into

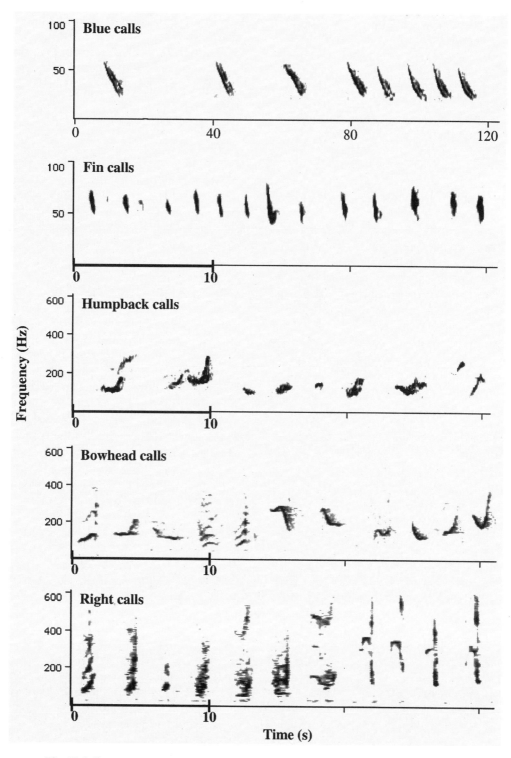

Fig. 73.1. Spectrographic examples of calls from five baleen whale species. The blue whale calls were from a continuous recording of a surface active group off San Nicolas Island, Calif., in October 1995. The fin whale calls are from multiple samples recorded off the British Isles in July–August 1997. The humpback calls are from multiple samples recorded in Frederick's Sound, Alaska, in 1984. The bowhead calls are from multiple samples recorded off Barrow, Alaska, in May 2000. The right whale calls were recorded in Gulfo San Jose, Argentina, in 1977. For fin, humpback, bowhead, and right whale call examples, time gaps between calls are uninformative and a result of editing.

phrases, which are repeated to form themes. In bow-heads, the most complex songs have two themes, each of which has a single phrase. Songs change annually, and there is a high degree of inter-individual variation. Humpback song organization is more extensive than that of bowheads, with up to 7–8 themes composed from multiple phrases. As with bowheads, the humpback song changes annually, and there is a great deal of intra- and inter-individual variability. In both species, there is a high degree of improvisation such that no two songs from the same animal are ever identical.

Blue, fin, and minke whales also produce patterned sequences of sounds (Clark and Fristrup 1997; Mellinger, Carson, and Clark 2000; Watkins et al. 1987). For fin whales, the seasonal occurrence of these signals in mid-latitudes from mid-fall through mid-spring coincides with the assumed breeding season, suggesting these sequences of sounds are male reproductive displays (Watkins et al. 1987). This suggestion, by default, is usually extended to blue and minke whales. However, for all the balaenoptera species there are no reports presenting direct evidence for the sex of the animal producing the sound. For blue and fin whales, this single functional interpretation of patterned sequences as a male reproductive display could be too simple. There is evidence that both species produce long, patterned sequences during the summer feeding season in high latitudes (Croll et al. 2001a; Ljungblad, Clark, and Shimada 1998). In any case, these long, patterned sequences are composed of repetitions of nearly identical 2–3 note phrases. Variability for blue and fin whales at the inter-annual, inter-individual, or intra-individual levels is low relative to bowhead or humpback whales. There are potential differences in acoustic features between populations of both blue and fin whales, but better measures of this variability are needed (Moore et al. 1998; Stafford, Nieukirk, and Fox 1999; Thompson, Findley, and Vidal 1992; Watkins et al. 1987). Fig. 73.2 shows spectrographic representations of songs for two pelagic species (blue and fin whales) and two coastal species (humpback and bowhead whales).

The Sonar Equation Approach

The sonar equation (Urick 1983) is a term applied to a generic family of highly simplified logarithmic equations used for describing and evaluating a wide range of acoustic propagation features. This form is amenable to understanding acoustic propagation and for measuring important variables, and the logarithmic scale is well suited to the very large dynamic range of most of the terms. Because many readers are likely unfamiliar with some of the more arcane forms of acoustic terminology, the sonar equation approach will be presented in terms of simple topical examples that will hopefully communicate the basics of underwater sound propagation.

First consider a single point source of sound, a single FM sweep from frequency f_0 to frequency f_1, with no

harmonics. This sweep occurs over duration T. Units are usually Hz (cycles/second) for frequency and seconds for time. The product of frequency multiplied by time is nondimensional. The difference between the upper and lower frequencies is the overall signal bandwidth (W) where:

$$W = |f_1 - f_0| \qquad (73.1)$$

The product of signal duration and bandwidth is referred to as the time-bandwidth product, TW.

The acoustic output of the underwater source is defined in terms of the source level, SL, typically described as the pressure level, P, at 1 m from the source, and presented in a decibel format relative to a standard pressure of 1 μPa.

$$SL = 20 \log(P) \text{ at } 1\text{m}/1\mu\text{Pa} \qquad (73.2)$$

As the sound emanates from the source, its intensity decreases due to spreading losses. There are two basic spreading losses: spherical and cylindrical. Spherical spreading loss is inversely proportional to the distance (R) from the source, while cylindrical spreading loss is inversely proportional to $(R)^{1/2}$ from the source. Spherical spreading usually persists from the source out to a range approximately equal to the depth of water, after which cylindrical spreading occurs. A secondary type of loss, referred to as absorption, occurs in water due to chemical effects. The absorption loss factor, α, is usually given in units of dB/unit distance. The combination of spreading and absorption losses is referred to as transmission loss (TL), and can be written in equation form as:

$$TL = 20 \log[R(\text{m})] + \alpha R, \text{ for } R \le 2000 \text{ m} \qquad (73.3a)$$

and

$$TL = TL_0 + 10 \log[R(\text{m})] + \alpha R \qquad (73.3b)$$

for $R > 2000$ m, and where $66 < TL_0 < 70$.

These are highly simplified forms of actual transmission losses but adequate for the intent of this paper. In particular, we have ignored TL effects very near the air/water interface of the ocean surface. (See Urick 1983 and Jensen 1981 for more detailed treatments on this subject.)

There are enough terms identified at this point to evaluate the received pressure level (RL) at some distance from a whale producing a sound in terms of the source level (SL) and TL. That is,

$$RL = SL - TL \qquad (73.4)$$

If it is assumed that the receiver is a listening whale, then several issues must be addressed to determine an-

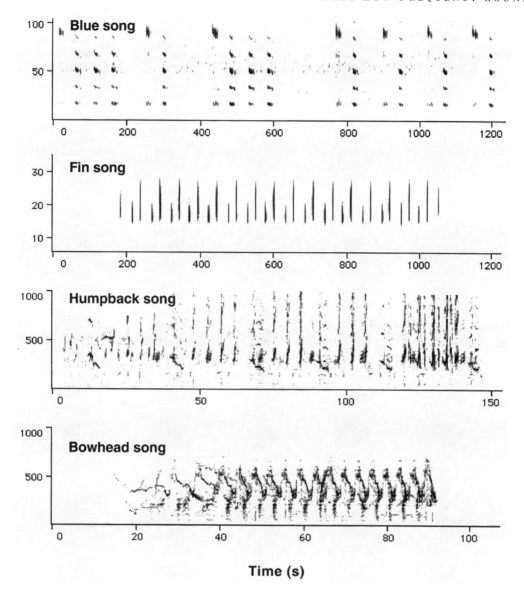

Time (s)

Fig. 73.2. Spectrographic examples of songs from four baleen whale species. The blue and fin whale samples were recorded off San Nicolas Island, Calif., in September 1994 and October 1995, respectively. The humpback whale was recorded off north Kauai, Hawaii, in March 1989. The bowhead song was recorded off Barrow, Alaska, in April 1993. The blue and humpback whale samples show only a portion of the songs to illustrate repetitions of phrases.

swers to the following questions: Can the sound be perceived or detected above the background noise, and can the sound be decoded and recognized above the background noise?

To address the first question, we assume the background noise is uniform over the frequency band of the sound and has a spectrum noise level (NL) in dB re 1 μPa^2/Hz. To compare a sound of received pressure level (RL) and bandwidth (W) to a uniform ambient noise given as a spectrum level requires that the spectrum level be converted to a pressure level. This is most

simply accomplished by summing up the individual 1 Hz noise levels in the detection band, W, from f_0 to $f_1 - f_0$, (eq. 73.1). In dB format, this is simply NL + 10 log (W). The comparison of sound level to noise level is called the signal-to-noise ratio (SNR) and the relationship is given by

$$SNR = RL - [NL + 10 \log(W)] + DI \qquad (73.5)$$

In eq. 73.5 an additional term called the directivity index (DI) has been introduced. For an animal, this can be viewed as a measure of its binaural hearing ability (Au 1993). A conservative estimate for the minimum gain against noise provided by two ears is given by

$$DI = 10 \log(2) \qquad (73.6)$$

Thus, DI = 3 dB for two receivers spaced less than one wavelength apart.

In reality, all listening mechanisms, both biological and humanmade, require some SNR > 0 for signal detection. This level above a SNR of zero is called the detection threshold (DT). The SNR excess above this detection threshold is called the signal excess (SE). SE in equation forms is

$$SE = (SL - TL - [NL + 10 \log(W)] + DI) - DT \qquad (73.7a)$$

and

$$SE = (SL - TL - [NL + 10 \log(W)] + DI) - RT \qquad (73.7b)$$

and SE must be positive for detection or recognition to occur.

Eqs. 73.7a and 73.7b are nearly identical. The difference is that in eq. 73.7b the term DT is replaced by RT, a recognition threshold. We assume that the bandwidths for detection and recognition are equal, although this is not necessarily true. We also assume that the amount of SE needed for detection is equal to or greater than the level of SE needed for recognition (i.e., RT ≥ DT), and that different thresholds exist for different tasks. For example, with an FM upsweep (e.g., contact call), the detection and recognition thresholds are almost identical. In contrast, with a broadband FM signal with harmonics and amplitude modulation (AM) (e.g., a female estrous call), the difference between detection and recognition thresholds could be as high as 20 dB.

The relationship of most of the various terms in eqs. 73.7a and 73.7b seems to be intuitive. However, the equations appear to imply that greater bandwidth increases the noise floor, and thereby apparently reduces the success of detection. In fact, the opposite is true, and the widespread use of broadband signals by most baleen whales suggests that there is some property of bandwidth that provides a gain in communication effectiveness. If we examine a signal that has duration and bandwidth features, then a simple processor used in both sonar and radar applications and well adapted to such a signal type is a "matched-filter" processor (Johnson and Johnson 1993). The output of a matched filter is the ratio of signal energy to noise energy, whereas the output of a processor that emulates eqs. 73.7a and 73.7b is the ratio of signal power to noise power. Thus, a matched filter provides a powerful technique for detecting broad-

band signals in both noise and reverberation, the theoretical gain being equivalent to 10 log (TW).

Bandwidth thus becomes an enormously valuable factor in a wide range of acoustic functions once some form of broadband processing has been implemented. One benefit of a broadband signal is that it offers the possibility for a receiver to successfully detect and recognize the signal in environments where portions of the signal are lost due to such factors as frequency-dependent multipath effects or masking. This benefit of broadband signals was expounded in a very simple relationship by Harrison and Harrison (1995). In essence, a signal of mean frequency (f_0) and bandwidth (W) can be viewed as if its received level (RL) was sampled at a number of points over a distance interval (R_{INT}) relative to a distance (R_0) such that

$$R_{INT} = R_0[W/f_0] \qquad (73.8)$$

This is tantamount to averaging the signal over the entire frequency band at a single distance point. Bandwidth, perforce, removes peaks and nulls that would otherwise be present in a pure tone transmission. The result is a more well behaved signal, and one that is easier to recognize.

In biological terms, selection should favor animals with sensory perception and processing mechanisms that take advantage of signal bandwidth. Although animals will not have a matched filter as strictly defined by engineering terms, there are clear cases (e.g., dolphins and bats) in which animal sensory systems have evolved to take advantage of signal bandwidth properties. To date, there is no direct evidence that baleen whales utilize a "matched filter" type of perception processing. Recently, Frazer and Mercado (2000) implied bandwidth gain in their attempt to model humpback song as a sonar function. An elegantly rigorous treatment of matched-field processing techniques was presented by Thode, D'Spain, and Kuperman (2000) to describe details of blue whale signal structures and the underwater movement of an animal producing sounds. Such approaches, if done vigilantly, coupled with careful integration of evolutionary principles offer a realistic paradigm for elucidating physical acoustic constraints on signal structure and function.

Baleen whale sounds in shallow water

What are the implications of eqs. 73.7a and 73.7b in terms of the relationship between the acoustic environment and baleen whale sounds? These equations state that acoustic detection and recognition can be increased by (1) higher source level, (2) lower transmission loss, (3) lower ambient noise, (4) greater directivity index, and (5) lower detection thresholds. Transmission loss (TL), as constrained by physical propagation, and ambient noise (NL) are determined by the environment. In contrast, the biological variables upon which selection can act include source level (SL), source features that potentially minimize TL, and physiological and neurophysiological adaptations that increase signal detection and recognition.

In biological terms, what are the most efficient, least costly signal features to change to increase detection and recognition? In general, since the detection and recognition components of communication (see eqs. 73.7a and 73.7b) are enhanced by matching signal characteristics to minimize TL, one would predict that selection should favor signals with frequencies in the band with minimum TL and signals with bandwidths less than or equal to the bandwidth of minimum TL. To address this question in more biological detail, let us take an evolutionary approach and make the following assumptions. We assume that: (1) early mysticetes lived in a very shallow-water, coastal environment (depths < 40 m); (2) these animals produced broadband sounds with some natural variations in frequency content and intensity; and (3) selection favored sounds that increased the contact range between individuals engaged in foraging and mating activities. Given these conditions, one needs to know about propagation (TL) and ambient noise (NL) factors in a shallow-water, coastal environment, and specifically about how sound variables that are biologically determined (e.g., source level, frequency, and bandwidth) influence detection range and recognition.

As shown above in eq. 73.3b, TL is strongly influenced by the absorption term (α) such that the lower the frequency, the less the transmission loss. This implies that selection should favor sounds in a frequency band somewhere below 1000 Hz, a band we will refer to as the intermediate frequency band. Both empirical and modeled physical acoustic evidence for shallow water indicate that the lowest TL occurs in the 100–500 Hz band (Jensen et al. 1994). Below 100 Hz there is very high attenuation (Jensen 1981; Jensen and Kuperman 1983), primarily because of low-frequency sound coupling into the bottom (see Premus and Spiesberger 1997). Above 500 Hz, volume attenuation and scattering loss lead to dramatically higher transmission losses. Overall, in shallow water over ranges of several tens of kilometers, TL is approximately 10 dB lower for sounds in the 100–500 Hz band compared to sounds outside this band, and frequency-dependent TL benefits increase with increasing range.

The major natural sources of noise in the coastal habitat come from local wind, waves impacting the shore, rain, and biological sounds primarily from invertebrates and fish. In coastal habitats these sources of sound are often in the immediate vicinity of a receiver, and more distant sources are cut off by the natural underwater boundaries of shoals, capes, and headlands. Local wind speed and bottom conditions are dominant factors ef-

fecting shallow-water ambient noise (Wille and Geyer 1984; Urick 1983). There is some empirical evidence on shallow-water environments inhabited by whales indicating that the region of lowest ambient noise spectra occurs in the 100–400 Hz band (Clark 1983, table 1). Thus, for the shallow-water environment, considerations of transmission loss and ambient noise predict that selection should favor sounds in the 100–500 Hz band. This suggests that in terms of sound characteristics that optimize communication range, selection should have initially favored changes in frequency band and less strongly favored changes in source level.

Baleen whale sounds in deep water

Throughout this discussion, we assume that the original ancestral mysticete lived in shallow coastal waters and from there moved out into deeper and deeper water to exploit new food resources. The most beneficial feeding opportunities would occur seasonally along upsloping edges as found around seamounts and along continental shelves. At such locations, the ocean temperature varies dramatically with depth, which varies from hundreds to thousands of meters, and sound transmission is not necessarily bounded by the air-water interface or the ocean bottom. In deep water, differences in ocean temperature and pressure can have a strong influence on sound velocity with the result that sound is refracted away from either the surface or bottom, or both. Under these conditions of upward and/or downward refraction, low-frequency sounds enter a sound channel in which most of the sound energy propagates within a limited depth regime. This effect of refraction, in combination with extremely low levels of absorption, can lead to exceptionally low levels of transmission loss and extremely long ranges of acoustic transmission. The depth of the sound channel varies dramatically with latitude and season (Jensen et al. 1994; Urick 1983). In temperate waters, the sound channel rises to the surface as it approaches shelf breaks from the deep, so animals positioned off separate shelf breaks could communicate without the need to dive deeply. In high latitudes where intensive feeding occurs during summer months, surface ducting confines the sound channel to the upper 100 m layer, an environmental feature that also facilitates long-range communication. In both an acoustic and evolutionary sense, the shelf edge was the interface between the shallow and deep-water habitats. It was the region where animals adapted for shallow water could exploit rich food resources and gain direct access to a novel mechanism for very long-range communication.

What are the implications of the shelf edge and deep-sea environment on signal features? Some answers to this question relative to the abyssal ocean become evident by consideration of eqs. 73.7a and 73.7b, especially in regard to TL and ambient noise. However, several new considerations are required to fully appreciate the different acoustic niche of the deep ocean. One of these considerations involves TL, while a second involves concepts of reverberation and reflection.

As mentioned above, our initial considerations of TL as expressed in eq. 73.3 were simple, when in fact TL is a complex phenomena (see Jensen et al. 1994). However, for purposes of biological acoustic considerations, we identify and apply only those more complex components that we believe are the most critical to the present discussion.

In the deep-sea environment TL is lower for low-frequency sounds, defined here as sounds below 100 Hz. A competing factor is ambient noise. The naturally dominant source of low-frequency ambient noise in the deep ocean is probably wind from high latitudes (Bannister 1986), while the present-day dominant source of low-frequency ambient noise is from shipping (Dyer 1997). Bannister (1986) estimates 50 Hz ambient spectral levels of 65–75 dB re 1 μPa^2/Hz with highest levels in the winter and lowest levels in the summer. In general, ambient noise level in the deep ocean is inversely related to frequency (Urick 1983; Wenz 1962); noise increases at a rate of about 6–10 dB/octave as frequency drops from 1000 Hz to 1 Hz. However, there appears to be an interesting exception to this relationship within the low-frequency band. Based on Wenz's (1962) ambient noise spectra for usual deep-water conditions and average wind speeds, there is a plateau in ambient noise level in the 10–100 Hz band. Curtis, Howe, and Mercer (1999) provided empirical evidence of a window of low ambient noise in the 10–30 Hz band based on seafloor sensors distributed in the deep sound channel of the North Pacific Ocean, and accounted for both the wind and shipping noise contributions. If we accept these empirical data as representative of average ancient ambient noise levels, and take into account TL, one would predict that for the deep ocean, selection should favor very long-range communication signals in the approximately 10–30 Hz band.

Comparison of baleen whale acoustic features for long-range sound transmission

Some acoustic characteristics of the mysticete group are shown in figs. 73.1 and 73.2. and listed in table 73.1. As described above, consideration of the sonar equation predicts that the acoustic characteristics of frequency, bandwidth, and duration are important variables constraining the range of effective one-way communication. Therefore, we would predict that differences in sound propagation for the shallow and deep environments would lead to differences in the frequency, bandwidth, and duration characteristics for species that inhabit the two environments.

Fig. 73.3 plots peak frequency versus bandwidth for blue, fin, humpback, and bowhead whale songs. Fig. 73.3 indicates that songs for animals that primarily sing in a

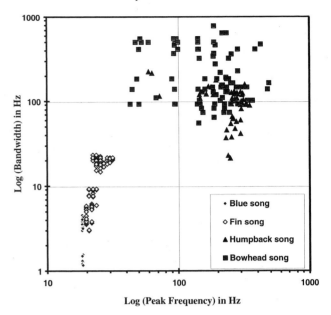

Fig. 73.3. Plot of peak frequency versus time-bandwidth product for songs from four baleen species. Samples for blue (North Atlantic), humpback (western North Atlantic), and bowhead (Beaufort Sea) whales are from three different dates, with three to four songs measured per date. Samples for fin whales are from four different locations (Gulf of California, Mediterranean Sea, British Isles, and Southern California Bight) with 3 songs per location.

coastal environment (humpback and bowhead) have greater bandwidths and higher peak frequencies than songs for pelagic species. Fig. 73.4 matches average spectra for different species sounds with the average ambient noise spectra interpreted for either the pelagic or coastal environment. Fig. 73.4 indicates that the overall frequency bands of baleen whale sounds are generally matched to the lowest ambient noise in their respective environments. For blue and fin whales, the major energy in their infrasonic songs is within a 15–35 Hz band coincident with an acoustic window of low ambient noise, thereby reducing transmission loss and increasing long-range detection. For bowhead, humpback, and right whales, their contact calls and songs are matched to a 100–400 Hz, low-noise window that occurs in shallow water. Given the benefits of increasing both range and effectiveness of communication, the advantage of matching sound production frequency to transmission properties of the environment is expected to be a powerful selective force. In the shallow-water environment, the combination of lower TL at moderate ranges and lower ambient noise in the 100–400 Hz band should have selected for sounds in this frequency band. In contrast, for the deep-ocean environment, the combination of lower TL over very long ranges and a lower average ambient noise window in the 10–30 Hz band should have selected for sounds in this low-frequency band.

Two other acoustic characteristics that strongly in-

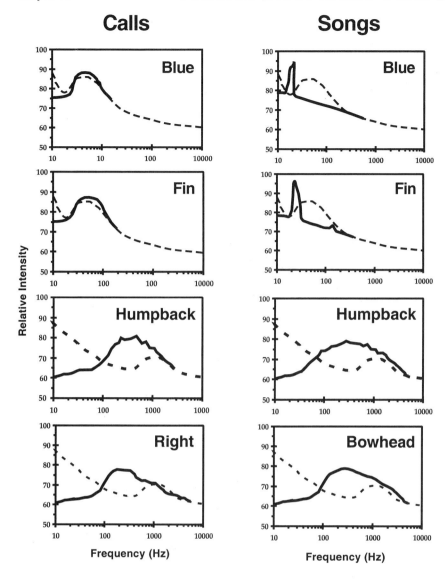

Fig. 73.4. *Left:* Average spectra for different species' calls coincident with the ambient noise for the pelagic (blue and fin whales) and coastal (humpback and right whales) environments. *Right:* Average spectra for different species' songs coincident with the ambient noise for the pelagic (blue and fin whales) and coastal (humpback and right whales) environments.

crease signal excess (SE) are signal redundancy and stereotypy and signal bandwidth. Differences in the duration of repetitions within the songs of the two coastal and two pelagic species are noteworthy in this regard. In the case of blue and fin whales, the largest portion of the song with nearly identical features is the entire song itself; their songs last approximately 900–1500 s with highly redundant and stereotypic phrases lasting on the order of 15–130 s and 10–35 s, respectively. In contrast, for humpback and bowhead whales, the largest portion of a song with similar features is the phrase, and phrases last approximately 10–30 s and 3–5 s, respectively (see fig. 73.2). Assuming that stereotypy and redundancy are directly proportional to detection and recognition, pelagic species could achieve as much as an order-of-magnitude gain (approximately 10 dB) from stereotypy and redundancy as compared to coastal species. (This gain is estimated by taking 10 log of the ratio between pelagic phrase duration and coastal phrase duration.) We stress this point concerning stereotypy because this

feature is a necessary, but not sufficient, characteristic of echolocation signals.

Another point we emphasize is the importance of bandwidth. On the first order, bandwidth alone predicts rates and complexities of communication. Although high acoustic bandwidth does not necessarily translate into high communication rate, it is true that high communication rates cannot be achieved without high bandwidth. Bandwidth also provides a mechanism for estimating range between conspecific communication. In the ocean, this is particularly true because of the frequency-dependent effect of absorption. A very simple example of the impact of absorption on bandwidth as a function of range is illustrated in fig. 73.5. Here we show spectrograms of blue and humpback whale songs as recorded over ranges of approximately 1 km and 100 km. As is obvious from these spectrograms, the higher frequency features that are prominent in humpback song at 1 km are totally absent at 100 km. In contrast, there is little difference in the received frequency

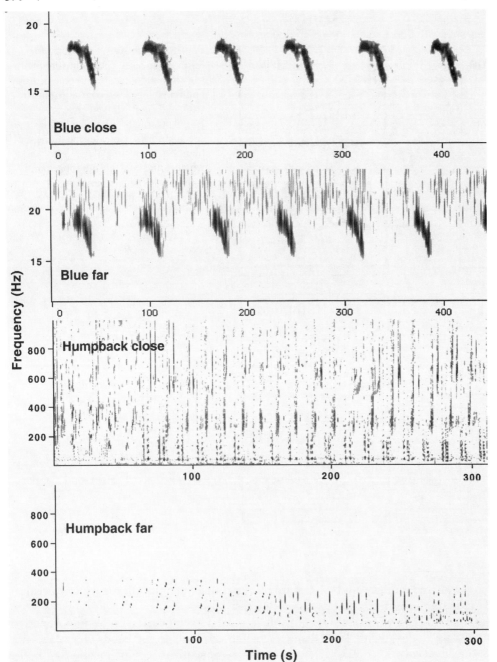

Fig. 73.5. Spectrographic examples for blue and humpback songs showing the differences between songs recorded at close range (approximately 1 km) and far range (approximately 100 km). Blue and humpback whale samples are from western North Atlantic during 1993 and 1996, respectively.

content of the blue whale songs for the two ranges, because all song energy is in the low-frequency band, which is not affected by absorption.

Consequences of long-range communication in the deep ocean

As shown above, the frequency band and bandwidth predictions from the sonar equations for shallow and deep water are well matched to the characteristics of baleen whale sounds. What are the costs and benefits to adopting a coastal frequency band and bandwidth versus a pelagic frequency band and bandwidth? The cost for the very shallow water environment is modest com-

munication range, while the benefit is higher communication rate. Thus, coastal species, simply on the basis of environmental considerations, are restricted by the extent to which individuals can remain in contact, but the availability of greater bandwidth provides the opportunity for a richer acoustic exchange. Two of the most obvious trade-offs when moving into a deep-water environment and shifting into the low-frequency band are higher ambient noise levels and loss of bandwidth with its implied lowering of information rate. However, the potential benefit of shifting into the lower frequency band in the deep-sea environment is a two order-of-magnitude increase in communication range. Based on this logic, we conclude that for pelagic species the decrease in potential communication rate resulting from the lower bandwidth is offset by the benefits of increased potential communication range. This is further improved by increasing (1) source level, (2) the duration of individual sounds, (3) the effective signal duration through multiple, periodic repetitions of the same sounds, (4) stereotypy, and (5) redundancy. There is evidence that these features are equal or greater in pelagic than in coastal species (fig. 73.2 and table 73.1).

Payne and Webb (1971) first formally hypothesized that the loud, infrasonic sounds of certain balaenoptera whales were adapted to take advantage of this physical acoustic niche, and that, prior to modern shipping, whales might have communicated across ocean basins. Based on the intense, low-frequency characteristics of these sounds, several authors have also suggested that whales might sense the echoes from ocean features as an aid for orientation and navigation (Norris 1967, 1969; Patterson and Hamilton 1964; Payne and Webb 1971; Thompson, Winn, and Perkins 1979).

We now further explore this idea. In particular, we extend the broadband sonar equation to include a term for signal reflection. We modify the notion of long-range communication to propose that, under certain circumstances, sounds with signal characteristics under strong selection pressure for very long-range communication are also used for sensing reverberations and reflections in a form of echo-ranging. In the next section, we modify the one-way eq. 73.7 to a two-way equation to account for the situation in which a sound emanating from a whale reflects off an underwater surface and propagates back to the whale that produced the sound.

This notion of signal use for detecting reverberations and reflections has particular salience in the ocean environment where the functional distance of the visual modality is severely limited. Thus, the argument can be made that there would be a selective advantage for those who could sense reflections of their sounds off ocean features. Here it is understood that acoustic signaling for communication is the primary function driving the selection of acoustic features, with reverberation and reflection sensing a possible secondary feature. However, we have no strong basis for this assumption, and the order could be switched. Although the evolutionary histories and derivations of present acoustic features and functions remain uncertain, we would argue that the assumption of communication as the primary function does not detract from the considerations of whether or not the use of sounds for echo-ranging is possible.

Using signals for assessment of the environment and the detection of objects

The representation of sound propagation in eqs. 73.2 and 73.3, although simple and effective for illustrating net TL along ducted paths, is insufficient for estimating transmission loss in regions where there is significant interaction with the ocean bottom. Highly accurate propagation prediction methods include both ray and wave modeling, with modal analysis and the use of parabolic equation (PE) methods that are beyond the scope of this chapter (Jensen et al. 1994). However, we will discuss the implications of some propagation model predictions and include several key empirical examples specifically related to marine mammal communication.

For this topic, we expand the three basic equations into a form that addresses active (i.e., two-way) transmission of signals for the purpose of detecting and evaluating the reflection of that signal from features in the environment. Eq. 73.7a thus becomes

$$SE = SL + 10 \log(W) + TS - 2TL - NL + DI - DT \tag{73.9}$$

where TS is target strength, the logarithmic ratio of incident to reflected signal power at a reflective boundary in dB re 1 m^2, and the transmission loss of the signal occurs both going to and returning from the target.

Five associated factors relevant to detection using active transmissions also must be introduced here. These are blanking range (R_{BLNK}), the effect of intersignal time delay (T_{INT}) on maximum search range (R_{MAX}), theoretical target size or range resolution (R_{RES}), and theoretical Doppler or range-rate resolution (V_{RES}).

R_{BLNK} is a function of the time during which the signal is being emitted. For an animal listening for echoes from a sound it produced, it is assumed that it cannot listen until it has stopped producing the sound. This means that the animal cannot receive the echo from an object within a certain range from its position—that is, a blanking range. If the duration of the transmitted signal is T, then the blanking range is given by

$$R_{BLNK} = C \cdot T/2 \tag{73.10}$$

where C = speed of sound, nominally 1500 m/s in seawater.

T_{INT} defines the likely maximum search range (R_{MAX}) being investigated with the transmitted signal. Here the underlying principle is that an animal waits to listen as long as reasonable until sending out a new querying signal.

$$R_{MAX} = C \cdot T_{INT}/2 \tag{73.11}$$

The last two terms, R_{RES} and V_{RES}, are dependent on the type of signal used. For this study we consider only the two most common signal types, tonelike signals (TW ≈ 1) that are most effective in velocity resolution, and broadband signals (TW > 1) that are most effective in range resolution. These relationships are shown in table 73.2.

TABLE 73.2. Velocity and range resolution characteristics of broadband and narrowband signals.

Active sonar terms	$TW \approx 1$	$TW > 1$
R_{RES}	$CT/2$	$C/2W$
V_{RES} (Note 2)	$C/2Tf_0$	(Note 1)

Note 1: Broadband pulses ($TW > 1$) also can be used for range (target velocity) evaluation if the arrival time of multiple successive signals are compared against a set baseline of the original transmitted delays. If the time difference between the first and last signal is T_{INT}, then the range rate resolution is $V_{RES} = C/(2WT_{INT})$.

Note 2: Velocity and range rate resolution are restricted to the vector component lying along the path of the signal. Thus, a target moving normal to this path would inherently demonstrate zero Doppler or range rate. In calculating Doppler shift, the term (f_0) is the transmit frequency.

The range resolution inherent to a broadband signal provides two immediate benefits: (1) it increases the potential target density of a reflective object by closely approximating the object's own resolution at the expense of other nearby and unwanted reflectors; and (2) it provides sizeable processing gain in the presence of distributed reverberation. Thus, a CW signal of duration T integrates reverberant returns over a span of reverberators that extend over a range equal to $C \cdot T/2$. However, an FM signal of bandwidth W integrates only over a region of $C/2W$. The net gain is about 10 log (TW). In an active sense, this means that for an FM signal, a target's range can be resolved to $R_{RES} = C/2W$ and its range rate to a velocity of $V_{RES} = C/(2W \cdot T_{INT})$.

In a passive arrangement, such as one-way interspecific communication, bandwidth may play an analogous role. Thus, if a whale is producing a repetitive series of sounds, a nearby listening whale can tell if the calling animal is approaching or retreating by changes in the intercall, T_{INT}, value. This ability will be significantly enhanced if the range resolution of that signal is improved by a high bandwidth—that is, if $C/(2W) < C \cdot T/2$.

The importance of broadband gain against multipath has already been discussed, but there are additional benefits available with an expanded frequency range: (1) increased signal diversity reduces signal overlap with other animal sounds; and (2) shifting center frequency reduces self-interference from reverberation. The latter is an active sonar technique used to reduce the chances of an early signal reflection returning from a distant object (e.g., from 1000 km away) being confused with a late signal reflection returning from a close object (e.g., 100 km away).

AMBIENT NOISE AND HEARING THRESHOLDS

Unlike biological variables related to the production of sound (e.g., SL, frequency, and bandwidth), eq. 73.7 includes two terms (DI and DT) related to the perception of sound. So far in these considerations of long-range communication, we have ignored detection threshold (DT). Does the natural environment provide any insights concerning thresholds of auditory perception?

Whales reside within a largely auditory environment. In this context, we hypothesize that ambient noise has likely played a major role in shaping the types of sounds produced by animals and the limitations of their sensory perception mechanisms. For the baleen whales, we assume there was an advantage to producing sounds in frequency bands with low levels of ambient noise. Likewise, it is reasonable to assume that a whale's auditory sensitivity is adapted to ambient noise conditions and that ambient noise has played an evolutionary role in defining their hearing thresholds. One of the major shortfalls in our understanding of the acoustic behavior of the baleen whales is our lack of measured hearing thresholds. By assuming an evolutionary role for ambient noise in shaping whale hearing threshold, as well as by knowing the approximate frequency ranges of sounds they produce, a generic hearing threshold for mysticete whales can be estimated. The process by which we estimate this threshold includes the following assumptions: (1) the frequency band of lowest hearing threshold includes the frequency band of sound production; and (2) the frequency of best hearing and the bounds on either side can be extrapolated by comparison with other mammalian underwater specialists with measured thresholds.

The absolute threshold at best hearing is estimated from two factors: (1) low ambient noise level (i.e., 10th percentile) at best hearing, and (2) auditory critical ratio. The logic is that evolutionary pressures should select for the most efficient use of the dynamic range experienced by the auditory mechanism. As a result, hearing thresholds would not be expected to occur significantly below the lowest background noise levels. Furthermore, all measured mammalian hearing systems work within a relatively narrow range of critical ratios, on the order of 16–24 dB re 1 Hz. Therefore, we propose that the threshold of best hearing can be estimated as the sum of the low ambient noise level (within the best hearing range) and the critical ratio.

A first-order evaluation of this hypothetical approach can be obtained by applying it to selected mammalian species with good listening capabilities, and for which both threshold and critical ratios are already established from direct measurements. The two examples chosen for this evaluation are human hearing thresholds in air and beluga whale (*Delphinapterus leucas*) hearing thresholds in water.

The calibrated measurements of human hearing thresholds are well founded, and for this example we have chosen the ISO MAF threshold that has been established for normal listeners and young people for whom presbycusis is not a factor (Harris 1998). The measurement of critical ratio (and the related value critical bandwidth) is also well established (Harris 1998). What has not been as well established is the lower bounds on ambient noise in the human habitat. There are, however, some well-documented ambient measurements in the wild that provide an acceptable estimate of this quantity. Fidell et al. (1996) report one-third octave band measurements at three wilderness sites based on 46 h of broadband recordings. The analyzed portions were based on samples free of all anthropogenic noise (especially aircraft fly-overs and other vehicular activity) or wind artifacts. The site measurements with the lowest level in the spectra of best human hearing were made at Superstition Wilderness area in Arizona and resulted in an A-weighted level of 29 dBA. This data point can be compared with a series of A-weighted measurements at the Grand Canyon (Harris 1998) of mean value 25 dBA. This latter measurement is of some importance as it is also reported with 90th and 10th percentiles, with a nominal 10th percentile of 25 dBA ± 5 dB at this location. Thus, the Grand Canyon value overlaps the Superstition measurement, and it places an estimated 10th percentile lower bound of 20 dBA for ambient at best hearing. These results are summarized in fig. 73.6A, where we have made a composite noise spectrum by combining the ambient levels from Superstition Wilderness for frequencies below 250 Hz (Fidell et al. 1996) with those from the Grand Canyon for 500–4000 Hz (Harris 1998). As can be seen in fig. 73.6A, the thresholds predicted from the sum of lowest ambient noise level (measured composite spectrum) and the human critical ratios are very similar to the measured ISO MAF thresholds (Harris 1998) for the frequency band of lowest threshold. This similarity is quite remarkable and does not contradict the hypothesis over the full range of best human hearing from 1000 to 4000 Hz.

In a very similar manner comparable results were compiled for the beluga, as displayed in fig. 73.6B. In performing this analysis, measured thresholds (Awbrey, Thomas, and Kastelein 1988) and critical ratios (Johnson and McManus 1989) for *Delphinapterus leucas* were used and combined with the lower ambient noise bounds displayed by Urick (1983) in his classic compendium of ambient noise mechanisms. As the beluga has incredibly good hearing and is also an inhabitant of the Arctic, we have also included some threshold estimates based on the quite low levels of ambient noise found under certain conditions in the Arctic (Milne and Ganton 1964). As with the human results, the proposed model of low ambient level plus critical ratio provides a remarkably good estimate of beluga hearing threshold in the region of best hearing sensitivity (20–60 kHz).

To extrapolate this approach to the baleen whales we have made two further assumptions: (1) the lower bound of the Urick (1983) compendium of ambient noise (absent of shipping noise effects) is valid in the region below 1 kHz; and (2) the range of mammalian critical ratios is reliably approximated by 16–24 dB re 1 Hz. Given these assumptions, the range of thresholds for baleen whales in the frequency regime below 1 kHz can be pre-

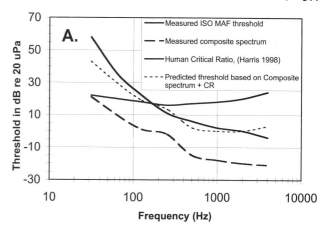

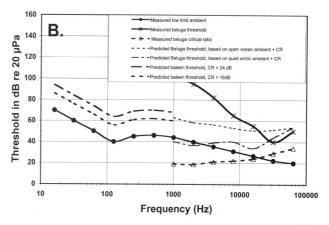

Fig. 73.6. Comparisons of measured and/or predicted auditory thresholds for (*A*) humans and (*B*) beluga whales and baleen whales. *A:* Measured ISO MAF human thresholds from Harris (1998) and predicted thresholds based on the sum of (i.e., composite) measured low ambient noise data (Fidell et al. 1996; Harris 1998) and measured human critical ratios (Harris 1998). *B:* Measured beluga whale thresholds from Awbrey et al. (1988) and predicted thresholds based on the sum of measured ambient noise levels from either the open ocean (Urick 1983) or the Arctic (Milne and Ganton 1964), and measured beluga critical ratios from Johnson and McManus (1989); predicted auditory threshold for baleen whale based on the sum of measured lower ambient spectrum from Urick (1983) and either a 16 dB or 24 dB critical ratio.

dicted and is provided in fig. 73.6C. Thus, for example, the expected hearing threshold for a species with best hearing in the 200–400 Hz band (e.g., right whales) would be expected to be 60–70 dB re 1 μPa.

Echo-Ranging Hypothesis

Simple rules, such as those provided in this chapter, can provide important insights into the potential utility of whale signals for echo-ranging. To illustrate, we present three specific examples of potential active interrogation of the environment by a mysticete whale.

The first case is an example for a bowhead whale migrating within the Arctic sea ice on its way to the summer feeding grounds. This migration starts in late winter

Reverberation Levels of Bowhead Call from Ice Keel and Thin Ice.

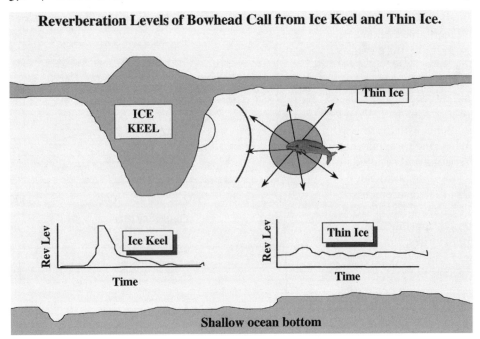

Fig. 73.7. Reverberation of a bowhead sound off an ice keel

when there is little to no sunlight, there is often no open water, and the migration route is impeded by winter pack ice. This is a hazardous environment in which animals can be trapped by and die in the ice. All indications are that the migration continues throughout the 24 h day and does not stop during darkness. Thus, there should be strong selective pressure to successfully navigate through the ice floes without the use of visual cues. Fig. 73.7 illustrates the difference in reverberation level of a bowhead call as reflected off thin ice and off a pressure ridge with 5–10 m keels of ice. In this instance, the reverberation time history of the return from the vicinity of the deep ice keel is substantially different from that of the adjacent thin ice cover. This difference predicts that bowheads could discriminate between areas of thin ice (for which the reverberation levels are comparable to that of open water) and the more dangerous areas of heavy ice. Indeed, there is some anecdotal evidence to support this prediction. In 1985, during the spring bowhead census off Pt. Barrow, Alaska, a thick (5–10 m) multiyear ice floe was located within the path of the bowhead migration. Acoustic tracks of bowhead singers and callers diverted around the multiyear floe, and some call rates for animals at the front of the migrating herd increased as they approached the floe (George et al. 1989; Ellison, Clark, and Bishop 1987; Tyack and Clark 2000).

The second case is for a male humpback whale singing along an island shelf break. An example of the TL sound field for a singer off the coast of the island of Hawaii is shown in fig. 73.8. At this location the only reverberations of his song come from reflections off the island's underwater slope. The region seaward of the singer returns no reverberation because the sound is entering directly into the deep sound channel. If one of the pri-

mary functions of singing is to communicate to as large an audience as possible, this propagation scenario predicts selection should favor males that sing along the shelf edge. If part of the message is directed at rival males and part at potential mates (both local and distant), then selection should favor animals that sing songs with high bandwidth and high intensity, and produce FM sounds of modest duration. To date almost all research on humpback singers has been directed at animals that sing in nearshore, shallow-water areas rather than offshore near the slope edge, so the functional proportion of humpbacks that sing near the shelf break remains to be determined. Frankel et al. (1995), using a hydrophone array to locate and track singers, found that half of the singers were beyond the 100 fathom (200 m) contour, and the offshore singer density was approximately one third that of the nearshore density. Recent acoustic monitoring for whales using the U.S. Navy's Sound Surveillance System (SOSUS) arrays has consistently found humpbacks singing in the deep ocean throughout the same periods of the year when they are singing in shallow water (Clark unpublished data; Charif, Clapham, and Clark 2001; Watkins et al. 2000). The subjective impression is that the deep-water singing behavior of these humpbacks is no different than that of humpbacks in shallow water.

The third case, for a blue whale singing or calling in the deep ocean, presents the potential situation of echoranging in a pelagic context. This example was motivated from observations by Christopher W. Clark during our SOSUS work. In 1993, a blue whale was tracked for 43 days as it traveled over 3500 km in the western North Atlantic. During this period the whale sang almost continuously on a regular basis throughout the 24 h day, and

Reverberant Field for Humpback Singing off Island of Hawaii

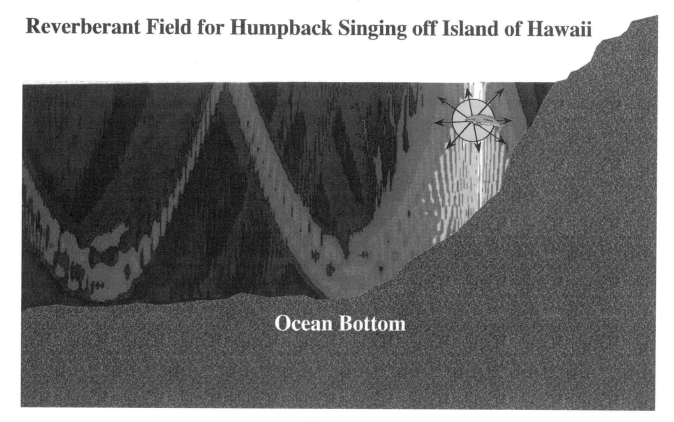

Ocean Bottom

Fig. 73.8. Reverberation of humpback song for a singer along the edge of a volcanic island

the time-of-arrival differences of its sounds at different arrays were used to position and track the animal (approximately 4–6 positions/day). It was first positioned approximately 650 km northeast of the Island of Bermuda, and the details of its daily movements were recorded and studied by navy analysts (Lt. Comdr. G. "Chuck" Gagnon, declassified list of track data, July 1993, pers. comm.). It swam on a steady course and speed (220–230° T, 2–4 knots). During the first three days, its track headed toward and passed just south of an underwater seamount, at which point it turned more to the west and headed toward Bermuda. For the next four days it swam on a line to Bermuda and passed just south of the island before turning in a more southerly direction. This behavior, together with the recordings on which echoes of its sounds off Bermuda were detected, suggested that the whale might have echo-ranged first off the seamount and then off Bermuda as a mechanism of orienting and navigating. The two-way travel distance for a 20 s blue whale FM signal is approximately 30 km, or a one-way distance of 15 km. Blue whales typically have 60–120 s intersignal intervals equivalent to a one-way distance of 45–90 km. Is it possible that the sounds from a blue whale (20 s, 17–20 Hz, 185 dB re 1 μPa) could reflect off Bermuda and be detectable at the whale when it is 45–90 km distant? Blue whale source levels are on the order of 180–185 dB re 1 μPa^2/Hz or 180 dB

band level (Thode, D'Spain, and Kuperman 2000), and typical dive depths when traveling are on the order of 30–100 m (Croll et al. 2001b; Fiedler et al. 1998). The time-bandwidth gain from a typical call is approximately 13 dB. NL for the lowest (10th percentile) in the band is 60 dB (Richardson et al. 1995; Urick 1983; Wenz 1962). TS for Bermuda in the low-frequency band of interest is 55 dB (WTE, unpubl. data). For the reflected sound to be detectable at the whale under the best conditions (see eq. 73.10), SE must be greater than 0 dB. Fig. 73.9 shows predicted TL sound fields for this situation and illustrates one-way propagation fields for a 20 Hz blue whale sound relative to the seamount at which the whale turned (top panel), the open ocean to the south of the seamount (middle panel), and Bermuda after it reached the seamount (bottom panel). For DI we assume the minimum value of 3 dB. For DT we assume a conservative value of −6 dB. In the direction of Bermuda at approximately 45 km from the base of the island, the two-way TL is 160–170 dB. Given these values, eq. 73.9 estimates that SE ≥ 15dB. According to this scenario, the echo returning off the base of Bermuda would easily be above background noise at the whale under these conditions.

In the above estimation of SE, the intersignal value (60–120 s) was used to determine the maximum distance out to which an animal could echo-range. If a blue

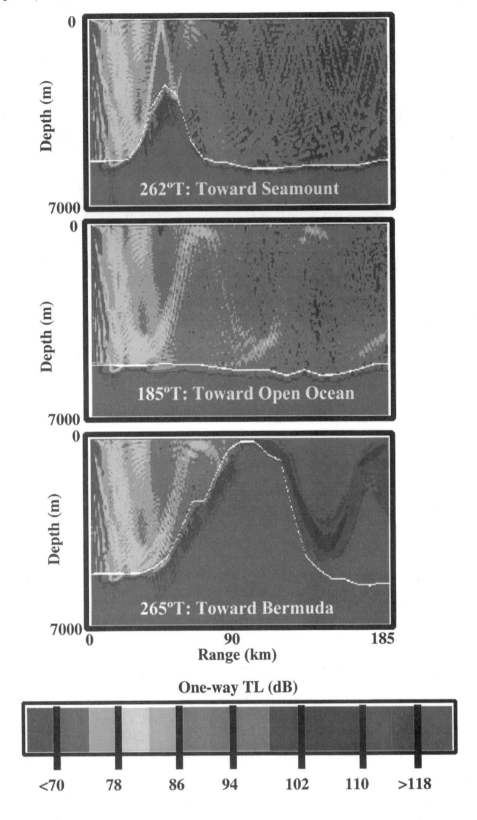

Fig. 73.9. Acoustic propagation of a blue whale's 20 Hz sounds relative to a seamount northeast of Bermuda (*top*), the open ocean (*middle*), and the island of Bermuda (*bottom*) as a function of range. Model parameters for whale: depth, 100 m; frequency, 20 Hz; location (*top* and *middle*), 33°78′ N, 62°00′ W; location (*bottom*), 32°35′ N, 63°85′ W. The white line represents the bathymetry.

whale does echo-range, one might predict that it would decrease its intersignal interval as it approached Bermuda. In fact, this is not the case, and we have never observed a blue or fin whale changing this interval as it approached an island or seamount. This ranging hypothesis assumes that the animal has the ability to de-termine the direction of the returning echo and measure the time difference between the production of the sound and the reception of the sound's echo. A more primitive form of perception does not include ranging and is not constrained by this ability to sense the time difference between the transmitted and echo signals. In this second case, the perception task is reduced to determining the approximate direction of the returning echo, and navigation is not constrained by two-way travel time as determined from the intersignal interval. Given the results of the above exercise in which SE ≥ 15 dB for a blue whale at a range of 45 km from Bermuda, predictably a whale could detect the reflection of its sound off a bathymetric feature at much greater distances, and orient and move toward it well before it could determine its distance from that feature.

Discussion

The expressed characteristics of communication sounds are a balance between multiple, often conflicting, selective pressures (Bradbury and Vehrencamp 1998). In the case of underwater acoustic communication, there are cases in which the environment can have a significant influence on sound propagation, which can lead to dramatic improvements in communication range. Throughout this chapter, a basic assumption has been that there is an advantage to greater communication range (larger audience, greater inter-individual spacing) and that selection should favor those individuals whose sounds are optimized for the transmission properties of the environment.

To evaluate this notion, the broadband one-way sonar equation was applied and used to predict that bandwidth and frequency band are two acoustic characteristics of adaptive advantage for optimizing communication range independent of environment. Empirical evidence supports this prediction. The bandwidths and peak frequencies for songs of the two example coastal species (humpback and bowhead whales) are between 25–600 Hz and 150–400 Hz, respectively, while bandwidths and peak frequencies for two pelagic species (blue and fin whales) are between 3–25 Hz and 18–35 Hz, respectively. These characteristics are well matched to both the acoustic transmission and ambient noise conditions of the respective environments. Further pronounced differences between songs for species in these two environments are found in signal redundancy and stereotypy. We conclude that many baleen whale sounds are well adapted for long-range communication within their predominant breeding and feeding environments.

Propagation considerations predict that communication ranges for coastal and pelagic species are different by at least an order of magnitude, with communication ranges for coastal species on the order of tens of kilometers, and pelagic species on the order of hundreds of kilometers. Given the pronounced differences between the coastal and pelagic species, there are several implications regarding behavioral ecology and the potential impact of humanmade activities on ambient noise.

Throughout this study, we have assumed that the original ancestral mysticete lived in shallow coastal waters and from there moved offshore to exploit feeding opportunities along shelf breaks and offshore seamounts. This assumption coincides with the existing molecular evidence indicating that right whales, the coastal group, are more ancestral than the rorquals, a group that contains the most pelagic of the mysticete species (Messenger and McGuire 1998). Another assumption is that a fundamental factor driving the scale of the communication range is the spatial distribution of resources. If acoustic communication is an important part of baleen whale feeding success, then these results predict that the distribution of food resource for the coastal species are more closely distributed on average than food resources for the pelagic species.

Noteworthy acoustic features of the pelagic species are high signal intensity, redundancy, and stereotypy. As selection favored the production of louder, more redundant low-frequency signals for communication, was there some other feature of this acoustical niche that whales might have exploited? Could the echoes returning from distant ocean features be exploited as an aid in navigation and orientation in the deep sea?

The broadband two-way sonar equation was used to explore the possibility that blue and fin whale sounds could be used as a form of echo-ranging. An empirical example for a blue whale approaching Bermuda (fig. 73.9) was presented to show that in this scenario physical acoustics is not a limiting factor in the detection of a reflection off the base of Bermuda. We conclude that there is no a priori reason, based on transmission properties of the environment and other physical acoustic considerations, to reject the hypothesis that certain baleen whales use sounds for echo-ranging. However, it must be made clear that there are no direct, unequivocal data to support this hypothesis. Despite this lack of evidence, we propose the following mechanism for the evolution of an echo-ranging function. As low-frequency signals became increasingly better adapted for very long-range communication within the deep-sea environment, selection favored signals that were infrasonic, intense, stereotypic, and redundant. These last two features are also advantageous for detection of reverberation and reflection. Thus, there was a secondary selective advantage to these signals as an aide in long-range navigation and orientation along the shelf edge and in the deep ocean. The bowhead example of possible echo-ranging off ice keels suggests that the possible use of reflection and reverberation does not appear to be confined to deep-water species. It is not clear whether the bowhead example of echo-ranging off ice keels is simply a special case of exploitation of reverberation in the unique coastal habitat of the Arctic, or whether echo-ranging is common to the baleen group.

Although early whales evolved from a land-based mammal, it is unclear whether all archaic mysticetes were restricted to shallow-water environments. The current distribution of the less derived balaenid species supports this assumption. However, it is difficult to prove that the common ancestor to existing right whales and rorquals was coastal. Ketten (1992a, 1992b) provides anatomical evidence that right whale ears are well adapted for intermediate frequency (<1000 Hz) hearing, while fin whale ears are well adapted for low-frequency hearing, but this does not preclude a common ancestor with echo-ranging capabilities.

Regardless of whether or not baleen whales echo-range, there is no doubt that species throughout the baleen group produce sounds with characteristics that are well matched to the species habitat. This result strongly supports the well-accepted, but not necessarily well-documented, assumption that many of these sounds are for long-range communication. What is especially compelling, however, is the match between signal characteristics and the habitat's frequency band with low transmission loss and low ambient noise. These features have important implications, especially regarding ambient noise, as originally implicated by Payne and Webb (1971). As ambient noise increases, especially the ubiquitous ambient noise from commercial shipping, the potential acoustic space of an individual decreases. If long-range communication is the norm for whales, then every 3 dB increase in ambient noise in the signal band halves the effective communication range for both the signaler and the receiver. The same is true in the case of a possible echo-ranging function: an increase in ambient noise reduces an animal's ability to navigate and orient over long distances.

Although propagation models and physical acoustic considerations help form predictions about how selection might act on bioacoustic signals, they do not answer why animals produce those sounds. Oceanographic models of circulation and productivity are beginning to be used to predict the availability of marine food resources that ultimately drive the distribution and abundance of top predators such as whales. The issue is now one of gathering the biological evidence to demonstrate and test whether or not baleen whales acoustically behave in ways that are consistent with sound propagation constraints and ecological conditions. To do this one could use oceanographic and acoustic models to predict acoustic communication differences in different coastal habitats or in different pelagic habitats. Evaluative testing of these predictions could then be done through comparisons of conspecific populations that inhabit different environments (e.g., inshore versus offshore populations of blue whales or minke whales), or comparisons of different species that inhabit similar environments (e.g., blue whales and fin whales, right whales and coastal minke whales).

This situation emphasizes the need for long-term research integrating the fundamentals of acoustic propagation, environmental conditions, prey productivity and distribution, and marine animal distribution and behavior over multiple spatial and temporal scales. The importance of baleen whale acoustic behaviors will only become evident through such a long-term, integrated approach. Knowledge of how and why whales use sound is critical for our deeper appreciation of sound in the marine environment and for a realistic understanding of the potential impacts of humanmade sound on those animals.

Acknowledgments

This synthesis would not have been possible without the support of many agencies and creative discussions with many individuals. Research support was provided by the National Geographic Society, New York Zoological Society, National Science Foundation, Naval Research Lab, North Slope Borough Department of Wildlife Management, and the Office of Naval Research. Jack Bradbury, Donald Croll, Kurt Fristrup, and Leila Hatch all provided helpful insights and stimulating comments. Christopher W. Clark is indebted to all the staff in the Bioacoustics Research Program for assistance with data collection and analysis.

Part Six / Literature Cited

AIRAPET'YANTS, E. SH. 1974. Echolocation in rodents and insectivores. Pp. 300–312 in *Echolocation in nature*, part 2. Arlington, Va.: Joint Publication Research Service (JPRS-63328-2).

AIRAPET'YANTS, E. SH. and A. I. KONSTANTINOV. 1970. Echolocation in birds and small land mammals. Pp. 207–214 in *Echolocation in animals*. Academy of Sciences USSR, Joint Scientific Council on Physiology of Man and Animals.

ALDERTON, D. 1996. *Rodents of the world.* London: Cassell.

ANDREW, R. J. 1964. The displays of primates. Pp. 227–309 in *Evolutionary and genetic biology of primates*, vol. 2, ed. J. Buettner-Janusch. New York: Academic Press.

AU, W. W. L. 1993. *The sonar of dolphins.* New York: Springer-Verlag.

AWBREY, F. T., J. A. THOMAS, and R. A. KASTELEIN. 1988. Low-frequency underwater hearing sensitivity in belugas, *Delphinapterus leucas. Journal of the Acoustical Society of America* 84:2273–2275.

BANNISTER, R. W. 1986. Deep sound channel noise from high-latitude winds. *Journal of the Acoustical Society of America* 79:41–48.

BARRETT-LENNARD, L. G., J. K. B. FORD, and K. A. HEISE. 1996. The mixed blessing of echolocation: Differences in sonar use by fish-eating and mammal eating killer whales. *Animal Behaviour* 51:553–565.

BATESON, G. 1966. Problems in cetacean and other mammalian communication. Pp. 565–579 in *Whales, dolphins and porpoises.* Berkeley and Los Angeles: University of California Press.

BELL, R. W. 1979. Ultrasonic control of maternal behavior: Developmental implications. *American Zoologist* 19:413–418.

BELL, R., W. NOBLE, and F. DAVES. 1971. Echolocation in the blinded rat. *Perception and Psychophysics* 10:112–114.

BERTA, A., and J. L. SUMICH. 2000. *Marine mammals: Evolutionary biology.* San Diego, Calif.: Academic Press.

BÉRUBÉ, M., and A. AQUILAR. 1998. A new hybrid between a blue whale, *Balaenoptera musculus,* and a fin whale, *B. physalus:* Frequency and implications of hybridization. *Marine Mammal Science* 14:82–98.

BLAKE, B. H. 1992. Ultrasonic vocalization and body temperature maintenance in infant voles of three species (Rodentia: Arvicolidae). *Developmental Psychobiology* 25:581–596.

BLUMBERG, M. S., and J. R. ALBERTS. 1990. Ultrasonic vocalizations by rat pups in the cold: An acoustic by-product of laryngeal braking? *Behavioral Neuroscience* 194:808–817.

BRADBURY, J. W., and S. L. VEHRENCAMP. 1998. *Principles of animal communication.* Sunderland, Mass.: Sinauer Associates, Inc.

BUCHLER, E. R. 1976. The use of echolocation by wandering shrews (*Sorex vagrans*). *Animal Behavior* 24:858–873.

BURGESS, W. C., P. L. TYACK, B. J. LEBOEUF, and D. P. COSTA. 1998. A programmable acoustic recording tags and first results from free-ranging northern elephant seals. *Deep-Sea Research II* 45:1327–1351.

CHABOT, D. 1988. A quantitative technique to compare and classify humpback whale (*Megaptera novaeangliae*) sounds. *Ethology* 77:89–102.

CHARIF, R. A., P. J. CLAPHAM, and C. W. CLARK. 2001. Acoustic detections of singing humpback whales in deep waters off the British Isles. *Marine Mammal Science* 17(4): 751–768.

CHARIF, R. A., D. K. MELLINGER, K. J. DUNSMORE, K. M. FRISTRUP, and C. W. CLARK. 2002. Estimated source levels of fin whale (*Balaenoptera physalus*) vocalizations: Adjustments for surface interference. *Marine Mammal Science* 18(1): 81–98.

CHASE, J. 1980. Rat echolocation: Correlations between object detection and click production. Pp. 875–878 in *Animal sonar systems,* ed. R. G. Busnel and J. F. Fish. New York: Plenum Press.

CHURCHFIELD, S. 1990. *The natural history of shrews.* Ithaca, N.Y.: Cornell University Press.

CLARK, C. W. 1982. The acoustical repertoire of the southern right whale: A quantitative analysis. *Animal Behavior* 30:1060–1071.

———. 1983. Acoustical communication and behavior of the southern right whale (*Eubalaena australis*). Pp. 163–198 in *Communication and behavior of whales,* ed. R. S. Payne. Boulder, Colo.: Westview Press.

CLARK, C. W., and J. M. CLARK. 1980. Sound playback experiments with southern right whales (*Eubalaena australis*). *Science* 207:663–665.

CLARK, C. W., and K. M. FRISTRUP. 1997. Whales '95: A combined visual and acoustical survey of blue and fin whales off Southern California. *Report of the International Whaling Commission* 47:583–600.

CLARK, C. W., and J. H. JOHNSON. 1984. The sounds of the bowhead whale, *Balaena mysticetus,* during the spring migrations of 1979 and 1980. *Canadian Journal of Zoology* 62:1436–1441.

CRANE, N. L., and K. LASHKARI. 1996. Sound production of gray whales, *Eschrichtius robustus,* along their migration route: A new approach to signal analysis. *Journal of the Acoustical Society of America* 100:1878–1886.

CROLL, D. A., B. TERSHY, R. HEWITT, D. DEMER, P. FIEDLER, S. SMITH, W. ARMSTRONG, J. POPP, T. KIEKHEFER, V. LOPEZ, J. URBAN, and D. GENDRON. 1998. An integrated approach to the foraging ecology of marine birds and mammals. *Deep-Sea Research II* 45:1353–1371.

CROLL, D. A., C. W. CLARK, J. CALAMBOKIDIS, W. T. ELLISON, and B. R. TERSHY. 2001a. Effect of anthropogenic low frequency noise on the foraging ecology of *Balaenoptera* whales. *Animal Conservation* 4:13–27.

CROLL, D. A., A. ACEVEDO-GUTIERREZ, B. TERSHY, and J. URBAN-RAMIREZ. 2001b. The diving behavior of blue and fin whales: Is dive duration shorter than expected based on oxygen stores? *Comparative Biochemistry and Physiology—Molecular and Integrative Physiology* 129A (4): 797–809.

CUMMINGS, W. C. 1985. Bryde's whale—*Balaenoptera edeni.* Pp. 137–154 in *Handbook of marine mammals,* vol. 3: *The sirenians and baleen whales,* ed. S. H. Ridgeway, and R. Harrison. New York: Academic Press.

CUMMINGS, W. C., and P. O. THOMPSON. 1971. Underwater sounds from the blue whale, *Balaenoptera musculus. Journal of the Acoustical Society of America* 50:1193–1198.

CURTIS, K. R., B. M. HOWE, and J. A. MERCER. 1999. Low-frequency ambient sounds in the North Pacific: Long time series observations. *Journal of the Acoustical Society of America* 106:3189–3200.

DAHLHEIM, M. E., H. D. FISHER, and J. D. SCHEMPP. 1984. Sound production by the gray whale and ambient noise levels in Laguna San Ignacio, Baja California Sur, Mexico. Pp. 511–541 in *The gray whale* Eschrichtius robustus, ed. M. L. Jones, S. L. Swartz, and S. Leatherwood. New York: Academic Press.

DAVIS, R. W., L. A. FUIMAN, T. M. WILLIAMS, S. O. COLLIER, W. P. HAGEY, S. G. KANATOUS, and M. HORNING. 1999. Hunting behavior of a marine mammal beneath the Antarctic fast ice. *Science* 283:993–996.

DAWBIN, W. H., and D. H. CATO. 1992. Sounds of pygmy right whale (*Caperea marginata*). *Marine Mammal Science* 8:213–219.

DEHNHARDT, G., B. MAUCK, and H. BLECKMANN. 1998. Seal whiskers detect water movements. *Nature* 394:235–236.

DIJKGRAAF, S. 1932. Über lautäusserungen der Elritze. *Zietschr. fur. vergleichende Physiologie* 17:802–805.

DUDZINSKI, K. 1995. Communication in Atlantic Spotted dolphins, *Stenella frontalis,* relationship between vocal and behavioral activity. Proceedings of the European Association of Aquatic Mammals, 23rd Annual Symposium, Nuremberg, Germany, 23 March 1993.

DYER, I. 1997. Ocean ambient noise. Pp. 549–557 in *Encyclopedia of acoustics,* ed. M. J. Crocker. New York: John Wiley and Sons.

EDDS, P. L. 1988. Characteristics of finback *Balaenoptera physalus* vocalizations in the St. Lawrence estuary. *Bioacoustics* 1:131–149.

EDDS, P. L., D. K. ODELL, and B. R. TERSHY. 1993. Vocalizations of a captive juvenile and free-ranging adult-calf pairs of Bryde's whales, *Balaenoptera edeni. Marine Mammal Science* 9:269–284.

EDDS-WALTON, P. L. 2000. Vocalizations of minke whales (*Balaenoptera acutorostrata*) in the St. Lawrence estuary. *Bioacoustics* 11:31–50.

EISENBERG, J. F. 1975. Tenrecs and solenodons in captivity. *International Zoo Yearbook* 15:6–12.

EISENBERG, J. F., and E. GOULD. 1966. The behaviour of *Solenodon paradoxus* in captivity with comments on the behavior of other Insectivora. *Zoologica* 51:49–58.

ELLISON, W. T., C. W. CLARK, and G. C. BISHOP. 1987. Potential use of surface reverberation by bowhead whales, *Balaena mysticetus,* in under-ice navigation: Preliminary considerations. *Report of the International Whaling Commission* 37:329–332.

ELWOOD, R. W., and F. KEELING. 1981. Temporal organization of ultrasonic vocalizations in infant mice. *Developmental Psychobiology* 15:221–227.

EVANS, W. E., and R. HAUGEN. 1963. An experimental study of the echolocation ability of the California sea lion, *Zalophus californianus* (Lesson). *Bulletin of the Southern California Academy of Science* 62:165–175.

FAY, R. R., and A. N. POPPER. 1994. *Comparative hearing in mammals.* New York: Springer-Verlag.

FELDHAMER, G. A., L. C. DRICKAMER, S. H. VESSEY, and J. F. MERRITT. 1999. *Mammalogy: Adaptation, diversity and ecology.* Dubuque, Iowa: McGraw-Hill.

FENTON, M. B. 1975. Acuity of echolocation in *Collocalia hirundinacea* (Aves: Apodidae) with comments on distribution of echolocating swiftlets and molossid bats. *Biotropica* 7:1–7.

———. 1983. *Just bats.* Toronto: University of Toronto Press.

FIDELL, S., L. SILVATI, R. HOWE, K. S. PEARSONS, B. TABACHNICK, R. C. KNOPF, J. GRAMANN, and T. BUCHANAN. 1996. Effects of aircraft overflights on wilderness recreationists. *Journal of the Acoustical Society of America* 100:2909–2918.

FIEDLER, P. C., S. REILLY, R. P. HEWITT, D. DEMER, A. PHILBRICK, S. SMITH, W. ARMSTRONG, D. A. CROLL, B. R. TERSHY, and B. MATE. 1998. Blue whale habitat and prey in the Channel Islands. *Deep-Sea Research II* 45:1781–1801.

FORSMAN, K. A., and M. G. MALMQUIST. 1988. Evidence for echolocation in the common shrew, *Sorex araneus. Journal of Zoology* (London) 216:655–662.

FRANKEL, A. S., C. W. CLARK, L. M. HERMAN, and C. M. GABRIELE. 1995. Spatial distribution, habitat utilization, and social interactions of humpback whales, *Megaptera novaeangliae,* off Hawai'i, determined using acoustical and visual techniques. *Canadian Journal of Zoology* 73:1134–1146.

Literature Cited   PART SIX | **585**

FRAZER, L. N., and E. MERCADO III. 2000. A sonar model for humpback song. *IEEE Journal of Ocean Engineering* 25:160–182.

GEORGE, J. C., C. W. CLARK, G. M. CARROLL, and W. T. ELLISON. 1989. Observations on the ice-breaking and ice navigation behavior of migrating bowhead whales (*Balaena mysticetus*) near Point Barrow, Alaska, spring 1985. *Arctic* 42:24–30.

GEORGE, S. R., J. R. CHOAT, and H. H. GENOWAYS. 1986. Short-tailed shrew, *Blarina brevicauda. Mammalian Species* 261:1–9.

GORMAN, M. L., and R. D. STONE. 1990. *The natural history of moles.* Ithaca, N.Y.: Comstock.

GOULD, E. 1965. Evidence for echolocation in the Tenrecidae of Madagascar. *Proceedings of the American Philosophical Society* 109:352–360.

———. 1969. Communication in three genera of shrews (Soricidae): *Suncus, Blarina* and *Cryptotis. Behavioral Biology* A3:11–31.

GOULD, E., N. NEGUS, and A. NOVICK. 1964. Evidence for echolocation in shrews. *Journal of Experimental Zoology* 156:19–38.

GOULD, S. J., and R. C. LEWONTIN. 1979. The spandrels of San Marco and the Panglossian paradigm. *Proceedings of the Royal Society of London B* 205:581–598.

GRIFFIN, D. R. 1954. Acoustic orientation in the oil bird, *Steatornis. Proceedings of the National Academy of Sciences* 39:884–893.

GRIFFIN, D. R., and R. A. SUTHERS. 1970. Sensitivity of echolocation in cave swiftlets. *Biological Bulletin* 139:495–501.

GRANT, P. R., and B. R. GRANT. 1992. Demography and the genetically effective sizes of two populations of Darwin's finches. *Ecology* 73:766–784.

GRUNWALD, A. 1969. Investigation on orientation in white-toothed shrews (Soricidae-Crocidurinae). *Zeitschrift Vergl. Physiol.* 65:191–217.

HAMILTON, W. J., JR. 1940. The biology of the smokey shrew (*Sorex fumeus fumeus* Miller). *Zoologica* 25:473–491.

HARRIS, C. M. 1998. *Handbook of acoustical measurements and noise control,* 3rd ed. Woodbury, N.Y.: Acoustical Society of America.

HARRISON, C., and J. HARRISON. 1995. A simple relationship between frequency and range averages for broadband sonar. *Journal of the Acoustical Society of America* 97:1314.

HARTMAN, G. D., and T. L. YATES. 1985. *Scapanus orarius. Mammalian Species Account* 253:1–5.

HENSON, O. W., JR. 1961. Some and functional aspects of certain structures of the middle ear in bats and insectivores. *University of Kansas Science Bulletin* 423:151–255.

HOFFMEISTER, D. F. 1989. *Mammals of Illinois.* Champaign: University of Illinois Press.

JACOBS, G. H., J. F. DEEGAN, M. A. CROGNALE, and J. A. FENWICK. 1993. Photopigments of dogs and foxes and their implications for canid vision. *Visual Neuroscience* 10:173–180.

JENSEN, F. B. 1981. Sound propagation in shallow water: A detailed description of the acoustical field close to surface and bottom. *Journal of the Acoustical Society of America* 70:1397–1406.

JENSEN, F. B., and W. A. KUPERMAN. 1983. Optimum frequency of propagation in shallow water environments. *Journal of the Acoustical Society of America* 73:813–819.

JENSEN, F. B., W. A. KUPERMAN, M. B. PORTER, and H. SCHMIDT. 1994. *Computational ocean acoustics.* New York: American Institute of Physics.

JOHNSON, C. S., and M. W. McMANUS. 1989. Masked tonal thresholds in the beluga whale. *Journal of the Acoustical Society of America* 85:2651–2654.

JOHNSON, D. H., and D. E. JOHNSON. 1993. *Array signal processing.* Englewood Cliffs, N.J.: Prentice-Hall.

JOHNSON, G. L. 1893. Observations on the refraction and vision of the seal's eye. Pp. 719–723 in the *Proceedings of the Zoological Society of London.*

KAHMANN, H., and K. OSTERMANN. 1951. Wahrenhmen und Hervorbringen hoher tone bei kleinen saugetieren. *Experientia* 7:268–269.

KASTAK, D., and R. J. SCHUSTERMAN. 1998. Low-frequency amphibious hearing in pinnipeds: Methods, measurements, noise, and ecology. *Journal of the Acoustical Society of America* 103:2216–2228.

———. 1999. In-air and underwater hearing sensitivity of a northern elephant seal (*Mirounga angustirostris*). *Canadian Journal of Zoology* 77:1751–1758.

KETTEN, D. R. 1992a. The cetacean ear: Form, frequency, and evolution. Pp. 53–75 in *Marine mammal sensory systems,* ed. J. A. Thomas, R. A. Kastelein, and A. Ya. Supin. New York: Plenum Press.

———. 1992b. The marine mammal ear: Specializations for aquatic audition and echolocation. Pp. 717–750 in *The Evolutionary biology of hearing: Springer handbook of auditory research,* ed. D. B. Webster, R. R. Fay, and A. N. Popper. Berlin: Springer-Verlag.

KING, JUDITH E. 1983. *Seals of the world.* Ithaca, N.Y.: Cornell University Press.

KINGDON, J. 1974. *East African mammals: Insectivores and bats,* vol. 2A. Chicago: University of Chicago Press.

KOOYMAN, G. 1968. An analysis of some behavioral and physiological characteristics related to diving in the Weddell seal. *Antarctic Research Series* 11:227–261.

———. 1981. *Weddell seal, consummate diver.* Cambridge: Cambridge University Press.

LANDAU, D., and W. W. DAWSON. 1970. The histology of retinas from the pinnipedia. *Vision Research* 10: 691–702.

LAVIGNE, D. M., and K. RONALD. 1975. Pinniped visual pigments. *Comparative Biochemistry and Physiology* 52B:325–329.

LE BOEUF, B. J., and R. M. LAWS, eds. 1994. *Elephant seals.* Berkeley and Los Angeles: University of California Press.

LEVENSON, D. H., and R. J. SCHUSTERMAN. 1997. Pupillometry in seals and sea lions: Ecological implications. *Canadian Journal of Zoology* 75:2050–2057.

———. 1999. Dark adaptation and visual sensitivity in shallow and deep-diving pinnipeds. *Marine Mammal Science* 15:1303–1313.

LJUNGBLAD, D. K., C. W. CLARK, and H. SHIMADA. 1998. A comparison of sounds attributed to pygmy blue whales (*Balaenoptera musculus musculus*) recorded south of the Madagascar Plateau and those attributed to "True" blue whales (*Balaenoptera musculus*) recorded off Antarctica. *Report of the International Whaling Commission* 48:439–442.

LJUNGBLAD, D. K., P. O. THOMPSON, and S. E. MOORE. 1982. Underwater sounds recorded from migrating bowhead whales, *Balaena mysticetus,* in 1979. *Journal of the Acoustical Society of America* 71:477–482.

LYTHGOE, J. N., and H. J. A. DARTNALL. 1970. A "deep sea rhodopsin" in a mammal. *Nature* 227:955–956.

MANN, D. A., A. LU, and A. N. POPPER. 1996. A clupeid fish can detect ultrasound. *Nature* 389:333.

MAYO, C. A., and M. K. MARX. 1990. Surface foraging behavior of the North Atlantic right whale, *Eubalaena glacialis,* and associated zooplankton characteristics. *Canadian Journal of Zoology* 68:2214–2220.

McDONALD, M. A., J. CALAMBOKIDIS, A. M. TERANISHI, and J. A. HILDEBRAND. 2001. The acoustic calls of blue whales off California with gender data. *Journal of the Acoustical Society of America* 109:1728–1735.

MELLINGER, D. K., C. D. CARSON, and C. W. CLARK. 2000. Characteristics of minke whale (*Balaenoptera acutorostrata*) pulse trains recorded near Puerto Rico. *Marine Mammal Science* 16:739–756.

MESSENGER, S. L., and J. A. McGUIRE. 1998. Morphology, molecules, and the phylogenetics of cetaceans. *Systematic Biology* 47:90–124.

MILNE, A. R., and J. H. GANTON. 1964. Ambient noise under Arctic-sea ice. *Journal of the Acoustical Society of America* 36:855–864.

MOBLEY, J. R., L. M. HERMAN, and A. S. FRANKEL. 1988. Responses of wintering humpback whales (*Megaptera novaeangliae*) to playback of recordings of winter and summer vocalizations and of synthetic sounds. *Behavioral Ecology and Sociobiology* 23:211–223.

MOORE, J. C. 1943. A contribution to the natural history of the Florida short-tailed shrew. *Proceedings of the Florida Academy of Science* 6:155–166.

MOORE, S. E., K. M. STAFFORD, M. E. DAHLHEIM, C. G. FOX, H. W. BRAHAM, J. POLOVINA, and D. BAIN. 1998. Seasonal variation in reception of fin whale calls at five geographic areas in the North Pacific. *Marine Mammal Science* 14:617–627.

MORTON, E. S. 1975. Ecological sources of selection on avian sounds. *American Naturalist* 109:17–34.

———. 1977. On the occurrence and significance of motivational-structural rules in some bird and mammal sounds. *American Naturalist* 111:855–869.

———. 1982. Grading, discreteness, redundancy and motivational-structural rules. Pp. 183–212 in *Evolution and ecology of acoustical communication in birds,* ed. D. E. Kroodsma and E. H. Miller. New York: Academic Press.

NEWMAN, J. D. 1988. *The physiological control of mammalian vocalization.* New York: Plenum Press.

NIKAIDO, M., A. P. ROONEY, and N. OKADA. 1999. Phylogenetic relationships among cetartiodactyls based on insertions of short and long interspersed elements: Hippopotamuses are the closest extant relatives of whales. *Proceedings of the National Academy of Sciences* 96:10261–10266.

NORRIS, K. S. 1967. Some observations on the migration and orientation of marine mammals. Pp. 101–125 in *Animal orientation and navigation: Proceedings of the twenty-seventh animal biology colloquium, May 6–7, 1966,* ed. R. M. Storm. Corvallis: Oregon State University Press.

———. 1969. The echolocation of marine mammals. Pp. 391–423 in *The biology of marine mammals,* ed. H. T. Andersen. New York: Academic Press.

NORRIS, K. S., and B. MØHL. 1983. Can odontocetes stun prey with sound? *American Naturalist* 122:85–104.

NOVICK, A., and E. GOULD. 1964. Comparative study of echolocation in Tenrecidae of Madagascar and other Old World insectivores. Final Report No. AFOSR. 64-0245 Yale University.

NOWAK, R. M. 1991. *Walker's mammals of the world,* vols. 1 and 2. 5th ed. Baltimore: Johns Hopkins University Press.

OLIVER, G. 1978. Navigation in mazes by a grey seal, *Halichoerus grypus. Behavior* 67:97–114.

PALMEIRIM, J. M., and R. S. HOFFMANN. 1983. *Gaemys pyrenaicus.* Mammalian Species, 207:1–5.

PANIGADA, S., M. ZANARDELLI, S. CANESE, and M. JAHODA. 1999. How deep can baleen whales dive? *Marine Ecology Progress Series* 187:309–311.

PATTERSON, B., and G. R. HAMILTON. 1964. Repetitive 20 cycle per second biological hydroacoustical signals at Bermuda. Pp. 125–145 in *Marine bio-acoustics,* ed. W. N. Tavolga. Oxford: Pergamon Press.

PAYNE, K. B., W. R. LANGBAUER JR., and E. M. THOMAS. 1986. Infrasonic calls of the Asian elephant (Elephas maximus). *Behavioral Ecology and Sociobiology* 18:297–301.

PAYNE, R., and S. McVAY. 1971. Songs of humpback whales. *Science* 173:587–597.

PAYNE, R., and D. WEBB. 1971. Orientation by means of long range acoustical signaling in baleen whales. *Annals of the New York Academy of Sciences* 188:110–141.

PILLERI, G. 1990. Adaptation to water and the evolution of echolocation in the Cetacea. *Ethology, Ecology and Evolution* 2:135–163.

POULTER, T. C. 1963. Sonar signals of the sea lion. *Science* 139:753–755.

———. 1966. The use of active sonar by the Califonia sea lion, *Zalophus californianus* (Lesson). *Journal of Auditory Research* 6:165–173.

———. 1967. Systems of echolocation. Pp. 157–185 in *Animal sonar systems: Biology and bionics,* ed. R. G. Busnel. Jouy-in-Josas, France: Laboratoire de Physiologie Acoustique.

PREMUS, V., and J. L. SPIESBERGER. 1997. Can acoustical multipath explain finback (*B. physalus*) 20-Hz doublets in shallow water? *Journal of the Acoustical Society of America* 101:1127–1138.

RADO, R., N. LEVI, H. HAUSER, J. WITCHER, N. ADLER, N. INTRATOR, Z. WOLLBERG, and J. TERKEL. 1987. Seismic signalling as a means of communication in a subterranean mammal. *Animal Behaviour* 35:1249–1266.

RENOUF, D., and DAVIS, M. B. 1982. Evidence that seals may use echolocation. *Nature* (London) 300:635–637.

RENOUF, D., G. GALWAY, and L. GABORKO. 1980. Evidence for echolocation in harbour seals. *Journal of the Marine Biology Association of the United Kingdom* 60:1039–1042.

REPENNING, C. A. 1976. Adaptive evolution of sea lions and walruses. *Systematic Zoology* 25:375–390.

RICE, C. R., S. H. FEINSTEIN, and R. J. SCHUSTERMAN. 1965. Echo-detection ability of the blind: Size and distance factors. *Journal of Experimental Psychology* 70:246–251.

RICHARD, P. B. 1973. Capture, transport, and husbandry of the Pyrenean desman, *Galemys pyrenaicus. Int. Zoo Yearbook* 13:175–177.

RICHARDSON, W. J., C. R. GREENE, JR., C. I. MALME, and D. H. THOMSON. 1995. *Marine mammals and noise.* New York: Academic Press.

RIDGWAY, S. H., and S. R. HARRISON. 1985. *Handbook of marine mammals,* vol. 3: *The sirenians and baleen whales.* New York: Academic Press.

RIEDMAN, M. 1990. *The pinnipeds: Seals and sea lions.* Berkeley and Los Angeles: University of California Press.

RILEY, D. A., and M. ROSENZWEIG. 1957. Echolocation in rats. *Journal of Comparative Physiology and Psychology* 50:323–328.

ROGERS, T., D. H. CATO, and M. M. BRYDEN. 1995. Underwater vocal repertoire of the leopard seal (*Hydruga leptonyx*) in Prydz Bay, Antarctica. Pp. 223–236 in *Sensory systems of aquatic mammals,* ed. R. A. Kastelein, J. A. Thomas, and P. E. Nachtigall. Woerden, The Netherlands: De Spil Publishers.

ROOD, J. 1958. Habits of the short-tailed shrew in captivity. *Journal of Mammalogy* 39:499–507.

ROSENBAUM, H. C., R. L. BROWNELL JR., M. W. BROWN, C. SCHAEFF, V. PORTWAY, B. N. WHITE, S. MALIK, L. A. PASTENE, N. J. PATENAUDE, C. S. BAKER, M. GOTO, P. B. BEST, P. J. CLAPHAM, P. HAMILTON, M. MOORE, R. PAYNE, V. ROWNTREE, C. T. TYNAN, and R. DESALLE. 2000. Worldwide genetic differentiation of Eubalaena: Questioning the number of right whale species. *Molecular Ecology* 9:1793–1802.

ROSENZWEIG, M., D. RILEY, and K. KRECH. 1955. Evidence for echolocation in the rat. *Science* 121:600.

ROSSBACH, K. A., and D. L. HERZING. 1997. Underwater observations of benthic feeding bottlenose dolphins

(*Tursiops truncatus*) near Grand Bahama Island, Bahamas. *Marine Mammal Science* 13:498–504.

SALES, G., and D. PYE. 1974. *Ultrasonic communication in animals.* London: Chapman and Hall.

SCHEVILL, W. E., and W. A. WATKINS. 1972. Intense low-frequency sounds from an Antarctic minke whale, *Balaenoptera acutorostrata. Breviora* 388:1–7.

SCHLUTER, D., and J. D. MCPHAIL. 1992. Ecological character displacement and speciation in sticklebacks. *American Naturalist* 140:85–108.

SCHUSTERMAN, R. J. 1966. Underwater click vocalization by a California sea lion: Effects of visibility. *Psychological Record* 16:129–136.

———. 1967. Perception and determinants of underwater vocalizations in the California sea lion. In *Animal sonar systems,* ed. R. G. Busnel and J. F. Fish. New York: Plenum Press.

———. 1981. Behavioral capabilities of seals and sea lions: A review of their hearing, visual, learning, and diving skills. *Psychological Record* 31:125–143.

SCHUSTERMAN, R. J., and R. F. BALLIET. 1971. Aerial and underwater visual acuity in the California sea lion (*Zalophus californianus*) as a function of luminance. *Annals of the New York Academy of Science* 188:37–46.

SCHUSTERMAN, R. J., and D. KASTAK. 1996. Pinniped acoustics: Habituation and sensitization to anthropogenic signals. *33rd Annual Meeting of the Animal Behavior Society,* Flagstaff, Arizona.

SCHUSTERMAN, R. J., R. GENTRY, and J. SCHMOOK. 1967. Underwater sound production by captive California sea lions, *Zalophus californianus. Zoologica* 52:21–24.

SCRONCE, B. L., and S. H. RIDGWAY. 1980. Grey seal, *Halichoerus:* Echolocation not demonstrated. In *Animal sonar systems,* ed. R. G. Busnel and J. F. Fish. New York: Plenum Press.

SERGEEV, V. E. 1973. Characteristics of the orientation of shrews in water. *Ekologia* 4:87–90.

SILBER, G. K. 1986. The relationship of social vocalizations to surface behavior and aggression in the Hawaiian humpback whale (*Megaptera novaeangliae*). *Canadian Journal of Zoology* 64:2075–2080.

SINIFF, D. B. 1991. An overview of the ecology of antarctic seals. *American Zoologist* 31:143–147.

SPEAKMAN, J. 1993. Evolution of echolocation for predation. *Symposium of the Zoological Society of London* 65:39–63.

SPRINGER, S. 1937. Observations on *Cryptotis floridana* in captivity. *Journal of Mammalogy* 18:237–238.

STAFFORD, K. M., S. L. NIEUKIRK, and C. G. FOX. 1999. Low-frequency whale sounds recorded on hydrophones moored in the eastern tropical Pacific. *Journal of the Acoustical Society of America* 106:3687–3698.

STIRLING, I., and D. SINIFF. 1979. Underwater vocalizations of leopard seals (*Hydrurga Leptonyx*) and crabeater seals (*Lobodon carcinophagus*) near the South Shetland Islands, Antarctic. *Canadian Journal of Zoology* 57:1244–1248.

THODE, A. M., G. L. D'SPAIN, and W. A. KUPERMAN. 2000. Matched-field processing, geoacoustic inversion, and source signature recovery of blue whale vocalizations. *Journal of the Acoustical Society of America* 107:1286–1300.

THOMAS, J. A. 1991. The sounds of seal society. *Natural History* 3 (March): 46–54.

THOMAS, J. A., and D. P. DEMASTER. 1982. An acoustic technique for determining the haulout of leopard (*Hydrurga leptonyx*) and crabeater (*Lobodon carcinophagus*) seals. *Canadian Journal of Zoology* 60:2028–2031.

THOMAS, J. A., and C. L. GOLLADAY. 1995. Geographic variation in leopard seal (*Hydrurga leptonyx*) underwater vocalizations. Pp. 201–222 in *Sensory systems of aquatic mammals,* ed. R. A. Kastelein, J. A. Thomas, and P. E. Nachtigall. Woerden, The Netherlands: De Spil Publishers.

THOMAS, J. A., and V. B. KUECHLE. 1982. Quantitative analysis of the underwater repertoire of the Weddell seal (*Leptonychotes weddelli*). *Journal of the Acoustical Society of America* 72:1730–1738.

THOMAS, J. A., and I. STIRLING. 1983. Geographic variations in the underwater vocalizations of Weddell seals (*Leptonychotes weddelli*) from Palmer Peninsula and McMurdo Sound, Antarctica. *Canadian Journal of Zoology* 61:2203–2212.

THOMAS, J. A., K. ZINNEL, and L. FERM. 1983. Investigation of Weddell seal (*Leptonychotes weddelli*) underwater calls using playback techniques. *Canadian Journal of Zoology* 61:1448–1456.

THOMAS, J. A., S. R. FISHER, W. E. EVANS, and F. T. AWBREY. 1983. Ultrasonic vocalizations of leopard seals (*Hydrurga leptonyx*). *Antarctic Journal of the United States* 17:186–187.

THOMAS, J. A., R. A. PUDDICOMBE, M. GEORGE, and D. LEWIS. 1988. Variations in underwater vocalizations of Weddell seals (*Leptonhychotes weddelli*) at the Vestfold Hills as a measure of breeding population discreetness. *Hydrobiologica* 165:279–284.

THOMPSON, P. O., L. T. FINDLEY, and O. VIDAL. 1992. 20-Hz pulses and other vocalizations of fin whales, *Balaenoptera physalus,* in the Gulf of California, Mexico. *Journal of the Acoustical Society of America* 92:3051–3057.

THOMPSON, T. J., H. E. WINN, and P. J. PERKINS. 1979. Mysticete sounds. Pp. 403–431 in *Behavior of marine mammals: Current perspectives in research,* ed. H. E. Winn and B. L. Olla. New York: Plenum Press.

TOMASI, T. E. 1979. Echolocation by the short-tailed shrew *Blarina brevicauda. Journal of Mammalogy* 60:751–759.

TYACK, P. 1983. Differential response of humpback whales (*Megaptera novaeangliae*) to playback of song or social sounds. *Behavioral Ecology and Sociobiology* 13:49–55.

TYACK, P. L., and C. W. CLARK. 2000. Communication and Acoustical behavior in dolphins and whales. Pp. 156–224 in *Hearing by whales and dolphins,* ed. W. W. L. Au, A. N. Popper, and R. R. Fay. New York: Springer-Verlag.

URICK, R. J. 1983. *Principles of underwater sound.* 3rd ed. New York: McGraw-Hill.

WALLS, G. 1942. *The vertebrate eye and its adaptive radiation.* New York: Hafner.

WARTZOK, D., R. J. SCHUSTERMAN, and J. GAILEY-PHIPPS. 1984. Seal echolocation? *Nature* 308:53.

WARTZOK, D., R. ELSNER, H. STONE, B. P. KELLY, and R. W. DAVIS. 1992a. Under-ice movements and the sensory basis of hole finding by ringed and Weddell seals. *Canadian Journal of Zoology* 70:1712–1722.

WARTZOK, D., S. SAYEIGH, H. STONE, J. BARCHAK, and W. BARNES. 1992b. Acoustic tracking system for monitoring under-ice movements of polar seals. *Journal of the Acoustical Society of America* 92:682–687.

WATKINS, W. A. 1981. Activities and underwater sounds of fin whales. *Scientific Reports of the Whales Research Institute* 33:83–117.

WATKINS, W. A., P. TYACK, K. E. MOORE, and J. E. BIRD. 1987. The 20-Hz signals of finback whales (*Balaenoptera physalus*). *Journal of the Acoustical Society of America* 82:1901–1912.

WATKINS, W. A., M. A. DAHER, G. M. REPPUCCI, J. E. GEORGE, D. L. MARTIN, N. A. DIMARZIO, and D. P. GANNON. 2000. Seasonality and distribution of whale calls in the North Pacific. *Oceanography* 13:62–67.

WENZ, G. M. 1962. Acoustic ambient noise in the ocean: Spectra and sources. *Journal of the Acoustical Society of America* 34:1936–1956.

WHITAKER, J. O., JR. 1974. The least shrew, *Cryptotis parva. Mammalian Species* 43:1–8.

WILLE, P. C., and D. GEYER. 1984. Measurements on the origin of the wind-dependent ambient noise variability in shallow water. *Journal of the Acoustical Society of America* 75:73–185.

WILSON, D. E., and D. M. REEDER. 1993. *Mammal species of the world.* 2nd ed. Washington, D.C.: Smithsonian Institution Press.

WINN, H. E., and P. J. PERKINS. 1976. Distribution and sounds of the minke whale, with a review of mysticete sounds. *Cetology* 19:1–12.

WÜRSIG, B., and C. CLARK. 1993. Behavior. Pp. 157–199 in *The bowhead whale,* ed. J. Burns, J. Montague, and C. J. Cowles. Lawrence, Kans.: The Society for Marine Mammalogy.

Author Index

Subject Index

L

laboratory rat, 562–563
Lagenorhynchus obliquidens. See Pacific white-sided dolphin
larynx: in bats, 5–6, 9–10, 32, 251: in dolphins, 28
Lasionycteris noctivagans. See silver-haired bat
Lasiurus cinereus, 351, 354. *See also* hoary bat
latency shift, in bats, 127
lateral lemniscus, in bats, 15, 149
learning: associative, 174, 284; by biomimetic sonar, 505; in echolocation, 273–277, 490
least shrew, 551, 558
leopard seal, 531, 535–541
Lepidoptera, hearing, 321, 327–331
Leptonychotes weddellii. See Weddell seal
lesser spearnosed bat, 260
Lipotes vexillifer. See Baiji
listening, passive, 391, 401
little brown bat, 5, 25, 116, 119, 141, 163, 185–189, 222–225, 319
Lobodon carcinophagus. See crabeater seal
low duty cycle, bats, 311, 317, 352–353, 355

M

Macrocephala sp. *See* pilot whale
magnetic resonance (MR), of dolphins, 65
maps (organizational) in auditory cortex, in animals, 174
masked shrew, 553
masking: backwards, 256, 315; forwards, 253, 256, 266–268, 315
matched-filter receiver, in bats, 252–253
McGurk effect, 281–283
medial division (MD), in bats, 144
medial geniculate body (MGB), in bats, 13, 118, 142, 203, 211, 213, 215
Megaderma lyra, 106, 260, 266–268. *See also* gleaning bat
Megadermatidae, 315
Megaptera novaengliae. See humpback whale
melon, in dolphins, 31, 66, 70
memory, spatial, in bats, 255
Mesoplodon carlhubbsi. See Hubbs' beaked whale
metabolic rate, in bats, 363
Microgale dobsoni, 555
microphone: air-dielectric, 477; condenser, 469, 477; electret, 477; piezoelectric transducers, 469, 477
Microtus spp. *See* voles
midbrain: in bats, 7, 123–128, 196–200, 214–221; in dolphins, 163
middle ear: of bats, 92; of dolphins, 92–93; reflex, 253, 335
mimicry, in bats, 254, 324, 330
minke whale, 565
Mirounga angustirostris. See elephant seal
modal action pattern (MAP), in dolphins, 419–425
model, of auditory perception in bats, 459–468
monauralism, in bats, 20
monkey lip dorsal bursae (MLDB), in dolphins, 29, 36, 65

Monodon monoceros. See narwhal
Mormoopidae, 315
moths. *See by family*
multimodal processing: in bats, 207, 260–262; in dolphins, 257–258, 283–287, 288–298
multiple target hypothesis, in bats, 370
multivariate analysis of variance (MANOVA), 488
mustached bat, xxiv, 3, 11, 93, 95–96, 100–102, 104, 141, 146, 176–184, 190, 201, 211, 214–221, 251, 316, 350
Myotis adversus, 318
Myotis alcathoe, 346
Myotis austroriparius, xxx, 8
Myotis bechsteini, 490
Myotis blythii, 346
Myotis brandtii, 346
Myotis californicus, 480–481
Myotis ciliolabrum, 480–481
Myotis daubentonii, xxx, 319–320, 356–361, 380
Myotis evotis, 346, 355
Myotis lucifugus, 319, 350, 356, 371, 480–481. *See also* little brown bat
Myotis myotis, 260, 346, 490
Myotis mystacinus, 346
Myotis natterei, 315, 489
Myotis septentrionalis, 355. *See also* northern long-eared bats
Myotis volans, 480–481
Myotis yumanensis, 346, 350, 356, 480–481
mysticetes, 564–582

N

narrow-space bats, 335–336
narwhal, 386
nasal apparatus: in bats, 35; in dolphins, 28–35
National Science Foundation (NSF), ix
neocortex: in bats, 208–209; in dolphins, 161–172
Neomys fodiens. See water shrew
neuroarchitecture: in bats, 208; in dolphins, 161
neurons: combination-sensitive, 148, 176, 201–202; constant-latency, 370; delay-sensitive, 187–188; delay-tuned, 151, 178; exciting, 125, 162, 199, 224; facilitated, 143, 148, 174, 178; immunoreactive, 162–172; inhibition, 125, 131, 142–143, 162, 196, 199, 224; latency, 138, 148; parallel-hierarchical processing, 207; T-cells, 372–380
Neurotrichus gibbsii. See shrew-mole
Noctilio leporinus. See fishing bats
noctuid moth, peri-stimulus-time histogram (PST), 205
Noctuoidea. See noctuid moth
noise: ambient, 112–113, 380, 475, 576; target detection in, 253
northern long-eared bats, 22–26, 346
northern right whale, 565
nose leaf, in bats, 352
nucleus ambiguus (NA), in bats, 7, 9
nucleus of central acoustic tract (NCAT), in bats, 13–15
Nyctalus leisleri, 314, 354

Nyctalus noctula, 341, 344, 487
Nycteridae, 315

O

Office of Naval Research (ONR), ix
open-space bats, 335
Orcinus orca. See killer whale
organ of corti: in bats, 101; in dolphins, 93–94
oscilloscope, 473, 477
otoacoustic emissions (OAE), in bats, 104–103
Otomops martiensseni, 340, 353–354
outer hair cells: in bats, 107; in dolphins, 94

P

Pacific white-sided dolphin, 36–39, 52–58, 93, 95, 163, 166
pallid bat, 129–135
paralemniscal area (PLA), in bats, 12, 15
parti-colored bat, xxiii, 341, 344
passive biosonar, 533, 564–582
perception, in dolphins, 288–298
performance in experiments: by bats, xv, xvii, xix, xxii, 249–259, 273–275, 383–384; by dolphins, xv–xxii, 249–259, 286
periaqueductal gray (PAG), in bats, 7–10, 13
period meter, 474, 477
phantom echo: with bats, xxv, 253, 255, 269, 273, 367; with dolphins, xxv, 514–518
Phoca hispida. See ringed seal
Phoca vitulina. See harbor seal
Phocoena phocoena. See harbour porpoise
Phocoenidae, 561
Phocoenoides dalli. See Dall's porpoise
phonic lips, in dolphins, 29
Phyllostomatidae, 315
phyllostomus discolor. *See* lesser spearnosed bat
Phyllostomus hastatus, 355
phylogeny, in bats, 347
Physeter catadon. See sperm whale
Physeter macrocephalus. See sperm whale
pilot whale, 163, 415
pinnae: of bats, 12, 91–92, 352, 391; of insectivores, 550; of rodents, 552
pinnipeds, 351–353, 535–541
Pipistrellus nathusii, 341
Pipistrellus pipistrellus, 318, 321, 341, 344, 350, 361, 381
Pipistrellus pygmaeus, 341, 361
Pipistrellus rueppellii, 355
Pipistrellus stenopterus, 353
playback techniques, in bats, 18, 20, 377
Plecotus auritus, 345
Plecotus austriacus, 345
pontine gray (PG), in bats, 12
power spectra displays: of bat signals, xv, 471; of dolphin signals, xi, xiv, 37, 51, 56, 71, 396, 429; of least shrew signals, 559
predator/prey coevolution, 251, 254, 324–326, 327–331, 366, 372–380, 390
prey evasive flight, 331
prey selection, in bats, 320, 339–345
prey size, for bats, 340, 342–344, 360
prey swamping, 330
prey temporal avoidance, 328